Handbook of Research on Nanoemulsion Applications in Agriculture, Food, Health, and Biomedical Sciences

Karthikeyan Ramalingam
B.S. Abdur Rahman Crescent Institute of Science and Technology, India

IGI Global
PUBLISHER of TIMELY KNOWLEDGE

A volume in the Advances in Chemical and Materials Engineering (ACME) Book Series

Published in the United States of America by
IGI Global
Engineering Science Reference (an imprint of IGI Global)
701 E. Chocolate Avenue
Hershey PA, USA 17033
Tel: 717-533-8845
Fax: 717-533-8661
E-mail: cust@igi-global.com
Web site: http://www.igi-global.com

Library of Congress Cataloging-in-Publication Data

Names: Ramalingam, Karthikeyan, 1977- editor.
Title: Handbook of Research on nanoemulsion applications in agriculture, food, health, and biomedical sciences / Karthikeyan Ramalingam, editor.
Description: Hershey PA : Engineering Science Reference, [2022] | Includes bibliographical references and index. | Summary: "This book focuses on the basic aspects of nanoemulsion, synthesis and characterization, and focuses on the different types of applications for nanoemulsions in agriculture, food, health, and biomedical sciences"-- Provided by publisher.
Identifiers: LCCN 2021021798 (print) | LCCN 2021021799 (ebook) | ISBN 9781799883784 (hardcover) | ISBN 9781799883807 (ebook)
Subjects: LCSH: Emulsions. | Nanobiotechnology. | Nanochemistry.
Classification: LCC TP156.E6 N36 2022 (print) | LCC TP156.E6 (ebook) | DDC 660/.294514--dc23
LC record available at https://lccn.loc.gov/2021021798
LC ebook record available at https://lccn.loc.gov/2021021799

This book is published in the IGI Global book series Advances in Chemical and Materials Engineering (ACME) (ISSN: 2327-5448; eISSN: 2327-5456)

British Cataloguing in Publication Data
A Cataloguing in Publication record for this book is available from the British Library.

For electronic access to this publication, please contact: eresources@igi-global.com.

Advances in Chemical and Materials Engineering (ACME) Book Series

J. Paulo Davim
University of Aveiro, Portugal

ISSN:2327-5448
EISSN:2327-5456

Mission

The cross disciplinary approach of chemical and materials engineering is rapidly growing as it applies to the study of educational, scientific and industrial research activities by solving complex chemical problems using computational techniques and statistical methods.

The **Advances in Chemical and Materials Engineering (ACME) Book Series** provides research on the recent advances throughout computational and statistical methods of analysis and modeling. This series brings together collaboration between chemists, engineers, statisticians, and computer scientists and offers a wealth of knowledge and useful tools to academics, practitioners, and professionals through high quality publications.

Coverage

- Thermo-Chemical Treatments
- Multifuncional and Smart Materials
- Statistical methods
- Heat Treatments
- Ductility and Crack-Resistance
- Numerical Techniques
- Toxicity of Materials
- Metallic Alloys
- Nanomaterials
- Industrial Chemistry

IGI Global is currently accepting manuscripts for publication within this series. To submit a proposal for a volume in this series, please contact our Acquisition Editors at Acquisitions@igi-global.com or visit: http://www.igi-global.com/publish/.

The Advances in Chemical and Materials Engineering (ACME) Book Series (ISSN 2327-5448) is published by IGI Global, 701 E. Chocolate Avenue, Hershey, PA 17033-1240, USA, www.igi-global.com. This series is composed of titles available for purchase individually; each title is edited to be contextually exclusive from any other title within the series. For pricing and ordering information please visit http://www.igi-global.com/book-series/advances-chemical-materials-engineering/73687. Postmaster: Send all address changes to above address.

Titles in this Series

For a list of additional titles in this series, please visit: http://www.igi-global.com/book-series/advances-chemical-materials-engineering/73687

Handbook of Research on Advancements in the Processing, Characterization, and Application of Lightweight Materials
Kaushik Kumar (Birla Institute of Technology, India) B. Sridhar Babu (CMR Institute of Technology, India) and J. Paulo Davim (University of Aveiro, Portugal)
Engineering Science Reference • © 2022 • 458pp • H/C (ISBN: 9781799878643) • US $285.00

Emerging Applications and Implementations of Metal-Organic Frameworks
Shimaa Mohamed Elsaeed (Egyptian Petroleum Research Institute, Egypt) Elsayed Zaki (Egyptian Petroleum Research Institute, Egypt) and Abdel-Azim Abdel-Azim (Egyptian Petroleum Research Institute, Egypt)
Engineering Science Reference • © 2021 • 254pp • H/C (ISBN: 9781799847601) • US $225.00

Normal Partitions and Hierarchical Fillings of N-Dimensional Spaces
Gennadiy Vladimirovich Zhizhin (Russian Academy of Natural Sciences, Russia)
Engineering Science Reference • © 2021 • 280pp • H/C (ISBN: 9781799867685) • US $195.00

Applications of Nanomaterials in Agriculture, Food Science, and Medicine
Mohd Amin Bhat (Higher Education Department, Government of Jammu and Kashmir, India) Irshad Ahmad Wani (Government Degree College for Boys, Anantnag, India) and Shah Ashraf (National Institute of Technology Srinagar (NIT Srinagar), India)
Engineering Science Reference • © 2021 • 442pp • H/C (ISBN: 9781799855637) • US $215.00

Technology Development for Adsorptive Heat Energy Converters Emerging Research and Opportunities
Kostyantyn M. Sukhyy (Ukrainian State University of Chemical Engineering, Ukraine) Elena A. Belyanovskaya (Ukrainian State University of Chemical Engineering, Ukraine) and Mikhailo P. Sukhyy (Ukrainian State University of Chemical Engineering, Ukraine)
Engineering Science Reference • © 2021 • 305pp • H/C (ISBN: 9781799844327) • US $195.00

Nanotechnologies and Clusters in the Spaces of Higher Dimension Emerging Research and Opportunities
Gennadiy Vladimirovich Zhizhin (Independent Researcher, Russia)
Engineering Science Reference • © 2021 • 286pp • H/C (ISBN: 9781799837848) • US $195.00

Advanced Surface Coating Techniques for Modern Industrial Applications
Supriyo Roy (Haldia Institute of Technology, India) and Goutam Kumar Bose (Haldia Institute of Technology, India)
Engineering Science Reference • © 2021 • 342pp • H/C (ISBN: 9781799848707) • US $215.00

IGI Global
PUBLISHER of TIMELY KNOWLEDGE

701 East Chocolate Avenue, Hershey, PA 17033, USA
Tel: 717-533-8845 x100 • Fax: 717-533-8661
E-Mail: cust@igi-global.com • www.igi-global.com

List of Contributors

Table of Contents

Detailed Table of Contents

surfactants) to the processing conditions during emulsification. The current options of emulsification methods, processing parameters, and optimization process of nanoemulsion are highlighted and elaborated in this chapter. This allows readers to understand the fundamental principles of nanoemulsion preparation and encourage future studies and applications in the related field.

Chitosan is a natural polysaccharide and emulsifier that can ensure a significant emulsion stability at suitable pH, ionic strength, composition, concentration, or thermal processing. The evaluation of the electrokinetic properties is a key factor in investigation of the stability of the nanoemulsions with a view to their potential applications in bionanotechnology. Consequently, the precise control over the physicochemical characteristics of chitosan (degree of acetylation, DA and molecular weight, Mw) can provide a high stability and specific biological properties of the developed functional structures. The chapter is focused on the interpretation of the electrokinetic response from nanoemulsion stabilized by adsorption of chitosan (as a polyelectrolyte or uncharged polymer) by using appropriate theoretical approaches.

Essential oil, endorsed as a natural cum powerful ingredient, encompasses incredible properties, and various approaches have been proposed and spurred over the last decades. One among them is the nano-emulsification of the essential oils, and the sub-micron size provides prolonged shelf life, ease of formulation and handling, nullification of the stated obstacles, as well as the augmentation of the essential oil antimicrobial activity. This chapter primarily reinforces and is well-articulated on (1) the components and its precise concentration adequate to fabricate nanoemulsion; (2) critical analysis on the experimentally inferred antimicrobial mechanism and the extent of contribution granted by the optimal concentration of the components in the nanoemulsion; (3) scrutinization of the possible impact rendered by the components on the other characteristics of nanoemulsion aside from antimicrobial potency. Altogether, the importance of careful formulation of the essential oil-loaded nanoemulsion is emphasized extensively.

Administration of drugs through the oral route is considered the simplest and most convenient way to offer greater patient compliance than other routes. Most active drugs discovered in the past and those being discovered in recent times are inadequate because of their inherent limitations in physicochemical properties such as low solubility and permeability, resulting in poor bioavailability, especially after oral administration in the form of tablet or capsule. Pharmaceutical nanoemulsion is the most promising, safer, and multimodal technique for delivering poorly soluble drugs and gaining more attention due to its characteristics such as higher solubilisation capacity, smaller size, surface charge, and site-specific drug targeting. This chapter focuses on the biological fate of nanoemulsion after oral administration and a few case studies related to the oral application of nanoemulsion in delivering poorly soluble drugs. In addition, the anatomy and physiology of the GI tract, components of nanoemulsion, and methods of preparation are addressed.

Nanoemulsions are an isotropical mixture of oil, surfactant, and water with droplet diameter approximately in the range of 10-100 nm. They are being exponentially used for drug delivery systems for the influential administration of therapeutical agents because of their potential advantages over other approaches. Nanoemulsions can be used to design delivery systems that have increased drug loading, enhanced drug solubility, increased bioavailability, controlled drug release, and enhanced protection against chemical or enzymatic degradation. Moreover, nanoemulsions have better thermodynamical stability to flocculation, sedimentation, and creaming than conventional emulsions. Their small droplet dimensions and large droplet surface area positively influence drug transport and delivery, along with allowing targeting to specific sites. This chapter focuses on recent applications of nanoemulsions in the area of drug delivery.

Section 2
Antimicrobial Nanoemulsion

An enhancement of antibiotic resistance in bacteria is associated with increased morbidity, mortality, and health infrastructure and hospital care charges. The Infectious Diseases Society of America (IDSA) has highlighted a section of antibiotic resistant bacteria termed as ESKAPE pathogens. These pathogens are proficient in 'escaping' the biocidal effect of antibiotics and mutually representing new paradigms in transmission of diseases, pathogenesis, and resistance in their genetic materials. Essential oil-based nanoemulsions (NEs) have great interest towards the "natural" therapies as potential antimicrobial agents. Thermodynamic properties and kinetically stable potential of biphasic system of nanoemulsion enable them to be used as an effective nano-carrier with controlled release at the targeted point. This chapter describes the mechanisms of ESKAPE pathogens and the mode of the mechanisms of antimicrobial action of nanoemulsions for the treatment of MDR human pathogens.

Tuberculosis is commonly called TB and considered to be the most contagious disease. This disease is caused by a causative agent known as Mycobacterium tuberculosis within the host body cells. Emergence incidence of XDR and MDR of tuberculosis are due to high dose intake and prolonged treatment of antibiotics. In this era, only one vaccine called as BCG is available which is ineffective against infected tuberculosis adults. Though several antibiotics have been produced to overcome drug resistance and even enhance the patient compliance towards treatment and reduce the treatment regimen, we require a novel strategy that can improve the potency of therapeutic synthetic drugs. This approach can be achieved by the application of nanotechnology associated with conventional therapy. Nanotechnology has attributed the promising effects associated with optimized treatment against chronic diseases. This novel technology has the ability to reduce the dose frequency and even resolves the poor patient compliance.

Nanoemulsions are an attractive approach for the delivery of antiviral drugs in the treatment of various viral infections. Nanoemulsions are easy to plan and develop, and their components exhibit high variability. Nanoemulsion system and its components have certain biophysical properties which could increase the efficacy of drug therapy. Pulmonary surfactant (PS)-assisted antiviral drug delivery by nanoemulsion system could be another effective approach for the treatment of COVID-19. Antiviral drug delivery of nebulization using an animation system could increase the efficacy of antiviral drug against COVID-19. Ginkgo biloba polyprenol nanoemulsion was also found to be stable, non-toxic, and had strong antiviral activity against influenza A H3N2 and hepatitis B virus in vitro. Nanoemulsion systems possess certain properties that make their system suitable for drug delivery by mobilization and hence would be promising systems for therapeutics in the future.

Chapter 10
Mushroom Bioactive Compounds: Potential Source for the Development of Antibacterial Nanoemulsion 213
Deepali Koreti, Pandit Ravishankar Shukla University, Raipur, India
Anjali Kosre, Pandit Ravishankar Shukla University, Raipur, India
Ashish Kumar, Sant Gahira Guru Vishwavidyalaya, Sarguja Ambikapur, India
Nagendra Kumar Chandrawanshi, Pandit Ravishankar Shukla University, Raipur, India

Mushrooms are widely utilized as a nutritional and functional food, and they are also appreciated for their medicinal as well as therapeutic applications. Bioactive compounds are isolated from mushrooms possess medicinal properties such as anti-inflammatory, antimicrobial, antitumor, immunomodulatory, and antioxidant. Mushroom bioactive compounds having antibacterial potential and can be used in medical sectors for the cure of several pathogenic disease. Nanoemulsions are one such example of nano-technique that has a very promising application in various fields. The use of nanoemulsion in encapsulation for food-grade ingredients in their compounds such as vitamins, lipids, antioxidants, and antimicrobial agents is a new technique. Bioactive compounds from mushrooms can be a good alternative source for antibacterial nanoemulsions development. This chapter discusses all bioactive compounds and the production and purification of these bioactive compounds. Another hand discusses the nanoemulsion formation by bioactive compounds and nanoemulsion used as antimicrobial agents.

Section 3
Nanoemulsion Applications in Cancer Research

Chapter 11
Application of Nanoemulsion in Cancer Treatment 237
Sumira Malik, Amity University, Ranchi, India
Shristi Kishore, Amity University, Ranchi, India
Manisha Kumari, Amity University, Ranchi, India
Archna Dhasmana, School of Applied and Life sciences, Uttaranchal University, Dehradun, India

Nanoemulsions are pharmaceutical-based nanometres ranged nanoformulated particles with significant and valuable contribution in field of the nanotechnology. In cancer treatment, the treatment through drugs fails primarily due to multidrug resistance (MDR), poor solubility, and unspecific toxicity. Nanoemulsions have the remarkable properties of non-immunogenicity, biodegradability, sustained encapsulation of low water solubility drugs, sustained regulated release of drug, stable and safe carrying tendency to deliver

Nanoemulsion is the major vehicle for delivering different types of drugs, nucleic acids, and imaging agents. Due to their attractive properties, it has been extensively used for diagnostics of cancer therapy and imaging. However, nanoemulsion is designed through multiple functions by modifications in surface and encapsulation of active compounds against cancer. In nanoemulsion, the surface alteration can be changed by targeting the surface charge, a targeting ligand. The core of the emulsion can be loaded with drugs, imaging agents, and contrast agents. In this chapter, the application of nanoemulsion against specifically liver and gastric cancer is explored briefly. The major focuses on the severity of cancer, multifunctional nature of respective drug-loaded nanoemulsions, how to defeat the physiological hurdles, targeted and non-targeted delivery of nanoemulsion, clinical and preclinical studies are discussed with trending examples from the review of the literature and future perspectives.

Oral cancer is one of the most common types of cancer, and lifestyle factors like extensive consumption of tobacco, betel quid, and alcohol are the major etiological factors of oral cancer. Treatment of oral cancer includes surgery, radiation therapy, and chemotherapy, but this treatment possesses lots of side effects. Therefore, scientists and medical experts are utilizing natural products and medicinal plants for new drug development. Natural products and phytochemicals showed better efficacy with less toxicity. However, most of the phytochemicals showed poor permeability and less bioavailability. To combat this problem, scientists developed nanosized nanoemulsions of phytochemicals to treat various ailments. Nanoemulsions of phytochemicals exhibited better efficacy than their free form due to increased permeability and bioavailability. Numerous phytopharmaceuticals have been formulated for nanoemulsions to date and tested for their anticancer potential against various cancers, including oral cancer and oral health management.

Cancer is a class of disorders that is characterized by the abnormal growth of cells in an uncontrolled manner. In cancer progression, tumor cells have become highly heterogeneous, and they create a mixed

agriculture and forestry are necessary. In this chapter, the synthesis, characterization, and application of nanoemulsion in pesticides formulation are reviewed. In addition, the effect of antifungal or insecticidal and antimicrobial enhancement activity of nanoemulsions with size, composition, and stability are also discussed.

Mushrooms belonging to basiodmycetes with their high nutritional value and biologically active compounds of medicinal importance can be developed into potential food products. They have been used as a traditional food, and their medicinal property is also appreciable all over the world. Naturally occurring active compounds such as polysaccharides, proteins, lipids, and glucans, etc. are obtained from various sources including plants, animals, bacteria, algae, and fungi. The efficiency of naturally derived compounds in food industry, as well as factors influencing its effectiveness, has been reported by researchers. Mushrooms produce a diversity of biologically active compounds such as proteoglucans, polysaccharides, phenolic compounds, lectins, steroids β-glucan, chitosan, and terpenoids, etc. The bioactive compounds and their concentration differ from species to species. Thus, these bioactives can be effectively used in the fabrication of fungal (mushroom)-derived nanoemulsions applicable for the food industry.

Nanoemulsions are increasingly used in the food industry for the encapsulation of lipophilic compounds such as nutraceuticals, flavors, vitamins, antioxidants, and colors. The advantages of nanoemulsion-based encapsulation include increased bioavailability, higher solubility, control release, protection from chemical degradation, and incorporation of different ingredients into food products. Food industries are interested in the preparation of natural polymer-based nanoemulsions to prepare clean label products. A variety of polymers such as soya lecithin, WPI, GA, and modified starches have been used to fabricate nanoemulsion-based delivery systems for nutraceuticals and micronutrients. Hence, nanoemulsion-fortified food products have received great attention due to consumer demand for safer and healthier food products. Additionally, nanoemulsions can be used for the texture modification of food products and encapsulation as well as delivery of antimicrobial agents. The main challenge for the preparation of nanoemulsions is the selection of appropriate ingredients.

Attention to nanoemulsions has significantly grown in recent years as a result of their unique features like better stability, special appearance, higher performance, and sensorial merits. Chronic injuries are the consequence of a disturbance in the extremely coordinated cataract of wound healing actions. Nevertheless, correlated with variations in the timescales of various physical methods embroiled in tissue renewal, the aggression of the tumor microenvironment, rich in decaying enzymes, as well as its increased pH, demands the use of efficient drug delivery applications. This chapter summarizes that the various stages of wound healing include four phases: hemostatic stage, inflammation, proliferation, and remodeling process, respectively. Moreover, the major reported classes of lipid-based elements were either vesicular (liposome, permeation increased vesicle, etc.), emulsion-based behavior (nano-emulsion and micro-emulsion), or comprise a solid-based liquid matrix in the wound-healing process.

Vaccination is one of the most effective approaches for the prevention of deadly and highly contagious diseases. One of the important biomedical applications of nanoemulsions is in the process of vaccination. Nanoemulsions are made from nano-sized safe, well-considered ingredients, amalgamated in a protective way to bring forth a stable emulsion. They have provided ways for vaccine delivery using intranasal or mucosal oil-based emulsions rather than using needles. Also, some nanoemulsions have effectively shown anti-pathogenic activities against several germs. Nanoemulsions are also used as vaccine adjuvants and are used to boost the effectiveness of vaccines. Nanoemulsion-based adjuvants put forward the possibility of non-irritating, needle-free vaccines, handed out as nose drops or as a simple nasal sprayer. The chapter aims to discuss the applications of nanoemulsions in the process of vaccination.

The area of healthcare needs new innovative methods and tools for improvisation and to impart better efficiencies. Nanoemulsions are pharmaceutical formulations containing nanometre-sized particles used for controlled and systemic delivery of bioactive pharmaceuticals. Various advantageous properties of nanoemulsions such as the presence of hydrophobic core region, higher stability, and smaller size have made them useful to a large extent in the biomedical field. They have been employed in transdermal drug delivery, intranasal drug delivery, pulmonary drug delivery, parenteral drug delivery, and improvised delivery of hydrophobic drugs. This chapter aims to discuss various applications of nanoemulsions in healthcare including cosmetics, antimicrobials, vaccine delivery, targeted drug delivery, gene delivery, cancer therapy, and many more in detail.

Preface

Having been inspired by the recent developments in nanotechnology, the scientists are interested in the identification of various applications of nanotechnology which include industrial, medicinal, and energy uses. The nanoemulsions are produced by mixing an oil phase with an aqueous phase under shear pressure; this procedure yields uniform populations of oil droplets ranging in diameter from 200 to 800 nm and are kinetically stable colloidal substances with enhanced properties compared to the conventional emulsion substances. Nanoemulsions have broad potential applications in the agriculture, Food, Health, and Biomedical Sciences. This book focuses on the fundamentals of nanoemulsions, such as preparation and characterization, plus numerous types of applications and current advances in nanoemulsions.

In nanoemulsion science, I have experienced and am well trained at the University of Texas Health Science Center at San Antonio, Texas, USA on the different aspects of nanoemulsion preparation, characterization, and application for antimicrobial development and I have used nanoemulsion against battlefield multi-resistant microorganisms to prevent battlefield related infections. In US Naval Medical Research Unit, San Antonio, Texas, USA, I had worked in different nanoemulsion and metal nano-particles against infectious disease causing microorganisms and I have developed antimicrobial nanoemulsion treatment for multi-resistant Acinetobacter baumannii. In India, I have obtained 23 Lakh funding from DST-SERB for the antimicrobial nanoemulsions formulation for the treatment of multi-drug resistant ESKAPE pathogens. Currently from the standardized formulation developing nano-ointment for the treatment of hospital infection-associated multi-drug resistant Klebsiella pneumoniae pathogens through the funds of Indian Council of Medical Research (ICMR), Government of India.

This book will focus on the various permutations of application nanoemulsions and give necessary theoretical knowledge. It will be written for researchers who want to improve their understanding of the basics of nanoemulsion and the wide applications of different nanoemulsion formulations. Also, the importance of the emulsification process, since without an emulsification process product development will be not possible in many fields. Hence this book will have lots of impact and values at the completion. This book is divided into 4 sections comprising 23 chapters overall. This book comprises contributions from authors worldwide- Nigeria, Pakistan, Italy, Brazil, Turkey, Bulgaria, China, Malaysia, and India.

The first section covers the fundamentals of nanoemulsions, such as what are nanoemulsions, their kinds, qualities, benefits, compositions, and so on. In the first chapter, the essential properties of nanoemulsions are discussed. Unique properties of nanoemulsions such as small droplet size, exceptional stability, transparent appearance, and tunable rheology are detailed in this chapter with specific examples.

The second chapter deals with the preparation of nanoemulsion. Its processing parameters and emulsification process are described in it. Nanoemulsions can be prepared by either high-energy or low-energy methods. Its size is dependent on its constituents, operating conditions, and preparatory methods. The emulsification process involves the break up of droplets into smaller ones, adsorption of surfactants, and collision of droplets.

Chapter 3 deals with chitosan stabilized oil-in-water nanoemulsions. A detailed note about these nanoemulsions is discussed along with their electrokinetic properties. The physicochemical properties of chitosan particles and O/W emulsions produced at the same time from different process conditions are discussed here in detail.

In Chapter 4, essential oil nanoemulsions are described. A detailed note on its composition along with its efficacy rate is also explained. Encapsulation of essential oils in nanoemulsions has been proved as a new method for enhancing their efficacy, stability, and utilization.

Chapter 5 deals with improved oral delivery of drugs using nanoemulsions. The nanosized droplets have led to the considerable attraction for this formulation, for the delivery of hydrophilic as well as hydrophobic drugs as drug carriers due to their improved drug solubilization capacity, long shelf life, ease of preparation, and improvement of bioavailability of drugs.

Chapter 6 provides an overview of current perspectives on nanoemulsions in targeted drug delivery. Nanoemulsion formulation offers several advantages such as delivery of drugs, biological or diagnostic agents. Nowadays, nanoemulsions are used for targeted drug delivery of various anticancer drugs, photosensitizers, or therapeutic agents. Nanoemulsion can also provide prolonged action of the medicaments.

Section 2 describes the various antimicrobial properties and applications of nanoemulsions. Chapter 7 explains the antimicrobial mechanisms of nanoemulsion against drug-resistant ESKAPE pathogens. The ESKAPE pathogens (Enterococcus faecium, Staphylococcus aureus, Klebsiella pneumoniae, Acinetobacter baumannii, Pseudomonas aeruginosa, and Enterobacter species) are the leading cause of nosocomial infections throughout the world. Most of them are multidrug-resistant isolates, which is one of the greatest challenges in clinical practice. Application of nanoemulsions against these ESKAPE pathogens would be a boon in medicine.

In Chapter 8, applications of nanoemulsion in tuberculosis treatment are described. Nanoemulsions are considered as one of the promising alternatives for augmenting the bioavailability of the antitubercular drugs through the oral route for improving their therapeutic efficacy.

Chapter 9 deals with antiviral drug therapy using nanoemulsions. Its insights and applications are discussed in detail. Nanoemulsions have significant potential to target microorganisms, especially viruses. They act as a vehicle for delivering antiviral drugs and vaccines. The emergence of newer viral illnesses, particularly coronavirus disease, has created a niche for using NEs to augment currently existing treatment choices because of the constriction of the drug discovery pipeline and the introduction of new viral diseases.

Chapter 10 discusses mushrooms' bioactive compounds as a potential source for the development of antibacterial nanoemulsion. Mushroom bioactive compounds have antibacterial potential and can be used in medical sectors for the cure of several pathogenic diseases. The bioactive compounds of mushrooms and the production, purification of these bioactive compounds are explained in detail here.

Section 3 deals with the application of nanoemulsions in cancer treatment. Chapter 11 elucidates the various options by which nanoemulsions can be used for anti-cancer therapy due to their remarkable properties of non-immunogenicity, biodegradability, sustained encapsulation of low water solubility drugs, sustained regulated release of the drug, stable and safe carrying tendency to deliver drugs. It highlights the current status and applications of nanoemulsions as anti-cancer therapeutics and their commercial importance.

Chapter 12 deals with applications of nanoemulsion in prostate cancer treatment. The importance of nanoemulsions as potential drug delivery mechanisms in prostate cancer therapy is discussed. This chapter focuses on reviewing different ways by which nanoemulsion can bring the solution to water-solubility problems and also target specific cancer cells.

In Chapter 13, applications of nanoemulsion in breast cancer treatment are explained. A summary of different NEs and their documented applications in cancer therapeutics, with emphasis on breast cancer, is presented in this chapter.

Chapter 14 deals with applications of nanoemulsion in the liver and gastric cancer. The major focus is on the severity of both the cancers, multifunctional nature of respective drug-loaded nanoemulsions, how to defeat the physiological hurdles, targeted and non-targeted delivery of nanoemulsion, clinical and preclinical studies, discussed with trending examples from the review of the literature and future perspectives.

Chapter 15 explains the applications of phytochemical nanoemulsions in oral cancer and oral health. Nanoemulsions of phytochemicals exhibited better efficacy than their free form due to increased permeability and bioavailability. Numerous phytopharmaceuticals have been formulated for nanoemulsions to date and tested for their anticancer potential against various cancers, including oral cancer and oral health management.

Chapter 16 deals with mushroom-based bioactive compounds nanoemulsion used for cancer therapy. The chapter focuses on mushroom-derived bioactive compounds and possible implications in nanotechnology, and further will be utilized for new advanced nanoemulsion techniques for the promising treatment of cancer.

Section 4 gives an account of various other applications of nanoemulsions. Chapter 17 explains nanoemulsions in the delivery of anti-hypertensive drugs. This chapter highlights various aspects of hypertension including its pathophysiology and potential approaches to combat high blood pressure. In addition, it discusses nanoemulsions and their utility in the oral delivery of anti-hypertensive drugs.

Chapter 18 deals with nanoemulsion in the pesticide formulation. In this chapter, the synthesis, characterization, and application of nanoemulsion in pesticides formulation are reviewed. In addition, the effect of antifungal or insecticidal and antimicrobial enhancement activity of nanoemulsions with size, composition, and stability are also discussed.

Chapter 19 explains Nanoemulsion based on Mushroom Bioactive compounds and their Application in Food Preservation. Naturally occurring active compounds such as polysaccharides, proteins, lipids, and glucans, etc. are obtained from various sources including plants, animals, bacteria, algae, and fungi. The efficiency of naturally derived compounds in the food industry, as well as factors influencing their effectiveness, is reported in this chapter.

Chapter 20 deals with Applications of Nanoemulsions in Food Manufacturing. Food industries are interested in the preparation of natural polymers-based nanoemulsions to prepare clean label products. A variety of polymers such as soya lecithin, WPI, GA, and modified starches have been used to fabricate nanoemulsions-based delivery systems for nutraceuticals and micronutrients. Hence, nanoemulsions fortified food products have received great attention due to consumers' demand for safer and healthier food products.

Chapter 21 describes Applications of Nano-emulsion for the Wound Healing Process. This chapter summarizes the various stages of wound healing includes four phases: hemostatic stage, inflammation, proliferation, and remodeling process, respectively. Moreover, the major reported classes of lipid-based elements were either vesicular (liposome, permeation increased vesicle, etc.) along with emulsion-based behavior (nano-emulsion and micro-emulsion) or comprise a solid-based liquid matrix in the wound healing process were also discussed.

In Chapter 22, applications of nanoemulsions in vaccination are described. Nanoemulsions are also used as vaccine adjuvants and are used to boost the effectiveness of vaccines. Nanoemulsion-based adjuvants put forward the possibility of non-irritating, needle-free vaccines, handed out as nose drops or as a simple nasal sprayer. The current chapter aims to discuss the applications of nanoemulsions in the process of vaccination.

Chapter 23 discusses the Healthcare application of nanoemulsions. This chapter aims to discuss various applications of nanoemulsions in healthcare including cosmetics, antimicrobials, vaccine delivery, targeted drug delivery, gene delivery, cancer therapy, and many more in detail.

The rapid emergence of emerging illnesses, the possible use of multidrug-resistant pathogens in bioweapons, and a scarcity of new antimicrobials from global pharmaceutical corporations have all prompted concerns about the creation of new antimicrobials with various target mechanisms using nanotechnology.

The book target audience of this book will be composed of students, teachers, scientists, and technical professionals working in the field of nanoemulsion science, agriculture, food, drug delivery, health, cosmetics, pharmaceuticals, materials technology, and other allied biomedical sciences. Moreover, the book will provide insights and support researchers concerned with the nanoemulsion sciences, and also this book will kindle new concepts for more effective nanoemulsion formulation for an application.

I strongly believe that this book would be a useful guide to students and researchers working in the field of nanoemulsion. The comprehensive chapter resources would greatly advance the reader's knowledge regarding the vast variety, properties, and applications of nanoemulsions.

Section 1
Basics of Nanoemulsion

Chapter 1
The Essential Properties of Nanoemulsions:
Basics of Nanoemulsion

Mohamed Akif S.
B.S. Abdur Rahman Crescent Institute of Science and Technology, India

Jahangir Ahmed
https://orcid.org/0000-0002-1380-939X
B.S. Abdur Rahman Crescent Institute of Science and Technology, India

Karthikeyan Ramalingam
https://orcid.org/0000-0002-9334-427X
B.S. Abdur Rahman Crescent Institute of Science and Technology, India

ABSTRACT

Nanoemulsions are a submicron with colloidal particulate systems ranging from 10 to 1,000 nm in size. Nanoemulsions hold enormous scope in the field of cosmetics, diagnostics, food, and paint. Moreover, nanoemulsions are ubiquitously regarded as superior drug carriers for the infusion of lipophilic cytotoxic antineoplastic agents on a particular target criterion. Nanoemulsions are prepared from two immiscible liquids that are mixed by employing surfactants and co-surfactants. It also encompasses some significant benefits like biocompatibility, non-immunogenicity, low toxicity, drug entrapment, nanoscale size, large surface area, long-term and restrained release, uncomplicated mode of formulation, as well as thermodynamic stability. Nanoemulsion drug delivery can address the major challenge of effective drug formulation due to its instability and poor solubility in the vehicle. The primary objective of this chapter is to provide a quick overview of various physico-properties of nanoemulsion, with a special emphasis on its various applications in various fields.

DOI: 10.4018/978-1-7998-8378-4.ch001

INTRODUCTION

Nanoemulsions (NE) are heterogeneous submicron-sized mixtures of lipid and aqueous phases stabilised by emulsifying agents which are also recognised as ultrafine emulsions (Khaleel *et al.*, 2020; Sharma *et al.*, 2010; Gurpret *et al.*, 2018; Patel *et al.*, 2012; Gupta *et al.*, 2016; Eral *et al.*, 2014; Helgeson *et al.*, 2012; Wu *et al.*, 2013; An *et al.*, 2013; Chen *et al.*, 2011; Khaled *et al.*, 2016; Trujillo*et al.*, 2016; Pathak*et al.*, 2017; Zhao *et al.*, 2021; Aswathanarayan *et al.*, 2019; Zhang *et al.*, 2018; Hamed *et al.*, 2015; Ganta *et al.*, 2014; Bainun *et al.*, 2015; Jaiswal *et al.*, 2015; Pandey *et al.*, 2018; Zhang *et al.*, 2013; Zhang *et al.*, 2011; Aboofazeli *et al.*, 2010; Solans *et al.*, 2005; Burapapadh *et al.*, 2010). Nanoemulsions are formulated when two immiscible liquids have been emulsified with emulsifiers, leading to the formation of colloidal dispersive systems which are thermodynamically stable (Helgeson*et al.*, 2016; Jaiswal*et al.*, 2015; Sarker *et al.*, 2005) and are fine oil-in-water dispersions with droplets ranging in diameter from 100 to 600 nm (Sharma *et al.*, 2010; Bouchemal*et al.*, 2004; Pagar *et al.*, 2019; Sutradhar*et al.*, 2013). Nanoemulsions are dispersed particle collections used in pharmaceuticals, biomedical aids, and vehicles with promising contribution in cosmetics, diagnostics, drug therapies, and biotechnologies (Gurpret *et al.*, 2018; Pagar *et al.*, 2019; Sutradhar*et al.*, 2013; Sivakumar *et al.*, 2014; Sharma*et al.*, 2012; Sonneville*et al.*, 2018; Khanna*et al.*, 2018) and are furthermore renowned as multiphase colloidal dispersion, which is perceived by its stability and clarity (Sheth*et al.*, 2020 ; Calderó*et al.*, 2016; Demisli*et al.*, 2020; Ren*et al.*, 2018; Pavoni *et al.*, 2020; Rehman *et al.*, 2017; Nam*et al.*, 2012; Streck *et al.*, 2016; Lococo*et al.*, 2012; Cheng *et al.*, 2020). Gupta et al. (2016) revealed that nanoemulsions have very high kinetic stability (Sarheed*et al.*, 2020; Liu*et al.*, 2019; Zhang *et al.*, 2014; Sadeghpour*et al.*, 2015; Barres*et al.*, 2017; Hashtjin *et al.*, 2015; Yu *et al.*, 2012). As a consequence, nanoemulsions can be kinetically stable for longer periods of time. The interfacial tension between oil and water appears to be exceptionally low in nanoemulsions, and they appear to be transparent due to droplet sizes that are less than 25% of the wavelength of visible light (Gupta *et al.*, 2016; Gaba*et al.*, 2019; Namratha*et al.*, 2021). It forms quickly and, in some cases, spontaneously, with little or no high-energy input. The research work focused on the formulation of nanoemulsions utilizing various approaches, which are broadly categorised into two essential classes: high-energy techniques and low-energy techniques (Gupta *et al.*, 2016; Kumar*et al.*, 2019; Kotta*et al.*, 2013; McClements*et al.*, 2013]. Low energy approaches to nanoemulsion development, on the other hand, entail Phase Inversion Temperature (PIT) and Emulsion Inversion Point (EIP) (Liu*et al.*, 2019; Kumar*et al.*, 2019; Salem*et al.*, 2019; Choradiya*et al.*, 2021; Mishra*et al.*, 2018). High-energy approaches like High-Pressure Homogenization (HPH) and ultrasonication requisites ginormous energy to yield small droplets (Khaleel *et al.*, 2020; Calligaris*et al.*, 2016). In addition to the surfactant, the oil phase, and the water phase, a co-surfactant or co-solvent is prevalently used (Kumar*et al.*, 2019). Emulsion droplet coalescence or coagulation is effectively repressed due to the smaller droplet size of emulsions. Additionally, it also facilitates the suppression of emulsion precipitation and also helps to deliver the active agents (Patel *et al.*, 2012). Eral *et al.* (2014), Gupta *et al.* (2016) and Li *et al.* (2021) had similar thoughts on illustrating that many of the problems encountered in prevailing pharmaceutical crystallisation practises could be eliminated by using nanoemulsions. Droplet size range and stability are the predominant significant variations among conventional emulsions (or macroemulsions), nanoemulsions, and microemulsions (Gupta *et al.*, 2016; Mcclements*et al.*, 2012). A research conducted by Helgeson *et al.*, (2012) reported the utilization of nanoemulsions to be building blocks for the conglomeratisation of complex substances, including compartmentalised nanoparticles as well as oil droplets capsidated in a protein shell which was in accordance with (Wu *et al.*, 2013) and (An *et* *al.,* 2013). Macroemulsions, as contrasted to nanoemulsions, are thermodynamically stable, equilibrium systems that are temperature and composition sensitive (Gupta*et al.,* 2016). Since nanoemulsions are considered to be less vulnerable to physicochemical changes than conventional emulsions, they are a desirable tool for the aforementioned applications (Gupta *et al.,* 2016; Azeem*et al.,* 2009). Owing to the convergence of polymer and perhaps even electrostatic layers encircling the small nanoemulsion droplets, gel-like attributes and possibly high viscosity can indeed be incorporated at abated droplet concentrations (Chen *et al.,* 2011; Alliod*et al.,* 2019; Fathordoobady*et al.,* 2021). Nanoemulsions could be engineered to improve the biological accessibility of bioactive molecules that are either entombed within them or maybe even consumed alongside them (Khaled *et al.,* 2016; Trujillo*et al.,* 2016). This chapter presents an analysis of the chemical and physical properties of nanoemulsions, with an emphasis on rheological attributes and its significance. Furthermore, the correlation among both the structure and function of nanoemulsions is succinctly summarized.

Physicochemical Characteristics of Nanoemulsions in General

Nanoemulsions were universally acknowledged as colloidal dispersions. It embodies an immiscible liquid in which another fluid is evenly disseminated as tiny droplets (Demisli*et al.,* 2020; Streck*et al.,* 2016; Sánchez*et al.,* 2019; Mountain*et al.,* 2014). The two immiscible liquids implicated throughout most industrial applications are oil and water (Ashaolu *et al.,* 2021). A nanoemulsion can be broadly categorized as oil in water (O/W) as well as water in oil (W/O) based on the relative spatial organisation of the oil and water phases. In figure 1, nanoemulsion is created by gently combining water and oil together with an emulsifier at a specific temperature. Both kinds of nanoemulsions are thermodynamically stable and being exploited for various industrial applications (Bernardi*et al.,* 2021; Niknam*et al.,* 2011). The composition, concentration, size, and interfacial properties of the droplets in nanoemulsions ascertain their physical and chemical characteristics. As a result, there appears to be a significant increase in research into the relationship between macroscopic and microscopic attributes of nanoemulsions.

Figure 1. Nanoemulsions are the thermodynamically stable isotropic system in which two immiscible liquids are mixed to form a single phase by help of emulsifying agent (surfactants)

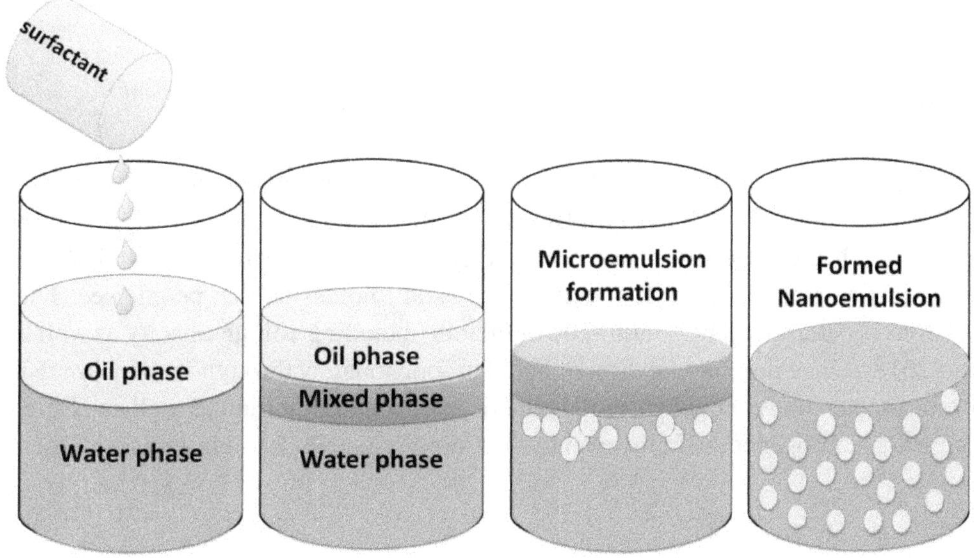

SIGNIFICANCE OF PHYSICAL AND CHEMICAL CHARACTERISTICS

Permanence

The expression "stability," according to Pathak et al., (2017), alludes to a nanoemulsions potential to withstand alterations in its physical and chemical attributes over time. Since the free energy of both the segregated oil and water phases is significantly smaller than those of the emulsified ones, nanoemulsions appear to be thermodynamically unstable, because some nanoemulsions need significant energy to be formed. As a result, given enough time, they always tend to degrade, with respect to the rate of modification evaluated by the elevation of the system's kinetic energy impediments and true nanoemulsion represent in figure 2. Nanoemulsions degrade through a variety of mechanisms, including coagulation, coalescence as well as gravitational estrangement. Owing to the physicochemical stability of nanoemulsions, many commercial implementations necessitate products to be hinged on nanoemulsion under specific circumstances during fabrication, interpretation, transportation, and utilization. Nanoemulsion, on the contrary, may need to degrade under different set of environmental conditions (Zhao *et al.,* 2021).

Figure 2. Nanoemulsions Appealing Principles

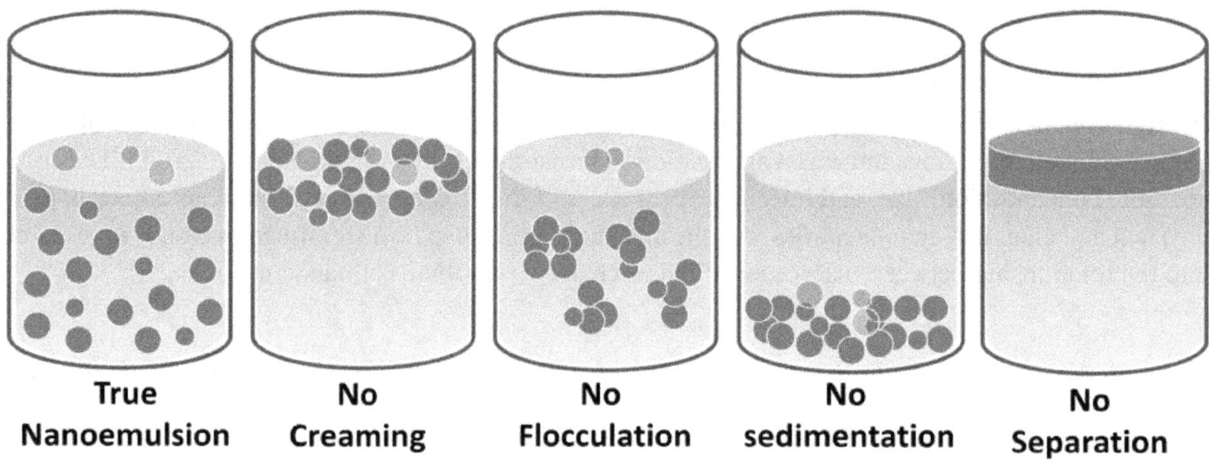

Demeanour

As said by Aswathanarayan and Vittal (2019), the optical properties of nanoemulsions are critical for their use in the food industry, and nanoemulsions are either optically clear or slightly turbid, depending on droplet size. The demeanour of the commodity based on nanoemulsion is vital in determining customer gratification since it provides the primary visual impression that people see. The semblance of nanoemulsions is determined by a multitude of factors, including colour, opacity as well as similitude (Zhang *et al.,* 2018). Once droplets are disseminated proportionately throughout nanoemulsion, it shows up to be homogeneous; however, when substantial droplet conglomeration as well as oiling off tends to happen, it appears heterogeneous. In the event of extreme scattering of light impact of the oil droplets, conventional emulsions possess a cloudy and perhaps even opaque appearance (McClements *et al.,* 2011). Nanoemulsions, on the other hand, can indeed be clear or cloudy, which depends primarily on the droplet size in relation to the wavelength of light. The size, number, and droplets' refractive index in a nanoemulsion determine the extent of light scattering. The palette of nanoemulsions has always been determined by the existence of any chromophores that intentionally absorb light in the visible region of the electromagnetic spectrum. This section of the study seeks to identify the factors that have a significant impact on the appearance of nanoemulsions, which are then thoroughly discussed.

Rheology

A study conducted by Jaworska *et al.* (2015) revealed the rheological parameters of nanoemulsions which were hinged on three sorts of oil phases namely, Crodamol GTCC, Crodamol PC, and oleic acid. For a myriad of purposes, understanding the rheology of nanoemulsions is prerequisite. The efficacy of droplet disruption in a homogenizer is primarily determined by the viscidity of the segregated water and oil phases, as well as the physicochemical characteristic of the processed nanoemulsion. Then, the shelf life of nanoemulsion commodities is influenced by the rheological characteristic of the discrete phases; for example, as the aqueous phase viscosity increases, the macerating rate of O/W nanoemulsions greatly diminishes. Consequently, the flow of a nanoemulsion has a significant impact on the prototype and implementation of several critical production methods, such as stirring within a container, flowing via tube, passing via heat exchanger, and maybe even processing into industrial output receptacle. Finally, the rheology of nanoemulsion-based products affects some of their sensory properties, such as creamy texture, width, pourability, and presumably perhaps streamability. As a result, nanoemulsion-based products should be designed and manufactured in such a way that they have the desired rheological properties for a specific application (Hamed *et al.,* 2015).

Release Attributes

Encapsulating compounds that have an affinity for water, do not have an affinity for water, or may even be amphiphilic biologically active compounds agents such as flavours and colours, is a crucially significant industrial application of nanoemulsions (Choi *et al.,* 2020; Das*et al.,* 2020). It is necessary to recognize the retention and release properties of biologically dynamic components from delivery systems which are based on nanoemulsion (Odriozola *et al.,* 2014). Nanoemulsions can be developed to keep biologically active compounds in an eatable while controlling their discharge once those that interacted with distinctive environmental conditions (Liu*et al.,* 2019).

CORRELATIONS AMONG BOTH STRUCTURE AND FUNCTION

The physical and chemical characteristics of goods produced from nanoemulsions are mainly governed by the droplets they comprise (Choi *et al.,* 2020; Li*et al.,* 2018). As a consequence, this portion describes the various ways wherein droplets within a nanoemulsion could perhaps vary.

Composition of the Droplets

Nanoemulsions can have different droplet compositions depending on the ingredients used in their production. The droplets of O/W nanoemulsion are composed of a hydrophobic core encased in a shell of surface-dynamic components. As a matter of fact, the components of nanoemulsion droplets could be modified by fabricating them with various hydrophobic and surface-active materials. The viscosities, phase behaviour of hydrophobic materials, densities, and polarities of nanoemulsions vary considerably, which has a significant impact on and functional attributes of nanoemulsions. Owing to the obvious difference in droplet dimensions, the comprehensive make-up of the droplets in a nanoemulsion is usually distinct from those of a standard emulsion made with alike additives (Choi *et al.,* 2020). The surface of the shell thickness in traditional emulsions is much infinitesimal than the radius of the hydrophobic core which is in juxtaposition to nanoemulsions, which has the surface of the shell thickness to be without prejudice analogous to the aquaphobic core (Tadros *et al.,* 2004). Changes in droplet composition could have an effect on the physicochemical properties of nanoemulsions (Zhang*et al.,* 2018). The values corresponding to mass to volume counterpoint between the oil and aqueous phases, for illustration, recognise the flocculation steadiness of droplets in a nanoemulsion, whilst the refractive index comparison among these phases assesses the looks (McClements*et al.,* 2004). Changes in droplet composition can also affect the release properties of nanoemulsions [Choi *et al.,* 2020).

Concentration of the Droplets

Droplet agglomeration in a nanoemulsion can be articulated using the numeral, mass and area of the surface in nanoemulsion (McClements *et al.,* 2011). The volume fraction (ϕ) undoubtedly the most popular means of describing droplet concentration which encompasses droplet quantity for every unit of volume with nanoemulsion. Modifications in the preliminary quantity of liquid phase used to concoct nanoemulsions, and perhaps even trying to dilute them after functionalization, could indeed significantly change droplet concentration (Choi*et al.,* 2020). In case of conventional emulsions, the impactful volume ratio of the droplets is comparable to the volume ratio of the oil phase. Due to the perceived presence of the shell layer, the effective volume ratio of oil droplets in nanoemulsions could be significantly greater than the liquid phase (Zhang*et al.,* 2018).

Size of the Droplets

Nanoemulsions frequently come with a variety of round droplets which come up with various ranges in diameter; as a result, their sizes may be interpreted by a PSD, which exemplifies the significant portion of droplets that fall into one of several dimensions (Sadeghpour*et al.,* 2015). The particle size distribution of a nanoemulsion is primarily regulated by changing the structure constitution, homogenization of the nanoemulsion prerequisites as well as additives characteristics. In high-energy homogenization methods, for illustration, the input energy potency and time interval, the nature and amount of the emulsifying agent, as well as the O/W interfacial tension, and the oil and water-phase viscosities seem to have a consequence on the PSD. Furthermore, Temperature vs. time interval heritage, stirring velocities, surfactant O/W proportion, nature of the surfactant, as well as influence of ionic potential, all directly impact the PSD of low-energy homogenization methodology. Based on the mode of preparation, distinctive interventions are designed to enhance the PSD of a nanoemulsion. Distinctive emulsifier types produce shell veneer of various diameters; consequently persuasive particle diameter could even vary considerably on the nature of the emulsifying agent. Likewise, the adsorptive tendency of additional things on the particle interfaces, such as biologically charged polymers and possibly supplementary charge substances that can attach onto the interfaces of counter charged with oil droplets coated with emulsifying agents, can indeed boost the diameter of the shell veneer (Guzey *et al.,* 2006).

Charge of the Droplets

Particles in nanoemulsions customarily have an electric veneer capability due to the attachment of charged components such as biodegradable polymers and possibly even ionic emulsifying agents (Zhang*et al.,* 2018). Surface potential has been essential in assessing nanoemulsion physical and chemical characteristics such as excipient interrelations, thermal stability, agglomeration consistency, and exterior attachment attributes to the surface. According to McClements (2010), potential is being used to delineate the electrostatic attributes of nanoemulsion particles (McClements*et al.,* 2010). Utilising cationic, non-ionic, and perhaps even anionic emulsifiers, the potency of nanoemulsion drops can recurrently be made constructive, unbiased, or even disavowal (Wu *et al.,*2019). Aside from that, the attachment of certain other electrified materials, which includes some group of polysaccharides, can significantly change the electric properties of emulsifier-coated droplets (McClements*et al.,* 2010).

Physical State of the Droplets

Numerous O/W nanoemulsions encompass absolutely liquid oil globules; consequently, by manipulating and depending on the oil phase heat capacity, completely or moderately crystalline oil globules could perhaps establish (Aswathanarayan*et al.,*2019; Weiss *et al.,*2008; Tan *et al.,* 2021) demonstrated that a nanoemulsion containing fluid globules can indeed be formed at temperatures beyond the melting point of the aqueous environment and then cooled below the melting point to facilitate droplet crystallisation (Tan*et al.,* 2021). Employing nanoemulsions as a template, two kinds of lipid nanoparticles could be formulated: solid lipid nanoparticles (SLNs) and nanostructured lipid carriers (NLCs) and furthermore, based on a study done by Gonullu et al. (2015) and Naseri et al. (2015) revealed that at various temperatures, SLN, NLC, and NE were found to be physically stable and in addition to it was being ascertained that nanoformulations increased drug penetration significantly and can indeed be considered for the relief of traumatic and inflammatory skin conditions (Naseri*et al.,* 2015; Gonullu*et al.,* 2015). Several experimental data have unveiled that SLNs are absolutely crystallised oil droplets with a crystalline framework that is immensely organised, whilst nanostructured lipid carriers are partially crystallised oil globules with a much more disordered solid structure (Jores *et al.,* 2005; Khater *et al.,* 2021; Qian *et al.,* 2012; Uner *et al.,*2005).

RHEOLOGICAL ATTRIBUTES

Rheology attribute is the systematic study of matter's deformation and flow, and it's a core component of many nanoemulsion-based materials (Brito-de *et al.,*2017; Vlachopoulos*et al.,* 2011). Consequently, the flow of a nanoemulsion has a significant impact on the prototype and implementation of several critical production methods, such as stirring within a container, flowing via tube, passing via heat exchanger and maybe even processing into industrial output receptacle (Goyal*et al.,* 2015). It also uncovers its sensory attributes (Choi *et al.,*2020). Bjorn et al. (2012) scrutinised the rheological behaviour of a nanoemulsion by establishing the link among consequent shear strain rate and application of shear stress. The apparent shear viscosity and perhaps even dynamic shear modulus (G) are amongst the most common parameters evaluated on nanoemulsions, but some numerous different specifications may be even more pertinent for some specific functions (Aswathanarayan *et al.,* 2019). Assessments of the rheological attributes of nanoemulsions can expose essential aspects about the structural organisation as well as interactions of the droplets contained within them (Erramreddy*et al.,* 2017). The particle density, droplet interrelations, as well as prolonged phase viscosity are perhaps the most significant factors influencing the rheology of nanoemulsions (Maphosa*et al.,* 2018).

Diluted Nanoemulsion

In a dilute nanoemulsion, the stream behaviour around its solitary drop of fluid seems to have no impact on the flow profile around another droplet. The subsequent formula could be used to interpolate the nature of its viscosity on a dilute nanoemulsion, which had been extricated by Albert Einstein (Hunter*et al.,* 2001).

In this, is the nanoemulsion viscosity, seems to be the continuous phase viscosity, and is thus the dispersed phase volume fraction. The extrapolation of this equation makes several conjecturations, with the exception of the assertion that the continuous phase is an absolutely perfect liquid. Amongst assertions made above, the Einstein equation predicts the dilute nanoemulsions nature of viscosity composed of particles that may or may not have a firm interaction among each other. Nonetheless, for nanoemulsions comprising tiny particles with significant attractive or repulsive interrelations, there could be significant differences among model in figure 3 and empirical results (McClement *et al.,*2004). Precipitated nanoemulsions tend to possess an elevated viscosity unlike non-precipitated nanoemulsions, since they retain some water between the tiny droplets, consequently enhancing the colloidal particle shape.

Figure 3. Illustrates the reliance of colloidal dispersions' viscosity ratio on the disperse-phase volume fraction, which gives a comprehensive report on the Einstein model and effective medium model.

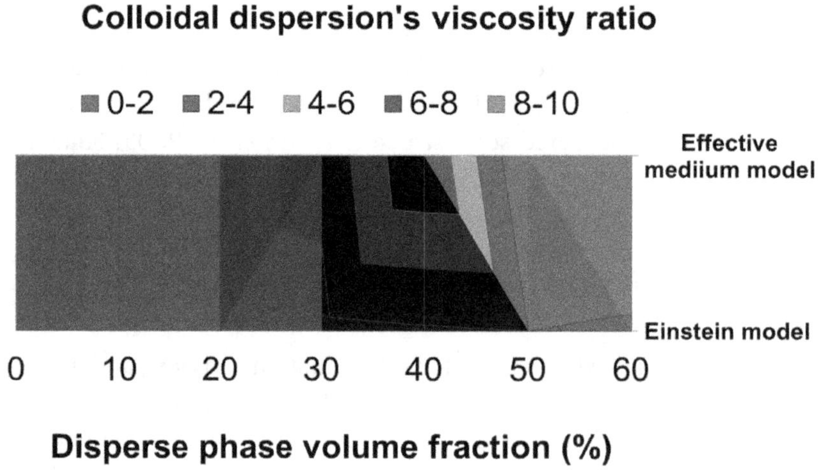

Concentrated Nanoemulsion

Only as the agglomeration of tiny nano droplets in an emulsion increases, will these droplets begin to interact with one another via hydrodynamic interactions, which have a substantial impact on shear viscosity (Helgeson *et al.,*2016; Tadros *et al.,*2011). Based on the constitution of the tiny nano droplet colloidal interactions, the shear viscosity is likely to be significantly higher than anticipated by the Einstein equation under such conditions. The humongous majority of real food emulsions are semi-dilute, if not concentrated, systems with remarkable droplet-droplet interactions. To put it another way, the perturbation of the fluid flow exacerbated by one droplet has a direct effect on the fluid flow around another droplet. The disperse phase volume fraction of colloidal dispersions had been strongly correlated with their rheological characteristics. In the event of emulsions, at large volume fractions, the fluid droplets could become disordered, attributing to the formation of a hexagonal close packed framework wherein the dispersed phase may inhabits literally the entire space, asserting that (ϕ) is close to unity (Zhu*et al.,* 2020).

Absence of Long-Range Colloidal Interactions

The shear viscosity of a nanoemulsion in its truancy is dominated by flow of fluid's interactions fraternized mostly with the stream of fluid around the tiny globules. The mere existence of several other droplets in concentrated nanoemulsions could perhaps significantly influence the fluid flow around one droplet, leading to increased degree of energy dissipation. In such systems, the aphorism below could be utilized to interpret the nature of viscosity of the tiny droplets in the emulsion over a wide range of droplet concentrations.

In this scenario, is a critical packing parameter related to the volume fraction at which the droplets become close-packed. This equation anticipates that as droplet agglomeration rises more towards the point at which globules tend to pack in a firm manner, the viscosity of a nanoemulsion will increase dramatically (Jafari*et al.,* 2018). Owing to the shift in the perceived importance of hydrodynamic impacts as well as Brownian motion, the viscosity of agglomerated nano-sized globules of the emulsion frequently exhibits remarkable decrease in viscosity under shear strain (Derkach*et al.,* 2009; Pal *et al.,*2016). At low shear stresses, Brownian motion endorses an arbitrary pattern of globules throughout the framework, culminating in the nano-sized globules of the emulsion with a significantly higher friction viscosity (McClements*et al.,* 2004; Larson *et al.,* 1999). Thence, fluid flow resistance may indeed be diminished, culminating in less energy dissipation as well as a relatively low nanoemulsion viscosity represent in figure 4.

Figure 4. Illustrates the apparent velocity's reliance on shear stress, which provides a systematic study on Brownian motion and hydrodynamic forces.

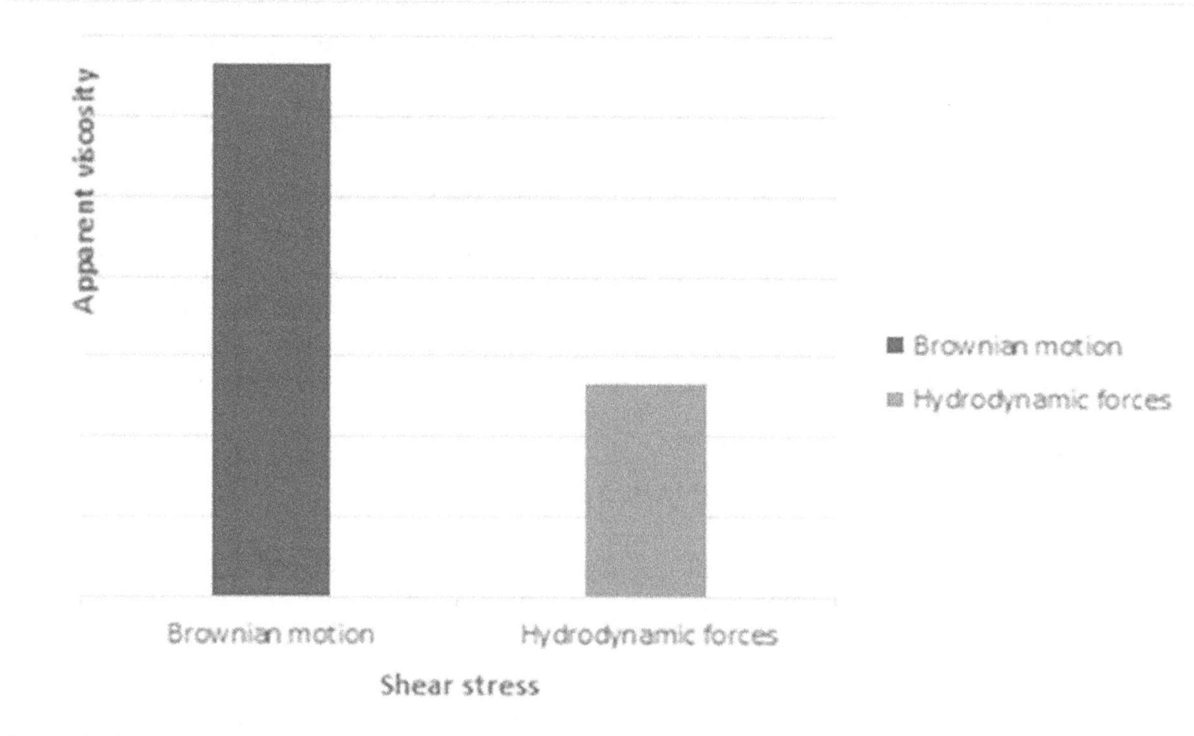

Repellent Interactions

Because of the predominance of an electrically charged and perhaps even thick interfacial layer, nanoemulsions typically contain droplets with relatively powerful electrostatic and steric repulsion between them (McClementset *al.,* 2011). Such abhorrent interactions retain the droplets from getting too close to each other when they come into contact, promoting their efficacious volume fraction (Tadros T. F. 2011).

In this case, roughly one-half of the interspace between the proximate separations of the two globules can be presumed to be an appropriate and efficient "shell thickness." The width of the interfacial veneer for globules retained together by steric aversion can be presumed to be the measure of delta. The relative magnitudes of shell thickness or even just droplet dimensions determine the influence of repugnant interactions on nanoemulsion rheology (Erramreddy*et al.,*2014). Under circumstances where the shell thickness relatively small than that of the droplet radius (r), including high ionic strengths and perhaps even humongous droplets the influence is virtually nil. In contra-distinction, the repercussions could be noticeable for frameworks where the width of the shell is analogous to the radius of the globule. The potential impact of revolting interrelationships in the nano-sized globules of nanoemulsion, as illustrated by Weiss and McClements (2000), is exemplified in figure. 5, which portrays the reliance of shear modulus as well as phase angle on droplet size. Sodium dodecyl-sulphate was used to stabilise the droplets in this system, resulting in a massive negative charge (Vatanparast*et al.,* 2018). Meng *et al.* (2020) demonstrated that larger initial droplet dimensions are detrimental to nanoemulsion stabilisation, and that a higher bending modulus may impede droplet size. This type of behaviour is especially noticeable in nano-sized globules of emulsions stabilised by surfactants that carries an ionic group, which then develop thick veneers.

Figure 5. Presents the Droplet radius's interconnection on Complex shear modulus that further enables for a scientific assessment of divergent globule radius.

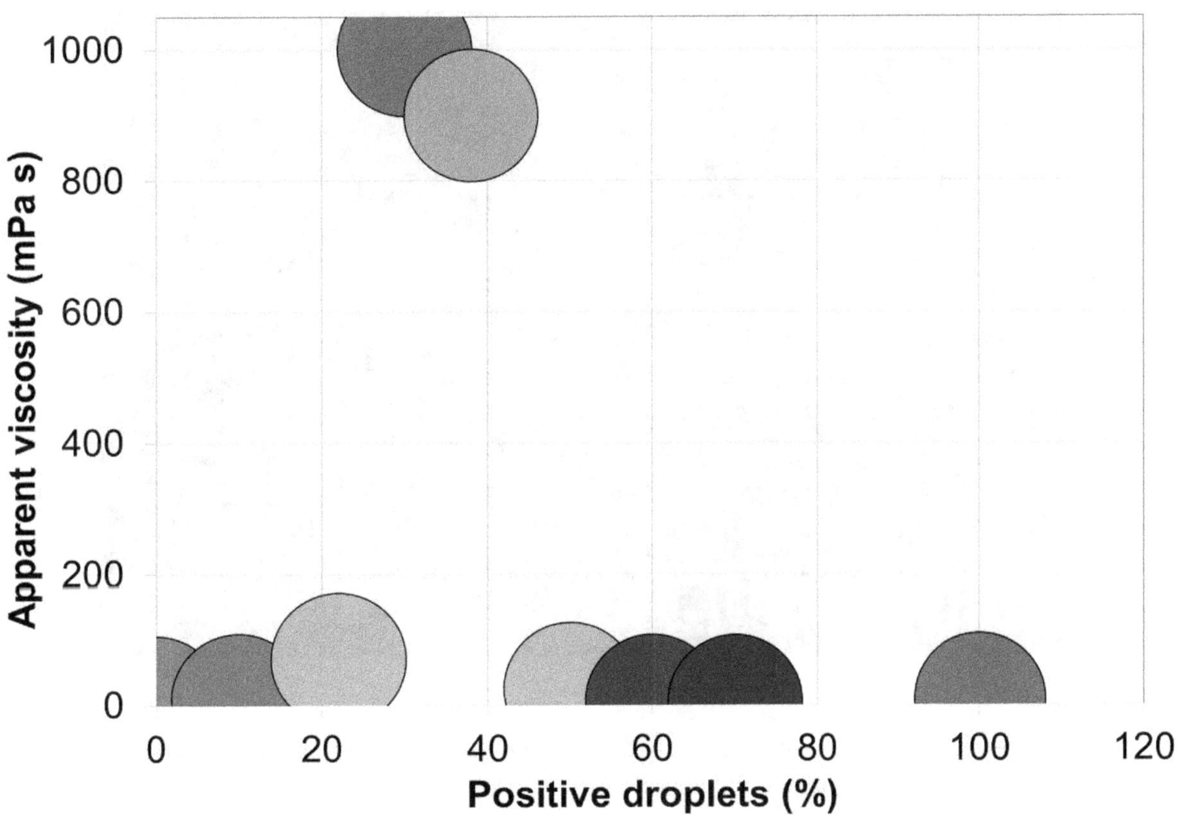

Alluring Interactions

The alluring interactions among globules in certain nanoemulsions do seem to be capable of overcoming the repugnant interactions, likely to result in droplet aggregation and perhaps even flocculation (Costa*et al.,* 2019). The viscosity of coagulated nano-sized globules of emulsion is significantly higher than that of a non-coagulated one due to the general vicinity of the continuous phase stranded in between accumulated globules. The viscosity of coagulated tiny globules of emulsion can be estimated by utilising alike formulae as non-flocculated nanoemulsion, assuming the granules are flocs instead of solitary oil globules (Chanamai *et al.,*2000). As a consequence, Quemada *et al.* (2002) developed an equation for the apparent viscosity of agglomerated coagulated nanoemulsions, which is represented as follows.

Figure 6. Illustrates an explanation of the droplet coagulation impact on the rheology of relatively agglomerated O/W nanoemulsions. Nonetheless, distinguished cationic/anionic oil globule proportions were evaluated (Mao et al., 2013). A broad spectrum of emulsifiers has been used to manipulate the charge (cationic or anionic) of the droplets in order to stabilise them (Sarathchandraprakash et al., 2013). Due to the obvious strong electrostatic repugnance between the droplets, the viscosity of nanoemulsions comprising just negative or only positive droplets had been significantly lower. Nanoemulsions composed of cationic and anionic globules; nevertheless, the viscosity was significantly higher due to electrostatic repulsion, which resulted in the establishment of large flocs.

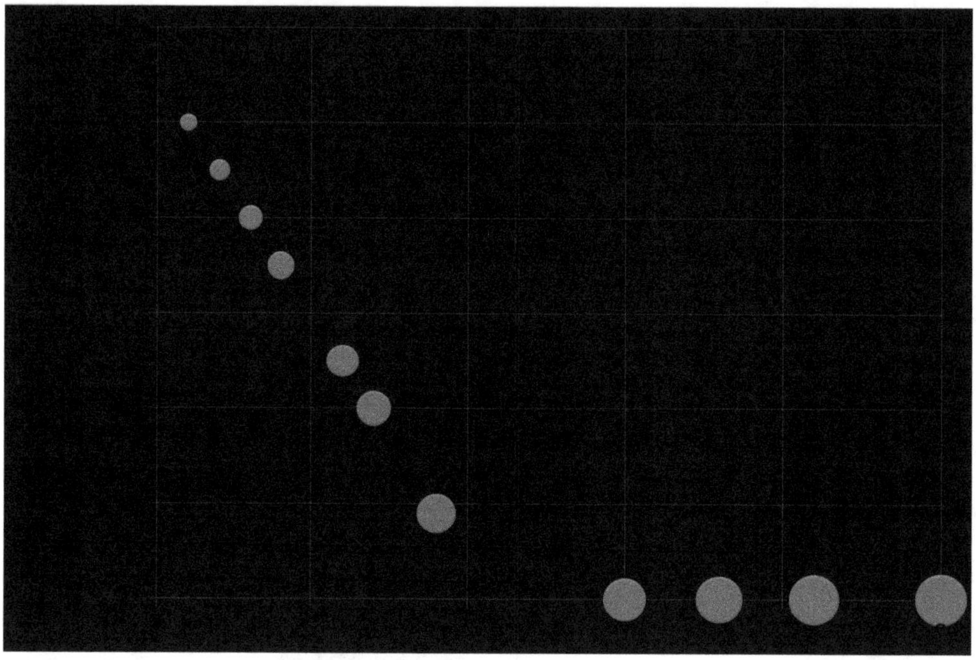

The effective volume fraction of the flocs in the nanoemulsion is expressed as $\Phi_{eff}(=\Phi/\Phi_{floc})$ whereas the volume fraction of oil droplets in the nanoemulsion and is possibly the volume fraction of oil droplets within the flocs. The figure 6 below portrays the interrelationship of positive droplets percentage on apparent viscosity.

DEMEANOUR

Several application domains of nanoemulsions rely on their optical characteristics [Aswathanarayan et al., 2019]. Since some consumer commodities are ought to be clear, the implementation of nano-sized globules from emulsions should not cause them to look cloudy (McClements *et al.*,2011; Kumar*et al.*, 2012). Numerous different commodities, on the other end of the spectrum, have always been anticipated to be cloudy (Włodarska 2019; 122). A correlation was noted among nanoemulsion composition/structure as well as optical characteristics would therefore be extremely valuable for the logical prototype of pharmaceutical formulations with ideal demeanour (Choi*et al.*, 2020; komaiko 2016). The optical characteristics of nano sized emulsions are primarily influenced by their interrelationships with the vibrations between an electric field and a magnetic field (Komaiko*et al.,* 2016). Due to their transparency, clear nanoemulsions with tiny droplets can be easily degraded by UV or visible light (Dickinson *et al.,*1992). The droplet size, concentration, as well as refractive index ascertain the dispersion of light waves in an angular orientation. The most indispensable factors which influence the appearance of nanoemulsions are taken into consideration in this section. The figure 6 represents distinctive formulations of oil in water nanoemulsions produced by employing various surfactants.

To calculate the optical characteristics of nanoemulsions, a wide range of analytical instruments are readily accessible (Malacara *et al.,*2002). These devices are categorized based on their operating principles: spectrophotometric colorimeters as well as trichromatic colorimeters (McClements *et al.,*2004). UV-visible spectrophotometers, which compute the transmission and/or reflection of light waves beyond a spectrum of wavelengths in the visible region, could be used to interpret the optical characteristics of nanoemulsions (Sharma *et al.,* 2021). The UV–Vis–NIR reflectance spectra of emulsions pre and post laser irradiation divulged that the physical properties of the emulsions modified throughout radiation exposure, with the dimensions of the droplets diminishing, culminating in an emulsion with better overall stabilisation (Dinache*et al.,*2020). Using a series of filters, trichromatic colorimeters calculate the potency of the "red," "green," and "blue" light constituents to which the human eye is intuitive (MacDougall*et al.,* 2010). Assessments of reflectance are highly recommended for assessing the optical characteristics of opaque nanoemulsions, whereas transmission observations are preferable for ascertaining the optical properties of clear nanoemulsions.

Crucial Cogitations Concerning Nanoemulsion Colour

Interrelationships with light waves evaluate the optical properties of nanoemulsions. Light scattering concept, could be used to interpret the transmission of light waves in particular which could be expressed simply as,

In this case, the wavelength of light is λ, the scattering angle is ϑ, the droplet radius is r, the dispersed phase volume fraction is ϕ, the refractive index juxtaposition and the absorption spectrum is α (λ).

These equation indicates that the globule diameter as well as globule agglomeration to be the most significant in impacting the optical characteristics of nano-sized globules of emulsion.

CONCLUSION

Nanoemulsions have promising future as well as economic benefits to be used as bioactive compound delivery systems. They are much more reliable than conventional emulsions as well as macroemulsions. The physicochemical characteristics of nanoemulsions seem to be key factors affecting several of their functional as well as sensory characteristics. Owing to the fairly insignificant globule diameter of nano-sized emulsions, it possesses greater reliability to gravitational separation as well as aggregation unlike conventional emulsions that may be worthwhile for industrial applications necessitating longer shelf life. The structure and morphology of nanoemulsions can indeed be modified to facilitate the production of rheological characteristics ranging from low-viscosity fluids to semisolid materials. In nanoemulsions, in which the oil droplets possesses a fairly strong attraction and perhaps even repulsion, highly viscous and semisolid materials could often be fabricated at much subsidiary globule agglomerations than that in traditional emulsions. This chapter has scrutinized nanoemulsions and their general categorisation, stability, and the effects of constituent phase, physicochemical properties and rheological properties on nanoemulsions. For enhanced durability and stability, highly concentrated emulsifiers are mandated. But nevertheless, owing to taste restrictions and enforced legislation, the use of highly concentrated emulsifiers is constrained. However, further research should be conducted to perform a comprehensive analysis of the stabilisation performance of various emulsifiers and also to develop standard criteria to evaluate their efficacy in specific requirements.

REFERENCES

Aboofazeli, R. (2010). Nanometric-scaled emulsions (nanoemulsions). *IJPR*, *9*(4), 325–326. PMID:24381596

Alliod, O., Messager, L., Fessi, H., Dupin, D., & Charcosset, C. (2019). Influence of viscosity for oil-in-water and water-in-oil nanoemulsions production by SPG premix membrane emulsification. *Chemical Engineering Research & Design*, *142*, 87–99. doi:10.1016/j.cherd.2018.11.027

An, H. Z., Safai, E. R., Eral, H. B., & Doyle, P. S. (2013). Synthesis of biomimetic oxygen-carrying compartmentalized microparticles using flow lithography. *Lab on a Chip*, *13*(24), 4765–4774. doi:10.1039/c3lc50610j PMID:24141406

Ashaolu, T. J. (2021). Nanoemulsions for health, food, and cosmetics: A review. *Environmental Chemistry Letters*, *19*(4), 1–15. Advance online publication. doi:10.100710311-021-01216-9 PMID:33746662

Aswathanarayan, J. B., & Vittal, R. R. (2019). Nanoemulsions and Their Potential Applications in Food Industry. *Frontiers in Sustainable Food Systems*, *3*, 95. doi:10.3389/fsufs.2019.00095

Azeem, A., Rizwan, M., Ahmad, F. J., Iqbal, Z., Khar, R. K., Aqil, M., & Talegaonkar, S. (2009). Nanoemulsion components screening and selection: A technical note. *AAPS PharmSciTech*, *10*(1), 69–76. doi:10.120812249-008-9178-x PMID:19148761

Bainun, I. N., Alias, N. H., & Syed-Hassan, S. S. A. (2015). Nanoemulsion: Formation, Characterization, Properties and Applications - A Review. *Advanced Materials Research*, *1113*, 147–152. . doi:10.4028/www.scientific.net/AMR.1113.147

Barres, A. R., Wimmer, M. R., & Mecozzi, S. (2017). Multicompartment Theranostic Nanoemulsions Stabilized by a Triphilic Semifluorinated Block Copolymer. *Molecular Pharmaceutics*, *14*(11), 3916–3926. doi:10.1021/acs.molpharmaceut.7b00624 PMID:28945386

Basha, S. K., Muzammil, M. S., Dhandayuthabani, R., Kumari, V. S., & Kaviyarasu, K. (2019). Nanoemulsion as Oral Drug Delivery-A Review. *Current Drug Research Reviews*, *12*(1), 4–15. doi:10.2174/2589977511666191024173508 PMID:31774040

Bernardi, D. S., Pereira, T. A., Maciel, N. R., Bortoloto, J., Viera, G. S., Oliveira, G. C., & Rocha-Filho, P. A. (2011). Formation and stability of oil-in-water nanoemulsions containing rice bran oil: In vitro and in vivo assessments. *Journal of Nanobiotechnology*, *9*(1), 44. doi:10.1186/1477-3155-9-44 PMID:21952107

Björn, A., de La Monja, P. S., Karlsson, A., Ejlertsson, J., & Svensson, B. H. (2012). Rheological characterization. *Biogas, 1*, 63-76. doi:10.5772/32596

Bouchemal, K., Briançon, S., Perrier, E., & Fessi, H. (2004). Nano-emulsion formulation using spontaneous emulsification: Solvent, oil and surfactant optimisation. *International Journal of Pharmaceutics*, *280*(1–2), 241–251. doi:10.1016/j.ijpharm.2004.05.016 PMID:15265563

Brito-de la Fuente, E., Turcanu, M., Ekberg, O., & Gallegos, C. (2017). Rheological aspects of swallowing and dysphagia: shear and elongational flows. In *Dysphagia* (pp. 687–716). Springer. org/ doi:10.1007/174_2017_119

Burapapadh, K., Kumpugdee-Vollrath, M., Chantasart, D., & Sriamornsak, P. (2010). Fabrication of pectin-based nanoemulsions loaded with itraconazole for pharmaceutical application. *Carbohydrate Polymers*, *82*(2), 384–393. doi:10.1016/j.carbpol.2010.04.071

Calderó, G., Montes, R., Llinàs, M., García-Celma, M. J., Porras, M., & Solans, C. (2016). Studies on the formation of polymeric nano-emulsions obtained via low-energy emulsification and their use as templates for drug delivery nanoparticle dispersions. *Colloids and Surfaces. B, Biointerfaces*, *145*, 922–931. doi:10.1016/j.colsurfb.2016.06.013 PMID:27341306

Calligaris, S., Plazzotta, S., Bot, F., Grasselli, S., Malchiodi, A., & Anese, M. (2016). Nanoemulsion preparation by combining high pressure homogenization and high power ultrasound at low energy densities. *Food Research International*, *83*, 25–30. doi:10.1016/j.foodres.2016.01.033

Chanamai, R., & McClements, D. J. (2000). Dependence of creaming and rheology of monodisperse oil-in-water emulsions on droplet size and concentration. *Colloids and Surfaces. A, Physicochemical and Engineering Aspects*, *172*(1-3), 79–86. doi:10.1016/S0927-7757(00)00551-3

Chen, E., Cao, L., McClements, D. J., Liu, S., Li, B., & Li, Y. (2018). Enhancement of physicochemical properties of whey protein-stabilized nanoemulsions by interfacial cross-linking using cinnamaldehyde. *Food Hydrocolloids*, *77*, 976–985. doi:10.1016/j.foodhyd.2017.11.047

Chen, H., Khemtong, C., Yang, X., Chang, X., & Gao, J. (2011). Nanonization strategies for poorly water-soluble drugs. *Drug Discovery Today*, *16*(7-8), 354–360. doi:10.1016/j.drudis.2010.02.009 PMID:20206289

Cheng, L. C., Kuei Vehusheia, S. L., & Doyle, P. S. (2020). Tuning Material Properties of Nanoemulsion Gels by Sequentially Screening Electrostatic Repulsions and Then Thermally Inducing Droplet Bridging. *Langmuir*, *36*(13), 3346–3355. doi:10.1021/acs.langmuir.0c00199 PMID:32216359

Choi, S. J., & McClements, D. J. (2020). Nanoemulsions as delivery systems for lipophilic nutraceuticals: Strategies for improving their formulation, stability, functionality and bioavailability. *Food Science and Biotechnology*, *29*(2), 149–168. doi:10.100710068-019-00731-4 PMID:32064124

Choradiya, B., & Patil, S. (2021). A comprehensive review on nanoemulsion as an ophthalmic drug delivery system. *Journal of Molecular Liquids*, *339*, 116751. doi:10.1016/j.molliq.2021.116751

Costa, C., Medronho, B., Filipe, A., Mira, I., Lindman, B., Edlund, H., & Norgren, M. (2019). Emulsion Formation and Stabilization by Biomolecules: The Leading Role of Cellulose. *Polymers*, *11*(10), 1570. doi:10.3390/polym11101570 PMID:31561633

Das, A. K., Nanda, P. K., Bandyopadhyay, S., Banerjee, R., Biswas, S., & McClements, D. J. (2020). Application of nanoemulsion-based approaches for improving the quality and safety of muscle foods: A comprehensive review. *Comprehensive Reviews in Food Science and Food Safety*, *19*(5), 2677–2700. doi:10.1111/1541-4337.12604 PMID:33336977

Demisli, S., Mitsou, E., Pletsa, V., Xenakis, A., & Papadimitriou, V. (2020). Development and Study of Nanoemulsions and Nanoemulsion-Based Hydrogels for the Encapsulation of Lipophilic Compounds. *Nanomaterials (Basel, Switzerland)*, *10*(12), 2464. doi:10.3390/nano10122464 PMID:33317080

Derkach, S. R. (2009). Rheology of emulsions. *Advances in Colloid and Interface Science*, *151*(1-2), 1-23. doi:org/10.1016/j.cis.2009.07.001

Dickinson, E. (1992). *Introduction to food colloids*. Oxford University Press.

Dinache, A., Tozar, T., Smarandache, A., Andrei, I. R., Nistorescu, S., Nastasa, V., & Romanitan, M. O. (2020). Spectroscopic characterization of emulsions generated with a new laser-assisted device. *Molecules (Basel, Switzerland)*, *25*(7), 1729. doi:10.3390/molecules25071729 PMID:32283754

Eral, H. B., Mahony, M. O., Shaw, R., Trout, B. L., Myerson, A. S., & Doyle, P. S. (2014). Composite Hydrogels Laden with Crystalline Active Pharmaceutical Ingredients of Controlled Size and Loading. *Chemistry of Materials*, *26*(21), 6213–6220. doi:10.1021/cm502834h

Erramreddy, V. V., & Ghosh, S. (2014). Influence of emulsifier concentration on nanoemulsion gelation. *Langmuir*, *30*(37), 11062–11074. doi:10.1021/la502733v PMID:25137632

Erramreddy, V. V., Tu, S., & Ghosh, S. (2017). Rheological reversibility and long-term stability of repulsive and attractive nanoemulsion gels. *RSC Advances*, *7*(75), 47818–47832. doi:10.1039/C7RA09605D

Fathordoobady, F., Sannikova, N., Guo, Y., Singh, A., Kitts, D. D., & Pratap-Singh, A. (2021). Comparing microfluidics and ultrasonication as formulation methods for developing hempseed oil nanoemulsions for oral delivery applications. *Scientific Reports*, *11*(1), 72. doi:10.103841598-020-79161-w PMID:33420136

Gaba, B., Khan, T., Haider, M. F., Alam, T., Baboota, S., Parvez, S., & Ali, J. (2019). Vitamin E Loaded Naringenin Nanoemulsion via Intranasal Delivery for the Management of Oxidative Stress in a 6-OHDA Parkinson's Disease Model. *BioMed Research International*, *2382563*, 1–20. Advance online publication. doi:10.1155/2019/2382563 PMID:31111044

Ganta, S., Talekar, M., Singh, A., Coleman, T. P., & Amiji, M. M. (2014). Nanoemulsions in translational research-opportunities and challenges in targeted cancer therapy. *AAPS PharmSciTech*, *15*(3), 694–708. doi:10.120812249-014-0088-9 PMID:24510526

Gönüllü, Ü., Üner, M., & Yener, G., Fatma Karaman, E. C. E. M., & Aydoğmuş, Z. (2015). Formulation and characterization of solid lipid nanoparticles, nanostructured lipid carriers and nanoemulsion of lornoxicam for transdermal delivery. *ActaPharmaceutica*, *65*(1), 1–13. PMID:25781700

Goyal, A., Sharma, V., Upadhyay, N., Singh, A. K., Arora, S., Lal, D., & Sabikhi, L. (2015). Development of stable flaxseed oil emulsions as a potential delivery system of ω-3 fatty acids. *Journal of Food Science and Technology*, *52*(7), 4256–4265. doi:10.100713197-014-1370-2 PMID:26139890

Gupta, A., Eral, H. B., Hatton, T. A., & Doyle, P. S. (2016). Nanoemulsions: Formation, properties and applications. *Soft Matter*, *12*(11), 2826–2841. doi:10.1039/C5SM02958A PMID:26924445

Gupta, A., Eral, H. B., Hatton, T. A., & Doyle, P. S. (2016). Nanoemulsions: Formation, Properties and Applications. *Soft Matter*, *12*(11), 2826–2841. Advance online publication. doi:10.1039/C5SM02958A PMID:26924445

Gurpret, K., & Singh, S. K. (2018). Review of Nanoemulsion Formulation and Characterization Techniques. *Indian Journal of Pharmaceutical Sciences*, *80*(5). Advance online publication. doi:10.4172/pharmaceutical-sciences.1000422

Guzey, D., & McClements, D. J. (2006). Formation, stability and properties of multilayer emulsions for application in the food industry. *Advances in Colloid and Interface Science*, *128–130*, 227–248. doi:10.1016/j.cis.2006.11.021 PMID:17223060

Hamed, R., Basil, M., AlBaraghthi, T., Sunoqrot, S., & Tarawneh, O. (2015). Nanoemulsion-based gel formulation of diclofenac diethylamine: Design, optimization, rheological behavior and in vitro diffusion studies. *Pharmaceutical Development and Technology*, *21*(8), 980–989. doi:10.3109/10837450.2015.1086372 PMID:26369621

Hashtjin, A. M., & Abbasi, S. (2015). Optimization of ultrasonic emulsification conditions for the production of orange peel essential oil nanoemulsions. *Journal of Food Science and Technology*, *52*(5), 2679–2689. doi:10.100713197-014-1322-x PMID:25892765

Helgeson, M. E. (2016). Colloidal behavior of nanoemulsions: Interactions, structure, and rheology. *Current Opinion in Colloid & Interface Science*, *25*, 39–50. doi:10.1016/j.cocis.2016.06.006

Helgeson, M. E., Moran, S. E., & Doyle, H. Z. (2012). Mesoporous organohydrogels from thermogelling photocross linkable nanoemulsions. *Nature Materials*, *11*(4), 344–352. doi:10.1038/nmat3248 PMID:22327746

Hunter, R. J. (2001). *Foundations of colloid science*. Oxford university press.

Jafari, S. M., & McClements, D. J. (Eds.). (2018). *Nanoemulsions: Formulation, applications, and characterization*. Academic Press.

Jaiswal, M., Dudhe, R., & Sharma, P. K. (2015). Nanoemulsion: An advanced mode of drug delivery system. *Biotech*. doi:org/10.1007/s13205-014-0214-0

Jaworska, M., Sikora, E., & Ogonowski, J. (2015). Rheological Properties of Nanoemulsions Stabilized by Polysorbate 80. *Chemical Engineering & Technology*, *38*(8), 1469–1476. doi:10.1002/ceat.201500069

Jha, D., & Mazumder, P. M. (2018). Biological, chemical and pharmacological aspects of Madhucalongifolia. *Asian Pacific Journal of Tropical Medicine*, *11*(1), 9. doi:10.4103/1995-7645.223528

Jores, K., Haberland, A., Wartewig, S., Mäder, K., & Mehnert, W. (2005). Solid lipid nanoparticles (SLN) and oil-loaded SLN studied by spectrofluorometry and Raman spectroscopy. *Pharmaceutical Research*, *22*(11), 1887–1897. doi:10.100711095-005-7148-5 PMID:16132349

Khaled, A., Soonham, Y., Taha, K., & David, M. (2016). Utilization of nanoemulsions to enhance bioactivity of pharmaceuticals, supplements, and nutraceuticals: Nanoemulsion delivery systems and nanoemulsion excipient systems. *Expert Opinion on Drug Delivery*, *13*(9), 1327–1336. Advance online publication. doi:10.1517/17425247.2016.1162154 PMID:26984045

Khaleel Basha, S., Syed Muzammil, M., & Dhandayuthabani, R. (2020). Nanoemulsion as Oral Drug Delivery - A Review. *Current Drug Research Reviews*, *12*(1), 4–15. doi:10.2174/2589977511666191024173508 PMID:31774040

Khanna, V., Singh, S., Baghel, D., Kaur, B., Nayar, A., & Ghimire, R. (2018). *Nanoemulsion present aand future perspective – An Overview*. Academic Press.

Khater, D., Nsairat, H., Odeh, F., Saleh, M., Jaber, A., Alshaer, W., & Mubarak, M. S. (2021). Design, Preparation, and Characterization of Effective Dermal and Transdermal Lipid Nanoparticles: *A Review. Cosmetics*, *8*(2), 39. doi:10.3390/cosmetics8020039

Komaiko, J. (2016). *Optimization of the fabrication, stability, and performance of food grade nanoemulsions with low and high energy methods*. Academic Press.

Kotta, S., Khan, A., Ansari, S., Sharma, R., & Ali, J. (2013). Formulation of nanoemulsion: A comparison between phase inversion composition method and high-pressure homogenization method. *Drug Delivery*, *22*(4), 455–466. doi:10.3109/10717544.2013.866992 PMID:24329559

Kumar, M., Bishnoi, R. S., Shukla, A. K., & Jain, C. P. (2019). Techniques for Formulation of Nanoemulsion Drug Delivery System: A Review. *Preventive Nutrition and Food Science*, *24*(3), 225–234. doi:10.3746/pnf.2019.24.3.225 PMID:31608247

Kumar, S. H., & Singh, V. (2012). Nanoemulsification-a novel targeted drug delivery tool. *Journal of Drug Delivery and Therapeutics*, *2*(4). Advance online publication. doi:10.22270/jddt.v2i4.185

Larson, R. G. (1999). *The structure and rheology of complex fluids* (Vol. 150). Oxford university press.

Li, G., Zhang, Z., Liu, H., & Hu, L. (2021). Nanoemulsion-based delivery approaches for nutraceuticals: Fabrication, application, characterization, biological fate, potential toxicity and future trends. *Food & Function*, *12*(5), 1933–1953. doi:10.1039/D0FO02686G PMID:33596279

Li, Y., Jin, H., Sun, X., Sun, J., Liu, C., Liu, C., & Xu, J. (2018). Physicochemical Properties and Storage Stability of Food Protein-Stabilized Nanoemulsions. *Nanomaterials (Basel, Switzerland)*, *9*(1), 25. doi:10.3390/nano9010025 PMID:30585224

Liu, Q., Huang, H., Chen, H., Lin, J., & Wang, Q. (2019). Food-Grade Nanoemulsions: Preparation, Stability and Application in Encapsulation of Bioactive Compounds. *Molecules (Basel, Switzerland)*, *24*(23), 4242. doi:10.3390/molecules24234242 PMID:31766473

Lococo, D., Mora-Huertas, C. E., Fessi, H., Zaanoun, I., & Elaissari, A. (2012). Argan oil nanoemulsions as new hydrophobic drug-loaded delivery system for transdermal application. *Journal of Biomedical Nanotechnology*, *8*(5), 843–848. doi:10.1166/jbn.2012.1445 PMID:22888756

Ma, D., Tu, Z. C., Wang, H., Zhang, Z., & McClements, D. J. (2017). Fabrication and characterization of nanoemulsion-coated microgels: Electrostatic deposition of lipid droplets on alginate beads. *Food Hydrocolloids*, *71*, 149–157. doi:10.1016/j.foodhyd.2017.05.015

MacDougall, D. B. (2010). Colour measurement of food: principles and practice. In *Colour Measurement* (pp. 312–342). Woodhead Publishing. doi:10.1533/9780857090195.2.312

Malacara, D. (2002). Color vision and colorimetry: Theory and applications. *SPIE Press*, *7*, 164.

Mao, Y., & McClements, D. J. (2013). Modulation of food texture using controlled heteroaggregation of lipid droplets: Principles and applications. *Journal of Applied Polymer Science*, *130*(6), 3833–3841. doi:10.1002/app.39631

Maphosa, Y., & Jideani, V. A. (2018). Factors affecting the stability of emulsions stabilised by biopolymers. *Science and Technology Behind Nanoemulsions*, *65*. Advance online publication. doi:10.5772/intechopen.75308

McClements, D. J. (2002). Theoretical prediction of emulsion color. *Advances in Colloid and Interface Science*, *97*(1-3), 63–89. doi:10.1016/S0001-8686(01)00047-1 PMID:12027025

McClements, D. J. (2004). *Food Emulsions: Principles, Practices, and Techniques* (2nd ed.). CRC Press., org/ doi:10.1201/9781420039436

McClements, D. J. (2010). Emulsion design to improve the delivery of functional lipophilic components. *Annual Review of Food Science and Technology*, *1*(1), 241–269. doi:10.1146/annurev.food.080708.100722 PMID:22129337

McClements, D. J. (2011). Edible nanoemulsions: Fabrication, properties, and functional performance. *Soft Matter*, *7*(6), 2297–2316. doi:10.1039/C0SM00549E

Mcclements, D. J. (2012). Nanoemulsions versus microemulsions: Terminology, differences, and similarities. *Soft Matter*, *8*(6), 1719–1729. doi:10.1039/C2SM06903B

McClements, D. J. (2013). Nanoemulsion-based oral delivery systems for lipophilic bioactive components: Nutraceuticals and pharmaceuticals. *Therapeutic Delivery*, *4*(7), 841–857. doi:10.4155/tde.13.46 PMID:23883127

McClements, D. J., & Rao, J. (2011). Food-Grade Nanoemulsions: Formulation, Fabrication, Properties, Performance, Biological Fate, and Potential Toxicity. *Critical Reviews in Food Science and Nutrition*, *51*(4), 285–330. doi:10.1080/10408398.2011.559558 PMID:21432697

Meng, R., Wang, C., & Shen, Z. (2020). Optimization and Characterization of Highly Stable Nanoemulsion for Effective Oil-Based Drilling Fluid Removal. *SPE Journal*, *25*(03), 1259–1271. doi:10.2118/199904-PA

Mishra, P., Balaji, A., Mukherjee, A., & Chandrasekaran, N. (2018). Bio-Based Nanoemulsions: An Eco-Safe Approach Towards the Eco-Toxicity Problem. Handbook of Ecomaterials, 1-23. doi:org/10.1007/978-3-319-48281-1_94-1

Mountain, G. A., Jelier, B. J., Bagia, C., Friesen, C. M., & Janjic, J. M. (2014). Design and formulation of nanoemulsions using 2-(poly(hexafluoropropylene oxide)) perfluoropropyl benzene in combination with linear perfluoro(polyethylene glycol dimethyl ether). *Journal of Fluorine Chemistry*, *162*, 38–44. doi:10.1016/j.jfluchem.2014.03.007 PMID:24976645

Nam, Y. S., Kim, J. W., Park, J., Shim, J., Lee, J. S., & Han, S. H. (2012). Tocopheryl acetate nanoemulsions stabilized with lipid-polymer hybrid emulsifiers for effective skin delivery. *Colloids and Surfaces. B, Biointerfaces*, *94*, 51–57. doi:10.1016/j.colsurfb.2012.01.016 PMID:22326341

Namratha Vinaya, K., Mary John, A., Mangsatabam, M., & Anna Philip, M. (2021). Ultrasound-assisted synthesis and characterization of sesame oil based nanoemulsion. *IOP Conference Series. Materials Science and Engineering*, *1114*(1), 012085. doi:10.1088/1757-899X/1114/1/012085

Naseri, N., Valizadeh, H., & Zakeri-Milani, P. (2015). Solid Lipid Nanoparticles and Nanostructured Lipid Carriers: Structure, Preparation and Application. *Advanced Pharmaceutical Bulletin*, *5*(3), 305–313. doi:10.15171/apb.2015.043 PMID:26504751

Niknam, S. M., Escudero, I., & Benito, J. M. (2020). Formulation and Preparation of Water-In-Oil-In-Water Emulsions Loaded with a Phenolic-Rich Inner Aqueous Phase by Application of High Energy Emulsification Methods. *Foods (Basel, Switzerland)*, *9*(10), 1411. doi:10.3390/foods9101411 PMID:33027911

Odriozola-Serrano, I., Oms-Oliu, G., & Martín-Belloso, O. (2014). Nanoemulsion-based delivery systems to improve functionality of lipophilic components. *Frontiers in Nutrition*, *1*, 24. doi:10.3389/fnut.2014.00024 PMID:25988126

Pagar, K.R., & Darekar, A.B. (2019). Nanoemulsion: A new concept of Delivery System. *Nanoemulsion: A New Concept of Delivery System*, *9*(1), 39-46. doi:org/10.5958/2231-5659.2019.00006.7

Pal, R. (2016). Modeling the viscosity of concentrated nanoemulsions and nanosuspensions. *Fluids*, *1*(2), 11. doi:10.3390/fluids1020011

Pandey, M., Choudhury, H., Yeun, O. C., Yin, H. M., Lynn, T. W., Tine, C., Wi, N. S., Yen, K., Phing, C. S., Kesharwani, P., Bhattamisra, S. K., & Gorain, B. (2018). Perspectives of Nanoemulsion Strategies in the Improvement of Oral, Parenteral and Transdermal Chemotherapy. *Current Pharmaceutical Biotechnology*, *19*(4), 276–292. doi:10.2174/1389201019666180605125234 PMID:29874994

Patel, R. B., Thakore, S. D., & Patel, M. R. (2016). Recent Survey on Patents of Nanoemulsions. *Current Drug Delivery*, *13*(6), 857–881. doi:10.2174/1567201812666150901111930 PMID:26324230

Patel, R. P., & Joshi, J. R. (2012). An overview on nanoemulsion: A Novel Approach. *International Journal of Pharmaceutical Sciences and Research*, *3*(12), 4640–4650. org/ doi:10.13040/IJPSR.0975-8232.3(12).4640-50

Pathak, M. (2017). Nanoemulsions and Their Stability for Enhancing Functional Properties of Food Ingredients. Nanotechnology Applications in Food, 87-106. doi:org/10.1016/B978-0-12-811942-6.00005-4

Pavoni, L., Perinelli, D. R., Bonacucina, G., Cespi, M., & Palmieri, G. F. (2020). An Overview of Micro- and Nanoemulsions as Vehicles for Essential Oils: Formulation, Preparation and Stability. *Nanomaterials (Basel, Switzerland)*, *10*(1), 135. doi:10.3390/nano10010135 PMID:31940900

Qian, C., Decker, E. A., Xiao, H., & McClements, D. J. (2012). Solid lipid nanoparticles: Effect of carrier oil and emulsifier type on phase behavior and physical stability. *Journal of the American Oil Chemists' Society, 89*(1), 17-28. doi:10.1007/s11746-011-1882-0

Quemada, D., & Berli, C. (2002). Energy of interaction in colloids and its implications in rheological modeling. *Advances in Colloid and Interface Science, 98*(1), 51–85. doi:10.1016/S0001-8686(01)00093-8 PMID:12061712

Rehman, F. U., Shah, K. U., Shah, S. U., Khan, I. U., Khan, G. M., & Khan, A. (2017). From nanoemulsions to self-nanoemulsions, with recent advances in self-nanoemulsifying drug delivery systems (SNEDDS). *Expert Opinion on Drug Delivery, 14*(11), 1325–1340. doi:10.1080/17425247.2016.1218462 PMID:27485144

Ren, J. N., Dong, M., Hou, Y. Y., Fan, G., & Pan, S. Y. (2018). Effect of olive oil on the preparation of nanoemulsions and its effect on aroma release. *Journal of Food Science and Technology, 55*(10), 4223–4231. doi:10.100713197-018-3358-9 PMID:30228421

Sadeghpour Galooyak, S., & Dabir, B. (2015). Three-factor response surface optimization of nanoemulsion formation using a microfluidizer. *Journal of Food Science and Technology, 52*(5), 2558–2571. doi:10.100713197-014-1363-1 PMID:25892755

Salem, M., & Ezzat, S. (2019). Nanoemulsions in Food Industry. *Some New Aspects of Colloidal Systems in Foods.* doi:org/10.5772/intechopen.79447

Sánchez-López, E., Guerra, M., Dias-Ferreira, J., Lopez-Machado, A., Ettcheto, M., Cano, A., Espina, M., Camins, A., Garcia, M. L., & Souto, E. B. (2019). Current Applications of Nanoemulsions in Cancer Therapeutics. *Nanomaterials (Basel, Switzerland), 9*(6), 821. doi:10.3390/nano9060821 PMID:31159219

Sarathchandraprakash, N. K., Mahendra, C., Prashanth, S. J., Manral, K., Babu, U. V., & Gowda, D. (2013). Emulsions and emulsifiers. *Asian Journal of Experimental Chemistry, 8*, 30–45.

Sarheed, O., Shouqair, D., Ramesh, K., Khaleel, T., Amin, M., Boateng, J., & Drechsler, M. (2020). Formation of stable nanoemulsions by ultrasound-assisted two-step emulsification process for topical drug delivery: Effect of oil phase composition and surfactant concentration and loratadine as ripening inhibitor. *International Journal of Pharmaceutics, 576*, 118952. doi:10.1016/j.ijpharm.2019.118952 PMID:31843549

Sarker, D. K. (2005). Engineering of Nanoemulsions for Drug Delivery. *Current Drug Delivery, 2*(4), 297–310. doi:10.2174/156720105774370267 PMID:16305433

Sharma, N., Bansal, M., Visht, S., Sharma, P., & Kulkarni, G. (2010). Nanoemulsion: A new concept of delivery system. *Chronicles of Young Scientists, 1*(2), 2.

Sharma, N., Kaur, G., & Khatkar, S. K. (2021). Optimization of emulsification conditions for designing ultrasound assisted curcumin loaded nanoemulsion: Characterization, antioxidant assay and release kinetics. *Lebensmittel-Wissenschaft + Technologie, 141*, 110962. doi:10.1016/j.lwt.2021.110962

Sharma, S., & SaranSgdevot, K. (2012). Nanoemulsions for cosmetics. *International Journal of Advanced Research in Pharmaceuticals Bio Sciences, 1*(4), 408.

Sheth, T., Seshadri, S., Prileszky, T., & Helgeson, M. E. (2020). Multiple nanoemulsions. *Nature Reviews. Materials*, *5*(3), 214–228. doi:10.103841578-019-0161-9

Sivakumar, M., Tang, S. Y., & Tan, K. W. (2014). Cavitation technology – A greener processing technique for the generation of pharmaceutical nanoemulsions. *Ultrasonics Sonochemistry*, *21*(6), 2069–2083. doi:10.1016/j.ultsonch.2014.03.025 PMID:24755340

Solans, C., Izquierdo, P., Nolla, J., Azemar, N., & García-Celma, M. J. (2005). Nano-emulsion. *Current Opinion in Colloid & Interface Science*, *10*(3-4), 102–110. doi:10.1016/j.cocis.2005.06.004

Sonneville-Aubrun, O., Yukuyama, M. N., & Pizzino, A. (2018). Application of Nanoemulsions in Cosmetics. Nanoemulsions, 435-475. doi:org/10.1016/B978-0-12-811838-2.00014-X

Streck, L., Sarmento, V. H., Machado, P. R., Farias, K. J., Fernandes-Pedrosa, M. F., & da Silva-Júnior, A. A. (2016). Phase Transitions of Isotropic to Anisotropic Biocompatible Lipid-Based Drug Delivery Systems Overcoming Insoluble Benznidazole Loading. *International Journal of Molecular Sciences*, *17*(7), 981. doi:10.3390/ijms17070981 PMID:27376278

Sutradhar, K., & Amin, M. (2013). Nanoemulsions: Increasing possibilities in drug delivery. *European Journal of Nanomedicine*, *5*(2), 97–110. doi:10.1515/ejnm-2013-0001

Tadros, T., Izquierdo, P., Esquena, J., & Solans, C. (2004). Formation and stability of nano-emulsions. *Advances in Colloid and Interface Science*, *108–109*, 303–318. doi:10.1016/j.cis.2003.10.023 PMID:15072948

Tadros, T. F. (2011). *Rheology of dispersions: principles and applications*. John Wiley & Sons.

Tan, C., & McClements, D. J. (2021). Application of Advanced Emulsion Technology in the Food Industry: A Review and Critical Evaluation. *Foods (Basel, Switzerland)*, *10*(4), 812. doi:10.3390/foods10040812 PMID:33918596

Trujillo, L. S., Belloso, O. M., & McClements, D. J. (2016). Excipient Nanoemulsions for Improving Oral Bioavailability of Bioactives. *Nanomaterials (Basel, Switzerland)*, *6*(1), 17. doi:10.3390/nano6010017 PMID:28344274

Uner, M., Wissing, S. A., Yener, G., & Müller, R. H. (2005). Solid lipid nanoparticles (SLN) and nanostructured lipid carriers (NLC) for application of ascorbylpalmitate. *Die Pharmazie*, *60*(8), 577–582. PMID:16124399

Vatanparast, H., Shahabi, F., Bahramian, A., Javadi, A., & Miller, R. (2018). The role of electrostatic repulsion on increasing surface activity of anionic surfactants in the presence of hydrophilic silica nanoparticles. *Scientific Reports*, *8*(1), 1–11. doi:10.103841598-018-25493-7 PMID:29740036

Vlachopoulos, J., Polychronopoulos, N., & Kontopoulou, M. (2011). Basic concepts in polymer melt rheology and their importance in processing. *Applied Polymer Rheology: Polymeric Fluids with Industrial Applications,* 1-27. doi:org/10.1002/9781118140611.ch1

Wang, L., Tabor, R., Eastoe, J., Li, X., Heenan, R. K., & Dong, J. (2009). Formation and stability of nanoemulsions with mixed ionic–nonionic surfactants. *Physical Chemistry Chemical Physics*, *11*(42), 9772–9778. doi:10.1039/b912460h PMID:19851556

Weiss, J., Decker, E. A., McClements, D. J., Kristbergsson, K., Helgason, T., & Awad, T. (2008). Solid lipid nanoparticles as delivery systems for bioactive food components. *Food Biophysics*, *3*(2), 146–154. doi:10.100711483-008-9065-8

Weiss, J., & McClements, D. J. (2000). Influence of Ostwald ripening on rheology of oil-in-water emulsions containing electrostatically stabilized droplets. *Langmuir*, *16*(5), 2145–2150. doi:10.1021/la9909392

Włodarska, K., Pawlak-Lemańska, K., Górecki, T., & Sikorska, E. (2019). Factors influencing consumers' perceptions of food: A study of apple juice using sensory and visual attention methods. *Foods*, *8*(11), 545. doi:10.3390/foods8110545 PMID:31684206

Wu, D., Lu, J., Zhong, S., Schwarz, P., Chen, B., & Rao, J. (2019). Influence of nonionic and ionic surfactants on the antifungal and mycotoxin inhibitory efficacy of cinnamon oil nanoemulsions. *Food & Function*, *10*(5), 2817–2827. doi:10.1039/C9FO00470J PMID:31049507

Wu, S. H., Hung, Y., & Mou, C. Y. (2013). Compartmentalized Hollow Silica Nanospheres Templated from Nanoemulsions. *Chemistry of Materials*, *25*(3), 352–364. doi:10.1021/cm303116u

Yu, L., Li, C., Xu, J., Hao, J., & Sun, D. (2012). Highly Stable Concentrated Nanoemulsions by the Phase Inversion Composition Method at Elevated Temperature. *Langmuir*, *28*(41), 14547–14552. doi:10.1021/la302995a PMID:22985401

Zhang, L. W., Al-Suwayeh, S. A., Hung, C. F., Chen, C. C., & Fang, J. Y. (2011). Oil components modulate the skin delivery of 5-aminolevulinic acid and its ester prodrug from oil-in-water and water-in-oil nanoemulsions. *International Journal of Nanomedicine*, *6*, 693–704. doi:10.2147/IJN.S17524 PMID:21556344

Zhang, P., Kopechek, J. A., & Porter, T. M. (2013). The impact of vaporized nanoemulsions on ultrasound-mediated ablation. *Journal of Therapeutic Ultrasound*, *1*(1), 2. doi:10.1186/2050-5736-1-2 PMID:24761223

Zhang, Y., Shang, Z., Gao, C., Du, M., Xu, S., Song, H., & Liu, T. (2014). Nanoemulsion for solubilization, stabilization, and in vitro release of pterostilbene for oral delivery. *AAPS PharmSciTech*, *15*(4), 1000–1008. doi:10.120812249-014-0129-4 PMID:24831090

Zhang, Z., & McClements, D. J. (2018). Overview of nanoemulsion properties: stability, rheology, and appearance. In *Nanoemulsions* (pp. 21–49). Academic Press. org/ doi:10.1016/B978-0-12-811838-2.00002-3

Zhao, A., Lee, J., & Emelianov, S. (2021). Formulation and Acoustic Modulation of Optically Vaporized Perfluorocarbon Nanodroplets. *Journal of Visualized Experiments*, *62814*(173). doi:10.3791/62814 PMID:34338676

Zhu, Y., Gao, H., Liu, W., Zou, L., & McClements, D. J. (2020). A review of the rheological properties of dilute and concentrated food emulsions. *Journal of Texture Studies*, *51*(1), 45–55. doi:10.1111/jtxs.12444 PMID:31107991

Chapter 2
Preparation of Nanoemulsion: The Processing Parameters and Emulsification Techniques in Nanoemulsion Preparation

Elaine Elaine
UCSI University, Malaysia

Kar Lin Nyam
UCSI University, Malaysia

ABSTRACT

Nanoemulsion is a versatile emulsion-based delivery system that can be structured or prepared with different compositions, methods, or processing variables. Although single nanoemulsion has been a promising delivery carrier in numerous industries, the innovation of double nanoemulsion is introduced to provide different benefits. However, the nanoemulsion must be optimized in terms of the composition and emulsification methods to maintain the integrity of the delivery system. Therefore, the optimization of nanoemulsion can range from deciding the type and concentration of compositions (aqueous, lipid, surfactants) to the processing conditions during emulsification. The current options of emulsification methods, processing parameters, and optimization process of nanoemulsion are highlighted and elaborated in this chapter. This allows readers to understand the fundamental principles of nanoemulsion preparation and encourage future studies and applications in the related field.

INTRODUCTION

Nanoemulsion is a complex colloidal system that consists of nano-sized droplets with a high surface area. Based on Figure 1, nanoemulsion is favored in many applications due to its kinetic stability, desired optical transparency, and rheological characteristics (Borthakur, Boruah, Sharma, and Das, 2016; Rehman, Akram, Seralin, Vandamme, and Anton, 2020). The typical emulsion is vulnerable to destabilization processes such as gravitational separation, flocculation, and coalescence. However, as

DOI: 10.4018/978-1-7998-8378-4.ch002

compared to conventional emulsion, nanoemulsion has a higher bioavailability and also great physical and kinetic stability against the destabilization process (Goodarzi and Zendehboudi, 2018; Teo, Goh, and Lee, 2014). The preparation concept of nanoemulsion is almost similar for food, health, agriculture, and biomedicine fields. The distinctive point will be the composition selection and emulsification method based on purpose of encapsulation. In the biomedicine field, low energy emulsification methods such as spontaneous emulsification, phase inversion methods are often used to encapsulation sensitive compounds that degrade under high heat, light, etc. Meanwhile, food and agriculture often incorporate large scale-based emulsification instrument for production efficiency and stability (Niknam et al., 2020; Saffarionpour, 2019). Most nanoemulsions are fabricated with food-grade nanomaterials and can incorporate functional composition within the formulation (Azmi, Elgharbawy, Motlagh, Samsudin, and Salleh, 2019; Bhushani and Anandharamakrishnan, 2017; Dasgupta and Ranjan, 2018; Liu, Huang, Chen, Lin, and Wang, 2019; Pathak, 2017).

Table 1. The advantages and disadvantages of nanoemulsion encapsulation

Advantages	Disadvantages
Good kinetic stability: bioavailability, shelf life	Surfactant is required to reduce interfacial tension and osmotic pressure
Optical transparency (opacity, turbidity)	Thermodynamically unstable
Rheological characteristics (viscosity, fluidity)	Sensitivity to pH and ionic strength changes*
Functional modification (composition, active compounds)	Optimization is required to ensure storage stability

*when sensitive composition such as biopolymers are used.

In general, nanoemulsion consists of aqueous phase, lipid phase, and surfactants. The two immiscible liquids (lipid and water) formed an interface that is generally stabilized by surfactants. Surfactants are required in nanoemulsion because the nanodroplets have higher interfacial tension and stress. Furthermore, the presence of both oil and aqueous phases within emulsion contributes to different densities which made the system thermodynamically unstable (Abdelrasoul, Eid, and Badawy, 2020; Maphosa and Jideani, 2018; Sheth, Seshadri, Prileszky, and Helgeson, 2020). Different preparation methods produced different functionalities and stability of the nanoemulsion. Although nanoemulsion can be produced in different charges, sizes, or compositions, the preparation process has always been strictly controlled and optimized to ensure emulsion stability under environmental stress. Ultimately, nanoemulsion has been developed for different applications by utilizing a wide range of natural ingredients and emulsification technologies. Due to its versatility, nanoemulsion can be applied in numerous industries such as food and beverages, cosmetics, pharmaceutical, and biomedical fields (A, Kovooru, Behera, Kumar, and Srivastava, 2021; Dasgupta and Ranjan, 2018). It is also being considered as one of the efficient and safer cancer treatments due to its therapeutic effect and target release capability (Chen and Hu, 2020; Jang, Jeong, and Lee, 2020; Sanchez-Lopez et al., 2019; Teo et al., 2014).

This chapter summarizes the recent literature on emulsification techniques, nanoemulsion preparation, and optimization methods. The understanding of the fundamental concept of emulsion preparation is crucial to design a nanoemulsion delivery system according to the desired functionality, stability, and physicochemical characteristics. In general, each emulsification technique has its advantages and disadvantages which will be discussed below. Besides, several publications regarding the nanoemulsion preparation were analyzed and highlighted to provide essential knowledge of the emulsification mechanism. Regarding the optimization methods, the basic principles of common experimental design derived from mathematical modeling were discussed in this chapter.

THE EMULSIFICATION TECHNOLOGIES USED IN PREPARING NANOEMULSION

In general, emulsification involves the disruption of initial droplets to form new smaller droplets. The emulsification technologies can be classified into high and low energy methods (Bhushani and Anandharamakrishnan, 2017). Generally, low energy methods are preferred to encapsulate sensitive compounds such as flavor compounds, probiotics, heat-labile compounds, etc. Both emulsification methods are still being used in producing nanoemulsion according to the objectives of the research. The innovation of developing or improving existing nanoemulsions can be efficiently aided with the optimization process. The benefits of optimizing nanoemulsion can discover the effect of the certain parameter or the interaction between factor variables on the output of nanoemulsion (Faghmous, Bouzid, Boumaza, Touati, and Boyron, 2020; Pongsumpun, Iwamoto, and Siripatrawan, 2020; Samiun, Ashari, Salim, and Ahmad, 2020).

Based on the Gibbs free energy (ΔG) equation, nanoemulsion formation requires high energy input and/or the presence of surfactants to disperse one phase into another phase consistently. This is because the presence of nano-sized droplets in nanoemulsion often consists of high interfacial tension between the inner and outer phases (Borthakur et al., 2016). The interfacial tension is proportional to Laplace pressure. Laplace pressure is the holding forces of droplet that must be overcome to break the droplet to nanometer size. Therefore, stabilizing the nanoemulsion requires the addition of surfactants and/or co-surfactants (Borthakur et al., 2016; Pathak, 2017; Sheth et al., 2020).

In general, the coarse emulsion must be produced by mixing the individual phase with suitable surfactants. A high-speed homogenizer (HSH) is normally used to prepare the coarse emulsion before subjecting to the nano emulsification process. The purpose of producing coarse emulsion is to allow surfactant to be blended well into respective phases (Hidajat, Jo, Kim, and Noh, 2020). Therefore, HSH has been a complementary feature in producing nanoemulsion either in high energy or low energy emulsification methods. Afterward, the coarse emulsion is subjected to high-energy and/or low energy emulsification methods to form the fine nanoemulsion. The common nano emulsification methods were briefly discussed below.

Figure 1. The illustrative emulsification method used in nanoemulsion production

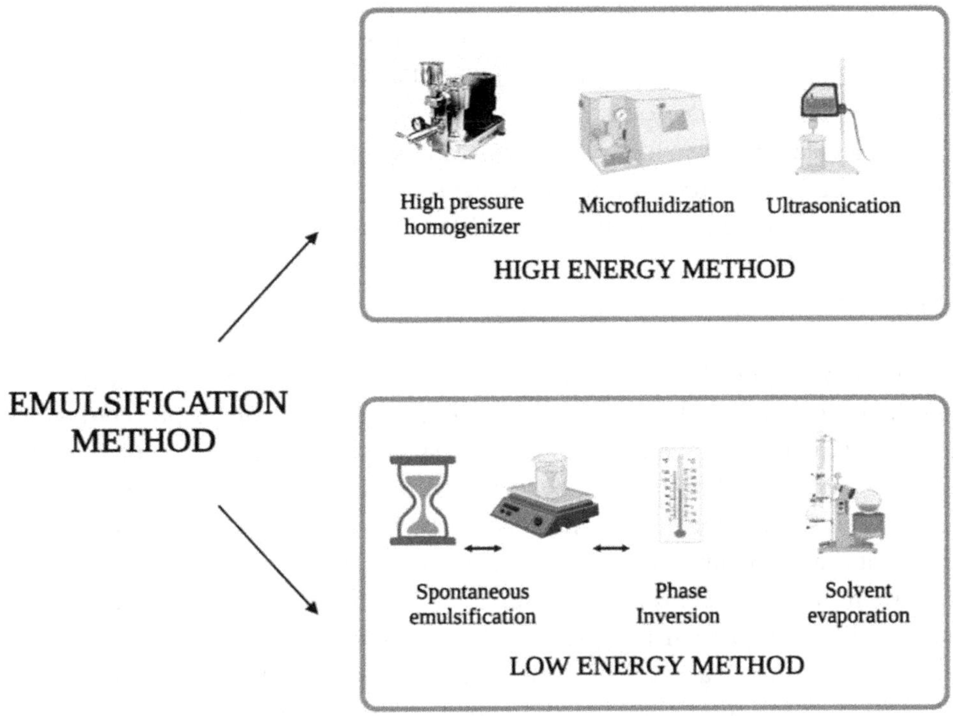

High Energy Emulsification Methods

High energy emulsification methods refer to the formation of emulsion by utilizing mechanical force to bring the immiscible liquids (oil and water) together forming homogenous droplets. Additionally, surfactants are added to stabilize the newly formed nanodroplet against destabilization forces. Nevertheless, the concentration of surfactant used in high-energy emulsification methods is lower than low energy methods (Jasmina, Džana, Alisa, Edina, and Ognjenka, 2017). The common high-energy methods are high-pressure homogenization, microfluidization, and ultrasonication.

High-pressure Homogenization

High-pressure homogenization (HPH) produces nanoemulsion by applying pressure (500-5000 psi or 3.5-35 MPa) which generates hydraulic shear and intense turbulence. Besides pressure, the type of nozzle used in the instrument can determine the fluid flow in the system. As a result, the mechanism of droplet breakdown can be different (McClements, 2015; Pathak, 2017; Rehman et al., 2020). HPH is the most widely used method to prepare nanoemulsion due to its simplicity, scalability, and consistency (Teo et al., 2014). High-pressure homogenizer devices are available in benchtop, pilot plant, and industrial-scale which is convenient to reduce the variation between the procedure in research lab and manufacturing.

The main components of the instrument consist of a pump, homogenization chamber, pressure valve, and interaction chamber (Borthakur et al., 2016). The basic concept of homogenization is classified into piston-gap and jet-stream flow direction. The piston gap homogenization concept is similar to the colloid mill principle which occurs by passing the sample through the narrow gap to break down the sample. The narrow gap lies between the moving rotor and fixed stator which produces high shear, stress, and grinding forces and leads to size reduction of droplets (Rehman et al., 2020). Piston gap homogenization is a commonly used homogenization design for mass production in food industries due to its simplified operation, scalability, and consistency (Pathak, 2017). Meanwhile, jet stream homogenization happened by the collision of samples from two high-velocity streams which reduces the droplet in size. Numerous research studies applied the high-pressure homogenizer to form nanoemulsion (Ding et al., 2021; Hidajat et al., 2020; H. D. Silva et al., 2019; W. Silva et al., 2018; Teo, Lee, and Goh, 2017).

Microfluidization

Microfluidization is the application of a high-velocity stream of fluid with relatively high pressure in the interaction chamber (Rehman et al., 2020). It usually complies with an air-powered intensifier pump which provides air turbulence in the chamber at a constant rate. With the pumping inlet, the sample passed through the channel within the chamber with the increment of pressure and number of cycles. The collision between fluids generates intense disruptive forces that produce smaller droplet sizes as the desirable output (Sadeghpour Galooyak and Dabir, 2015; Sanchez-Lopez et al., 2019). Controlling processing conditions like homogenization pressure, number of passes and surfactant concentration are important optimization parts when using microfluidization (McClements, 2015; Sadeghpour Galooyak and Dabir, 2015). The main advantages of using a microfluidizer will be the high-processing volume and high shear rates. Besides, it can form uniform droplet size (<100 nm) and distribution with lesser or no contamination risk because all of the processes are within the chamber (Pathak, 2017; Rehman et al., 2020). It can be used in small-scale production up to large-scale industrial production. In general, there are two types of microfluidizers which are single inlet and double inlet microfluidizers. A single inlet microfluidizer is mainly used in the bench setting to reduce the droplet size of the coarse emulsion into nanoemulsion (Artiga-Artigas, Lanjari-Perez, and Martin-Belloso, 2018). On the other hand, a double inlet microfluidizer is capable to produce fine nanoemulsion directly from the water and oil phase input. Therefore, it is highly advantageous than single inlet microfluidization and HPH because the pre-mixing of coarse emulsion is not required (Azmi et al., 2019).

Microfluidizers are typically suitable for homogenizing low to intermediate viscosity liquids as the fluid is required to flow at high-velocity stream in the chamber (McClements, 2015). The cost of instruments and energy consumption is relatively expensive (Jasmina et al., 2017). The energy generated by collisions dissipates as heat which is considered as energy wastage and destabilizing factor for nanoemulsion. As a result, cooling systems are required within the chamber to maintain the stability and size distribution of the nanoemulsions. Besides, excessive energy might leads to overprocessing conditions within nanoemulsion (Koubaa et al., 2018).

Ultrasonication

Ultrasonication is the application of sound energy by ultrasonicator to agitate the fluid sample. In detail, the basic principle of emulsification using ultrasonication can be classified into two stages: i) the interfacial wave is introduced to fluid which aids in droplet formation; ii) the applied ultrasound produces sound cavitation which forms the microbubble and collapses continuously due to pressure fluctuation of the sound wave. This induces highly localized turbulence and subsequently breaks the droplet into a smaller size (A et al., 2021). The functionality of ultrasonicators differs depending on the available dimensions of the instrument. In general, there are bath ultrasonication and probe ultrasonication, respectively. The former has lower sonication intensity as the sample is loaded in the container and subjected to bath ultrasonication. The container can act as the barrier for the wave to reach the sample. Meanwhile, the latter which has an indirect contact wall that acts as a barrier of the wave to reach the sample while the latter has the ultrasonic device contact directly with the sample via the ultrasonic probe. The latter showed better consistency of the nano emulsification process in terms of physicochemical properties and storage stability (Pongsumpun et al., 2020).

In terms of probe ultrasonicator, the source of the wave came from the piezo-electric crystal within ultrasonic probe which converts the electrical waves into intense pressure waves. When the probe is dipped into the sample, bubbles are produced and diminished which pressurizes the droplets to size reduction. The combination of cavitation, turbulence, and interfacial waves are the main contributing factors to nanoemulsion formation. The minimum frequency to form nanoemulsion is equal to or higher than 20 kHz (Borthakur et al., 2016). For the optimal condition of ultrasonication, the depth of the dipped probe into the liquid sample should be carefully monitored and avoid direct contact with the solid surface. The improper position of the probe can influence the pressure of sample distribution negatively. One of the main issues when applying probe ultrasonication is scalability. The sound wave distribution of the probe to the sample in a bigger vessel might not be as effective as batch emulsification in smaller scales (Bhushani and Anandharamakrishnan, 2017). Furthermore, the amplitude intensity and sonication time significantly affected the overall stability of emulsion. Leong et al. (2017) highlighted the increased amplitude intensity leads to higher shear forces to reduce droplet diameter size. The ultrasonication allows the greater exposure of hydrophobic groups of emulsifier to interface that leads to stronger film (Silva et al., 2020). This can be an alternative emulsification method to high pressure homogenization and microfluidization due to its effectiveness in generating shear forces through acoustic cavitation. In terms of energy consumption, the former method consumes lesser electrical energy than the latter. However, the amplitude and time of sonication did not have linear relationship with final droplet size. The ultrasonication might induce overprocessing condition which produces broader dispersion if it is not strictly controlled (Taha et al., 2020).

The main advantages of ultrasonication are the lower possibility of overprocessing as compared to other high-energy techniques. Despite lesser energy was consumed for the whole process, ultrasonication also can produce stable nanoemulsions with smaller droplet size and lower polydispersity index (Carpenter, George, and Saharan, 2020; Gharehbeglou, Jafari, Homayouni, Hamishekar, and Mirzaei, 2019; Rodriguez-Burneo, Busquets, and Estelrich, 2017). Besides, ultrasonication is suitable to prepare double nanoemulsions with fewer surfactant concentrations. This is considered as advantageous as the high concentration of surfactants might alter the physicochemical properties and exceed the permissible limit of surfactants (Jang et al., 2020; McClements, 2015; T. B. Tan et al., 2016). Research has compared high energy methods (HPH, microfluidization, and ultrasonication) with identical processing variables showed that ultrasonication requires less amount surfactant to produce comparable stable nanoemulsion as others (Pongsumpun et al., 2020). There are many research studies utilized ultrasonication to form nanoemulsion (Abdelrasoul et al., 2020; Alzorqi, Ketabchi, Sudheer, and Manickam, 2016; Ghazy, Fouad, Saleh, Kholif, and Morsy, 2021; Pongsumpun et al., 2020; Taha et al., 2019; Velderrain-Rodríguez, Acevedo-Fani, González-Aguilar, and Martín-Belloso, 2019).

Low energy Methods

Low-energy methods are mainly dependent on the internal energy from the chemical reaction of the sample to form nanoemulsion. Examples of low energy methods are spontaneous emulsification, phase inversion, solvent displacement techniques, etc. These techniques do not require any expensive equipment which made the emulsification process appealing on both benchtop and manufacturing scale (Bhushani and Anandharamakrishnan, 2017). However, a higher concentration of surfactants and co-surfactants are required to allow low energy emulsification to occur (Jasmina et al., 2017; McClements, 2015).

Spontaneous Emulsification

Spontaneous emulsification or commonly known as the self-emulsification process because it does not involve any phase transition and high mechanical forces to form nanoemulsion. Therefore, the method can be applied without special instruments (A et al., 2021). Numerous research studies have formed nano-emulsion with the self-emulsification method (Faghmous et al., 2020; Hosny, Al Nahyah, and Alhakamy, 2020; Xu et al., 2019). The chemical composition of the lipid phase, aqueous phase, and surfactants is important in this method. The coarse emulsion must be prepared beforehand to blend well respective phase with surfactants under magnetic stirring or rotor-stator homogenization (Teo et al., 2017). To form oil-in-water nanoemulsion, the dispersed phase consists of oil and water-soluble surfactant. The dispersed phase is then titrating into a water continuous phase under continuous stirring. When the surfactant moves from dispersed phase to continuous phase, the nanodroplets are formed (McClements, 2015).

On the other hand, spontaneous emulsification can be similar to the solvent evaporation method. After the dispersed phase is added into the continuous phase, the aqueous phase is evaporated under reduced pressure. As a result, the surfactant-covered dispersed phase in the continuous phase causes turbulence which forms small microemulsions. The turbulence at the interface between two phases must be induced and co-surfactants played a crucial role in this case (Azmi et al., 2019). When the microemulsion was diluted with water, the co-surfactant can diffuse to the aqueous phase and disrupt the formation of microemulsion. Therefore, the nanoemulsion can be formed at the same time.

Phase Inversion Emulsification

Phase inversion is defined as the transition of phase which takes place due to the chemical energy that existed in the system (Rehman et al., 2020). The phase transition happens when the alteration of spontaneous surfactant curvature occurs from positive to negative or vice versa. This is due to the changes of solubility or affinity of the nonionic surfactant in the respective phase (A et al., 2021; Azmi et al., 2019). The phase inversion method can be influenced by the temperature and composition of the emulsification process. In such a manner, phase inversion methods can be differentiated into phase inversion composition (PIC) and phase inversion temperature (PIT) (Maurya and Aggarwal, 2019).

Phase Inversion Temperature (PIT)

Phase Inversion Temperature (PIT) utilizes the changes of physicochemical properties of nonionic surfactant along with temperature changes. The formation of nanoemulsion through phase inversion temperature (PIT) can be conducted in three steps. Firstly, the mixing of oil, water, and surfactant at room temperature to prepare the coarse emulsion. Secondly, the emulsion is heated to temperature close to the PIT of the surfactant. PIT is the temperature where the interaction between emulsions and surfactant occurs (Goodarzi and Zendehboudi, 2018). The initial non-ionic surfactant has a well-hydrated group which contributes to high solubility in the water phase. However, the water solubility decreases at high temperature because the head group of surfactant becomes dehydrated (Borthakur et al., 2016). At the PIT condition, the barrier to resist the coalescence process is low and nanoemulsion can be formed. Afterward, the emulsion is rapidly cooled under continuous magnetic stirring which enables the formation of stable nanoemulsions (Bhushani and Anandharamakrishnan, 2017). Rapid cooling favors the production of small oil droplets which allows the surfactant in the oil phase to move to the aqueous phase (McClements, 2015).

The PIT of nanoemulsion is mainly affected by the type of non-ionic surfactants used in the emulsion. The ionic surfactants are not advisable to be used in the PIT emulsification method. Through the electrical conductivity measurement, PIT can be determined by visualizing the cloud points with the rise of temperature (Azmi et al., 2019; Pham and Nguyen, 2020; Sarheed et al., 2020). This method is not suitable for heat-sensitive compound encapsulation and food industries as most food processing require heating process. (Azmi et al., 2019). The temperature control must be carefully monitored as the droplet has a higher tendency to coalesce above the PIT. Nevertheless, the formation of nanoemulsions with PIT method is more preferable over PIC due to the lower surfactant usage in the formulation (Pham and Nguyen, 2020).

Phase Inversion Composition (PIC)

Phase Inversion composition is one of the isothermal techniques to form nanoemulsion (Sanchez-Lopez et al., 2019). In brief, the formation of nanodroplets is due to the existing bi continuous phase zone in the system. To develop a bi-continuous phase, the dispersed phase is blended with one portion of the continuous phase. After mixing, the remaining continuous phase is added into the former mixture to obtain the final composition of nanoemulsion. This causes the rearrangement of the dispersed phase into small droplets because the former continuous phase favors the latter added continuous phase. This trapping phenomenon resulted in phase transition and droplet formation (Azmi et al., 2019). The PIC can produce an average droplet size as small as 30 nm (Nantarat, Chansakaow, and Leelapornpisid, 2015). For example, ginger oil-in-water nanoemulsion was produced with this technique with the effect of stirring and water addition rates as their processing variables (Farshbaf-Sadigh, Jafarizadeh-Malmiri, Anarjan, and Najian, 2019).

Solvent Evaporation Method

The solvent evaporation method can form nanoemulsion by incorporating the organic solvent into the dispersed phase. As the common nanoemulsion formed is O/W configuration, the organic solvent is normally mixed with the lipid phase. However, the organic solvent is typically chosen with water-miscible characteristics and is added into the formulation which causes turbulence and supersaturation at the interface. Once the organic solvent is evaporated by a rotational evaporator, it causes shrinkage in size and forming nanodroplets (Azmi et al., 2019; Teo et al., 2014). This causes the nucleation of supersaturated oil droplets which are slightly miscible in water. As a result, the diffusion of oil droplets to neighboring droplets within the interface might occur. By diffusion, the level of supersaturation decreases and forming a metastable nanoemulsion. The condition of continuous phases and surfactant played a major role in the formation of nanoemulsion in this method (Saffarionpour, 2019). The presence of surfactant is crucial in this state to adsorb at the oil-water interface and lowering down the interfacial tension.

Examples of organic solvents that have been used for nanoemulsion production are ethyl acetate (Min, Ahn, Lee, Kwak, and Chang, 2018), acetone (Firoozy and Anarjan, 2019), and ethanol. The main advantages of this method are energy efficiency, cost-effectiveness, and scalability. Meanwhile, the obvious disadvantage will be the utilization of hazardous solvents such as ethanol and acetone (Azmi et al., 2019; Rehman et al., 2020). The application of organic solvents generally must have a high solvent-to-oil ratio to form small-sized droplets. The organic solvent must be water-miscible solvents that have lipophilic functional compounds at the same time (Sonawane, Bhanvase, and Sivakumar, 2020).

The Types of Nanoemulsion and Its Preparation

In general, nanoemulsion can be classified into single nanoemulsion and double nanoemulsion based on the number of interfacial layers (Figure 2). Moreover, the nanoemulsions can be also be differentiated based on the nature of dispersion, number of interfacial layers, and functionalized structure (A et al., 2021; Goodarzi and Zendehboudi, 2018; Pal et al., 2020; Sheth et al., 2020). The current research showed that the multilayer nanoemulsion is limited to double layers. This is because the additional layer can contribute to the increase of droplet size which therefore cannot be considered as nanoemulsion. Furthermore, the complex multilayer nanoemulsion still possesses considerable numbers of challenges in terms of stabilizing the internal structure and limitation of characterization techniques. However, overcoming these issues can bring numerous opportunities for a wide range of encapsulating applications (Koubaa et al., 2018; Sheth et al., 2020).

Figure 2. The schematic diagram of the structure of nanoemulsion

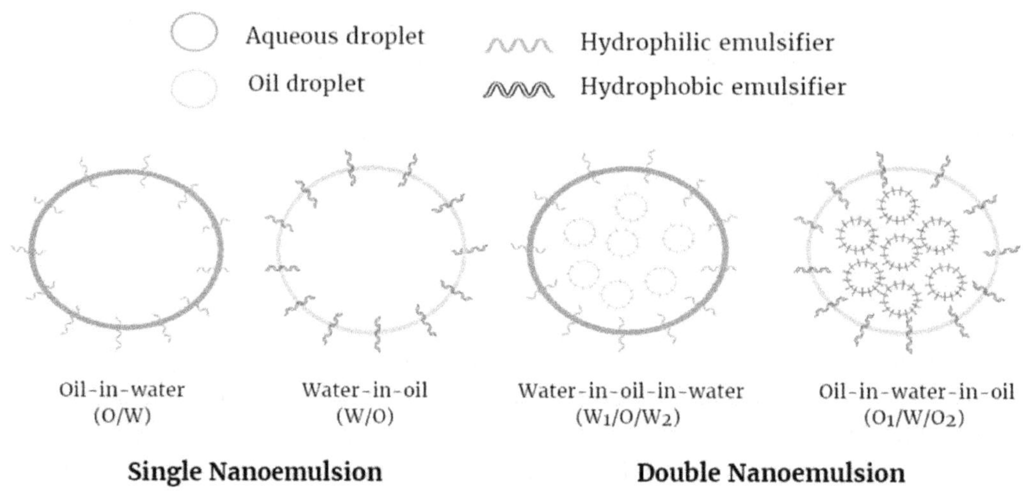

The Preparation of Single Nanoemulsion

Table 2. The recent literature publications about the preparation of single nanoemulsion

Active compound	Dispersed phase	Continuous phase	Emulsification method	Optimization method	References
Monoterpenes	Mixed monoterpenes (R-carvone, cinnamaldehyde, citral, geraniol, and pulegone), Dimethyl sulfoxide	Distilled water, Tween-80	Ultrasonication	N/A	(Abdelrasoul et al., 2020)
Flaxseed oil	Flexseed oil	Tween-80, sodium alginate	Ultrasonication	N/A	(Abbasi, Samadi, Jafari, Ramezanpour, and Shams Shargh, 2019)
Carp oil	Carp oil, Tween-80	Chitosan, gelatin, distilled water	HSH	N/A	(Esquerdo, Silva, Dotto, and Pinto, 2018)
Tocopherol	Acetone, tocopherol	Maltodextrin, distilled water, polysorbate 20	Solvent evaporation	N/A	(Firoozy and Anarjan, 2019)
Coconut oil and green tea seed oil	Coconut oil, green tea seed oil	WPI, soy lecithin, tea saponin, Tween-80 in phosphate buffer	Ultrasonication	N/A	(Gao et al., 2020)
Capsaicin	Capsaicin, peanut oil, co-surfactant (ethanol)	Distilled water, tocopheryl polyethylene glycol	HPH	N/A	(Han, Zhang, Liu, and Xiao, 2020)
Limonene	Limonene, soybean oil	Distilled water, Tween-80	HPH	OFAT	(Hidajat et al., 2020)
Methotrexate	Medium chain triglyceride (MCT), ethanol	Distilled water	HSH, Microfluidization	Pseudo-ternary phase diagram	(Jang et al., 2020)
Geranium oil	Geranium essential oil, Tween-80, ethanol	Distilled water	Ultrasonication	OFAT	(Jesser et al., 2020)
Anise extract	Anise extract	Distilled water, Tween-80	Ultrasonication	N/A	(Ghazy et al., 2021)
Vitamin D_3	Caprylic-/capric triglyceride	NaCl, distilled water, Kolliphor HS 15	Phase Inversion Temperature	Ternary diagram	(Maurya and Aggarwal, 2019)
Tamanu oil	Tamanu oil, nonionic surfactants (Tween 20,40,80)	Distilled water	Phase Inversion Temperature	N/A	(Pham and Nguyen, 2020)
Lutein	Lutein, acetone	Surfactants (Tween-80/SDS/ sodium caseinate/ SDS–Tween-80	Solvent evaporation	N/A	(T. B. Tan et al., 2016)
Cinnamon oil	Cinnamon essential oil	Tween-80, deionized water	Ultrasonication	RSM-BBD	(Pongsumpun et al., 2020)
Curcumin	Corn oil, curcumin	Sodium alginate, distilled water, surfactants (Tween-20/Lecithin// sucrose monopalmitate)	HSH, Microfluidization	N/A	(Artiga-Artigas et al., 2018)
Curcumin	Palm olein, curcumin in ethanol solution	Distilled water, Tween-80	HSH, Microfluidization	RSM-BBD	(Raviadaran, Chandran, Shin, and Manickam, 2018)
Aripiprazole	Palm Kernel Oil Esters, lecithin, Tween-80	Glycerol, deionized water	HPH	RSM-CCRD	(Samiun et al., 2020)
Clove oil	Clove oil	Surfactants (Span 80/Tween-80), distilled water	Ultrasonication	N/A	(Shahavi, Hosseini, Jahanshahi, Meyer, and Darzi, 2019)
nystatin	Paraffin oil, nystatin	Exopolysaccharides, distilled water, PEG-400	Ultrasonication	RSM, six-factor	(Song et al., 2020)
Green tea seed oil	Green tea seed oil, mixed surfactant (Tween 80 and Span 80)	Distilled water	Spontaneous emulsification	N/A	(Xu et al., 2019)
Eucommia ulmoides seed oil	Tween 80, Ethanol, *Eucommia ulmoides* seed oil	Distilled water	Microfluidization	Manual	(S. Wang et al., 2020)
Curcumin	Clove oil, Tween-80 (surfactant), PEG-400 (co-surfactant)	Distilled water	Ultrasonication	RSM-CCD, Five-factor, three-level	(Ahmad et al., 2019)
Valproic acid	MCT, safflower seed oil, valproic acid in MS, and lecithin solution	Distilled water	Ultrasonication	RSM-CCD, three-factor	(S. F. Tan et al., 2016)

(Abbreviations are given in the table: HSH – High-speed homogenization; HPH – High-pressure homogenization; N/A – Not Applicable; RSM – Response surface methodology; BBD – Box Behnken Design; CCD – Central composite design; OFAT – One factor at time; O/W – oil-in-water; W/O – water-in-oil; CCRD -)

Single nanoemulsion contains one continuous phase and dispersed phase, namely are water-in-oil and oil-in-water nanoemulsions. In recent years, oil-in-water nanoemulsion is more prevalent than water-in-oil nanoemulsion due to their lipophilic compound encapsulation and water compatibilities (Pongsumpun et al., 2020). From the oil-in-water emulsion, people have been creative in structuring the continuous phase with the complex interface for stability and functional improvement. The functionalized complex structure can be achieved by applying multilayer, mixed, or covalent complex structures (M. Li, McClements, Liu, and Liu, 2020). On the other hand, the water-in-oil emulsion is considered a cost-ineffective system as it consumed a higher volume of lipid and more susceptible to oxidation due to its lipid coating ingredients. All emulsification methods are suitable to form a single nanoemulsion. Spontaneous emulsification is widely used to form nanoemulsions in pharmaceutical industries while high energy methods such as high-pressure homogenization and microfluidization are commonly used in the food and cosmetic field. Table 2 showed the summary of recent literature about the preparation of single nanoemulsion.

The Preparation of Double Nanoemulsion

Double nanoemulsions are a complex colloidal dispersion in which emulsion is coated in another outer continuous phase (Faghmous et al., 2020). It can protect both hydrophilic and hydrophobic compounds at the same time. The emerging innovation of double nanoemulsion improves the bioavailability of emulsion carriers and protects the active compounds under environmental conditions. Typically, double emulsions have two oil-water interfaces which provide another protection layer for bioactive compounds, therefore, increasing the resistance to degradation (Li et al., 2021). Water-in-oil-in-water (W/O/W) and oil-in-water-in-oil (O/W/O) double emulsions are examples of double emulsions (J. Wang, Shi, Agyei, and Wang, 2017). W/O/W double emulsions are preferred over the latter due to lower usage of lipid. This improves the overall lipid profile of the emulsion carrier and allows good physicochemical and sensory properties. Besides, W/O/W nanoemulsion solves the oxidation susceptibility of the lipid coating phase in the conventional water-in-oil emulsion. Table 2 showed the recent literature studies conducted by numerous researchers about the preparation method of double nanoemulsion.

However, the main issues for the stability of double nanoemulsions are the stabilization of internal structure within the system. The unstable structure of double nanoemulsion might invert into the simple inverse emulsion. Therefore, the existed multiple phases might have a higher tendency of coalescence due to the inter interaction between the inner and outer aqueous phases. The type and concentration of surfactants are crucial in stabilizing the double nanoemulsion specifically (Faghmous et al., 2020; Koubaa et al., 2018). With the two interfaces that exist in the system, two types of surfactants are required to produce stable nanoemulsions. Critical micelle concentration (CMC) is the concentration point of surfactant where it self-associates and forms micelles. When the concentration of surfactant is less than CMC, the double emulsions are stable (Rehman et al., 2020). Once it reached the CMC, the increase of surfactant concentration does not influence the adsorption of droplets anymore. The surfactants self-associate and form micelles (McClements, 2015; Saffarionpour, 2019; Velderrain-Rodríguez et al., 2019).

The preparation of double nanoemulsion can be prepared in a single-step or two-step approach (Rehman et al., 2020). Single-step double nanoemulsion has the same composition for both the inner and outer aqueous phases of the system. It has a relatively low encapsulation efficiency and not suitable for high added value active compounds. Meanwhile, the two-step method has different inner and outer aqueous phases which potentially has higher encapsulation efficiency (W. Silva et al., 2018). It is preferable to apply low-energy emulsification options to prepare a secondary emulsion. This is because the high-energy method has a higher possibility to damage the integrity of encapsulated ingredients of the primary emulsion (Faghmous et al., 2020). However, studies showed that high-energy methods such as ultrasonication are still desirable to form the secondary emulsion (Gharehbeglou et al., 2019).

On the other hand, biopolymers are normally incorporated as a secondary coating for double nanoemulsions due to their stabilization properties. However, the concentration of biopolymers used in double nanoemulsions must be relatively low. This is because most biopolymers have a relatively high molecular weight which can increase the droplet size. Research conducted by Faghmous et al. (2020) used as low as ~0.07% of sodium alginate to form double nanoemulsion. In general, the biopolymer can impart electrostatic interaction and/or steric interaction depending on the types of ingredients used in the formulation. It has also been reported the blending of protein and polysaccharides biopolymers contribute to overall better stability and functionality of nanoemulsion.

Table 3. The recent literature publications about the preparation of double nanoemulsion

Active Compound	Inner Phase	Primary Continuous Phase	Outer Continuous Phase	Emulsification method	Optimization method	Size (nm)	References
Flaxseed oil	Flaxseed oil	Whey protein, distilled water	Sodium alginate, distilled water	Ultrasonication	N/A	< 500	(Abbasi et al., 2019)
Insulin	Phosphate-buffered saline	Paraffin oil, Span 80	Sodium alginate, Tween-80, Chitosan, calcium chloride	Spontaneous emulsification	RSM-BBD	< 200	(Faghmous et al., 2020)
Gallic acid	Sodium acetate buffer, electrolytes (NaCl/CaCl2/ $MgSO_4$/glucose)	Canola oil, sorbitan monooleate and PGPR	Polysorbate 80 / Polaxamer 407, polysaccharides (sodium alginate/chitosan/xanthan gum)	Phase Inversion	N/A	< 500	(Martins et al., 2020)
Oleuropein	Ultrapure water, propyl paraben	Soybean oil, Span 80,	Distilled water, Pectin, WPC	HSH, ultrasonication	RSM	< 200	(Gharehbeglou et al., 2019)
Thyme oil	Thyme oil	Tween-80, distilled water	Chitosan-alginate	Ultrasonication	N/A	50 – 165	(Hamedi, Moradi, Tonelli, and Hudson, 2019)
Gallic acid, quercetin	Gallic acid/Quercetin, ethanol, Miglyol 812	Olive oil, linseed oil, PGPR	Sodium caseinate, distilled water	Solvent evaporation, HPH	RSM- three-level	> 1000	(W. Silva et al., 2018)
Curcumin	Medium chain triglyceride	WPI, distilled water	Chitosan	HPH, spontaneous emulsification	OFAT	< 200	(H. D. Silva et al., 2019)
Mango Peel Extract (MPE)	NaCl, distilled water	Corn oil, PGPR, glycerol	NaCl, distilled water, hydrophilic surfactants (Tween-20/Tween-80/ Lecithin)	Ultrasonication	OFAT	400–800	(Velderrain-Rodríguez et al., 2019)
Emulsifier (WPH, SC, GA)	Corn oil and emulsifier solution (WPH/SC, WPH/ SC/GA)	Phosphatee buffer	N/A	HPH	N/A	140–200	(X. Li et al., 2019)
Thyme oil	Thyme oil and MCT	Whey protein isolate, distilled water	Chitosan hydrochloride, distilled water	HSH, HPH	N/A	100–300	(S. Li et al., 2021)
Red gingseng extract	Ethyl acetate	MCT, PGPR	Maltodextrin	HSH, solvent displacement technique	RSM-CCD	100–500	(Min et al., 2018)
N/A	Corn oil, ethyl acetate	WPI, distilled water	Lactoferrin, distilled water	HPH, solvent evaporation	N/A	80–2000	(Teo et al., 2017)

(Abbreviations are given in the table: HSH – High-speed homogenization; HPH – High-pressure homogenization; N/A – Not Applicable; RSM – Response surface methodology; BBD – Box Behnken Design; CCD – Central composite design; OFAT – One factor at time)

OPTIMIZATION OF NANOEMULSION PREPARATION

The preparation of nanoemulsion has included an optimization process to bring the most effective and efficient production of the final nanoemulsion. The stability of the emulsion is depending on the interactions, size, charge, composition, and method induced during preparation (Sonawane et al., 2020). The common experimental design model for optimization is the one-factor-at-time (OFAT) and factorial design model (W. Silva et al., 2018). OFAT method is suitable to determine the effect of one single factor as the independent variable with the other processing conditions are fixed variables. This study is beneficial when the researchers want to discover the influence of certain variables only (J. Wang et al., 2017). Meanwhile, factorial design models are involving in two or more factors at the same time in the experimental run. The most commonly used factorial design models are response surface methodology (RSM) with central composite design (CCD) (Min et al., 2018) or Box Behnken Design (BBD) (Faghmous et al., 2020). CCD is the typical RSM design experiment that can be created from a two-level or five-level factorial design with the center point and a group of axial points. These points allowed researchers to estimate the curvature and gave excellent prediction capability based on the design space. Meanwhile, BBD only utilizes three-level for each factor and it provides strong coefficient estimates based on center design space. However, BBD has lesser design points and thus will be having fewer experimental tests as compared to CCD. The factorial design allows the operator to run the experiments effectively and the interaction trend between the factors can be evaluated.

Many parameters must be investigated when optimizing nanoemulsions which require numerical methods such as RSM to reduce the difficulties of experimental runs (Pongsumpun et al., 2020). Numerous statistical online software is available to aid the optimization process namely SPSS, Minitab (Faghmous et al., 2020), and Design Expert (Ahmad et al., 2019; Song et al., 2020; Wahgiman, Salim, Abdul Rahman, and Ashari, 2019). Regardless of the optimization design model, the independent and dependent variables are the important parts in optimizing the emulsion system.

Factor Variables (Independent Variables)

The factor variables are the independent variables that normally modified to evaluate the response variables of final emulsion. In general, the experimental runs were conducted at a different level of all factors which are set manually or aided by software. Afterward, the response of experimental runs can be compared with the predicted responses produced by the software to determine the relevance of the study. In preparing nanoemulsion, independent variables can be divided into the preparation and/or composition variables. These independent variables are the main contributing factor which determines the desired final emulsion products (Niknam et al., 2020).

The composition variable is often the most important in optimizing nanoemulsion produced with low energy methods. Composition variables are normally prioritizing the type, concentration, and ratio of ingredients used in the formulation (Azmi et al., 2019). The concentration of ingredients is crucial to reduce the rate of flocculation within the emulsion system. The excessive amount of emulsifier might flocculate by overpassing the CMC point (Faghmous et al., 2020). Meanwhile, the preparation variable is often being the contributing factor for nanoemulsion produced by high-energy methods. Preparation variables refer to the parameters of emulsification methods. For example, the homogenization pressure, temperature and time, number of passes, etc are the main considerations when applying high-pressure homogenization and microfluidization. On the other hand, the variables for the ultrasonication process can be considered based on the sonication frequency, time, temperature, the depth of the ultrasonic probe, etc. However, numerous research studies have included both composition and preparation variables to get the best result (Bhushani and Anandharamakrishnan, 2017; Sheth et al., 2020). Multifactorial design is preferred because the interaction between several independent variables on the response variable can be evaluated.

Response Variables (Dependent Variables)

Response variables are often evaluated to determine the optimal formulation and operating conditions. The nanoemulsion is normally evaluated based on the physicochemical, functional, and target compound properties. The microstructure, encapsulation efficiencies, size, and distribution of droplets are the crucial response parameters to evaluate the overall integrity of nanoemulsion. Besides, physicochemical properties such as viscosity and creaming index can be complemented in experimental studies. The response variables are often associated with the emulsion characterization.

FUTURE TRENDS

The innovation of biomedical and pharmaceutical industries in utilizing nanoemulsion as a drug delivery system can be both functional and safer alternatives to existing drug treatments. The versatility of nanoemulsion preparation allows numerous enticing opportunities in developing a personalized system (Chen and Hu, 2020). Although double nanoemulsion has better bioavailability and storage stability, the stabilization of the multilayer structure within the system remains challenging (Faghmous et al., 2020). Moreover, the current characterization method to determine the microstructure of double nanoemulsion is still in the infancy stage. Therefore, the prospects in developing effective characterization of double nanoemulsion can be beneficial for the optimization process. The design of nano delivery system is very fast-moving trends which are considered as sustainable and efficient for delivering the functional properties of active ingredients. With the fast-moving trend of nanotechnologies, the transition to utilizing natural ingredients as the coating for nanoemulsion can be beneficial for sustainable application in different aspects. Therefore, the discovery of natural surfactants to stabilize nanoemulsion will open numerous ideas of nanoemulsion preparation (Sanchez-Lopez et al., 2019). Besides, ultrasonication has been shown as an efficient nano emulsification method with considerable low energy consumption and lesser usage of surfactants. Therefore, the future innovation in upscaling the ultrasonication emulsification process is appealing to both producers' and consumers' benefits.

CONCLUSION

The potential applications of nanoemulsion in numerous industries are boosting the research of emulsion delivery systems. The suitable operating conditions in preparing nanoemulsions with existing instruments are equally important to form a stable nanoemulsion delivery system. Current research breakthrough has allowed researchers to develop nanoemulsions based on desired composition, stability, and functional properties. The importance of understanding fundamental concepts and mechanisms of nanoemulsion preparation enables the researcher to develop the emulsion delivery system according to the needs. Despite the flexibility and versatility of method preparation, the careful selection of composition, emulsification, processing variables is crucial to design stable nanoemulsions. The optimization method accelerates the process of developing nanoemulsions with validation through experimental data analysis. Although recent publications discussed the innovative achievement in nanoemulsion technologies, further innovation in improving the molecular structures of nanoemulsion is encouraged.

REFERENCES

A, N., Kovooru, L., Behera, A. K., Kumar, K. P. P., & Srivastava, P. (2021). A critical review of synthesis procedures, applications and future potential of nanoemulsions. *Adv Colloid Interface Sci, 287*, 102318. doi:10.1016/j.cis.2020.102318

Abbasi, F., Samadi, F., Jafari, S. M., Ramezanpour, S., & Shams Shargh, M. (2019). Ultrasound-assisted preparation of flaxseed oil nanoemulsions coated with alginate-whey protein for targeted delivery of omega-3 fatty acids into the lower sections of gastrointestinal tract to enrich broiler meat. *Ultrasonics Sonochemistry, 50*, 208–217. doi:10.1016/j.ultsonch.2018.09.014 PMID:30249371

Abdelrasoul, M. A., Eid, A. R., & Badawy, M. E. I. (2020). Preparation, characterizations and antibacterial activity of different nanoemulsions incorporating monoterpenes: In vitro and in vivo studies. *Archiv für Phytopathologie und Pflanzenschutz, 53*(7-8), 310–334. doi:10.1080/03235408.2020.1744977

Ahmad, N., Ahmad, R., Al-Qudaihi, A., Alaseel, S. E., Fita, I. Z., Khalid, M. S., & Pottoo, F. H. (2019). Preparation of a novel curcumin nanoemulsion by ultrasonication and its comparative effects in wound healing and the treatment of inflammation. *RSC Advances, 9*(35), 20192–20206. doi:10.1039/C9RA03102B

Alzorqi, I., Ketabchi, M. R., Sudheer, S., & Manickam, S. (2016). Optimization of ultrasound induced emulsification on the formulation of palm-olein based nanoemulsions for the incorporation of antioxidant beta-D-glucan polysaccharides. *Ultrasonics Sonochemistry, 31*, 71–84. doi:10.1016/j.ultsonch.2015.12.004 PMID:26964925

Artiga-Artigas, M., Lanjari-Perez, Y., & Martin-Belloso, O. (2018). Curcumin-loaded nanoemulsions stability as affected by the nature and concentration of surfactant. *Food Chemistry, 266*, 466–474. doi:10.1016/j.foodchem.2018.06.043 PMID:30381213

Azmi, N. A. N., Elgharbawy, A. A. M., Motlagh, S. R., Samsudin, N., & Salleh, H. M. (2019). Nanoemulsions: Factory for Food, Pharmaceutical and Cosmetics. *Processes (Basel, Switzerland), 7*(9), 617. doi:10.3390/pr7090617

Bhushani, A., & Anandharamakrishnan, C. (2017). Food-Grade Nanoemulsions for Protection and Delivery of Nutrients. *Food-Grade Nanoemulsions for Protection and Delivery of Nutrients., 24*, 99–139. doi:10.1007/978-3-319-53112-0_3

Borthakur, P., Boruah, P. K., Sharma, B., & Das, M. R. (2016). *Nanoemulsion: Preparation and its application in food industry*. doi:10.1016/B978-0-12-804306-6.00005-2

Carpenter, J., George, S., & Saharan, V. K. (2020). Ultrasonic cavitation assisted synthesis of multilayer emulsions as encapsulating and delivery systems for bioactive compounds. *Encapsulation of Active Molecules and Their Delivery System, 23*.

Chen, J., & Hu, L. (2020). Nanoscale Delivery System for Nutraceuticals: Preparation, Application, Characterization, Safety, and Future Trends. *Food Engineering Reviews*, *12*(1), 14–31. doi:10.100712393-019-09208-w

Dasgupta, N., & Ranjan, S. (2018). Food Nanoemulsions: Stability, Benefits and Applications. *An Introduction to Food Grade Nanoemulsions*, 19-48.

Ding, M., Liu, L., Zhang, T., Tao, N., Wang, X., & Zhong, J. (2021). Effect of interfacial layer number on the storage stability and in vitro digestion of fish oil-loaded multilayer emulsions consisting of gelatin particle and polysaccharides. *Food Chemistry*, *336*, 127686. doi:10.1016/j.foodchem.2020.127686 PMID:32763735

Esquerdo, V. M., Silva, P. P., Dotto, G. L., & Pinto, L. A. A. (2018). Nanoemulsions From Unsaturated Fatty Acids Concentrates of Carp Oil Using Chitosan, Gelatin, and Their Blends as Wall Materials. *European Journal of Lipid Science and Technology*, *120*(2), 1700240. doi:10.1002/ejlt.201700240

Faghmous, N., Bouzid, D., Boumaza, M., Touati, A., & Boyron, O. (2020). Optimization of chitosan-coated W/O/W multiple emulsion stabilized with Span 80 and Tween 80 using Box–Behnken design. *Journal of Dispersion Science and Technology*, 1–13. doi:10.1080/01932691.2020.1774387

Farshbaf-Sadigh, A., Jafarizadeh-Malmiri, H., Anarjan, N., & Najian, Y. (2019). Preparation of Ginger Oil in Water Nanoemulsion Using Phase Inversion Composition Technique: Effects of Stirring and Water Addition Rates on their Physico-Chemical Properties and Stability. *Zeitschrift für Physikalische Chemie*, *0*(0). Advance online publication. doi:10.1515/zpch-2019-1427

Firoozy, M., & Anarjan, N. (2019). Preparation of maltodextrin stabilized alpha-tocopherol nanoemulsions using solvent-displacement technique. *Food Science & Technology International*, *25*(5), 404–413. doi:10.1177/1082013219825893 PMID:30704297

Gao, W., Jiang, Z., Du, X., Zhang, F., Liu, Y., Bai, X., & Sun, G. (2020). Impact of Surfactants on Nanoemulsions based on Fractionated Coconut Oil: Emulsification Stability and in vitro Digestion. *Journal of Oleo Science*, *69*(3), 227–239. doi:10.5650/jos.ess19264 PMID:32051356

Gharehbeglou, P., Jafari, S. M., Homayouni, A., Hamishekar, H., & Mirzaei, H. (2019). Fabrication of double W1/O/W2 nano-emulsions loaded with oleuropein in the internal phase (W1) and evaluation of their release rate. *Food Hydrocolloids*, *89*, 44–55. doi:10.1016/j.foodhyd.2018.10.020

Ghazy, O. A., Fouad, M. T., Saleh, H. H., Kholif, A. E., & Morsy, T. A. (2021). Ultrasound-assisted preparation of anise extract nanoemulsion and its bioactivity against different pathogenic bacteria. *Food Chemistry*, *341*(Pt 2), 128259. doi:10.1016/j.foodchem.2020.128259 PMID:33068847

Goodarzi, F., & Zendehboudi, S. (2018). A Comprehensive Review on Emulsions and Emulsion Stability in Chemical and Energy Industries. *Canadian Journal of Chemical Engineering*, *97*(1), 281–309. doi:10.1002/cjce.23336

Hamedi, H., Moradi, S., Tonelli, A. E., & Hudson, S. M. (2019). Preparation and Characterization of Chitosan–Alginate Polyelectrolyte Complexes Loaded with Antibacterial Thyme Oil Nanoemulsions. *Applied Sciences (Basel, Switzerland)*, *9*(18), 3933. doi:10.3390/app9183933

Han, J., Zhang, S., Liu, X., & Xiao, C. (2020). Fabrication of capsaicin emulsions: Improving the stability of the system and relieving the irritation to the gastrointestinal tract of rats. *Journal of the Science of Food and Agriculture*, *100*(1), 129–138. doi:10.1002/jsfa.10002 PMID:31452202

Hidajat, M. J., Jo, W., Kim, H., & Noh, J. (2020). Effective Droplet Size Reduction and Excellent Stability of Limonene Nanoemulsion Formed by High-Pressure Homogenizer. *Colloids and Interfaces*, *4*(1), 5. doi:10.3390/colloids4010005

Hosny, K. M., Al Nahyah, K. S., & Alhakamy, N. A. (2020). Self-Nanoemulsion Loaded with a Combination of Isotretinoin, an Anti-Acne Drug, and Quercetin: Preparation, Optimization, and In Vivo Assessment. *Pharmaceutics*, *13*(1), 46. Advance online publication. doi:10.3390/pharmaceutics13010046 PMID:33396942

Jang, J. H., Jeong, S. H., & Lee, Y. B. (2020). Enhanced Lymphatic Delivery of Methotrexate Using W/O/W Nanoemulsion: In Vitro Characterization and Pharmacokinetic Study. *Pharmaceutics*, *12*(10), 978. Advance online publication. doi:10.3390/pharmaceutics12100978 PMID:33081266

Jasmina, H., Džana, O., Alisa, E., Edina, V., & Ognjenka, R. (2017). *Preparation of nanoemulsions by high-energy and lowenergy emulsification methods CMBEBIH 2017*. Springer.

Jesser, E., Lorenzetti, A. S., Yeguerman, C., Murray, A. P., Domini, C., & Werdin-Gonzalez, J. O. (2020). Ultrasound assisted formation of essential oil nanoemulsions: Emerging alternative for Culex pipiens pipiens Say (Diptera: Culicidae) and Plodia interpunctella Hubner (Lepidoptera: Pyralidae) management. *Ultrasonics Sonochemistry*, *61*, 104832. doi:10.1016/j.ultsonch.2019.104832 PMID:31675660

Koubaa, M., Roohinejad, S., Sharma, P., Nikmaram, N., Hashemi, S. S., Abbaspourrad, A., & Greiner, R. (2018). Multiple Emulsions. *Emulsion-based Systems for Delivery of Food Active Compounds: Formation, Application, Health and Safety*, 69.

Leong, T. S. H., Zhou, M., Kukan, N., Ashokkumar, M., & Martin, G. J. O. (2017). Preparation of water-in-oil-in-water emulsions by low frequency ultrasound using skim milk and sunflower oil. *Food Hydrocolloids*, *63*, 685–695. doi:10.1016/j.foodhyd.2016.10.017

Li, M., McClements, D. J., Liu, X., & Liu, F. (2020). Design principles of oil-in-water emulsions with functionalized interfaces: Mixed, multilayer, and covalent complex structures. *Comprehensive Reviews in Food Science and Food Safety*, *19*(6), 3159–3190. doi:10.1111/1541-4337.12622 PMID:33337043

Li, S., Sun, J., Yan, J., Zhang, S., Shi, C., McClements, D. J., Liu, X., & Liu, F. (2021). Development of antibacterial nanoemulsions incorporating thyme oil: Layer-by-layer self-assembly of whey protein isolate and chitosan hydrochloride. *Food Chemistry*, *339*, 128016. doi:10.1016/j.foodchem.2020.128016 PMID:33152858

Li, X., Wu, G., Qi, X., Zhang, H., Wang, L., & Qian, H. (2019). Physicochemical properties of stable multilayer nanoemulsion prepared via the spontaneously-ordered adsorption of short and long chains. *Food Chemistry*, *274*, 620–628. doi:10.1016/j.foodchem.2018.09.002 PMID:30372987

Liu, Q., Huang, H., Chen, H., Lin, J., & Wang, Q. (2019). Food-Grade Nanoemulsions: Preparation, Stability and Application in Encapsulation of Bioactive Compounds. *Molecules (Basel, Switzerland), 24*(23), 4242. Advance online publication. doi:10.3390/molecules24234242 PMID:31766473

Maphosa, Y., & Jideani, V. A. (2018). *Factors Affecting the Stability of Emulsions Stabilised by Biopolymers*. doi:10.5772/intechopen.75308

Martins, C., Higaki, N. T. F., Montrucchio, D. P., Oliveira, C. F., Gomes, M. L. S., Miguel, M. D., Miguel, O. G., Zanin, S. M. W., & Dias, J. F. G. (2020). Development of W1/O/W2 emulsion with gallic acid in the internal aqueous phase. *Food Chemistry, 314*, 126174. doi:10.1016/j.foodchem.2020.126174 PMID:31981885

Maurya, V. K., & Aggarwal, M. (2019). A phase inversion based nanoemulsion fabrication process to encapsulate vitamin D3 for food applications. *The Journal of Steroid Biochemistry and Molecular Biology, 190*, 88–98. doi:10.1016/j.jsbmb.2019.03.021 PMID:30923018

McClements, D. J. (2015). *Food emulsions: Principles, practices, and techniques*. CRC Press. doi:10.1201/b18868

Min, J.-Y., Ahn, S.-I., Lee, Y.-K., Kwak, H.-S., & Chang, Y. H. (2018). Optimized conditions to produce water-in-oil-in-water nanoemulsion and spray-dried nanocapsule of red ginseng extract. *Food Science and Technology (Campinas), 38*(3), 485–492. doi:10.1590/fst.09517

Nantarat, T., Chansakaow, S., & Leelapornpisid, P. (2015). Optimization, characterization and stability of essential oils blend loaded nanoemulsions by PIC technique for anti-tyrosinase activity. *International Journal of Pharmacy and Pharmaceutical Sciences, 7*, 308–312.

Niknam, S. M., Escudero, I., & Benito, J. M. (2020). Formulation and preparation ofwater-in-oil-in-water emulsions loaded with a phenolic-rich inner aqueous phase by application of high energy emulsification methods. *Foods, 9*(10), 13–28. doi:10.3390/foods9101411 PMID:33027911

Pal, K., Banerjee, I., Sarkar, P., Kim, D., Deng, W. P., Dubey, N. K., & Majumder, K. (2020). *Biopolymer-Based Formulations: Biomedical and Food Applications*. Academic Press.

Pathak, M. (2017). *Nanoemulsions and Their Stability for Enhancing Functional Properties of Food Ingredients*. doi:10.1016/B978-0-12-811942-6.00005-4

Pham, D. H., & Nguyen, T. T. (2020). Preparation of Tamanu Oil Nanoemulsions by Phase Inversion Temperature. *IOP Conference Series. Materials Science and Engineering, 991*, 012116. doi:10.1088/1757-899X/991/1/012116

Pongsumpun, P., Iwamoto, S., & Siripatrawan, U. (2020). Response surface methodology for optimization of cinnamon essential oil nanoemulsion with improved stability and antifungal activity. *Ultrasonics Sonochemistry, 60*, 104604. doi:10.1016/j.ultsonch.2019.05.021 PMID:31539730

Raviadaran, R., Chandran, D., Shin, L. H., & Manickam, S. (2018). Optimization of palm oil in water nano-emulsion with curcumin using microfluidizer and response surface methodology. *Lebensmittel-Wissenschaft + Technologie, 96*, 58–65. doi:10.1016/j.lwt.2018.05.022

Rehman, A. U., Akram, S., Seralin, A., Vandamme, T., & Anton, N. (2020). *Lipid nanocarriers: Formulation, properties, and applications*. doi:10.1016/B978-0-12-816770-0.00021-6

Rodriguez-Burneo, N., Busquets, M. A., & Estelrich, J. (2017). Magnetic Nanoemulsions: Comparison between Nanoemulsions Formed by Ultrasonication and by Spontaneous Emulsification. *Nanomaterials (Basel, Switzerland)*, *7*(7), 190. Advance online publication. doi:10.3390/nano7070190 PMID:28737673

Sadeghpour Galooyak, S., & Dabir, B. (2015). Three-factor response surface optimization of nano-emulsion formation using a microfluidizer. *Journal of Food Science and Technology*, *52*(5), 2558–2571. doi:10.100713197-014-1363-1 PMID:25892755

Saffarionpour, S. (2019). Preparation of Food Flavor Nanoemulsions by High- and Low-Energy Emulsification Approaches. *Food Engineering Reviews*, *11*(4), 259–289. doi:10.100712393-019-09201-3

Samiun, W. S., Ashari, S. E., Salim, N., & Ahmad, S. (2020). Optimization of Processing Parameters of Nanoemulsion Containing Aripiprazole Using Response Surface Methodology. *International Journal of Nanomedicine*, *15*, 1585–1594. doi:10.2147/IJN.S198914 PMID:32210553

Sanchez-Lopez, E., Guerra, M., Dias-Ferreira, J., Lopez-Machado, A., Ettcheto, M., Cano, A., Espina, M., Camins, A., Garcia, M. L., & Souto, E. B. (2019). Current Applications of Nanoemulsions in Cancer Therapeutics. *Nanomaterials (Basel, Switzerland)*, *9*(6), 821. Advance online publication. doi:10.3390/nano9060821 PMID:31159219

Sarheed, O., Shouqair, D., Ramesh, K., Khaleel, T., Amin, M., Boateng, J., & Drechsler, M. (2020). Formation of stable nanoemulsions by ultrasound-assisted two-step emulsification process for topical drug delivery: Effect of oil phase composition and surfactant concentration and loratadine as ripening inhibitor. *International Journal of Pharmaceutics*, *576*, 118952. doi:10.1016/j.ijpharm.2019.118952 PMID:31843549

Shahavi, M. H., Hosseini, M., Jahanshahi, M., Meyer, R. L., & Darzi, G. N. (2019). Evaluation of critical parameters for preparation of stable clove oil nanoemulsion. *Arabian Journal of Chemistry*, *12*(8), 3225–3230. doi:10.1016/j.arabjc.2015.08.024

Sheth, T., Seshadri, S., Prileszky, T., & Helgeson, M. E. (2020). Multiple nanoemulsions. *Nature Reviews. Materials*, *5*(3), 214–228. doi:10.103841578-019-0161-9

Silva, H. D., Beldíková, E., Poejo, J., Abrunhosa, L., Serra, A. T., Duarte, C. M. M., Brányik, T., Cerqueira, M. A., Pinheiro, A. C., & Vicente, A. A. (2019). Evaluating the effect of chitosan layer on bioaccessibility and cellular uptake of curcumin nanoemulsions. *Journal of Food Engineering*, *243*, 89–100. doi:10.1016/j.jfoodeng.2018.09.007

Silva, M., Anh Bui, T. H., Dharmadana, D., Zisu, B., & Chandrapala, J. (2020). Ultrasound-assisted formation of double emulsions stabilized by casein-whey protein mixtures. *Food Hydrocolloids*, *109*(June), 106143. doi:10.1016/j.foodhyd.2020.106143

Silva, W., Torres-Gatica, M. F., Oyarzun-Ampuero, F., Silva-Weiss, A., Robert, P., Cofrades, S., & Gimenez, B. (2018). Double emulsions as potential fat replacers with gallic acid and quercetin nanoemulsions in the aqueous phases. *Food Chemistry*, *253*, 71–78. doi:10.1016/j.foodchem.2018.01.128 PMID:29502846

Sonawane, S., Bhanvase, B. A., & Sivakumar, M. (2020). *Encapsulation of Active Molecules and Their Delivery System*. Elsevier.

Song, R., Yan, F., Cheng, M., Dong, F., Lin, Y., Wang, Y., & Song, B. (2020). Ultrasound-Assisted Preparation of Exopolysaccharide/Nystatin Nanoemulsion for Treatment of Vulvovaginal Candidiasis. *International Journal of Nanomedicine*, *15*, 2027–2044. doi:10.2147/IJN.S241134 PMID:32273700

Taha, A., Ahmed, E., Hu, T., Xu, X., Pan, S., & Hu, H. (2019). Effects of different ionic strengths on the physicochemical properties of plant and animal proteins-stabilized emulsions fabricated using ultrasound emulsification. *Ultrasonics Sonochemistry*, *58*, 104627. doi:10.1016/j.ultsonch.2019.104627 PMID:31450289

Taha, A., Ahmed, E., Ismaiel, A., Ashokkumar, M., Xu, X., Pan, S., & Hu, H. (2020). Ultrasonic emulsification: An overview on the preparation of different emulsifiers-stabilized emulsions. *Trends in Food Science & Technology*, *105*, 363–377. doi:10.1016/j.tifs.2020.09.024

Tan, S. F., Masoumi, H. R., Karjiban, R. A., Stanslas, J., Kirby, B. P., Basri, M., & Basri, H. B. (2016). Ultrasonic emulsification of parenteral valproic acid-loaded nanoemulsion with response surface methodology and evaluation of its stability. *Ultrasonics Sonochemistry*, *29*, 299–308. doi:10.1016/j.ultsonch.2015.09.015 PMID:26585010

Tan, T. B., Chu, W. C., Yussof, N. S., Abas, F., Mirhosseini, H., Cheah, Y. K., Nehdi, I. A., & Tan, C. P. (2016). Physicochemical, morphological and cellular uptake properties of lutein nanodispersions prepared by using surfactants with different stabilizing mechanisms. *Food & Function*, *7*(4), 2043–2051. doi:10.1039/C5FO01621E PMID:27010495

Teo, A., Goh, K., & Lee, S. J. (2014). *Nanoparticles and nanoemulsions. Functional foods and dietary supplements: Processing effects and health benefits*. John Wiley & Sons, Ltd. Publication.

Teo, A., Lee, S. J., & Goh, K. K. T. (2017). Formation and stability of single and bi-layer nanoemulsions using WPI and lactoferrin as interfacial coatings under different environmental conditions. *Food Structure*, *14*, 60–67. doi:10.1016/j.foostr.2017.06.005

Velderrain-Rodríguez, G. R., Acevedo-Fani, A., González-Aguilar, G. A., & Martín-Belloso, O. (2019). Encapsulation and stability of a phenolic-rich extract from mango peel within water-in-oil-in-water emulsions. *Journal of Functional Foods*, *56*, 65–73. doi:10.1016/j.jff.2019.02.045

Wahgiman, N. A., Salim, N., Abdul Rahman, M. B., & Ashari, S. E. (2019). Optimization of nanoemulsion containing gemcitabine and evaluation of its cytotoxicity towards human fetal lung fibroblast (MRC5) and human lung carcinoma (A549) cells. *International Journal of Nanomedicine*, *14*, 7323–7338. doi:10.2147/IJN.S212635 PMID:31686809

Wang, J., Shi, A., Agyei, D., & Wang, Q. (2017). Formulation of water-in-oil-in-water (W/O/W) emulsions containing trans-resveratrol. *RSC Advances*, *7*(57), 35917–35927. doi:10.1039/C7RA05945K

Wang, S., Wang, X., Liu, M., Zhang, L., Ge, Z., Zhao, G., & Zong, W. (2020). Preparation and characterization of Eucommia ulmoides seed oil O/W nanoemulsion by dynamic high-pressure microfluidization. *Lebensmittel-Wissenschaft + Technologie*, *121*, 108960. doi:10.1016/j.lwt.2019.108960

Xu, W., Zhu, D., Xu, B., Huang, L., Xiong, Y., Ge, P., & Wang, Z. (2019). Preparation, antibacterial and antioxidant properties of green tea seed oil nanoemulsions by self-emulsification method. *Micro & Nano Letters*, *14*(12), 1219–1222. doi:10.1049/mnl.2019.0185

Chapter 3
Chitosan-Stabilized Oil-in-Water Nanoemulsions:
Electrokinetic Properties

Viktoria Milkova
https://orcid.org/0000-0002-6737-1936
Institute of Physical Chemistry, Bulgarian Academy of Sciences, Bulgaria

ABSTRACT

Chitosan is a natural polysaccharide and emulsifier that can ensure a significant emulsion stability at suitable pH, ionic strength, composition, concentration, or thermal processing. The evaluation of the electrokinetic properties is a key factor in investigation of the stability of the nanoemulsions with a view to their potential applications in bionanotechnology. Consequently, the precise control over the physico-chemical characteristics of chitosan (degree of acetylation, DA and molecular weight, Mw) can provide a high stability and specific biological properties of the developed functional structures. The chapter is focused on the interpretation of the electrokinetic response from nanoemulsion stabilized by adsorption of chitosan (as a polyelectrolyte or uncharged polymer) by using appropriate theoretical approaches.

INTRODUCTION

Chitosan is a natural polysaccharide with unique structure and properties, highly sophisticated functionality and a wide range of application in biotechnology. It is almost the one polycationic polysaccharide in nature with remarkable potential in medicine and pharmacy. The polymer is biocompatible, biodegradable and non-toxic to living tissues as well as it has an antibacterial, antifungal, anti-inflammatory and anti-tumour activity. Therefore it is relevant for various bio-medical and pharmaceutical applications (wound-healing, drug delivery, vaccine delivery, gene therapy, tissue engineering, bio-imagining) (Dash *et al.*, 2011).

Because of the pronounced capacity of chitosan to stabilize nano- and microstructures suitable for tissue engineering and drug encapsulation, the polymer is used for design of various functional delivery platforms. The stability of the produced nano- or microformulations is a key factor for their applications.

DOI: 10.4018/978-1-7998-8378-4.ch003

Since the significant part of the developed structures are result of an electrostatic assembly process of charged components (polyelectrolytes, proteins, dyes, enzymes, bio-active substances, aptamers etc.), the stability of the systems is govern predominantly by the electrical properties of the components and the electrostatic interactions between them.

The ability of chitosan to stabilize emulsions is due mainly to the structural heterogeneity of the molecules. Therefore the variations in the molecular structure of polymer through control on the degree of acetylation or molecular weight may entail different charge distribution, electrical properties of molecules and emulsified capacity.

Concerning on the nanoemulsions, the mixtures of oil in water are thermodynamically instable systems and require presence of surface active substances (emulsifiers) in order to improve the stability of the produced emulsion droplets. The stability against flocculation, coalescence, creaming or Ostwald ripening depends on the surface properties of the droplets, stability of the thin liquid film between them and rheological characteristics of the system. The adsorption of emulsifier on the droplet surface decreasing the surface tension. The stability of the dispersion is due to the repulsive steric interactions between the interface surfactant layers and repulsive electrostatic interactions between similarly charged droplets.

The present chapter is addressed to the ability of chitosan as a polyelectrolyte or uncharged polymer to stabilize nanoemulsions through formation thick adsorption monolayer or chitosan-based multilayer film on droplets.

The evaluation of the electrokinetic properties of the nanoemulsion is a key point in the investigation of the stability of the produced systems. Therefore, the precise control over the physicochemical characteristics of chitosan can provide a high stability and specific biological properties of the developed functional structures. The interpretation of the experimental data for electrokinetic potential or electrophoretic mobility of the nanoemulsion droplets stabilized by adsorbed polymer mono- or multilayer requires application of appropriate theoretical approaches. In the present chapter are pointed out two scenarios for interpretation of the electrokinetic response from nanoemulsion stabilized with chitosan.

BACKGROUND

Effect of Chitosan in Stabilization of Oil-in-Water Nanoemulsion

Chitosan is a natural polysaccharide, copolymer of *β-(1-4)-2-acetamido-2-deoxy-β-D-glucopyranose* and *2-amino-2-deoxy- β-D-glucopyranose*. The polymer has been used as an emulsifier for stabilization of convectional emulsions through increasing the viscosity of the continuous phase or improving the emulsion stability by electrostatic adsorption on the oppositely charged surface of surfactant-stabilized droplets (electrosteric stabilization) (Jumaa and Muller, 1999; Klinkesorn, 2013; Wang and Heuzey, 2016). The value of hydrophilic / lipophilic balance (HLB) of chitosan that express the ability of emulsifier to stabilize the emulsions is ca. 34-36 (Schulz *et al.*, 1998).

The ability of chitosan to stabilize the emulsion is govern predominantly by its structural heterogeneity (Klinkesorn, 2013). The chitosan molecule consists of block-wise or alternating sequence of hydrophilic (deacetylated) and hydrophobic (acetylated) D-glucosamine residues and the degree of acetylation define the hydrophilic / hydrophobic character of the polymer chain (Figure 1).

Figure 1. Schematic chemical structure of chitosan: acetylated (A) and deacetylated (D) monomers.

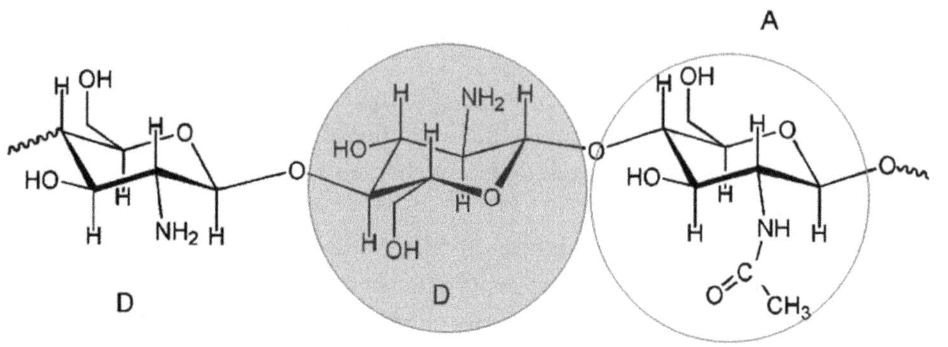

Recently has been reported general findings about the correlation in the stability of chitosan-coated nanoemulsion droplets and physicochemical characteristic of polymer (molecular weight or DA) or experimental conditions (salt concentration, pH of the dispersion, polymer and droplet concentration) (Del Blanco *et al.*, 1999; Mun *et al.*, 2005; Li and Xia, 2011; Santander-Ortega *et al.*, 2011). Hence, the variation in the molecular structure of chitosan correlate with the physicochemical, bio-active and pharmacological properties of its molecules, and polymer capacity for development and improving the stability of many soft functional nanostructures (emulsions, liposomes, capsules, etc.). Moreover, chitosan-covered nanoemulsion droplets are shown remarkable stability in biological medium (Santander-Ortega *et al.*, 2010; Goycoolea *et al.*).

Electrokinetic Properties of Nanoemulsion

The extensive experimental and theoretical studies have been focussed on the characterization of electrokinetic behaviour of emulsion droplets and their stability. Recently, the interest has focused on the investigation of the electrokinetic response of nanoemulsion, in order to evaluate the stability of the emulsion-based drug delivery platforms suitable for various applications in bio-nano-technology (Makvandi *el al.*, 2021).

The electrokinetic potential of the surfactant-stabilized droplets in nanoemulsion has conventionally measured by using electrophoresis. However, the analysis of electrophoretic mobility or electrokinetic ζ-potential of the droplets is more complicated compare to the rigid particles because of the droplet hydrodynamics, stabilizing agents or droplet deformation. The main reason is that liquid flow may also occur within the droplets because the momentum transfer across the interface and the classical notion for slip plane loses its meaning (Delgado *et al.*, 2007). As the result from the flow inside the droplets, the tangential velocity of the liquid surrounding droplet does not have to become zero at the particle surface. Moreover, the droplet hydrodynamic is complicated by the presence of applied electric field in electro-kinetic measurements. Therefore, the experimental value of the electrophoretic mobility for droplets is higher than for a corresponding rigid particle. In order to ensure stability of the droplets, surface active components have to be adsorb on the droplet surface. The presence of emulsifiers (polymers, surfactants or proteins) adsorbed on the surface further complicate the droplet hydrodynamics but because of presence of these adsorbed layers, the droplet surface will behaves as rigid and the momentum transfer to the liquid droplet will be very slight (Washington *et al.*, 1989). If the flow of the surrounding liquid is not conveyed into the particle interior so that there occurs no fluid flow inside the particle core, then the particle core behaves like a rigid unreformed spherical particle (Barchini and Saville, 1996).

Moreover, the general question is if the electrokinetic models developed for emulsion are applicable for nanoemulsion. Because of the small droplet size and large surface/volume ratio, it is comprehensible to expect different surface properties of the nano-droplets. Therefore, on the basis of the standard electrokinetic model have been proposed theoretical approaches suitable for characterization of the electrokinetic behavior of nano-emulsions. (De Aguiar *et al.*, 2010; Zdrali *et al.*, 2019; Hill and Afuwape, 2020; Hill, 2020).

The adsorption of polyelectrolyte on the droplet surface make the analysis of the electrokinetic behaviour of the system more complicated. In this case, the interpretation of the electrophoretic mobility of the droplets with adsorbed polyelectrolyte-surfactant membrane has performed on the basis of suggestion that the electrical properties of the adsorbed polymer define the properties of coated particles (McClements, 2012).

Counterion Condensation at Fully Charged Chitosan Polycation

In solution of highly charged polyelectrolytes some of the counterions undergo "condensation" near to the polyion chain (Manning, 1978). This phenomena occurs when the average distance b between the two elementary backbone charges e is less than the Bjerrum length l_b:

$$l_b = \frac{e^2}{\varepsilon kT} \tag{1}$$

The Manning's dimensionless charge parameter, related to the charge density of polyelectrolyte is defined as

$$\xi = l_b / b \tag{2}$$

When $\xi > 1$, monovalent counterions "condense" along the polyion chain to yield an effective charge density one electron per Bjerrum length. This means that the "real" charge density of the polyion is reduced and becomes close to that of the low charged polyelectrolyte.

Upon the polyelectrolyte adsorption on an oppositely charged surface, full release of the condensed counterions from the molecules is not always observed (Sens and Joanny, 2000). Previously have been shown that upon the adsorption of high charged polyelectrolyte on weak charged surface, the molecules retain a part of condensed counterions and the fraction of released counterions, β, is increasing with the charge density of the surface (Cheng and Olvera de la Cruz, 2003):

$$\log\left[8\xi\left(1-\beta\right)\right] = \log\left[\left(\kappa r\right)^2\right] - 2\xi\beta\log\left[\frac{rh}{L_k\left(h\right)^2}\right] + 2\log\left[\frac{\kappa\left(h+\lambda\right)}{2}\right] \tag{3}$$

where κ^{-1} is the Debye screening length, r is a chain radius, h is a distance between the polymer and surface (or the thickness of the adsorbed layer), σ is the surface charge density λ is a Gouy-Chapmann length and L_k is a local screening length in the Gouy-Chapmann layer, respectively

$$\lambda = 1/\left(2\pi l_B \sigma\right) \tag{4}$$

$$L_k = \lambda / \sqrt{2} \tag{5}$$

These results are in line with the previous experimental and theoretical studies on polyelectrolyte adsorption on rigid particles surface which predict that in the electrokinetic measurements is estimated the "effective" charge of the particles or emulsion droplets covered by polyelectrolyte (Milkova and Radeva, 2006).

In the presence of a weak electric field (like in the electrokinetic measurements) the polymer coated macroion (particle or emulsion droplet) and its counterions (diffuse and condensed), drifts together along the electric field in the direction implied by the inverted charge, and the charge inversion is not altered until the electric field exceeds a critical value.

Chitosan is a weak polyelectrolyte with reactive ($-NH_2$) functional groups that are responsible for the positive charge of the molecules under acidic conditions (pKa~6.5) (Rinaudo *et al.*, 1993). Therefore, the degree of substitution with acetic residues ($-NH\text{-}CH_3$) is in correlation to the charge density of the polymer. At high degree of acetylation (high DA), when the significant part of the chitosan monomers along the polymer chain are acetylated (or the number of functional NH_2-groups is low), the polymer charge density is low. Whereas at low DA, the high number of NH_2-groups define a typical behaviour of chitosan as a polycation. The distance between two charges along the chitosan chain, b, is related to the DA by

$$b=0.515/(1-DA) \tag{6}$$

where 0.515 nm is the length per monosaccharide unit (Arguelles-Mondal *et al.*, 2000). At very low charge density (DA<28%), the average distance b becomes smaller than the Bjerrum length. Consequently, for chitosan with high charge density it is can expect the condensation of the part of counterions (Schatz *et al.*, 2004).

STABILIZATION OF OIL-IN-WATER NANOEMULSION DROPLETS BY ADSORPTION OF CHITOSAN MONOLAYER OR CHITOSAN-BASED MULTILAYER FILMS

Deposition of Chitosan Monolayer on Oil-in-Water Nanoemulsion Droplets

In order to explain the process of stabilization of the emulsion through adsorption of chitosan, in the present chapter is considered a stable nanoemulsion from negatively charged oil droplets dispersed in water. The charge of the droplets is govern by the presence of emulsifier adsorbed on the surface. The nanoemulsion is stable because of strong repulsive steric interactions between the interface surfactant layers on the droplet surface and the participation of repulsive electrostatic interactions between similarly charged droplets. In order to prevent the unfavourable interactions and improve the stability of the system, the droplet concentration in the dispersion is very low ($C_{droplet}$<C*, C* is the concentration of droplets in suspension if the double electric layers of the two adjacent droplets are touched).

Therefore, the electrokinetic behaviour and stability of the nanoemulsion in the presence of chitosan can be evaluated as a function of the concentration and physicochemical characteristics (DA and Mw).

At very low concentrations of chitosan added to the dispersion, all molecules are adsorb on the droplet surface due to the strong electrostatic attraction between the positively charged monomers and negatively charged moiety of the surfactant layer on the surface (Figure 2). Consequently, the electrokinetic potential of the droplets will gradually decreases with increasing the concentration of polymer as a result of partial neutralization of the surface charge upon the chitosan adsorption. Simultaneously, the size of the droplets increases due to the formation of adsorption layer onto the droplet surface and / or flocculation in the emulsion. The full aggregation in the dispersion is corresponding to the presence of neutral polymer-droplets complexes stick to each other due to the strong short-range attraction. The peak value of the size of the droplets is indicator for this aggregation. The coincidence of the concentration for maximum aggregation with the isoelectric points (IEP) is indicator that the interaction between chitosan molecules and droplets is mainly electrostatic, otherwise the participation of other no-electrostatic or/ and hydrophobic interactions have to be taken into account.

Figure 2. Dependence of the electrokinetic potential, ζ, and diameter of nanoemulsion droplets, D, as a function of concentration of chitosan (DA 15% and Mw 50 kDa) in dispersion.

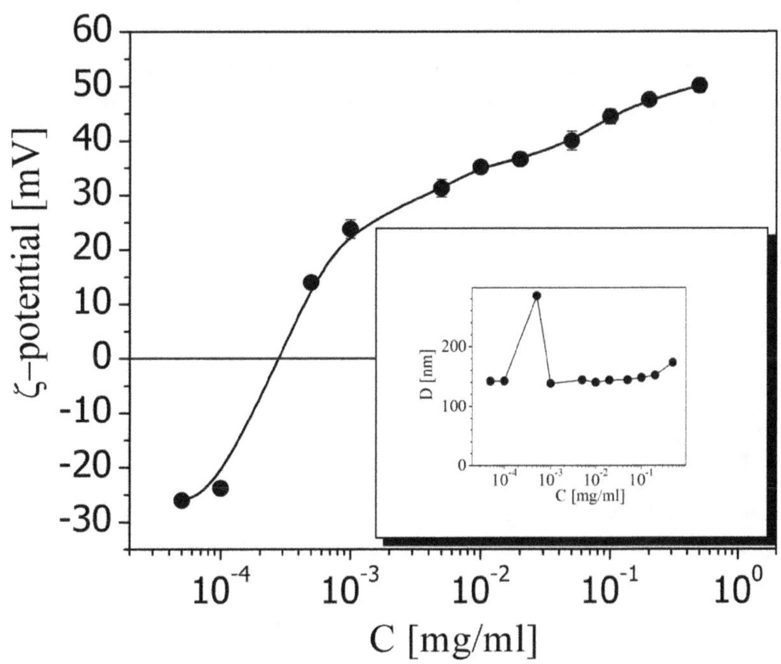

The further increasing of the polymer concentration leads to increase in the electrokinetic charge (in absolute value) of the polymer-coated droplets, but with opposite sign. This is indicator for the over-compensation of the surface charge and re-stabilization of the system. The nanoemulsion stabilization is result from the increased electrostatic and steric repulsion between isolated droplets covered with chitosan layer (electrosteric stabilization) (Klinkesorn, 2013).

The presented adsorption scenarios illustrates a typical polyelectrolyte behaviour of chitosan in stabilization of nanoemulsion (or particles) but the organization of the molecules and structure of the deposited layer on the surface depends on the Mw and DA of chitosan (Quemeneur et al., 2007; Quemeneur et al., 2010). Recent studies have been shown that the adsorption of chitosan with low Mw or high DA promoted more efficient packing of the molecules on the surface, while the deposition of chitosan with high Mw or low DA results in more heterogeneous distribution of the chains (Goycoolea *et al.*, 2012).

In order to evaluate the effect of chitosan charge density on emulsion stabilization could be distinguish three DA-dependent regimes (Figure 3.). At very low DA of chitosan, the polyelectrolyte adsorption on negatively charged droplet surfaces is govern by the strong attractive electrostatic interactions (Sonvico *et al.*, 2006). Moreover, upon the adsorption of fully charged chains (DA<28%) on weakly charged surfaces, the molecules could retain a part from the condensed counterions that results in decreasing in the "real" polyelectrolyte charge density. That is the reason for observed by electrokinetics almost the same values of the electrokinetic potential of the nanoemulsion droplets with adsorbed chitosan with different charge density (Goycoolea and Milkova, 2017; Milkova, 2021).

The subsequent increasing of DA of chitosan molecules results in increasing of the fraction of uncharged monomers and the polymer backbone becomes more hydrophobic. Therefore, the adsorption of chitosan is govern by the contribution of electrostatic and hydrophobic interactions. The decreasing of the polymer charge density correlates with the decreasing of the electrokinetic potential of the chitosan-coated particles, but the organization of the molecules on the surface and structure of the adsorbed layer depends on the balance of electrostatic / hydrophobic interactions and molecular weight (length of the polymer chain, respectively). At very high DA of the chitosan, the hydrophobic interaction between the polymer molecules and surface are dominate (Chandler, 2005).

In the present consideration, the analysis of the electrical properties of nanoemulsion with droplets covered by surfactant /chitosan layer require the application of different electrokinetic models.

Figure 3. Chitosan adsorption on surfactant-stabilized nanoemulsion droplets.

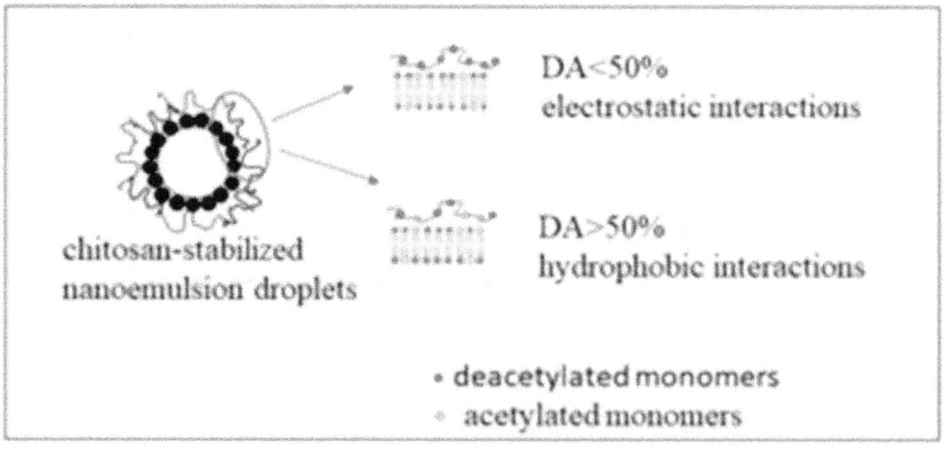

First, the ζ-potential of droplets with adsorbed fully charged chitosan molecules (low DA) can be estimated with assumption that the droplet behaves as a rigid particle. The approximation is correct if the flow of the surrounding liquid is not conveyed into the particle interior (because of the presence of surfactant layer) so that there occurs no fluid flow inside the particle core (Delgado et al., 2007).

Moreover, the estimation is correct if the adsorption of chitosan on the droplets surface is govern by the electrostatic interactions. Therefore, according to the theory, for diluted suspension of weak charged rigid particles with radius r, in the range between low and high κr regime, in such conditions that the concentration polarizability and surface conductivity are negligible, the ζ-potential can be calculated by using Henry's formula (O'Brien and White, 1978):

$$U_{ef} \frac{\varepsilon \varepsilon_0 \zeta}{\eta} f(\kappa r) \tag{7}$$

where κ is the Debye-Hückel parameter, ε is the relative permittivity of electrolyte solution, ε_0 is the permittivity of vacuum, η is viscosity of solution an $f(\kappa r)$ is Henry's function. Consequently, the ζ-potential of the chitosan-coated droplets can be calculated by using standard electrokinetic model by using the experimental values for the electrophoretic mobility, U_{ef}.

Second, at the very high degree of acetylation of the chitosan (DA > 50%) is expected the adsorption of almost uncharged polymer chains on the negatively charged emulsion droplets. In this case, the process is govern predominantly by the attractive hydrophobic interactions between the hydrophobic moiety of chains and hydrophobic patches of the surfactant layer on droplets. In this a point of view, the chitosan oligosaccharides (COS, DA > 90%) give unique opportunity for precisely characterization of and distinguishing of the contribution of the electrostatic and hydrophobic interactions in analysis of the adsorption or electrokinetic behaviour of dispersion in the presence of chitosan (Milkova, 2021).

The elecrokinetic response from dispersion of rigid particles in the presence of uncharged polymer has been shown the decreasing of the experimental value of ζ-potential. Generally, this observation is addressed to an outward shift of the "plane of shear" of the double electric layer of the particles and defines the "effective electrokinetic layer thickness", δ_E (Cohen Stuart *et al.*, 1984; Delgado *et al.*, 2007). In the limit of very low electrolyte concentration, the calculated value of δ_E is identical to hydrodynamic thickness of the adsorbed polymer layer, δ_H and can be estimated by using the relation derived by Cohen Stuart et. al, (Cohen Stuart *et al*, 1984).

$$\delta_E = \frac{1}{\kappa a} ln \frac{\zeta_0}{\zeta}, \tag{8}$$

where ζ_o is the electrokinetic potential of uncovered surface, ζ is the potential of surface with adsorbed polymer, I is the ionic strength of the suspension, a is corresponding to the cross-section of the polymer chain.

Stabilization of Oil-in-Water Nanoemulsion Droplets by Adsorption of Chitosan-based Multilayer Film

Among the huge number methods for developing of innovative structures suitable for drug delivery, the systems based on nanoemulsion offered exceptional opportunities for encapsulation of different hydrophobic bio-active molecules (Rosas-Durazo *et al.*, 2011). Moreover, the development of oil-core nanocapsules produced through adsorption of mono-or multilayer hydrophilic polymer film on nano-emulsion droplets allows to co-encapsulate hydrophobic and hydrophilic drugs. The drug in the produced nanostructures is protected by the polymer shell and drug release occurs upon the destruction of the shell (Shchukina and Shchukin, 2012; Szczepanowicz *et al.*, 2015). The specific properties of the polymers in the shell determine the bio-distribution, selectivity and stability of the nanostructures, whereas the internal structure of the shell effects on the drug release.

The multilayer oil-core nanocapsules can be formed through subsequent electrostatic adsorption of oppositely charged polyelectrolytes (Layer-by-Layer method, LbL method) (Decher *et al.*, 1992). The classical LbL method is an electrostatic assembly process, where the charged substrate is alternately exposed into solutions of two oppositely charged polyelectrolytes. The overcompensation of the surface charge after each adsorption step ensures the adsorption of the next layer from polyelectrolyte with opposite charge. The subsequent deposition process finally results in multilayered polyelectrolyte complex, stabilized by strong electrostatic forces. Because the driving force of the film formation is the electrostatic interaction between the oppositely charged components, the method is one of the most versatile techniques for design of nanostructures functional coatings on charged colloidal core with different nature, properties, shape and size. The procedure allows to incorporate various small charged components (inorganic or polymer nanoparticles, proteins, dyes, enzymes and etc.) into the polymer shell results in multicomponent shell with specific properties (thickness, permeability, morphology, water content, ion content) The drug encapsulation in the capsules can allow achieving fine control of the processes of transport and drug release through the membrane. In this case, the properties of the multilayer membrane will define the mechanisms and kinetics of the process.

Figure 4. Procedure for preparation of polymer-coated nanoemulsion droplets in o/w emulsion by adsorption of chitosan monolayer or chitosan-based multilayer film.

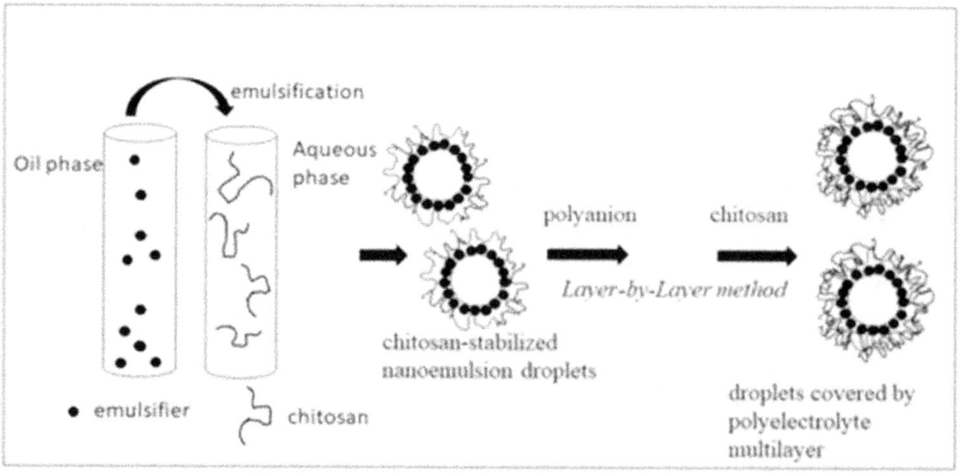

Chitosan as a natural polyelectrolyte is preferable component in development of oil-core capsules. In series of investigations, McClements *et al.* have been reported the application of the LbL method for development of emulsions stabilized by deposition of chitosan-based multilayer film. The authors have been studied the influence of different factors on the properties and stability of the produced emulsions (pH, ionic strength, concentration of droplets and chitosan, droplet size, thermal proceedings) (Ogawa *et al.*, 2004; McClements, 2014). In Figure 4 is presented the procedure used for the development of much diluted nanoemulsion consists of oil droplets covered by polyelectrolyte multilayer (Milkova and Goycoolea, 2020). The system has been used for the investigation of the dependence of the encapsulation efficiency and stability of caffeine loaded oil-in-water nanoemulsion droplets as a function of the number of adsorbed chitosan/alginate bi-layers and physicochemical characteristics of chitosan. The procedure is developed on the basis of the original method for preparation of stable nanoemulsions proposed by Calvo *et.al.* (Calvo *et al.*, 1997).

The interpretation of the electrokinetic response of the multi-layered nanoemulsion is similar to the behaviour of the particles covered with polymer film because the general assumption is that the droplets behaves as a rigid particles. The experimental results indicate the overcompensation of the surface charge (oscillation) after each deposition step formed from oppositely charged polyelectrolytes. The electrical properties of the component in the last deposited layer define the electrical properties of the produced structure.

FUTURE RESEARCH DIRECTIONS

The direction of the fundamental studies focused on the characterization of the properties and stability of polymer-stabilized nanoemulsion are in correlation with opportunity for their application in biotechnology, medicine or pharmacy.

CONCLUSION

In the present chapter are presented two scenarios for interpretation of the electrokinetic response from nanoemulsion stabilized by chitosan-surfactant membrane. The evaluation of the electrokinetic properties and stability of these system is very important with respect to their potential in the development of drug delivery platforms suitable for application in medicine or pharmacy.

The ability of chitosan to stabilize the emulsion is govern by the structural heterogeneity of the molecules and corresponding variation in hydrophilic / hydrophobic character of the polymer chain. At very low charge density of the polymer chain (when DA<50%), chitosan evinces a typical polyelectrolyte behaviour and the capacity to stabilize nanoemulsion is due to the strong electrostatic interactions between the positively charged monomers and negatively charged surface of surfactant-stabilized droplets. The registered electrical properties of nanoemulsion can be interpreted by using a standard electrokinetic model.

Respectively, at very high charge density (when DA>50%), because of the pronounced hydrophobic character of the polymer chain, the ability of chitosan to stabilize the nanoemulsion results predominantly from the attractive hydrophobic interactions between the hydrophobic moiety of chains and hydrophobic patches of the surfactant layer on droplets. Therefore, the electrical properties of nanoemulsion can be explained by using the approximation for the outward shift of the "plane of shear" of the double electric layer of the droplets.

In the other point of view, the development of oil-core nanocapsules produced through adsorption of mono-or multilayer hydrophilic polymer film on nanoemulsion droplets allows to encapsulate various (hydrophilic and hydrophobic) bio-active molecules. According to the presented comparative analysis of the nanoemulsion stabilized by absorption of chitosan-based film, the precise control over the physicochemical characteristics of polymer can provide a high stability and specific biological properties of the developed functional structures.

REFERENCES

Arguelles-Mondal, W., Cabrera, G., Peniche, C., & Rinaudo, M. (2000). Conductimetric study of the interpolyelectrolyte reaction between chitosan and polygalacturonic acid. *Polymer*, *41*(7), 2373–2378. doi:10.1016/S0032-3861(99)00396-1

Barchini, R., & Saville, D. A. (1996). Electrokinetic Properties of Surfactant-Stabilized Oil Droplets. *Langmuir*, *12*(6), 1442–1445. doi:10.1021/la950693a

Calvo, P., Remufin-Ldpez, C., Vila-Jato, J. L., & Alonso, M. J. (1997). Development of positively charged colloidal drug carriers: Chitosan-coated polyester nanocapsules and submicron-emulsions. *Colloid & Polymer Science*, *275*(1), 46–53. doi:10.1007003960050050

Chandler, D. (2005). Interfaces and the driving force of hydrophobic assembly. *Nature*, *437*(7059), 640–647. doi:10.1038/nature04162 PMID:16193038

Cheng, H., & Olvera de la Cruz, M. (2003). Adsorption of rod-like polyelectrolytes onto weakly charged surfaces. *The Journal of Chemical Physics*, *119*(23), 12635–12644. doi:10.1063/1.1626630

Cohen Stuart, M. A., Waajen, F. H. W. H., & Dukhin, S. S. (1984). Electrokinetic effects of adsorbed neutral polymers. *Colloid & Polymer Science*, *262*(5), 423–426. doi:10.1007/BF01410263

Dash, M., Chiellini, F., Ottenbrite, E., & Chiellini, E. (2011). Chitosan-A Versatile Semi-Synthetic Polymer in Biomedical Applications. *Progress in Polymer Science*, *36*(8), 981–1014. doi:10.1016/j.progpolymsci.2011.02.001

De Aguiar, H. B., De Beer, A. G. F., Strader, M. L., & Roke, S. (2010). The interfacial tension of nanoscopic oil droplets in water is hardly affected by SDS surfactant. *Journal of the American Chemical Society*, *132*(7), 2122–2123. doi:10.1021/ja9095158 PMID:20121125

Decher, G., Hong, J. D., & Schmitt, J. (1992). Buildup of ultrathin multilayer films by a self-assembly process: III. Consecutively alternating adsorption of anionic and cationic polyelectrolytes on charged surfaces. *Thin Solid Films*, *210/211*, 831–835. doi:10.1016/0040-6090(92)90417-A

Del Blanco, L. F., Rodriguez, M. S., Schulz, P. C., & Agguló, E. (1999). Influence of the deacetylation degree on chitosan emulsification properties. *Colloid & Polymer Science*, *277*(11), 1087–1092. doi:10.1007003960050495

Delgado, A. V., Gonzalez-Caballero, F., Hunter, R. J., Koopal, L. K., & Lyklema, J. (2007). Measurement and interpretation of electrokinetic phenomena. *Journal of Colloid and Interface Science*, *309*(2), 194–224. doi:10.1016/j.jcis.2006.12.075 PMID:17368660

Goycoolea, F. M., & Milkova, V. (2017). Electrokinetic behaviour of chitosan adsorbed on o/w nanoemulsion droplets. *Colloids and Surfaces. A, Physicochemical and Engineering Aspects*, *519*, 205–211. doi:10.1016/j.colsurfa.2016.05.093

Goycoolea, F. M., Valle-Gallego, A., Stefani, R., Menchicchi, B., David, L., Rochas, C., Santander-Ortega, M. G., & Alonso, M. J. (2012). Chitosan-based nanocapsules: Physical characterization, stability in biological media and capsaicin encapsulation. *Colloid & Polymer Science*, *290*(14), 1423–1434. doi:10.100700396-012-2669-z

Hill, R. J. (2020). Electrokinetic spectra of dilute surfactant-stabilized nano-emulsions. *Journal of Fluid Mechanics*, *902*, A15. doi:10.1017/jfm.2020.579

Hill, R. J., & Afuwape, G. (2020). Dynamic mobility of surfactant-stabilized nano-drops: Unifying equilibrium thermodynamics, electrokinetics and Marangoni effects. *Journal of Fluid Mechanics*, *895*, A14. doi:10.1017/jfm.2020.256

Jumaa, M., & Muller, B. W. (1999). Physicochemical properties of chitosan-lipid emulsions and their stability during autoclaving process. *International Journal of Pharmaceutics*, *183*(2), 175–184. doi:10.1016/S0378-5173(99)00086-1 PMID:10361168

Klinkesorn, U. (2013). The role of chitosan in emulsion formation and stabilization. Food Reviews International. *Taylor & Francis Group*, *29*(4), 371–393. doi:10.1080/87559129.2013.818013

Li, X., & Xia, W. (2011). Effect of concentration, degree of deacetylation and molecular weigh on emulsifying properties of chitosan. *International Journal of Biological Macromolecules*, *48*(5), 768–772. doi:10.1016/j.ijbiomac.2011.02.016 PMID:21382402

Makvandi, P., Josic, U., Delfi, M., Pinelli, F., Jahed, V., Kaya, E., Ashrafizadeh, M., Zarepour, A., Rossi, F., Zarrabi, A., Agarwal, T., Zare, T. N., Ghomi, M., Maiti, T. K., Breschi, L., & Tay, F. R. (2021). Drug Delivery (Nano)Platforms for Oral and Dental Applications: Tissue Regeneration, Infection Control, and Cancer Management. *Advanced Science (Weinheim, Baden-Wurttemberg, Germany)*, *8*(8), 2004014. doi:10.1002/advs.202004014 PMID:33898183

Manning, G. S. (1978). Limiting laws and counterion condensation in polyelectrolyte solutions V. Further development of the chemical model. *Biophysical Chemistry*, *9*(1), 65–70. doi:10.1016/0301-4622(78)87016-1 PMID:753405

McClements, D. J. (2012). Nanoemulsions versus microemulsions: Terminology, differences, and similarities. *Soft Matter*, *8*(6), 1719–1729. doi:10.1039/C2SM06903B

McClements, D. J. (2014). Advances in fabrication of emulsions with enhanced functionality using structural design principles. *Current Opinion in Colloid & Interface Science*, *17*(5), 235–245. doi:10.1016/j.cocis.2012.06.002

Milkova, V. (2021). Electrosteric stabilization of oil/water emulsions by adsorption of chitosan oligosaccharides – An electrokinetic study. *Carbohydrate Polymers*, *265*, 118072–1180820. doi:10.1016/j.carbpol.2021.118072 PMID:33966836

Milkova, V., & Goycoolea, F. M. (2020). Encapsulation of caffeine in polysaccharide oil-core nanocapsules. *Colloid & Polymer Science*, *298*(8), 1035–1041. doi:10.100700396-020-04653-0

Milkova, V., & Radeva, Ts. (2006). Counterion release from adsorbed highly charged polyelectrolyte – an electrooptical study. *Journal of Colloid and Interface Science*, *298*(2), 550–555. doi:10.1016/j.jcis.2006.01.020 PMID:16480999

Mun, S., Decker, E. A., & McClements, D. J. (2005). Influence of droplet characteristics on the formation of oil-in water emulsions stabilized by surfactant-chitosan layers. *Langmuir*, *12*(14), 6228–6234. doi:10.1021/la050502w PMID:15982024

O'Brien, R., & White, L. R. (1978). Electrophoretic mobility of a spherical colloidal particle. *Journal of the Chemical Society. Faraday Transactions II*, *43*, 1607–1626. doi:10.1039/f29787401607

Ogawa, S., Decker, E. A., & McClements, D. J. (2004). Production and characterization of O/W emulsions containing droplets stabilized by lechithin-chitosan-pectin multilayered membranes. *Journal of Agricultural and Food Chemistry*, *52*(11), 3595–3600. doi:10.1021/jf034436k PMID:15161236

Quemeneur, F., Rammal, A., Rinaudo, M., & Pepin-Donat, B. (2007). Large and giant vesicles "decorated" with chitosan: Effects of pH, salt or glucose stress, and surface adhesion. *Biomacromolecules*, *8*(8), 2512–2519. doi:10.1021/bm061227a PMID:17658883

Quemeneur, F., Rinaudo, M., Maret, G., & Pepin-Donat, B. (2010). Decoration of lipid vesicles by polyelectrolytes: Mechanism and structure. *Soft Matter*, *6*(18), 4471–4481. doi:10.1039/c0sm00154f

Rinaudo, M., Milas, M., & Le Dung, P. (1993). Characterization of chitosan. Influence of ionic strength and degree of acetylation on chain expansion. *International Journal of Biological Macromolecules*, *15*(5), 281–285. doi:10.1016/0141-8130(93)90027-J PMID:8251442

Rosas-Durazo, A., Lizardi, J., Higuera-Ciapara, I., Arduelles-Monal, W., & Goycoolea, F. M. (2011). Development and characterization of nanocapsules comprising dodecyltrimethylammonium chloride and κ-carrageenan. *Colloids and Surfaces. B, Biointerfaces*, *86*(1), 242–246. doi:10.1016/j.colsurfb.2011.03.020 PMID:21497069

Santander-Ortega, M. J., Peula-Garcia, J. M., Goycoolea, F. M., & Ortega-Vinuesa, J. L. (2011). Chitosan nanocapsules: Effect of chitosan molecular weight and acetylation degree on electrokinetic behavior and colloidal stability. *Colloids Surf. B. Biointerphases*, *82*(2), 571–580. doi:10.1016/j.colsurfb.2010.10.019 PMID:21071187

Santander-Ortega, M. J., Stauner, T., Loretz, B., Ortega-Vinuesa, J. L., Bostos-Gonzalez, D., Wenz, G., Schaefer, U. F., & Lehr, C. M. (2010). Nanoparticles made from novel starch derives for transdermal drug delivery. *Journal of Controlled Release*, *141*(1), 85–92. doi:10.1016/j.jconrel.2009.08.012 PMID:19699771

Schatz, Ch., Lucas, J.-M., Viton, Ch., Domard, A., Pichot, Ch., & Delair, Th. (2004). Formation and properties of positively charged colloids based on polyelectrolyte complexes of biopolymers. *Langmuir, 20*(18), 7766–7778. doi:10.1021/la049460m PMID:15323530

Schulz, P. C., Rodriguez, M. S., Del Blanco, L. F., Pistonesi, M., & Agullo, E. (1998). Emulsification properties of chitosan. *Colloid & Polymer Science, 267*(12), 1159–1165. doi:10.1007003960050359

Sens, P., & Joanny, J.-F. (2000). Counterion release and electrostatic adsorption. *Physical Review Letters, 84*(21), 4862–4265. doi:10.1103/PhysRevLett.84.4862 PMID:10990817

Shchukina, E. M., & Shchukin, D. (2012). Layer-by-layer coated emulsion microparticles as storage and delivery tool. *Current Opinion in Colloid & Interface Science, 17*(5), 281–289. doi:10.1016/j.cocis.2012.06.003

Sonvico, F., Cagnani, A., Rossi, A., Motta, S., Di Bari, T. T., Cavatorta, F., Alonso, M. J., Deriu, A., & Colombo, P. (2006). Formation of self-organized nanoparticles by lecithin/chitosan ionic interaction. *International Journal of Pharmaceutics, 324*(1), 67–73. doi:10.1016/j.ijpharm.2006.06.036 PMID:16973314

Szczepanowicz, K., Bazylinska, U., Pietkiewicz, J., Szyk-Warszynska, L., Wilk, K. A., & Warszynski, P. (2015). Biocompatible long-sustained release oil-core polyelectrolyte nanocarriers: From controlling physical state and stability to biological impact. *Advances in Colloid and Interface Science, 222*, 678–691. doi:10.1016/j.cis.2014.10.005 PMID:25453660

Wang, X. Y., & Heuzey, M. C. (2016). Chitosan-based conventional and pickering emulsions with long-term stability. *Langmuir, 32*(4), 924–936. doi:10.1021/acs.langmuir.5b03556 PMID:26743171

Washington, C., Chawla, A., Christy, N., & Davis, S. S. (1989). Electrokinetic properties of phospholipid-stabilized fat emulsions. *Intertational J. of Pharmaceutics, 54*(3), 191–197. doi:10.1016/0378-5173(89)90096-3

Zdrali, E., Etienne, G., Smolentsev, N., Amstad, E., & Roke, S. (2019). The interfacial structure of nano- and micron-sized oil and water droplets stabilized with SDS and Span80. *The Journal of Chemical Physics, 150*(20), 204704. doi:10.1063/1.5083844 PMID:31153210

ADDITIONAL READING

Chaudharya, S., Kumara, S., Kumarb, V., & Sharmac, R. (2020). Chitosan nanoemulsions as advanced edible coatings for fruits and vegetables: Composition, fabrication and developments in last decade. *International Journal of Biological Macromolecules, 152*(1), 154–170. doi:10.1016/j.ijbiomac.2020.02.276 PMID:32109479

Choi, A. J., Kim, C. J., Cho, Y. J., Hwang, J. K., & Kim, C. T. (2011). Characterization of Capsaicin-Loaded Nanoemulsions Stabilized with Alginate and Chitosan by Self-assembly. *Food and Bioprocess Technology, 4*(6), 1119–1126. doi:10.100711947-011-0568-9

Giacone, D.V., Dartora, V.F.M.C., de Matos. J.K.R., Passos. J.S., Miranda, D.A.G., de Oliveira, E.A., Silveira, E.R., Costa-Lotufo, L.V., Maria-Engler, S.S., & Lopes, L.B. (2020). Effect of nanoemulsion modification with chitosan and sodium alginate on the topical delivery and efficacy of the cytotoxic agent piplartine in 2D and 3D skin cancer models. *International Journal of Biological Macromolecules, 165*(Pt A), 1055-1065. doi:10.1016/j.ijbiomac.2020.09.167

Gupta, A., Eral, H. B., Hattona, T. A., & Doyle, P. S. (2016). Nanoemulsions: Formation, properties and applications. *Soft Matter, 12*(11), 2826–2841. doi:10.1039/C5SM02958A PMID:26924445

Jaiswal, M., Dudhe, R., & Sharma, P. K. (2015). Nanoemulsion: an advanced mode of drug delivery system. *3 Biotech., 5*(2), 123–127. doi:10.1007/s13205-014-0214-0

Langaroudi, N. F., & Kazemi, N. M. (2019). Preparation and characterization of O/W nanoemulsion with Mint essential oil and Parsley aqueous extract and the presence effect of chitosan. *Nanomed. Res. J., 4*(1), 48–55. doi:10.22034/nmrj.2019.01.008

Liu, C., Tan, Y., Xu, Y., McCleiments, D. J., & Wang, D. (2019). Formation, characterization and application of chitosan/pectin-stabilized multilayer emulsions as astaxanthin delivery systems. *International Journal of Biological Macromolecules, 140*, 985–997. doi:10.1016/j.ijbiomac.2019.08.071 PMID:31401274

Yang, E., Lee, J. W., Chang, P. S., & Park, I. K. (2021). Development of chitosan-coated nanoemulsions of two sulfides present in onion (*Allium cepa*) essential oil and their nematicidal activities against the pine wood nematode, *Bursaphelenchus xylophilus. Environmental Science and Pollution Research International.* Advance online publication. doi:10.100711356-021-15451-8 PMID:34291413

Chapter 4
Composition and Efficacy of Essential Oil Nanoemulsions

Sowbarnika Arul Senthil
https://orcid.org/0000-0002-6354-1774
B.S. Abdur Rahman Crescent Institute of Science and Technology, India

Sneha Unnikrishnan
B.S. Abdur Rahman Crescent Institute of Science and Technology, India

Karthikeyan Ramalingam
https://orcid.org/0000-0002-9334-427X
B.S. Abdur Rahman Crescent Institute of Science and Technology, India

ABSTRACT

Essential oil, endorsed as a natural cum powerful ingredient, encompasses incredible properties, and various approaches have been proposed and spurred over the last decades. One among them is the nano-emulsification of the essential oils, and the sub-micron size provides prolonged shelf life, ease of formulation and handling, nullification of the stated obstacles, as well as the augmentation of the essential oil antimicrobial activity. This chapter primarily reinforces and is well-articulated on (1) the components and its precise concentration adequate to fabricate nanoemulsion; (2) critical analysis on the experimentally inferred antimicrobial mechanism and the extent of contribution granted by the optimal concentration of the components in the nanoemulsion; (3) scrutinization of the possible impact rendered by the components on the other characteristics of nanoemulsion aside from antimicrobial potency. Altogether, the importance of careful formulation of the essential oil-loaded nanoemulsion is emphasized extensively.

INTRODUCTION

Paracelsus von Hohenheim, who entitled "Quinta essential" for the effective component present in the drug, was the first person ever to use the term "ESSENTIAL OIL", in the 16th century (Guenther, 1952). Essential oils are defined as a complex, but diverse group of volatile aromatic compounds extracted

DOI: 10.4018/978-1-7998-8378-4.ch004

from non-woody plant materials extensively (Edris, 2007). In general, essential oil resides in specialized structures present in the surfaces of plants such as secretory glandules or internal compartments like large vacuoles, owing to their elevated molecular reactivity (Dima & Dima., 2015). It's regarded to acquire a wide range array of biological activities, like a pool of phytochemicals that could be isolated from different parts of the plant (Burt., 2004). In particular, it exhibits a high level of antibacterial as well as antifungal activities. This is because, the major component of EO comprises a wide range array of aldehyde, alcohols, esters, hydrocarbons (aliphatic) which are low in molecular weight, followed by terpenoids, especially monoterpenes (C10), sesquiterpenes (C15), and diterpenes (C20) (Dorman & Deans, 2000; Jones, 1996; LisBalchin & Deans, 1997) (***FIG 1***). However, their composition might vary as per their biological and chemical inception and to the diverse processing methods which could be subjected, for extraction and purification process (Sharma & Tripathi, 2008).

Figure 1. Representative structures typical for oregano essential oil present in the nanoemulsion. It comprises Carvacrol, Thymol, Gamma terpinene & beta-fenchyl alcohol (from clockwise)

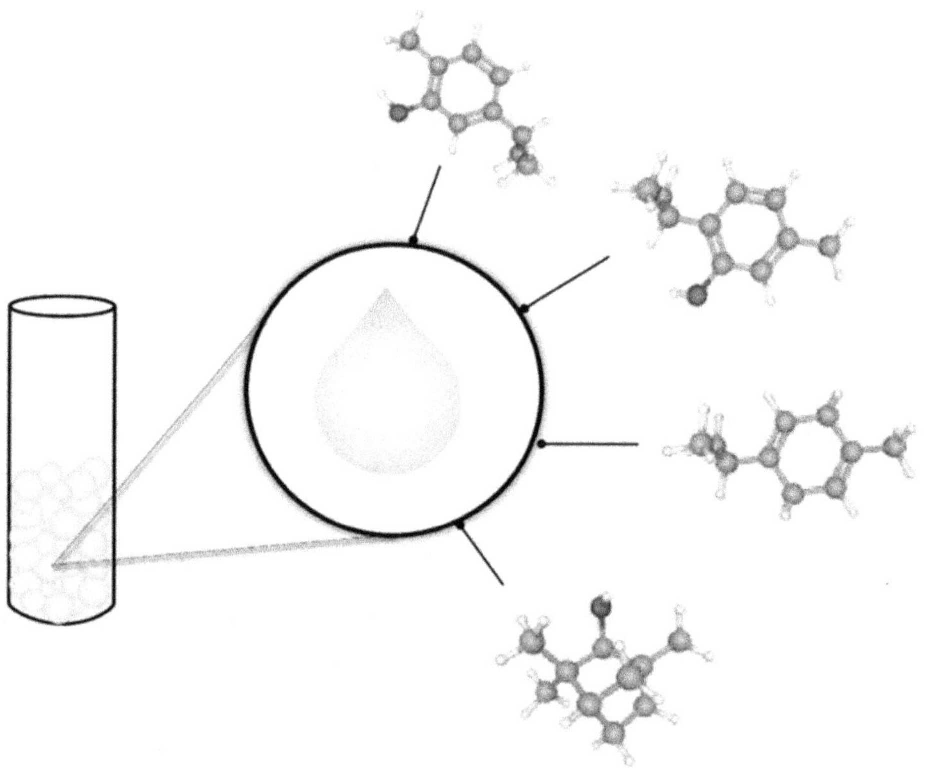

The antibacterial attributes of essential oil are attributed to the phenolic compounds present in it (Cosentino *et al.*, 1999). In particular, the acquisition of various compounds by the essential oil are characterized by a distinct mode of activity, which relies on the nature of the specific bio-active molecule (Di Pasqua *et al.*, 2007). Additionally, several pieces of literature proposed the antagonistic mechanisms carried out by these EO's which are: (a) Cell wall degradation (b) Vandalization of the plasma membrane (c) Creating deterioration to the proteins of the membrane (d) Exudation of cellular components (e) Cytoplasm condensation and Proton motive force exhaustion (Burt, 2004) (***FIG 4***). For an instance, carvacrol, a phenolic compound, likely enacts as protonophore, i.e., a proton transporter across the phospholipid bilayer, thereby the proton motive force is dispersed (Lambert *et al.*, 2001; Ultee *et al.*, 2002). Additionally, the synergistic action between various components present in the essential oil is said to improve the bio-efficacy and preventing the progress of aversion in microorganisms (Basak & Guha, 2018). Essential oils have their own huge set of commercial applications, since consumer's demand for natural in preference to the synthesized additives are on a steep rise, provided that the essential oils are regarded as "Natural" components (Chang *et al.*, 2011 & 2012). About 3000 Essential oils are estimated to be present and well-recognized, of which 300 oils are traded extensively (Freires, Denny, Benso, de Alencar, Rosalen., 2015). Although essential oils exhibit superior antibacterial, antifungal, insecticidal, antioxidant & therapeutic properties, practical application is greatly limited and remains challenging since these EO's are highly volatile & exhibits a frequent tendency to degrade as a result of oxidation via treatment to heat, light, oxygen, humidity, directly (Bilia *et al.*, 2014) (***FIG 2 & 3***).

Figure 2. Determinants of the essential oil's stability

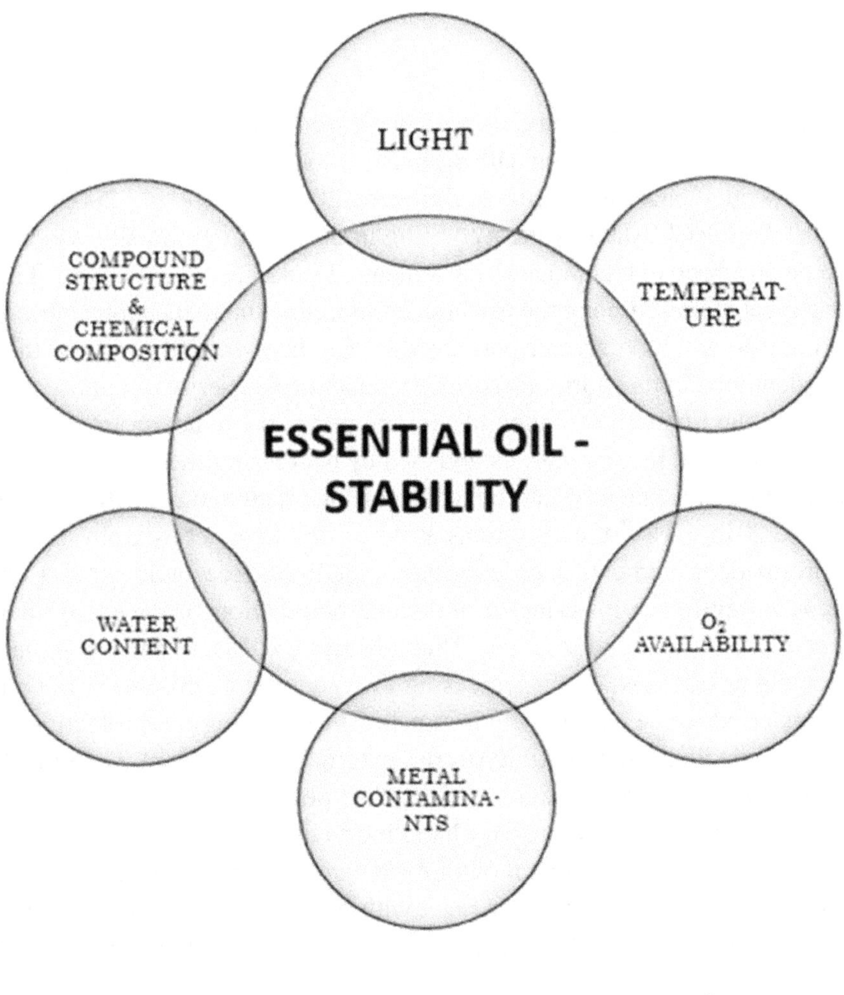

Figure 3. Proposed conversion reactions takes place in essential oil, on exposure to heat/air/light

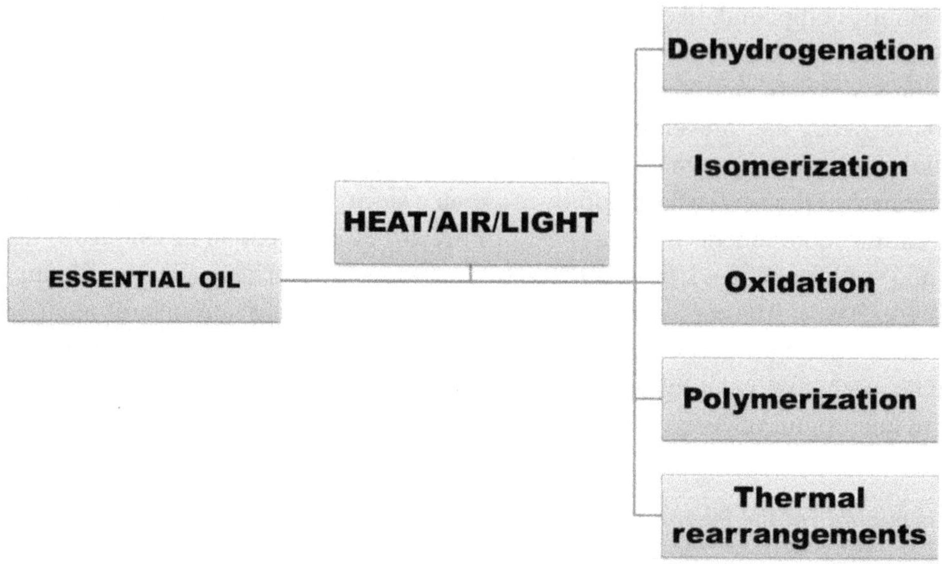

The bactericidal action of the EO's could be greatly delayed due to the property of hydrophobicity of essential oils, which restricts the interaction with the bacteria in the aqueous environment, thus decreasing the toxicity to the respective bacterium (Burt & Reinders, 2003; Ultee *et al.*, 2002). Protection of such compounds could be well-established by incorporating in oil-in-water emulsion systems as a carrier & it is considered to be an effective technique (McClements, Decker & Weiss, 2007). This encapsulation system is attracting great interest among the research community due to its appreciable level of stability, in comparison to emulsions with a greater particle size (Qadir *et al.*, 2016) where the nanoemulsions steadiness were credentialed to its enormous surface area and free energy of nanoscale droplets (Jemaa *et al.*, 2019). Typically, the lipophilic compounds, (entrapped in the nanoemulsions) could be very well transported through biological membranes, established by decreasing the droplet size & increasing the surface area/volume ratio, subsequently leading to the enhancement in the reactivity of such compounds (Salvia-Trujillo *et al.*, 2015). Also, these systems are regarded to enact as a physical-chemical barrier that counteracts pro-oxidant elements such as ultra-violet light, free radicals, and oxygen, which in turn enables the circumvention or lowering of the oxidation reactions upon encapsulating oil in these systems (Gonnet *et al.*, 2010; Fernández-Lopez, Viuda-Martos, 2018). In particular, nanoencapsulation promotes the following characteristics: 1)Decreasing the rate of evaporation, or the pace at which the core material is transferred to the external environment 2) Conferring the protection for core material from degradation via reducing the receptivity to the external environment 3) Controlling the release rate of the core material, either gradually or at a specified point in time 4) The physical properties of the original material are altered to facilitate easy handling 5) dilution of core materials when only small amounts are required for the accomplishment of uniform dispersion in the host material 6) Separation of components in a mixture that might otherwise react with one other 7) Concealment of the undesirable taste or the flavor of the core material (Fang & Bhandari, 2010). In short, encapsulation of the bioactive oil provides a cost-effective and practical method for modifying oil release which imparts protection against oxidation processes in the environment, increment of physical stability, reduce the chances of volatility & toxicity, enhancement in the bioactivity (Rodríguez *et al.*, 2016). In this regard, EO-loaded nanoemulsions are designated by entrapment of the essential oil, which enacts as the dispersed phase, on a nanoscale range. Such nanoemulsions based on essential oil offer several advantages and some of them include a) Possession of high kinetic stability b) Solubilization of hydrophobic bioactive molecule and enhancement of bioavailability c) Can be utilized for taste-making d) Non-toxic and non-irritant. However, gathered data demonstrated that special consideration must be provided to the formulation stage during the production of essential oil-based nanoemulsion as unsuitable formulation, owing to the misconception of the essential oil entrapment process, might paves way for instability and/or the inefficiency of the gained nanoemulsion (Jemaa et al., 2019). It's because, the composition of the nano-emulsion is reported to be the central factor affecting the antimicrobial property (Yildirium *et al.*, 2017). Formulation of such components accounts for reducing the size of the droplet, determining a plethora of NE's properties with their surface and biological activity (Donsì and Ferrari, 2016).

IMPORTANCE OF ESSENTIAL OIL NANOEMULSIONS IN ANTIMICROBIAL THERAPY

Antimicrobial resistance (AMR) has transpired as one of the most serious public health issues of the twenty-first century, posing a threat to the effective prevention and treatment of an ever-widening range of infections caused by bacteria, fungi, parasites, virus that are no longer susceptible to common medicines which are employed for treatment (Prestinaci et al.,2015). Despite the development of numerous new generation antibiotics, many bacteria strains continue to acquire resistance to some antibiotics (Joyce and Ramamurthy, 2017) and hence the resistance of bacteria towards antibiotics makes the problem of AMR even more emergent (Prestinaci et al.,2015). Furthermore, medicine only achieves a limited efficacy of the prescribed dosage concentration (Joyce and Ramamurthy, 2017). Due to rising microbe resistance to conventional antimicrobials, a general rise in the usage of "natural" products, and the need for anti-microbial agents with improved safety and toxicity profiles, there emerges a demand for unconventional agents and hence the application of essential oil has been dramatically rekindled the interest (Thormar, 2010). Antibacterial, antiviral, antifungal, anti-helmintic, anti-tumor, anti-oxidant, anti-lice, antidandruff, spasmodic, hormonal, and insect repellent properties are all well-known properties of essential oils (Joyce and Ramamurthy, 2017) and the same has been clarified in the previous section. Secondly, nanoemulsions can reduce the size of droplets to the nanometric scale. In the case of antibacterial activity, this results in increased surface area for easier contact with microbial pathogens, resulting in lysis and cell death. Thus, Essential oil-impregnated nanoemulsions provide an increment in the aqueous solubility of lipophilic components, enhancement in the oral bioavailability at lower dose concentrations, confer-ment of transparency along with stability at an appreciable rate, reduction of hazardous side effects with a better bio-compatibility, thereby attracting great importance in the medicinal field. Furthermore, the Nanoemulsions' antibacterial action is not as specific as that of conventional antibiotics. They have an antibacterial effect against bacteria, enveloped viruses, and spores in quantities that aren't toxic to animals. Resistance strains are less likely to arise thanks to their physical kill-on-contact mechanism. In particular, they connect with the cell membrane, causing the pathogen to die (Joyce and Ramamurthy, 2017). The above-said fact has been magnified further in the upcoming sections to well elucidate and validate the tremendous antimicrobial potency demonstrated by these nanoemulsions and the routes they employ to render those incredible effects.

FUNDAMENTAL ANTIMICROBIAL MECHANISM OF EO AIDED NANOEMULSIONS

The plethora of authors in their studies, state that the nanoemulsion might magnify the EO's antimicrobial potential, upon employing at a concentration below their solubilities in water. The broad spectrum of antimicrobial potential of essential oil-based NE's was stronger, in comparison with the bulk-free form of EO's. Upon the application of EO-NE, measurable reduction in the microbial count heavily relies on the components of the particular essential oil, microbial strain screened for the study, and the emulsion formulation size. The surface charge & the mean droplet size of nanoemulsion governs the penetration of EO's to the cellular membrane so that it well interacts with numerous molecular sites embedded in the membrane of microbial cells. This particular interaction of EO's in emulsion systems with the microbial cell is likely to be promoted through the following routes (Donsì and Ferrari., 2016):

Figure 4. Schematic depiction on the fate of the bacteria upon treatment with essential oil which leads to leakage of ions & metabolites (1), Alteration of Proton motive force (2) & Fatty acids (3), Coagulation of essential oil (in yellow) with the cytoplasm (4), Increase in membrane permeability (5).

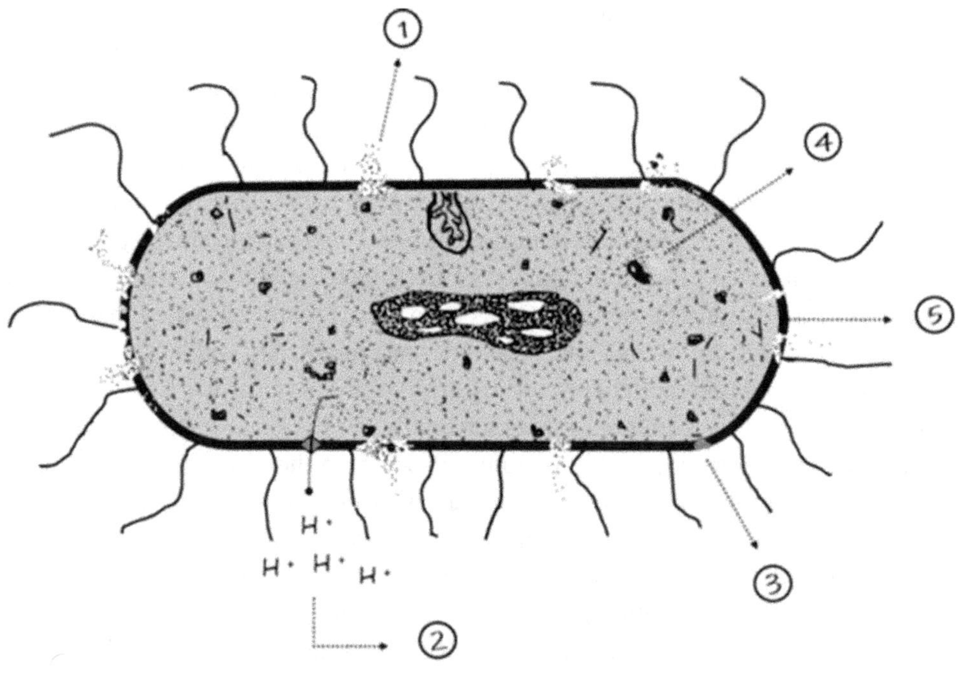

[1] Improvisation in the interaction with the cytoplasmic membrane furthermore is driven by transference in passive mode through the outer cytoplasmic membrane as well as the increased surface area

(Donsì *et al.*, 2012). NE droplets push the EO molecules to the active sites for interaction in the case of bacteria that are gram-positive (Majeed et al., 2016). Acceleration of penetration of droplets due to their size and the hydrophilic surface they possess into the cytoplasmic membrane via hydrophilic transmembrane channels known as porin proteins embedded and it's reported in gram-negative bacteria (Nazzaro *et al.*, 2013). On account of the interplay taking place in the middle of EO and the microbial cells, active transport proteins which are arranged in the phospholipid bilayer are intervened extensively thereby the integrity of the bilayer was distorted (Moghimi *et al.*, 2016).

[2] An increment in the concentration of EO at the target site is managed by the positively charged NE droplets that interact electrostatically with microbial cell walls which are negatively charged. (Chang *et al.*, 2015).

[3] Perpetual release of the EO's from the nano-sized droplets, accelerated by the partition taking place in between the aqueous phase & droplets of oil, paves way for the prolonged EO activity. Nanodroplets which are the nano-tanks bears the EO molecules which were in a dynamic equilibrium between the dispersed oil phase and the aqueous phase (Donsì *et al.*, 2012).

[4] The intended release of essential oil at the desired sites is likely to get promoted by the droplets of the emulsifier & its integration with the phospholipid bilayer of the cell membrane (Donsì and Ferrari, 2016).

Therefore, in various aspects, the implementation of lipophilic functional components into nanoemulsion-based systems serves several advantages which are specifically true in the case of essential oil (Donsì and Ferrari., 2016).

FUNCTIONAL COMPONENTS OF NANOEMULSION

Essential Oil - Its Associated Antimicrobial Potentiality

1. Thyme Essential Oil (TEO)

Belonging to the family of Lamiaceae, Thyme is an aromatic and medicinal plant, having 928 species which is being identified in Europe, Asia, the Northern part of Africa, Southern America, and Australia so far. *T. serpyllum* and *T. vulgaris*, also known as wild thyme and common thyme, are the representative species of the thymus genus (Nabavi *et al.*, 2015; Badi *et al.*, 2004). *Thymus vulgaris* comprises Thymol (47.59%), gamma-terpinene (30.90%), para-cymene (8.41%) as its principal constituents, as per the analysis via Gas chromatography (Borugă *et al.*, 2014).

1.1 Antimicrobial Activities of Thyme Essential Oil

Thyme essential oil presented a superior anti-microbial activity against a wide range array of fungal species & bacteria among various essential oil types (Ziani *et al.*, 2011; Šegvić *et al.*, 2007). TEO-based NE was observed to produce an antagonistic effect, at a concentration of 10% (TEO), against *Streptococcus aureus*, *Klebsiella pneumoniae*, *Enterococcus faecalis*, which are regarded as serious foodborne pathogens. The diameter of inhibitory zones was reached highest for *S. aureus* (26.1 mm) and TEO achieved the diameter of 25.0 mm. *K. pneumoniae* and *E. faecalis* were unable to withstand the inhibitory effect exerted by thyme EO encapsulated NE. It is noticeable that a remarkable difference in microbial count decline was created when compared to the anti-microbial activities of individual thyme EO, thereby suggesting that the encapsulation technique improvised the EO's anti-microbial potency to a better extent. Unlike the foodborne pathogens, thyme EO-NE was less counteractive with an array of fish spoilage bacteria, viz., *E. faecalis, V. vulnificus, P. mirabilis, P. luteola,* yet *Photobacterium damselae and Serratia liquefaciens* stood as an exceptional case. The antimicrobial activity of TEO was attributed to carvacrol, which was observed to be the principal component of thyme (Ozogul *et al*., 2020). In addition to the tremendous potency, thyme EO, when loaded in nanoemulsion along with chitosan films, displayed a successful deterioration of the biofilms formed by *S. aureus* and *E. coli* and the values are 60.35 and 55.50% respectively. However, it was a little less than the activity of another candidate, thymol-chitosan (T-CS) loaded nanoemulsion, as the antibiofilm activity was observed to be significant, where the values are 83.78% and 83.64% for the same pathogenic strains respectively. A key highlight of the nanostructure of T-CS is that it facilitates the reduction of mean particle size, which is recorded to be 123.30 nm. Moreover, the diameter of inhibitory zones was also measured to be higher for the T-CS loaded NE, against the sample strains (*S. aureus* and *E. coli*) during research. These results observed for T-CS based NE was conferred by the enrichment of thymol content in the nanoemulsion (Liu & Liu, 2020). *Haemophilus influenzae, Pseudomonas aeruginosa* was also highly sensitive to the action of Thyme oil-impregnated NE when it interacts with the liquid phase of NE, at EO's concentration of 2w/w% & this factual data was evident through the MIC values recorded during the study (31.2μg/mL, 3.9μg/mL, 2000μg/mL respectively). Meanwhile, contradistinction was observed in the performance of the vapor phase of the same Thyme oil-NE, where it presented no bacterial inactivation of *H. influenzae* and *P. aeruginosa*. MIC value of 0.293 μL/L was recorded only against *S. pneumoniae* in the vapor phase. The reason for this phenomenon is that the movement of thymol is intervened by a designated nanoemulsion in the vapor phase. Nevertheless, entrapment in emulsion-based systems proved to be advantageous in contributing to the enhancement of thyme oil's antimicrobial potentiality in both of the phases (Lida Ghaderi *et al*., 2017). In addition to the above-described superior effects, Thymus essential oil nanoemulsions were reported to have the capability in altering the functionality of fungal cell walls including fluidity or permeability as such (Wan et al.,2019). For an instance, 2.5 wt% of thyme EO in nanoemulsion was observed to be an effective fungicidal concentration for *Fusarium graminaearum* isolates, which is determined when it underwent assays such as the agar dilution method. Thyme EO entrapped NE emerged successfully among the other EO's, i.e., Clove, Cinnamon, Peppermint, Lemongrass, and its nanoemulsion eventually. In comparison with the NE's of other EO's, the highest EC90 values were observed only for thyme oil, which falls at a range of 7.61 $\pm$ 0.09 mg/g & these results were credentialed to the presence of constituents such as thymol on large scale (Wan et al., 2019). Thyme oil encapsulated – nanoemulsion also displayed an enhanced mycotoxin inhibitory activity, in comparison with the bulk essential oil (Wan et al.,2019).

2. Anise Essential Oil (AEO)

Being a flowering plant, Anise (*Pimpinella anisum*) has been extensively utilized in folk medicine for the treatment of anorexia, digestive problems, cough (Ernst, 2004). Anise comprises a wide range of compounds such as Pseudoisoeugenol Anethole, Eugenol, Estragole, Methyl chavicol, Anisaldehyde, Estrols, Scopoletin, Polenes, Umbelliferone, Terpene hydrocarbons, Coumarins, Polyacetylenes, which was determined in chemical studies (Tepe *et al*., 2006). Anethole is enacting as an antimicrobial compound in Anise EO, whereas its total content in Anise EO is quantified as 97.3% (Alvessi *et al.*, 2004; Tepe *et al.*, 2006).

2.1 Antimicrobial potency of Anise Essential Oil:

Anise oil, once it is encapsulated, significantly tends to decrease the mean diameter of nanoemulsions and achieved the same, by increasing the levels of Anise EO up to 75% during the formulation of nanoemulsion. A gradual increase in the viscosity of the Nanoemulsion could also be observed, in addition. Nano-emulsification of anise oil, when applied at a minute quantity of 1% (v/v), both *E. coli* & *L. monocytogenes* counts were drastically dropped down, as a result of the bactericidal investigation via Agar diffusion assay (Topuz *et al.*, 2016). Moreover, AEO based NE was made to encounter a series of foodborne pathogens (*E. coli*, *Typhimurium*, *Y. entrerocolitica*, *B. cereus*, *P. aeruginosa*, *S. aureus*, *L. monocytogenes*) & diameters of inhibitory zones were measured for evaluation. This NE scored the highest inhibitory value for *Y. enterocolitica* (25 mm), *E. coli* (20 mm) & *Typhimurium* (15mm), on comparing with the bulk extract. Yet, no effect was observed for *P. aeruginosa* (0 mm) & these values were attributed to the presence of AEO (Ghazy *et al.*, 2021). Conversely, author Nirmal *et al.*, (2018) show that the AEO and its nanoemulsion had failed to exert bactericidal effect when it encountered *E. coli*, *L. monocytogenes*, *S. aureus*, and *P. aeruginosa*. Henceforth, further studies are needed to be performed for the analysis of various parameters which are found to have a direct influence on the antimicrobial activities of EO's nanoemulsions.

3. Eucalyptus Essential Oil (EEO)

Eucalyptus (*Eucalyptus globulus*), being a tall evergreen tree, had Australia as its native (Bruneton, 1995). This vast genus belongs to the family of Myrtaceae & that comprises 900 species and subspecies (Brooker & Kleinig, 2004). Eucalyptol, otherwise known as 1,8 cineole, is reported to be the central compound of *Eucalyptus globulus* oil with a concentration of 63.81%. Other chemical constituents include α-pinene (16.06%), Aroma dendrene (3.68%), and o-cymene (2.35%) (Luís *et al.*, 2015). Entrapment of Eucalyptus EO in a nano-emulsified structure is observed to be the best solution for the circumvention of degradative processes such as thermal degradation and oxidation which is mainly associated with the utilization of free EEO (Shahi *et al.*, 2000).

3.1 Antimicrobial Activity of Eucalyptus Essential Oil:

Eucalyptus oil, in association with the nanostructure, significantly tends to antagonize various bacterial as well as fungal species. Chitosan-based nanoemulsion (NE-CH; an oil in water model) showed remarkable antibacterial activity, when on treatment against *S. aureus*, with the increasing concentration of EEO (5% NE-CH), owing to the diameter inhibitory zones values with a reach of 15mm (recorded via Agar disc diffusion assay). Typically, on treatment with 5% NE-CH, turbidity of the bacterium cell was also well reduced. In addition to this, curtailment of moisture content of the chitosan films was presented due to the incorporation of the EEO (Sugumar *et al.*, 2015) & this was in coincidence with the result of Shojaee-Aliabadi *et al.*, (2014) where the films tend to behave hydrophobically to a greater degree which attributed to the inclusion of EEO to the edible films. EEO-NE was also effective against *E. coli*, *S. aureus*, and *B. cereus* at 50 – fold dilution when in comparison with the oil in free form &

viable cell count was also attenuated to a better extent than the positive controls, although microbial proliferation was interpreted through cell viability curve over time (Sugumar *et al.*, 2013). Eucalyptus EO in free form showed a better inhibitory effect than the nano-emulsified form in the case of *P. aeruginosa* and the same was reported by Quatrin *et al.*, (2017). The work also defined the fungicidal potency of Eucalyptus EO-NE against 3 selected fungal species, i.e., *Candida albicans*, *Candida tropicalis* & *Candida galbrata* through MFC values & was found to be 0.219, 0.885, 0.219 mg/mL respectively, thus achieved in a lower concentration of EO. Yet, implication through MIC values suggests that the infusion of 5% EEO in nanoemulsion is inadequate for the subsequent reduction of *P. aeruginosa* & only attained 114.125 mg/mL. Data from biofilm formation assay of the fungal species suggested that EEO loaded NE successfully deteriorated biofilms of 2 out of 3 species, i.e *C. albicans* and *C. glabrata* where the inhibitory percentage falls at 84.5 & 84% at the concentration of 22.5 mg/mL respectively. Yet, it doesn't present a remarkable difference for *C. tropicalis*. These results corroborate with the data from the literature of Hendry *et al.*, (2009) which examined the anti-bio-film potency of nanoemulsion bearing Eucalyptus essential oil. In addition to this, Damjanovic'-Vratnica *et al.*, (2011) perceived the profile which is coincidental to work by Quatrin *et al.*, (2017), where EEO-based NE inhibited *Candida* species remarkably concerning *P. aeruginosa* which was evident through the MIC and Disc diffusion assays.

4. Clove Essential Oil (CEO)

Clove (*Syzygium aromaticum*), belonged to the family of Myrtaceae (Cui *et al.*, 2018). This comprises about 30-35 wide variety of species having their origin in North America, Asia minor Europe, and the Near East (Khaleque *et al.*, 2016). According to the previously conducted studies, the central components of Clove EO are Eugenol acetate, β-caryophyllene, Eugenol (Toledano-Medina *et al.*, 2016; Teixerira *et al.*, 2014). The antimicrobial activity is hugely credited to Phenylpropanoid eugenol, a phytochemical compound present in the CEO, conferring the property to act as a natural preservative, which could be utilized in leading food industries (Bentayeb *et al.*, 2014). In particular, the antimicrobial activity is greatly imputed to the presence of the phenyl group which has the potency of altering the permeability of the bacterial population by denaturing the protein molecules of the cell membrane, thus resulting in the death of the cell subsequently (Hossain *et al.*, 2012; Sharif *et al.*, 2017).

4.1 Antimicrobial Activities of Clove Essential Oil:

Clove essential oil has been observed to be a potent candidate in inhibiting a broad range of bacterial as well as fungal species. Antibacterial influence of clove essential oil entrapped NE on an array of serious foodborne pathogens, i.e., *Escherichia coli, Bacillus subtilis, Bacillus cereus, Listeria monocytogenes, Salmonella typhimurium, Staphylococcus aureus* was heightened, having the highest inhibitory zone diameter at a range of 5.12 – 14.34 mm respectively. Moreover, it was also reported to be effective against probiotic bacteria, i.e., *L. acidophilus, L. reuteri, L. casei, L. rhamnosus* where their inhibitory range falls at 2.57 – 4.44 mm. Regarding fungal species, CEO-NE exhibited a better inhibitory effect when tested against *Aspergillus niger* and *Candida albicans* where their inhibition zones are 2.13 and 3.19 mm respectively (Shahbazi *et al.*, 2019). MIC values also suggest that the Clove EO incorporated NE demonstrated elevated bactericidal action at 7 mg/ml as its concentration than the oil in the free form where the values reached 856 μg/ml and 999 μg/ml corresponding to the population of gram-positive (*Staphylococcus aureus, Streptococcus mutans, Staphylococcus epidermis*) and gram-negative bacteria

(*K. pneumonia, E. coli, S. choleraesuis*) screened for the study (DE Meneses *et al.*, 2019). Also, nano-emulsified CEO reached an MBC value of 16 µg/ml, 32 µg/ml for *E. coli* and *B. cereus* correspondingly, thus suggesting its better efficacy. Clove oil contribution also, had a significant effect on NE's performance indicators, particularly in the reduction of mean droplet size that had superior effects (Shahavi *et al.*, 2015). Moreover, enhanced bacterial inactivation was demonstrated within a short period by nanoemulsion containing CEO followed by Lemongrass oil- encapsulated NE which was reconfirmed via kinetics of killing assay performed by Salvia Trujillo et al., (2015), against *E. coli*, a gram-negative bacterium.

5. Citrus Essential Oil (CEO)

An oily aromatic liquid extracted from various parts such as flowers, peel, leaves, and twigs of lime & are aromatic (Ruberto, 2002). There is a wide range of lime species such as *Citrus aurantifolia* (Key lime), *Citrofortunella microcarpa* (Calamansi lime), *Citrus latifolia* (Persian lime), *Citrus glauca* (Australian desert lime), *Citrus hystrix* (Kaffir lime), *Citrus limetta* (Sweet lime) and much more (Cruz- Valenzuela *et al.*, 2017). The drawback associated with the lime EO is the frequent susceptibility of bioactive compounds (such as monoterpene hydrocarbons) to degradation, on exposure to humidity, oxygen, or heat directly leading to the alteration in the functionality of its own (Perez- Cacho & Rouseff, 2008). The designation of a nanoemulsion is observed to be a feasible method for the effective preservation of the bioactive compounds present in EO (Liew *et al.*, 2020).

5.1 Antimicrobial Activity of Citrus Essential Oil

Nano-emulsified EO of lemon (*Citrus limonum*) (contains 10% w/w of EO) was observed to produce a noticeable bactericidal performance equivalent to the effect of pure lemon oil alone (100%), against food-borne pathogens (*S. aureus, E. faecalis, S. Para typhi A*) with the highest inhibitive activity observed for *Salmonella* para typhi A, as per the MBC report. *K. pneumoniae* was highly resistant to the effect exerted by CEO-NE, but it was sensitive when it interacts with pure lemon oil which produced an inhibitory zone diameter of 19.73 ± 0.50 mm. MIC values suggested that, under the same concentration of EO, the nano-emulsified form inhibited the growth but doesn't present a remarkable difference in its action against fish spoilage bacteria (*V. vulnificus, P. luteola, P. damselae, S. liquefaciens, E. faecalis, P. mirabilis*) comparatively. The free form of lemon oil (at a concentration of 10%) also displayed its role in the inhibition of spoilage bacterial population but not the pathogenic bacteria as significant as the nano-emulsified structure. These outcomes, indirectly suggested that the antimicrobial activity of the CEO was rapidly elevated when it gets converted to its nano-structure (Yazgan *et al.*, 2019). Recently, nanoemulsion of EO extracted from Calamansi lime (*Citrofortunella macrocarpa*) was observed to emerge successfully in inhibiting *S. aureus*, where this factual data was confirmed through disc diffusion assay & the diameter of the inhibitory zone reached 9.98 mm for the respective bacterium. Moreover, *Salmonella* spp. and *E. coli* were also inhibited effectively by the nanoemulsion. Overall, the performance of calamansi lime was far better, than those of the other 2 EO's of Kaffir lime (*Citrus hystrix*) & Key lime (*Citrus aurantifolia*) and its nanoemulsion (Liew *et al.*, 2020). EO of Calamansi lime contains limonene, which dominated its entire composition (94%) which is in contrast to kaffir lime & key lime (36 & 39% respectively) (Othman *et al.*, 2016) which conferred the anti-microbial efficacy to the nanoemulsion fabricated with this EO. Also, Limonene displayed inhibitory effects on both pathogenic as well as on spoilage bacteria (Cruz-Valenzuela *et al.*, 2017). Furthermore, this highest percentage of limonene in Calamansi lime oil accounts for a larger mean particle size of the nanoemulsion (60 nm) followed by a high viscosity rate (4.33 mPa.s), in comparison with the NE's of key & Kaffir lime EO (Liew *et al.*, 2020). Another EO, obtained from *Citrus medica L. var. sarcodactylis*, when introduced to a nanoemulsion, paved the way for a 100% reduction in the count of *S. aureus* cells than the pure essential oil, owing to its MIC value with a reach of 0.48 mg/mL. (Concentration of EO in NE- 50%). A significant inhibitory rate was also displayed by this NE against *E. coli*, indeed. Apart from the antimicrobial efficacy, this essential oil has created an impression on the formation & stability of nanoemulsion which was revealed through the result of the Z-average diameter of the nanoemulsion (Lou *et al.*, 2017).

b. Stabilizers

When the oil segment and an aqueous segment are homogenized together, the framework will swiftly break down through various unique mechanisms, inclusive of Ostwald ripening, droplet flocculation, gravitational separation, coalescence. Therefore, it's extremely vital to feature a wide variety of stabilizers available to enhance the nanoemulsions' long-standing stability (McClements and Rao, 2011). Few examples of stabilizers employed are Ripening inhibitors, Weighing agents, Emulsifiers, Texture modifiers (Aswathanarayan & Vittal, 2019).

Emulsifiers

Emulsifiers, which are amphiphilic, are the surface-active molecules that significantly decrease interfacial tension & are observed to be exceptionally dissoluble in one of the liquid segments, at least (Mason *et al.*, 2006). These could be categorized as anionic, cationic, zwitter ionic & non-ionic ones. Small molecule surfactants, phospholipids, proteins, polysaccharides stand as some notable examples of emulsifiers (McClements, 2005; Kralova & Sjöblom, 2009). Quaternary ammonium halide is widely used as both cationic as well as zwitterions surfactants, whereas prominently utilized anionic surfactants were the potassium laurate, sodium lauryl sulphate. Examples of non-ionic surfactants were sorbitan esters and polysorbates (Singh and Gurpreet, 2018).

Selection of a Surfactant

The selection of an optimal surfactant blend is a key factor in producing stable nanoemulsion due to repulsive force, thereby averting flocculation and coalescence between the nanodroplets (Sugumar *et al.*, 2016). Several factors are needed to be considered for a better nomination of a desirable surfactant, its concentration concerning the oil phase. These factors comprise of: i) Surface coverage which is requisite for emulsion formation with utmost stability ii) Interfacial tension iii) Hydrophilic-Lipophilic balance (HLB) of the surfactant molecules or its concentration and the location of hydrophobic moieties in case of emulsifiers where the molecular weight is high iv) Kinetics of adsorption taking place at the oil-in-water interface and the rearrangement of molecules especially for emulsifiers of high molecular weight, such as polysaccharides and proteins (Donsì *et al.*, 2012).

Properties and Comparison of Different Surfactants

The surfactant's concentration imparts a crucial role in ascertaining the characteristics of emulsion systems to favor stability (Ghazy *et al.*, 2021). Structure Composition & Charge possessed by the interfacial layer wrapping the drops of oil can be influenced only by the various categories of surfactants of distinct Hydrophilic Lipophilic Balance (HLB) values for emulsion formulation (McClements *et al.,* 2016). In general, a decrease in interfacial tension results in a) Increment in the efficacy of breaking the droplets (Walstra, 1993) b) Droplet coalescence scale gets elevated (Rao & McClements, 2012) c) The emulsifiers' absorption rate onto the newly created o/w interfaces will be decreased (Chanamai *et al.*, 2002). The concentration of surfactant at a higher rate predominantly decreases the interfacial tension at the oil-water interface which results in the production of smaller droplets (Davidov-Pardo & McClements, 2015). To sum up, surfactant concentration in the emulsion should be sufficient for coverage of the droplets of essential oil & retain them in the dispersed phase (Nirmal *et al.*, 2018). For an instance, thyme oil loaded nano emulsion's mean particle diameter was diminished drastically via increment in the concentration of emulsifiers, viz., BSA (180 nm), Q-Naturale (199 nm), and Lysolecithin (181 nm) from 0.1 to 1.0% (w/v) and the values are observed to be consistent for the emulsifiers listed. Also, when the concentration was raised furthermore (say BSA) to 1.5% (w/v), the mean droplet size reduced at 177 nm. Also, in terms of stability, Q-Naturale or BSA containing thyme essential oil NE formulations were highly stable for about 28 days rather than the Soy lecithin-based ones. It was attributed to the vulnerability of the interfacial layer to the disruption that led to the rapid elevation in the mean droplet diameter when it's stored for a day (Wan *et al.*, 2019). Another example, which is about Tween 80, a low mass non-ionic surfactant, was portrayed by intense activity on the surface that results in a translation of quick absorption at o/w interfaces, thereby efficiently deterring the droplets from coalescing as efficiently as possible (Li et al., 2015). Others have reported that Tween 80 was an effective surface-active agent due to its high HLB value (15) and recognized to be satisfactory for o/w NE formulation (Gehan Marei *et al.*, 2018). Tween 80 was investigated to be a little effective in producing smaller droplets (130.95±5.40 nm) when it's utilized for the fabrication of clove oil nanoemulsions, than the other emulsifiers BSA and Quillaja saponin (152.78±4.36nm & 179.82±4.36 nm respectively) (Wan *et al.*, 2020) & the interpretation was in agreement with another study, where the same surfactant was found to produce finer droplet size of the nanoemulsions of Citral EO, significantly than the other emulsifier, whey isolate, independent of the formulation (Ribes *et al.*, 2017). Unfortunately, Tween 80 stabilized nanoemulsion showed sudden heightening of mean droplet radii when stored for about a week to 406.33nm, but, BSA and Quillaja saponin fabricated nanoemulsions exhibited a high degree of stability & quite consistent throughout 28 days. at 23°C. The possible reason observed for the phenomenon exhibited by tween 80 might be the higher concentration than its critical micelle concentration where the excess molecules of tween 80 might arrange themselves into micelles spontaneously (Wan et al., 2020). Incorporation of Lauric arginate, a cationic surfactant was indicated to enhance the stability of Thyme oil nanoemulsion and loading capacity of the EO, once emulsified in combination with tween 80 (Chang *et al.*, 2015). Even so, there are possibilities that higher levels of surfactant might lead to the acceleration of droplet growth via Ostwald ripening or Coalescence, thus stimulating the mechanisms favoring instability. This was witnessed in the case of Tween 80, upon emulsification in cinnamon oil-based nanoemulsion at a concentration of 20 wt%, during its storage at both 4 & 25°C (Chuesiang *et al.*, 2017). There are numerous physicochemical phenomena responsible for impacting the stability, some of which are: a) Transportation of oil molecules from one droplet to another by the surfactant micelles, thereby driving Ostwald ripening (Peña

& Miller, 2006; Weiss *et al.*, 2000) b) Generation of osmotic stress in the occupancy of surfactant micelles in the aqueous phase, which drags the droplets together and enhancing coalescence (McClements, 1994). Nevertheless, these differences (droplet growth) don't seem to have any ramifications on the optical characteristics of the nanoemulsion (Chuesiang *et al.*, 2017). On contrary, eucalyptus EO impregnated nanoemulsion containing Tween 80 was reported to be efficient in decreasing droplet size and conferring stability than the tween 20 based nanoemulsions (Sugumar *et al.*, 2015). Nanoemulsion based on Anise EO stabilized by molecules of Tween 80, at a concentration of 5% was also found to be sufficient in decreasing the extent of a collision via steric hindrance (Ghazy *et al.*, 2021). Polysorbate 80 was also reported to possess a negative surface density which is attributed to the oxygen atoms present in the molecules (Marchiori *et al.*,2010). Due to this nature, surfactant enables to provide a surface charge, which was monitored when this was incorporated into the nanoemulsion of eucalyptus essential oil & the ZP values were averaged in -9.42 mV, at a pH around 5 (Quatrin *et al.*, 2017). Furthermore, the average zeta potential measured for non-ionic tween 80 stabilized emulsion systems were slightly negative which was around -8.0mV for Cinnamon leaf EO & Bergamot essential oil nanoemulsions and -16.4 mV for lemon oil nanoemulsions. Whey isolates based on Citral nanoemulsions also presented a ZP value of ranged from -30 to -43 mV and it's observed that the droplet stability may be aided by electrostatic repulsive forces. (Ribes *et al.*, 2017). In general, emulsifiers stabilize the nanoemulsion through the effect of either steric hindrance electrostatic repulsion (Ozturk and McClements, 2016). Clove oil nanoemulsion formulations of Pluronic F 127, a non-ionic surfactant, presented a Zeta Potential value which was around -12mV (DE Meneses *et al.*, 2019) and this profile find agreement with the other study, where lemongrass oil-nanoemulsion was stabilized with Tween 80 (Gündel *et al.*, 2018). Moreover, other works have indicated that the droplet size of the emulsions also has a remarkable influence on the loading efficacy of different core materials (Soottitantawat *et al.*, 2003; Meynier et al., 2005; Jafari et al., 2007). In one study, Citral nanoemulsions were formulated and stabilized with a surfactant mix composed of Brij 97 and Span 95 (under different HLB values). The encapsulation ratio was heightened to 82.8%. It's noticeable that the encapsulation ratio increased with the increasing value of HLB. A lower droplet size of 28 nm was also established by the surfactant blend at an HLB value of 12 (Lu et al., 2018). However, high surfactant concentration might lead to the impairment of the antimicrobial properties of the extract's active components through their infusion in their micelles. Poly sorbate-based surfactants such as Tween 80 stood as a paradigm in the reduction of essential oil's antimicrobial activity (El Sayed *et al.*, 2017). This was evident during the application of Tween 80 in the cinnamon oil nanoemulsions where the antimicrobial activity was drastically impacted through the increasing surfactant concentration. Also, it was mentioned that even the MIC values of essential oils against fungal as well as bacterial species were heightened by higher amounts of tween 80 (Inouya *et al.*, 2001; Ma *et al.*, 2016). Nevertheless, the elevation of antibacterial efficacy of nanoemulsion encapsulated with cinnamon essential oil (formulated via spontaneous emulsification method) was pronounced upon the increment in the concentration of Tween 80, from 10-20 wt% against *Escherichia coli* (Yildirium *et al.*, 2017). Cationic surfactants such as Cetyl Pyridinium Chloride (CPC), characterized to possess a negative charge, was also noticed to contribute to the deterioration of microbial membranes efficiently than tween 80 (Simões *et al.*, 2008) even though it contains a high HLB value (26) compared to Tween 80 (Huang *et al.*, 2012). Moreover, research works demonstrated the emulsifier's ascendancy on the essential oil-based nano emulsion's antimicrobial activity via determining the degree of diffusion of essential oil-based nanoemulsion betwixt the cell membrane of bacteria. A substantial rate of diffusion was noted in nanoemulsion of thyme EO ladened with Q-Naturale than Tween 80 as it forms smaller micelles (Ryu *et al.*,

2018). This positive impact was parallel with another study, where it presented that the Q-Naturale stabilized thyme oil-NE, dramatically paved the way in pulling out a slightly high MGI (Mycelium Growth Inhibition) rate than the other natural emulsifiers, BSA and lysolecithin. In addition to this, significant changes in the morphological structure of hyphae of *F. Graminaearum* isolates were also reported upon the exposure to Q-Naturale or BSA loaded TEO nanoemulsion which encloses modification in the thickness of the hyphal wall, disruption or appeared to evanesce still, in few hyphae as well as the precipitates appeared on the cellular membrane's outermost layer. The reasons justified for such phenomenal activities are stated as either the difference in the emulsifier's immanent fungicidal action or variation in the rate of dissemination by the driblet of nanoemulsion in the interior of the cellular membrane, thus promoting an intended release of TEO at suitable targeted sites. Contradictorily, the hyphal wall's integrity was only lost after the treatment of lysolecithin-based nanoemulsion of thyme essential oil but hypha wasn't observed to be destructed. Another highlight of the natural emulsifiers, BSA or lysolecithin, is that they had a direct influence on inducing morphological changes in *F. graminaearum* isolates. Particularly, on interaction with the spores, an irregularly wrinkled shape was presented followed by possession of rough surfaces (Wan *et al.*, 2019). Reduction in cytoplasmic contents of the spores was pinpointed as the factor behind such alterations (Yamamoto-Ribeiro *et al.*, 2013). Ultimately, Thyme oil nanoemulsions stabilized with natural emulsifiers are substantially proved in bringing about an impact on morphological structures of spores, which might ascribe to their antifungal activity (Wan *et al.*, 2019). Furthermore, the effectiveness of thyme oil nanoemulsions was reported to increase against *Zygosaccharomyces bailii*, upon the inclusion of lauric arginate which was evident in comparison with the TEO-based nanoemulsion which is devoid of lauric arginate at the same initial TEO level. The influence was attributed to the inherent antimicrobial activity which is thought to interact with the yeast cells directly, thus contributing to the anti-microbial potency of the nanoemulsion (Chang *et al.*, 2015). Whey Protein Isolate (WPI) based nanoemulsions of CEO, BEO, and LEO resulted in a better intervention in the mycelium growth of *Aspergillus niger* than the tween 80 based ones, at a concentration of 1 wt%. Two potential mechanisms appeared to play a critical role in the inhibitory action which is: a) Prevalence of negative charge in WPI was envisaged in stimulating a notable rate of electrostatic repulsion from the highly anionic surface of the fungi, then tween 80 b) Emulsion droplets integrate with the cellular membrane of the fungi. WPI based NE's had a better influence on the spore germination which was facilitated by the availability of a substantial number of sites appropriate for the Whey Protein Isolate to interact with the surface of the spore (Ribes *et al.*, 2017). Typically, the complex fungal surface composition contains multiple sites for interaction with the emulsifier, thus leading to the targeted EO release (Ziani *et al.*, 2011).

OTHER COMPONENTS REQUIRED FOR FORMULATION OF ESSENTIAL OIL BASED NANOEMULSION

a. Co-surfactants

Co-surfactants are known as surface-active molecules, hold both hydrophilic & lipophilic properties, possess a hydrocarbon chain and a hydroxyl group often, but are unable to stabilize the nanoemulsion systems on their own as the polar head group's size is limited (McClements & Rao, 2011). The ability of Co-surfactant in facilitating the formation of nanoemulsion is enclosed by various physicochemical mechanisms that are listed hereunder:

i. Interface fluidization.
ii. Optimizing the ratio of dispersed to continuous phase viscosity.
iii. Deduction in the electrical repulsion occurring between the head groups of ionic surfactants at an interface by acting as spacers.
iv. Triggering a suitable curvature of the interfacial zone (Gradzielski, 1998; Garti *et al.*, 2001; Shafiq-un-Nabi *et al.*, 2007).

The most intriguing co-surfactants comprise alcohols or glycols having a low molecular weight which presents a carbon chain between 2-10 carbon atoms. Examples of such components are ethanol, propylene glycol, n-butanol (Leite *et al.*,2018).

Characterization of Individual Co-Surfactants and Their Mixtures

Various categories of co-surfactants have been investigated for the formulation of nanoemulsions (Handa *et al.*, 2021). When thyme oil nanoemulsion was formulated with co-surfactants such as lecithin and Cetyl alcohol where Tween 80 acts as a surfactant, lecithin incorporation proved itself to be effective extensively in the stabilization of droplet diameter under 200 nm at all the concentrations fixed. Nevertheless, Cetyl alcohol imposed a similar profile but induced the growth of droplets upon the increment in its concentration (Ghaderi *et al.*, 2017). Ethanol, when mimics as a co-surfactant, enables the reduction of droplet size to 211 nm at a concentration of 5%, followed by the enhanced dispersion of tween and dill essential oil when formulated in nanoemulsion systems. In addition, the implementation of glycerol as a co-surfactant was ascribed to be a novel approach for obtaining nanoemulsions of extreme stability through relatively simple, low cost, and scalable technologies (Schreiner *et al.*, 2020). In another case, Ethanol & Tween 80 acts as co-surfactant and surfactant correspondingly, was incorporated (at an equal ratio of 1:1) with a view of producing *Pimpinella anisum* EO (14%) based nanoemulsion by employing low energy methods (Hashem *et al.*, 2018). Also, ethanol was found to contribute to producing transparent and clearer nanoemulsion of pine oil than Poly Ethylene Glycol 400 (PEG) during investigation alongside, and the phenomenon was ascribed to the inclined rate of formation of hydrogen bonds with the water molecules yet PEG was unable to do so. In contrast, increment in the clear nanoemulsion regions was witnessed in the case of PEG 400 when formulated with Tween 80, only at higher surfactant concentrations (Handa *et al.*, 2020).

b. Ripening retardants

In general, the essential oil's water solubility is adequately high to permit the quick transfer of oil molecules from smaller to bigger droplets, because the variations in chemical potentials are related to the curvature of the droplets (Taylor, 1998). To circumvent the process to occur, ripening retarders are incorporated that facilitate the generation of compositional ripening effect due to the entropy of creating an effect, which in turn counteracts the mechanism of Ostwald ripening (Peña & Miller, 2006). Determining the nature and the concentration of the ripening inhibitor which is intended to be loaded, is regarded to be a crucial parameter during the formulation of nanoemulsion bearing essential oils (Chang *et al.*, 2015). A ripening retarder can be Long Chain Triglyceride (LCT), ester gum, or mineral oil which are hydrophobic, i.e., dispersible in an oil phase, not in water (Kabalnov *et al.*, 1987; Sonneville-Aubrun *et al.*, 2004).

Impact of Ripening Inhibitors in the Properties of EO Aided Nanoemulsions:

(i) Particle size:

Generally, LCT is known to be sparingly soluble in water than the MCT that makes them enact as strong ripening inhibitors (Chang *et al.*, 2012; Wooster *et al.*, 2008). However, supporting evidence suggests that when LCT, as a carrier oil (between 0-50% concentration), was amalgamated in the Origanum oil-based nanoemulsion, the size of the droplet remained small initially, but a steep rise was observed when utilized at a higher concentration (Ryu *et al.*, 2019). This phenomenon was likely promoted because, LCT-surfactant system was incompetent to generate flimsy bi-continuous microemulsions & that's a prerequisite for NE to form (Komaiko & McClements, 2016). Regardless of the factual data stated above, the presence of adequate MCT (Middle Chain Triglycerides) levels in the oil phase of nanoemulsion leads to the impediment of Ostwald ripening which is facilitated by compositional ripening effect, i.e., upon the mobilization of molecules of essential oil from smaller to larger droplets, it creates the droplets to possess a different internal composition that begets osmotic pressure, thereby opposing additional alterations in the size of the droplet (McClements et al., 2012). Studies have also demonstrated a successful stabilization of carvacrol-loaded nanoemulsion through the inclusion of MCT (Chang *et al.*, 2013). Moreover, MCT was found to decrease the mean droplet diameter, achieved up to 50±1.8 nm (at a concentration of 60%) when employed in nanoemulsion containing low thyme oil levels (40%) (Ryu, 2017).

(ii) Stability:

No incidence of visible creaming was observed in MCT prepared nanoemulsions as well as in the nanoemulsions containing both LCT and MCT (at equal ratio), when these were stashed for 30 days at both 4 and 20°C. Although, larger droplet size was prompted by LCT containing Origanum oil nanoemulsion than the MCT ones, both of these types encouraged a higher rate of stability eventually, which could be sufficient enough for commercial applications (Ryu *et al.*, 2019).

(iii) Partitioning of antimicrobial compounds:

Studies illustrate that the type and the concentration of ripening inhibitors have a pronouncing regulation in the partitioning of compounds containing antimicrobial activity (Ryu *et al.*, 2018). Furthermore, it was hypothesized that the subsequent transference of antimicrobial compounds between the droplets of an emulsion and delivered to the cytoplasmic membrane of bacterial cells is navigated by the micelles present in the nanoemulsion (Ryu *et al.*, 2018). Henceforth, the antimicrobial efficacy of nanoemulsions having different ripening retarder compositions is adamantly increased which is marked by the distribution of those components both by micelles as well as by aqueous segment (Ryu *et al.*, 2019).

(iv) Antimicrobial activity:

The composition of ripening inhibitors holds a striking impression on the nano emulsion's antimicrobial efficacy which is indicative through the MIC values. This is evident in a study, when LCT, on loading to the nanoemulsion of Origanum oil, reduces the MIC values from 4000 to 1375 ppm, thereby indicating a considerable rise in the antimicrobial efficacy against *Salmonella sp*. On the other hand, when the same antimicrobial efficacy was monitored just by increasing the proportion of LCT to ≥66.6%, the potential of nanoemulsion was declined which is indicated by the increased MIC values (Ryu *et al.*, 2019). The possible reason which accounts for the initial decrease in the MIC value might be the antimicrobial components and the variation in their equilibrium partitioning taking place in between the nanoemulsion and the membrane of the bacterial cell (Ryu *et al.*, 2018). Similarly, as for the MCT-prepared nanoemulsion (containing the same oil), the presence of MCT makes the oil less effective in inactivating *Salmonella sp.* (Ryu *et al.*, 2019). Contradictorily, another study suggests that the antimicrobial efficacy of the TEO-nanoemulsion is well enhanced which is attributed to the increased proportion of corn oil (acts as LCT) in the nanoemulsion (having high LCT-MCT ratio) and the MIC values were reduced remarkably (Ryu, 2017). In a nutshell, it reveals that the antimicrobial activity in addition to the physical stability of nanoemulsion must be optimized by controlling the composition of carrier oil (Ryu *et al.*, 2019).

c. Texture modifiers

A texture modifier is a substance included with an aim of thickening or gelling the continuous phase (Imeson, 2010). These substances are frequently embodied with a view of conferring stability to the emulsion by retarding the droplet movement. Furthermore, it might be utilized to provide various expedient attributes such as thickness, creaminess, richness, or gel strength. Biopolymers like proteins (vegetable proteins, egg, milk) or polysaccharides such as starch, alginate, xanthan, pectin, carrageenan, guar gum are some of the commonly employed texture modifiers, particularly in the food industry (McClements and Rao, 2011). Evidence of such inclusion could be found in the study conducted by Guerra-Rosas *et al.*, (2017) who formulated 4 EO-based nanoemulsions (Thyme, Oregano, Mandarin, and Lemongrass) entrapping High Methoxyl Pectin (HMP) solution as its modifier followed by tween 80 as an emulsifier. HMP prepared nanoemulsion of lemongrass EO yielded the smallest droplet size (11±1 nm) and displayed a high degree of stability, on storage among the other nanoemulsions characterized. Moreover, it was reported that the incorporation of food-grade biopolymers such as pectin is observed to impart certain desirable characteristics such as discrete interfacial behavior (rheology, electrostatic forces, and steric repulsion) (Chen, Remondetto & Subirade,2006) because of its anionic nature (Dickinson *et al.*, 1998). A study by Trujillo-Cayado *et al.*, (2020), suggested that Aerosil COK84 was effective in promoting the occurrence of gel behavior and the presence of a yield point upon entrapment in the nanoemulsion of thyme oil. Additionally, the ingredient was reported to diminish the mechanism of Ostwald ripening as they wrapped the droplets preventing them to diffuse with each other.

Figure 5. Hierarchal classification on overall composition of an essential oil loaded nanoemulsion

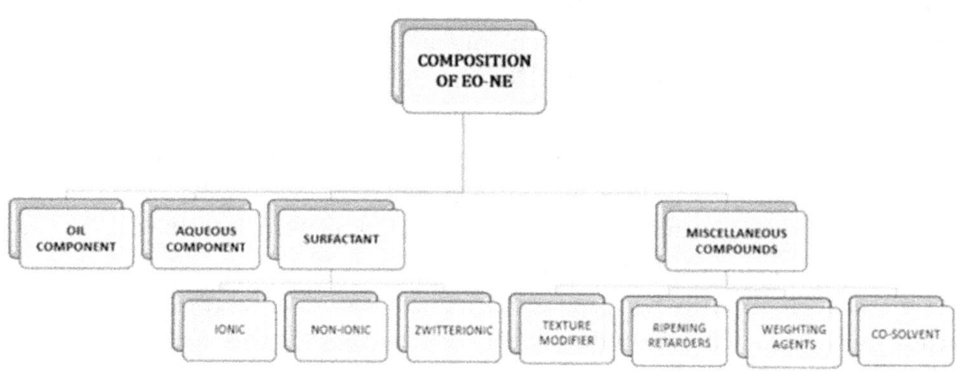

d. Weighting agents

Generally, sedimentation & creaming are some types of instability found in emulsion and this could be eliminated by the inclusion of "Weighting Agents". Weighting agents are the substances that are incorporated into the oil droplets of oil/water nanoemulsion with a view of matching the density with the surrounding aqueous phase. These are observed to decrease the driving force which is necessary for gravitational separation, leading to the formation of stable emulsion (McClements, 2005; McClements & Rao, 2011). The "hydrophobic" weighting agents are highly denser in comparison with the water molecules. Rosin gum, ester gum, sucrose acetate isobutyrate, brominated vegetable oil was a few paragons for weighting agents. (McClements & Gumus, 2016).

Table 1. List of nanoemulsions entrapped with various essential oils and their associated antimicrobial efficacy against different microorganisms

NAME OF THE ESSENTIAL OIL (EO)	FORMULATION OF NANOEMULSION	TARGET MICRO-ORGANISM	IMPACT/ACTIVITY OF NANOEMULSION	ASCRIPTION	REFERENCE(S)
Basil	*O. basilicum EO* + Sorbitan monooleate + Polysorbate 80 + Ultrapure water.	*C. albicans* *C. tropicalis* *E.coli* *P. mirabilis* *S. aureus*	Basil oil-based NE shoots up the antimicrobial potential about 2-fold for *C. albicans*, 4 times for *P. mirabilis*, *S. aureus*, and 9 times for *E. coli*.	The reduced size of the designed nanoemulsion and the increased surface area led to such tremendous effects.	Silva Gündel et al., (2018).
Cinnamon	Cinnamon oil + Tween 80 + Sterile DI water	*L. monocytogenes* & *S. typhimurium*	MIC values for *L. monocytogenes* & *S. typhimurium* is 0.039% (v/v) & 0.78% (v/v) respectively. Also, Tween 80 is found to have no effect on bacterial growth.	Enrichment of the oil with cinnamaldehyde than eugenol is believed to explain the lower MIC value for gram-negative bacteria.	Paudel et al., (2019).
Cleome viscosa	*Cleome viscosa* EO + Triton X- 100 + Distilled water.	*C. albicans*	Potentiated remarkable anti-fungal activity. Reduction in the formed bio-film was observed at 70% MFC value.	The presence of bio-active alkaloids, flavonoids, tannins in EO might account for such activity.	Krishnamoorthy, Gaseem, Athinarayanan et al., (2021); Bala et al., (2010).
Ginger	Ginger EO + Tween 20 + Arabic gum + Deionized water	*E. coli* & *S. aureus*	Exertion of relatively high antibacterial activity was observed in Gram-negative than the Gram-positive bacterium.	Specific active compound chemical structure made the nanoemulsion display bactericidal effect.	Firoozi et al., (2020)
Neem	Neem EO + Polysorbate 20.	*Saprolegnia parasitica*	The EO-loaded NE results in 100% parasitic mortality when employed at a concentration of 200mg/mL.	The presence of tween 20 was also attributed to the activity apart from the nanoemulsion.	Rabelo et al., (2021)

Continued on following page

Table 1. Continued

NAME OF THE ESSENTIAL OIL (EO)	FORMULATION OF NANOEMULSION	TARGET MICRO-ORGANISM	IMPACT/ACTIVITY OF NANOEMULSION	ASCRIPTION	REFERENCE(S)
Origanum vulgare	Oregano oil + Sunflower oil + Cremophor RH 40 + Span 80 + Deionized water.	*E. coli & S. aureus*	MBC values were significant for S. aureus but changed (increased) in the case of *E. coli* before storage for 30 days.	Difference in the values was related to the difference in the cell wall structure of E. coli & S. aureus. Deposition of oregano EO components in the cytoplasmic membrane induced the loss of integrity, thereby causing more permeability to ions & protons which in turn imparts anti-staphylococcal activity.	Moraes-Lovison et al., (2017) Souza, Barros, Oliveira & Conceição, (2010)
Peppermint	Peppermint EO + MCT + Purity Gum 2000 + Starch	*L. monocytogenes & S. aureus*	Peppermint EO formulated NE exhibited prolonged inhibition of growth of both bacteria which was revealed via time-kill kinetics studies.	Emulsion system found to be useful in enhancing the solubility & stability of peppermint EO in culture medium for extension of antimicrobial activity.	Liang et al., (2012)
Sage	Sage EO + Span 80 + Tween 80 + Water	*E. coli, Shigella dysenteriae, S. typhi*	Potent antibacterial activity was exhibited against *E. coli & S. typhimurium*, owing to MIC values being 4 times lower than the pure EO. Deterioration of *E. coli*'s cell membrane was observed, which paves the way to the nucleic acid leakage & protein in the culture medium.	Encapsulation of EO increased the effective mass transfer to the microbial membrane, thereby the antibacterial activity is enhanced.	Moghimi et al., (2016) Pan, Chen, Davidson & Zhong, (2014)
Satureja montana L.	EO (either of 3 varieties) + Tween 20/ Tween 80 + HEPES buffer	*L.monocytogenes, S. aureus, S. haemolyticus, E. coli, K. pneumoniae, P. aeruginosa, S. marcescens*	NE's formulated with EO (3rd type) were found to reach tremendous antimicrobial activity. Induction of decrease in the MIC values was achieved for *S. aureus, E. coli, S. marcescens* (0.78,0.39,0.78 mg/mL)	Lipophilic compounds (Terpenes) might be concerning the alteration stimulated in the structural and functional integrity of cell membrane of gram-positive, gram-negative as well in fungi.	Maccelli et al., (2019)

Continued on following page

Table 1. Continued

NAME OF THE ESSENTIAL OIL (EO)	FORMULATION OF NANOEMULSION	TARGET MICRO-ORGANISM	IMPACT/ACTIVITY OF NANOEMULSION	ASCRIPTION	REFERENCE(S)
Zataria multiflora Boiss	EO + Polysorbate 20 + Sodium Phosphate Buffer	*L. monocytogenes*, *S. typhimurium*	Nanoemulsion bearing ZEO exerted a better inhibitory effect against both the sample strains in milk than in the BHI broth. *L. monocytogenes* was observed to behave highly resistant towards ZEO-NE in BHI broth.	Nanoemulsion performs differently in BHI broth compared to milk where the interactions with other compounds seem to be positively regulating the activity. Resistance of *L. monocytogenes* in BHI broth is presumably because of its slow rate of growth in the system.	Shahabi et al., (2017)

e. Aqueous component

Water forms the central component of the aqueous phase but, in addition, it comprises components that include proteins, carbohydrates, minerals, co-solvents (polyols, simple alcohols), bases, acids & are polar. The production, stability, and physicochemical attributes of the nanoemulsions are governed by refractive index, polarity, rheology, density, interfacial tension, phase behavior, pH, aqueous phase's ionic strength and these parameters rely on the type and concentration of the components in the aqueous phase (McClements and Rao, 2011). The predominant criterion to ascertain the efficacy of the EO constituent appears to be the amount of the essential oil in the aqueous component, instead of the delivery system's droplet size (Donsì *et al.*, 2012). Hence, the composition of the aqueous phase must be carefully monitored to hone the generation or enhance the stability of nanoemulsion (McClements and Rao, 2011).

CONCLUSION

The increased concern in the utilization of the essential oil's biological attributes has been enacting as the driving force of the scientific community to develop an appropriate formulation for their mobilization. Nanoemulsions are a versatile platform for creating soft materials that may be customized in terms of composition as well as optical, rheological, and stability characteristics. Nanoemulsions confer a wide array of desirable chemical compositions as well as physical properties which in turn offer remarkable advantages instead of microemulsions. Nevertheless, to prevent distinct physicochemical phenomena that might significantly influence the stability and biological effectiveness of the generated nanoemulsion, particular attention should be paid to the formulation stage of essential oil nanoemulsion. With the progress in time the weaknesses and advances in science and technology, dosage composition of oil based nanoemulsion forms have evolved from simple mixtures to highly sophisticated systems; which are the known as novel EO-NE systems to develop commercially successful delivery system. The chapter broadly elucidates the role of nanoemulsions in imparting the protection of essential oil from harsh conditions followed by the co-relation established between the nanoemulsions' composition & the enhancement of the antibacterial efficacy of essential oil. Furthermore, to throw light on the contribution of other additional components during the formulation stage of nanoemulsion as well its impression on the entire properties of nanoemulsion, investigations demonstrated by several authors has been precisely portrayed as suitable illustrations, aside from essential oil. Desirably, the technical research should ideally focus on an in-depth analysis of probable co-relationships between the system's composition and its stability.

REFERENCES

Aswathanarayan, J. B., & Vittal, R. R. (2019). Nanoemulsions and Their Potential Applications in Food Industry. *Frontiers in Sustainable Food Systems*, *3*, 95. doi:10.3389/fsufs.2019.00095

Avlessi, F., Dangou, J., Wotto, V., Alitonou, G., Sohounhloue, D., & Menut, C. (2004). Antioxidant activities of essential oil from Clausena anisata (Wild) Hook. *Comptes Rendus. Chimie*, *7*, 1057–1061. doi:10.1016/j.crci.2003.11.015

Badi, H., Yazdani, D., Ali, S. M., & Nazari, F. (2004). Effects of spacing and harvesting time on herbage yield and quality/quantity of oil in thyme, Thymus vulgaris L. *Industrial Crops and Products*, *19*(3), 231–236. doi:10.1016/j.indcrop.2003.10.005

Bala, A., Kar, B., Haldar, P. K., Mazumder, U. K., & Bera, S. (2010). Evaluation of anticancer activity of Cleome gynandra on Ehrlich's Ascites Carcinoma treated mice. *Journal of Ethnopharmacology*, *129*(1), 131–134. doi:10.1016/j.jep.2010.03.010 PMID:20307641

Basak, S., & Guha, P. (2018). A review on antifungal activity and mode of action of essential oils and their delivery as nano-sized oil droplets in food system. *Journal of Food Science and Technology*, *55*(12), 4701–4710. doi:10.100713197-018-3394-5 PMID:30482966

Bentayeb, K., Vera, P., Rubio, C., & Nerín, C. (2014). The additive properties of Oxygen Radical Absorbance Capacity (ORAC) assay: The case of essential oils. *Food Chemistry*, *148*, 204–208. doi:10.1016/j.foodchem.2013.10.037 PMID:24262547

Bilia, A. R., Guccione, C., Isacchi, B., Righeschi, C., Firenzuoli, F., & Bergonzi, M. C. (2014). Essential oils loaded in nanosystems: a developing strategy for a successful therapeutic approach. *Evidence-Based Complementary and Alternative Medicine*.

Borugă, O., Jianu, C., Mişcă, C., Goleţ, I., Gruia, A. T., & Horhat, F. G. (2014). Thymus vulgaris essential oil: chemical composition and antimicrobial activity. *Journal of Medicine and Life*, *3*(Spec Iss 3), 56–60.

Brooker, M. I. H., & Kleinig, D. A. (2004). Field guide to Eucalypts (2nd ed.). Bloomings Books.

Bruneton, J. (1995). Pharmacognosy, phytochemistry, medicinal plants. Andover: Intercept / Paris: Lavoisier.

Burt, S. (2004). Essential oils: Their antibacterial properties and potential applications in foods—a review. *International Journal of Food Microbiology, 94*(3), 223–253. doi:10.1016/j.ijfoodmicro.2004.03.022 PMID:15246235

Burt, S. A., & Reinders, R. D. (2003). Antibacterial activity of selected plant essential oils against Escherichia coli O157:H7. *Letters in Applied Microbiology, 36*(3), 162–167. doi:10.1046/j.1472-765X.2003.01285.x PMID:12581376

Chanamai, R., Horn, G. R., & Mcclements, D. (2002). Influence of oil polarity on droplet growth in oil-in-water emulsions stabilized by a weakly adsorbing biopolymer or a nonionic surfactant. *Journal of Colloid and Interface Science, 247*(1), 167–176. doi:10.1006/jcis.2001.8110 PMID:16290453

Chang, Y., McLandsborough, L., & McClements, D. J. (2011). Interactions of a cationic antimicrobial (ε-polylysine) with an anionic biopolymer (pectin): An isothermal titration calorimetry, microelectrophoresis, and turbidity study. *Journal of Agricultural and Food Chemistry, 59*(10), 5579–5588. doi:10.1021/jf104299q PMID:21462961

Chang, Y., McLandsborough, L., & McClements, D. J. (2011). Physicochemical properties and antimicrobial efficacy of electrostatic complexes based on cationic ε-polylysine and anionic pectin. *Journal of Agricultural and Food Chemistry, 59*(12), 6776–6782. doi:10.1021/jf201162g PMID:21574554

Chang, Y., McLandsborough, L., & McClements, D. J. (2012). Cationic antimicrobial (ε-polylysine)-anionic polysaccharide (pectin) interactions: Influence of polymer charge on physical stability and antimicrobial efficacy. *Journal of Agricultural and Food Chemistry, 60*(7), 1837–1844. doi:10.1021/jf204384s PMID:22268773

Chang, Y., McLandsborough, L., & McClements, D. J. (2012). Physical properties and antimicrobial efficacy of thyme oil nanoemulsions: Influence of ripening inhibitors. *Journal of Agricultural and Food Chemistry, 60*(48), 12056–12063. doi:10.1021/jf304045a PMID:23140446

Chang, Y., McLandsborough, L., & McClements, D. J. (2013). Physicochemical properties and antimicrobial efficacy of carvacrol nanoemulsions formed by spontaneous emulsification. *Journal of Agricultural and Food Chemistry, 61*(37), 8906–8913. doi:10.1021/jf402147p PMID:23998790

Chang, Y., McLandsborough, L., & McClements, D. J. (2015). Fabrication, stability and efficacy of dual-component antimicrobial nanoemulsions: Essential oil (thyme oil) and cationic surfactant (lauric arginate). *Food Chemistry, 172*, 298–304. doi:10.1016/j.foodchem.2014.09.081 PMID:25442557

Chen, L., Remondetto, G., & Subirade, M. (2006). Food protein-based materials as nutraceutical delivery systems. *Trends in Food Science & Technology, 17*(5), 272–283. doi:10.1016/j.tifs.2005.12.011

Chuesiang, P., Siripatrawan, U., Sanguandeekul, R., McLandsborough, L., & McClements, D. J. (2018). Optimization of cinnamon oil nanoemulsions using phase inversion temperature method: Impact of oil phase composition and surfactant concentration. *Journal of Colloid and Interface Science, 514*, 208–216. doi:10.1016/j.jcis.2017.11.084 PMID:29257975

Cosentino, S., Tuberoso, C. I., Pisano, B., Satta, M., Mascia, V., Arzedi, E., & Palmas, F. (1999). In-vitro antimicrobial activity and chemical composition of Sardinian Thymus essential oils. *Letters in Applied Microbiology, 29*(2), 130–135. doi:10.1046/j.1472-765X.1999.00605.x PMID:10499301

Cruz-Valenzuela, M. R., Tapia-Rodríguez, M. R., Vazquez-Armenta, F. J., Silva-Espinoza, B. A., & Ayala-Zavala, J. F. (2017). Lime (Citrus aurantifolia) Oils. *Essential Oils in Food Preservation. Flavor and Safety*, *61*(3), 532–537. doi:10.1016/B978-0-12-416641-7.00061-4

Cui, H., Bai, M., Rashed, M., & Lin, L. (2018). The antibacterial activity of clove oil/chitosan nanoparticles embedded gelatin nanofibers against *Escherichia coli* O157:H7 biofilms on cucumber. *International Journal of Food Microbiology*, *266*, 69–78. doi:10.1016/j.ijfoodmicro.2017.11.019 PMID:29179098

Damjanović Vratnica, B., Djakov, T., Sukovic, D., & Damjanovic, J. (2011). Antimicrobial Effect of Essential Oil Isolated from Eucalyptus globulus Labill. from Montenegro. *Czech Journal of Food Sciences*, *29*(3), 277–284. doi:10.17221/114/2009-CJFS

Davidov-Pardo, G., & McClements, D. J. (2015). Nutraceutical delivery systems: Resveratrol encapsulation in grape seed oil nanoemulsions formed by spontaneous emulsification. *Food Chemistry*, *167*, 205–212. doi:10.1016/j.foodchem.2014.06.082 PMID:25148980

de Meneses, A. C., Sayer, C., Puton, B. M. S., Cansian, R. L., Araújo, P. H. H., & de Oliveira, D. (2019). Production of clove oil nanoemulsion with rapid and enhanced antimicrobial activity against gram-positive and gram-negative bacteria. *Journal of Food Process Engineering*, *6*(42),. doi:10.1111/jfpe.13209

de Souza, E. L., de Barros, J. C., de Oliveira, C. E., & da Conceição, M. L. (2010). Influence of Origanum vulgare L. essential oil on enterotoxin production, membrane permeability and surface characteristics of Staphylococcus aureus. *International Journal of Food Microbiology*, *137*(2-3), 308–311. doi:10.1016/j.ijfoodmicro.2009.11.025 PMID:20015563

Di Pasqua, R., Betts, G., Hoskins, N., Edwards, M., Ercolini, D., & Mauriello, G. (2007). Membrane toxicity of antimicrobial compounds from essential oils. *Journal of Agricultural and Food Chemistry*, *55*(12), 4863–4870. doi:10.1021/jf0636465 PMID:17497876

Dickinson, E., Semenova, M. G., Antipova, A. S., & Pelan, E. (1998). Effect of high-methoxy pectin on properties of casein-stabilized emulsions. *Food Hydrocolloids*, *12*(4), 425–432. doi:10.1016/S0268-005X(98)00057-5

Dima, C., & Dima, S. (2015). Essential oils in foods: Extraction, stabilization, and toxicity. *Current Opinion in Food Science*, *5*, 29–35. doi:10.1016/j.cofs.2015.07.003

Donsì, F., Annunziata, M., Vincensi, M., & Ferrari, G. (2012). Design of nanoemulsion-based delivery systems of natural antimicrobials: Effect of the emulsifier. *Journal of Biotechnology*, *159*(4), 342–350. doi:10.1016/j.jbiotec.2011.07.001 PMID:21763730

Donsì, F., & Ferrari, G. (2016). Essential oil nanoemulsions as antimicrobial agents in food. *Journal of Biotechnology*, *233*, 106–120. doi:10.1016/j.jbiotec.2016.07.005 PMID:27416793

Donsì, F., Sessa, M., & Ferrari, G. (2012). Effect of Emulsifier Type and Disruption Chamber Geometry on the Fabrication of Food Nanoemulsions by High Pressure Homogenization. *Industrial & Engineering Chemistry Research*, *51*(22), 7606–7618. doi:10.1021/ie2017898

Dorman, H. J., & Deans, S. G. (2000). Antimicrobial agents from plants: Antibacterial activity of plant volatile oils. *Journal of Applied Microbiology*, *88*(2), 308–316. doi:10.1046/j.1365-2672.2000.00969.x PMID:10736000

Edris, A. E. (2007). Pharmaceutical and therapeutic potentials of essential oils and their individual volatile constituents: A review. *Phytotherapy Research*, *21*(4), 308–323. doi:10.1002/ptr.2072 PMID:17199238

El-Sayed, H. S., Chizzola, R., Ramadan, A. A., & Edris, A. E. (2017). Chemical composition and antimicrobial activity of garlic essential oils evaluated in organic solvent, emulsifying, and self-microemulsifying water based delivery systems. *Food Chemistry*, *221*, 196–204. doi:10.1016/j.foodchem.2016.10.052 PMID:27979186

Ernst, E. (2004). Phytotherapy: A Quick Reference to Herbal Medicine. *Focus on Alternative and Complementary Therapies*, *9*(2), 164–164. doi:10.1111/j.2042-7166.2004.tb04305.x

Fang, Z. X., & Bhandari, B. (2010). Encapsulation of polyphenols – a review. *Trends in Food Science & Technology*, *21*(10), 510–523. doi:10.1016/j.tifs.2010.08.003

Fernández-López, J., & Viuda-Martos, M. (2018). Introduction to the Special Issue: Application of Essential Oils in Food Systems. *Foods (Basel, Switzerland)*, *7*(4), 56. doi:10.3390/foods7040056 PMID:29621143

Firoozi, M., Rezapour-Jahani, S., Shahvegharasl, Z., & Anarjan, N. (2020). Ginger essential oil nanoemulsions: Preparation and physicochemical characterization and antibacterial activities evaluation. *Journal of Food Process Engineering*, *43*(8), 43. doi:10.1111/jfpe.13434

Freires, I. A., Denny, C., Benso, B., de Alencar, S. M., & Rosalen, P. L. (2015). Antibacterial Activity of Essential Oils and Their Isolated Constituents against Cariogenic Bacteria: A Systematic Review. *Molecules (Basel, Switzerland)*, *20*(4), 7329–7358. doi:10.3390/molecules20047329 PMID:25911964

Garti, N., Yaghmur, A., Leser, M. E., Clement, V., & Watzke, H. J. (2001). Improved oil solubilization in oil/water food grade microemulsions in the presence of polyols and ethanol. *Journal of Agricultural and Food Chemistry*, *49*(5), 2552–2562. doi:10.1021/jf001390b PMID:11368635

Ghaderi, L., Moghimi, R., Aliahmadi, A., McClements, D. J., & Rafati, H. (2017). Development of antimicrobial nanoemulsion-based delivery systems against selected pathogenic bacteria using a thymol-rich Thymus daenensis essential oil. *Journal of Applied Microbiology*, *123*(4), 832–840. doi:10.1111/jam.13541 PMID:28714250

Ghazy, O. A., Fouad, M. T., Saleh, H. H., Kholif, A. E., & Morsy, T. A. (2021). Ultrasound-assisted preparation of anise extract nanoemulsion and its bioactivity against different pathogenic bacteria. *Food Chemistry*, *341*, 128259. doi:10.1016/j.foodchem.2020.128259 PMID:33068847

Gonnet, M., Lethuaut, L., & Boury, F. (2010). New trends in encapsulation of liposoluble vitamins. *Journal of Controlled Release*, *146*(3), 276–290. doi:10.1016/j.jconrel.2010.01.037 PMID:20600399

Gradzielski, M. (1998). Effect of the cosurfactant structure on the bending elasticity in nonionic oil-in-water microemulsions. *Langmuir*, *14*(21), 6037–6044. doi:10.1021/la980074c

Guenther, E. (1952). The essential oils. *Journal of the American Pharmaceutical Association*, *41*(6), 339–340. doi:10.1002/jps.3030410627

Guerra-Rosas, M. I., Morales-Castro, J., Cubero-Márquez, M. A., Salvia-Trujillo, L., & Martín-Belloso, O. (2017). Antimicrobial activity of nanoemulsions containing essential oils and high methoxyl pectin during long-term storage. *Food Control, 77*, 131–138. doi:10.1016/j.foodcont.2017.02.008

Gündel, S., de Souza, M. E., Quatrin, P. M., Klein, B., Wagner, R., Gündel, A., Vaucher, R. A., Santos, R., & Ourique, A. F. (2018). Nanoemulsions containing Cymbopogon flexuosus essential oil: Development, characterization, stability study and evaluation of antimicrobial and antibiofilm activities. *Microbial Pathogenesis, 118*, 268–276. doi:10.1016/j.micpath.2018.03.043 PMID:29581028

Gündel, S., Velho, M. C., Diefenthaler, M. K., Favarin, F. R., Copetti, P. M., Fogaça, A. O., Klein, B., Wagner, R., Gündel, A., Sagrillo, M. R., & Ourique, A. F. (2018). Basil oil-nanoemulsions: Development, cytotoxicity and evaluation of antioxidant and antimicrobial potential. *Journal of Drug Delivery Science and Technology, 46*, 378–383. doi:10.1016/j.jddst.2018.05.038

Gurpreet, K., & Singh, S. K. (2018). Review of Nanoemulsion Formulation and Characterization Techniques. *Indian Journal of Pharmaceutical Sciences, 80*, 781–789. doi:10.4172/pharmaceutical-sciences.1000422

Handa, M., Ujjwal, R., Vasdev, N., Flora, S., & Shukla, R. (2021). Optimization of Surfactant- and Cosurfactant-Aided Pine Oil Nanoemulsions by Isothermal Low-Energy Methods for Anticholinesterase Activity. *ACS Omega, 6*(1), 559–568. doi:10.1021/acsomega.0c05033 PMID:33458508

Hashem, A. S., Awadalla, S. S., Zayed, G. M., Maggi, F., & Benelli, G. (2018). Pimpinella anisum essential oil nanoemulsions against Tribolium castaneum-insecticidal activity and mode of action. *Environmental Science and Pollution Research International, 25*(19), 18802–18812. doi:10.100711356-018-2068-1 PMID:29713977

Hendry, E. R., Worthington, T., Conway, B. R., & Lambert, P. A. (2009). Antimicrobial efficacy of eucalyptus oil and 1,8-cineole alone and in combination with chlorhexidine digluconate against microorganisms grown in planktonic and biofilm cultures. *The Journal of Antimicrobial Chemotherapy, 64*(6), 1219–1225. doi:10.1093/jac/dkp362 PMID:19837714

Hossain, M., Al-Hashmi, R. A., Weli, A. M., Al-Riyami, Q., & Al-Sabahib, J. N. (2012). Constituents of the essential oil from different brands of Syzigium caryophyllatum L by gas chromatography–mass spectrometry. *Asian Pacific Journal of Tropical Biomedicine, 2*(3), S1446–S1449. doi:10.1016/S2221-1691(12)60435-3

Huang, J., Xiong, Y., Zeng, G., Guo, S., Xie, G., Zhang, D., Tang, X., & Liu, Z. (2012). Separation of phenol from various micellar solutions using MEUF. *Separation and Purification Technology, 98*, 1–6. doi:10.1016/j.seppur.2012.06.037

Imeson, A. (2010). *Food Stabilizers, Thickeners and Gelling Agents*. JohnWiley & Sons.

Inouye, S., Tsuruoka, T., Uchida, K., & Yamaguchi, H. (2001). Effect of sealing and Tween 80 on the antifungal susceptibility testing of essential oils. *Microbiology and Immunology, 45*(3), 201–208. doi:10.1111/j.1348-0421.2001.tb02608.x PMID:11345529

Jafari, S., He, Y., & Bhandari, B. (2007). Production of sub-micron emulsions by ultrasound and microfluidization techniques. *Journal of Food Engineering, 82*(4), 478–488. doi:10.1016/j.jfoodeng.2007.03.007

Jemaa, M.B., Falleh, H., & Ksouri, R. (2019). *Encapsulation of Natural Bioactive Compounds: Nano-emulsion Formulation to Enhance Essential Oils Activities*. Academic Press.

Jones, F. A. (1996). Herbs- useful plants. Their role in history and today. *European Journal of Gastroenterology & Hepatology*, *8*(12), 1227–1231. doi:10.1097/00042737-199612000-00018 PMID:8980946

Joyce, N., & Ramamurthy, N. (2017). Recent Research Trends in Fabrication and Applications of Plant Essential Oil Based Nanoemulsions. *Journal of Nanomedicine & Nanotechnology*, *08*(02). Advance online publication. doi:10.4172/2157-7439.1000434

Kabal'nov, A. S., Pertzov, A. V., & Shchukin, E. D. (1987). Ostwald ripening in two-component disperse phase systems: Application to emulsion stability. *Colloids and Surfaces*, *24*(1), 19–32. doi:10.1016/0166-6622(87)80258-5

Khaleque, M., Keya, C. A., Hasan, K. N., Hoque, M., Inatsu, Y., & Bari, M. (2016). Use of cloves and cinnamon essential oil to inactivate Listeria monocytogenes in ground beef at freezing and refrigeration temperatures. *Lebensmittel-Wissenschaft + Technologie*, *74*, 219–223. doi:10.1016/j.lwt.2016.07.042

Komaiko, J., & Mcclements, D. (2016). Formation of Food-Grade Nanoemulsions Using Low-Energy Preparation Methods: A Review of Available Methods. *Comprehensive Reviews in Food Science and Food Safety*, *15*(2), 331–352. doi:10.1111/1541-4337.12189 PMID:33371595

Králová, I., & Sjöblom, J. (2009). Surfactants Used in Food Industry: A Review. *Journal of Dispersion Science and Technology*, *30*(9), 1363–1383. doi:10.1080/01932690902735561

Krishnamoorthy, R., Gassem, M. A., Athinarayanan, J., Periyasamy, V. S., Prasad, S., & Alshatwi, A. A. (2021). Antifungal activity of nanoemulsion from Cleome viscosa essential oil against food-borne pathogenic Candida albicans. *Saudi Journal of Biological Sciences*, *28*(1), 286–293. doi:10.1016/j.sjbs.2020.10.001 PMID:33424308

Lambert, R. J., Skandamis, P. N., Coote, P. J., & Nychas, G. J. (2001). A study of the minimum inhibitory concentration and mode of action of oregano essential oil, thymol and carvacrol. *Journal of Applied Microbiology*, *91*(3), 453–462. doi:10.1046/j.1365-2672.2001.01428.x PMID:11556910

Leite, C. B. S., Coelho, J. M., Muehlmann, L. A., Azevedo, R. B., & Sousa, M. H. (2018). Microemulsions as Platforms for Transdermal Delivery of Hydrophilic Drugs - A Review. *Current Nanoscience*, *14*(3), 170–178. Advance online publication. doi:10.2174/1573413714666171218145416

Li, W., Chen, H., He, Z., Han, C., Liu, S., & Li, Y. (2015). Influence of surfactant and oil composition on the stability and antibacterial activity of eugenol nanoemulsions. *Lebensmittel-Wissenschaft + Technologie*, *62*(1), 39–47. doi:10.1016/j.lwt.2015.01.012

Liang, R., Xu, S., Shoemaker, C. F., Li, Y., Zhong, F., & Huang, Q. (2012). Physical and antimicrobial properties of peppermint oil nanoemulsions. *Journal of Agricultural and Food Chemistry*, *60*(30), 7548–7555. doi:10.1021/jf301129k PMID:22746096

Liew, S., Utra, U., Alias, A. K., Tan, T. B., Tan, C. P., & Yussof, N. S. (2020). Physical, morphological and antibacterial properties of lime essential oil nanoemulsions prepared via spontaneous emulsification method. *Lebensmittel-Wissenschaft + Technologie*, *128*, 109388. Advance online publication. doi:10.1016/j.lwt.2020.109388

Lis-Balchin, M., & Deans, S. G. (1997). Bioactivity of selected plant essential oils against Listeria monocytogenes. *Journal of Applied Microbiology*, *82*(6), 759–762. doi:10.1046/j.1365-2672.1997.00153.x PMID:9202441

Liu, T., & Liu, L. (2020). Fabrication and characterization of chitosan nanoemulsions loading thymol or thyme essential oil for the preservation of refrigerated pork. *International Journal of Biological Macromolecules*, *162*, 1509–1515. doi:10.1016/j.ijbiomac.2020.07.207 PMID:32738320

Lou, Z., Chen, J., Yu, F., Wang, H., Kou, X., Ma, C., & Zhu, S. (2017). The antioxidant, antibacterial, antibiofilm activity of essential oil from Citrus medica L. var. sarcodactylis and its nanoemulsion. *Lebensmittel-Wissenschaft + Technologie*, *80*, 371–377. doi:10.1016/j.lwt.2017.02.037

Lu, W. C., Huang, D. W., Wang, C. R., Yeh, C. H., Tsai, J. C., Huang, Y. T., & Li, P. H. (2018). Preparation, characterization, and antimicrobial activity of nanoemulsions incorporating citral essential oil. *Yao Wu Shi Pin Fen Xi*, *26*(1), 82–89. doi:10.1016/j.jfda.2016.12.018 PMID:29389592

Luís, Â., Duarte, A., Gominho, J., Domingues, F., & Duarte, A. P. (2016). Chemical composition, antioxidant, antibacterial and anti-quorum sensing activities of Eucalyptus globulus and Eucalyptus radiata essential oils. *Industrial Crops and Products*, *79*, 274–282. doi:10.1016/j.indcrop.2015.10.055

Ma, Q., Davidson, P. M., & Zhong, Q. (2016). Antimicrobial properties of microemulsions formulated with essential oils, soybean oil, and Tween 80. *International Journal of Food Microbiology*, *226*, 20–25. doi:10.1016/j.ijfoodmicro.2016.03.011 PMID:27016636

Maccelli, A., Vitanza, L., Imbriano, A., Fraschetti, C., Filippi, A., Goldoni, P., Maurizi, L., Ammendolia, M. G., Crestoni, M. E., Fornarini, S., Menghini, L., Carafa, M., Marianecci, C., Longhi, C., & Rinaldi, F. (2019). Satureja montana L. Essential Oils: Chemical Profiles/Phytochemical Screening, Antimicrobial Activity and O/W NanoEmulsion Formulations. *Pharmaceutics*, *12*(1), 7. doi:10.3390/pharmaceutics12010007 PMID:31861717

Majeed, H., Liu, F., Hategekimana, J., Sharif, H. R., Qi, J., Ali, B., Bian, Y.Y., Ma, J., Yokoyama, W., & Zhong, F. (2016). Bactericidal action mechanism of negatively charged food grade clove oil nanoemulsions. *Food Chemistry*, *197*(Prt A), 75-83. . doi:10.1016/j.foodchem.2015.10.015

Marchiori, M. L., Lubini, G., Dalla Nora, G., Friedrich, R. B., Fontana, M. C., Ourique, A. F., Bastos, M. O., Rigo, L. A., Silva, C. B., Tedesco, S. B., & Beck, R. C. (2010). Hydrogel containing dexamethasone-loaded nanocapsules for cutaneous administration: Preparation, characterization, and in vitro drug release study. *Drug Development and Industrial Pharmacy*, *36*(8), 962–971. doi:10.3109/03639041003598960 PMID:20590450

Marei, G. I. K., Rabea, E. I., & Badawy, M. E. I. (2018). Preparation and Characterizations of Chitosan/Citral Nanoemulsions and their Antimicrobial Activity. *Applied Food Biotechnology*, *5*(2), 69–78. doi:10.22037/afb.v5i2.19005

Mason, T. G., Wilking, J., Meleson, K., Chang, C. B., & Graves, S. (2006). Nanoemulsions: Formation, structure, and physical properties. *Journal of Physics Condensed Matter*, *18*(41), 079001. doi:10.1088/0953-8984/18/41/R01

McClements, D. J. (1994). Ultrasonic determination of depletion flocculation in oil-in-water emulsions containing a non-ionic surfactant. *Colloids and Surfaces. A, Physicochemical and Engineering Aspects*, *90*(1), 25–35. doi:10.1016/0927-7757(94)02881-8

McClements, D. J. (2004). *Food Emulsions: Principles, Practices, and Techniques*. CRC Press. doi:10.1201/9781420039436

McClements, D. J., Decker, E. A., & Weiss, J. (2007). Emulsion-based delivery systems for lipophilic bioactive components. *Journal of Food Science*, *72*(8), R109–R124. doi:10.1111/j.1750-3841.2007.00507.x PMID:17995616

McClements, D. J., & Gumus, C. E. (2016). Natural emulsifiers - Biosurfactants, phospholipids, biopolymers, and colloidal particles: Molecular and physicochemical basis of functional performance. *Advances in Colloid and Interface Science*, *234*, 3–26. doi:10.1016/j.cis.2016.03.002 PMID:27181392

McClements, D. J., Henson, L., Popplewell, L. M., Decker, E. A., & Choi, S. J. (2012). Inhibition of Ostwald ripening in model beverage emulsions by addition of poorly water soluble triglyceride oils. *Journal of Food Science*, *77*(1), C33–C38. doi:10.1111/j.1750-3841.2011.02484.x PMID:22133014

McClements, D. J., & Rao, J. (2011). Food-grade nanoemulsions: Formulation, fabrication, properties, performance, biological fate, and potential toxicity. *Critical Reviews in Food Science and Nutrition*, *51*(4), 285–330. doi:10.1080/10408398.2011.559558 PMID:21432697

McClements, D. J., Saliva-Trujillo, L., Zhang, R., Zhang, Z., Zou, L., Yao, M., & Xiao, H. (2016). Boosting the bioavailability of hydrophobic nutrients, vitamins, and nutraceuticals in natural products using excipient emulsions. *Food Research International*, *88*(Pt A), 140–152. doi:10.1016/j.foodres.2015.11.017

Md Othman, S., Hassan, M. A., Nahar, L., Basar, N., Jamil, S., & Sarker, S. D. (2016). Essential Oils from the Malaysian Citrus (Rutaceae) Medicinal Plants. *Medicines (Basel, Switzerland)*, *3*(2), 13. doi:10.3390/medicines3020013 PMID:28930124

Meynier, A., Lecoq, C., & Genot, C. (2005). Emulsification enhances the retention of esters and aldehydes to a greater extent than changes in the droplet size distribution of the emulsion. *Food Chemistry*, *93*(1), 153–159. doi:10.1016/j.foodchem.2004.10.018

Moghimi, R., Aliahmadi, A., Mcclements, D., & Rafati, H. (2016). Investigations of the effectiveness of nanoemulsions from sage oil as antibacterial agents on some food borne pathogens. *Lebensmittel-Wissenschaft + Technologie*, *71*, 69–76. doi:10.1016/j.lwt.2016.03.018

Moghimi, R., Ghaderi, L., Rafati, H., Aliahmadi, A., & McClements, D. J. (2016). Superior antibacterial activity of nanoemulsion of Thymus daenensis essential oil against E. coli. *Food Chemistry*, *194*, 410–415. doi:10.1016/j.foodchem.2015.07.139 PMID:26471573

Moraes-Lovison, M., Marostegan, L., Peres, M., Menezes, I., Ghiraldi, M., Rodrigues, R., Fernandes, A., & Pinho, S. (2016). Nanoemulsions encapsulating oregano essential oil: Production, stability, antibacterial activity and incorporation in chicken pâté. *Lebensmittel-Wissenschaft + Technologie*, *77*, 233–240. Advance online publication. doi:10.1016/j.lwt.2016.11.061

Nabavi, S. M., Marchese, A., Izadi, M., Curti, V., Daglia, M., & Nabavi, S. F. (2015). Plants belonging to the genus Thymus as antibacterial agents: From farm to pharmacy. *Food Chemistry*, *173*, 339–347. doi:10.1016/j.foodchem.2014.10.042 PMID:25466031

Nazzaro, F., Fratianni, F., De Martino, L., Coppola, R., & De Feo, V. (2013). Effect of essential oils on pathogenic bacteria. *Pharmaceuticals (Basel, Switzerland)*, *6*(12), 1451–1474. doi:10.3390/ph6121451 PMID:24287491

Nirmal, N. P., Meredy, R., Li, L., & Sultanbawa, Y. (2018). Formulation, characterisation and antibacterial activity of lemon myrtle and anise myrtle essential oil in water nanoemulsion. *Food Chemistry*, *254*, 1–7. doi:10.1016/j.foodchem.2018.01.173 PMID:29548427

Özoğul, Y., Boğa, E. K., Akyol, I., Durmuş, M., Uçar, Y., Regenstein, J., & Kosker, A. R. (2020). Antimicrobial activity of thyme essential oil nanoemulsions on spoilage bacteria of fish and food-borne pathogens. *Food Bioscience*, *36*, 100635.

Ozturk, B., & McClements, D. J. (2016). Progress in natural emulsifiers for utilization in food emulsions. *Current Opinion in Food Science*, *7*, 1–6. https://doi.org/10.1016/j.cofs.2015.07.008

Pan, K., Chen, H., Davidson, P. M., & Zhong, Q. (2014). Thymol nanoencapsulated by sodium caseinate: Physical and antilisterial properties. *Journal of Agricultural and Food Chemistry*, *62*(7), 1649–1657. https://doi.org/10.1021/jf4055402

Paudel, S., Bhargava, K., & Kotturi, H. (2019). Antimicrobial activity of cinnamon oil nanoemulsion against Listeria monocytogenes and Salmonella spp. on melons. *LWT- Food Science and Technology*. doi: . doi:10.1016/j.lwt.2019.05.087

Peña, A. A., & Miller, C. A. (2006). Solubilization rates of oils in surfactant solutions and their relationship to mass transport in emulsions. *Advances in Colloid and Interface Science*, *123-126*, 241–257. https://doi.org/10.1016/j.cis.2006.05.005

Perez-Cacho, P. R., & Rouseff, R. (2008). Processing and storage effects on orange juice aroma: A review. *Journal of Agricultural and Food Chemistry*, *56*(21), 9785–9796. https://doi.org/10.1021/jf801244j

Prestinaci, F., Pezzotti, P., & Pantosti, A. (2015). Antimicrobial resistance: A global multifaceted phenomenon. *Pathogens and Global Health*, *109*(7), 309–318. https://doi.org/10.1179/2047773215Y.0000000030

Qadir, A., Faiyazuddin, M. D., Talib Hussain, M. D., Alshammari, T. M., & Shakeel, F. (2016). Critical steps and energetics involved in a successful development of a stable nanoemulsion. *Journal of Molecular Liquids*, *214*, 7–18. https://doi.org/10.1016/j.molliq.2015.11.050

Quatrin, P. M., Verdi, C. M., de Souza, M. E., de Godoi, S. N., Klein, B., Gundel, A., Wagner, R., de Almeida Vaucher, R., Ourique, A. F., & Santos, R. (2017). Antimicrobial and antibiofilm activities of nanoemulsions containing Eucalyptus globulus oil against Pseudomonas aeruginosa and Candida spp. *Microbial Pathogenesis*, *112*, 230–242. https://doi.org/10.1016/j.micpath.2017.09.062

Rabelo, A. S., Sutili, F. K., Meneses, J. O., Severino, P., Souto, E. B., Fujimoto, R. Y., & Cardoso, J. C. (2021). 23 central composite rotatable design for the production of neem oil nanoemulsion for antifungal and antiparasitic applications. *Journal of Chemical Technology and Biotechnology (Oxford, Oxfordshire)*, *96*, 2159–2167. https://doi.org/10.1002/jctb.6623

Rao, J., & Mcclements, D. (2012). Impact of lemon oil composition on formation and stability of model food and beverage emulsions. *Food Chemistry*, *134*(2), 749–757.

Ribes, S., Fuentes, A., Talens, P., Barat, J., Ferrari, G., & Donsì, F. (2017). Influence of emulsifier type on the antifungal activity of cinnamon leaf, lemon and bergamot oil nanoemulsions against Aspergillus niger. *Food Control*, *73*, 784–795.

Rodríguez, J., Martín, M. J., Ruiz, M. A., & Clares, B. (2016). Current encapsulation strategies for bioactive oils: From alimentary to pharmaceutical perspectives. *Food Research International*, *83*, 41–59. doi:10.1016/j.foodres.2016.01.032

Ruberto, G. (2002). Analysis of Volatile Components of Citrus Fruit Essential Oils. In J. F. Jackson & H. F. Linskens (Eds.), Analysis of Taste and Aroma. Molecular Methods of Plant Analysis, 21. Springer. https://doi.org/10.1007/978-3-662-04857-3_7

Ryu, V. (2017). *Use Of Different Ripening Inhibitors To Enhance Antimicrobial Activity Of Essential Oil Nanoemulsion* (Thesis). University of Massachusetts Amherst.

Ryu, V., Corradini, M. G., McClements, D. J., & McLandsborough, L. (2019). Impact of ripening inhibitors on molecular transport of antimicrobial components from essential oil nanoemulsions. *Journal of Colloid and Interface Science*, *556*, 568–576. https://doi.org/10.1016/j.jcis.2019.08.059

Ryu, V., McClements, D. J., Corradini, M. G., & McLandsborough, L. (2018). Effect of ripening inhibitor type on formation, stability, and antimicrobial activity of thyme oil nanoemulsion. *Food Chemistry*, *245*, 104–111. https://doi.org/10.1016/j.foodchem.2017.10.084

Ryu, V., McClements, D. J., Corradini, M. G., Yang, J. S., & McLandsborough, L.National Center for Biotechnology Information. (2021). *PubChem* Natural antimicrobial delivery systems: Formulation, antimicrobial activity, and mechanism of action of quillaja saponin-stabilized carvacrol nanoemulsions. *Food Hydrocolloids*, *82*, 442–450. doi:10.1016/j.foodhyd.2018.04.017

Salvia-Trujillo, L., Rojas-Graü, A., Soliva-Fortuny, R., & Martín-Belloso, O. (2015). Physicochemical characterization and antimicrobial activity of food-grade emulsions and nanoemulsions incorporating essential oils. *Food Hydrocolloids*, *43*, 547–556. https://dx.doi.org/10.1016/j.foodhyd.2014.07.012

Schreiner, T. B., Santamaria-Echart, A., Ribeiro, A., Peres, A. M., Dias, M. M., Pinho, S. P., & Barreiro, M. F. (2020). Formulation and Optimization of Nanoemulsions Using the Natural Surfactant Saponin from Quillaja Bark. *Molecules (Basel, Switzerland)*, *25*(7), 1538. doi:10.3390/molecules25071538

Segvić Klarić, M., Kosalec, I., Mastelić, J., Piecková, E., & Pepeljnak, S. (2007). Antifungal activity of thyme (Thymus vulgaris L.) essential oil and thymol against moulds from damp dwellings. *Letters in Applied Microbiology*, *44*(1), 36–42. https://doi.org/10.1111/j.1472-765X.2006.02032.x

Shafiq-un-Nabi, S., Shakeel, F., Talegaonkar, S., Ali, J., Baboota, S., Ahuja, A., Khar, R. K., & Ali, M. (2007). Formulation development and optimization using nanoemulsion technique: A technical note. *AAPS PharmSciTech*, *8*(2), E12–E28. https://doi.org/10.1208/pt0802028

Shahabi, N., Tajik, H., Moradi, M., Forough, M., & Ezati, P. (2017). Physical, antimicrobial and antibiofilm properties of Zataria multiflora Boiss essential oil nanoemulsion. *International Journal of Food Science & Technology*, *52*, 1645–1652.

Shahavi, M. H., Hosseini, M., Jahanshahi, M., Meyer, R. L., & Darzi, G. N. (2015). Clove oil nanoemulsion as an effective antibacterial agent: Taguchi optimization method. *Desalination and Water Treatment*, *57*, 18379–18390. doi:10.1080/19443994.2015.1092893

Shahbazi, Y. (2019). Antioxidant, antibacterial, and antifungal properties of nanoemulsion of clove essential oil. *Nanomedicine Research Journal*, *4*(4), 204–208. doi:10.22034/nmrj.2019.04.001

Shahi, S. K., Shukla, A. C., Bajaj, A. K., Banerjee, U., Rimek, D., Midgely, G., & Dikshit, A. (2000). Broad spectrum herbal therapy against superficial fungal infections. *Skin Pharmacology and Applied Skin Physiology*, *13*(1), 60–64. https://doi.org/10.1159/000029909

Sharif, H. R., Abbas, S., Majeed, H., Safdar, W., Shamoon, M., Khan, M. A., Shoaib, M., Raza, H., & Haider, J. (2017). Formulation, characterization and antimicrobial properties of black cumin essential oil nanoemulsions stabilized by OSA starch. *Journal of Food Science and Technology*, *54*(10), 3358–3365. https://doi.org/10.1007/s13197-017-2800-8

Sharma, N., & Tripathi, A. (2008). Effects of Citrus sinensis (L.) Osbeck epicarp essential oil on growth and morphogenesis of Aspergillus niger (L.) Van Tieghem. *Microbiological Research*, *163*(3), 337–344. https://doi.org/10.1016/j.micres.2006.06.009

Shojaee-Aliabadi, S., Hosseini, H., Mohammadifar, M. A., Mohammadi, A., Ghasemlou, M., Hosseini, S. M., & Khaksar, R. (2014). Characterization of κ-carrageenan films incorporated plant essential oils with improved antimicrobial activity. *Carbohydrate Polymers*, *101*, 582–591. https://doi.org/10.1016/j.carbpol.2013.09.070

Simões, M., Simões, L. C., Cleto, S., Pereira, M. O., & Vieira, M. J. (2008). The effects of a biocide and a surfactant on the detachment of Pseudomonas fluorescens from glass surfaces. *International Journal of Food Microbiology*, *121*(3), 335–341. https://doi.org/10.1016/j.ijfoodmicro.2007.11.041

Sonneville-Aubrun, O., Simonnet, J. T., & L'Alloret, F. (2004). Nanoemulsions: A new vehicle for skincare products. *Advances in Colloid and Interface Science*, *108-109*, 145–149. https://doi.org/10.1016/j.cis.2003.10.026

Soottitantawat, A., Yoshii, H., Furuta, T., Ohkawara, M., & Linko, P. (2003). Microencapsulation by Spray Drying: Influence of Emulsion Size on the Retention of Volatile Compounds. *Journal of Food Science*, *68*, 2256–2262. https://doi.org/10.1111/j.1365-2621.2003.tb05756.x

Sugumar, S., Mukherjee, A., & Chandrasekaran, N. (2015). Eucalyptus oil nanoemulsion-impregnated chitosan film: Antibacterial effects against a clinical pathogen, Staphylococcus aureus, in vitro. *International Journal of Nanomedicine*, *10*(Suppl 1), 67–75. https://doi.org/10.2147/IJN.S79982

Sugumar, S., Nirmala, J., Ghosh, V., Anjali, H., Mukherjee, A., & Chandrasekaran, N. (2013). Bio-based nanoemulsion formulation, characterization and antibacterial activity against food-borne pathogens. *Journal of Basic Microbiology*, *53*(8), 677–685. https://doi.org/10.1002/jobm.201200060

Sugumar, S., Singh, S., Mukherjee, A., & Chandrasekaran, N. (2016). Nanoemulsion of orange oil with non ionic surfactant produced emulsion using ultrasonication technique: Evaluating against food spoilage yeast. *Applied Nanoscience*, *6*, 113–120. https://doi.org/10.1007/s13204-015-0412-z

Taylor, P. (1998). Ostwald ripening in emulsions. *Advances in Colloid and Interface Science*, *75*(2), 107–163. https://doi.org/10.1016/S0001-8686(98)00035-9

Teixeira, B., Marques, A., Pires, C., Ramos, C., Batista, I., Saraiva, J., & Nunes, M. (2014). Characterization of fish protein films incorporated with essential oils of clove, garlic and origanum: Physical, antioxidant and antibacterial properties. *Lebensmittel-Wissenschaft + Technologie*, *59*, 533–539.

Teixeira, B., Marques, A., Ramos, C., Serrano, C., Matos, O., Neng, N. R., Nogueira, J. M. F., Saraiva, J. A., & Nunes, M. L. (2013). Chemical composition and bioactivity of different oregano (Origanum vulgare) extracts and essential oil. *Journal of the Science of Food and Agriculture*, *93*, 2707–2714. https://doi.org/10.1002/jsfa.6089

Tepe, B., Akpulat, H. A., Sokmen, M., Daferera, D., Yumrutas, O., Aydin, E., Polissiou, M., & Sokmen, A. (2006). Screening of the antioxidative and antimicrobial properties of the essential oils of Pimpinella anisetum and Pimpinella flabellifolia from Turkey. *Food Chemistry*, *97*(4), 719–724. https://doi.org/10.1016/j.foodchem.2005.05.045

Thormar, H. (2010). *Lipids and Essential Oils as Antimicrobial Agents*. John Wiley & Sons.

Toledano-Medina, M. A., Pérez-Aparicio, J., Moreno-Rojas, R., & Merinas-Amo, T. (2016). Evolution of some physicochemical and antioxidant properties of black garlic whole bulbs and peeled cloves. *Food Chemistry*, *199*, 135–139. https://doi.org/10.1016/j.foodchem.2015.11.128

Topuz, O. K., Özvural, E. B., Zhao, Q., Huang, Q., Chikindas, M., & Gölükçü, M. (2016). Physical and antimicrobial properties of anise oil loaded nanoemulsions on the survival of foodborne pathogens. *Food Chemistry*, *203*, 117–123. https://doi.org/10.1016/j.foodchem.2016.02.051

Trujillo-Cayado, L. A., Santos, J., Calero, N., Alfaro-Rodríguez, M. C., & Muñoz, J. (2020). Strategies for reducing Ostwald ripening phenomenon in nanoemulsions based on thyme essential oil. *Journal of the Science of Food and Agriculture*, *100*(4), 1671–1677. https://doi.org/10.1002/jsfa.10181

Turek, C., & Stintzing, F. C. (2013). Stability of Essential Oils: A Review. *Comprehensive Reviews in Food Science and Food Safety*, *12*, 40–53. https://doi.org/10.1111/1541-4337.12006

Ultee, A., Bennik, M. H., & Moezelaar, R. (2002). The phenolic hydroxyl group of carvacrol is essential for action against the food-borne pathogen Bacillus cereus. *Applied and Environmental Microbiology*, *68*(4), 1561–1568. https://doi.org/10.1128/AEM.68.4.1561-1568.2002

Walstra, P. (1993). Principles of emulsion formation. *Chemical Engineering Science*, *48*(2), 333–349. https://doi.org/10.1016/0009-2509(93)80021-H

Wan, J., Jin, Z., Zhong, S., Schwarz, P., Chen, B., & Rao, J. (2020). Clove oil-in-water nanoemulsion: Mitigates growth of Fusarium graminearum and trichothecene mycotoxin production during the malting of Fusarium infected barley. *Food Chemistry*, *312*, 126120. https://doi.org/10.1016/j.foodchem.2019.126120

Wan, J., Zhong, S., Schwarz, P., Chen, B., & Rao, J. (2019). Enhancement of antifungal and mycotoxin inhibitory activities of food-grade thyme oil nanoemulsions with natural emulsifiers. *Food Control*, *106*, 106709. https://doi.org/10.1016/j.foodcont.2019.106709

Wan, J., Zhong, S., Schwarz, P., Chen, B., & Rao, J. (2019). Physical properties, antifungal and mycotoxin inhibitory activities of five essential oil nanoemulsions: Impact of oil compositions and processing parameters. *Food Chemistry*, *291*, 199–206. https://doi.org/10.1016/j.foodchem.2019.04.032

Weiss, J., Cancelliere, C., & Mcclements, D. J. (2000). Mass Transport Phenomena in Oil-in-Water Emulsions Containing Surfactant Micelles: Ostwald Ripening. *Langmuir*, *16*, 6833–6838. https://doi.org/10.1021/la991477v

Wooster, T. J., Golding, M., & Sanguansri, P. (2008). Impact of oil type on nanoemulsion formation and Ostwald ripening stability. *Langmuir*, *24*(22), 12758–12765. https://doi.org/10.1021/la801685v

Yamamoto-Ribeiro, M. M., Grespan, R., Kohiyama, C. Y., Ferreira, F. D., Mossini, S. A., Silva, E. L., Filho, B. A., Mikcha, J. M., & Machinski, M. Jr. (2013). Effect of Zingiber officinale essential oil on Fusarium verticillioides and fumonisin production. *Food Chemistry*, *141*(3), 3147–3152. https://doi.org/10.1016/j.foodchem.2013.05.144

Yazgan, H., Ozogul, Y., & Kuley, E. (2019). Antimicrobial influence of nanoemulsified lemon essential oil and pure lemon essential oil on food-borne pathogens and fish spoilage bacteria. *International Journal of Food Microbiology*, *306*, 108266. https://doi.org/10.1016/j.ijfoodmicro.2019.108266

Yildirim, S. T., Oztop, M. H., & Soyer, Y. (2017). Cinnamon oil nanoemulsions by spontaneous emulsification: Formulation, characterization and antimicrobial activity. *Lebensmittel-Wissenschaft + Technologie*, *84*, 122–128. https://doi.org/10.1016/j.lwt.2017.05.041

Ziani, K., Chang, Y., McLandsborough, L., & McClements, D. J. (2011). Influence of surfactant charge on antimicrobial efficacy of surfactant-stabilized thyme oil nanoemulsions. *Journal of Agricultural and Food Chemistry*, *59*(11), 6247–6255. https://doi.org/10.1021/jf200450m

Chapter 5
Improved Oral Delivery of Drugs Using Nanoemulsion

Subramanian Natesan
National Institute of Pharmaceutical Education and Research, Kolkata, India

Victor Hmingthansanga
National Institute of Pharmaceutical Education and Research, Kolkata, India

Nidhi Singh
National Institute of Pharmaceutical Education and Research, Kolkata, India

Pallab Datta
National Institute of Pharmaceutical Education and Research, Kolkata, India

Sivakumar Manickam
Universiti Teknologi Brunei, Brunei

V. Ravichandiran
National Institute of Pharmaceutical Education and Research, Kolkata, India

ABSTRACT

Administration of drugs through the oral route is considered the simplest and most convenient way to offer greater patient compliance than other routes. Most active drugs discovered in the past and those being discovered in recent times are inadequate because of their inherent limitations in physicochemical properties such as low solubility and permeability, resulting in poor bioavailability, especially after oral administration in the form of tablet or capsule. Pharmaceutical nanoemulsion is the most promising, safer, and multimodal technique for delivering poorly soluble drugs and gaining more attention due to its characteristics such as higher solubilisation capacity, smaller size, surface charge, and site-specific drug targeting. This chapter focuses on the biological fate of nanoemulsion after oral administration and a few case studies related to the oral application of nanoemulsion in delivering poorly soluble drugs. In addition, the anatomy and physiology of the GI tract, components of nanoemulsion, and methods of preparation are addressed.

DOI: 10.4018/978-1-7998-8378-4.ch005

1. INTRODUCTION

Administration of drugs through the oral route is one of the most ideal and convenient routes that has been successfully employed in treating several diseases (Thiagarajan et al., 2011). Oral drug delivery has been the most favoured or accommodated drug delivery system in the pharmaceutical field because of its numerous impending advantages, including patient-friendly, cost-effective, convenient, non-invasiveness, and well-established (Muheem et al., 2016). Administration of drugs via oral route mainly holds three goals. First is the local delivery of drugs to treat gastrointestinal (GI) disease. The drugs are normally absorbed into the GI mucosa but will not undergo systemic absorption or inadequately be absorbed. Second is systemic drug delivery, where the absorbed drugs cross the mucosal wall and enter the systemic circulation. Third, mainly focusing on enhancing the dissolution rate of poorly soluble drugs (Martinez and Amidon, 2002). Related to the absorption of drugs in the GI tract, several factors govern the process, such as concentration of the drug at the site of absorption, blood flow to the site of absorption, surface area for absorption, the physical state of the drug, and its solubility (Brunton et al., 2018). The entire GI tract consists of epithelial cells that form the innermost layer and the continuous lining of the GI tract, which assist absorption and the drugs have to penetrate these epithelial cells. These cells act as a barrier but selectively regulate the transport of drugs from the lumen to the essential tissue compartment (Brunton et al., 2018). The mechanisms by which drug molecules are transported through the epithelial cells include passive transport through paracellular diffusion (between the cells), transcellular transport (through the cells), and active transport through carrier-mediated transport or receptor-mediated transport through endocytosis. Among these, transcellular transport is considered the major mechanism of drug absorption along the GI tract and is generally comparable to the solubility of the drug in lipids. Hence, absorption is much more favoured when the drug molecule is in the non-ionised form or more lipophilic than the ionised form (Homayun et al., 2019).

Conventional dosage forms such as tablets or capsules, when administered orally, exhibit severe negative aspects which impede their applicability in some of the definite populations, explicitly, pediatric and geriatric, and oral dosing to these populations is sometimes limited as the patients are not capable of swallowing the product (Chime et al., 2014). In addition, numerous pharmaceutical agents are not stable in diverse environmental conditions (from the acidic stomach to basic intestinal conditions). A requirement for oral delivery is adequate aqueous solubility of the drug under GI conditions, which, when not accomplished, leads to drug amassing or instability *in-vivo* and ultimately pave a way to low bioavailability and partial treatment of the disease (Muheem et al., 2016; Jaiswal, Dudhe, and Sharma, 2015). This phenomenon is usually encountered in hydrophobic drugs which have very poor solubility. Numerous nano-based drug delivery systems such as liposomes, solid lipid nanoparticles, nanostructured lipid carriers, microemulsions, and nanoemulsions have been exploited extensively to overcome the aforementioned drawbacks. However, nanoemulsion has emerged as a superior formulation among these nano-based drug delivery systems than the other formulations for oral drug delivery because of its capability to improve the bioavailability of the drug. Since nanoemulsion is considered a colloidal liquid formulation, it is expediently administered to diverse populations of all ages. It can also encapsulate the drugs, which can protect them from the external environment of the GI tract, thereby imparts stability to the drugs leading to the efficient treatment of targeted disease. Nanoemulsion can also be used for the targeted drug delivery by conjugating a specific ligand on the surface of its droplets. By definition, nanoemulsion is an isotropic, transparent/translucent, heterogeneous system of two immiscible liquids, thermodynamically unstable but kinetically stable. It typically possesses nanodroplets of 200 to 700 nm

(Thakore and Patel, 2014). Nanoemulsion is composed of an oil phase and an aqueous phase, mixed in the presence of an emulsifier commonly known as a surfactant (co-emulsifier/co-surfactant), forming a coarse emulsion, which is subsequently broken down into oil droplets into nanosize range by employing high energy (McClements et al., 2012). Generally, nanoemulsion is classified into three types, i.e. O/W (oil phase is dispersed into an aqueous phase), W/O (aqueous phase is dispersed into an oil phase), and multiple emulsions where microdomains of oil and water phases are inter-dispersed within the system. Furthermore, nanoemulsion is classified based on its surface charge, i.e. neutral, anionic, and cationic (Koroleva and Yurtov, 2012).

A brief outline regarding the oral administration of drugs and the possible applications of nanoemulsion in oral drug delivery is emphasised in this particular section. Following this, anatomy and physiology, excipients used in the preparation of nanoemulsion, methods of preparation, modes of nanoemulsion transport through oral route, and application of nanoemulsions in improving oral delivery will be discussed thoroughly.

2. ANATOMY AND PHYSIOLOGY OF GI TRACT

2.1. Anatomy of GI Tract

The GI tract is approximately 9 m in length and has a tube-like structure with varying diameters along its course. The pictorial depiction of the GI tract is shown in Figure 1. Ingestion of food, nutrients absorption, and secretion of dissipating products are the most important functions carried out by the GI tract (Marieb et al., 2007; Reinus et al., 2014). After oral administration, the transfer of food and pharmaceuticals from the oesophagus to the stomach is assisted by peristaltic contraction. Subsequently, a major portion of the digestion process occurs in the stomach with the assistance of various enzymes and its acidic environment (Reinus et al., 2014).

The stomach presents itself as a short-term reservoir holding the ingested food and pharmaceuticals before their delivery to the duodenum at a controlled rate. Smaller amounts of the drugs are absorbed in the stomach because of their small surface area and physiological conditions. Several functional units are responsible for regulating gastric secretion and gastric emptying; among them, local nervous plexus, hormones, mechanoreceptors sensitive to GI wall tension, and chemoreceptors are the most common ones. Gastric glands in the stomach are accountable for gastric juices, which cover up approximately the whole wall of the gastric body. A gastric gland comprises three types of cells: mucosal cells of the neck, parietal cells, and peptic cells (Hall et al., 2015). Briefly, when the ingested food arrives at the stomach, it causes gastric wall distension, resulting in gastrin release; it also stimulates the production of both acid and pepsinogen while enhancing the motor functions of the stomach. When the gastric secretion and the food have been mixed properly, the resultant product is called chyme, which is required for the pyloric sphincter to enter the small intestine during gastric emptying (GE).

The small intestine is the most convoluted and longest (approximately 6 m in length) component of the GI tract, where the ingestion process completes utilising enzymes secreted from the pancreas and liver, and it is also responsible for the absorption of nutrients (Marieb et al., 2007; Reinus et al., 2014). Owing to its large surface area, it is considered the major site of drug absorption. In an adult, the surface area of the small intestine increases due to the presence of villi (finger-like protrusion that projects into the intestinal lumen) and microvilli which are well furnished with blood vessels (Marieb et al., 2007;

Reinus et al., 2014). The initial part of the small intestine is the duodenum having a length of 20-30 cm, to which most of the ingested food enters where the pancreas and the gallbladder drain their contents. Optimal pH must digest the food by the pancreatic enzymes (proteases, amylases, lipases). Therefore, the duodenal pH is maintained in the range of 6-6.5 due to bicarbonate that counterbalances the acid chyme dump by the stomach (Ruiz et al., 2018). The lipase enzymes degrade the fatty components of the food. However, before this step, there was a requirement for the dispersion of fatty components. It is carried out by the surfactants present in the bile. They are also responsible for the dissolution of hydrophobic compounds. The duodenum part of the small intestine also does another critical function: converting an inactive form of a drug into its active form via enzymatic hydrolysis. The other parts of the small intestine are the jejunum and ileum, presenting an increased pH value (8 in the most distal area), decreasing internal diameter and number of folds. Still, the mucous membrane preserves its absorptive ability.

The final part of the GI tract is the large intestine. Its function is to absorb the residual nutrients and water for maintaining homeostasis. It also processes the waste products (Reinus et al., 2014). The residual dissipation is subsequently transferred to the rectum, where it is cleared from the body in the form of a stool. A specific part of the large intestine known as the colon is exploited extensively for systemic and local drug delivery. Based on the anatomy of the colon, it is classified into four distinct parts: ascending, transverse, descending, and sigmoid. In contrast to the small intestine, the colon does not possess specialised villi; consequently, the surface area is immensely smaller. Nevertheless, the surface area of the large intestine is increased because of its structural presentation, i.e. tomb-like structure. A wide number and an array of bacteria permanently colonise the colon and form the microbiome (Consortium, 2012).

Figure 1. Anatomy of the GI tract

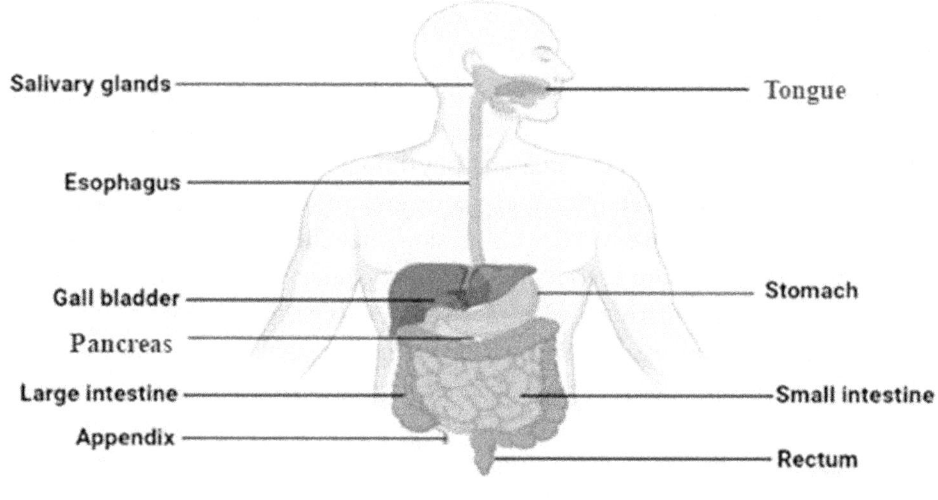

2.2. Physiology of GI Tract

Drug delivery via the oral route may be the most suitable route of administration. Still, due to the inherent physiological properties of the GI tract, in some instances, it hinders its applicability. On the other hand, it may also offer a suitable environment for site-specific drug delivery. The physiology of the GI tract may vary within the healthy individual. As a result, while designing a formulation intended for oral delivery, importance should be given to physiological factors of the GI tract. Several physiological factors govern the functionality of the GI tract, such as its transit time, pH, mucous, intestinal fluid volume, enzymes and microbiome.

2.2.1. GI transit time

Dosage forms and drugs designed to have site-specific targeting and GI transit time plays a prime factor. The GI transit time can vary from several minutes to several hours depending on physiological factors, including age, gender, food intake, osmolarity, and body posture (Kagan & Hoffman, 2018). For instance, the GI transit time in a fasted state can vary from 0 to 2 h, while in a fed state, it can be long-lasting up to 6 h. The GI time for the small intestine is comparatively invariable, approximately 3 to 4 h (Hu et al., 2000). Nevertheless, it can also be prolonged in healthy individuals from 2 to 6 h (Reinus et al., 2014).

2.2.2. GI pH

The gastrointestinal environment presents a diverse range of pH and has been extensively exploited with the intention of delayed-release therapies. The pH of the gastric environment at the initial part of the GI tract is highly acidic (pH 1.5-2 in the fasted state and pH 3-6 in the fed state), and the pH quickly increases to 6 when it reaches the duodenum, and increases down the small intestine to pH 7.4 at the extreme of the ileum (Fallingborg et al., 1993; Braten and Jones, 2006). However, several important factors control the variation in pH in an individual. The most important ones are dietary intake (food and fluids) and microbial metabolism (Ibekwe et al., 2008). The pH of the GI tract also influences the ionisation process of drugs, which in succession affects the drug absorption (Brunton et al., 2018).

2.2.3. GI mucosas

The GI mucous is continuously secreted in the GI tract and serves as a lubricant for transferring food and other pharmaceutical compounds. Besides, it also shields the epithelium from foreign pathogens and motorised stress (Atuma et al., 2001). The GI mucosa collects water and mucin protein covered with proteoglycans bearing a negative charge (Homayun et al., 2019). Depending upon the anatomy of the GI tract, the concentration or thickness of the mucous layer varies. There is no distinct mucous layer in the mouth and oesophagus, but they are continuously bathed by the mucous produced from the salivary glands (Johansson et al., 2013). There is only one type of mucous system in the small intestine that is unattached, while the colon and the stomach are composed of two types of mucous system that is attached and unattached (Atuma et al., 2001). The mucous is constantly produced by the goblet cells down the GI tract, and later it is emptied from the tissues because of the turnover of cells. Concerning drug delivery aimed for systemic absorption, the mucous presents itself as a significant barricade for the drug molecules when penetration is intended. In addition, it can also lessen the dwelling time of the designed dosage form.

2.2.4. GI Enzymes and Microbiome

Enzymes present in different parts of the GI tract carry out the digestion process of food and degradation of dosage forms and drugs. Majority of the enzymes involved in the digestion process of food act on the stomach and small intestine (Reinus et al., 2014). GI enzymes can influence the stability of the drugs susceptible to enzymatic degradation; however, they can also be used as an asset for designing a formulation intended for regional delivery of the drugs.

The intestinal microbiome consists of more than 500 discrete bacterial species, and it is the most integral part of the GI tract for accomplishing digestion and maintaining intestinal health (Sartor et al., 2008). The greater numbers of the intestinal microbiome inhabit the anaerobic colon, and the most important source of nutrition for these inhabitants is the fermentation of carbohydrates (Macfarlane et al., 2011). Both environmental and genetic aspects contribute to the substantial difference in the microbiome structure evident among the individuals (Sartor et al., 2010). Nevertheless, the main species such as bacteroidetes, firmicutes, proteobacteria, and bacteroidetes emerge to be consistent and correspond to the mainstream of the colonic flora (Frank et al., 2007).

3. NANOEMULSION COMPONENTS

Nanoemulsion is composed of oils, surfactants, and co-surfactants as the main components. In addition, nanoemulsion may contain excipients like polymers, preservatives, buffers, stabilisers, and antioxidants, which are essential for the drug or oil depending on their physicochemical properties (McClements et al., 2013). Excipients screening in drug solubility is essential in nanoemulsion formulation since it helps select the best oil and surfactant. However, the sensible choice of excipients depends on the compatibility with the individual route and intended application.

3.1. Oil

Oil is one of the most important excipients in nanoemulsion formulation. It serves as a solvent system for water-insoluble drugs (hydrophobic drugs) to encapsulate the requisite amount by rapid emulsification. The selection of oils is achieved by making the emulsification region (ER) in phase studies. However, the chain length of the oil and the molecular volume also influence the selection of oil. Vegetable oils and triglycerides of long-chain fatty acids produce a small ER; consequently, it presents difficulties in the emulsification process (Warisnoicharoen et al., 2000; Sapra et al., 2013). The hydrophobicity of the oil phase also has an immense impact on the generation of the emulsification region. A positive effect is encountered in the ER when there is a decline in the hydrophobicity of the oil phase. In addition to this, a decline in the oil phase's hydrophobicity negatively affects the solubility of a lipophilic drug. As a general rule, the solubilising capability of different classes of oils ought to increase in the following manner: medium-chain mono- and diglycerides < medium-chain triglycerides < vegetable oils (Sapra et al., 2013). A fine balance between the emulsification tendency and solubilising capacity is the final choice of the oil phase. In search of enhanced solubilising capacity and emulsification tendency, formulation scientists employ a combination of oils that meets the above conditions. Oils such as vitamin E (d-α-tocopherol) enhances the solubility of poorly soluble drugs like itraconazole, saquinavir, and paclitaxel (Constantinides et al., 2004). Moreover, the viscosity or the consistency of the selected oil might be beneficial in selecting the preparation in specific the requirement of high-energy methods (Narang et al., 2007). Examples of oils that are employed for nanoemulsion are castor oil, corn oil, coconut oil, evening primrose oil, olive oil, linseed oil, peanut oil, mineral oil, triglycerides, Myglyol 812R, fatty acids, as well as acid, and esters of saccharose (such as mono-palmitate of saccharose, di- palmitate of saccharose or tri-palmitate of saccharose) and isopropyl myristate (IPM). Nevertheless, the oils mentioned above are not suitable for all the routes of administration. Therefore before using the nanoemulsion system, a compatibility study of oil with individual routes is mandatory (Azeem et al., 2009).

3.2. Surfactant

Surfactants are important excipients incorporated in the nanoemulsion system to promote stability by preserving the size of the globules within the nano range; else, the individual globules amalgamate simultaneously, resulting in increased size, which eventually directs to phase separation. The prime concern in nanoemulsion formulation is stability; it depends on the fine selection of surfactants and their concentration. Some of the required properties of surfactants are quick absorption onto the surface of globules to form an inclusive and articulate film so that the globules do not coalesce, the capability to decrease the surface tension to less than 10 dynes/cm, ability to give requisite viscosity and zeta potential in the nanoemulsion system to give optimal stability, and lastly efficient at the minimum amount (Gupta et al., 2015). Since most of the active drugs are less soluble, the preferred surfactant should have the capacity to enhance the solubility and facilitate high drug loading into the nanoemulsion system. Concurrently, the selected surfactant should also present excellent emulsification to allow high oil inclusion (Sapra et al., 2013). The hydrophilic-lipophilic balance (HLB) value is of great importance for any surfactant since it is useful in deciding the appropriate surfactant for a specific application. In general, lipophilic surfactants or water-insoluble surfactants have a low HLB value, i.e. below 9 and hydrophilic surfactants or water-soluble surfactants have a high HLB value i.e. above 11. Intermediate surfactants usually have an HLB value between 9 and 11. Non-ionic surfactants generally have an HLB value in the range of 1-20. Still, for ionic surfactants, due to the greater strength of the hydrophilic regions, the HLB value considers the weight and the strength of the hydrophilic and lipophilic regions (Davies et al., 1957). Normally for the development of oil-in-water nanoemulsion, a surfactant having a high HLB value is required. As a common practice, a combination of surfactants is preferred when adopting low-energy phase inversion temperature to prepare nanoemulsion. The combination consists of hydrophobic and hydrophilic surfactants to form a multifaceted network at the interface. On the other hand, it may bring a stability concern for the partially trapped drugs in the oil and water phase.

The commonly used surfactants in nanoemulsion preparation are non-ionic, and owing to the presence of an ether functional group, the water solubility of the surfactant increases. Besides the emulsification property, a surfactant should also possess certain characteristics like non-toxic, suitable taste and odour, and optimum chemical stability. Ionic surfactants are also used to prepare nanoemulsions, but they are not preferred due to their toxic property. From a positive standpoint, various surfactants employed for nanoemulsion preparation possess the GRAS (generally regarded as safe) status that is fit for human consumption. Nevertheless, a maximum of the surfactants are toxic. Therefore, surfactants derived from natural sources such as phospholipid and lecithin have gained much importance due to their low toxicity profile and high biocompatibility (Musa et al., 2013; Dordevic et al., 2013; Hu et al., 2014). Lecithin might propose a very bright prospect, but due to the affinity to form lamellar liquid crystalline phases, an auxiliary substance such as polysorbates are included to slow down this problematic event (Moreno et al., 2003). In general, the concentration of surfactants used in nanoemulsion preparation is moderately high (10% w/v) as compared to traditional emulsion preparation (1% w/v) owing to the huge interfacial area in nanoemulsion (Vandamme et al., 2002). In addition, the nanoemulsion preparation is favoured with a lesser amount of surfactant producing low toxicity and low interaction with oral routes. However, this should be carefully assessed of quick emulsification and sufficient thermodynamic stability at large-scale preparation. The preferred surfactants for human applications include polysorbates, sorbitan monoesters, polyethylene glycol (PEG), phospholipid derivatives tyloxapol, lecithins, poloxamers, polyethoxylated castor oil derivatives such as polyoxyethylene 10-oleyl ether (Brij 96), and Solutol® HS 15 (PEG-660-12-hydroxystearate).

3.3. Co-surfacant

Co-surfactant is usually added in combination with surfactant molecules to provide structural elasticity to the nanoemulsion system. A rigid surfactant film is formed when a single surface-active agent is used in the nanoemulsion system, which cannot reduce the oil-water interfacial tension at a satisfactory level in the emulsification process. The combination of co-surfactant and surfactant in the nanoemulsion system produces a more active surfactant system in which the co-surfactant infiltrates the rigid layer of the surfactant, disturbs the interfacial film, provides the necessary flexibility, lessens the interfacial tension, and ultimately assists in the formation of nanoemulsion (Azeem et al., 2009). Based upon the comparative significance of their polar groups, co-surfactants also participate in modifying the curvature of the oil-water interface. In addition, they also help in the solubilisation of oil. Co-surfactants generally used in nanoemulsion formulation include short-chain alcohols, such as ethanol, glycerol propylene glycol, PEG 400 benzyl alcohol, and glycofurol (tetrahydrofurfuryl alcohol PEG ether or tetraglycol). Alcohols such as hexanol and pentanol are not recommended in nanoemulsion formulation due to their irritating character. Long-chain alcohols are more efficient than alcohols with an increased hydroxyl group (Sapra et al., 2013).

4. METHODS OF PREPARATION

Nanoemulsions are thermodynamically unstable systems; therefore, energy and a high amount of surfactants are necessary to obtain proper nanosized emulsions. The development and stability of nanoemulsions are certain by choice of the preparation method. Nanoemulsions are generally prepared by high or low-energy techniques (Thiagarajan et al., 2011; Mason et al., 2006).

4.1. Low-Energy Methods

It employs low energy for the formation of nanoemulsions. They are also often addressed as the condensation process of nanoemulsification due to phase transitions. The processes that fall under low-energy methods are phase transition, phase inversion temperature, and spontaneous emulsification (Wang et al., 2007). The processes mentioned above mainly depend on the modulation of interfacial properties of excipients like oil, surfactant, and the ratio of surfactant mixture to provide a suitable formulation. The inherent physicochemical characteristics of the ternary phase and interfacial activities play a crucial part in preparing nanoemulsions (Lamaallam et al., 2005; Solans et al., 2002). Other factors such as the HLB value of the surfactants, temperature, and ratio of the excipients are known to be of prime importance (Sonneville-Aubrun et al., 2004; Solans et al., 2005). Low-energy methods are now gaining much interest in nanoemulsion preparation due to their broad applicability and stability features. The in-depth discussion regarding these techniques have been illustrated below.

4.1.1. Phase Inversion Temperature Method

Phase inversion temperature is a temperature-dependent method used to solubilise oil and water molecules with the help of non-ionic surfactant molecules. It is one of the most common methods employed in the preparation of nanoemulsions and is considered the basis of nanoemulsion formulations. It is carried out in two simple steps; firstly, a nanoemulsion (O/W) is prepared by mixing an appropriate amount of oil, water, and non-ionic surfactant at ambient temperature. Secondly, the prepared nanoemulsion (O/W) is heated at a specific temperature. It initiates changing into W/O type nanoemulsion due to the surfactant's lipophilic nature, which holds a monolayer of negative curvature (Izquierdo et al., 2005). The intermediate HLB value and the experimental temperature drive the negative curvature to zero at low interfacial tension and involve bicontinuous emulsions and lamellar liquid crystalline phase (Sole et al., 2006, 2010). This method also tends to form nanodroplets upon rapid cooling above the phase inversion temperature or under normal circumstances.

4.1.2. Solvent Displacement Method

The solvent displacement technique is a spontaneous nanoemulsification developed from the nano-precipitation concept. In this technique, the oil phase is dispersed into an organic solvent, including alcohol, acetone, diethyl ether, etc. Subsequently, it is added slowly into a mixture of surfactant and co-surfactant representing the aqueous phase accompanied by an agitation which results in the formation of nanodroplets. The excess solvent is removed in the last stage by employing a rotary evaporator (Porras et al., 2008).

4.1.3. Phase Inversion Composition Method

The phase inversion composition technique is the self-nanoemulsification process that can generate nanoassemblies without using solvent and heat at room temperature. Nanoemulsions are a kinetically stable system at the nanosize (around 200 nm) range. An explicit preparation method engages the addition of surfactant holding an aqueous phase into the oil phase accompanied with agitation at a steady temperature. The phase inversion composition technique involves phase transition throughout the formation of a bicontinuous microemulsion. The rapid alterations of droplets size from micron to nano is known as the spontaneous nanoemulsification process, which may be due to phase inversion composition (Forgiarini et al., 2001).

4.2. High-energy Methods

High energy emulsification methods are considered as one of the most efficient methods for the preparation of nanoemulsion. Fabrication of nanoemulsion is accomplished by employing high kinetic energy generated from sophisticated mechanical equipment. The high energy generated is required to reduce micro-droplets to nanodroplets to obtain the primary emulsion (Anton and Vandamme, 2009). The high-energy sources used for the production of nanoemulsion are high-pressure homogeniser, ultrasonicator, and microfluidiser. The working and operating conditions of the equipment affect the droplet size of the nanoemulsion. The most commonly used methods for the preparation of nanoemulsion are homogenisation and ultrasonication. Ultrasonication generates ultrasonic waves (cavitation forces) with the help of probes required for the size-reduction of micron-size droplets of an emulsion into tiny droplets. The droplet's size of nanoemulsion can be tuned by varying the time of application and input energy. In contrast to ultrasonication, the homogenisation method employs high pressure (500 to 5000 psi) to produce high energy through hydraulic force and shear, requisite for converting micron-size droplets into nano-size. Ultrasound has been well exploited in generating nanoemulsion encapsulating active constituents, including pharmaceuticals and natural products. Notably, detailed optimisation studies demonstrate the critical parameters to obtain the smallest droplet to realise enhanced properties (Alzorqi et al., 2016; Sivakumar et al., 2014; Tang et al., 2012a, Tang et al., 2012b). Microfluidization is another technique that is employed for the production of nanoemulsion. The working principle of microfluidization is based on generating high energy by applying high pressure of about 20,000 psi through a displacement pump. In the microfluidization process, the micron-size droplets are reduced into fine nanosized droplets, resulting in nanoemulsion formation. High-pressure homogeniser, ultrasonicator, and microfluidiser are well-accepted equipment for nanoemulsions production in laboratories and industries (Graves et al., 2005; Jafari et al., 2007; Qian and McClements, 2011).

5. POTENTIAL BIOLOGICAL FATE OF NANOEMULSION AFTER ORAL ADMINISTRATION

The ability to increase a hydrophobic component's bioavailability is one of the primary reasons for using nanoemulsions as a drug delivery system. Several mechanisms account for this effect which includes physicochemical and physiological mechanisms. First, the solubility of a hydrophobic component encapsulated within the oil phase surrounded by an aqueous phase increases as the droplet size decreases due to a thermodynamic consequence related to particle curvature (Acosta et al., 2009). As a result, the concentration gradient of the hydrophobic component increases from the surface of the lipid droplets to the epithelium cell surfaces leads to an elevated driving force for mass transport and absorption. Second, the rate of lipid digestion increases as the surface area of the lipid droplets within the nanoemulsion system increases with a decrease in the particle size (Li et al., 2011). Third, the small particle size may enter through the mucous layer's pores (≈ 400 nm), coating the epithelium cells; thus, they may get absorbed by the epithelial cells (Cone et al., 2009). In contrast, the same phenomenon might not occur in larger particle sizes, and this size-dependent penetration through the epithelial cells of the GI tract is illustrated in Figure 2.

Figure 2. Size-dependent penetration through the epithelial cells of the GI tract

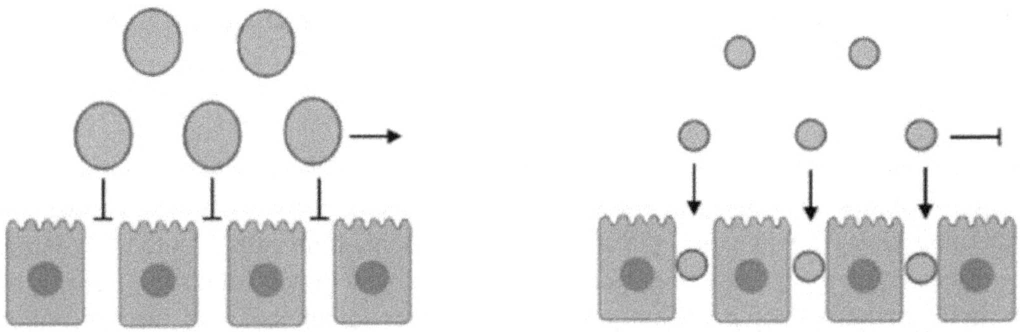

5.1. Influence GI Conditions on Nanoemulsion

The GI tract consists of various regions such as the mouth, stomach, small intestine, and large intestine with a diverse physiological environment that may have the capability to alter the physicochemical properties of the nanoemulsion when administered by the oral route (Golding et al., 2011). Nanoemulsions pass through very complex physiological environments within the GI tract, with changes in pH, temperature, ionic composition mechanical forces, fluid flow, enzyme activities, surface-active agents, and biological surfaces. As a result, alterations in their size, charge, composition, and concentration occur, affecting the overall bioavailability of the encapsulated hydrophobic components. For instance, the size of the lipid droplets present within a nanoemulsion system may either decrease or increase as they travel through the GI tract owing to processes such as flocculation, coalescence, digestion, and solubilisation (McClements et al., 2012). The surface charge of nanoemulsion may be altered due to the digestion of emulsifiers, co-adsorption, and free fatty acid production. The interfacial composition may alter due to the digestion of absorbed surface-active components, co-adsorption, and free fatty acid or monoacylglycerol production. The lipid-core composition might also alter lipase digestion and subsequent solubilisation (McClements et al., 2011).

5.2. Bioaccessibility

Bioaccessibility for conventional emulsion is defined as the fraction of hydrophobic drug released from the oil droplets into the GI tract (Porter et al., 2007). After the hydrophobic drug is released from the oil droplets, it integrates into the mixed micelles inside the lumen of the small intestine (Nike et al., 2011). Conventional emulsions usually have large droplet sizes too large to enter into the mucous layer of the GI tract. The digestion of the lipid phase is required to integrate hydrophobic drugs into the mixed micelles to promote the release of hydrophobic drugs and facilitate the formation of mixed micelles supplemented with a high solubilisation capacity (Porter et al., 2008). Furthermore, a small amount of hydrophobic drug leaves from the oil droplets and enters the mixed micelles through diffusion (Wang et al., 2012). A similar mechanism can be assumed for nanoemulsions; however, they may be directly absorbed by the epithelium cells due to their small droplet size. In addition, nanoemulsion may penetrate the mucous layer and enter the epithelial cells, where they could be absorbed by active or passive transport mechanisms (Ensign et al., 2012). However, this prospect will rely on digestion time and the corresponding time required for penetration into the mucus layer. Some of the key factors recognised to impact the bioaccessibility of the encapsulated hydrophobic drug is discussed in the following sections.

5.3.1. Absorption

The absorption patterns of hydrophobic bioactive components present inside the lumen of the small intestine are as follows. First, they are transferred through the digestive fluids, and second, they pass through the mucous layer, where they get absorbed by the epithelium cells (Pouton and Porter, 2008). Digestible lipids demonstrate great significance when employed in nanoemulsion preparation as they can form mixed micelles upon digestion, resulting in greater solubilisation of the hydrophobic drug. The schematic representation of this process is shown in Figure 3. A study was conducted on Caco-2 cell lines using curcumin as a model drug to support this hypothesis, demonstrating that the main absorption mechanism is mixed micelle rather than nanoemulsion droplets (Yu et al., 2012). In general, a hydrophobic drug encapsulated within the lipid droplets of nanoemulsion using indigestible lipids or partially digestible lipids may be transported to the epithelium cells. However, no mechanistic studies of the absorption of the hydrophobic drug from indigestible nanoemulsions have been reported.

Figure 3. Schematic representation of digestible lipids promoting the formation of mixed micelles and solubilisation of hydrophobic drug

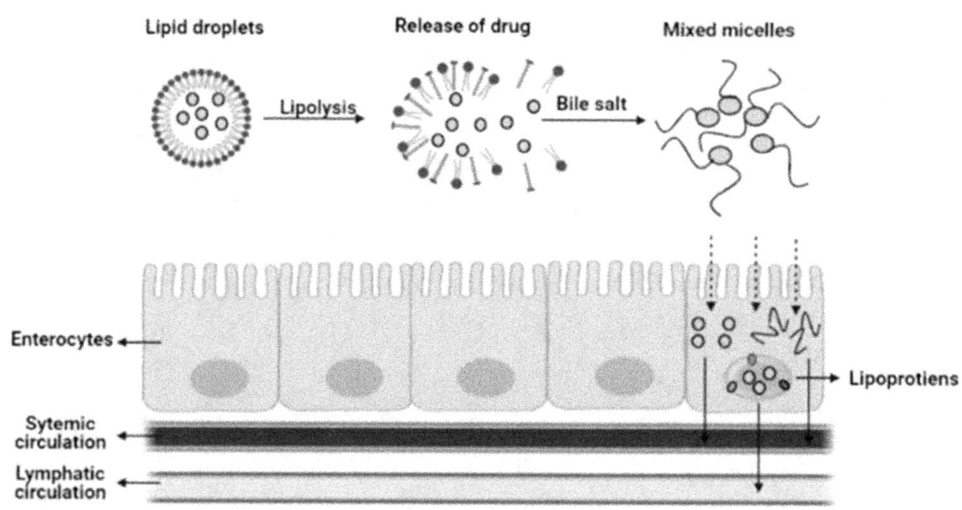

The physicochemical properties of nanoemulsion in the small intestine may affect the absorption process. The speed at which the lipid droplets travel through the mucous layer relies on their size compared to the pore sizes of the thick biopolymer layer. Compared to larger particle sizes, lipid droplets with a relatively small particle size (~400 nm) may penetrate well through the mucous layer (Florence et al., 2012). The ability to travel through the mucous layer by the lipid droplets depends on mass transport rate and lipid digestion rate. However, digestible lipid droplets may be totally digested before they arrive at the epithelium cells. Owing to their small size, the retention of lipid droplets within the small intestine increases, resulting in greater lipid digestion and an enhanced bioavailability (Ensign et al., 2012). The surface charge also affects the transfer of lipid droplets into and through the mucous layer. For instance, cationic lipid droplets may bind with the anionic biopolymer molecules, thus increasing their retention time but inhibiting their transport to the epithelium cells, which may be considered as a hindrance. Once the hydrophobic drug reaches the apical surface of the epithelium cells, it may be absorbed by active or passive-transport mechanisms depending on its molecular structure and any of its integrating nanoparticles (Cai et al., 2010; Florence and Hussain, 2001). At this stage, the absorption of the hydrophobic drug may be restricted by its permeation through the cell membrane (Metcalfe and Thomas, 2011). However, any undigested lipid droplets penetrating through the mucous layer and reaching the apical layer of epithelium cells may be absorbed through several different mechanisms, including paracellular-transported through the tight junctions between the epithelium cells; transcellular-transported straight across the epithelium cell walls by a variety of endocytosis processes, and persorption-transported through holes created in the epithelium layer due to cell turnover, which is shown in Figure 4.

Figure 4. Schematic representation of undigestible lipids promoting paracellular, transcellular, and persorption transport

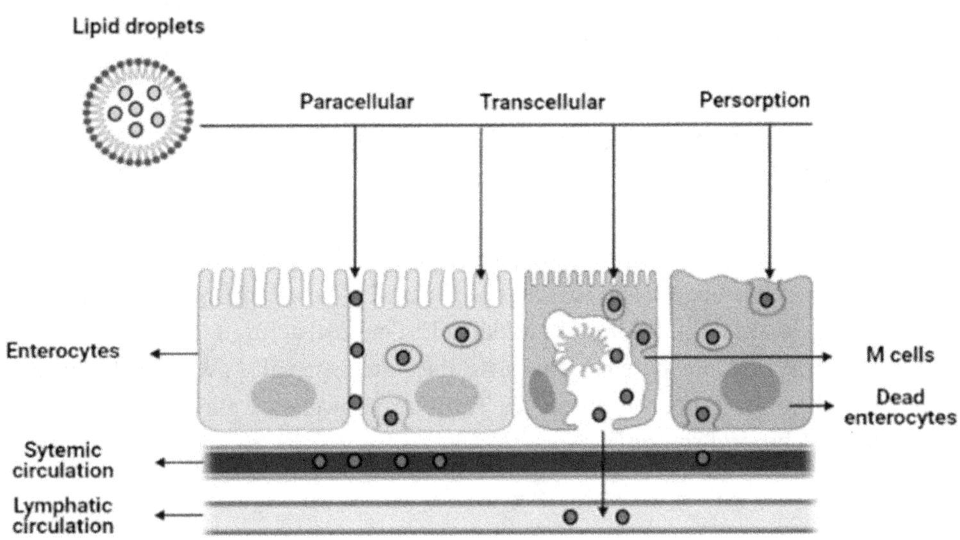

5.3.2. Metabolism

The GI tract, epithelium cells, and the systemic circulation consist of various chemicals and enzymes that metabolise orally administered hydrophobic components and undergo modification in their chemical structure (Xie et al., 2012). The degree and nature of modifications can be controlled by the cautious design of a delivery system. For instance, by changing the lipid-phase composition, it is possible to direct the uptake of a delivery system either through the lymphatic or the portal vein route. The delivery systems that travel through the lymphatic route bypass the first-pass metabolism inside the liver and avoid the biotransformation (Trevaskis et al., 2008). Additional research is needed to ascertain the function of lipid droplet properties on the metabolism of hydrophobic components encapsulated inside the nanoemulsion-based delivery systems.

6. LIMITATIONS OF NANOEMULSION IN ORAL DRUG DELIVERY

As discussed in the previous sections, nanoemulsion has numerous advantages, such as improving the drug's bioavailability and expediently administering it to diverse populations of all ages due to its colloidal nature. In addition, it can also protect the drug from the external environment of the GI tract and imparts stability to the drugs. Regardless of its advantages, nanoemulsion also showcases some limitations because of its small size. The first limitation of nanoemulsion in terms of oral drug delivery is the cost of manufacturing because nanoemulsion requires some sophisticated instrument and procedure to get the desired size. For example, high-pressure homogeniser, microfluidiser, and probe sonicator require a higher investment. The second limitation of nanoemulsion is its inadequate stability and a major problem during the storage for a long period. The main reason for this instability is the Ostwald ripening. The elevated curvature of lipid droplets presents another noteworthy solubility when distinct, with large lipid droplets having a small radius of curvature. The third but not the least limitation is that the accessibility of the surfactant and co-surfactant necessary for nanoemulsion preparation is less, which is a prime factor in the fabrication of nanoemulsion.

7. EXISTING PATENTS ON THE USE OF NANOEMULSION FOR ORAL DRUG DELIVERY

The drug's having low oral bioavailability is one of the greatest hurdles faced by the formulation scientist, and a significant effort has been made to enhance the oral bioavailability through nanoformulations such as nanoemulsion, liposomes, solid lipid nanoparticles, nanostructured lipid carriers, etc. The novel inventions in this specific area are patented because they are generally awarded for ground-breaking inventions. Several patents have been reported in Table 1, showcasing the significance of delivering drug's efficiently as NEs by oral route.

Table 1. Specific patents related to oral nanoemulsion

Patent number	Title	Description	References
WO016664A2	Composition and methods for treating cancer with dacarbazine nanoemulsions	A uniform microfluidized nanoemulsion of dacarbazine improves the cell membrane's permeability and significantly increases the intracellular concentration of anticancer agents. As a nanoemulsion, dacarbazine has greater anticancer efficacy than when applied as a free solution	Nicolosi et al., 2008
WO035311A2	Antioxidant synergy nanoemulsion formulations to treat cancer	The microfluidized nanoemulsion improves the cell membrane's permeability by at least fourfold, significantly increasing the intracellular concentration of typically cell-impermeant antioxidants and/or systemic bioavailability. This synergistic combination has greater anticancer efficacy than the same combination applied as a free solution	Nicolosi et al., 2007
US0026988A1	Composition and methods for human immunodeficiency virus vaccination	This invention relates to methods and composition for the stimulation of immune responses. It provides methods of inducing an immune response to human immunodeficiency virus in a subject and composition useful in such methods (e.g., a nanoemulsion comprising HIV or antigenic portion thereof)	Baker et al., 2008
101700224A	Method for preparing oral chondroitin sulfate nanoemulsion	The proportion of the oil phase to the surfactant is 45:1 - 1:7, and the proportion of the surfactant to the cosurfactant is 4:1- 1:4. It improves the bioavailability, reduces the toxic and side effects and fully exerts the efficacy of the chondroitin sulfate	Sha et al., 2010
WO105052A1	Novel oral formulations of ospemifene	The invention provides an improved drug formulation containing ospemifene, where the absorption of the drug is essentially increased, and the variability in plasma level is essentially decreased	Lehtola et al., 2005
200810150354	Ivermectin nanoemulsion drug combination and preparation method thereof	O/w nanoemulsion consisting of ethyl oleate, tween 80, propylene glycol and Ivermectin in double-distilled water. It enhanced the safety, dissolubility and bioavailability	Jiyu et al., 2010

8. APPLICATION OF NANOEMULSION IN IMPROVING ORAL DELIVERY

Oral administration of drugs is highly preferred owing to its numerous advantages. Most of the drugs used clinically to treat various diseases are hydrophobic, having extremely low bioavailability. Therefore, it is necessary to enhance the bioavailability of hydrophobic drugs. In recent times, nanoemulsion has been used as a potential carrier for hydrophobic drugs, presenting encouraging results to enhance the bioavailability and alter drug release. Generally, drugs are integrated into the oil phase of nanoemulsion;

therefore, its lipid solubility is extremely important as it governs the quality of the formulation (Khani et al., 2016). Nanoemulsion as a system for the oral delivery of poorly soluble drugs has various applications reliant on the major drug effects.

Numerous studies have been reported to enhance the oral bioavailability of hydrophobic drugs. For instance, a nanoemulsion formulation was developed to increase the oral bioavailability of mebudipine, indicating that the relative bioavailability of mebudipine nanoemulsion could be increased by 2.6-, 2.0- and 1.9-fold as compared to suspension, ethyl oleate solution, and the micellar solution, respectively. In addition, the optimum droplet size of mebudipine nanoemulsion was 22.28 ± 4.0 nm, and the formulation was found to be stable for the required period (Mahour et al., 2015). Cilostazol, an antithrombotic drug, has a very low oral bioavailability limiting its application clinically. To address this issue, cilostazol nanoemulsion formulation was developed using triacetin combined with capryol 90 in a ratio of 1:1 as an oil phase with tween 80 as surfactant and transcutol P as co-surfactant. The droplet size of cilostazol nanoemulsion was 93.72 nm, and the pharmacokinetics study indicated that the relative bioavailability of cilostazol nanoemulsion was increased 3.29 times compared with cilostazol suspension (Wulff-Perez et al., 2014). N-oleoylethanolamine (OEA) is a lipid mediator that acts as a satiety factor. Like other hydrophobic drugs, OEA also has poor water solubility, a main limiting factor. Hence, a stable and non-toxic nanoemulsion delivery system was designed for OEA. The obtained results support nanoemulsion's effectiveness in delivering highly hydrophobic bioactive drugs such as OEA (Harwansh et al., 2017). Betulinic acid (BA), a potent hepatoprotective drug, has very poor bioavailability due to its inherent low solubility that restricts its potential. Therefore, BA loaded nanoemulsion was developed to increase its bioavailability using olive oil as an oil phase, labrasol as a surfactant, and plurol as a co-surfactant. The droplet size and zeta potential of the developed BA loaded nanoemulsions were 150.3 ± 0.56 nm and -10.2 ± 0.11 mV, respectively. The pharmacokinetic parameter was also improved in terms of BA loaded nanoemulsion compared with BA. BA loaded nanoemulsion enhanced the hepatoprotective effect due to its increased solubility and enhanced bioavailability (Groo et al., 2017). An anticancer drug such as pyridoclax, generally used in chemoresistant cancers, falls under the BCS Class II drug usually has low water solubility. Two approaches were considered and compared to surmount this undesirable property: formulation of pyridoclax loaded nanoemulsion and synthesis of pyridoclax dihydrochloride salt. The results indicated that pyridoclax loaded nanoemulsion was more stable, and the solubility was increased by 1000-fold compared to salt form. Thus, it has been concluded that a nanoemulsion is a promising option for delivering hydrophobic drugs (Choudhury et al., 2014). Paclitaxel, another promising anticancer drug, is restricted from oral use due to its poor aqueous solubility. Therefore, to improve its oral bioavailability, paclitaxel loaded nanoemulsion was developed consisting of capryol 90 and Tween 80 as oil and aqueous phases, respectively. The developed paclitaxel loaded nanoemulsion was characterised for its droplet size, zeta potential, morphology, viscosity, thermodynamic stability, and *in-vitro* drug release. The paclitaxel loaded nanoemulsion exhibited small droplet size, excellent thermodynamic stability, low viscosity, and sustained release. Furthermore, the absolute oral bioavailability of the paclitaxel loaded nanoemulsion was assessed in a mouse model, and improvements up to 55.9% were noted (Sessa et al., 2014). Natural compounds have a wide range of therapeutic activity but possess a poor aqueous solubility and experience the first-pass effect, restricting its application. Resveratrol is one such natural compound that falls under this category. Different nanoemulsion formulations were developed to optimise oral bioavailability. The developed formulation could transport resveratrol across the cell monolayers, indicating a better protecting potential with sustained release. In addition, fluorescence spectroscopy also confirmed that resveratrol was encapsulated in the inner core of the nanoemulsion (Sessa et al., 2014). A parallel study was conducted by encapsulating resveratrol in O/W food-grade nanoemulsion. The results indicated that resveratrol encapsulated nanoemulsion has a droplet size of 180 nm with maximum physical and chemical stability. Consequently, the developed formulation also exhibited higher antioxidant activity than its free form, signifying that encapsulated resveratrol was not metabolised in the GI tract and can be likely absorbed through the intestinal wall in active form (Sessa et al., 2011). Another study was carried out for improving the bioavailability of silymarin through silymarin loaded nanoemulsion. In this study, silymarin loaded nanoemulsion was obtained using capryol 90 as an oil phase, solutol as a surfactant, and transcutol HP as a co-surfactant along with high-pressure homogenisation for size reduction. The droplet size, zeta potential, polydispersity index (PDI), percentage transmittance, and percentage *in-vitro* drug release was 50.02 ± 4.5 nm, 0.45 ± 0.02, −31.49 mV, 100.00 ± 2.21% and 90.00 ± 1.83%, respectively. The pharmacokinetic studies indicated that the relative oral bioavailability of silymarin loaded nanoemulsion was enhanced significantly ($p < 0.05$) in contrast to oral suspension (Nagi et al., 2017). Breviscapine loaded nanoemulsion was developed to improve its oral bioavailability and was characterised for droplet size, PDI, encapsulation efficiency (%), *in-vitro* drug release, and stability in the GI tract. The droplet size and PDI were 45 nm and 0.105, respectively, and the encapsulation efficiency (%) was 95.2%. *In-vitro* drug release fitted to first-order kinetics and was stable in the GI tract. The permeation study suggested that nanoemulsion assists in increasing the apparent permeability coefficient (Papp) from the apical side to the basal side in contrast to the free drug. Furthermore, the relative bioavailability of breviscapine was increased significantly when it was given in NE form (Ma et al., 2015). The effects of encapsulated tocotrienols (TRF) and caffeic acid (CA) in stable water-in-oil-in-water (W/O/W) multiple nanoemulsions with cisplatin towards cancer cells were examined (Raviadran et al., 2021). It has been observed that encapsulated TRF and CA in W/O/W with cisplatin synergistically improved therapeutic efficacy towards cancer cells, as well as lowered the toxicity effects towards normal cells. Numerous researches have been conducted for improving the oral bioavailability of various drugs by using nanoemulsion. Table 1 lists out some case studies related to the pharmacokinetic parameters of drugs after oral administration.

Table 2. Summary of pharmacokinetic parameters obtained through the administration of oral nanoemulsion

Drug	Dose (mg/kg)	AUC			C_{max}	T_{max} (h)	BA enhancement compared to pure drug	References
		AUC_{t0-24}	AUC_{t0-48}	AUC_{t0-}				
CoenzymeQ10	30	1888	-	-	128±20 ngmL^{-1}	2	1.8 fold	Nui et al., 2020
CoenzymeQ10	60	-	26.14±3.24	-	1.21±0.12 μgmL^{-1}	6±1.5	5 fold	Belhaj et al., 2012
CoenzymeQ10	30	5070±2770	-	-	480±347 ngmL^{-1}	6±2.1	4.6 fold	Nepal et al., 2010
CoenzymeQ10	45	68.42±12.35	-	95.22±24.56	8.51±1.23 μgmL^{-1}	2	2.2 fold	Khattab et al., 2016
CoenzymeQ10	60	71.3±10.8	-	-	3.2±0.6 μgmL^{-1}	1.5±0.5	1.7 fold	Hatakana et al., 2008
Piplartine	10	-	-	33055.994	1208.67 ngmL^{-1}	10	1.5 fold	Fofaria et al., 2015
Amlodipine Besilate	4	4.25±0.97	-	15.19±1.26	0.202±0.042 μgmL^{-1}	6±0.32	4.7 fold	Chhabra et al., 2011
Simvastatin	20	-	-	61.75±8.03	$p < 0.001$	-	3.7 fold	Chavan et al., 2013
Efavirenz	18	43.53±3.134	-	101.59±3.584	3.12±0.0596 μgmL^{-1}	4	2.6 fold	Kotta et al., 2014
Artemether	4	308.707	-	430.959	0.986 μgmL^{-1}	2	2.6 fold	Laxmi et al., 2014

CONCLUSION

Nanoemulsions are extensively explored for improving the oral bioavailability of poorly soluble drugs. These systems provide numerous advantages in comparison to conventional drug delivery systems. Their main features include improving oral bioavailability by enhancing the solubility and permeability and are regarded as a safe and efficient drug delivery system. The poorly soluble drug resides inside the lipid nanodroplets that form mixed micelles after lipid digestion and promote the solubilisation of poorly soluble drugs. Due to their nanosize, they immensely increase the surface area, aiding improved bioavailability. Furthermore, it is highly acknowledged due to its thermodynamic stability and prolongation of drug retention inside the bio-system. More research is required to translate this delivery system from bench to bedside with proven efficacy. Therefore, more controlled studies in animals and humans are needed to prove their safety and efficacy. Moreover, systematically optimised formulae and preparation method is required to produce nanoemulsion with optimum properties.

REFERENCES

Acosta, E. (2009). Bioavailability of nanoparticles in nutrient and nutraceutical delivery. *Current Opinion in Colloid & Interface Science*, *14*(1), 3–15. doi:10.1016/j.cocis.2008.01.002

Alzorqi, I., Ketabchi, M. R., Sudheer, S., & Manickam, S. (2016). Optimisation of ultrasound induced emulsification on the formulation of palm-olein based nanoemulsions for the incorporation of antioxidant β-D-glucan polysaccharides. *Ultrasonics Sonochemistry*, *31*, 71–84. doi:10.1016/j.ultsonch.2015.12.004 PMID:26964925

Anton, N., & Vandamme, T. F. (2009). The universality of low-energy nano-emulsification. *International Journal of Pharmaceutics*, *377*(1-2), 142–147. doi:10.1016/j.ijpharm.2009.05.014 PMID:19454306

Atuma, C., Strugala, V., Allen, A., & Holm, L. (2001). The adherent gastrointestinal mucus gel layer: Thickness and physical state in vivo. *American Journal of Physiology. Gastrointestinal and Liver Physiology*, *280*(5), G922–G929.

Azeem, A., Rizwan, M., Ahmad, F. J., Iqbal, Z., Khar, R. K., Aqil, M., & Talegaonkar, S. (2009). Nanoemulsion components screening and selection: A technical note. *AAPS PharmSciTech*, *10*(1), 69–76.

Baker, J.R., Bielinska, A., & Myc, A. (2008). *Compositions and methods for human immunodeficiency virus vaccination*. US0026988A1.

Belhaj, N., Dupuis, F., Arab-Tehrany, E., Denis, F. M., Paris, C., Lartaud, I., & Linder, M. (2012). Formulation, characterisation and pharmacokinetic studies of coenzyme Q10 PUFA's nanoemulsions. *European Journal of Pharmaceutical Sciences*, *47*(2), 305–312.

Bratten, J., & Jones, M. P. (2006). New directions in the assessment of gastric function: Clinical applications of physiologic measurements. *Digestive Diseases (Basel, Switzerland)*, *24*(3-4), 252–259.

Brunton, L. L., Hilal-Dandan, R., & Knollmann, B. C. (Eds.). (2018). *Goodman & Gilman's The Pharmacological Basis of Therapeutics* (p. 7). McGraw-Hill Education.

Cai, Z., Wang, Y., Zhu, L. J., & Liu, Z. Q. (2010). Nanocarriers: A general strategy for enhancement of oral bioavailability of poorly absorbed or pre-systemically metabolised drugs. *Current Drug Metabolism*, *11*(2), 197–207.

Chavhan, S. S., Petkar, K. C., & Sawant, K. K. (2013). Simvastatin nanoemulsion for improved oral delivery: Design, characterisation, in vitro and in vivo studies. *Journal of Microencapsulation*, *30*(8), 771–779.

Chhabra, G., Chuttani, K., Mishra, A. K., & Pathak, K. (2011). Design and development of nanoemulsion drug delivery system of amlodipine besylate for improvement of oral bioavailability. *Drug Development and Industrial Pharmacy*, *37*(8), 907–916.

Chime, S. A., Kenechukwu, F. C., & Attama, A. A. (2014). Nanoemulsions-Advances in Formulation. *Characterisation and Applications in Drug Delivery*, *3*, 77–126.

Choudhury, H., Gorain, B., Karmakar, S., Biswas, E., Dey, G., Barik, R., & Pal, T. K. (2014). Improvement of cellular uptake, in vitro antitumor activity and sustained release profile with increased bioavailability from a nanoemulsion platform. *International Journal of Pharmaceutics*, *460*(1-2), 131–143.

Cone, R. A. (2009). Barrier properties of mucus. *Advanced Drug Delivery Reviews*, *61*(2), 75–85.

Consortium, T. H. M. P. (2012). Structure, Function and Diversity of the Healthy Human Microbiome. *Nature*, *486*, 207–214.

Constantinides, P. P., Tustian, A., & Kessler, D. R. (2004). Tocol emulsions for drug solubilisation and parenteral delivery. *Advanced Drug Delivery Reviews*, *56*(9), 1243–1255.

Davies, J. T. (1957). A quantitative kinetic theory of emulsion type. I. Physical chemistry of the emulsifying agent. In *Gas/Liquid and Liquid/Liquid Interface. Proceedings of the International Congress of Surface Activity* (*Vol. 1*, pp. 426-438). Academic Press.

Dordevic, S. M., Radulovic, T. S., Cekic, N. D., Randelovic, D. V., Savic, M. M., Krajisnik, D. R., & Savic, S. D. (2013). Experimental design in formulation of diazepam nanoemulsions: Physicochemical and pharmacokinetic performances. *Journal of Pharmaceutical Sciences*, *102*(11), 4159–4172.

Ensign, L. M., Cone, R., & Hanes, J. (2012). Oral drug delivery with polymeric nanoparticles: The gastrointestinal mucus barriers. *Advanced Drug Delivery Reviews*, *64*(6), 557–570.

Fallingborg, J., Christensen, L. A., Jacobsen, B. A., & Rasmussen, S. N. (1993). Very low intraluminal colonic pH in patients with active ulcerative colitis. *Digestive Diseases and Sciences*, *38*(11), 1989–1993.

Florence, A. T. (2012). "Targeting" nanoparticles: The constraints of physical laws and physical barriers. *Journal of Controlled Release*, *164*(2), 115–124.

Florence, A. T., & Hussain, N. (2001). Transcytosis of nanoparticle and dendrimer delivery systems: Evolving vistas. *Advanced Drug Delivery Reviews*, *50*, S69–S89.

Fofaria, N. M., Qhattal, H. S. S., Liu, X., & Srivastava, S. K. (2016). Nanoemulsion formulations for anticancer agent piplartine - Characterisation, toxicological, pharmacokinetics and efficacy studies. *International Journal of Pharmaceutics*, *498*(1-2), 12–22.

Forgiarini, A., Esquena, J., Gonzalez, C., & Solans, C. (2001). Formation of nanoemulsions by low-energy emulsification methods at constant temperature. *Langmuir*, *17*(7), 2076–2083.

Frank, D. N., Amand, A. L. S., Feldman, R. A., Boedeker, E. C., Harpaz, N., & Pace, N. R. (2007). Molecular-phylogenetic characterisation of microbial community imbalances in human inflammatory bowel diseases. *Proceedings of the National Academy of Sciences of the United States of America*, *104*(34), 13780–13785.

Golding, M., Wooster, T. J., Day, L., Xu, M., Lundin, L., Keogh, J., & Clifton, P. (2011). Impact of gastric structuring on the lipolysis of emulsified lipids. *Soft Matter*, *7*(7), 3513–3523.

Graves, S., Meleson, K., Wilking, J., Lin, M. Y., & Mason, T. G. (2005). Structure of concentrated nanoemulsions. *The Journal of Chemical Physics*, *122*(13), 134703.

Groo, A. C., De Pascale, M., Voisin-Chiret, A. S., Corvaisier, S., Since, M., & Malzert-Freon, A. (2017). Comparison of 2 strategies to enhance pyridoclax solubility: Nanoemulsion delivery system versus salt synthesis. *European Journal of Pharmaceutical Sciences*, *97*, 218–226.

Gupta, R., Shea, J., Scaife, C., Shurlygina, A., & Rapoport, N. (2015). Polymeric micelles and nano-emulsions as drug carriers: Therapeutic efficacy, toxicity, and drug resistance. *Journal of Controlled Release*, *212*, 70–77.

Hall, J. E. (2015). Secretory Functions of the Alimentary Tract. In J. E. Hall & A. C. Guyton (Eds.), *Guyton and Hall Textbook of Medical Physiology* (pp. 817–832). Elsevier, Inc.

Harwansh, R. K., Mukherjee, P. K., & Biswas, S. (2017). Nanoemulsion as a novel carrier system for improvement of betulinic acid oral bioavailability and hepatoprotective activity. *Journal of Molecular Liquids*, *237*, 361–371.

Hatanaka, J., Kimura, Y., Lai-Fu, Z., Onoue, S., & Yamada, S. (2008). Physicochemical and pharmacokinetic characterisation of water-soluble Coenzyme Q10 formulations. *International Journal of Pharmaceutics*, *363*(1-2), 112–117.

Homayun, B., Lin, X., & Choi, H. J. (2019). Challenges and recent progress in oral drug delivery systems for biopharmaceuticals. *Pharmaceutics*, *11*(3), 129.

Hu, J., Chen, D., Jiang, R., Tan, Q., Zhu, B., & Zhang, J. (2014). Improved absorption and in vivo kinetic characteristics of nanoemulsions containing evodiamine–phospholipid nanocomplex. *International Journal of Nanomedicine*, *9*, 4411.

Hu, Z., Mawatari, S., Shibata, N., Takada, K., Yoshikawa, H., Arakawa, A., & Yosida, Y. (2000). Application of a biomagnetic measurement system (BMS) to the evaluation of gastrointestinal transit of intestinal pressure-controlled colon delivery capsules (PCDCs) in human subjects. *Pharmaceutical Research*, *17*(2), 160–167.

Ibekwe, V. C., Fadda, H. M., McConnell, E. L., Khela, M. K., Evans, D. F., & Basit, A. W. (2008). Interplay between intestinal pH, transit time and feed status on the in vivo performance of pH responsive ileo-colonic release systems. *Pharmaceutical Research*, *25*(8), 1828–1835.

Izquierdo, P., Feng, J., Esquena, J., Tadros, T. F., Dederen, J. C., Garcia, M. J., & Solans, C. (2005). The influence of surfactant mixing ratio on nanoemulsion formation by the pit method. *Journal of Colloid and Interface Science*, *285*(1), 388–394.

Jafari, S. M., He, Y., & Bhandari, B. (2007). Optimisation of nano-emulsions production by microfluidization. *European Food Research and Technology*, *225*(5), 733–741.

Jaiswal, M., Dudhe, R., & Sharma, P. K. (2015). Nanoemulsion: an advanced mode of drug delivery system. *3 Biotech, 5*(2), 123-127.

Jiyu, Z., Jianyong, L., & Xuzheng, Z. (2010). *Ivermectin nanoemulsion drug combination and preparation method thereof.* Academic Press.

Johansson, M. E., Sjövall, H., & Hansson, G. C. (2013). The gastrointestinal mucus system in health and disease. *Nature Reviews. Gastroenterology & Hepatology*, *10*(6), 352.

Kagan, L., & Hoffman, A. (2008). Systems for region selective drug delivery in the gastrointestinal tract: Biopharmaceutical considerations. *Expert Opinion on Drug Delivery*, *5*(6), 681–692.

Khani, S., Keyhanfar, F., & Amani, A. (2016). Design and evaluation of oral nanoemulsion drug delivery system of mebudipine. *Drug Delivery*, *23*(6), 2035–2043.

Khattab, A., Hassanin, L., & Zaki, N. (2017). Self-nanoemulsifying drug delivery system of coenzyme (Q10) with improved dissolution, bioavailability, and protective efficiency on liver fibrosis. *AAPS PharmSciTech*, *18*(5), 1657–1672.

Koroleva, M. Y., & Yurtov, E. V. (2012). Nanoemulsions: The properties, methods of preparation and promising applications. *Russian Chemical Reviews*, *81*(1), 21.

Kotta, S., Khan, A. W., Ansari, S. H., Sharma, R. K., & Ali, J. (2014). Anti HIV nanoemulsion formulation: Optimisation and in vitro–in vivo evaluation. *International Journal of Pharmaceutics*, *462*(1-2), 129–134.

Lamaallam, S., Bataller, H., Dicharry, C., & Lachaise, J. (2005). Formation and stability of miniemulsions produced by dispersion of water/oil/surfactants concentrates in a large amount of water. *Colloids and Surfaces. A, Physicochemical and Engineering Aspects*, *270*, 44–51.

Laxmi, M., Bhardwaj, A., Mehta, S., & Mehta, A. (2015). Development and characterisation of nanoemulsion as carrier for the enhancement of bioavailability of artemether. *Artificial Cells, Nanomedicine, and Biotechnology*, *43*(5), 334–344.

Lehtola & Anttila. (2005). *Novel oral formulations of ospemifene.* Academic Press.

Li, Y., Hu, M., & McClements, D. J. (2011). Factors affecting lipase digestibility of emulsified lipids using an in vitro digestion model: Proposal for a standardised pH-stat method. *Food Chemistry*, *126*(2), 498–505.

Ma, Y., Li, H., & Guan, S. (2015). Enhancement of the oral bioavailability of breviscapine by nanoemulsions drug delivery system. *Drug Development and Industrial Pharmacy*, *41*(2), 177–182.

Macfarlane, G. T., & Macfarlane, S. (2011). Fermentation in the human large intestine: Its physiologic consequences and the potential contribution of prebiotics. *Journal of Clinical Gastroenterology*, *45*, S120–S127.

Mahour, R., Sahni, J. K., Sharma, S., Kumar, S., Ali, J., & Baboota, S. (2015). Nanoemulsion as a tool for improvement of cilostazol oral bioavailability. *Journal of Molecular Liquids*, *212*, 792–798.

Marieb, E. N., & Hoehn, K. (2007). *Human Anatomy & Physiology*. Pearson education.

Martinez, M. N., & Amidon, G. L. (2002). A mechanistic approach to understanding the factors affecting drug absorption: A review of fundamentals. *Journal of Clinical Pharmacology*, *42*(6), 620–643.

Mason, T. G., Graves, S. M., Wilking, J. N., & Lin, M. Y. (2006). Extreme emulsification: Formation and structure of nanoemulsions. *Condensed Matter Physics*, *9*, 193–199.

McClements, D. J. (2012). Nanoemulsions versus microemulsions: Terminology, differences, and similarities. *Soft Matter*, *8*(6), 1719–1729.

McClements, D. J. (2013). Nanoemulsion-based oral delivery systems for lipophilic bioactive components: Neutraceuticals and pharmaceuticals. *Therapeutic Delivery*, *4*(7), 841–857.

McClements, D. J., & Rao, J. (2011). Food-grade nanoemulsions: Formulation, fabrication, properties, performance, biological fate, and potential toxicity. *Critical Reviews in Food Science and Nutrition*, *51*(4), 285–330.

McClements, D. J., & Xiao, H. (2012). Potential biological fate of ingested nanoemulsions: Influence of particle characteristics. *Food & Function*, *3*(3), 202–220.

Metcalfe, P. D., & Thomas, S. (2010). Challenges in the prediction and modeling of oral absorption and bioavailability. *Current Opinion in Drug Discovery & Development*, *13*(1), 104–110.

Moreno, M. A., Ballesteros, M. P., & Frutos, P. (2003). Lecithin-based oil-in-water microemulsions for parenteral use: Pseudoternary phase diagrams, characterisation and toxicity studies. *Journal of Pharmaceutical Sciences*, *92*(7), 1428–1437.

Muheem, A., Shakeel, F., Jahangir, M. A., Anwar, M., Mallick, N., Jain, G. K., & Ahmad, F. J. (2016). A review on the strategies for oral delivery of proteins and peptides and their clinical perspectives. *Saudi Pharmaceutical Journal*, *24*(4), 413–428.

Musa, S. H., Basri, M., Masoumi, H. R. F., Karjiban, R. A., Abd Malek, E., Basri, H., & Shamsuddin, A. F. (2013). Formulation optimisation of palm kernel oil esters nanoemulsion-loaded with chloramphenicol suitable for meningitis treatment. *Colloids and Surfaces. B, Biointerfaces*, *112*, 113–119.

Nagi, A., Iqbal, B., Kumar, S., Sharma, S., Ali, J., & Baboota, S. (2017). Quality by design based silymarin nanoemulsion for enhancement of oral bioavailability. *Journal of Drug Delivery Science and Technology*, *40*, 35–44.

Narang, A. S., Delmarre, D., & Gao, D. (2007). Stable drug encapsulation in micelles and microemulsions. *International Journal of Pharmaceutics*, *345*(1-2), 9–25.

Nepal, P. R., Han, H. K., & Choi, H. K. (2010). Preparation and in vitro–in vivo evaluation of Witepsol® H35 based self-nanoemulsifying drug delivery systems (SNEDDS) of coenzyme Q10. *European Journal of Pharmaceutical Sciences*, *39*(4), 224–232.

Nicolosi R & Jean T. (2008). *Compositions and methods for treating cancer with dacarbazine nanoemulsions*. Academic Press.

Nicolosi R & Shea T. (2007). *Antioxidant synergy formulation nanoemulsions to treat cancer.* Academic Press.

Nik, A. M., Corredig, M., & Wright, A. J. (2011). Release of lipophilic molecules during in vitro digestion of soy protein-stabilised emulsions. *Molecular Nutrition & Food Research*, *55*(S2), S278–S289.

Niu, Z., Acevedo-Fani, A., McDowell, A., Barnett, A., Loveday, S. M., & Singh, H. (2020). Nanoemulsion structure and food matrix determine the gastrointestinal fate and in vivo bioavailability of coenzyme Q10. *Journal of Controlled Release*, *327*, 444–455.

Porras, M., Solans, C., Gonzalez, C., & Gutierrez, J. M. (2008). Properties of water-in-oil (W/O) nanoemulsions prepared by a low-energy emulsification method. *Colloids and Surfaces. A, Physicochemical and Engineering Aspects*, *324*(1-3), 181–188.

Porter, C. J., Pouton, C. W., Cuine, J. F., & Charman, W. N. (2008). Enhancing intestinal drug solubilisation using lipid-based delivery systems. *Advanced Drug Delivery Reviews*, *60*(6), 673–691.

Porter, C. J., Trevaskis, N. L., & Charman, W. N. (2007). Lipids and lipid-based formulations: Optimising the oral delivery of lipophilic drugs. *Nature Reviews. Drug Discovery*, *6*(3), 231–248.

Pouton, C. W., & Porter, C. J. (2008). Formulation of lipid-based delivery systems for oral administration: Materials, methods9 and strategies. *Advanced Drug Delivery Reviews*, *60*(6), 625–637.

Qian, C., & McClements, D. J. (2011). Formation of nanoemulsions stabilised by model food-grade emulsifiers using high-pressure homogenisation: Factors affecting particle size. *Food Hydrocolloids*, *25*(5), 1000–1008.

Raviadaran, R., Ng, M. H., Chandran, D., Ooi, K. K., & Manickam, S. (2021). Stable W/O/W multiple nanoemulsion encapsulating natural tocotrienols and caffeic acid with cisplatin synergistically treated cancer cell lines (A549 and HEP G2) and reduced toxicity on normal cell line (HEK 293). *Materials Science and Engineering C*, *121*, 111808.

Reinus, J., & Simon, D. (2014). *Gastrointestinal Anatomy and Physiology: The essentials*. Wiley-Blackwell Publishers.

Rowe, R. C., Sheskey, P., & Quinn, M. (2009). *Handbook of Pharmaceutical Excipients*. Pharmaceutical Press.

Ruiz, M. E., & Montoto, S. S. (2018). Routes of drug administration. In *ADME Processes in Pharmaceutical Sciences* (pp. 97–133). Springer.

Sapra, B., Thatai, P., Bhandari, S., Sood, J., Jindal, M., & Tiwary, A. (2013). A critical appraisal of microemulsions for drug delivery: Part I. *Therapeutic Delivery*, *4*(12), 1547–1564.

Sartor, R. B. (2008). Microbial influences in inflammatory bowel diseases. *Gastroenterology*, *134*(2), 577–594.

Sartor, R. B. (2010). Genetics and environmental interactions shape the intestinal microbiome to promote inflammatory bowel disease versus mucosal homeostasis. *Gastroenterology*, *139*(6), 1816–1819.

Sessa, M., Balestrieri, M. L., Ferrari, G., Servillo, L., Castaldo, D., D'Onofrio, N., & Tsao, R. (2014). Bioavailability of encapsulated resveratrol into nanoemulsion-based delivery systems. *Food Chemistry*, *147*, 42–50.

Sessa, M., Tsao, R., Liu, R., Ferrari, G., & Donsì, F. (2011). Evaluation of the stability and antioxidant activity of nanoencapsulated resveratrol during in vitro digestion. *Journal of Agricultural and Food Chemistry*, *59*(23), 12352–12360.

Sha, L., & Rongmin, Y. (2010). *Method for preparing oral chondroitin sulfate nanoemulsion*. Academic Press.

Sivakumar, M., Tang, S. Y., & Tan, K. W. (2014). Cavitation technology - A greener processing technique for the generation of pharmaceutical nanoemulsions. *Ultrasonics Sonochemistry*, *21*, 2069–2083.

Solans, C., Esquena, J., Forgiarini, A. M., Uson, N., & Izquierdo, P. (2002). Nanoemulsions: formation and properties. In K. L. Mittal & D. O. Shah (Eds.), *Surfactants in Solution: Fundamentals and Applications*. Marcel Dekker.

Solans, C., Izquierdo, P., Nolla, J., Azemar, N., & Garcia-Celma, M. J. (2005). Nano-emulsions. *Current Opinion in Colloid & Interface Science*, *10*(3-4), 102–110.

Sole, I., Maestro, A., Pey, C. M., Gonzalez, C., Solans, C., & Gutierrez, J. M. (2006). Nano-emulsions preparation by low energy methods in an ionic surfactant system. *Colloids and Surfaces. A, Physicochemical and Engineering Aspects*, *288*(1-3), 138–143.

Sole, I., Pey, C. M., Maestro, A., Gonzalez, C., Porras, M., Solans, C., & Gutierrez, J. M. (2010). Nanoemulsions prepared by the phase inversion composition method: Preparation variables and scale up. *Journal of Colloid and Interface Science*, *344*(2), 417–423.

Sonneville-Aubrun, O., Simonnet, J. T., & L'alloret, F. (2004). Nanoemulsions: A new vehicle for skincare products. *Advances in Colloid and Interface Science*, *108*, 145–149.

Tang, S. Y., Manickam, S., Wei, T. K., & Nashiru, B. (2012). Formulation development and optimisation of a novel cremophor EL-based nanoemulsion using ultrasound cavitation. *Ultrasonics Sonochemistry*, *19*, 330–345.

Tang, S.Y., Sivakumar, M., Ng, A.M., & Shridharan, P. (2012). Anti-inflammatory and analgesic activity of novel oral aspirin-loaded nanoemulsion and nano multiple emulsion formulations generated using ultrasound cavitation. *International Journal of Pharmaceutics, 430*(1-2), 299-306.

Thakore, S., Patel, R., & Patel, M. (2014). Nanoemulsion or Microemulsion? Understanding the Differences and Similarities. *The Pharma Review*, 136-142.

Thiagarajan, P. (2011). Nanoemulsions for drug delivery through different routes. *Research in Biotechnology*, *2*(3), 1–13.

Trevaskis, N. L., Charman, W. N., & Porter, C. J. (2008). Lipid-based delivery systems and intestinal lymphatic drug transport: A mechanistic update. *Advanced Drug Delivery Reviews*, *60*(6), 702–716.

Vandamme, T. F. (2002). Microemulsions as ocular drug delivery systems: Recent developments and future challenges. *Progress in Retinal and Eye Research*, *21*(1), 15–34.

Wang, L., Li, X., Zhang, G., Dong, J., & Eastoe, J. (2007). Oil-in-water nanoemulsions for pesticide formulations. *Journal of Colloid and Interface Science*, *314*(1), 230–235.

Wang, P., Liu, H. J., Mei, X. Y., Nakajima, M., & Yin, L. J. (2012). Preliminary study into the factors modulating β-carotene micelle formation in dispersions using an in vitro digestion model. *Food Hydrocolloids*, *26*(2), 427–433.

Warisnoicharoen, W., Lansley, A. B., & Lawrence, M. J. (2000). Nonionic oil-in-water microemulsions: The effect of oil type on phase behaviour. *International Journal of Pharmaceutics*, *198*(1), 7–27.

Wulff-Perez, M., Pavon, F. J., Martin-Rodriguez, A., De Vicente, J., Alen, F., De Fonseca, F. R., & Serrano, A. (2014). Preparation, characterisation and in vivo evaluation of nanoemulsions for the controlled delivery of the antiobesity agent N-oleoylethanolamine. *Nanomedicine (London)*, *9*(18), 2761–2772.

Xie, G., Zhao, A., Zhao, L., Chen, T., Chen, H., Qi, X., & Jia, W. (2012). Metabolic fate of tea polyphenols in humans. *Journal of Proteome Research*, *11*(6), 3449–3457.

Yu, H., & Huang, Q. (2012). Improving the oral bioavailability of curcumin using novel organogel-based nanoemulsion. *Journal of Agricultural and Food Chemistry*, *60*(21), 5373–5379.

Chapter 6
Current Perspectives on Nanoemulsions in Targeted Drug Delivery:
An Overview

Ali Zeytunluoglu
Pamukkale University, Turkey

Idris Arslan
Bülent Ecevit University, Turkey

ABSTRACT

Nanoemulsions are an isotropical mixture of oil, surfactant, and water with droplet diameter approximately in the range of 10-100 nm. They are being exponentially used for drug delivery systems for the influential administration of therapeutical agents because of their potential advantages over other approaches. Nanoemulsions can be used to design delivery systems that have increased drug loading, enhanced drug solubility, increased bioavailability, controlled drug release, and enhanced protection against chemical or enzymatic degradation. Moreover, nanoemulsions have better thermodynamical stability to flocculation, sedimentation, and creaming than conventional emulsions. Their small droplet dimensions and large droplet surface area positively influence drug transport and delivery, along with allowing targeting to specific sites. This chapter focuses on recent applications of nanoemulsions in the area of drug delivery.

INTRODUCTION

Nanotechnology manipulates the molecular structure of materials to change their intrinsic properties and obtain the optimal materials with revolutionary applications. The use of nanotechnology techniques in the medical and pharmaceutical industry has gained momentum in recent years. Traditonally, nanoemulsions have been commonly used in clinics for over five decades as parenteral nutrition formula. The

DOI: 10.4018/978-1-7998-8378-4.ch006

nanotechnology-driven pharmaceuticals are usually termed as "nanopharmaceuticals". Nanoemulsions are oil-in-water (o/w) or water in oil (w/o) emulsions with mean droplet diameters usually ranging from approximately 50 to 500 nm. Submicron emulsions, ultrafine emulsions and miniemulsions can be also used to refer the nanoemulsions (McClements 2012).

Nanoemulsions possess the following advantages;

- Nanoemulsions possess a considerably higher free energy and surface area than that of other emulsions which allow them an effective transport system.
- Nanoemulsions is a primary method of encapsulation.
- When compared to dispersal systems, nanoemulsions are thermodynamically stable systems. Nanoemulsions for the delivery of bioactive molecules have a high kinetic stability.
- Nanoemulsions are biocompatible. Therefore, they are suitable for human and veterinary therapeutic purposes.
- The problems of inherent creaming, flocculation, coalescence and sedimentation are not issues for nanoemulsions.
- Nanoemulsions can be formulated in various formulations e.g. liquids, foams, creams and sprays.
- Nanoemulsions are non-toxic, non-irritant and non-allergic.
- Once nanoemulsions are formulated with surfactants, approved for human consumption (GRAS), they can be taken by enteric route (Shah et al., 2010)

Besides the advantages mentioned above, nanoemulsions also have some disadvantages:

- Advanced (=expensive) equipment and privite procedures are needed to prepare smaller size nanoemulsions.
- High concentrations of surfactant and cosurfactant for stability are needed.
- Surfactants can have toxic effect.
- Prepared nanoemulsions have storage problems.
- It is affected by environmental parameters such as temperature and pH.
- It has limited solubility for substances with high melting points.
- Organic solvents are difficult to remove from this formulation.
- Little is known about interfacial chemistry in nanoemulsion preparation.

It is estimated that the portion of nanoemulsions in the pharmaceutical market will increase by 8.6% until 2024 (Global Market, 2019). In order to actualize this expectation in the nanoemulsion market, the problems that appears during the nanoemulsion production process and storage stages should be eliminated and the biosafety of the product should be increased.

It is recommended to use Plan Do Check At (PDCA) and seven quality control (seven QC) tools together for impeccable production and protection of real-time product quality in industrial scale production of nanoemulsions (Fig 1).

Figure 1. PDCA cycle, B. Seven quality control tools (Rai and Sharma, 2021).

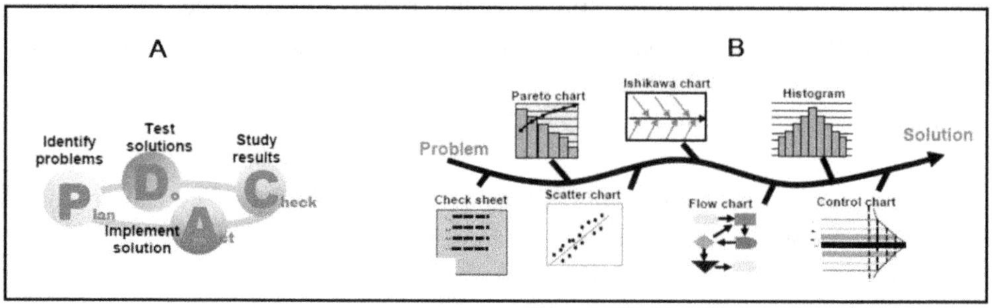

With seven QC tools, data is collected and analyzed throughout the all production process.The main reasons of the problem is identified, and activities are carried out to solve or prevent the problem. Thus, the seven QC tools play a more active role in decision making in each part of the PDCA cycle (Rai and Sharma, 2021). This application in the production process utilizer in the identification of key parameters affecting nanoemulsion formation, in the design of suitable equipment for large-scale production, the development and scaling of control methods. Thus, nanoemulsions with better stability and higher biosafety can be prepared at the end of the production process.

The main factor determining nanoemulsions interacting with the human body using different administration routes safety in therapeutic uses is that it does not have a negative toxic effect on human health of nanoemulsions and is biocompatible at the maximum level . The prepared nanoemulsion components, structure and application route can cause toxic effects. There's no simple method developed yet for the determination of the toxic effect at the nanoscale (Wani et al., 2018).

Figure 2. Nanotoxicological classification system (NCS) (Müller et al., 2011).

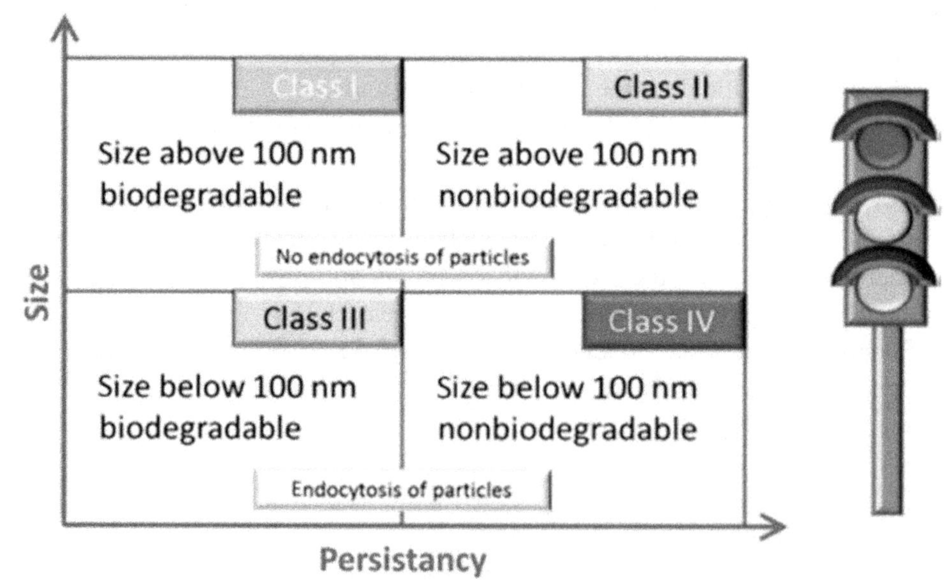

Müller and friends (Müller et al., 2011) formed a classification table of nanoemulsions showing the effect of particle size and biodegradability on toxicity. It is stated that the nanoparticle prepared according to this classification table, the toxicity increases and the usability of the nanoemulsion decrease when below 100 nm and nonbiodegradable (Fig 2).

The interest have increased gradually more efficient on nanoemulsion preparation methods and stabilization mechanisms to provide more efficiency in the target area with the increase of pharmacological uses of nanoemulsions in drug delivery. There are to used mainly several techniques like microfluidization, high pressure homogenization, emulsion inversion point (EIP), phase inversion temperature (PIT) and sonication for the preparation of nanoemulsion.

Nanoemulsions have higher thermodynamical stability compared to dispersed systems. Ostwald ripening is most common destabilization mechanism which is defined as the growth of the dispersed phase larger droplets at the expense of the smaller droplet.

In addition to the chemical nature of nanoemulsions, their physical characteristic properties, viscosity, zeta potential and type of surfactant play an important role in its stability.

Naturally occurring products are found in many natural sources (Arslan and Celik 2008). The term nanopharmaceutical is used for nanotechnology-driven pharmaceuticals. The kinetic stability characteristics of nanoemulsions and particularly nanopharmaceuticals are an important factor that determines their shelf life. Therefore, in order for the bulk and droplet properties of nanoemulsions to remain stable for a longer time in terms of electrostatic, steric and mechanical stability, it is necessary to define the factors affecting stability very well and to introduce new strategies to improve stability. Presence of repulsive electrostatic force between droplets in the stability of nanoemulsions; It prevents droplets from coalescence and agglomeration by separating the droplets at appropriate distances (Helgeson, 2016). The formation of an elastic film by non-absorbent molecules around the droplets, the formation of a voluminous surface by the branched molecules, the existance of hydrophobic interactions between absorbent and surfactant molecules play an effective role in the steric stabilization of nanoemulsions (Cardosa-Ugarte, 2016). The presence of gelling polymers and viscosity agents is effective in the mechanical stability of nanoemulsions and creates a barrier that prevents the collection of droplets. Conditions of use and proportions must be well adjusted, as they affect droplet size, encapsulation efficiency and optical appearance (Barradas and Silva, 2021).

Flocculation, coalescence, sedimentation, Ostwald ripening, creaming and phase separation are phenomena major responsible from instability of nanoemulsions (Fig 3).

Flocculation can be defined as a reversible phenomenon in which droplets dispersed in an emulsion clustered after random collision. In flocculation, layer formation (Creaming) take place at the surface when the droplet aggregation is less dense, and sedimentation occurs at the bottom when the aggregation is very dense. Other destability phenomena such as Ostwald ripening, coalescence and phase separation are the irreversible circumstances that occur in the form of increase in droplet size and phase separation as a result of the coalescence of droplets. It is considered to play a vital role of droplet sizes in elimination of effects in instability of nanoemulsion of mentioned above all active phenomena. Therefore, reducing the droplet size and minimizing the density difference will also increase the stability of the nanoemulsion as it will prevent the creaming, sedimentation and coalescence accompanying flocculation (McClements, 2011, McClements and Jafari, 2018). When compared to dispersed systems, nanoemulsions possess a considerably high thermodynamical stability.

The design and development of new drug delivery systems in order to provide more effective bioavailability from the drugs used in the treatment are among the important studies carried out in pharmacological area researches. It is necessary to develop suitable formulations for the delivery of lipophilic drugs with high hydrophobic character to their required places in the human body. Because reasons such as low solubility of lipophilic drugs in biological fluids and slow absorption in oral delivery reduce the bioavailability rate of drugs in this group and reduce the therapeutic effectiveness of this drugs (Campos et al, 2012).

Figure 3. Schematic representation of major factors lead to instabilities of nanoemulsions

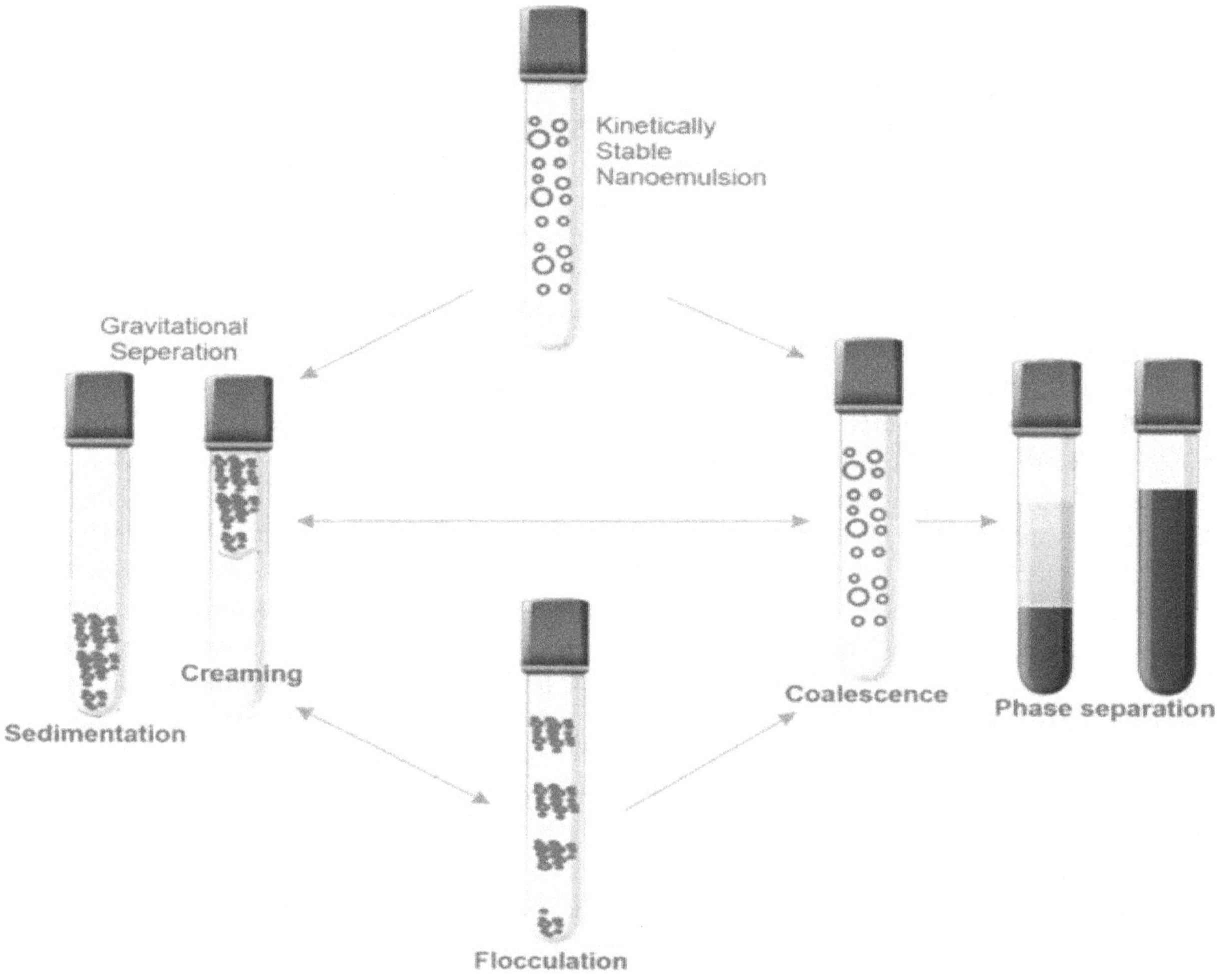

Nanostructured drug delivery systems have been the subject of most interest in lipid-based formulation studies and this systems have been used to effectively delivery many drugs at higher concentrations than the lymphatic system (Depnath et al, 2013).

Danazol is a synthetic steroid which have with low oral bioavailability due to its low solubility in water, slow dissolution rate. Danazol-containing oil-in-water nanoemulsions to enhance bioavailability from danazol has been prepared by Devalapally and colleagues (Devalapally et al., 2013). Stiripentol is a drug highly insoluble in water that used in the treatment Severe Myoclonic Epilepsy of Infancy (SMEI).

The stiripentol-nanoemulsion has been prepared using solvent-diffusion/ultrasonication technique for enhancement of stability and bioavailability of stiripentol by Lu and colleagues (Lu et al., 2015).

CHARACTERIZATION PARAMETERS OF NANOEMULSIONS

The characterization steps of nanoemulsions are carried out various methods such as electron microscopy (SEM/TEM) analyses, zeta potential measurements, droplet size analyses, polydispersity index analyses, drug content analyses, fluorescence tests, viscosity measurements, conductance measurements, dye dilution and filter paper tests. These methods using in the characterization of nanoemulsions are described below:

Electron Microscopy Analyses

Scanning and/or Transmission Electron Microscopy techniques (SEM/TEM) are widely used to determine the morphological and structural characteristics of nanoemulsion-based formulas. The most preferred method is to use three-dimensional images obtained from SEM to evaluate the surface morphology of nanoemulsions. However, TEM can also be used in the droplet size analysis. In the last two decades, atomic force microscope technique has been used together with SEM to comprehensive examine surface roughness of nanoemulsions (Klang, 2012).

Zeta Potential Measurement

Zeta potential is generally used to determine the surface charge property and their long-term physical stability of nanoemulsion systems. For this purpose, a ZetaPALS tool is a widely used instrument for the measure of zeta potential (Yilmaz and Borchert, 2005). The measurements are performed with diluted nanoemulsion formulations and its values are determined from the electrophoretic mobility of the oil droplets. As a rule of thumb zeta potential values $> +30$ mV or < -30 mV are considered as good indicators of long term stability (Singh, 2017). Doubtlessly, thermodynamically stability of nanoemulsion-assisted drug substances are so important in drug delivery.

Droplet Size and Polydispersity Index (PDI) Analysis

The polydispersity index value gives information about the droplet size dispersion in the nanoemulsion. A higher polydispersity index value indicates a large difference in nanoemulsion droplet sizes, while a lower polydispersity index value (<0.2) indicates a uniform size of nanoemulsion droplets and referred to as 'monodisperse'. The photon correlation spectroscopy (PCS), dynamic light scattering (DLS) or laser diffraction is used in droplet size analysis. Data obtained with PCS alone can be misleading because it misses very small nanoemulsions, nanoemulsions are not always perfectly spherical, unable to distinguish between nanoemulsions loaded with drug molecules and those without. The nanoemulsion structure changes due to sample dilutions, so it needs to be confirmed by other techniques such as SEM, TEM, X-Ray (Singh, 2017).

Drug Content Analysis

The drug content in nanoemulsions is determined using high performance liquid chromatography or UV spectrophotometer. In order to determine the drug content in the nanoemulsion, the nanoemulsion is firstly dissolved with the appropriate organic solvent and then the drug content is taken into the solvent phase and analyzed.

Fluorescence Test

Nanoemulsions prepared with some oils show fluorescent properties when exposed to UV light. Thanks to these features, it is possible to define the nanoemulsion type.

Viscosity Measurement

Viscosity is an important parameter in characterisation process of the nanoemulsions and it is determined by a viscometer instrument through measuring the concentrations of water, oil, surfactants in the nanoemulsion content. Viscosity influences both bioavailability and delivery of nanoemulsion-based drug substances.

Conductance Measurement

Conductance measurements are applied to determine the electrical conductivity of nanoemulsion systems. After the electrodes connected to the lamp are immersed in the nanoemulsion solution the current is supplied to the nanoemulsion solution from the electrical source. While the lamp connected to the electrodes lights in oil-in-water nanoemulsions, does not light in water-in-oil type emulsions.

Dye Dilution and Filter Paper Tests

Dye test is used in assesing the type of nanoemulsion. The nanoemulsions prepared by dissolving oils in water (continuous phase) are dyed homogeneously on all sides with water soluble paint while the nanoemulsions prepared by dissolving water in oil (dispersed phase) are seen that the paint becomes soluble only in the dispersed phase of the nanoemulsion. Filter paper test is used to determine the nanoemulsion type according to the migration of nanoemulsions at different rates on the filter paper surface. The use of water or oil in nanoemulsion dilution plays a role in determining the nanoemulsion type. A water-in-oil nanoemulsion can be diluted by adding oil only while A oil-in-water nanoemulsion can be diluted an by adding water only.

APPLICATION OF NANOEMULSION IN DRUG DELIVERY

Nanoemulsions are important structures that play a role in easily delivering the water-insoluble pharmaceutical drug component to the target area due to their very small size. Nanoemulsion-based drug delivery systems characteristically include a colloid of oil and aqueous phases.

Table 1. Several studies on the use of nanoemulsions in different drug delivery routes

Drug	Composition	Method of Preparation	Route of administration	Contribution	Therapeutic use	References
Meloxicam	Rhodamine, Caprylic acid, Tween 80, Propylene glycol, Carbopol 940	Magnetic stirring	Topical	Enhanced transdermal permeation and penetration	Anti-inflammatory therapy	Khurana et al., 2013
Tributyl phosohate	Soybean oil, Triton X-100	Spontaneous emulsification	Topical	enable to use for biocidal application of a toxic compound	Biocidal activity	Hamouda et al., 2001
Ceramide	Lipoid E-80, Tween 80, Glycerol, Potassium sorbate	Ultraturrax followed High pressure homogenization (HPH) method	Topical	Increased permeability of the skin	Transdermal delivery	Yilmaz and Borchert, 2005
Transanethole	Oleic acid,Tween 20, Propylene glycol	Ultrasonication	Topical	ensure prolonged antidiabetic action	Antidiabetic activity	Mostofa et al., 2005
Clotrimazole	Capryol 90, Solutol HS 15 Gelucire 44/14	Spontaneous emulsification	Oral	Increased bioavailability	Malaria therapy	Borhade et al., 2012
Thalidomide	Castor oil Soybean Lecithin, Glycerol, Polysorbate 80	Spontaneous emulsification	Oral	Overcome poor solubility	Cancer theraphy	Araújo et al., 2011
Cilostazol	Triacetin, Capryol, Tween 80, Transcutol P	Aqueous titration technique	Oral	Increased bioavailability	Antithrombotic activity	Mahour et al., 2015
Curcumin	Carboxymethyl cellulose, Chitosan, Octenyl-succinic-anhydride (OSA)-modified starch, Medium chain triglycerides	Ultrasonic homogenization	Oral	Stability and shelf improvement, minimum aggregation	Antitumour activity	Abbas et al., 2015
Ezetimibe	Capryol 90, Crempophor EL, Transcutol P	Aqueous titration technique	Oral	Higher release and bioavailability	Lowering cholesterol and triglyceride levels	Bali et al., 2011
Risperidone	Capmul MCM, Tween 80, Transcutol, propylene glycol	Magnetic stirring	İntranasal	Rapid and larger amount of transmission	Antipsychotic drug	Kumar et al., 2008
Saquinavir mesylate	Capmul MCM, Tween 80, PEG 400	Vortex and Magnetic stirring	İntranasal	Rapid and larger amount of transmission	Anti-HIV activity	Mahajan et al., 2013
Paliperidone	Labrasol, Labrafil M 1944 CS, Transcutol HP 241, Tween 80, Tween 20, PEG 400, Brij 35, Poloxamer 188	Orbital shaker	İntranasal	Allows dose reduction for desired therapeutic activity	Antipsychotic drug	Pidaparthi and Suares, 2016
Selegiline	grape seed oil, Sefsol 218, Tween 80, Unitop FFT-40, Lauroglycol 90	High energy emulsification	İntranasal	Increased bioavailability of the brain	Parkinson and major depressive disorder	Kumar et al., 2016
Dexamethasone acetate and polymyxin B sulfate	Eutanol G, Lipoid S100, Tween 80, Cetylpyridinium chloride, glycerol	HPH method	Ocular	Increased bioavailability	Ophthalmic infection treatment	Li et al., 2016
Paclitaxel	Poly(ethylene oxide)-copoly(D,L-Lactide), Perfluoro-15-crown-5-ether	Sonication	Parenteral	high therapeutic efficacy, lower systemic toxicity, prevention of drug resistance	Antitumour activity	Gupta et al., 2015
Fisetin	Labrasol, Tween 80, Miglyol® 812, Lipoid E80,	Ultraturrax followed Sonication	Parenteral	enable to use for antitumour application of a toxic compound	Antitumour activity	Ragelle et al., 2012
Docetaxel	Oleic acid, Stearyl amine, Egg lecithin	Ultrasonication and homogenization	Parenteral	Increasing efficacy against poor solubility and hydrolytic instability, reducing side effects	Antitumour activity	Venkateshwarlu et al., 2010

The main advantage of using nanoemulsions as drug carriers are redound the level of bioavailability from drug with deliver much higher concentrations of the drug to the target site in compared to conventional delivery systems. Besides, other advantages of nanoemulsions are that they can be prepared as cream, foam, spray or liquid, they do not show any flocculation or sedimentation problems, they can be fastly penatrate due to their small size, they are non-toxic and biocompatible. The cost of the nanoemulsion preparation process is a high and they cannot be stored for a long time due to their varying stability over time.

Drugs are delivered to the target area using nanoemulsions, topical, oral, ocular, pulmonary intravenous, intranasal and parenteral delivery routes. Several studies on the use of nanoemulsions in different drug delivery routes are showed Table 1.

Transdermal and Topical Delivery

Skin is most preferred an application methods for topical drug delivery. The skin and mucous membranes (eye, nose, ears, urethra, vagina, colon) are considered a route of administration for the topical delivery of drugs prepared in the form of drops, sprays, ointments, creams, lotions and powders. There are various advantages of drug application to with skin route. Interestingly, transdermally applied drugs might penetrate the skin by hair follicles, sweat ducts and then reach the blood circulation. However; stratum corneum acts as a protective barrier which preventing intake of drug into the body by diffusion and therefore limits their bioavailability. The structural properties of nanoemulsions play a important role in overcoming the obstacle to drug absorption by skin. Another advantage of drug administration in this way is that it creates a low risk of side effects. However, its disadvantage is that the absorption of the drug by the deeper layers of skin is prevent since the human skin acts as a protective layer for the human body. The inherent properties of o/w emulsions can enhance the absorption of lipophilic drugs when applied topically to the skin. Also, the use components (surfactants and costabilizers) within the interface can enhance penetration activity of the skin outer layer. Nanoemulsions are accepted as an alternative approaches in the delivery of drugs have poor aqueous solubility, penetration and low adsorption through the skin. Numerous studies have been published on the delivery skin route of drugs through encapsulation, some of which have been in clinical trial stage (Tayeb and Sainsbury, 2018; Talegaonkar and Negi, 2015).

There has been a plethora of distinct drug substances delivered to or through the skin using nanoemulsion-driven formulations, including betamethasone, clobetasol, corticosterone, flufenamic acid, flurbiprofen, tocopheryl acetate, tolterodine tartrate and amlodipine (Sivaramakrishnan et al., 2004; Jain et al., 2005, Kumar et al., 2009; Elshafeey et al., 2009; Zhao et al., 2009; Senyigit et al., 2010; Jensen et al., 2010; Luengo et al., 2006). (Fig 4)

Topical lipophilic drug delivery to the eye are improved by the use of o/w emulsion systems. Lipophilic drug loaded oil in water ocular emilsions provide equivocally a better balance between patient comfort topical instillation *e.g.* indomethacin and piroxicam (Fig.5) to the eye and ocular bioavailablity improvement (El-Leithy, Inrahim and Sorour 2015).

Figure 4. Molecular structures of Betamethasone, clobetasol, corticosterone and flufenamic acid

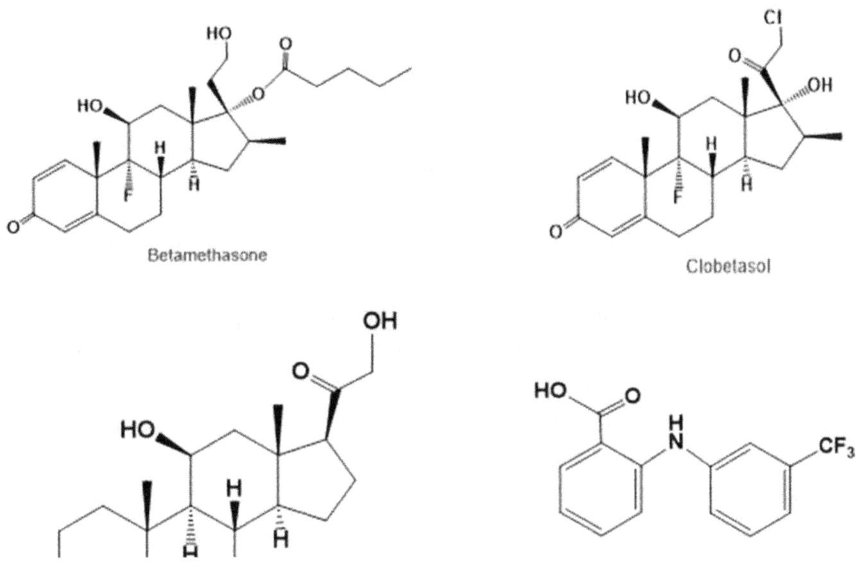

Figure 5. Molecular structure of indomethacin and piroxicam

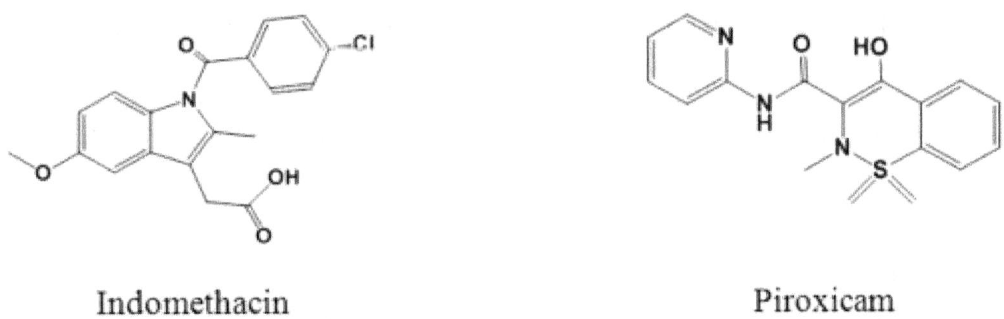

Oral Delivery

Currently, nanoemulsion-driven platforms have obtained increasing attention as a convenient drug carrier system for insoluble bioactive substances to enhance their bioavailability and modify drug release properties (Han et al., 2008; Kotta et al., 2013; Sharma et al., 2014; Verma et al., 2014). For example, improved oral bioavailability of amlodipine and felodipine nanoemulsions by comparison the classic preparations has been revelaed (Veerareddy et al., 2012; Chhabra et al., 2011). Doubtlessly, oral route of delivery is the most appropriate, easiest, cost-effective drug delivery system. However, some health conditions such as geriatric, pediatric, trauma epileptic patients and some inherently difficulties restrict the oral route of delivery. Peptide-based drugs are generally hydrolyzed through enzymatic degradation in gastrointestinal track which limits their absorption and bioavailability. On the other hand, it is well-known that some therapeutics have poor oral bioavailability due to variable absorption and high drug effluxing through P-glycoprotein transporters in the lumen (Mei et al. 2013).

In recent years, many scientists have reported considerable oral bioactivity development of unsoluble drugs. Yoo and collaegues has reported that nanoparticles are swallowed by the M-cells, after that they are fastly assimilated and "shuttled" to the lumphocytes (Yoo et al. 2010). In the last decades, administration of nanoemulsions as oral drug delivery systems has been demonstated to rise the efficiency of the drug at the action site (Khatri and Shao, 2017).

Self-nanoemulsifying systems can enhance the bioavailability by bypassing the hepatic portal route, protecting drug against enzymatic degradation in the hard gastrointestinal surroundings, facilitating intestinal lymphatic transport of drugs, reducing cytochrome P450-triggered metabolism, and arresting the P-glycoprotein mediated efflux due to intrinsic property of excipients used in formulation (Ahn and Park 2016).

Many of the drugs prepared in the form of nanoemulsions taken orally are transported throughout the gastrointestinal tract and are by being absorbed the small intestines and released into the blood or created directly a local effect in the intestines. The pH of the intestinal ambiance, the thickness of the intestinal mucosa and the diversity of the bacterial fauna are limiting factors that affect the retention and absorption time of the drug by the intestines. (Homayun et al., 2019).

However, drug formulation, whether the drug is targeted, solubility ratio, hydrophilicity, pore density and size, surface area, enzymatic degradation status, food-drug interaction depending on the satiety-hunger variability of the patient given the drug, depending on in the gastrointestinal tract regional differences of the patients are shown as other limiting factors affecting the bioavailability of the drug (Viswanathan et al, 2017; Buya et al, 2020 ; Vinarov et al, 2021).

Patent applications on nanoemulsions have increased tremendously in the last decades and continue to increase. The information about some important patents on using nanoemulsions in drug delivery through oral route are summarized in Table 2.

Table 2. Patents summary on oral drug delivery

No	Patent No	Patent name	References
1	US 05961970	Submicron emulsions as vaccine adjuvants	Lowel et al., 1999
2	US 0216751	Liquid Bevirimat Dosage Forms for Oral Administration	Jacob et al., 2010
3	US 0294900	Self emulsifying drug delivery system for a curcuminoid based composition	Kohli et al., 2011
4	US 0269792	Eutectic based self-nanoemulsified drug delivery system	Khan and Nazzal, 2012
5	US 0052235	Herpes simplex virus nanoemulsion vaccine	Fattom et al, 2013
6	WO 205226	Self-nanoemulsion of poorly soluble drugs	Desai et al., 2014
7	US 0281573	Medicinal Composition for enhanced delivery of triterpenes	Schwarz and Weisspapir, 2017
8	US 0087120	Composition for improving boavailibility and efficacy of taxane	Sachdeva et al, 2017
9	US 0311146	Composition for intraoral delivery of biologically active peptides and proteins	Schwarz and Weisspapir, 2018
10	US 0009067	Formulation and method for increasing oral bioavailability of drugs	Hoffman et al, 2020
11	US 0186870	Improved cannobinoid bioavailability	Garabagi et al., 2021
12	US 0290586	Oral drug delivery composition containing oxaliplatin and method for preparing same	Park et al., 2021

Aerosol Applications

Pulmonary delivery route; Besides having a thin epithelial barrier, it also creates a large surface area for the absorption of aerosolized therapeutics. In addition, therapeutics encounter low enzymatic activity in this pathway and this pathway has a connection with the vascular system (Azarmi, 2008; Gaspar, 2008). For this reason, the use of this pathway in noninvasive drug delivery seems promising for the treatment of respiratory tract diseases. In the last 20 years, many nano drug carriers have been developed to be used in this way (Beija, 2012). The type and characteristics of the device to be used in pulmonary drug administration are a decisive factor to nanoemulsion formulation. Apart from pulmonary route, the other way of delivery aerosol therapeutics into the body painless by non-invasive way is intranasal administration. The nasal route is generally used for the local treatment of nasal-related ailments such as sinusitis and nasal congestion. However, there are studies on the administration of drugs for the treatment of various diseases such as influenza hypertension, migraine through the nasal mucosa (Abdou, 2017; Das, 2012; Jain, 2009).

Ingestible Applications

Oral drug delivery is the easiest and most cost-effective in all drug administration routes. The drug delivery via oral route are faced with many obstacles in the body: (pH changes, digestive enzymes, with high affinity the presence of structures to drugs). Nanoemulsions by protecting drugs from degradation by digestive enzymes affect the absorption durations of the them in the gastrointestinal region and their effectiveness are enhanced. The main purpose of drug delivery is ensuring to reach to maximum performance levels in the drug's efficacy area. It has been observed that the bioavailability from drug increased in the nanoemulsion prepared in as Oil-in-water of paclitaxel (Fig 3) used in the treatment of many types of cancer (Tiwari, 2006).

Parenteral Applications

Parenteral drug delivery demonstrates the effectiveness of this route compared to other delivery routes by passing of the adsorption step the rate-limiting step in drug delivery and increasing bioavailability levels. However, the problems encountered in other drug delivery routes, such as poor aqueous solubility and stability conditions, also apply to parenteral administration. This problems are limiting factors for drug delivery by the parenteral application (Patel, 2018). Different strategies are described in the literature for this purpose e.g. modification of vehicle pH, addition of cosolvent systems, or inclusion complexes with cyclodextrins (Mallick et al. 2007). However, these approaches are restricted with some limiting factors such as immunoadjuvant biocompatibility, triggering of pain, and also precipitation of drugs during the administration. Nanoemulsions are considered as tools to help overcome existing problems in parenteral drug delivery due to their property protect drugs against enzymatic degradation and hydrolysis (Prakash and Thiagarajan, 2011).

A recent paper has shown that the highly hydrophobic sorafenib molecule, which is used in renal cell carcinomas and, to some extent, liver and thyroid carcinomas, can be applied peranetal as a nanoemulsion (Izadiyan, 2017). The obstacles such as short life time and low solubility in front of peranetal route delivery of therapeutic molecules in different types of cancer such as lung and multiple myeloma have been overcome with nanoemulsion application (Zhao, 2013, Araújo, 2011).

In recent years, scientists have shown that the thermodynamical stability of nanoemulsions used for parenteral delivery largely depends upon their profiles, preparation techniques, and storage conditions (Cheng et al., 2015). However, the potential of nanoemulsions to dissolve large quantities of hydrophobic bioactive substances, as well as their biocompatibility, and the ability to conserve drugs from chemical and/or enzymatic degradation perform them ideal vehicles for parenteral delivery (Gue et al., 2016; Hörmann and Zimmer, 2016).

Even though natural surfactants are considerably preferred to synthetic ones, having appropriate emulsification finding, adjuvant emulsifying substances have been employed to deliver greater emulsion stability. For example; a polysorbate-type nonionic surfactants from the poloxamers and polyoxyethylene sorbitans have been revealed to yield encouraging result when used in combination with phospholipids because they cause the formation of compact mixed films, giving greater stability to the formulation. Moreover, hemolytic effects and changes in the droplet diameter of nanoemulsions stabilized by Tween 80 during autoclaving significantly restrict their use. Researchers have shown that polyoxyethylated castor oil derivatives (Cremophor EL) have also received noticeable attention for parenteral use, but the existance of hemolysis and undesired toxicity has been characterized for some of these substances (Lu et al., 2008; Söderlind et al., 2003).

Intranasal Delivery

The management of neurological diseases is challenging, due to the need of drug substances to cross the blood-brain barrier and reach the brain. Normally, the blood-brain barrier acts a basic role in protecting the brain against toxic agents and thus controls and limits the entry of drug molecules. A recent paper has shown that the US Food and Drug Administration (FDA) have not approved more than 90% of new drugs used to treat central nerveous system diseases, due to the difficulty of crossing the blood-brain barrier (Islam et al. 2020).

The nasal administration of drug molecules using the nose-to-brain route leads to direct drug targeting into brain bypassing the blood-brain barrier at a higher rate and extent than other traditional routes. The olfactory region of the nasal mucosa mediates a direct connection between the nose and the brain. Therefore, the intranasal route was suggested as a promising approach to improve the delivery of drugs to the brain avoiding the need to cross the blood-brain barrier. Nasal nanoemulsions appear to be influential, non-invasive and safe drug delivery platforms to archive brain targeting for the treatment of various neurological disorders. Nanoemulsion-based drugs have been used for the treatment of some health conditions such as metoclopraminde (motion sickness condition), estradiol (hormone therapy), olanzapine (schizophreia), desmopressin (enuresis) (Deshpande et al. 2014; Elsheikh et al. 2012).

Intranasal route has various advantages, including high drug bioavailability, non-invasiveness, high blood flow, large surface area available for drug absorption, practicality of application, fast onset of action, avoidance of gastrointestinal and hepatic metabolism and, possibility to bypass the blood-brain barrier (Ali et al 2010; Sabir et al. 2020; Alavian and Shams 2019).

A lot of studies have been published as topical drug delivery via encapsulation of drugs within nanoemulsions. The encapsulated nanoemulsions which contains diclofenac, testosterone and aminolevulinic acid as active pharmaceutical ingredient are used in clinical trials.

Nanomulsion formulations also reveal promise in targeted cancer therapy as carriers for prolonging the drug release after intra-muscular and intratumoral injection to improve the transport of antineoplastic agents by the lymphatic system.

Camptothecin is a topoisomerase I inhibitor which is used for a broad spectrum of cancer therapy. However, its clinical trial is restricted by its insolubility, poor stability, and toxicity. It is noteworthly to mention that the nanoemulsions were prepared using liquid perfluorocarbons and coconut oil as the cores of the inner phase. Nanoemulsions were stabilized by phospholipids and/or Pluronic F68. The nanoemulsions were prepared at high drug loading of approximately 100% with a mean droplet diameter of 220-420 nm. Interestingly, camptothecin have showed retarded drug release in this system. It was confirmed by confocal laser scanning microscopy that nanoemulsions uptake into cells. Using a 1 MHz ultrasound, an increased release of camptothecin from the system with lower oil concentration could be established, illustrating a drug-targeting effect (Fig 6).

Another study has reported the formulation of filter sterilizable nanoemulsion of paclitaxel using α-tocopherol as the oil phase and α-tocopherylpolyethyleneglycol-1000 succinate and poloxamer 407 as emulsifiers. The nanoemulsion system has showed better efficacy and has been more tolerable when it has studied in B16 melanoma tumor models in mice (Constantinite et al. 2000).

Figure 6. Molecular structure of camptothecin and paclitaxel

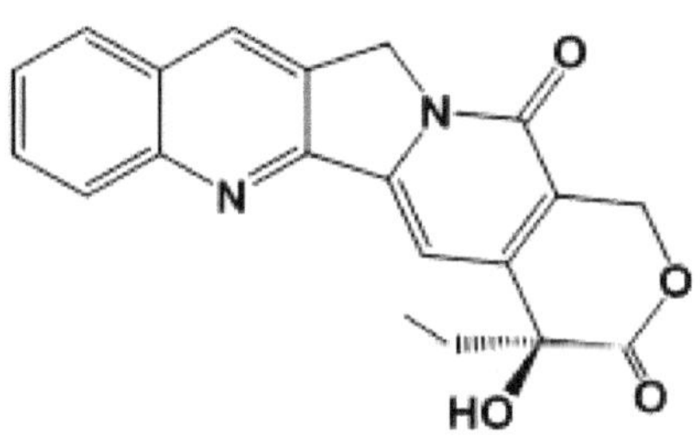

Camptothecin

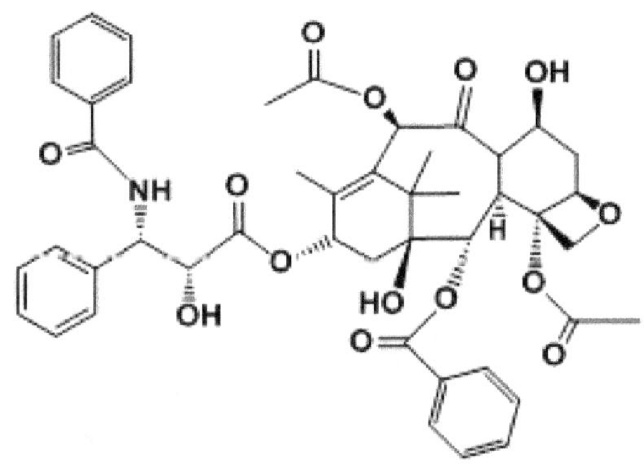

Paclitaxel

CONCLUSION

Nanoemulsions have opened a new door in the field of drug therapy by overcoming some limitations in drug delivery. This chapter focuses on structures, avantages and disavantages preparation methods, stabilization, characterization, applications in different drug delivery systems of nanoemulsions as drug delivery vehicle (Fig 7).

Nanoemulsion-driven formulations are being exponentially improved and utilized as drug delivery systems for nearly all types of drug administration. They offer various advantages fort he delivery of pharmaceuticals, drugs, biologicals, and/or diagnostic agents. A comprehensive study was performed with diverse applications in drug delivery, with the most investigated field being the advanced solubilization and delivery poorly soluable actives. Nanoemulsion-based platforms might supply with a solution to the delivery of 'difficult-to-formulate' pharmaceuticals in which stability is a so critical.

The management of nanoemulsion-based drugs have a few disadvantegous. The utilization of a large concentration of surfactant and co-surfactants is fundamental for stabilizing the nanodroplets.

Figure 7. The avantages and disavantages, preparation methods, instabilization mechanisms, characterization parameters, route of administration in different drug delivery systems of nanoemulsions as drug delivery vehicle.

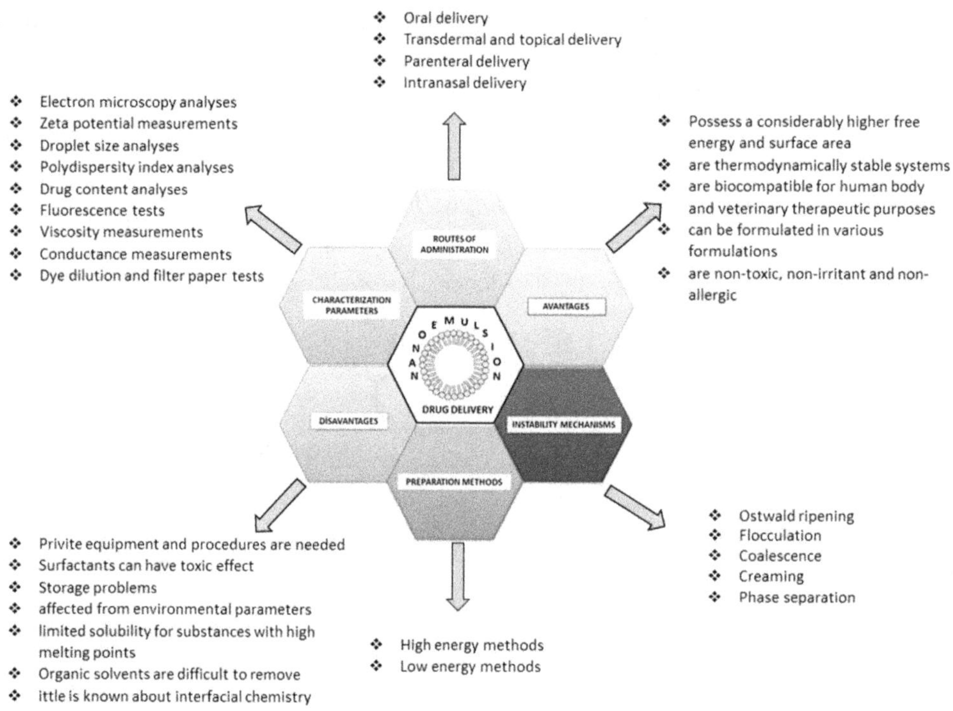

The preperation of nanoemulsions is a costly procedure because of size decrease of droplets requiring a unique sort of instruments and procedure techniques. Storage of nanoemulsion formulations is one of the other problems. Despite all these disadvantages nanoemulsion-based drug delivery is still significantly superior to other approaches. When considered broadly, nanoemulsion-based drug delivery looks like it will be a breakthrough in the near future.

REFERENCES

Abbas, S., Karangwa, E., Bashari, M., Hayat, K., Hong, X., Sharif, H. R., & Zhang, X. (2015). Fabrication of polymeric nanocapsules from curcumin-loaded nanoemulsion templates by self-assembly. *Ultrasonics Sonochemistry*, *23*, 81–92. doi:10.1016/j.ultsonch.2014.10.006 PMID:25453208

Abdou, E. M., Kandil, S. M., & Miniawy, H. M. F. E. (2017). Brain targeting efficiency of antimigrain drug loaded mucoadhesive intranasal nanoemulsion. *International Journal of Pharmaceutics*, *529*(1-2), 667–677. doi:10.1016/j.ijpharm.2017.07.030 PMID:28729175

Ahn, H., & Park, J. H. (2016). Liposomal delivery systems for intestinal lymphatic drug transport. *Biomaterials Research*, *20*(1), 36. doi:10.118640824-016-0083-1 PMID:27895934

Alavian, F., & Shams, N. (2019). Oral and intranasal administration of nanoparticles in the cerebral ischemia treatment; Considering the advantages and disadvantages. *Current Clinical Pharmacology*, *14*, 1–10. doi:10.2174/1574884714666190704115345 PMID:31272358

Ali, J., Ali, M., Baboota, S., Kaur Sahni, J., Ramassamy, C., Dao, L., & Bhavna. (2010). Potential of Nanoparticulate Drug Delivery Systems by Intranasal Administration. *Current Pharmaceutical Design*, *16*(14), 1644–1653. doi:10.2174/138161210791164108 PMID:20210751

Araújo, F. A., Kelmann, R. G., Araújo, B. V., Finatto, R. B., Teixeira, H. F., & Koester, L. S. (2011). Development and characterization of parenteral nanoemulsions containing thalidomide. *European Journal of Pharmaceutical Sciences*, *42*(3), 238–245. doi:10.1016/j.ejps.2010.11.014 PMID:21130164

Arslan, I., & Celik, A. (2008). Chemical composition and antistaphylococcal activity of Salvia chrysophylla Stapf. Naturally distributed in Denizli province (Turkey) and its vicinity. *Pakistan Journal of Botany*, 401799–401803.

Azarmi, S., Roa, W. H., & Löbenberg, R. (2008). Targeted delivery of nanoparticles for the treatment of lung diseases. *Advanced Drug Delivery Reviews*, *60*(8), 863–875. doi:10.1016/j.addr.2007.11.006 PMID:18308418

Bali, V., Ali, M., & Ali, J. (2011). Nanocarrier for the enhanced bioavailability of a cardiovascular agent: In vitro, pharmacodynamic, pharmacokinetic and stability assessment. *International Journal of Pharmaceutics*, *403*(1-2), 46–56. doi:10.1016/j.ijpharm.2010.10.018 PMID:20969935

Beija, M., Salvayre, R., Lauth-de Viguerie, N., & Marty, J. D. (2012). Colloidal systems for drug delivery: From design to therapy. *Trends in Biotechnology*, *30*(9), 485–496. doi:10.1016/j.tibtech.2012.04.008 PMID:22673692

Borhade, V., Pathak, S., Sharma, V., & Patravale, V. (2012). Patravale, Clotrimazole nanoemulsion for malaria chemotherapy. Part I: Preformulation studies, formulation design and physicochemical evaluation. *International Journal of Pharmaceutics*, *431*(1-2), 138–148. doi:10.1016/j.ijpharm.2011.12.040 PMID:22227344

Buya, A. B., Beloqui, A., Memvanga, P. B., & Préat, V. (2020). Self-Nano-Emulsifying Drug-Delivery Systems: From the Development to the Current Applications and Challenges in Oral Drug Delivery. *Pharmaceutics*, *12*(1194), 1–55. doi:10.3390/pharmaceutics12121194 PMID:33317067

Campos, V. E. B., Ricci-Júnior, E., & Mansur, C. R. E. (2012). Nanoemulsions as Delivery Systems for Lipophilic Drugs. *Journal of Nanoscience and Nanotechnology*, *12*(3), 2881–2890. doi:10.1166/jnn.2012.5690 PMID:22755138

Cheng, Y., Liu, M., Hu, H., Liu, D., & Zhou, S. (2015). Development, Optimization, and Characterization of PEGylated Nanoemulsion of Prostaglandin E1 for Long Circulation. *AAPS PharmSciTech*, *17*(2), 409–417. doi:10.120812249-015-0366-1 PMID:26195071

Chhabra, G., Chuttani, K., Mishra, A. K., & Pathak, K. (2011). Design and development of nanoemulsion drug delivery system of amlodipine besilate for improvement of oral bioavailability. *Drug Development and Industrial Pharmacy*, *37*(8), 907–916. doi:10.3109/03639045.2010.550050 PMID:21401341

Constantinides, P. P., Lambert, K. J., Tustian, A. K., Schneider, B., Lalji, S., Ma, W., & Quay, S. C. (2000)... *Pharmaceutical Research*, *17*(2), 175–182. doi:10.1023/A:1007565230130 PMID:10751032

Das, S. C., Hatta, M., Wilker, P. R., Myc, A., Hamouda, T., Neumann, G., & Kawaoka, Y. (2012). Nanoemulsion W805EC improves immune responses upon intranasal delivery of an inactivated pandemic H1N1 influenza vaccine. *Vaccine*, *30*(48), 6871–6877. doi:10.1016/j.vaccine.2012.09.007 PMID:22989689

Debnath, S., And, S., & Kumar, G.V., (2010), Nanoemulsion-a method to improve the solubility of lipophilic drugs. *International Journal of Advances in Pharmaceutical Sciences, 2*(2-3), 72-83.

Desai, D., Vaka, S., Shah, N., Jain, A., & Phuapradit,W. (2014). *Self-nanoemulsion of poorly soluble drugs*. Academic Press.

Deshpande, D., Kethireddy, S., Gattacceca, F., & Amiji, M. (2014). Comparative pharmacokinetics and tissue distribution analysis of systemically administered 17-β-estradiol and its metabolites in vivo delivered using a cationic nanoemulsion or a peptide-modified nanoemulsion system for targeting atherosclerosis. *Journal of Controlled Release*, *180*, 117–124. doi:10.1016/j.jconrel.2014.02.009 PMID:24556419

Devalapally, H., Silchenko, S., Zhou, F., McDade, J., Goloverda, G., Owen, A., & Hidalgo, I. J. (2013). Evaluation of a Nanoemulsion Formulation Strategy for Oral Bioavailability Enhancement of Danazol in Rats and Dogs. *Journal of Pharmaceutical Sciences*, *102*(10), 3808–3815. doi:10.1002/jps.23667 PMID:23878097

El-Leithy, E. S., Ibrahim, H. K., & Sorour, R. M. (2013). In vitroandin vivoevaluation of indomethacin nanoemulsion as a transdermal delivery system. *Drug Delivery*, *22*(8), 1010–1017. doi:10.3109/10717544.2013.844742 PMID:24188427

Elshafeey, A. H., Kamel, A. O., & Fathallah, M. M. (2009). Utility of Nanosized Microemulsion for Transdermal Delivery of Tolterodine Tartrate: Ex-Vivo Permeation and In-Vivo Pharmacokinetic Studies. *Pharmaceutical Research*, *26*(11), 2446–2453. doi:10.100711095-009-9956-5 PMID:19697106

Elsheikh, M. A., Elnaggar, Y. S. R., Gohar, E. Y., & Abdallah, O. Y. (2012). Nanoemulsion liquid preconcentrates for raloxifene hydrochloride: Optimization and in vivo appraisal. *International Journal of Nanomedicine*, *3787*. Advance online publication. doi:10.2147/IJN.S33186 PMID:22888234

Fattom, A.I., Simon, J., & Baker, J.R. (2013). *Herpes simplex virus nanoemulsion vaccine*. Academic Press.

Garabagi, F., Morello, G., & Wagner, C. (2021). *Improved cannobinoid bioavailability*. Academic Press.

Gaspar, M. M., Bakowsky, U., & Ehrhardt, C. (2008). Inhaled Liposomes–Current Strategies and Future Challenges. *Journal of Biomedical Nanotechnology*, *4*(3), 245–257. doi:10.1166/jbn.2008.334

Gué, E., Since, M., Ropars, S., Herbinet, R., Le Pluart, L., & Malzert-Fréon, A. (2016). Evaluation of the versatile character of a nanoemulsion formulation. *International Journal of Pharmaceutics*, *498*(1-2), 49–65. doi:10.1016/j.ijpharm.2015.12.010 PMID:26685727

Gupta, R., Shea, J., Scaife, C., Shurlygina, A., & Rapoport, N. (2015). Polymeric micelles and nano-emulsions as drug carriers: Therapeutic efficacy, toxicity, and drug resistance. *Journal of Controlled Release*, *212*, 70–77. doi:10.1016/j.jconrel.2015.06.019 PMID:26091919

Hamouda, T., Myc, A., Donovan, B., Shih, A. Y., Reuter, J. D., & Baker, J. R. (2001). A novel surfactant nanoemulsion with a unique non-irritant topical antimicrobial activity against bacteria, enveloped viruses and fungi. *Microbiological Research*, *156*(1), 1–7. doi:10.1078/0944-5013-00069 PMID:11372645

Han, M., He, C.-X., Fang, Q.-L., Yang, X.-C., Diao, Y.-Y., Xu, D.-H., & Gao, J.-Q. (2008). A Novel Camptothecin Derivative Incorporated in Nano-Carrier Induced Distinguished Improvement in Solubility, Stability and Anti-tumor Activity Both In Vitro and In Vivo. *Pharmaceutical Research*, *26*(4), 926–935. doi:10.100711095-008-9795-9 PMID:19048358

Hoffman, A., Domb, A.J., & Cherniakov, A.E. (2020). *Formulation and method for increasing oral bioavailability of drugs*. Academic Press.

Homayun, B., Lin, X., & Choi, H. J. (2019). Challenges and Recent Progress in Oral Drug Delivery Systems for Biopharmaceuticals. *Pharmaceutics*, *11*(129), 1–29. doi:10.3390/pharmaceutics11030129 PMID:30893852

Hörmann, K., & Zimmer, A. (2016). Drug delivery and drug targeting with parenteral lipid nano-emulsions— A review. *Journal of Controlled Release*, *223*, 85–98. doi:10.1016/j.jconrel.2015.12.016 PMID:26699427

Islam, S. U., Shehzad, A., Ahmed, M. B., & Lee, Y. S. (2020). Intranasal Delivery of Nanoformulations: A Potential Way of Treatment for Neurological Disorders. *Molecules (Basel, Switzerland)*, *25*(8), 1929. doi:10.3390/molecules25081929 PMID:32326318

Izadiyan, Z., Basri, M., Fard Masoumi, H.R., Abedi Karjiban, R., Salim, N., & Shameli, K. (2017). Modeling and optimization of nanoemulsion containing Sorafenib for cancer treatment by response surface methodology. *Chemistry Central Journal*, *11*(21), 1-9. doi:10.1186/s13065-017-0248-6

Jacob, J. S., Richards, J., Augustine, J. G., & Milea, J. S. (2010). Liquid Bevirimat Dosage Forms for Oral Administration. Academic Press.

Jain, R., & Patravale, V. B. (2009). Development and Evaluation of Nitrendipine Nanoemulsion for Intranasal Delivery. *Journal of Biomedical Nanotechnology*, *5*(1), 62–68. doi:10.1166/jbn.2009.031 PMID:20055107

Jain, S. K., Chourasia, M. K., Masuriha, R., Soni, V., Jain, A., Jain, N. K., & Gupta, Y. (2005). Solid Lipid Nanoparticles Bearing Flurbiprofen for Transdermal Delivery. *Drug Delivery*, *12*(4), 207–215. doi:10.1080/10717540590952591 PMID:16036715

Jensen, L. B., Magnussson, E., Gunnarsson, L., Vermehren, C., Nielsen, H. M., & Petersson, K. (2010). Corticosteroid solubility and lipid polarity control release from solid lipid nanoparticles. *International Journal of Pharmaceutics*, *390*(1), 53–60. doi:10.1016/j.ijpharm.2009.10.022 PMID:19836439

Joseph Schwarz, J., & Michael Weisspapir, M. (2017). *Medicinal Composition for enhanced delivery of triterpenes*. Academic Press.

Joseph Schwarz, J., & Michael Weisspapir, M. (2018). *Composition for intraoral delivery of biologically active peptides and proteins*. Academic Press.

Khan, M., & Nazzal, S. (2012). *Eutectic based self-nanoemulsified drug delivery system*. Academic Press.

Khatri, P., & Shao, J. (2017). Transport of lipid nano-droplets through MDCK epithelial cell monolayer. *Colloids and Surfaces. B, Biointerfaces*, *153*, 237–243. doi:10.1016/j.colsurfb.2017.02.024 PMID:28267668

Khurana, S., Jain, N. K., & Bedi, P. M. S. (2013). Nanoemulsion based gel for transdermal delivery of meloxicam: Physico-chemical, mechanistic investigation. *Life Sciences*, *92*(6-7), 383–392. doi:10.1016/j.lfs.2013.01.005 PMID:23353874

Klang, V., Matsko, N. B., Valenta, C., & Hofer, F. (2012). Electron microscopy of nanoemulsions: An essential tool for characterisation and stability assessment. *Micron (Oxford, England)*, *43*(2-3), 85–103. doi:10.1016/j.micron.2011.07.014 PMID:21839644

Kohli, K., Chopra, S., Khar, R.K., Pillai, K.K., & Arora, S. (2011). *Self emulsifying drug delivery system for a curcuminoid based composition*. Academic Press.

Kotta, S., Khan, A. W., Ansari, S. H., Sharma, R. K., & Ali, J. (2013). Formulation of nanoemulsion: A comparison between phase inversion composition method and high-pressure homogenization method. *Drug Delivery*, *22*(4), 455–466. doi:10.3109/10717544.2013.866992 PMID:24329559

Kumar, D., Aqil, M., Rizwan, M., Sultana, Y., & Ali, M. (2009). Investigation of a nanoemulsion as vehicle for transdermal delivery of amlodipine. *Die Pharmazie*, *64*, 80–85. doi:10.1691/ph.2009.8200 PMID:19320278

Kumar, M., Misra, A., Babbar, A. K., Mishra, A. K., Mishra, P., & Pathak, K. (2008). Intranasal nanoemulsion based brain targeting drug delivery system of risperidone. *International Journal of Pharmaceutics*, *358*(1-2), 285–291. doi:10.1016/j.ijpharm.2008.03.029 PMID:18455333

Kumar, S., Ali, J., & Baboota, S. (2016). Design Expert®supported optimization and predictive analysis of selegiline nanoemulsion via the olfactory region with enhanced behavioural performance in Parkinson's disease. *Nanotechnology*, *27*(43). doi:10.1088/0957-4484/27/43/435101

Li, X., Müller, R. H., Keck, C. K., & Bou-Chacra, N. A. (2016). Mucoadhesive dexamethasone acetate-polymyxin B sulfate cationic ocular nanoemulsion – novel combinatorial formulation concept. *Die Pharmazie*, *71*, 327–333. doi:10.1691/ph.2016.5190 PMID:27455551

Lowel, G., Amselem, S., Friedman, D., & Aviv, H. (1999). *Submicron emulsions as vaccine adjuvants*. Academic Press.

Lu, R., Liu, S., Wang, Q., & Li, X. (2015). Nanoemulsions as novel oral carriers of stiripentol: Insights into the protective effect and absorption enhancement. *International Journal of Nanomedicine*, *10*, 4937–4946. PMID:26261418

Lu, Y., Wang, Y., & Tang, X. (2008). Formulation and thermal sterile stability of a less painful intravenous clarithromycin emulsion containing vitamin E. *International Journal of Pharmaceutics*, *346*(1-2), 47–56. doi:10.1016/j.ijpharm.2007.06.003 PMID:17640833

Luengo, J., Weiss, B., Schneider, M., Ehlers, A., Stracke, F., König, K., & Schaefer, U. F. (2006). Influence of Nanoencapsulation on Human Skin Transport of Flufenamic Acid. *Skin Pharmacology and Physiology*, *19*(4), 190–197. doi:10.1159/000093114 PMID:16679821

Mahajan, H. S., Mahajan, M. S., Nerkar, P. P., & Agrawal, A. (2013). Nanoemulsion-based intranasal drug delivery system of saquinavir mesylate for brain targeting. *Drug Delivery*, *21*(2), 148–154. doi:10.3109/10717544.2013.838014 PMID:24128122

Mahour, R., Sahni, J. K., Sharma, S., Kumar, S., Ali, J., & Baboota, S. (2015). Nanoemulsion as a tool for improvement of Cilostazol oral bioavailability. *Journal of Molecular Liquids*, *212*, 792–798. doi:10.1016/j.molliq.2015.10.027

Mallick, S., Pattnaik, S., Swain, K., & De, P. K. (2007). Current Perspectives of Solubilization: Potential for Improved Bioavailability. *Drug Development and Industrial Pharmacy*, *33*(8), 865–873. doi:10.1080/03639040701429333 PMID:17729104

McClements, D. J. (2012). Nanoemulsions versus microemulsions: Terminology, differences, and similarities. *Soft Matter*, *8*(6), 1719–1729. doi:10.1039/C2SM06903B

Mei, L., Zhang, Z., Zhao, L., Huang, L., Yang, X.-L., Tang, J., & Feng, S.-S. (2013). Pharmaceutical nanotechnology for oral delivery of anticancer drugs. *Advanced Drug Delivery Reviews*, *65*(6), 880–890. doi:10.1016/j.addr.2012.11.005 PMID:23220325

Mostafa, D. M., Abd El-Alim, S. H., Asfour, M. H., Al-Okbi, S. Y., Mohamed, D. A., & Awad, G. (2015). Transdermal nanoemulsions of Foeniculum vulgare Mill. essential oil: Preparation, characterization and evaluation of antidiabetic potential. *Journal of Drug Delivery Science and Technology*, *29*, 99–106. doi:10.1016/j.jddst.2015.06.021

Muller, R. H., Gohla, S., & Keck, C. M. (2011). State of the art of nanocrystals—Special features, production, nanotoxicology aspects and intracellular delivery. *European Journal of Pharmaceutics and Biopharmaceutics*, *78*(1), 1–9. doi:10.1016/j.ejpb.2011.01.007 PMID:21266197

Park, J.W., Byun, Y.R., Choi, Y.K., Chang, K.Y., & Lee, J.B. (2021). *Current Oral drug delivery composition containing oxaliplatin and method for preparing same*. Academic Press.

Patel, M.R., Patel, R.B., & Thakore, S.D. (2018). Nanoemulsion in drug delivery. Applications of Nanocomposite. *Materials in Drug Delivery*, 667–700. doi:10.1016/B978-0-12-813741-3.00030-3

Pidaparthi, K., & Suares, D. (2016). Comparison of Nanoemulsion and Aqueous Micelle Systems of Paliperidone for Intranasal Delivery. *AAPS PharmSciTech*, *18*(5), 1710–1719. doi:10.120812249-016-0640-x PMID:27714701

Prakash, R.T., & Thiagarajan, P. (2011). Nanoemulsions for drug delivery through different routes. *Research in Biotechnology*, *2*(3), 1-13.

Ragelle, H., Crauste-Manciet, S., Seguin, J., Brossard, D., Scherman, D., Arnaud, P., & Chabot, G. G. (2012). Nanoemulsion formulation of fisetin improves bioavailability and antitumour activity in mice. *International Journal of Pharmaceutics*, *427*(2), 452–459. doi:10.1016/j.ijpharm.2012.02.025 PMID:22387278

Rai, V. K., Sharma, A., & Thakur, A. (2021). Quality Control of Nanoemulsion: By PDCA Cycle and 7QC Tools. *Current Drug Delivery*, *18*(9), 1–12. doi:10.2174/1567201818666210203180516 PMID:33538674

Sabir, F., Ismail, R., & Csoka, I. (2020). Nose-to-brain delivery of antiglioblastoma drugs embedded into lipid nanocarrier systems: Status quo and Outlook. *Drug Discovery Today*, *25*(1), 185–194. doi:10.1016/j.drudis.2019.10.005 PMID:31629966

Sachdeva, M.S., Patel, K., & Godugu, C. (2017). *Composition for improving boavailibility and efficcacy of taxane*. Academic Press.

Şenyiğit, T., Sonvico, F., Barbieri, S., Özer, Ö., Santi, P., & Colombo, P. (2010). Lecithin/chitosan nanoparticles of clobetasol-17-propionate capable of accumulation in pig skin. *Journal of Controlled Release*, *142*(3), 368–373. doi:10.1016/j.jconrel.2009.11.013 PMID:19932722

Shah, P., Bhalodia, D., & Shelat, P. (2010). Nanoemulsion: A pharmaceutical review. *Systematic Reviews in Pharmacy*, *1*(1), 24–32. doi:10.4103/0975-8453.59509

Sharma, S., Sahni, J. K., Ali, J., & Baboota, S. (2014). Effect of high-pressure homogenization on formulation of TPGS loaded nanoemulsion of rutin – pharmacodynamic and antioxidant studies. *Drug Delivery*, *22*(4), 541–551. doi:10.3109/10717544.2014.893382 PMID:24625264

Singh, Y., Meher, J. G., Raval, K., Khan, F. A., Chaurasia, M., Jain, N. K., & Chourasia, M. K. (2017). Nanoemulsion: Concepts, development and applications in drug delivery. *Journal of Controlled Release*, *252*, 28–49. doi:10.1016/j.jconrel.2017.03.008 PMID:28279798

Sivaramakrishnan, R., Nakamura, C., Mehnert, W., Korting, H. C., Kramer, K. D., & Schäfer-Korting, M. (2004). Glucocorticoid entrapment into lipid carriers—Characterisation by parelectric spectroscopy and influence on dermal uptake. *Journal of Controlled Release*, *97*(3), 493–502. doi:10.1016/S0168-3659(04)00169-5 PMID:15212881

Söderlind, E., Wollbratt, M., & Corswant, C. (2003). The usefulness of sugar surfactants as solubilizing agents in parenteral formulations. *International Journal of Pharmaceutics*, *252*(1-2), 61–71. doi:10.1016/S0378-5173(02)00599-9 PMID:12550781

Talegaonkar, S., & Negi, L. M. (2015). Nanoemulsion in drug targeting. In P. V. Devarajan & S. Jain (Eds.), *Targeted Drug Delivery: Concepts and Design* (pp. 433–459). Springer. doi:10.1007/978-3-319-11355-5_14

Tayeb, H. H., & Sainsbury, F. (2018). Nanoemulsions in drug delivery: Formulation to medical application. *Nanomedicine (London)*, *13*(19), 2507–2525. doi:10.2217/nnm-2018-0088 PMID:30265218

The Future of the Global Nanoemulsion Market to 2024: Rising Adoption of Targeted Therapeutics & Image-Guided Therapies. (2019). https://www.prnewswire.com/news-releases/the-future-of-the-global-nanoemulsion-market-to-2024-rising-adoption-of-targeted-therapeutics--image-guided-therapies-300873615.html

Tiwari, S. B., & Amiji, M. M. (2006). Improved Oral Delivery of Paclitaxel Following Administration in Nanoemulsion Formulations. *Journal of Nanoscience and Nanotechnology*, *6*(9), 3215–3221. doi:10.1166/jnn.2006.440 PMID:17048539

Veerareddy, P.R., Poluri, K., Sistla, R., Chaganty, S., & Venishetty, V.K. (2012). Formulation development and comparative pharmacokinetic evaluation of felodipine nanoemulsion in SD rats. *American Journal of PharmTech Research, 2*(3), 931–45.

Venkateshwarlu, I., Prabhakar, K., Ali, M., & Kishan, V. (2010). Development and In Vitro Cytotoxic Evaluation of Parenteral Docetaxel Lipid Nanoemulsions for Application in Cancer Treatment. *PDA Journal of Pharmaceutical Science and Technology*, *64*(3), 233–241. PMID:21502023

Verma, P., Meher, J. G., Asthana, S., Pawar, V. K., Chaurasia, M., & Chourasia, M. K. (2014). Perspectives of nanoemulsion assisted oral delivery of docetaxel for improved chemotherapy of cancer. *Drug Delivery*, *23*(2), 479–488. doi:10.3109/10717544.2014.920430 PMID:24901205

Vinarov, Z., Abrahamsson, B., Artursson, P., Batchelor, H., Berben, P., Bernkop-Schnürch, A., Butler, J., Ceulemans, J., Davies, N., Dupont, D., Flaten, G. E., Fotaki, N., Griffin, B. T., Jannin, V., Keemink, J., Kesisoglou, F., Koziolek, M., Kuentz, M., Mackie, A., ... Augustijns, P. (2021). Current challenges and future perspectives in oral absorption research: An opinion of the UNGAP network. *Advanced Drug Delivery Reviews*, *171*, 289–331. doi:10.1016/j.addr.2021.02.001 PMID:33610694

Viswanathan, P., Muralidaran, Y., & Ragavan, G. (2017). Challenges in oral drug delivery: a nano-based strategy to overcome. *Nanostructures for Oral Medicine*, 173–201. doi:10.1016/B978-0-323-47720-8.00008-0

Wani, T.A., & Masoodi, F.A., MahdiJafari, S., & McClements, D.J. (2018). *Safety of Nanoemulsions and Their Regulatory Status, Nanoemulsions, Formulation, Applications, and Characterization*. Academic Press.

Yoo, M. K., Kang, S. K., Choi, J. H., Park, I. K., Na, H. S., Lee, H. C., & Cho, C. S. (2010). Targeted delivery of chitosan nanoparticles to Peyer's patch using M cell-homing peptide selected by phage display technique. *Biomaterials*, *31*(30), 7738–7747. doi:10.1016/j.biomaterials.2010.06.059 PMID:20656343

Zhao, H., Lu, Gong, T., & zhang. (2013). Nanoemulsion loaded with lycobetaine–oleic acid ionic complex: Physicochemical characteristics, in vitro, in vivo evaluation, and antitumor activity. *International Journal of Nanomedicine*, *8*, 1959–1973. doi:10.2147/IJN.S43892 PMID:23723698

Zhao, Y., Moddaresi, M., Jones, S. A., & Brown, M. B. (2009). A dynamic topical hydrofluoroalkane foam to induce nanoparticle modification and drug release in situ. *European Journal of Pharmaceutics and Biopharmaceutics*, *72*(3), 521–528. doi:10.1016/j.ejpb.2009.03.002 PMID:19303438

KEY TERMS AND DEFINITIONS

Chemical Stability: It refers to the potential changes in composition in the product due to chemical reactions that may occur, such as oxidation, polymerization, and hydrolysis.

Drug Delivery: It refers to drug preparation, route of administration, site-specific targeting, metabolism, and toxicity are used to optimize efficacy and safety, and to improve patient convenience and compliance.

Flocculation: It is a reversible phenomenon in which droplets dispersed in an emulsion clustered after random collision.

Nanoemulsions: A colloidal particulate system in the submicron size and their sizes varied from 10 to 1.000 nm.

Polydispersity Index: A numeric value gives information about the droplet size dispersion in the nanoemulsion.

Surfactant: A compound that lower the interfacial tension between two liquids, between a gas and a liquid, or between a liquid and a solid.

Zeta Potential: It is the electrical potential at the slipping plane.

Section 2
Antimicrobial Nanoemulsion

Chapter 7
Antimicrobial Mechanisms and Mode of Actions of Nanoemulsion Against Drug-Resistant ESKAPE Pathogens

Karthikeyan Ramalingam
https://orcid.org/0000-0002-9334-427X
B.S. Abdur Rahman Crescent Institute of Science and Technology, India

Mohd Hashim Khan
https://orcid.org/0000-0002-0537-7902
B.S. Abdur Rahman Crescent Institute of Science and Technology, India

ABSTRACT

An enhancement of antibiotic resistance in bacteria is associated with increased morbidity, mortality, and health infrastructure and hospital care charges. The Infectious Diseases Society of America (IDSA) has highlighted a section of antibiotic resistant bacteria termed as ESKAPE pathogens. These pathogens are proficient in 'escaping' the biocidal effect of antibiotics and mutually representing new paradigms in transmission of diseases, pathogenesis, and resistance in their genetic materials. Essential oil-based nanoemulsions (NEs) have great interest towards the "natural" therapies as potential antimicrobial agents. Thermodynamic properties and kinetically stable potential of biphasic system of nanoemulsion enable them to be used as an effective nano-carrier with controlled release at the targeted point. This chapter describes the mechanisms of ESKAPE pathogens and the mode of the mechanisms of antimicrobial action of nanoemulsions for the treatment of MDR human pathogens.

INTRODUCTION

In recent years, antimicrobial resistance (AMR) is one of the major health concerns around the world. According to the World Health Organization (WHO), AMR poses a major threat to the treatment of

DOI: 10.4018/978-1-7998-8378-4.ch007

bacterial infections because the greater intake of drugs has led to an increase in the incidence of incorrect use of antibiotics leads to increased morbidity and mortality. Rapid increases in resistance and a downturn in the introduction of new antibiotic groups have severely limited existing therapeutic options for multidrug-resistant (MDR) pathogens (Gelband et al., 2015). The increase in the rate of drug resistance in bacterial infections and prevalence of MDR bacteria have been closely linked to the use of broad-spectrum antibiotics. Some Gram-positive and Gram-negative bacteria have developed resistance to not only commonly used antibiotics, but also to whole classes of antibiotics (Nikaido, 2009). A growing category of multidrug resistant human pathogenic bacteria mentioned as ESKAPE pathogens (*Enterococcus faecium*, *Staphylococcus aureus*, *Klebsiella pneumoniae*, *Acinetobacter baumannii*, *Pseudomonas aeruginosa*, *Enterobacter* spp), are capable of "escaping" the biocidal action of antimicrobial agents and commonly recognized as hospital acquired pathogens (Rice, 2008). Persistent use of antibiotics has provoked the emergence of MDR in pathogenic microorganisms including extended spectrum b-lactamase (ESBL) and carbapenemase producing bacteria, which render even the effectual drugs unsuccessful. According to the urgent need of new antimicrobials, these pathogens are describing three categories as critical, high and medium priority. Carbapenem-resistant *A. baumannii* and *P. aeruginosa*, as well as extended spectrum ESBL and carbapenem-resistant *K. pneumoniae* and *Enterobacter* spp are on the critical priority catalog of pathogenic bacteria, whereas vancomycin-resistant *E. faecium* (VRE) and methicillin- and vancomycin-resistant *S. aureus* (MRSA and VRSA) are in the list of high priority group of the need. Under the research investigation, ESKAPE pathogens showed mechanisms of multidrug resistance are broadly grouped into three evidences as, modification of the target site where the antibiotic may bind, drug inactivation by an irreversible cleavage that is catalyzed by the presence of enzyme, reduced drug absorption due to lower permeability or increased drug efflux system (Santajit and Indrawattana, 2016). The current therapeutic options for the treatment of multidrug-resistant pathogens have been severely limited. Collectively, due to the danger infectious nature of these pathogens resulted in the strong demand of new antimicrobials as alternative therapies that are currently in practice (Gelband et al., 2015).

A special interest has been developed by the scientific community to treat drug resistant infections, particularly those caused by ESKAPE pathogens through the use of nanoemulsions. Essential oils (EOs) are aromatic oily natured liquids obtained from plant material. The most active components of EOs include flavonoids, monoterpenes, phenyl propanoids and polyunsaturated omega-6-fatty acids (Franklyne et al., 2016). The medicinal properties of essential oil based nanodroplets usually in the range of 10–100 nm size, have received a great interest for the reason that of their low toxicity, kinetically stable properties, isotropic, transparent, pharmacokinetic activities and economic feasibility (Chouhan et al., 2017). The attempt to formulate a potential nanoemulsion with multiple potential targets or complex mechanisms as well as desirable biological and physicochemical properties, advanced the area of research to develop innovative antimicrobials. The cell membrane of bacteria made-up of phospholipid bilayers contains charged lipids with high binding efficiency and the potential for electrostatic attraction (Krishnamoorthy et al., 2018).

In this context, nanoemulsions have broad spectrum biocidal effect against bacteria particularly by disrupting their charged lipid layers of outer membranes through electrostatic attraction (Hamouda and Baker, 2000). Antibacterial nanodroplets of nanoemulsions are made up with charged lipids in their outer shell layers and a core stratum containing health-promoting bioactive compounds. The hydrophobic nature of the essential oils and their components is another important factor, which enables them to partition with the lipids present in the cell membrane of bacteria, making them more permeable by disrupting/ disturbing cell structures and their matrix arrangements. Consequently, the leakage of molecules, enzymes, ions and critical components of the cell ultimately results in the death of bacterial cell (Krishnamoorthy et al., 2018). The present chapter describes the drug resistant traits in ESKAPE pathogens, mode of the mechanisms of action of antimicrobial nanoemulsions and their role as delivery vehicles against the drug resistant pathogens. Therefore, nanoemulsions acquires antibacterial, antifungal, antiviral, antiseptic, antioxidant and insecticidal activities and have been screened on a global scale as potential sources of novel antimicrobial agents and alternatives to treat infectious diseases in humans caused by drug resistant ESKAPE pathogens.

PROCESS OF EMULSIFICATION

An emulsion is a colloidal suspension of two or more immiscible liquids stabilized by surfactants or emulsifier. The process of making emulsions is termed as emulsification. The combination of crude oil and water can shape an *oil-in-water emulsion*, wherein the dispersed phase is the oil, and water is the dispersion medium or a *water-in-oil emulsion*, wherein the dispersed phase is the water, and the oil is the dispersion medium, Figure 1 represents the process of emulsifications.

Figure 1. Schematic representation of two major types of emulsions: oil in water (O/W) and water in oil (W/O) emulsions

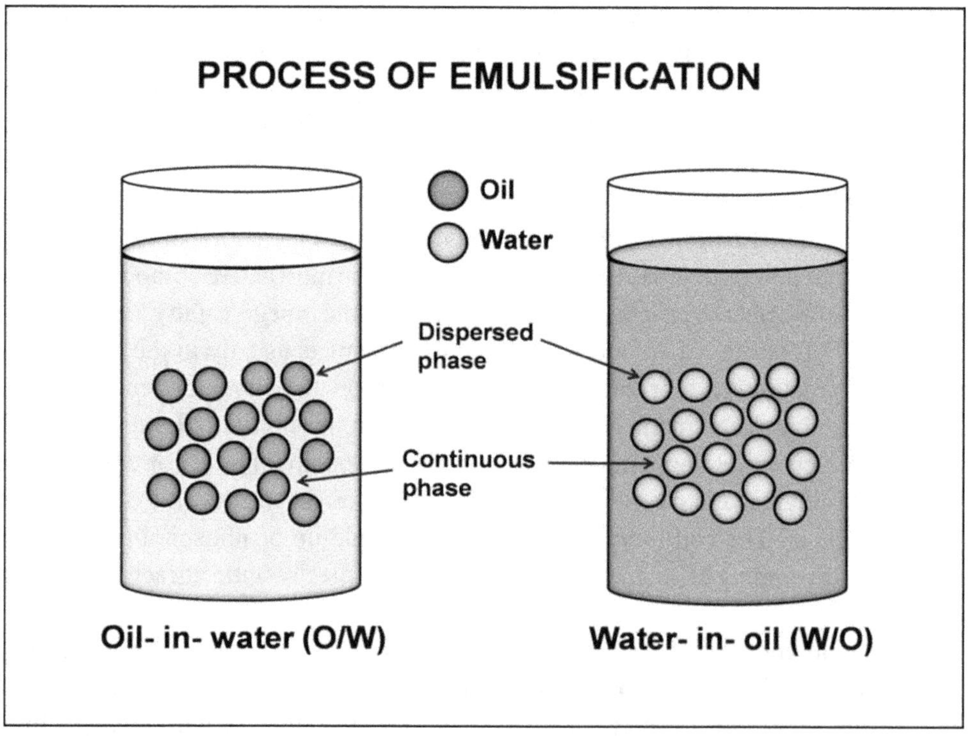

However on the basis of the size of the liquid droplets the emulsions can be divided into three categories as Macroemulsions, Microemulsion and Nanoemulsion. The difference in their properties between these emulsion types are illustrated and summarized in Table1.

Table 1. The differences between macroemulsion, microemulsion and nanoemulsion

Properties	Macroemulsions	Microemulsion	Nanoemulsion
Size range	1-100 μM	20-500 nm	10-100 nm
Shape	Spherical	Spherical, laminar	Spherical
Appearance	Turbid/Opaque	Clear	Clear/turbid
Preparation method	High and low energy	Low energy	High and low energy
Formation	Application of flow	Spontaneous	Application of extreme flow
Polydispersity	High (>40%)	Very low (<10%)	Low (10-20%)
Stability	Unstable	Thermodynamically stable, kinetically unstable	Thermodynamically unstable, kinetically stable

NANOEMULSIONS

Nanoemulsions are colloidal, transparent, kinetically stable nanocarrier systems of nanometric size (50 nm- 500 nm) that contain water and oil domains which is stabilized by a surfactant or co-surfactant. These are also known as miniemulsion, sub-micron emulsions or fine emulsions. A nanoemulsion thermodynamically unstable biphasic system having both the property hydrophilic and lipophillic heads stabilized with the addition of emulsifying agents which are generally crude substances (Rai et al., 2018). Essential and natural oil based nanoemulsions show higher antimicrobial properties due to non-phospholipid-based, stable, inexpensive and non-toxic effect in clinical relevance. They are widely used as effective antimicrobial agents against multi drug resistant human pathogens (Singh et al., 2019).

THE "ESKAPE" PATHOGENS

Nosocomial infections are in fact appropriately being recognized as a major patient safety matter for years but manageable problem. The Infectious Diseases Society of America (IDSA) highlighted a faction of pathogenic bacteria capable to 'escaping' the biocidal action of available antibiotics as a result acronymically- termed as 'ESKAPE' pathogens together representing new paradigms in resistance, transmission, pathogenesis and hospital acquired infections (Rice, 2008). The problem of antimicrobial resistance is not equivalently extended across all bacteria that infect hospitalized patients but among the pathogenic bacteria like ESKAPE pathogens are the main concern about the AMR now and the coming years (Rice, 2008). These pathogens rank among the most prevalent causes of life threatening infections and the ability to acquire antimicrobial resistance for longer persistence According to the updated data available from the National Healthcare Safety Network, ESKAPE pathogens are responsible for significantly more than 40% of infections in ICU patients (Weiner et al., 2016). These microorganisms have become more and more common and there are currently very few drugs under development that are efficiently inhibiting their growth. Immediate alert towards the AMR in pathogenic microorganisms help to researchers to focus in finding the solution and permit efficient serious evaluation of unmarked antimicrobial agents. The problem of virulence, resistance, transmission and pathogenicity in these microbes may escort to alternative strategies for the development of new antimicrobials based on nanotherapy. According to the most recent data from the Centers for Disease Control and Prevention (CDC), two-thirds of all health care-associated infections (HAIs) are caused by only this group of six assemblies of ESKAPE bacteria.

ANTIMICROBIAL RESISTANCE TRAITS IN "ESKAPE" PATHOGENS

Antimicrobial resistance (AMR) in human pathogens is one of the most health concerns in present scenario around the world. Innovative drug as a blessing to the mankind penicillin, started the era of antibiotics in 1928 and since then it has immensely extended modern medicine to the generations. Persistent use with greater intake of drugs, self-medication, and exposure to infections in hospitals has led to an increase in the incidence of incorrect use of antibiotics (World Health Organization 2014) and provoked the emergence of multidrug resistant (MDR) bacteria responsible for 15.5% Hospital Acquired Infections in the world (Rice, 2008; Allegranziet al., 2011; Pendleton et al., 2013). Antimicrobial resistance makes it difficult to provide successful health-care services to manage infectious diseases in the community. Drug-resistant pathogens negatively impact to the patients suffering from infectious disease as a rise in incidence of human infections by the same pathogens, the increase in AMR among the same strains with significant outcome; they cause infections with worse clinical results than sensitive strains, development of persister strains responsible for recurrent bacterial infections on antibiotic treatment to eradicate them (Michiels et al., 2016). The term "ESKAPE" introduced as the biocidal effect of antibiotics escaping group of bacteria responsible for the majority of nosocomial infections (Rice, 2008). These pathogens rank among the most prevalent causes of life-threatening infections due to the persistence of drug resistance genes that disseminate antimicrobial resistance in these bacterial cells. For the treatment of ESKAPE pathogens, the misuse and overdose of antibacterial agents such as antibiotics are the major cause behind the development of AMR. The ESKAPE pathogens have evolved resistance mechanisms against synthetic drugs and multiple antibiotics such as oxazolidinones, tetracyclines, macrolides, lipopeptides, fluoroquinolones, β- lactams, combination of β-lactam-β-lactamase inhibitor, as well as antibiotics which are the last line of protection including glycopeptides, polymyxins, carbapenems and lipoglycopeptides (Naylor et al., 2018). The novel β-lactamases inhibitor agents are capable to inhibit the effects of a wide range of antibiotics including, clavulanic acid, sulbactam, avibactam, vaborbactam and relebactam (Santajit and Indrawattana, 2016).The constitutive expression of the drug resistance mechanisms in these bacteria required a special attention to the physicians to prescribe the antibiotics with the background information of the drug for the patient. The collaborative endeavors and the knowledge of local susceptibility patterns in bacteria will require sustainable stewardship practices to reduce the improper utilization of antibiotics in both the human health and agricultural divisions (Holmes et al., 2016). Although ESKAPE pathogens represent the paradigm for disease transmission in both the community and clinical settings with increasing resistance trade, pathogenesis, required immediate attention at the genetic level and mechanisms surrounding ESKAPE pathogen emergence and broadly persistence in society.

The frequency of drug resistance in ESKAPE pathogens is increasingly encountered in the community as well as clinical settings. These pathogens are able to carry resistance genes, may be passed through the bacterial chromosome, transposons or plasmids leading to next generations. Mechanisms of drug resistance in ESKAPE pathogens (De Oliveira et al., 2020) fall into numerous categories and can be broadly classified into four groups:

[1] Inactivation or alteration of drugs/ antimicrobial agents
[2] Modifications of bacterial target site to escape drug bindings
[3] Reduction of intracellular drug accumulation/antimicrobial penetration
[4] The ability to form biofilms

ANTIMICROBIAL ACTIVITY OF NANOEMULSIONS AGAINST "ESKAPE" PATHOGENS

The progress in the formulation of novel antibacterial agent as nanoemulsions against ESKAPE pathogens is the urgent need of time to combat the drug resistance in human pathogens. Several researches have attempted to formulate the nanoemulsions with multiple potential targets and complex mechanisms. Nanoemulsions synthesized using natural oils have demonstrated potent antibacterial, antifungal, and antiviral efficacy against pathogenic microbes (Krishnamoorthy et al., 2021; Hobson et al., 2018; Sugumar et al., 2014). The nanoemulsions formulated with sunflower oil, clove oil, black seed oil, sage oil, thyme oil, glycerol monolaurate, monolaurin, carvacrol, ethyl oleate, tea tree oil and eucalyptus oil, reported highly susceptible to drug resistant ESKAPE bacteria, fungal spores and viral pathogens (Franklyne et al., 2016). Nanoemulsions have broad biocidal activities against drug resistant bacteria by disrupting the outer membranes of both the Gram positive and Gram-negative bacteria. Bacterial cell membrane comprises charged lipids with a high binding efficiency and electrostatic attraction potential while the droplets of nanoemulsions as antibacterial agent have charged lipids in their outer shell layers and a core portion that encloses health promoting bioactive edible complexes (Shaker et al., 2019).

Antimicrobial Nanoemulsions Against *Enterococcus faecium*

The human pathogen, *E. faecium* developed characteristics that enabled it to become a competent nosocomial pathogen. For successful prevention and reduction of enterococcal bacteremia with vancomycin resistant *E. faecium* (VRE) there is an urgent need to focus on reducing VRE and find the effective antimicrobial nano drugs to combat the hospital acquired infections.

The stable cinnamon oil nanoemulsions (CON) prepared by using soy protein and lecithin as natural emulsifiers with droplet size (141 nm) by using ultrasonic emulsification method. The minimum inhibitory concentration (MIC) and the minimum bactericidal concentration (MBC) of cinnamon oil nanoemulsions were determined and the results of (MICs/MBCs) were reported against *E. faecium* (600/2500 ppm), *S. aureus* (600/3500 ppm), *P. aeruginosa* (600/3500 ppm) and *E. coli* (700/3500 ppm). Investigation of the kinetics of microbial deactivation indicating that the synthesized nanoemulsions are significantly effective in deactivating *E. faecium* after 24 hrs incubation and completely inhibited the growth at concentrations and time periods (Ghani et al., 2019).

Antimicrobial nanoemulsions (NE-1, NE- 2, NE-3, NE-7, NE-9, NE-10, NE-12, NE-13, NE-14, and NE-15) were synthesized with the special combinations of surfactant (Triton X-100 Brij 30, Tween 60 and Span 20) Cetylpyridinium chloride, Mahua oil and ultra pure water to analyze the antimicrobial activities against ESKAPE pathogens. The results of antimicrobial activities of ten formulated NEs were reported as MIC and MBC. The maximum inhibitory action was reported at the dilution of 1:128 against *E. faecium* MCC 2763. Among all formulated NEs a marked NE-12, exhibited (19.07%) inhibition of pre formed biofilm by *E. faecium* MCC 2763 and showed a significant antibiofilm action to eradicate the matrix (Khan and Ramalingam, 2019).

The most important antimicrobial mechanisms of nanoemulsions is to damage the cell membrane of bacteria and rise in cell membrane permeability, thus Increased permeability decreases the integrity of the membrane and consequently increasing the leakage of intercellular contents to come out from the cell. When the antibacterial activity of CON against *E. faecium* and *E. coli* were evaluated and treated with nanoemulsion at the increasing concentration, a significant increase in the cell components release was observed compared to the control. Approximately 2- fold rise in the optical density of the bacterial cell culture indicating the greater damage of membrane integrity leading to the death of the bacterial cells. The efficacies of CON were also access as potassium leakage and nucleic acid from the bacterial cells (Moghimi et al., 2016).

It was observed that the increasing the concentration of CON increasing the leakage of potassium representing the maximum damage to the wall and increases cell permeability to potassium and nucleic acid leakage (Ghani et al., 2019). When the droplet of CON entered into the cell membrane, they interfere with the phospholipid bilayers integrity and active transports ultimately rupture the cell membrane. In parallel study, the antimicrobial activities of Peppermint oil nanoemulsion (PON) were access by using Agar well diffusion method. The 5% PON was dissolved in dimethyl sulphoxide (DMSO) and tested against *E. faecalis* NCDC 115. The standards of zone of inhibitions reported at three different concentrations 50, 75 and 100 µl with the values 7.4±0.32, 11.8±0.35, and 18.0±0.24 mm respectively (Mann et al., 2020).

In similar study, antimicrobial activities of lemon oil based nanoemulsion displayed strong antimicrobial effect and compared in terms of disc diffusion, MIC and MBC against E. faecalis and the results were reported as zone of inhibition 24.25 ± 0.5mm, and 12.50 and 25 mg/l respectively (Yazgan et al., 2019).

Antibacterial effect of a newly formulated Chx HCl nanoemulsion composed of 2% Labrafil, 12% Tween 80 and6% Propylene glycol with the particle size 12.18 nm. In terms of viable count, Chx.HClnanoemulsion1.6% shows the best cleansing effect and penetration ability due to the smaller particle size. The results of the high concentrations of Chx HCl nanoemulsion showed better cleansing ability and antibacterial effect by means of high efficacy against *E. faecalis* with high reduction rate or complete bacterial cells eradication (Abdelmonem et al., 2019).

Likewise, the Lemon essential oil has strong bactericidal activity against *E. faecium* with the MIC and MBC values reported as the same concentration 0.5µl/ml (Espina et al., 2011). Lemon, sweet orange, bergamot essential oils were also tested for their antimicrobial activity against vancomycin-resistant and vancomycin-sensitive strains of *E. faecium* and *E. faecalis* and found effective with the minimum inhibitory dose (MID) of 50 mg/l under a range of conditions for inhibiting the growth and survival of the pathogens (Fisher and Phillips, 2009).

Formulations of Nanoemulsions Against *Staphylococcus aureus*

S. aureus is common opportunistic pathogens that reside in skin abrasions, burns and wounds infection may lead to septicemia. Methicillin-resistant *Staphylococcus aureus* (MRSA) infections are common among immunocompromised patients and can appear surgical wounds or invasive devices including catheters or implanted feeding tubes. For the treatment of these drug resistant life threatening pathogens, nanoemulsion is important which possess a broad spectrum of activity against bacterial pathogens due to the reduced droplet diameter and hydrophobic nature of the oil system.

Essential oil based nanoemulsion comprising of eucalyptus oil, Tween 80, and water as continuous phase with the mean droplet diameter of 3.8 nm showed broad spectrum of antimicrobial activities against *S. aureus* (MTCC 96). The kinetics of killing experiment demonstrated complete loss of viability within15 min of interaction with the formulated NE. There was immediate log reduction (0.324 logcfu/ml) within a min when compare to control cells. Kinetics of killing experiment demonstrated that even the dilutions (10, 100, 1000 fold) of NE represent significant antibacterial activity against *S. aureus*. The cytoplasmic leakage of *S. aureus* was also studied between the range of 0 to 60 min and found that leakage of cell contents occurs immediately within a min (83.32%) compared to control bacteria (0.03%) on interaction with NE. The morphological changes were also observed with the help of SEM in treated cells of *S. aureus* with NE, the cells were remarkably disintegrated. After being treated with 10-fold diluted NE for a minute, the structure of cells and integrity were altered (Sugumar et al., 2014).

In similar studies, stable nanoemulsion was formulated with eucalyptus oil, Tween 80 and water content with the particle size 9.4 nm. The nanoemulsion-impregnated chitosan (NE-CH) film showed higher antibacterial activity than chitosan (CH) film alone. To access the antibacterial activities Agar disc diffusion method was employed. The elevated content of NE-CH (1%, 3%, and 5%) showed increasing inhibitory effect as zone of inhibition (7 mm, 11 mm, and 15 mm respectively) against *S. aureus*. The hydrophilic nature of NE-CH found to have an enhanced inhibitory zone of inhibitions (Sugumar et al. 2015). The standard plate count assay conducted with the varying percentages of NE-CH (1%, 3%, and 5%) against the wound isolate *S. aureus*. There were no viable cells were observed with 5% emulsion-impregnated films whereas 80% viable cells were observed with 1% and 3%, after 24-hour treatments. Turbidity test of the bacterial cell showed highly reduction in turbidity when treated with 5% NE-CH and found to have the highest morbidity and mortality agent. As a result the increasing concentration of essential oil mentioned effective antibacterial activity (Sugumar et al., 2015).

Non-toxic linoleic acid nanoemulsion (LA-NE) formulated to enhance the antibacterial activity against *S. aureus* and methicillin-resistant *S. aureus* (MRSA) and exhibited a 205-fold increase in the antibacterial activity than plain linoleic acid (LA). The MIC values of LA NE with a 1: 1 (w/w) ratio reported greater activity against *S. aureus* and MRSA as of 3 mg/ml and 1.5 mg/ml while the MIC values of LA found as 625 mg/ml and 312.5 mg/ml respectively. The TEM analysis of LA-NE treaded *S. aureus* and MRSA cells showed the disruption of the bacterial cell membrane and pointed out the main target site of the bacterial membrane (Sonawane et al., 2015).

Essential oil (*Minthostachys verticillata*) based nanoemulsion synthesized and evaluated for its antibacterial activity against *S. aureus* (ATCC, 209213). In plating serial dilutions assay, the interaction of NE with *S. aureus* shows a bacterial growth inhibition of 58.87%. A time killing kinetics assay was performed to estimate the viability of *S. aureus* upon interaction with EO and NE for 24 hrs. The significant ($p < 0.05$) reduction in bacterial growth was observed after 3 hrs (1.28 log UFC/mL) but a greatest effect was observed after 24 hrs of incubation as 5 log reduction growth survival (Cecchini et al., 2021).

A novel microbicidal nanoemulsion (NB-201) developed against methicillin-resistant *S. aureus* infection tested in *in vitro* and *in vivo* murine and porcine models infected with MRSA. Topical treatment with NB- 201 significantly decreased bacterial load with no toxic effects on healthy skin tissues. Antimicrobial activity of NB- 201 greatly reduces inflammation and effectively kills MRSA (Cao et al., 2017).

The formulation of NE from *C.viscosa* essential oil represented broad-spectrum antimicrobial activity against MRSA, which exhibited high bacterial cell penetration capability due to nanoscale bioactive droplets containing amphiphilic nature. The results of zone of inhibition assay demonstrated largest zone of inhibition (15.6 mm) during the treatment of NE against MRSA among other treated Gram-positive bacteria. The MIC of 25 µg/mL and the MBC of 40 µg/mL were observed for drug resistant pathogens. Fluorescence microscopic analysis revealed efficiently killing of NE-treated MRSA bacterial cells and exposed a potent bacterial efflux pump inhibitor (Krishnamoorthy et al., 2018).

The citral nanoemulsions (CNE) showed significant antimicrobial activities against *S. aureus*. The zone of inhibition assay screened the antimicrobial activity with zone size 19.2a ± 2.3 and represents the potency of the CNE against drug resistant *S. aureus* (Lu et al., 2018).

Garlic oil nanoemulsion (GON) with diameter size 36.3 nm was prepared with Tween-80 surfactant, propylene glycol as co-surfactant and Garlic oil used for the formulations. The antibacterial activity of GON investigated as the zone of inhibition assay and found effective against S. *aureus* as 12 mm zone of inhibition (Hassan and Mujtaba, 2019).

Antimicrobial Nanoemulsions for *Klebsiella pneumoniae*

K. pneumoniae is a pathogenic bacterium having the ability to form a thick layer of extracellular biofilm and representing one of the main agents in hospital arise infections. The aggregated *K. pneumoniae* as biofilm can attach to medical devices including endotracheal tubes and catheters, leading to the resistance and spreading infections. For the explorations of new strategies for possible treatment, researchers are focused on oils based nanoemulsions to fight against multi drug resistant bacterial infections.

Natural oil based nanoemulsion comprised of *Cleome viscosa* essential oil, Tween 80 surfactant and water content with droplet size 7 nm. The antimicrobial potential of the NEs were tested *in vitro* against different pathogens including *K. pneumonia*. Influence of structural factors on the antimicrobial activity accessed through the zone of inhibition assay against drug resistant pathogens. The results of the inhibition assay indicate that the NE formulation from *C.viscosa* essential oil showed broad-spectrum inhibition activity against *K. pneumonia* and produced the largest zone (12.4 mm) of inhibition. The MIC and MBC values of same NEs against *K. pneumonia* reported as 30 µg/mL and an MBC of 40 µg/mL respectively (Krishnamoorthy et al., 2018).

Essential oil based nanoemulsions developed with 5% (w/v) Polysorbate 80 at a total mass of 50 g, 5% (w/v) essential oil and 90% (w/v) water contents with the average droplet size range within 200 nm. The MIC carried out to access the antimicrobial activity of nanoemulsion to various bacterial pathogens including *K. pneumoniae*. Microdilution method utilized to evaluate antimicrobial efficacy and the MIC value 10µg/ml of the nanoemulsion was obtained against *K. pneumoniae* (MTCC-432) (Sundararajan et al., 2018).

In a similar study, nanoemulsion was developed with globule size of 151.7 nm, containing green tea catechins known as polyphenon 60 (P60) and ciprofloxacin (CF) to treat urinary tract infection. Ultrasonication technique used to encapsulate the Polyphenon 60 and ciprofloxacin in a single nanoemulsion system. Antibacterial activity evaluation against extended spectrum beta lactamase strains of *K. pneumoniae* revealed that nanoemulsions containing 4 and 10 mg/ml each of polyphenon 60 and ciprofloxacin effectively inhibited the growth of bacteria. The effect of nanoemulsions on *K. pneumoniae* was assayed by agar well diffusion method to determine the MBC. The findings with encapsulated nanoemulsions demonstrated that ESBL-producing *K. pneumoniae* showed the inhibition of growth at 1:5 and 1:2 dilutions containing 4 and 10 mg/ml of each P60 and CF, respectively (Atinderpal et al., 2018).

The antimicrobial activities of lemon oil based nanoemulsion (LON) and lemon essential oil (100% and 10%) were estimated in terms of disc diffusion, MIC and MBC against *K. pneumoniae* (ATCC700603). LON and 10% lemon essential oil exhibited higher inhibition zones with values of 17.88 and 24.25 mm, respectively while100% lemon essential oil showed strong resistance against *K. pneumoniae* and no activates demonstrated. Both of essential lemon oil and its nanoemulsion showed strong antimicrobial activities when compared to positive control group of tetracycline neomycin and streptomycin. Nanoemulsion displayed antimicrobial effect as growth inhibition MIC value 6.25 mg/mL and bactericidal effect as MBC values 25 mg/mL against *K. pneumoniae*. The result of paper disc diffusion assay reported as14.38 mm for *K. pneumoniae* and confirms the enhanced antimicrobial effect of LON (Yazgan et al., 2019).

Oil-in-water nanoemulsions containing lauryl acrylate (10% w/w), clove oil (5% w/w) and varying degrees of 1,6-hexanediol diacrylate (0, 1, 2% w/w) were prepared as nanocapsules by photopolymerisation method. Nanocapsules loaded with clove oil with varying degree of cross linking were assessed for their ability to inhibit growth of *K. pneumoniae*. To assess the ability of surface-coated nanocapsules for the reduction of bacterial (*K. pneumoniae*) binding, and treated with nanocapsules for 24 h, stained and were fixed. A fluorescence imaging analysis showed a clear reduction of binding on surfaces with the nanocapsules when compared to untreated controls. Clove oil encapsulated nanocapsules produced the best bacteria repelling surfaces of coverslips coated with nanoemulsions (Venkateswaran, et al., 2016).

Thymus vulgaris (TV) and Syzygium aromaticum (SA) essential oils based emulsion were selected for the preparation of chitosan coated nanoemulsions (NEs). The formulated NEs were evaluated for their *in vitro* antimicrobial efficacy. Chitosan coated NEs showed a potential and effect intranasal against multi-drug resistant *K. pneumoniae* strains. The TV-NE and SA-NE were tested for their antimicrobial efficacy by means of MIC and MBC which showed same MIC and MBC values (expressed in % v/v) against *K. pneumoniae* 0.03% v/v and same for both the nanoemulsions. The Time-Kill Studies of TV-NEs showed a remarkable anti-bacterial activity with a concentration-dependent effect and found rapidly bactericidal after 2 hrs of incubation at the concentration of 0.06% v/v whereas SA-NEs was bactericidal after 2 and 6 hrs at the concentrations 0.125% and 0.06% v/v, respectively (Rinaldi et al., 2020).

The nanoemulsion 0.25% aqueous-based chlorhexidine nanoemulsion (NM-Cl 0.25% w/v) developed to evaluate the antiseptic efficacy against *K. pneumoniae* carbapenemase (clinical isolate). Antimicrobial action of NM-Cl, evaluated by the broth microdilution method and the MIC reported (0.015 µg/mL) against *K. pneumoniae.* The time killing result demonstrates the residual effect of the NM-Cl. The 2x MIC concentrations were more effective to reduce the CFUs at first evaluation and the 4xMIC concentrations were more effective to maintain lower CFUs counting over 6 and 18hrs time (Risso et al., 2020).

Nanoemulsion containing a mixture of Tween 80(1% w/w), sage essential oil (10% w/w), and aqueous phase (89% of the total volume) with the mean particle diameter 204.4 nm, manufactured by an ultrasonic homogenization method. The sage essential oil nanoemulsion (SEN) displayed inhibitory impact with the MIC and MBC at the same concentration of 25 mg/ml against *K. pneumoniae,* which indicates inhibitory effect. It was also observed that SEN produces an effective zone of inhibition as 11.50 mm and revealed that the antimicrobial efficacy of SEN is strongly related to essential oil compounds, tested pathogens, size and their formulation (Yazgan, 2020).

Antimicrobial Action of Nanoemulsions Against *Acinetobacter baumannii*

The pathogen, *A.baumannii* is an important MRD opportunistic bacterium, documented as the causative agent of nosocomial infections including pneumonia, bacteremia, UTI, surgical site infection and wound/

skin infections in hospitalized patients. Increasing antibiotic resistance in pathogen demands a novel broad-spectrum antimicrobial therapeutic alternative. The developments of nanoemulsions have broad antimicrobial activity against MRD bacteria to defeat the MRD pathogens.

The combination of five oil-in-water nanoemulsions (N1, N2, N3, N4, N5) were prepared with different concentrations of surfactants (Triton X-100, Tween 60, Tween 80) chelating agents, CPC and soybean oil. The solution mixtures were passes two times for emulsification at 20,000 lb/in^2 by using a high energy Microfluidization method. Nanoemulsions were serially diluted to evaluate the antimicrobial activities as MIC and MBC against four strains of *A. baumannii* ATCC (strains ATCC 19606, ATCC 19003, ATCC 17961 and ATCC BAA-1605) by using microdilution methods. The nanoemulsions N1, N2, N3 and N4 not reported as significant antibacterial agent but N5 showed excellent antibacterial effect against all four strains at higher dilutions with MIC value 1:2,187 for all, and MBC dilutions values ranges 1:729 to 1:2,187. The kinetics of killing activity of N5 evaluated at dilution ranges of 1:500 to 1:3,333 for the time duration 1, 15, 30, and 60 mins. The viability of all bacterial strains reduced by 2 logs on dilution 1:3333, at the completion of 30 mins time exposure with N5. The increasing concentrations (0, 10%, 20%, 30%, 50%, and 70%) of ethanol mixed with the dilutions (1:1,333, 1:2,000, 1:4,000, and 1:8,000) of N5. Evaluation of killing effect (CFU/ml) of N5 with 1:4,000 dilutions and 30% ethanol resulted in a 1 to 3-log increase in killing effect, against all four bacterial strains tested. In order to evaluate the antibiofilm activity of N5 with the dilutions (1:400, 1:1,000, and 1:2,000) against *A. baumannii* were measured utilizing an alamar Blue cell proliferation assay kit. The sound effects of N5 (1:1,000 dilution) reported as 80% lost of metabolic activity and 90% loss of metabolic activity at increasing concentration (1:400 dilution) within 1 hr. SEM analysis of treated biofilms of antibiotic-resistant strain (*A. baumannii* ATCC BAA-1605) showed disruption and dispersion of the bacterial biofilms with a clear reduction (Hwang et al., 2013).

Alhagi maurorum essential oil based nanoemulsion containing chitosan, prepared with the mean droplet size 172 ± 4 nm and Zeta potentials +28.6 mv by using the ionotropic gelation method. The antimicrobial activity of the NEs evaluated through methods. The MIC and MBC of nanoemulsion were determined by using the macro-broth dilution method and compared with *A. maurorum* essential oil against *A. baumannii* MCCKM345. The MIC and MBC values of nanoemulsion and essential oil were reported as 12.5, 25mg/ ml and 25, 50mg ml respectively and accomplished the NEs inhibited bacterial growth at a lower concentration with no colony formation. The Zone of Inhibition assay evaluates the sensitivity of the NEs and essential oil, the values of zones of inhibition calculated as 29±0.6mm 15±0.3mm, respectively. Chitosan loaded NEs showed great antibacterial activity than free essential oil against *A. baumannii.* The inhibitory effects of NEs assessed for prevention of biofilm formation, its destruction and inhibition of biofilm metabolic activities. A great destruction of biofilms happened when treated with increasing concentrations (12.5mg/ml) of NEs and reported as 90% loss of the *A. baumannii* biofilm formation. The effect of NEs on curing and deletion of plasmid also reported with two different plasmids of *A. baumannii* and completely cured after treatment (Hassanshahian et al., 2020).

The o/w nanoemulsion contain surfactant Brij 30(10% v/v), EDTA (65% v/v), soybean oil (25% v/v) and water prepared by using high energy microfluidization method. The particle sizes of nanoemulsion measured as 295.3 nm and 5.5 nm for deferiprone (DF) and 293.0 nm for combined (DF/NE). The antimicrobial activities of deferiprone (NE/DF) recoded as bactericidal concentrations (MBCs) against four strains S1, S2, S3, S4 of *A.baumannii* (BAA- 1605, ATCC 17961, ATCC 19606, ATCC 19003) reported as (320 dilution / 3.13mM), (213 dilution / 5.21 mM), (427 dilution / 2.6 mM), (320 dilution / 3.13 mM) respectively. The synergistic effect of nanoemulsion and deferiprone demonstrated as an increased efficacy up to 2- to 3.3-fold. The inhibitory effect of NE/DF on adherence of *A. baumannii* strains (S1, S2, S3, and S4) to glass surfaces demonstrated in higher inhibition of the growing cells and the adhesion of all strains significantly (p<0.05) affected with combined one (NE/DF). A range of inhibition of adhesion of all *A. baumannii* strains (S1, S2, S3, and S4) by combined NE/DF reported as 21.5 to 83.7%, 27.9 to 82.9%, 48.4 to 84.9% and 29.1 to 83.6% respectively. Biofilm viability as Live and dead assay with NE/DF showed a significantly increased dead cell ($p < 0.05$). The combined effect of NE and DF reported as 1.69-fold increase in killing of *A. baumannii* strains. SEM observations reported the morphological changes in pili-bonded intact cells and surfaces were disintegrated with highly reduced cell numbers. A direct effect of exposure to NE, DF and NE/DF on biofilms collapsed the matrix and causes the irregulation in the boundaries of nucleus and cell wall of *A. baumannii* strains (Ramalingam and Lee, 2019).

Cinnamon oil nanoemulsion (CNEs) prepared by mixing cinnamon essential oil, 5% tween 80 as a surfactant and deionized water through ultrasonication method. They investigated the antimicrobial activity of CNEs alone and in combination with different antibiotics against MRD strain of *A. baummannii* H72721. The percentage of polydispersity and average particles diameter size of the CNEs were measured as 25.73±5 nm and 9.35±2.5 nm, respectively. The MIC of gentamicin, kanamycinc and ampicillin alone were reported as 150 µg/ml, 200 µg/ml and 700 µg/ml respectively. The antimicrobial properties of the CNEs were examined between the concentration ranges of 0% to 0.15%, the total eradication of the bacterial growth found at the concentration of 0.09% against *A. baumannii*. The synergistic effect of CNEs with either gentamycin or kanamycin, but not ampicillin, showed a significant increase in their antimicrobial action in comparison of alone effect. Interestingly, the results of combined effect of gentamycin or kanamycin with CNEs suggested the strong synergistic potential as increased to 22% and 72% respectively, (Sabir and Sidiq, 2019).

The composition of nanoemulsion formulated by high energy Microfluidization method with (25% v/v) soybean oil, (10% v/v), Brij 30, and (65% v/v) 60mM EDTA added with deionized water for the peroration. The distribution of droplets size measured with a mean diameter of 269 nm for the synthesized NEs. Antimicrobial activity of nanoemulsions against *A. baumannii* evaluated MBC by serial dilution methods. The MBC of NEs for *A. baumannii* S1 (225 dilutions), S2 (158 dilutions), S3 (125 dilution) and S4 (175 dilutions) were recorded with no visible colonies observation. For adherence activity, the plates were incubated aerobically for 24 h with slanted glass cover slips at different concentrations of NEs with *A. baumannii* strains (S1, S2, S3 and S4). The inhibition of adherence reported as S1- ATTC BAA-1605 (82.8–99.8%), S2- ATTC17961 (95.9–99.9%), S3- ATTC19606 (61.8–92.9%), S4- ATTC19003 (74.1–93.9%) on different dilutions (10, 50, 100, 500, 1000 and 2000) used as anti-adherence agents. In biofilm viability test, the glass coverslips with grown biofilm treated with NE at 1, 10,100, 250 and 500-fold dilution for 1 hr and the inhibition of biofilm formation recorded against *A. baumannii* S1, S2, S3 and S4 as 59.9–92.0%, 63.3–82.0%, 58.1–74.1% and 56.2–87.7% respectively. SEM examination represented significant morphological changes in *A. baumannii* caused by the exposure of NE treatment. Biofilms of *A. baumannii* treated with NE demonstrated clear view of the disintegration and collapse of the matrix (Ramalingam and Lee, 2018).

Thymus daenensis essential oil based nanoemulsion prepared containing a non-ionic surfactant (Tween 80) and lecithin as co-surfactant by using ultrasonication method. The particle size of the formulated nanoemulsion measured as 171.88 nm. The antibacterial activities of nanoemulsion and pure oil analyzed against 10 MDR clinical isolates of *A. baumannii*. MICs and MBCs of both were determined by using broth microdilution method in a 96-well microtiter plate. Microdilution assay represented the MIC values of nanoemulsion and pure oil as 50 µg/mL and 125 µg/mL, respectively and MBC values of nanoemulsion and pure oil as 200g/mL and 2000 µg/mL, respectively against all10 MDR clinical isolates of *A.baumannii*. The antibiofilm activity of *T. daenensis* nanoemulsion and pure oil screened against most resistant isolate of MDR *A. baumannii.* The sub lethal concentration *T. daenensis* nanoemulsion and pure oil (1/2MIC) showed the best anti-biofilm activity as 56.43% inhibition and 20.65% inhibition respectively, after 24 h of incubation (Moghimi et al., 2018).

Synthesis of Antimicrobial Nanoemulsions for *Pseudomonas aeruginosa*

P. aeruginosa is common opportunistic pathogens implicated in a broad range of infections and are able to form biofilms which resist multiple antibiotics leading to serious infections in hospitals settings. Serious infections associated with including, endocarditis, meningitis, keratitis, UTI, folliculitis, acute/chronic respiratory infections and wound infections causing a high mortality rate in patients. The development of resistance to the multiple antimicrobial agents has been the major concern, in human pathogens; nanoemulsions can be an effective alternative to the MDR *P. aeruginosa*, as they are efficient and safer to the human health.

A novel formulation of eugenol nanoemulsion (ENE) containing 4% eugenol, triglyceride, 8% Tween 80, and 88% saline buffer as aqueous phase, prepared through titration. The particle-size distribution measured in the ranges of 10.59 to 11.12 nm as a stable emulsion. The antimicrobial efficacy of ENE access through zone of inhibition assay and the diameter of the inhibition zone measured as 1.20 cm. For expression of QS genes, *P. aeruginosa* investigated at various concentrations of ENE. The highest concentration (0.2 mg/mL) eugenol and ENE inhibited 3-oxo-C12-HSL production approx 33% and 50%, respectively, whereas the inhibition of C4-HSL production reported as 30% and 40%, respectively in *P. aeruginosa*. The inhibitive effect of eugenol and ENE on the expression of the QS synthase genes responsible for the production of 3-oxo-C12-HSL and C4-HSL reported as concentration reliant. At 0.2 mg/mL concentration of eugenol and ENE, the effect on the expression level of the *lasI* synthase gene calculated as 65% and 52%respectively, whereas the *rhlI* synthase gene expression calculated as 61% and 45%, respectively. The effect of eugenol and ENE on pyocyanin production by *P. aeruginosa* investigated at different concentrations. At a concentration of 0.2 mg/ mL of eugenol and ENE inhibited pyocyanin production by 28 and 48%, respectively. *P. aeruginosa* grown in the presence of eugenol and ENE showed a significant reduction in QS-mediated swarming motility demonstrated as 45% and 68%, respectively at 0.2 mg/ mL concentration. Eugenol and ENE inhibited QS mediated rhamnolipid production and its related gene (*rhlA*) expression in *P. aeruginosa* at concentration dependent manner. At the concentration of 0.05 mg/mL, eugenol and ENE, the expression level of the rhlA gene reported as 83% and 74%, respectively. The effect of eugenol and ENE on the formation of biofilm by *P. aeruginosa* assessed and inhibited biofilm formation at concentration-dependent manner with the greatest inhibition 36 and 63%, respectively at a concentration of 0.2 mg/mL. Confocal laser scanning microscopy analysis results for the biofilm inhibition proved a potential candidate for the development of novel agents against QS dependent bacteria (Lou et al., 2019).

Neem oil nanoemulsion (NOE) formulated with Tween 20 (surfactant), neem oil (1:3 ratio) and deionized water as aqueous phase by using sonication method. Antimicrobial activity of NOE carried out by using well diffusion and disc diffusion methods. In well diffusion 40 µg/well and disc diffusion 40µg/disc analysis NOE added to check the sensitivity of nanoemulsion against *P. aeruginosa*. The results showed that NOE treated wells and loaded disc produced zone of inhibitions as 25 mm and 35 mm respectively. The histopathological examination of fish organs (*L. rohita*) infected by *P. aeruginosa*. Histopathological examination revealed a severe damage due to microbial pathogenicity in the normal structure of lamellae in fishes. As a result, 90% survival rate of infected fishes observed when intramuscularly, 25 μL CON injected for the treatment of *P. aeruginosa* infection in fishes (Mishra et al., 2014).

Argentinean oregano essential oil (OEO) based four nanoemulsions (NEs) developed as EONE, EONE2, EONE3 and CNE (control) with composition of 3% soy lecithin, 16% oil phase and 81% water phase and the mean particle sizes measured as 131.7, 89.17, 74.7 and150.77 respectively. Antimicrobial activity of NEs accessed against *P. aeruginosa* ATCC 14213 as MIC by using a broth microdilution assay. Among all EONE3 showed greater inhibitory effect against *P. aeruginosa* at the concentration 2-mg/ml after the completion of 9 hrs time period. However the demonstration of the results with NEs revealed that highly lipophilic compounds penetrate easily through the outer membrane of several pathogenic bacteria (Asensio et al., 2020).

The nanoemulsions (EONE) prepared containing 5% *E. globulus* oil, 2% polysorbate 80 (Tween 80) and 25 mL ultrapure water employing the high stirring emulsification method. The EONE presented mean values of droplets 76 ± 0, 77 nm. Antimicrobial activity of EONE determined as MIC by using the broth microdilution technique and disc diffusion test against *P. aeruginosa* PA0. Inhibitory effect as MIC evaluated 114.125 mg/mL and zone of inhibition reported as 24 mm against *P. aeruginosa* (Quatrin et al., 2017).

Non-cationized nanoemulsion (NCN) formulated with the help of analogous method, containing Capmul MCM C8 (CMCM8), Labrasol (LBS) and Cremophor RH40 (CRH40) in the ratio of (1:1:2). Similarly, the cationized nanoemulsion (CNE) also formulated with the inclusion of 0.5% w/w of Oleyl amine (OA). The antimicrobial activities of NCN and CNE evaluated by Agar diffusion and cytoplasmic release studies against *P. aeruginosa*. Zones of inhibition exhibited by NCNs and CNEs as 10.3 ± 0.6 and 14.3 ± 1.15 mm respectively against *P. aeruginosa*. The MIC of NCNs and CNEs evaluated using microdilution method for *P. aeruginosa* and reported as 11.37 and 5.70 mg/mL, respectively. Killing kinetics of CNEs (at 60 min) represented higher decline as compared to NCN (15 min) with no observed viable colonies of *P. aeruginosa* (Singh et al., 2015).

Oil-in-water nanoemulsions with different combinations formulated, using non-ionic surfactant Tween 20, neem oil and water by using Ultrasonication method with the mean droplet size 31.3nm. The formulated neem oil nanoemulsion tested on *Pseudomonas aeruginosa* by agar dilution plate count method and kinetics for killing. Antibacterial results indicate the effective zone of inhibition and killing rate against bacterial growth (Deeksha et al., 2019).

A stable black pepper oil nanoemulsion (BPON) prepared with the mixing of black pepper oil and surfactant (Tween 80) at the ratio of 1:2 by using high-pressure homogenization method. Average mean droplet size of the nanoemulsion measured as 30.3±2 nm. The antimicrobial susceptibility of BPON tested against the pathogenic isolate *P. aeruginosa* by micro-broth dilution and well diffusion methods. Treated bacterial strain exhibited pronounced inhibition on the growth at the MIC concentration of 0.234μl/ml and the zone of inhibition diameter of 16.6±1.2 mm for *P. aeruginosa*. SEM analysis confirmed the BPON induced membrane damage when interacted with cells. The treated cells showed pronounced morphological changes and appeared to disintegrate with irregular arrangement as a result cells exhibited significant damage on the cell surface. Qualitative estimation of biofilm formation performed for the detection of biofilm inhibition potential at different concentrations (0.014, 0.058, 0.234, 0.937, 3.75,15 and 60 μl/mL) of BPON against *P. aeruginosa*. A significant reduction ($p<0.05$) in biofilm formation in the presence of nanoemulsion observed at the concentration of 0.234 μl/mL, indicates the strong inhibition of biofilm formation (Swathy et al., 2018).

The nanoemulsions prepared from 3% Satureja khuzistanica essential oils (SKEO), 9% surfactants (Tween80 and Span80), 10%hydrophilic lipophilic balance (HLB) and 88% water with two EO constituents (carvacrol and 1,8-cineol) prepared by mixing at high speed homogenizer. Mean droplet size of SKEO measured by DLS as 84.87 ± 1.20 nm. Antimicrobial efficacy of NEs and the pure EOs determined against *P. aeruginosa* PAO1 as MIC by a broth micro-dilution method and reported as 8µg/mL and 64µg/mL respectively represented greater antibacterial activity than bulk oil. The anti-biofilm properties of SKEO evaluated by applying 1/2 MIC concentrations, the results of biofilm inhibition assay revealed the best anti-biofilm effect as 89.53±4.88% for *P. aeruginosa* PAO1 biofilm inhibition. For the evaluation of biofilm eradication, biofilms were grown in 96-well plates for 24 hrs and treated with SKEO, reported as 56.65% eradication in established biofilms formed by *P. aeruginosa* PAO1. Another biofilm generated on the glass pieces in the presence of sub-MIC concentration of SKEO. The results of SEM analysis showed a great disruption in the condensed matrix of *P. aeruginosa* PAO1 biofilms. Pyocyanin producing *P. aeruginosa* PAO1 cells exposed to concentration of 1/2 and 1/4 MIC of SKEO and incubated for 24 and 48 hrs. The maximum inhibitory effect of SKEO reported as 89.44 ± 3.41% inhibition of pyocyanin production at concentration of 1/2 MIC in 24 hrs with statistically ($P < 0.05$) significant data (Ghaderi et al., 2020).

Antimicrobial Nanoemulsions Against *Enterobacter* Species

The Enterobacter species are considered opportunistic pathogens and increasing cause of nosocomial infections belonging to the family Enterobacteriaceae. The species of *E. aerogenes* and *E. cloacae* are the complex pathogens associated with intensive care unit patients to cause bacteremia, UTIs, osteomyelitis, endocarditis, lower respiratory tract infections and may be transmitted via contaminated infusion of blood products as well as linked with hospital required instruments or devices. Natural oil based nanoemulsions possess bioactive compounds that act to protect from chronic infections, although it can be exploited and employed by humans to increase the bioavailability of drugs as well as nanomedicine to treat the drug resistant human pathogens including Enterobacter species.

Ten antimicrobial nanoemulsions (NEs) formulated for the treatment of multi drug resistant ESKAPE pathogens. All the NEs prepared as oil-in-water by using different combination of (3%) surfactant (Tween 60, Brij 30, Triton X-100, Span 20), (10%) Mahua oil, (1%) Cetylpyridinium chloride (CPC), and sterile ultra pure water. For emulsification, the mixed phase solutions subsequently passed with three rounds at 20,000 lb/in^2 by using a high energy Microfluidization method. The droplet size of each synthesized NE1, NE2, NE3, NE7, NE9, NE10, NE12, NE13, NE14 and NE15 measured as an average size of 142.00 nm, 106.00 nm, 106.00 nm, 7.53 nm, 106.00 nm,37.80 nm, 106.00 nm, 78.80 nm, 91.30 nm and 106.00 nm respectively. Synthesized NEs evaluated for their antimicrobial activity through micro-dilution method. Antimicrobial efficacy of NEs determined by means of MIC, MBC, and antibiofilm studies against ESKAPE pathogens including *E. aerogenes* MTCC 111. The formulation of NE1, NE2, NE3, NE7, NE9, NE10, NE12, NE13, NE14 and NE15 presented effective growth inhibition as MIC at dilutions 1:4, 1:4, 1:6, 1:4, 1:4, 1:4, 1:4, 1:4, 1:4, 1:12 and MBCs at dilutions1:3, 1:3, 1:4, 1:2, 1:2, 1:2, 1:2, 1:2, 1:2, 1:8 respectively, against *E. aerogenes* MTCC 111. Based on the preliminary results of antimicrobial effect, four NEs tested to estimate the antibiofilm activities formed by ESKAPE pathogens including *E. aerogenes* MTCC 111. Selected NEs showed significant antibiofilm effect during post treatment of NEs to the biofilm formed for 72 hrs. The NE2, NE7, NE12 and NE13 remarkably expressed biofilm inhibition and reported as 89.23%, 54.48%, 53.12% and 54.26% respectively against *E. aerogenes*. The synthesized NEs disrupted the integrity of existing biofilms and lowering the metabolic activities of pathogens which indicates the antibiofilm potential of the nanoemulsions (Khan and Ramalingam, 2019).

Nanoemulsion (o/w) formulated with oil phase (olive oil, squalene, Triton X100, and polysorbate 80) and aqueous phase (0.1%) polyvinyl alcohol or (0.1%) polyethylene glycol) prepared by ultrasonication method. Finally obtained NESH fabricated by using clarithromycin, olive oil, polysorbate 80, polyvinyl alcohol and water. The droplet size of prepared NEs ranged from 68 to380 nm. For all six different formulated NEs as NESH 01, NESH 02, NESH 03, NESH 04, NESH 05 and NESH 06 evaluated for their droplet size as 68, 125, 100,220,380 and 293 nm respectively. Antibacterial activity of the fabricated nanoemulsion determined by the zone of inhibition assay against MDR human pathogens including, *Enterobacter aerogenes* MTCC 111. The zones of inhibitions reported as 2.2 cm against *E. aerogenes* MTCC 111. Synthesized NESH demonstrated great solubility of antibiotics clarithromycin hence, increased bioavailability of the antibiotic in a nanoemulsion system (Vatsraj et al., 2014).

The study with Acnano (a polyherbal nanoemulsion) procured from Venus Medicine Research Centre, Baddi (India) and Adapalene used as a control to compare the efficacies against acne causing microbes including *E. cloacae* MTCC 441 and *E. coli* MTCC 739. The estimation of Acnano nanoemulsion and Adapalene as antimicrobial agents estimated as zone of inhibition against *E. cloacae* (23.39±0.23 and 11.23±0.10 mm) and *E. coli* (19.29±0.16 and 13.99±0.17 mm) respectively. A polyherbal nanoemulsion against *E. cloacae* found to be two times effective than that of Adapalene (a positive control) during the same applications (Naithani et al., 2010).

A nanoemulsion containing an aqueous phase including (4.5% w/w) Tween 80 with distilled water, (6% w/w) Z. multiflora essential oil and fortified with (6% w/w) cinnamaldehyde and cinnamaldehyde, mixed with the homogenization method. The particles size of fortified Z. multiflora essential oil nanoemulsion (ZEON) measured as 184.7 nm. Antimicrobial efficacy estimated against Enterobacteriaceae bacteria through Plate count agar method for the total viable counts and counts reported as log10 CFU/g. In initial count for the microbial flora of chicken meat counted as 4.30 log10 CFU/g with no statistically significant difference but the count increased to the final value of 10.94 log10 CFU/g on 20th day of storage chicken meat. Eventually the chicken treated with NZEO reduced the Enterobacteriaceae count about 2.8 log10 cycles at the end of storage (Abbasi et al., 2020).

Thymol nanoemulsion (TNE) composed of thymol essential oil, sodium alginate, surfactant-Tween 80 and ultra pure water homogenized by using Ultrasonication method. Formulated TNE accessed for microbiological analyses, the birds slaughtered and cecal contents were aseptically removed, weighed, homogenized and serially diluted for the count of *Enterobacteriaceae* bacteria by Plate Count Agar method. Birds meat challenged with TNE, significantly reduced ($p < 0.05$) *Enterobacteriaceae* population by 2- 2.6 log10CFU/g and also it increased the number of lactobacillus copies by 2 log10CFU/g after 22 days of storage (Ibrahim et al., 2021).

A microemulsion (ME) composed of 23.5% Cremophor EL, 30% ethyl decanoate, 12.5% transcutol and 34% distilled water. The morphological structures of the ME characterized as spherical and their droplet diameter measured as 15.55 ± 3.17 nm. Antibacterial activity of ME estimated by using the agar well diffusion assay and reported as 12.6 ± 0.9 mm against *Enterobacter* spp human pathogenic bacteria (Alkhatib et al., 2016).

The chitosan-loaded nanoemulsion prepared from a mixture of (4% w/v) *Zataria Multiflora* Boiss (*ZEO*) and *Bunium persicum* Boiss (BEO) with (4% w/v) Tween 80, (0.01% w/v) lecithin and homogenized by using high-intensity sonication method. *In vitro* evaluations performed to evaluate the effect of prepared nanoemulsion using microdilution method to count the bacteria. The highest reduction rate of total viable count of *Enterobacteriaceae* bacteria reported as 2.51log CFU/g in meat sample stored for 15-18 days and significantly increases the shelf life of treated meat (Keykhosravy et al., 2020).

MECHANISMS OF ANTIMICROBIAL ACTION OF NANOEMULSION

Nanoemulsions containing natural/essential oils generally cause the damage of bacterial cell wall or cell membrane resulting in complete cell disruption and disorganization. The increase in cell membrane permeability and partial solubilization of the cell membrane by fatty acids, leading to release of membrane proteins which leads to uncoupling of oxidative phosphorylation or enzyme and nutrient uptake inhibition causing bacteriostatic phenomenon in pathogens (Desbois and Smith, 2010). The nanoemulsion system promotes interaction with the microbial cell membranes via four key connections:

(i) The increased surface area of nanodroplets increases passive transport across the outer cell membrane leads to enhance contact with cytoplasmic membranes.
(ii) The fusion of the high energy loaded nanodroplets with the phospholipid bilayers of the microbial cell membrane, most likely enhances the targeted discharge of the essential oils with high energy release at the specific sites.
(iii) Constantly release of essential oils preserved by nanodroplets adds the prolonged effect on the targeted sites at viable cells. The driven force of detachment among aqueous phase and oil phase of nanodroplets persisted over time to leave their effect on microbial cells.
(iv) The electrostatic interaction of negatively charged microbial cell wall/membranes with positively charged nanoemulsion droplets enhances the absorption of essential oils at the site of stroke.

To understand the mechanism of action of emulsions, the investigation for cell wall damage have been determined by the analysis of release of UV-absorbing substances, protein release, changes in intracellular potassium ion concentrations (Sugumar et al., 2014). Ultimately the visualization of these events through electron microscopy demonstrates the mode of the mechanisms of actions of antimicrobial nanoemulsions in microbes (Moghimi et al., 2018). There are two major modes of antimicrobial action of nanoemulsions have been hypothesized and described in previous investigations for the possible mechanisms: **(1)** Charge based antimicrobial action of nanoemulsions **(2)** High pressure explosion based action of nanoemulsions.

(1) Charge Based Antimicrobial Action of Nanoemulsions

The microbicidal NEs are surfactant-containing emulsions which have been shown to be non-toxic to animals but most effective against many pathogenic bacteria. The mechanism of action of nanoemulsions has been proposed that due to the charged based interactions, the nanodroplets come with the contact of outer membrane of the microorganism, destabilizing the organism's lipid envelope and initiating its disruption.Charge based interactions of positively charged polymeric materials with teichoic acid and LPS have been reported in previous studies in Gram positive and negative bacteria respectively. The cell wall of Gram positive bacteria possesses teichoic acid, due to which more occurrence of phosphate groups confers the polyanionic properties of its structures. In case of Gram negative bacteria the outer membrane possesses lipopolysaccharides (LPS), due to which abundance of phospholipids found which provides the net negative charge to bacteria (Dillen et al., 2008). To detect the charged supported interaction mechanisms, negatively charged purity gum ultra (PGU) based nanoemulsion prepared. The study revealed that the size of *S. aureus* and *L. monocytogenes* increase on adhesion of negatively charged droplets which suggested the interaction with bacterial cell wall components. Adhesion of droplets probably occurs through self-assembly or chemical linkage rather than electrostatic interaction because both bacteria and nanoemulsion droplets have negative surface charge (Majeed et al., 2016). At the same time bacteria like *E. coli*, *S. aureus* and *L. monocytogenes* treated with Tween 80 containing nanoemulsion (with neutral charge) which showed no interaction with bacterial cells and pointed out the event probably due to no charge on nanodroplets (Majeed et al., 2016). In this regard, the schematic representation in Figure 3 which is based on the previous results and hypothesis depicted the charge based synthesis of nanoemulsions and the mechanism of action of positive charge containing nanodroplets on negatively charged cell wall/ membrane of bacteria. The self-assembly or chemical interaction of nanodroplets with the microbial cell wall generally demonstrated through microscopic analysis (Ramalingam and Lee., 2018) and found significantly effective against both gram negative and positive bacterial populations .

Figure 2. Schematic representation shows the synthesis of nanoemulsions and charge based mechanism of action of positive charge containing nanodroplets on negatively charged cell wall/ membrane of individual bacterial cell.

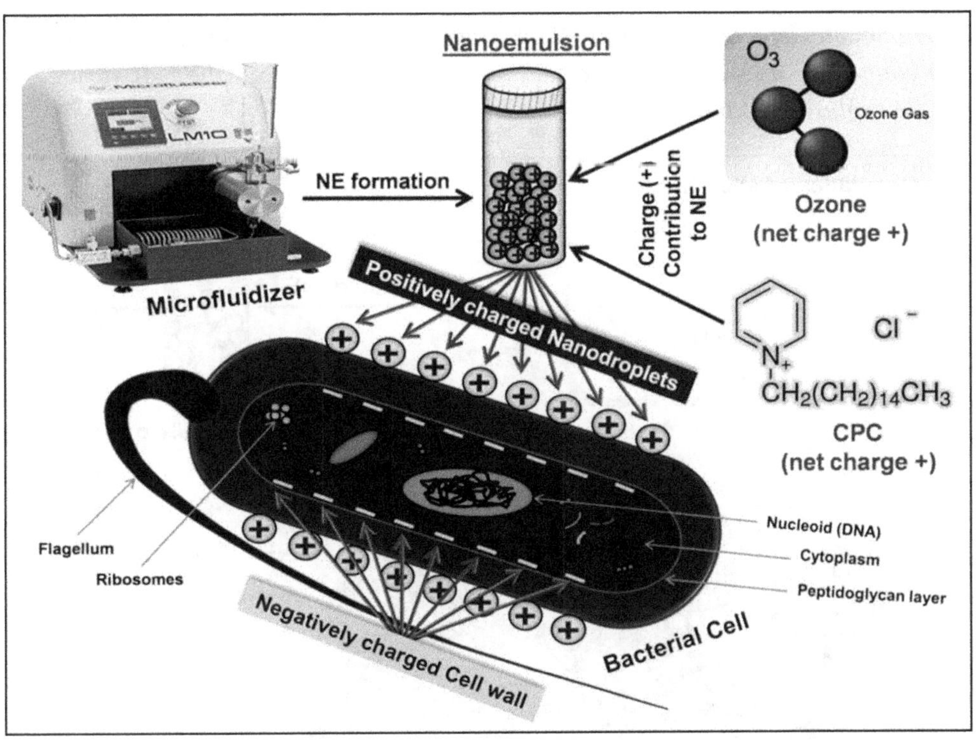

(2) High Pressure Explosion Based Action of Nanoemulsions

The antimicrobial mechanism function as the attraction of nanoemulsions with microbial cells due to the acquired surface charges of nanodroplets. Most of the nanoemulsions are having the ability to fuse with the outer membranes of the bacteria promoted by the electrostatic interactions. Cationic charges of nanodroplets and the anionic charges on the bacterial cells, generating destabilization of the lipid bilayers of the membrane, cellular permeability and Efflux pumps regulation, ultimately leading to the disfunction of the bacterial cell (Krishnamoorthy et al., 2018). The fusion of the emulsified nanodroplets with the phospholipid bilayers eventually stimulate the sustained release of the oils from the nanoemulsion droplets at the targeted site amplify the action of the oils (Li et al., 2015). Here, the diagrammatic depiction in Figure 3 represents as the nanodroplets produced at high pressure (10,000 or 20,000 psi), once hit on the surface of the pathogens causing bombing effect on the cell wall/membrane with high energy release leading the disruption and ultimately death of the microorganisms (Hwang et al., 2013). Nanodroplets are thermodynamically driven to fuse with lipid containing organism's membrane. When a sufficient number of nanoparticles bind to pathogens, they release most of the energy trapped inside the nanodroplets. The active components as well as the energy released; disrupt the pathogen lipid membrane resulting in cell lysis (Chime et al., 2014). Hence the nanoemulsions represent the broad spectrum activity trough their mode of actions on bacteria.

Figure 3. Diagrammatic depiction shows the Microfluidization based depletion of high pressure (20000 or 40000 psi) containing nanodroplets and its mechanism of action on the cell surface of bacterial population while explosion with high pressure bombardment.

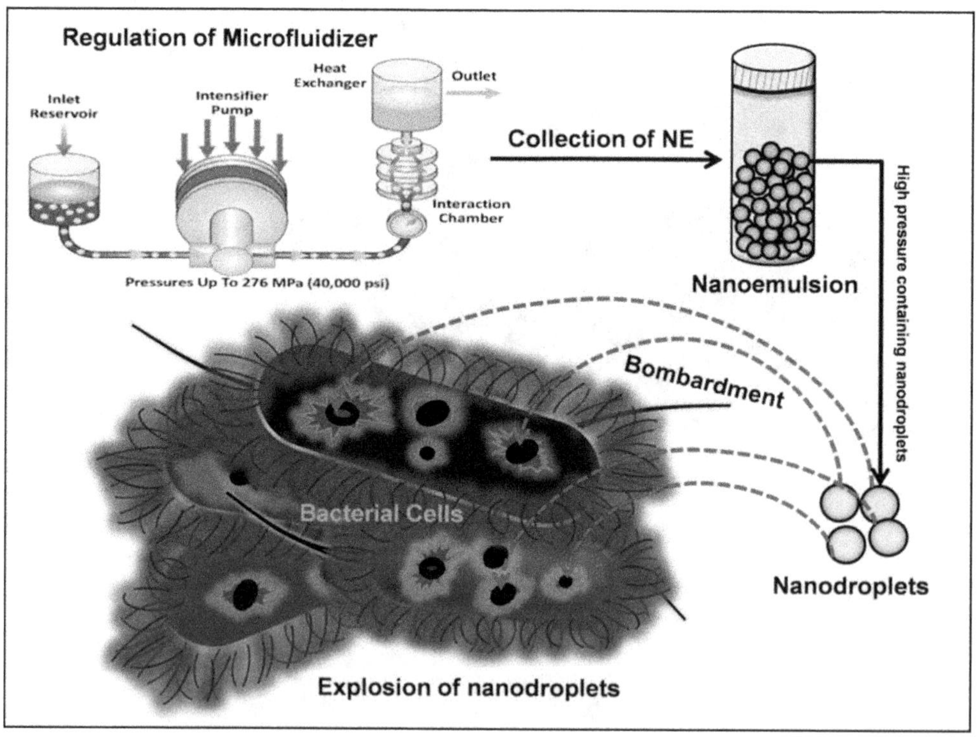

CONCLUSION

Nanoemulsions potentially represents a suitable alternative to standard antibiotic and antimicrobial compounds for the treatment of multi drug resistant group of bacteria termed as ESKAPE pathogens. The ESKAPE pathogens represent the paradigm for resistance, pathogenesis and the emergence of disease transmission in both the community and clinical settings are broadly shared. NE formulations offer numerous promising properties such as high surface area, transparent appearance, high kinetic stability, low viscosity, controlled drug release, increased bioavailability and ability to dissolve non-polar active substances. Increasing multidrug resistance in human pathogens with high rates of morbidity and mortality is a great challenge faced by society, clinicians and researchers. To tackle the existing problem, NE containing oils possess important bioactive volatile compounds with diverse antimicrobial potential to inhibit the microbial growth. Bioactive phytochemicals in NEs are able to block the drug efflux mechanism and induce broad-spectrum metabolic enzyme inhibition in drug resistant pathogens leading to combat the resistance mechanisms in bacteria. NEs are effectively able to reduce the metabolic activities of ESKAPE pathogens in biofilms by consistent antimicrobial activity after exposure to the bacterial populations. The mechanism and mode of actions of NE are depicted in the present chapters which are based on electrostatic forces/charge interaction and high pressure containing nanodroplets explosion with the microbial cell surfaces. Close interactions and controlled release of oils/drugs from nanodroplets disrupt the cell membrane of the targeted pathogens by altering membrane permeability. The interruption in the cellular metabolism as well as enzyme kinetics leading to the leakage of intracellular constituent such as nucleic acids, proteins, and potassium ultimately death of the bacteria takes place. The current study suggested a prominent anti-biofilm activity, inhibitions of adherence, and morphological disruption with pili destruction potencies of antimicrobial nanoemulsions. These considerations could be used for the rational design of nanoemulsions as alternative antimicrobial agents for the treatment of drug resistant ESKAPE pathogens.

FUNDING

This work was supported by the DST Science and Engineering Research Board (SERB), India [SERB/LS-267/2014] and Extra Mural Research Funding of AYUSH (Ayurveda, Yoga and Naturopathy, Unani, Siddha and Homoeopathy), India [28015/209 /2015-HPC] for providing fellowship and funds to perform this research.

ACKNOWLEDGMENT

The authors are pleased to express thanks to ICMR (Indian Council of Medical Research) for awarding Senior Research Fellowship [File No: AMR/Fellowship/6/2019/ECD-II] to complete this task of study.

REFERENCES

Abbasi, Z., Aminzare, M., Azar, H. H., & Rostamizadeh, K. (2020). Effect of corn starch coating incorporated with nanoemulsion of Zatariamultiflora essential oil fortified with cinnamaldehyde on microbial quality of fresh chicken meat and fate of inoculated Listeria monocytogenes. *Journal of Food Science and Technology*, 1–11. doi:10.100713197-020-04774-y

Abdelmonem, R., Younis, M. K., Hassan, D. H., & El-Sayed, M. A. E. G. (2019). Formulation and characterization of chlorhexidineHCl nanoemulsion as a promising antibacterial root canal irrigant: In-vitro and ex-vivo studies. *International Journal of Nanomedicine*, *14*, 4697. doi:10.2147/IJN.S204550

Alkhatib, M. H., Aly, M. M., Saleh, O. A., & Gashlan, H. M. (2016). Antibacterial activity of a microemulsion loaded with cephalosporin. *Biologia*, *71*(7), 748–756. doi:10.1515/biolog-2016-0105

Allegranzi, B., Nejad, S. B., Combescure, C., Graafmans, W., Attar, H., Donaldson, L., & Pittet, D. (2011). Burden of endemic health-care-associated infection in developing countries: Systematic review and meta-analysis. *Lancet*, *377*(9761), 228–241. doi:10.1016/S0140-6736(10)61458-4

Asensio, C. M., Quiroga, P. R., Al-Gburi, A., Huang, Q., & Grosso, N. R. (2020). Reological behaviour, Antibacterial and Quorum sensing inhibition study of an *Argentinean oregano* essential oil nanoemulsion. *Frontiers in Nutrition*, *7*, 193. doi:10.3389/fnut.2020.569913

Atinderpal, K., Kapoor, N., Gupta, S., Tyag, A., Sharma, R. K., Ali, J., & Dang, S. (2018). Development and characterization of green tea catechins and ciprofloxacin-loaded nanoemulsion for intravaginal delivery to treat urinary tract infection. *Indian Journal of Pharmaceutical Sciences*, *80*(3), 442–452. doi:10.4172/pharmaceutical-sciences.1000377

Bassetti, M., Vena, A., Croxatto, A., Righi, E., & Guery, B. (2018). How to manage Pseudomonas aeruginosa infections. *Drugs in Context*, *7*. Advance online publication. doi:10.7573/dic.212527

Brink, A. J., Coetzee, J., Clay, C. G., Sithole, S., Richards, G. A., Poirel, L., & Nordmann, P. (2012). Emergence of New Delhi metallo-beta-lactamase (NDM-1) and Klebsiellapneumoniaecarbapenemase (KPC-2) in South Africa. *Journal of Clinical Microbiology*, *50*(2), 525–527. doi:10.1128/JCM.05956-11

Bush, K. (2010). Alarming β-lactamase-mediated resistance in multidrug-resistant Enterobacteriaceae. *Current Opinion in Microbiology*, *13*(5), 558–564. doi:10.1016/j.mib.2010.09.006

Cao, Z., Spilker, T., Fan, Y., Kalikin, L. M., & Ciotti, S., LiPuma, J. J., & Wang, S. H. (2017). Nanoemulsion is an effective antimicrobial for methicillin-resistant Staphylococcus aureus in infected wounds. *Nanomedicine (London)*, *12*(10), 1177–1185. doi:10.2217/nnm-2017-0025

Castanheira, M., Deshpande, L. M., Mathai, D., Bell, J. M., Jones, R. N., & Mendes, R. E. (2011). Early dissemination of NDM-1-and OXA-181-producing Enterobacteriaceae in Indian hospitals: Report from the SENTRY Antimicrobial Surveillance Program, 2006-2007. *Antimicrobial Agents and Chemotherapy*, *55*(3), 1274–1278. doi:10.1128/AAC.01497-10

Cecchini, M. E., Paoloni, C., Campra, N., Picco, N., Grosso, M. C., Perez, M. S., & Bellingeri, R. (2021). Nanoemulsion of Minthostachysverticillata essential oil, In-vitro evaluation of its antibacterial activity. *Heliyon*, *7*(1), e05896. doi:10.1016/j.heliyon.2021.e05896

Chime, S. A., Kenechukwu, F. C., & Attama, A. A. (2014). Nanoemulsions—Advances in formulation, characterization and applications in drug delivery. In A. D. Sezer (Ed.), *Application of Nanotechnology in Drug Delivery* (Vol. 3, pp. 77–126). Intech.

Cholley, P., Ka, R., Guyeux, C., Thouverez, M., Guessennd, N., Ghebremedhin, B., & Hocquet, D. (2014). Population structure of clinical Pseudomonas aeruginosa from West and Central African countries. *PLoS One*, *9*(9), e107008. doi:10.1371/journal.pone.0107008

Chouhan, S., Sharma, K., & Guleria, S. (2017). Antimicrobial activity of some essential oils present status and future perspectives. *Medicines (Basel, Switzerland)*, *4*(3), 58. doi:10.3390/medicines4030058

Codjoe, F. S., & Donkor, E. S. (2018). Carbapenem resistance: A review. *Medical Science*, *6*(1), 1. doi:10.3390/medsci6010001

Davin-Regli, A. (2015). Enterobacteraerogenes and Enterobacter cloacae; versatile bacterial pathogens confronting antibiotic treatment. *Frontiers in Microbiology*, *6*, 392. doi:10.3389/fmicb.2015.00392

De Oliveira, D. M., Forde, B. M., Kidd, T. J., Harris, P. N., Schembri, M. A., Beatson, S. A., & Walker, M. J. (2020). Antimicrobial resistance in ESKAPE pathogens. *Clinical Microbiology Reviews*, *33*(3), e00181–e19. doi:10.1128/CMR.00181-19

Deeksha, K., & Hari, S. (2019). Synthesis, Characterization and Antibacterial effect of Neem Oil Nanoemulsion. *Research Journal of Pharmacy and Technology*, *12*(9), 4400–4404. doi:10.5958/0974-360X.2019.00757.1

Desbois, A. P., & Smith, V. J. (2010). Antibacterial free fatty acids: Activities, mechanisms of action and biotechnological potential. *Applied Microbiology and Biotechnology*, *85*(6), 1629–1642. doi:10.100700253-009-2355-3

Dillen, K., Bridts, C., Van der Veken, P., Cos, P., Vandervoort, J., Augustyns, K., Stevens, W., & Ludwig, A. (2008). Adhesion of PLGA or Eudragit/PLGA nanoparticles to Staphylococcus and Pseudomonas. *International Journal of Pharmaceutics*, *349*(1-2), 234–240. doi:10.1016/j.ijpharm.2007.07.041

Espina, L., Somolinos, M., Lorán, S., Conchello, P., García, D., & Pagán, R. (2011). Chemical composition of commercial citrus fruit essential oils and evaluation of their antimicrobial activity acting alone or in combined processes. *Food Control*, *22*(6), 896–902. doi:10.1016/j.foodcont.2010.11.021

Fisher, K., & Phillips, C. (2009). In vitro inhibition of vancomycin-susceptible and vancomycin-resistant *Enterococcus faecium* and E. faecalis in the presence of citrus essential oils. *British Journal of Biomedical Science*, *66*(4), 180–185. doi:10.1080/09674845.2009.11730270

Franklyne, J. S., Mukherjee, A., & Chandrasekaran, N. (2016). Essential oil micro-and nanoemulsions: Promising roles in antimicrobial therapy targeting human pathogens. *Letters in Applied Microbiology*, *63*(5), 322–334. doi:10.1111/lam.12631

Gelband, H., Molly Miller, P., Pant, S., Gandra, S., Levinson, J., Barter, D., & Laxminarayan, R. (2015). The state of the world's antibiotics. *Wound Healing Southern Africa, 8*(2), 30-34. doi.org/hdl.handle.net/10520/EJC180082

Ghaderi, L., Aliahmadi, A., Ebrahimi, S. N., & Rafati, H. (2021). Effective Inhibition and eradication of Pseudomonas aeruginosa biofilms by Saturejakhuzistanica essential oil nanoemulsion. *Journal of Drug Delivery Science and Technology*, *61*, 102260. doi:10.1016/j.jddst.2020.102260

Ghani, S., Barzegar, H., Noshad, M., & Hojjati, M. (2019). Development of nanoemulsion-based antimicrobial activity of cinnamon prepared with soy protein isolate-lecithin. *Journal of Food and Bioprocess Engineering*, *2*(2), 147–154.

Hamouda, T., & Baker, J. R. Jr. (2000). Antimicrobial mechanism of action of surfactant lipid preparations in enteric Gram-negative *bacilli. Journal of Applied Microbiology*, *89*(3), 397–403. doi:10.1046/j.1365-2672.2000.01127.x

Hassan, K. A., & Mujtaba, M. A. (2019). Antibacterial efficacy of garlic oil nano-emulsion. *AIMS Agriculture and Food*, *4*(1), 194–205. doi:10.3934/agrfood.2019.1.194

Hassanshahian, M., Saadatfar, A., & Masoumipour, F. (2020). Formulation and characterization of nanoemulsion from Alhagimaurorum essential oil and study of its antimicrobial, antibiofilm, and plasmid curing activity against antibiotic-resistant pathogenic bacteria. *Journal of Environmental Health Science & Engineering*, *18*(2), 1015–1027. doi:10.100740201-020-00523-7

Hobson, J. J., Edwards, S., Slater, R. A., Martin, P., Owen, A., & Rannard, S. P. (2018). Branched copolymer-stabilised nanoemulsions as new candidate oral drug delivery systems. *RSC Advances*, *8*(23), 12984–12991.

Hoiby, N., Bjarnsholt, T., Givskov, M., Molin, S., & Ciofu, O. (2010). Antibiotic resistance of bacterial biofilms. *International Journal of Antimicrobial Agents*, *35*(4), 322–332. doi:10.1016/j.ijantimicag.2009.12.011

Howard, A., O'Donoghue, M., Feeney, A., & Sleator, R. D. (2012). Acinetobacter baumannii: An emerging opportunistic pathogen. *Virulence*, *3*(3), 243–250. doi:10.4161/viru.19700

Hwang, Y. Y., Ramalingam, K., Bienek, D. R., Lee, V., You, T., & Alvarez, R. (2013). Antimicrobial activity of nanoemulsion in combination with cetylpyridinium chloride in multidrug-resistant Acinetobacter baumannii. *Antimicrobial Agents and Chemotherapy*, *57*(8), 3568–3575. doi:10.1128/AAC.02109-12

Ibrahim, D., Abdelfattah-Hassan, A., Badawi, M., Ismail, T. A., Bendary, M. M., Abdelaziz, A. M., & Abd El-Hamid, M. I. (2021). Thymol nanoemulsion promoted broiler chicken's growth, gastrointestinal barrier and bacterial community and conferred protection against Salmonella Typhimurium. *Scientific Reports*, *11*(1), 1–20. doi:10.103841598-021-86990-w

Keykhosravy, K., Khanzadi, S., Hashemi, M., & Azizzadeh, M. (2020). Chitosan-loaded nanoemulsion containing ZatariaMultifloraBoiss and BuniumpersicumBoiss essential oils as edible coatings: Its impact on microbial quality of turkey meat and fate of inoculated pathogens. *International Journal of Biological Macromolecules*, *150*, 904–913. doi:10.1016/j.ijbiomac.2020.02.092

Khan, M. H., & Ramalingam, K. (2019). Synthesis of antimicrobial nanoemulsions and its effectuality for the treatment of multi-drug resistant ESKAPE pathogens. *Biocatalysis and Agricultural Biotechnology*, *18*, 101025. doi:10.1016/j.bcab.2019.101025

Krishnamoorthy, R., Athinarayanan, J., Periasamy, V. S., Adisa, A. R., Al-Shuniaber, M. A., Gassem, M. A., & Alshatwi, A. A. (2018). Antimicrobial activity of nanoemulsion on drug-resistant bacterial pathogens. *Microbial Pathogenesis*, *120*, 85–96. doi:10.1016/j.micpath.2018.04.035

Krishnamoorthy, R., Gassem, M. A., Athinarayanan, J., Periyasamy, V. S., Prasad, S., & Alshatwi, A. A. (2021). Antifungal activity of nanoemulsion from Cleome viscosa essential oil against food-borne pathogenic Candida albicans. *Saudi Journal of Biological Sciences*, *28*(1), 286–293. doi:10.1016/j.sjbs.2020.10.001

Li, W., Chen, H., He, Z., Han, C., Liu, S., & Li, Y. (2015). Influence of surfactant and oil composition on the stability and antibacterial of eugenol nanoemulsions. *Lebensmittel-Wissenschaft + Technologie*, *62*(1), 39–47. doi:10.1016/j.lwt.2015.01.012

Livermore, D. M. (2002). Multiple mechanisms of antimicrobial resistance in Pseudomonas aeruginosa: Our worst nightmare. *Clinical Infectious Diseases*, *34*(5), 634–640. doi:10.1086/338782

Lou, Z., Letsididi, K. S., Yu, F., Pei, Z., Wang, H., & Letsididi, R. (2019). Inhibitive effect of eugenol and its nanoemulsion on quorum sensing–mediated virulence factors and biofilm formation by Pseudomonas aeruginosa. *Journal of Food Protection*, *82*(3), 379–389. doi:10.4315/0362-028X.JFP-18-196

Lu, W. C., Huang, D. W., Wang, C. C., Yeh, C. H., Tsai, J. C., Huang, Y. T., & Li, P. H. (2018). Preparation, characterization, and antimicrobial activity of nanoemulsions incorporating citral essential oil. *Yao Wu Shi Pin Fen Xi*, *26*(1), 82–89. doi:10.1016/j.jfda.2016.12.018

Majeed, H., Liu, F., Hategekimana, J., Sharif, H.R., Qi, J., Ali, B., Bian, Y.Y., Ma, J., Yokoyama, W. & Zhong, F. (2016). Bactericidal action mechanism of negatively charged food grade clove oil nanoemulsions. *Food Chem*, *197*(Pt A), 75-83. doi:10.1016/j.foodchem.2015.10.015

Manchanda, V., Sanchaita, S., & Singh, N. P. (2010). Multidrug resistant acinetobacter. *Journal of Global Infectious Diseases*, *2*(3), 291. doi:10.4103/0974-777X.68538

Mann, B., Sonu, K. S., Sharma, R., Kumar, R., & Singh, R. (2020). Formulation of peppermint oil nanoemulsion using conjugates of whey proteins with maltodextrin and its characterization. *Indian Journal of Traditional Knowledge*, *19*(2), 394–400.

Michiels, J. E., Van den Bergh, B., Verstraeten, N., Fauvart, M., & Michiels, J. (2016). In vitro emergence of high persistence upon periodic aminoglycoside challenge in the ESKAPE pathogens. *Antimicrobial Agents and Chemotherapy*, *60*(8), 4630–4637. doi:10.1128/AAC.00757-16

Mishra, P., R. S, S. K., Jerobin, J., Thomas, J., Mukherjee, A., & Chandrasekaran, N. (2014). Study on antimicrobial potential of neem oil nanoemulsion against P seudomonasaeruginosa infection in L abeorohita. *Biotechnology and Applied Biochemistry*, *61*(5), 611–619. doi:10.1002/bab.1213

Moghimi, R., Aliahmadi, A., Rafati, H., Abtahi, H. R., Amini, S., & Feizabadi, M. M. (2018). Antibacterial and anti-biofilm activity of nanoemulsion of Thymus daenensis oil against multi-drug resistant Acinetobacterbaumannii. *Journal of Molecular Liquids*, *265*, 765–770. doi:10.1016/j.molliq.2018.07.023

Moghimi, R., Ghaderi, L., Rafati, H., Aliahmadi, A., & McClements, D. J. (2016). Superior antibacterial activity of nanoemulsion of Thymus daenensis essential oil against E. coli. *Food Chemistry*, *194*, 410–415. doi:10.1016/j.foodchem.2015.07.139

Naithani, V., Kumar, S., Srivastava, S. M., Pathak, N., & Chaudhary, M. (2010). An evaluation of antimicrobial efficacy of acano against some acne causing microorganisms. *Int. J. Drug Dev. Res*, *2*(1), 134–140.

Naylor, N. R., Atun, R., Zhu, N., Kulasabanathan, K., Silva, S., Chatterjee, A., & Robotham, J. V. (2018). Estimating the burden of antimicrobial resistance: A systematic literature review. *Antimicrobial Resistance and Infection Control*, *7*(1), 1–17. doi:10.118613756-018-0336-y

Newman, J. W., Floyd, R. V., & Fothergill, J. L. (2017). The contribution of Pseudomonas aeruginosa virulence factors and host factors in the establishment of urinary tract infections. *FEMS Microbiology Letters*, *364*(15). Advance online publication. doi:10.1093/femsle/fnx124

Nikaido, H. (2003). Molecular basis of bacterial outer membrane permeability revisited. *Microbiology and Molecular Biology Reviews*, *67*(4), 593–656. doi:10.1128/MMBR.67.4.593-656.2003

Nikaido, H. (2009). Multidrug resistance in bacteria. *Annual Review of Biochemistry*, *78*, 119–146. doi:10.1146/annurev.biochem.78.082907.145923

Pendleton, J. N., Gorman, S. P., & Gilmore, B. F. (2013). Clinical relevance of the ESKAPE pathogens. *Expert Review of Anti-Infective Therapy*, *11*(3), 297–308. doi:10.1586/eri.13.12

Quatrin, P. M., Verdi, C. M., de Souza, M. E., de Godoi, S. N., Klein, B., Gundel, A., & Santos, R. C. V. (2017). Antimicrobial and antibiofilm activities of nanoemulsions containing Eucalyptus globulus oil against Pseudomonas aeruginosa and Candida spp. *Microbial Pathogenesis*, *112*, 230–242. doi:10.1016/j.micpath.2017.09.062

Rai, V. K., Mishra, N., Yadav, K. S., & Yadav, N. P. (2018). Nanoemulsion as pharmaceutical carrier for dermal and transdermal drug delivery: Formulation development, stability issues, basic considerations and applications. *Journal of Controlled Release*, *270*, 203–225. doi:10.1016/j.jconrel.2017.11.049

Ramalingam, K., & Lee, V. (2019). Synergistic Effects of Nanoemulsion and Deferiprone (1, 2 Dimethyl-3-Hydroxypyrid-4-One) on Multi-Drug Resistant Acinetobacterbaumannii. *Nano Biomedicine and Engineering*, *11*(3), 226–237. doi:10.5101/nbe.v11i3.p226-237

Ramalingam, K. & Lee, V. A. (2018). Antibiofilm activity of an EDTA-containing nanoemulsion on multidrug-resistant Acinetobacterbaumannii. *Artificial Cells, Nanomedicine, and Biotechnology*, *46*(sup2), 737-743. doi:10.1080/21691401.2018.1468771

Rice, L. B. (2008). Federal funding for the study of antimicrobial resistance in nosocomial pathogens, no ESKAPE. doi.org/ doi:10.1086/533452

Rinaldi, F., Oliva, A., Sabatino, M., Imbriano, A., Hanieh, P. N., Garzoli, S., & Carafa, M. (2020). Antimicrobial Essential Oil Formulation: Chitosan Coated Nanoemulsions for Nose to Brain Delivery. *Pharmaceutics*, *12*(7), 678. doi:10.3390/pharmaceutics12070678

Risso, N. H., Stopiglia, C. D. O., Oliveira, M. T., Haas, S. E., Maciel, T. R., Lazzari, N. R., & Beckmann, D. V. (2020). Chlorhexidine Nanoemulsion: A New Antiseptic Formulation. *International Journal of Nanomedicine*, *15*, 6935. doi:10.2147/IJN.S228280

Sabir, D. K., & Sidiq, K. R. (2020). Antimicrobial Activity of Combined Cinnamon Nano Emulsions-Antibiotics against AcinetobacterBaumannii. *Passer Journal*, *1*(2), 2. doi:10.24271/psr.03

Santajit, S., & Indrawattana, N. (2016). Mechanisms of antimicrobial resistance in ESKAPE pathogens. *BioMed Research International*. Advance online publication. doi:10.1155/2016/2475067

Shaker, D. S., Ishak, R. A., Ghoneim, A., & Elhuoni, M. A. (2019). Nanoemulsion: A review on mechanisms for the transdermal delivery of hydrophobic and hydrophilic drugs. *Scientia Pharmaceutica*, *87*(3), 17. doi:10.3390cipharm87030017

Singh, N., Verma, S. M., Singh, S. K., & Verma, P. R. P. (2015). Evidence for bactericidal activities of lipidic nanoemulsions against Pseudomonas aeruginosa. *Antonie van Leeuwenhoek*, *107*(6), 1555–1568. doi:10.100710482-015-0449-8

Singh, S., Hussain, A., Shakeel, F., Ahsan, M. J., Alshehri, S., Webster, T. J., & Lal, U. R. (2019). Recent insights on nanomedicine for augmented infection control. *International Journal of Nanomedicine*, *14*, 2301–2325. doi:10.2147/IJN.S170280

Sonawane, S. J., Kalhapure, R. S., Jadhav, M., Rambharose, S., Mocktar, C., & Govender, T. (2015). Transforming linoleic acid into a nanoemulsion for enhanced activity against methicillin susceptible and resistant Staphylococcus aureus. *RSC Advances*, *5*(110), 90482–90492.

Sugumar, S., Ghosh, V., Nirmala, M. J., Mukherjee, A., & Chandrasekaran, N. (2014). Ultrasonic emulsification of eucalyptus oil nanoemulsion: Antibacterial activity against Staphylococcus aureus and wound healing activity in Wistar rats. *Ultrasonics Sonochemistry*, *21*(3), 1044–1049. doi:10.1016/j.ultsonch.2013.10.021

Sugumar, S., Mukherjee, A., & Chandrasekaran, N. (2015). Eucalyptus oil nanoemulsion-impregnated chitosan film: Antibacterial effects against a clinical pathogen, Staphylococcus aureus, in vitro. *International Journal of Nanomedicine*, *10*(Suppl 1), 67. doi:10.2147/IJN.S79982

Sundararajan, B., Moola, A. K., Vivek, K., & Kumari, B. R. (2018). Formulation of nanoemulsion from leaves essential oil of Ocimumbasilicum L. and its antibacterial, antioxidant and larvicidal activities (Culexquinquefasciatus). *Microbial Pathogenesis*, *125*, 475–485. doi:10.1016/j.micpath.2018.10.017

Swathy, J. S., Mishra, P., Thomas, J., Mukherjee, A., & Chandrasekaran, N. (2018). Antimicrobial potency of high-energy emulsified black pepper oil nanoemulsion against aquaculture pathogen. *Aquaculture (Amsterdam, Netherlands)*, *491*, 210–220. doi:10.1016/j.aquaculture.2018.03.045

Vatsraj, S., Chauhan, K., & Pathak, H. (2014). Formulation of a novel nanoemulsion system for enhanced solubility of a sparingly water soluble antibiotic, clarithromycin. *Journal of Nanoscience*. Advance online publication. doi:10.1155/2014/268293

Venkateswaran, S., Dos Santos, O. D. H., Scholefield, E., Lilienkampf, A., Gwynne, P. J., Swann, D. G., & Bradley, M. (2016). Fortified interpenetrating polymers–bacteria resistant coatings for medical devices. *Journal of Materials Chemistry. B, Materials for Biology and Medicine*, *4*(32), 5405–5411. doi:10.1039/c6tb01110a

Vila, J., Martí, S., & Sanchez-Céspedes, J. (2007). Porins, efflux pumps and multidrug resistance in Acinetobacterbaumannii. *The Journal of Antimicrobial Chemotherapy*, *59*(6), 1210–1215. doi:10.1093/jac/dkl509

Weiner, L. M., Webb, A. K., Limbago, B., Dudeck, M. A., Patel, J., Kallen, A. J., & Sievert, D. M. (2016). Antimicrobial-resistant pathogens associated with healthcare-associated infections: Summary of data reported to the National Healthcare Safety Network at the Centers for Disease Control and Prevention, 2011–2014. *Infection Control and Hospital Epidemiology*, *37*(11), 1288–1301. doi:10.1017/ice.2016.174

World Health Organization. (2014). *World health statistics 2014*. World Health Organization. https://apps.who.int/iris/handle/10665/112738

Wyres, K. L., Hawkey, J., Hetland, M. A., Fostervold, A., Wick, R. R., Judd, L. M., & Holt, K. E. (2019). Emergence and rapid global dissemination of CTX-M-15-associated *Klebsiella pneumoniae* strain ST307. *The Journal of Antimicrobial Chemotherapy*, *74*(3), 577–581. doi:10.1093/jac/dky492

Yayan, J., Ghebremedhin, B., & Rasche, K. (2015). Antibiotic resistance of Pseudomonas aeruginosa in pneumonia at a single university hospital center in Germany over a 10-year period. *PLoS One*, *10*(10), e0139836. doi:10.1371/journal.pone.0139836

Yazgan, H. (2020). Investigation of antimicrobial properties of sage essential oil and its nanoemulsion as antimicrobial agent. *Lebensmittel-Wissenschaft + Technologie*, *130*, 109669. doi:10.1016/j.lwt.2020.109669

Yazgan, H., Ozogul, Y., & Kuley, E. (2019). Antimicrobial influence of nanoemulsified lemon essential oil and pure lemon essential oil on food-borne pathogens and fish spoilage bacteria. *International Journal of Food Microbiology*, *306*, 108266. doi:10.1016/j.ijfoodmicro.2019.108266

Zarrilli, R., Pournaras, S., Giannouli, M., & Tsakris, A. (2013). Global evolution of multidrug-resistant Acinetobacterbaumannii clonal lineages. *International Journal of Antimicrobial Agents*, *41*(1), 11–19. doi:10.1016/j.ijantimicag.2012.09. 008

Chapter 8
Application of Nanoemulsion in Tuberculosis Treatment

Sonia Singh
https://orcid.org/0000-0003-1503-2745
GLA University, India

Yogesh Murti
https://orcid.org/0000-0002-5278-0681
GLA University, India

Nitin Agrawal
https://orcid.org/0000-0003-4637-0608
Raja Balwant Singh Engineering Technical Campus, India

ABSTRACT

Tuberculosis is commonly called TB and considered to be the most contagious disease. This disease is caused by a causative agent known as Mycobacterium tuberculosis within the host body cells. Emergence incidence of XDR and MDR of tuberculosis are due to high dose intake and prolonged treatment of antibiotics. In this era, only one vaccine called as BCG is available which is ineffective against infected tuberculosis adults. Though several antibiotics have been produced to overcome drug resistance and even enhance the patient compliance towards treatment and reduce the treatment regimen, we require a novel strategy that can improve the potency of therapeutic synthetic drugs. This approach can be achieved by the application of nanotechnology associated with conventional therapy. Nanotechnology has attributed the promising effects associated with optimized treatment against chronic diseases. This novel technology has the ability to reduce the dose frequency and even resolves the poor patient compliance.

1. INTRODUCTION

Tuberculosis is a chronic persistent communicable disease (Kaufmann & McMichael, 2005; Gaspar, 2008; Mitnick, 2008). It is caused by *Mycobacterium tuberculosis* and *Mycobacterium bovis*, mainly in lungs to be called as pulmonary tuberculosis; however this lung infection also infect the other organs named

DOI: 10.4018/978-1-7998-8378-4.ch008

as extra pulmonary tuberculosis (Patil, 2018; Padayatchi & Friedland, 2008). In 1993, WHO announced this chronic infectious disease as a global emergency. This disease contained the highest mortality rate as compared to other communicable disease. Moreover, the potentially significant pharmacotherapies are being available for over many 50 years, still TB considered as the main cause of fatal deaths. This infectious disease has multidrug resistant strains which can kill immunosuppressed patients within very short time duration (Singh, 2007; Mehta, 2018; Shivangi, 2018; Bisht, 2019).

2. FEW KEY FACTS OF TUBERCULOSIS

2.1 Causative Agent: Tuberculosis

Figure 1. Pathogenesis of tuberculosis

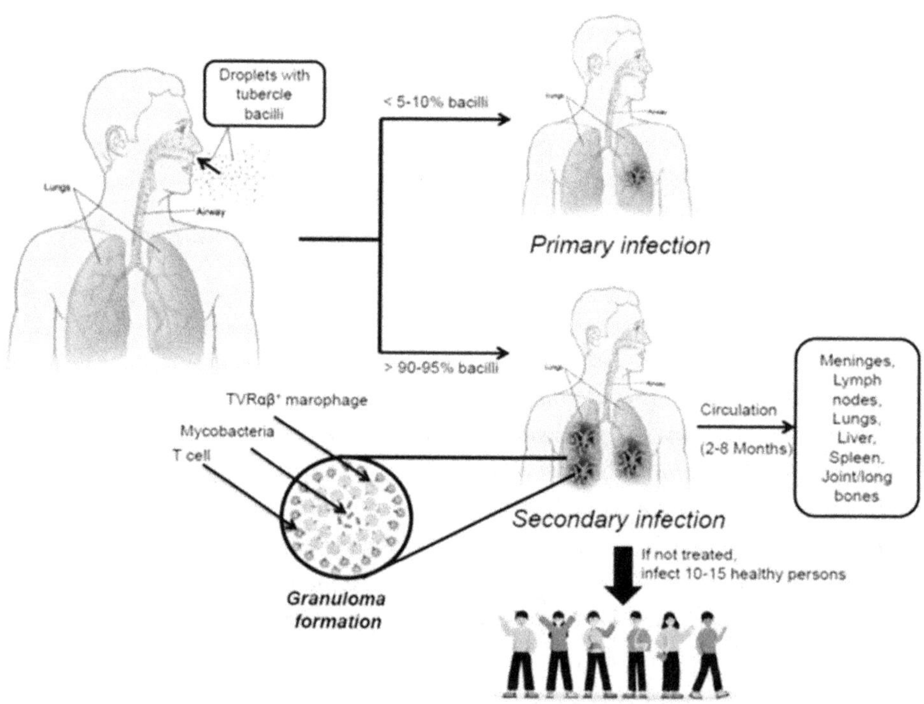

Mycobacterium tuberculosis is a gram-positive micro-organism. It is aerobic rod-shaped acid fast bacteria; thereby produces acid stable complexes with arylmethane dyes (Comstock, 1974; Singh, 2004). These bacteria remain in viable condition in the air for longer time and eventually get inhaled by human being *via* lungs. And they are phagocytosed by alveolar macrophages and then start their replication process within duration of two to three weeks. The infected macrophages develop a granuloma, a hard shell within the bacilli. The bacilli able to replicate and grow inside the granuloma and then rupture to start a new cycle into another host body. The development of infection rate of disease may depend on several factors including time period of latent, immunity of an individual and age of the patient (Figure

1). Several other species of *M. tuberculosis* have adapted their genetic makeup accordingly as able to infect the living populations (Skeiky & Sadoff, 2006; du Toit, 2006; Aagaard, 2009). It was observed that in almost 95% of infected cases, the alveolar macrophages able to carry the infectious agent without any noticeable signs and symptoms. However, in case if the micro-organisms are not destroyed completely, they may remain in dormant phase for many days and become reactivate after years (Schatz, 2005; Kaufmann, 2001).

2.2 Drug Resistanti- Tuberculosis Drugs

Table 1. Anti-tubercular drugs with adverse side effects

Anti-Tubercular Drugs	Adverse Side Effects	References
Group 1: First-line Anti-TB Drugs		
Isoniazid	Hepatitis, seizures, hypersensitivity and peripheral neuropathy	Drew, 2020
Rifampicin	Anorexia, hepatitis, thrombocytopenia, renal failure, multiple drug interaction, cutaneous reaction	Somoskovi, 2001; Floss & Yu, 2005; Wehrli, 1983
Rifabutin	Skin discoloration, leucopaenia, arthralgia, uveitis	Jamieson, 2014; Kunin, 1996
Pyrazinamide	Hepatitis, hyperuricemia, nausea, anorexia, rash	Zhang, 2003
Ethambutol	Arthralgia, optic neuritis, gastrointestinal discomfort	Wolucka, 1994
Group 2: Fluoroquinolone Anti-TB Drugs		
Ofloxacin Levofloxacin Moxifloxacin	Vomiting, nausea, abdominal pain, diarrhea, drowsiness, pruritus, insomnia, photosensitivity reactions, hepatotoxicity, tremor	Von Groll, 2009; Vacher, 1999; Willby, 2015
Group 3: Injectable Anti-TB Drugs		
Streptomycin, Kanamycin Amikacin, Capreomycin	Vestibular dysfunction, ototoxicity, dizziness, rash, nephrotoxicity, hypersensitivity reaction	Georghiou, 2012
Group 4: Second-Line Anti-TB Drugs		
Ethionamide or Prothionamide	Loss of appetite, nausea, abdominal pain, vomiting, neurotoxicity, hepatotoxicity, photosensitivity	Morlock, 2003
Cycloserine or Terizidone	Vertigo, headache, dysarthria, convulsion, Stevens-Johnson syndrome, skin rashes	Almeida Da Silva, 2011
p-Aminosalicylic acid	Diarrhea, anorexia, nausea, hypothyroidism, allergic reactions, vomiting	Zhang, 2015; Zhao, 2014
Group 5: Drugs with clinical data claimed		
Linezolid	Lactic acidosis, myelosuppression, serotonin syndrome, peripheral neuropathy	Narita, 2007
Clofazimine	Abdominal pain, hyperpigmentation, bleeding, bowel obstruction	Cholo, 2012; Zhang, 2015
Amoxicillin with clavulanate	Nausea, vomiting, headache, diarrhea, allergic reactions	Li, 2018; Gresser, 2001; Horita, 2014
Imipenem	Skin rash, nausea, vomiting, pruritus, diarrhea	Hornik, 2013
Clarithromycin	Indigestion, mild diarrhea, nausea, vomiting abdominal pain, skin rash, unpleasant taste	Van der Paardt, 2017
Thiacetazone	Diarrhea, nausea, vomiting, loss of appetite, skin rashes, clumsiness severe cutaneous hypersensitivity	Ipuge, 1995

The drug resistant tuberculosis becomes a major problem as to control the infection throughout the world. This drug resistant TB generally belongs to MDR and XDR. The term" MDR-TB" is the strains of *M. tuberculosis* which are resistant to the first-line anti-TB drugs namely rifampicin and isoniazid (Nasiri, 2017; Trauner, 2014). And XDR-TB is defined as the strains of *M. tuberculosis* resistant to one fluroquinolone and second-line anti-TB drugs along with first-line drugs such as isoniazid and rifampicin. It is estimated that more than four lakh cases of MDR-TB occurs every year worldwide. However, out of them around 9% gets affected by XDR-TB. The treatment of MDR-TB requires about 2 years with the application of second-line anti-TB drugs. Moreover, the XDR-TB treatment becomes little bit complicated as the options of using drugs becomes limited because of resistant against fluoroquinolone and second-line anti-TB drugs. The treatment of XDR-TB is costly, time- consuming and develops severe adverse effects as discussed in Table 1. It has been found that XDR-TB have high occurrence of infection in HIV patients. Around 36 million populations with HIV gets infected with TB (Chan, 2008; Millard & Moore, 2015; D'Ambrosio, 2015; Caminero, 2010; Sun, 2010; Skeiky & Sadoff, 2006).

2.3 Anti-Tuberculosis Drugs/Therapy

The first-line drugs are highly effective against TB whereas the second-line drugs are effective only when the first-line therapy are failed to show significant recovery due to the development of drug resistance. Generally, the course treatment of TB consists of two phases involving four antibiotics for about six to nine months.

In the first intensive phase, the treatment occurs with four oral first-line anti-TB drugs for about two months and it is found that during this drug regimen therapy, most of the bacilli destroyed. During the second phase i-e called as continuation phase, the treatment includes the intake of the first-line anti-TB drugs, rifampicin and isoniazid either three times a week for around 4-6 months to eradicate some of the survived bacilli from the first phase treatment. These treatment strategies may be associated with long term adverse side effects. Novel approaches with some advanced therapy are required to get overcome from this epidemic infectious disease (Zhang, 2015).

In the current era, nanotechnology has emerged as one of the advanced and novel technologies 5concerned with nano-sized particles showed improved physicochemical properties of drug components that have been formulated into nano-forms. Moreover, with the application of nanoformualtion, it may become easier to eradicate these life threatening diseases such as HIV, TB and so on. Table 2 describes the various nanoformulations of anti-TB drugs developed by several researchers against TB.

Table 2. Nanotechnology employed in the management of tuberculosis

Drug delivery system	Drug component	Advantages	References
Nanoparticles: poly lactic-co-glycolic-acid	Rifampicin	Improved therapeutic index of the drug, slow and continuous drug release	Tripathi, 2010
Nanoparticles: thioridazine in poly(lactic-*co*-glycolic) acid	Rifampicin	Reduced the mycobacterial infection	Vibe, 2016
Nanoparticles: poly(butyl cyanoacrylate	Moxifloxacin	Improved therapeutic efficacy	Kisich, 2007
Nanoparticles: poly(butyl-2-cyanoacrylate)	Moxifloxacin	Reduced lung infection	Shipulo, 2008
Nanoparticles: poly (DL-lactide-co-glycolide)	Rifampicin	Improved therapeutic efficacy	Pandey, 2003
Nanoparticles: poly lactide-co-glycolides	Rifampicin	Improved antibacterial efficacy	Esmaeili, 2007
Nanoparticles: poly(d-L-lactide-co-glycolide)	Rifampin, Azithromycin	Reduced the microbial burden	Toti, 2011
Liposomes: egg phosphatidylcholine and cholesterol	Rifampicin	Improved therapeutic efficacy	Vyas, 2004
Solid lipid particles	Rifampicin, Isoniazid, Pyrazinamide	Improved drug bioavailability, reduced the drug dosing frequency	Pandey & Khuller, 2005
Flower-like polymeric micelles: poly(*ε*-caprolactone)–*b*-poly(ethylene glycol)–poly(*ε*-caprolactone)	Rifampicin	Improved therapeutic efficacy	Moretton, 2012
Micelles: poly(epsilon-caprolactone)-*b*-poly(ethylene glycol)-poly(epsilon-caprolactone)	Rifampicin	Enhanced physical stability	Moretton, 2010
'flower-like' polymeric micelles: poly(ε-caprolactone)-*b*-PEG-*b*-poly(ε-caprolactone)	Rifampicin, Isoniazid	Improved therapeutic efficacy	Moretton, 2014
Micelle: poly(ethylene glycol)-poly(aspartic acid)	Pyrazinamide	Improved therapeutic efficacy	Silva, 2006
Dendrimer: 4th-generation poly(amidoamine)	Rifampicin	Suitable drug delivery system	Bellini, 2015
Mannosylated dendrimers	Rifampicin	Suitable drug delivery system	Kumar, 2006
Nanoparticles: mesoporous silica	Isoniazid, Rifampicin	Improved therapeutic efficacy	Clemens,2012
Gelatin nanoparticles	Rifampicin	Increased drug target response, reduced dose frequency	Saraogi, 2010
Nanoparticles: poly-(dl-lactide-co-glycolide)	Econazole, Moxifloxacin	Proved as intermittent therapy of TB	Ahmad, 2008
Microemulsion: oleic acid, phosphate buffer, Tween 80, ethanol	Isoniazid	Showed sustained and controlled drug release	Mehta, 2008
Nanosuspensions	Clofazimine	Promised as a new drug formulation for intravenous therapy of TB	Peters, 2000
Micro/ nanoparticles	Rifampicin	Improved therapeutic efficacy	Reverchon, 2002
Microparticles: poly(DL-lactide-co-glycolide)	Isoniazid, Rifampin	Exhibited equivalent clearance of CFU in organs	Dutt & Khuller, 2001
Microparticles: poly (DL-lactide-co-glycolide)	Rifampicin	Promised as a sustained release carrier systems	Dutt & Khuller, 2001
Niosome	Rifampicin	Improved therapeutic efficacy	Mullaicharam & Murthy, 2004
Nanoparticles: poly (dl-lactide-co-glycolide)	Isoniazid, Rifampicin Pyrazinamide	Reduced in dosing frequency	Pandey, 2003
Nanoparticles	Isoniazid, Rifampin, Streptomycin	Improved therapeutic efficacy	Anisimova, 2000
Niosomes	Rifampicin	Improved *in vitro* drug release	Jain & Vyas, 1995
Niosomes	Ceftriaxone sodium, Rifampicin	Exhibited high drug entrapment efficiencies, improved drug release profiles.	Khan, 2020
Niosomes	Rifampicin, Gatifloxacin	Exhibited significant inhibition rate, reduced growth index	Rani, 2010
Niosomes	Rifampicin	Improved therapeutic efficacy	Jatav, 2011
Niosomes	Rifampicin	Reduced drug dose, dose frequency, toxicity	Singh, 2010
Nanosuspension	Rifampicin	Improved drug bioavailability	Schianti, 2011
Nanosuspensions	Rifampicin	Increased drug dissolution	Melo, 2020
Nanocomposites: carbohydrate-based spray-dried	Rifampicin	Enhanced drug inhalation performance, reduced cellular toxicity	Mehanna, 2019

Continued on following page

Table 2. Continued

Drug delivery system	Drug component	Advantages	References
Nanosuspensions: Povacoat ®	Rifampicin	Improved drug bioavailability	Melo, 2015

3. NANOEMULSION: NOVEL DRUG DELIVERY EXPLORED IN TUBERCULOSIS CHEMOTHERAPY

Emulsions are the dispersion of two immiscible solvents forming dispersed phase and continuous phase. Most commonly employed liquids are water and oil that are used in the formation of emulsion. When oil dispersed in water then it referred as oil-in-water emulsion and are generally used in the delivery of hydrophobic substances, whereas the water droplets dispersed in oil phase, called as water-in-oil (w/o) emulsions and such type of emulsions are for the delivery of hydrophilic substances (Tadros, 2004; McClements, 2007; Acosta, 2009). However, multiple emulsions may also be developed such as oil-in-water-in-oil (o/w/o), water-in-in-water (w/o/w) emulsions. In w/o/w type of emulsions, the large size of oil droplets along with water droplets is dispersed in an aqueous phase. And in o/w/o emulsion, the water/oil droplets are dispersed in an oil phase (Garti & Benichou, 2003, Weiss, 2006).

The emulsions are broadly classified on the basis of droplet size and the stability as nanoemulsions, microemulsions and coarse emulsions. In Figure 2, the different types of emulsions have been shown with their general properties (Komaiko & McClements, 2016; Anton & Vandamme, 2011; McClements & Rao, 2011).

Table 3. Advantages of Nanoemulsion

S. No	Advantages of Nanoemulsion as Novel Drug Delivery System
1	Higher surface area as well as free energy which make them an effective drug delivery system
2	Non-irritant, non-toxic
3	Formulated in many forms creams, liquids foams and sprays
4	Reduces the problems associated with flocculation, coalescence, sedimentation and creaming
5	Increased the rate of absorption
6	Declined the absorption variability
7	Applicable for lipophilic drug
8	Enhances bioavailability
9	Can be deliver by tropical, intravenous, oral route of administration
10	Thermodynamically stable
11	Controlled drug release and drug targeting
12	Protects the drug from hydrolysis, oxidation
13	Deliver both lipophilic and hydrophilic drugs

Nanoemulsions are employed as a promising drug delivery system in cosemotology, drug therapies, vaccines adjuvant, pharmaceutical, biomedical and biotechnological field (Velikov & Pelan, 2008; Ines, 2015). It is oil-in-water formulation with an average droplet size range from 100 to 500 nm. It has been called by different synonyms named as mini-emulsion, sub-micron emulsion and ultra-fine emulsion. They contain small droplet size range with highly stable colloidal system. This novel drug delivery system contained several enhanced properties over conventional emulsion formulations (Table 3).

The nanoemulsion contains two immiscible liquid with an emulsifying agent. The immiscible liquid must contain oil and water phase which contributes as dispersed and continuous phase. Both type of nanoemulsions i-e o/w and w/o comprises of a core-shell (Komaiko & McClements, 2016; Anton & Vandamme, 2011; McClements & Rao, 2011; Bouchemal, 2004; Kim, 2001; Wagner, 1996).

3.1 Components of nanoemulsions

Different components are involved in the preparation of nanoemulsion such as lipids, oils, surfactants, water and co-solvents. The oil phase in this system may contain triglycerides such as glycerols, mineral oils, fatty acids, vegetables oils and this selection of oil component is usually based on the solubility of drug component (Gasco, 1991; Kriwet & Müller-Goymann, 1995). The oil phases that contain high drug leading efficiency are generally employed for nanoemulsion development. Commonly used surfactants used are tweens, cremophor EL, spans, phospholipids, proteins, polysaccharides and lauroyl macrogo-glycerides. However, the combination of surfactants and co-surfactants are used for the nanoemulsion system development. Some of the co-surfactants which are commonly utilized in the development and formulation of propylene glycol, ethanol, polyethylene glycol, ethylene glycol, glycerine, propanol and transcutol-P (Trotta, 1999; Wooster, 2008; Jaiswal, 2015).

The oils employed in the formulation of nanoemulsions may include corn oil, coconut oil, castor oil, mineral oil, olive oil, linseed oil, evening primrose oil etc. Generally, the combination of water and oil may develop a temporary phase emulsion, which on prolonged standing gets separate into two separate phases because of coalescence of dispersed globules. However, to get overcome from this unstability of emulsion, emulgents/emulsifying agents are incorporated. Some of the emulgents are broadly catego-rized as surface active agents such as tweens, spans, hydrophilic colloids like acacia, and even finely divided solid particles such as veegum, bentonite (Table 4). An emulgent should be safe, non-toxic and their physical and chemical stabilities should be compatible with the drug sample. Although some other desirable properties of emulsifying agent are as follow: (1) it is able to decrease the surface ten-sion; (2) it is able to prevent globule coalescence; (3) it is able to produce significant zeta potential as well as viscosity which ultimately improve the system's stability; (4) it is effective at low concentration. The emulsifying agents develop mono-, multi-molecular or solid particulate films around the dispersed globules (Jaiswal, 2015).

Table 4. Classification of Emulsifier/Emulsifying agent with examples (Sarathchandraprakash, 2013)

Classification of Emulsifying agents
Based on Chemical structure
A. Synthetic Emulsifers
(a) Anionic emulsifiers
- Carboxylates(soaps, amino carboxylates, N-acyl-sarcosinates, acylated protein hydrolysates); sulfonates, sulfates and sulfated products 9alkyl benzene sulfonates, petroleum sulfonates, dialkyl sulfosuccinates, naphthalene sulfonates, olefin sulfonates); sulfated natural fats and oils (olive oil, polyunsaturated fatty acid)
(b) Cationic emulsifiers
- Employed as preservatives and disinfectants - Used on skin for cleansing wounds and burns - Quaternary ammonium components, cetrimide, benzalkonium chloride, cetyl pyridinium chloride
(c) Non-ionic emulsifiers
- Less irritant than anionic/cationic surfactants - Excellent grease or oil remover emulsifier - Polyolesters, polyoxyethylene esters, poloxamers, glycol, glycerol esters, sorbitan derivatives, polyethylene glycol (PEG-40, PEG-50, PEG-55), ethers of fatty alcohol, glycerine fatty acid esters (mono, di and tri-glyceride), sorbitan esters of fatty acids
(d) Amphoteric/Zwitter ionic
- Suitable for personal use - Especially for sensitive skins - Employed in shampoo, dish washing solutions
B. Natural Emulsifying agents
- Obtained from plant/animal tissues - Forms hydrated lyophilic colloids which exert protective colloidal properties and would reduce the risk of coalescence
(a) Vegetable derivative
- Non-toxic - Easy to handle, non-expensive - Limited its application in o/w emulsion - Acacia, pectin,carrageenan
(b) Animal source
- Form w/o emulsions - Lecithin, lanolin, cholesterol
(c) Semi-synthetic emulsifier
- Strong emulsifier, non-toxic - Less prone to microbial growth - Methylcellulose, carboxymethyl cellulose
(d) Synthetic agents
- Strongest emulsifier - Non-toxic, do not support microbial growth - Carbopol
C. Finely divided solid particle emulsifier
- Form a particulate layer around the dispersed globules - Enhancing the viscosity and reducing the interaction between dispersed particles - Bentonite, veegum, magnesiu hydroxide, trisilicate of magnesium, aluminium hydroxide, hectorite
D. Auxillary agents – Tween
Based on Mechanism of action
(a) Monomolecular adsorption

Continued on following page

Table 4. Continued

Classification of Emulsifying agents
- On the addition of emulsifying agent, the system forms a monolayer film at o/w interface
(b) Multi-molecular adsorption
- This includes hydrophilic colloids which help in stabilizing the emulsions by developing multi-molecular adsorption at o/w emulsion interface - Usually they are known to be called as emulsion stabilizers by enhancing the viscosity of the dispersed mixture
(c) Solid particle adsorption
- In this type of emulsion, finely divided solid particles are adsorbed at the emulsion interface and thereby help in stabilizing it. Such particles are wetted by either water or oil; thereby used to form a particulate film which would reduce the risk of coalescence

3.1.1 Oils/lipids

The application of lipids/oils used in the formulation of nanoemulsions are usually depends on the solubility of drug components. However, nanoemuslions contain about 5-20% of lipid or oil droplets especially to develop o/w type of emulsion. Some of the examples of re-esterified derivatives are sesame oil, soya-bean oil, safflower oil, rice bran oil, coconut oil, cotton seed oil are employed to formulate the novel drug delivery system. D-α-tocopherol, celeic acid, ethyl oleate has been extensively employed in the formulation of oral, tropical and parenteral based nanoemulsions. Suitable marine oils such as salmon oil used for several emulsification purposes.

3.1.2 Surfactants/co-surfactants

These are usually amphiphilic components used to stabilize nanoemulsions by reducing the interfacial surface tension and helps in preventing droplet aggregration. Commonly employed surfactants are lecithin obtained soyabean and egg yolk, cremophor EL, sodium deoxychoelate, tween 20, 40, 60, 80, solutol HS-15, sodium dodecyl sulfate, casein, polysaccharides such as starch gums, β-lactoglobulin, PEG. The cosurfactants are generally used to complement surfactants as to make the system more stabilize inducing ethanol, glycerine, propanol, transcutol IP, ethylene glycol, propylene glycol (Singh, 2017).

3.1.3 Antioxidants, Chemoprotectants, Preservatives

These components should be of low toxic, thermostable, low cost, ease of availability, acceptability in taste, odour, colour, and chemical/physical compatibility. Benzoic acid, sorbic acid, propionic acid, acetic acid may be used as antifungal agents. Phenoxy-2-ethanol and chlorobutanol are used in ophthalmic nanoformulation. Antioxidants such as ascorbic acid, citranoic acid, phosphoric acid, tartaric acid, citric acid, metabisulfite, thiourea, sodium formaldehyde, sodium bisulfate, tocopherols, butyl hydroxytoluene, citranoic acid, phosphoric acid are employed to formulate nanoemulsion and offer oxidative stability. Addition of pH stabilizers, chelating agents and UV protectants to counter the environmental degradation (Sarathchandraprakash, 2013; Singh, 2017; Lante & Friso, 2013).

Characterization of nanoemulsion can be analyzed by evaluating the stability of emulsions as shown in Table 5.

Table 5. Stability of emulsions (Jaiswal, 2015)

Instability condition	Observations
Flocculation and creaming	-Floccules lead to creaming.
	-Joining of globules to form floccules that settle down in the emulsion.
Cracking	- Separation of dispersed phase as one of the layer. - Permanent instability. - Addition of an emulgent of opposite properties. - Precipitation of emulsifying agent. - Addition of common solvent where both water and oil phase are miscible. - Temperature, micro-organisms, creaming
Phase immersion	- Change in the type of emulsion either o/w to w/o or vice-versa. - Addition of electrolytes. - Phase volume ratio. - Temperature change.

Table 6. Evaluation parameters of Nanoemulsion (Jaiswal, 2015; Alka, 2007; Farhan, 2008; Singh & Vingkar, 2008; Chen, 2008)

Examination parameters	Characteristics
Droplet size analysis	Diffusion method using a light scattering particle size analyzer counter, LS230, TEM
Viscosity	Brookfield viscometer at different temperature
Dilution test	This test can be based to find out the type of emulsion o/w or w/o
Electrical conductivity test	The water is considered as a good conductor of electricity. Therefore, an emulsion with water as a continuous phase will immediately conduct electricity than with oil as in continuous phase.
Staining test/dye-solubility test	This test is carried out with scarlet dye. If the system is w/o type, then the water phase consumed red stain and the background would remain colourless. This test will remain in vice-versa with o/w type.
Fluorescence test	o/w type of emulsion shows fluorescence whereas w/o emulsion shows spotty appearance.
Refractive index	Abbes refractometer
pH	pH meter
Percentage transmittance	UV-visible spectrophotometer.
Conductance measurement	Conductometer

3.2 Methods of Nanoemulsion Formulation Aspects

It involves two effective techniques, using low and high energy processes. In case of high energy method, high pressure homogenization and ultrasonication are involved whereas in low energy methods involve self nanoemulsification and phase-inversion emulsification techniques.

1. High Energy Methods

a. **High energy homogenization**: It provides strong distruptive forces to break down large droplets into nanosized molecules with high kinetic energy. For this, mechanical devices such as microfluidizer, high pressure homogenizers and ultrsonicators are employed. It controls for rheology, stability and colour of the emulsion. High-pressure homogenizers provide high energy to develop small sized droplet upto 1 mm. However, several other forces including hydraulic shear, turbulence are applied in combination with this technique to generate smallest particle sized droplets.
b. **Microfluidization:** It is a mixing technique to generate microsize particle droplets with the application of microfluidizer. During this process, the fluids are allowed to pass through the micro-sized channels of microfluidizer under high pressure ranged from 500 to 20, 000 psi; thereby inside the interaction chamber, the two streams of macroemulsion colloid with each other at very high velocity that develop nanoemulsions.
c. **Ultrasonication:** This is one of the best methods over other two high energy methods. Here, ultrasonic waves form cavitation forces which may break the macro-sized emulsion into nano-sized emulsion.

2. Low energy methods

It requires low energy for the formulation of nanoemulsions.

a. **Phase inversion emulsification:** During this method, phase transition can occur by the spontaneous curvature of surface-active agents.
b. **Transitional phase inversion:** This technique may occur because of changes involved in the spontaneous affinity/curvature of surfactants.
    i. **Phase inversion temperature:** In this, change in spontaneous curvature of surfactant can be inversed due to change in the temperature.
    ii. **Phase inversion composition:** In this system, there may be a change in the system composition than in the temperature of the system.
c. Catastrophic phase inversion

Emulsion inversion point: This may take place due to catastrophic phase inversion as there may be change in the fractioned volume of dispersed phase.

d. **Self nanoemulsification:** This can be done without causing change in the spontaneous curvature of surfactant (Jaiswal, 2015; Tiwari & Amiji, 2006; Banker, 2002; Anton, 2008; Pouton, 1997).

The phase behavior of novel drug delivery system, nanoemulsion consisting of oil, water and surfactant/cosurfactant can be examined using ternary phase diagram where each corner of the phase diagram represents 100% of that particulate component. In case of four or more than four components are examined, each corner may represent a binary mixture of two components such as oil/drug, water/drug and surfactant/co-surfactant. It should be taken into consideration that not every combination of components develops nanoemulsions over a possible range of composition. The multiple phases may develop at low concentration of surfactant and co-surfactant as described below and shown in figure.2

Figure 2. Ternary phase diagram of novel drug delivery systems

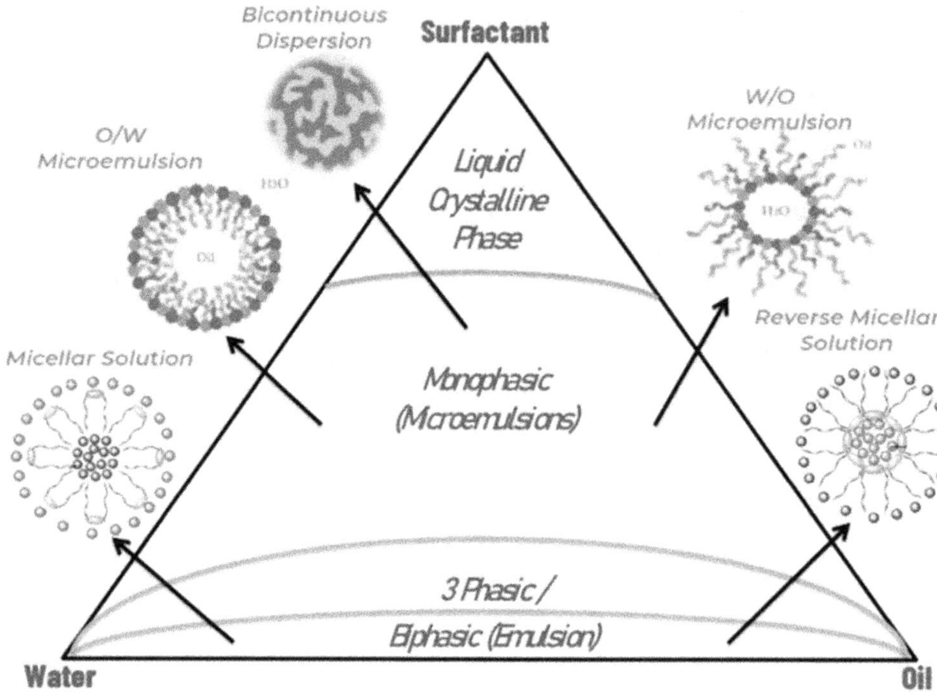

3.2 Literature Survey Based on Nanoemulsion

Researchers developed carvone-schiff base isoniazid nanoemulsions with low concentration of 12% w/w of triacetin by using aqueous phase titration and were evaluated for droplet size, zeta potential, refractive index, % transmittance, surface morphology, in-vitro drug release, thermodynamic stability, emulsification efficiency and polydispersity index. The optimized nanoemulsion formulation should enhance rate of absorption in case of pharmacokinetic studies studied in rat models, when compared to suspension formulation.

Lopes *et al* prepared different colloidal formulations including Nanocapsules, nanoemulsions and nanospheres in association with ethionamide. These formulations were developed using spontaneous emulsification and interfacial deposition methods. In another research work, the investigators developed oil-in-water nanoemulsion containing clarithromycin as antibiotic drug (Lopes, 2000). The developed optimized formulation showed a particle size of 30 nm with improved physical stability and enhanced bioavailability at room temperature.

Mehata *et al* formulated microemulsion of anti-TB drug, rifampicin with oleic acid, phosphate buffer, tween 80 and ethanol. The formulation showed the changes in the structure as from water-in-oil micro-emulsion to oil-in-water phase and controlled drug release in *in-vitro* dissolution studies (Mehta, 2007).

Reddy *et al* developed phospholipid phosal 53 MCT nanoemulsions containing capuramycin such as SQ641, SQ64taua, SQ641-2aua and SQ997-3aua. These nanoemulsion formulations were assessed against M.tuberculosis using J774A.1 mouse macrophage cell line. The nanoemulsion with an emulsifier, d-tocopheryl polyethylene glycol 1000 succinate revealed significant enhanced intracellular activity against free capuramycin (Reddy, 2011).

Table 7. Summary of some of the nano drug delivery formulated anti-TB drugs

Novel Drug Delivery	Drugs	References
Cationic nanoemulsion: chitosan and polymyxin B	Rifampicin	Hussain, 2020
Cationic nanoemulsion gel	Rifampicin	Shobo, 2018
o/w Nanoemulsion	Pretomanid	Alshehri, 2020
Cationic nanoemulsion	Rifampicin	Kramer, 2018
Nanoemulsion adjuvanted vaccine	-	Orr, 2014
Nanoemulsion adjuvanted vaccine	-	Ahmed, 2017
Nanoemulsion vaccine	-	Kaur, 2015
Microemulsion	Rifampicin, Isoniazid, Pyrazinamide	Kaur, 2014
Microemulsion	Rifampicin, Isoniazid, Pyrazinamide	Baboota, 2007
Microemulsion	Terbinafine	Glass, 2005
Microemulsion	Rifampicin, Isoniazid and Pyrazinamide	Baker, 2008

Table 8. Some of the effective anti-TB drugs in nanoemulsion form in clinical aspects

Title of Patent	Application number	Country	References
Nanoemulsion therapeutic compositions and methods of using the same	US8747872B2	U.S	Baker, 2008
An antibacterial rifampicin nanoemulsion and preparation method thereof	CN100528160C	China	Chen, 2006
Moxifloxacin nanoemulsion and its preparation method	CN102920656A	China	Makidon, 2011
Gelled oral pharmaceutical compositions	WO/2011/128628	London	Drageet, 2011
Compositions for inactivating pathogenic microorganisms, methods of making the compositions, and methods of use thereof	US20150150244A9	U.S	Baker, 2004
Polypeptide vaccine and vaccination strategy against mycobacterium	EP2818178A1	European Patent Office	Sable, 2009
Nanoemulsion compositions having anti-inflammatory activity	EP1951200A2	European Patent Office	Baker, 2006

CONCLUSION

This manuscript provides the detail information of various applications of nano drug delivery system in the management of anti-TB drugs. The nanotechnology can overcome from the formulation problems associated with the drug bioavailability, dose frequency, route of administration, stability and side effects. Nanoemulsions are becoming centre of research and development in nanotechnology field. Nanoemulsions are defined as a biphasic two immiscible solvent which can able to produce o/w or w/o emulsion system that has been stabilized using suitable emulsifying agents. The lipidic properties of nanoemulsion usually facilitates the drug target effects within the lymph nodes and thereby with this it ultimately enhances the drug bioavailability and reducing their dose frequency. Incorporation of anti-tubercular drug(s) within the nanoemulsion system can make them to cross the biological barriers more effectively so as to reach the systemic circulations and thus help in declining the loads of Mycobacterium tuberculosis. Novel drug delivery systems not only contain the potential of enhancing the biopharmaceutic parameters but also have the effective potency to deliver the active drug molecule/drug component. This current manuscript has explored various excipients including oils, lipids, surfactants, emulsifying agents/emulsifier, antioxidants, preservatives and so on. Nanoemulsion has become more promising novel drug delivery technology employed in the management of tuberculosis. Nanoemulsion contains various beneficiary effects over other drug delivery systems which include chemical/physical properties, stability, reduced drug dose frequency, bioavailability and side effects. Nanotechnology-based rational targeting may be able to enhance the therapeutic property of a drug component through many ways by reducing adverse side effects and administration drug regimens, either alone or in combinations. This would effectively results in higher patient compliance. Now-a –days the potential novel treatments and strategies consists of several drug components in randomized placebo trials or used in various preclinical studies with promising and effective activities in the management of tuberculosis against various strains of Mycobacterium tuberculosis. The encapsulation of existing available anti-tuberculosis drugs into novel drug delivery systems and moreover the introduction of some active newer drugs either alone or in combination therapies against all different forms of tuberculosis would have revealed in the novel and effective treatments along with reduced side effects. The present book chapter relies on the basic information of nanoemulsion system along with detailed account on the application in the delivery of anti-tubercular drugs as to enhance the immune system against tuberculosis. Additionally, the literature on patent also included in this chapter. Moreover, the book chapter would provide the various nanotechnology applications with respect to the treatment of TB.

REFERENCES

Aagaard, C., Dietrich, J., Doherty, M., & Andersen, P. (2009). TB vaccines: Current status and future perspectives. *Immunology and Cell Biology*, *87*(4), 279–286. doi:10.1038/icb.2009.14 PMID:19350048

Acosta, E. (2009). Bioavailability of nanoparticles in nutrient and nutraceutical delivery. *Current Opinion in Colloid & Interface Science*, *14*(1), 3–15. doi:10.1016/j.cocis.2008.01.002

Ahmed, M., Smith, D. M., Hamouda, T., Rangel-Moreno, J., Fattom, A., & Khader, S. A. (2017). A novel nanoemulsion vaccine induces mucosal Interleukin-17 responses and confers protection upon *Mycobacterium tuberculosis* challenge in mice. *Vaccine*, *35*(37), 4983–4989. doi:10.1016/j.vaccine.2017.07.073 PMID:28774560

Ahmad, Z., Pandey, R., Sharma, S., & Khuller, G. K. (2008). Novel chemotherapy for tuberculosis: Chemotherapeutic potential of econazole- and moxifloxacin-loaded PLG nanoparticles. *International Journal of Antimicrobial Agents*, *31*(2), 142–146. doi:10.1016/j.ijantimicag.2007.10.017 PMID:18155883

Alka, A. J. A., Baboota, S., Shakeel, F., & Shafiq, S. (2007). Design development and evaluation of novel nanoemulsion formulations for transdermal potential of Celecoxib. *Acta Pharmaceutica (Zagreb, Croatia)*, *57*(3), 315–332. doi:10.2478/v10007-007-0025-5 PMID:17878111

Almeida Da Silva, P. E., & Palomino, J. C. (2011). Molecular basis and mechanisms of drug resistance in *Mycobacterium tuberculosis*: Classical and new drugs. *The Journal of Antimicrobial Chemotherapy*, *66*(7), 1417–1430. doi:10.1093/jac/dkr173 PMID:21558086

Alshehri, S., Altamimi, M. A., Hussain, A., Imam, S. S., Singh, S. K., & Faruk, A. (2020). Morphological transition of *M. tuberculosis* and modulation of intestinal permeation by food grade cationic nanoemulsion: *In vitro-ex vivo-in silico* GastroPlus™ studies. *Journal of Drug Delivery Science and Technology*, *60*, 101971. doi:10.1016/j.jddst.2020.101971

Anisimova, Y. V., Gelperina, S. I., Peloquin, C. A., & Heifets, L. B. (2000). Nanoparticles as antituberculosis drugs carriers: Effect on activity against *Mycobacterium tuberculosis* in human monocyte-derived macrophages. *Journal of Nanoparticle Research*, *2*(2), 165–171. doi:10.1023/A:1010061013365

Anton, N., Benoit, J. P., & Saulnier, P. (2008). Design and production of nanoparticles formulated from nano-emulsion templates-a review. *Journal of Controlled Release*, *128*(3), 185–199. doi:10.1016/j.jconrel.2008.02.007 PMID:18374443

Anton, N., & Vandamme, T. F. (2011). Nano-emulsions and micro-emulsions: Clarifications of the critical differences. *Pharmaceutical Research*, *28*(5), 978–985. doi:10.100711095-010-0309 1 PMID:21057856

Baboota, S., Al-Azaki, A., Kohli, K., Ali, J., Dixit, N., & Shakeel, F. (2007). Development and evaluation of a microemulsion formulation for transdermal delivery of terbinafine. *PDA Journal of Pharmaceutical Science and Technology*, *61*(4), 276–285. PMID:17933209

Banker, G. S., Lieberman, H. A., & Rieger, M. M. (2002). Pharmaceutical dosage forms. *Disperse Systems Marcel Dekker*, *2*(3), 339–340.

Baker, J. R. (2006). *Nanoemulsion compositions having anti-inflammatory activity*. https://patents.google.com/patent/EP1951200A2/en

Baker, J. R., Annis, T. C., Hamouda, T., & Vengroff, L. (2004). *Compositions for inactivating pathogenic microorganisms, methods of making the compositions, and methods of use thereof*. https://patents.google.com/patent/US20150150244

Baker, J. R., Rathinavelu, S., Makidon, P. E., LiPuma, J.J., & Nigavekar, S. (2008). *Nanoemulsion therapeutic compositions and methods of using the same*. https://patents.google.com/patent/US8747872B2/en

Bellini, R. G., Guimarães, A. P., Pacheco, M. A., Dias, D. M., Furtado, V. R., de Alencastro, R. B., & Horta, B. A. (2015). Association of the anti-tuberculosis drug rifampicin with a PAMAM dendrimer. *Journal of Molecular Graphics & Modelling*, *60*, 34–42. doi:10.1016/j.jmgm.2015.05.012 PMID:26093506

Bisht, D., Sharma, D., Sharma, D., Singh, R., & Gupta, V. K. (2019). Recent insights into *Mycobacterium tuberculosis* through proteomics and implications for the clinic. *Expert Review of Proteomics*, *16*(5), 443–456. doi:10.1080/14789450.2019.1608185 PMID:31032653

Bouchemal, K., Briancon, S., Fessi, H., & Perrier, E. (2004). Nano-emulsion formulation using spontaneous emulsification: Solvent, oil and surfactant optimization. *International Journal of Pharmaceutics*, *280*(1-2), 242–259. doi:10.1016/j.ijpharm.2004.05.016 PMID:15265563

Caminero, J. A., Sotgiu, G., Zumla, A., & Migliori, G. B. (2010). Best drug treatment for multidrug-resistant and extensively drug-resistant tuberculosis. *The Lancet. Infectious Diseases*, *10*(9), 621–629. doi:10.1016/S1473-3099(10)70139-0 PMID:20797644

Chan, E. D., Strand, M. J., & Iseman, M. D. (2008). Treatment outcomes in extensively resistant tuberculosis. *The New England Journal of Medicine*, *359*(6), 657–659. doi:10.1056/NEJMc0706556 PMID:18687654

Chen, F. J. (2006). *An antibacterial rifampicin nanoemulsion and preparation method thereof.* https://patents.google.com/patent/CN100528160C/en

Chen, H., Du, D., Mao, C. M. D., Wan, J., Xu, H., & Yang, X. (2008). Hydrogel-thickened nanoemulsion system for topical delivery of lipophilic drugs. *International Journal of Pharmaceutics*, *353*(1-2), 270–276. doi:10.1016/j.ijpharm.2007.11.051 PMID:18215479

Cholo, M. C., Steel, H. C., Fourie, P. B., Germishuizen, W. A., & Anderson, R. (2012). Clofazimine: Current status and future prospects. *The Journal of Antimicrobial Chemotherapy*, *67*(2), 290–298. doi:10.1093/jac/dkr444 PMID:22020137

Clemens, D. L., Lee, B. Y., Xue, M., Thomas, C. R., Meng, H., Ferris, D., Nel, A. E., Zink, J. I., & Horwitz, M. A. (2012). Targeted intracellular delivery of antituberculosis drugs to *Mycobacterium tuberculosis*-infected macrophages via functionalized mesoporous silica nanoparticles. *Antimicrobial Agents and Chemotherapy*, *56*(5), 2535–2545. doi:10.1128/AAC.06049-11 PMID:22354311

Comstock, G. W., Livesay, V. T., & Woolpert, S. F. (1974). The prognosis of a positive tuberculin reaction in childhood and adolescence. *American Journal of Epidemiology*, *99*(2), 131–138. doi:10.1093/oxfordjournals.aje.a121593 PMID:4810628

D'Ambrosio, L., Centis, R., Sotgiu, G., Pontali, E., Spanevello, A., & Migliori, G. B. (2015). New anti-tuberculosis drugs and regimens: 2015 update. *ERJ Open Research*, *1*(1), 00010–02015. doi:10.1183/23120541.00010-2015 PMID:27730131

Drew, R. H. (2020). Isoniazid: An overview. *UpToDate*. https://www.uptodate.com/contents/isoniazid-an-overview

du Toit, L. C., Pillay, V., & Danckwerts, M. P. (2006). Tuberculosis chemotherapy: Current drug delivery approaches. *Respiratory Research*, *7*(1), 118–125. doi:10.1186/1465-9921-7-118 PMID:16984627

Dutt, M., & Khuller, G. K. (2001). Therapeutic efficacy of poly(DL-lactide-co-glycolide)-encapsulated antitubercular drugs against *Mycobacterium tuberculosis* infection induced in mice. *Antimicrobial Agents and Chemotherapy*, *45*(1), 363–366. doi:10.1128/AAC.45.1.363-366.2001 PMID:11121000

Dutt, M., & Khuller, G. K. (2001). Chemotherapy of *Mycobacterium tuberculosis* infections in mice with a combination of isoniazid and rifampicin entrapped in Poly (DL-lactide-co-glycolide) microparticles. *The Journal of Antimicrobial Chemotherapy*, *47*(6), 829–835. doi:10.1093/jac/47.6.829 PMID:11389115

Esmaeili, F., Hosseini-Nasr, M., Rad-Malekshahi, M., Samadi, N., Atyabi, F., & Dinarvand, R. (2007). Preparation and antibacterial activity evaluation of rifampicin-loaded poly lactide-co-glycolide nanoparticles. *Nanomedicine (London)*, *3*(2), 161–167. doi:10.1016/j.nano.2007.03.003 PMID:17468055

Farhan, A. J., Mushir, A., Faiyaz, S., Cushman, T., Roop, K. K., & Sheikh, S. (2008). Investigation of nanoemulsion system for transdermal delivery of domperidone: *Ex-vivo* and *in vivo* studies. *Current Nanoscience*, *4*(4), 381–390. doi:10.2174/157341308786306071

Floss, H. G., & Yu, T. W. (2005). Rifamycin-mode of action, resistance, and biosynthesis. *Chemical Reviews*, *105*(2), 621–632. doi:10.1021/cr030112j PMID:15700959

Garti, N., & Benichou, A. (2003). Recent developments in double emulsions for food applications. *Sematic Scholar*, *132*, 353–412. doi:10.1201/9780203913222.ch10

Gasco, M. R., Gallarate, M., & Pattarino, F. (1991). *In vitro* permeation of azelaic acid from viscosized microemulsions. *International Journal of Pharmaceutics*, *69*(3), 193–196. doi:10.1016/0378-5173(91)90361-Q

Gaspar, M. M., Cruz, A., Fraga, A. G., Castro, A. G., Cruz, M. E., & Pedrosa, J. (2008). Developments on drug delivery systems for the treatment of mycobacterial infections. *Current Topics in Medicinal Chemistry*, *8*(7), 579–591. doi:10.2174/156802608783955629 PMID:18473884

Georghiou, S. B., Magana, M., Garfein, R. S., Catanzaro, D. G., Catanzaro, A., & Rodwell, T. C. (2012). Evaluation of genetic mutations associated with *Mycobacterium tuberculosis* resistance to amikacin, kanamycin and capreomycin: A systematic review. *PLoS One*, *7*(3), e33275. doi:10.1371/journal.pone.0033275 PMID:22479378

Glass, B. D., Agatonovic-Kustrin, S., & Wisch, M. H. (2005). Artificial neural networks to optimize formulation components of a fixed-dose combination of rifampicin, isoniazid and pyrazinamide in a microemulsion. *Current Drug Discovery Technologies*, *2*(3), 195–201. doi:10.2174/1570163054866864 PMID:16472228

Gresser, U. (2001). Amoxicillin-clavulanic acid therapy may be associated with severe side effects - review of the literature. *European Journal of Medical Research*, *6*, 139–149. PMID:11309226

Horita, Y., Maeda, S., Kazumi, Y., & Doi, N. (2014). *In vitro* susceptibility of *Mycobacterium tuberculosis* isolates to an oral carbapenem alone or in combination with β-lactamase inhibitors. *Antimicrobial Agents and Chemotherapy*, *58*(11), 7010–7014. doi:10.1128/AAC.03539-14 PMID:25224000

Hornik, C. P., Herring, A. H., Benjamin, D. K., Capparelli, E. V., Kearns, G. L., van den Anker, J., Cohen-Wolkowiez, M., Clark, R. H., & Smith, P. B. (2013). Best pharmaceuticals for children act-pediatric trials network. Adverse events associated with meropenem versus imipenem/cilastatin therapy in a large retrospective cohort of hospitalized infants. *The Pediatric Infectious Disease Journal*, *32*(7), 748–753. doi:10.1097/INF.0b013e31828be70b PMID:23838776

Hussain, A., Altamimi, M. A., Alshehri, S., Imam, S. S., Shakeel, F., & Singh, S. K. (2020). Novel approach for transdermal delivery of rifampicin to induce synergistic antimycobacterial effects against cutaneous and systemic tuberculosis using a cationic nanoemulsion gel. *International Journal of Nanomedicine*, *15*, 1073–1094. doi:10.2147/IJN.S236277 PMID:32103956

Ines, G. R. M., Juliana, M. C., Araceli, O. M., Laura, S. T., & Olga, M. B. (2015). Long-term stability of food-grade nanoemulsions from high methoxyl pectin containing essential oils. *Food Hydrocolloids*, *52*, 438–446. doi:10.1016/j.foodhyd.2015.07.017

Ipuge, Y. A. I., Rieder, H. L., & Enarson, D. A. (1995). Adverse cutaneous reactions to thiacetazone for tuberculosis treatment in Tanzania. *Lancet*, *346*(8976), 657–660. doi:10.1016/S0140-6736(95)92278-4 PMID:7544858

Jain, C. P., & Vyas, S. P. (1995). Preparation and characterization of niosomes containing rifampicin for lung targeting. *Journal of Microencapsulation*, *12*(4), 401–407. doi:10.3109/02652049509087252 PMID:8583314

Jaiswal, M., Dudhe, R., & Sharma, P.K. (2015). Nanoemulsion: an advanced mode of drug delivery system. *3 Biotech, 5*(2), 123-127. doi:10.1007/s13205-014-0214-0

Jamieson, F. B., Guthrie, J. L., Neemuchwala, A., Lastovetska, O., Melano, R. G., & Mehaffy, C. (2014). Profiling of rpoB mutations and MICs for rifampin and rifabutin in *Mycobacterium tuberculosis. Journal of Clinical Microbiology*, *52*(6), 2157–2162. doi:10.1128/JCM.00691-14 PMID:24740074

Jatav, V. S., Singh, S. K., Khatri, P., Sharma, A. K., & Singh, R. (2011). Formulation and *in-vitro* evaluation of rifampicin-loaded niosomes. *Journal of Chemical and Pharmaceutical Research*, *3*(2), 199–203.

Kaufmann, S.H., & McMichael, A.J. (2005). Annulling a dangerous liaison: Vaccination strategies against AIDS and tuberculosis. *Nature Medicine, 4*, S33-44. . doi:10.1038/nm1221

Kaufmann, S. H. E. (2001). How can immunology contribute to the control of tuberculosis? *Nature Reviews. Immunology*, *1*(1), 20–30. doi:10.1038/35095558 PMID:11905811

Kaur, G., & Mehta, S. K. (2014). Probing location of anti-TB drugs loaded in Brij 96 microemulsions using thermoanalytical and photophysical approach. *Journal of Pharmaceutical Sciences*, *103*(3), 937–944. doi:10.1002/jps.23857 PMID:24425102

Kaur, G., Mehta, S. K., Kumar, S., Bhanjana, G., & Dilbaghi, N. (2015). Coencapsulation of hydrophobic and hydrophilic antituberculosis drugs in synergistic Brij 96 microemulsions: A biophysical characterization. *Journal of Pharmaceutical Sciences*, *104*(7), 2203–2212. doi:10.1002/jps.24469 PMID:25951802

Khan, D. H., Bashir, S., Khan, M. I., Figueiredo, P., Santos, H. A., & Peltonen, L. (2020). Formulation optimization and *in vitro* characterization of rifampicin and ceftriaxone dual drug loaded niosomes with high energy probe sonication technique. *Journal of Drug Delivery Science and Technology*, *58*, 101763. doi:10.1016/j.jddst.2020.101763

Kim, C. K., Cho, Y. J., & Gao, Z. G. (2001). Preparation and evaluation of biphenyl dimethyl dicarboxylate microemulsions for oral delivery. *Journal of Controlled Release*, *70*(1-2), 149–155. doi:10.1016/S0168-3659(00)00343-6 PMID:11166415

Kisich, K.O., Gelperina, S., Higgins, M.P., Wilson, S., Shipulo, E., Oganesyan, E., & Heifets, L. (2007). Encapsulation of moxifloxacin within poly (butyl cyanoacrylate) nanoparticles enhances efficacy against intracellular *Mycobacterium tuberculosis. International Journal of Pharmaceutics, 10*(345), 154-162. . doi:10.1016/j.ijpharm.2007.05.062

Komaiko, J. S., & McClements, D. J. (2016). Formation of food-grade nanoemulsions using low-energy preparation methods: A review of available methods. *Comprehensive Reviews in Food Science and Food Safety*, *15*(2), 331–352. doi:10.1111/1541-4337.12189 PMID:33371595

Kramer, R. M., Archer, M. C., Orr, M. T., Dubois, C. N., Beebe, E. A., Huang, P. D., Dowling, Q. M., Schwartz, A. M., Fedor, D. M., Vedvick, T. S., & Fox, C. B. (2018). Development of a thermostable nanoemulsion adjuvanted vaccine against tuberculosis using a design-of-experiments approach. *International Journal of Nanomedicine*, *26*(13), 3689–3711. doi:10.2147/IJN.S159839 PMID:29983563

Kriwet, K., & Müller-Goymann, C. (1995). Diclofenac release from phospholipid drug systems and permeation through excised human stratum corneum. *International Journal of Pharmaceutics*, *125*(2), 231–242. doi:10.1016/0378-5173(95)00130-B

Kumar, P. V., Asthana, A., Dutta, T., & Jain, N. K. (2006). Intracellular macrophage uptake of rifampicin loaded mannosylated dendrimers. *Journal of Drug Targeting*, *14*(8), 546–556. doi:10.1080/10611860600825159 PMID:17050121

Kunin, C. M. (1996). Antimicrobial activity of rifabutin. *Clinical Infectious Diseases*, *22*(1, Supplement_1), S3–S13. doi:10.1093/clinids/22.Supplement_1.S3 PMID:8785253

Lante, A., & Friso, D. (2013). Oxidative stability and rheological properties of nanoemulsions with ultrasonic extracted green tea infusion. *Food Research International*, *54*(1), 269–276. doi:10.1016/j.foodres.2013.07.009

Li, F., Wan, L., Xiao, T., Liu, H., Jiang, Y., Zhao, X., Wang, R., & Wan, K. (2018). *In vitro* activity of β-Lactams in Combination with β-Lactamase Inhibitors against *Mycobacterium tuberculosis* Clinical Isolates, *BioMed Research International*. *Article ID*, *3579832*, 1–8. Advance online publication. doi:10.1155/2018/3579832 PMID:30065936

Lopes, E., Pohlmann, A. R., Bassani, V., & Guterres, S. S. (2000). Polymeric colloidal systems containing ethionamide: Preparation and physico-chemical characterization. *Die Pharmazie*, *55*(7), 527–530. PMID:10944782

Makidon, P. E. (2011). *Moxifloxacin nanoemulsion and its preparation method*. https://patents.google.com/patent/CN102920656A/en

McClements, D. J., Decker, E. A., & Weiss, J. (2007). Emulsion-based delivery systems for lipophilic bioactive components. *Journal of Food Science*, *72*(8), 109–124. doi:10.1111/j.1750-3841.2007.00507.x PMID:17995616

McClements, D. J., & Rao, J. (2011). Food-grade nanoemulsions: Formulation, fabrication, properties, performance, biological fate, and potential toxicity. *Critical Reviews in Food Science and Nutrition*, *51*(4), 285–330. doi:10.1080/10408398.2011.559558 PMID:21432697

Mehanna, M.M., Mohyeldin, S.M., & Elgindy, N.A. (2019). Rifampicin-carbohydrate spray-dried nanocomposite: A futuristic multiparticulate platform for pulmonary delivery. *International Journal of Nanomedicine, 14*, 9089-9112. Doi: org/ doi:10.2147/IJN.S211182

Mehta, P., Bothiraja, C., Kadam, S., & Pawar, A. (2018). Potential of dry powder inhalers for tuberculosis therapy: Facts, fidelity and future. *ArtifCells Nanomedicine and Biotechnology, 46*(3), 791-806. Doi: org/ doi:10.1080/21691401.2018.1513938.3

Mehta, S. K., Kaur, G., & Bhasin, K. K. (2007). Analysis of Tween based microemulsion in the presence of TB drug rifampicin. *Colloids and Surfaces. B, Biointerfaces*, *60*(1), 95–104. doi:10.1016/j.colsurfb.2007.06.012 PMID:17646089

Mehta, S. K., Kaur, G., & Bhasin, K. K. (2008). Incorporation of antitubercular drug isoniazid in pharmaceutically accepted microemulsion: Effect on microstructure and physical parameters. *Pharmaceutical Research*, *25*(1), 227–236. doi:10.100711095-007-9355-8 PMID:17577642

Melo, K. J., Souza, N., Ferraz, H., Fotaki, N., Lobenberg, R., & Bou-Chacra, N. (2015). Formulation with Povacoat® to Prevent Aggregation of Rifampin Nanocrystal Using High-Pressure Homogenization. *AAPS Annual Meeting*.

Melo, K. J. C., Henostroza, M. A. B., Löbenberg, R., & Bou-Chacra, N. A. (2020). Rifampicin nanocrystals: Towards an innovative approach to treat tuberculosis. *Materials Science and Engineering C*, *112*, 110895. doi:10.1016/j.msec.2020.110895 PMID:32409052

Millard, J., & Moore, D. A. J. (2015). Multidrug resistant tuberculosis. *BMJ (Clinical Research Ed.)*, *350*(feb26 10), h882. doi:10.1136/bmj.h882 PMID:25721508

Mitnick, C. D., Appleton, S. C., & Shin, S. S. (2008). Epidemiology and treatment of multidrug resistant tuberculosis. *Seminars in Respiratory and Critical Care Medicine*, *29*(5), 499–524. doi:10.1055-0028-1085702 PMID:18810684

Moretton, M. A., Chiappetta, D. A., & Sosnik, A. (2012). Cryoprotection-lyophilization and physical stabilization of rifampicin-loaded flower-like polymeric micelles. *Journal of the Royal Society, Interface*, *9*(68), 487–502. doi:10.1098/rsif.2011.0414 PMID:21865255

Moretton, M. A., Glisoni, R. J., Chiappetta, D. A., & Sosnik, A. (2010). Molecular implications in the nanoencapsulation of the anti-tuberculosis drug rifampicin within flower-like polymeric micelles. *Colloids and Surfaces. B, Biointerfaces*, *79*(2), 467–479. doi:10.1016/j.colsurfb.2010.05.016 PMID:20627665

Moretton, M. A., Hocht, C., Taira, C., & Sosnik, A. (2014). Rifampicin-loaded 'flower-like' polymeric micelles for enhanced oral bioavailability in an extemporaneous liquid fixed-dose combination with isoniazid. *Nanomedicine (London)*, *9*(11), 25–30. doi:10.2217/nnm.13.154 PMID:24410279

Morlock, G. P., Metchock, B., Sikes, D., Crawford, J. T., & Cooksey, R. C. (2003). ethA, inhA, and katG loci of ethionamide-resistant clinical *Mycobacterium tuberculosis* isolates. *Antimicrobial Agents and Chemotherapy*, *47*(12), 3799–3805. doi:10.1128/AAC.47.12.3799-3805.2003 PMID:14638486

Mullaicharam, A., & Murthy, R. (2004). Lung accumulation of niosome-entrapped rifampicin following intravenous and intratracheal administration in the rat. *Journal of Drug Delivery Science and Technology, 14*(2), 99–104. doi:10.1016/S1773-2247(04)50020-5

Narita, M., Tsuji, B. T., & Yu, V. L. (2007). Linezolid-associated peripheral and optic neuropathy, lactic acidosis, and serotonin syndrome. *Pharmacotherapy, 27*(8), 1189–1197. doi:10.1592/phco.27.8.1189 PMID:17655517

Nasiri, M. J., Haeili, M., Ghazi, M., Goudarzi, H., Pormohammad, A., Fooladi, A. A. I., & Feizabadi, M. M. (2017). New insights in to the intrinsic and acquired drug resistance mechanisms in mycobacteria. *Frontiers in Microbiology, 8*, 681. Advance online publication. doi:10.3389/fmicb.2017.00681 PMID:28487675

Orr, M. T., Kramer, R. M., Barnes, L., Dowling, Q. M., Desbien, A. L., Beebe, E. A., Laurance, J. D., Fox, C. B., Reed, S. G., Coler, R. N., & Vedvick, T. S. (2014). Elimination of the cold-chain dependence of a nanoemulsion adjuvanted vaccine against tuberculosis by lyophilization. *Journal of Controlled Release, 177*, 20–26. doi:10.1016/j.jconrel.2013.12.025 PMID:24382398

Padayatchi, N., & Friedland, G. (2008). Decentralised management of drug-resistant tuberculosis (MDR- and XDR-TB) in South Africa: An alternative model of care. *The International Journal of Tuberculosis and Lung Disease, 12*, 978–980. PMID:18647461

Pandey, R., Zahoor, A., Sharma, S., & Khuller, G. K. (2003). Nanoparticle encapsulated antitubercular drugs as a potential oral drug delivery system against murine tuberculosis. *Tuberculosis (Edinburgh, Scotland), 83*(6), 373–378. doi:10.1016/j.tube.2003.07.001 PMID:14623168

Pandey, R., & Khuller, G. K. (2005). Solid lipid particle-based inhalable sustained drug delivery system against experimental tuberculosis. *Tuberculosis (Edinburgh, Scotland), 85*(4), 227–234. doi:10.1016/j.tube.2004.11.003 PMID:15922668

Pandey, R., Sharma, A., Zahoor, A., Sharma, S., Khuller, G. K., & Prasad, B. (2003). Poly (DL-lactide-co-glycolide) nanoparticle-based inhalable sustained drug delivery system for experimental tuberculosis. *The Journal of Antimicrobial Chemotherapy, 52*(6), 981–986. doi:10.1093/jac/dkg477 PMID:14613962

Patil, K., Bagade, S., Bonde, S., Sharma, S., & Saraogi, G. (2018). Recent therapeutic approaches for the management of tuberculosis: Challenges and opportunities. *Biomedicine and Pharmacotherapy, 99*, 735–745. doi:10.1016/j.biopha.2018.01.115 PMID:29710471

Peters, K., Leitzke, S., Diederichs, J. E., Borner, K., Hahn, H., Müller, R. H., & Ehlers, S. (2000). Preparation of a clofazimine nanosuspension for intravenous use and evaluation of its therapeutic efficacy in murine *Mycobacterium avium* infection. *The Journal of Antimicrobial Chemotherapy, 45*(1), 77–83. doi:10.1093/jac/45.1.77 PMID:10629016

Pouton, C. W. (1997). Formulation of selfemulsifying drug delivery systems. *Advanced Drug Delivery Reviews, 25*(1), 47–58. doi:10.1016/S0169-409X(96)00490-5

Rani, N. P., Suriyaprakash, T. N. K., & Senthamarai, R. (2010). Formulation and evaluation of rifampicin and gatifloxacin niosomes on logarithmic-phase cultures of *Mycobacterium tuberculosis. International Journal of Pharma and Bio Sciences, 1*(4), 379–387.

Reddy, V. M., Bogatcheva, E., Einck, L., & Nacy, C. A. (2011). Nanoemulsion formulation enhances intracellular activity of capuramycin analogues against Mycobacterium tuberculosis. *Drug Delivery Letters*, *1*(2), 150–158. doi:10.2174/2210304x11101020150

Reverchon, E., De Marco, I., & Della Porta, G. (2002). Rifampicin microparticles production by supercritical antisolvent precipitation. *International Journal of Pharmaceutics*, *243*(1-2), 83–91. doi:10.1016/S0378-5173(02)00261-2 PMID:12176297

Sable, S., Plikaytis, B. B., Shinnick, T. M., Amara, R. R., & Cheruvu, M. (2009). *Polypeptide vaccine and vaccination strategy against Mycobacterium*. https://patents.google.com/patent/EP2818178A1/en

Saraogi, G. K., Gupta, P., Gupta, U. D., Jain, N. K., & Agrawal, G. P. (2010). Gelatin nanocarriers as potential vectors for effective management of tuberculosis. *International Journal of Pharmaceutics*, *385*(1-2), 143–149. doi:10.1016/j.ijpharm.2009.10.004 PMID:19819315

Sarathchandraprakash, N. K., Mahendra, C., Prashanth, S. J., Manral, K., Babu, U. V., & Gowda, D. V. S. (2013). Emulsions and emulsifiers. *Asian Journal of Experimental Chemistry*, *8*(1&2), 30–45.

Schatz, A., Bugie, E., Waksman, S. A., Hanssen, A. D., Patel, R., & Osmon, D. R. (2005). Streptomycin, a substance exhibiting antibiotic activity against gram-positive and gram-negative bacteria. *Clinical Orthopaedics and Related Research*, (437), 3–6. doi:10.1097/01.blo.0000175887.98112.fe PMID:16056018

Schianti, J. N., Cerize, N. N. P., Oliveira, A. M., Derenzo, S., & Gingora-Rubio, M. R. (2011). Rifampicin nanosuspension production using microfluidic device. *NSTI-Nanotech*, *1*, 229–232.

Shipulo, E. V., Lyubimov, I. I., Maksimenko, O. O., Vanchugova, L. V., Oganesyan, E. A., Sveshnikov, P. G., Biketov, S. F., Severin, E. S., Heifets, L. B., & Gel'perina, S. E. (2008). Development of a nanosomal formulation of moxifloxacin based on poly(butyl-2-cyanoacrylate). *Pharmaceutical Chemistry Journal*, *42*(3), 43–47. doi:10.100711094-008-0073-2

Shivangi, & Meena, L.S. (2018). A novel approach in treatment of tuberculosis by targeting drugs to infected macrophages using biodegradable nanoparticles. *Applied Biochemistry and Biotechnology, 185,* 815–821. Doi: doi:org/10.1007/s12010-018-2695-5

Shobo, A., Pamreddy, A., Kruger, H. G., Makatini, M. M., Naicker, T., Govender, T., & Baijnath, S. (2018). Enhanced brain penetration of pretomanid by intranasal administration of an oil-in-water nanoemulsion. *Nanomedicine (London)*, *13*(9), 997–1008. doi:10.2217/nnm-2017-0365 PMID:29790418

Silva, M., Ricelli, N. L., El Seoud, O., Valentim, C. S., Ferreira, A. G., Sato, D. N., Leite, C. Q., & Ferreira, E. I. (2006). Potential tuberculostatic agent: Micelle-forming pyrazinamide prodrug. *Archiv der Pharmazie*, *339*(6), 283–290. doi:10.1002/ardp.200500039 PMID:16688684

Singh, G., Raghuvanshi, H. K., Anand, A., Pundir, R., & Dwivedi, H. (2010). Targeted delivery of rifampicin by niosomal drug delivery system. *Journal of Pharmacy Research*, *3*(5), 1152–1154.

Singh, J. A., Upshur, R., & Padayatchi, N. (2007). XDR-TB in South Africa: No time for denial or complacency. *PLoS Medicine*, *4*(1), e50. doi:10.1371/journal.pmed.0040050 PMID:17253901

Singh, K. K., & Vingkar, S. K. (2008). Formulation, antimalarial activity and biodistribution of oral lipid nanoemulsion of primaquine. *International Journal of Pharmaceutics*, *347*(1&2), 136–143. doi:10.1016/j.ijpharm.2007.06.035 PMID:17709216

Singh, S. (2004). Tuberculosis. *Current Anaesthesia and Critical Care*, *15*(3), 165–171. doi:10.1016/j.cacc.2004.05.007

Singh, Y., Meher, J. G., Raval, K., Khan, F. A., Chaurasia, M., Jain, N. K., & Chourasia, M. K. (2017). Nanoemulsion: Concepts, development and applications in drug delivery. *Journal of Controlled Release*, *28*(252), 28–49. doi:10.1016/j.jconrel.2017.03.008 PMID:28279798

Skeiky, Y. A., & Sadoff, J. C. (2006). Advances in tuberculosis vaccine strategies. *Nature Reviews. Microbiology*, *4*(6), 469–476. doi:10.1038/nrmicro1419 PMID:16710326

Somoskovi, A., Parsons, L. M., & Salfinger, M. (2001). The molecular basis of resistance to isoniazid, rifampin, and pyrazinamide in *Mycobacterium tuberculosis*. *Respiratory Research*, *2*(3), 164–168. doi:10.1186/rr54 PMID:11686881

Sun, Z., Zhang, J., Song, H., Zhang, X., Li, Y., Tian, M., Liu, Y., Zhao, Y., & Li, C. (2010). Concomitant increases in spectrum and level of drug resistance in *Mycobacterium tuberculosis* isolates. *The International Journal of Tuberculosis and Lung Disease*, *14*(11), 1436–1441. PMID:20937184

Tadros, T., Izquierdo, P., Esquena, J., & Solans, C. (2004). Formation and stability of nano-emulsions. *Advances in Colloid and Interface Science*, *108*, 303–318. doi:10.1016/j.cis.2003.10.023 PMID:15072948

Tiwari, S. B., & Amiji, M. M. (2006) Nanoemulsion formulations for tumor-targeted delivery. In Nanotech Cancer Therapy (pp. 723-739). Taylor and Francis Group.

Toti, U. S., Guru, B. R., Hali, M., McPharlin, C. M., Wykes, S. M., Panyam, J., & Whittum-Hudson, J. A. (2011). Targeted delivery of antibiotics to intracellular chlamydial infections using PLGA nanoparticles. *Biomaterials*, *32*(27), 6606–6613. doi:10.1016/j.biomaterials.2011.05.038 PMID:21652065

Trauner, A., Borrell, S., Reither, K., & Gagneux, S. (2014). Evolution of drug resistance in tuberculosis: Recent progress and implications for diagnosis and therapy. *Drugs*, *74*(10), 1063–1072. doi:10.100740265-014-0248-y PMID:24962424

Tripathi, A., Gupta, R., & Saraf, S. A. (2010). PLGA nanoparticles of antitubercular drug: Drug loading and release studies of a water-in-soluble drug. *International Journal of Pharm Tech Research*, *2*(3), 2116–2123.

Trotta, M. (1999). Influence of phase transformation on indomethacin release from microemulsions. *Journal of Controlled Release*, *60*(2-3), 399–405. doi:10.1016/S0168-3659(99)00094-2 PMID:10425344

Vacher, S., Pellegrin, J. L., Leblanc, F., Fourche, J., & Maugein, J. (1999). Comparative antimycobacterial activities of ofloxacin, ciprofloxacin and grepafloxacin. *The Journal of Antimicrobial Chemotherapy*, *44*(5), 647–652. doi:10.1093/jac/44.5.647 PMID:10552981

Van der Paardt, A. L., Akkerman, O. W., Gualano, G., Palmieri, F., Davies Forsman, L., Aleksa, A., Tiberi, S., de Lange, W. C., Bolhuis, M. S., Skrahina, A., van Soolingen, D., Kosterink, J. G., Migliori, G. B., van der Werf, T. S., & Alffenaar, J. C. (2017). Safety and tolerability of clarithromycin in the treatment of multidrug-resistant tuberculosis. *The European Respiratory Journal*, *49*(3), 1601612. doi:10.1183/13993003.01612-2016 PMID:28331034

Velikov, K. P., & Pelan, E. (2008). Colloidal delivery systems for micronutrients and nutraceuticals. *Soft Matter*, *4*(10), 1964–1980. doi:10.1039/b804863k

Vibe, C. B., Fenaroli, F., Pires, D., Wilson, S. R., Bogoeva, V., Kalluru, R., Speth, M., Anes, E., Griffiths, G., & Hildahl, J. (2016). Thioridazine in PLGA nanoparticles reduces toxicity and improves rifampicin therapy against mycobacterial infection in zebrafish. *Nanotoxicology*, *10*(6), 680–688. doi:10.3109/17435390.2015.1107146 PMID:26573343

Von Groll, A., Martin, A., Jureen, P., Hoffner, S., Vandamme, P., Portaels, F., Palomino, J. C., & da Silva, P. A. (2009). Fluoroquinolone resistance in *Mycobacterium tuberculosis* and mutations in gyrA and gyrB. *Antimicrobial Agents and Chemotherapy*, *53*(10), 4498–4500. doi:10.1128/AAC.00287-09 PMID:19687244

Vyas, S. P., Kannan, M. E., Jain, S., Mishra, V., & Singh, P. (2004). Design of liposomal aerosols for improved delivery of rifampicin to alveolar macrophages. *International Journal of Pharmaceutics*, *269*(1), 37–49. doi:10.1016/j.ijpharm.2003.08.017 PMID:14698575

Wagner, J. G., Gerrard, E. S., & Kaiser, D. G. (1996). The effect of the dosage form on serum levels of indoxole. *Clinical Pharmacology and Therapeutics*, *7*(5), 610–619. doi:10.1002/cpt196675610 PMID:5957166

Wehrli, W. (1983). Rifampin: Mechanisms of action and resistance. *Reviews of Infectious Diseases*, *5*(3), S407–S411. doi:10.1093/clinids/5.Supplement_3.S407 PMID:6356275

Weiss, J., Takhistov, P., & McClements, D. J. (2006). Functional materials in food nanotechnology. *Journal of Food Science*, *71*(9), 107–116. doi:10.1111/j.1750-3841.2006.00195.x

Willby, M., Sikes, R. D., Malik, S., Metchock, B., & Posey, J. E. (2015). Correlation between GyrA substitutions and ofloxacin, levofloxacin, and moxifloxacin cross-resistance in Mycobacterium tuberculosis. *Antimicrobial Agents and Chemotherapy*, *59*(9), 5427–5434. doi:10.1128/AAC.00662-15 PMID:26100699

Wolucka, B. A., McNeil, M. R., de Hoffmann, E., Chojnacki, T., & Brennan, P. J. (1994). Recognition of the lipid intermediate for arabinogalactan/arabinomannan biosynthesis and its relation to the mode of action of ethambutol on mycobacteria. *The Journal of Biological Chemistry*, *269*(37), 23328–23335. doi:10.1016/S0021-9258(17)31657-5 PMID:8083238

Wooster, T. J., Golding, M., & Sanguansri, P. (2008). Impact of oil type on nanoemulsion formation and Ostwald ripening stability. *Langmuir ACS Journal of Surface Colloids*, *24*(22), 12758–12765. doi:10.1021/la801685v PMID:18850732

Zhang, S., Chen, J., Cui, P., Shi, W., Zhang, W., & Zhang, Y. (2015). Identification of novel mutations associated with clofazimine resistance in *Mycobacterium tuberculosis*. *The Journal of Antimicrobial Chemotherapy*, *70*(9), 2507–2510. doi:10.1093/jac/dkv150 PMID:26045528

Zhang, X., Liu, L., Zhang, Y., Dai, G., Huang, H., & Jin, Q. (2015). Genetic determinants involved in *p*-aminosalicylic acid resistance in clinical isolates from tuberculosis patients in northern China from 2006 to 2012. *Antimicrobial Agents and Chemotherapy*, *59*(2), 1320–1324. doi:10.1128/AAC.03695-14 PMID:25421465

Zhang, Y., Wade, M. M., Scorpio, A., Zhang, H., & Sun, Z. (2003). Mode of action of pyrazinamide: Disruption of *Mycobacterium tuberculosis* membrane transport and energetics by pyrazinoic acid. *The Journal of Antimicrobial Chemotherapy*, *52*(5), 790–795. doi:10.1093/jac/dkg446 PMID:14563891

Zhao, F., Wang, X. D., Erber, L. N., Luo, M., Guo, A. Z., Yang, S. S., Gu, J., Turman, B. J., Gao, Y. R., Li, D. F., Cui, Z. Q., Zhang, Z. P., Bi, L. J., Baughn, A. D., Zhang, X. E., & Deng, J. Y. (2014). Binding pocket alterations in dihydrofolate synthase confer resistance to para-aminosalicylic acid in clinical isolates of *Mycobacterium tuberculosis. Antimicrobial Agents and Chemotherapy*, *58*(3), 1479–1487. doi:10.1128/AAC.01775-13 PMID:24366731

Chapter 9
Nanoemulsion-Based Antiviral Drug Therapy:
Insight and Applications

Halima M.
B.S. Abdur Rahman Crescent Institute of Science and Technology, India

Ihsana Banu Ishthiaq
B.S. Abdur Rahman Crescent Institute of Science and Technology, India

Sneha Unnikrishnan
B.S. Abdur Rahman Crescent Institute of Science and Technology, India

Karthikeyan Ramalingam
https://orcid.org/0000-0002-9334-427X
B.S. Abdur Rahman Crescent Institute of Science and Technology, India

ABSTRACT

Nanoemulsions are an attractive approach for the delivery of antiviral drugs in the treatment of various viral infections. Nanoemulsions are easy to plan and develop, and their components exhibit high variability. Nanoemulsion system and its components have certain biophysical properties which could increase the efficacy of drug therapy. Pulmonary surfactant (PS)-assisted antiviral drug delivery by nanoemulsion system could be another effective approach for the treatment of COVID-19. Antiviral drug delivery of nebulization using an animation system could increase the efficacy of antiviral drug against COVID-19. Ginkgo biloba polyprenol nanoemulsion was also found to be stable, non-toxic, and had strong antiviral activity against influenza A H3N2 and hepatitis B virus in vitro. Nanoemulsion systems possess certain properties that make their system suitable for drug delivery by mobilization and hence would be promising systems for therapeutics in the future.

DOI: 10.4018/978-1-7998-8378-4.ch009

INTRODUCTION

The world has advanced in numerous domains, however viral infections proceed to exist and add to the mortality of mankind alongside its fluctuated financial signs. As of late, there have been outbreaks of a few viral infections brought about by COVID- 19, Nipah infection, Ebola infection, Zika infection, dengue infection, chikungunya infection and various strains of influenza virus H5N1 (avian influenza), H1N1 and H3N2 (pig influenza). As of late, the novel Covid (nCoV) has caused extreme pandemic asserting existences of roughly 3 million people approximately until now, with high effect on socioeconomic implications throughout the world. In 2018, nineteen Nipah virus cases were accounted for in India, 17 of which brought about mortality. Since 2001, the casualty rate because of Nipah virus has been accounted in the range of 68 and 100% in India (Chauhon at al. 2018). Significant episode of Ebola infection illness in West Africa during 2014–2016 guaranteed 11,315 lives out of 28,616 detailed cases. In Australia, in 2019, in the primary quarter itself, 27,540 warnings of influenza were received. Although diminished influenza action is accounted for in different continents across the world, various strains of the influenza infection are found in the different parts of the world with the occasional influenza A virus prevailing. Zika virus transmission has assumed control over epidemic extent in different parts of the world in the course of recent years. By and by, dengue apparently afflicts South East Asia multiple times more when contrasted with other viral infections, subsequently heightening the expense of dengue treatment to about $950 million. Further, in May 2018, around 164,000 dengue rates had struck internationally. Along these lines, the financial implications related with viral sicknesses have been very high. Different danger factors recognized for viral infections incorporate natural dangers including water supply, disinfection office and environment, way of life including smoking and liquor abuse, specific geological region, different operations like blood bonding, medical procedure, transmission from vectors, and so on. While from among these, a couple of variables are unavoidable and safety measures should be practiced to keep away from them, endeavours can be guided toward the others to get positive reaction (Lopez-Diez et al. 2017; Fuller et al. 2017; Reker et al. 2014). The significant challenge that stays in the advancement of successful antiviral agents is the capacity of the virus to increase in the host cell by liberating its own DNA or RNA. The host's immune system is exceptionally undermined if there should arise an occurrence of viral infections and backslides are normal. Likewise, because of the compexities related with viruses, treatment is for the most part symptomatic and complete destruction of the virus may not be conceivable. Recognizing and diagnosing specific sort of viral disease is very difficult. Now and again, due to past exposure, viral antibodies present in the host may get activated, delivering it hard to distinguish accidental infection. Difficulties in avoidance, identification, or treatment are viewed as a red sign by the research community (Charavarty et al. 2020). With the arrival of nanotechnology, it's been possible to understand the cellular mechanisms of the living cells and to develop technologies which facilitate early diagnosis and treatment of viral infections. A number of its applications comprise:

- Drug and gene delivery
- Usage of fluorescent biological labels,
- Detection of proteins, pathogens, and tumours
- Purification of biological molecules and cells
- Tissue engineering
- MRI contrast heightening
- Pharmacokinetic studies (Joo et al. 2008; Saini et al. 2010; Blecher et al. 2011).

Thus, it displayed an enormous field of research and application with its ability to cater to viral diseases in an efficient manner and address the issues posed by traditional antiviral medicines. Nano formulations haven't only helped to beat the issues associated with drug solubility and bioavailability but also by themselves has acted as antiviral agents through various mechanisms.

2. ANTIVIRAL THERAPY OF NANOEMULSION

Table 1. Distinct compositions of nanoemulsions used for antiviral treatments.

S.NO.	DISTINCT COMPOSITION OF NANOEMULSION	RESULTS	REFERENCE
1	8N8 Nanoemulsion	*Herpes simples*, influenza A and vaccinia were sensitive to 8N8 nanoemulsion (8 volumes Tributy phosohate, 64 volumes soybean oil and 8 volumes Triton X-100). 8N8 nanoemulsion is virucidal to HIV virus. Nonenveloped viruses, such as adenovirus are resistant to 8N8 nanoemulsion.	Hamouda et al. 2001
2	Nanoemulsion ATB	Virucidal against ebola virus (strain Zaire) from vero cell culture fluid (EBO-zc) and from blood of infected monkeys (EBO-zb).	Chepurnov et al. 2003
4	Curcumin Nanoemulsion (nanocurcumin)	Improved curcumin physicochemical properties are antiviral against Dengue virus. Exhibited higher cell cytotoxicity compared to curcumin solution. Suppressed DENV growth. Greater virus reduction was observed for DENV-1 and DENV-2.	Nabila et al. 2020.
5	Respiratory Syncytial virus (RSV) inactivated by and adjuvanted with W805EC Nanoemulsion	Elicits robust humoral and cellular immune response thereby protecting from RSV. Intranasal immunization resulted in high antibody titter and robust Th-1 skewed cellular response. Vaccination provided sterilizing cross protection immunity against heterologous RSV challenges. During viral challenge there was no marked or severe histological effects in lungs. Safe and effective for immunization in multiple animal models.	O'konek et al. 2015.
6	Recombinant HIV gp 120 and nanoemulsion adjuvants.	Nasal immunization of gp 120/ nanoemulsion results in a systemic and mucosal antibody response along with significant, cross-reactive neutralizing immunity and a Th1 polarized CTL response. Nanoemulsion could be evaluated as mucosal adjuvant for multiple HIV vaccine.	Bielinska et al. 2008.
7	HSV-2 surface glycoproteins gD2 and gB2 formulated in nanoemulsion adjuvants (NE01-Gd2/gB2).	Induced complete protection against genital herpetic lesion, dorsal root ganglia (DRG) infection. Prevention of chronic recurrent infection. Reduced recurrent lesions and virus shedding in previously infected animals.	Bernstein et al. 2019.
8	Metadichol (nanoemulsion of long chain lipid alcohol.	In humans, it acts as the most safe and effective inhibitory against enveloped virus. Metadicol is largely based on its safety and activity against parasites, bacteria and viruses Serves as preventive agent for many tropical diseases, by strengthening innate immunity through VDR binding Could be used as nutritional supplement in countries where viral infections are prevalent.	Raghavan et al. 2016.
9	Acyclovir- loaded chitosan nanospheres from nanoemulsion	Enhanced drug loading amount, prolonged release kinetics, no cytotoxicity, improved in vitro antiviral efficacy. Improved topical delivery of anti-infective drugs. Higher antiviral activity of the acyclovir loaded chitosan nanospheres in both HSV-1 and HSV-2 strains compared to free acyclovir.	Donalisio et al. 2018.
10	Nanoemulsion adjuvanted H5N1 vaccine.	Elicited significantly higher and longer lasting antibody responses. Significantly improved the immunogenicity of H5N1 vaccine. Enhanced protection against homologous clade 1 and heterologous clade 2 H5N1 virus challenge in young and aged mice respectively.	Cao et al. 2016.
11	Achyrocline satureioides extract incorporate into topical nanoemulsion	Improved antiherpes activity of hydroethanoic extract. Detection of higher flavonoid retention in the porcine ear skin epidermis and upper layers of oesophageal mucosa. Increase in retention during tissue injury.	Bidone et al. 2015.
12	Berberine nanoemulsion	Induced phototoxcity in photodynamic therapy Increase in ROS Increase in autophagia Could be used as photosensitizing agents in PDT to treat cervical carcinoma.	Floriano et al. 2021.

Decreasing drug discovery and increasing viral disease created a niche to use nanoemulsion for improving currently available therapeutic options. Emulsification process significantly enhances the efficacy of active ingredients which in turn improves antiviral drug delivery. Table 1 classifies distinct compositions of nanoemulsions in antiviral therapy.

3. EFFECTIVE APPROACHES FOR TREATING COVID 19: ANTIVIRAL DRUG DELIVERY USING NANOEMULSION SYSTEM

3.1 Pulmonary Surfactant (PS) Assisted Antiviral Drug Delivery by Nanoemulsion

For the treatment of COVID 19, pulmonary surfactant (PS) assisted antiviral drug delivery by nanoemulsion could be an effective approach. It is synthesized by type II alveolar cells and act as lung surfactants. It is generally denoted as functional excipient in the nanoemulsion system because they are naturally composed of proteins and lipids. Main compositions of PS are:

i. Saturated and unsaturated lipids: 92%.
ii. Surfactant proteins (SPs), hydrophilic surfactant protein A (SP-A) and surfactant protein D (SP-D) and hydrophobic surfactant protein B (SP-B) and surfactant protein C (SP-C): 8% (Kumar et al. 2020).

A lipid and lipophilic portions of PS forms stable surface active film at the water interface for normal expiration process and could be explored as oil phase of the biphasic nanoemulsion system for loading antiviral drug respectively (Kumar et al. 2020; Parra et al. 2015; Pilcer et al. 2010).

On the other hand, surfactants proteins (SPs) play potent role in protecting the lungs from infections. SP-A and its recombinant derivatives have the ability to inhibit virus attachment, inhibit entry in a host cell and prevent the transmission of infection (Guagliardio et al. 2018; Hartshorne et al. 2010). As the COVID-19 infection is caused by binding to the ACE2 receptors (Fig. 1A), in previous research, it has been reported that when SP-D binds along glycosylated spike protein (S-protein) of SARS-CoV (severe acute respiratory syndrome coronavirus), virus destruction and clearance occurs (Fig. 1B). Due to the efficient binding of carbohydrate recognition domains and carbohydrate structures present on glycosyated S-protein of SARS-CoV, SP-D could facilitate targeted antiviral drug delivery against SARS- CoV- 2 infection (Leth- Larson et al. 2007). SP-B and SP-C are capable for transferring and spreading the antiviral drug loaded nanoemulsion system along the respiratory epithelial surface. Thus, nanoemulsion system with antiviral drugs and functional excipients (PS and associated SPs) will enable targeted and effective pulmonary drug delivery along with synergizing antiviral action against the sever SARS-C0V 2 infection (Fig. 1B) (Kumar et al. 2020).

3.2 Curcumin: Nanoemulsion Based Antiviral Drug

Curcumin is extracted from the roots of rhizome plant *Curcuma longa.* It is a natural polyphenolic compound that exhibits anti-oxidant, anti-microbial, anti-proliferative and anti- inflammatory therapeutic properties (Pang et al. 2015). It was reported that curcumin possessed anti-viral activity against

HIV, HSV-2, HPV viruses, Influenza, virus, Zikavirus, Hepatitis virus and adenovirus (Das et al. 2020; Praditya et al. 2019). Topical application of curcumin in emulsion form could predominantly disturb the viral entry site of ACE2 receptors at the nasal cells and mucosal surface of respiratory tract thereby preventing the SARS –CoV2 infection in humans (Fig. 1B). Large scale clinical trials should be conducted to study the benefits of curcumin for their pharmaceutical applications in nanoemulsion system (Manoharan et al. 2020).

Figure 1. A) COVID-19 infection in nasal passage by binding to the ACE2 receptor B) Function of various antiviral nanoemulsions agains COVID-19

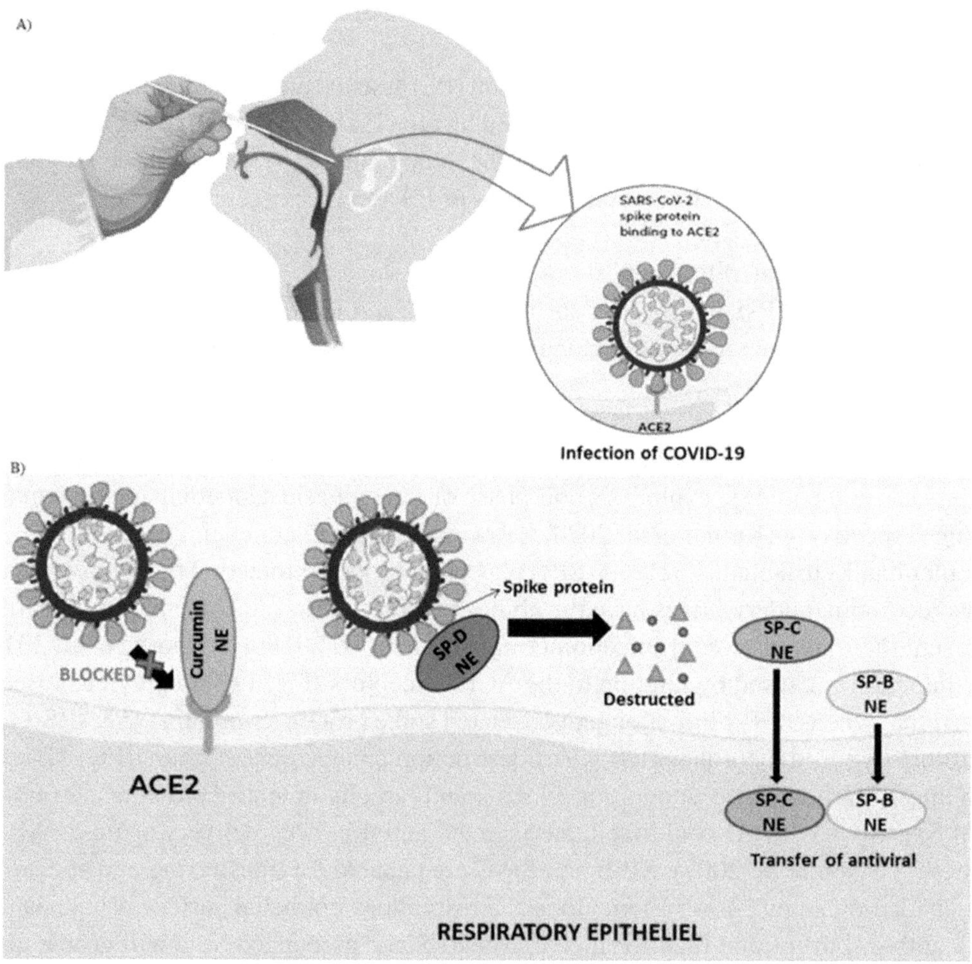

3.3 Vitamin D Nanoemulsion Treatment for COVID 19

Vitamin D supplement is a cost-effective functional method to enhance the immune system through dietary sources that are easily available among all the countries (Aguilar et al. 2020). Vitamin D and bioactivity arises from high hydrophobicity, chemical degradation and variable oral bioactivity either to make Vitamin D miscible in aqueous environment or to complete the needs of vitamin D against inactivation and degradation, nanoparticle encapsulation in colloidal dispersion is followed. Nanocapsulation of vitamin D have successfully enhanced the taste, shelf-life, appearance and the absorption level (Khalid et al. 2015; Haham et al. 2012). Vitamin D has the ability to increase its bioavailability, when it is co-administrated along with lipids. Thus, compositions of vitamin D nanoemulsion, when administrated in human body could make it more bio-available and biocompatible. Co-administration of vitamin along with lipids could be carried out by using lipids as carrier or in the form of a nanoemulsion (McClements et al. 2012). Research studies have reported that vitamin D nanoemulsions have successfully improved its bioavailability, reinforced its potency and enhanced its shelf- stability in the absence of harmful synthetics. Enhancement of biocompatibility takes place based on the chosen components of nanoemulsion that protect vitamin D from being degraded or rejected by enzymes present in the body. Better shelf- life can be obtained by embedding vitamin D droplets in nanoemulsion. Increase in solubility level occurs when nanoemulsion of vitamin D is combined with natural materials, thereby making it miscible with water. Clinical studies ensures that vitamin D have the ability to regulate the overwhelming host immune response to COVID 19, due to their inflammatory response during viral infection (Aguilar et al. 2020).

4. NANOEMULSION IMMUNIZATION FOR INFLUENZA VIRUS

Nowadays vaccination for influenza virus is generally focused on humoral immunity along with the limitation of short duration of protection, narrow cross-strain efficacy and suboptimal immunogenicity. Wong et al. analyzed whether the diverse mechanisms of two chemically and biologically different adjuvants could improve the immunogenicity and breadth of protection against influenza virus. Results from these combinations of adjuvants, oil in water nanoemulsion and RNA based agonists of RIG-I demonstrated the enhanced immunogenicity of the adjuvant combination were synergistic and also highlighted the potential value of a combined adjuvant approach for improving the vaccination against the influenza virus (Wong et al. 2020).

From the previous preliminary studies, it is concluded that nanoemulsion has the ability of inhibiting pathogenic enveloped virus infections and simultaneously provide the impetus for testing as prophylactic barriers to enveloped virus infection in vivo. The murine model of pneumonitis is well suitable model for analyzing antiviral agents however there may be vast difference between these models and humans. In mice infection develops from alveoli and spreads to upper respiratory tract whereas in humans it develops from lower respiratory tract and spreads to the lung parenchyma (Donovan et al. 2000). Unique immune response was produced with live viruses and emulsion was administrated. Administration of nanoemulsion along with live virus through intranasal medication provided full protection of animal body against influenza pneumonitis. Immunized animals remained healthy, maintained normal temperature range for 14 days, maintained grossly normal lungs and only during histologic examination resulted with minimal inflammatory changes compared with influenza pneumonitis. In the absence of nanoemulsion, non lethal or sub lethal doses of live virus were unable to elicit the immune response there by indicating reduced cytokine production (Myc et al. 2003).

Among few adjuvant formulations in vaccine, squalene-based oil in water emulsion is approved for human use (Fox et al. 2013). Fox et al. compared the adjuvant properties of nanoemulsion that were made up of different classes of oil structures to squalene nanoemulsion. In their study, they introduced nanoemulsion prepared with polyprenols (derived from species of Pinaceace family) as novel vaccine adjuvant compositions. This experiment demonstrated long term physical stability, stimulation of cytokine production from human cell in vitro and promoting antigen specific immune responses in various animal models, specifically during the formulation of TLR4 ligand glucopyranosyl lipid adjuvant (GLA). Using plant-based isoprenoid substances in nanoemulsion adjuvants could enhance the sustainability by reducing the reliance on animals to obtain squalene in pharmaceutical preparations (Fox et al. 2019).

Figure 2. Function of various plant derived polyprenol in antiviral nanoemulsion.

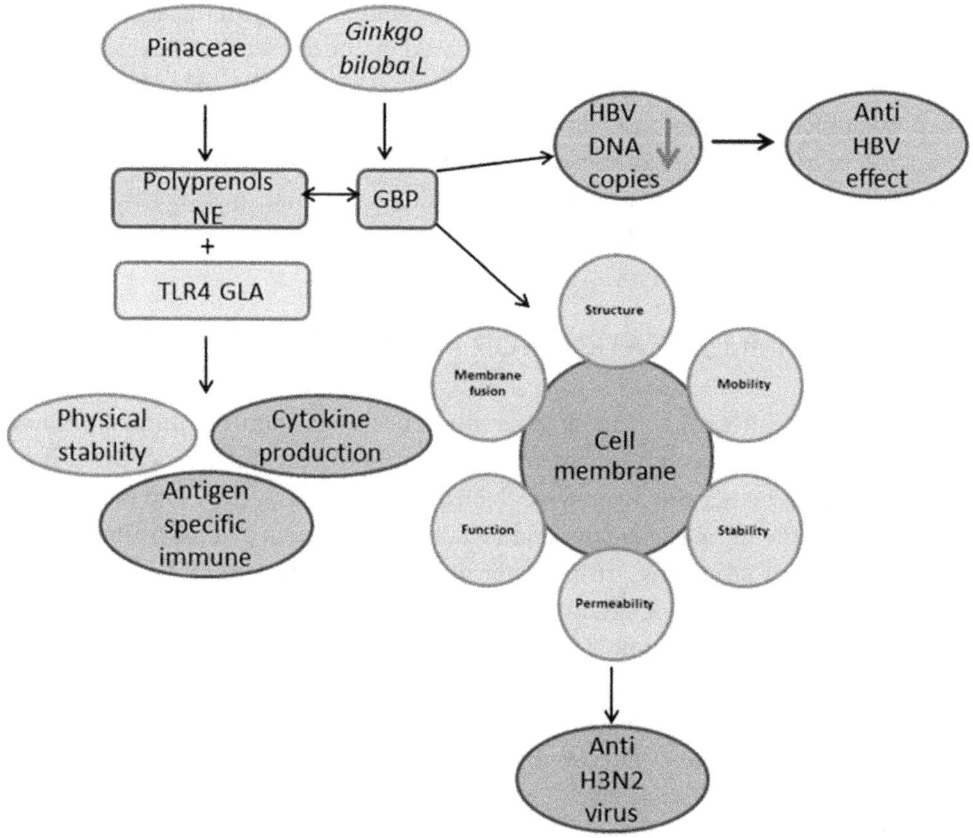

Due to the rapid mutation of genes and frequent variations, influenza virus causes considerable disease epidemic. As the virus has the ability to cause epidemic pathogens frequently and becomes common across the world, influenza A virus have been denoted as the influenza H3N2 subtype. Surface markers of the influenza virus, hemagglutinin (HA) and nerve neuraminidase (NA) undergo frequent mutation and thus no effective vaccines have been developed (Zhong et al. 2012). *Ginkgo biloba L.* is well known for its bioactive substance such as flavonoids, terpene lactones and polyprenols. Pharmacological effect of flavonoids and terpene lactones of *Ginkgo biloba L.* extracts and its preparation have been globally used in the treatment of cardiovascular and cerebrovascular disease. Polyprenols are non-toxic in general especially in their action as antiviral agents (Fig 2). They take part during the biosynthesis of glycoprotein. From the animal experiments it was concluded that plant polyprenols has the best inhibitory and therapeutic effects of 60%-90% approximately on canine infection hepatitis, murine hepatitis, cat infectious enteritis, bovine leukaemia, rabies and distemper virus (Danilov et al. 1996).

Zhong et al. concluded that nanoemulsion of polyprenols from Ginkgo Leaves (GBP), a natural active lipid is non toxic to normal cells and concluded with better protective effects on H3N2 virus infected MDCK cells at a concentration of 100 μ g/mL. Hence GBP is the promising potential antiviral drug. Impact of the polyprenol structure towards cell membranes has been considered as the potent factor (Danilov et al. 1996; Bergamini et al. 2004; Walinska 2004 and Walinsa et al. 2001). The nanoemulsion of GBP can reinforce membrane fusion and regulate the structure and function of biological membranes by improving the mobility, stability and also the permeability of the cell membrane.

5. HEPATITIS

After HIV, hepatitis B and C are the other diseases caused by the use of contaminated needles. This urges scientists to come up with orally administered vaccines and medications than needles and syringes (Simonsen et al., 1999; Chircov et al., 2019). Many in vitro and in vivo studies have been published to show that nano emulsions have antiviral properties against the hepatitis virus A, B, C and E stereotypes. Nano emulsions are used as carriers for delivering various types of antiviral drugs that are associated with low dissolvability and bioavailability (Fig. 3). Evidence for nanoemulsions increasing the solubility and bioavailability has been proven with both in vitro and in vivo studies (Franklyne et al., 2021).

Figure 3. Nanoemulsion based anti-hv drugs and vaccines are derived from various sources besides can be administered in different forms.

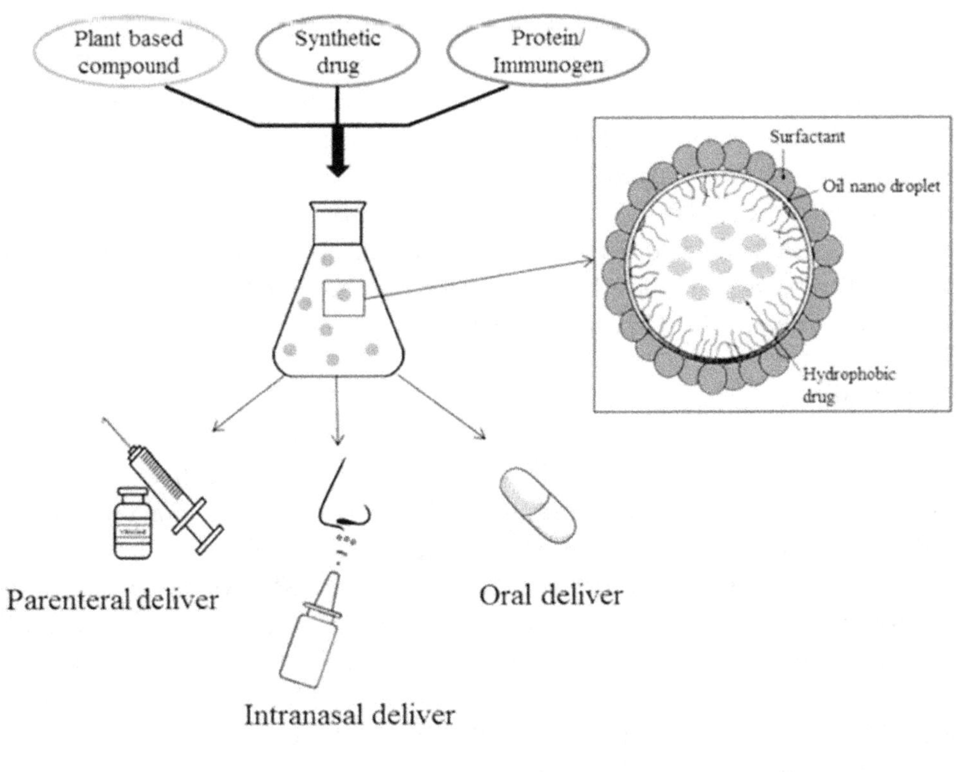

5.1 Drug Mediated Treatment

Adefovir dipivoxil, used to particularly treat HBV is an orally administering nucleoside reverse transcriptase inhibitor with a significant antiviral effect but low bioavailability (Hadziyannis et al., 2003; Stephanos et al., 2003). However, the same drug was administered to male albino rats via a different nanoemulsion carrier thereby, resulting in a significant 3.5 fold higher bioavailability of the drug than the native form of the drug during oral administration (Durai, 2015). In an in vivo study performed on Swiss Albino mice, the drug adefovir dipivoxil was tested to have enhanced stability and bioavailability when administered orally using nanoemulsion. It was experimented to be effective in delivering the compound to the mice liver by the emulsion composed of Capmul MCM, precirol ATO-5 and cremophor RH40 or Pluronic F68 which are the glycerolized fats and oils, solid lipid and surfactant respectively (El-Halim et al., 2020).

5.2 Plant-based Nanoemulsions

Baicalin nanoemulsion, meanwhile, has been shown to be an effective delivery mechanism for chronic HBV infection. Baicalin is a glycosyloxyflavone, a flavonoid obtained from Scutellaria plants. The researchers discovered that in another in vivo study, murine models given a lipid-based nanoemulsion of baicalin had an 11.5-fold greater Cmax value in their lymph nodes. This research was designed to target the HBV that hid in the lymphatic system (Xu et al., 2019). Similarly, silibinin, a plant-based compound with potential antiviral activity against hepatitis C, showed improved solubility, bioavailability, liver accumulation and antioxidant activity against HCV when administered in nanoemulsion form with polyvinylpyrrolidone as the carrier. The crystalline structure of the drug was transformed to an amorphous phase in the nanoemulsified form, allowing drug release only at alkaline pH 7.4 at a rate of more than 75% after 20 minutes of administration, in the intestine facilitating better targeting (Ellah et al., 2019). GPB was testified to have an anti HBV activity in an in vitro study conducted on HepG 2215 cells. As GBP was difficult to disperse in water a nanoemulsion was prepared to increase its bioavailability. The cells were treated with GBP at various concentrations, when the incubation time and GBP concentration were increased. GPB had an inhibitory effect and reactive procedure on the HBV antigen and DNA as the positive drug 3TC. GBP inhibited HBsAg by 52.11 percent at 50 g/mL, and by 67.32 percent at 100 g/mL on Day 9, and reduced copies on HBV DNA when the concentration was between 12.5 and 25 g/mL. Thus, it can be concluded that GBP has a sound anti HBV effect (Wang et al., 2015).

5.3 Immunization

Other than drug delivery systems, nanoemulsions have also been used to deliver immunizing agents to provoke antibody synthesis against hepatitis viruses (Table 2). For providing oral immunization as study designed solid fat nanoemulsion (SFN) made up of monophosphoryl lipid A (MPLA) for targeted release of HBV antigen (HBsAg) into the colon. MPLA is an adjuvant in the SNF which together 1,2-dipalmitoyl-sn-glycero-3-phosphocholine (DPPC) with mimics the structure of the actual pathogen (Shahu et al., 2019). TLR7, an oral agonist, was tested to have an active anti-HBV effect in woodchuck and chimpanzee models. Besides, it has been proven to increase the response of HBV specific immune cells in a clinical trial (Boni et al., 2018). MLPA has been listed to be the first TLR agonist category adjuvant which is safe and compatible with most antigens and used in human mass vaccination. The vaccine was prepared by emulsifying the antigen in nano form followed by solvent evaporation, encapsulation and lyophilization to obtain the oral administered vaccine for immunization. The in vivo studies presented that the HBsAg was proven to be targeted to 7.4 pH in the colon with 17.5% ± 1.45% release percentage after 5 to 6hours of administration (Shahu et al., 2019). Likewise, a related, comparative study was performed in murine model to understand the immunization effect of SFN of MPLA against HBV. A proper comparison was done between response created by Genvac B, a commercial intramuscular vaccine used to treat HBV and 1μg of rHBsAg entrapped in MPLA adjuvanted SFN. The results revealed the nanoemulsion mediated delivery prompted a significantly higher cellular and humoral response in the subjects. Moreover, it was suggested that the chylomicrons mimicking structure of SFN has thereby encouraged the lymphatic cells in better uptake and processing (Minz and Pandey, 2017).

Attempting to make a nasal delivered nanoemulsion based parenteral vaccine against HBV; scientists synthesized a novel intranasal vaccine with recombinant hepatitis B surface antigen- HBsAg adw serotype which was effective with little number of administrations (two in the study). In the pre-clinical trial, mice, rats and guinea pigs models were tested to display a protective immunity ≥1000 IU/ml after administration of intranasal nanoemulsion based vaccine. It was also depicted that the HBsAg-NE vaccine had lesser toxicity and was tolerated better by many animal models. The immunized animal models presented a vigorous and sustainable immune response in the form of IgG, mucosal IgA and strong antigen-specific cellular immunity. The serum IgG levels were detected to be $\geq 10^6$ specific to the antigen, post immunization along with, IFN-γ and TNF-α cytokine as an indication of Th1 polarized immune reaction. The thus produced HBsAg-NE vaccine showed both an immunizing effect and a therapeutic effect for chronic hepatitis B. Moreover, the vaccine has a long-term stability in storage condition where, the complete anti-HBV immunogenicity of the HBsAg-NE vaccine was reserved for 6 weeks at 40°C, half a year at 25°C and a whole year at 4°C (Makidon et al., 2008).

As the next step of immunization, another study was conducted for the detection of antibodies against the HBsAg, S protein and E2 protein using chimeric HBV (adw genotype)-HCV (1a genotype) envelope subviral particles delivered in a nanoemulsion form to stretch a protective immunity against hepatitis C virus. In this study nanoemulsion was used to deliver and immunize the rabbits with the corresponding immunogens subcutaneously (Beaumont and Roingeard, 2015). Recently, a relatively similar research was performed in young female BALB/c mice endeavoring to immunize against hepatitis C virus. HBV small surface antigen (sHBsAg) and Chimeric virus-like particles (VLPs) derived from HCV E2 glycoprotein (HCV 412–425 and 523–535 epitopes) were acquired in the Leishmania tarentolae expression system to construct an HBV/HCV particle, along with Addavax as the adjuvant these were combined in squalene-based oil-in-water nanoemulsion. This mixture was subcutaneously injected to mice with a dosage of 15μg protein on 0th day and 10μg protein on 14th, 28th day. The administration provoked the immune system to produce efficient cross-neutralizing anti-HCV therapeutic antibodies (Czarnota et al., 2020).

Besides HBV and HCV, HEV is also a cause of quite large number of fatalities including still births. The two truncated ORF2 proteins (54 kDa and 26 kDa) are found to be effective immunogens and can be efficiently used as nanoemulsion-based vaccine candidate. Chitosan nanoemulsion composed of olive oil and surfactant (PluronicF-127) was used to deliver the two truncated ORF2 proteins in encapsulated form as a mode of oral administration. The study was conducted using in vitro models. The purity of the protein and cytotoxicity of the vaccine was tested using affinity chromatography, western blot using anti-HEV antibody and by cell viability in cell lines HeLa and THP1 respectively. It was witnessed that 100 μg/mL nanoemulsion was analyzed to be safe in both cell lines. The encapsulated proteins were targeted to be released at a pH of 7.4. A percentage release of 26–59% and 9.7–40% was seen for 26 kDa and 54 kDa protein one hour post administration which further progressively increased along the time (Rani et al., 2018).

Table 2. Immunization against Hepatitis virus

Sl. No.	Drug/ protein delivered	Nanoemulsion used	Mode of administration	Benefit	Reference
1.	HBVAg	SFN of MLA	Oral	Targeted release into colon with 17.5% ± 1.45% release percentage after 5 to 6hrs of administration, TLR agonist	Shahu et al., 2019
2.	rHBsAg	SFN of MPLA	Oral	Higher cellular and humoral response, better uptake and processing by lymphatic cells	Minz and Pandey, 2017
3.	HBsAg	oil-in-water nanoemulsion	Nasal	Fewer number of administrations, lesser toxicity, IgG, mucosal IgA release, better shelf life	Makidon et al., 2008
4.	Two truncated ORF2 proteins (54 kDa and 26 kDa)	olive oil nanoemulsion	Oral	Targeted released at pH of 7.4, percentage release of 26–59% and 9.7–40% for 26 kDa and 54 kDa protein in 1hr post administration	Rani et al., 2018
5.	S and E2 protein in chimeric HBV, HCV envelope subviral particles	Squalene based oil-in-water nanoemulsion	Subcutaneously	Antibodies against the HBsAg, immunity against hepatitis C virus.	Beaumont and Roingeard, 2015
6.	sHBsAg VLPs	Oil-in-water nanoemulsion	Subcutaneously	cross-neutralizing anti-HCV therapeutic antibodies release	Czarnota et al., 2020

6. ADVANTAGES OF NANOEMULSIONS

Anti-viral treatment using Nanoemulsions have affluent advantages over other conventional treatments. Most drugs and components that are vulnerable such as proteins or DNA or mRNA can be efficiently delivered through nanoemulsified form as these ensure drug safety against harsh environmental conditions, biocompatibility, and biodegradability, making them a safer lipid-based system alternative (Aguilera-Correa et al., 2021). Moreover, nanoemulsions provide the ease of large-scale processing using a high-pressure homogenization process which makes it is easier to commercialize drugs and vaccines in larger quantities. When it comes to vaccination, the presence of surfactants and lipophilic compounds nanoemulsion makes it easier and enhances the transdermal absorption of the therapeutic compounds. Furthermore, nanoemulsions have a higher degree of skin permeation and drug aggregation than lipid nanoparticles. Higher stability, mild nature, lesser use of surfactants and its ease of formulation through simple methods make nanoemulsions a better candidate for dermal drug delivery systems. Using nanoemulsions in antiviral therapy eliminates the limitation of low drug loading or irritating side effects (Chircov et al., 2019). In orally administered capsules, nanoemulsion increase the bioavailability of the drug delivered up to several folds, this benefit also provides the advantage of administering the drug in a lover dose and with better effectiveness. Thereby, will reduce the toxicity of the drug and lessen the number of administrations of the booster dose (Makidon et al., 2008). Besides drugs, vitamins such as D are also delivered via nanoemulsion with higher bioavailability as an antiviral treatment (Aguilar et al., 2020). Along with better bioavailability and solubility the compounds antihypertensive effect is also seen to be enhanced (Manoharan et al., 2020). In addition, nanoemulsion based drugs and vaccines also display greater stability with no loss of anti-viral therapeutic effect even at raised temperatures which ensure the effectiveness even when appropriate cold chain maintenance is lost (Makidon et al., 2008, Donovan et al., 2000). This is possible due to slow down chemical degradation reactions which enhances the stability of encapsulated lipophilic components by lowering the rate of molecular diffusion through the lipid process (McClements, 2012).

Besides, a nanoemulsified drug component poses few biophysical properties, thus reformulation the conventional drugs using nanoemulsion with functional excipients could possibly elaborate the efficacy of drug therapy (Kumar and Jain, 2021). Modified drug formulation with nanoemulsion has also been witnessed to increase the solubility addition to improvement in maintenance of higher permeability and low drug efflux enhancing bioavailability especially in orally administered treatment (Manyarara et al., 2018). Polyunsaturated fatty acids (PUFA) rich oil based nanoemulsions were tested to offer efficiently solubility of the drug in the oil droplet and thereby, significantly enhance drug absorption across the gastro-intestinal tract following oral administration. Similarly, nanoemulsions made from pine-nut oil rich in alpha and gamma-linolenic acid, which is an omega-3 fatty acid displayed the efficiency of delivering across the blood-brain barrier (Vyas et al., 2007). Due to the small size, lipid nanoemulsions present the flexibility of crossing the endothelial lining (Shaker et al., 2019). In addition to the advantages offered by the lipid components the emulsifying agent such as tween 80 used in nanoemulsion offers the benefit of expand the permeability of several drugs (Prabhakar et al., 2013).

7. FUTURE PROSPECTIVE

Unlike antibiotics, formulating an antiviral drug requires more study. The poor solubility and bioavailability of most antiviral drugs is a huge concern to be solved. Thus, reformulations these drugs with nanoemulsion and adding suitable surfactants, emulsifiers would surely augment the efficiency of the drug or compound. Many drugs and compound which are water soluble and insoluble have still not been completely studied and tested in nanoemulsified form against many viral infections (Hobson et al., 2018). This would seal the gap in formulation an antiviral drug or vaccine with good bioavailability, solubility by enhancing their delivery, availability and implementing those in antiviral therapy would have potential effect for currently known viruses and novel viruses in the future. Many recent researches have tested commercial drugs by reformulating with nanoemulsion and have proven to be more effective in nanoemulsified form, as well as, many drugs and compounds are yet to be tested in nanoemulsified form. Nanoemulsion based drugs have a potential future for oral administration for targeted delivery along with increasing bioavailability such as in accurately releasing the drug at the intestine.

As most oils with antiviral activity in nanoemulsion are derived from natural compounds, they would have lower toxicity, while also provide promising solution in antiviral therapy. Considering the nano size, these can be used as flexible carriers to delivery compound that is difficult to be delivered across the barriers enabling better penetration. Nanoemulsions are capable of delivering compound through the skin by intensifying the diffusion and absorption of the compounds. Although the mechanism of penetration is not completely understood, this provides opportunity for further research in this area. Moreover, due to the higher chemical stability, vaccines with better heat tolerance and biocompatibility can be developed. Drugs and vaccines that can be resistant to disrupted cold chain management would offer a great scope during transportation. Though nanoemulsion based antiviral drugs and vaccine are being used, still additional research is required in testing nanoemulsions as antiviral therapeutics.

CONCLUSION

Virus-related deaths continue to be a significant cause of death, despite advances in health-care technology. As a result, a growing number of researchers are turning to nanotechnology to tackle the spread of viral diseases. Several nanoemulsions have been developed for drug delivery, diagnosis, and antiviral treatment. A void has opened up for nanoemulsion to be used to boost currently available therapeutic options due to a decline in drug production and a rise in viral disease. The emulsification process improves the bioavailability, biocompatibility, solubility, absorbability, antihypertensive effect, and durability of active ingredients, resulting in better antiviral drug delivery. Many other strategies to treating viral diseases have been attempted, including using nanoemulsion to treat the global pandemic COVID-19. Nanoemulsions were used to deliver a possible pulmonary surfactant, curcumin, and vitamin D to achieve anti-COVID-19 activity. Another viral infection with faster communicability and contagiousness is influenza; to treat it, boosting humoral immunity by delivering viral surface antigens in nanoemulsions with adjuvants was seen as a reliable line of defence. After HIV, hepatitis is the next disease to cause death by blood and sexually transmitted infection. Hepatitis has a significant mortality rate. Effective results have been obtained after delivering drugs, active compounds and surface proteins against various serotypes of hepatitis virus. Nanoemulsion has the benefit of being able to be delivered in a variety of ways, including dermally, intranasally, orally, offering users a wide range of options while maintaining a consistent therapeutic effect. This field, on the other hand, emboldens scientists to conduct further research into nanoemulsion reformulation and its mechanism of action. Figure 4 represents the overview of antiviral drug delivery system.

Figure 4. A. Drug delivered in nanoemulsified form.

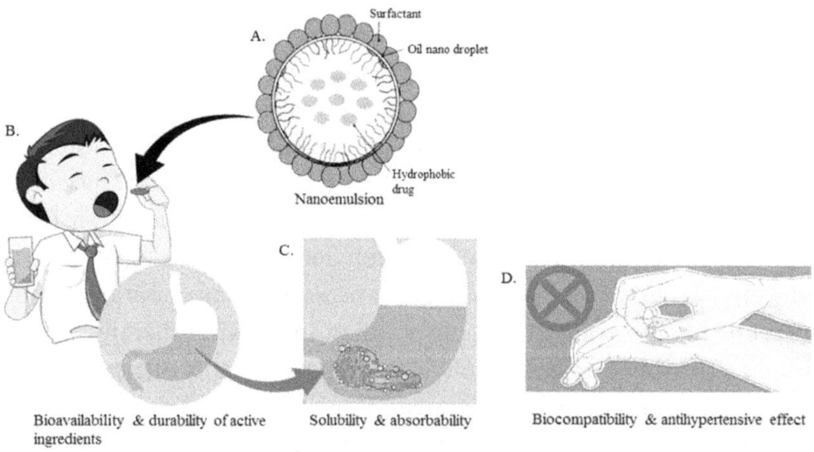

B. Improves bioavailability and durability of the active ingredient.
C. Has higher solubility and absorbability.
D. Is more biocompatible and has antihypertensive effect making it a better antiviral drug delivery system.

REFERENCES

Abd Ellah, N. H., Tawfeek, H. M., John, J., & Hetta, H. F. (2019). Nanomedicine as a future therapeutic approach for Hepatitis C virus. *Nanomedicine (London)*, *14*(11), 1471–1491. doi:10.2217/nnm-2018-0348 PMID:31166139

Aguilar, Z. P., Alhashim, E., & Benedict, V. H. R. (2020). COVID-19 pathology and vitamin d nanoemulsion as preventive and alternative remediation for severity: A review. *European Journal of Biomedical*, *7*(6), 488–499.

Aguilera-Correa, J. J., Esteban, J., & Vallet-Regí, M. (2021). Inorganic and Polymeric Nanoparticles for Human Viral and Bacterial Infections Prevention and Treatment. *Nanomaterials (Basel, Switzerland)*, *11*(1), 137. doi:10.3390/nano11010137 PMID:33435597

Beaumont, E., & Roingeard, P. (2015). Chimeric hepatitis B virus (HBV)/hepatitis C virus (HCV) subviral envelope particles induce efficient anti-HCV antibody production in animals pre-immunized with HBV vaccine. *Vaccine*, *33*(8), 973–976. doi:10.1016/j.vaccine.2015.01.007 PMID:25596457

Bergamini, E., Bizzarri, R., Cavallini, G., Cerbai, B., Chiellini, E., Donati, A., Gori, Z., Manfrini, A., Parentini, I., Signori, F., & Tamburini, I. (2004). Ageing and oxidative stress: A role for dolichol in the antioxidant machinery of cell membranes? *Journal of Alzheimer's Disease*, *6*(2), 129–135. doi:10.3233/JAD-2004-6204 PMID:15096696

Bernstein, D. I., Cardin, R. D., Bravo, F. J., Hamouda, T., Pullum, D. A., Cohen, G., Bitko, V., & Fattom, A. (2019). Intranasal nanoemulsion-adjuvanted HSV-2 subunit vaccine is effective as a prophylactic and therapeutic vaccine using the guinea pig model of genital herpes. *Vaccine*, *37*(43), 6470–6477. doi:10.1016/j.vaccine.2019.08.077 PMID:31515143

Bidone, J., Argenta, D. F., Kratz, J., Pettenuzzo, L. F., Horn, A. P., Koester, L. S., ... Teixeira, H. F. (2015). Antiherpes activity and skin/mucosa distribution of flavonoids from achyrocline satureioides extract incorporated into topical nanoemulsions. *BioMed Research International*, •••, 2015. PMID:26101767

Bielinska, A. U., Janczak, K. W., Landers, J. J., Markovitz, D. M., Montefiori, D. C., & Baker, J. R. Jr. (2008). Nasal immunization with a recombinant HIV gp120 and nanoemulsion adjuvant produces Th1 polarized responses and neutralizing antibodies to primary HIV type 1 isolates. *AIDS Research and Human Retroviruses*, *24*(2), 271–281. doi:10.1089/aid.2007.0148 PMID:18260780

Blecher, K., Nasir, A., & Friedman, A. (2011). The growing role of nanotechnology in combating infectious disease. *Virulence*, *2*(5), 395–401. doi:10.4161/viru.2.5.17035 PMID:21921677

Boni, C., Vecchi, A., Rossi, M., Laccabue, D., Giuberti, T., Alfieri, A., Lampertico, P., Grossi, G., Facchetti, F., Brunetto, M. R., Coco, B., Cavallone, D., Mangia, A., Santoro, R., Piazzolla, V., Lau, A., Gaggar, A., Subramanian, G. M., & Ferrari, C. (2018). TLR7 agonist increases responses of hepatitis B virus–specific T cells and natural killer cells in patients with chronic hepatitis B treated with nucleos (t) ide analogues. *Gastroenterology*, *154*(6), 1764–1777. doi:10.1053/j.gastro.2018.01.030 PMID:29378197

Cao, W., Davis, W. G., Kim, J. H., Juan, A., Taylor, A., Hendrickson, G. R., ... Sambhara, S. (2016). An oil-in-water nanoemulsion enhances immunogenicity of H5N1 vaccine in mice. *Nanomedicine; Nanotechnology, Biology, and Medicine*, *12*(7), 1909–1917. doi:10.1016/j.nano.2016.04.005 PMID:27112307

Chakravarty, M., & Vora, A. (2020). Nanotechnology-based antiviral therapeutics. *Drug Delivery and Translational Research*, 1–40. PMID:32748035

Chepurnov, A. A., Bakulina, L. F., Dadaeva, A. A., Ustinova, E. N., Chepurnova, T. S., & Baker, J. R. Jr. (2003). Inactivation of Ebola virus with a surfactant nanoemulsion. *Acta Tropica*, *87*(3), 315–320. doi:10.1016/S0001-706X(03)00120-7 PMID:12875924

Chircov, C., & Grumezescu, A. M. (2019). Nanoemulsion preparation, characterization, and application in the field of biomedicine. In *Nanoarchitectonics in biomedicine* (pp. 169–188). William Andrew Publishing. doi:10.1016/B978-0-12-816200-2.00019-0

Czarnota, A., Offersgaard, A., Pihl, A. F., Prentoe, J., Bukh, J., Gottwein, J. M., Bieńkowska-Szewczyk, K., & Grzyb, K. (2020). Specific Antibodies Induced by Immunization with Hepatitis B Virus-Like Particles Carrying Hepatitis C Virus Envelope Glycoprotein 2 Epitopes Show Differential Neutralization Efficiency. *Vaccines*, *8*(2), 294. doi:10.3390/vaccines8020294 PMID:32532076

Danilov, L. L., Maltsev, S. D., Deyeva, A. V., Narovlyansky, A. N., Sanin, A. V., Ozherelkov, S. V., & Pronin, A. V. (1996). Phosprenyl: A novel drug with antiviral and immunomodulatory activity. *Archivum Immunologiae et Therapiae Experimentalis*, *44*(5-6), 395–400. PMID:9017158

Das, S., Sarmah, S., Lyndem, S., & Singha Roy, A. (2020). An investigation into the identification of potential inhibitors of SARS-CoV-2 main protease using molecular docking study. *Journal of Biomolecular Structure & Dynamics*, 1–11. doi:10.1080/07391102.2020.1763201 PMID:32362245

Donalisio, M., Leone, F., Civra, A., Spagnolo, R., Ozer, O., Lembo, D., & Cavalli, R. (2018). Acyclovir-loaded chitosan nanospheres from nano-emulsion templating for the topical treatment of herpesviruses infections. *Pharmaceutics*, *10*(2), 46. doi:10.3390/pharmaceutics10020046 PMID:29642603

Donovan, B. W., Reuter, J. D., Cao, Z., Myc, A., Johnson, K. J., & Baker, J. R. Jr. (2000). Prevention of murine influenza A virus pneumonitis by surfactant nano-emulsions. *Antiviral Chemistry & Chemotherapy*, *11*(1), 41–49. doi:10.1177/095632020001100104 PMID:10693653

Durai, R. D. (2015). Drug delivery approaches of an antiviral drug: A comprehensive review. *Asian Journal of Pharmaceutics*, *9*(1), 1-12.

Floriano, B. F., Carvalho, T., Lopes, T. Z., Takahashi, L. A. U., Rahal, P., Tedesco, A. C., & Calmon, M. F. (2021). Effect of berberine nanoemulsion Photodynamic therapy on cervical carcinoma cell line. *Photodiagnosis and Photodynamic Therapy*, *33*, 102174. doi:10.1016/j.pdpdt.2020.102174 PMID:33401021

Fox, C. B., & Haensler, J. (2013). An update on safety and immunogenicity of vaccines containing emulsion-based adjuvants. *Expert Review of Vaccines*, *12*(7), 747–758. doi:10.1586/14760584.2013.811188 PMID:23885820

Fox, C. B., Van Hoeven, N., Granger, B., Lin, S., Guderian, J. A., Hartwig, A., Marlenee, N., Bowen, R. A., Soultanov, V., & Carter, D. (2019). Vaccine adjuvant activity of emulsified oils from species of the Pinaceae family. *Phytomedicine*, *64*, 152927. doi:10.1016/j.phymed.2019.152927 PMID:31465981

Franklyne, J. S., Gopinath, P. M., Mukherjee, A., & Chandrasekaran, N. (2021). Nanoemulsions: The rising star of antiviral therapeutics and nano-delivery system-current status and prospects. *Current Opinion in Colloid & Interface Science*, *54*, 101458. doi:10.1016/j.cocis.2021.101458 PMID:33814954

Fuller, T. L., Calvet, G., Genaro Estevam, C., Rafael Angelo, J., Abiodun, G. J., Halai, U. A., De Santis, B., Carvalho Sequeira, P., Machado Araujo, E., Alves Sampaio, S., Lima de Mendonça, M. C., Fabri, A., Ribeiro, R. M., Harrigan, R., Smith, T. B., Raja Gabaglia, C., Brasil, P., Bispo de Filippis, A. M., & Nielsen-Saines, K. (2017). Behavioral, climatic, and environmental risk factors for Zika and Chikungunya virus infections in Rio de Janeiro, Brazil, 2015-16. *PLoS One*, *12*(11), e0188002. doi:10.1371/journal.pone.0188002 PMID:29145452

Guagliardo, R., Pérez-Gil, J., De Smedt, S., & Raemdonck, K. (2018). Pulmonary surfactant and drug delivery: Focusing on the role of surfactant proteins. *Journal of Controlled Release*, *291*, 116–126. doi:10.1016/j.jconrel.2018.10.012 PMID:30321577

Hadziyannis, S. J., Tassopoulos, N. C., Heathcote, E. J., Chang, T. T., Kitis, G., Rizzetto, M., Marcellin, P., Lim, S. G., Goodman, Z., Wulfsohn, M. S., Xiong, S., Fry, J., & Brosgart, C. L. (2003). Adefovir dipivoxil for the treatment of hepatitis B e antigen–negative chronic hepatitis B. *The New England Journal of Medicine*, *348*(9), 800–807. doi:10.1056/NEJMoa021812 PMID:12606734

Haham, M., Ish-Shalom, S., Nodelman, M., Duek, I., Segal, E., Kustanovich, M., & Livney, Y. D. (2012). Stability and bioavailability of vitamin D nanoencapsulated in casein micelles. *Food & Function*, *3*(7), 737–744. doi:10.1039/c2fo10249h PMID:22569895

Hamouda, T., Myc, A., Donovan, B., Shih, A. Y., Reuter, J. D., & Baker, J. R. (2001). A novel surfactant nanoemulsion with a unique non-irritant topical antimicrobial activity against bacteria, enveloped viruses and fungi. *Microbiological Research*, *156*(1), 1–7. doi:10.1078/0944-5013-00069 PMID:11372645

Hartshorn, K. L. (2010). Role of surfactant protein A and D (SP-A and SP-D) in human antiviral host defense. *Frontiers in Bioscience (Scholar Edition)*, *2*(1), 527–546. doi:10.274183 PMID:20036966

Hobson, J. J., Edwards, S., Slater, R. A., Martin, P., Owen, A., & Rannard, S. P. (2018). Branched copolymer-stabilised nanoemulsions as new candidate oral drug delivery systems. *RSC Advances*, *8*(23), 12984–12991. doi:10.1039/C8RA01944D

Khalid, N., Kobayashi, I., Wang, Z., Neves, M. A., Uemura, K., Nakajima, M., & Nabetani, H. (2015). Formulation of monodisperse oil-in-water emulsions loaded with ergocalciferol and cholecalciferol by microchannel emulsification: Insights of production characteristics and stability. *International Journal of Food Science & Technology*, *50*(8), 1807–1814. doi:10.1111/ijfs.12832

Kumar, M., & Jain, C. P. (2020). Possible benefits of reformulating antiviral drugs with nanoemulsion system in the treatment of novel coronavirus infection. *Current Drug Therapy, 2*, 3.

Leth-Larsen, R., Zhong, F., Chow, V. T., Holmskov, U., & Lu, J. (2007). The SARS coronavirus spike glycoprotein is selectively recognized by lung surfactant protein D and activates macrophages. *Immunobiology, 212*(3), 201–211. doi:10.1016/j.imbio.2006.12.001 PMID:17412287

Lopez-Diez, E., Perez, S., Carballo, M., Iñarrea, A., de la Orden, A., Castro, M., Rodríguez, M., Almuster, S., Montero, R., Perez, M., Sanchez, J., & Ojea, A. (2017). Lifestyle factors and oncogenic papillomavirus infection in a high-risk male population. *PLoS One, 12*(9), e0184492. doi:10.1371/journal.pone.0184492 PMID:28898279

Makidon, P. E., Bielinska, A. U., Nigavekar, S. S., Janczak, K. W., Knowlton, J., Scott, A. J., Mank, N., Cao, Z., Rathinavelu, S., Beer, M. R., Wilkinson, J. E., Blanco, L. P., Landers, J. J., & Baker, J. R. Jr. (2008). Pre-clinical evaluation of a novel nanoemulsion-based hepatitis B mucosal vaccine. *PLoS One, 3*(8), e2954. doi:10.1371/journal.pone.0002954 PMID:18698426

Manoharan, Y., Haridas, V., Vasanthakumar, K. C., Muthu, S., Thavoorullah, F. F., & Shetty, P. (2020). Curcumin: A wonder drug as a preventive measure for COVID19 management. *Indian Journal of Clinical Biochemistry, 35*(3), 373–375. doi:10.100712291-020-00902-9 PMID:32641876

Manyarara, T. E., Khoza, S., Dube, A., & Maponga, C. C. (2018). Formulation and characterization of a paediatric nanoemulsion dosage form with modified oral drug delivery system for improved dissolution rate of nevirapine. *MRS Advances, 3*(37), 2203–2219. doi:10.1557/adv.2018.320

McClements, D. J. (2012). Advances in fabrication of emulsions with enhanced functionality using structural design principles. *Current Opinion in Colloid & Interface Science, 17*(5), 235–245. doi:10.1016/j.cocis.2012.06.002

Minz, S., & Pandey, R. S. (2018). Lipid a adjuvanted chylomicron mimicking solid fat nanoemulsions for immunization against hepatitis B. *AAPS PharmSciTech, 19*(3), 1168–1181. doi:10.120812249-017-0932-9 PMID:29243216

Myc, A., Kukowska-Latallo, J. F., Bielinska, A. U., Cao, P., Myc, P. P., Janczak, K., Sturm, T. R., Grabinski, M. S., Landers, J. J., Young, K. S., Chang, J., Hamouda, T., Olszewski, M. A., & Baker, J. R. Jr. (2003). Development of immune response that protects mice from viral pneumonitis after a single intranasal immunization with influenza A virus and nanoemulsion. *Vaccine, 21*(25-26), 3801–3814. doi:10.1016/S0264-410X(03)00381-5 PMID:12922114

Nabila, N., Suada, N. K., Denis, D., Yohan, B., Adi, A. C., Veterini, A. S., Anindya, A. L., Sasmono, R. T., & Rachmawati, H. (2020). Antiviral action of curcumin encapsulated in nanoemulsion against four serotypes of dengue virus. *Pharmaceutical Nanotechnology, 8*(1), 54–62. doi:10.2174/2211738507666191210163408 PMID:31858909

O'Konek, J. J., Makidon, P. E., Landers, J. J., Cao, Z., Malinczak, C. A., Pannu, J., Sun, J., Bitko, V., Ciotti, S., Hamouda, T., Wojcinski, Z. W., Lukacs, N. W., Fattom, A., & Baker, J. R. Jr. (2015). Intranasal nanoemulsion-based inactivated respiratory syncytial virus vaccines protect against viral challenge in cotton rats. *Human Vaccines & Immunotherapeutics*, *11*(12), 2904–2912. doi:10.1080/21645515.2015.1075680 PMID:26307915

Pang, X. F., Zhang, L. H., Bai, F., Wang, N. P., Garner, R. E., McKallip, R. J., & Zhao, Z. Q. (2015). Attenuation of myocardial fibrosis with curcumin is mediated by modulating expression of angiotensin II AT1/AT2 receptors and ACE2 in rats. *Drug Design, Development and Therapy*, *9*, 6043. PMID:26648693

Parra, E., & Pérez-Gil, J. (2015). Composition, structure and mechanical properties define performance of pulmonary surfactant membranes and films. *Chemistry and Physics of Lipids*, *185*, 153–175. doi:10.1016/j.chemphyslip.2014.09.002 PMID:25260665

Pilcer, G., & Amighi, K. (2010). Formulation strategy and use of excipients in pulmonary drug delivery. *International Journal of Pharmaceutics*, *392*(1-2), 1–19. doi:10.1016/j.ijpharm.2010.03.017 PMID:20223286

Prabhakar, K., Afzal, S. M., Surender, G., & Kishan, V. (2013). Tween 80 containing lipid nanoemulsions for delivery of indinavir to brain. *Acta Pharmaceutica Sinica. B*, *3*(5), 345–353. doi:10.1016/j.apsb.2013.08.001

Praditya, D., Kirchhoff, L., Brüning, J., Rachmawati, H., Steinmann, J., & Steinmann, E. (2019). Anti-infective properties of the golden spice curcumin. *Frontiers in Microbiology*, *10*, 912. doi:10.3389/fmicb.2019.00912 PMID:31130924

Raghavan, P. R. (2016). Inhibition of Dengue and other enveloped viruses by Mctadichol®, a novel Nano emulsion Lipid. *Journal of the science of Healing Outcomes, 8*, 19-25.

Rani, D., Saxena, R., Nayak, B., & Srivastava, S. (2018). Cloning and expression of truncated ORF2 as a vaccine candidate against hepatitis E virus. *3 Biotech, 8*(10), 1-5.

Saini, R., Saini, S., & Sharma, S. (2010). Nanotechnology: The future medicine. *Journal of Cutaneous and Aesthetic Surgery*, *3*(1), 32–33. doi:10.4103/0974-2077.63301 PMID:20606992

Shaker, D. S., Ishak, R. A., Ghoneim, A., & Elhuoni, M. A. (2019). Nanoemulsion: A review on mechanisms for the transdermal delivery of hydrophobic and hydrophilic drugs. *Scientia Pharmaceutica*, *87*(3), 17. doi:10.3390cipharm87030017

Simonsen, L., Kane, A., Lloyd, J., Zaffran, M., & Kane, M. (1999). Unsafe injections in the developing world and transmission of bloodborne pathogens: A review. *Bulletin of the World Health Organization*, *77*(10), 789. PMID:10593026

Soni, H., & Sharma, S. (2021). Current Update on Nanoemulsion: A Review. *Sch Int J Anat Physiol*, *4*(1), 6–13.

Vyas, T. K., Shahiwala, A., & Amiji, M. M. (2008). Improved oral bioavailability and brain transport of Saquinavir upon administration in novel nanoemulsion formulations. *International Journal of Pharmaceutics*, *347*(1-2), 93–101. doi:10.1016/j.ijpharm.2007.06.016 PMID:17651927

Walinska, K. (2004). Comparison of the influence of the polyprenol structure on model membranes. *Desalination*, *163*(1-3), 239–245. doi:10.1016/S0011-9164(04)90195-6

Walinska, K., & Janas, T. (2001). The effect of phosphate ester derivatives of polyprenol on model membrane stability. *Cellular & Molecular Biology Letters*, *6*(2A).

Wang, C. Z., Li, W. J., Tao, R., Ye, J. Z., & Zhang, H. Y. (2015). Antiviral activity of a nanoemulsion of polyprenols from ginkgo leaves against influenza A H3N2 and hepatitis B virus in vitro. *Molecules (Basel, Switzerland)*, *20*(3), 5137–5151. doi:10.3390/molecules20035137 PMID:25808155

Wang, C. Z., Li, W. J., Tao, R., Ye, J. Z., & Zhang, H. Y. (2015). Antiviral activity of a nanoemulsion of polyprenols from ginkgo leaves against influenza A H3N2 and hepatitis B virus in vitro. *Molecules (Basel, Switzerland)*, *20*(3), 5137–5151. doi:10.3390/molecules20035137 PMID:25808155

Wong, P. T., Goff, P. H., Sun, R. J., Ruge, M. J., Ermler, M. E., Sebring, A., ... Baker, J. R. Jr. (2020). Combined Intranasal Nanoemulsion and RIG-I Activating RNA Adjuvants Enhance Mucosal, Humoral, and Cellular Immunity to Influenza Virus. *Molecular Pharmaceutics*. PMID:32491861

Xu, Q., Zhou, A., Wu, H., & Bi, Y. (2019). Development and in vivo evaluation of baicalin-loaded W/O nanoemulsion for lymphatic absorption. *Pharmaceutical Development and Technology*, *24*(9), 1155–1163. doi:10.1080/10837450.2019.1646757 PMID:31342830

Zhong, J., Huang, P., Wen, M., Liang, L., Zhang, X., Tan, S., & Zhu, X. (2012). Antigenic epitope peptides of influenza H3N2 virus neuraminidase gene based on experiments. *Chinese Science Bulletin*, *57*(22), 2908–2913. doi:10.100711434-012-5266-0

Chapter 10
Mushroom Bioactive Compounds:
Potential Source for the Development of Antibacterial Nanoemulsion

Deepali Koreti
Pandit Ravishankar Shukla University, Raipur, India

Anjali Kosre
Pandit Ravishankar Shukla University, Raipur, India

Ashish Kumar
Sant Gahira Guru Vishwavidyalaya, Sarguja Ambikapur, India

Nagendra Kumar Chandrawanshi
https://orcid.org/0000-0002-9093-5365
Pandit Ravishankar Shukla University, Raipur, India

ABSTRACT

Mushrooms are widely utilized as a nutritional and functional food, and they are also appreciated for their medicinal as well as therapeutic applications. Bioactive compounds are isolated from mushrooms possess medicinal properties such as anti-inflammatory, antimicrobial, antitumor, immunomodulatory, and antioxidant. Mushroom bioactive compounds having antibacterial potential and can be used in medical sectors for the cure of several pathogenic disease. Nanoemulsions are one such example of nano-technique that has a very promising application in various fields. The use of nanoemulsion in encapsulation for food-grade ingredients in their compounds such as vitamins, lipids, antioxidants, and antimicrobial agents is a new technique. Bioactive compounds from mushrooms can be a good alternative source for antibacterial nanoemulsions development. This chapter discusses all bioactive compounds and the production and purification of these bioactive compounds. Another hand discusses the nanoemulsion formation by bioactive compounds and nanoemulsion used as antimicrobial agents.

DOI: 10.4018/978-1-7998-8378-4.ch010

INTRODUCTION

Mushrooms are a rich source of primary and secondary metabolites. These metabolites are participated various biological activities *in vitro* as well as *in vivo*. Compounds that can perform biological activities are called bioactive compounds. Major bioactive compounds in fungal groups are known as mycochemicals, can be found in the form of their cell wall components (proteins, polysaccharides) or as secondary metabolites (phenolic terpenes, compounds, and steroids), etc. Mushrooms are well known for rich bioactive compounds and therefore also known as a functional food. Various natural sources have been utilized as therapeutic agents since ancient times like there are more than 10,000 plants identified for their medicinal purposes (Takala *et al.*, 2011). The medicinal value of the plant or other foods depends on the active compounds within them.

These natural active compounds in plants and foods tend to be poorly absorbed in the body. Recently scientists discovered various techniques for the extraction, purification, and characterization of its biological potentials. Extracted compounds are chemically unstable by nature, and their high doses and repeated consumption may cause other therapeutic effects. The bioactive compound of foods can degrade by internal factors and other environmental factors like sunlight, UVB light, and high temperatures (Marsup *et al.*, 2020). Improving the absorption of bioactive compounds from the plant and other foods requires suitable delivery systems. Nanotechnology is now a trading way for solving many problems in the medical sector; it can be effective and efficient for targeted drug delivery. In this way, nanoemulsions-based bioactive compounds delivery systems represent the most suitable approach. Nanoemulsions are nano-sized (size varies from 10 to 1,000 nm) emulsions designed for increasing the delivery rate and surface area of active components. Formation of bioactive nanoemulsions initially lipid droplets of nanometric size are dispersed in an aqueous phase by using a suitable emulsifier at the oil/water interface and produced through high-pressure homogenization. Immobilization of bioactive compounds can enhance it's the dispensability in aqueous solutions. It's also minimized the phase's separation and protects bioactive compounds from interactions with food ingredients. Nanoemulsions can maintain structural-functional properties, improve their absorption rate and bioavailability inside the body. One possible mechanism given by researchers for its high efficiency is that the enhanced passive transport across the cell membrane (Porter *et al.*, 2008). Nanoemulsions work as a carrier for the delivery of active compounds and increase their absorption. Nanoemulsions can be effectively utilized as therapeutic agents in, food, cosmetic, and pharmaceutical industries. Currently, extensive bioactive compounds are isolated and utilized in the nanoemulsions development from plants (Donsi *et al.*, 2010).

This chapter mainly focuses on the nanoemulsion in the field of antibacterial agents developments. Recently there are various antibacterial agents available for different bacterial species, but there is some limitation associated with a conventional antibacterial drug such as the generation of drug-resistant strain. The number of drug-resistant bacterial strains has evolved worldwide that generates economic losses to patients and countries. Nanoemulsion was recently applicable in the field of antibacterial development and helpful for overcoming the problem of drug resistance strain generation. Mushroom-based bioactive compound and their nanoemulsions are new areas of research. Many bioactive compounds possess bioactivity like antioxidants, anti-inflammatory, antibiotics, anticancerous, immunomodulatory, and so on, but their bioavailability is very low therefore nanoemulsions are the best option for efficient delivery and enhance bioavailability (Wenzel *et al.*, 2005). Mushrooms are a rich source of bioactive compounds, hence they can be the most alternative and probable source for making bioactive nanoemulsions for antibacterial agent development.

BIOACTIVE COMPOUNDS OF MUSHROOMS

Mushrooms' bioactive compounds are found in their cell wall components like; proteins and polysaccharides or as secondary metabolites like; phenolic compounds, terpenes, and steroids. (Patel *et al.*, 2012; Gupta *et al.*, 2019). Their concentration and efficacy depend on the type of mushroom, type of substrate, substrate composition, growth conditions, developmental stage, culture or postharvest conditions, storage, and cooking procedures (Guillamon *et al.*, 2010; Enshasy *et al.*, 2013). Besides their pharmacological features, and are consumed as a functional food that is beneficial in our diet (Khatun *et al.*, 2012). Researchers investigating the preparation and structural characterizations methods of bioactive compounds from mushrooms. Mushroom bioactive compounds are categorized into two categories such as primary and secondary metabolites. Secondary metabolites include; vitamins, polyphenols, alkaloids, terpenoids, sesquiterpenes, lactones, sterols, nucleotide analogs, glycoprotein's and polysaccharides (*β*-glucans). Some identified bioactive compounds are; laccase, lectins, lignin-degrading enzymes, proteases, protease inhibitors, ribosome-inactivating proteins, and hydrophobins, etc. (Erjavec e*t al.*, 2012). These mushroom bioactive compounds can contribute to the development of potential functional foods source with high nutritional values (Valverde e*t al.*, 2015). Mushrooms contain all essential amino acids required for the human body; also they are an excellent resource of vitamins and minerals (Barros e*t al.*, 2008).

Polysaccharides

Polysaccharide is the chief bioactive substance in edible mushrooms, which are biological activities and received attention worldwide in the medicine and pharmaceutical sector (Bisen *et al.*, 2010). Various studies reported that mushroom polysaccharides are used as potential pharmacological agents with different bioactivities (Bai *et al.*, 2018; Rodrigues Barbosa *et al.*, 2019; Cheng *et al.*, 2019). Polysaccharides are formed in numerous types in the mushrooms cells, either can be simpler ones like (monosaccharides) or they can be a complex form (oligo and polysaccharides). Some edible mushrooms polysaccharides reported for the developed as functional food substances such as the schizophyllan from *Schizophyllum commune*, lentinan from *L. edodes,* pleuran from *Pleurotus* species, calocyban from *Calocyb*e species, all of these are proved and considered as the excellent representatives of D-glucans. Simple polysaccharides such as glucose, fructose, galactose, xylose, mannose, fucose, rhamnose, arabinose, trehalose, and mannitol, are the most common polysaccharide found in mushrooms. Mushrooms polysaccharides having antitumor, anti-inflammatory, and immunomodulatory activities and provide health benefits for the human body. The mushroom polysaccharide can activate dendritic cells, monocytes, natural killer cells, neutrophils, cytotoxic macrophages, and different cytokines (Wasser, 2015). *β*-Glucans are biologically active molecules or glucose polymers found in yeast, grain, alga, and fungus. *β*-glucans are polysaccharides composed of multiple linear and branched D-glucose units linked by β-1-3 and 1-6, bonds. The bioactivity of polysaccharides is significantly influenced by physicochemical modifications such as the degree of branching and addition of substituent groups (sulfates, selenates) (Li *et al.*, 2015). *Pleurotus* species mushrooms are an excellent source of crude fiber (10.2%) and particularly for β-glucans content (25.9%), which can be widely associated with prebiotic and anti-cholesterol, anti-cancer, and immunomodulatory properties (Aida *et al.*, 2009; Zhu *et al.*, 2015). *P. citrinopileatus* and *P. ostreatus* have been reported for their highest crude fiber and β-glucans content which are 20.7 and 50%, respectively. It can be used as a drug and is also known as a biological response modifier. β-glucans, being pleuran the more studied and recognized in *Pleurotus* genus mushrooms having bioactivity in humans and natural immunostimulant

(Imunoglukan P4H®) (Jesenak *et al.*, 2013; Bergendiova *et al.*, 2011). Sulfated polysaccharide shows increased solubility, antiviral, anticancer, anticoagulant, and antioxidant activities. It has anti-infective, antitumor, antiviral, and immunomodulatory activities as also wound healing properties (Chandrawanshi *et al.*, 2017; Majtan *et al.*, 2018). As a form of a drug, it was reported that their intestinal permeability is very low, because of its macromolecular structure (Matsuda *et al.*, 2012). Utilization as an oral biological substance, polysaccharides are damaged by the acidic environment of the stomach and reduce their desired effect (Koning *et al.*, 2012; Pu *et al.*, 2016). It was poorly soluble in water. The extracted polysaccharide can aggregate with storage time, therefore, forms larger droplets in water. Therefore it's important to find an efficacious delivery system for polysaccharides by reducing the particle size and improving the solubility and stability. Li *et al.* (2019) prepared and optimized water in oil nanoemulsion by *Lentinus edodes* polysaccharide and evaluates its antitumor activity. These studies demonstrate that enhanced antitumor potential in the form of nano emulsified polysaccharides.

Proteins and Peptides

Numerous proteins and peptides are identified, characterized and their mechanisms of action are well established by researchers. Edible mushrooms are also an excellent source of bioactive proteins and peptides, which possess different biological activities. But there are many other unidentified proteins, so mushrooms represent a relatively unexplored source of novel proteins. Similarly, it's also an endless source of novel peptides. Studies revealed that the dry mushrooms content 228 to 249g (g/kg of mushroom) protein, which is much higher than other protein sources (Petrovska, 2020). Proteins and peptides are a large group of bioactive compounds with multiple health benefits such as; enhances the digestion and absorption of exogenous nutritional components and boost up the immune system. Mushrooms proteins with pharmaceutical potential have been found and named lectins, ribonucleases, ribosome-inactivating proteins, immunomodulatory proteins, and laccase (Ma *et al.*, 2018). Bhandari *et al.* (2020) reported another group of a peptide called bioactive peptides (BAPs) is small fragments (contain 2-20 amino acids) of proteins that provide health benefits (Ryan *et al.*, 2011). Depending on their amino acid composition and confirmation they perform various biological activities (Bechaux *et al.*, 2019). Because of their size, these BAPs could be easily absorbed by the intestine and directly produce local effects in the digestive tract, or they also entered the circulatory system and show their physiological effects (Erdmann *et al.*, 2009). According to Daliri *et al.* (2017) and Montesano *et al.* (2020), the biological function of the bioactive proteins has been credited to their encoded BAPs and can be released without losing the bioactivities. Peptide shows improved bioactivity as compared to their parent proteins (Udenigwe *et al.*, 2012).

Lectin

Lectins are bioactive proteins of varying complexity that reversibly bind to carbohydrates and are found throughout nature. It's played a significant role in cellular interaction, cell adhesion to glycoprotein synthesis, controlling the protein level in the blood; regulate differentiation, in host defense mechanisms, inflammation, drug targeting, and organ formation, metastasis of cancer cells (Sharon *et al.*, 1989; Wang *et al.*, 1998; Singh *et al.*, 1999; Tiwary *et al.*, 1999; Hassan *et al.*, 2015). Lectins were found and isolated from plants seeds and roots, bacteria, insects, algae, body fluid of vertebrates, mammalian cell membranes, and fungi (Singh, 1999). 10% of the mushroom species have been reported for potentially bioactive substances, including lectins (Chang, 1992; Wang *et al.*, 1998). Approximately 80% of fungal lectins are isolated from mushrooms (Singh *et al.*, 2010; Varrot *et al.*, 2013), whereas edible mushrooms are reported for various lectins (Pohleven *et al.*, 2009; Liu *et al.*, 2013; S. Kim *et al.*, 2018). In mushrooms, lectins are widely distributed in the fruiting body and vegetative mycelia. Different chromatographic techniques and affinity chromatography were widely used for the isolation and purification of mushroom lectins. Mostly intracellular and surface mycelial associated lectins have been reported *by Pleurotus* species (Sarup Singh *et al.*, 2019). Mushroom lectins having various physicochemical characteristics such as carbohydrate content, number of subunits and stability, molecular weight, pH, and temperature range. Generally, mushroom lectins range from 12 to190 kDa and content 2 or more subunits (identical or non-identical) (Khan and Khan, 2010). *Pleurotus* species mushrooms have been reported for various lectins with a range of bioactivity such as antihepatoma and antisarcoma (Yoshida *et al.*, 1994; Wang *et al.*, 2000; Wang *et al.*, 2001; Wang *et al.*, 2003; Sumisa *et al.*, 2004). Lectins play a role in molecular recognition in various biological processes, furthermore, making them valuable structural probes for use in future biotechnology research.

Phenols

Phenols also called phenolic compounds they have one aromatic ring (C6) at least and one or more (OH) groups. It is found in the form of small simple molecules to complicated polymers (Michalak, 2006). Phenolic compounds are found in various biological plants as well as fungus such as mushrooms. It's a group of compounds that shows various biological activities. It's reported as anti-microbial, anti-inflammatory, anti-thrombotic, and anti-allergenic properties (Assadpour *et al.*, 2018). Phenolic compounds are lipophilic and hydrophilic nature compounds. Polyphenols are aromatic molecules with phenolic structural units that have low *in vivo* bioavailability and therefore their appliance in chronic disease treatment and prevention are limited. Curcumin, tannin, stilbenes and flavonoids are the some oplyphenols (Scalbert and Williamson, 2000). Different kinds of phenolic compounds are identified in edible mushroom extract such as phenolic acids, flavonoids, tannins, and tocopherols, etc. (Abdelshafy *et al.*, 2021). Studies revealed that the antioxidant activity is the most significant bioactivity shows by phenolic compounds (Chandrawanshi *et al.,* 2015; Chandrawanshi et al., 2018). It's well known as free radical inhibitors, metal inactivators, peroxide decomposers, and oxygen scavengers (Heleno *et al.*, 2012). It is well known that phenolic compounds are active bimolecular and protect against several degenerative disorders such as; brain dysfunction, cardiovascular diseases, and aging (Finimundy *et al.*, 2013).

Terpenes

Chemically terpenes are the cluster of volatile unsaturated hydrocarbons and are classified as monoterpenoids, diterpenoids, triterpenoids, and sesquiterpenoids. Researchers have found and isolate numerous types of sesquiterpenoids from mushrooms, including aristolane, bisabolene, cuparene, drimane, fomannosane, lactarane, nordasinane, spiro, sterpurane. Diterpenoids are mainly detected as cyathane types in the mushroom. Most of the triterpenoid compounds are lanostane type which is isolated from mushrooms. Moreover, the mushroom terpenes have been proved for numerous health benefits such as antioxidant, antiviral, and anticancer activities (Klaus *et al.*, 2017).

MUSHROOMS BIOACTIVE COMPOUNDS: AS ANTIBACTERIAL AGENTS

A variety of compounds have been isolated from various species of edible mushrooms that have been shown to possess medicinal properties such as anti-inflammatory, antihyperglycemic, antimicrobial, antiviral, antitumor, immunomodulatory, and antioxidant (Periasamy 2005; Hu 2006; Xu et al., 2016; Liu et al., 2019). Bioactive compounds extraction, identification, and utilization in the medicinal application have been increased significantly in past years. Various mushroom crude extract and isolated bioactive compounds have been reported for their antimicrobial potential. In the medical sector, there are numerous antibiotic drugs are available for the treatment of several bacterial diseases. During past decades several drug-resistant varieties or strains of microbes are developed. These multi-drug resistant strains are developed due to the extensive use of antibiotics in common infectious diseases. Therefore researchers are searching and working for the development and formulation of a novel antibiotic from a natural source that has a new mechanism of action. Hence mushroom-based bioactive for antimicrobial products development has received great interest in currents years. Some reported mushrooms and their bioactive compounds are given in table 1 and used as antibacterial agents. Various bioactive phenolic compounds which having antimicrobial activities are reported in mushrooms (Ozen *et al.,* 2011). Their mode of action on microbes is dependent upon the concentration and type of active compound as well as mushroom species. Chowdhury et al. (2015) screened the antimicrobial and antioxidant potential of three edible mushrooms. They are also estimated bioactive constituents such as phenols, flavonoids, and ascorbic acid in mushroom extract and observed that phenols are the major bioactive constituent among others. Gebreyohannes et al. (2019) reported antimicrobial activity of 35 indigenous wild mushrooms against gram-positive and gram-negative bacteria species. Bach *et al.* (2019) reported antimicrobial activity of the phenolic extracts of five edible mushrooms; *Agaricus bisporus, Lentinula edodes, Flammulina velutipes,* and *Agaricus brasiliensis* were observed against some pathogenic bacteria, *including E. coli, S. enteritidis, S. aureus,* and *B. cereus*. Recently Angelini *et al.* (2020) reported antimicrobial activity of *Tricholosporum goniospermum* mushroom extract (methanol, ethyl acetate, and n-hexane extraction) was tested antimicrobial activity against Gram-positive bacteria, including *B. cereus, B. subtilis, and S. aureus*, and Gram-negative bacteria, including *E.coli, P. aeruginosa*, and *S. typhy* and yeasts, including *C. albicans, C. albicans, C. tropicalis,* and *C. parapsilopsis*. HPLC analysis in this study reveals that mushroom phenolic extracts were richest in Gallic Acid as well as a low quantity of Catechin. As novel bioactive compounds of mushrooms are being discovered, the possibilities of its use in the medicinal sector are continuously increasing.

Table 1. Some mushroom species and their reported antibacterial activity.

S. No.	Mushroom species name	Extraction solvent	Purification method	Bioactive compounds	Bacterium strain	References
1.	*Agaricus cf.nigrecentulus, Irpex lacteus, Leucoagaricus cf. cinereus, Marasmius cf. bellus, Marasmius sp., Nothopanus hygrophanus, Pycnoporus sanguineus,* and *Tyromyces duracinus*	Ethyl acetate extraction	-	-	*Bacillus* species, *Enterococcus faecalis, E.coli, Listeria monocytogenes, Pseudomonas aeruginosa, Salmonella typhimurium, Staphylococcus aureus, S. epidermidis, S.saprophyticu, Streptococcus pyogenes* and *S. pneumoniae*	Rosa *et al.*, 2003
2.	*Pleurotus sajor-caju*	Aqueous extract	Chromatography	Ribonuclease	*P. aeruginosa* and *S. aureus*	Ngai *et al.*, 2004
3.	*Cordyceps sinensis*	HPLC	HPLC	CSAP isolated	*Bacillus subtilis, E. coli Proteus vulgaris S. aureus Salmonella typhi*	Zheng *et al.*, 2006
4.	*Agaricus bisporus* and *Pleurotus sajor caju*	Methanol, ethanol, ether, xylene, benzene, acetone extraction	-	-	*E. aerogenes, E.coli, P. aeruginosa, Klebsiella pneumonia*	Tambekar *et al.*, 2006
5.	*Agaricus species, Boletus edulis, Calocybe gambosa, Cantharellus cibarius, Craterellus cornucopioides,* and *Marasmius oreades*	Methanol extraction	-	Phenols, flavonoids, ascorbic acid and carotenoids	*Bacillus* species, *S.s aureus, E.coli, P. aeruginosa CECT and K. pneumoniae*	Barros *et al.*, 2008
6.	*Lentinus edodes*	Ethyl acetate extract and aqueous extract	-	-	*Bacillis* species, *Cupriavidis, E. coli, Klebsiella* species, *E. faecalis, Listeria monocytogenes Pseudomonas* species, *Salmonella poona, Serratia marcescens, Serratia, Staphylococcus* species and Wild type hand bacteria 1,2,3	Ishikawa *et al.*, 2001; Hearst *et al.*, 2009
7.	*Lentinus cladopus* LC4	Methanol extraction	Chromatography	-	*Bacillus subtilis*	Sudirman, 2010
8.	*Armillaria mellea, Meripilus giganteus, Morchella costata, Morchella elata, Morchella esculenta var. vulgaris, Morchella hortensis, Morchella rotunda, Paxillus involutus, Pleurotus* species	Ethanol extraction	-	-	*Bacillus* species *Enterobacter* species, *E. coli, Proteus vulgaris, Salmonella typhimurium, Sarcina lutea*, and *S. aureus*	Kalyoncu *et al.*, 2010
9.	*Jahnoporus hirtus* and *Albatrellus flettii*	MeOH in CH_2Cl_2	TLCs, HPLC	Terpenes, Confluentin, Grifolin and Neogrifolin	*B. cereus* and *Enterococcus faecalis*	Liu *et al.*, 2010
10.	*Ganoderma pfeifferi*	Dichloromethane extraction	Sephadex LH 20 column	Ganomycin A and B	*S. aureus, B. subtilis, E.coli, Proteus mirabilis, and Pseudomonas aeruginosa, Serratia marcescens*, and *M. flavus*	Mothana *et al.*, 2010
11.	Mushrooms of genus *Cortinarius*	Ethyl acetate and aqueous fraction	TLC	Anthraquinones derivatives	*Staphylococcus aureus* and *Pseudomonas aeruginosa*	Beattie *et al.*, 2010
12.	*Ganoderma lucidum*	Ethyl alcohol, methanol, acetone and distilled water	-	-	*E. coli, S. aureus, K. pneumoniae, Bacillus subtilis, Salmonella typhi* and *P. aeruginosa*	Quereshi *et al.*, 2010

Continued on following page

Table 1. Continued

S. No.	Mushroom species name	Extraction solvent	Purification method	Bioactive compounds	Bacterium strain	References
13.	*Lactarius* species, *Chanterellus cibarius, A. bisporus, Ramaria flava, Macrolepiota procera, Leatiporus sulphureus, Boletus edulis, Hydnum repandum, Cortinarius sp.*	Methanol extraction	-	-	*Staphylococcus* species, *Bacillus cereus, E. coli,* and *P. aeruginosa*	Ozen *et al.*, 2011
14.	*Lentinus edodes*	Aqueous extracts	-	-	*P. intermedia*	Signoretto *et al.*, 2011
15.	*Trametes* species, *Trametes* species, and *Microporus* species	Chloroform, ethanol and hot water extraction	-	-	*S. aureus,* and *K. pneumonia*, and *E. coli*	Gebreyohannes *et al.* 2019

ANTIBACTERIAL NANOEMULSION AND THE DEVELOPMENT OF BIOACTIVE BASED ANTIBACTERIAL NANOEMULSION

During the earlier period, many phytochemicals have been encapsulated within nanoemulsion and used for medicinal as well industrial applications. It has been already well documented that these nano-sized bioactive compounds have enhanced bioavailability (Huang *et al.*, 2010; Arunkumar *et al.*, 2013). Nanoemulsion technology has various advantages like their targetability, biocompatibility, and biodegradability that makes it easily applied for several administration routes like; intravenous, ocular, and transdermal (Ezhilarasi *et al.*, 2013; Fathi *et al.*, 2014). Nanoemulsions can help for dispersibility of poorly soluble compounds in the aqueous phase, increases their bioavailability as well as protect the bioactive compounds from interaction with other compounds. Nanoemulsions reduce the force on the organoleptic properties of food and enhance their absorption and bioavailability through the passive transport across the cell membrane (Gupta *et al.*, 2013). Research data showed that fruit, vegetables, and mushrooms, (they contain bioactive compounds) in diet can have a health-protective role and reduces the risk of chronic diseases.

Antimicrobial nanoemulsions are emulsified mixtures of oil, detergent and have shown a broad range of antimicrobial activity against pathogenic microbes like; bacteria, viruses, and fungi (Hwang *et al.*, 2013). Antimicrobial nanoemulsion works on lipid bilayers of the cell membrane when nanoemulsions fused and released the compounds stored inside the emulsion and destabilize the lipid membrane of the microbes. The nonspecific activity of nanoemulsions provides its broad-spectrum antibiotic activity without generating resistance strains, therefore nanoemulsion can be a suitable candidate for used as an effective antimicrobial agent and used in a broad range of products development (Hamouda and Baker, 2000). Prakash *et al.* (2020) studied the effect of nano emulsification of some spices and their major compound on the antibacterial and antibiofilm activities against the common food-borne pathogen such as *S. enterica*. This experiment observed that lemongrass and its major constituent like citral exhibited higher antibacterial and anti-biofilm activities. Further encapsulation of citral through nanoemulsion has improved the antibacterial and anti-biofilm activities against a pathogen.

Many bioactive compounds such as polycarbohydrates, lipids, essential oils, vitamins, polyphenols, flavor compounds, carotenoids are isolated from plants and used for nanoemulsion development. Mushrooms contain bioactive compounds including; tannins, polyphenols, and anthocyanins, *etc.* (da Silva *et al.*, 2016; Aditya *et al.*, 2017). These isolated compounds are less soluble in water and stability criteria make it challenging in food industrial application (Weiss *et al.*, 2006; Salvia-Trujillo *et al.*, 2017). Nanoemulsions are formulated to retain their bioactive compounds during storage within a product or a food product as well as effectively release in the target area (Soukoulis *et al.*, 2017). Recently various bioactive compounds from mushrooms were commercially produced and applied in different sectors. Isolation of individual compounds from mushrooms and its use in nanoemulsion development is a new research area in which not much work has been done yet. Encapsulations of bioactive natural products are very useful for novel nutraceuticals development (Sarkar *et al.*, 2015). *Cordyceps militaris*, also known as parasitic fungus and was extensively used as a traditional medicine in China. It contains bioactive compounds such as; cordycepin, D-mannitol (cordycepic acid), and polysaccharides. These compounds were identified for their antiaging, whitening, antitumor, anti-inflammatory, immunomodulatory, and/or blood glucose and cholesterol-lowering activities (Kiho *et al.*, 1992; Yu *et al.*, 2015). Rupa *et al.* (2020) produces and optimized a *Cordyceps militaris* fungus-based oil-in-water nanoemulsion encapsulated in sea buckthorn oil, by using an ultrasonication process. This fungus-based nanoemulsion exhibited good antioxidant activity and potent antimicrobial activity against *E. coli* and *S. aureus* bacteria.

Table 2. Some bioactive nanoemulsions are reported as an antimicrobial agent.

S. No.	Source of the compounds	Active compounds	Formation method	Applications	References
1.	Curcumin	Curcumin	High pressure homogenization	Antimicrobial properties	Dons *et al.*, 2011
2.	Peppermint oil	-	Pressure homogenization	Prolonged antibacterial Activities	Liang *et al.*, 2012
3.	Thyme oil	-	High pressure homogenization	Antimicrobial activities against an spoilage yeast, *Zygosaccharomyces bailii*	Chang *et al.*, 2012
4.	Thyme oil	-	Homogenization	Antibacterial activity against *E. coli*	Xue *et al.*, 2015
5.	Cinnamaldehyde	-	Spontaneous emulsification	Provide a long-term inhibition on the bacterial growth	Tian *et al.*, 2016
6.	Eugenol oil	-	High speed homogenization	Antimicrobial activity of eugenol against *Listeria monocytogenes* and *Salmonella Enteritidis*	Hu *et al.*, 2016
7.	Oregano, thyme, lemongrass or mandarin essential oils	-	Microfluidizer	antimicrobial activity against *E. coli* and *Listeria innocua*	Guerra-Rosas *et al.*, 2017
8.	Cinnamon oil	-	Spontaneous emulsification, microfluidization and ultrasonication	Antimicrobial	Yildirim *et al.*, 2017
9.	D-limonene oil	-	Ultrasonication	*Bacillus cereus, E. coli, Enterococcus faecalis* and *Salmonella typhi*	Sonu *et al.*, 2018
10.	Thymus daenensis oil	lecithin and Tween 80 as emulsifiers	Ultrasonication	Antibacterial activity against *Acinetobacter baumannii*	Moghimi *et al.*, 2018
11.	Propolis	Phenolic acids, flavonoids, and artepillin C	Phase Inversion Emulsification method	Antimicrobial and antioxidant activities	Seibert *et al.*, 2019
12.	*Lentinus edodes*	Polysaccharide	ND	Improved the stability, higher anti-tumor activity	Li *et al.*, 2019
13.	Spice essential oil	cuminaldehyde, anethole, linalool, b-caryophyllene	Ultrasonication	Antibacterial activities against *Salmonella enterica Typhimurium*	Prakash *et al.*, 2020
14.	*Cordyceps militaris* Fungus	-	Ultrasonication	Improvement antioxidant, antimicrobial, and anti-inflammatory activities	Rupa *et al.*, 2020

METHODS OF NANOEMULSION FORMATION

The emulsion is a mixture of two or more liquids that are immiscible where the dispersed phase is dispersed throughout the continuous phase (Kennedy *et al.*, 2000; Naseema *et al.,* 2020). Nanoemulsions droplet sizes generally range between 20 to 200 nm diameter (Solans *et al.*, 2005), they are heterogeneous dispersions of two immiscible phases of oil and water that can thermodynamically be stabilized by surfactants and co-surfactants. Brownian motion is the main reason for its stability. Ostwald's ripening is the process that makes nanoemulsion unstable by smaller droplets combined and forms larger ones (Tesch *et al.*, 2002; Anjana *et al.*, 2012). Ostwald's ripening is controlled by surfactants and co-surfactants (Garti *et al.*, 2004). In the process of nanoemulsion formation, there are some components requires such as oil (for solubilization of the lipophilic molecules/ drug), Surfactant, Water, Co surfactant (enhance the action of surfactant). Some reported methods are used for making nanoemulsions are; High-pressure homogenization, Ultrasonication, Phase inversion, and Micro fluidization (Solans *et al.*, 2005). This chapter deals with methods applied for the bioactive base nanoemulsion formation that can be used as an antibacterial agent.

High-pressure homogenizers (HPH)

HPH is commonly used in industrial-level nanoemulsion formation (Schubert *et al.*, 2003). Devices for HPH are based on the same principle as those employed for the beverages amalgamation. In this process, macroemulsions are push through the piston pump to the downstream situated narrow valve. Extreme hydraulic shear and turbulence in the valve break the macro scale droplets into smaller ones and by repeating this process many times forms of nanoemulsions at bulk level (McClements *et al.*, 2011; Espitia *et al.*, 2019). High pressure by external source deforms the interface between the water and oil is deformed to such an extent that the formed droplets are broken up into smaller ones (up to 1 nm) (Aqil *et al.*, 2016). The external energy is given in the form of mechanical or chemical energy (Shakeell *et al.*, 2008). Piston homogenizer used with very high pressure up to 500 to 5000 psi for nanoemulsion formation. After high-pressure application products were subjected to intense turbulence and hydraulic shear that makes enormously fine nanoparticles. Nanoemulsion formation can be mathematically calculated by the equation. High energy consumption and sometimes temperature rising is the drawback associated in this technique. This high-pressure homogenizer is generally used in milk industries to reduce the size of fat globules and also to homogenize the milk (Scherze *et al.*, 1999). Sometimes that can be used as equipment for disrupting the bacterial cell walls and extracting protein in the cell.

Ultrasonication

For small-scale nanoemulsion formation, ultrasonication is one of the best methods. Agitation of molecules created by the sound waves (ultrasonic frequency of 20 kHz) is basic of ultrasonication. This technique is mainly adopted when a droplet size less than 0.2 μ is required (Gurpreet and Singh, 2018; Naseema *et al.*, 2020). Formed nanoemulsions morphology is affected by the power supply, the frequency or amplitude of the ultrasound waves, and treatment time. They also influence by hydrostatic pressure, dissolved gas concentration, apparatus configuration, and temperature of Ultrasound devices (Abbas *et al.*, 2013). Laboratory-scale ultrasonic homogenizer has many advantages like a simple operation, excellent energy efficiency, ease to optimized, low emulsifier requirements, dispersion stability high and a less prone to microbial contaminants in the processing stage (Abbas *et al.*, 2013). This cannot suitable for a large scale for the reason that the waves have a good impact on the molecules near it. There are two proposed mechanisms. The first one is that the acoustic wave generated helps the oil phase disperse in the water phase and the second mechanism is that the sound waves generate local turbulence that helps the oil droplets break down into sub-micron-sized droplets and helping it to remain stable. The Ultrasonication method was mostly used industrial method for natural active compounds-based nanoemulsion formation. Several advantages are associated with this method such as better encapsulation, control of delivery, higher yields, and smaller droplet size obtained by using less energy (Flory *et al.*, 1942).

Phase Inversion

The emulsion phase inversion method was also known as the catastrophic phase inversion method, it's typically referred for the creation of water oil nanoemulsions by using traditional high-speed mixers. After that, these were transformed into reverse oil/water nanoemulsions by the temperature or the composition modifications (Espitia *et al.*, 2019). This technique doesn't require any type of external force, it involves the formation of fine dispersions when phase transitions occur. Phase transition was created by changing the temperature or composition though keeping the other alternate parameter was constant. The phase inversion temperature method is based on the changes of solubility of polyoxyethylene- type surfactant with temperature (Shinoda *et al.*, 1969). In this method, fine dispersion is obtained by chemical energy resulting from phase transitions taking place through emulsification (Izquierdo *et al.*, 2002). Adequate phase transitions are produced by varying the composition at constant temperature or by varying the temperature at constant composition. This surfactant becomes lipophilic with an increase in temperature due to dehydration of the polymer chain.

Micro Fluidization

This method is based on high shear force on substance inside the microfluidizer, that utilizes a high-pressure positive dislocation pump (500-20 000 psi) that pushes the product out through the interaction chamber that forms very small particles of sub-micron range (Ganesan *et al.*, 2018). In this technique, high pressure around 40,000 psi/ 270Mpa the jets of crude emulsions were allowed to collide with each other under high pressure around 40,000 psi/ 270Mpa (Mahdi *et al.*, 2006). In the Y type, the crude emulsion jet comes together and collides. In the microfluidizer, the mixture is repeatedly circulated until the required particle size was achieved further the resultant product is passed through the filter for the separation of smaller droplets from larger ones for obtaining uniform nanoemulsion (Gurpreet and Singh, 2018). The sheer pressure created was proportional to the pressure and the velocity of the jets. It also determined the droplet size, greater the pressure smaller the droplet size (Mahdi *et al.*, 2006; Serdaroglu *et al.*, 2015). Scientists recommended micro fluidization, for the production of multiple nanoemulsions and dispersed phase's production in organized droplet size. This technique did not apply to bulk production of nanoemulsion (Hadziabdi *et al.*, 2015). Some of the examples of micro fluidization-based nanoemulsions are curcumin-loaded nanoemulsions that can be used for the treatment of chronic diseases by enhancing curcumin bioavailability. This technique can be useful for nutraceutical development or food-grade bioactive compounds based nanoemulsions formation (Ganesan *et al.*, 2018).

Spontaneous Emulsification

Spontaneous emulsification is also known as emulsification by the solvent diffusion process and it's requiring low energy input for nanoemulsion formation. This method is developed at room temperature, therefore, does not need any special equipment. Addition of water to a solution of oil and surfactant in step by step manner at fixed temperature stir gently to create nano-emulsions. This technique was firstly formed of an organic solution, comprising of oil and lipophilic surfactant in a water-miscible solvent. Then this oil/water emulsion is formed by injecting the organic phase into the aqueous phase in the magnetic stirring. Lastly, the organic solvent was removed by evaporation. The spontaneity of the process of emulsification depends on phase transition area, surfactant concentration and structure, interfacial and bulk viscosity, and interfacial tension. The spontaneous formation of nanoemulsion facilitates by specific temperatures and the utilized emulsifier (Bouchemal *et al.*, 2004; Francois *et al.*, 2005; Komaiko *et al.*, 2015). The limitation in this method is fewer amounts of the oil phase and their solvent presence (Hadziabdi *et al.*, 2017). From the food industrial point of view solvents' uses are problematic due to their cost, flavor, and safety circumstances.

DELIVERY SYSTEM FOR ANTIMICROBIAL NANOEMULSION

There are some researcher works with the nanoemulsion delivery for its antimicrobial activity (Donsi *et al.*, 2012). The delivery method depends on the open type and process utilized for nanoemulsion formation as well as the main component of nanoemulsion. The antimicrobial activity of nanoemulsions is reported for different microorganisms and its effectivity depends on its formulation. Researchers reported that the efficiency of delivery methods of antimicrobial activity was correlated with the concentration of the compound and the emulsifier used. The essential oil contains various bioactive compounds such as; phenols, terpenes, and aldehydes (Ceylan and Fung, 2004), and the antimicrobial activities of different essential oils and components have been well worked for different microorganisms (Gill and Holley, 2006; Di Pasqua *et al.*, 2007). Mushroom extract and its extracted compounds are used as an antimicrobial agent for a long time. Their utilization in form of nanoemulsion is the new way for enhancing its biological activity. Due to their low solubility in water, mushroom extract can be encapsulated in suitable delivery systems. The suitable delivery system for antimicrobial compounds enhances the transfer rates in the sites of action that maximize the antimicrobial activity and can be sufficient at the low amount of application. Nano-encapsulation of bioactive mushroom compounds leads to an increase in their physical stability and protects them from the outer environment (Weiss *et al.*, 2009). Several encapsulation methods are available for oil-based antimicrobial nanoemulsion formation such as; solid-fat micrometric emulsion and entrapped into the crystallized-fat structure (Gavini *et al.*, 2005) some biocompatible compounds also used such as chitosan (Pedro *et al.*, 2009) or Ca-alginate (Wang *et al.*, 2009). Some of the plant-based nanoemulsions are; *Origanum dictamnus* L. (Liolios *et al.*, 2009), Citrus limon extracts (Gortzi *et al.*, 2007), Myrtus extracts (Gortzi *et al.*, 2008), and *Artemisia arborescens* L. essential oil (Sinico *et al.*, 2005). It was reported by Burt (2004), phenolic compounds, in the essential oil constituents display the highest antimicrobial activity.

PROSPECT FOR UTILIZATION OF MUSHROOM BIOACTIVE COMPOUNDS FOR THE GENERATION OF ANTIBACTERIAL NANOEMULSION

Figure 1. Prospect for utilization of mushroom bioactive compounds for the generation of antibacterial nanoemulsion

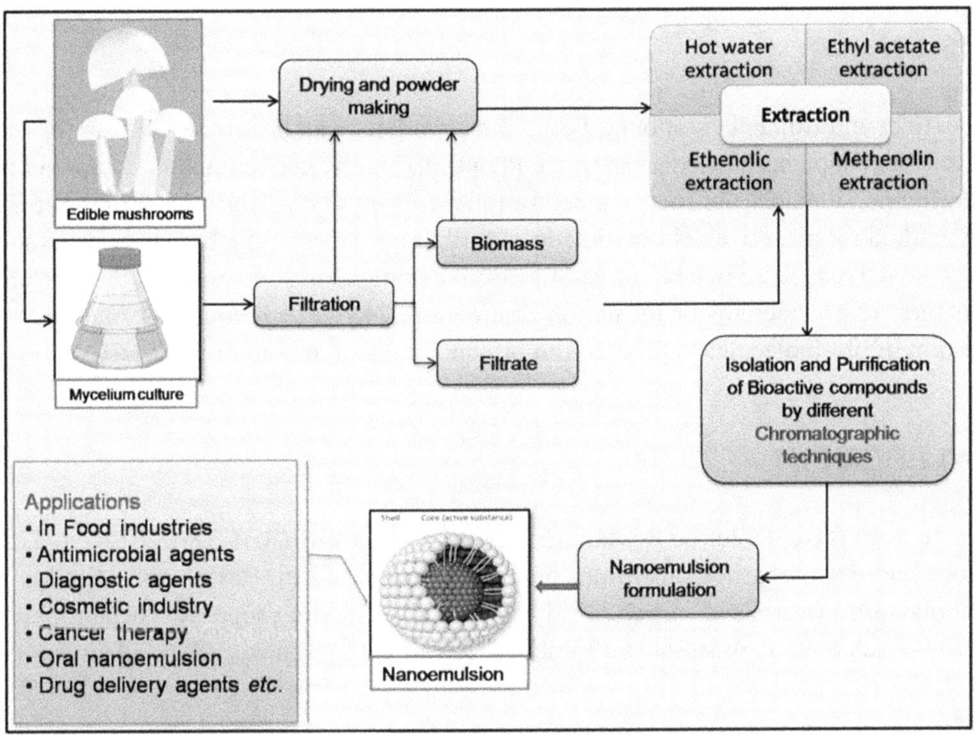

Mushrooms are a healthy food source having various health benefits. Recently mushrooms cultivation was widely appreciated for its increasing demand in the food sector as well as in the pharmaceutical sector. The mushroom was easily cultivated by using various agro-wastes and it can also apply for bioremediation purposes. Bulk production of mushrooms offers both advantages for farmers, small entrepreneurs, and pharmaceutical sectors. Mushrooms having various metabolites that have known biological activities therefore it's the best alternative source for isolation of bioactive compounds. The development of new techniques and well-designed types of equipment make cultivation and compounds isolation from mushrooms are very easy. Isolated and purified compounds can be further use for the development of other products. Nanoemulsions are one such option for the effective utilization of these bioactive compounds. Nanoemulsion technology has its advantages like it protects compounds from the outer environment, enhance their absorption, bioavailability, and is easy to use. Mushroom technology and nanoemulsion technology (shown in figure 1), both have a high potential for the growth of income and employment generating sectors. Mushrooms isolated compounds can be used in various types of nanoemulsion development. In the medical field, different antibiotics were used against different bacterial strains but there was a problem associated with it that is the generation of drug-resistant strains. The generation of drugs resistance strains in the environments requires new antibacterial agents continually. Mushrooms can be good and easily available natural sources for bioactive compound isolation. Mushroom-based bioactive compounds are the best and can produce at a large scale. Overall mushroom-based bioactive compounds can be an effortless source for antibacterial antibiotic development with other biomedical properties. It was given the future opportunity of research in the efficient bioactive compounds isolation from mushrooms and its different types of nanoemulsions development.

CONCLUSION

Mushroom extract and isolated compound-based nanoemulsion can be a better option for future nano-drug for biomedical applications. By changing and optimizing the nanoemulsion components and their variables like temperature, pH, and ionic concentration can easily control the release of the core materials. It's the best method for bioactive delivery inside as well as out-sides of the human body. Nanoemulsion that contains bioactive is increases effective of bioactive compounds and protects them from the outer. The efficient bioactive nanoemulsion formation requires an effective technique for isolation, purification, characterization of bio-molecules, and selection of nanoemulsion making method.

ACKNOWLEDGMENT

The authors are thankful to the Junior Research Fellowship (DBT/JRF/BET-18/I/2018/AL/123), Department of Biotechnology, Biotech Consortium of India Limited and Pt. Ravishankar Shukla University Research scholarship award (797/Fin/Sch./2021) for providing funding support. The authors are also are thankful to the Head, School of Studies in Biotechnology, Pt. Ravishankar Shukla University, Raipur.

REFERENCES

Abbas, S., Hayat, K., Karangwa, E., Bashari, M., & Zhang, X. (2013). An overview of ultrasound-assisted food-grade nanoemulsions. *Food Engineering Reviews*, *5*(3), 139–157. doi:10.100712393-013-9066-3

Angelini, P., Venanzoni, R., Flores, G. A., Tirillini, B., Orlando, G., Recinella, L., Chiavaroli, A., Brunetti, L., Leone, S., Di Simone, S. C., Ciferri, M. C., Zengin, G., Gunes, A., Menghini, L., & Ferrante, C. (2020). Evaluation of antioxidant, antimicrobial and tyrosinase inhibitory activities of extracts from *Tricholosporum goniospermum,* an edible wild mushroom. *Antibiotics (Basel, Switzerland)*, *9*(8), 9. doi:10.3390/antibiotics9080513 PMID:32823710

Anjana, D., Ravichandran, S., Parmar, H., & Upadhyay, R. V. (2014). Development of curcumin based ophthalmic formulation development of Curcumin. *American Journal of Infectious Diseases*, *8*, 41–49.

Anke, T., Kupka, J., Schramm, G., & Steglich, W. (1980). Antibiotics from basidiomycetes. X. Scorodonin, a new antibacterial and antifungal metabolite from Marasmius scorononius (Fr.). *The Journal of Antibiotics*, *33*(5), 463–467. doi:10.7164/antibiotics.33.463 PMID:7191846

Aqil, M., Kamran, M., Ahad, A., & Imam, S. S. (2016). Development of clove oil based nanoemulsion of olmesartan for transdermal delivery: Box–Behnken design optimization and pharmacokinetic evaluation. *Journal of Molecular Liquids*, *214*, 238–248. doi:10.1016/j.molliq.2015.12.077

Assadpour, E., & Jafari, S. M. (2018). A systematic review on nanoencapsulation of food bioactive ingredients and nutraceuticals by various nanocarriers. *Critical Reviews in Food Science and Nutrition*, *59*(19), 3129–3151. doi:10.1080/10408398.2018.1484687 PMID:29883187

Barros, L., Cruz, T., Baptista, P., Estevinho, L. M., & Ferreira, I. C. F. R. (2008). Wild and commercial mushrooms as source of nutrients and nutraceuticals. *Food and Chemical Toxicology*, *46*(8), 2742–2747. doi:10.1016/j.fct.2008.04.030 PMID:18538460

Beattie, K. D., Rouf, R., Gander, L., May, T. W., Ratkowsky, D., Donner, C. D., Gill, M., Grice, I. D., & Tiralongo, E. (2010). Antibacterial metabolites from Australian macrofungi from the genus *Cortinarius*. *Phytochemistry*, *71*(8-9), 948–955. doi:10.1016/j.phytochem.2010.03.016 PMID:20392467

Campani, V., Biondi, M., Mayol, L., Cilurzo, F., Pitaro, M., & de Rosa, G. (2016). Development of nanoemulsions for topical delivery of vitamin K1. *International Journal of Pharmaceutics*, *511*(1), 170–177. doi:10.1016/j.ijpharm.2016.07.004 PMID:27386792

Ceylan, E., & Fung, D. Y. C. (2004). Antimicrobial activity of spices. *Journal of Rapid Methods and Automation in Microbiology*, *12*(1), 1–55. doi:10.1111/j.1745-4581.2004.tb00046.x

Chandrawanshi, N. K., Jadhav, S. K., Tiwari, K. L., & Quraishi, A. (2015). In vitro tuberization and Colchicine content analysis of *Gloriosa superba L. Biotechnology*, *14*, 142–147. doi:10.3923/biotech.2015.142.147

Chandrawanshi, N. K., Tandia, D., & Jadhav, S. K. (2017). Nutraceutical properties evaluation of *schizophyllum commune. Indian Journal of Scientific Research*, *13*, 57–62.

Chandrawanshi, N.K., Tandia, D.K. & Jadhav, S.K. (2018). Determination of antioxidant and antidiabetic activities of polar solvent extracts of *Daedaleopsis confragosa* (Bolton) J. Schrot. *Research Journal of Pharmacy and Technology, 11,* 5623–5630.

Chang, Y., McLandsborough, L., & McClements, D. J. (2012). Physical properties and antimicrobial efficacy of thyme oil nanoemulsions: Influence of ripening inhibitors. *Journal of Agricultural and Food Chemistry*, *60*(48), 12056–12063. doi:10.1021/jf304045a PMID:23140446

Chowdhury, M. M. H., Kubra, K., & Ahmed, S. R. (2015). Screening of antimicrobial, antioxidant properties and bioactive compounds of some edible mushrooms cultivated in Bangladesh. *Annals of Clinical Microbiology and Antimicrobials*, *14*(1), 2–6. doi:10.118612941-015-0067-3 PMID:25858107

Di Pasqua, R., Betts, G., Hoskins, N., Edwards, M., Ercolini, D., & Mauriello, G. (2007). Membrane toxicity of antimicrobial compounds from essential oils. *Journal of Agricultural and Food Chemistry*, *55*(12), 4863–4870. doi:10.1021/jf0636465 PMID:17497876

Donsi, F., Annunziataa, M., Vincensia, M., & Ferrari, G. (2012). Design of nanoemulsion-based delivery systems of natural antimicrobials: Effect of the emulsifier. *Journal of Biotechnology*, *159*(4), 342–350. doi:10.1016/j.jbiotec.2011.07.001 PMID:21763730

Donsi, F., Wang, Y., Li, J., & Huang, Q. (2010). Preparation of curcumin sub-micrometer dispersions by high-pressure homogenization. *Journal of Agricultural and Food Chemistry*, *58*(5), 2848–2853. doi:10.1021/jf903968x PMID:20146520

Espitia, P. J. P., Fuenmayor, C. A., & Otoni, C. G. (2019). Nanoemulsions: Synthesis, characterization, and application in bio-based active food packaging. *Comprehensive Reviews in Food Science and Food Safety*, *18*(1), 264–285. doi:10.1111/1541-4337.12405 PMID:33337016

Flory, P. J. (1942). Thermodynamics of high polymer solutions. *The Journal of Chemical Physics*, *10*(1), 51–61. doi:10.1063/1.1723621

Ganesan, P., Karthivashan, G., Park, S. Y., Kim, J., & Choi, D. K. (2018). Microfluidization trends in the development of nanodelivery systems and applications in chronic disease treatments. *International Journal of Nanomedicine*, *13*, 6109–6121. doi:10.2147/IJN.S178077 PMID:30349240

Ganesan, P., Karthivashan, G., Park, S. Y., Kim, J., & Choi, D. K. (2018). Microfluidization trends in the development of nanodelivery systems and applications in chronic disease treatments. *International Journal of Nanomedicine*, *13*, 6109–6121. doi:10.2147/IJN.S178077 PMID:30349240

Garti, N., Yuli, I. A., Spernath, A., & Hoffman, R. E. (2004). Transition and loci of solubilization of nutraceuticals in U-type nonionic microemulsions studied by self-diffusion NMR. *Physical Chemistry*, *6*, 2968–2976.

Gavini, E., Sanna, V., Sharma, R., Juliano, C., Usai, M., Marchetti, M., Karlsen, J., & Giunchedi, P. (2005). Solid lipid microparticles (SLM) containing juniper oil as anti-acne topical carriers: Preliminary studies. *Pharmaceutical Development and Technology*, *10*(4), 479–487. doi:10.1080/10837450500299727 PMID:16370177

Gebreyohannes, G., Nyerere, A., Bii, C., & Sbhatu, D. B. (2019). Determination of antimicrobial activity of extracts of indigenous wild mushrooms against pathogenic organisms. *Evidence-Based Complementary and Alternative Medicine*, *2019*, 1–7. doi:10.1155/2019/6212673 PMID:30906415

Gill, A. O., & Holley, R. A. (2006). Disruption of *Escherichia coli, Listeria monocytogenes and Lactobacillus sakei* cellular membranes by plant oil aromatics. *International Journal of Food Microbiology*, *108*(1), 1–9. doi:10.1016/j.ijfoodmicro.2005.10.009 PMID:16417936

Gortzi, O., Lalas, S., Tsaknis, J., & Chinou, I. (2007). Enhanced bioactivity of Citrus limon (Lemon Greek cultivar) extracts, essential oil and isolated compounds before and after encapsulation in liposomes. *Planta Medica*, *73*(09), 184. doi:10.1055-2007-986965

Guerra-Rosas, M., Morales-Castro, J., Cubero-Marquez, M., Salvia-Trujillo, L., & Martín-Belloso, O. (2017). Antimicrobial activity of nanoemulsions containing essential oils and high methoxyl pectin during long-term storage. *Food Control*, *77*, 131–138. doi:10.1016/j.foodcont.2017.02.008

Gurpreet, K., & Singh, S. K. (2017). Review of nanoemulsion formulation and characterization techniques. *Indian Journal of Pharmaceutical Sciences*, *80*, 781–789.

Hadziabdi, J., Dzana, O., Alisa, E., Edina, V., & Ognjenka, R. (2017). Preparation of nanoemulsions by high-energy and low- energy emulsification methods. *IFMBE Procesing*, *62*, 317–322. doi:10.1007/978-981-10-4166-2_48

Hamouda, T., & Baker, J. R. Jr. (2000). Antimicrobial mechanism of action of surfactant lipid preparations in enteric Gram-negative bacilli. *Journal of Applied Microbiology*, *89*(3), 397–403. doi:10.1046/j.1365-2672.2000.01127.x PMID:11021571

Hearst, R., Nelson, D., McCollum, G., Millar, B. C., Maeda, Y., Goldsmith, C. E., Rooney, P. J., Loughrey, A., Rao, J. R., & Moore, J. E. (2009). An examination of antibacterial and antifungal properties of constituents of Shiitake (*Lentinula edodes*) and Oyster (*Pleurotus ostreatus*) mushrooms. *Complementary Therapies in Clinical Practice*, *15*(1), 5–7. doi:10.1016/j.ctcp.2008.10.002 PMID:19161947

Hemmila, M. R., Mattar, A., Taddonio, M. A., Arbabi, S., Hamouda, T., Ward, P. A., Wang, S. C., & Baker, J. R. Jr. (2010). Topical nanoemulsion therapy reduces bacterial wound infection and inflammation after burn injury. *Surgery*, *148*(3), 499–509. doi:10.1016/j.surg.2010.01.001 PMID:20189619

Hu, Q., Gerhard, H., Upadhyaya, I., Venkitanarayanan, K., & Luo, Y. (2016). Antimicrobial eugenol nanoemulsion prepared by gum arabic and lecithin and evaluation of drying technologies. *International Journal of Biological Macromolecules*, *87*, 130–140. doi:10.1016/j.ijbiomac.2016.02.051 PMID:26902894

Hu, S. H., Liang, Z. C., Chia, Y. C., Lien, J. L., Chen, K. S., Lee, M. Y., & Wang, J. C. (2006). Antihyperlipidemic and antioxidant effects of extracts from *Pleurotus citrinopileatus. Journal of Agricultural and Food Chemistry*, *54*(6), 2103–2110. doi:10.1021/jf052890d PMID:16536582

Hwang, Y. Y., Ramalingam, K., Bienek, D. R., Lee, V., You, T., & Alvareza, R. (2013). Antimicrobial activity of nanoemulsion in combination with Cetylpyridinium Chloride in multidrug-resistant *Acinetobacter baumannii. Antimicrobial Agents and Chemotherapy*, *57*(8), 3568–3575. doi:10.1128/AAC.02109-12 PMID:23669390

Ishikawa, N. K., Kasuya, M. C. M., & Vanetti, M. C. D. (2001). Antibacterial activity of *Lentinula edodes* grown in liquid medium. *Brazilian Journal of Microbiology*, *32*(3), 206–210. doi:10.1590/S1517-83822001000300008

Izquierdo, P., Esquena, J., Tadros, T. F., Dederen, C., Garcia, M. J., Azemar, N. S. C., & Solans, C. (2002). Formation and stability of nano-emulsions prepared using the phase inversion temperature method. *Langmuir*, *18*(1), 26–30. doi:10.1021/la010808c

Kalyoncu, F., & Oskay, M. (2008). Antimicrobial activities of four wild mushroom species collected from Turkey. *Proceeding of the 6th International Conference on Mushroom Biology and Mushroom Products*.

Kalyoncu, F., Oskay, M., Saglam, H., Erdogan, T. F., & Tamer, A. U. (2010). Antimicrobial and antioxidant activities of mycelia of 10 wild mushroom species. *Journal of Medicinal Food*, *13*(2), 415–419. doi:10.1089/jmf.2009.0090 PMID:20192845

Kennedy, J. F., & Thorley, M. (2000). Introduction to soft matter: polymers, colloids, amphiphiles and liquid crystals-IW Hamley. Carbohydrate Polymere, 1, 78.

Kiho, T., Shiose, Y., Nagai, K., & Ukai, S. (1992). Polysaccharides in fungi. XXX. Antitumor and immunomodulating activities of two polysaccharides from the fruiting bodies of *Armillariella tabescens. Chemical & Pharmaceutical Bulletin*, *40*(8), 2110–2114. doi:10.1248/cpb.40.2110 PMID:1423764

Kumar, D. D., Mann, B., Pothuraju, R., Sharma, R., Bajaj, R., & Minaxi, M. (2016). Formulation and characterization of nanoencapsulated curcumin using sodium caseinate and its incorporation in ice cream. *Food Functinal*, *7*(1), 417–424. doi:10.1039/C5FO00924C PMID:26501983

Lee, S. J., Yeo, W. H., Yun, B. S., & Yoo, I. D. (1999). Isolation and sequence analysis of new Peptaibol, Boletusin, from Boletus spp. *Journal of Peptide Science*, *5*(8), 374–378. doi:10.1002/(SICI)1099-1387(199908)5:8<374::AID-PSC211>3.0.CO;2-X PMID:10507687

Li, M., & Fogler, H. (1978). Acoustic emulsification- II Break-up of the large primary oil droplets in a water medium. *Journal of Fluid Mechanics*, *88*(3), 513–528. doi:10.1017/S0022112078002244

Li, T., Han, X., Bao, R., Hao, Y., & Li, S. (2019). Preparation and properties of water-in-oil shiitake mushroom polysaccharide nanoemulsion. *International Journal of Biological Macromolecules*, *140*, 343–349. doi:10.1016/j.ijbiomac.2019.08.134 PMID:31425759

Liang, R., Xu, S., Shoemaker, C. F., Li, Y., Zhong, F., & Huang, Q. (2012). Physical and antimicrobial properties of peppermint oil nanoemulsions. *Journal of Agricultural and Food Chemistry*, *60*(30), 7548–7555. doi:10.1021/jf301129k PMID:22746096

Liolios, C. C., Gortzi, O., Lalas, S., Tsaknis, J., & Chinou, I. (2009). Liposomal incorporation of carvacrol and thymol isolated from the essential oil of Origanum dictamnus L. and in vitro antimicrobial activity. *Food Chemistry*, *112*(1), 77–83. doi:10.1016/j.foodchem.2008.05.060

Liu, Q., Huang, H., Chen, H., Lin, J., & Wang, Q. (2019). Food-grade nanoemulsions: Preparation, stability and application in encapsulation of bioactive compounds. *Molecules (Basel, Switzerland)*, *24*(23), 4242. doi:10.3390/molecules24234242 PMID:31766473

Liu, X. T., Winkler, A. L., Schwan, W. R., Volk, T. J., Rott, M. A., & Monte, A. (2010). Antibacterial compounds from mushrooms I: A Lanostane-type triterpene and prenylphenol derivates from *Jahnoporus hiritus* and *Albatrellus flettii* and their activities against *Bacillus cereus* and *Enterococcus faecalis. Planta Medica*, *76*(02), 182–185. doi:10.1055-0029-1186001 PMID:19644795

Lohith Kumar, D. H., & Sarkar, P. (2017). Encapsulation of bioactive compounds using nanoemulsions. *Environmental Chemistry Letters*, *16*(1), 59–70. doi:10.100710311-017-0663-x

Lv, S., Gu, J., Zhang, R., Zhang, Y., Tan, H., & McClements, D. J. (2018). Vitamin E encapsulation in plant-Based nanoemulsions fabricated using dual-channel microfluidization: Formation, stability, and bioaccessibility. *Journal of Agricultural and Food Chemistry*, *66*(40), 10532–10542. doi:10.1021/acs.jafc.8b03077 PMID:30240207

Mahdi Jafari, S., He, Y., & Bhandari, B. (2006). Nano-Emulsion production by sonication and microfluidization - A aomparison. *International Journal of Food Properties*, *9*(3), 475–485. doi:10.1080/10942910600596464

Marsup, P., Yeerong, K., Neimkhum, W., Sirithunyalug, J., Anuchapreeda, S., To-Anun, C., & Chaiyana, W. (2020). Enhancement of chemical stability and dermal delivery of *Cordyceps militaris* Extracts by Nanoemulsion. *Nanomaterials (Basel, Switzerland)*, *10*(8), 1565. doi:10.3390/nano10081565 PMID:32784892

McClements, D. J. (2011). Edible nanoemulsions: Fabrication, properties, and functional performance. *Soft Matter*, *7*(6), 2297–2316. doi:10.1039/C0SM00549E

Moghimi, R., Aliahmadi, A., Rafati, H., Abtahi, H. R., Amini, S., & Feizabadi, M. M. (2018). Antibacterial and anti-biofilm activity of nanoemulsion of Thymus daenensis oil against multi-drug resistant *Acinetobacter baumannii*. *Journal of Molecular Liquids*, *265*, 765–770. doi:10.1016/j.molliq.2018.07.023

Mothana, R. A. A., Jansen, R., Julich, W.-D., & Lindequist, U. (2000). Ganomycins A and B, new antimicrobial Farnesyl hydroquinones from the Basidiomycete *Ganoderma pfeifferi*. *Journal of Natural Products*, *63*(3), 416–418. doi:10.1021/np990381y PMID:10757736

Nakalembe, I., Kabasa, J. D., & Olila, D. (2015). Comparative nutrient composition of selected wild edible mushrooms from two agro-ecological zones, Uganda. *SpringerPlus*, *4*(1), 433. doi:10.118640064-015-1188-z PMID:26306295

Naseema, A., Kovooru, L., & Behera, A. K. (2020). A critical review of synthesis procedures, applications and future potential of nanoemulsions. *Advances in Colloid and Interface Science*. Advance online publication. doi:10.1016/j.cis.2020.102318 PMID:33242713

Ngai, P. H. K., & Ng, T. B. (2004). A ribonuclease with antimicrobial, antimitogenic and antiproliferative activities from the edible mushroom *Pleurotus sajor-caju*. *Peptides*, *25*(1), 11–17. doi:10.1016/j.peptides.2003.11.012 PMID:15003351

Ozen, T., Darcan, C., Aktop, O., & Turkekul, I. (2011). Screening of antioxidant, antimicrobial activities and chemical contents of edible mushrooms wildly grown in the Black Sea region of Turkey. *Combinatorial Chemistry & High Throughput Screening*, *14*(2), 72–84. doi:10.2174/138620711794474079 PMID:20958253

Ozturk, M., Duru, M. E., Kivrak, S., Mercan-Dogan, N., Turkoglu, A., & Ozler, M. A. (2011). In vitro antioxidant, anticholinesterase and antimicrobial activity studies on three Agaricus species with fatty acid compositions and iron contents: A comparative study on the three most edible mushrooms. *Food and Chemical Toxicology*, *49*(6), 1353–1360. doi:10.1016/j.fct.2011.03.019 PMID:21419821

Palafox-Carlos, H., Ayala-Zavala, J. F., & Gonzalez-Aguilar, G. A. (2011). The role of dietary fiber in the bioaccessibility and bioavailability of fruit and vegetable antioxidants. *Journal of Food Science*, *76*(1), 6–15. doi:10.1111/j.1750-3841.2010.01957.x PMID:21535705

Park, H., Mun, S., & Kim, Y. R. (2019). UV and storage stability of retinol contained in oil-in-water nanoemulsions. *Food Chemistry*, *272*, 404–410. doi:10.1016/j.foodchem.2018.08.076 PMID:30309562

Pedro, A. S., Cabral-Albuquerque, E., Ferreira, D., & Sarmento, B. (2009). Chitosan: An option for development of essential oil delivery systems for oral cavity care? *Carbohydrate Polymers*, *76*(4), 501–508. doi:10.1016/j.carbpol.2008.12.016

Periasamy, K. (2005). Novel antibacterial compounds obtained from some edible mushrooms. *International Journal of Medicinal Mushrooms, 7*, 443–444. doi:10.1615/IntJMedMushrooms.v7.i3.820

Porter, C., Pouton, C. W., Cuine, J. F., & Charman, W. N. (2008). Enhancing intestinal drug solubilisation using lipid-based delivery systems. *Advanced Drug Delivery Reviews, 60*(6), 673–691. doi:10.1016/j.addr.2007.10.014 PMID:18155801

Prakash, A., Baskaran, R., Nithyanand, P., & Vadivel, V. (2020). Effect of nanoemulsification on the antibacterial and anti-biofilm activities of selected spice essential oils and their major constituents against *Salmonella enterica Typhimurium. Journal of Cluster Science, 31*(5), 1123–1135. doi:10.100710876-019-01720-7

Quereshi, S., Pandey, A. K., & Sandhu, S. S. (2010). Evaluation of antibacterial activity of different *Ganoderma lucidum* extracts. *Journal of Science and Research, 3*, 9–13.

Ribeiro, B., de Pinho, P. G., Andrade, P. B., Oliveira, C., Silva Ferreira, A. C., Baptista, P., & Valentao, P. (2011). Do bioactive carotenoids contribute to the color of edible mushrooms? *The Open Chemical and Biomedical Methods Journal, 4*(1), 14–18. doi:10.2174/1875038901104010014

Roger, K. (2016). Nanoemulsification in the vicinity of phase inversion: Disruption of bicontinuous structures in oil / surfactant / water systems. *Current Opinion in Colloid & Interface Science, 25*, 120–128. doi:10.1016/j.cocis.2016.09.015

Rosa, L. H., Machado, K. M. G., Jacob, C. C., Capelari, M., Rosa, C. A., & Zani, C. L. (2003). Screening of Brazilian Basidiomycetes for antimicrobial activity. *Memorias do Instituto Oswaldo Cruz, 98*(7), 967–974. doi:10.1590/S0074-02762003000700019 PMID:14762527

Rupa, E. J., Li, J. F., Arif, M. H., Yaxi, H., Puja, A. M., Chan, A. J., Hoang, V., Kaliraj, L., Yang, D. C., & Kang, S. C. (2020). *Cordyceps militaris* fungus extracts-mediated nanoemulsion for improvement antioxidant, antimicrobial, and anti-inflammatory activities. *Molecules (Basel, Switzerland), 25*(23), 5733. doi:10.3390/molecules25235733 PMID:33291776

Salvia-Trujillo, L., Soliva-Fortuny, R., Rojas-Grau, M. A., McClements, D. J., & Martin-Belloso, O. (2017). Edible Nanoemulsions as carriers of active ingredients: A Review. *Annual Review of Food Science and Technology, 8*(1), 439–466. doi:10.1146/annurev-food-030216-025908 PMID:28125342

Santos, D. T., & Meireles, M. A. A. (2010). Carotenoid pigments encapsulation: Fundamentals, techniques and recent trends. *The Open Chemical Engineering Journal, 4*(1), 42–50. doi:10.2174/1874123101004010042

Scherze, I., Marzilger, K., & Muschiolik, G. (1999). Emulsification using micro porous glass (MPG): Surface behaviour of milk protein. *Colloids and Surfaces. B, Biointerfaces, 12*(3-6), 213–221. doi:10.1016/S0927-7765(98)00076-9

Schubert, H., Ax, K., & Behrend, O. (2003). Product engineering of dispersed systems. *Trends in Food Science & Technology, 14*(1-2), 9–16. doi:10.1016/S0924-2244(02)00245-5

Seibert, J. B., Bautista-Silva, J. P., Amparo, T. R., Petit, A., Pervier, P., Almeida, J. C. D. S., Azevedo, M. C., Silveira, B. M., Brandao, G. C., de Souza, G. H. B., & ... . (2019). Development of propolis nanoemulsion with antioxidant and antimicrobial activity for use as a potential natural preservative. *Food Chemistry, 287*, 61–67. doi:10.1016/j.foodchem.2019.02.078 PMID:30857719

Serdaroglu, M., Ozturk, B., & Kara, A. (2015). An overview of food emulsions: Description, classification and recent potential applications. *Turkish Journal of Agriculture -. Food Science and Technology (Campinas), 3*, 430–438.

Shakeell, F., Baboota, S., Ahuja, A., Ali, J., Faisal, M. S., & Shafiq, S. (2008). Stability evaluation of celecoxib nanoemulsion containing Tween 80. *Thaiphesatchasan, 32*, 4–9.

Sharma, A., Sharma, N. K., Srivastava, A., Kataria, A., Dubey, S., Sharma, S., & Kundu, B. (2018). Clove and lemongrass oil based non-ionic nanoemulsion for suppressing the growth of plant pathogenic *Fusarium oxysporum f.sp. lycopersici. Industrial Crops and Products, 123*, 353–362. doi:10.1016/j.indcrop.2018.06.077

Shinoda, K., & Saito, H. (1969). The stability of O / W Type emulsions as functions of temperature and the HLB of emulsifiers: The emulsification by PIT method. *Journal of Colloid and Interface Science, 30*(2), 258–263. doi:10.1016/S0021-9797(69)80012-3

Signoretto, C., Marchi, A., Bertoncelli, A., Burlacchini, G., Tessarolo, F., Caola, I., Pezzati, E., Zaura, E., Papetti, A., Lingstrom, P., Pratten, J., Spratt, D. A., Wilson, M., & Canepari, P. (2011). Effects of mushroom and chicory extracts on the physiology and shape of *Prevotella intermedia*, a Periodontopathogenic bacterium. *Journal of Biomedicine & Biotechnology, 2011*, 1–8. Advance online publication. doi:10.1155/2011/635348 PMID:21918594

Sinico, C., De Logu, A., Lai, F., Valenti, D., Manconi, M., Loy, G., Bonsignore, L., & Fadda, A. M. (2005). Liposomal incorporation of Artemisia arborescens L. essential oil and in vitro antiviral activity. *European Journal of Pharmaceutics and Biopharmaceutics, 59*(1), 161–168. doi:10.1016/j.ejpb.2004.06.005 PMID:15567314

Solans, C., Izquierdo, P., Nolla, J., Azemar, N., & Garcia-Celma, M. J. (2005). Nanoemulsions. *Current Opinion in Colloid & Interface Science, 10*(3-4), 102–110. doi:10.1016/j.cocis.2005.06.004

Sonu, K. S., Mann, B., Sharma, R., Kumar, R., & Singh, R. (2018). Physico-chemical and antimicrobial properties of d-limonene oil nanoemulsion stabilized by whey protein-maltodextrin conjugates. *Journal of Food Science and Technology, 55*(7), 2749–2757. doi:10.100713197-018-3198-7 PMID:30042591

Soukoulis, C., Tsevdou, M., Andre, C. M., Cambier, S., Yonekura, L., Taoukis, P. S., & Homann, L. (2017). Modulation of chemical stability and in vitro bioaccessibility of beta-carotene loaded in kappa-carrageenan oil-in-gel emulsions. *Food Chemistry, 220*, 208–218. doi:10.1016/j.foodchem.2016.09.175 PMID:27855891

Sudirman, L. I. (2010). Partial purification of antimicrobial compounds isolated from mycelia of tropical *lentinus cladopus* lc4. *Hayati Journal of Biosciences, 2*(2), 63–67. doi:10.4308/hjb.17.2.63

Takala, P., Vu, K., Salmieri, S., & Lacroix, M. (2011). Effect of antimicrobial coatings on the radiosensitization of *Escherichia coli, Salmonella Typhimurium,* and *Listeria monocytogenes* in fresh broccoli. *Journal of Food Protection*, *74*(7), 1065–1069. doi:10.4315/0362-028X.JFP-10-520 PMID:21740707

Tambekar, D. H., Sonar, T. P., Khodke, M. V., & Khante, B. S. (2006). The Novel antibacterials from two edible mushrooms: *Agaricus bisporus and Pleurotus sajor caju. International Journal of Pharmacology*, *2*(5), 584–587. doi:10.3923/ijp.2006.584.587

Tesch, S., Gerhards, C., & Schubert, H. (2002). Stabilization of emulsions by OSA starches. *Journal of Food Engineering*, *54*(2), 167–174. doi:10.1016/S0260-8774(01)00206-0

Tian, W.-L., Lei, L.-L., Zhang, Q., & Li, Y. (2016). Physical stability and antimicrobial activity of encapsulated Cinnamaldehyde by self-emulsifying nanoemulsion. *Journal of Food Process Engineering*, *39*(5), 462–471. doi:10.1111/jfpe.12237

Wang, Q., Gong, J., Huang, X., Yu, H., & Xue, F. (2009). In vitro evaluation of the activity of microencapsulated carvacrol against Escherichia coli with K88 pili. *Journal of Applied Microbiology*, *107*(6), 1781–1788. doi:10.1111/j.1365-2672.2009.04374.x PMID:19645769

Weiss, J., Gaysinksy, S., Davidson, M., & Mcclements, J. (2009). Nanostructured encapsulation systems: food antimicrobials. In *IUFoST World Congress Book: Global Issues in Food Science and Technology*. Elsevier Inc.

Weiss, J., Takhistov, P., & McClements, D. J. (2006). Functional materials in food nanotechnology. *Journal of Food Science*, *71*(9), 107–116. doi:10.1111/j.1750-3841.2006.00195.x

Wenzel, E., & Somoza, V. (2005). Metabolism and bioavailability of trans-resveratrol. *Molecular Nutrition & Food Research*, *49*(5), 472–481. doi:10.1002/mnfr.200500010 PMID:15779070

Xu, D., Wang, H., Zheng, W., Gao, Y., Wang, M., Zhang, Y., & Gao, Q. (2016). Charaterization and immunomodulatory activities of polysaccharide isolated from Pleurotus eryngii. *International Journal of Biological Macromolecules*, *92*, 30–36. doi:10.1016/j.ijbiomac.2016.07.016 PMID:27392770

Xue, J., Davidson, P. M., & Zhong, Q. (2015). Antimicrobial activity of thyme oil co-nanoemulsified with sodium caseinate and lecithin. *International Journal of Food Microbiology*, *210*, 1–8. doi:10.1016/j.ijfoodmicro.2015.06.003 PMID:26082324

KEY TERMS AND DEFINITIONS

Antibiotic: It is an agent that kills or inhibits bacterial growth and widely used for various bacterial infections. Antibiotics are substances that are produced by living organisms like bacteria and fungi.

Bioactive Compounds: Compounds or molecules that give specific biological activity like antioxidant, antibacterial, immunomodulatory, etc. It can be found and isolated in various biological objects such as plants, algae, bacteria, and fungus.

Drug Resistance: Microbes that can't kill or are inhibited by specific drugs are known as drug resistant strains. Drug resistance is formed when microbes like; bacteria, fungi, viruses, and parasites are changing themselves over time and increase the risk of disease or illness.

Functional Food: Functional foods are the dietary stuff that provides energy and nutrients, besides this they can give additional targeted health benefits. It can be natural or precede form and contained various bioactive ingredients.

Immobilization: Entrapments of substance or biologically active molecules in a defined region of space that can be other biological or chemical substances. It's basically used for the protection of the substances from outer environmental changes as well as for enhancing their affectivity.

Nutraceutical: Nutraceuticals are natural food products that give additional health benefits or protect from disease or illness. It's available in supplemented or fortified food.

Pathogenic: Pathogenic are the microbes that cause or can cause infection in the living organism. It's also known as the infectious agent. Some species of bacteria and fungus are pathogenic in nature.

Section 3
Nanoemulsion Applications in Cancer Research

Chapter 11
Application of Nanoemulsion in Cancer Treatment

Sumira Malik
https://orcid.org/0000-0001-5077-1493
Amity University, Ranchi, India

Shristi Kishore
https://orcid.org/0000-0002-8851-5951
Amity University, Ranchi, India

Manisha Kumari
Amity University, Ranchi, India

Archna Dhasmana
https://orcid.org/0000-0002-1120-2009
School of Applied and Life sciences, Uttaranchal University, Dehradun, India

ABSTRACT

Nanoemulsions are pharmaceutical-based nanometres ranged nanoformulated particles with significant and valuable contribution in field of the nanotechnology. In cancer treatment, the treatment through drugs fails primarily due to multidrug resistance (MDR), poor solubility, and unspecific toxicity. Nanoemulsions have the remarkable properties of non-immunogenicity, biodegradability, sustained encapsulation of low water solubility drugs, sustained regulated release of drug, stable and safe carrying tendency to deliver such drugs, and specificity in targeting only cancer cells to overcome multidrug resistance through for clinical and therapeutic application. They excellently address the noncompliance issues associated with the conventional anti-cancerous chemotherapeutic dosage issues. Currently multifunctional nanoemulsions are under experimentation for the treatment of various types of cancer. The chapter highlights the current status and applications of nanoemulsions as anti-cancer therapeutics and their commercial importance.

DOI: 10.4018/978-1-7998-8378-4.ch011

1. INTRODUCTION

Nanoemulsions (NEs) are types of colloidal dispersions which are used as drug carriers, mainly for the molecules that are poorly water soluble or have low water solubility composed of safe guard excipients. These drug formulations lead to high solubility and stability because they are constituted of heterogeneous dispersion of the nanometer droplets in different liquid (Ganta et al., 2014c, McClements D.J., 2012). NEs have the capacity to encapsulate the drugs and these encapsulations safeguard the drug by increasing its half-life in the plasma and also prevent it from degradation (Maeda et al., 2000). NEs are heterogeneous system in which the oil phase is dispersed in an aqueous phase as a droplet and stabilized with the help of emulsifying agents. Emulsifying agents are surfactants or amphiphilic surface-active molecules that lower the interfacial tension present between two phase of immiscible liquids of water and oil by absorbing them at their interfaces. Emulsifying agents are composed of charged or polar head that tends to keep themselves in polar liquids and the non-polar or hydrophobic bicarbonate tail resides themselves in non-polar liquids (Tadros et al., 2004). A schematic diagram of an oil-in-water NE droplet is represented in Figure 1.

Figure 1. An oil-in-water nanoemulsion droplet

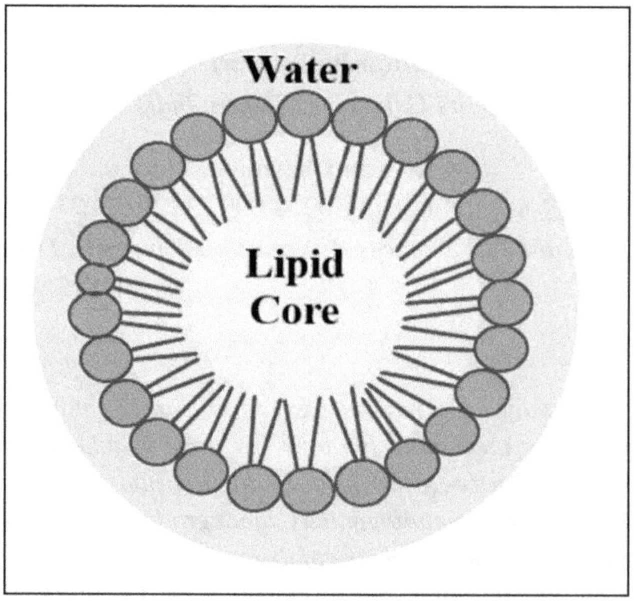

In the pharmaceutical industry number of drugs that contain NEs dosage forms have been found and some other drugs are under clinical and periclinical round of development. Moreover, NEs are used as excellent carriers for drug encapsulation, besides this they also help in removing the toxicities that are associated with the surfactant and formulation based on essential oils. Currently, for parental drug delivery there has been a rush for the exploration of NEs in the treatment of complex diseases such as in the treatment of cancer. NEs have the ability to protect the drug from hydrolysis and enzymatic degradation and their capacity to solubilise the larger amounts of hydrophobic drug make them perfect platform for parental drug delivery (Ganta et al., 2014c).

The major problem that arises in the cancer therapy is the effective and target drug delivery to the agent sites, while avoiding the adverse damage that results from the systematic administration. Although systemic drug delivery already depends mainly on physiochemical properties of the drug like binding property of plasma protein, size and diffusivity, tumors have an outward net connective flow and a dense heterogeneous vasculature that mainly acts as a hurdle in the path of efficient drug delivery at target sites. NEs have come out as best strategies to intensify the imaging agents and the therapeutic to tumors, thereby boosting the potential for the purpose of diagnosis for therapeutic success and thus they are useful in the case of target drug delivery to cancer (Kumar and Divya., 2015). In cancer therapy, the research has been more concentrated on NEs because they hold characteristics essential features that are important to achieve vital therapeutic effects such as, specific targeting, superficial charge, half-life of circulation, large surface area and the capacity of imaging for the formulation. Due to the small size of the oil droplet, they are easily targeted to the tumor causing tissue with the help of targeting constituent on the surface of NEs. Besides this NEs are being exploited for the purpose of image-guided drug delivery by imaging components and utilizing targeting (Tiwari, Tan and Amiji, 2006).

The present chapter aims to discuss the current and forthcoming applications of NEs in the cancer therapy in detail.

2. NANOEMULSIONS IN CANCER THERAPY: OPPURTUNITIES AND CHALLENGES

Nanotechnology has primarily improved the effectiveness and safety of cancer treatment by advancing the drug delivery system such as nanocarriers. Nanocarriers because of their small size are perfectly suitable for the chemotherapeutic passive targeting through the enhanced permeability retention (EPR) effect. Furthermore, they can achieve active targeting through receptor-mediated uptake to the host tissues and specific cell types (Senapati et al., 2018). In addition, they also solve the problem related to water solubility faced by hydrophobic drugs, enhance drug stability, and provides a controlled release (Neubi et al., 2018). To achieve the drug delivery for cancer therapy NEs, liposomes, polymetric nanoparticles, and nanostructured lipid carriers have shown a greater approach among all the nanostructured systems (Senapati et al., 2018). NEs have various advantages because of their composition. They are formulated mainly with biocompatible components and generally recognized as safe (GRAS) and they are easy to manufacture and scale-up (Ma et al., 2014). Moreover, to the advantages that are mentioned above commonly shown by most of the nanosystem, NEs possess great physiochemical stability, high encapsulation capacity for the hydrophobic drugs, enhanced bioavailability and low inter and intra-individual variability for the drug pharmacokinetics. NEs have stability similar to microspheres, liposomes, and ethosomes but they have the advantages of increased absorption and solubility of molecules that are poorly bioavailable (Verma et al., 2014). In an *in vivo* study, the accumulation of NEs and lipid nanoparticles in the brain were been compared. Through this it was shown that NEs increase the retention time in a proper manner compared to lipid nanoparticles (Hörmann and Zimmer., 2016). Furthermore, comparative studies between NEs, liposomes and solid lipid nanoparticles about the effect of skin permeation have indicated that NEs and liposomes have the ability to permeate to deeper layers of the skin, while solid lipid nanoparticles are only able to release the drug in superficial layers of the skin (Clares et al., 2014). Accordingly, a nebulized lipid-based NEs for the treatment of lung cancer was also carried out to study the possibility to enhance the resistance towards the enzymatic degradation and hydrolysis and of dissolving a higher amount of hydrophobic drugs (Asmawi et al., 2019).

In the delivery of the drugs to cancer cells, NEs can be of great use because of the fact these systems are found to be safe, enhancing the drug effects, avoiding toxicity and are capable of delivering the drug at the site of target tissue. Currently, there are no formulation of this type has been accepted by the FDA and various studies related to this study are decisive and therefore, they can enhance the system of delivery of therapeutics. In the production of these formulations, there might be involvement of high pressure and temperature conditions, depending on the excipients and drugs and thus, for several manufacturing process not all types of the starting materials are suitable. It is important to guarantee that the liable drugs are applicable and can be generated at a larger scale. To conceive the multifunctional NEs in large scale production can be quite hard as there are fair number of the variables to be considered (McClements D.J., 2012). Thus, to investigate the suitable method used for a particular nanoformulations, scale–up, material safety, and all the parameters related to quality control are needed to be taken into account (Ganta et al., 2014c).

2.1 Exploitation of Tumor Microenvironment

Tumor microenvironment is made up of glial cells, endothelial cells, extra cellular matrix, immune cells, proteins, adipocytes, activated fibroblasts, vascular cells and pericytes. The following components of the tumor microenvironment encourage the tumor to proceed into a malignant phenotype by supporting the function, structure and physiology of tumor that nurtures an environment (Ganta et al., 2014c). As we know, the tumor microenvironment promotes cancer growth but in the case of delivery of therapeutic agents to the site of tumor tissues there are important differences that can be exploited such as oxygenation, vascular abnormalities, metabolic states, perfusion, pH and mainly angiogenesis. Angiogenesis is a type of physical mechanism which encourages the formation of a new blood vessels from the one that exists. Angiogenesis is performed by cancer cells when the size of the tumor reaches greater than $2mm^3$ because in this case the core of the tumor are sensitive to hypoxic condition therefore to survive in this condition cancer cells have to perform angiogenesis to continue the presence of nutrients and the oxygen for their further growth while in contrary the small-sized tumors maintain the supply of nutrients and oxygen to the tumor cells through simple diffusion (Danhier, Feron and Préat., 2010).

Judah Folkman, a biomedical pioneer and an American medical scientist in 1971, hypothesized that as for the supply of nutrients and oxygen the cancer cells depend on the process of angiogenesis, therefore if this process could somehow be stopped, there might be a chance in the cancer treatment (Folkman J., 1971). Following his work and observations, some of the synthetic as well as natural angiogenesis inhibitors were discovered, and many of them were clinically used for the treatment of cancer such as everolimus, srafenib, sunitinib, bevacizumab and pazopanib. Among these angiogenesis inhibitors some of them are reported to face problems in delivery and toxicity (Kamba and McDonald., 2007). Therefore, NEs delivery system can be used to improve the therapeutic efficiency and reducing the toxicity by encapsulating these angiogenesis inhibitors. Betulinic acid that functions as an angiogenesis inhibitor is a pentacyclic triterpene that has the potential to target the vascular endothelial cells and also the tumor cells. According to the work, it was discovered that the betulinic acid can be used in a nanoemulsion delivery system to perform the strong anti-inflammatory activity to decrease the exposure of vital organs with betulinic acid (Dehelan et al., 2011). Similarly, a mycotoxin fumagillin has exhibited the antiangiogeneic properties, although the use of fumagillin is mainly associated with the neurotoxicity at a systemic dose (Winter et al., 2008). Thus this indicates that NEs-based delivery system could be used to overcome several shortcomings of the capable angiogenesis inhibitors utilizing a phenomenon that is highly exhibited by the tumor microenvironment in comparison with normal tissues. The another unique feature of tumor microenvironment which can be exploited for the purpose of target drug delivery is pH of the tumor microenvironment (Ganta et al., 2014bc).

2.2 Passive Targeting

Passive targeting of NEs is helpful in targeting the cancer cells by the means of increased retention effect and permeability, where the delivery system is of crucial size. The phenomenon of increased retention and permeability effect arises due to the dependency of tumor tissues on the process of angiogenesis that leads to the growth of poorly grown tumor vasculatures with the deficiency of pericytes, fenestrations irregular basement membrane. Thus, it increases the permeability of vessels at the site of tumor formation. Similarly, tumor tissues also have non-functional or absent lymphatic vessels that lead to deficient lymphatic drainage from the site of tumor. The union of deficient lymphatic drainage and increased vascular permeability is a phenomenon which has been named as the retention (EPR) effect and have increased permeability. Thus, it used in the delivery of some nano-therapeutic agents to the tumor tissues (Maeda et al., 2000).

The vascular pore size of the tumor mainly depends on the microenvironment, growth rate and the type of tumor, however in majority the vascular pore size of solid tumor exists between 380 and 780 nm. To achieve the maximum therapeutic accumulation at the tumor mass with the help of EPR effect the optimal size of drug delivery system exists between 10 and 100 nm (Danhier, Feron and Préat., 2010). Nanoparticles exhibits different size, particles with size less than 400 nm can easily eructate the tumor mass, however the particles of size less than 10 nm can be cleared with the help of rapid renal filtration and the particles of size greater than 100 nm get easily engulfed by the immune cells or by mononuclear phagocyte system (MPS) they can easily be acknowledged (Caron et al., 2013). In nano-delivery system the charges present on the surface of the nanoparticles is also an important component and should be of negative charge or no charge to elude the immune system (Ganta et al., 2014c).

Some authors in their published work have found that the minimization of adverse effect and enhanced therapeutic activity is due to the formulation of therapeutic agents in the nanoparticles that takes the benefit of retention effect. Among them some of the systems are already been translated successfully for clinical practices (Gabizon et al., 2003, Allen and Martin., 2004). In studies, researchers typically found that NEs negative charges (Tagne et al., 2008, Huynh et al., 2009) present on the surface and the particle size less than 200 nm is acceptable. During the formulation development if the negative charge on the surface of NEs increases it is known as prevent droplet coalescence and in the case of *in vivo* it is found that it effects the permeability of blood-brain barrier (BBB) (Wang, Wang and Huang., 2009). In nanoemulsion delivery system apart from the surface charge and size of the nanoparticles circulation time also plays an important role in it. Due to fast opsonization by the mononuclear phagocyte system the circulating half-life of the NEs are short. Therefore,

Figure 2. Passive targeting strategy of NEs drug delivery systems

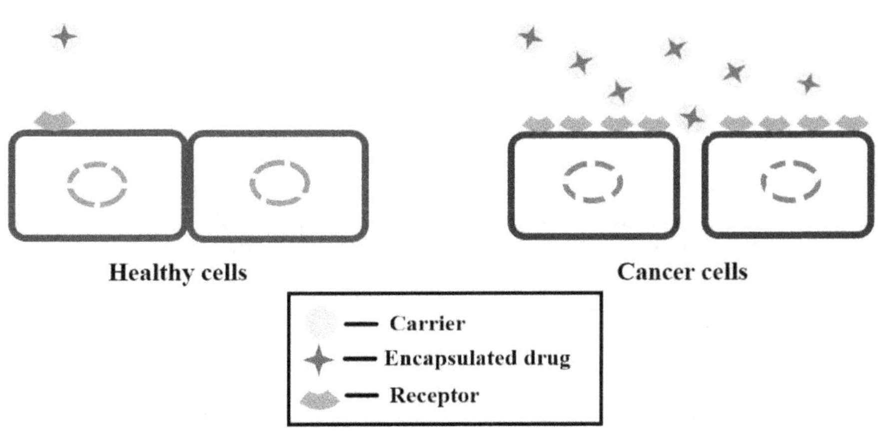

to increase the circulating half life and therapeutic efficacy of the NEs, the surface of the NEs are been coated with the help of hydrophilic polymers like poloxamine, PEG, and poloxamers which are helpful in decreasing the recognition by the MPS cells and absorption of blood protein (Allen and Martin., 2004). Hence, by passive targeting of the NEs the combination of parameters such as surface charge, size and NEs hydrophilicity can be improved to achieve the efficacy of NEs at the site of tumor tissues (Ganta et al., 2014c). The passive targeting strategy of NEs drug delivery system has been shown in Figure 2.

2.3. Active Targeting

The active targeting process of NEs could be achieved through the attachment of components on the surface of the NEs that have the capability to recognize the target throughout the tumor effected cells, tissue and organ, or in intra cellular organelle that leads to the accumulation of the NEs. In contrast to passive targeting, active targeting of the NEs not only utilizes the physiological benefits that are provides by the tumor microenvironment but also helps in delivery to tumor tissues specifically and therefore minimize the interaction between tumor tissue and healthy tissues (Ganta et al., 2014c). Active targeting is mainly beneficial for those therapeutic agents that require facilitation by fusion, endocytosis is not efficiently internalized. Therapeutic agents following active targeting also allow higher delivery of payloads intra-cellularly with the help of receptor-mediated uptake present in NEs system. The conjugation of ligands with the carrier lead drugs to remain unmodified and thus enables the drug to show its therapeutic activity by intracellular uptake.

At the NEs surface, varieties of targeting ligands are attached to enable identification by suitable receptors that are expressed at the target sites. For the selectivity of ideal receptors related to tumors, it is essential to select those receptors that are not expressed on normal cells but are overexpressed on tumor cells. Receptors such as, epidermal growth factor (EGF), transferring and folate are overexpressed in some type of cancers such as breast and ovarian cancers (Milane, Duan, Amiji., 2011). For the purpose of targeting, specific antigenic site the targeting element that are conjugated to NEs include synthetic polymers, proteins, vitamins, antibodies, aptamers and peptides (Zeng et al., 2013; Tayeb at al., 2017; Talekar et al., 2012; Mulik et al., 2010). In active tumor targeting, monoclonal antibodies and the fragments derived from them are commonly used because of their high affinity, specificity and scFv antibodies and Fab fragments are used due to their small size (Ahmad et al., 2012).

In literature, many authors have published their works of NEs on active targeting of cancer to deliver the wide range of drugs that are insoluble in water, including this concept to explore more about the ways to

Figure 3. Active targeting strategy of NEs drug delivery systems

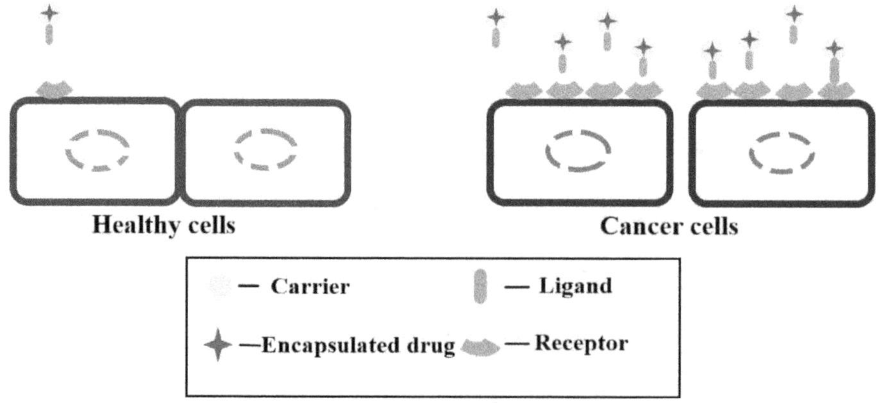

overcome the multi drug resistance (MDR) (Meng et al., 2016). In comparison to an untargeted formulation, NEs that are functionalized externally with folate to target the nasopharyngeal tumor cells and NEs that are loaded with aclacinomycin A exhibits better efficiency of the loaded drug (Ohguchi et al., 2008). The active targeting strategy of NEs drug delivery systems is shown in Figure 3.

2.4 Imaging

In cancer therapy, imaging has an important characteristic that allows real time monitoring of the cancer with less destruction of tissue and invasiveness. Biomedical imaging plays an important role in biopsy guidance, therapy guidance and planning, screening, prediction, progmography, and staging. In anatomical imaging, the techniques that are used traditionally are ultrasound, computed X-ray tomography, and imaging by magnetic resonance (Ganta et al., 2014c). Among them, the most frequently used technique is ultrasound that is used as an external incitement because it is cost-effective, easily assessable, and also has the ability to be used in co-existence with the multi-modal system that can incorporate microbubbles which is an ultrasound contrast agent. The development of these microbubbles are in the form of perfluocarbon (PFC) NEs which is a type of synthetic organic compound in which almost all of the hydrogen atoms are been replaced with fluorine atoms. In the isotopes of PFC's 19F isotope gives great sensitivity *in vivo* because it is chemically and biologically inert. A wide range of therapeutic and contrast agent can be combined on or encapsulated in PFCs due to which PFC-NEs are suitable in delivery or multi-modal imaging (Kaneda et al., 2008). NEs accumulate inside the tumor with the help of EPR effect when the formulation of microbubbles is done in the form of PFC-NEs upon the intravenous administration. Nano-droplets get converted to bubbles when the accumulation takes place in tumor, under the actions shown by ultrasound. In NEs, when cytotoxic agents are incorporated, the encapsulated drug and post-droplet-bubble-transition are released with ability to regress the tumor effectively and increase the intracellular uptake (Rapoport et al., 2009, Gao et al., 2008). Besides this, these NEs system can also be used to target the tumor tissues actively that achieves higher rate of tumor accumulation. In recent studies Bae *et al.* reported the development of rhodamine/PFC NEs with the MRI, bimodal imaging contrast agent, and the capabilities of optical imaging (Bae et al., 2013). Furthermore, NEs also pervade with folate that allowed them to get localized into the tumors that express FR which improves the tumor detection, provides high specificity, and great signal sensitivity. It has been seen that folate-PFC NEs can also be used for accurate diagnosis, targeted tumor therapy in early detection of disease for clinical applications.

PFC-NEs are mainly used for image-guided therapy but apart from this Gianella *et al.* recently described the development of Cy7 for imaging of near infrared fluorescence, prednisolone acetate valerate foe the therapeutic purpose and for MRI multi-functional NEs with nanocrystals of iron oxide. The unique theranostic properties of multi-functional delivery system makes it suitable and can be applied for the purpose of image-guided therapy for cancer, and therefore indicates the abilities of multimode NEs system for tissue imaging and therapeutic delivery (Gianella et al., 2011).

3. APPLICATIONS OF NANOEMULSIONS TO OVERMCOME MULTI DRUG RESISTANCE (MDR)

Multi Drug Resistant (MDR) tumors acts as a crucial hurdle in the pathway of effective cancer treatment and in addition with metastasis they are found to be main contributors for the reason of death caused by cancer. In MDR of the tumor cells, the expressions of the multifunctional efflux transporters from the family of ABC gene have been known to play a major role in it (Dean M., 2009). ABCs genes are mainly responsible for the efflux that are generated by different endogenous ligands including metabolic products, proteins, lipids, and also cytotoxic antibiotic drugs by utilizing the energy generated during the ATP hydrolysis (Mohammad, He and Yin et al., 2018).

The multi drug resistance of various anti-tumor drugs are caused due to some of the transporters that are expressed by the family of ABC genes. The first ABC transporters that was identified is MDR-1 also cited as ABCB-1 or P-gp, encoded with the ABC-1 gene and it has ability to pump etoposide, colchicines, pacitaxel and vinblastine. Along with ABCB-1 transporters, there are other applicable transporters such as ABCA-2, encoded with ABCA-2 gene, which are responsible for estramustine resistance. In case of tumor cells, the most widely used transporter is known as MRP-1, encoded with ABCC1 gene and is responsible for colchine, vincristine, methotrexate and etoposide resistance. On the cell membrane of some polarized cells such as liver, intestinal epithelium, and kidney one more transporter of this family, MRP-2 is located, which is responsible for resistance from cisplantin, methotrexate, doxorubicin, and vinblastine. In addition, MRP-3 encoded with ABCC-3 gene is liable for the transportations of organic anions and in pumping anti-tumor drugs such as etoposide and methotrexate (Dean M., 2009, Mohammad, He, and Yin et al., 2018). MRP-4 is encoded with ABCC-4 pumps 6-TG-6-thioguanine and 6- mercaptopurine. In several cases, it has been seen that for one ABC transporters more than one drug can also be substrate for example, 6-TG-6-thioguanine and 6-mercaptopurine are also pumped by the MPR-5 transporter and etoposide is also pumped by MPR-6 along with MPR-3 and MPR-1 (Dean M., 2009). In the field of oncology hunt for important compounds for the blocking of these hyperactive ABC pumps is a growing field of interest in order to enhance the chemotherapeutic effects. In this contrast, some of the ABC inhibitor functionalizing NEs have been found to address the MDR associated with cancer (Mohammad, He and Yin et al., 2018).

Ganta *et al.* developed the NEs which is functionalized with folate in order to deliver the docetaxel efficiently to ovarian cancer cells to overcome the docetaxel multi drug resistance. NEs are targeted with the folate because the folate receptors have the tendency to be expressed poorly in majority of the normal tissues, but in epithelial ovarian cancer it is overexpressed. Thus, the NEs that are targeted with folate are capable of delivering the docetaxel by receptor mediated endocytosis which shows increased cytotoxicity capable of overcoming the ABC transporter mediated with taxane resistance (Ganta et al., 2014b). However, overcoming MDR is an important issue, but currently, the taxanes including docetaxel and paclitaxel have been found as fundamental drugs in breast cancer therapy (Crown, O'Leary and Ooi., 2004). Thus, to get rid of this obstacle Meng *et al.* used the baicalein to inhibit the P-gp and also to enhance oxidative stress. To increase the cell sensitivity to paclitaxel, increasing the oxidative stress is found to be most suitable strategy due to the fact that glutathione and cellular reactive oxygen species are very important for redox reactions shown by cells. Therefore, to treat the breast cancer the researchers used this strategy and co-encapsulated baicalein and paclitaxel in NEs. The evolved NEs have the capabilities to decrease the cellular GSH, increase ROS and improve caspase-3 activity in MCF-7 cells. More notably, it has been seen in an antitumor study that NEs co-encapsulated with baicalein and paclitaxel show greater antitumor potency than other formulated paclitaxel. So, to overcome MDR, the co-delivery of baicalein and paclitaxel in NEs can be used as possible combine therapeutic strategy (Meng et al., 2016). After that, Zheng *et al.* developed a different strategy to overcome MDR by aiming to prepare a NEs that have the ability to change the levels of Bcl-2 and Bax expression and also inhibit the transport function carried by P-gp. In order to achieve their goals, they used a derivative of vitamin E, which is an antioxidant and its mechanism comprise of eliminating the chain reaction carried by fatty acid radical propagation and reducing peroxyl radicals (Zheng et al., 2016). It has also been indicated that the derivative of vitamin E used in this study, TPGS, is one of the most commercially available and strong surfactants that can reverse the process of MDR in cancer and also serves as a P-gp inhibitor (Tang et al., 2013). Vitamin E has the ability to disturb the Bcl-xL-Bax interaction, which activates the Bax and therefore intervene mitochondrial-centered apoptotic cell death. Thus, the suitable way found for the study of cell lines of paclitaxel-resistan human ovarian carcinoma can be a vitamin E based NEs containing paclitaxel (Zheng et al., 2016).

4. APPLICATION OF NANOEMULSIONS IN THE TREATMENT OF DIFFERENT TYPES OF CANCERS

Multiple advantageous properties of NEs have made them explored in the treatment of various types of cancers. They:

1. have high surface-to-volume ratio than macro-emulsions that makes them efficient vehicles for the delivery of therapeutics,
2. can easily be formulated in creams, foams, sprays, liquids, etc.,
3. are generally non-toxic to the mucous membranes, thus can easily be applied on the skin,
4. can be administered via the oral route, as the surfactants used in NE formulations are approved for human use, and
5. are safe for practical healthcare applications as they are generally non-toxic to the healthy cells (Souto, Nayak, & Murthy, 2011).

For the safe treatment of different types of cancers, NEs have been widely explored. Several applications of NEs in the therapy of different types of cancers and their anticancer activities are discussed in this section.

4.1 Nanoemulsions in the Treatment of Lung Cancer

A number of anti-cancer drugs are used in the treatment of different types of cancers like lung cancer, prostate cancer, breast cancer, etc. These drugs enhance the apoptotic activities of the cancer cells, inhibit their cellular functions, arrest their cell division, and lead to the breakdown of microtubules (Favero et al., 2018). One of such anti-cancer drugs is Paclitaxel (PTX). But one limitation that restricts the use of PTX in therapeutics is that it has very low solubility in water. For this reason, some NEs formulations containing PTX have been developed. One of the example of such formulations is Taxol® that contains ethanol and Cremophol-EL®. However, some evidences have shown toxic effects of Cremophol-EL® (Monge-Fuentes et al., 2017). To overcome this problem of PTX's water-insolubility, hyaluronic acid-complexed PTX Nanoemulsions (HPNEs) were prepared. These HPNEs were developed to target small lung cancer cells. Further *in vivo* investigations in mouse showed that HPNEs had exhibited more antitumor activities than Taxol®. Conjugated hyaluronic acid in HPNEs inhibited the growth of lung tumor cells due to their affinity for CD44 overexpressed cancer cells. This study suggests that HPNEs can be effectively be used as nano-vehicles for the delivery of drugs having low solubility to target the small-lung cancer cells overexpressing CD44 (Kim and Park, 2017).

Docetaxel is another drug that exhibits anti-cancer activities against lung cancer. But due to its poor aqueous solubility and low bioavailability, the practical application of this drug in the clinical practice is restricted. Development of Docetaxel NEs (DNEs) has overcome these limitations of docetaxel. DNEs have shown high stability, controlled drug release, and high selectivity for A549 lung cancer cells than normal MRC-5 cells. Their aerodynamic characterization shows that they can be ideally used for the targeting of docetaxel in lung

Table 1. Some NE formulations and their anti-cancer activities against lung cancer cells

Nanoemulsion formulation	Observed anti-lung cancer activities	References
Hyaluronic acid-complexed PTX NEs	Inhibition of small-lung cancer cells overexpressing CD44	Kim and Park, 2017
Neobavailsoflavone NEs	Suppression of lung cancer by regulating the tumor microenvironment.	Ye, He, & Feng, 2020
Epigallocatechin-3-gallate NEs	Inhibition of H1299 lung cancer cells growth and proliferation by activating AMP-activated protein kinase signal pathway.	Chen et al., 2020a
Natural Burneol NEs	Stimulation of apoptotic activities in A549 non-small lung cancer cells by regulating EGFR/EHD1-mediated apoptotic pathways.	Yuan et al., 2020
Naringenin NEs	Stimulation of apoptotic activities of the A549 cancer cells by increasing activities of Bax and Caspase-3 and reducing the Bcl2 gene expression.	Md et al., 2020
Tanshinone NEs	Cell cycle arrest of A549 lung cancer cells at G0/G1 phase.	Lee, Liang, and Chen, 2016
Citrus lemon essential oil NEs	Stimulation of apoptotic activities of the A549 cancer cells by increasing Caspase-3 gene expression.	Yousefian Rad et al., 2020
Arachis hypogaaea essential oil NEs	Stimulation of apoptotic activities by SubG1 and Cas-3 overexpression in A549 lung cancer cells.	Fazelifar et al., 2020
Curcuminoid extracts NEs	Stimulation of apoptosis in A549 and H460 lung cancer cells.	Chang and Chen, 2015
Docetaxel-NEs (DNE4)	Selective toxicity against A549 lung carcinoma cells than normal MRC-5 cells, ideal to be administered via pulmonary routes.	Asmawi et al., 2019
Diferuloylmethane Nanoemulsion-hybrid nanocarriers	Increased inhibitory effects on A549 lung adenocarcinomic cells than free Diferuloylmentane alone.	Sun et al., 2016

cancer treatment via pulmonary routes (Asmawi et al., 2019). Some NE formulations containing essential oils such as *Citrus lemon* essential oil and *Arachis hypogaaea* essential oil have also shown potent anti-cancer activities against A549 lung cancer cells by stimulating their apoptotic pathways (Yousefian Rad et al., 2020; Fazelifar et al., 2020). These NE formulations lead to the overexpression of pro-apoptotic genes such as Cas-3 and SubG-1. For the effective targeting of non-small lung cancer cells natural burneol nanoemulsions (NBNEs) were formulated. The NBNEs enhanced the solubility and distribution of natural burneol. It was further found by the proteomics study that they exhibit strong anti-cancer activities against A549 non-small lung cancer cells by regulating their EGFR/EHD1-mediated apoptotic pathways (Yuan et al., 2020). These studies show that NE formulations have vigorous efficiency to act as therapeutic agents in advanced chemotherapy. Some other NE formulations having anti-lung cancer activities are summarized in Table 1.

It has been found that both of the two types of lung cancer cells- small lung cancer cells and non-small lung cancer cells can develop resistance to certain anti-cancer drugs. Overexpression of some specific proteins like lung-resistance associated proteins and superficial membrane proteins, deregulation or inhibition of apoptotic pathway of cells' apoptotic pathways, etc. contribute to the development of drug resistance in these tumor cells. To resolve this problem, surfactant-based drug delivery carriers like nanocarriers (NEs, nanoparticles, nanogels, liposomes), microspheres, novel powders, etc. have been developed (Kaur et al., 2016).

4.2 Nanoemulsions in Ovarian Cancer Therapy

Ovarian cancer is one of the top ten diagnosed cancers and among top five fatal cancers in most of the countries that causes gynaecologic malignancy globally (Knutson et al., 2006). To treat ovarian cancer, several strategies have been employed to date. *Linum usitatissimum* seed essential oil (LSEO) has been found to have anti-cancer activities. LSEO was formulated with NEs and their anti-cancerous activities were evaluated. Formulation of LSEO with NEs lead to the increase in bio-compatibility of the LSEO. The LSEO-NEs showed significant cytotoxic activities against A2780 human ovarian cancer cells and led to their death. They were also seen to show considerable anti-angiogenic activities (Keykhasalar et al., 2020). These cytotoxic and anti-angiogenic activities of LSEO-NEs make them novel therapeutics in the ovarian cancer chemotherapy.

In another study, the efficiency of Pyridoclax-loaded NEs (PNEs) was tested for treatment of ovarian cancer. Pyridoclax is an anti-cancerous drug that has been found effective in the treatment of chemo-resistant ovarian cancers. However, the PNEs showed 2.5 times more anti-cancerous activities on chemo-resistant ovarian carcinoma cells than the free Pyridoclax *in vivo*. This study demanded for permitting the use of PNEs in the treatment of ovarian cancer by administration through oral routes (Groo et al., 2020). In another experiment, NEs formulations containing a platinum prodrug, myriplastin, and C6 ceramide (a pro-apoptotic agent) were developed for platinum-based chemotherapy in the treatment of ovarian cancer. When these myriplastin/C6-ceramide NEs were tested for ovarian cancer cells A2780, A2780CP, and SKOV3, 50 times increased cytotoxicity was observed for SKOV3 cells than that of anti-cancerous cisplastin in solo. These observations indicate that these NEs formulation have the potential to be applied as theranostics in the treatment of ovarian cancer (Ganta et al., 2014a).

4.3 Nanoemulsions in Prostate Cancer Therapy

Even in the today's modern-scientific world, the therapeutic options for the treatment of cancer are very limited in numbers. Also, the existing options have their own limitations. In comparison with other types cancer, the current decade has evidenced a significant increase in the death rates of patients suffering from prostate cancer (Zhou et al., 2016). According to a report on global cancer statistics, it has been found that more than 70% patients suffer from post-treatment reappearance and transition of the disease to an untreatable state (Jemal et al., 2011). Like other cancers, the root causes of prostate cancers are the Cancer Stem Cells (CSCs) or the Tumor-Initiating Cells (TICs), that are responsible for the tumor development, metastasis, and development of drug-resistance (Zheng et al., 2016). Several studies have shown that CDC markers expressing cancer cells such as CD4 and CD133 not only display resistance to certain therapeutics but also have the ability to proliferate even after the therapy (Konrad et al., 2017). Several underlying mechanisms such as deactivation of apoptotic pathways, activation of anti-apoptotic pathways, quiescent state, enhanced DNA repair mechanisms, etc. can contribute to the development of this therapeutic-resistance.

The current approaches for the therapy of prostate cancer only rely on targeting bulk and rapidly-propagating carcinoma cells but do not focus on inner subpopulations of CDCs. For this reason, the rate of survival and success rate of anti-cancerous therapeutics in case of prostate cancer therapy is comparatively lower than other types of cancers (Zhu et al., 2009). In a study, the therapeutic effects of NEs conjugated with DHA-SBT-1214, which is a toxoid prodrug was evaluated against stem cells of prostate cancer. These NE formulations lead to complete cell death of PPT2 cancer cells *in vitro*. They showed higher efficiency in the suppression of tumor than Abraxane® (a paclitaxel prodrug) when administered weekly to a NOD/SCID mouse having subcutaneous PPT2 tumors. This study shows that DHA-SBT-1214 NEs have the potential to supress prostate cancer stem cells CD133/CD44+ both *in vivo* and *in vitro*.

Another chemical that is known for its anti-prostate cancerous activities is curcumin. But due to its poor solubility and low bioavailability, its therapeutic potentials have been limited. In a study, curcumin was conjugated with oil-in-water NEs and its anticancer activities were explored for the treatment of prostate cancer. The curcumin NEs showed higher cytotoxicity, greater cellular uptake, and absorption rates as compared to the free curcumin. They enhanced the apoptotic activities of the cancer cells indicating their potentialities to be used in the effective delivery of therapeutics in the treatment of prostate cancer (Guan et al., 2017). Similar to curcumin, the therapeutic applications of rutin bioactive are limited due to its low bioavailability and degradation tendency in aqueous environments. To overcome these limitations, rutin NEs were developed. These rutin NEs showed significant anti-cancer activities against human prostate cancer cell line (P3C). Further investigations indicated that they stimulated the apoptotic pathways by inducing ROS (Ahmad et al., 2017).

4.4 Nanoemulsions in the Treatment of Breast Cancer

Breast cancer is the most prevalent tumor in women that accounts for an incidence of 11.6% and a mortality rate of 6.6% (Bray et al., 2018). Although the most common approach for the treatment of breast cancer is chemotherapy, several limitations including low accumulation of therapeutics, drug resistance, accumulation of drugs at other organs, etc. decrease its efficiency (Fidler and Kripke, 2015). To improve the current treatment procedures, several strategies have been adopted.

Lapachol (LAP), that belongs to napthoquinones class, is a well-known bioactive compound with anti-tumor properties. But due to its negative side-effects and poor aqueous solubility, its use in the clinical treatments is limited. In a latest study, the anti-cancer activities of LAP loaded NEs were explored in a breast cancer (4T1) tumor model. The anti-tumor activities of these stable LAP-NEs were found to be more than the free LAP, and no signs of negative-side effects were observed. Hence, LAP-NEs can effectively be used in the treatment of breast cancer (Mendes Miranda et al., 2021). In another study, sour cherry pit oil (SCPO) having anti-inflammatory and antioxidant activities was formulated with NEs. The formulated SCPO-NEs showed considerable apoptotic and cytotoxic activities against MCF-7 breast cancer cells than normal HFF cells. They also led to the reduction of tumor size in murine breast cancer model. Thus, they are capable of treating breast cancer effectively by inducing apoptotic pathways (Maragheh et al., 2019).

For the effective treatment of breast cancer, combination of more than one anti-cancer therapeutic can prove beneficial. In a study, two anti-cancer drugs docetaxel and thymoquinone were formulated in borage oil-based NEs and evaluated for their anti-tumor activities to treat breast cancer. The formulated NEs showed increased cytotoxicity against MCF-7 and MDA-MB-231 cancer cells by inducing apoptosis and stimulating autophagy simultaneously. These NEs enhance the efficiency of docetaxel and reduce its associated toxicity, thus can effectively be used in the breast cancer chemotherapy in future (Alkhatib et al., 2020).

4.5 Nanoemulsions in the Treatment of Leukemia

In children, leukemia is the chief cause of cancer-associated deaths. To treat leukemia, several approaches have been employed till now. The use of NEs in delivering anti-cancer therapeutics can be an effective approach in the treatment of leukemia. Quercetin loaded NEs have shown effective anti-cancer activities against MDR and non-MDR human leukemia cells. These NEs were shown to maintain the cytotoxic effects of quercetin that might lead to an increase in their bioavailability. The quercetin NEs have efficiency to serve as potent anti-cancer therapeutic agent for the treatment of leukemia *in vivo* (Marques et al., 2021).

In a study, gambogic acid nanoemulsions (GA-NEs) were formulated to exploit their activities against acute myeloid leukemia both *in vitro* and *in vivo*. The formulated GA-NEs were found to have 4000 fold more solubility than the free gambogic acid and 318.2% more bioavailability than its suspension. Also, they exhibited better anti-cancer activities against the acute myeloid leukemia both *in vitro* and *in vivo* by decreasing half inhibitory concentration values in HL-60, Jurket, and MV4-11 cells, and by increasing the time and percentage of endurance. Thus, for the treatment of acute myeloid leukemia, GA-NEs can prove as promising therapeutic agents (Feng et al., 2018). Some other NEs such as tucumã extract NEs have also shown better antileukemic activities than the free tucumã extract. This shows their potential to be used as alternative medicines in the treatment of leukemia (Copetti et al., 2020).

4.6 Nanoemulsions in the Treatment of Melanoma

Melanoma is the most serious type of skin cancer, which when treated with traditional therapies, results in an average survival time of the patient with six to twelve months (in case of metastatic melanoma). In most of the patients, the five-year survival rates are found to be lower than 10% (Balch et al., 2009). Increasing chemo resistance in the cancer cells has also become a major issue of concern. Thus, there is a crucial need for the development of fully-effective and cost-worthy therapeutics agents having explicit potentials to treat melanoma. Nowadays, in the treatment of melanoma, apart from conventional clinical treatments, some advanced targeted therapies are also being employed.

In some recent studies, cold atmospheric plasma (CAP) has been well established of having promising applications in cancer therapy (Yan, Sherman, & Keidar, 2017). In an experiment, silyamarin nanoemulsions (SNEs) along with air CAP were tested for their anti-cancer activities against G-361 human melanoma cells. Melonama cells treated with SNEs and CAP showed increased cellular toxicity, considerable ROS and RNS (Reactive Nitrogen Species) generation, significant DNA damage, and an increased Caspase 8,9,3,7 activities. These cells groups were observed with blocked HGF/c-MET pathways. These SNEs along with CAP were seen to reduce the tumor size and weight in mice xenografts (Adhikari et al., 2019). Thus, they have a huge potential to be applied in the clinical chemotherapy for the treatment of melanoma in future. In another study, *Tectona grandis* (TG) leaf extract NEs were formulated for their applications as photosensitizer in the therapy for melanoma B16 cells. The formulised TG-NEs were found to have cytotoxic effects on B6F10 and murine fibroblast NIH3T3 cells but were comparatively too less toxic for the normal cells in the dark than the free TG extracts. This makes them compatible for treating melanoma in a safe as well as effective manner (de Menezes Furtado et al., 2017). In another study, NEs were prepared by formulating coffee oil, algal oil, surfactants (Tween: Span=4:1), and deionised water. *In vivo* experiments in animals showed that these NEs were able to reduce trans-epidermal water loss, melanin production, skin erythema, and sub-cutaneous flow of blood. They inhibited the growth of melanoma tumor cells B16-F10 by arresting them at G2/M phase. They also induced the apoptotic pathways by rising Caspase-3,8,9 activities (Yang, Hung, and Chen, 2017).

4.7 Nanoemulsions in Colon Cancer Therapy

According to a statistical report published in 2020, colon cancer is most common cause of death due to cancers in the United States. In 2020 alone, 53,200 individuals died due to colon cancer in the USA (Siegel et al., 2020). Today, the most effective way of treating colon cancers is by jointly using operations with conventional chemotherapy, immunotherapy, radiotherapy, herbs, etc. However, even after the therapy, the chances of survival are still lower than 20-40% after five years, and there are risks of tumor migrations and recurrences (Huang et al., 2015). The major cause behind the death due to colon cancers is attributed to migration of the cancer cells, not the cancer itself.

Nano-therapeutics, particularly NEs hold huge potential to be explored in the treatment of cancers. In a study, NEs containing cinnamon and algal oils were incorporated with epirubicin (EPI). The EPI-NEs showed potent inhibitory actions against human colon cancer cells HCT116. The half-minimum inhibitory concentrations of EPI=NEs were found to lower than that of the free EPI. Also, they significantly increased the ROS generation inside the cells and stimulated their apoptotic pathways. Thus, it was concluded that the NEs had increased the therapeutic anti-cancer activities of the EPI (Alkhatib, Aljadani, and Mahassni, 2020).

In another study, NEs carrying lycopene and gold nanoparticles were tested against HT29 colon cancer cell lines. Incorporation of lycopene (an important bioactive compound having anti-tumor properties found in tomatoes) into NEs overcame its problem of poor stability and low bioavailability. The use of Au-Lycopene nanoemulsions (AL-NEs) led to an increase in the apoptotic activities by fifteen times in the HT29 colon cancer cells. Also, the migration capabilities of these colon cancer cells was found to be decreased by five times by treating with these AL-NEs. Thus, in future, they can potentially be used in the colon cancer therapy (Huang et al., 2015).

4.8 Nanoemulsions in the Treatment of Other Types of Cancers

Apart from these cancers, NEs have also been explored for the treatment of other types of cancers. Some of the studies exploring the anti-tumor effects of NEs in the treatment of other types of cancers are summarized in Table 2.

Table 2. Some nanoemulsion formulations and their activities against different types of cancer cells

Name of the cancer	Nanoemulsion formulations	Observed anti-cancer activity	References
Cervical cancer	Curcumin-NEs	Phototoxic effects on cervical cancer cell lines, stimulates apoptotic pathways by increasing Cas-3/7 activities.	de Matos et al., 2018
Cervical cancer	Berberine-NEs	Increasing the cervical carcinomic cell deaths when treated jointly with photodynamic therapy.	Floriano et al., 2021
Oral cancer	Spearmint Oil NEs	Cytotoxic effects against oral carcinoma cell lines (KON).	Tubtimsri et al., 2018
Bladder cancer	Cisplatin and gemcitabine NEs	Deep intravesical deposition of chemotherapeutics for the treatment of bladder cancer cells both *in vitro* and *in vivo.*	Chen et al., 2020b
Bladder cancer	Anti-EGFR indocyanine green mitomycin C-perfluorocarbon NEs	Eradication of bladder cancer cells five-times more than mitomycin-C alone, low chemotoxicity on heathy cells	Lee and Lin, 2018
Cervical cancer	Mitomycin C-Essential oil NEs	Increased cytotoxicity on HeLa and MCF-7 cells.	Al-Otaibi et al., 2018
Pancreatic cancer	Propofol NEs	Improvement of immune functions of patients with pancreatic cancer during surgery and decrement in the level of inflammatory factors.	Wang and Zhou, 2021
Liver cancer	5-Fluorouracil/Curcumin NEs.	Synergistic treatment of liver cancer by enhanced efficiency and cytotoxicity of NE formulations against liver cancer cells.	Guo et al., 2020
Nasopharyngeal cancer	Folate decorated doxorubicin lipid NEs	Inhibition of growth of nasal cancer cell lines CN1. Immense necrosis and haemorrhage of CN1 cancer cells.	Liu et al., 2017

5. CLINICAL TRIALS AND FUTURE PROSPECTIVES

NEs have emerged out as potent and reliable tools in the treatment of different types of cancer. However, as compared to studies and experiments exploring potentials of NE formulations in cancer treatment, the number of clinical trials performed till date are very less in numbers. Some of the clinical trials that used NEs in the treatment of cancer and cancer-related symptoms are listed in Table 3.

Table 3. Clinical trials involving the use of NEs in the treatment of cancer and cancer-related symptoms

ClinicalTrial.gov Identifier	Condition or disease	Intervention/Treatment	Description	References
NCT02685592	Lentigo Maligna	Drug: 5-aminolevulinic acid NEs	Investigation of photodynamic therapy to treat Lentigo Malinga.	https://clinicaltrials.gov/ct2/show/NCT02685592?term=Nanoemulsion&cond=Tumor&draw=2&rank=6
NCT01975363	-Atypical Ductal Breast Hyperplasia -BRCA1 Gene Mutation -BRCA2 Gene Mutation -Ductal Breast Carcinoma in Situ -Lobular Breast Carcinoma in Situ	Dietary Supplement: curcumin Other: Biomarker analysis Other: Assessment of Dietary Intake Other: Daily Log	Investigation of curcumin NEs in inducing inflammatory changes in breast tissues in obese women at high risk of breast cancer.	https://clinicaltrials.gov/ct2/show/NCT01975363?term=Nanoemulsion&cond=Tumor&draw=2&rank=3
NCT03865992	-Breast Cancer -Joint Pain	Dietary Supplement: Curcumin Other: Placebo Other: Nanoemulsion Other: Quality-of-Life Assessment Behavioral: Questionnaire	Assessment of Curcumin NEs versus Placebo in reducing joint pain in breast cancer survivors having Aromatase-inhibitor induced joint diseases.	https://clinicaltrials.gov/ct2/show/NCT03865992?term=Nanoemulsion&cond=Tumor&draw=2&rank=2
NCT01326078	Leukemia	Drug: propofol	Comparison of safety and efficiency of propofol NEs and lipid emulsions for sedation in non-invasive ambulatory procedures.	https://clinicaltrials.gov/ct2/show/NCT01326078?term=Nanoemulsion&cond=Tumor&draw=2&rank=4

The emergence of NEs as novel drug delivery system has provided alternatives for enhancing the solubility and bioavailability of drugs. The ability of NEs to encapsulate different varieties of drugs and bioactive compounds has proven helpful in the delivery of therapeutics via desired routes in an efficient manner. One of the major forthcoming challenges in the development of NEs is to continuously find dynamic mechanisms for the improvement of NE formulations and their efficiency in the drug delivery. To maintain the drug stability and uptake while increasing the therapeutic competence is also a challenging task for the researchers.

In future, studies should focus on exploring the mechanisms behind the NEs-target cells interaction. Also, the interaction of the NEs with the conjugated drug and surfactants should further be explored. In the recent years, the application of NEs in imaging cancers is also increasing. The use of NEs to monitor tumors provides ways for merest destruction and invasion. Thus, this application of NEs in the imaging of tumors also needs to be explored more (Sánchez-López et al., 2019). At last, NEs need to be applied more frequently and practically in the cancer therapy as very few clinical trials have been performed to date.

CONCLUSION

NEs are heterogeneous system of particles that have arisen as forthcoming strategic tools in the treatment of cancer. The presence of a hydrophobic core region in the NEs have helped us to overcome the problem of delivery of lipophilic cancer therapeutics. NE-formulations have been effectively used to eliminate the cancer cells without damaging the healthy tissues. Moreover, some NE-formulations are found to have the capability to target tumor cells either by active or passive targeting. Also, they are found to stop the migration of metastatic cancer cells and avoid multi-drug resistance. To date, many NEs have been tested both *in vitro* and *in vivo* on animal models for the treatment of cancers such as breast cancer, lung cancer, colon cancer, ovarian cancer, prostate cancer, lymphoma, melanoma, and many more. Some clinical trials have also been performed to test their efficiency in the treatment of human cancers. The current chapter discusses the mechanisms of targeting NEs against cancer cells and overcoming multi-drug resistance. Furthermore, the chapter concisely summarizes the recent studies exploring the applications of NEs in the therapy of different types of cancers.

REFERENCES

Adhikari, M., Kaushik, N., Ghimire, B., Adhikari, B., Baboota, S., Al-Khedhairy, A. A., Wahab, R., Lee, S. J., Kaushik, N. K., & Choi, E. H. (2019). Cold atmospheric plasma and silymarin nanoemulsion synergistically inhibits human melanoma tumorigenesis via targeting HGF/c-MET downstream pathway. *Cell communication and signaling. CCS, 17*(1), 52. doi:10.118612964-019-0360-4

Ahmad, M. Sahabjada, -., Akhtar, J., Hussain, A., Badaruddeen, -., Arshad, M., & Mishra, A. (2017). Development of a new rutin nanoemulsion and its application on prostate carcinoma PC3 cell line. *EXCLI Journal, 16*, 810–823. doi:10.17179/excli2016-668

Ahmad, Z. A., Yeap, S. K., Ali, A. M., Ho, W. Y., Alitheen, N. B., & Hamid, M. (2012). scFv antibody: Principles and clinical application. Clinical & Developmental Immunology, 2012, 980250 doi:10.1155/2012/980250

Al-Otaibi, W. A., Alkhatib, M. H., & Wali, A. N. (2018). Cytotoxicity and apoptosis enhancement in breast and cervical cancer cells upon coadministration of mitomycin C and essential oils in nanoemulsion formulations. *Biomedicine & pharmacotherapy = Biomedecine & pharmacotherapie, 106*, 946–955. doi:10.1016/j.biopha.2018.07.041

Alkhatib, M. H., Aljadani, M. A., & Mahassni, S. H. (2020). Carrying epirubicin on nanoemulsion containing algae and cinnamon oils augments its apoptotic and anti-invasion effects on human colon cancer cells. *American Journal of Translational Research, 12*(6), 2463–2472.

Alkhatib, M. H., Bawadud, R. S., & Gashlan, H. M. (2020). Incorporation of docetaxel and thymoquinone in borage nanoemulsion potentiates their antineoplastic activity in breast cancer cells. *Scientific Reports, 10*(1), 18124. https://doi.org/10.1038/s41598-020-75017-5

Allen, T. M., & Martin, F. J. (2004). Advantages of liposomal delivery systems for anthracyclines. [doi:10.1053/j.seminoncol.2004.08.001]. *Seminars in Oncology, 31*(6, Suppl. 13), 5–15.

Asmawi, A. A., Salim, N., Ngan, C. L., Ahmad, H., Abdulmalek, E., Masarudin, M. J., & Abdul Rahman, M. B. (2019). Excipient selection and aerodynamic characterization of nebulized lipid-based nanoemulsion loaded with docetaxel for lung cancer treatment. *Drug Delivery and Translational Research, 9*(2), 543–554. https://doi.org/10.1007/s13346-018-0526-4

Bae, P. K., Jung, J., Lim, S. J., Kim, D., Kim, S. K., & Chung, B. H. (2013). Bimodal perfluorocarbon nanoemulsions for nasopharyngeal carcinoma targeting. *Molecular Imaging & Biology*, *15*(4), 401–410.

Balch, C. M., Gershenwald, J. E., Soong, S. J., Thompson, J. F., Atkins, M. B., Byrd, D. R., Buzaid, A. C., Cochran, A. J., Coit, D. G., Ding, S., Eggermont, A. M., Flaherty, K. T., Gimotty, P. A., Kirkwood, J. M., McMasters, K. M., Mihm, M. C. Jr, Morton, D. L., Ross, M. I., Sober, A. J., & Sondak, V. K. (2009). Final version of 2009 AJCC melanoma staging and classification. *Journal of Clinical Oncology*, *27*(36), 6199–6206. https://doi.org/10.1200/JCO.2009.23.4799

Bray, F., Ferlay, J., Soerjomataram, I., Siegel, R. L., Torre, L. A., & Jemal, A. (2018). Global cancer statistics 2018: GLOBOCAN estimates of incidence and mortality worldwide for 36 cancers in 185 countries. *CA: a Cancer Journal for Clinicians*, *68*(6), 394–424. https://doi.org/10.3322/caac.21492

Caron, W. P., Lay, J. C., Fong, A. M., La-Beck, N. M., Kumar, P., Newman, S. E., ... Zamboni, W. C. (2013). Translational studies of phenotypic probes for the mononuclear phagocyte system and liposomal pharmacology. *The Journal of Pharmacology and Experimental Therapeutics*, *347*(3), 599–606.

Chang, H. B., & Chen, B. H. (2015). Inhibition of lung cancer cells A549 and H460 by curcuminoid extracts and nanoemulsions prepared from Curcuma longa Linnaeus. *International Journal of Nanomedicine*, *10*, 5059–5080. https://doi.org/10.2147/IJN.S87225

Chen, B. H., Hsieh, C. H., Tsai, S. Y., Wang, C. Y., & Wang, C. C. (2020a). Anticancer effects of epigallocatechin-3-gallate nanoemulsion on lung cancer cells through the activation of AMP-activated protein kinase signaling pathway. *Scientific Reports*, *10*(1), 5163. https://doi.org/10.1038/s41598-020-62136-2

Chen, T. Y., Tsai, M. J., Chang, L. C., & Wu, P. C. (2020b). Co-Delivery of Cisplatin and Gemcitabine via Viscous Nanoemulsion for Potential Synergistic Intravesical Chemotherapy. *Pharmaceutics*, *12*(10), 949. https://doi.org/10.3390/pharmaceutics12100949

Clares, B., Calpena, A. C., Parra, A., Abrego, G., Alvarado, H., Fangueiro, J. F., & Souto, E. B. (2014). Nanoemulsions (NEs), liposomes (LPs) and solid lipid nanoparticles (SLNs) for retinyl palmitate: Effect on skin permeation. *International Journal of Pharmaceutics*, *473*(1–2), 591–598.

Copetti, P. M., Gündel, S., de Oliveira, P., Favarin, F. R., Ramos, A. P., Pintos, F. G., Pappis, L., Gündel, A., Machado, A. K., Ourique, A. F., & Sagrillo, M. R. (2020). Development, characterisation, stability study and antileukemic evaluation of nanoemulsions containing *Astrocaryum aculeatum* extract. *Natural product research*, 1–6. Advance online publication. https://doi.org/10.1080/14786419.2020.1862830

Crown, J., O'Leary, M., & Ooi, W. S. (2004). Docetaxel and paclitaxel in the treatment of breast cancer: A review of clinical experience. *The Oncologist*, *9*(Suppl. 2), 24–32.

ClinicalTrials.gov. (2021). *Curcumin in Reducing Joint Pain in Breast Cancer Survivors With Aromatase Inhibitor-Induced Joint Disease.* Retrieved from https://clinicaltrials.gov/ct2/show/NCT03865992?term=Nanoemulsion&cond=Cancer&draw=2&rank=2

Danhier, F., Feron, O., & Préat, V. (2010). To exploit the tumor microenvironment: Passive and active tumor targeting of nanocarriers for anti-cancer drug delivery. *Journal of Controlled Release*, *148*(2), 135–146.

de Matos, R., Calmon, M. F., Amantino, C. F., Villa, L. L., Primo, F. L., Tedesco, A. C., & Rahal, P. (2018). Effect of Curcumin-Nanoemulsion Associated with Photodynamic Therapy in Cervical Carcinoma Cell Lines. *BioMed research international, 2018*, 4057959. https://doi.org/10.1155/2018/4057959

de Menezes Furtado, C., de Faria, F. S., Azevedo, R. B., Py-Daniel, K., Dos Santos Camara, A. L., da Silva, J. R., de Holanda Oliveira, E., Rodriguez, A. F., & Degterev, I. A. (2017). Tectona grandis leaf extract, free and associated with nanoemulsions, as a possible photosensitizer of mouse melanoma B16 cell. *Journal of Photochemistry and Photobiology. B, Biology*, *167*, 242–248. https://doi.org/10.1016/j.jphotobiol.2017.01.004

Dean, M., & ABC. (2009). ABC Transporters, drug resistance, and cancer stem cells. Mammary Gland Boil. *Journal of Mammary Gland Biology and Neoplasia, 14*(1), 3–9. doi:10.1007/s10911-009-9109-9

Dehelean, C. A., Feflea, S., Gheorgheosu, D., Ganta, S., Cimpean, A. M., Muntean, D., & Amiji, M. M. (2013). Anti-angiogenic and anti-cancer evaluation of betulin nanoemulsion in chicken chorioallantoic membrane and skin carcinoma in Balb/c mice. *Journal of Biomedical Nanotechnology*, *9*(4), 577–589.

Favero, G. M., Paz, J. L., Otake, A. H., Maria, D. A., Caldini, E. G., de Medeiros, R., Deus, D. F., Chammas, R., Maranhão, R. C., & Bydlowski, S. P. (2018). Cell internalization of 7-ketocholesterol-containing nanoemulsion through LDL receptor reduces melanoma growth *in vitro* and *in vivo*: A preliminary report. *Oncotarget*, *9*(18), 14160–14174. https://doi.org/10.18632/oncotarget.24389

Fazelifar, P., Tabrizi, M. H., & Rafiee, A. (2020). The *Arachis hypogaea* Essential Oil Nanoemulsion as an Efficient Safe Apoptosis Inducer in Human Lung Cancer Cells (A549). *Nutrition and cancer*, 1–9. Advance online publication. https://doi.org/10.1080/01635581.2020.1783330

Feng, Z., Wang, Z., Yang, Y., Du, Y., Cui, S., Zhang, Y., Tong, Y., Song, Z., Zeng, H., Zou, Q., Peng, L., & Sun, H. (2018). Development of a safety and efficacy nanoemulsion delivery system encapsulated gambogic acid for acute myeloid leukemia in vitro and in vivo. *European journal of pharmaceutical sciences: official journal of the European Federation for Pharmaceutical Sciences, 125*, 172–180. https://doi.org/10.1016/j.ejps.2018.10.001

Fidler, I. J., & Kripke, M. L. (2015). The challenge of targeting metastasis. *Cancer and Metastasis Reviews*, *34*(4), 635–641. https://doi.org/10.1007/s10555-015-9586-9

Floriano, B. F., Carvalho, T., Lopes, T. Z., Takahashi, L., Rahal, P., Tedesco, A. C., & Calmon, M. F. (2021). Effect of berberine nanoemulsion associated with Photodynamic therapy in cervical carcinoma cell line. *Photodiagnosis and photodynamic therapy*, 102174. Advance online publication. https://doi.org/10.1016/j.pdpdt.2020.102174

Folkman, J. (1971). Tumor angiogenesis: Therapeutic implications. *The New England Journal of Medicine*, *285*(21), 1182–1186.

Gabizon, A., Shmeeda, H., & Barenholz, Y. (2003). Pharmacokinetics of pegylated liposomal doxorubicin: Review of animal and human studies. *Clinical Pharmacokinetics*, *42*(5), 419–436.

Ganta, S., Singh, A., Patel, N. R., Cacaccio, J., Rawal, Y. H., Davis, B. J., Amiji, M. M., & Coleman, T. P. (2014a). Development of EGFR-targeted nanoemulsion for imaging and novel platinum therapy of ovarian cancer. *Pharmaceutical Research*, *31*(9), 2490–2502. https://doi.org/10.1007/s11095-014-1345-z

Ganta, S., Singh, A., Rawal, Y., Cacaccio, J., Patel, N., Kulkarni, P., & ... . (2014b). Formulation development of a novel targeted theranostic nanoemulsion of docetaxel to overcome multidrug resistance in ovarian cancer. *Drug Delivery*, *23*(3), 958–970. doi:10.3109/10717544.2014.923068

Ganta, S., Talekar, M., Singh, A., Coleman, T. P., & Amiji, M. M. (2014c). Nanoemulsions in translational research—Opportunities and challenges in targeted cancer therapy. *AAPS PharmSciTech*, *15*(3), 694–708.

Gao, Z., Kennedy, A. M., Christensen, D. A., & Rapoport, N. Y. (2008). Drugloaded Nano/microbubbles for combining ultrasonography and targeted chemotherapy. *Ultrasonics*, *48*(4), 260–270.

Gianella, A., Jarzyna, P., Mani, V., Ramachandran, S., Calcagno, C., Tang, J., & … . (2011). Multifunctional Nanoemulsion Platform for Imaging Guided Therapy Evaluated in Experimental Cancer. *ACS Nano*, *5*(6), 4422–4433. doi:10.1021/nn103336a

Groo, A. C., Hedir, S., Since, M., Brotin, E., Weiswald, L. B., Paysant, H., Nee, G., Coolzaet, M., Goux, D., Delépée, R., Freret, T., Poulain, L., Voisin-Chiret, A. S., & Malzert-Fréon, A. (2020). Pyridoclax-loaded nanoemulsion for enhanced anticancer effect on ovarian cancer. *International Journal of Pharmaceutics*, *587*, 119655. https://doi.org/10.1016/j.ijpharm.2020.119655

Guan, Y. B., Zhou, S. Y., Zhang, Y. Q., Wang, J. L., Tian, Y. D., Jia, Y. Y., & Sun, Y. J. (2017). Therapeutic effects of curcumin nanoemulsions on prostate cancer. *Journal of Huazhong University of Science and Technology. Medical Sciences*, *37*(3), 371–378. https://doi.org/10.1007/s11596-017-1742-8

Guo, P., Pi, C., Zhao, S., Fu, S., Yang, H., Zheng, X., Zhang, X., Zhao, L., & Wei, Y. (2020). Oral co-delivery nanoemulsion of 5-fluorouracil and curcumin for synergistic effects against liver cancer. *Expert Opinion on Drug Delivery*, *17*(10), 1473–1484. https://doi.org/10.1080/17425247.2020.1796629

Hörmann, K., & Zimmer, A. (2016). Drug delivery and drug targeting with parenteral lipid nanoemulsions—A review. *Journal of Controlled Release*, *223*, 85–98.

Huang, R. F., Wei, Y. J., Inbaraj, B. S., & Chen, B. H. (2015). Inhibition of colon cancer cell growth by nanoemulsion carrying gold nanoparticles and lycopene. *International Journal of Nanomedicine*, *10*, 2823–2846. https://doi.org/10.2147/IJN.S79107

Huynh, N. T., Passirani, C., Saulnier, P., & Benoit, J. P. (2009). Lipid nanocapsules: A new platform for nanomedicine. *International Journal of Pharmaceutics*, *379*(2), 201–209.

Jemal, A., Bray, F., Center, M. M., Ferlay, J., Ward, E., & Forman, D. (2011). Global cancer statistics. *CA: a Cancer Journal for Clinicians*, *61*(2), 69–90. https://doi.org/10.3322/caac.20107

Kamba, T., & McDonald, D. M. (2007). Mechanisms of adverse effects of anti-VEGF therapy for cancer. *British Journal of Cancer*, *96*(12), 1788–1795.

Kaneda, M. M., Caruthers, S., Lanza, G. M., & Wickline, S. A. (2009). Perfluorocarbon nanoemulsions for quantitative molecular imaging and targeted therapeutics. *Annals of Biomedical Engineering*, *37*(10), 1922–1933.

Kaur, P., Garg, T., Rath, G., Murthy, R. S., & Goyal, A. K. (2016). Surfactant-based drug delivery systems for treating drug-resistant lung cancer. *Drug Delivery*, *23*(3), 727–738. https://doi.org/10.3109/10717544.2014.935530

Keykhasalar, R., Tabrizi, M. H., Ardalan, P., & Khatamian, N. (2020). The Apoptotic, Cytotoxic, and Anti-angiogenic Impact of *Linum usitatissimum* Seed Essential Oil Nanoemulsions on the Human Ovarian Cancer Cell Line A2780. *Nutrition and cancer*, 1–9. Advance online publication. https://doi.org/10.1080/01635581.2020.1824001

Kim, J. E., & Park, Y. J. (2017). Improved Antitumor Efficacy of Hyaluronic Acid-Complexed Paclitaxel Nanoemulsions in Treating Non-Small Cell Lung Cancer. *Biomolecules & Therapeutics*, *25*(4), 411–416. https://doi.org/10.4062/biomolther.2016.261

Knutson, K. L., Krco, C. J., Erskine, C. L., Goodman, K., Kelemen, L. E., Wettstein, P. J., Low, P. S., Hartmann, L. C., & Kalli, K. R. (2006). T-cell immunity to the folate receptor alpha is prevalent in women with breast or ovarian cancer. *Journal of Clinical Oncology, 24*(26), 4254–4261. https://doi.org/10.1200/JCO.2006.05.9311

Konrad, C. V., Murali, R., Varghese, B. A., & Nair, R. (2017). The role of cancer stem cells in tumor heterogeneity and resistance to therapy. *Canadian Journal of Physiology and Pharmacology, 95*(1), 1–15. https://doi.org/10.1139/cjpp-2016-0079

Kumar, G. P., & Divya, A. (2015). Nanoemulsion based targeting in cancer therapeutics. *Medicinal Chemistry (Shariqah, United Arab Emirates), 5*(5), 272–284.

Lee, W. D., Liang, Y. J., & Chen, B. H. (2016). Effects of tanshinone nanoemulsion and extract on inhibition of lung cancer cells A549. *Nanotechnology, 27*(49), 495101. https://doi.org/10.1088/0957-4484/27/49/495101

Lee, Y. H., & Lin, Y. C. (2018). Anti-EGFR Indocyanine Green-Mitomycin C-Loaded Perfluorocarbon Double Nanoemulsion: A Novel Nanostructure for Targeted Photochemotherapy of Bladder Cancer Cells. *Nanomaterials (Basel, Switzerland), 8*(5), 283. https://doi.org/10.3390/nano8050283

Liu, Y., Yu, X. M., Sun, R. J., & Pan, X. L. (2017). Folate-Functionalized Lipid Nanoemulsion to Deliver Chemo-Radiotherapeutics Together for the Effective Treatment of Nasopharyngeal Carcinoma. *AAPS PharmSciTech, 18*(4), 1374–1381. https://doi.org/10.1208/s12249-016-0595-y

Ma, Y., Liu, D., Wang, D., Wang, Y., Fu, Q., Fallon, J., & .... (2014). Combinational Delivery of Hydrophobic and Hydrophilic Anticancer Drugs in Single Nanoemulsions To Treat MDR in Cancer. *Molecular Pharmaceutics, 11*(8), 2623–2630. doi:10.1021/mp400778r

Maeda, H., Wu, J., Sawa, T., Matsumura, Y., & Hori, K. (2000). Tumor vascular permeability and the EPR effect in macromolecular therapeutics: A review. *Journal of Controlled Release, 65*(1–2), 271–284.

Maragheh, A. D., Tabrizi, M. H., Karimi, E., Seyedi, S., & Khatamian, N. (2019). Producing the sour cherry pit oil nanoemulsion and evaluation of its anti-cancer effects on both breast cancer murine model and MCF-7 cell line. *Journal of Microencapsulation, 36*(4), 399–409. https://doi.org/10.1080/02652048.2019.1638460

Marques, M. B., Machado, A. P., Santos, P. A., Carrett-Dias, M., Araújo, G. S., Alves, B., de Oliveira, B. S., da Silva Júnior, F., Dora, C. L., Cañedo, A. D., Filgueira, D., & Fernandes, E. Silva, E., & Votto, A. (2021). Anti-MDR Effects of Quercetin and its Nanoemulsion in Multidrug Resistant Human Leukemia Cells. *Anticancer Agents in Medicinal Chemistry*. Advance online publication. doi:10.2174/1871520621999210104200722

McClements, D. J. (2012). Nanoemulsions versus microemulsions: Terminology, differences, and similarities. *Soft Matter, 8*(6), 1719–1729.

Md, S., Alhakamy, N. A., Aldawsari, H. M., Husain, M., Kotta, S., Abdullah, S. T. A., Fahmy, U., Alfaleh, M. A., & Asfour, H. Z. (2020). Formulation Design, Statistical Optimization, and In Vitro Evaluation of a Naringenin Nanoemulsion to Enhance Apoptotic Activity in A549 Lung Cancer Cells. *Pharmaceuticals (Basel, Switzerland), 13*(7), 152. https://doi.org/10.3390/ph13070152

Mendes Miranda, S. E., Alcântara Lemos, J., Fernandes, R. S., Silva, J. O., Ottoni, F. M., Townsend, D. M., Rubello, D., Alves, R. J., Cassali, G. D., Ferreira, L., & de Barros, A. (2021). Enhanced antitumor efficacy of lapachol-loaded nanoemulsion in breast cancer tumor model. *Biomedicine & pharmacotherapy, 133*, 110936. https://doi.org/10.1016/j.biopha.2020.110936

Meng, L., Xia, X., Yang, Y., Ye, J., Dong, W., Ma, P., & … . (2016). Co-encapsulation of paclitaxel and baicalein in nanoemulsions to overcome multidrug resistance via oxidative stress augmentation and P-glycoprotein inhibition. *International Journal of Pharmaceutics*, *513*(1-2), 8–16. doi:10.1016/j.ijpharm.2016.09.001

Milane, L., Duan, Z., & Amiji, M. (2011). Development of EGFR-targeted polymer blend nanocarriers for combination paclitaxel/lonidamine delivery to treat multi-drug resistance in human breast and ovarian tumor cells. *Molecular Pharmaceutics*, *8*(1), 185–203.

Mohammad, I. S., He, W., & Yin, L. (2018). Understanding of human ATP binding cassette superfamily and novel multidrug resistance modulators to overcome MDR. *Biomedicine and Pharmacotherapy*, *100*, 335–348.

Monge-Fuentes, V., Muehlmann, L. A., Longo, J. P., Silva, J. R., Fascineli, M. L., de Souza, P., Faria, F., Degterev, I. A., Rodriguez, A., Carneiro, F. P., Lucci, C. M., Escobar, P., Amorim, R. F., & Azevedo, R. B. (2017). Photodynamic therapy mediated by acai oil (Euterpe oleracea Martius) in nanoemulsion: A potential treatment for melanoma. *Journal of Photochemistry and Photobiology. B, Biology*, *166*, 301–310. https://doi.org/10.1016/j.jphotobiol.2016.12.002

Mulik, R. S. Mönkkönen, J., Juvonen, R.O., Mahadik, K. R., Paradkar, A. R. (2010). Transferrin mediated solid lipid nanoparticles containing curcumin: enhanced in vitro anticancer activity by induction of apoptosis. *International Journal of Pharmaceutics, 398(1),* 190–203. doi:10.1016/j.ijpharm.2010.07.021

Neubi, G. M. N., Opoku-Damoah, Y., Gu, X., Han, Y., Zhou, J., Ding, Y., & Damoah-Opoku, Y. (2018). Bio-inspired drug delivery systems: An emerging platform for targeted cancer therapy. [doi:10.1039/C8BM00175H]. *Biomaterials Science*, *6*(5), 958–973.

Ohguchi, Y., Kawano, K., Hattori, Y., & Maitani, Y. (2008). Selective delivery of folate–PEG-linked, nanoemulsion-loaded aclacinomycin A to KB nasopharyngeal cells and xenograft: Effect of chain length and amount of folate–PEG linker. *Journal of Drug Targeting*, *16*(9), 660–667.

Photodynamic Therapy for Lentigo Maligna Using 5-aminolevulinic Acid Nanoemulsion as a Light Sensitizing Cream - Full Text View - ClinicalTrials.gov. (2021). Retrieved 14 March 2021, from https://clinicaltrials.gov/ct2/show/NCT02685592?term=Nanoemulsion&cond=Cancer&draw=2&rank=6

Pilot Study of Curcumin for Women With Obesity and High Risk for Breast Cancer - Full Text View - ClinicalTrials. gov. (2021). Retrieved 14 March 2021, from https://clinicaltrials.gov/ct2/show/NCT01975363?term=Nanoemulsion&cond=Cancer&draw=2&rank=3

Rapoport, N. Y., Kennedy, A. M., Shea, J. E., Scaife, C. L., & Nam, K. H. (2009). Controlled and targeted tumor chemotherapy by ultrasound activated nanoemulsions/microbubbles. [doi:10.1016/j.jconrel.2009.05.026]. *Journal of Controlled Release*, *138*(3), 268–276.

Sánchez-López, E., Guerra, M., Dias-Ferreira, J., Lopez-Machado, A., Ettcheto, M., Cano, A., Espina, M., Camins, A., Garcia, M. L., & Souto, E. B. (2019). Current Applications of Nanoemulsions in Cancer Therapeutics. *Nanomaterials (Basel, Switzerland)*, *9*(6), 821. https://doi.org/10.3390/nano9060821

Senapati, S., Mahanta, A. K., Kumar, S., & Maiti, P. (2018). Controlled drug delivery vehicles for cancer treatment and their performance. *Signal Transduction and Targeted Therapy, 3*, 7. doi:10.1038/s41392-017-0004-3

Siegel, R. L., Miller, K. D., Goding Sauer, A., Fedewa, S. A., Butterly, L. F., Anderson, J. C., Cercek, A., Smith, R. A., & Jemal, A. (2020). Colorectal cancer statistics, 2020. *CA: a Cancer Journal for Clinicians, 70*(3), 145–164. https://doi.org/10.3322/caac.21601

Souto, E. B., Nayak, A. P., & Murthy, R. S. (2011). Lipid nanoemulsions for anti-cancer drug therapy. *Die Pharmazie, 66*(7), 473–478.

Sun, L., Wan, K., Hu, X., Zhang, Y., Yan, Z., Feng, J., & Zhang, J. (2016). Functional nanoemulsion-hybrid lipid nanocarriers enhance the bioavailability and anti-cancer activity of lipophilic diferuloylmethane. *Nanotechnology, 27*(8), 085102. https://doi.org/10.1088/0957-4484/27/8/085102

Tadros, T., Izquierdo, P., Esquena, J., & Solans, C. (2004). Formation and stability of nano-emulsions. *Advances in Colloid and Interface Science*, 108–109, 303–318, 303–318. doi:10.1016/j.cis.2003.10.023

Tagne, J. B., Kakumanu, S., Ortiz, D., Shea, T., & Nicolosi, R. J. (2008). A nanoemulsion formulation of tamoxifen increases its efficacy in a breast cancer cell line. *Molecular Pharmaceutics, 5*(2), 280–286.

Talekar, M., Ganta, S., Singh, A., Amiji, M., Kendall, J., Denny, W. A., & Garg, S. (2012). Phosphatidylinositol 3-kinase inhibitor (PIK75) containing surface functionalized nanoemulsion for enhanced drug delivery, cytotoxicity and pro-apoptotic activity in ovarian cancer cells. [doi:10.1007/s11095-012-0793-6]. *Pharmaceutical Research, 29*(10), 2874–2886.

Tang, J., Fu, Q., Wang, Y., Racette, K., Wang, D., & Liu, F. (2013). Vitamin E reverses multidrug resistance in vitro and in vivo. [doi:10.1016/j.canlet.2013.04.020]. *Cancer Letters, 336*(1), 149–157.

Tayeb, H., Piantavigna, S., Howard, C., Nouwens, A., Mahler, S., Middelberg, A., & … . (2017). Insights into the interfacial structure–function of poly(ethylene glycol)-decorated peptide-stabilised nanoscale emulsions. *Soft Matter, 13*(43), 7953–7961. doi:10.1039/c7sm01614j

Tiwari, S., Tan, Y.-M., & Amiji, M. (2006). Preparation and in vitro characterization of multifunctional nanoemulsions for simultaneous MR imaging and targeted drug delivery. *Journal of Biomedical Nanotechnology, 2*(3), 217–224.

Tubtimsri, S., Limmatvapirat, C., Limsirichaikul, S., Akkaramongkolporn, P., Inoue, Y., & Limmatvapirat, S. (2018). Fabrication and characterization of spearmint oil loaded nanoemulsions as cytotoxic agents against oral cancer cell. *Asian journal of pharmaceutical sciences, 13*(5), 425–437. https://doi.org/10.1016/j.ajps.2018.02.003

Use of Lipid Emulsion or Nanoemulsion of Propofol on Children Undergoing Ambulatory Invasive Procedures. - Full Text View - ClinicalTrials.gov. (2021). Retrieved from https://clinicaltrials.gov/ct2/show/NCT01326078?term=Nanoemulsion&cond=Tumor&draw=2&rank=4

Verma, P., Meher, J. G., Asthana, S., Pawar, V. K., Chaurasia, M., & Chourasia, M. K. (2014). Perspectives of nanoemulsion assisted oral delivery of docetaxel for improved chemotherapy of cancer. *Drug Delivery, 23*, 1–10.

Wang, X., Wang, Y.-W., & Huang, Q. (2009). Enhancing stability and oral bioavailability of polyphenols using nanoemulsions. Micro/Nanoencapsulation Act Food Ingredients: Am Chem. *Società*, 198–212.

Wang, Z., & Zhou, S. (2021). Effect of Compound Propofol Nanoemulsion on Immune Function in Patients with Pancreatic Cancer. *Journal of Nanoscience and Nanotechnology, 21*(2), 1390–1396. https://doi.org/10.1166/jnn.2021.18661

Winter, P. M., Schmieder, A. H., & Caruthers, S. D., Keene et al. (2008). Minute dosages of αvβ3-targeted fumagillin nanoparticles impair Vx-2 tumor angiogenesis and development in rabbits. *The FASEB Journal*, *22*(8), 2758–2767.

Yan, D., Sherman, J. H., & Keidar, M. (2017). Cold atmospheric plasma, a novel promising anti-cancer treatment modality. *Oncotarget*, *8*(9), 15977–15995. https://doi.org/10.18632/oncotarget.13304

Yang, C. C., Hung, C. F., & Chen, B. H. (2017). Preparation of coffee oil-algae oil-based nanoemulsions and the study of their inhibition effect on UVA-induced skin damage in mice and melanoma cell growth. *International Journal of Nanomedicine*, *12*, 6559–6580. https://doi.org/10.2147/IJN.S144705

Ye, H., He, X., & Feng, X. (2020). Developing neobavaisoflavone nanoemulsion suppresses lung cancer progression by regulating tumor microenvironment. *Biomedicine & pharmacotherapy = Biomedecine & pharmacotherapie, 129*, 110369. https://doi.org/10.1016/j.biopha.2020.110369

Yousefian Rad, E., Homayouni Tabrizi, M., Ardalan, P., Seyedi, S., Yadamani, S., Zamani-Esmati, P., & Haghani Sereshkeh, N. (2020). *Citrus lemon* essential oil nanoemulsion (CLEO-NE), a safe cell-depended apoptosis inducer in human A549 lung cancer cells with anti-angiogenic activity. *Journal of Microencapsulation*, *37*(5), 394–402. https://doi.org/10.1080/02652048.2020.1767223

Yuan, R., Huang, Y., Chan, L., He, D., & Chen, T. (2020). Engineering EHD1-Targeted Natural Borneol Nanoemulsion Potentiates Therapeutic Efficacy of Gefitinib against Nonsmall Lung Cancer. *ACS Applied Materials & Interfaces*, *12*(41), 45714–45727. https://doi.org/10.1021/acsami.0c08069

Zeng, B. J., Chuan, Y. P., O'Sullivan, B., Caminschi, I., Lahoud, M. H., Thomas, R., & Middelberg, A. P. (2013). Receptor-specific delivery of protein antigen to dendritic cells by a nanoemulsion formed using top-down non covalent click self-assembly, *Small*, *9*(22), 3736–3742.

Zheng, N., Gao, Y., Ji, H., Wu, L., Qi, X., Liu, X., & Tang, J. (2016). Vitamin E derivative-based multifunctional nanoemulsions for overcoming multidrug resistance in cancer. *Journal of Drug Targeting*, *24*(7), 663–669. https://doi.org/10.3109/1061186X.2015.1135335

Zhou, C. K., Check, D. P., Lortet-Tieulent, J., Laversanne, M., Jemal, A., Ferlay, J., Bray, F., Cook, M. B., & Devesa, S. S. (2016). Prostate cancer incidence in 43 populations worldwide: An analysis of time trends overall and by age group. *International Journal of Cancer*, *138*(6), 1388–1400. https://doi.org/10.1002/ijc.29894

Zhu, L., Gibson, P., Currle, D. S., Tong, Y., Richardson, R. J., Bayazitov, I. T., Poppleton, H., Zakharenko, S., Ellison, D. W., & Gilbertson, R. J. (2009). Prominin 1 marks intestinal stem cells that are susceptible to neoplastic transformation. *Nature*, *457*(7229), 603–607. https://doi.org/10.1038/nature07589

Chapter 12
Nanoemulsion Applications:
Application of Nanoemulsion in Prostate Cancer Treatment

Ayodeji Ojo Oteyola
Federal University of Minas Gerais, Belo-Horizonte, Brazil

Raffaele Pilla
St. John of God – Fatebenefratelli Hospital, Benevento, Italy

Folasade Adesola Ola-Oladimeji
https://orcid.org/0000-0002-6422-9213
Ekiti State University, Ado-Ekiti, Nigeria

Omotayo Fagbuaro
Ekiti State University, Ado-Ekiti, Nigeria

ABSTRACT

Prostate cancer (PCa) is a global issue with increasing rise in morbidity and mortality. PCa treatment has been a global challenge for many years because drugs designed to combat this disease might show low efficacy as a result of low solubility. Limitations of chemo-drugs for treating PCa give birth to the use of nanomedicine which helps to improve drug delivery systems. Nanoemulsions are particles that are pharmaceutically formulated and comprised within the range of a nanometer (10-200nm). Nanoemulsions are thermodynamically stable and made up of safe gradient agents. This chapter elucidates the physiological, biological, and molecular barriers affecting drug delivery in PCa. The authors discussed the importance of nanoemulsions as potential drug delivery mechanisms in PCa therapy. This chapter focuses on reviewing different ways by which nanoemulsion can bring solution to water-solubility problems and also target specific cancer cells. Limitations of nanoemulsions in the drug delivery field were also highlighted.

DOI: 10.4018/978-1-7998-8378-4.ch012

INTRODUCTION

Prostate is an accessory gland associated with the male sexual system in mammals (Roehrborn, 2006; Luke and Coffey, 1994). The prostate helps in the production of fluid that helps in sperm functioning (Luke and Coffey, 1994). In humans, it surrounds men urethra; it is differentiated into different glandular regions such as the peripheral zone and the transition zone. Peripheral zone encircles the proximal prostate urethra where 70-85% of adenocarcinoma occurs (Noguchi et al., 2000) while the transition zone surrounds the distal prostate urethra and benign hyperplasia lesions occurs in this area. In a normal condition, the prostate grows and mature during puberty as the androgen levels increases. Its growth is primarily regulated by testosterone (Prins and Lindgren, 2014).

Globally, Prostate cancer (PCa) is the most common non-cutaneous cancer with an estimation of one million, six hundred thousand cases and three hundred and sixty-six thousand deaths in a year (Torre et al., 2015). PCa is characterized by aberrant dividing cells in the prostate gland which leads to abnormal growth of the prostate gland. Prostate malignant transformation follows a multiple process, starting from prostatic intraepithelial neoplasia (PIN) to localized prostate cancer and followed by the advanced prostate adenocarcinoma in which with the primary invasion leads to metastatic prostate cancer. Following histological patterns of prostate cancer, the Gleason grading system (Gleason and Mellinger, 1974) has been used to identify prostate cancer stages and it has been refined over the years (Epstein et al., 2014; 2016). Death from PCa normally occurs as a result of metastasis when the cancer cells spread to other part of the body. Regions in which the cancer cell can spread include the pelvic, lymph nodes, bladder, rectum, brain and bone.

Presently, there are no preventive measures for prostate cancer, however, nutrition and lifestyle changes have been suggested as ways of prevention but research studies are still ongoing and scientific recommendations are not available at the moment. Pomegranate, green tea, broccoli, crocetin, flaxseed, soy extracts, vegan diets and physical exercise have been proposed to lower the risk of PCa development. Majority of the studies associated with prevention and inhibition of PCa have been done in animal models but it is challenging how these results can be interpreted and applied in human PCa (Schatten, 2018).

Hormone responsiveness is an important characteristic of prostate cancer (Huggins and Hodges, 1941). Huggins and Hodges reported that castration led to decrease in tumor in patient with prostate cancer (Huggins and Hodges, 1941). Androgen deprivation therapy (ADT) which involves the use of agents that obstruct the androgen pathway is used in the treatment of prostate cancer. However, there can be resistant to ADT which can result in primary castration-resistant prostate cancer (CRPC) or the metastatic type (mCRPC) (Zou et al., 2017). The use of drugs (Pharmacotherapy) in PCa depends on the stage of the disease. In one perspective, early detection and use of drugs is encouraged (Armstrong and Gao, 2015; Malarvizhi et al., 2014). In another perspective, drugs are used after surgical removal and radiation (Maeda, 2012). The use of drugs in treating prostate cancer is challenging as a result of little vascularization as the disease progresses (Maeda, 2012). Some of the chemo drugs (chemotherapy) for treating prostate cancer include docetaxel (taxotere), cabazitaxel (jevtana), mitoxantrone (novantrone) and estramustine (emcyt). These drugs are associated with limited efficacy and the associated risk or side effects on patients outweigh the efficacy (Armstrong and Gao, 2015). Armstrong and Gao (2015) reported the association between the metabolism of cancer tissue and drug resistance, this causes a challenge in the effectiveness in the use of drugs. In order to solve the problem of drug resistance, there is need to subsequently increase the dose of the drug which in turn leads to systemic accumulation of the drug and toxicity Therefore, there is an urgent need to improve drug therapy and drug delivery system by increasing specificity with less systemic toxicity.

Targeted drug delivery using controlled released (CR) – nanocarriers are gaining attention in improving drug tolerance (Torchilin, 2000; Danhier et al., 2010). CR-Nanocarriers together with chemotherapy help in the reduction of loading dose and systemic toxicity (Ho et al., 2015; Tsukihara et al., 2015). The development of nanocarriers with improved therapeutic efficacy and safety leading to reduction in tumor burden is the newly identified clinical need. Therefore, in this chapter, we focus on the design of nanocarriers using nanoemulsion for targeted delivery to prostate cancer.

Physiological, Biological and Molecular barriers affecting drug delivery in prostate cancer

It is very important to examine blood fluid and urine flow to have an understanding of how drug delivery is affected by the prostate transport process. Normal prostate has a blood flow of 0.21 mL/g/min; there is an increase in the blood flow in benign prostate hyperplasia (BPH) (0.28 mL/g/min). However, the blood flow in PCa is much higher (0.64 mL/g/min) (Vaupel and Kelleher, 2013). Normal prostate has a mean oxygen pressure of 26 mmHg, there is reduction in the mean oxygen pressure in prostate cancer (6 mmHg) (Inaba, 1992). The increase in the prostate blood volume follows the order: PCa (6.1%) < BPH (8.9%) < normal prostate (8.1%) (Inaba, 1992). In spite of the high blood flow in prostate cancer, the tumor normally reveals hypoxia, a condition in which the tissue is denied of adequate oxygen supply (Vaupel and Kelleher, 2013). Normal prostate is more vascularized than PCa and this lower vascularization in PCa affects the nutrients transportation (Maeda, 2012). Considering the urine flow, the advancement of prostate diseases significantly affects it, because the prostate surrounds the urethra. In physiological conditions, the prostate volume increases while the flow of urine decreases with age and disease (Trumbeckas et al., 2011). The differences in blood fluid and urine flow reveals changes in mass transport, especially nutrients and drugs, and this is greatly altered as the disease progresses as a result of prostate tissue remodeling (Trumbeckas et al., 2011). Nacey and colleagues (1995) suggested that prostate tissue remodeling in BPH can be suppressed by the use of specific drugs but this becomes more complicated during prostate cancer as a result of abnormal tissue remodeling.

In the prostate tumor microenvironment, hormone composition, enzymes and cytokines increase and this intensifies the metabolism of prostate cancer (Higgins et al., 2009). Li and Bluth (2011) reported that elevation of the hormone, metabolic enzymes and cytokines at the tumor site increase the degradation of drugs which further minimizes its effectiveness at the target site. Furthermore, drug resistance mechanisms that aid the survival of tumors also exist (Wang et al., 2011; Armstrong and Gao, 2015). Sfanos and De Marzo (2012) reported that the "enabling cause" of prostate cancer is the chronic inflammation of the prostate. In the advanced state, the intra acinar T cells are greatly influenced by the cytotoxic T cells which suggest adaptive immune responses (John et al., 2000). Regardless of the immune dysregulation, prolonged use of drugs seems to be effective, but results in systemic toxicity. In order to reduce this toxicity, there is need to consider the development of nanocarriers using nanoemulsion for effective delivery to target site (prostate tumor).

Pharmacotherapeutic Protocols and Clinical Approach to Prostate Cancer

The three main approaches involved in treating PCa include radiation, surgery and pharmacotherapy (NCI, 2020). The classes of the pharmacological agents include anti-neoplastics, chemotheraupeutic modulating agents, anti-fungals and bisphosphonate. Anti-neoplastics are used during castration, it helps to reduce the production of testosterone in such a way that the testosterone level is lower than the castrate threshold (50 ng/dL) (NCI, 2020). Chemotherapy modulating agents are Sipuleucel-T, which enhance the immune system against prostatic acid phosphatase (PAP). The anti-fungals react by obstructing steroidal synthesis (e.g. ketoconazole). Bisphosphonates, which include zoledronic acid, are generally used in reducing bone metastasis effects. The hormone therapy inhibits testosterone production and activity. There are three classifications of hormone therapy, based on the methodology of the treatment; (1) androgen depletion therapy (ADT), (2) combined androgen blockade (CAB), (3) triple androgen blockade (TAB). Testosterone blockers, estrogens and LH-RH antagonists are generally used in hormone-dependent prostate cancer (Henry and O'mahony, 1999). However, anti-neoplastics which include Liarazole (van Ginckel et al., 1990), Eramustine and Vinblastine combination (Newling, 1997) are used in treating hormone refractory prostate cancer. Anti-neoplastic molecules are generally used in advanced stages of prostate cancer. Some of them are gonadotropin-releasing hormone (GnRH) agonists, which include Leuprolide, Goserelin, Histrelin and Triptorelin. Some are anti-androgenic, such as Abiraterone, Degarelix, Flutamide and Nilutamide, and others are anti-microtubular, which include Docetaxel and Cabazitaxel. Table 1 shows a list of drugs for treating PCa and their relative side effects. Hormone therapy is mostly used, but other agents are administered when the hormone therapy seems ineffective or less effective, but many of these agents have systemic side effects that sometimes outweigh their functions.

Table 1. Drugs used for the treatment of prostate cancer and its side effects

Drugs	Route and mode of action	Related side effects
Anti-neoplastics: Hormone therapy		
Abiraterone Acetate (Zytiga)	Taken orally, used in treating metastatic PCa, helps in reducing testosterone production.	Breathing difficulty, edema, hypertension.
Bicalutamide (Casodex)	Taken orally, used in treating metastatic PCa, helps to suppress cancer cells growth by shutting androgen receptor.	Diarrhea.
Cabazitaxel	Taken through injection, used in treating metastatic PCa, helps in inducing apoptosis.	Anemia.
Degarelix	Taken through subcutaneous injection, used in treating metastatic PCa by inhibiting the GnRH.	Headache, dim vision.
Enzalutamide	Taken through subcutaneous injection, used in treating metastatic PCa by inhibiting the testosterone.	Fatigue, seizures, reproductive dysfunctions.
Flutamide (Drogenil)	Taken orally, used in treating metastatic PCa.	Gastrointestinal side effects.
Goserelin Acetate	Taken through subcutaneous injection, used in treating PCa.	Diarrhea.
Leuprolide Acetate (Lupron)	Taken through injection, used in treating PCa.	Redness, pain, memory loss, swelling of the breast.
Chemotherapy		
Docetaxel (Taxotere)	Taken through injection, used in treating metastatic PCa, helps inducing apoptosis.	Liver dysfunction, hypersensitivity reactions.
Mitoxantrone Hydrochloride (Novatrone)	Taken through intravenous infusion, used in treating PCa, is a type II topoisomerase inhibitor, helps in inhibiting DNA synthesis and RNA repair in cancer cell.	Heart muscle damage.
Immunotherapy		
Sipuleucel-T (Provenge)	Taken through intravenous injection, used in preventing and treating prostate cancer.	Anemia, hair loss, diarrhea, fertility issues.

Source: (Sasikumar and Kamalasanan, 2017)

After radiation or elimination of tumor through surgery (Maeda, 2012) and also in castration resistant patients (Henry and O'mahony, 1999), anti-neoplastic drugs are encouraged. Tumor growth progression determines the length of pharmacotherapy application; pharmacological agents can be used as the main therapy or in combination with other chemotherapeutic agents. Nevertheless, this combination approach is relatively linked to high systemic toxicity (Kummar et al., 2006). Considering the various side effects associated with the mentioned therapy, it is important to develop effective drug delivery systems for efficient treatment of prostate cancer.

Consideration of Nanomedicine to Overcome the Barriers in Prostate Cancer Therapy

Li and Bluth (2011) reported that the change in fluids, drug resistance and the tumor microenvironment causes alterations in the drug delivery parameters in PCa. Progress in nanotechnology and nanocarriers which includes nanoparticles, liposomes, nanocapsules and nanoemulsions in the treatment of tumor, is an encouraging approach because it improves drug delivery (Zhang et al., 2019). Chen and colleagues (2018) reported that nanocarriers based drug-delivery is crucial in the production of nanomedicine, which helps increasing the time of circulation, improves targeted delivery and lowers the systemic toxicity in pharmacotherapy. Nanoemulsion also helps in preventing drug from degradation (McCarron and Faheem, 2010). Bouchaala and group (2016) reported that nanoemulsions can be targeted to the tumor site and therefore releasing their contents. The first production of nanoemulsions for treating PCa has a reservoir and a targeting ligand that help in the delivery systems (Desai et al., 2008; Almeida et al., 2010). The main importance of nanomedicine is the ability to develop therapeutic molecules in the absence of dose-limiting toxic excipients that are present in marketed formulations (Hare et al., 2017). However, challenges with drug delivery mechanisms, and notably to the prostates, are anatomical barriers, physiological barriers, cellular and molecular level barriers; hence, there is a need to develop a nanocarrier based targeted drug delivery system. As a result of the systemic toxicity caused by prolonged chemotherapy, the new trending strategy in the development of new therapeutic is the drug repurposing (Antonarakis et al 2013). Antonarakis and colleagues (2013) defined drug repurposing as the use of established drug information for newer pharmaceutical applications. Drug repurposing sometimes involves the combination of drugs. In prostate cancer treatment, examples of approved drug repurposing are metformin and itraconazole. Metformin is an anti-diabetic drug and it has been repurposed as a treatment for PCa. Itraconazole is an anti-fungal drug that has been repurposed for the treatment of PCa (Leone et al., 2014). Considering the effectiveness of nanoemulsion, it is evident that, when these repurposed drugs are loaded into nanocarriers, they might be used to target cancer tissues in the prostate. When preparing a nanoemulsion to avoid the barrier of delivery, it is important to put into consideration, the composition of drug loading compartment (Rapoport et al., 2011), the inactivation of the targeted ligands (Ganta et al., 2014) and the stabilization of nanoemulsion interface (Ye et al., 2013).

Importance of Nanoemulsions as a Potential Drug Delivery Mechanism in Cancer Therapy

Nanoemulsions are heterogeneous colloidal systems that have two immiscible phases which are oil and water and one phase is dispersed in another which are stabilized by the addition of emulsifying agents (Gorain et al., 2014). They are translucent water in oil (w/o) or oil in water (o/w) droplets which is normally 10-200 nm. Ganta et al. (2009) reported that nanoemulsions are kinetically stable but they are thermodynamically unstable. The emulsifying agents are the surfactants and the co-surfactants and the macromolecules that are adsorbed to the interface. The oily phase immersed the hydrophobic tail while the aqueous phase immersed the hydrophilic groups. With respect to the nature of nanoemulsion, nanoemulsion delivery systems can be used in correcting the non-targeted delivery, systemic toxicity and stability challenges that are associated with the conventional chemotherapy (Choudhury et al., 2014). Choudhury et al. (2014) reported that nanoemulsion is a better delivery mechanism for chemotherapeutics compared with the conventional delivery mechanism in relation to specificity, stability and delivery to the targeted site. It has also been established that the ability to design formulation with tumor-specific ligands makes nanoemulsion delivery mechanism unique (Choudhury et al., 2014). Nanoemulsion shows a very unique delivery system because (1) it helps in improving the bioavailability of the drug as a result of its increase surface area (Khan et al., 2018), (2) it improves the stability of the drug by protecting the drug from degradation and UV light (Singh et al., 2007), (3) it protects the drug from bacteria, fungi and viruses (Hwang et al., 2013), (4) the small size of the droplets can facilitates penetration (Singh et al., 2007). In vaccine delivery, nanoemulsion are purported to play an important role because it is in the circulation for a longer time. It is interesting to also know that in nanoemulsion preparation, it is composed of GRAS (generally recognized as safe) constituents and it is approved by FDA. Therefore, formulation scientists make use of nanoemulsions (Choudhury et al., 2014; Ganta et al., 2014; Praveen-Kumar and Divya, 2015).

Nanoemulsion Composition and Formulation

The main constituents of nanoemulsions are oil, surfactants, co-surfactants and aqueous phase (McClements, 2013). In addition to the main components, other excipients such as buffers, preservatives, stabilizers, polymers and antioxidants can also be incorporated.

Oil acts as a medium for the hydrophobic drug of interest to capture the needed amount of loading by emulsification. Emulsification region (ER) generation helps in the selection of oils, and the selection is generally influenced by the chain length of the oil. Sapra et al. (2013) reported difficulties in the emulsification process as a result of small ERs of oils with long chain fatty acids. It was reported that, when reduction in the hydrophobicity of the oil occurs, this could have a positive effect on the ER but this could also have a negative effect on the solubility of the drugs inside it. Generally, the nature of solubility of different oils increases in the order; medium-chain mono- and diglycerides < medium-chain triglycerides < vegetable oils (Sapra et al., 2013). The choice of choosing the oil depends on the balance between the solubility with the drug and the emulsification. Oils that can be used for nanoemulsions include; castor oil, soybean oil, corn oil, coconut oil, olive oil, triglycerides, peanut oil and others. Nevertheless, individual oils are route specific and the compatibility of each oil to a specific route should be tested before nanoemulsion formulation for that route (Vandamme, 2002).

Surfactants are used in stabilizing the emulsion system and also maintaining the globule sizes. The basis for the selection of surfactants is that (1) it must have the potential to reduce the surface tension (< 10 dynes/cm), (2) it must also possess the ability to form a complete and cogent film so as to avoid coalescence, (3) it must have the potential to provide needed viscosity in the emulsion system so as to bring about stability (4) even at low concentration, it must be effective (Gupta et al., 2015) (5). It must have the ability to dissolve high amount of drugs, as well as be able to have high drug loading in the nanoemulsion system. In addition to the emulsification nature of a surfactant, other characteristics that a surfactant must possess are that it must be non-toxic, have an acceptable taste and be chemically stable. Surfactants used in nanoemulsion preparations must have the GRAS (generally recognized as safe) status which means they are safe for human consumption. Examples of surfactants used for human application include; lecithins, phospholipid derivatives, polysorbates, sorbitan monoesters, tyloxapol, polyethylene glycol (PEG) and poloxamers (Sapra et al., 2013).

Co-surfactants play a vital role by uniting with the surfactants to provide the needed structural flexibility to the nanoemulsion system. The use of surfactants alone results in rigidity in the surfactant film which cannot reduce the oil-water interface tension. However, when a co-surfactant is used in combination with surfactants, it enters the surfactant layer and disrupts the interfacial film and this helps to provide the needed fluidity and help the whole process of emulsification (Azeem et al., 2009). Co-surfactants also aid oil solubility. In nanoemulsion systems, examples of co-surfactants that can be used include, short-chain alcohols (ethanol, propylene glycol, glycerol, benzyl alcohol, PEG 400). Techniques such as high pressure homogenization, ultrasonication, microfluidization, phase-inversion and emulsification are means by which nanoemulsion can be prepared.

Controlled Drug Delivery System of Nanomedicine for Prostate Cancer Using Nanoemulsion

Nanotechnology has enhanced the safety and the efficacy of prostate cancer therapy through the development of delivery mechanisms like nanocarriers. When considering the development of nanomedicine based on controlled released technologies, the main step is to outline an appropriate controlled drug delivery system (CDDS). The traditional way of CDDS are classified based on reservoir type, activation based (Liu et al., 2014), feedback based and targeted drug delivery system (Wong and Choi, 2015). Considering the several ways of targeting prostate cancer, drug polymer conjugates and nanoemulsions are primarily used (Tai et al., 2011).

The suitability of developing nanoemulsion is very high because it can be prepared with adequate size range and multiple drugs can be load into it. On-going research shows that nanoemulsion can be functionalized with target ligands (Tiwari et al., 2006) in order to target prostate cancer. Bouchaala and colleagues reported that nanoemulsion shows low leakage of drugs in bloodstream. The group confirmed that nanoemulsion can be targeted to the site of the tumor and thereby releasing its content (Bouchaala et al., 2016). The initial nanoemulsion produced for targeting prostate cancer are reservoir type targeted delivery systems which has a reservoir and a targeting ligand (Almeida et al., 2010). Mura et al. (2013) reported that the present research is now to develop multifunctional nanoemulsions with activation components.

The following are indispensable in the composition of nanoemulsions (1) drug loading compartment (Liu et al., 2014), (2) released properties and reduction in drug leakage in the blood (Irby et al., 2017), (3) immobilizing targeting ligands (Ganta et al., 2014) and (4) stabilizing the nanoemulsion interface (Liu et al., 2014). Conventional drugs can be loaded into nanoemulsions. Generally recognized as safe (GRAS) grade excipients are used in the preparation of nanoemulsion. In order to increase the stability of nanoemulsion, many surfactants have been developed. Nevertheless, all the surfactants that were shown to be safe in low-risk condition such as cream are not safe for systemic applications. For instance, Cremophor EL, which is a derivative of castor oil, has been used as a channel for paclitaxel but it is found to be toxic clinically later on (Gelderblom et al., 2001). The use of Cremophor EL leads to abnormal lipoprotein pattern, hyperlipidemia, erythrocytes assemblage and delayed seizures. Therefore, consideration of clinical importance is necessary in the selection of excipients. Presently, triglycerides surfactants are being used in the development of nanoemulsion (Ganta et al., 2014). The use of long chain triglycerides bring about larger droplet size when compared to short chain triglycerides (Ganta et al., 2014). Bouchemal et al. (2004) reported that the nature of oil affects the average droplet size of nanoemulsion. The more viscous the oil is, the smaller the droplet size with long term stability.

The drug loading is realized by mixing the hydrophilic drug with water and hydrophobic drug with oil phase prior to the preparation of NE (Desai et al., 2008). The immobilization of the targeting group is achieved by coupling (Ganta et al., 2014). The use of nanoemulsion has been suggested in cancer therapy as a medium of drug targeting (Constantinides et al., 2008). Oil in water nanoemulsion is used in intravenous applications. Nanoemulsion has the potential to deliver multiple drugs loaded specifically to the targeted tumor without damaging or impairing normal cell. Anticancer drugs that are lipophilic can be loaded into oil phase of nanoemulsion while hydrophilic drugs can be loaded into the water phase of the nanoemulsion (Khatri et al., 2013). Polymeric coatings are used in stabilizing nanoemulsion interface. Polyethlene glycol (PEG) is usually used for coating the surface nanoparticles (Kohli et al., 2014). Polyoxazolines, Poly (Amino acids), N-(2-hydroxypropyl) methacrylamide (HPMA), Poly-betaines, Poly-glycerols and Poly-saccharides (Kohli et al., 2014) and proteins (Kaladhar and Sharma, 2006) are alternative to be used in coating. Conventionally, folate and transferin are examples of ligands that are generally explored for cellular targeting in cancer. Recently, antibodies, aptamers, peptide ligands, anisamide are other targeting ligands that are being explored (Duskey and Rice, 2014).

The main challenge in the development of nanoemulsion is the stability challenge, Ostwald ripening (OR). Ostwald ripening results in coalescence of the particles of nanoemulsion which in turn leads to phase separation (Gupta et al., 2016). Gupta et al. (2016) reported that the potency of phase separation is higher in w/o when compared to o/w. Weiss et al. (1999) suggested that amphiphilic nonionic polysaccharides and polyethylene glycol derivatives can be used for stabilizing the interface whereas oil with greater viscosity enhances the stability of the nanoemulsion by inhibiting the OR from approximately 3 hours to many months. The effect of the oil on the OR depends on the density of the oil. In targeting prostate cancer, coalescence must be avoided and the primary advantage of nanoemulsion over other controlled released-nanocarriers is the simple drug loading and it is industrially feasible.

A range of controlled released-nanocarriers and drug combination can broaden the boarder of nanomedicine for prostate cancer. The primary aim and focus is to enhance the efficacy of pharmacotherapy without causing toxicity to surrounding tissues.

Advanced Drug Delivery System (ADDS) of Nanomedicine for Prostate Cancer

Advanced Drug Delivery Systems (ADDS) can be developed through nanotechnology (Parhi et al., 2012) and can be used to target the different stages of prostate cancer such as node or metastasis. The focus in advanced drug delivery system is not the release at a predetermined rate but it is the target to tumor, node and cancer cells. Considering clinical advancement from a controlled drug delivery to advanced drug delivery system, many biological barriers need to be overcome. First and foremost, the time of delivery to the targeted site needs to be assessed (Li et al., 2013; Anton et al., 2016). Secondly, there is need to keep track of its course of action (Maeda, 2012). Thirdly, the effect on nonspecific targeting tissues needs to be monitored (Li et al., 2013; Anton et al., 2016).

At the moment, many of the preclinical studies are revealing reduced tumor growth and cancer cells death. However, this is the end point after lessen the burden with the help of radiation and surgery. Moreover, there is need to consider disease stabilization end points and explore it for studying the therapeutic effect of controlled released-nanocarriers in disease stabilization (Kummar et al., 2006). Nevertheless, current research is tending towards this direction.

Several factors that influence the development of ADDS include blood compatibility (Attia et al., 2014), the effect of serum interaction on nanoemulsion (Attia et al., 2014), cancer cells targetability (Goldstein et al., 2007), immune responses after administration (Abu Lila et al., 2013), nanoemulsion stability in bloodstream (Li et al., 2013), the effects of extravagation on nanoemulsion (Li et al., 2013). Nanoemulsions can be targeted to cancer cells depending on the ligand, research has shown that nanoemulsions have good stability nature in the blood and remain for a long time in circulation. The polyethylene glycol (PEG) layer aids the prolonged circulation in blood (Bouchaala et al., 2016).

In prostate cancer, it is evident that fluid dynamics and mass transport are reduced (Trumbeckas et al., 2011). Therefore, ADDS targets from the systemic circulation to the tumor microenvironment. From a drug delivery perspective, fluid dynamics reduction in vasculature aids targeting, this is as a result of the availability of the required time for the interaction of enhanced permeability and retention (EPR) and ligand-receptor (Maeda, 2012). Neovascularization is greater during the initial stages after surgical removal or radiation therapy or metastasis of prostate cancer, but lesser in larger tumors (Weidner et al., 1993). However, these are the stages where EPR targeting can be efficient. When metastasis occur, microvasculature increases near the tumor host tissues (Barth et al., 1996). Fukumura et al. (2010) reported that tumor blood vessels are generally chaotic and that leaky vasculature is common in many tumors. Mass transport limitations by nanomedicine are still challenging.

From a technological perspective, reports show that 0.7% of administered dose reach the targeted site (Wilhelm et al., 2016). The remaining reach non-targeted organ such as the kidney, liver and spleen (Li et al., 2013). Maeda (2012) reported that controlled released nanocarriers are usually achieved by passive targeting. In fact, available nanocarriers in the market today are working based on the principles of passive targeting (Bae and Park, 2011). Intracellular active targeting such as ligand-receptor interaction and pathways need to be explore (Sahay et al., 2010). Research is on-going in this regard. Nevertheless, FDA is yet to approve any product based on active targeting principles (van der Meel et al., 2017). Sahay et al. (2010) reported that cell targeting by nanoparticles is either based on charge or cell type. In endocytosis and pinocytosis pathways, nanoparticles that are lower than 100 nm are internalized by clathrin and calveolae. In some polarized epithelial cells that do not possess calveolae in the surface, nanoparticles solely depend on clathrin mediated endocytosis. In case of non-polarised cells, endocytosis is always through multiple pathways with clathrin and calveolae (Sahay et al., 2010). Cancer cells are known to be non-polarized (Harush-Frenkel et al., 2007).

Nanoparticles that are lower than 100nm are efficient through endocytosis. Sahay et al. (2010) classified the routes as Arf6-dependent, Cdc42-dependent, MHC-1 dependent. Another routes explored for intracellular targeting is the receptor mediated endocytosis and micropinocytosis (Prokop and Davidson, 2008). Many of the nanocarriers use more than a pathway for cellular entry (Sahay et al., 2010). Understanding the various pathways through which nanoparticles can be delivered and in relation to their intracellular processes could help in the advanced drug delivery systems in prostate cancer.

FUTURE RESEARCH DIRECTIONS

Conventional chemotherapy plays an important role in cancer treatment. However, limitations associated with chemotherapy are a great challenge. The use of nanoemulsion as a drug delivery system in biomedical and pharmaceutical industries can be an interesting alternative to the existing treatments. Currently, there is an approval of several cancer nanomedicines such as, liposomal Irinotecan and liposomal Vincristine, which is used for treating metastatic pancreatic cancer and acute lymphoblastic leukemia respectively (Davis and Farag, 2013). Kundranda and Niu (2015) reported that in 2005, albumin-bound paclitaxel was approved for treating breast cancer and in 2013, it was approved for treating metastatic pancreatic cancer. Clinical trials outcome of nanomedicines involving single drugs reveal mixed results. For instance, the phase II results of nanomedicines using polymeric nanoparticles for prostate targeting seems to be effective in castrate-resistant prostate cancers (CRPC) and non-small cell lung cancer (NSCLC) but not effective against cervical cancer. Studies are also on-going on the effectiveness of combination therapy of nanomedicines and it will be interesting to have more studies in this direction as this may prove to have a lasting solution to the challenge of solubility and targetability. Future studies should also address the long-term stability and safety of nanoemulsion.

CONCLUSION

Globally, Prostate cancer (PCa) is the most common non-cutaneous cancer. However, current advancement in nanotechnology has revealed the importance in prevention, treatment and early diagnosis. Nanoemulsion is found to be the suitable drug-delivery tool which helps in the improvement of drug therapy and drug delivery systems. In this chapter, the authors discuss the underlying mechanisms to improve pharmacokinetics and pharmacodynamics characteristics. Modification of droplet surface targeting some specific ligands for prostate cancer can also improve the delivery tool for intracellular active targeting, so as to minimize toxicity associated with conventional chemotherapy. Generally, in this chapter, the authors analyzed nanoemulsion as a promising nanocarrier for effective delivery of drug therapy against prostate cancer.

Today, nanoemulsions are gaining increasing attention in cancer therapy. However, considering the transition from laboratory bench to patient's bed side, many challenges need to be looked in to ensure efficiency and effectiveness. These challenges include accuracy in targeting cancer cell-specific ligands, excipients safety to avoid unfavorable side effects. Thus, these challenges need to be overcome before marketing them.

REFERENCES

Abu Lila, A. S., Kiwada, H., & Ishida, T. (2013). The accelerated blood clearance (ABC) phenomenon: Clinical challenge and approaches to manage. *Journal of Controlled Release*, *172*(1), 38–47. doi:10.1016/j.jconrel.2013.07.026 PMID:23933235

Almeida, C. P., Vital, C. G., Contente, T. C., Maria, D. A., & Maranhao, R. C. (2010). Modification of composition of a nanoemulsion with different cholesteryl ester molecular species: Effects on stability, peroxidation, and cell uptake. *International Journal of Nanomedicine*, *5*, 679–686. doi:10.2147/IJN.S12293 PMID:20957219

Anton, N., Hallouard, F., Attia, M. F., & Vandamme, T. F. (2016). Nano-emulsions for drug delivery and biomedical imaging. In A. Propkop & V. Weissig (Eds.), *Intracellular delivery III* (pp. 273–300). Springer. doi:10.1007/978-3-319-43525-1_11

Antonarakis, E. S., Heath, E. I., Smith, D. C., Rathkopf, D., Blackford, A. L., Danila, D. C., King, S., Frost, A., Ajiboye, A. S., Zhao, M., Mendonca, J., Kachhap, S. K., Rudek, M. A., & Carducci, M. A. (2013). Repurposing itraconazole as a treatment for advanced prostate cancer: A noncomparative randomized phase II trial in men with metastatic castration-resistant prostate cancer. *The Oncologist*, *18*(2), 163–173. doi:10.1634/theoncologist.2012-314 PMID:23340005

Armstrong, C. M., & Gao, A. C. (2015). Drug resistance in castration resistant prostate cancer: Resistance mechanisms and emerging treatment strategies. *American Journal of Clinical and Experimental Urology*, *3*(2), 64–76. PMID:26309896

Attia, M. F., Anton, N., Chiper, M., Akasov, R., Anton, H., Messaddeq, N., Fournel, S., Klymchenko, A. S., Mély, Y., & Vandamme, T. F. (2014). Biodistribution of X-ray iodinated contrast agent in nanoemulsions is controlled by the chemical nature of the oily core. *ACS Nano*, *8*(10), 10537–10550. doi:10.1021/nn503973z PMID:25284066

Azeem, A., Rizwan, M., Ahmad, F. J., Iqbal, Z., Khar, R. K., Aqil, M., & Talegaonkar, S. (2009). Nanoemulsion components screening and selection: A technical note. *AAPS PharmSciTech*, *10*(1), 69–76. doi:10.120812249-008-9178-x PMID:19148761

Bae, Y. H., & Park, K. K. (2011). Targeted drug delivery to tumours: Myths, reality and possibility. *Journal of Controlled Release*, *153*(3), 198–205. doi:10.1016/j.jconrel.2011.06.001 PMID:21663778

Barth, P. J., Weingärtner, K., Köhler, H. H., & Bittinger, A. (1996). Assessment of the vascularization in prostatic carcinoma: A morphometric investigation. *Human Pathology*, *27*(12), 1306–1310. doi:10.1016/S0046-8177(96)90342-1 PMID:8958303

Bouchaala, R., Mercier, L., Andreiuk, B., Mély, Y., Vandamme, T., Anton, N., Goetz, J. G., & Klymchenko, A. S. (2016). Integrity of lipid nanocarriers in bloodstream and tumor quantified by near-infrared ratiometric FRET imaging in living mice. *Journal of Controlled Release*, *236*, 57–67. doi:10.1016/j.jconrel.2016.06.027 PMID:27327767

Bouchemal, K., Briancon, S., Perrier, E., & Fessi, H. (2004). Nano-emulsion formulation using spontaneous emulsification: Solvent, oil and surfactant optimization. *International Journal of Pharmaceutics*, *280*(1-2), 241–251. doi:10.1016/j.ijpharm.2004.05.016 PMID:15265563

Chen, T., Gong, T., Zhao, T., Fu, Y., Zhang, Z., & Gong, T. (2018). A comparison study between lycobetaine-loaded nanoemulsion and liposome using nRGD as therapeutic adjuvant for lung cancer therapy. *European Journal of Pharmaceutical Sciences*, *111*, 293–302. doi:10.1016/j.ejps.2017.09.041 PMID:28966099

Choudhury, H., Gorain, B., Karmakar, S., Biswas, E., Dey, G., Barik, R., Mandal, M., & Pal, T. K. (2014). Improvement of cellular uptake, in vitro antitumor activity and sustained release profile with increased bioavailability from a nanoemulsion platform. *International Journal of Pharmaceutics*, *460*(1-2), 131–143. doi:10.1016/j.ijpharm.2013.10.055 PMID:24239580

Constantinides, P. P., Chaubal, M. V., & Shorr, R. (2008). Advances in lipid nanodispersions for parenteral drug delivery and targeting. *Advanced Drug Delivery Reviews*, *60*(6), 757–767. doi:10.1016/j.addr.2007.10.013 PMID:18096269

Danhier, F., Feron, O., & Préat, V. (2010). To exploit the tumor microenvironment: Passive and active tumor targeting of nanocarriers for anti-cancer drug delivery. *Journal of Controlled Release, 148*(2), 135–146. doi:10.1016/j.jconrel.2010.08.027 PMID:20797419

Davis, T., & Farag, S. S. (2013). Treating relapsed or refractory Philadelphia chromosome-negative acute lymphoblastic leukemia: Liposome-encapsulated vincristine. *International Journal of Nanomedicine, 8*, 3479–3488. doi:10.2147/IJN.S47037 PMID:24072970

Desai, A., Vyas, T., & Amiji, M. (2008). Cytotoxicity and apoptosis enhancement in brain tumor cells upon coadministration of paclitaxel and ceramide in nanoemulsion formulations. *Journal of Pharmaceutical Sciences, 97*(7), 2745–2756. doi:10.1002/jps.21182 PMID:17854074

Duskey, J. T., & Rice, K. G. (2014). Nanoparticle ligand presentation for targeting solid tumors. *AAPS PharmSciTech, 15*(5), 1345–1354. doi:10.120812249-014-0143-6 PMID:24927668

Epstein, J. I., Amin, M. B., Beltran, H., Lotan, T. L., Mosquera, J. M., Reuter, V. E., Robinson, B. D., Troncoso, P., & Rubin, M. A. (2014). Proposed morphologic classification of prostate cancer with neuroendocrine differentiation. *The American Journal of Surgical Pathology, 38*(6), 756–767. doi:10.1097/PAS.0000000000000208 PMID:24705311

Epstein, J. I., Egevad, L., Amin, M. B., Delahunt, B., Srigley, J. R., Humphrey, P. A., & Grading, C. (2016). The 2014 International Society of Urological Pathology (ISUP) Consensus Conference on Gleason Grading of Prostatic Carcinoma: Definition of grading patterns and proposal for a new grading system. *The American Journal of Surgical Pathology, 40*(2), 244–252. doi:10.1097/PAS.0000000000000530 PMID:26492179

Fukumura, D., Duda, D. G., Munn, L. L., & Jain, R. K. (2010). Tumor microvasculature and microenvironment: Novel insights through intravital imaging in pre-clinical models. *Microcirculation (New York, N.Y.), 17*(3), 206–225. doi:10.1111/j.1549-8719.2010.00029.x PMID:20374484

Ganta, S., & Amiji, M. (2009). Coadministration of paclitaxel and curcumin in nanoemulsion formulations to overcome multidrug resistance in tumor cells. *Molecular Pharmaceutics, 6*(3), 928–939. doi:10.1021/mp800240j PMID:19278222

Ganta, S., Singh, A., Patel, N. R., Cacaccio, J., Rawal, Y. H., Davis, B. J., Amiji, M. M., & Coleman, T. P. (2014). Development of EGFR targeted nanoemulsion for imaging and novel platinum therapy of ovarian cancer. *Pharmaceutical Research, 31*(9), 2490–2502. doi:10.100711095-014-1345-z PMID:24643932

Gelderblom, H., Verweij, J., Nooter, K., & Sparreboom, A. (2001). Cremophor EL: The drawbacks and advantages of vehicle selection for drug formulation. *European Journal of Cancer, 37*(13), 1590–1598. doi:10.1016/S0959-8049(01)00171-X PMID:11527683

Gleason, D. F., & Mellinger, G. T. (1974). Prediction of prognosis for prostatic adenocarcinoma by combined histological grading and clinical staging. *The Journal of Urology, 111*(1), 58–64. doi:10.1016/S0022-5347(17)59889-4 PMID:4813554

Goldstein, D., Gofrit, O., Nyska, A., & Benita, S. (2007). Anti-HER2 cationic immunoemulsion as a potential targeted drug delivery system for the treatment of prostate cancer. *Cancer Research, 67*(1), 269–275. doi:10.1158/0008-5472.CAN-06-2731 PMID:17210707

Gorain, B., Choudhury, H., Kundu, A., Sarkar, L., Karmakar, S., Jaisankar, P., & Pal, T. K. (2014). Nanoemulsion strategy for olmesartan medoxomil improves oral absorption and extended antihypertensive activity in hypertensive rats. *Colloids and Surfaces. B, Biointerfaces*, *115*, 286–294. doi:10.1016/j.colsurfb.2013.12.016 PMID:24388859

Gupta, A., Eral, H. B., Hatton, T. A., & Doyle, P. S. (2016). Nanoemulsions: Formation, properties and applications. *Soft Matter*, *12*(11), 2826–2841. doi:10.1039/C5SM02958A PMID:26924445

Gupta, R., Shea, J., Scaife, C., Shurlygina, A., & Rapoport, N. (2015). Polymeric micelles and nanoemulsions as drug carriers: Therapeutic efficacy, toxicity, and drug resistance. *Journal of Controlled Release*, *212*, 70–77. doi:10.1016/j.jconrel.2015.06.019 PMID:26091919

Hare, J. I., Lammers, T., Ashford, M. B., Puri, S., Storm, G., & Barry, S. T. (2017). Challenges and strategies in anti-cancer nanomedicine development: An industry perspective. *Advanced Drug Delivery Reviews*, *108*, 25–38. doi:10.1016/j.addr.2016.04.025 PMID:27137110

Harush-Frenkel, O., Debotton, N., Benita, S., & Altschuler, Y. (2007). Targeting of nanoparticles to the clathirin mediated endocytic pathway. *Biochemical and Biophysical Research Communications*, *353*(1), 26–32. doi:10.1016/j.bbrc.2006.11.135 PMID:17184736

Henry, R. Y., & O'mahony, D. (1999). Treatment of prostate cancer. *Journal of Clinical Pharmacy and Therapeutics*, *24*(2), 93–102. doi:10.1046/j.1365-2710.1999.00207.x PMID:10380061

Higgins, L. H., Withers, H. G., Garbens, A., Love, H. D., Magnoni, L., Hayward, S. W., & Moyes, C. D. (2009). Hypoxia and the metabolic phenotype of prostate cancer cells. *Biochimica et Biophysica Acta*, *1787*(12), 1433–1443. doi:10.1016/j.bbabio.2009.06.003 PMID:19524545

Ho, C. H., Hsu, J. L., Liu, S. P., Hsu, L. C., Chang, W. L., Chao, C. C., & Guh, J. H. (2015). Repurposing of phentolamine as a potential anticancer agent against human castration-resistant prostate cancer: A central role on microtubule stabilization and mitochondrial apoptosis pathway. *The Prostate*, *75*(13), 1454–1466. doi:10.1002/pros.23033 PMID:26180030

Huggins, C., & Hodges, C. V. (1941). Studies on prostatic cancer. I. The effect of castration, of estrogen and of androgen injection on serum phosphatases in metastatic carcinoma of the prostate. *Cancer Research*, *1*, 293–297.

Hwang, Y. Y., Ramalingam, K., Bienek, D. R., Lee, V., You, T., & Alvarez, R. (2013). Antimicrobial activity of nanoemulsion in combination with cetylpyridinium chloride in multidrug-resistant Acinetobacter baumannii. *Antimicrobial Agents and Chemotherapy*, *57*(8), 3568–3575. doi:10.1128/AAC.02109-12 PMID:23669390

Inaba, T. A. (1992). Quantitative measurements of prostatic blood flow and blood volume by positron emission tomography. *The Journal of Urology*, *148*(5), 1457–1460. doi:10.1016/S0022-5347(17)36939-2 PMID:1279212

Irby, D., Du, C., & Li, P. (2017). Lipid–drug conjugate for enhancing drug delivery. *Molecular Pharmaceutics*, *14*(5), 1325–1338. doi:10.1021/acs.molpharmaceut.6b01027 PMID:28080053

John, H., Barghorn, A., Funke, G., Sulser, T., Hailemariam, S., Hauri, S., & Joller-Jemelka, H. M. (2000). Noninflammatory chronic pelvic pain syndrome: Immunological study in blood, ejaculate and prostate tissue. *European Urology*, *39*(1), 72–78. doi:10.1159/000052415 PMID:11173942

Kaladhar, K., & Sharma, C. P. (2006). *Novel albumin self-assembled liposomes for drug delivery applications.* Annual Meeting-Society For Biomaterials In Conjunction With The International Biomaterials Symposium, Annual Meeting: Biomaterials- The enabling Technology, Pittsburgh, PA.

Khan, I., Bahuguna, A., Kumar, P., Bajpai, V. K., & Kang, S. C. (2018). *In vitro* and *in vivo* antitumor potential of carvacrol nanoemulsion against human lung adenocarcinoma A549 cells via mitochondrial mediated apoptosis. *Scientific Reports*, *8*(1), 144. doi:10.103841598-017-18644-9 PMID:29317755

Khatri, S., Lohani, P., & Gandhi, S. (2013). Nanoemulsions in cancer therapy. Indo Global. *Journal of Pharmaceutical Sciences*, *3*(2), 124–133.

Kohli, A. G., Kierstead, P. H., Venditto, V. J., Walsh, C. L., & Szoka, F. C. (2014). Designer lipids for drug delivery: From heads to tails. *Journal of Controlled Release*, *190*, 274–287. doi:10.1016/j.jconrel.2014.04.047 PMID:24816069

Kummar, S., Gutierrez, M., Doroshow, J. H., & Murgo, A. J. (2006). Drug development in oncology: Classical cytotoxics and molecularly targeted agents. *British Journal of Clinical Pharmacology*, *62*(1), 15–26. doi:10.1111/j.1365-2125.2006.02713.x PMID:16842375

Kundranda, M. N., & Niu, J. (2015). Albumin-bound paclitaxel in solid tumors: Clinical development and future directions. *Drug Design, Development and Therapy*, *9*, 3767–3777. doi:10.2147/DDDT.S88023 PMID:26244011

Leone, A., Di Gennaro, E., Bruzzese, F., Avallone, A., & Budillon, A. (2014). New perspective for an old antidiabetic drug: Metformin as anticancer agent. *Cancer Treatment and Research*, *159*, 355–376. doi:10.1007/978-3-642-38007-5_21 PMID:24114491

Li, J., & Bluth, M. H. (2011). Pharmacogenomics of drug metabolizing enzymes and transporters: Implications for cancer therapy. *Pharmacogenomics and Personalized Medicine*, *4*, 11–33. doi:10.2147/PGPM.S18861 PMID:23226051

Li, X., Anton, N., Zuber, G., Zhao, M., Messaddeq, N., Hallouard, F., Fessi, H., & Vandamme, T. F. (2013). Iodinated α-tocopherol nano-emulsions as non-toxic contrast agents for preclinical X-ray imaging. *Biomaterials*, *34*(2), 481–491. doi:10.1016/j.biomaterials.2012.09.026 PMID:23083930

Liu, F., Lin, S., Zhang, Z., Hu, J., Liu, G., Tu, Y., Yang, Y., Zou, H., Mo, Y., & Miao, L. (2014). pH responsive nanoemulsions for controlled drug release. *Biomacromolecules*, *15*(3), 968–977. doi:10.1021/bm4018484 PMID:24527876

Luke, M. C., & Coffey, D. S. (1994). Human androgen receptor binding to the androgen response element of prostate specific antigen. *Journal of Andrology*, *15*(1), 41–51. PMID:7514587

Maeda, H. (2012). Vascular permeability in cancer and infection as related to macromolecular drug delivery, with emphasis on the EPR effect for tumor-selective drug targeting. *Proceedings of the Japan Academy. Series B, Physical and Biological Sciences*, *88*(3), 53–71. Advance online publication. doi:10.2183/pjab.88.53 PMID:22450535

Malarvizhi, G. L., Retnakumari, A. P., Nair, S., & Koyakutty, M. (2014). Transferrin targeted core-shell nanomedicine for combinatorial delivery of doxorubicin and sorafenib against hepatocellular carcinoma. *Nanomedicine (London), 10*(8), 1649–1659. doi:10.1016/j.nano.2014.05.011 PMID:24905399

McCarron, P. A., & Faheem, A. M. (2010). Nanomedicine-based cancer targeting: A new weapon in an old war. *Nanomedicine (London), 5*(1), 3–5. doi:10.2217/nnm.09.89 PMID:20025458

McClements, D. J. (2013). Nanoemulsion-based oral delivery systems for lipophilic bioactive components: Nutraceuticals and pharmaceuticals. *Therapeutic Delivery, 4*(7), 841–857. doi:10.4155/tde.13.46 PMID:23883127

Mura, S., Nicolas, J., & Couvreur, P. (2013). Stimuli-responsive nanocarriers for drug delivery. *Nature Materials, 12*(11), 991–1003. doi:10.1038/nmat3776 PMID:24150417

Nacey, J. N., Meffan, P. J., & Delahunt, B. (1995). The effect of finasteride on prostate volume, urinary flow rate and symptom score in men with benign prostatic hyperplasia. *The Australian and New Zealand Journal of Surgery, 65*(5), 35–39. doi:10.1111/j.1445-2197.1995.tb01744.x PMID:7529489

National Cancer Institute. (2020). *Prostate cancer treatment.* Retrieved January 16, 2020, from http://www.cancer.gov/types/prostate/patient/prostate-treatment-pdq#section/_120

Newling, D. W. (1997). The palliative therapy of advanced prostate cancer, with particular reference to the results of recent European clinical trials. *British Journal of Urology, 79*(S1), 72–81. doi:10.1111/j.1464-410X.1997.tb00805.x PMID:9088277

Noguchi, M., Stamey, T. A., Mc Neal, J. E., & Yemoto, C. E. (2000). An analysis of 148 consecutive transition zone cancers: Clinical and histological characteristics. *The Journal of Urology, 163*(6), 1751–1752. doi:10.1016/S0022-5347(05)67535-0 PMID:10799175

Parhi, P., Mohanty, C., & Sahoo, S. K. (2012). Nanotechnology-based combinational drug delivery: An emerging approach for cancer therapy. *Drug Discovery Today, 17*(17-18), 1044–1052. doi:10.1016/j.drudis.2012.05.010 PMID:22652342

Praveen-Kumar, G., & Divya, A. (2015). Nanoemulsion based targeting in cancer therapeutics. *Medicinal Chemistry (Sunnyvale, Calif.), 5*(6), 272–284. doi:10.4172/2161-0444.1000275

Prins, G. S., & Lindgren, M. (2014). Accessory sex glands in the male. In Physiology of Reproduction (pp. 773–804). Elsevier Inc.

Prokop, A., & Davidson, J. M. (2008). Nanovehicular intracellular delivery systems. *Journal of Pharmaceutical Sciences, 97*(9), 3518–3590. doi:10.1002/jps.21270 PMID:18200527

Rapoport, N. Y., Nam, K. H., Gupta, R., Gao, Z., Mohan, P., Payne, A., Todd, N., Liu, X., Kim, T., Shea, J., Scaife, C., Parker, D. L., Jeong, E. K., & Kennedy, A. M. (2011). Ultrasound mediated tumor imaging and nanotherapy using drug loaded, block copolymer stabilized perfluorocarbon nanoemulsions. *Journal of Controlled Release, 153*(1), 4–15. doi:10.1016/j.jconrel.2011.01.022 PMID:21277919

Roehrborn, C. (2006). Insights into the relationships between prostatic disorders and their potential impact on future urologic practice. *European Urology Supplements, 5*(12), 698–703. doi:10.1016/j.eursup.2006.06.007

Sahay, G., Alakhova, D. Y., & Kabanov, A. V. (2010). Endocytosis of nanomedicines. *Journal of Controlled Release*, *145*(3), 182–195. doi:10.1016/j.jconrel.2010.01.036 PMID:20226220

Sapra, B., Thatai, P., Bhandari, S., Sood, J., Jindal, M., & Tiwary, A. (2013). A critical appraisal of microemulsions for drug delivery: Part I. *Therapeutic Delivery*, *4*(12), 1547–1564. doi:10.4155/tde.13.116 PMID:24304251

Sasikumar, A., & Kamalasanan, K. (2017). Nanomedicine for prostate cancer using nanoemulsion: A review. *Journal of Controlled Release*, *260*, 111–123. doi:10.1016/j.jconrel.2017.06.001 PMID:28583444

Schatten, H. (2018). Brief overview of prostate cancer statistics, grading, diagnosis and treatment strategies. *Advances in Experimental Medicine and Biology*, *1095*, 1–14. doi:10.1007/978-3-319-95693-0_1 PMID:30229546

Sfanos, S., & De Marzo, A. M. (2012). Prostate cancer and inflammation: The evidence. *Histopathology*, *60*(1), 199–215. doi:10.1111/j.1365-2559.2011.04033.x PMID:22212087

Singh, Y., Meher, J. G., Raval, K., Khan, F. A., Chaurasia, M., Jain, N. K., & Chourasia, M. K. (2007). Nanoemulsion: Concepts, development and applications in drug delivery. *Journal of Controlled Release*, *252*, 28–49. doi:10.1016/j.jconrel.2017.03.008 PMID:28279798

Tai, W., Shukla, R. S., Qin, B., Li, B., & Cheng, K. (2011). Development of a peptide-drug conjugate for prostate cancer therapy. *Molecular Pharmaceutics*, *8*(3), 901–912. doi:10.1021/mp200007b PMID:21510670

Tiwari, S., Tan, Y. M., & Amiji, M. (2006). Preparation and *in vitro* characterization of multifunctional Nanoemulsions for simultaneous MR imaging and targeted drug delivery. *Journal of Biomedical Nanotechnology*, *2*(3), 217–224. doi:10.1166/jbn.2006.038

Torchilin, V. P. (2000). Drug targeting. *European Journal of Pharmaceutical Sciences*, *11*(Suppl 2), S81–S91. doi:10.1016/S0928-0987(00)00166-4 PMID:11033430

Torre, L. A., Bray, F., Siegel, R. L., Ferlay, J., Lortet-Tieulent, J., & Jemal, A. (2015). Global cancer statistics, 2012. *CA: a Cancer Journal for Clinicians*, *65*(2), 87–108. doi:10.3322/caac.21262 PMID:25651787

Trumbeckas, D., Milonas, D., Jievaltas, M., Matjosaitis, A. J., Kincius, M., Grybas, A., & Kopustinskas, V. (2011). Importance of prostate volume and urinary flow rate in prediction of bladder outlet obstruction in men with symptomatic benign prostatic hyperplasia. *Central European Journal of Urology*, *64*(2), 75–79. doi:10.5173/ceju.2011.02.art5 PMID:24578868

Tsukihara, H., Nakagawa, F., Sakamoto, K., Ishida, K., Tanaka, N., Okabe, H., Uchida, J., Matsuo, K., & Takechi, T. (2015). Efficacy of combination chemotherapy using a novel oral chemotherapeutic agent, TAS-102, together with bevacizumab, cetuximab, or panitumumab on human colorectal cancer xenografts. *Oncology Reports*, *33*(5), 2135–2142. doi:10.3892/or.2015.3876 PMID:25812794

van der Meel, R., Lammers, T., & Hennink, W. E. (2017). Cancer nanomedicines: Oversold or underappreciated? *Expert Opinion on Drug Delivery*, *14*(1), 1–5. doi:10.1080/17425247.2017.1262346 PMID:27852113 van Ginckel, R., De Coster, R., Wouters, W., Vanherck, W., Veer, V., Der, D., Goeminne, N., Jagers, E., Van Cauteren, H., Wouters, L., & Distelmans, W. (1990). Antitumoral effects of R 75251 on the growth of transplantable R3327 prostatic adenocarcinoma in rats. *The Prostate*, *16*(4), 313–323. doi:10.1002/pros.2990160406 PMID:2371176

Vandamme, T. F. (2002). Microemulsions as ocular drug delivery systems: Recent developments and future challenges. *Progress in Retinal and Eye Research*, *21*(1), 15–34. doi:10.1016/S1350-9462(01)00017-9 PMID:11906809

Vaupel, P., & Kelleher, D. K. (2013). Blood flow and oxygenation status of prostate cancers. *Advances in Experimental Medicine and Biology*, *765*, 299–305. doi:10.1007/978-1-4614-4989-8_42 PMID:22879048

Wang, F., Wang, Y. C., Dou, S., Xiong, M. H., Sun, T. M., & Wang, J. (2011). Doxorubicin-tethered responsive gold nanoparticles facilitate intracellular drug delivery for overcoming multidrug resistance in cancer cells. *ACS Nano*, *5*(5), 3679–3692. doi:10.1021/nn200007z PMID:21462992

Weidner, N., Carroll, P. R., Flax, J., Blumenfeld, W., & Folkman, J. (1993). Tumor angiogenesis correlates with metastasis in invasive prostate carcinoma. *American Journal of Pathology*, *143*(2), 401–409. PMID:7688183

Weiss, J., Herrmann, N., & McClements, D. J. (1999). Ostwald ripening of hydrocarbon emulsion droplets in surfactant solutions. *Langmuir*, *15*(20), 6652–6657. doi:10.1021/la981739d

Wilhelm, S., Tavares, A. J., Dai, Q., Ohta, S., Audet, J., Dvorak, H. F., & Chan, W. C. W. (2016). Analysis of nanoparticle delivery to tumours. *Nature Reviews. Materials*, *1*(5), 16014. doi:10.1038/natrevmats.2016.14

Wong, P. T., & Choi, S. K. (2015). Mechanisms of drug release in nanotherapeutic delivery systems. *Chemical Reviews*, *115*(9), 3388–3432. doi:10.1021/cr5004634 PMID:25914945

Ye, A., Zhu, X., & Singh, H. (2013). Oil-in-water emulsion system stabilized by protein-coated nanoemulsion droplets. *Langmuir*, *29*(47), 14403–14410. doi:10.1021/la403493y PMID:24175702

Zhang, J., Wang, L., You, X., Xian, T., Wu, J., & Pang, J. (2019). Nanoparticle therapy for prostate cancer: Overview and perspectives. *Current Topics in Medicinal Chemistry*, *19*(1), 57–73. doi:10.2174/1568026619666190125145836 PMID:30686255

Zou, M., Toivanen, R., Mitrofanova, A., Floch, N., Hayati, S., Sun, Y., Le Magnen, C., Chester, D., Mostaghel, E. A., Califano, A., Rubin, M. A., Shen, M. M., & Abate-Shen, C. (2017). Transdifferentiation as a mechanism of treatment resistance in a mouse model of castration-resistant prostate cancer. *Cancer Discovery*, *7*(7), 736–749. doi:10.1158/2159-8290.CD-16-1174 PMID:28411207

Chapter 13
Application of Nanoemulsions in Breast Cancer Treatment

M Joyce Nirmala
Indian Institute of Technology, Madras, India

Uma Kizhuveetil
Jayoti Vidyapeeth Womens University, India

Shiny P. J.
Indian Institute of Technology, Madras, India

Uppada Sumanth Raj
Indian Institute of Technology, Madras, India

Sindhu Priya Dhas
Jaya College of Arts and Science, India

R. Nagarajan
Indian Institute of Technology, Madras, India

ABSTRACT

A new, efficient, and secure clinical approach is increasingly being sought for the treatment of cancer. Nanoemulsions (NE) are projected to have a profound effect on delivering improved healthcare services with significant implications on forthcoming healthcare policies. In contrast to other drug carriers, the key value of NEs is that they can be engineered to target tumor cells and overcome the major challenge of multi-drug resistance. Multifunctional NEs are being investigated by researchers in various fields of study, primarily in the treatment of different forms of cancer. The congruent presence of NEs with contrast agents or certain dyes increases the accuracy of cancer status identification by enhancing the responsiveness of the agents; thus, they are finding application as nanotheranostics. A summary of different NEs and their documented applications in cancer therapeutics, with emphasis on breast cancer, is presented in this chapter.

1. INTRODUCTION

Nanoemulsions (NEs) are kinetically-stable nanodroplet dispersions of two immiscible liquids, oil, and water, of size ranging from 20–200 nm (Gupta et al., 2016; McClements, 2012). They can be either oil-in-water (o/w) or water-in-oil (w/o) nanodroplet systems, with one being distributed in the other through a surfactant/ co-surfactant mixture (Mkandawire and Aryee, 2018). The inclusion of an emulsifier (surfactant/ co-surfactant) is essential to form the tiny droplets (Handa et al., 2021). It reduces the interfacial tension between the oil and water phases of the emulsion, i.e., it lowers the surface energy

DOI: 10.4018/978-1-7998-8378-4.ch013

per unit area. The emulsifier also helps to stabilize nanoemulsions due to its effect on the repulsive electrostatic interactions and steric hindrance (Mason et al., 2006).

Nanoemulsions exhibit high surface area, enhanced stability, optical transparency, spontaneous emulsification, and tunable rheological properties. Furthermore, nanoemulsions can stabilize chemically unstable compounds by blocking their oxidative and light-induced degradation (Tayeb and Sainsbury, 2018; Chime et al., 2014). Thus, NEs are widely used in the food industry, pharmaceuticals, and cosmetics due to their enhanced functional properties.

In the field of pharmaceuticals, NEs are being used in marketed drugs, with many others in the clinical development stage. NE's remarkable characteristics such as the ease of manufacture, increased loading capacities, high rate of drug solubility, long-term reliability, increased bioavailability, safety, and regulated cargo discharge, make them promising drug carriers (Kotta et al., 2012). NEs are usually formulated with surfactants considered to be GRAS (generally recognized as safe), and they can be consumed orally. As a result, nanoemulsions are typically used in oral drug delivery. In addition, they are widely exploited for ophthalmic, transdermal, pulmonary, and image-guided drug delivery. In recent years, NEs have been researched for parenteral drug delivery in cancer therapy (Ganta et al., 2014).

NEs are promising candidates for future diagnostics, medication therapies, dermatology, and biotechnology. NEs have many applications such as in cancer therapy, drug targeting, mucosal vaccines, transdermal drug delivery mediums, lipophilic drug delivery mechanisms, etc. If the Oswald-ripening process is constrained by highly insoluble oils, there is enormous scope for nano-emulsion applications (Sharma et al., 2013). In addition, they can be quickly targeted to the tumor sites specifically due to their size characteristics. Furthermore, the ability to surface functionalize with targeting molecules is advantageous for selective delivery of medications, proteins, photosensitizers, etc. to the tumor region. Perfluorochemical NEs have demonstrated positive effects in cancer therapy when combined with other treatment methods and delivered to the neovasculature (Shah et al., 2010).

Researchers must overcome material protection, scale-up, and quality management concerns to realize the promise of nanoemulsion dosage forms as anticancer therapeutics in clinical settings. Furthermore, before being used in therapeutic applications, the *in-vivo* delivery and metabolism of NEs of novel compounds must be thoroughly evaluated.

This Chapter will focus on applications of nanoemulsions in breast cancer, the most common tumor among women and commonly treated using chemotherapy. However, chemo-treatment efficacy is poor owing to side effects, multidrug resistance (MDR), and low specificity to tumor cells. Nanoemulsions are suitable alternatives for breast cancer treatment because of their unique functional properties. The objectives of the review are as follows:

- To explore the use of nanoemulsions in therapeutics
- To present the current role of nanoemulsions in cancer nanomedicine, and
- To outline the treatment procedures for breast cancer using nanoemulsions

2. NANOEMULSIONS IN THERAPEUTICS

The development of advanced screening methods as well as advances in combinatorial chemistry have led to the development of many novel therapeutic candidates. However, the majority of these potential medicinal compounds are poorly soluble in water, due to which it is estimated that nearly 40% of newly discovered therapeutic compounds are not usable in practice (Ganta et al., 2014). Nanoemulsions have been extensively studied due to their extensive use in therapeutics and pharmaceutics. The flexible characteristics of NEs provide improved drug loading properties as well as adequate drug safety in biological environments. Furthermore, their non-toxic and non-irritant properties make NEs the perfect therapeutic system (Aboofazeli, 2010; Prakash & Thiagarajan, 2011). This Section focuses on nanoemulsions as therapeutic agents in multiple fields of medicine, such as cancer medicine, drug delivery, and antimicrobials.

Cancer develops at a molecular level as several subsets of genes undergo genetic changes, later resulting in a malignant growth of cancer cells, tissue invasion, and organ dysfunction (Sarkar et al., 2007). The intrinsic complexity of the tumor environment and the presence of P-glycoprotein typically serve as obstacles to conventional chemotherapy by blocking the effective targeting of the medication to the tumor site (Wiradharma et al., 2009). The big downside of anticancer medicines is their lack of selectivity for the tumor site, which causes extreme side effects and poor recovery rates. As a result, it is a major challenge to kill dysfunctional cells using the traditional drug delivery mechanism (Bayón-Cordero et al., 2019). Surface morphology of nanocarriers such as nanoemulsions can lead to improved imaging, screening, and delivery functions, enhance the bioavailability of drugs, minimize undesirable side effects, and increase selective absorption, enabling specific targeting of the tumor sites (Yu & Zhang, 2009). The identification of an effective delivery mechanism, such as NEs, is intended to selectively target cancer cells, allowing drugs and imaging agents to be administered specifically to the tumor site, thus improving clinical effectiveness. Specific alterations of NE properties have been investigated over the years in order to enhance and meet the precise treatment objectives (Sareen et al., 2012). Besides that, NEs containing fatty acids such as omega-3 and omega-6 fatty acids, linoleic acid, and vitamins E and K are often used as carriers for various diagnostic and imaging agents for chemotherapy (Tiwari et al., 2006). These are usually linked with multifunctional molecules that aid in image-guided diagnosis and drug targeting. The following Sections discuss advancements of various nanoemulsions in the field of cancer therapy.

2.1. Antimicrobial Nanoemulsions

Antimicrobial NEs, emulsions of antibiotics or natural products such as essential oils with reported antimicrobial activity and a surfactant or cosurfactant in water, have been extensively studied in microbiology for a variety of applications. These applications include treating bacterial and fungal infections, as well as against parasites such as *Leishmania spp.* and *Plasmodium spp.* (Hakemi-Vala et al., 2017). These NE particles are lipophilic and therefore bind to lipid molecules on the pathogen. Once a sufficient number of nanoparticles bind to pathogens, some of the energy stored within the emulsion is release. The activity of the antimicrobial component and the energy released destabilize the pathogen lipid membrane, leading to cell lysis and death (Subhashis et al., 2011). A few examples of such antimicrobial nanoemulsions are given below.

- A NE obtained by combining a heated oil phase (tributyl phosphate, soybean oil) and Triton X-100 in a 1:8:1 mixture with deionized water was reported to be bactericidal against most Gram-positive bacteria, a few Gram-negative bacteria, including *H. Influenzae* and *N. Gonorrhoeae* and to varying degrees, many enveloped viruses including Herpes simplex, vaccinia virus and HIV (Hamouda et al., 2011; Hamouda et al., 1999).

- NB-002 is an oil-in-water emulsion containing highly filtered oil, ethanol, polysorbate 20, CPC, and water that is used to treat skin, hair, and nail infections. It displays a have reported fungicidal activity against many dermatophytes and leads to the formation of fewer stable drug resistant mutants as compared to other cutaneous infection drugs (Pannu et al., 2009).
- A palatable lipid nanoemulsion of primaquine with a small particle size range of 50–200 nm that was stabilized with lecithins and poloxamer, upon oral administration, demonstrated effective antimalarial activity at a 25% lower dose than the plain drug solution (Singh & Vingkar, 2008).
- Topical nanoemulsions of 5-aminolevulinic acid (ALA) and temoporfin have been reported to be effective for photodynamic therapy of skin conditions with an extended-release time of around 24-48 hours, which compares favorably to aqueous solutions of ALA (Zhang et al., 2011; Primo et al., 2008).

2.2. Nanoemulsions as Vaccine and Drug Delivery Systems

NE-based mucosal surface vaccines delivering inactivated species to mucosal surfaces have been developed. These vaccines are administered nasally to elicit an immune response (Charles & Attama, 2011). Clinical trials for an influenza vaccine, and an HIV vaccine are under progress. Intranasal administration of *Mycobacterium tuberculosis* immunodominant antigen-mediated potent IL-17 T-cell responses and conferred defense against tuberculosis threat in mice (Ahmed et al., 2017). Some vaccines, such as Anthrax and Hepatitis B, are undergoing further studies in laboratory trials. The NE acts as an adjuvant for proteins, when applied to the mucosal surface and aids their uptake via antigen-presenting cells. As a result of the synthesis of distinct IgG and IgA antibodies, a strongly systemic and mucosal immune response occurs (Chime et al., 2014).

NEs are highly effective means for ocular drug delivery because the sustained steady-state release of the drug from the NE allows the drug to penetrate deeper into the ocular tissue. NE has a low formulation cost and has a lower viscosity. This colloidal dispersion improves ocular residence time and increases opioid permeation across the cornea. The interaction of NE with the lipid layer of the tear film allows the drug to remain in the target area for a longer period of time and function as a drug depot (Alany et al., 2006). Brinzolamide (Azopt) loaded in NE was developed by a group of researchers for the treatment of glaucoma. The formulation was stable and had a greater penetrating effect than the ocular suspension of Azopt (Mahboobian et al., 2019).

Since the upper layer of the epidermis prevents certain medications from passing through, chemical permeation enhancers are used for topical drug distribution (Jampilek & Brychtova, 2012). Since many of these permeation enhancers have side effects such as skin irritation and allergies (Kovacik et al., 2020), novel methods for topical drug delivery needs to be explored. Nanoscale drug delivery devices, such as NEs, are commonly used to improve epidermal and dermal drug deposition. The fact that NEs are vehicles that often function as transdermal permeation enhancers is a beneficial aspect (Prow et al., 2011, Shakeel et al., 2010). Water-in-oil NEs of caffeine, a compound with known anti-cancer activity against oral cancers, have revealed a large improvement in permeability, thereby removing the need for added chemical permeability enhancers (Shakeel & Ramadan, 2010).

NEs are suitable vehicles for parenteral administration as well. The ability of NEs to shield drugs from hydrolysis and enzymatic degradation, allows them to dissolve and deliver large amounts of hydrophobic drugs. Their lack of flocculation, sedimentation, and creaming, along with their large surface area and free energy, present clear advantages for this route of administration. Major clinical and preclinical experiments of parenteral nanoemulsion-based carriers have shown that these emulsions ensure regulated and sustained release of medications over long periods of time (Prakash and Thiagarajan 2011). An example in case is carbamezepine, an anticonvulsant medication that previously could not be used intravenously, owing to its poor water solubility. However, NEs of carbamezepine have shown favorable kinetics upon parenteral administration (Kelmann et al., 2007).

Nanocarrier structures (nanoemulsions) that can be inhaled have a number of benefits. As particle size decreases, the surface area increases, resulting in increased dissolution rate and relatively uniform distribution of drug dose among the alveoli. Nanocarrier systems have the advantage of allowing for prolonged release in lung tissue, resulting in lower dosing frequency and increased patient compliance (Mansour et al., 2009). The quercetin-loaded O/W NE, prepared using palm oil ester/ricinoleic acid as the oil step, showed strong selective cytotoxicity against A549 lung cancer cells while having no effect on normal cells. Hence, NEs could be used as a carrier mechanisms for pulmonary delivery of molecules with poor water solubility (Arbain et al., 2019).

Taking into account the broad interest in the delivery of nasal drugs and the possible advantages of intranasal administration, the industry will soon begin to market new nasal products. They will not only provide medications for acute and chronic illnesses, but also new nasal vaccines that exhibit improved local or systemic defense against infections. It is possible to produce medications that specifically affect the brain in order to have a successful clinical result in the CNS, even while minimizing systemic side effects (Pires et al., 2009). A synergistic emulsion with 2 antioxidants was formulated to treat cancer (Nicolosi & Shea, 2015). To overcome multi-drug resistance caused by cancer cells, an unsaturated fatty acid as an oil-in water emulsion has been prepared to target cancer stem cells (Egan et al., 2018). The classification of nanoemulsions in therapeutics is given in Figure 1.

Figure 1. Nanoemulsion in therapeutics
Few nanoemulsions used in therapeutics approved for clinical trials have been listed in Table 1 (https://clinicaltrials.gov/).

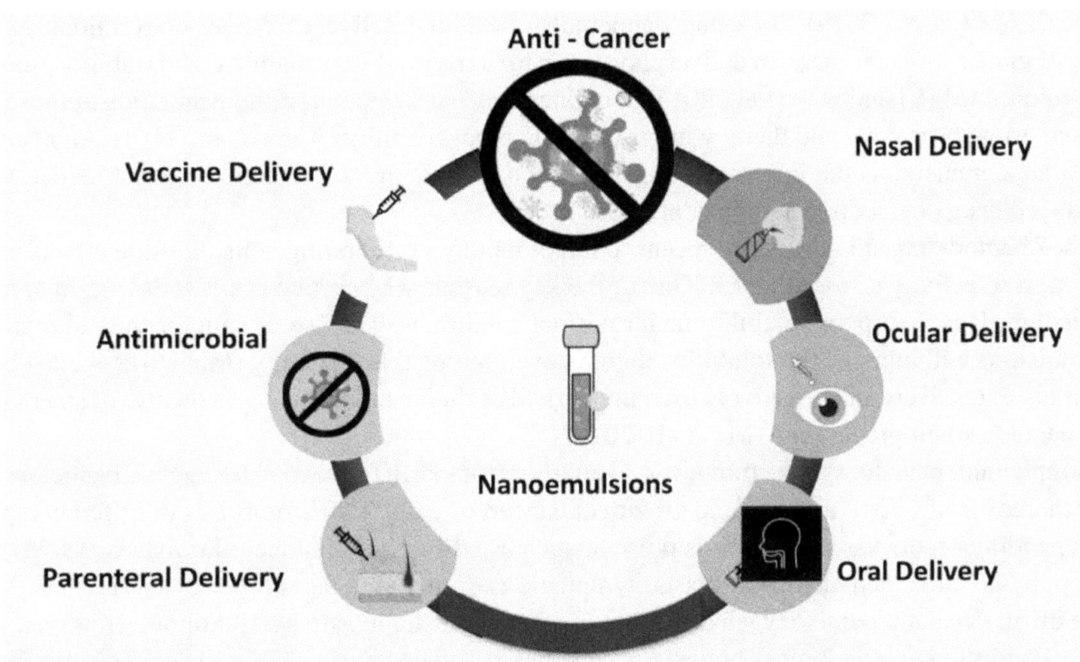

Table 1. Nanoemulsion used in therapeutics approved for clinical trials

Active Compound	Disease	Phase	Status	NCT Number
5-aminolevulinic acid	Lentigo Maligna	Phase IV	Completed	NCT02685592
Clobetasol propionate	Cataract	Phase III	Completed	NCT04249076
Testosterone	Menopause	Phase II	N/A	NCT02445716
3%-Diclofenac	Osteoarthritis of the Knee	Phase II	Completed	NCT00484120
Brimonidine Tartrate	Dry Eye	Phase III	Completed	NCT03785340
Brimonidine Tartrate	Ocular Graft-vs-Host Disease	Phase III	Completed	NCT03591874
Aminolevulinic Acid	Carcinoma, Basal Cell	Phase II	Active, not recruiting	NCT02367547
Curcumin	Breast Cancer	N/A	Completed	NCT01975363
BW-1010	Anthrax	Phase I	Active, not recruiting	NCT04148118
Curcumin	Breast Cancer, Joint Pain	N/A	Recruiting	NCT03865992
14C-cholesteryl oleate, 3H-cholesterol	Impaired Glucose Tolerance	-	Completed	NCT01020578

3. NANOEMULSIONS IN CANCER NANOMEDICINE

3.1. Nanoemulsions as Carriers for Cancer Drugs

Nanoemulsions have become a priority of cancer therapy research since they possess essential characteristics for achieving an adequate therapeutic efficacy: high surface area, superficial charge, enhanced circulation half-life, specificity of drug delivery to the target site, and potential in imaging techniques (Sánchez-López et al., 2019). By using nanoemulsions as the delivery system, conventional dosage forms associated with non-targeted delivery, systemic toxicity, low bioavailability, and stability problems can be minimized (Choudhury et al., 2014). The increased surface area of the nanoemulsion results in increased drug absorption, and thereby improved drug bioavailability (Khan et al., 2018). Another benefit of a nanoemulsion is the improved drug stability achieved due to protection against oxidative and UV-induced drug degradation (Singh et al., 2017).

Since chemotherapeutic drugs are meant to target rapidly proliferating cells, the side effects can be fatal if high selectivity is not achieved. One of the key reasons why certain drugs fails to gain approval in clinical trials is their poor solubility under normal conditions. The use of nanoformulations such as nanoemulsions will improve the solubility, thereby overcoming this difficulty (Mahato et al., 2011). NE systems are extensively and effectively used in the field of therapeutics, owing to the use of safer GRAS components for their preparation (Ma et al., 2014).

A simple understanding of the tumor microenvironment (TME) is needed to create a nanocarrier for cancer therapy as the TME aids in rapid growth of cancerous cells. TME comprises of different types of cells depending on the tissue where it is present, such as adipocytes, extracellular matrix (ECM), glial cells, epithelial cells, activated fibroblasts, lymphatic endothelial cells and pericytes (Lawler, 2009). Simple diffusion can provide oxygen and other essential cell supplies to small tumors. However, if the tumor becomes larger than 2 mm^3, hypoxia occurs due to reduced permeability for oxygen, resulting in angiogenesis, i.e., formation of new blood vessels (Danhier et al., 2010). Angiogenesis inhibitors such as bevacizumab, sorafenib, and sunitinib have been clinically used as anticancer therapeutics. The concern, however, is that these angiogenesis inhibitors have significant toxicity, as well as an elevated risk of developing therapeutic tolerance and transmission problems (Kamba, McDonald, 2007; Carmeliet and Jian, 2011). Encapsulation of these angiogenesis inhibitors using nanoemulsion delivery systems is an efficient method to reduce toxicity and improve therapeutic effectiveness. Betulinic and perfluorocarbon fumagillin-loaded nanoemulsions have been shown to have a significant inhibitory effect on angiogenesis, with minimal to nil toxic side effects (Dehelean et al., 2013; Winter et al., 2008). A patent on use of an oxysterol or a similar compound was used to create a nanoemulsion for treating cancer by targeting (Kreutz, 2019).

Another result of hypoxic condition in cancer cell mass is tumor acidification. Due to hypoxic condition, lactate, the final product of anaerobic glycolysis, is removed by a monocarboxilate transporter repeating to H+(protons). Furthermore, as hypoxia persists, the carbonic anhydrase IX levels increase, resulting in the formation of bicarbonate from carbon dioxide. Owing to this, a pH gradient develops between the external and internal TME. pH-responsive lipids, which are stable at the physiological pH of 7.4 but can release therapeutic load under acidic conditions, gain significance as drug carriers to passively target tumor sites in such conditions (Mahato, 2017).

The Enhanced Permeability and Retention (EPR) effect can also be used as a method for passive targeting of anti-cancer therapeutics to the tumor site. EPR is most beneficial to macromolecular and hydrophobic drugs, and their key function is dependent on retention ability, since small molecules are filtered and reverted to blood circulation (Maeda et al., 2000). The surface of the nanoemulsion, when coated with hydrophilic polymers, prevents opsonization by the Mononuclear Phagocytic System (MPS) resulting in improved circulation half-life and therapeutic efficacy. However, this could also lead to reduced levels of interactions with tumor cells (Allen et al., 2004; Romberg et al., 2008).

Active targeting, like passive targeting, exploits the TME, but it also discovers a way to directly deliver the drug directly to the target cells. This mechanism is only effective if the target cells have unique receptors distributed uniformly on the surface (Peer et al., 2007). The resulting bond may be analogous to ligand-receptor or antigen-antibody. Targeting molecules, such as folate, transferrin, or epidermal growth factor (EFGR), bind to overexpressed receptors in cancer cells (Ganta et al., 2014).

The expression of multifunctional efflux transporters from the ABC gene family has been linked to cancer cell MDR, which is a significant barrier to successful cancer therapy. ATP-binding cassettes (ABCs) use the energy produced from ATP hydrolysis to efflux a variety of endogenous ligands such as lipids, proteins, metabolic products, and drugs such as cytotoxic antibiotics (Dean, 2009; Mohammad et al., 2018). P-glycoprotein (MDR1 or ABCB1), multidrug resistance-associated proteins (MRPs), of which MRP1 or ABCC is the most examined, and breast cancer resistance protein (ABCG2) are some of the most intensively-researched MDR transporters. These proteins have different shapes, but they all perform the same function: they eliminate chemotherapy drugs from cells (Peer et al., 2007). Small interfering RNA (siRNA) molecules were designed to help with chemotherapy because they suppress MDR protein expression and control anti-apoptotic genes. Nanoemulsions, in combination with P-gp modulators or Bcl-2 inhibitors, might be quite helpful to overcome MDR (Yu et al., 2014).

Another means of targeting NEs would be to conjugate them with antibodies or antibody fragments, since antigen-antibody (Ab) binding is precise and selective. The monoclonal antibodies-- trastuzumab, bevacizumab, rituximab, and cetuximab have all been approved by the FDA for cancer treatment (McClements, 2010; Yu et al., 2012). Oligonucleotides can also be conjugated to nanoemulsions, but are unstable, have a low half-life in biological fluids, and have poor intracellular penetration, making them an unlikely alternative for nanoemulsion conjugation (Toub et al., 2006). To resolve these concerns, researchers have used a variety of techniques, including conjugating siRNA with poly(ethylene glycol), developing cationic liposomes, polyelectrolyte complex (PEC) micelles, pH prone nanocarriers, and lipid and plasmid DNA complexes (Mahato, 2017).

A library of ssDNA and ssRNA can be selected to shape DNA or RNA oligonucleotides, ensuing in aptamers, using a procedure called SELEX (Ganta et al., 2008). Aptamers fulfill their function by binding to the target compound and folding into secondary and tertiary DNA/RNA structures. SELEX generates aptamers against tumor cells dependent on the identification of receptors and biomarkers (Wang and Huang, 1987). A significant number of aptamers have been produced against well-known cancer proteins such as VEGF, tenascin-C, HER3, prostate-specific membrane antigen (PSMA), and platelet-derived growth factor (PDGF) (Li et al., 2011). Folic acid is an ideal target site because folate receptors and folic acid have a high affinity and can easily be conjugated with nanoemulsions (Modi and Anderson, 2013). Further, the high levels of folate receptor expression in lung, colon, brain, pancreas, ovary, endometrium, cervix, prostate, and breast cancers, emphasizes the advantages of the use of folate receptors as a target site (Qian and McClements, 2011). Microfluidized nanoemulsion have been found to have a higher cell permeability containing anticancer drug (Nicolosi & Tagne., 2010).

3.2. Nanoemulsions for Various Cancers

3.2.1. Lung Cancer

Lung cancer is the most common form of cancer in males and third most common form in females (Barta et al., 2019). Lung cancer, based on biological characteristics and clinical performance, can be classified into small cell lung and non-small cell lung cancer. Small-cell lung tumors develop and spread faster to the nervous, skeletal and lymphatic systems than non-small-cell lung tumors, and are more sensitive to chemical and radiation therapies (Herbst et al., 2008).

There have been various *in vitro* studies on the anti-cancer effects of NEs on lung cancer cell lines. Citrus lemon essential oil nanoemulsion (CLEO-NE) has a reported cytotoxic and apoptotic effect in A549 human lung cancer cells. It also has an anti-angiogenic effect (Yousefian et al., 2020;). Curcuminoid extracts and nanoemulsions, when used to investigate the mechanism of inhibition of lung cancer cells A549 and H460, gave promising results. (Chang & Chen, 2015).

3.2.2. Cervical Cancer

Cervical cancer is the fourth most-commonly reported form of cancer in women (Jemal et al., 2011). It has been shown that combining essential oils with the chemotherapeutic agent mitomycin C (MMC) in NEs may be helpful in cervical cancer treatment. Naturally-occurring substances with anticancer effects have received focused attention in the last two decades to tackle the poor therapeutic index of traditional anticancer drugs (Goldberg et al., 1994). The effect of O/W nanoemulsions based on two essential oils of Chamomile (Ch) and Garlic (Gar) with mitomycin C have been reported to be effective in reducing the viability of HeLa cells (Alkhatib et al., 2018). When HeLa cells were exposed to the blank and MMC-loaded NE formulas, the viability percentages were significantly reduced. However, when HeLa cells were exposed to Ch-NE and Ch-MMC, the nuclei of the cells shrank more than when they were exposed to Gar-NE and Gar-MMC (Alkhatib et al., 2018). Another study used curcumin-nanoemulsion as a photosensitizing drug on cervical carcinoma cell lines to examine the impact of photodynamic therapy on cellular viability. In all cell lines tested, the empty nanoemulsion had very low cytotoxicity, whereas incubation with curcumin-nanoemulsion at 20 μM curcumin resulted in cytotoxicity of more than 80% for the cell lines. The nanoemulsion-curcumin was capable of producing an effective photodynamic response on carcinoma cervical cell lines, resulting in death only in those cells that were irradiated (de Matos et al., 2018).

3.2.3. Colon Cancer

Colon cancer accounts for a significant number of cancer-related deaths worldwide, and shows the highest occurrence rate. *Carum Carvi* essential oil NE (CCONE) has shown robust anticancer activity by inducing greater caspase3 expression and thus apoptotic death in the HT29 cell line compared to normal human umbilical vein endothelial cells (HUVEC) (Hsu et al., 2017). Similarly, a carotenoid NE of *L. barbarum*, was found to be more effective than the *L. barbarum* extract in its anticancer activity in colon cancer cell line HT-29 (Khatamian et al., 2019).

3.2.4. Prostate Cancer

Fatalities due to prostate cancer have increased in the last decade, with a substantial number of treated patients experiencing relapse and progressing to an untreatable condition. A catechin NE fabricated using catechin extract, lecithin, Tween 80, and deionized water showed significant improvement over the catechin extract alone in inhibiting prostate cancer cell PC-3 proliferation (Chrastina et al., 2018). Another study used oleic acid dispersions with polyoxyethylene sorbitan monooleate as a surfactant to create plumbagin-loaded nanoemulsions. When compared to free plumbagin, the antiproliferative activity of plumbagin-loaded nanoemulsion was higher in PTEN-P2 cells. The increased cytotoxicity of plumbagin in the NE may be due to increased cellular absorption of plumbagin in the nanoparticulate form, a stabilizing effect of nanoemulsions on plumbagin, or both. The oleic-acid-based nanoemulsion formulation has a high drug-loading capability and increased antiproliferative activity, indicating a sound plumbagin delivery system (Tsai & Chen, 2016).

3.2.5. Ovarian Cancer

Ovarian cancer is also a common cancer diagnosed in women worldwide. Pyridoclax, an Mc11 inhibitor, used as anticancer drug, when encapsulated in NE, has been reported to have improved efficacy on ovarian cancer cells *in vitro* at a 2.5-fold lower concentration as compared to the drug alone. Also, NE formulation showed better bioavailability of the drug upon oral administration in mice (Groo et al., 2020). The distribution of paclitaxel-loaded hyaluronan solid nanoemulsions (PTX-HSNs) to boost ovarian cancer treatment via active tumor targeting against SK-OV-3 and OVCAR-3 (CD44) cells revealed that PTX-HSN had a hundredfold higher targeting potential than PTX-loaded solid nanoemulsions without hyaluronan. PTX-HSN's are, therefore, a highly efficient nano system that enhances the targeting of PTX in the ovarian tumor (Kim & Park., 2017)

4. NANOEMULSIONS IN BREAST CANCER TREATMENT

4.1. Conventional Breast Cancer Therapy

4.1.1. Radiation Therapy

Breast-conserving treatment (BCT), including surgery and adjuvant radiotherapy, is one of the most notable evidence-based advances in clinical cancer care for early-stage breast cancer and DCIS (Castaneda and Strasser, 2017). Adjuvant radiation therapy has been shown to increase local regulation and breast cancer mortality in trials assessing breast-conserving treatment. In contrast, post-mastectomy radiation therapy has been shown to improve longevity in properly-chosen individuals in trials. Regardless of stage, patients with breast cancer were given the same radical surgical treatment alternative for decades until the landmark trials contrasting adapted radical mastectomy and breast conservation (Valente and Shah, 2020). In addition, RT has been shown to reduce local-regional recurrence and remote metastases and increase long-term survival following BCT or mastectomy (Harris, 2014). However, there is an increasing emphasis on shifting away from treatment recommendations focused exclusively on clinical and pathologic aspects and more towards personalized treatment recommendations. Studies are required to help understand which patients benefit from regional nodal irradiation in patients with reduced nodal involvement. Finally, while prior research looked at omitting RT in favor of endocrine therapy for low-risk patients, recent experiments are looking at missing endocrine therapy instead of RT, partially due to shorter treatment durations, increased compliance, and different toxicity profiles (Ward et al., 2019; Woodward et al., 2020; Hughes et al., 2013).

4.1.2. Endocrine Therapy

Endocrine therapy is a practical approach in the treatment of hormone-positive breast cancer, both early and advanced. Endogenous estrogen, which contains estrone, estradiol, and estriol, causes the ER to dimerize and facilitates estrogen-regulated gene transcription in HR + breast cancer. Endocrine therapy works by inhibiting estrogen synthesis or by suppressing the estrogen's action at the receptor level (Reinbolt et al., 2015). Adjuvant endocrine therapy (AET) reduces the risk of relapse and mortality dramatically. Unfortunately, although AET is successful, 31-73 percent of breast cancer survivors do not choose to use it. Improved symptom control and implementing supportive treatment interventions that respect the principles and attitudes maintained by breast cancer survivors when emphasizing the advantages of AET, and exploring women's motivations for non-persistence are all needed to increase AET persistence (Lambert et al., 2018). However, recent research shows that prolonged endocrine therapy and ovarian suppression have substantial benefits in some patients. The MA-17 trial recommended moving from Tamoxifen to Letrozole in postmenopausal women, while the ATLAS and aTTom trials recommend a 10-year Tamoxifen therapy. The TEXT and SOFT trials show that an AI with ovarian suppression is consistent with substantial survival advantages in some premenopausal patients(Glassman et al., 2017). The advancement and acceptance of targeted treatments, such as everolimus in conjunction with AET for patients with advanced breast cancer, has resulted from a better understanding of the pathways of resistance to endocrine therapy. Since the addition of selective agents to endocrine therapy is intended to enhance toxicity, assessing which patients are at greater risk of recurrence and more likely to recover would be crucial to improving long-term results for HR-positive breast cancer patients in the early stages (Zelnak and O'Regan, 2015).

4.1.3. Chemotherapy

According to recent studies, neoadjuvant chemotherapy is widely used in early-stage breast cancer (EBC) and locally-advanced breast cancer (LBC) to improve the odds of breast-conserving surgery (BCS) by downstaging the disease and determining tumor response to treatment (Kümmel et al., 2014). In addition, chemotherapy advances include developing and using new cytotoxic agents, experimental therapeutic methods, and neoadjuvant therapy, all of which have been shown to improve treatment outcomes such as overall mortality and aseptic survival (Houssami et al., 2012). Furthermore, randomized prospective studies have shown that patients with early-stage breast cancer who favor breast conservation benefit from neoadjuvant chemotherapy, with around a 25% complete pathologic response and more than an 80% partial pathologic response. For patients with Her 2-neu–positive tumors, neoadjuvant trastuzumab seems to be an outstanding choice (Buzdar et al., 2015). Taxanes increase CT effectiveness regardless of age, nodal status, tumor size or level, steroid receptor expression, or tamoxifen application, but at the expense of increased non-cardio toxicity. Anthracycline and taxane-based CT regimens decrease breast cancer mortality by nearly one-third (EBCTCG et al., 2012). According to a study, anthracycline-containing regimens cut the annual breast cancer mortality risk by 38% for patients under 50 years old, and by 20% for those aged 50 to 69 years. The addition of a taxane to an anthracycline-based regimen has been shown to improve the treatment of early-stage breast cancer (EBCTCG, 2005).

4.2. Modern Strategies Used in Breast Cancer Therapy

4.2.1. Active and Passive Targeting

Antitumor drugs delivered by targeted drug delivery systems have been shown to have a high potential for reducing cytotoxicity while increasing therapeutic effects. Improved cell targeting can be passive or aggressive. The therapeutic agent is incorporated into a nanocarrier that passively approaches the target organ, resulting in passive targeting. It may be enhanced by the concentration of active ingredients at a given location due to various physicochemical, pathophysiological, and anatomical factors (Dhankhar et al., 2010; Vasir and Labhasetwar, 2005). Active targeting was suggested as a means of improving targeting effectiveness. This method is focused on ligand-receptor and antibody-antigen interactions. In cancer, overexpression of receptors or antigens can potentially target efficient drug uptake through receptor-mediated endocytosis.

Substantial progress has been achieved in this field of study, both preclinical and clinical, with the approval of a range of (primarily passively tumor-targeted) nanomedicine formulations for clinical usage (Lammers et al., 2012).

4.2.2. Nanoparticles

Although nanoparticles in drug delivery systems are not commonly used in clinical procedures, several observational trials are currently being performed to capitalize on the possible benefits of NPs in systems for cancer therapy. NPs have gained popularity as nanocarriers due to properties such as water dispersibility, biocompatibility, and biodegradability. The use of nanoparticles in cancer therapy improves drug solubility, improving the bioavailability of certain chemotherapeutic drugs (Peer et al., 2007; Wang et al., 2008). Recent research has primarily focused on the targeting, visualization, and diagnosing of primary breast cancer and its metastasis using multiple multifunctional nanoparticles. Accruing preclinical data suggest that nanoparticles can be used to target and illuminate primary breast cancer and metastasis. Active and passive targeting NPs for conventional imaging methods, such as magnetic resonance imaging, computer topographies, etc., can be aimed towards breast cancers (Mu et al., 2017). A better understanding of the individual tumor, tumor location, and host conditions, which affect nanomaterial distribution, especially at sites of metastatic breast cancer, would be essential to realize the potential of nanoparticle strategies (Grobmyer et al., 2012). Natural and inorganic particles are often used as drug carriers. Natural particles include dendrimers, carbon nanotubes, and liposomes. Magnetic and gold nanoparticles are examples of inorganic nanoparticles. Anticancer inhibitors can be either consciously or passively introduced with nanostructures in treating and diagnosing breast cancer (Bamrungsap et al., 2012).

4.2.3. Liposomes

Liposome NPs (LNPs) have proven advantageous as a platform for targeted delivery of oligonucleotides, peptides, and siRNA-based gene therapy for breast cancer. The use of LNP-encapsulation prevents peptides and nucleotides from degrading in the blood vessels, and allows for targeted delivery through target ligands (Hortobagyi et al., 2001). The combination drug delivery of quercetin and vincristine via liposomes was shown to have synergistic effects in treating ER-negative breast cancer (Wong and Chiu, 2010). For the treatment of metastatic breast cancer, non-PEGylated liposomes containing doxorubicin and cyclophosphamide, are used (Dhankhar et al., 2010). Furthermore, a liposome-conjugated antibody that overexpresses HER-2 reportedly provided 22-fold calcein to mammary epithelial cells (Kullberg et al., 2010).

4.2.4. Polymers and Dendrimers

Recent research suggested a novel drug delivery mechanism for HER-2-overexpressing metastatic breast cancer using HER-2-targeted HPMA copolymer conjugates in conjunction with a tyrosine kinase inhibitor. Their study on targeting HER-2 receptors through extracellular TRZ binding and intracellular kinase domains indicated that a drug conjugate delivery mechanism might synergistically affect anticancer activity. Various tests have shown that medications encapsulated with polymeric micelles are more effective against breast cancer than other drugs (Lee and Nan, 2012). The doxorubicin-containing polyion complex micelle accumulates in the nucleus of drug-resistant MCF-7 cells and is thought to have a powerful antiproliferative effect on targeted tumors (Lu et al., 2011). Several trials have shown that medications encapsulated with polymeric micelles are more effective against breast cancer than non-encapsulated drugs (Zhang et al., 2017).

4.2.5. Inorganic Nanoparticles

As reported in to a recent review, gold nanoparticles were targeted to the EGFR/VEGFR-2 signaling pathway for breast cancer. Their function in the inhibition of angiogenesis, migration, and metastasis of breast cancer cells was investigated. As a result, AuNPs-Qu-5 nanoparticles were found to inhibit various proteins in these cells, including PI3K, snail, slug, vimentin, Akt, N-cadherin, and p-GSK3β (Balakrishnan et al., 2016). When Superparamagnetic Iron Oxide NPs (SPIONs) are synthesized using laser pyrolysis, antitumor activity on MCF-7 cells is observed after coating the NPs with the anthracyclinic antibiotic violamycine B1 (Marcu et al., 2013). Another study compared the effects of three different SPIONs on BC cells, which differed in scale, form, zeta potential, and surface coating. The impacts of SPIONs on cellular absorption, magnetic properties, and cytotoxicity were analyzed in addition (Poller et al., 2017).

4.2.6. Carbon-based Nanoparticles

Multi-walled carbon nanotubes are coming into prominence in breast cancer therapy. A study suggested using trastuzumab-functionalized multiwalled carbon nanotubes and trastuzumab–diphtheria toxin conjugate to ablate breast cancer cells. Another intriguing technique employed carbon nanotubes as a preferential magnetic targeting method for non-invasive breast cancer monitoring (Oraki Kohshour et al., 2014; Al Faraj et al., 2015). Later, another study was published on the BC treatment using Carbon dots as the nanocarrier. Carbon dots and Dox were electrostatically conjugated, and the conjugate achieved higher cellular absorption and anti-tumor efficacy on MCF-7 cells than free Dox (Kong et al., 2018). The various strategies used in breast cancer therapy are listed in Figure 2.

Figure 2. Classification of Breast Cancer Therapy Methods

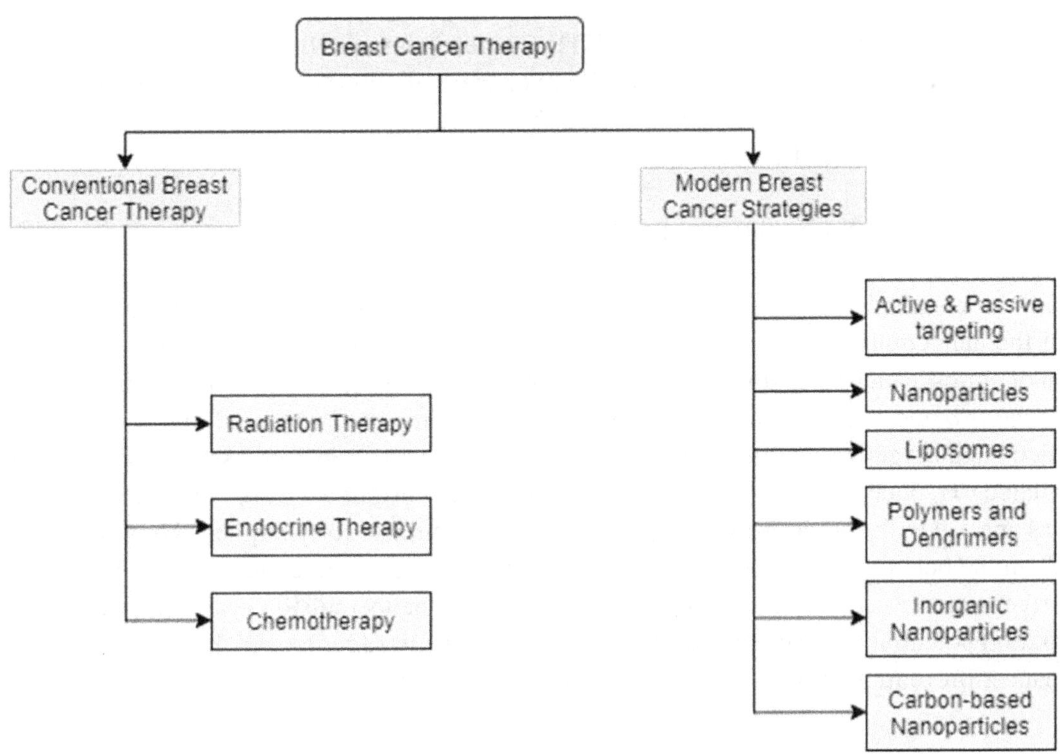

4.3. Synthetic Nanoemulsions for Breast Cancer Therapy

4.3.1. Docetaxel and Thymoquinone in Borage Oil-based NE

Studies on borage oil-based nanoemulsion (B-NE) and the combination (docetaxel + thymoquinone) B-NE showed that the (docetaxel + thymoquinone) B-NE combination inhibits MCF-7 and MDA-MB-231 cells better than the free combination drugs (docetaxel + thymoquinone). In addition, the cell death study of the (DTX + TQ) B-NE revealed a potential synergistic role of the combined drugs in simultaneously inducing autophagy and apoptosis (Alkhatib et al., 2020).

4.3.2. Paclitaxel Vitamin-E Nanoemulsion

Incorporation of paclitaxel (PTX) using high-pressure homogenization into a vitamin E NE resulted in more significant cytotoxicity in a breast cancer cell line (MCF-7). Furthermore, incorporating vitamin E in nanoemulsions resulted in a cascade of complementary molecular effects that amplified the primary antiproliferative function of PTX, and separately demonstrated efficiency in preserving the aggressive nature of the sluggish chronic immune response to breast cancer. In addition, the *in vivo* anticancer activity of PTX packed nanoemulsion was considerably higher than that of Taxol and free PTX (Pawar et al., 2014).

4.3.3. Tamoxifen nanoemulsion

A tamoxifen NE solution induced apoptosis in HTB-20 breast cancer cells. The *in vivo* effectiveness when assessed by inoculating mice with viable HTB-20 cancer cells followed by treatment with tamoxifen formulation after cancer had developed, showed that mortality due to cancer was reduced in animals treated with tamoxifen NE (Tagne et al., 2008).

4.3.4. Multifunctional C6 Ceramide NE

C6 ceramide was administered intraductal using a cationic bioadhesive nanoemulsion. The oil components of the nanoemulsion were monoolein, tricaprylin, and tributyrin. The aqueous phase included a solution of chitosan and poloxamer 407, with Polysorbate 80 acting as a surfactant. NBD ceramide was used for *in vitro* and *in vivo* experiments, while the pure compound was used to test the nanocarrier's cytotoxicity against cells in culture. The concentration of C6 ceramide decreased by 4.5-fold when nano encapsulated to reduce the cell viability of MCF-7 to 50%; a further decrease was observed when tributyrin was applied to the nanoemulsion, a phenomenon attributable to synergism. The findings obtained here affirm the benefit of the bioadhesive nanoemulsion for C6 ceramide localization in mammary tissue. Furthermore, because of the higher cytotoxicity of the nanoemulsion against cancer cells and the low irritation potential of the unloaded nanocarrier, it may be used for local delivery of active agents for breast cancer prevention in high-risk patients (Migotto et al., 2018).

4.3.5. Cholesterol-rich – nanoemulsion

A lipid mixture of cholesteryl oleate, egg phosphatidylcholine, triolein, and cholesterol was used to make a nanoemulsion, combined with paclitaxel oleate. The tumoral absorption, pharmacokinetics, and toxicity of paclitaxel associated with nanoemulsions were studied in breast cancer patients in a clinical trial. Two protocols were developed, one using paclitaxel oleate and the other using paclitaxel oleate combined with a nanoemulsion. The half-life of paclitaxel oleate is longer, and the overall plasma clearance is lower than paclitaxel dissolved in cremophor EL. Since the pharmacokinetic parameters are increased, the drug is concentrated in the neoplastic tissue, and the toxicity of paclitaxel is decreased. Thus, the paclitaxel-nanoemulsion preparation may be beneficial for use in the treatment of breast cancer (Pires et al., 2009).

4.3.6. Docetaxel-loaded Ligand - Solid Fat Nanoemulsions

Tristearin and soya PC-based solid fat nanoemulsions with a folate receptor-specific ligand (folate-PEG-cholesterol) have been shown to be effective for breast cancer therapy. The cytotoxicity of folate-conjugated PEG-solid fat nanoemulsions was compared to plain solid fat nanoemulsions and free DTX in vitro. In folate receptor-positive cells, the IC50 value of targeted folate-PEG-cholesterol nanoemulsions was lower than that of plain solid fat nanoemulsions. According to the findings, folate effectively facilitated the internalization of solid fat nanoemulsions encapsulating DTX to folate receptor-positive target tumor cells, and thereby provided the most effective killing of cancer cells through receptor-mediated targeting (Yadav and Gupta, 2015).

4.3.7. Coencapsulated DOX and W198 Lipid Nanoemulsions

Doxorubicin (DOX) and Bromotetrandrine (W198), two clinically approved chemotherapeutic agents, encapsulated in Lipid Nanoemulsion (DOX/W198-LNs) were shown to successfully reverse multidrug resistance in MCF-7/ADR breast cancer cells with substantially improved DOX cytotoxicity. This preparation reported an enhanced cellular uptake and intra-tumoral dissemination in mouse xenograft models based on the inhibitory action of W198 against P-glycoproteins (P-gp), and on the nanoscale LN mechanism that bypasses P-gp efflux via endocytosis. Furthermore, DOX/W198-LNs showed significantly reduced cardiac or GI tract toxicity, proving it to be a successful method for delivering cytotoxic chemotherapeutics in MDR tumors (Cao et al., 2015).

4.3.8. Edelfosine Nanoemulsions

Edelfosine nanoemulsions (ET-NEs), with edelfosine, Miglyol 812, and phosphatidylcholine as components, are preferred due to their high safety profile. ET-NEs' ability to stop tumor development in triple-negative breast cancer was shown *in vitro* and *in vivo* using zebrafish embryos and a highly aggressive and invasive triple-negative breast cancer cell line. Importantly, ET-NEs can cross the skin layer of MDA-MB 231 xenografted zebrafish embryos into the yolk sac, resulting in an effective decrease in the proliferation of highly aggressive and parasitic tumoral cells. Overall, the findings show that ET-NEs can constitute alternative treatment options for triple-negative breast cancer (Saraiva et al., 2021).

4.4. Natural-based Nanoemulsions for Breast Cancer Therapy

4.4.1. Sour Cherry Pit Oil

Sour cherry pit oil (SCPO)-NE has shown promising cytotoxic, anti-apoptotic, and anti-tumor activity in MCF7 cells and in a breast cancer model. Tumor size reduction in the treated BCMM model and MCF-7 cancer cells indicated an increased apoptotic function in SCPO nanoemulsions compared to SCPO alone. However, the findings showed that the cytotoxic behavior of SCPO-NE is cell-dependent, requiring further studies (Maragheh et al., 2019).

4.4.2. Lapachol

Lapachol NE was prepared using hot homogenization method, demonstrating optimal conditions for intravenous administration. In addition, biodistribution and blood clearing tests confirm an improvement in the blood circulating time, leading to preferential tumor absorption. Thereby, it is confirmed to contribute to the higher antitumor activity of Lapachol NE compared to Lapachol. Furthermore, in vitro experiments confirmed that encapsulation has little effect on LAP's cytotoxicity function. These favorable results, in the absence of toxicity *in vivo* (in mice), lead to the authors' recommendation of Lapachol NE as an efficient cancer treatment technique (Mendes Miranda et al., 2021).

4.4.3. Curcumin

Interfacial pre-polymer deposition and spontaneous nano-emulsification methods were used to formulate the curcumin-nanoemulsion (CNE). In vitro breast cancer model, MCF-7 cells were used as a photosensitizing agent in photodynamic therapy. The use of CNE, synergistically with photodynamic therapy, increased caspase 3/7 activity in the studied MCF-7 cell model, signaling that this therapy initiates a series of events contributing to cellular apoptosis. Furthermore, CNE was successful as a photosensitizing drug in photodynamic treatment, as evidenced by phototoxic effects, increased ROS in both cell lines, and increased caspase-mediated apoptosis in the breast cancer line (Machado et al., 2019).

4.4.4. Camptothecin

Using the solvent evaporation technique, camptothecin (CPT)-loaded polymer-stabilized nanoemulsion (PSNE) has been formulated for passive targeted transmission to breast cancer. Drug release, *in-vivo* targeting ability, cytotoxicity, genotoxicity, and in-vivo biodistribution were all improved when PSNEs were used as compared to the extract. In addition, PSNE demonstrated potent cytotoxicity and genotoxicity in cancer cells but low DNA damage potential in human peripheral blood lymphocytes. In 4T1-breast tumor xenografted BALB/c mice, the efficacy of PSNEs for the targeted transmission of CPT into tumor regions was demonstrated. CPT was passively targeted from PSNE to breast cancer tissue, according to an *in-vivo* bio-distribution analysis (Sugumaran et al., 2018).

4.4.5. Citrus-Pectin Nanoemulsion Zataria Essential Oil

Citrus-Pectin nanoemulsion Zataria Essential Oil (CP/ZEONE) has been reported to be effective in a z-dose-dependent and time-dependent manner on breast cancer cells with no substantial effect on normal cells. According to the findings, CP/ZEO NE therapy resulted in ROS generation and MMP disruption in MDA-MB-231 cells, causing activation of the mitochondrial intrinsic apoptosis pathway. Furthermore, this study shows that CP/ZEO NE contributed to cell death by interfering with DNA, halting the cell cycle, and destroying DNA. The use of CP nanoemulsion thus protects or increases the potential of ZEO's anti-cancer activity by increasing its stability (Salehi et al., 2020).

4.4.6. *Nigella Sativa* Essential Oil

Table 2. Recent Nanoemulsions formulated and their functions

Nanoemulsion	Active Compound	Formulation Technique	Type of Cancer	Nanoemulsion Type	Reference
[10]-gingerol-loaded Nanoemulsion	[10]-gingerol	Ultrasonication	Breast Cancer	Natural	Zanesco-Fontes et al., 2021
Paclitaxel-loaded Nanoemulsion with Satureja Khuzestanica	Paclitaxel	High-speed Homogenizer	Breast Cancer	Synthetic	Attari et al., 2021
Edelfosine Nanoemulsions with Miglyol 812 And Phosphatidylcholine	Edelfosine	Ethanol Injection Method	Breast Cancer	Synthetic	Saraiva et al., 2021
Organogel Based Nanoemulsion	Paclitaxel	Emulsification Through Ultrasonication	Melanoma	Synthetic	Fardous et al., 2021
Date Palm Extract (Phoenix Dactylifera) Pegylated Nanoemulsion	Polyphenol (Date Palm Extract)	Twenty-three Full Factorial Design	Liver Cancer & Breast Cancer	Natural	Khalil et al., 2021
Protein-olive Oil-in-water Nanoemulsions	Curcumin	Ultrasonication	Breast Cancer	Natural	Bharmoria et al., 2021
Lipid-oil-nanodroplets	Combretastatin A4	Two-step High-pressure Homogenization	Colorectal Cancer	Natural	Charalambous et al., 2021
Cholesterol-rich Nanoemulsion	Aluminum Phthalocyanine Chloride	Emulsification	Brain Cancer (Glioblastoma)	Synthetic	Tedesco et al., 2021
Agarose-polyvinylpyrrolidone -hydroxyapatite in Double Nanoemulsion	Quercetin	Emulsification	Breast Cancer	Natural	Samadi et al., 2021
Nanemulsion Loaded Paclitaxel And BEZ235	Paclitaxel	Emulsification	Colon Cancer	Synthetic	Hu et al., 2021
W/O/W Nanoemulsion Encapsulating Tocotrienols, Caffeic Acid and Cisplatin	Tocotrienols, Caffeic Acid And Cisplatin	High-speed Homogenizer & High-speed Homogenizer	Lung Cancer & Liver Cancer	Natural	Ravindaran et al., 2021
Biomimetic Nanoemulsion Phd@Pm	Pd-1 Protein	Sonication	Breast Cancer	Synthetic	Zhang et al., 2021
Irinotecan-lipiodol Nanoemulsion	Irinotecan	Tessari Technique (Emulsification)	Colorectal Cancer	Synthetic	Melancon et al., 2021
Disulfiram Loaded Ion-sensitive Nanoemulsion	Disulfiram	Three-factor And Three-level Optimal Design.	Brain Cancer (Glioblastoma)	Synthetic	Qu et al., 2021
Berberine Nanoemulsion	Berberine	Spontaneous Emulsification	Cervical Cancer	Natural	Floriano et al., 2021
Celastrol Nanoemulsion	Celastrol	Ultrasonic Emulsification	Melanoma	Natural	Qiu et al., 2021
Ir780-perfluoropolyether Nanoemulsion	IR780	Emulsion Inversion Point	Ovarian Cancer	Synthetic	Hong et al., 2021
Lapachol-loaded Nanoemulsion	Lapachol	Hot Homogenization Method	Breast Cancer	Natural	Mendes Miranda et al., 2021
Arachis Hypogaea Oil Nanoemulsion	*Arachis Hypogaea*	Ultrasonication	Lung Cancer	Natural	Fazelifar et al., 2021

In this research, ultrasonic emulsification, polysorbate 80, and water were used to develop a highly-stable *Nigella Sativa* essential oil nanoemulsion (NSEO-NE). NSEO-NE dramatically reduces the survival of MCF-7 breast cancer cells. It has been shown that NSEO nanoemulsion decreases cell viability and changes nuclear morphology in a dose and time-dependent manner. The findings explicitly show that NSEO-NE mediated apoptosis in MCF-7 cells. These results support the possible use of NSEO-NE in breast cancer treatment, with a lot of scope for future translational studies (Periasamy et al., 2016). Some recently-formulated nanoemulsions have been listed in Table 2.

5. FUTURE PERSPECTIVES

Nanoemulsions have emerged as a visible and superior option in the field of therapeutics, especially in cancer therapy, over the last few decades. A steady increase is anticipated in the need for and demand for nanoemulsions due to their unique properties that make drug delivery much simpler. Until now, various nanoemulsions have been studied in animal models to determine their efficacy in drug delivery and encapsulation, as well as their toxicity. Large-scale industrial production of nanoemulsions is needed after clinical trials to open new avenues for improving and unlocking the full potential of nanoemulsions. During large-scale development, drug stability and the effect on formulation processes in nanoemulsions should be the prime factors of importance and studied with care.

Along with manufacturing, the association of nanoemulsions with target cells is a key research focus in drug development, with researchers investigating various methods for drug release and absorption (Sánchez-López et al., 2019). Recent studies have shown that combining traditional as well as novel treatments with nanoemulsions improve their performance. The advancement in nanoemulsion technology would necessitate a thorough understanding of the relationship between NE formulation and the myriad physical, chemical, and physiological variables associated with various pathologies and administration routes (Tayeb and Sainsbury, 2018). Advancements in the field of nanoemulsion in cancer therapy are in the early stages and are focused on fundamental research in laboratories, with results likely to be of very little clinical significance. As a result, it is critical to take more realistic and long-term strategic approaches to tackle the scale-up limitations in cancer therapy using nanoemulsions.

6. CONCLUSION

Nanoemulsions have the ability to be targeted to specific tumor sites, displaying enhanced permeability, as well as retention and sustained release of drugs. Various studies at clinical and preclinical stages have shown promising results in terms of enhanced cytotoxic and apoptotic effects on the cancer cells, with reduced side effects on the normal cells surrounding the region. A low but effective dose of drug because of the specificity and possible synergistic effect might give a high therapeutic potential. It has been proven to overcome the major problems of multi drug resistance and the ability of encapsulate opens up varied options for tailored treatment modalities. The freedom to use several plant-derived components and custom made polymers are innumerable in this system with superior properties.

These characteristics, along with the added advantage of versatility in terms of drug delivery routes, make nanoemulsions an extremely promising candidate for cancer therapy. This review has highlighted on the recent progress in methodology to treat breast cancer using nanoemulsions of varied types. However, more research needs to be done to overcome the production challenges and a fully effective methodology. Further clinical trials and related studies would give a better understanding of the in-vivo behavior of the nanoemulsions, which might be a potential cure for the dreaded cancer. With constant efforts, it is anticipated that several nanoemulsion formulations will be available in the market in the coming years.

REFERENCES

Aboofazeli, R. (2010). Nanometric-scaled emulsions (nanoemulsions). *Iranian Journal of Pharmaceutical Research*, *9*(4), 325–326.

Ahmed, M., Smith, D. M., Hamouda, T., Rangel-Moreno, J., Fattom, A., & Khader, S. A. (2017). A novel nanoemulsion vaccine induces mucosal Interleukin-17 responses and confers protection upon Mycobacterium tuberculosis challenge in mice. *Vaccine*, *35*(37), 4983–4989.

Al Faraj, A., Shaik, A. S., & Al Sayed, B. (2015). Preferential magnetic targeting of carbon nanotubes to cancer sites: Noninvasive tracking using MRI in a murine breast cancer model. *Nanomedicine (London)*, *10*(6), 931–948.

Alany, R. G., Rades, T., Nicoll, J., Tucker, I. G., & Davies, N. M. (2006). W/O microemulsions for ocular delivery: Evaluation of ocular irritation and precorneal retention. *Journal of Controlled Release*, *111*(1-2), 145–152.

Alkhatib, M. H., Al-Otaibi, W. A., & Wali, A. N. (2018). Antineoplastic activity of mitomycin C formulated in nanoemulsions-based essential oils on HeLa cervical cancer cells. *Chemico-Biological Interactions*, *291*, 72–80.

Alkhatib, M. H., Bawadud, R. S., & Gashlan, H. M. (2020). Incorporation of docetaxel and thymoquinone in borage nanoemulsion potentiates their antineoplastic activity in breast cancer cells. *Scientific Reports*, *10*(1), 18124.

Allen, T. M., & Martin, F. J. (2004). Advantages of liposomal delivery systems for anthracyclines. *Seminars in Oncology*, *31*(6, Suppl 13), 5–15.

Arbain, N. H., Salim, N., Masoumi, H., Wong, T. W., Basri, M., & Abdul Rahman, M. B. (2019). In vitro evaluation of the inhalable quercetin loaded nanoemulsion for pulmonary delivery. *Drug Delivery and Translational Research*, *9*(2), 497–507.

Attari, F., Hazim, H., Zandi, A., Mazarei, Z., & Rafati, H. (2021). Circumventing paclitaxel resistance in breast cancer cells using a nanoemulsion system and determining its efficacy via an impedance biosensor. *Analyst (London)*, *146*(10), 3225–3233.

Balakrishnan, S., Bhat, F. A., Raja Singh, P., Mukherjee, S., Elumalai, P., Das, S., Patra, C. R., & Arunakaran, J. (2016). Gold nanoparticle-conjugated quercetin inhibits epithelial-mesenchymal transition, angiogenesis and invasiveness via EGFR/VEGFR-2-mediated pathway in breast cancer. *Cell Proliferation*, *49*(6), 678–697.

Bamrungsap, S., Zhao, Z., Chen, T., Wang, L., Li, C., Fu, T., & Tan, W. (2012). Nanotechnology in therapeutics: A focus on nanoparticles as a drug delivery system. *Nanomedicine (London)*, *7*(8), 1253–1271. https://doi.org/10.2217/nnm.12.87

Barta, J. A., Powell, C. A., & Wisnivesky, J. P. (2019). Global Epidemiology of Lung Cancer. *Annals of Global Health*, *85*(1), 8.

Bayón-Cordero, L., Alkorta, I., & Arana, L. (2019). Application of Solid Lipid Nanoparticles to Improve the Efficiency of Anticancer Drugs. *Nanomaterials (Basel, Switzerland)*, *9*(3), 474.

Bharmoria, P., Bisht, M., Gomes, M. C., Martins, M., Neves, M. C., Mano, J. F., Bdikin, I., Coutinho, J., & Ventura, S. (2021). Protein-olive oil-in-water nanoemulsions as encapsulation materials for curcumin acting as anticancer agent towards MDA-MB-231 cells. *Scientific Reports*, *11*(1), 9099.

Buzdar, A. U., Ibrahim, N. K., Francis, D., Booser, D. J., Thomas, E. S., Theriault, R. L., Pusztai, L., Green, M. C., Arun, B. K., Giordano, S. H., Cristofanilli, M., Frye, D. K., Smith, T. L., Hunt, K. K., Singletary, S. E., Sahin, A. A., Ewer, M. S., Buchholz, T. A., Berry, D., & Hortobagyi, G. N. (2005). Significantly higher pathologic complete remission rate after neoadjuvant therapy with trastuzumab, paclitaxel, and epirubicin chemotherapy: Results of a randomized trial in human epidermal growth factor receptor 2-positive operable breast cancer. *Journal of Clinical Oncology*, *23*(16), 3676–3685.

Cao, X., Luo, J., Gong, T., Zhang, Z. R., Sun, X., & Fu, Y. (2015). Coencapsulated doxorubicin and bromotetrandrine lipid nanoemulsions in reversing multidrug resistance in breast cancer in vitro and in vivo. *Molecular Pharmaceutics*, *12*(1), 274–286.

Carmeliet, P., & Jain, R. K. (2011). Molecular mechanisms and clinical applications of angiogenesis. *Nature*, *473*(7347), 298–307.

Castaneda, S. A., & Strasser, J. (2017). Updates in the Treatment of Breast Cancer with Radiotherapy. *Surgical Oncology Clinics of North America*, *26*(3), 371–382.

Chang, H. B., & Chen, B. H. (2015). Inhibition of lung cancer cells A549 and H460 by curcuminoid extracts and nanoemulsions prepared from Curcuma longa Linnaeus. *International Journal of Nanomedicine*, *10*, 5059–5080.

Charalambous, A., Mico, V., McVeigh, L. E., Marston, G., Ingram, N., Volpato, M., Peyman, S. A., McLaughlan, J. R., Wierzbicki, A., Loadman, P., Bushby, R. J., Markham, A. F., Evans, S. D., & Coletta, P. L. (2021). Targeted microbubbles carrying lipid-oil-nanodroplets for ultrasound-triggered delivery of the hydrophobic drug, Combretastatin A4. *Nanomedicine; Nanotechnology, Biology, and Medicine*, 102401. Advance online publication.

Charles, L., & Attama, A.A. (2011) Current state of nanoemulsions in drug delivery. *Journal of Biomaterials amd Nanobiotechology, 2*, 626-639.

Chime, S. A., Kenechukwu, F. C., & Attama, A. A. (2014). *Nanoemulsions — Advances in Formulation*. In *Characterization and Applications in Drug Delivery, Application of Nanotechnology in Drug Delivery*. IntechOpen.

Choudhury, H., Gorain, B., Karmakar, S., Biswas, E., Dey, G., Barik, R., Mandal, M., & Pal, T. K. (2014). Improvement of cellular uptake, in vitro antitumor activity and sustained release profile with increased bioavailability from a nanoemulsion platform. *International Journal of Pharmaceutics*, *460*(1-2), 131–143.

Chrastina, A., Baron, V. T., Abedinpour, P., Rondeau, G., Welsh, J., & Borgström, P. (2018). Plumbagin-Loaded Nanoemulsion Drug Delivery Formulation and Evaluation of Antiproliferative Effect on Prostate Cancer Cells. *BioMed Research International.*

Peto, R., Davies, C., Godwin, J., Gray, R., Pan, H. C., Clarke, M., Cutter, D., Darby, S., McGale, P., Taylor, C., Wang, Y. C., Bergh, J., Di Leo, A., Albain, K., Swain, S., Piccart, M., & Pritchard, K. (2012). Comparisons between different polychemotherapy regimens for early breast cancer: Meta-analyses of long-term outcome among 100,000 women in 123 randomised trials. *Lancet*, *379*(9814), 432–444.

Danhier, F., Feron, O., & Préat, V. (2010). To exploit the tumor microenvironment: Passive and active tumor targeting of nanocarriers for anti-cancer drug delivery. *Journal of Controlled Release*, *148*(2), 135–146.

de Matos, R., Calmon, M. F., Amantino, C. F., Villa, L. L., Primo, F. L., Tedesco, A. C., & Rahal, P. (2018). Effect of Curcumin-Nanoemulsion Associated with Photodynamic Therapy in Cervical Carcinoma Cell Lines. *BioMed Research International*, *2018*, 4057959.

Dean, M. (2009). ABC transporters, drug resistance, and cancer stem cells. *Journal of Mammary Gland Biology and Neoplasia*, *14*(1), 3–9.

Dehelean, C. A., Feflea, S., Gheorgheosu, D., Ganta, S., Cimpean, A. M., Muntean, D., & Amiji, M. M. (2013). Anti-angiogenic and anti-cancer evaluation of betulin nanoemulsion in chicken chorioallantoic membrane and skin carcinoma in Balb/c mice. *Journal of Biomedical Nanotechnology*, *9*(4), 577–589.

Dhankhar, R., Vyas, S. P., Jain, A. K., Arora, S., Rath, G., & Goyal, A. K. (2010). Advances in novel drug delivery strategies for breast cancer therapy. *Artificial Cells, Blood Substitutes, and Immobilization Biotechnology*, *38*(5), 230–249.

Early Breast Cancer Trialists' Collaborative Group (EBCTCG). (2005). Effects of chemotherapy and hormonal therapy for early breast cancer on recurrence and 15-year survival: An overview of the randomised trials. *Lancet*, *365*(9472), 1687–1717.

Egan, J.E., Ojima, I., Amiji, M.M., & Botchkina, G.I. (2018). *Nanoemulsion compositions of taxoid drugs, and methods for the use thereof to target cancer cells and cancer stem cells*. Academic Press.

Fardous, J., Omoso, Y., Joshi, A., Yoshida, K., Patwary, M., Ono, F., & Ijima, H. (2021). Development and characterization of gel-in-water nanoemulsion as a novel drug delivery system. *Materials Science and Engineering C*, *124*, 112076.

Fazelifar, P., Tabrizi, M. H., & Rafiee, A. (2021). The *Arachis hypogaea* Essential Oil Nanoemulsion as an Efficient Safe Apoptosis Inducer in Human Lung Cancer Cells (A549). *Nutrition and Cancer*, *73*(6), 1059–1067.

Floriano, B. F., Carvalho, T., Lopes, T. Z., Takahashi, L., Rahal, P., Tedesco, A. C., & Calmon, M. F. (2021). Effect of berberine nanoemulsion Photodynamic therapy on cervical carcinoma cell line. *Photodiagnosis and Photodynamic Therapy*, *33*, 102174.

Ganta, S., Devalapally, H., Shahiwala, A., & Amiji, M. (2008). A review of stimuli-responsive nanocarriers for drug and gene delivery. *Journal of Controlled Release*, *126*(3), 187–204.

Ganta, S., Talekar, M., Singh, A., Coleman, T. P., & Amiji, M. M. (2014). Nanoemulsions in translational research-opportunities and challenges in targeted cancer therapy. *AAPS PharmSciTech*, *15*(3), 694–708.

Glassman, D., Hignett, S., Rehman, S., Linforth, R., & Salhab, M. (2017). Adjuvant Endocrine Therapy for Hormone-positive Breast Cancer, Focusing on Ovarian Suppression and Extended Treatment: An Update. *Anticancer Research*, *37*(10), 5329–5341.

Goldberg, M. A., & Schneider, T. J. (1994). Similarities between the oxygen-sensing mechanisms regulating the expression of vascular endothelial growth factor and erythropoietin. *The Journal of Biological Chemistry*, *269*(6), 4355–4359.

Grobmyer, S. R., Zhou, G., Gutwein, L. G., Iwakuma, N., Sharma, P., & Hochwald, S. N. (2012). Nanoparticle delivery for metastatic breast cancer. *Nanomedicine; Nanotechnology, Biology, and Medicine*, *8*(Suppl 1), S21–S30.

Groo, A. C., Hedir, S., Since, M., Brotin, E., Weiswald, L. B., Paysant, H., Nee, G., Coolzaet, M., Goux, D., Delépée, R., Freret, T., Poulain, L., Voisin-Chiret, A. S., & Malzert-Fréon, A. (2020). Pyridoclax-loaded nanoemulsion for enhanced anticancer effect on ovarian cancer. *International Journal of Pharmaceutics*, *587*, 119655.

Gupta, A., Eral, H. B., Hatton, T. A., & Doyle, P. S. (2016). Nanoemulsions: Formation, properties and applications. *Soft Matter*, *12*(11), 2826–2841.

Hakemi-Vala, M., Rafati, H., Aliahmadi, A., & Ardalan, A. (2017). Nanoemulsions: A Novel Antimicrobial Delivery System. *Nano- and Microscale Drug Delivery Systems*, 245–266.

Hamouda, T., Hayes, M. M., Cao, Z., Tonda, R., Johnson, K., Wright, D. C., Brisker, J., & Baker, J. R. Jr. (1999). A novel surfactant nanoemulsion with broad-spectrum sporicidal activity against Bacillus species. *The Journal of Infectious Diseases*, *180*(6), 1939–1949.

Hamouda, T., Myc, A., Donovan, B., Shih, A. Y., Reuter, J. D., & Baker, J. R. Jr. (2001). A novel surfactant nanoemulsion with a unique non-irritant topical antimicrobial activity against bacteria, enveloped viruses and fungi. *Microbiological Research*, *156*(1), 1–7.

Handa, M., Ujjwal, R. R., Vasdev, N., Flora, S. J. S., & Shukla, R. (2021). Optimization of Surfactant- And Cosurfactant-Aided Pine Oil Nanoemulsions by Isothermal Low-Energy Methods for Anticholinesterase Activity. *ACS Omega*, *6*(1), 559–568.

Harris, J. R. (2014). Fifty years of progress in radiation therapy for breast cancer. *American Society of Clinical Oncology educational book. American Society of Clinical Oncology Annual Meeting*, 21–25.

Herbst, R. S., Heymach, J. V., & Lippman, S. M. (2008). Lung cancer. *The New England Journal of Medicine*, *359*(13), 1367–1380.

Hong, L., Wang, J. L., Geng, J. X., Zhao, Y. H., Zhou, G. X., Zhang, J., Liu, L. W., & Qu, J. L. (2021). Rational design of an oxygen-enriching nanoemulsion for enhanced near-infrared laser activatable photodynamic therapy against hypoxic tumors. *Colloids and Surfaces. B, Biointerfaces*, *198*, 111500.

Hortobagyi, G. N., Ueno, N. T., Xia, W., Zhang, S., Wolf, J. K., Putnam, J. B., Weiden, P. L., Willey, J. S., Carey, M., Branham, D. L., Payne, J. Y., Tucker, S. D., Bartholomeusz, C., Kilbourn, R. G., De Jager, R. L., Sneige, N., Katz, R. L., Anklesaria, P., Ibrahim, N. K., ... Hung, M. C. (2001). Cationic liposome-mediated E1A gene transfer to human breast and ovarian cancer cells and its biologic effects: A phase I clinical trial. *Journal of Clinical Oncology*, *19*(14), 3422–3433.

Houssami, N., Macaskill, P., von Minckwitz, G., Marinovich, M. L., & Mamounas, E. (2012). Meta-analysis of the association of breast cancer subtype and pathologic complete response to neoadjuvant chemotherapy. *European Journal of Cancer*, *48*(18), 3342–3354.

Hsu, H. J., Huang, R. F., Kao, T. H., Inbaraj, B. S., & Chen, B. H. (2017). Preparation of carotenoid extracts and nanoemulsions from Lycium barbarum L. and their effects on growth of HT-29 colon cancer cells. *Nanotechnology*, *28*(13), 135103.

https://clinicaltrials.gov/

Hu, Y., Zhang, K., Zhu, X., Zheng, X., Wang, C., Niu, X., Jiang, T., Ji, X., Zhao, W., Pang, L., Qi, Y., Li, F., Li, L., Xu, Z., Gu, W., & Zou, H. (2021). Synergistic Inhibition of Drug-Resistant Colon Cancer Growth with PI3K/mTOR Dual Inhibitor BEZ235 and Nano-emulsioned Paclitaxel via Reducing Multidrug Resistance and Promoting Apoptosis. *International Journal of Nanomedicine*, *16*, 2173–2186.

Hughes, K. S., Schnaper, L. A., Bellon, J. R., Cirrincione, C. T., Berry, D. A., McCormick, B., Muss, H. B., Smith, B. L., Hudis, C. A., Winer, E. P., & Wood, W. C. (2013). Lumpectomy plus tamoxifen with or without irradiation in women age 70 years or older with early breast cancer: Long-term follow-up of CALGB 9343. *Journal of Clinical Oncology*, *31*(19), 2382–2387.

Jaiswal, M., Dudhe, R., & Sharma, P. K. (2015). Nanoemulsion: An advanced mode of drug delivery system. *Biotech*, *5*(2), 123–127.

Jampilek, J., & Brychtova, K. (2010). Azone analogues: Classification, design, and transdermal penetration principles. *Medicinal Research Reviews*, *32*(5), 907–947.

Jemal, A., Bray, F., Center, M. M., Ferlay, J., Ward, E., & Forman, D. (2011). Global cancer statistics. *CA: a Cancer Journal for Clinicians*, *61*(2), 69–90.

Kamba, T., & McDonald, D. M. (2007). Mechanisms of adverse effects of anti-VEGF therapy for cancer. *British Journal of Cancer*, *96*(12).

Kelmann, R. G., Kuminek, G., Teixeira, H. F., & Koester, L. S. (2007). Carbamazepine parenteral nanoemulsions prepared by spontaneous emulsification process. *International Journal of Pharmaceutics*, *342*(1-2), 231–239.

Khalil, H. E., Alqahtani, N. K., Darrag, H. M., Ibrahim, H. M., Emeka, P. M., Badger-Emeka, L. I., Matsunami, K., Shehata, T. M., & Elsewedy, H. S. (2021). Date Palm Extract (*Phoenix dactylifera*) PEGylated Nanoemulsion: Development, Optimization and Cytotoxicity Evaluation. *Plants*, *10*(4), 735.

Khan, I., Bahuguna, A., Kumar, P., Bajpai, V. K., & Kang, S. C. (2018). In vitro and in vivo antitumor potential of carvacrol nanoemulsion against human lung adenocarcinoma A549 cells via mitochondrial mediated apoptosis. *Scientific Reports*, *8*(1), 144.

Khatamian, N., Homayouni Tabrizi, M., Ardalan, P., Yadamani, S., & Darchini Maragheh, A. (2019). Synthesis of Carum Carvi essential oil nanoemulsion, the cytotoxic effect, and expression of caspase 3 gene. *Journal of Food Biochemistry*.

Kim, J. E., & Park, Y. J. (2017). Paclitaxel-loaded hyaluronan solid nanoemulsions for enhanced treatment efficacy in ovarian cancer. *International Journal of Nanomedicine*, *12*, 645–658.

Kong, T., Hao, L., Wei, Y., Cai, X., & Zhu, B. (2018). Doxorubicin conjugated carbon dots as a drug delivery system for human breast cancer therapy. *Cell Proliferation*, *51*(5), e12488.

Kotta, S., Khan, A. W., Pramod, K., Ansari, S. H., Sharma, R. K., & Ali, J. (2012). Exploring oral nanoemulsions for bioavailability enhancement of poorly water-soluble drugs. *Expert Opinion on Drug Delivery*, *9*(5), 585–598.

Kováčik, A., Kopečná, M., & Vávrová, K. (2020). Permeation enhancers in transdermal drug delivery: Benefits and limitations. *Expert Opinion on Drug Delivery*, *17*(2), 145–155. https://doi.org/10.1080/17425247.2020.1713087

Kreutz, F.T. (2019). *Nanoemulsions and methods for cancer therapy*. Academic Press.

Kullberg, M., Owens, J. L., & Mann, K. (2010). Listeriolysin O enhances cytoplasmic delivery by Her-2 targeting liposomes. *Journal of Drug Targeting*, *18*(4), 313–320.

Kümmel, S., Holtschmidt, J., & Loibl, S. (2014). Surgical treatment of primary breast cancer in the neoadjuvant setting. *British Journal of Surgery*, *101*(8), 912–924.

Lambert, L. K., Balneaves, L. G., Howard, A. F., Chia, S. K., & Gotay, C. C. (2018). Understanding adjuvant endocrine therapy persistence in breast Cancer survivors. *BMC Cancer*, *18*(1), 732.

Lammers, T., Kiessling, F., Hennink, W. E., & Storm, G. (2012). Drug targeting to tumors: Principles, pitfalls and (pre-) clinical progress. *Journal of Controlled Release*, *161*(2), 175–187.

Lawler, J. (2009). Introduction to the tumour microenvironment review series. *Journal of Cellular and Molecular Medicine*, *13*(8A), 1403–1404.

Lee, J. H., & Nan, A. (2012). Combination drug delivery approaches in metastatic breast cancer. *Journal of Drug Delivery*, *2012*.

Li, N., Nguyen, H. H., Byrom, M., & Ellington, A. D. (2011). Inhibition of cell proliferation by an anti-EGFR aptamer. *PLoS One*, *6*(6), e20299. https://doi.org/10.1371/journal.pone.0020299

Lu, H. L., Syu, W. J., Nishiyama, N., Kataoka, K., & Lai, P. S. (2011). Dendrimer phthalocyanine-encapsulated polymeric micelle-mediated photochemical internalization extends the efficacy of photodynamic therapy and overcomes drug-resistance in vivo. *Journal of Controlled Release*, *155*(3), 458–464.

Ma, Y., Liu, D., Wang, D., Wang, Y., Fu, Q., Fallon, J. K., Yang, X., He, Z., & Liu, F. (2014). Combinational delivery of hydrophobic and hydrophilic anticancer drugs in single nanoemulsions to treat MDR in cancer. *Molecular Pharmaceutics*, *11*(8), 2623–2630.

Machado, F. C., Adum de Matos, R. P., Primo, F. L., Tedesco, A. C., Rahal, P., & Calmon, M. F. (2019). Effect of curcumin-nanoemulsion associated with photodynamic therapy in breast adenocarcinoma cell line. *Bioorganic & Medicinal Chemistry*, *27*(9), 1882–1890.

Maeda, H., Wu, J., Sawa, T., Matsumura, Y., & Hori, K. (2000). Tumor vascular permeability and the EPR effect in macromolecular therapeutics: A review. *Journal of Controlled Release*, *65*(1-2), 271–284.

Mahato, R. (2017). Nanoemulsion as targeted drug delivery system for cancer therapeutics. *Journal of Pharmaceutical Sciences and Pharmacology*, *3*, 82–97.

Mahato, R., Tai, W., & Cheng, K. (2011). Prodrugs for improving tumor targetability and efficiency. *Advanced Drug Delivery Reviews*, *63*(8), 659–670.

Mahboobian, M. M., Seyfoddin, A., Aboofazeli, R., Foroutan, S. M., & Rupenthal, I. D. (2019). Brinzolamide-loaded nanoemulsions: Ex vivo transcorneal permeation, cell viability and ocular irritation tests. *Pharmaceutical Development and Technology*, *24*(5), 600–606.

Mansour, H. M., Rhee, Y. S., & Wu, X. (2009). Nanomedicine in pulmonary delivery. *International Journal of Nanomedicine*, *4*, 299–319.

Maragheh, A. D., Tabrizi, M. H., Karimi, E., Seyedi, S., & Khatamian, N. (2019). Producing the sour cherry pit oil nanoemulsion and evaluation of its anti-cancer effects on both breast cancer murine model and MCF-7 cell line. *Journal of Microencapsulation*, *36*(4), 399–409.

Marcu, A., Pop, S., Dumitrache, F., Mocanu, M., Niculite, C. M., Gherghiceanu, M., Lungu, C. P., Fleaca, C., Ianchis, R., Barbut, A., Grigoriu, C., & Morjan, I. (2013). Magnetic iron oxide nanoparticles as drug delivery system in breast cancer. *Applied Surface Science*, 281.

Mason, T. G., Wilking, J. N., Meleson, K., Chang, C. B., & Graves, S. M. (2006). Nanoemulsions: Formation, structure, and physical properties. *Journal of Physics Condensed Matter*, *18*, R635–R666.

McClements, D. J. (2010). Emulsion design to improve the delivery of functional lipophilic components. *Annual Review of Food Science and Technology*, *1*, 241–269.

McClements, D. J. (2012). Nanoemulsions versus microemulsions: Terminology, differences, and similarities. *Soft Matter*, *8*, 1719–1729.

Melancon, M. P., Yevich, S., Avritscher, R., Swigost, A., Lu, L., Tian, L., Damasco, J. A., Dixon, K., Cortes, A. C., Munoz, N. M., Liang, D., Liu, D., & Tam, A. L. (2021). A novel irinotecan-lipiodol nanoemulsion for intravascular administration: Pharmacokinetics and biodistribution in the normal and tumor bearing rat liver. *Drug Delivery*, *28*(1), 240–251.

Mendes Miranda, S. E., Alcântara Lemos, J., Fernandes, R. S., Silva, J. O., Ottoni, F. M., Townsend, D. M., Rubello, D., Alves, R. J., Cassali, G. D., Ferreira, L., & de Barros, A. (2021). Enhanced antitumor efficacy of lapachol-loaded nanoemulsion in breast cancer tumor model. *Biomedicine and Pharmacotherapy*, *133*, 110936.

Mkandawire, M., & Aryee, A. N. (2018). Resurfacing and modernization of edible packaging material technology. *Current Opinion in Colloid & Interface Science*, *19*, 104–112.

Modi, S., & Anderson, B. D. (2013). Determination of drug release kinetics from nanoparticles: Overcoming pitfalls of the dynamic dialysis method. *Molecular Pharmaceutics*, *10*(8), 3076–3089.

Mohammad, I. S., He, W., & Yin, L. (2018). Understanding of human ATP binding cassette superfamily and novel multidrug resistance modulators to overcome MDR. *Biomedicine and Pharmacotherapy*, *100*, 335–348.

Mu, Q., Wang, H., & Zhang, M. (2017). Nanoparticles for imaging and treatment of metastatic breast cancer. *Expert Opinion on Drug Delivery*, *14*(1), 123–136.

Nicolosi, R., & Shea, T. (2015). *Anti-oxidant synergy formulation nanoemulsions to treat cancer*. Academic Press.

Nicolosi, R., & Tagne, J. (2010). *Compositions and methods for treating cancer with dacarbazine nanoemulsions*. Academic Press.

Oraki Kohshour, M., Mirzaie, S., Zeinali, M., Amin, M., Said Hakhamaneshi, M., Jalili, A., Mosaveri, N., & Jamalan, M. (2014). Ablation of breast cancer cells using trastuzumab-functionalized multi-walled carbon nanotubes and trastuzumab-diphtheria toxin conjugate. *Chemical Biology & Drug Design*, *83*(3), 259–265.

Pannu, J., Mccarthy, A., Martin, A., Hamouda, T., Ciotti, S., Fothergill, A., & Sutcliffe, J. (2009). NB-002, a novel nanoemulsion with broad antifungal activ-ity against dermatophytes, other filamentous fungi, and Candida albicans. *Antimicrobial Agents and Chemotherapy*, *53*, 3273–3279.

Patel, R. B., Patel, M. R., Thakore, S. D., & Patel, B. G. (2017). Nanoemulsion as a valuable nanostructure platform for pharmaceutical drug delivery. In *Nano-and Microscale Drug Delivery Systems* (pp. 321–341). Elsevier.

Pawar, V. K., Panchal, S. B., Singh, Y., Meher, J. G., Sharma, K., Singh, P., Bora, H. K., Singh, A., Datta, D., & Chourasia, M. K. (2014). Immunotherapeutic vitamin E nanoemulsion synergies the antiproliferative activity of paclitaxel in breast cancer cells via modulating Th1 and Th2 immune response. *Journal of Controlled Release*, *196*, 295–306.

Peer, D., Karp, J. M., Hong, S., Farokhzad, O. C., Margalit, R., & Langer, R. (2007). Nanocarriers as an emerging platform for cancer therapy. *Nature Nanotechnology*, *2*(12), 751–760.

Periasamy, V. S., Athinarayanan, J., & Alshatwi, A. A. (2016). Anticancer activity of an ultrasonic nanoemulsion formulation of Nigella sativa L. essential oil on human breast cancer cells. *Ultrasonics Sonochemistry*, *31*, 449–455.

Pires, A., Fortuna, A., Alves, G., & Falcão, A. (2009). Intranasal drug delivery: How, why and what for? *Journal of Pharmacy & Pharmaceutical Sciences*, *12*(3), 288–311.

Pires, L. A., Hegg, R., Valduga, C. J., Graziani, S. R., Rodrigues, D. G., & Maranhão, R. C. (2009). Use of cholesterol-rich nanoparticles that bind to lipoprotein receptors as a vehicle to paclitaxel in the treatment of breast cancer: Pharmacokinetics, tumor uptake and a pilot clinical study. *Cancer Chemotherapy and Pharmacology*, *63*(2), 281–287.

Poller, J. M., Zaloga, J., Schreiber, E., Unterweger, H., Janko, C., Radon, P., Eberbeck, D., Trahms, L., Alexiou, C., & Friedrich, R. P. (2017). Selection of potential iron oxide nanoparticles for breast cancer treatment based on in vitro cytotoxicity and cellular uptake. *International Journal of Nanomedicine*, *12*, 3207–3220.

Prakash, R.T., & Thiagarajan, P. (2011). Nanoemulsions for drug delivery through different routes. *Research in Biotechnology, 2*(3), 1-13.

Primo, F. L., Bentley, M. V., & Tedesco, A. C. (2008). Photophysical studies and in vitro skin permeation/retention of foscan/ nanoemulsion (NE) applicable to photodynamic therapy skin cancer treatment. *Journal of Nanoscience and Nanotechnology*, *8*, 340–347.

Prow, T. W., Grice, J. E., Lin, L. L., Faye, R., Butler, M., Becker, W., Wurm, E. M., Yoong, C., Robertson, T. A., Soyer, H. P., & Roberts, M. S. (2011). Nanoparticles and microparticles for skin drug delivery. *Advanced Drug Delivery Reviews*, *63*(6), 470–491.

Qian, C., & McClements, D. J. (2011). Formation of nanoemulsions stabilized by model food-grade emulsifiers using high-pressure homogenization: Factors affecting particle size. *Food Hydrocolloids*, *25*, 1000–1008.

Qiu, N., Liu, Y., Liu, Q., Chen, Y., Shen, L., Hu, M., Zhou, X., Shen, Y., Gao, J., & Huang, L. (2021). Celastrol nanoemulsion induces immunogenicity and downregulates PD L1 to boost abscopal effect in melanoma therapy. *Biomaterials*, *269*, 120604.

Qu, Y., Li, A., Ma, L., Iqbal, S., Sun, X., Ma, W., Li, C., Zheng, D., Xu, Z., Zhao, Z., & Ma, D. (2021). Nose-to-brain delivery of disulfiram nanoemulsion in situ gel formulation for glioblastoma targeting therapy. *International Journal of Pharmaceutics*, *597*, 120250.

Ravindaran, R., Ng, M. H., Chandran, D., Ooi, K. K., & Manickam, S. (2021). Stable W/O/W multiple nanoemulsion encapsulating natural tocotrienols and caffeic acid with cisplatin synergistically treated cancer cell lines (A549 and HEP G2) and reduced toxicity on normal cell line (HEK 293). *Materials Science and Engineering C*, *121*, 111808.

Reinbolt, R. E., Mangini, N., Hill, J. L., Levine, L. B., Dempsey, J. L., Singaravelu, J., Koehler, K. A., Talley, A., & Lustberg, M. B. (2015). Endocrine therapy in breast cancer: The neoadjuvant, adjuvant, and metastatic approach. *Seminars in Oncology Nursing*, *31*(2), 146–155.

Romberg, B., Hennink, W. E., & Storm, G. (2008). Sheddable coatings for long-circulating nanoparticles. *Pharmaceutical Research*, *25*(1), 55–71.

Salehi, F., Jamali, T., Kavoosi, G., Ardestani, S. K., & Vahdati, S. N. (2020). Stabilization of Zataria essential oil with pectin-based nanoemulsion for enhanced cytotoxicity in monolayer and spheroid drug-resistant breast cancer cell cultures and deciphering its binding mode with gDNA. *International Journal of Biological Macromolecules*, *164*, 3645–3655.

Samadi, A., Pourmadadi, M., Yazdian, F., Rashedi, H., Navaei-Nigjeh, M., & Eufrasio-da-Silva, T. (2021). Ameliorating quercetin constraints in cancer therapy with pH-responsive agarose-polyvinylpyrrolidone -hydroxyapatite nanocomposite encapsulated in double nanoemulsion. *International Journal of Biological Macromolecules*, *182*, 11–25. Advance online publication.

Sánchez-López, E., Guerra, M., Dias-Ferreira, J., Lopez-Machado, A., Ettcheto, M., Cano, A., Espina, M., Camins, A., Garcia, M. L., & Souto, E. B. (2019). Current Applications of Nanoemulsions in Cancer Therapeutics. *Nanomaterials (Basel, Switzerland)*, *9*(6), 821.

Saraiva, S. M., Gutiérrez-Lovera, C., Martínez-Val, J., Lores, S., Bouzo, B. L., Díez-Villares, S., Alijas, S., Pensado-López, A., Vázquez-Ríos, A. J., Sánchez, L., & de la Fuente, M. (2021). Edelfosine nanoemulsions inhibit tumor growth of triple negative breast cancer in zebrafish xenograft model. *Scientific Reports*, *11*(1), 9873.

Sareen, S., Mathew, G., & Joseph, L. (2012). Improvement in solubility of poor water-soluble drugs by solid dispersion. *International Journal of Pharmaceutical Investigation*, *2*(1), 12–17.

Sarkar, F. H., Banerjee, S., & Li, Y. (2007). Pancreatic cancer: Pathogenesis, prevention and treatment. *Toxicology and Applied Pharmacology*, *224*(3), 326–336.

Shah, P., Bhalodia, D., & Shelat, P. (2010). Nanoemulsion: A pharmaceutical review. *Systematic Review in Pharmacy*, *1*, 24–32.

Shakeel, F., & Ramadan, W. (2010). Transdermal delivery of anticancer drug caffeine from water-in-oil nanoemulsions. *Colloids and Surfaces. B, Biointerfaces*, *75*(1), 356–362.

Shakeel, F., Ramadan, W., Faisal, M., Rizwan, M., Faiyazuddin, M., Mustafa, G., & Shafiq, S. (2010). Transdermal and Topical Delivery of Anti-inflammatory Agents Using Nanoemulsion/Microemulsion: An Updated Review. *Current Nanoscience*, *6*(2), 184–198.

Sharma, N., Mishra, S., Sharma, S., Desphande, R. D., & Sharma, R. K. (2013). Preparation and Optimization of Nanoemulsions for targeting Drug Delivery. *Journal of Drug Development and Research*, *5*(4), 37–48.

Singh, K. K., & Vingkar, S. K. (2008). Formulation, antimalarial activity and biodistribution of oral lipid nanoemulsion of primaquine. *International Journal of Pharmacology*, *347*, 136–143.

Singh, Y., Meher, J. G., Raval, K., Khan, F. A., Chaurasia, M., Jain, N. K., & Chourasia, M. K. (2017). Nanoemulsion: Concepts, development, and applications in drug delivery. *Journal of Controlled Release*, *252*, 28–49.

SMigotto, A., Carvalho, V., Salata, G. C., da Silva, F., Yan, C., Ishida, K., Costa-Lotufo, L. V., Steiner, A. A., & Lopes, L. B. (2018). Multifunctional nanoemulsions for intraductal delivery as a new platform for local treatment of breast cancer. *Drug Delivery*, *25*(1), 654–667.

Subhashis, D., Satayanarayana, J., & Gampa, V. K. (2011). Nanoemulsion-a method to improve the solubility of lipophilic drugs. *International Journal of Advanced Pharmacology and Science*, *2*(2-3), 72–83.

Sugumaran, A., Ponnusamy, C., Kandasamy, P., Krishnaswami, V., Palanichamy, R., Kandasamy, R., Lakshmanan, M., & Natesan, S. (2018). Development and evaluation of camptothecin loaded polymer stabilized nanoemulsion: Targeting potential in 4T1-breast tumour xenograft model. *European Journal of Pharmaceutical Sciences*, *116*, 15–25.

Tagne, J. B., Kakumanu, S., Ortiz, D., Shea, T., & Nicolosi, R. J. (2008). A nanoemulsion formulation of tamoxifen increases its efficacy in a breast cancer cell line. *Molecular Pharmaceutics*, *5*(2), 280–286.

Tayeb, H. H., & Sainsbury, F. (2018). Nanoemulsions in drug delivery: Formulation to medical application. *Nanomedicine (London)*, *13*(19), 2507–2525.

Tedesco, A. C., Silva, E., Jayme, C. C., Piva, H. L., & Franchi, L. P. (2021). Cholesterol-rich nanoemulsion (LDE) as a novel drug delivery system to diagnose, delineate, and treat human glioblastoma. *Materials Science and Engineering C*, *123*, 111984.

Tiwari, S., Tan, Y. M., & Amiji, M. (2006). Preparation and In Vitro Characterization of Multifunctional Nanoemulsions for Simultaneous MR Imaging and Targeted Drug Delivery. *Journal of Biomedical Nanotechnology*, *2*, 217–224.

Toub, N., Malvy, C., Fattal, E., & Couvreur, P. (2006). Innovative nanotechnologies for the delivery of oligonucleotides and siRNA. *Biomedicine and Pharmacotherapy*, *60*(9), 607–620.

Tsai, Y. J., & Chen, B. H. (2016). Preparation of catechin extracts and nanoemulsions from green tea leaf waste and their inhibition effect on prostate cancer cell PC-3. *International Journal of Nanomedicine*, *11*, 1907–1926.

Valente, S. A., & Shah, C. (2020). The Landmark Series: Adjuvant Radiation Therapy for Breast Cancer. *Annals of Surgical Oncology*, *27*(7), 2203–2211.

Vasir, J. K., & Labhasetwar, V. (2005). Targeted drug delivery in cancer therapy. *Technology in Cancer Research & Treatment*, *5*(5), 363–374.

Wang, C. Y., & Huang, L. (1987). pH-sensitive immunoliposomes mediate target-cell-specific delivery and controlled expression of a foreign gene in mouse. *Proceedings of the National Academy of Sciences of the United States of America*, *84*(22), 7851–7855.

Wang, X., Yang, L., Chen, Z. G., & Shin, D. M. (2008). Application of nanotechnology in cancer therapy and imaging. *CA: a Cancer Journal for Clinicians*, *58*(2), 97–110.

Ward, M. C., Vicini, F., Chadha, M., Pierce, L., Recht, A., Hayman, J., Thaker, N. G., Khan, A., Keisch, M., & Shah, C. (2019). Radiation Therapy Without Hormone Therapy for Women Age 70 or Above with Low-Risk Early Breast Cancer: A Microsimulation. *International Journal of Radiation Oncology, Biology, Physics*, *105*(2), 296–306.

Winter, P. M., Schmieder, A. H., Caruthers, S. D., Keene, J. L., Zhang, H., Wickline, S. A., & Lanza, G. M. (2008). Minute dosages of alpha(nu)beta3-targeted fumagillin nanoparticles impair Vx-2 tumor angiogenesis and development in rabbits. *The FASEB Journal*, *22*(8), 2758–2767.

Wiradharma, N., Zhang, Y., Venkataraman, S., Hedrick, J. L., & Yang, Y. Y. (2009). Self-assembled polymer nanostructures for delivery of anticancer therapeutics. *Nano Today*, *4*(4), 302–317.

Wong, M. Y., & Chiu, G. N. (2010). Simultaneous liposomal delivery of quercetin and vincristine for enhanced estrogen-receptor-negative breast cancer treatment. *Anti-Cancer Drugs*, *21*(4), 401–410.

Woodward, W. A., Barlow, W. E., Jagsi, R., Buchholz, T. A., Shak, S., Baehner, F., Whelan, T. J., Davidson, N. E., Ingle, J. N., King, T. A., Ravdin, P. M., Osborne, C. K., Tripathy, D., Livingston, R. B., Gralow, J. R., Hortobagyi, G. N., Hayes, D. F., & Albain, K. S. (2020). Association between 21-Gene Assay Recurrence Score and Locoregional Recurrence Rates in Patients With Node-Positive Breast Cancer. *JAMA Oncology*, *6*(4), 505–511.

Yadav, S., & Gupta, S. (2015). Development and in vitro characterization of docetaxel-loaded ligand appended solid fat nanoemulsions for potential use in breast cancer therapy. *Artificial Cells, Nanomedicine, and Biotechnology*, *43*(2), 93–102.

Yousefian Rad, E., Homayouni Tabrizi, M., Ardalan, P., Seyedi, S. M. R., Yadamani, S., Zamani-Esmati, P., & Haghani Sereshkeh, N. (2020). Citrus Lemon essential oil nanoemulsion (CLEO-NE), a safe cell-depended apoptosis inducer in human A549 lung cancer cells with anti-angiogenic activity. *Journal of Microencapsulation*, 1–24.

Yu, H., Xu, Z., Chen, X., Xu, L., Yin, Q., Zhang, Z., & Li, Y. (2014). Reversal of lung cancer multidrug resistance by pH-responsive micelleplexes mediating co-delivery of siRNA and paclitaxel. *Macromolecular Bioscience*, *14*(1), 100–109.

Yu, M. K., Park, J., & Jon, S. (2012). Targeting strategies for multifunctional nanoparticles in cancer imaging and therapy. *Theranostics*, *2*(1), 3–44. https://doi.org/10.7150/thno.3463

Yu, W., & Zhang, N. (2009). Surface Modification of Nanocarriers for Cancer Therapy. *Current Nanoscience*, *5*(2), 123–134.

Zanesco-Fontes, I., Silva, A., da Silva, P. B., Duarte, J. L., Di Filippo, L. D., Chorilli, M., Cominetti, M. R., & Martin, A. (2021). [10]-Gingerol-Loaded Nanoemulsion and its Biological Effects on Triple-Negative Breast Cancer Cells. *AAPS PharmSciTech*, *22*(5), 157.

Zelnak, A. B., & O'Regan, R. M. (2015). Optimizing Endocrine Therapy for Breast Cancer. *Journal of the National Comprehensive Cancer Network: JNCCN*, *13*(8), e56–e64.

Zhang, J., Kinoh, H., Hespel, L., Liu, X., Quader, S., Martin, J., Chida, T., Cabral, H., & Kataoka, K. (2017). Effective treatment of drug resistant recurrent breast tumors harboring cancer stem-like cells by staurosporine/epirubicin co-loaded polymeric micelles. *Journal of Controlled Release*, *264*, 127–135.

Zhang, L. W., Al-Suwayeh, S. A., Hung, C. F., Chen, C. C., & Fang, J. Y. (2011). Oil components modulate the skin delivery of 5-aminolevulinic acid and its ester pro drug from oil-in-water and water-in-oil nanoemulsions. *International Journal of Nanomedicine*, *6*, 693–704.

Zhang, Y., Liao, Y., Tang, Q., Lin, J., & Huang, P. (2021). Biomimetic Nanoemulsion for Synergistic Photodynamic-Immunotherapy Against Hypoxic Breast Tumor. *Angewandte Chemie*, *60*(19), 10647–10653.

Chapter 14
Nanoemulsion:
A Boon to Liver and Gastric Cancer Treatment

Priyadarshini Mohapatra
Vellore Institute of Technology, India

Natarajan Chandrasekaran
Vellore Institute of Technology, India

Amitava Mukherjee
Vellore Institute of Technology, India

ABSTRACT

Nanoemulsion is the major vehicle for delivering different types of drugs, nucleic acids, and imaging agents. Due to their attractive properties, it has been extensively used for diagnostics of cancer therapy and imaging. However, nanoemulsion is designed through multiple functions by modifications in surface and encapsulation of active compounds against cancer. In nanoemulsion, the surface alteration can be changed by targeting the surface charge, a targeting ligand. The core of the emulsion can be loaded with drugs, imaging agents, and contrast agents. In this chapter, the application of nanoemulsion against specifically liver and gastric cancer is explored briefly. The major focuses on the severity of cancer, multifunctional nature of respective drug-loaded nanoemulsions, how to defeat the physiological hurdles, targeted and non-targeted delivery of nanoemulsion, clinical and preclinical studies are discussed with trending examples from the review of the literature and future perspectives.

1. INTRODUCTION

According to IARC GLOBOCAN 2020 report: cancer is the leading cause of death, representing 19.3 million new cases and 10 million deaths in 2019. Cancer is the loss of normal control of growth while compared with normal cells. It has disrupted the balance between new cell growth and old cell death(Gorain et al., 2020). This disruption of cells is the reason behind the abnormal growth of the cells.

DOI: 10.4018/978-1-7998-8378-4.ch014

The basic reason behind cancer is a mutation in the genetic code. 7.1% of millions of people died of liver cancer, and 1.01% died of gastric cancer (Mahato, 2017). Gastric cancer is public health burden in developing countries. It is the second-highest reason behind death compared to others (Sánchez-López et al., 2019). Even though so many therapies are delineated to cure cancer, but the success is very limited. There are different types of treatments noted down (chemotherapy, immunotherapy, radiotherapy). Among all these treatments, chemotherapy is the most used therapy and is the primary treatment for malignant disease(Ashaolu, 2021). Targeting the obstacles, the Nano system has great potential to take over against cancer treatment(Plaza-Oliver et al., 2021). Different nanocarriers are being explored due to their efficient drug delivery, such as Nanoemulsion, Nanoparticle, nanoshells, nanocapsule, core-shell nanoparticle, nanotube, lipid nanoparticle, dendrimers etc. (Gorain et al., 2020). Some of these nanomaterial's toxicity is not explored, and to get clear from this study clinical, and preclinical studies are required(Sánchez-López et al., 2019). There are different types of cancer, out of which Liver and gastrointestinal cancer are two major and known as major global death issues(Upadhyay et al., 2020). During the past few decades, public health care efforts and the incidence of age-standardized cases and new cases of liver cancer are increased globally. Different studies suggested that the patterns used in the occurrence of liver cancer are different across the regions; it is mainly occurred by the currency of risk factors(Sánchez-López et al., 2019). The reasons are different for the disease in different countries. In China, the Hepatitis B virus (HBV) is the crucial reason for Liver cancer(Yousefian Rad et al., 2020), whereas, in South Korea and Japan, the Hepatitis C virus (HCV) is found to be the major reason behind hepatocellular carcinoma(Praveen Kumar, 2015). Getting the occurrence pattern of respective cancer and trends will optimize more beneficial and curious prevention strategies(Ganta et al., 2014). Globally it may help to decrease the cases of liver cancer. According to the GBD (Global Burden of Disease) study, the main three reasons behind liver cancer are Hepatitis B, Hepatitis C and alcohol consumption(Choudhury et al., 2019). The mortality rate of liver cancer is highest in Hepatitis B (43.4%), and the Hepatitis C mortality rate (18.7%) is more than the alcohol consumption mortality rate (14.7%)(Ganta et al., 2014). Due to these many reasons, lung cancer is very much effective and difficult to cure. The mortality rate of liver cancer is presented in **Figure 1.**

Figure 1. According to the GBD Study, Liver cancer mortality rate(Upadhyay et al., 2020).

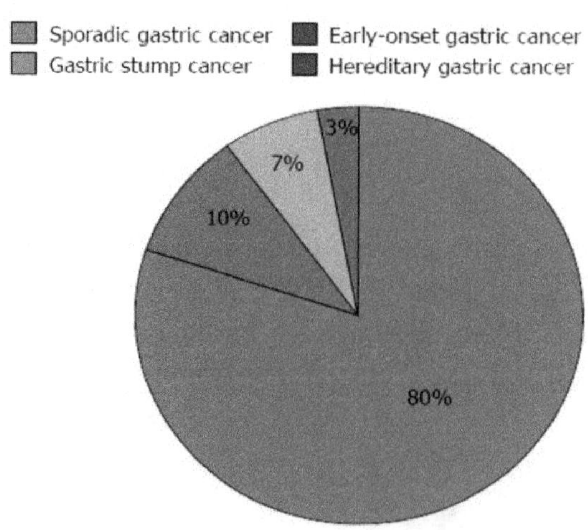

Gastric cancer is known as the fourth highest health problem in the cancer world. It is also known as the highest rate of global burden in recent years(Gorain et al., 2020). It is a major problem in North America and Western European countries. This is caused by Helicobacter pylori, which induce chronic inflammation, carcinogenesis, and various factors, including bacterial host (Ganta et al., 2014). Cell proliferation and apoptosis may lead to inflammation linked to oncogenesis. Another pathogen linked with gastric cancer is the Epstein-Bar virus, which is mostly found in malignant cells compared to normal epithelial cells(Plaza-Oliver et al., 2021). There are some specific reasons behind it, the first reason is hereditary, and the second is gastric adenocarcinoma and intestinal gastric cancer. In hereditary cases, CDH1 has mutations found in around 45% of families. Apart from this, mutation CTNNA1 is also identified as the genetic reason behind gastric cancer. There are many environmental factors associated with this, and day to day lifestyle has a great impact on it(Choudhury et al., 2019). People with gastric oesophageal reflux and obesity also contribute to the increasing incidence of gastric cancer(Sánchez-López et al., 2019). The severity rate of different types of gastric cancer is presented in **Figure 2.**

Figure 2. Different types of Gastric cancer and severity rate(Sánchez-López et al., 2019).

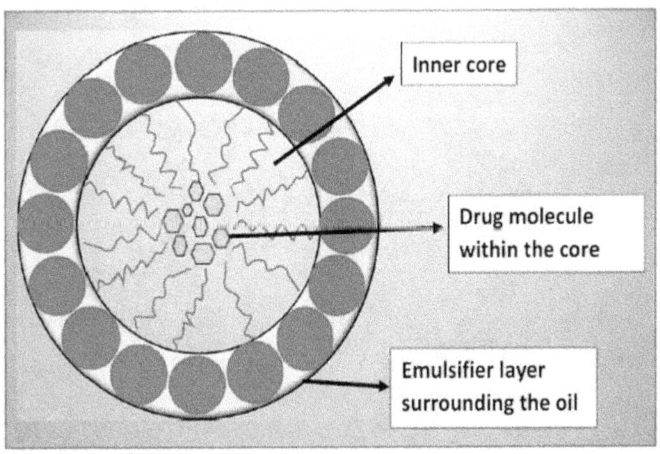

There are so many treatment therapies available against both these cancers. Surgeries are preferred if the disease is detected in an early stage. Nevertheless, where the patient's condition is critical and cannot go for surgery, they carry out chemotherapy drugs(Guo et al., 2018). The major hurdle in the liver and gastric cancer is targeting and efficacious delivery of agents to the particular site while avoiding the side effects of systematic administration. Moreover, this is very important to develop a more efficient drug delivery system with less toxic effects on the normal cells(Tang et al., 2020). So, nanotechnology is very powerful for the release and targeting of Hepatocellular carcinoma and Gastric cancer. Nanosystem will help conquer the toxic effects of systematic administration of chemotherapeutics by getting better pharmacokinetics, the combination of cytotoxic agents in the tumour's location, and the efficiency of treatment via nanosystem(Li et al., 2020). Specifically for HCC, there are delivery strategies explored for nanoparticles out of which liposomes and micelles are passively and receptor-mediated dynamic targeting pathways (Aravalli et al., 2008). Depending upon the mechanism, it is allowed for novel therapeutic agents and rational design of methodology. The emergence of combinatorial and high-output methods are mentioned for treatment(Syed, 2020).

The drug will be delivered through a nanosystem and may reach the tumour through vascular tissue.

Moreover, the cancer patient is almost developed based on prolonged inflammatory processes(Date et al., 2010), which emerged due to genetic alterations(Jerobin et al., 2012). Compared to all the nanosystem techniques, nanoemulsion is on the trend because of its many pharmacokinetics properties against cancer cells(Journal & Issue, 2013). Nanoemulsion is a heterogeneous system where the oil phase will be dispersed in the water phase as a droplet, and it will stabilize by emulsifying agent(Journal & Issue, 2013). Emulsifying agents play a great role to reduce the interfacial surface tension between the immiscible liquid phase of the aqueous phase and oil (Ghosh et al., 2014). Emulsifying agents are amphiphilic surface-active molecules or surfactants in nature. The droplet size of the nanoemulsion is presented in **Figure. 3.**

Figure 3. Structural representation of nanoemulsion droplet size (Plaza-Oliver et al., 2021).

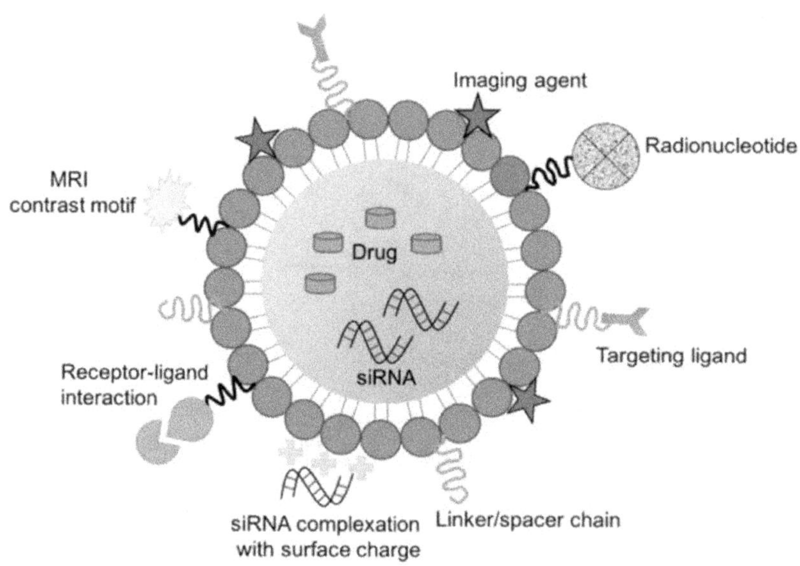

Nanoemulsion is the delivery system because of its greater stability and bioavailability in chemotherapeutic than conventional drug delivery(Saranya Sugumar et al., 2012). Apart from all the nanosystems, nanoemulsion is a better one because of its increased surface, and so it is used to help in the higher adsorption of the drug and help improve the drug's bioavailability(Kentish et al., 2008). The second important benefits of nanoemulsion are to protect the drug against UV light and degradation; these are used to enhance the stability of the drug(Choudhury et al., 2019). The globule nanoemulsion size may increase their ability to penetration, so the small size of droplet nanoemulsion is marked as low skin irritation due to penetration of droplets through skin and hair follicles without disrupting the healthy tissues(Praveen Kumar, 2015). Compared with conventional emulsions, nanoemulsion has a greater surface area, free energy, and is more stable in creaming, cracking, and phase separation(S Sugumar et al., 2014). It also plays a major part in vaccine delivery because it stays in circulation longer (El-readi, 2019). They used to take up highly antigen-presenting cells. Due to all these important characteristics of nanoemulsion, nowadays, it is used rapidly in pharma sectors, specifically in the case of cancer-targeting therapy, nanocarrier, drug delivery etc. (Jaiswal et al., 2015).

2. Risk Factors of Liver And Gastric Cancer

Every disease has certain reasons; due to the lack of some major reasons and lack of some particular biological molecules, it is used to make a biological difference in the cell growth, this is specifically known as cancer risk factor and reason behind cancer(Zakaria, 2015). A risk factor is used to increase the chance of severity.

2.1 In Liver Cancer

There is no. risk factors used to help in the severity of the disease. Such as HBV(Hepatitis B Virus), HCV(Hepatitis Chronic Virus), Alcohol Intake, NAFLD(Non-Alcoholic Fatty Liver disease), Exposure to Aflatoxin(Bangaru et al., 2020).

HBV(Hepatitis B Virus)

HBV virus is contagious and used to increase the risk factors in Hepatocellular carcinoma. It used to spread one to the other. The disease is the first phase called acute hepatitis B infection(Huynh et al., 2003). The virus holds double standard DNA and overlaps with viral DNA polymerase, two structural proteins(LeCluyse et al., 2005). The molecular components in HBV are used to play a major role in inactivating and targeting the pathways.

HCV(Hepatitis Chronic Virus)

In the development of hepatocellular carcinoma, HCV infection is also a major risk factor. It contains single-stranded positive RNA with a single open reading frame and encodes all structural proteins(Wu & Li, 2012).HCV infection according to viral proteins with infected hepatocytes without damage to the host genome(Shin & Chung, 2013). In HCV infections, the nucleus is used to help in the activation of the signalling pathway.

Alcohol intake

It used to more risk full fatty liver through alcoholic hepatitis, and more intake of alcohol cause alcoholic hepatitis, hepatic fibrosis to cirrhosis and ultimately HCC(Wu & Li, 2012).In the experiment of invivo, male rats used to treat with 37% ethanol for 8-10 weeks. In the rat model, after an overtime period, the experiment was used to test the nuclear and cytoplasmic expression of β catenin used to decrease in the liver(Ching et al., 2003).

Non-Alcoholic Fatty Liver Disease

This is a major risk factor in the progression of HCC. This disease ranges from simple steatosis to non-alcoholic steatohepatitis(NASH)(Shin & Chung, 2013). The increased burden of fat in hepatocytes, oxidative stress, and lipid peroxidation induce hepatic inflammation.

Exposure of aflatoxin

Aflatoxin is used to produce fungi. It is mostly known as carcinogenic mycotoxin. It is used to speed up the risk of HCC. It is used to induce the G: C to T: A transversion at the third base in codon 249(Bangaru et al., 2020). It is used to regulate the Wnt expression.

2.2 In gastric cancer

Over recent years, Gastric cancer has been getting attention from epidemiologic investigators. There is no. of factors is the reason behind the severity of gastric cancer. Out of all the risk factors H.Pylori and Family history of gastric cancer. Apart from this, other factors include ageing, tobacco, sun exposure, overweight, chemical substances, certain hormones, poor diet, alcohol intake, etc. (Słotwiński et al., 2018).

Helicobacter pylori

Infection with H.Pylori is the major reason behind stomach cancer. It used to affect the lower part of the stomach. Long time infection with H.Pylori used to cause atrophic gastritis in the inner linings of the stomach(Zhou et al., 1996). Some people used to get more infections, and it used to get more chance of stomach cancer.

Pernicious anemia

In the stomach, some linings are used to make substances known as the Intrinsic Factor. So, that body needs to absorb vitamin B12 from food. The body used to be vitamin B12 deficiency, which is normally used to affect the ability to make new red blood cells. It used to come by after stomach surgery in some of these cases. So, it is also a reason behind stomach cancer(Tricker & Preussmann, 1991).

HDGC(Hereditary diffuse gastric cancer)

The chance of getting this cancer is 70%, according to research. In the case of a woman, It used to affect as well in breast cancer. This is a syndrome of mutated CDH1 gene in expression(Ching et al., 2003).

Diet

It is the major reason and important risk factor for gastric cancer. People have a bad diet in daily life, and it may be a reason behind this. The food is salty, fermented more time, fridged food, pickle, and all make stomach ulcers (Zhou et al., 1996). Eating more grilled, packaged food continuous will increase the severity of stomach cancer.

Alcohol and Tobacco Use

The risk of gastric cancer used to double in the people habituated with a daily intake of alcohol and smoking(Gedaly et al., 2014). The upper part near to oesophagus used to affect earliest in the people with this habit(Pianetti et al., 2001).

3. PERSPECTIVES OF LIVER AND GASTRIC CANCER: ITS MOLECULAR MECHANISM AND BIOLOGICAL BEHAVIOURS

The characterization of liver and gastric cancer, such as the advancement of the genomic and clinical tools, brought the unveiling of minute molecular alterations within the tumour cell and environment(Roskams et al., 1998). Pointing out molecular alterations of heterogeneity within the epithelial cells could lead the recent research with developing the strategies of detection and improvement of patient's quality of life(Fausto & Campbell, 2003). During carcinogenesis, a cellular adaptation of the mutated liver and gastric cancer cells manages the oncogenic changes due to the changes in genetic and epigenetic during carcinogenesis(El-Zahaby et al., 2019). According to investigational data and clinical reports, the enzymes and receptors have a potential role in the development and proliferation of cancer, and they are used to help different stages of cancer development(Thomas & Abbruzzese, 2005). It has been noticed that the cancer cell captures the biological functions of the cells at the molecular level, so it is used to help malignant cells to multiply easily and improve the supply of nutrients and oxygen(Efferth, 2017).

These molecular alterations play an important role in angiogenesis during cancer growth and propagation of malignant cells to distant organs (Efferth et al., 2007). The lack of oxygen and nutrients within the cancer microenvironment promotes the propagation of new vasculature(Ždralević et al., 2018). The cancer cells and Fibroblast cells stimulate the secretion of angiogenic factors. The stimulation will affect the expression of matrix metalloproteinase by endothelial cells while allowing the formation of sprouting of parent blood pressure values(Li et al., 2020). In tumour cells, basement membranes are specialized in the extracellular matrix structure, which plays a major function in signal transduction actions via pathways initiated by integrin-mediated cell-matrix adhesions, leading to cell proliferation and survival(Tang et al., 2020). In the early-stage detection of cancer cells within the capsule of fibronectin, the cancer patient's mortality is a consequence of the dissemination of tumour cells to distant sites(Sharma et al., 2012). The cancer spreads to distant organs includes disruption and tight junction surrounding cancer origin, key signalling pathways, transforming growth factor beta-catenin, focal adhesion kinase accompanied by inactivation of suppression MMP-9 and activate protein(Alkreathy et al., 2012). The functional approaches for metastasis and tissue invasion are phytochemicals of natural products with fatty acids in the diet(Lee et al., 2006). Identifying the targeted genes and pathways of disease progression regulate its expression in specific liver and gastric cancer, which could help the disease molecularly and biologically identify, use them to discover novel diagnostic and therapeutic pathways through the research(S. Darvesh et al., 2011).

4. DELIBERATION OF NANOEMULSION TO DEFEAT THE PHYSIOLOGICAL HURDLES IN LIVER AND GASTRIC CANCER

Major changes in nanotechnology and nanocarriers such as liposomes, nanoparticles, dendrimers, nanocapsules and nanoemulsion have been explored to extend the half-life, the targeted delivery and reduce the toxicity in chemotherapy(Guo et al., 2018). The stability of nanoemulsion is a major challenge due to its large surface area and the characteristics of nanoemulsion completely understood in optimum efficacy(Saranya Sugumar & Singh, 2015). Nanoemulsion-based drug delivery system has contributed to developing nanomedicine with more efficacy, safety, and stability(Zhang et al., 2020). The nanoemulsions vary with characteristics features such as shape, size, surface charge and modified surface pave, which are the major part for achieving desirable nanomedicine with biopharmaceuticals advantages(Saranya Sugumar et al., 2012). Apart from other advantages, nanoemulsion can hold drugs without any refusal of stability, so the drug-loaded nanoemulsion will formulate therapeutic molecules using dose limit toxic excipients(S Sugumar et al., 2014).

During the delivery of drugs through the nanosystem into the liver and stomach, the major problems faced include disease and biological barriers because of its liver anatomy and physiology(Ghosh et al., 2014). Due to its variation in physicochemical properties of the nanoemulsion, the circulation time is reduced. Hence, the valuable knowledge of molecular and biological interaction between nanoemulsion and the disease environment is essential to design a safer nanoemulsion-based drug delivery system(Shamseddine et al., 2014). The delivery of nanoemulsion offers the advantages of delivering the therapeutics molecules to specifically the site of action, so it is used to enhance the delivery of a drug(Jerobin et al., 2012). There are so many nano-based systems available with toxicity, and this toxicology part is not studied completely. Through the help of clinical and preclinical studies, the toxicity of nanomaterials may be understood. The two things that must be noticed very carefully are the stability and toxicity of materials before trial in the human body. Because exposure to those particles may increase the various types of side effects such as irritation, inflammatory response, phagocytosis etc. (El-readi, 2019), focusing all the hurdles in the physical way of both these cancers, ingredients for nanoemulsion will be thoroughly investigated before the formulation process at the nanoscale(Praveen Kumar, 2015). In recent years, nanoemulsion is getting more attention towards pharmaceutical industries apart from other nanosystems because the formulation safety rules have already been established(Gorain et al., 2020).

5. NANOEMULSION USING AS DIFFERENT APPROACHES IN BOTH OF THE CANCER TREATMENT

Scientists are currently trying to discover the potential of nanoemulsion in the liver and gastric cancer to deliver different approaches against cancer. The schematic representation of multifunctional nanoemulsion application against cancer treatment is presented in **Figure 4**. Thermodynamic stability, large surface area, stability, droplet size, and desired drug release study make nanoemulsion major and attractive approaches towards hepatocellular carcinoma and gastric cancer(Jaiswal et al., 2015). The formulations of nanoemulsion against cancers studies found that these are a safe grade of excipients (GRAS) in which the encapsulation efficiency of nanoemulsion hydrophobic components is a high potential of physiochemical stability, supper efficacy with enhanced bioavailability (Efferth, 2017). In treatment purpose of liver and gastric cancer, nanoemulsion approaches in a wide range of routes of administration in different fields such as:

Figure 4. Structural representation of Multifunctional nanoemulsion application against cancer

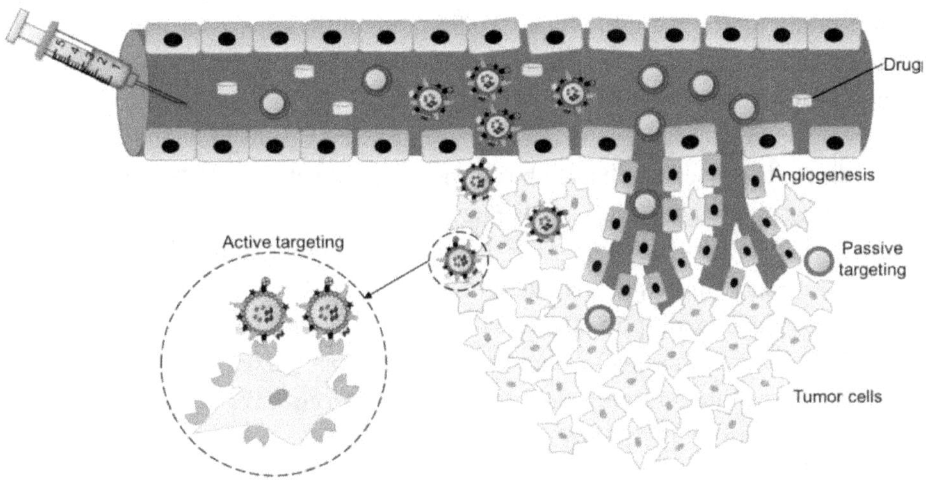

Cancer Molecular Imaging

Imaging cancer in an early stage is a very important aspect of cancer treatment. It provides the tumour microenvironment and monitors the appropriate with less tissue destruction in particular cancer(Mahato, 2017). In real-time, the novel molecular imaging helps in screening, biopsy, diagnosis and treatment progression (Gorain et al., 2020). The molecular imaging of cancer cells helps in liver cancer in in-vivo visualization and characterization of biological processes at both molecular and cellular levels. The different techniques are used in multiple imaging criteria such as molecular imaging (PET, SPECT, MRI, CEU and optical fluorescence), CT, BLI, MRI, ultrasound(Sánchez-López et al., 2019). Nanoemulsion formulation is already successfully progressed against cancer imaging of tumour microenvironment as hydrophobic dyes(Mahato, 2017). For example, the scientist penetrates the gold nanosphere in oil-in-water nanoemulsion to increase the nonlinear contrast while analyzing the molecular images(Yousefian Rad et al., 2020). This core nanoemulsion coated with gold nanosphere generates photoacoustic signals by converting light energy to sound energy(Ganta et al., 2014). The development of microtubules forms as PFC (perfluorocarbon) nanoemulsions, a synthetic organic compound where all the hydrogen compounds will be replaced with fluorine atoms(Upadhyay et al., 2020). So, these microtubules provide very good sensitivity, and it is used as a multipurpose imaging and delivery system by encapsulating varieties of contrast agents(Mahato, 2017). In clinical application studies also, nanoemulsion can be used for early detection purposes and targeted cancer therapy. In the platform for imaging and targeting, the iron oxide nanoparticle is encapsulated with any nanoemulsion in the core part, which helps detect MRI in optical imaging as a fluorescence signal(Ganta et al., 2014).

Nanoemulsion Conjugated Monoclonal Antibodies

The major advantage of nanoemulsion using monoclonal antibodies or specific fragments as targeting agents in the antigen-binding region is highly beneficial among antibodies(Islam et al., 2017). The different specificity can be imparted with different monoclonal antibodies(Mcclements & Jafari, 2018). Single monoclonal antibodies have two epitope binding sites that offer more selectivity and high binding affinity for the target with interest(Mahato, 2017). Sometimes, the whole molecule is used in Abs, or only the part of the binding site is used, such as Fab fragments(Ganta et al., 2014). The various immune-nanoemulsion has been investigated back to some years.

In Gastric cancer, it was studied that anti -HIF-1_ antibody conjugated triblock copolymer nano micelles filled with paclitaxel for targeting the MGC-803 cells. In the study, they found that Pluronic (polyethene oxide-poly propylene oxide-polyethene oxide) nano micelles specifically bound with MGC-803 cancer cells and HIF-1 (overexpressing) were used to release paclitaxel inside the cells(Choudhury et al., 2019). This study proves that antibody-targeted nanoemulsions have strong potential in clinical tumours towards imaging and targeting therapy.

Another study involved gastric cancer cell line treatment. They developed a Her2 antibody fab fragment conjugated with immune-nano micelles entrapping with doxorubicin and QD fluorescence probes which included both the properties of thermosensitive and antigen antibody-mediated targeted delivery to the individual tumour cells(Mahato, 2017).

Nanoemulsions Encapsulated with Oligonucleotide

Due to major obstacles such as short half-life and instability in biological fluids, oligonucleotides cannot be used in therapeutics. Chemical modifications of oligonucleotides in phosphodiester backbones partially used to increase the chemical stability for degradation of enzymes but unable to improve the intracellular penetration(Sánchez-López et al., 2019). To overcome these major problems, different strategies have been applied to the oligonucleotide, such as pH-sensitive nanocarriers developing verities of delivery systems such as cationic liposomes etc. (Islam et al., 2017).

In the study, they showed that PEG-DOPE (polyethene glycol-dioleoyl phosphatidylethanolamine), a copolymer system modification with G(4)-PAMAM dendrimer nanocarrier, was used to deliver a drug si-RNA to tumour cells (Periasamy et al., 2016). The modification of lipid in cationic polymers increased transfection efficiency, and the micellar dendrimers system yielded higher stability and si-RNA against degradation in enzymatic(Zhang et al., 2020).

Nanoemulsion Conjugated in Aptamers

Through an in-vitro selection process, aptamers are selected from an initial library containing random single-stranded DNA or RNA and the process is known as SELEX(Gorain et al., 2020). Though aptamers and antibodies possess similar properties, aptamers have some critical advantages over antibodies, such as being small in size, lack of immunogenicity, rapid penetration in tissues(Mahato, 2017). Aptamers bind to their target molecules with more affinity and selectively fold in their secondary and tertiary structures. Several aptamers are selected against various no. of cancer proteins such as VEGF, PDGF, HER3 and nuclear factor -B(Praveen Kumar, 2015). Aptamers can be selected against the cancer cell lines by SELEX, which is based on the biomarkers and receptor proteins on the surface of cancer cells(Plaza-Oliver et al., 2021). In the study of aptamer conjugated, doxorubicin-loaded acoustic droplets comprise a core of lipids and lipid-based compounds. In leukaemia, DNA aptamers sgc8c provide the ability to specifically target cells for both imaging and therapy studies(Kathe & Kathpalia, 2017). Through the study of drug release, high intensity of ultrasound has interacted with trigger targeted acoustic droplet vaporization, resulting in cell destruction and chemical treatment(Mcclements & Jafari, 2018).

6. DELIVERY OF VARIOUS THERAPEUTIC AGENTS BASED NANOEMULSION AGAINST LIVER AND GASTRIC CANCER: -

Chemotherapy is one of the major therapeutic processes against cancer treatment, but it is associated with some limitations in use(Iqbal et al., 2021). Different novel approaches-based nanotechnology brought better therapeutic approaches in delivering chemotherapeutics(Shamseddine et al., 2014). Apart from chemotherapeutic drugs, there are different drugs available through nanoemulsions, specifically in liver and Gastric cancer, such as synthetic, combinational, and natural drugs(Xu et al., 2019).

In liver cancer

Targeting the severity of liver cancer, there are some therapeutics drugs available as natural drugs-

Curcumin

Since immemorial time, curcumin has been used as traditional medicine and a very common spice in the family. Turmeric is another name for curcumin. In the in-vivo study, they found that after treatment, only 40% curcumin is in the blood and 60% curcumin in the liver and stomach of rats(S. Darvesh et al., 2011). It proves that very poor absorption of curcumin through oral administration. When it is loaded with nanoemulsion, the bioavailability and lipid concentration are increased significantly in liver cancer rats (Tang et al., 2020).

Eugenol

Polyphenols based compounds have low water solubility, such as Eugenol. The oral administration of Eugenol is low in water bioavailability. After loading with nanoemulsion, it is used to enhance the bioavailability(Press, 2015b). In the study of Eugenol nanoemulsion against HB8065 cell lines, it is observed that the cancer cells are induced in apoptosis and Reactive oxygen species(Upadhyay et al., 2020). Eugenol nanoemulsion showed higher apoptosis in cell lines compared to bulk Eugenol.

Nigella Sativa

The other name of *Nigella Sativa is* Black cumin seed. This is the form of oil-in-water nanoemulsion consisting of oils from medicinal plants. This form of nanoemulsion of oils is used to enhance the bioavailability of medications(Periasamy et al., 2016). Another major property of the nanoemulsion is highly stable over six months duration. In the study, it was found that Black cumin seed oil nanoemulsion showed a strong effect on liver cancer cell lines (HepG2) with reduced cell viability and increased Reactive Oxygen Species(ROS)(Farshi et al., 2017).

Sorafenib

It is a synthetic drug against liver cancer. It is used to inhibit tyrosine, serine kinase and cell proliferation in Liver cancer. While using this, there are some hurdles, like 39-49% low bioavailability because of its poor water solubility and hepatic metabolism(Monga et al., 2001). Through the help of high energy based nanoemulsion, it showed potential safe and effective parenteral drug delivery than conventional drugs form as it contained a lesser amount drug efficacy(Date et al., 2010). It was studied from MTT assay; optimal formulation did not affect the normal cells but still could remove cancer cells.

Curcumin and Sorafenib

This are known as a combinational drug of both natural drugs and synthetic drugs. Sorafenib is a standard medicine for liver cancer and the first molecular targeting medicine that got permission against advanced hepatocellular carcinoma(Nejak-Bowen et al., 2009). In using curcumin, it used to act as an agent to reduce drug resistance and enhance the drug's efficacy. The help of nanoparticles showed enhanced cytotoxicity and apoptosis in HepG2 cells(Petkov et al., 2004). The co-delivery of through nanoemulsion-based delivery system could promise results.

In Gastric Cancer

Zingiber Officinale

The other name of **Zingiber Officinale is** Ginger. This is the form of oil-in-water nanoemulsion consisting of medicinal oils from the medicinal plants with surfactant and water. Compared to bulk oil, nanoemulsion formulation is used to enhance the bioavailability of medications(Jaiswal et al., 2015). The major benefit of the nanoemulsion is the high rates of stability over the months. The study was done using **Zingiber Officinale oil** and nanoemulsion against HGC-27 gastric cancer cell lines. As a result, nanoemulsion-based studies showed better results than bulk oil(Efferth, 2017).

Ginkgo biloba

This plant is one of the most beneficial medicinal plants having the capability of treating gastric cancer. The emulsion is water-in-oil ultrasonic emulsification to further develop nanoparticles out of them using sodium tripolyphosphate(Roskams et al., 1998). The formulated nanoemulsion showed better results against the gastric cancer cell line.

Gemcitabine

The study observed that Gemcitabine loaded nanoparticles showed the greatest improvement in the mice against gastric cancer(El-Zahaby et al., 2019). In the study of in-vivo anticancer activity, an isomer of vitamin E with Gemcitabine is incorporated with nanoemulsion (Espinoza et al., 2020). The drug-loaded nanoemulsion showed better and higher anticancer activity than drug-free nanoemulsion.

Doxorubicin

It is one of the most used chemotherapeutic drugs in the use of different cancers. It showed a positive result in some experiments done against gastric cancer. The antitumour activity of nanoemulsion loaded with doxorubicin is effective on the swiss albino mice bearing gastric cancer(Petkov et al., 2004). It was noticed that the levels of serum enzyme and other parts of tissues were less affected in the group of animals treated with nanoemulsion loaded drugs. Reduction in side effects is a major positivity towards nanoemulsion loaded drugs (Angsubhakorn et al., 1988).

Vitamin C

Vitamin C is the most common vitamin available for treating common diseases. Vitamin C based formulated nanoemulsion balances the pH. The citrus oil is the major source of vitamin C to evaluate the hepatoprotective effect(Yousefian Rad et al., 2020). The ulcer is the major issue nowadays, which causes *H. pylori* based gastric cancer.

7. TARGETED AND NON-TARGETED DELIVERY OF NANOEMULSION TO CANCER

In the pharmaceutical sector, nanoemulsion is highly applicable due to its major advantages and characteristics. These are described very briefly in this chapter and another literature part as well. Targeting any cancer through nanoemulsion-based drug delivery holding compounds like natural, synthetic and semisynthetic, is the most popular part of research(Kentish et al., 2008). Different types of antitumour compounds targeting different cancers are well established based upon active and passive targeting presented in **Figure 5**. The problem is the delivery of the compounds in respective sites because of their less stability and lipophilic characteristics(Salvia-trujillo et al., 2013). To target all the drug delivery issues, nanoemulsion is the most popular drug delivery system among other nanotechnology-based drug delivery systems for compound based lipophilic characteristics in treating specific liver and gastric cancer(Saranya Sugumar & Singh, 2015). In liver cancer, studies have already shown that, because the nanosized range and nanoemulsion droplets hold the drugs, it easily penetrates the tumour tissue and targets moiety attached to the surface(Fausto & Campbell, 2003).

Moreover, it ensured that droplets size is not needed to change after penetration in tumour tissue. This chapter delivers the possibilities of lipophilic compounds, including natural, synthetic, or semisynthetic occurring active ingredients for liver and gastric cancer. There are different types of drugs available according to the structure of cancer(Alkhatib et al., 2017). So, paclitaxel is mostly used semisynthetic drug against solid cancer due to its limited solubility and up potential by transporter proteins. While compared with conventional treatment strategies, a report on pharmacodynamic activity and the taxological approaches revealed better control of cancer without showing saviour toxicities(Thomas & Abbruzzese, 2005). Some studies show that the formulation is a mixture of different compounds. Some compounds derived from medicinal plants actively works against the anti-inflammatory, antioxidant, antiparasitic and antitumour activity. For example, diferuloylmethane is derived from turmeric, and when it is added with nanoemulsion, the bioavailability is enhanced, and it is worked as an anti-liver carcinoma agent(Huang et al., 2019). Another compound, Carvacrol, is known as a flavonoid compound, which is indicated as a potential compound to treat liver cancer by inducing apoptosis and leading to cell cycle arrest(Satyal et al., 2017). Nanoemulsion loaded with any drugs can induce apoptosis or show cytotoxicity in some drug-resistant liver cancer cells. In this study, the authors formulated the flavonoid compound based nanoemulsion using a different emulsifier. The drug-loaded nanoemulsion targets the cell cycle complex and results in the downregulation of cell proteins such as CDK2, CDK4, CDK6, cyclin D, cyclin E(Siegel et al., 2018). These are important biomarkers established as promoters of the cell cycle and cell growth. So, their downregulation leads to cell cycle inhibition(George & Kuttan, 1997). Quercetin is another flavonoid that has gained the attention of researchers due to its antiproliferative, growth factor suppression and the effects of cell signalling against different types of cancers such as liver, gastric, lung, colon cancer etc. (Ferlay et al., 2015). The Arabian research team developed the quercetin nanoemulsion strategy of the aqueous phase by high energy methods.

Figure 5. Schematic representation of drug-loaded nanoemulsions based on active and passive targeting(Mahato, 2017).

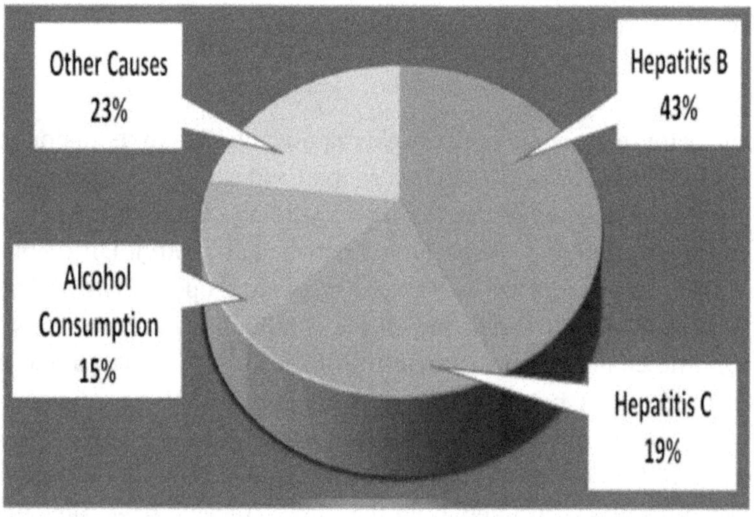

Like other techniques, gene therapy is also a major approach to treat any type of cancer. It is an effective genetic material at a cellular level without any hurdles. In nanocapsules, DNA multifunctional is prepared by layer-by-layer methods. Then the chitosan and DNA opposite charged are absorbed in nanoemulsion as core loaded with any marker(El-readi, 2019). Then finally nanoemulsion core loaded with nanocapsules are composed of DNA and chitosan against cancer cells. Nanoemulsions have many impending delivery systems in molecular imaging, theragnostic, quantum dots with unique fluorescence and exceptional photostability, narrow and symmetrical emission(Press, 2015a). Many researchers' effort has been towards the toxicity and surface medication.

8. MECHANISM OF NANOEMULSION LOADED DRUG DELIVERED TO TUMOUR CELL

The rapid proliferating of cancer cells by chemotherapeutic drugs is expected, but due to their toxicity and side effects, the chemotherapeutic drugs have limitations. Therefore, researchers have been trying their best for a few years to develop a specific drug for the respective tumours which can be selectively targeted and deliver the cytotoxic compounds to the cancer cells while controlling the toxicity and side effects to the healthy tissue(Fausto & Campbell, 2003). Over the last few years, the effort has been towards developing a nanoemulsion as a drug carrier for tumour targeting because nanoemulsion offers to deliver small molecules, macromolecules as drugs, protein humoral and cellular antigen peptides, antibodies etc. (Guo et al., 2018). Nanoemulsion showed various advantages in terms of delivery drugs, cellular uptake by antigen-presenting cells. In one of the studies, a nanoemulsion vaccine was developed to co-deliver immunostimulatory CpG and gastric cancer-specific antigen MG7(Sharma et al., 2012). They found that the tumour growth of mice pre-immunized with nanoemulsion encapsulated with MG7 and CpG were challenged later with an expression of MG7 cells(Alkreathy et al., 2012). This study suggested that nanoemulsion could be used for vaccine delivery to immunize against cancer. There are two types of targeting such as passive targeting and active targeting.

Passive targeting

In passive targeting, tumours vasculatures are abnormal in every step. Blood vessels are heterogeneous with branching structure and vessel lumen with enlarged gap junction of around 100 nm. In general, tumours have a longer drug retention period than normal tissues. The tumours lack a well-defined lymphatic system. So, the growing cancer cells compress the lymph vessels mainly in the centre of the tumour, which collapses and results in poor lymphatic drainage from tumours(Praveen Kumar, 2015). There is an effect called the EPR effect, which combines vascular permeability and poor lymphatic drainage. This effect plays a gold standard in anticancer drug delivery in several nanotherapeutic drugs, including gene delivery, molecular imaging, antibody therapy, and protein-polymer conjugates(Gorain et al., 2020). There are so many lists of nanoemulsions working as passive targeting by enhancing the permeability and retention effect.

Active targeting

It means nanoemulsion is attached with varieties of targeting molecules on its surface. So, it targets inside the tumour affected cells, tissue, organs or intracellular organelles(Ganta et al., 2014). Active targeting helps in the physiological benefits and offers delivery of drugs to the specific sites of tumours while binding with receptors overexpressed on target cancer(Bonferoni et al., 2017). Inactive targeting moieties are beneficial because they reduce or eliminate the toxic drugs to the healthy cells. Active targeting takes advantage of the overexpression of receptors such as transferrin, epidermal growth and prostate-specific membrane antigen(Sánchez-López et al., 2019). Targeted delivery results better in increased cytotoxicity to tumour cells and reduced side effects compared to non-targeted delivery. These effects help in targeting the cell membrane and internalization (Choudhury et al., 2019).

9. CLINICAL APPROACHES OF NANOEMULSION TOWARDS CANCER-

Nanoemulsion-based therapy of anticancer effect would be beneficial because of targeted drug delivery clinically. Through targeted drug delivery, the efficacy is improved and the side effects minimized (Praveen Kumar, 2015). According to its advantages, nanoemulsion-based therapy targets both liver and gastric cancer. In the clinical studies, along with stable emulsion, unstable emulsion also works against cancer cells, but sometimes patients show unwanted experience effects(Ganta et al., 2014). The preparation is challenging due to many side issues, the preparation methods and maintenance of the aseptic condition. The process of sterilization is difficult because dry heat or moist heat cannot be followed. An unstable, nonsterile and toxic ingredient in nanoemulsion is clinically undesirable and threatens patient health(Mahato, 2017). So many ligands as inactive addictive are now incorporated with the standard components of emulsions for targeting the specific cancer cells. The formulation stage of nanoemulsion holds organic compounds as well anticancer agents who are always challenged with the size of particles and appropriate drug loading desired tumour site; hence it goes through all the steps of clinical and preclinical studies with a series of challenges required to be passed(Yousefian Rad et al., 2020). There is no particular nanoemulsion currently available in the market against liver and gastric cancer. The major reasons behind this are the clinical aspects of delivery systems, characteristics and commercially applicable manufacturing process to adequate the methods complex(Plaza-Oliver et al., 2021).

10. FUTURE PERSPECTIVES AND CONCLUSION

Nanoemulsions are very much used for the drug's delivery system to encapsulate lipophilic, hydrophobic, and hydrophilic molecules to design in order. The main hurdles of nanoemulsions in future development are finding mechanisms to improve the nanoemulsion (Bonferoni et al., 2017). This research process has to keep in mind the interactions of the drug with other components in the delivery system, the manufacturing mechanism, and the drug's stability (Nácher et al., 2019). The nanoemulsion loaded drug can target the pathway of respective cancer and be the main study of future drug development (Ashaolu, 2021). There are different applications over nanoemulsions targeting different types of cancer. Liver and Gastric cancer are more focused in this study with exploring the traditional imaging, development of vaccine carriers, delivery of different types of molecules such as macromolecules and antigens etc. (Mahato, 2017). It allows long circulation time and uptake by cells with the particular antibody on the surface of the antigen or vice-versa(Gorain et al., 2020).

Liver and Gastric cancer are the two major death causes in the world. Moreover, nanotechnology has a major role in the prevention, treatment and diagnosis. Nanoemulsion is marked as the appropriate drug delivery through its explored treatments in cancer through different routes of administration. The conventional treatment strategies need to develop accordingly with drug delivery because it has less effect on the tumour and affects normal cells. The application of nanoemulsion against liver and gastric cancer cells seems to be the solution in these conditions. Nanoemulsion can be a good potential drug holder. For getting better solutions from nanoemulsion towards cancer, the clinical trials and mechanistic approaches must be strong and more explored. This chapter discusses different types of mechanisms to direct the nanoemulsion targeting the improved pharmacokinetics and pharmacodynamics characteristics.

ACKNOWLEDGMENT

Priyadarshini Mohapatra heartfully acknowledge Indian Council of Medical research (ICMR), the Government of India agency for the senior research fellowship grant [IRIS ID: 2020-8238] and the Vellore Institute of Technology for giving the laboratory facilities. The authors involved reveal no sources of conflict of interest.

11. REFERENCES

Alkhatib, M. H., Binsiddiq, B. M., & Backer, W. S. (2017). in Vivo Evaluation of the Anticancer Activity of a Water-In-Garlic Oil Nanoemulsion Loaded with Docetaxel. *International Journal of Pharmaceutical Sciences and Research, 8*(12), 5373–5379. doi:10.13040/IJPSR.0975-8232.8(12).5373-79

Alkreathy, H. M., Damanhouri, Z. A., Ahmed, N., Slevin, M., & Osman, A. M. M. (2012). Mechanisms of cardioprotective effect of aged garlic extract against doxorubicin-induced cardiotoxicity. *Integrative Cancer Therapies, 11*(4), 364–370. doi:10.1177/1534735411426726 PMID:22172987

Angsubhakorn, S., Bhamarapravati, N., Sahaphong, S., & Sathiropas, P. (1988). Reducing effects of rodent malaria on hepatic carcinogenesis induced by dietary aflatoxin B1. *International Journal of Cancer, 41*(1), 69–73. doi:10.1002/ijc.2910410114 PMID:3121525

Aravalli, R. N., Steer, C. J., & Cressman, E. N. K. (2008). *Molecular Mechanisms of Hepatocellular Carcinoma*. doi:10.1002/hep.22580

Ashaolu, T. J. (2021). Nanoemulsions for health, food, and cosmetics: A review. *Environmental Chemistry Letters*, (4), 3381–3395. Advance online publication. doi:10.100710311-021-01216-9 PMID:33746662

Bangaru, S., Marrero, J. A., & Singal, A. G. (2020). Review article: New therapeutic interventions for advanced hepatocellular carcinoma. *Alimentary Pharmacology & Therapeutics, 51*(1), 78–89. doi:10.1111/apt.15573 PMID:31747082

Bonferoni, M. C., Riva, F., Invernizzi, A., Dellera, E., Sandri, G., Rossi, S., Marrubini, G., Bruni, G., Vigani, B., Caramella, C., & Ferrari, F. (2017). Alpha-tocopherol loaded chitosan oleate nanoemulsions for wound healing. Evaluation on cell lines and ex vivo human biopsies, and stabilization in spray-dried Trojan microparticles. *European Journal of Pharmaceutics and Biopharmaceutics*. Advance online publication. doi:10.1016/j.ejpb.2017.11.008 PMID:29155053

Ching, Y. P., Wong, C. M., Chan, S. F., Leung, T. H. Y., Ng, D. C. H., Jin, D. Y., & Ng, I. O. L. (2003). Deleted liver cancer (DLC) 2 encodes a RhoGAP protein with growth suppressor function and is underexpressed in hepatocellular carcinoma. *The Journal of Biological Chemistry*, *278*(12), 10824–10830. doi:10.1074/jbc.M208310200 PMID:12531887

Choudhury, H., Pandey, M., Grain, B., Chatterjee, B., Madheswaran, T., Md, S., Mak, K.-K., Tambuwala, M., Chourasia, M. K., & Kesharwani, P. (2019). Nanoemulsions as Effective Carriers for the Treatment of Lung Cancer. In *Nanotechnology-Based Targeted Drug Delivery Systems for Lung Cancer*. Elsevier Inc. doi:10.1016/B978-0-12-815720-6.00009-5

Darvesh, Aggarwal, & Bishayee. (2011). Curcumin and Liver Cancer: A Review. *Current Pharmaceutical Biotechnology*, *13*(1), 218–228. doi:10.2174/138920112798868791

Date, A. A., Dixit, R., & Nagarsenker, M. (2010). *Self-nanoemulsifying drug delivery systems : formulation insights, applications and advances review.* Academic Press.

Efferth, T. (2017). Cancer combination therapies with artemisinin-type drugs. *Biochemical Pharmacology*, *139*(March), 56–70. doi:10.1016/j.bcp.2017.03.019 PMID:28366726

Efferth, T., Li, P. C. H., Konkimalla, V. S. B., & Kaina, B. (2007). From traditional Chinese medicine to rational cancer therapy. *Trends in Molecular Medicine*, *13*(8), 353–361. doi:10.1016/j.molmed.2007.07.001 PMID:17644431

El-readi, M. Z. (2019). *Review Article Cancer Nanomedicine : A New Era of Successful Targeted Therapy.* Academic Press.

El-Zahaby, S. A., Elnaggar, Y. S. R., & Abdallah, O. Y. (2019). Reviewing two decades of nanomedicine implementations in targeted treatment and diagnosis of pancreatic cancer: An emphasis on state of art. *Journal of Controlled Release*, *293*, 21–35. doi:10.1016/j.jconrel.2018.11.013 PMID:30445002

Espinoza, J. A., Zisi, A., Kanellis, D. C., Carreras-Puigvert, J., Henriksson, M., Hühn, D., Watanabe, K., Helleday, T., Lindström, M. S., & Bartek, J. (2020). The antimalarial drug amodiaquine stabilizes p53 through ribosome biogenesis stress, independently of its autophagy-inhibitory activity. *Cell Death and Differentiation*, *27*(2), 773–789. doi:10.103841418-019-0387-5 PMID:31285544

Farshi, P., Tabibiazar, M., Ghorbani, M., & Hamishehkar, H. (2017). Evaluation of Antioxidant Activity and Cytotoxicity of Cumin Seed Oil Nanoemulsion Stabilized by Sodium Caseinate- Guar Gum. Tabriz University of Medical Sciences, 23(4), 293–300. doi:10.15171/PS.2017.43

Fausto, N., & Campbell, J. S. (2003). The role of hepatocytes and oval cells in liver regeneration and repopulation. *Mechanisms of Development*, *120*(1), 117–130. doi:10.1016/S0925-4773(02)00338-6 PMID:12490302

Ferlay, J., Soerjomataram, I., Dikshit, R., Eser, S., Mathers, C., Rebelo, M., Parkin, D. M., Forman, D., & Bray, F. (2015). *Cancer incidence and mortality worldwide : Sources, methods and major patterns in GLOBOCAN 2012.* doi:10.1002/ijc.29210

Ganta, S., Talekar, M., Singh, A., Coleman, T. P., & Amiji, M. M. (2014). Nanoemulsions in translational research - Opportunities and challenges in targeted cancer therapy. *AAPS PharmSciTech*, *15*(3), 694–708. doi:10.120812249-014-0088-9 PMID:24510526

Gedaly, R., Galuppo, R., Daily, M. F., Shah, M., Maynard, E., Chen, C., Zhang, X., Esser, K. A., Cohen, D. A., Evers, B. M., Jiang, J., & Spear, B. T. (2014). Targeting the Wnt/β-catenin signalling pathway in liver cancer stem cells and hepatocellular carcinoma cell lines with FH535. *PLoS One*, *9*(6), e99272. Advance online publication. doi:10.1371/journal.pone.0099272 PMID:24940873

George, J., & Kuttan, R. (1997). CANCER shell liquid. *Group*, *112*, 5–10.

Ghosh, V., Mukherjee, A., & Chandrasekaran, N. (2014). *Colloids and Surfaces B : Biointerfaces Eugenol-loaded antimicrobial nanoemulsion preserves fruit juice against, microbial spoilage*. Academic Press.

Gorain, B., Choudhury, H., Nair, A. B., Dubey, S. K., & Kesharwani, P. (2020). Theranostic application of nanoemulsions in chemotherapy. *Drug Discovery Today*, *25*(7), 1174–1188. doi:10.1016/j.drudis.2020.04.013 PMID:32344042

Guo, Y., Wu, Z., Shen, S., Guo, R., Wang, J., Wang, W., Zhao, K., Kuang, M., & Shuai, X. (2018). Nanomedicines reveal how PBOV1 promotes hepatocellular carcinoma for effective gene therapy. *Nature Communications*, *9*(1), 3430. Advance online publication. doi:10.103841467-018-05764-7 PMID:30143633

Huang, Y., Zhou, B., Luo, H., Mao, J., Huang, Y., Zhang, K., Mei, C., Yan, Y., Jin, H., Gao, J., Su, Z., Pang, P., Li, D., & Shan, H. (2019). ZnAs@SiO2 nanoparticles as a potential anti-tumor drug for targeting stemness and epithelial-mesenchymal transition in hepatocellular carcinoma via SHP-1/JAK2/STAT3 signaling. *Theranostics*, *9*(15), 4391–4408. doi:10.7150/thno.32462 PMID:31285768

Huynh, H., Tuyen Nguyen, T. T., Pierce Chow, K. H., Tan, P. H., Soo, K. C., & Tran, E. (2003). Over-expression of the mitogen-activated protein kinase (MAPK) kinase (MEK)-MAPK in hepatocellular carcinoma: Its role in tumor progression and apoptosis. *BMC Gastroenterology*, *3*(1), 1–21. doi:10.1186/1471-230X-3-19 PMID:12906713

Iqbal, H., Razzaq, A., Uzair, B., Ain, N. U., Sajjad, S., Khan, N. U., & Menaa, F. (2021). *Breast Cancer Inhibition by Biosynthesized Titanium Dioxide Nanoparticles Is Comparable to Free Doxorubicin but Appeared Safer in BALB / c Mice*. Academic Press.

Islam, M. T., Streck, L., Vinícius, M., Barros, O., Wanessa, S., Silva, C., Machado, C., Machado, C., Luiz, A., Júnior, G., Fernanda, M., & Jardim, C. (2017). Evaluation of toxic, cytotoxic and genotoxic effects of phytol and its nanoemulsion. *Manuscript*, *177*, 93–101. Advance online publication. doi:10.1016/j.chemosphere.2017.02.145 PMID:28284120

Jaiswal, M., Dudhe, R., & Sharma, P. K. (2015). Nanoemulsion: an advanced mode of drug delivery system. *3 Biotech*, *5*(2), 123–127. doi:10.1007/s13205-014-0214-0

Jerobin, J., Sureshkumar, R. S., Anjali, C. H., Mukherjee, A., & Chandrasekaran, N. (2012). Biodegradable polymer-based encapsulation of neem oil nanoemulsion for controlled release of Aza-A. *Carbohydrate Polymers*, *90*(4), 1750–1756. doi:10.1016/j.carbpol.2012.07.064 PMID:22944443

Journal, A., & Issue, S. (2013). *Formulation and Characterization of Plant Essential Oil Based Nano-emulsion : Evaluation of its Larvicidal Activity Against Aedes aegypti*. Academic Press.

Kathe, K., & Kathpalia, H. (2017). Film Forming Systems for Topical and Transdermal Drug Delivery. *Asian Journal of Pharmaceutical Sciences*. doi:10.1016/j.ajps.2017.07.004

Kentish, S., Wooster, T. J., Ashokkumar, M., Balachandran, S., Mawson, R., & Simons, L. (2008). *The use of ultrasonics for nanoemulsion preparation*. doi:10.1016/j.ifset.2007.07.005

LeCluyse, E. L., Alexandre, E., Hamilton, G. A., Viollon-Abadie, C., Coon, D. J., Jolley, S., & Richert, L. (2005). Isolation and culture of primary human hepatocytes. *Methods in Molecular Biology (Clifton, N.J.)*, *290*, 207–229. doi:10.1385/1-59259-838-2:207 PMID:15361665

Lee, J. S., Heo, J., Libbrecht, L., Chu, I. S., Kaposi-Novak, P., Calvisi, D. F., Mikaelyan, A., Roberts, L. R., Demetris, A. J., Sun, Z., Nevens, F., Roskams, T., & Thorgeirsson, S. S. (2006). A novel prognostic subtype of human hepatocellular carcinoma derived from hepatic progenitor cells. *Nature Medicine*, *12*(4), 410–416. doi:10.1038/nm1377 PMID:16532004

Li, N., Chen, Y., Sun, H., Huang, T., Chen, T., Jiang, Y., Yang, Q., Yan, X., & Wu, M. (2020). Decreasing acute toxicity and suppressing colorectal carcinoma using Sorafenib-loaded nanoparticles. *Pharmaceutical Development and Technology*, *25*(5), 556–565. doi:10.1080/10837450.2020.1718704 PMID:31958240

Mahato, R. (2017). Nanoemulsion as Targeted Drug Delivery System for Cancer Therapeutics. *Journal of Pharmaceutical Sciences and Pharmacology*, *3*(2), 83–97. doi:10.1166/jpsp.2017.1082

Mcclements, D. J., & Jafari, S. M. (2018). General Aspects of Nanoemulsions and Their Formulation. In *Nanoemulsions*. Elsevier Inc. doi:10.1016/B978-0-12-811838-2.00001-1

Monga, S. P. S., Pediaditakis, P., Mule, K., Stolz, D. B., & Michalopoulos, G. K. (2001). Changes in wnt/β-catenin pathway during regulated growth in rat liver regeneration. *Hepatology (Baltimore, Md.)*, *33*(5), 1098–1109. doi:10.1053/jhep.2001.23786 PMID:11343237

Nácher, M. P. A., Jesús, M., Ofelia, H. M. A., & Busó, V. (2019). A novel lidocaine hydrochloride mucoadhesive films for periodontal diseases. *Journal of Materials Science. Materials in Medicine*, *30*(1), 14. Advance online publication. doi:10.100710856-018-6213-7 PMID:30635738

Nejak-Bowen, K. N., Zeng, G., Tan, X., Cieply, B., & Monga, S. P. (2009). β-catenin regulates vitamin C biosynthesis and cell survival in murine liver. *The Journal of Biological Chemistry*, *284*(41), 28115–28127. doi:10.1074/jbc.M109.047258 PMID:19690176

Periasamy, V. S., Athinarayanan, J., & Alshatwi, A. A. (2016). Anticancer activity of an ultrasonic nanoemulsion formulation of Nigella sativa L. essential oil on human breast cancer cells. *Ultrasonics Sonochemistry*, *31*, 449–455. doi:10.1016/j.ultsonch.2016.01.035 PMID:26964971

Petkov, P. M., Zavadil, J., Goetz, D., Chu, T., Carver, R., Rogler, C. E., Bottinger, E. P., Shafritz, D. A., & Dabeva, M. D. (2004). Gene Expression Pattern in Hepatic Stem/Progenitor Cells during Rat Fetal Development Using Complementary DNA Microarrays. *Hepatology (Baltimore, Md.)*, *39*(3), 617–627. doi:10.1002/hep.20088 PMID:14999680

Pianetti, S., Arsura, M., Romieu-Mourez, R., Coffey, R. J., & Sonenshein, G. E. (2001). Her-2/neu overexpression induces NF-κB via a PI3-kinase/Akt pathway involving calpain-mediated degradation of IκB-α that can be inhibited by the tumor suppressor PTEN. *Oncogene, 20*(11), 1287–1299. doi:10.1038j.onc.1204257 PMID:11313873

Plaza-Oliver, M., Santander-Ortega, M. J., & Lozano, M. V. (2021). Current approaches in lipid-based nanocarriers for oral drug delivery. *Drug Delivery and Translational Research, 11*(2), 471–497. doi:10.100713346-021-00908-7 PMID:33528830

Praveen Kumar, G. (2015). Nanoemulsion Based Targeting in Cancer Therapeutics. *Medicinal Chemistry (Shariqah, United Arab Emirates), 5*(6), 272–284. doi:10.4172/2161-0444.1000275

Press, D. (2015a). *Antibacterial activity of neem nanoemulsion and its toxicity assessment on human lymphocytes in vitro*. Academic Press.

Press, D. (2015b). *Eucalyptus oil nanoemulsion-impregnated chitosan film : antibacterial effects against a clinical pathogen, Staphylococcus aureus, in vitro*. Academic Press.

Roskams, T., De Vos, R., Van Eyken, P., Myazaki, H., Van Damme, B., & Desmet, V. (1998). Hepatic OV-6 expression in human liver disease and rat experiments: Evidence for hepatic progenitor cells in man. *Journal of Hepatology, 29*(3), 455–463. doi:10.1016/S0168-8278(98)80065-2 PMID:9764994

Salvia-trujillo, L., Rojas-graü, A., Soliva-fortuny, R., & Martín-belloso, O. (2013). *Physicochemical Characterization of Lemongrass Essential Oil – Alginate Nanoemulsions : Effect of Ultrasound Processing Parameters*. doi:10.1007/s11947-012-0881-y

Sánchez-López, E., Guerra, M., Dias-Ferreira, J., Lopez-Machado, A., Ettcheto, M., Cano, A., Espina, M., Camins, A., Garcia, M. L., & Souto, E. B. (2019). Current applications of nanoemulsions in cancer therapeutics. *Nanomaterials (Basel, Switzerland), 9*(6), 821. Advance online publication. doi:10.3390/nano9060821 PMID:31159219

Satyal, P., Craft, J. D., Dosoky, N. S., & Setzer, W. N. (2017). The chemical compositions of the volatile oils of garlic (Allium sativum) and wild garlic (Allium vineale). *Foods, 6*(8), 1–10. doi:10.3390/foods6080063 PMID:28783070

Shamseddine, A., Saleh, A., Charafeddine, M., Seoud, M., Mukherji, D., Temraz, S., & Sibai, A. M. (2014). Cancer trends in Lebanon: A review of incidence rates for the period of 2003-2008 and projections until 2018. *Population Health Metrics, 12*(1), 2–9. doi:10.1186/1478-7954-12-4 PMID:24593777

Sharma, N., Thomas, S., Golden, E. B., Hofman, F. M., Chen, T. C., Petasis, N. A., Schönthal, A. H., & Louie, S. G. (2012). Inhibition of autophagy and induction of breast cancer cell death by mefloquine, an antimalarial agent. *Cancer Letters, 326*(2), 143–154. doi:10.1016/j.canlet.2012.07.029 PMID:22863539

Shin, J. W., & Chung, Y. H. (2013). Molecular targeted therapy for hepatocellular carcinoma: Current and future. *World Journal of Gastroenterology, 19*(37), 6144–6155. doi:10.3748/wjg.v19.i37.6144 PMID:24115810

Siegel, R. L., Jemal, A., Wender, R. C., Gansler, T., Ma, J., & Brawley, O. W. (2018). An assessment of progress in cancer control. *CA: a Cancer Journal for Clinicians, 68*(5), 329–339. doi:10.3322/caac.21460 PMID:30191964

Słotwiński, R., Lech, G., & Słotwińska, S. M. (2018). MicroRNAs in pancreatic cancer diagnosis and therapy. *Central European Journal of Immunology, 43*(3), 314–324. doi:10.5114/ceji.2018.80051 PMID:30588176

Sugumar, S., Nirmala, J., Anjali, H., Mukherjee, A., & Chandrasekaran, N. (2012). *Bio-based nanoemulsion formulation, characterization and antibacterial activity against food-borne pathogens*. doi:10.1002/jobm.201200060

Sugumar, S., & Singh, S. (2015). *Nanoemulsion of orange oil with non ionic surfactant produced emulsion using ultrasonication technique : Evaluating against food spoilage yeast*. doi:10.1007/s13204-015-0412-z

Sugumar, S, Clarke, S. K., Nirmala, M. J., Tyagi, B. K., Mukherjee, A., & Chandrasekaran, N. (2014). *Nanoemulsion of eucalyptus oil and its larvicidal activity against Culex quinquefasciatus*. doi:10.1017/S0007485313000710

Syed, Y. Y. (2020). Ramucirumab : A Review in Hepatocellular Carcinoma. *Drugs*, (3), 315–322. Advance online publication. doi:10.100740265-020-01263-6 PMID:32034692

Tang, K., Du, S., Wang, Q., Zhang, Y., & Song, H. (2020). Traditional Chinese medicine targeting cancer stem cells as an alternative treatment for hepatocellular carcinoma. *Journal of Integrative Medicine, 18*(3), 196–202. doi:10.1016/j.joim.2020.02.002 PMID:32067923

Thomas, M. B., & Abbruzzese, J. L. (2005). Opportunities for targeted therapies in hepatocellular carcinoma. *Journal of Clinical Oncology, 23*(31), 8093–8108. doi:10.1200/JCO.2004.00.1537 PMID:16258107

Tricker, A. R., & Preussmann, R. (1991). Carcinogenic N-nitrosamines in the diet: occurrence, formation, mechanisms and carcinogenic potential. *Mutation Research/Genetic Toxicology, 259*(3–4), 277–289. doi:10.1016/0165-1218(91)90123-4

Upadhyay, T., Ansari, V. A., Ahmad, U., Sultana, N., & Akhtar, J. (2020). Exploring Nanoemulsion for Liver Cancer Therapy. *Current Cancer Therapy Reviews, 16*(4), 260–268. doi:10.2174/1573394716666200302123336

Wu, X., & Li, Y. (2012). *Signaling Pathways in Liver Cancer*. Liver Tumors., doi:10.5772/31381

Xu, T., Gao, C., Feng, X., Yang, Y., Shen, X., & Tang, X. (2019). PT SC College of Food Science and Engineering / Collaborative Innovation Center for. *International Journal of Biological Macromolecules*. Advance online publication. doi:10.1016/j.ijbiomac.2019.04.189 PMID:31047924

Yousefian Rad, E., Homayouni Tabrizi, M., Ardalan, P., Seyedi, S. M. R., Yadamani, S., Zamani-Esmati, P., & Haghani Sereshkeh, N. (2020). Citrus lemon essential oil nanoemulsion (CLEO-NE), a safe cell-depended apoptosis inducer in human A549 lung cancer cells with anti-angiogenic activity. *Journal of Microencapsulation, 37*(5), 394–402. doi:10.1080/02652048.2020.1767223 PMID:32400238

Zakaria, M. K. (2015). Hepatic Cancer Stem Cells and Signaling Pathways. *International Journal of Cancer and Clinical Research, 2*(2). Advance online publication. doi:10.23937/2378-3419/2/2/1017

Ždralević, M., Vučetić, M., Daher, B., Marchiq, I., Parks, S. K., & Pouysségur, J. (2018). Disrupting the 'Warburg effect' re-routes cancer cells to OXPHOS offering a vulnerability point via 'ferroptosis'-induced cell death. *Advances in Biological Regulation*, *68*(December), 55–63. doi:10.1016/j.jbior.2017.12.002 PMID:29306548

Zhang, Y., Lv, C., Dong, Y., & Yang, Q. (2020). Aspirin-targeted PD-L1 in lung cancer growth inhibition. *Thoracic Cancer*, *11*(6), 1587–1593. doi:10.1111/1759-7714.13433 PMID:32297484

Zhou, J., Mulshine, J. L., Unsworth, E. J., Scott, F. M., Avis, I. M., Vos, M. D., & Treston, A. M. (1996). Purification and characterization of a protein that permits early detection of lung cancer: Identification of heterogeneous nuclear ribonucleoprotein-A2/B1 as the antigen for monoclonal antibody 703D4. *The Journal of Biological Chemistry*, *271*(18), 10760–10766. doi:10.1074/jbc.271.18.10760 PMID:8631886

Chapter 15
Potential Phytochemical Nanoemulsions in the Treatment of Oral Cancer and Oral Health:
Recent Advancement in Cancer Therapy

Dharmeswar Barhoi
https://orcid.org/0000-0002-3184-990X
Assam University, Silchar, India

Sweety Nath Barbhuiya
Assam University, Silchar, India

Sarbani Giri
Assam University, Silchar, India

ABSTRACT

Oral cancer is one of the most common types of cancer, and lifestyle factors like extensive consumption of tobacco, betel quid, and alcohol are the major etiological factors of oral cancer. Treatment of oral cancer includes surgery, radiation therapy, and chemotherapy, but this treatment possesses lots of side effects. Therefore, scientists and medical experts are utilizing natural products and medicinal plants for new drug development. Natural products and phytochemicals showed better efficacy with less toxicity. However, most of the phytochemicals showed poor permeability and less bioavailability. To combat this problem, scientists developed nanosized nanoemulsions of phytochemicals to treat various ailments. Nanoemulsions of phytochemicals exhibited better efficacy than their free form due to increased permeability and bioavailability. Numerous phytopharmaceuticals have been formulated for nanoemulsions to date and tested for their anticancer potential against various cancers, including oral cancer and oral health management.

DOI: 10.4018/978-1-7998-8378-4.ch015

INTRODUCTION

Oral cancer is a severe public health issue and a major cause of morbidity and mortality in developed and developing countries (Borse et al., 2020). It is the sixth most common cancer globally (Upadhaya et al., 2019). Oral cancer occurs in the lips, cheeks, mouth, and tongue. Recently, International Agency for Research on Cancer (IARC) released the worldwide cancer burden estimated through GLOBOCAN 2020 estimates. In the report, new cancer cases were increased up to 3, 77,713 in which 1, 77,757 deaths have occurred in 2020 globally (IARC, Cancer fact sheet 2020 https://gco.iarc.fr/today/data/factsheets/cancers/1-Lip-oral-cavity-fact-sheet.pdf). Oral cancer is the third most widespread cancer in India and one of the major challenges for the nation affecting 20 individuals per 1,00,000 annually (Danaraddi et al., 2014; Scrobata et al., 2016). According to a recent report, around 77,000 new cases of oral cancer and 52,000 death cases annually were reported (Borse et al., 2020). Figure 1 showed the scenario of new cancer cases and their death rate in both Indian and global level. Therefore, the incidence of oral cancer and its related mortality rate is continuously increasing. The prevention of oral cancer and its treatment is a major challenging task for the scientist and health professional around the globe. Scientists and health professionals are adopting new and new strategies for cancer treatment to lower its burden and mortality rate.

Figure 1. Graphical representation of Indian and global cancer incidence (A) Incidence and death of new cancer (all sites) at global level as per the GLOBOCAN estimates 2020 (Sung et al., 2021). (B) Incidence and death of oral cancer (lip and oral cavity) at global level (GLOBOCAN estimates 2020; Sung et al., 2021). (C) Incidence and death of oral cancer in India (Borse et al., 2020).

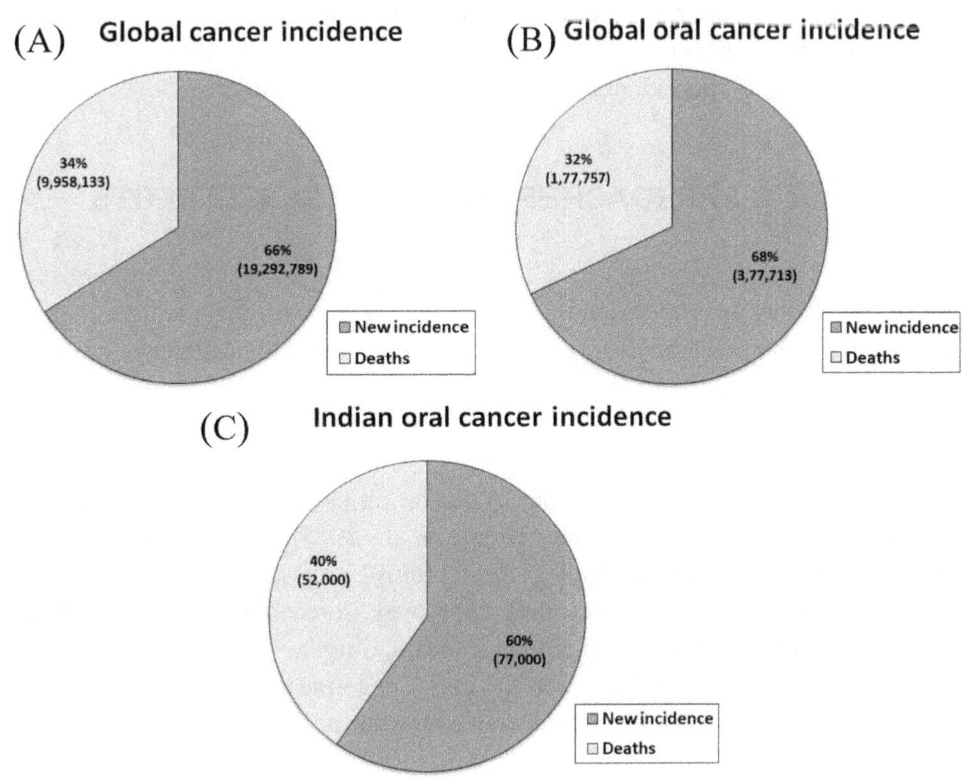

Oral health has a strong association with multiple cancer types, including oral cancer. The oral cavity is a niche of a variety of microorganism species that causes serious oral diseases. For example, a carcinogenic oral bacterium inhabits in the oral cavity and increases the risk of oral squamous cell carcinoma (OSCC) (Whitmore and Lamont, 2014). Apart from cancer, pathogenic microorganism accumulation in the oral cavity causes other diseases such as periodontal diseases, orodental trauma, dental caries, cleft lip and palate, etc. The global burden of oral diseases is escalating, and many populations of all ages suffer from the said disease (Ramalingam, 2011; Ramalingam and Amaechi, 2019). Therefore, intensive care should be taken to get rid of cancer risk and oral diseases. Phytochemicals also play a significant role as a disinfectant to microorganisms and help in promoting oral health. In this book chapter, phytochemicals and/or other plant-based products formulated to nanoemulsion in the treatment of oral cancer and other oral diseases were discussed.

RISK FACTORS OF ORAL CANCER AND ORAL DISEASES

A group of causative agents of oral cancer is there. Lifestyle factors such as indiscriminate tobacco consumption (both smoking and smokeless tobacco (SLT)) and chewing areca nut play a crucial role in oral carcinogenesis. Studies reported a positive correlation between SLT consumption and cancer incidence (Giri et al., 2021). Oral cancer is also associated with an unhealthy diet and poor dental care.

Viral infection also causes oral carcinogenesis. Several strains of viruses, such as the human papillomavirus (HPV), adenoviruses, hepatitis C virus (HPC), and herpes group viruses (HGV), are the potential risk factors for oral cancer initiation and its progression (Gupta and Metgud, 2013). HGV includes human herpes virus-8, Epstein-barr virus, and cryomegalovirus. Some viruses perform a synergistic role in oral carcinogenesis and are referred to as "synergistic viruses," and HPV and HGV viruses are classic examples (Chan et al., 2001). Bacterial and viral infections are the major etiological factors of oral diseases like- dental caries, periodontal diseases, and other oral ailments.

CANCER DIAGNOSTIC APPROACHES AND THEIR SIDE EFFECTS

In the last 2-3 decades, the cancer biology area developed more quickly, and the scientist and medical experts developed cancer diagnosis approaches. There are many conventional methods to treat cancer *viz.* surgery, radiation therapy, chemotherapy, surgery, targeted therapy, hormonal therapy and immunotherapy.

Surgery

Surgery is the crucial treatment for various types of cancer, in particular for solid tumors that include eliminating cancerous tissues from cancer site (Huang et al., 2017). It also involves exclusion of the entire organ affected with cancer. For example, "mastectomy" i.e., removal of entire breast in case of breast cancer or a part of the breast (lumpectomy). Moreover, in recent days, a range of methods have been devised out to treat premalignant and malignant cancers. Surgery is accompanied by chemotherapy in many cases for better consequences. However, the major negative aspect of surgery is that only solid tumors can be excluded and thus, this technique is not useful for non-solid cancers like lymphomas and leukemia.

Radiation therapy

Ionization radiation is used in the radiation therapy procedure that hinders the process of cell division by demolishing the genetic material (Symonds, 2001; Jackson and Bartek, 2009). Many a time, radiation therapy is accompanied by other treatment methods such as chemotherapy, surgery and immunotherapy in order to enhance the results (Baskar et al., 2012). The main limitation of this technique is the destruction of each and every cell indiscriminate of normal and cancerous cells upon exposure.

Chemotherapy

Cancer chemotherapy is the application of chemical compounds to prohibit the growth of cancerous cells. Previously it was thought that chemotherapy is limited to synthetic drugs only. But nowadays, it is extensively disseminated to employ chemical compounds from natural sources also (Alam et al., 2018). The term combined chemotherapy is employed in cancer chemoprevention that potray more than one chemotherapeutic drug for better results. Based on the mechanism of action, numerous anticancer chemotherapeutic drugs have been formulated so far. The chemotherapeutic drugs includes (a) antibiotics that interact with DNA replicating enzymes (b) alkylating agents damaging DNA materials, (c) antimetabolites that surrogate usual RNA and DNA molecules in cells, (d) topoisomerase inhibitors that prevent topoisomerase enzymes activity during replication and transcription (e) mitotic inhibitors that obstruct the mitotic cell division and (f) corticosteroids that are used to reprieve from the side effects of other drugs besides its application in cancer chemoprevention (Huang et al., 2017). Few chemotherapeutic drugs, their mechanism of action and the side effects are listed below-

Mitomycln C

Mitomycin C is an alkylating agent that inhibiting DNA synthesis by crosss linking with the DNA helix. It is an antineoplastic antibiotic isolated from *Streptomyces caespitosus* for the first time by Japanese microbiologists in 1950. Administration of this drug at higher concentration could also inhibit RNA and protein synthesis. It is used to treat a variety of cancer types after its discovery. Recently, it is also approved for the treatment of low-grade Upper Tract Urothelial Cancer. Besides, its potential to cure a variety of cancers, it has lots of side effects including nausea and vomiting, loss appetite, tumor site pain, abdominal pain, headache etc.

Methotrexate

Methotrexate is an antineoplastic agent widely used in the treatment of several cancer types and also in severe psoriasis, severe rheumatoid arthritis, and juvenile rheumatoid arthritis. It is a folate derivative that inhibits enzymes responsible for nucleotide synthesis that in turn leads to suppression of inflammation and inhibition of cell division. These effects of methotrexate led it to frequently use in the treatment of inflammation caused by arthritis or to control cell division in neoplastic diseases such as breast cancer and non-Hodgkin's lymphoma. Because of the adverse toxicity of methotrexate, it is used for the treatment of some forms of arthritis and severe psoriasis only if first line treatment has failed or patients are intolerant of those treatments.

Paclitaxel

Paclitaxel is a taxol derivative used as a chemotherapeutic agent for the treatment of advanced carcinoma of the ovary, and other various cancers including breast and lung cancer. It inhibits cancer cell proliferation by acting as a mitotic inhibitor. It was first isolated from the bark of Pacific yew tree in 1971 which contains endophytic fungi that synthesize it. Overdose includes bone marrow suppression, peripheral neurotoxicity, and mucositis. The side effects of paclitaxel include low blood counts, hair loss, pain in joints and muscles, nausea and vomiting, diarrhea etc.

Targeted therapy

Like chemotherapy, synthetic compounds or natural products are applied in targeted therapy as a medicine to decrease malignant cells' growth (Baudino, 2015). These drugs interact with particular proteins involved in the pathways associated with cancer development. Targeted cancer therapy is beneficial for numerous cancers, including lung, breast, colorectal, leukemia, and lymphoma. More than 15 drugs have been accepted by Food and Drug Administration agency (FDA, USA) since 2000. The major drugs used in targeted therapies are (a) Immunotoxins, (b) Small molecule inhibitors, and (c) monoclonal antibodies (Baudino, 2015; Zhong et al., 2021). These drugs obstruct with the molecules or a specific protein that triggers cancer development. Blocking these specific target protein compounds and restrict the carcinogenic pathways, thereby playing an indispensable role in the inhibition of tumorigenesis (Du et al., 2021).

Immunotherapy

Immunotherapy is a newly developed technique in cancer biology. William B Coley has exploited the power of immune system for the first time to treat various cancers (Kruger et al., 2019). In this therapeutic procedure, the immune system has been regulated by different immune agents (Esfahani et al., 2020). These immune agents include vaccines, cytokines, transfection agents, cell therapies, etc. (Stanculeanu et al., 2016). These agents promote the anti-tumor response in the body by intensifying the number of effector cells and construction of soluble mediators. These agents also influence the immune checkpoints (Esfahani et al., 2020) and stimulate a tumor-killing microenvironment. Altering the immune cells in immunotherapy elevates the autoimmunity of the immune cells that boosts the immune response against foreign pathogens and tumors (Esfahani et al., 2020). However, this may result in successive autoimmune reactions. Chemotherapy and radiotherapy both are utilized with immunotherapy for better assistance from the disease.

Hormonal therapy

Hormonal therapy is applied for the type of cancers that utilizes hormones for their augmentation and propagation and is very much effective in reducing growth and development of such type of cancers. Hormonal therapy broadly functions in two ways: (a) by reducing the body's capability for hormone generation and (b) interferes with the pathways of hormonal action and thus regulating the hormone metabolic activity. A number of drugs, interact with the hormone receptors, thereby, reducing tumorigenesis. For example, tamoxifen, is an effective drug candidate that communicates with the estrogen receptor and inhibits the progression of breast cancer (Fisher et al., 1998; Puhalla et al., 2012). Many other drugs are also available that use diverse receptors to demolish cancer progression by modulating hormone functions (Abdulkareem and Zurmi, 2011). Therefore, this treatment method has become a novel strategy for cancer treatment.

SIDE EFFECTS OF CONVENTIONAL CANCER THERAPEUTICS

The detrimental effects of contemporary cancer treatments remain a serious problem for patients and clinicians. Although the present cancer therapies enhanced the survival time of the patients, the side effects of the conservative treatments of cancer, especially chemotherapeutic drugs, are related with undesirable effects that confine the utilization of modern therapies. Medications used in conventional cancer treatment process are partly effective, contain life-long discarded toxicity, and do not meet the anticipated results (Choudhari et al., 2020). Patients even feel discomforts following treatments with current therapeutic drugs. The 5-year survival rate of cancer patients following cancer treatment is ~50% only (Barhoi et al., 2021b; Bhavana and Lakshmi, 2014). Therefore, scientists are more considerate in finding new approaches to improve the patient's survival time, the efficacy of the treatment approach, and to get rid of the harmful toxic effects of cancer treatment approaches.

A huge group of patient's undergone therapies with contemporary cancer intervention suffers from rigorous side effects. These side effects are often life-threatening. Among the side effects of conservative cancer therapies, nausea and vomiting are the most common and feared side effects for patients undergoing treatments. Both oral and gastrointestinal mucositis that might cause ulceration and pain are common in patients undergoing chemotherapy, radiotherapy, and targeted therapy (Cinausero et al., 2017). New cancer therapies cause cardiovascular toxicity, nephrotoxicity, and chronic subclinical skeletal muscle toxicity in patients (Nurgali et al., 2018). Recently available cancer therapies cause hazardous effects to the patients when drugs affect the normal cells. For example, chemotherapeutic drugs usually affect rapidly growing cells like cells with malignant properties. On the other hand, it also affects the body's healthy cells, but not limited to, hematopoietic cells, hair follicles, and cells in the mouth, reproductive, and digestive tract. In a study by Wochna-Loerzel (2015), it was observed that older patients (>65 years of age) are suffering from severe to moderately intense side effects of cancer therapy.

The center for disease control and prevention and the national cancer institute (NCI) has listed many ill effects of contemporary cancer therapies. The main side effects of conventional cancer therapies include neutropenia, lymphedema, hair loss, nausea and vomiting, cancer pain, and blood clots (Barhoi et al., 2021a; Widakowich et al., 2007; Kerns et al., 2014). Choudhari et al. (2020) reported that undesirable toxicity to non-specific organs, drug resistance, and cancer recurrence are also associated with traditional cancer therapies. Therefore, scientists are persistently searching for novel and promising anticancer therapeutics of enhanced effectiveness and lesser toxicity to overcome this deadly disease. Thus, scientists and medicinal experts are constantly searching for novel and substitute medicines with a reduced amount of toxicity and high efficacy to overcome this life-threatening obstacle.

PHYTOCHEMICALS AS AN ALTERNATIVE TO MODERN CANCER INTERVENTION

Medicinal plants are being used for the treatment of various diseases, including cancer. Using medicinal plants and/or natural products against cancer showed better results than conventional treatments. Therefore, experts are now using plants and plant products for drug development. To date, ~80% of the drugs are being developed using medicinal plants and other natural products to treat various ailments (Khan et al., 2019). Plant-based pharmaceuticals or phytochemicals have fewer side effects as compared to conventional cancer therapies.

Although natural products and medicinal plant extracts showed better results against various cancers, including oral cancer, it is still less than expected. Moreover, the results obtained in clinical trials are far away from the effects of preclinical studies. Thus, poor bioavailability and strategies adopted for phytochemical administration fail to show the expected outcome. In recent years, scientists and medical experts are trying to devise new techniques to overcome this problem. Scientists from different countries are now using nanotechnological methods for phytopharmaceutical/plant products to increase their efficacy.

PLANT-BASED PHYTOPHARMACEUTICALS IN ORAL CANCER

As mentioned above, medicinal plants are being widely used in drugs developed besides their traditional use. Several medicinal plants and their isolated products have been used in preclinical and clinical studies. We have previously documented some well-known phytochemicals used in treating oral cancer along with their action mechanism. We also reported that sulforaphane, apigenin, curcumin, piperine, ferulic acid, resveratrol, nimbolide, genistein, and lycopene are widely used phytopharmaceuticals used in oral cancer diagnosis (Barhoi et al., 2021b). These plant-based pharmaceuticals showed successful outcomes in the treatment of oral cancer in preclinical models. The phytopharmaceuticals reduced the non-targeted toxicity associated with modern cancer diagnosis approaches. However, the clinical trials do not show expected outcomes, mainly due to bioavailability and administration problems. Low passaging ability through the barriers is also one of the reasons for the unfortunate result in the clinical trial. The research findings of preclinical and clinical studies of photochemical and their nanoparticulate forms have shown very different results. For example, Epigallocatechin gallate (EGCG) is a polyphenol that reduces cell growth, induce apoptosis and inhibits angiogenesis (Ko et al., 2007). EGCG potentially inhibit cancer cell proliferation in cell culture studies and inhibits tumor growth in mice models (Chen et al., 2011). However, in clinical settings, the effect of EGCG was very less as found in cell lines and animal model studies. This is might be due to the inefficient systemic delivery and poor bioavailability. Therefore, scientist formulated and introduced the term "nanochemoprevention" where the phytochemicals were encapsulated to biocompatible nanoparticles with the hope of effective results in clinical trials. The encapsulated EGCG showed higher efficacy than the EGCG alone. Similarly, curcumin also showed potent anticancer activity against a variety of cancer cell lines and in animal studies also (Siddappa et al., 2017; Liao et al., 2018). This compound also showed unexpected results in clinical trials. Therefore, scientists have formulated liposome-encapsulated curcumin that showed effective anticancer activity against head and neck cancers (Wang et al., 2008).

Therefore, the phytochemicals have been incorporated/formulated to its nanoparticulate form to increase the bioavailability and solubility. Thus, nanotechnology based drug administration have now become very useful technique in the field of medicine for drug delivery.

NANOTECHNOLOGY: INNOVATIVE TECHNIQUE FOR DRUG CONSTRUCTION AND DELIVERY

A group of scientists from different fields has made successful innovative technologies for cancer drug constructs and systemic delivery. The collective effort of biologists, biomedical engineers, chemical scientists, pharmaceutical scientists, and oncologists contributed significantly to cancer treatment by implementing a novel approach to drug delivery called Nanotechnology. According to National Nanotechnology Initiative, "Nanotechnology" can be defined as "research and technology development at the atomic, molecular or macromolecular scale leading to control the creation and use of structures, devices, and systems with a length scale of ~100 nanometer (nm)" (Roco, 1999). In addition, nanoparticles of various phytopharmaceuticals used to treat oral cancer are being produced using the same technology. Nanoparticles of these phytopharmaceuticals have several advantages that increase the encapsulated drug's solubility and bioavailability and thus increase the efficacy of the encapsulated phytochemicals and/or drug compounds (Figure 2A).

Nanotechnology based drug delivery in the treatment of cancer have shown promising therapeutic efficacy than the conventional drugs. Nanoparticles upon delivery and exposure to tissues and fluids immediately absorbed onto their surface which depends surface characteristics of the nanoparticles. Nanoparticles can be designed with specific biomolecule linkers in their surface or within the vesicle and liposome that could assist in anchoring with the specific protein in the cancer cells of specific types. In this manner, the affinity of the nanoparticles/nanoemulsion can be improved to cancer cells/damaged cells even if they occur in small percentage of cells. Besides, the large surface area leads to accumulation of nanoparticles in organs. For example, Bailly et al. (2019) observed that administration of laser-synthesized dextran-coated gold particle is absorbed in the liver and spleen and did not causes any toxicity to the organs upon evaluation of its safety efficacy. These laser-synthesized nanoparticles can also be targeted to cancer cells either actively or passibly and their accumulation in cancer cells could be used either as contrast agents in cancer imaging or sensitizers of different local therapies under external stimuli. In this manner, nanotechnology based cancer delivery system and drug formulation helps in disease prevention and improved the efficacy by several folds.

NANOEMULSION: ADVANCED FORMULATION OF DRUGS

Nanoemulsions are nanotechnology-based colloidal dispersions of two merely immiscible liquids, oil, and water, where one is dispersed to the other with the help of surfactant or co-surfactant (Figure 2B) (Fofaria et al., 2016; Shaker et al., 2019). Nanoemulsions have several advantages over convention drug delivery systems due to their high stability, small-sized particulates (<100 nm), and large surface area (Chuan et al., 2015; Fofaria et al., 2016). In addition, encapsulation protects the therapeutic from being degraded in its nano-emulsified form and increases the life of the drug in the plasma (Sánchez-López et al., 2019). Formulation and utilization of nanoemulsions in cancer research and therapy are getting more focused. Nanoemulsions possess the essential features showing those as mentioned above promising therapeutic potential, including specific targeting.

Two or more drugs can be administered at the same time using nanoemulsion formulation. However, these combinations of two or multiple agents do not always indicate the upgraded efficacy than the sum of the drugs administered individually. Combination index of drug mixture is an important measure to assess the effect level. Co-encapsulation of two drugs and their delivery for disease treatment have been shown better efficacy, advantage of synchronizing the pharmacokinetics of the agents, and co-localization of the two drugs into the same tumor site maintain their pre-determined synergistic effects in many cancers (Gao et al., 2020). Co-encapsulation is the loading of two drug molecule in a single nanoparticle. For example, delivery of salinomycin and doxorubicin liposomal co-encapsulation exhibited 2-fold increased efficacy in eliminating cancer cells than their individual effect. Nevertheless, co-encapsulation of multi-agents at a predetermined dose is often risky due to its diverse pharmacokinetics. Therefore, before co-encapsulation, the interaction and effect level of the drugs to be combined should be thoroughly investigated to overcome the problem.

Figure 2. Function of nanoemulsion and schematic representation of their formulation (A) Functions of nanoemulsion (B) Representative diagram of phytochemical-encapsulated oil in water nanoemulsion

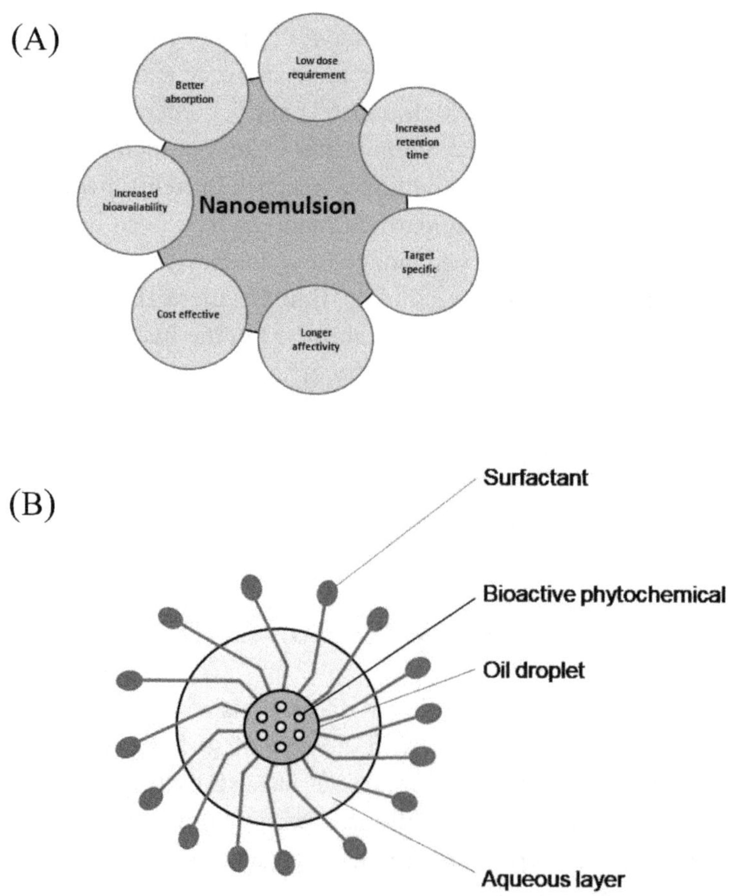

Nanoemulsion formulation drugs and their delivery in disease treatment improve the therapeutic efficacy of the drugs. Increased efficacy, solubility and stability of the nano-formulations are the main advantages of nanotechnology-based drug delivery system. Nanoemulsions can easily reach the cancer site as the blood vascular system surrounds cancer cells. Through this vascularised system, nanoemulsion can get the area of cancer. Besides, the other advantages of nanoemulsion are the smaller size that helps in easy passage through the plasma membrane and other barrier systems (Tiwari et al., 2006).

PHYTOCHEMICAL NANOEMULSIONS IN THE TREATMENT AND MANAGEMENT OF ORAL DISEASES

Several plant-based phytochemicals/products have been utilized in oral disease management. Phytochemicals are an excellent source of traditional medicines for treating toothache, dental caries, and other oral diseases. In recent times, plant products were also transformed into nanoemulsions, and their efficacy in oral disease management was investigated (Figure 3). The nanoemulsions formulated for oral disease management in the past few years were discussed below-

Soybean oil

Soybean oil is acquired from the soybean seeds (*Glycine max*). It is edible oil. A nanoemulsion of soybean oil was prepared, and their efficacy against adhesion and biofilm production in many cryogenic bacteria species (*Streptococus mutans, Lactobacillus casei, actinomyces viscosus, and Candida albicans*) was investigated (Ramalingam et al., 2011). It was observed that upon soybean oil nanoemulsion exposure, the adhesiveness of bacterial species was reduced by ~94.2 to 99.5%, and the bacterial count was also significantly reduced (Ramalingam et al., 2011). In a study, Lee et al. (2010) demonstrated that cetylpyridinium chloride containing nanoemulsion potentially prevented the dental carries in artificial mouth by using continuous flow dual-organisms namely, *S. mutans and L. casei* and also reduced the biofilm formation of the bacterial species tested.

Ascorbic Acid and α-Tocopherol

Ascorbic acid is a water-soluble vitamin (vitamin c) rich in citrus fruit and vegetables. It is a potent antioxidant that defends the body from reactive oxygen species. Nanoemulsion of ascorbic acid was formed, and its efficacy in oral disease treatment was evaluated. Chae et al. (2007) reported that ascorbic acid nanogel protects wound healing and ulcers in the oral cavity. Likewise, α-tocopherol also exhibited protective effects in the gingiva following the application of α-tocopherol nanogel paste (Chae et al., 2007). Apart from this, ascorbic acid and α-tocopherol showed an anti-inflammatory effect in the oral cavity's soft tissues (gingiva).

Figure 3. Chemical structure of some phytochemicals formulated to nanoemulsion in the treatment of oral cancer and other oral disease

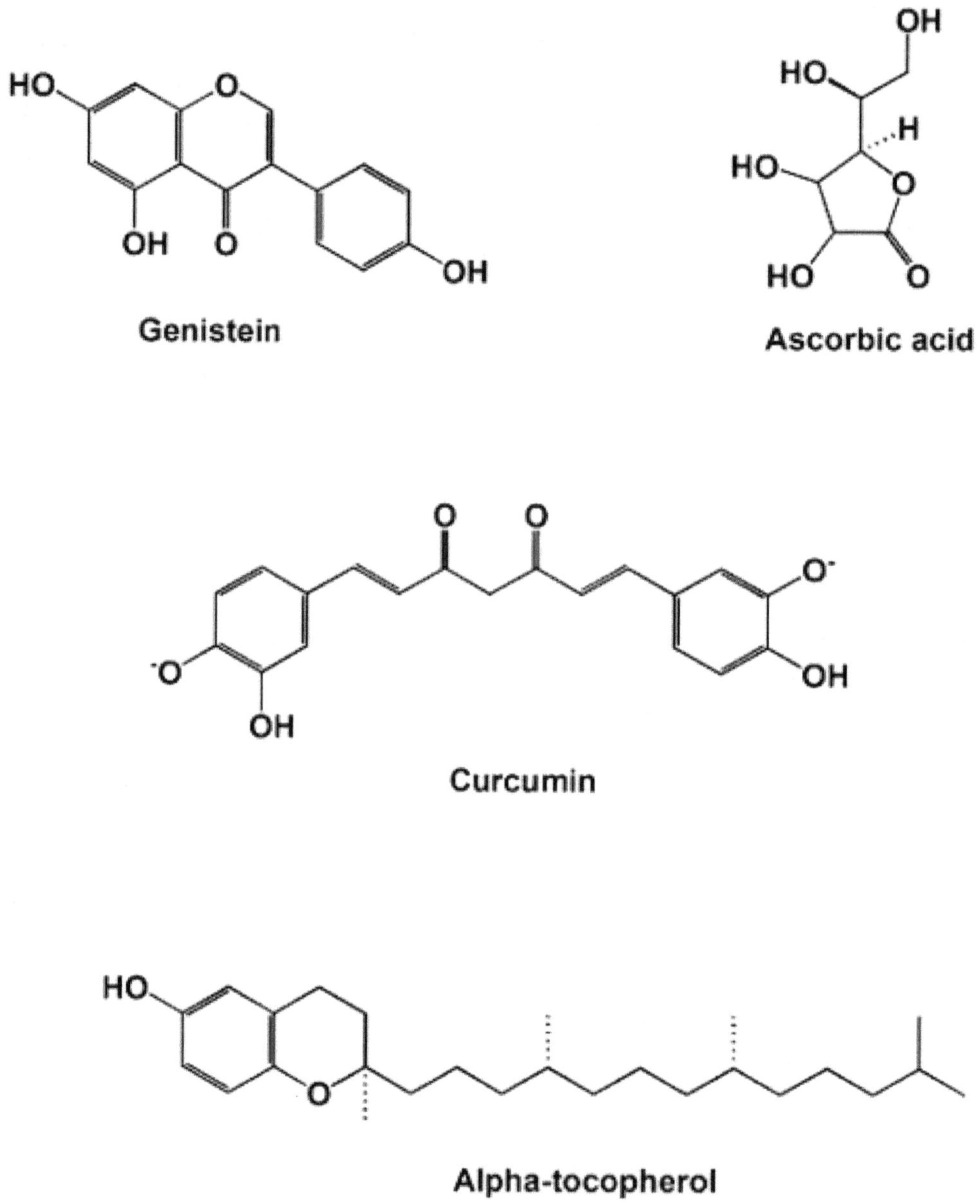

Cinnamon oil

Cinnamon oil is edible oil. The bark of cinnamon is extensively used as a spice. The nanoemulsions formulated from cinnamon showed interesting findings when tested against oral microorganisms. Treatment of cinnamon essential oil nanoemulsion exhibited a comparable decrease in bacterial biofilm production in established microsm biofilm model (Jiong et al., 2021). Moreover, cinnamon oil nanoemulsion showed protective effects against dental caries and periodontal disease (Jiong et al., 2021) (Table 1).

Table 1. Phytochemical nanoemulsions and their potential role in oral health management

Phytochemical	Nanoemulsion formulation	Application to oral health	Therapeutic potential	References (s)
Soybean oil	Soybean oil nanoemulsion	Dental caries	Treatment with soybean oil nanoemulsion exhibits a protective effect against dental caries. Biofilm production of oral microorganisms are also reduced following soybean oil nanoemulsion exposure	Lee et al., 2010
		Antimicrobial activity against cariogenic oral bacteria	Soybean oil nanoemulsion treatment significantly reduced >94% of the bacterial biofilm production	Ramalingam et al., 2011
Ascorbic acid (Vitamin C)	Ascorbic acid nanogel	Oral cavity ulcers	Treatment with nanoemulsion showed protection from ulcers in the oral cavity. Ascorbic acid nanoemulsion also potential wound healing activity	Chae et al., 2007
α- tocopherol (Vitamin E)	α-tocopherol nanogel	Oral cavity ulcer and inflammation in gingival tissues	α-tocopherol nanogel potentially reduced the inflammatory responses in the soft tissues of oral gingiva. Besides, α-tocopherol nanogel also protect from oral ulcers	Chae et al., 2007
Cinnamon oil	Cinnamon oil nanoemulsion	Antimicrobial potential against oral bacteria	Cinnamon essential oil nanoemulsion formulation has a potential role in the reduction of biofilm production against oral bacteria. Besides, it showed potent therapeutic efficacy against dental caries and orondontal disease	Jiang et al., 2021

PHYTOCHEMICAL NANOEMULSIONS IN ORAL CANCER THERAPY

Cancer pharmaceuticals from plant sources have been transformed into nanoemulsion formulation to achieve the best efficient effect of the drug. Studies reported that nanoemulsions of phytopharmaceuticals and other plant products had shown better results than phytochemicals alone in preclinical and clinical experimental settings. Therefore, phytochemicals nanoemulsions of apigenin, lycopene, genistein, curcumin, etc., have been investigated for their potential anticancer effect against various cancers, including oral cancer (Prasad et al., 2020; De Matos et al., 2017; Pham et al., 2015; Sánchez-López et al., 2019; Ashrafizadeh et al., 2020). The phytochemicals that have been evaluated for their anticancer potential against oral cancer (Figure 3) have been discussed herein.

Genistein

Genistein [5,7-dihydrox-3-(-4-hydroxyphenyl)-chromen-4-one] is an isoflavone. It is a major phytochemical in soybeans and other soy-based products and chickpeas. The antioxidant, anti-apoptotic, anti-metastatic, anti-proliferative, and anti-inflammatory activity of genistein has been well described (Ganai and Farooqi, 2015; Russo et al., 2016; Tuli et al., 2019). Epidemiological studies confirmed the negative correlation between dietary genistein intake and cancer occurrence. The anticancer properties of genistein against oral cancer in cell lines and animal models have been studied extensively (Yang et al., 2006; Ardito et al., 2017). However, genistein alone did not show anticipated results in clinical trials due to its lipophilic nature. Therefore, with the advancement of technology, nanoemulsions of genistein have been formulated that showed better outcomes than genistein alone.

Mucoadhesive nanoemulsions (NEs) have been prepared to increase the efficacy of the genistein in cancer therapy, and its anticancer potential has also been investigated. Gavin et al. (2015) designed mucoadhesive nanoemulsions of genistein by using chia seed oil [rich in polyunsaturated fatty acid (PUFA)] and α-tocopherol. Genistein nanoparticles (Gen-NPs) were incorporated into this oily phase that increased the mucoadhesive nanoemulsions. As a result, mucoadhesive nanoemulsions of Gen-NPs showed potent anticancer activity against human tongue squamous cell carcinoma (SCC-4 cell line) and pharyngeal squamous cell carcinoma (FaDu cell line) (Table 2). In addition, Gen-NPs-NE increased the cytotoxicity to both the SCC-4 cell line and FaDu cell line compared to Gen-NPs alone. Moreover, nanoemulsions of Gen-NPs facilities surmount the administration-associated obstacles.

Genistein protects from oral carcinogenesis by modulating multiple pathways in the cell. It increased the apoptosis rate in cancer cells by changing the expression of various apoptotic proteins like- Caspase-3, Bax, Bad, Bak, LC3-II, and p62. It also decreases PKL1 and MDM2 protein expression activity, resulting in increased apoptosis in cancer cells (Shafiee et al., 2016; Hsiao et al., 2019). In addition, genistein showed its anticancer activity by modulating the expression of the proteins associated with NF-κB signaling, PI3K/Akt signaling, Wnt/β-catenin signaling, ERK 1/2, and MAPK pathway (Tuli et al., 2019). Moreover, genistein exhibits its anticancer effect by inducing stress in the endoplasmic reticulum in cancer cells (Yang et al., 2016).

Table 2. Phytochemical nanoemulsions and their potential role in oral cancer treatment

Phytochemical	Nanoemulsion formulation	Oral cancer cell line	Therapeutic potential	References (s)
Genistein	Genistein-loaded nanoemulsion	Tongue squamous cell carcinoma (SCC-4 cell line) Pharyngeal cancer cell (FaDu cell line)	Genistein-loaded mucoadhesive nanoemulsion treatment led to dose-dependent cell death in SCC-4 and FaDu cell lines.	Gavin et al., 2015
Curcumin	Curcumin nanogel	Oral squamous cell carcinoma cell (OSCC cell line)	Curcumin naogel showed better solubility efficiency and high permeability. In addition, Curcumin nanogel exposure-induced mitochondrial-mediated programmed cell death. It also showed potential anticancer activity in the OSCC cell line by inducing cell death upon exposure.	Ganesh et al., 2019
α- tocopherol succinate	α-tocopherol succinate nanoemulsion	Oral cancer cell (KB cell line)	α-tocopherol succinate nanoemulsion showed cancer cell cytotoxicity and reduced the tumor growth in an animal model by activating intrinsic pathways of mitochondria-mediated apoptosis in KB cell lines	Gao et al., 2016
Spearmint oil	Spearmint oil nanoemulsion	Oral cancer cell (KON cell line)	Spearmint oil exposure to KON cell lines exhibited a dose-dependent decrease in viable cell number. The spearmint oil led to downregulation and upregulation of Bcl-2 and Bax protein expression, respectively, and induced mitochondrial-mediated apoptosis in the KON cell line	Tubtimsri et al., 2018
Celery oil	Celery oil nanoemulsion	Oral squamous cell carcinoma (SAS cell line)	Celery oil nanoemulsion showed a significant reduction in cell proliferation, inhibits colony formation, and induced programmed cell death in SAS cell line	Nirmala et al., 2020

α- Tocopherol

α-tocopherol (Vitamin E) is a phenolic compound with lipophilic. It has substantial antioxidant property and protects from cellular lipid peroxidation. It is abundant in vegetable oils like corn, cottonseed, sesame, and soybeans (Jiang et al., 2001). Due to the antioxidant ability of α-tocopherol, it exhibits potential anticancer activity against various cancers, including oral cancer. In a human intervention study based on 10 case-control studies, the association of dietary α-tocopherol intake and head and neck cancer (HNC) risk were investigated. It was observed that dietary ingestion of α-tocopherol reduced cancer jeopardy (Edefonti et al., 2015). Conversely, reports also indicated that the subordinate nutritional status of α-tocopherol is associated with the increased risk of various cancers (Taylor et al., 2003; Mahabir et al., 2008). However, large-scale human intervention studies showed inconsistent findings of the cancer therapeutic potential of α-tocopherol.

NEs of α-tocopherol have increased the efficacy of the compound, which is evident from experimental findings. The NE showed the most promising discoveries when the effectiveness of α-tocopherol nano-emulsions was compared with its free α-tocopherol. Gao et al. (2016) prepared α-tocopherol succinate nanoemulsions and investigated and compared their efficacy against KB cells. The authors reported that α-tocopherol nanoemulsions have more cytotoxic potential than free α-tocopherol. A consistent pharmacokinetic finding was also observed in the rat animal model (Gao et al., 2016). The anticancer properties of α-tocopherol nanoemulsions have been reported in various other types of cancers. However, very few studies have been conducted by taking oral cancer cell lines and/or oral cancer animal models.

The experts have proposed numerous mechanisms of action of α-tocopherol. It has been reported that α-tocopherol inhibits the cyclooxygenases (COX2) activity in cancer cells (Jiang et al., 2008). Downregulation of estrogen receptor α (ER- α), Akt signaling proteins, cyclin D1, and cyclin E are the pathways involved in α-tocopherol mediated anticancer potential (Ju et al., 2010; Huang et al., 2013). α-tocopherol also induces PPRAγ expression, p21 and p27 protein expression, and caspase-3 and caspase-9 protein expression that induced the apoptosis in cancer cells (Jiang et al., 2004a; Ju et al., 2010). Further, the synthesis of sphingolipids is also the mechanism of α-tocopherol showing cancer preventive efficacy (Jiang et al., 2004b).

Curcumin

Curcumin [(1E, 6E)-1,7-bis(4-hydroxy-3-methoxyphenyl)-1,6-heptadiene-3,5-dione] is a polyphenol isolated from the plant *Curcuma longa* (turmeric). It has promising therapeutic activity against multiple chronic diseases like inflammation, liver disease, neurodegenerative disease, metabolic syndrome, and cancer (Giordano and Tommonaro, 2019). Curcumin possesses strong antioxidant, anti-bacterial, anti-proliferative, antiviral, and pro-apoptotic activity evidenced by its preclinical studies (Hewlings and Kalman, 2017). The anticancer properties of curcumin in its free form have been evaluated in head and neck cancer cell lines, breast cancer cell lines, prostate cancer cell lines, and human colon cancer cell lines (Tomeh et al., 2019). However, poor bioavailability and less solubility in water and saline make it very difficult and tedious.

Like other phytochemicals, nanoemulsions of curcumin have several advantages over their free form and exhibited more satisfactory results. Curcumin nanoemulsion has been prepared, and its anticancer potential has been evaluated against oral squamous cell carcinoma (OSCC) cancer cell lines. Ganesh et al. (2019) designed and developed the nanogel formation of curcumin for OSCC therapy. Curcumin nanogel formulation using Carbopol 940 as gel matrix and polyethylene glycol (PEG) as a surfactant showed better transdermal permeability. Ganesh and co-workers have also reported that curcumin nanogel exhibited better permeability, biocompatibility, and solubility than its free form (Ganesh et al., 2019). Thus, curcumin nanogel showed promising therapeutic efficiency against OSCC.

Curcumin showed its anticancer effects by interfering with multiple mechanisms in the cell. Studies reported that curcumin affects and modulates the NF-κβ pathway in various human cancer cell lines (Wang et al., 2008). Thus, curcumin affects a variety of cellular signaling pathways and prevents carcinogenesis of multiple types. Curcumin also impacts the ROS generating enzymes decreased lipoxygenase /cyclo-oxygenase/cyclooxygenases and xanthine hydrogenase/oxidase in the number of cancer cells (Gheibi et al., 2019). In addition, PD-L1 and pSTAT3 protein expression was also downregulated upon curcumin treatment. It was also reported that curcumin treatment led to a change in the immunosuppressive micro-environment in tumor tissues or tongue squamous cell carcinoma (TSCC) cell lines (Liao et al., 2018).

Spearmint oil

Spearmint oil (SO) is commonly employed for oral care product manufacture. It is obtained from *Mentha sicata* leaves. SO have been reported to possess antioxidant, anti-inflammatory, anti-bacterial, antifungal, and antispasmodic activity. In addition, the anticancer potential of SO has been well documented (Manosroi et al., 2006). SO exhibited potential cytotoxicity against human mouth epidermal carcinoma cell lines, human epithelial type 2 cancer cell line, murine leukemia, and human breast adenocarcinoma cell lines. Therefore, SO is one of the plant-based oils having substantial anticancer properties against various cancer cell lines irrespective of its origin. However, SO oil in cancer treatment is limited because of water insolubility and inadequate bioavailability. Therefore, scientists have manufactured the nanoemulsions of SO to overcome the problems associated with hydrophilicity nature and bioavailability nature.

Nanoemulsions of SO and virgin coconut oils (VCO) were prepared to elucidate the anticancer activity of SO against the human oral squamous carcinoma (KON) cell line. Tubtimsri et al. (2018) reported that SO nanoemulsion preparations showed potential cytotoxicity in KON cell lines. SO-VCO nanoemulsion causes a dose-dependent decrease in viable cell number. Recently, Tubtimsri et al. (2021) reported that SO nanoemulsion induced apoptosis in KON cell lines, thus showing potential anticancer activity. The anticancer potential of SO can be attributed to the chemical compounds present in SO. GC-MS analysis revealed the various bioactive compounds from SO, and among them, the major bioactive compounds are phenolics, essential oils, and long hydrocarbon chains. In addition, SO possess flavonoids, caffeic acid, rosmaric acid, thymosin, limonene, and carvone as the major bioactive compound (Sharafi et al., 2010). Because of the presence of these bioactive compounds, SO exhibit several pharmacological activities. Reports are available that SO nanoemulsion treatment modulates the Bax and Bcl-2 protein expression leading to cytochrome c from mitochondria (Tubtimsri et al., 2018). The mitochondrial-mediated programmed cell death is a potent mechanism of action of SO-mediated anticancer activity.

Celery oil

Celery oil (CO) is obtained from the plant *Apium graveolens,* used as a traditional medicinal plant for decades (Suslick et al., 1999). CO is used to treat various chronic and sub-chronic ailments such as jaundice, rheumatism, and spleen and liver diseases (Hong et al., 2002). In addition, CO is also reported to carry cytotoxic and anti-inflammatory activity (Ahmedy, 2016). The pharmacological activity of CO could be attributed to the bioactive compounds present in it. The major bioactive compounds of CO include apigenin, caffeic acid, p-coumaric acid, tannins, and saponins. In addition, CO inhibits the growth of various cancers, including hepatocellular carcinoma (Seleh et al., 1998; Ahmedy, 2016). However, CO is insoluble in water due to its hydrophobicity. Besides, poor bioavailability of CO imposed the scientist to find the new techniques to increase its efficacy, solubility, and bioavaibility.

Recently, the nanoemulsions of CO were prepared, and CO nanoemulsions' anticancer potential has been investigated in SAS cell lines. SAS is an OSCC cell line. Exposure of CO nanoemulsions to SAS cells showed potential anticancer activity by inducing apoptosis. In addition, the anchorage-independent colony formation was reduced upon exposure to CO nanoemulsions (Nirmala et al., 2020). The authors also reported a significant reduction of cell proliferation rate following CO nanoemulsions treatment.

Previous reports revealed that CO induces apoptosis in cancer cell lines by up-regulating the protein expression involved in the pathway. For example, CO induces caspase-3 protease enzyme activity and thus activates the intrinsic pathways of apoptosis (Bhatia et al., 2013). In addition, CO modulates the expression of other proteins associated with cancer progression. It was also reported that CO potentially reduced the TNF-α, NF-κβ, and COX-2 proteins expression in the hepatic carcinoma model (Ahmedy, 2016).

CONCLUSION AND FUTURE DIRECTION OF NANOEMULSIONS AS THERAPEUTICS

Nanoemulsions-based drug delivery system shows better efficacy of the drugs due to increased solubility, bioavailability, and site-specific delivery of the drugs. Thus, this new approach is preferred by physicians and medical practitioners. Nano-sized nanoemulsions of phytochemicals made a significant contribution to oral cancer and oral health management and therapy. Phytochemical nanoparticles were encapsulated to form nanoemulsions. Upon investigation, it showed promising anticancer potential against various cancer cell lines, including breast cancer, colon cancer, prostate cancer, lung cancer, etc. The application of nanoemulsion is supposed to be increased in upcoming years. However, a few studies only conducted taking plant-based products to treat oral cancer and other oral manifestations included in the present report.

Based on the above discussion, we conclude that nanoemulsions-based drug delivery systems could be an excellent approach to curing and preventing cancer and other diseases. However, the toxicity of the nanoemulsion formulations and associated ingredients used in nanoemulsion formulations should be investigated for less toxicity and better efficacy.

REFERENCES

Abdulkareem, I. H., & Zurmi, I. B. (2012). Review of hormonal treatment of breast cancer. *Nigerian Journal of Clinical Practice*, *15*(1), 9. doi:10.4103/1119-3077.94088 PMID:22437080

Ahmedy, O. (2016). Study of the anticancer potential of celery seed oil against chemically induced hepatocellular carcinoma in rats: A mechanistic approach. *Al-Azhar Journal of Pharmaceutical Sciences*, *53*(1), 14–28. doi:10.21608/ajps.2016.6686

Alam, A., Farooq, U., Singh, R., Dubey, V. P., Kumar, S., Kumari, R., ... Dhar, K. L. (2018). Chemotherapy treatment and strategy schemes: A review. *Open Access J Toxicol*, *2*(5), 555600. doi:10.19080/OAJT.2018.02.555600

Ardito, F., Pellegrino, M. R., Perrone, D., Troiano, G., Cocco, A., & Muzio, L. L. (2017). In vitro study on anti-cancer properties of genistein in tongue cancer. *OncoTargets and Therapy*, *10*, 5405–5415. doi:10.2147/OTT.S133632 PMID:29180873

Ashrafizadeh, M., Bakhoda, M. R., Bahmanpour, Z., Ilkhani, K., Zarrabi, A., Makvandi, P., Khan, H., Mazaheri, S., Darvish, M., & Mirzaei, H. (2020). Apigenin as Tumor Suppressor in Cancers: Biotherapeutic Activity, Nanodelivery, and Mechanisms With Emphasis on Pancreatic Cancer. *Frontiers in Chemistry*, *8*, 8. doi:10.3389/fchem.2020.00829 PMID:33195038

Bailly, A. L., Correard, F., Popov, A., Tselikov, G., Chaspoul, F., Appay, R., Al-Kattan, A., Kabashin, A. V., Braguer, D., & Esteve, M. A. (2019). In vivo evaluation of safety, biodistribution and pharmacokinetics of laser-synthesized gold nanoparticles. *Scientific Reports*, *9*(1), 1–12. doi:10.103841598-019-48748-3 PMID:31501470

Barhoi, D., Giri, S., Upadhaya, P., & Barbhuiya, S. N. (2021b). Chemopreventive and Therapeutic Potential of Phytopharmaceuticals Against Oral Cancer: Evidence-Based Reports From Preclinical Studies in Animal Models. In Handbook of Research on Advancements in Cancer Therapeutics (pp. 541-569). IGI Global.

Barhoi, D., Upadhaya, P., Barbhuiya, S. N., Giri, A., & Giri, S. (2021a). Aqueous extract of Moringa oleifera exhibit potential anticancer activity and can be used as a possible cancer therapeutic agent: A study involving in vitro and in vivo approach. *Journal of the American College of Nutrition*, *40*(1), 70–85. doi:10.1080/07315724.2020.1735572 PMID:32191153

Baskar, R., Lee, K. A., Yeo, R., & Yeoh, K. W. (2012). Cancer and radiation therapy: Current advances and future directions. *International Journal of Medical Sciences*, *9*(3), 193–199. doi:10.7150/ijms.3635 PMID:22408567

Baudino, T. A. (2015). Targeted cancer therapy: The next generation of cancer treatment. *Current Drug Discovery Technologies*, *12*(1), 3–20. doi:10.2174/1570163812666150602144310 PMID:26033233

Bhatia, D., Thoppil, R. J., Mandal, A., Samtani, K. A., Darvesh, A. S., & Bishayee, A. (2013). Pomegranate bioactive constituents suppress cell proliferation and induce apoptosis in an experimental model of hepatocellular carcinoma: Role of Wnt/β-catenin signaling pathway. *Evidence-Based Complementary and Alternative Medicine*, *2013*, 2013. doi:10.1155/2013/371813 PMID:23606879

Bhavana, S. M., & Lakshmi, C. R. (2014). Oral oncoprevention by phytochemicals-A systematic review disclosing the therapeutic dilemma. *Advanced Pharmaceutical Bulletin*, *4*(Suppl 1), 413. PMID:25364656

Borse, V., Konwar, A. N., & Buragohain, P. (2020). Oral cancer diagnosis and perspectives in India. *Sensors International*, 100046.

Chae, C. H., Choi, D. J., Shim, H. Y., Byun, E. S., Hong, S. M., Park, Y. H., & Park, J. W. (2007). treatment of oral soft tissue lesions and wounds with high functional tooth paste made from nanoemulsion gel. *Journal of the Korean Association of Oral and Maxillofacial Surgeons*, *33*(6), 694–700.

Chan, P. K. S., Chan, M. Y. M., Li, W. W. H., Chan, D. P. C., Cheung, J. L. K., & Cheng, A. F. (2001). Association of human β-herpesviruses with the development of cervical cancer: Bystanders or cofactors. *Journal of Clinical Pathology*, *54*(1), 48–53. doi:10.1136/jcp.54.1.48 PMID:11271789

Chen, P. N., Chu, S. C., Kuo, W. H., Chou, M. Y., Lin, J. K., & Hsieh, Y. S. (2011). Epigallocatechin-3 gallate inhibits invasion, epithelial– mesenchymal transition, and tumor growth in oral cancer cells. *Journal of Agricultural and Food Chemistry*, *59*(8), 3836–3844. doi:10.1021/jf1049408 PMID:21388137

Choudhari, A. S., Mandave, P. C., Deshpande, M., Ranjekar, P., & Prakash, O. (2020). Phytochemicals in cancer treatment: From preclinical studies to clinical practice. *Frontiers in Pharmacology*, *10*, 1614. doi:10.3389/fphar.2019.01614 PMID:32116665

Chuan, L. I., Zhang, J., Yu-Jiao, Z. U., Shu-Fang, N. I. E., Jun, C. A. O., Qian, W. A. N. G., ... Shu, W. A. N. G. (2015). Biocompatible and biodegradable nanoparticles for enhancement of anti-cancer activities of phytochemicals. *Chinese Journal of Natural Medicines*, *13*(9), 641–652. doi:10.1016/S1875-5364(15)30061-3 PMID:26412423

Cinausero, M., Aprile, G., Ermacora, P., Basile, D., Vitale, M. G., Fanotto, V., Parisi, G., Calvetti, L., & Sonis, S. T. (2017). New frontiers in the pathobiology and treatment of cancer regimen-related mucosal injury. *Frontiers in Pharmacology*, *8*, 354. doi:10.3389/fphar.2017.00354 PMID:28642709

Danaraddi, S., Koneru, A., Hunasgi, S., Ramalu, S., & Vanishree, M. (2014). Natural ways to prevent and treat oral cancer. *Journal of Oral Research and Review*, *6*(1), 34. doi:10.4103/2249-4987.140213

De Matos, R. P. A., Calmon, M. F., Amantino, C. F., Villa, L. L., Primo, F. L., Tedesco, A. C., & Rahal, P. (2018). Effect of curcumin-nanoemulsion associated with photodynamic therapy in cervical carcinoma cell lines. *BioMed Research International*, *2018*, 2018. doi:10.1155/2018/4057959 PMID:29581972

Du, R., Wang, X., Ma, L., Larcher, L. M., Tang, H., Zhou, H., Chen, C., & Wang, T. (2021). Adverse reactions of targeted therapy in cancer patients: A retrospective study of hospital medical data in China. *BMC Cancer*, *21*(1), 1–13. doi:10.118612885-021-07946-x PMID:33639888

Edefonti, V., Hashibe, M., Parpinel, M., Ferraroni, M., Turati, F., Serraino, D., Matsuo, K., Olshan, A. F., Zevallos, J. P., Winn, D. M., Moysich, K., Zhang, Z.-F., Morgenstern, H., Levi, F., Kelsey, K., McClean, M., Bosetti, C., Schantz, S., Yu, G.-P., ... Decarli, A. (2015). Vitamin E intake from natural sources and head and neck cancer risk: A pooled analysis in the International Head and Neck Cancer Epidemiology consortium. *British Journal of Cancer*, *113*(1), 182–192. doi:10.1038/bjc.2015.149 PMID:25989276

Esfahani, K., Roudaia, L., Buhlaiga, N., Del Rincon, S. V., Papneja, N., & Miller, W. H. (2020). A review of cancer immunotherapy: From the past, to the present, to the future. *Current Oncology (Toronto, Ont.)*, *27*(12, s2), 87–97. doi:10.3747/co.27.5223 PMID:32368178

Fisher, B., Costantino, J. P., Wickerham, D. L., Redmond, C. K., Kavanah, M., Cronin, W. M., Vogel, V., Robidoux, A., Dimitrov, N., Atkins, J., Daly, M., Wieand, S., Tan-Chiu, E., Ford, L., & Wolmark, N. (1998). Tamoxifen for prevention of breast cancer: Report of the national surgical adjuvant breast and bowel project P-1 study. *Journal of the National Cancer Institute*, *90*(18), 1371–1388. doi:10.1093/jnci/90.18.1371 PMID:9747868

Fofaria, N. M., Qhattal, H. S. S., Liu, X., & Srivastava, S. K. (2016). Nanoemulsion formulations for anti-cancer agent piplartine—Characterization, toxicological, pharmacokinetics and efficacy studies. *International Journal of Pharmaceutics*, *498*(1-2), 12–22. doi:10.1016/j.ijpharm.2015.11.045 PMID:26642946

Ganai, A. A., & Farooqi, H. (2015). Bioactivity of genistein: A review of in vitro and in vivo studies. *Biomedicine and Pharmacotherapy*, *76*, 30–38. doi:10.1016/j.biopha.2015.10.026 PMID:26653547

Ganesh, G. N. K., Singh, M. K., Datri, S., & Karri, V. V. S. R. (2019). Design and Development of Curcumin Nanogel for Squamous Cell Carcinoma. *Journal of Pharmaceutical Sciences and Research*, *11*(4), 1683–1683.

Gao, Y., Qi, X., Zheng, Y., Ji, H., Wu, L., Zheng, N., & Tang, J. (2016). Nanoemulsion enhances α-tocopherol succinate bioavailability in rats. *International Journal of Pharmaceutics*, *515*(1-2), 506–514. doi:10.1016/j.ijpharm.2016.10.026 PMID:27746330

Gao, Y., Tang, M., Leung, E., Svirskis, D., Shelling, A., & Wu, Z. (2020). Dual or multiple drug loaded nanoparticles to target breast cancer stem cells. *RSC Advances*, *10*(32), 19089–19105. doi:10.1039/D0RA02801K

Gavin, A., Pham, J. T., Wang, D., Brownlow, B., & Elbayoumi, T. A. (2015). Layered nanoemulsions as mucoadhesive buccal systems for controlled delivery of oral cancer therapeutics. *International Journal of Nanomedicine*, *10*, 1569. PMID:25759580

Gheibi, S., Ghaleh, H. E. G., Motlagh, B. M., & Azarbayjani, A. F. (2019). Therapeutic effects of curcumin and ursodexycholic acid on non-alcoholic fatty liver disease. *Biomedicine and Pharmacotherapy*, *115*, 108938. doi:10.1016/j.biopha.2019.108938 PMID:31071511

Giordano, A., & Tommonaro, G. (2019). Curcumin and cancer. *Nutrients*, *11*(10), 2376. doi:10.3390/nu11102376 PMID:31590362

Giri, S., Barhoi, D., Barbhuiya, S. N., Giri, A., Das, S., Das, A., ... Singh, S. (2021). Consumption pattern and genotoxic potential of various smokeless tobacco products in Assam, India: A public health concern. *Mutation Research/Genetic Toxicology and Environmental Mutagenesis*, *866*, 503349. doi:10.1016/j.mrgentox.2021.503349 PMID:33985693

Gupta, K., & Metgud, R. (2013). Evidences suggesting involvement of viruses in oral squamous cell carcinoma. *Pathology Research International*, *2013*, 2013. doi:10.1155/2013/642496 PMID:24455418

Hewlings, S. J., & Kalman, D. S. (2017). Curcumin: A review of its effects on human health. *Foods*, *6*(10), 92. doi:10.3390/foods6100092 PMID:29065496

Hong, C., Kim, H. A., Firestone, G. L., & Bjeldanes, L. F. (2002). 3, 3′-Diindolylmethane (DIM) induces a G1 cell cycle arrest in human breast cancer cells that is accompanied by Sp1-mediated activation of p21WAF1/CIP1 expression. *Carcinogenesis*, *23*(8), 1297–1305. doi:10.1093/carcin/23.8.1297 PMID:12151347

Hsiao, Y. C., Peng, S. F., Lai, K. C., Liao, C. L., Huang, Y. P., Lin, C. C., Lin, M.-L., Liu, K.-C., Tsai, C.-C., Ma, Y.-S., & Chung, J. G. (2019). Genistein induces apoptosis in vitro and has antitumor activity against human leukemia HL-60 cancer cell xenograft growth in vivo. *Environmental Toxicology*, *34*(4), 443–456. doi:10.1002/tox.22698 PMID:30618158

Huang, C. Y., Ju, D. T., Chang, C. F., Reddy, P. M., & Velmurugan, B. K. (2017). A review on the effects of current chemotherapy drugs and natural agents in treating non–small cell lung cancer. *Biomedicine (Taipei)*, *7*(4), 23. doi:10.1051/bmdcn/2017070423 PMID:29130448

Huang, P. H., Chuang, H. C., Chou, C. C., Wang, H., Lee, S. L., Yang, H. C., Chiu, H.-C., Kapuriya, N., Wang, D., Kulp, S. K., & Chen, C. S. (2013). Vitamin E facilitates the inactivation of the kinase Akt by the phosphatase PHLPP1. *Science Signaling*, *6*(267), ra19–ra19. doi:10.1126cisignal.2003816 PMID:23512990

IARC. (2020). *Cancer fact sheet*. https://gco.iarc.fr/today/data/factsheets/cancers/1-Lip-oral-cavity-fact-sheet.pdf

Jackson, S. P., & Bartek, J. (2009). The DNA-damage response in human biology and disease. *Nature*, *461*(7267), 1071–1078. doi:10.1038/nature08467 PMID:19847258

Jeong, Y. J., Kim, H. E., Han, S. J., & Choi, J. S. (2021). Antibacterial and antibiofilm activities of cinnamon essential oil nanoemulsion against multi-species oral biofilms. *Scientific Reports*, *11*(1), 1–8. doi:10.103841598-021-85375-3 PMID:33723345

Jiang, Q., Christen, S., Shigenaga, M. K., & Ames, B. N. (2001). γ-Tocopherol, the major form of vitamin E in the US diet, deserves more attention. *The American Journal of Clinical Nutrition*, *74*(6), 714–722. doi:10.1093/ajcn/74.6.714 PMID:11722951

Jiang, Q., Wong, J., & Ames, B. N. (2004a). γ-Tocopherol Induces Apoptosis in Androgen-Responsive LNCaP Prostate Cancer Cells via Caspase-Dependent and Independent Mechanisms. *Annals of the New York Academy of Sciences*, *1031*(1), 399–400. doi:10.1196/annals.1331.056 PMID:15753180

Jiang, Q., Wong, J., Fyrst, H., Saba, J. D., & Ames, B. N. (2004b). γ-Tocopherol or combinations of vitamin E forms induce cell death in human prostate cancer cells by interrupting sphingolipid synthesis. *Proceedings of the National Academy of Sciences of the United States of America*, *101*(51), 17825–17830. doi:10.1073/pnas.0408340102 PMID:15596715

Jiang, Q., Yin, X., Lill, M. A., Danielson, M. L., Freiser, H., & Huang, J. (2008). Long-chain carboxychromanols, metabolites of vitamin E, are potent inhibitors of cyclooxygenases. *Proceedings of the National Academy of Sciences of the United States of America*, *105*(51), 20464–20469. doi:10.1073/pnas.0810962106 PMID:19074288

Ju, J., Picinich, S. C., Yang, Z., Zhao, Y., Suh, N., Kong, A. N., & Yang, C. S. (2010). Cancer-preventive activities of tocopherols and tocotrienols. *Carcinogenesis*, *31*(4), 533–542. doi:10.1093/carcin/bgp205 PMID:19748925

Kerns, S. L., Ostrer, H., & Rosenstein, B. S. (2014). Radiogenomics: Using genetics to identify cancer patients at risk for development of adverse effects following radiotherapy. *Cancer Discovery*, *4*(2), 155–165. doi:10.1158/2159-8290.CD-13-0197 PMID:24441285

Khan, F., Sarker, M., Rahman, M., Ming, L. C., Mohamed, I. N., Zhao, C., ... Rashid, M. A. (2019). Comprehensive review on phytochemicals, pharmacological and clinical potentials of Gymnema sylvestre. *Frontiers in Pharmacology*, *10*, 1223. doi:10.3389/fphar.2019.01223 PMID:31736747

Ko, S. Y., Chang, K. W., Lin, S. C., Hsu, H. C., & Liu, T. Y. (2007). The repressive effect of green tea ingredients on amyloid precursor protein (APP) expression in oral carcinoma cells in vitro and in vivo. *Cancer Letters*, *245*(1-2), 81–89. doi:10.1016/j.canlet.2005.12.029 PMID:16458426

Kruger, S., Ilmer, M., Kobold, S., Cadilha, B. L., Endres, S., Ormanns, S., Schuebbe, G., Renz, B. W., D'Haese, J. G., Schloesser, H., Heinemann, V., Subklewe, M., Boeck, S., Werner, J., & von Bergwelt-Baildon, M. (2019). Advances in cancer immunotherapy 2019–latest trends. *Journal of Experimental & Clinical Cancer Research*, *38*(1), 1–11. doi:10.118613046-019-1266-0 PMID:30606223

Lee, V. A., Karthikeyan, R., Rawls, H. R., & Amaechi, B. T. (2010). Anti-cariogenic effect of a cetylpyridinium chloride-containing nanoemulsion. *Journal of Dentistry*, *38*(9), 742–749. doi:10.1016/j.jdent.2010.06.001 PMID:20600554

Liao, F., Liu, L., Luo, E., & Hu, J. (2018). Curcumin enhances anti-tumor immune response in tongue squamous cell carcinoma. *Archives of Oral Biology*, *92*, 32–37. doi:10.1016/j.archoralbio.2018.04.015 PMID:29751146

Mahabir, S., Schendel, K., Dong, Y. Q., Barrera, S. L., Spitz, M. R., & Forman, M. R. (2008). Dietary α-, β-, γ-and δ-tocopherols in lung cancer risk. *International Journal of Cancer*, *123*(5), 1173–1180. doi:10.1002/ijc.23649 PMID:18546288

Manosroi, J., Dhumtanom, P., & Manosroi, A. (2006). Anti-proliferative activity of essential oil extracted from Thai medicinal plants on KB and P388 cell lines. *Cancer Letters*, *235*(1), 114–120. doi:10.1016/j.canlet.2005.04.021 PMID:15979235

Nirmala, M. J., Durai, L., Gopakumar, V., & Nagarajan, R. (2020). Preparation of Celery Essential Oil-Based Nanoemulsion by Ultrasonication and Evaluation of Its Potential Anticancer and Antibacterial Activity. *International Journal of Nanomedicine*, *15*, 7651–7666. doi:10.2147/IJN.S252640 PMID:33116493

Nurgali, K., Jagoe, R. T., & Abalo, R. (2018). Adverse effects of cancer chemotherapy: Anything new to improve tolerance and reduce sequelae? *Frontiers in Pharmacology*, *9*, 245. doi:10.3389/fphar.2018.00245 PMID:29623040

Pham, J., Grundmann, O., & Elbayoumi, T. (2015). Mitochondriotropic nanoemulsified genistein-loaded vehicles for cancer therapy. In *Mitochondrial Medicine* (pp. 85–101). Humana Press. doi:10.1007/978-1-4939-2288-8_7

Prasad, C., Bhatia, E., & Banerjee, R. (2020). curcumin encapsulated Lecithin nanoemulsions: An oral platform for Ultrasound Mediated Spatiotemporal Delivery of curcumin to the tumor. *Scientific Reports*, *10*(1), 1–15. doi:10.103841598-020-65468-1 PMID:32444829

Puhalla, S., Bhattacharya, S., & Davidson, N. E. (2012). Hormonal therapy in breast cancer: A model disease for the personalization of cancer care. *Molecular Oncology*, *6*(2), 222–236. doi:10.1016/j.molonc.2012.02.003 PMID:22406404

Ramalingam, K., & Amaechi, B. T. (2019). Phytochemicals: Their Therapeutic Potential Against Dental Caries. In Computational Techniques for Dental Image Analysis (pp. 238-275). IGI Global.

Ramalingam, K., Amaechi, B. T., Ralph, R. H., & Lee, V. A. (2012). Antimicrobial activity of nanoemulsion on cariogenic planktonic and biofilm organisms. *Archives of Oral Biology*, *57*(1), 15–22. doi:10.1016/j.archoralbio.2011.07.001 PMID:21807359

Roco, M. C. (1999). Nanoparticles and nanotechnology research. *Journal of Nanoparticle Research*, *1*(1), 1–6. doi:10.1023/A:1010093308079

Russo, M., Russo, G. L., Daglia, M., Kasi, P. D., Ravi, S., Nabavi, S. F., & Nabavi, S. M. (2016). Understanding genistein in cancer: The "good" and the "bad" effects: A review. *Food Chemistry*, *196*, 589–600. doi:10.1016/j.foodchem.2015.09.085 PMID:26593532

Saleh, M. M., Hashem, F. A., & Glombitza, K. W. (1998). Cytotoxicity and in vitro effects on human cancer cell lines of volatiles of Apium graveolens var filicinum. *Pharmaceutical and Pharmacological Letters*, *8*(2), 97–99.

Sánchez-López, E., Guerra, M., Dias-Ferreira, J., Lopez-Machado, A., Ettcheto, M., Cano, A., Espina, M., Camins, A., Garcia, M. L., & Souto, E. B. (2019). Current applications of nanoemulsions in cancer therapeutics. *Nanomaterials (Basel, Switzerland)*, *9*(6), 821. doi:10.3390/nano9060821 PMID:31159219

Scrobota, I., Bolfa, P., Filip, A. G., Catoi, C., Alb, C., Pop, O., ... Baciut, G. (2016). Natural chemopreventive alternatives in oral cancer chemoprevention. *Journal of Physiology and Pharmacology*, *67*(1), 161–172. PMID:27010905

Shafiee, G., Saidijam, M., Tavilani, H., Ghasemkhani, N., & Khodadadi, I. (2016). Genistein induces apoptosis and inhibits proliferation of HT29 colon cancer cells. *International Journal of Molecular and Cellular Medicine*, *5*(3), 178. PMID:27942504

Shaker, D. S., Ishak, R. A., Ghoneim, A., & Elhuoni, M. A. (2019). Nanoemulsion: A review on mechanisms for the transdermal delivery of hydrophobic and hydrophilic drugs. *Scientia Pharmaceutica*, *87*(3), 17. doi:10.3390cipharm87030017

Sharafi, S. M., Rasooli, I., Owlia, P., Taghizadeh, M., & Astaneh, S. D. A. (2010). Protective effects of bioactive phytochemicals from Mentha piperita with multiple health potentials. *Pharmacognosy Magazine*, *6*(23), 147. doi:10.4103/0973-1296.66926 PMID:20931070

Siddappa, G., Kulsum, S., Ravindra, D. R., Kumar, V. V., Raju, N., Raghavan, N., Sudheendra, H. V., Sharma, A., Sunny, S. P., Jacob, T., Kuruvilla, B. T., Benny, M., Antony, B., Seshadri, M., Lakshminarayan, P., Hicks, W. Jr, Suresh, A., & Kuriakose, M. A. (2017). Curcumin and metformin-mediated chemoprevention of oral cancer is associated with inhibition of cancer stem cells. *Molecular Carcinogenesis*, *56*(11), 2446–2460. doi:10.1002/mc.22692 PMID:28618017

Stanculeanu, D. L., Daniela, Z., Lazescu, A., Bunghez, R., & Anghel, R. (2016). Development of new immunotherapy treatments in different cancer types. *Journal of Medicine and Life*, *9*(3), 240. PMID:27974927

Sung, H., Ferlay, J., Siegel, R. L., Laversanne, M., Soerjomataram, I., Jemal, A., & Bray, F. (2021). Global cancer statistics 2020: GLOBOCAN estimates of incidence and mortality worldwide for 36 cancers in 185 countries. *CA: a Cancer Journal for Clinicians*, *71*(3), 209–249. doi:10.3322/caac.21660 PMID:33538338

Suslick, K. S., Didenko, Y., Fang, M. M., Hyeon, T., Kolbeck, K. J., McNamara III, W. B., ... Wong, M. (1999). Acoustic cavitation and its chemical consequences. *Philosophical Transactions of the Royal Society of London. Series A: Mathematical, Physical and Engineering Sciences, 357*(1751), 335-353.

Symonds, R. P. (2001). Radiotherapy. *BMJ (Clinical Research Ed.)*, *323*(7321), 1107–1110. doi:10.1136/bmj.323.7321.1107 PMID:11701579

Taylor, P. R., Qiao, Y. L., Abnet, C. C., Dawsey, S. M., Yang, C. S., Gunter, E. W., Wang, W., Blot, W. J., Dong, Z.-W., & Mark, S. D. (2003). Prospective study of serum vitamin E levels and esophageal and gastric cancers. *Journal of the National Cancer Institute*, *95*(18), 1414–1416. doi:10.1093/jnci/djg044 PMID:13130117

Tiwari, S., Tan, Y. M., & Amiji, M. (2006). Preparation and in vitro characterization of multifunctional nanoemulsions for simultaneous MR imaging and targeted drug delivery. *Journal of Biomedical Nanotechnology*, *2*(3-4), 217–224. doi:10.1166/jbn.2006.038

Tomeh, M. A., Hadianamrei, R., & Zhao, X. (2019). A review of curcumin and its derivatives as anticancer agents. *International Journal of Molecular Sciences*, *20*(5), 1033. doi:10.3390/ijms20051033 PMID:30818786

Tubtimsri, S., Limmatvapirat, C., Limsirichaikul, S., Akkaramongkolporn, P., Inoue, Y., & Limmatvapirat, S. (2018). Fabrication and characterization of spearmint oil loaded nanoemulsions as cytotoxic agents against oral cancer cell. *Asian Journal of Pharmaceutical Sciences, 13*(5), 425-437.

Tuli, H. S., Tuorkey, M. J., Thakral, F., Sak, K., Kumar, M., Sharma, A. K., Sharma, U., Jain, A., Aggarwal, V., & Bishayee, A. (2019). Molecular mechanisms of action of genistein in cancer: Recent advances. *Frontiers in Pharmacology*, *10*, 1336. doi:10.3389/fphar.2019.01336 PMID:31866857

Upadhaya, P., Barhoi, D., Giri, A., Bhattacharjee, A., & Giri, S. (2019). Joint detection of claudin-1 and junctional adhesion molecule-A as a therapeutic target in oral epithelial dysplasia and oral squamous cell carcinoma. *Journal of Cellular Biochemistry*, *120*(10), 18117–18127. doi:10.1002/jcb.29115 PMID:31161679

Wang, D., Veena, M. S., Stevenson, K., Tang, C., Ho, B., Suh, J. D., Duarte, V. M., Faull, K. F., Mehta, K., Srivatsan, E. S., & Wang, M. B. (2008). Liposome-encapsulated curcumin suppresses growth of head and neck squamous cell carcinoma in vitro and in xenografts through the inhibition of nuclear factor κB by an AKT-independent pathway. *Clinical Cancer Research*, *14*(19), 6228–6236. doi:10.1158/1078-0432.CCR-07-5177 PMID:18829502

Whitmore, S. E., & Lamont, R. J. (2014). Oral bacteria and cancer. *PLoS Pathogens*, *10*(3), e1003933. doi:10.1371/journal.ppat.1003933 PMID:24676390

Widakowich, C., de Castro, G. Jr, De Azambuja, E., Dinh, P., & Awada, A. (2007). Side effects of approved molecular targeted therapies in solid cancers. *The Oncologist*, *12*(12), 1443–1455. doi:10.1634/theoncologist.12-12-1443 PMID:18165622

Wochna Loerzel, V. (2015). Symptom Experience in Older Adults Undergoing Treatment for Cancer. *Oncology Nursing Forum*, *42*(3), E269–E278. doi:10.1188/15.ONF.E269-E278 PMID:25901389

Yang, Y., Zhou, Z. T., & Ge, J. P. (2006). Effect of genistein on DMBA-induced oral carcinogenesis in hamster. *Carcinogenesis*, *27*(3), 578–583. doi:10.1093/carcin/bgi234 PMID:16195236

Yang, Y. M., Yang, Y., Dai, W. W., Li, X. M., Ma, J. Q., & Tang, L. P. (2016). Genistein-induced apoptosis is mediated by endoplasmic reticulum stress in cervical cancer cells. *European Review for Medical and Pharmacological Sciences*, *20*(15), 3292–3296. PMID:27467006

Zhong, L., Li, Y., Xiong, L., Wang, W., Wu, M., Yuan, T., Yang, W., Tian, C., Miao, Z., Wang, T., & Yang, S. (2021). Small molecules in targeted cancer therapy: Advances, challenges, and future perspectives. *Signal Transduction and Targeted Therapy*, *6*(1), 1–48. doi:10.103841392-021-00572-w PMID:34054126

Chapter 16
Mushroom-Derived Bioactive-Based Nanoemulsion:
Current Status and Challenges for Cancer Therapy

Nagendra Kumar Chandrawanshi
https://orcid.org/0000-0002-9093-5365
Pt. Ravishankar Shukla University, India

Deepali Koreti
Pt. Ravishankar Shukla University, India

Anjali Kosre
Pt. Ravishankar Shukla University, India

Pramod Kumar Mahish
Government Digvijay Autonomous Post Graduate College, Rajnandgaon, India

ABSTRACT

Cancer is a class of disorders that is characterized by the abnormal growth of cells in an uncontrolled manner. In cancer progression, tumor cells have become highly heterogeneous, and they create a mixed population of cells with different molecular characteristics. The mushroom bioactive compounds have a rich biological activity including immunomodulatory, anticarcinogenic, antiviral, antioxidant, and anti-inflammatory, etc. Besides, conventional anticancer drugs and applied therapy have tremendous challenges and limitations such as poor solubility, narrow therapeutic window, cytotoxicity to normal tissues, etc., which may be the causes of treatment failure in cancer. A previous study reported mushroom bioactive compounds against cancer treatment. The chapter focuses on mushroom-derived bioactive compounds and possible implications in nanotechnology and, further, will be utilized for new advanced nanoemulsion techniques for the promising treatment of cancer.

DOI: 10.4018/978-1-7998-8378-4.ch016

INTRODUCTION

As a form of macrofungi, mushrooms are cultivated and consumed by the human race for dozens of centuries attributable to their enticing sensory quality and richness for nutritionary components Venturella et al. (2021). In general, mushrooms are categorized into two categories such as edible, non-edible and medicinal mushrooms. It is documented that higher Basidiomycetes have a good spectrum of therapeutic and prophylactic properties (Mau et al., 2002). Furthermore, the intention of the mushrooms redefined as a possible functional food with a less fat and high amount of protein, vitamin, mineral, glycoproteins, ergosterol, and lectins etc. (Thatoi and Singdevsachan 2014; Ma et al., 2018). Multitudinous research studies has been carried out to targeted to utilized of mushroom as health benefit and therapeutic uses of bioactive components. Moreover, mushrooms have additionally been proved to be indebted to the ability to attenuate the health vulnerability (Chandrawanshi et al., 2017; Rathore et al., 2017). The varieties of bioactive contain presence in mushrooms, which has exploit and used for new drugs development (Ma et al., 2018).

BRIEF DESCRIPTION OF CANCER

Cancer is a cluster of diseases that involve uncontrolled biological processes, replicative immortality and resistance to cell death. Cancer cells grow into an associate degree in abnormal cell mass, which is known as a tumor. Cancer processes are primarily originated by damage or mutation of proto-oncogene that code for proteins concerned within the induction of cell proliferation and differentiation. Alterations in each oncogene and tumor suppressor genes are necessary for tumor development and are favoured by mutations in the tumor status genes that code for a family of proteins involved within the control of deoxyribonucleic acid (DNA) damage. The mutations that initiate a tumor are clonally designated to favor aberrant and uncontrolled biological processes, the absence of inhibition of the excessive cell growth, avoidance of immune system, blockage of cell death, transmission and accumulation of genetic material errors (Perez-Herrero and Fernandez-Medarde, 2015). Globally, the morbidity and mortality rate due to cancer is increasing continuously. For the treatment of cancer generally prefer the chemotherapeutic and followed by radiotherapy approaches. However, chemotherapy has rigorous side effects. Therefore, biological scientists are continuously working with other interdisciplinary experts for developing a new effective management system for cancer treatment (Joseph et al., 2018).

To date, the majority of cancer patients are being treated by surgery, radiation therapy and chemotherapy. Generally, cancer tissue is removed by the surgery and then radiation therapy is given followed by chemotherapy. However, these treatment strategies possess potential disadvantages because of their non-selectivity, which produces harmful effects on normal tissues too. Hence, to overcome these serious drawbacks there would like novel targeting treatment approach (Choudhary et al., 2020). Stem cell therapy is recently focused on cancer treatment, having more promising therapeutic uses than the classical treatment systems (Chandrawanshi and Verma 2020). Medicinal mushrooms are reported to have diverse pharmacological and therapeutic actions like immunomodulatory, antidiabetic, cytotoxic, anti-inflammatory, antimicrobial, antiallergic, antioxidant and anticancer etc. (Chandrawanshi et al., 2018; Jeitler et al., 2020). The bioactive metabolites are produced from the fungal metabolic system. These activities are due to several bioactive metabolites found in the mycelium. Among the most important ones are immunomodulating and anti-tumor activities of various species of Basidiomycetes reported worldwide. Numerous studies conducted on anti-tumor activity possessed by natural compounds, as well as compounds of mushroom origin, contributed to the search for possibilities to resolve this global problem (Wasser, 2002). Within half a century, mycology has created a profound impact from the first the articles on research of anti-tumor activity of extracts from carpophores of various species of macromycetes in making "basidial" remedies and introducing them to the world pharmaceutical market. For the past decades, the strategy of research on the anti-tumor and immunomodulating properties in high-Basidiomycetes logically headed on a general course of complex research of therapeutic, prophylactic, nutritious and other benefits of macromycetes (Chang, 1999; Venturella et al. (2021).

However, many other species deserved for further research. Globally reported that variety of mushroom species has considered for immunomodulating and anti-tumor activities, such as: *Armillaria mellea, Bjerkandera adusta, Flammulina velutipes, Fomes fomentarius, Fomitopsis pinicola, Ganoderma lipsiense, Piptoporus betulinus, Pleurotus cornucopiae, P. ostreatus, Trametes suaveolens, Auricularia auricula-judae, Inonotus obliquus, Inonotus radiatus, Tremella foliacea, Tricholoma portentosum, Grifola frondosa* and *Lentinus edodes*. Foremost, the initial approach centered on the most common species in mushrooms industry, from genus *Agaricus, Pleurotus, Lentinus, Ganoderma, Grifola, Flammulina, Volvariella, Auricularia* and *Tremella* etc. (Mark et al., 2004; Ma et al., 2018; Venturella et al. (2021). Conventional therapies are still used for cancer treatment but they are associated with side effects, to overcome this many research studies have going on to find out new therapies for reducing the side effects.

Nanotechnology is an emerging area of research in health and therapeutic. Therefore, various methods and approaches previously reported by numerous scientists for cancer treatment. Among to cutting edge research on nanotechnology, nanoemulsion approaches has continuously developing and created a milestone for treating a cancer and other disease, due to having an efficient and targeted drug delivery. Thus present chapter confers about the status of nanoemulsion development in the field of cancer therapeutics from mushroom-derived bioactive compounds and their prospect for the treatment of cancer in the form of nanoemulsion formation.

MUSHROOM BIOACTIVE COMPOUNDS

Mushrooms known for their potential culinary and rich nutritional importance, present scenario medicinal valued of mushroom have been increased, along with sources for functional foods (Wasser, 2014; Venturella et al., 2021). Numerous macromolecules found on mushrooms, globally researcher involved in the investigated of bioactive compounds structure characterization, and validation of their biological functions. The major bioactive compounds presences in mushrooms such as are proteins, terpenes, unsaturated fatty acids and phenolic compounds etc. The components are directly process for nanoemulsion developments. Furthermore, modern medicinal studies have demonstrated that this mushroom possesses a broad range of bioactivities, including anti-inflammatory, anti-oxidant, anti-glycemic, anti-ulcer, anti-cancer, and immunostimulating effects etc. (Valverde et al., 2015; Venturella et al. (2021). Table 1, Graph 1 &2, exhibited and delineated concerning the anticancer properties of bioactive components of mushrooms.

Table 1. Mushroom bioactive components targeted to anticancer uses

SN	Mushroom species	Characteristics	Health benefits	References
1	*Agaricus bisporus*	Homogeneous polysaccharide	Antitumor activity	Ma et al., 2018
2	*Albatrellus confluens*	Grifolin	Osteosarcoma	Joseph et al., 2018
3	*A. confluens*	Neogrifolin	Osteosarcoma	Joseph et al., 2018
4	*Auricularia auricular*	Polysaccharide	Liver cancer HepG2	Joseph et al., 2018
5	*A. bisporus*	Hetero polysaccharide	Antitumor (Hela cells)	Ma et al., 2018
6	*Amauroderma rude*	Ergosterol	Breast cancer	Joseph et al., 2018
7	*Antrodia camphorata*	Polysaccharide	Hepatocellular carcinoma	Chang et al., 2011
8	*A. camphorata*	4-Acetylantroquinonol B	Colorectal cancer	Chang et al., 2017
9	*A. bisporus*	Lectin	Immuno-modulatory activity,	Ma et al., 2018
10	*Coprinus comatus*	Laccases	Anti-tumor and Antiviral activities (MCF7 and HepG2)	Zhao et al., 2014
11	*Cerrena unicolor*	Laccases	Anti-tumor (leukemic cell)	Matuszewsk et al., 2016
12	*Cordyceps militaris*	Cordycepin	NRK-52E cell line	Kadomatsu et al., 2012
13	*Coriolus versicolor*	PSK, Krestin (Protein bound polysaccharide)	Immunostimulant property and inhibit tumor growth	Joseph et al., 2018
14	*C. versicolor*	PSP(Polysaccharide peptide)	Immunostimulant property and inhibit tumor growth	Joseph et al., 2018
15	*Flammulina velutipes*	RIPs	Antiviral (HIV-1)	Ma et al., 2018
16	*F. velutipes*	Sesquiterpenoids	Anticancer (MCF-7, HepG2, A549 and SGC7901)	Wang et al., 2012
17	*Fomes fomentarius*	Polysaccharide (MFKFAP1β)	Lung cancer	Joseph et al., 2018
18	*Ganoderma applanatum*	Lectin	Anti-tumor (HT-29 colon adenocarcinoma cells)	Ma et al., 2018
19	*G. lucidum*	Homogeneous polysaccharide	Antitumor (HL-60 acute leukemia cells)	Ma et al., 2018
20	*G. atrum*	FIPs	Anti-tumor (Breast cancer cells)	Ma et al., 2018
21	*G. tsugae*	FIPs	Anti-tumor	Ma et al., 2018
22	*G. atrum*	FIPs	Anti-tumor (Breast cancer cells)	Ma et al., 2018
23	*G. microsporum*	FIPs	Anti-tumor (Lung cancer cells)	Ma et al., 2018
24	*G. lucidum*	Laccases	Antiviral (HIV-1)	Wang & Ng, 2006
25	*G. lucidum*	Triterpenoids	Anti-tumor (3T3-L1 cells)	Ma et al., 2018
26	*G. lucidum*	Polysaccharides (GLP) is composed of (1→3), (1→6)-α/β-glucans, glycoprotein's and water soluble heteropolysaccharides	Anti-proliferative, pro-apoptotic, anti-metastatic and anti-angiogenic	Sohretoglu& Huang,2018
27	*G. lucidum*	Polysaccharide	Anti-Liver cancer	Joseph et al., 2018

Continued on following page

Table 1. Continued

SN	Mushroom species	Characteristics	Health benefits	References
28	*Grifola frondosa*	Polysaccharide	Anti-tumor (Liver cancer Hep3B)	Joseph et al., 2018
29	*G. frondosa*	Grifolan (Polysaccharide peptide)	Antiviral and antitumor	Joseph et al., 2018
30	*G. frondosa*	Protein bound Polysaccharide	Anti-tumor (Breast cancer MCF7 cell line)	Joseph et al., 2018
31	*G. frondosa*	Sulfated polysaccharide	Anti-tumor (Liver cancer)	Joseph et al., 2018
32	*Hypsizigus marmoreus*	FIPs	Anti-proliferative (Human leukemia and hepatoma cells)	Ma et al., 2018
33	*I. obliquus*	Ergosterol peroxide	Anti-tumor (Colorectal cancer)	Joseph et al., 2018
34	*Lentinula edodes*	Homogeneous polysaccharide	Antitumor (HeLa cells)	Ma et al., 2018
35	*L. edodes*	Homogeneous polysaccharide	Antitumor (HepG2 cells)	Ma et al., 2018
36	*L. edodes*	Heterogeneous polysaccharide	Anti-tumor	Ma et al., 2018
37	*L. edodes*	Heterogeneous polysaccharide	Anti-tumor (Human colon cancer)	Ma et al., 2018
38	*L. edodes*	Homogeneous polysaccharide	Anti-tumor	Ma et al., 2018
39	*L. edodes*	Lentinan (Polysaccharide)	Immunostimulant	Joseph et al., 2018
40	*L. edodes*	Protein (Latcripin-15)	Anti-tumor (Lung cancer)	Joseph et al. 2018
41	*L. edodes*	Ploysaccharide	Anti-tumor (Hepatocarcinoma)	Joseph et al., 2018
42	*Lentinus crinits*	Polysaccharide	Panepoxydone (PP)	Joseph et al., 2018
43	*Lyophyllum shimeiji*	Ribonucleas Anti	Anti-tumor (Breast cancer MCF7 and Hepatoma HepG2)	Zhang et al., 2010.
44	*Macrolepiota procera*	Lectin	Lectin Anti-tumor activity, bind two glycoprotein's (amino peptidase N (CD13), integrin -31)	Ma et al., 2018
45	*Naematoloma fasciculare*	Triterpenoids	Antiproliferative activities (Human cancer cell lines (HCT-15, SK-MEL-2, SK-OV-3 and A549)	Ma et al., 2018
46	*Phellinus linteus*	Protein-bound polysaccharide	Anti-tumor (Colon cancer)	Joseph et al., 2018
47	*P. linteus*	Hispolon	Anti-tumor (Human hepatoma cells)	Joseph et al., 2018
48	*Pleurotus eryngii*	Hetero polysaccharide	Antitumor (HepG-2 cells)	Ma et al., 2018
49	*P. ostreatus*	Homogeneous polysaccharide	Antitumor (BGC-823 cells)	Ma et al., 2018

Continued on following page

Table 1. Continued

SN	Mushroom species	Characteristics	Health benefits	References
50	*P. eryngii*	Homogeneous polysaccharide	Antitumor (HepG-2 cells)	Ma et al., 2018
51	*P. ostreatus*	Lectin	Immuno-modulatory activity,	Ma et al., 2018
52	*P. eryngii*	Laccases	Antiviral	Wang & Ng, 2006
53	*P. cornucopiae*	Monoterpenoids	Inhibitory activity	Wang et al., 2014
54	*P. eryngii*	Diterpenoids	Cytotoxicity against two human cancer lines	Ma et al., 2018
55	*P. ferulae*	Terpenoids	Melanoma/Gastric cancer	Joseph et al., 2018
56	*P. ostreatus*	Polysaccharide	Inhibitory activity (Sarcoma Sarcoma 180 cells)	Joseph et al., 2018
57	*P. pulmonarius*	Protein-bound polysaccharide	Anti-tumor (Liver cancer)	Xu et al., 2012
58	*Ramaria formosa*	Ribonucleas Antiviral activity	Inhibitory activity (HIV-1)	Zhang et al., 2015
59	*Schizophyllum commune*	Sonifilan (Polysaccharide SPG)	Immunostimulant	Joseph et al., 2018
60	*Stachybotrys chlorohalonata*	FIPs	Immuno-modulatory activity,	Ma et al., 2018
61	*Termitomyces clypeatus*	Sugar entities	Anti-tumor activity (ovary, breast, and Brain)	Joseph et al., 2018
62	*Cordyceps militaris*	Based oil-in-water (O/W)	Anti-tumor activity (RAW 264.7 or HaCaT (human keratinocyte) Shown not toxic.	Esrat et al. (2020)

Graph 1.

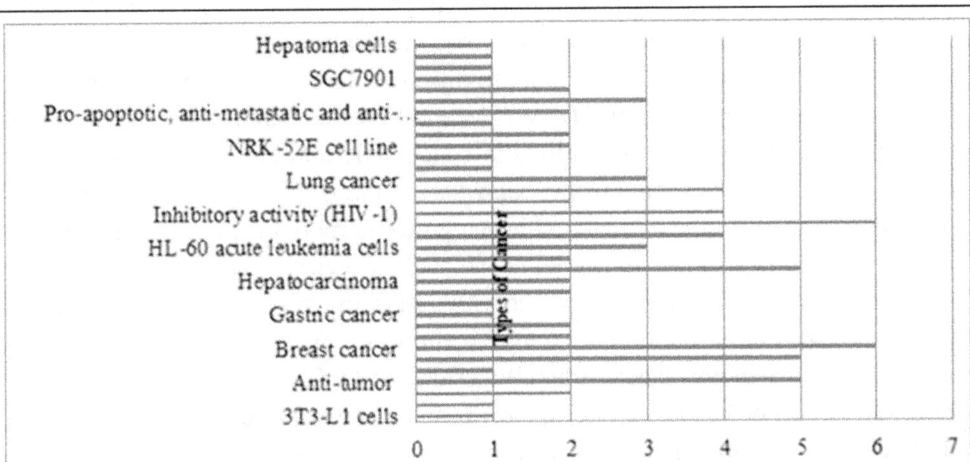

Graph 2.

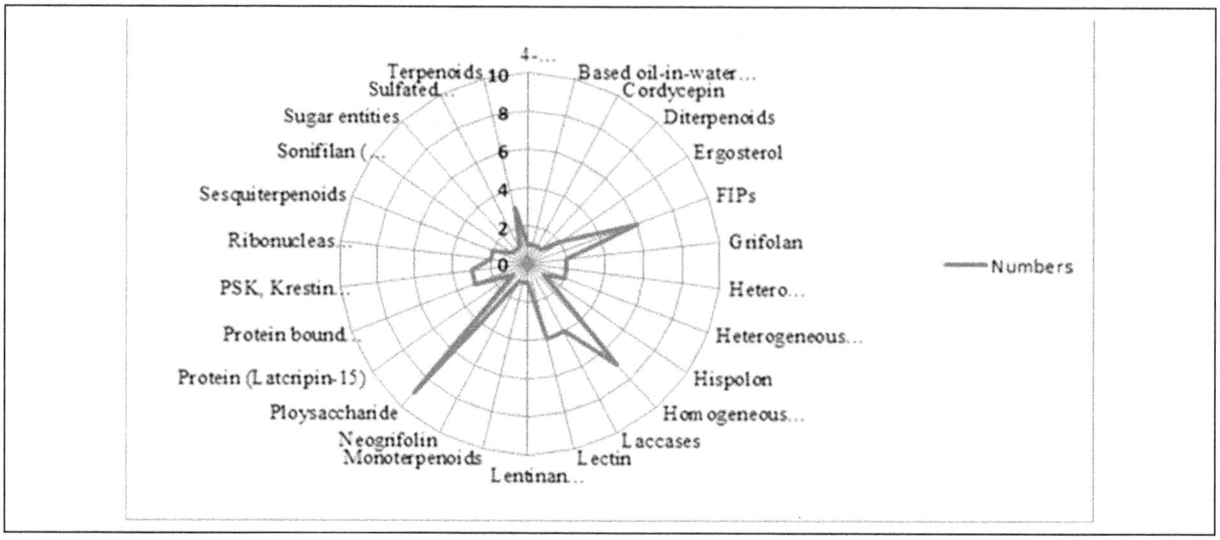

POLYSACCHARIDES

Polysaccharides are termed as the most potent and common compounds which have derived from mushrooms. It played important role in various biochemical pathways associated with bioactive protein molecules. Thus influence varied health advantages activities. Various bioactive polysaccharides has been extracted from mushroom, such as calocyban from *Calocybe indica*, schizophyllan from *Schizophyllum commune*, lentinan from *Lentinus edodes*, ganoderan from *Ganoderma lucidium* and pleuran from *Pleurotus* species etc. (Villares et al., 2012; Badalyan, 2014). The most common monosaccharide's found in mushrooms. They were characterized such as mannose, fucose, glucose, galactose, arabinose, fructose, xylose, rhamnose, trehalose and mannitol (Valverde et al., 2015). According to Venturella et al. (2021) *C. versicolor* have a potent immunomodulatory properties due to presence of two protein-bound polysaccharides present in the fungal extract, termed as the polysaccharide peptide (PSP), and therefore the conjugated glycoprotein PSK (krestin), derived from the strain CM101. They are mainly composed of β-glucans. β-glucan is recognized for activating to different immune cells expressing the appropriate receptors (like toll-like receptors) and complement receptors 3 etc. (Saleh et al., 2017). Polysaccharide peptides induce apoptosis in human promyelocytic cancer HL-60 cells (Chang et al., 2017). Water-soluble proteoglycan which has extracted from *Pleurotus ostreatus*, it's have potent immunomodulatory and antitumor activities against of sarcoma-180-bearing mouse model (Sarangi et al., 2006). Similarly, β-glucans which has extracted from *A. blazei,* it has the potent down regulated proliferation of malignant neoplasm (Murakawa et al., 2007). Some polysaccharides are primary constituents of polar and non polar extracts, respectively. It has been found that the malignant tumor properties of *G. lucidum* are primarily attributed due to the presence of polysaccharides (Wu et al., 2013). *G. lucidum* polysaccharides (GLP) are composed of (1→3), (1→6)-α/β-glucans, glycoproteins and water soluble heteropolysaccharides (Nie et al., 2013). GLP executes the anti-cancer actions through inhibiting tumor growth and metastasis (Weng & Yen 2010). *Auricularia auricular* polysaccharide (AAP) is a soluble polysaccharide extracted from the fruit bodies of *Auricularia auricular*. It is believed to be of high nutritional value due to the presence of high content of carbohydrates, amino acids, trace elements and vitamins. Also, since AAP is a water-soluble natural polysaccharide it has many favourable properties, such as excellent biodegradability and biocompatibility, sensible safety and anticancer activity. However, as a result of its water solubility, it cannot be used alone as a drug delivery carrier (Wang et al., 2017). Mushrooms polysaccharides biological function shown in Table 1. Throughout most of the revealed reports, additional studies are needed within the future disclose to the potent behaviour of mushrooms.

PROTEINS AND PEPTIDES

Protein and other related peptide is primary substrates of mushroom. Mushrooms have well known for the best natural resources for proteins since ancient periods (Petrovska, 2001). Proteins and peptides are significant bioactive components in mushrooms, with having a plentiful vigour benefits and thus enhanced of the digestion and absorption of exogenous nutritional components (Ma et al., 2018). Proteins regulated for the immune function to protect host defensive against the invasion of pathogens and management of other microbial infection (Valverde et al., 2015). Mushroom have a pharmaceutical potential due to presence of ribosome inactivating proteins (RIPs), fungal immunomodulatory proteins (FIPs), ribonucleases, lectins and laccases etc. (Xu et al., 2011). Lectins have been confirmed as for potent property of antitumor and other related immunomodulatory activities (Singh et al., 2014). FIPs are a novel family of bioactive proteins, playing a suppression of metastasis and invasion of tumor cells proliferation (Lin et al., 2010). RIPs are a group of catalyzes enzymes, they suppress polymerase activity of HIV-I (Puri et al., 2012). Similarly, cold-water extracts of *P. nebrodensis* and *P. eryngii* have demonstrated on *in vitro* anticancer property on human HCT116 carcinoma cell lines (Fontana et al., 2014). Similarly, *G. frondosa*, extracted glycoprotein known as GFG-3a. It have a displayed on programmed cell death in human SGC-7901 (Cui et al., 2016).

LACCASES

These are a widespread class of enzymes and are potent immunomodulatory and anticancer properties (Puri et al., 2012). Earlier studies has described of these function such as immune modulation and the anti-tumor activities against of human leukemic T cells, carcinoma MCF7 cells and hepatocarcinoma HepG2 cells (Beaumont et al., 2017). *Hericium erinaceus* derived single-band protein (HEP3), which was exhibited an immunomodulator against RAW 264.7 macrophages by declining the excess production of cytokines (Diling et al., 2017).

TERPENES

Terpenes are a group of volatile unsaturated hydrocarbons. They are various types such as monoterpenoids, triterpenoids, diterpenoids, and sesquiterpenoids. There are varieties of sesquiterpenoids extracted from mushrooms such as aristolane, bisabolane, cuparene, drimane, fomannosane, lactarane, nordasinane and spiroetc (Duru & Cayan, 2015; Ma et al. 2018). Previous research showed that the Terpenes have potent antioxidant, antitumor and anti inflammatory activities (Song et al., 2013; Klaus et al. 2017). Triterpenes are the most constituents of polar and non polar extracts, respectively. It has been found that the anticancer properties of *G. lucidum* are primarily attributed to its triterpenes. These properties are mainly because of two major groups of metabolites occur in *G. lucidum*, such as triterpenoids and triterpenes.Triterpene are derivatives from lanosterol, as well as ganodermic acid, ganoderic acids, lucidones, lucinedic acids and ganodermic alcohols etc. Due to they have remarkable property of antimetastatic, cytotoxic and antitumor (Bulam et al., 2019). Triterpenes butyl lucidenate A and N and butyl ganoderate A and B has shown a suppressed proliferation of adipogenesis in 3T3-L1 cells (Lee et al., 2010), Ganodermanondiol potent down regulated the cell proliferation for B16F10 skin cancer cells (Kim et al., 2016). Simillary, Bao et al. (2018) showed that out of 13 steroidal isolated from the mushrooms of *G. sinense*, (22E,24R)-6 β-methoxyergosta-7,9(11), 22-triene-3 β, 5 α-diol was assumed to have a high HKII binding ability and was validated by *in vitro* micro protein inhibition and cell-based assays. It absolutely was the primary natural compound that is reported to possess HKII inhibitory activity for cancer treatment. Lentinan was found to selectively inhibit or suppressing a some cancer proliferation such as skin cancer 2 (AIM2) colorectal carcinoma cell line CT26 and lung carcinoma cell line LAP0927 (Deng et al., 2018). *H. erinaceus* benzyl alcohol extractedcompotents such as erinacines (A-I), they could be move through the blood–brain barrier, thus have demonstrated property of neurotropic and neuroprotective effects. (Chong et al., 2020). *F. velutipes* having two capable terpenes such as flammulinolides andflammulinol, they are potent antitumor property against of cancer cell lines i.e. KB, Hela and HepG2 (Lee et al., 2011). Terpenesbiological function has shown on table 1.

PHENOLIC COMPOUNDS

Mushroom phenolic compounds, primarily consisted in the category such as hydroxycinnamic acids, flavonoids, hydroxybenzoic acids, phenolic acids, lignans, tannins, stilbenes, and modify polyphenols(DArchivio et al., 2010; Ma et al. 2018). They are structurally containing with a one or more hydroxyl groups with aromatic hydroxylated compoundwhich amid one or more aromatic rings. (DArchivio et al., 2010; Palacios et al., 2011). *Phellinus baumii* (Hymenochaetaceae) extracted three phenolic compounds shown the capacityto inhibit LPS-stimulated nitric oxide development in RAW264.7 cells (Lee et al., 2017). *Inonotus sanghuang* extracted phenolic compounds illustrated their potent property of anti-proliferative and anti-microbial activities, as well as the strongest anti-proliferative result against tumor cell PC3 (Liu et al., 2017). Hispolon isolated from *P. linteus*, having a potent role for edema, regulated the suppression of TNF-α and NO (Chang et al., 2011). Moreover, hispolon induce epidermis and gastric cancer-cell programmed cell death and suppresses the bladder and breast cancer cell growth (Patel & Goyal, 2012). The dichloromethane extract of *G. lucidum* composed of alkaloids, phenolic, terpenoids, and flavonoids, exhibited anti-human papillomavirus 16 (HPV 16) and E6oncoprotein activity (Lai et al., 2010). A liquid extract of polyphenols of shiitake mushroom have potent antiproliferative activity against Hep-2 and HeLa (Finimundy et al., 2013). The consumption of A. *bisporus* is suggested to prevent prostate cancer. The anti-proliferative and proapoptotic action of mushroom extracts by inhibiting the expansion of prostate cancer was successfully done in athymic mice (Adams et al., 2008).

NANOTECHNOLOGICAL DELIVERY SYSTEMS OF BIOACTIVE COMPOUNDS

The design of the delivery method is vital for numerous diseases such as neurodegenerative and cancer etc. The most effective nanotechnological strategies have the development of the delivery process. In consideration and getting high efficacy to deliver to a variety of compounds to target sites, routes have been developed. They are categorized as that solid macromolecule nanoparticles, liposomes, chemical compound nanoparticles, nanoemulsions, and nanoliposomes, etc. Methods are shown in Figure 1(Ganesan et al., 2015).

Figure 1. Schematic presentation of mushrooms bioactive nanoemulsion preparation and validation for cancer treatment

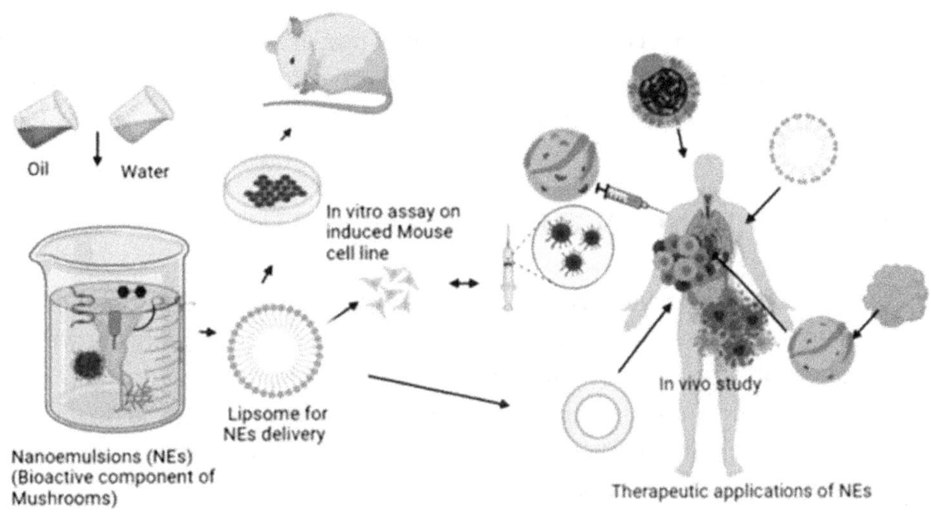

Nanoemulsions

Nanoemulsions are sub-micron dimension emulsions, variety from 5 to 200 nm. They are mostly prepared by using low or high energy techniques that consist of two immiscible phases with facilitation of an appropriate emulsifying agent (McClements, 2012). The emulsions are developed by using high-energy as well as low-energy techniques. In the high-energy development process such as employ physical force, homogenizer, to get the emulsion, and in addition, the low-energy techniques considered as impulsive creation with suitable water, oil, or another surfactant in an appropriate condition for effective preparation. A pomegranate seed oil nanoemulsion with one hundred thirty-five nm particles developed from sonication process, has successfully used for the treatment of the neurological disorder (Ganesan et al., 2015). Nanoemulsion formulations developed from the water-in-oil (W/O) or oil-in-water (O/W), frequently have a vice versa types. In O/W nanoemulsions oil and certain specific surfactant molecules are dispersed in an aqueous medium during emulsification. Thus surfactant has acts to decreasing the interfacial surface between in miscible liquid. The considering property depend upon nanoemulsions should have maximum lipophlic, stable for higher kinetic rate, smaller size of droplet and ability to increase drug bioavailability of encapsulated active ingredients in an aqueous environment. The present scenario use ultrasonication process to prepare an O/W nanoemulsion, which containing with anonionic surfactant, fungus extract, plant oil and co-surfactant etc. Due to ultrasound methodology implemented as primarily, because in this method, have an improved encapsulation, higher yields, smaller droplet size and control of delivery, of droplet then previous drug delivery methods (Palanivel et al., 2015).

MUSHROOM BIOACTIVE COMPOUND IN NANOEMULSION TECHNIQUE FOR CANCER TREATMENT

Recently, an oil and water nanoemulsion has been developed for *Cordyceps* mushroom. This O/W nanoemulsion developed from *Cordyceps*extract (COR-Ex), *Hippophae rhamnoides* oil, Tween 80 used as surfactants, and chitosan etc. The active constituents are as cordycepin (30-deoxyadenosine), polysaccharides, and cordycepic acid, etc. They have the potential property of immunomodulatory, antitumor, antiaging and anti-inflammatory, etc. (Yu et al., 2015). *Cordyceps militaris*-based oil-in-water (O/W) nanoemulsions that have shown on growth suppress of cancer i.e. RAW 264.7 or HaCaT (Yu et al., 2015).

CLASSIFICATION OF NANOENCAPSULATION SYSTEMS

According to Pateiro et al. (2021) nanoencapsulationshas containing various external materials (referred to as shell, wall or carrier material) and a coating target compounds (commonly known as core or internal phase, displayed on table two, discus with major mushroom bioactive compounds and their effects. Particular preparation of nanoemulsions can be combined both approach (up and down methods) to gain optimum results. This categorization is not used currently (Jafari, 2017). Various techniques are employing for the preparation of effective nanoencapsulation for drug carriers, such as liquid-based method, emulsification-solvent evaporation, critical fluid extraction of emulsion (O/W), ionotropic gelation, and electrochemical-based techniques, etc. (Panigrahi et al., 2018).

The benefits offered by these nanodelivery systems depend on the compatibility of nanoparticlespotential with the properties of the bioactive components and the desired application, which allow their usage in an exceedingly wide diversity of cancer treatment. It's essential to take under consideration several parameters for optimal nanoparticles and nanoemulsions preparations, such as surfactants, pH, suitable organic solvents, procedure duration, used polymer(s), interest compound(s) attached with carrier polymer(s), their preparation ratio and used technique (s) of preparation etc. Nanoparticles development for these encapsulation methods frequently vary in size, mechanical properties and shape, including with a composition that provide significantly maximum absorption, metabolism, distribution and excretion etc., future improvement will have considered like double layer of carrier molecules, increasing concentration of interest compounds, higher bio absorption rate etc. (Liang et al., 2016; Murthy et al., 2018).

Table 2. Summary of habitual food bioactive components and prospers mushroom bioactive contains (core substances) that are susceptible to be nanoencapsulation.

SN	Bioactive compounds and their encapsulation function	Mushroom bioactive compounds	References
1	Polysaccharides (Defence, improvement of their antioxidant)	Lentinan, schizophyllan, calocyban, fucose, glucose, galactose, arabinose, trehalose and mannitol	Villares et al., 2012; Badalyan, 2014; Pateiro et al., 2021
2	Phenolic compounds (Protection, improvement of their antioxidant and functional activities,	Hydroxycinnamic acids, Flavonoids, Hydroxybenzoic acids, Phenolic acids, Lignans, and Tannins	DArchivio et al., 2010; Pateiro et al., 2021
3	Peptides and enzymes (Improved antimicrobial and/or antioxidant activity)	Ribosome inactivating proteins (RIPs), Fungal immunomodulatory proteins (FIPs), Ribonucleases, Lectins and Laccases	Xu et al., 2011; Pateiro et al., 2021
4	Other function (Stabilization, controlled release, improved homogeneity)	Single-band protein (HEP3) Monoterpenoids, Triterpenoids, Diterpenoids and Sesquiterpenoids	Diling et al., 2017 Duru & Cayan, 2015 ; Pateiro et al., 2021

FUTURE PERSPECTIVE AND CHALLENGES

Nanocarriers are mixture nano scale systems capable of transporting anticancer agents, such as small molecular weight drugs or macromolecules, when they are transporting to the target tissue region, easily enable these anticancer agents to avoid the effect of normal tissues. But in addition, nanocarriers protect the drug from degradation and, reduce the nephritic clearance and increase its half-life in the bloodstream, augment the payload of cytotoxic drugs, allow the control of the release mechanics of the anticancer drugs, and improve the solubility of those insoluble. Angiogenesis in cancer generates new blood vessels to the tumor, but these new vessels have got increased permeability (enhanced permeability and retention or EPR effect), that together with the poor liquid body substance drainage of tumours allow the passive accumulation of the nanocarriers in tumoral tissues, releasing the therapy agents in the locality of the tumor. To exploit the singularities of the tumor vasculature, nanocarriers must have enough circulation half-life too passively (Perez-Herrero & Fernandez-Medarde, 2015; Verma et al., 2020). A lot of challenges for nanoemulsions technology such as appropriate drug delivery mechanism is needed to improve the absorption of bioactive compounds and also decreasing the treating dose on particular manner with minimized of unwanted side effects. Nanoemulsions will be effective carrier systems as therapeutic agents and have attracted attention in the food, cosmetic, and bioactive compounds, pharmaceutical industries and cancer research etc (Table 1, 2; Figure 1 and Graph 1). The current scientific studies has displayed an enhanced solubility, maximum permeability, and improved absorption of the active compound when has fabricated with nanoemulsion, compare with an only (bio-components) drug deliver (Saravana et al., 2019; Das et al., 2020).

Recently, several food and nutrition researchers have targeted on the fortification of foods with health-promoting bioactive ingredients, similar to nutrients or nutraceuticals, therefore as to enhance their ability to enhance human health and performance (Choi et al., 2011; McClements, 2007; Wang et al., 2008). A wide variety of bioactive ingredients are examined for this purpose, as well as macronutrients (proteins, carbohydrates, and lipids), micronutrients (vitamins and minerals), and nutraceuticals (carotenoids, geraniol, curcumin, and resveratrol) (de Vos, Faas et al., 2010; Dey et al., 2012; Syed et al., 2020; Wang et al., 2008; Xiong et al., 2020). Table 1 summarizes some of the bioactive ingredients that would be encapsulated inside nanoemulsion-based delivery systems and therefore the nutritionary benefits claimed for them. These kinds of delivery systems may be particularly useful for the incorporation of this type of bioactive ingredient into cancer tissue regions, as a result of they will overcome a number of the challenges usually related to the use of those ingredients, such as their poorwater solubility, chemical stability, and bioavailability, thus efficiently regulate the anticancer potency (McClements, 2007; Salminen et al., 2013; Salvia-Trujillo et al., 2017).

Food-grade nanoemulsions can lessen the dose and growth the efficacy of bioactive compounds, in addition to enhancing universal meals houses inclusive of texture, taste, and balance (Cenobio-Galindo et al., 2019). Despite that, nanoemulsions are thermodynamically risky structures that could breakdown through diverse bodily or chemical processes (Sainsbury et al., 2016).Reducing the scale of the droplets in nanoemulsions typically will increase their balance to gravitational separation and aggregation; however it could lessen their chemical balance due to the huge touch vicinity among the oil and water phases (Erdmann et al., 2017; Salminen et al., 2013).

Factors Affecting the Application of Nanoemulsion

There are different technical demanding situations that want to be addressed properly while formulating nanoemulsions. For example, many encapsulated bioactive additives degrade while uncovered to extended light, oxygen, or temperature degrees, thereby lowering their hobby (Liu, 2013). In such, cases it often to optimize the composition, processing operations, and storage conditions in order to maintain good physical and chemical stability (Sari et al., 2015). The large surface areas of nanoemulsions manner that a better degree of emulsifier is needed to formulate them than for traditional emulsions (Azeem et al., 2009; Gupta et al., 2019), which has cost, taste, and toxicity implications. Some of the emulsifiers used to formulate nanoemulsions, especially artificial surfactants, may also cause off-flavours or causes toxicity in final food products (McClements & Rao, 2011; Dasgupta et al., 2019). Furthermore, the decreased droplet length of nanoemulsions may also drastically extrude their gastrointestinal fate (McClements & Xiao, 2012).

Therefore, extra threat tests regularly want to be executed to assess the capacity toxicological outcomes of sure nanoemulsion development. Medicinal mushroom have wide variety of bioactive compounds, numerous previous researches has shown that, they used for treating to cancer, results has depiction, impact on suppression of growth pattern of tumor cells. Cordyceps oil and water-based nano emulsions have a potent down regulated against the HaCaT (human keratinocyte) (Esrat et al., 2020). The bioactive compounds would be a greater prosperity for development of nanomaterilas products that can easily be converted with nanoemulsions for targeted delivery in affected cells line and in *in vivo* clinical studies. In addition, it is going to be essential to deal with regulatory troubles surrounding the utility of nanotechnology in cancer treatment, as many nations have policies and rules touching on the usage of non-material's for cancer treatment. Finally, it is going to be essential to make certain regulatory guidelines for the development of nanoemulsion based cancer therapy for the treatment to various types of cancer

CONCLUSION

Nanoemulsions are being explored as delivery structures for hydrophobic bioactive substances such as nutrients, nutraceuticals, antimicrobials, and antioxidants, anticancer to improve the protection against uncontrolled cell growth patterns, excellent and dietary profile, etc. Numerous researches have already proven that nanoemulsions can improve the stability of the drugs for targeted approaches. Many of the useful outcomes of nanoemulsions are related to their small droplet length and large droplet surface vicinity. In addition, it is possible to incorporate a variety of different bioactive into a single delivery system, such as hydrophobic, hydrophilic, and amphiphilic ones, which can increase their potency through additive or synergistic effects. The worldwide researcher and worker to be intending that develop a bank of mushroom bioactive nanoemulsions for treating a variety of cancer might be controlled the global mortality on caused by cancer, another side will door open for arise a safe and significant treatment procedure.

ACKNOWLEDGMENT

The authors are thankful to the Junior Research Fellowship (DBT/JRF/BET-18/I/2018/AL/123), Department of Biotechnology, Biotech Consortium of India Limited, and Pt. Ravishankar Shukla University Research scholarship award (797/Fin/Sch./2021) for providing funding support. The authors are also thankful to the Head, School of Studies in Biotechnology, Pt. Ravishankar Shukla University, Raipur, for continuously encouraging.

REFERENCES

Adams, L. S., Phung, S., Wu, X., Ki, L., & Chen, S. (2008). White button mushroom (*Agaricus bisporus*) exhibits antiproliferative and proapoptotic properties and inhibits prostate tumor growth in athymic mice. *Nutrition and Cancer*, *60*(6), 44–56. doi:10.1080/01635580802192866 PMID:19005974

Azeem, A., Rizwan, M., Ahmad, F. J., Iqbal, Z., Khar, R. K., Aqil, M., & Talegaonkar, S. (2009). Nanoemulsion components screening and selection: A technical note. *AAPS PharmSciTech*, *10*(1), 69–76. doi:10.120812249-008-9178-x PMID:19148761

Badalyan, S. M. (2014). Potential of mushroom bioactive molecules to develop healthcare biotech products. *Proceedings of the 8th International Conference on Mushroom Biology and Mushroom Products*, 373-378.

Bao, F., Yang, K., Wu, C., Gao, S., Wang, P., Chen, L., & Li, H. (2018). New natural inhibitors of hexokinase 2 (HK2): Steroids from *Ganoderma sinense*. *Fitoterapia*, *125*, 123–129. doi:10.1016/j.fitote.2018.01.001 PMID:29305912

Beaumont, M., Portune, K. J., Steuer, N., Lan, A., Cerrudo, V., Audebert, M., Dumont, F., Mancano, G., Khodorova, N., Andriamihaja, M., Airinei, G., Tomé, D., Benamouzig, R., Davila, A.-M., Claus, S. P., Sanz, Y., & Blachier, F. (2017). Quantity and source of dietary protein influence metabolite production by gut microbiota and rectal mucosa gene expression: A randomized, parallel, double-blind trial in overweight humans. *The American Journal of Clinical Nutrition*, *106*(4), 1005–1019. doi:10.3945/ajcn.117.158816 PMID:28903954

Bulam, S., Ustun, N.S., & Peksen, A. (2019). Health benefits of *Ganoderma lucidum* as a medicinal mushroom. *TURJAF*, *7*(sp 1), 84–93.

Cenobio-Galindo, A. (2019). Development and incorporation of nanoemulsions in food. *International Journal of Food Studies*, *8*(2), 105–124. doi:10.7455/ijfs/8.2.2019.a10

Chandrawanshi, N. K., Tandia, D. K., & Jadhav, S. K. (2017). Nutraceutical properties evaluation of *Schizophyllum commune. Indian Journal of Scientific Research*, *13*(2), 57–62.

Chandrawanshi, N. K., Tandia, D. K. & Jadhav, S. K. (2018) Determination of antioxidant and antidiabetic activities of polar solvent extracts of *Daedaleopsis confragosa* (Bolton) J. Schrot. *Research J. Pharm. and Tech, 11*(12), 5623-5630. . doi:10.5958/0974-360X.2018.01020.X

Chandrawanshi, N. K., & Verma, S. (2020). *Recent research and development in stem cell therapy for cancer treatment: future promising and challenges*. IGI Global. doi:10.4018/978-1-7998-6530-8.ch018

Chang, H.Y., Sheu, M. J. & Yang, C.H. (2011) Analgesic effects and themechanisms of anti-inflammation of hispolon in mice. *Evid.-Based Compl.Alt. Med.*

Chang, S. T. (1999). Global impact of edible and medicinal mushrooms on human welfare in the 21st century: Nongreen revolution. *International Journal of Medicinal Mushrooms*, *1*(1), 1–7. doi:10.1615/IntJMedMushrooms.v1.i1.10

Chang, Y., Zhang, M., Jiang, Y., Liu, Y., Luo, H., Hao, C., Zeng, P., & Zhang, L. (2017). Preclinical and clinical studies of *Coriolus versicolor* polysaccharopeptide as an immunotherapeutic in China. *Discovery Medicine*, *23*, 207–219. PMID:28595034

Choi, A. J., Kim, C. J., Cho, Y. J., Hwang, J. K., & Kim, C. T. (2011). Characterization of capsaicin-loaded nanoemulsions stabilized with alginate and chitosan by self-assembly. *Food and Bioprocess Technology*, *4*(6), 1119–1126. doi:10.100711947-011-0568-9

Chong, P. S., Fung, M., Wong, K. H., & Lim, L. W. (2020). Therapeutic potential of *Hericium erinaceus* for depressive disorders. *International Journal of Molecular Sciences*, *2*, 163.

Choudhary, D., Goykar, H., Karanwad, T., Kannaujia, S., Gadekar, V. & Mishra, M. (2020). An understanding of mitochondria and its role in targeting nanocarriers for diagnosis and treatment of cancer. *Asian Journal of Pharmaceutical Sciences*. . doi:10.1016/j.ajps.2020.10.002

Cui, F., Zan, X., Li, Y., Sun, W., Yang, Y., & Ping, L. (2016). *Grifola frondosa* glycoprotein GFG-3a arrests S phase, alters proteome, and induces apoptosis in human gastric cancer cells. *Nutrition and Cancer*, *68*(2), 267–279. doi:10.1080/01635581.2016.1134599 PMID:27040446

D'Archivio, M., Filesi, C., Varì, R., Scazzocchio, B., & Masella, R. (2010). Bioavailability of the polyphenols: Status and controversies. *International Journal of Molecular Sciences*, *11*(4), 1321–1342. doi:10.3390/ijms11041321 PMID:20480022

Das, A. K., Nanda, P. K., Bandyopadhyay, S., Banerjee, R., Biswas, S., & McClements, D. J. (2020). Application of nanoemulsion-based approaches for improving the quality and safety of muscle foods: A comprehensive review. *Comprehensive Reviews in Food Science and Food Safety*, *19*(5), 2677–2700. doi:10.1111/1541-4337.12604 PMID:33336977

Dasgupta, N., Ranjan, S., & Gandhi, M. (2019). Nanoemulsion ingredients and components. *Environmental Chemistry Letters*, *17*(2), 917–928. doi:10.100710311-018-00849-7

de Vos, P., Faas, M. M., Spasojevic, M., & Sikkema, J. (2010). Encapsulation for preservation of functionality and targeted delivery of bioactive food components. *International Dairy Journal*, *20*(4), 292–302. doi:10.1016/j.idairyj.2009.11.008

Deng, S., Zhang, G., Kuai, J., Fan, P., Wang, X., Zhou, P., Yang, D., Zheng, X., Liu, X., Wu, Q., & Huang, Y. (2018). Lentinan inhibits tumor angiogenesis via interferon and in a T cell independent manner. *Journal of Experimental & Clinical Cancer Research*, *37*(1), 270. doi:10.118613046-018-0932-y PMID:30373628

Dey, T. K., Ghosh, S., Ghosh, M., Koley, H., & Dhar, P. (2012). Comparative study of gastrointestinal absorption of EPA & DHA rich fish oil fromnano and conventional emulsion formulation in rats. *Food Research International*, *49*(1), 72–79. doi:10.1016/j.foodres.2012.07.056

Diling, C., Chaoqun, Z., Jian, Y., Jian, L., Jiyan, S., Yizhen, X., & Guoxiao, L. (2017). Immunomodulatory activities of a fungal protein extracted from *Hericium erinaceus* through regulating the gut microbiota. *Frontiers in Immunology*, *8*, 666. doi:10.3389/fimmu.2017.00666 PMID:28713364

Duru, M. E., & Cayan, G. T. (2015). Biologically active terpenoids from mushroom origin:a review. *Records of Natural Products*, *9*, 456–483.

Erdmann, M. E., Lautenschlaeger, R., Schmidt, H., Zeeb, B., Gibis, M., Bruggemann, D. A., & Weiss, J. (2017). Influence of droplet size on the antioxidant efficacy of oil-in-water emulsions loaded with rosemary in rawfermented sausages. *European Food Research and Technology*, *243*(8), 1415–1427. doi:10.100700217-017-2852-z

Erdmann, M. E., Lautenschlaeger, R., Schmidt, H., Zeeb, B., Gibis, M., Bruggemann, D. A., & Weiss, J. (2017). Influence of droplet size on the antioxidant efficacy of oil-in-water emulsions loaded with rosemary in raw fermented sausages. *European Food Research and Technology*, *243*(8), 1415–1427. doi:10.100700217-017-2852-z

Esrat, J. R., Jin, F. L., Muhammad, H. A., Han, Y., Mitra, A. P., Ahn, J. C., Van-An, H., Lalitha, K., Deok, C. Y., & Se, C. K. (2020). *Cordyceps militaris* fungus extracts-mediatednanoemulsion for improvement antioxidant, antimicrobial, and anti-inflammatory activities. *Molecules (Basel, Switzerland)*, *25*(23), 5733. doi:10.3390/molecules25235733

Finimundy, T., Gambato, G., Fontana, R., Camassola, M., Salvador, M., Moura, S., Hess, J., Henriques, J. A. P., Dillon, A. J. P., & Roesch-Ely, M. (2013). Aqueous extracts of *Lentinula edodes* and *Pleurotus sajor-caju* exhibit high antioxidant capability and promising *in vitro* antitumor activity. *Nutrition Research (New York, N.Y.), 33*(1), 76–84. doi:10.1016/j.nutres.2012.11.005 PMID:23351413

Fontana, S., Flugy, A., Schillaci, O., Cannizzaro, A., Gargano, M. L., Saitta, A., De Leo, G., Venturella, G., & Alessandro, R. (2014). *In vitro* antitumor effects of the cold-water extracts of mediterranean species of genus *Pleurotus* (higher Basidiomycetes) on human colon cancer cells. *International Journal of Medicinal Mushrooms, 16*(1), 49–63. doi:10.1615/IntJMedMushr.v16.i1.50 PMID:24940904

Ganesan, P., Ko, H.-M., Kim, I.-S., & Choi, D.-K. (2015). Recent trends in the development of nanophytobioactive compounds and delivery systems for their possible role in reducing oxidative stress in Parkinson's disease models. *International Journal of Nanomedicine, 10*, 6757–6772. doi:10.2147/IJN.S93918 PMID:26604750

Gupta, P. K., Bhandari, N. N., Shah, H., Khanchandani, V., Keerthana, R., Nagarajan, V., & Hiremath, L. (2019). *An update on nanoemulsions using nanosized liquid in liquid colloidal systems*. Published by Intechopen. doi:10.5772/intechopen.84442

Jafari, S. M. (2017). An overview of nanoencapsulation techniques and their classification. In S. M. Jafari (Ed.), *Nanoencapsulation Technologies for the Food and Nutraceutical Industries* (pp. 1–34). Academic Press. doi:10.1016/B978-0-12-809436-5.00001-X

Jeitler, M., Michalsen, A., Frings, D., Hübner, M., Fischer, M., Koppold-Liebscher, D. A., Murthy, V., & Kessler, C. S. (2020). Significance of medicinal mushrooms in integrative oncology: A narrative review. *Frontiers in Pharmacology, 11*, 580656. doi:10.3389/fphar.2020.580656 PMID:33424591

Joseph, T. P., Chanda, W., Padhiar, A. A., Batool, S., Li Qun, S., Zhong, M., & Huang, M. (2018). Preclinical evaluation of the antitumor activities of edible and medicinal mushrooms: A molecular insight. *Integrative Cancer Therapies, 17*(2), 200–209. doi:10.1177/1534735417736861 PMID:29094602

Kadomatsu, M., Nakajima, S., Kato, H., Gu, L., Chi, Y., Yao, J., & Kitamura, M. (2012). Cordycepin as a sensitizer to tumour necrosis factor (TNF) - α-induced apoptosis through eukaryotic translation initiation factor 2α (eIF2 α) - and mammalian target of rapamycin complex 1 (mTORC1)-mediated inhibition of nuclear factor (NF)-κB. *Clinical and Experimental Immunology, 168*(3), 325–332. doi:10.1111/j.1365-2249.2012.04580.x PMID:22519596

Kim, J., Kim, H., Kim, J. H., Kwon, O., Son, E., Lee, C., & Park, Y. (2016). Effects of Ganodermanondiol, a new melanogenesis inhibitor from the medicinal mushroom *Ganoderma lucidum. Int. Mol. Sci, 17*(11), 1798. doi:10.3390/ijms17111798 PMID:27801787

Klaus, A., Kozarski, M., & Vunduk, J. (2017). Antibacterial and antifungal potential of wild basidiomycete mushroom *Ganoderma applanatum. Lek. Sirovine, 36*, 37–46.

Lai, L. K., Abidin, N. Z., Abdullah, N., & Sabaratnam, V. (2010). Anti-human papillomavirus (HPV) 16E6 activity of Ling Zhi or reishi medicinal mushroom, *Ganoderma lucidum* (W. Curt.: Fr.) P. Karst. (Aphyllophoromycetideae) extracts. *International Journal of Medicinal Mushrooms, 12*(3), 279–286. doi:10.1615/IntJMedMushr.v12.i3.70

Lee, I., Ahn, B., Choi, J., Hattori, M., Min, B. S., & Bae, K. H. (2011). Selective cholinesterase inhibition by lanostanetriterpenes from fruiting bodies of *Ganoderma lucidum. Bioorganic & Medicinal Chemistry Letters*, *21*(21), 6603–6607. doi:10.1016/j.bmcl.2011.04.042 PMID:21924611

Lee, I., Kim, J., Ryio, I., Kim, Y., Choo, S., Yoo, I., Min, B., Na, M., Hattori, M., & Bae, K. (2010). Lanostane triterpenes from *Ganoderma lucidum* suppress the adipogenesis in 3T3-L1 cells through down-regulation of SREBP-1c. *Bioorganic & Medicinal Chemistry Letters*, *20*(18), 5577–5581. doi:10.1016/j.bmcl.2010.06.093 PMID:20702092

Lee, S., Lee, D., Jang, T. S., Kang, K., Nam, J.-W., Lee, H.-J., & Kim, K. (2017). Anti-inflammatory phenolic metabolites fromthe edible fungus *Phellinus* baumii in LPS-stimulated RAW 264.7 cells. *Molecules (Basel, Switzerland)*, *22*(10), 1583. doi:10.3390/molecules22101583

Liang, J., Yan, H., Yang, H.-J., Kim, H. W., Wan, X., Lee, J., & Ko, S. (2016). Synthesis and controlled-release properties of chi-tosan/ Lactoglobulin nanoparticles as carriers for oral administration of epigallocatechin gallate. *Food Science and Biotechnology*, *25*(6), 1583–1590. doi:10.100710068-016-0244-y PMID:30263448

Lin, C.-H., Sheu, G.-T., Lin, Y.-W., Yeh, C.-S., Huang, Y.-H., Lai, Y.-C., Chang, J.-G., & Ko, J.-L. (2010). A new immunomodulatory protein from *Ganoderma microsporum* inhibits epidermal growth factor mediated migration and invasion in A549 lung cancer cells. *Process Biochemistry (Barking, London, England)*, *45*(9), 1537–1542. doi:10.1016/j.procbio.2010.06.006

Liu, K., Xiao, X., Wang, J., Chen, C.-Y. O., & Hu, H. (2017). Polyphenolic composition and antioxidant, antiproliferative, and antimicrobial activities of mushroom *Inonotus sanghuang*. *Lebensmittel-Wissenschaft + Technologie*, *82*, 154–161. doi:10.1016/j.lwt.2017.04.041

Liu, R. H. (2013). Dietary bioactive compounds and their health implications. *Journal of Food Science*, *78*(s1, Suppl), A18–A25. doi:10.1111/1750-3841.12101 PMID:23789932

Ma, G., Yang, W., Zhao, L., Pei, F., Fang, D., & Hu, Q. (2018). A critical review on the health promoting effects of mushrooms nutraceuticals. *Food Science and Human Wellness*, *7*(2), 125–133. doi:10.1016/j.fshw.2018.05.002

Mark, S., Valentina, K., Yulia, M., Alexander, G., Andrey, P., Andrey, S., Nikolay, P., & Nina, D. (2004). Immunomodulating and anti-tumor action of extracts of several mushrooms. *Journal of Biotechnology*, *113*(1-3), 77–83. doi:10.1016/j.jbiotec.2004.04.034 PMID:15380649

Matuszewska, A., Karp, M., Jaszek, M., Janusz, G., Osińska-Jaroszuk, M., Sulej, J., Stefaniuk, D., Tomczak, W., & Giannopoulos, K. (2016). Laccase purified from *Cerrena unicolor* exerts antitumor activity against leukemic cells. *Oncology Letters*, *11*(3), 2009–2018. doi:10.3892/ol.2016.4220 PMID:26998114

Mau, J. L., Lin, H. C., & Chen, C. C. (2002). Antioxidant properties of several medicinal mushrooms. *Journal of Agricultural and Food Chemistry*, *50*(21), 6072–6077. doi:10.1021/jf0201273 PMID:12358482

McClements, D. J. (2007). *Critical review of techniques and methodologies for characterization of emulsion stability*. Critical Reviews in Food Science and Nutrition. doi:10.1080/10408390701289292

McClements, D. J., & Rao, J. (2011). Food-grade nanoemulsions: Formulation, fabrication, properties, performance, biological fate, and potential toxicity. *Critical Reviews in Food Science and Nutrition, 51*(4), 285–330. doi:10.1080/10408398.2011.559558 PMID:21432697

McClements, D. J., & Xiao, H. (2012). Potential biological fate of ingested nanoemulsions: Influence of particle characteristics. *Food & Function, 3*(3), 202–220. doi:10.1039/C1FO10193E PMID:22105669

Murakawa, K., Fukunaga, K., Tanouchi, M., Hosokawa, M., Hossain, Z., & Takahashi, K. (2007). Therapy of myeloma *in vivo* using marine phospholip in combination with *Agaricus blazei* Murill as an immune respond activator. *Journal of Oleo Science, 56*(4), 179–188. doi:10.5650/jos.56.179 PMID:17898480

Murthy, K. N. C., Monika, P., Jayaprakasha, G. K., & Patil, B. S. (2018) Nanoencapsulation: An advanced nanotechnological approach to enhance the biological efficacy of curcumin. *Proceedings of the ACS Symposium Series*, 383-405. 10.1021/bk-2018-1286.ch021

Palacios, I., Lozano, M., Moro, C., D'Arrigo, M., Rostagno, M. A., Martínez, J. A., García-Lafuente, A., Guillamón, E., & Villares, A. (2011). Antioxidant properties of phenolic compounds occurring in edible mushrooms. *Food Chemistry, 128*(3), 674–678. doi:10.1016/j.foodchem.2011.03.085

Panigrahi, S. S., Syed, I., Sivabalan, S., & Sarkar, P. (2018). Nanoencapsulation strategies for lipid-soluble vitamins. *Chemical Papers, 73*(1), 1–16. doi:10.100711696-018-0559-7

Pateiro, M., Gomez, B., Munekata, P. E. S., Barba, F. J., Putnik, P., Bursa, D., Cevi, K., & Lorenzo, J. M. (2021). Nanoencapsulation of promising bioactive compounds to improve their absorption, stability, functionality and the appearance of the final food products. *Molecules (Basel, Switzerland), 26*(6), 1547. doi:10.3390/molecules26061547 PMID:33799855

Patel, S., & Goyal, A. (2012). Recent developments in mushrooms as anti-cancer therapeutics: A review. *Biotechnology, 2*, 1–15. PMID:22582152

Perez-Herrero, E., & Fernandez-Medarde, A. (2015). Advanced targeted therapies in cancer: Drug nanocarriers, the future of chemotherapy. *European Journal of Pharmaceutics and Biopharmaceutics, 93*, 52–79. Advance online publication. doi:10.1016/j.ejpb.2015.03.018 PMID:25813885

Petrovska, B. B. (2001). Protein fraction in edible Macedonian mushrooms. *European Food Research and Technology, 212*(4), 469–472. doi:10.1007002170000285

Puri, M., Kaur, I., Perugini, M. A., & Gupta, R. C. (2012). Ribosome-inactivating proteins: Currentstatus and biomedical applications. *Drug Discovery Today, 17*(13-14), 774–783. doi:10.1016/j.drudis.2012.03.007 PMID:22484096

Rathore, H., Prasad, S., & Sharma, S. (2017). Mushroom nutraceuticals forimproved nutrition and better human health: A review. *Pharma Nutr, 5*, 35–46.

Sainsbury, J., Grypa, R., Ellingworth, J., Duodu, K. G., & De Kock, H. L. (2016). The effects of antioxidants and shelf life conditions on oxidation markers in a sunflower oil salad dressing emulsion (SOSDE). *Food Chemistry, 213*, 230–237. doi:10.1016/j.foodchem.2016.06.081 PMID:27451176

Saleh, M. H., Rashedi, I., & Keating, A. (2017). Immunonodulatory properties of *Coriuolus versicolor*: The role of Polysaccharopeptide. *Frontiers in Immunology*, *8*, 1087. doi:10.3389/fimmu.2017.01087 PMID:28932226

Salminen, H., Herrmann, K., & Weiss, J. (2013). Oil-in-water emulsions as delivery systemfor n-3 fatty acids in meat products. *Meat Science*, *93*(3), 659–667. doi:10.1016/j.meatsci.2012.11.035 PMID:23273478

Salvia-Trujillo, L., Soliva-Fortuny, R., Rojas-Grau, M. A., McClements, D. J., & Martin-Belloso, O. (2017). Edible nanoemulsions as carriers of active ingredients: A review. *Annual Review of Food Science and Technology*, *8*(1), 439–466. doi:10.1146/annurev-food-030216-025908 PMID:28125342

Sarangi, I., Ghosh, D., Bhutia, S. K., Mallick, S. K., & Maiti, T. K. (2006). Anti-tumor and immunomodulating effects of *Pleurotus ostreatus* mycelia-derived proteoglycans. *International Immunopharmacology*, *6*(8), 1287–1297. doi:10.1016/j.intimp.2006.04.002 PMID:16782541

Saravana, P. S., Shanmugapriya, K., Gereniu, C. R. N., Chae, S.-J., Kang, H. W., Woo, H.-C., & Chun, B.-S. (2019). Ultrasound-mediated fucoxanthin rich oil nanoemulsions stabilized by κ -carrageenan: Process optimization, bio-accessibility and cytotoxicity. *Ultrasonics Sonochemistry*, *55*, 105–116. doi:10.1016/j.ultsonch.2019.03.014 PMID:31084784

Sari, T. P., Mann, B., Kumar, R., Singh, R. R. B., Sharma, R., Bhardwaj, M., & Athira, S. (2015). Preparation and characterization of nanoemulsion encapsulating curcumin. *Food Hydrocolloids*, *43*, 540–546. doi:10.1016/j.foodhyd.2014.07.011

Singh, S. S., Wang, H., Chan, Y. S., Pan, W., Dan, X., Yin, C., Akkouh, O., & Ng, T. (2014). Lectins from edible mushrooms. *Molecules (Basel, Switzerland)*, *20*(1), 446–469. doi:10.3390/molecules20010446 PMID:25558856

Sohretoglu, D., & Huang, S. (2018). *Ganoderma lucidum* Polysaccharides as an anti-cancer agent. *Anti-cancer Agents in Medicinal Chemistry*, *18*(5), 667–674. doi:10.2174/1871520617666171113121246 PMID:29141563

Song, F.-Q., Liu, Y., Kong, X.-S., & Chang, S. (2013). Progress on understanding the anticancer mechanisms of medicinal mushroom: *Inonotus obliquus*. *AsianPac. Journal of Cancer Prevention*, *14*, 1571–1578. PMID:23679238

Syed, I., Banerjee, P., & Sarkar, P. (2020). Oil-in-water emulsions of geraniol and carvacrol improve the antibacterial activity of these compounds on raw goat meat surface during extended storage at 4°C. *Food Control*, *107*, 106757. doi:10.1016/j.foodcont.2019.106757

Thatoi, H., & Singdevsachan, S. K. (2014). Diversity, nutritional composition andmedicinal potential of Indian mushrooms: A review. *African Journal of Biotechnology*, *13*(4), 523–545. doi:10.5897/AJB2013.13446

Valverde, M. E., Hernandez-Perez, T., & Paredes-Lopez, O. (2015). Edible mushrooms: Improving human health and promoting quality life. *International Journal of Microbiology*, *2015*, 1–14. doi:10.1155/2015/376387 PMID:25685150

Venturella, G., Ferraro, V., Cirlincione, F., & Gargano, M. L. (2021). Medicinal mushrooms: Bioactive compounds, use, and clinical trials. *International Journal of Molecular Sciences*, *22*(2), 634. doi:10.3390/ijms22020634 PMID:33435246

Verma, S., Jain, V., & Chandrawanshi, N. K. (2020). *Surfactant Based Anhydrous Nano Carrier System for Poorly Aqueous Soluble Anti-Cancer Drugs*. IGI Global. doi:10.4018/978-1-7998-6530-8.ch014

Villares, A., Mateo-Vivaracho, L., & Guillamon, E. (2012). Structural features and healthy properties of polysaccharides occurring in mushrooms. *Agriculture*, *2*(4), 452–471. doi:10.3390/agriculture2040452

Wang, H., & Ng, T. (2006). A laccase from the medicinal mushroom *Ganoderma lucidum*. *Applied Microbiology and Biotechnology*, *72*(3), 508–513. doi:10.100700253-006-0314-9 PMID:16636832

Wang, W., Chen, K., Liu, Q., Johnston, N., Ma, Z., Zhang, F., & Zheng, X. (2014). Suppression of tumor growth by *Pleurotus ferulae* ethanol extract through induction of cell apoptosis, and inhibition of cell proliferation and migration. *PLoS One*, *9*(7), e102673. doi:10.1371/journal.pone.0102673 PMID:25029345

Wang, X., Jiang, Y., Wang, Y. W., Huang, M. T., Ho, C. T., & Huang, Q. (2008). Enhancing anti-inflammation activity of curcumin through O/W nanoemulsions. *Food Chemistry*, *108*(2), 419–424. doi:10.1016/j.foodchem.2007.10.086 PMID:26059118

Wang, Y., Bao, L., Yang, X., Li, L., Li, S., Gao, H., Yao, X.-S., Wen, H., & Liu, H.-W. (2012). Bioactive sesquiterpenoids from the solidculture of the edible mushroom *Flammulina velutipes* growing on cookedrice. *Food Chemistry*, *132*(3), 1346–1353. doi:10.1016/j.foodchem.2011.11.117 PMID:29243621

Wang, Y., Li, P., Chen, F., Jia, L., Xu, Q., Gai, X., Yu, X., Di, Y., Zhu, Z., Liang, Y., Liu, M., Pan, W., & Yang, X. (2017). A novel pH-sensitive carrier for the delivery of antitumor drugs: Histidine-modified *Auricularia auricular* polysaccharide nanomicelles. *Scientific Reports*, *7*(1), 4751. doi:10.103841598-017-04428-8 PMID:28684737

Wasser, S. P. (2002). Medicinal mushrooms as a source of antitumor and immunomodulating polysaccharides. *Applied Microbiology and Biotechnology*, *60*(3), 258–274. doi:10.100700253-002-1076-7 PMID:12436306

Wasser, S. P. (2014). Medicinal mushroom science: Current perspectives, advances, evidences, and challenges. *Biomedical Journal*, *37*(6), 345–356. doi:10.4103/2319-4170.138318 PMID:25179726

Xiong, Y., Li, S., Warner, R. D., & Fang, Z. (2020). Effect of oregano essential oil and resveratrol nanoemulsion loaded pectin edible coating on the preservation of pork loin in modified atmosphere packaging. *Food Control*, *114*, 107226. doi:10.1016/j.foodcont.2020.107226

Xu, W., Huang, J. J., & Cheung, P. C. (2012). Extract of *Pleurotus pulmonarius* suppresses liver cancer development and progression through inhibition of VEGF-induced PI3K/AKT signalling pathway. *PLoS One*, *7*(3), e34406. doi:10.1371/journal.pone.0034406 PMID:22470568

Xu, X., Yan, H., Chen, J., & Zhang, X. (2011). Bioactive proteins from mushrooms. *Biotechnology Advances*, *29*(6), 667–674. doi:10.1016/j.biotechadv.2011.05.003 PMID:21605654

Yu, S.-H., Chen, S.-Y. T., Li, W.-S., Dubey, N. K., Chen, W.-H., Chuu, J.-J., Leu, S.-J., & Deng, W.-P. (2015). Hypoglycemic activity through a novel combination of fruiting body and mycelia of *Cordyceps militaris* in high-fat diet-induced type 2 diabetes mellitus mice. *Journal of Diabetes Research*. PMID:26258146

Zhang, R., Tian, G., Zhao, Y., Zhao, L., Wang, H., Gong, Z., & Ng, T. B. (2015). A novel ribonuclease with HIV-1 reverse transcriptase inhibitory activity purified from the fungus *Ramaria Formosa. Journal of Basic Microbiology*, *55*(2), 269–275. doi:10.1002/jobm.201300876 PMID:24523258

Zhang, Y., Xin, C., Cheng, C., & Wang, Z. (2020). Antitumor activity of nanoemulsion based on essential oil of *Pinus koraiensis* pinecones in MGC-803 tumor-bearing nude mice. *Arabian Journal of Chemistry*, *13*(11), 8226–8238. doi:10.1016/j.arabjc.2020.09.058

Zhao, S., Rong, C.-B., Kong, C., Liu, Y., Xu, F., Miao, Q.-J., Wang, S.-X., Wang, H.-X., & Zhang, G.-Q. (2014). A novel laccase with potentantil proliferative and HIV-1 reverse transcriptase inhibitory activities from mycelia of mushroom *Coprinus comatus. BioMed Research International*, *2014*, 1–8. doi:10.1155/2014/417461

KEY TERMS AND DEFINITIONS

Anticancer: Anticancer is a term defined as suppressing or control to abnormal cellular growth. The management of cancer used various chemical, physical and surgical approaches. The worldwide research outcomes and findings indicated that the varieties of bioactive components, that will be potential resources for cancer management, etc.

Bioactive Components: Bioactive compounds are a huge category found in plants and mushrooms such as polysaccharides, fatty acids, proteins, peptides, terpenes, polyphenols, and flavonoids, etc. These particular compounds were obtained by proper extraction with followed standards guidelines. The variety of compounds has to have an important role in cellular mechanism and they direly or indirectly influence the cellular metabolic patterns. The bioactive compounds played a remarkable role such as antitumor, anti metastasic, antidiabetic, antioxidant, and antiangiogenesis, etc.

Cancer Treatment Therapy: Cancer is a cluster of diseases in which abnormal cells grow and divide uncontrolled manner. Abnormal cells are also affected by normal body tissues. It has a lead factor that causes the higher mortality rate and continuously increasing constantly at the global level. The presently various methods used for cancer treatment such as a chemotherapeutic, chemical, and radiation, including surgery, etc. There are a lot of technical ambiguities in the management of cancer. Thus nanotechnological cutting edges research has achieved several milestones for health research. The available potential mushroom bioactive compounds would be possible intelligence drugs for cancer and existence methods such as nanoemulsion will be promising for drug delivery to targeted cancer treatment.

Immunomodulatory Activity: The potent bioactive compound (Flavonoids, Terpenes, polyphenols, and other compounds, etc.) will be capable of modifying or regulating the immune response or acted as an immunomodulator. It's an important mechanism for cancer management.

Liposome: The liposome is a spherical-shaped vesicle that is composed of one or more phospholipids bilayers. It has resembled the cell membrane structure. The liposome is a novel drug carrier; they are encapsulated with hydrophobic or hydrophilic targeted drugs (bioactive compounds), the containing vesicle will be capable of well-suited for drugs delivery mechanism.

Mushroom: A mushroom is a spore containing a fruiting body of a fleshy macro fungus. They are produced above on soil, ground, and waste agro substrate, etc. The numerous varieties of mushroom species have occurred in the biotic system. Accordingly, the edibility of mushrooms has been categorized, such as medicinal and edible mushrooms, etc. The mushrooms have a potent source of nutrients, rich in protein, carbohydrates, and fewer fats; furthermore, they are the main resources for bioactive compounds. The various mushroom directly prefers for culinary and many mushrooms used for treating human ailments etc.

Nanoemulsion: Nanoemulsion technique is categorized under the nano-technology and designed for quantifiable clinical and therapeutic application. The presently various nano-carriers-based nanoemulsions approach is widely targeted for a treat to various cancer. This therapy also overcomes various challenges such as low solubility, multidrug resistance, and toxicity, etc. Thus, nanoemulsions might be the solution to attain proficient and safe treatment of cancer. Nanoemuslison having several advantages such as nanometric size, ease of preparation, site-specific, drug encapsulation, non-immunogenic, biocompatibility, biodegradable, sustained and controlled release, large surface area, and thermodynamic stability, etc. Recently nanoemulsions are extensive investigated as drug carriers to improve the delivery rate of therapeutic agents or compounds. It will be formulated in a variety of forms such as liquids, sprays, foams, creams, ointments, and gels, etc. from mushroom bioactive components. All these potential developed nanoemulsions will be the great promising future of cosmetics, diagnostics, drug therapies, and biotechnologies.

Nanotechnology: It's defined as cutting edges research area. In the present scenario, various nanotechnological approaches have been developed for curing disease that provides compatible drug delivery approaches, maximum absorption on target sites, with high effectiveness of clinical studies.

Section 4

Other Applications of Nanoemulsion

Chapter 17
Nanoemulsions for the Delivery of Anti-Hypertensive Drugs

Subramanian Natesan
National Institute of Pharmaceutical Education and Research, Kolkata, India

Nidhi Singh
National Institute of Pharmaceutical Education and Research, Kolkata, India

Victor Hmingthansanga
National Institute of Pharmaceutical Education and Research, Kolkata, India

Pallab Datta
National Institute of Pharmaceutical Education and Research, Kolkata, India

Sai Hanuman Sagar Boddu
College of Pharmacy and Health Sciences, Ajman University, UAE

V. Ravichandiran
National Institute of Pharmaceutical Education and Research, Kolkata, India

ABSTRACT

Hypertension refers to an increase in the arterial blood pressure. Most commonly used antihypertensive drugs are available in conventional dosage forms as it offers superior patient compliance. A majority of anti-hypertensive drugs pose bioavailability issues as they belong to BCS class II and BCS class IV categories with poor solubility profile and rate limiting dissolution. Emerging drug delivery technologies like nanoemulsion are found to be promising and safer alternatives for the delivery of anti-hypertensive drugs. Nanoemulsion gained more attention due to favourable properties such as small size, good physical stability, rapid action, drug targeting, prevents photo-degradation, and improved bioavailability. This chapter highlights various aspects of hypertension including its pathophysiology and potential approaches to combat high blood pressure. In addition, the authors thoroughly discussed nanoemulsions and their utility in the oral delivery of anti-hypertensive drugs.

DOI: 10.4018/978-1-7998-8378-4.ch017

INTRODUCTION

Hypertension is a severe medical ailment in which arterial blood vessels show adamantly elevated pressure (Foëx & Sear, 2004). It substantially increases the risk for numerous cardiovascular diseases (CVDs) like congestive heart failure, stroke (hemorrhagic as well as ischemic), coronary artery disease, peripheral arterial disease, and renal failure (Weber et al., 2014). As per the World Health Organization 2015 data, an estimated 1.13 billion people globally are suffering from hypertension that leads to substantial morbidity, mortality, and monetary burden to the society. On average, 1 in 4 men and 1 in 5 women are supposed to be suffering from hypertension (Kumar, Mishra, Shrivastava, & Zaidi, 2020). The majority of the anti-hypertensive drugs belong to Biopharmaceutical Classification System (BCS) class II (low solubility, high permeability) or class IV (low solubility, low permeability) and are P-glycoprotein (Pgp) substrates thus undergo substantial pre-systemic metabolism resulting in low bioavailability (Alam et al., 2017). These obstacles prevent several effective anti-hypertensive drug molecules from entering the drug development pipeline.

Most of the antihypertensive drugs are currently available as conventional dosage forms as these are non-invasive, cost-effective, and convenient dosage forms. However, the orally administered conventional dosage form provides poor patient acceptability in geriatric, pediatric, and unconscious patients and is unstable under the extensive condition of the gastrointestinal tract (GIT) limiting their usefulness (Chime, Kenechukwu, & Attama, 2014). Nanotechnology-based drug delivery systems offer substantial technological advancements to overcome the limitations of conventional dosage forms. Various nanotechnology-based techniques used in drug delivery are nanoemulsions (submicron-sized colloidal dispersion system basically emulsion), nanospheres (drug encapsulated nanoparticles in a polymer matrix), nanosuspensions (nano-sized suspension), lipid nanoparticles (nanoparticles composed of lipids such as solid lipid nanoparticle, nanostructured lipid carrier, and polymer lipid hybrid nanoparticles) and liposomes (nano-sized spherical vesicles consisting of phospholipid). Amongst all these nanotechnology-based drug delivery systems nanoemulsion has gained more attention due to its potential to enhance the bioavailability of the orally administered drug by avoiding hepatic first-pass metabolism. It is a colloidal liquid dosage form, thus can easily be acceptable for patients of all age groups. In addition, both hydrophilic and lipophilic drugs can be formulated as nano-emulsions and the stability of the drug can be enhanced under the different conditions of GIT (Chircov & Grumezescu, 2019). In this section, a brief outline regarding the oral administration of anti-hypertensive drugs and their possible application in the oral administration of an anti-hypertensive drug is highlighted. Next to this, an overview of hypertension including its pathophysiology, nanoemulsions, mechanism of uptake of orally administered nanoemulsion and their utility in the oral delivery of anti-hypertensive drugs is discussed thoroughly.

AN OVERVIEW OF HYPERTENSION

Hypertension or high blood pressure is a medical condition in which systolic blood pressure (SBP) is more than 140 mmHg or diastolic blood pressure (DBP) is more than 90 mmHg (Delacroix, Chokka, & Worthley, 2014). It is a primary risk factor for cardiovascular-related mortality and approximately 8 million premature fatalities are caused due to uncontrolled hypertension (Hall et al., 2012). Cardiac output and peripheral resistance collectively give rise to blood pressure (BP). Cardiac output is defined as the amount of blood pumped by the heart throughout the body in a minute. It is calculated by heart rate and stroke volume whereas, peripheral resistance is defined as the resistance developing in by the arteries against the blood flow and can be calculated by physiological or anatomical alteration in tiny arteries and arterioles (Beevers, Lip, & O'Brien, 2001). Various stages of hypertension as per the American Society of Hypertension (ASH) are shown in Figure 1. In teenagers, raised cardiac output could lead to the development of hypertension, while in old age people it is mainly due to enhanced peripheral resistance and higher muscle stiffness. Is a substantial peril factor for numerous CVDs and upsurges the risk of congestive heart failure, stroke (hemorrhagic as well as ischemic), coronary artery disease, peripheral arterial disease, and renal failure. Despite numerous developments in this area, it is predicted that by the year 2025, approximately 1.56 billion people will be affected by hypertension (Delacroix et al., 2014).

Figure 1. Various stages of hypertension as per ASH guidelines

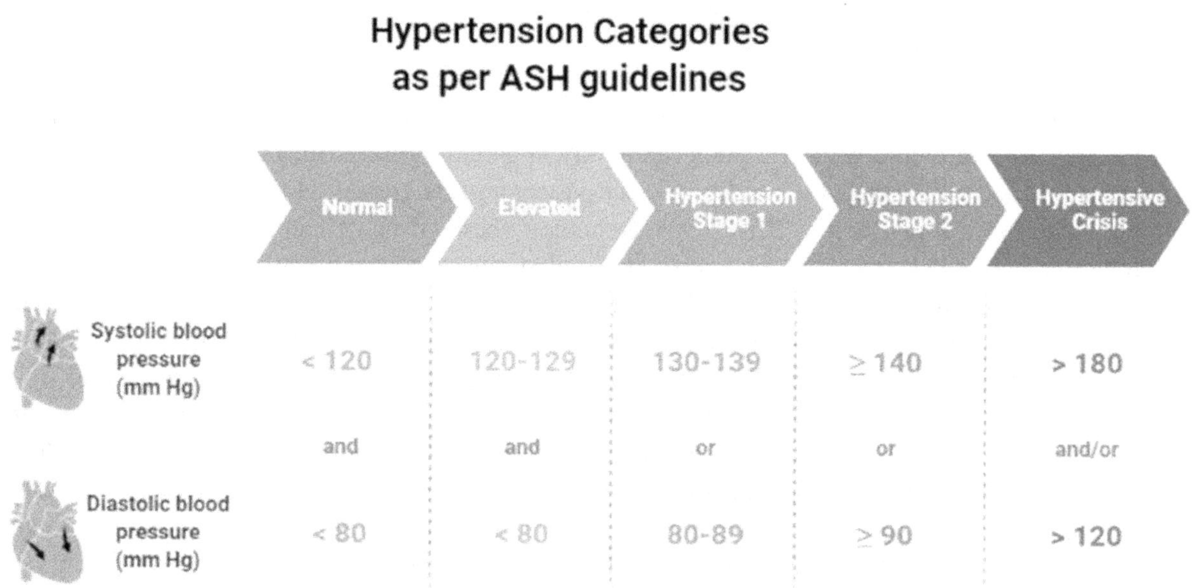

PATHOPHYSIOLOGY OF HYPERTENSION

Hypertension is a persistent rise in the arterial blood pressure and after prolong period it causes impairment of organs leading to severe illness and death. Hypertension is categorized as primary or secondary hypertension accounting for approximately 95% and 5% of total hypertensive cases respectively (Weber et al., 2014). The root cause of primary or essential hypertension is unidentified, but most probably it happens at the age of 50 or 60 years as a consequence of higher salt intake, obesity, and prior family history. In the case of secondary hypertension factors such as severe kidney ailment, sleep disorder, renal artery stenosis, endothelial dysfunction, and insulin resistance play a prominent role (Cain & Khalil, 2002) as shown in Figure 2. In both circumstances, a regulatory mechanism involved in the maintenance of normal blood pressure is affected, resulting in the development of disease.

Figure 2. The major organs are involved in the pathogenesis of hypertension.

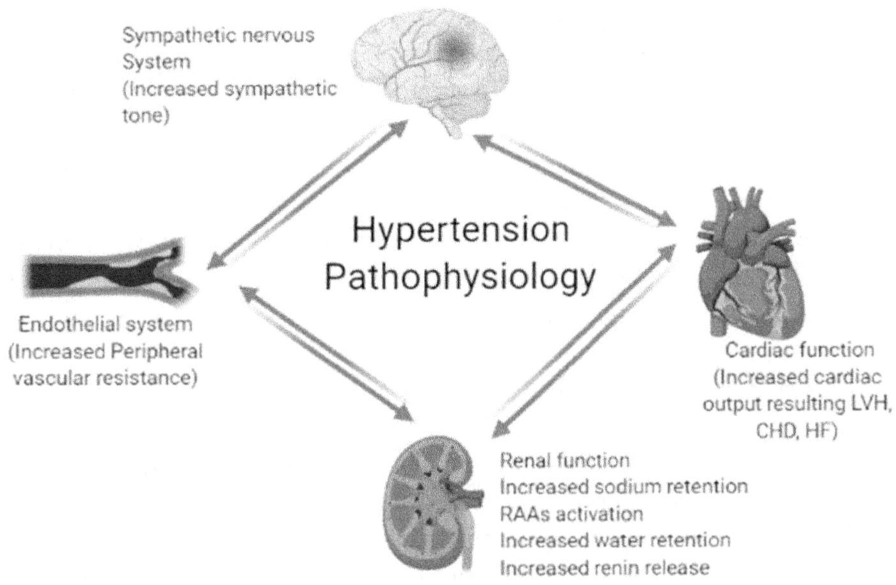

Cardiac output is defined as the amount of blood pumped by the heart per minute throughout the body whereas, vascular resistance is the extent of contraction accomplished by blood vessels (Weber et al., 2014). In the case of hypertension, the raised level of vascular resistance may be either due to the higher release of peptides like angiotensin and endothelins (vasoconstrictor) or by enhanced alpha-adrenoceptor stimulation (Carretero & Oparil, 2000). Along with this, another reason could be an elevated level of cellular calcium concentration in peripheral vascular smooth muscle that leads to vessel constriction (Pagore & Biyani). However, an exact mechanism behind the pathophysiology of hypertension is still unidentified but some physiological processes that play a key role in the development of hypertension are discussed thoroughly below section:

Cardiac Output and Peripheral Vascular Resistance

The equilibrium between cardiac output and peripheral vascular resistance is an important requirement for the normal functioning of blood pressure. Most patients with essential hypertension have normal cardiac output but elevated peripheral vascular resistance. This increased level causes contraction of vascular smooth muscle resulting in higher cytosolic calcium concentration. A calcium channel blocker works by blocking this mechanism and hence acts as a vasodilator. However, in the case of an initial stage of hypertension, the peripheral resistance remains the same and the rise in blood pressure occurs due to an elevated cardiac output, which is associated with the increased activity of the sympathetic system (Mayet & Hughes, 2003).

Renin-Angiotensin-Aldosterone System (RAAS)

It is the most prominent endocrine system that plays a major role in the regulation of blood pressure. The juxtaglomerular gland of the kidney releases renin in response to lower salt intake. Renin facilitates the conversion of angiotensinogen (a renin substrate) to angiotensin I. Further angiotensin I (an inactive substance inside the body) is quickly converted to angiotensin II by angiotensin-converting enzyme (ACE) in the lungs (Cody, 1997). Angiotensin II causes vasoconstriction property thus elevating blood pressure. Moreover, it triggers the release of aldosterone hormone from the adrenal gland resulting in increased blood pressure. The level of renin and angiotensin II is low in hypertensive patients (particularly in geriatric patients). In such conditions, the antagonist of the renin-angiotensin system is not therapeutically effective (Silva & Flynn, 2012).

Sympathetic Nervous System Overactivity

The overactivity of the sympathetic nervous system is connected with raised cardiac output, heart rate, peripheral resistance, and peripheral postganglionic sympathetic nerve discharge. The constriction and dilation of arteries are regulated by the sympathetic nervous system and helps in maintaining normal blood pressure. Additionally, a small alteration in blood pressure due to physical exercise or stress is also mediated by the sympathetic nervous system (Mark, 1996).

Endothelial cell dysfunction

Endothelial cells aid in maintaining the normal functioning of the heart by releasing vasoactive agents like nitric oxide (vasodilator) and endothelin (vasoconstrictor) (Cardillo, Kilcoyne, Quyyumi, Cannon III, & Panza, 1998). Modification of endothelial function could be a promising approach in decreasing the complications associated with hypertension (Panza, García, Kilcoyne, Quyyumi, & Cannon III, 1995).

Vasoactive agents

Vasoactive agents are defined as pharmaceutically active ingredients or intracellular substances that have capabilities either to enhance or reduce blood pressure. An excellent example of an endogenous vasoactive agent is bradykinin and nitric oxide. Bradykinin is a potent vasodilator that exhibits dilation of vascular smooth muscle and helps in maintaining normal blood pressure (Wagner et al., 1992). Nitric oxide is an admirable endothelial vasodilator, released from the endothelium of arteries and veins. It exhibits a vasodilation effect via diffusion from the blood vessel walls into the smooth muscle. Another vasoactive agent is endothelin, a powerful endothelial vasoconstrictor that causes constriction of vascular smooth muscle resulting in raised blood pressure (Vierhapper, Wagner, Nowotny, & Waldhäusl, 1990).

Figure 3. The reduced nitric oxide synthesis causes decreased renal pressure natriuresis resulting in hypertension

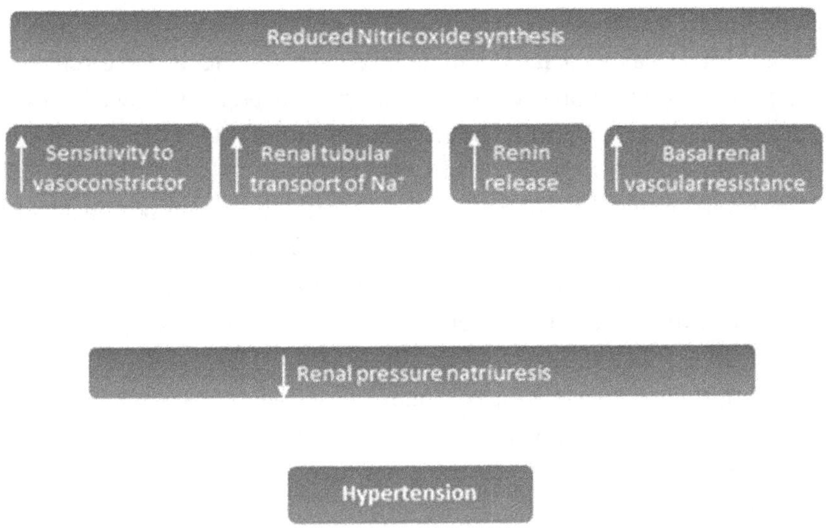

Atrial natriuretic peptides

The heart releases atrial natriuretic peptides in response to enhanced blood volume. It acts as a natural diuretic and promotes the enhanced excretion of water and sodium from the kidney. Abnormal functioning of this pathway causes fluid retention resulting in hypertension. The intracellular transportation of sodium through the walls of vascular smooth muscle may influence the rise in blood pressure due to its correlation with calcium transport (Brenner, Ballermann, Gunning, & Zeidel, 1990).

Genetic factors

Genetic factors play a significant role in the development of essential hypertension. Every gene has its significance and it is difficult to postulate which gene is primarily responsible for the development of hypertension. However, the consequences of hypertension are more in such situations where one or both parents have a prior history of hypertension. Many studies proposed that approximately 30% of alteration in blood pressure is due to genetic factors (Beevers et al., 2001).

Different randomized controlled trials showed that even a small reduction in blood pressure i.e., by 10 mmHg lessens patients' mortality risk by up to 25%. Likewise, cardiovascular ailments reduce the risk and stroke-associated death by up to 40%. A depth understanding of pathophysiological processes leading to cardiovascular diseases can help in the development of innovative therapies for effective treatment (Saxena, Ali, & Saxena, 2018).

ANTI-HYPERTENSIVE DRUG THERAPY

In 1946, Pentaquin was developed as the first anti-hypertensive drug for the effective treatment of hypertension but it exhibited low therapeutic efficacy and numerous side effects (Alam et al., 2017). Around the 1950s hexamethonium, a ganglionic blocking agent was introduced as a potent drug for treating hypertension. However, its use was associated with unpleasant side effects and circulatory complications from uncontrolled depressor action. Later, veratrum was introduced as it showed shorter onset of action than the previous drugs, but it was found to be toxic. After realizing the side effects associated with ganglionic blocking agents, hydralazine was developed but it was also not that effective and is rarely prescribed today. Then reserpine was introduced as the most effective antihypertensive drug but the efforts were not fruitful due to its adverse events such as impotency and depression (Alam et al., 2017). After 1960 the modern era of anti-hypertensive therapy started with the introduction of innovative drugs like beta-blockers and diuretics. In the year 1990, some other antihypertensive agents like calcium channel blockers, angiotensin receptor blockers, and ACE inhibitors were developed and are currently prescribed as the first-line antihypertensive drugs (Paulis & Unger, 2010). These agents are given either alone or in combination with other drugs. Recent advancements in discovering novel targets have unlocked a new door for the design of effective anti-hypertensive drugs which are currently in the development process.

Figure 4. The different categories of anti-hypertensive drugs

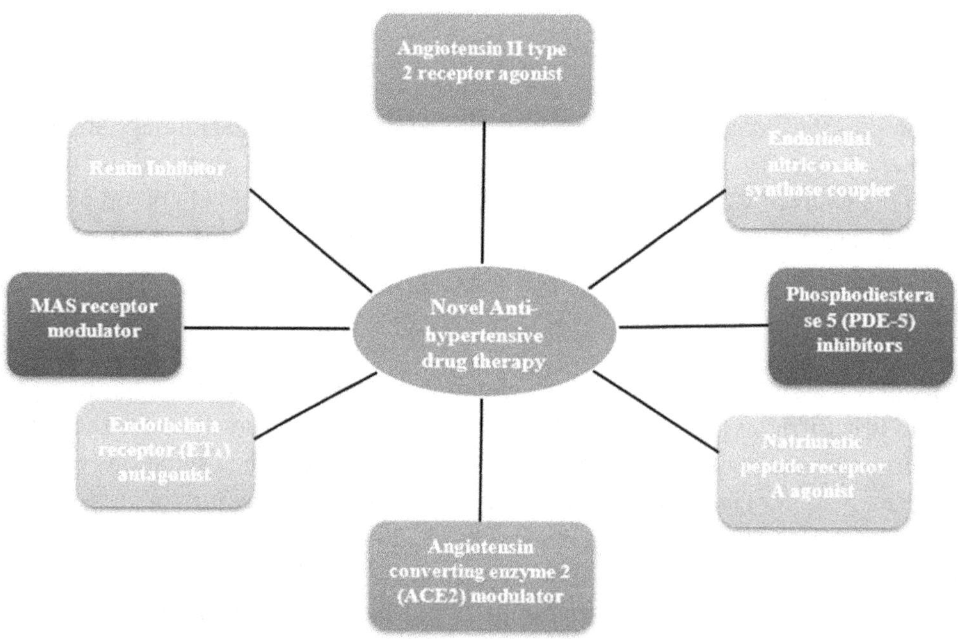

Renin Inhibitors: Renin, a potential target for anti-hypertensive drugs, is released from the kidney and engages in the body's renin-angiotensin-aldosterone system (RAAS). In 2007, the first renin inhibitor aliskiren was introduced into the market. Another new entity i.e. VTP2799 is currently undergoing phase II clinical trials of drug development (Paulis, Rajkovicova, & Simko, 2015).

Angiotensin II type 2 receptor agonists: Angiotensin II type 2 receptor agonists has vasodilation properties similar to bradykinin, cGMP, and nitric oxide. In addition, angiotensin II type 2 receptor agonists also mediate natriuresis. For example, compound 21 is known to exhibit natriuresis properties and is currently under the clinical development phase. Compound 21, mediates angiotensin II type 2 receptor agonist activity through its action on sodium/potassium ATPase and sodium/hydrogen exchanger 3 (NHE 3) (Paulis et al., 2015).

Phosphodiesterase 5 (PDE-5) inhibitors: Phosphodiesterase 5 promotes deprivation of cGMP (vasodilatory action). The inhibitors of PDE-5 prevent the degradation of cGMP and thus result in vasodilation. A recently approved PDE-5 inhibitor, named tadalafil is known to exhibit vasodilatory properties. Another PDE-5 inhibitor under phase II clinical development is KD027 (Paulis et al., 2015).

Angiotensin-converting enzyme 2 (ACE2) modulators: Angiotensin-converting enzyme 2 is responsible for the metabolism of angiotensin I and angiotensin II that are primary peptides and key regulators of the RAAS cascade. APN01 is currently in phase II clinical trials. APN01 is known to regulate high blood pressure through ACE2 modulation (Paulis et al., 2015).

MAS receptor modulator: MAS receptor is similar to angiotensin 2 receptor and is responsible for the release of nitrous oxide. Blockage of MAS receptors with hetero-dimerization results in obstruction of other angiotensin receptors. However, MAS receptor categories are still under the preclinical development phase (Paulis et al., 2015).

Endothelin a receptor (ET_A) antagonist: Endothelin receptors ET_A and ET_B are located on the cell membrane and are responsible for the pulmonary and systemic vasoconstriction. Endothelin receptor antagonists bind to these receptors and prevent vasoconstriction (Paulis et al., 2015). Currently approved endothelin receptor antagonists for pulmonary hypertension are ambrisentan and macitentan. Both drugs showed an excellent binding efficiency for ET_A and ET_B receptors (Iglarz et al., 2008).

Aldosterone synthase inhibitors: Recently aldosterone synthase (a cytochrome P450 enzyme responsible for aldosterone biosynthesis) has gained more attention in regulating blood pressure because mineralocorticoid receptor antagonists are not as much efficient in controlling aldosterone levels. ASILCI699 is an aldosterone synthase inhibitor, currently in phase II clinical trials (Niaz et al., 2016). This study showed control in high blood pressure by hindering RAAS.

Endothelial nitric oxide synthase (eNOS) coupler: In the case of pulmonary artery hypertension, the level of nitric oxide is low. Nitric oxide is a chemical substance responsible for the normal functioning of blood vessels due to its vasodilatory property. Endothelial nitric oxide synthase (eNOS) is an enzyme that plays a key role in the yielding of nitric oxide. In the case of uncoupled eNOS, nitric oxide production is reduced, meanwhile reactive oxygen species (ROS) generation increases and both these factors lead to higher vasoconstriction. Cicletanine is a thiazide-like diuretic that also works as an eNOS coupler, thus helps in enhancing the production of NO as well as reduction in ROS generation (Acierno & Labato, 2005).

Combined AT1 receptor blocker and NPE inhibitor: Neprilysin, enkephalinase, and vasopeptidase are the neutral endopeptidase (NEP) tending to deteriorate numerous peptide hormones (such as endothelin and angiotensin I, II) into its sedentary form. These peptide hormones show vasoconstriction properties. In addition, NEP also diminishes the vasodilatory properties of kinin and natriuretic peptides. The inhibition of NEP lowers the rise in high blood pressure by enhancing its vasodilatory action, meanwhile, the vasoconstriction is also increased. Thus, to overcome this situation a combined therapy of angiotensin I receptor blocker and neutral endopeptidase inhibitors were given. An example of this is LCZ699 and

Daglutril. Both agents are in phase III and II clinical trials and they are reported to exhibit both NEP inhibitory property as well as AT1 receptor blocking action (McMurray, 2015) (Paulis et al., 2015).

Natriuretic peptide receptor A (NPRA) agonist: Natriuretic peptides located in the brain and arteries exhibits vasodilation property. These peptides bind to the natriuretic peptide receptor A and exhibits dilation of a blood vessel via cGMP. A novel agent named PL3994 is an NPRA agonist which is currently in phase II clinical trial. This study reported a lowering in blood pressure and the initiation of natriuresis due to an increased cGMP level (Paulis et al., 2015).

LIMITATIONS OF ORAL DELIVERY OF ANTIHYPERTENSIVE DRUGS

Usually, solubility and permeability are the two essential parameters that affect the oral absorption of drug molecules as shown in figure. According to the biopharmaceutical classification system (BCS), the drugs that belong to the BCS class II (low solubility, high permeability) and class IV (low solubility, low permeability) are poor candidates for oral absorption. Therapeutically active substances that belong to BCS class II exhibits an irregular absorption pattern as well as lower bioavailability. A large number of the antihypertensive drugs belong to BCS class II and thus pose significant solubility and permeability issues resulting in low oral bioavailability. Some antihypertensives agents such as diltiazem, nifedipine, and nicardipine are metabolized by cytochrome P450 enzyme and are also substrates of P-glycoprotein (P-gp) efflux transporter located on the intestinal walls (Hetal, Bindesh, & Sneha, 2010) (El-Kattan & Varma, 2012).

The use of nanocarrier systems could be an effective approach to solve the limitations associated with orally delivered antihypertensive agents. Numerous nanocarrier systems like nanoemulsion, polymeric nanoparticles, solid lipid nanoparticles, nanostructured lipid carriers, and liposomes have been considered to overcome the drawbacks related to the oral delivery of antihypertensive agents (Chhabra, Chuttani, Mishra, & Pathak, 2011).

Figure 5. Schematic representation of various challenges to the oral delivery of anti-hypertensive drugs

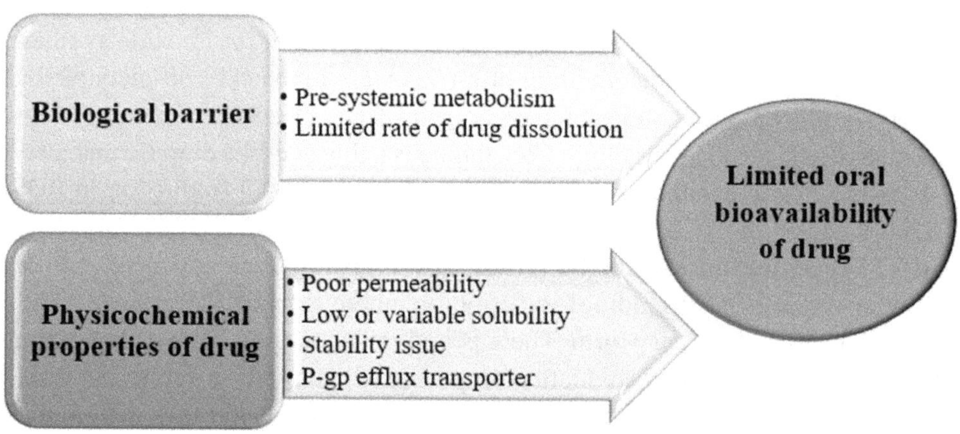

NANOEMULSIONS

Nanoemulsions are submicron-sized heterogeneous, transparent, or translucent colloidal dispersion of two azeotropic liquids having an average droplet size ranging from 200 to 700 nm (Thakore, Patel, & Patel, 2014). It comprises an oil and aqueous phase that coalesced together in the presence of surfactants (co-surfactants) usually called an emulsifier (co-emulsifier). Emulsifiers are the GRAS listed amphiphilic surface-active molecules (having lipophilic/non-polar hydrocarbon tails that prefer to be in the oil phase and polar/hydrophilic head groups that prefer to be in the aqueous phase) that offer stability to the system by reducing the interfacial tension between two immiscible phases (Y. Singh et al., 2017). Nanoemulsion can be administered via different routes as shown in Figure .6. In general, nanoemulsion exists in three different forms i.e. o/w type (oil phase is the dispersed phase and the aqueous phase is the dispersion medium), w/o type (aqueous phase is the dispersed phase and the oil phase is dispersion medium), and multiple emulsion (o/w/o or w/o/w) in which the microdomains of both oil and aqueous phases are inter-dispersed within the system (Tayeb & Sainsbury, 2018). Nanoemulsions can also be classified based on their surface charge into neutral, anionic, and cationic nanoemulsions (Koroleva & Yurtov, 2012). Emulsifiers are amphiphilic surface-active molecules that can reduce the interfacial tension between two immiscible liquid phases of oil and water by preferentially adsorbing at the interfaces. The kinetic stability of nanoemulsion is due to the emulsifier molecules that cover the dispersed oil droplet surface. Emulsifier selection is mainly done based on its solubility in oil and aqueous phases, HLB value, and toxicity profile. The oils that are used play a vital role in nanoemulsion formulation and help in the solubility of drugs with inherently low solubility. The amount of oil phase may vary from 5 to 20% w/w based on the site of administration (Sutradhar & Amin, 2013). Nanoemulsions are advantageous in the delivery of antihypertensive drugs over other nanocarriers systems due to their thermodynamic stability, rapid action, excellent shelf-life upon storage, and enhanced delivery of the drug near to the target site (Ganta, Talekar, Singh, Coleman, & Amiji, 2014). Following oral administration, the miniscule size droplets quickly absorb into enterocytes via internalization of droplets and are subsequently transported into the systemic circulation through the portal vein and lymphatic pathway. The tiny size of droplets in nanoemulsions and their ability to solubilize lipophilic drugs facilitate the rate of drug dissolution and increase the bioavailability (Lu, Qi, & Wu, 2012). An excellent example is an amlodipine-loaded nanoemulsion. As amlodipine is a BCS class IV drug with low aqueous solubility and permeability and thus low bioavailability. The drug-loaded nanoemulsion enhances the aqueous solubility, bioavailability, and payload of amlodipine to the target site (heart) (Chhabra et al., 2011). Nanoemulsion offers protection to the drug against rigorous environmental conditions (such as hydrolysis, pH, and oxidation) and possesses an organ-specific targeting via enhanced permeation and retention effects.

Figure 6. The various routes of the administration of nanoemulsions.

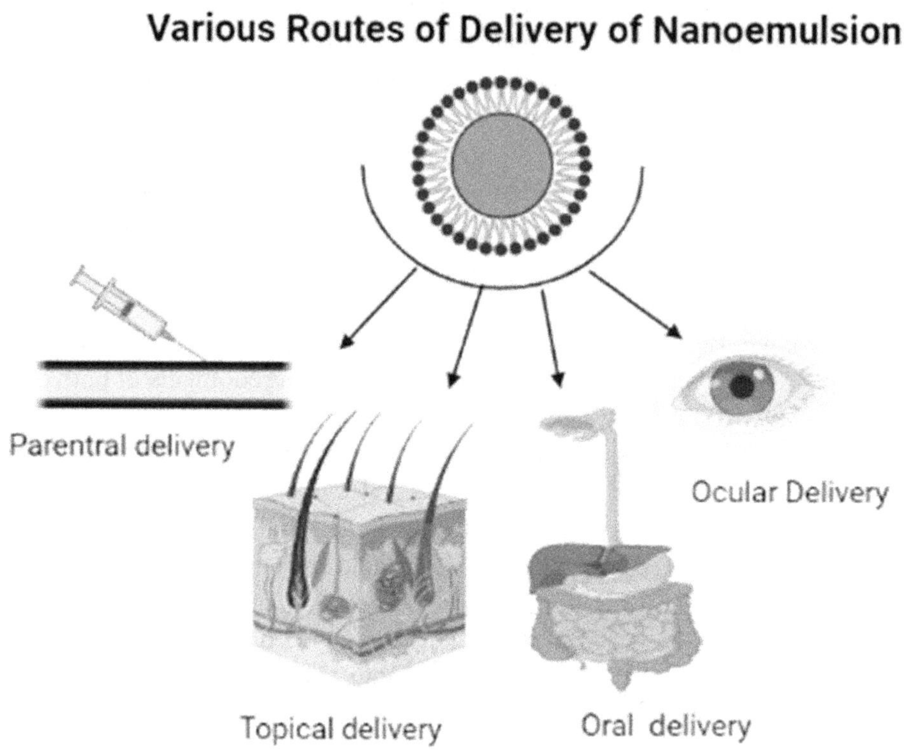

ADMINISTRATION OF NANOEMULSIONS VIA ORAL ROUTE

The oral route is the most preferred route for drug administration due to the ease of administration and patient compliance (Sastry, Nyshadham, & Fix, 2000). However, the drugs administered via the oral route encounter numerous hurdles such as harsh pH changes in the gastrointestinal tract (i.e., acidic pH in stomach and basic pH in intestine) and various digestive enzymes present in it. These enormous alterations in pH can hinder the therapeutic activity of the drug. For example, candesartan cilexetil an antihypertensive drug (angiotensin receptor blocker) endures chemical degradation at acidic pH. In addition to pH changes, numerous metabolic enzymes such as liver esterase and cytochrome P450 cause drug degradation in most antihypertensive drugs (Subramanian & Ghosal, 2004). In general, a huge number of antihypertensive drugs belong to BCS class II (low aqueous solubility and high permeability) and BCS class IV (low aqueous solubility and low permeability), and their delivery via the oral route is challenging as the drugs possess low bioavailability (Kotta et al., 2012). Nanoemulsions could be an effective approach to protect anti-hypertensive drugs from endogenous factors and enhance their bioavailability by increasing the drug permeability (Tayeb & Sainsbury, 2018). For instance, the olmesartan medoxomil (OM) dose was reduced by 3-folds from nanoemulsion due to enhanced solubility and bioavailability of the drug after oral administration. The in-vivo study data of OM nanoemulsion showed a 2.8-fold increase in AUC along with a sustained release profile (Gorain et al., 2014). Another example is the amlodipine besylate (AB) loaded nanoemulsion for the enhanced solubility and oral bioavailability of AB through the delivery of the drug at the target site. Pharmacokinetic study data showed 4.78-fold higher AUC, 2.2-fold higher $C_{max,}$ and a percent relative bioavailability of 475% as compared to the amlodipine besylate suspension (Chhabra et al., 2011). In addition to enhanced solubility nanoemulsions also extend the residence time of drugs in the GI tract and prevent pre-systemic metabolism. Orally administered nanoemulsions get absorbed into the systemic circulation via an intestinal lymphatic pathway. This can be achieved through the association of hydrophobic drugs to the chylomicrons and thus preventing the hepatic first-pass metabolism. Moreover, nanoemulsions prevent the hydrolysis and degradation of lipophilic drugs by encapsulating the drugs into the oil phase. This improves the overall stability of the drug (Sharma, Sharma, & Jain, 2016). For example, amlodipine and isradipine, suffer from low bioavailability and undergo photo-degradation. These drugs when loaded into nanoemulsion showed higher bioavailability and superior protection against light (DongJin, SungTae, HwaMi, BaeChan, & ChongKook, 2005).

Mechanism of Uptake of Nanoemulsion via Oral Administration

When administered orally, nanoemulsions enter into the GIT and are further exposed to an enormous GI environment (Subramanian & Ghosal, 2004). Initially, the stimulation of the lipid sensing mechanism (responsible for the secretion of gastric lipase) takes place in GIT resulting in the release of gastric lipase that causes lipid digestion. The digestion of lipids in the GIT ensues via three steps: (Carey, Small, & Bliss, 1983).

- The very first step is the formation of coarse emulsion along with fine dispersion of oil globules.
- Then lipid hydrolysis occurs at the oil/water interface.
- At last dispersion of products into a digestible form.

After oral administration, systemic absorption of the drug via the intestinal milieu is the primary concern for lipophilic drugs as these drugs have poor water solubility and a slow rate of dissolution. The formation of oil globules accelerates bile secretion and delay GI tract motility. The liver secretes bile fluids that are stored inside the gallbladder. The key components of bile are phospholipids, bile acids, and cholesterol. Components of bile act as an intrinsic surfactant and enhances the solubility profile of lipophilic drugs (Rautureau, Bisalli, & Rambaud, 1981) through the formation of mixed micelles. Bile salt mixed micelles help in solubilizing free drugs and offer its transportation across to the enterocytes as show in figure 7 (Carey & Small, 1970). Hydrophobic molecules are generally solubilized in the upper part of GIT, where the secretion of biliary lipids and pancreatic fluids takes place. Nanoemulsion globules are absorbed by either paracellular pathway, transcellular pathway, or by M-cells that reside in Peyer's patches. The transcellular pathway is the primary route of the entrance of lipophilic drugs through the intestinal lymphatic pathway. Gastric lipase promotes the digestion of triglycerides into a simple unit of di-glycerides, mono-glycerides, and free fatty acids in the stomach. Meanwhile, the sheer movement of the stomach and the passage of lipid molecules via pyloric sphincter results in emulsion formation (Abrams et al., 1988). The bile salts are absorbed over the oil/water interface and enhance the emulsion stability with abridged droplet size. The drug molecules are detached from micelles before absorption into enterocytes and once the drug enters into enterocytes, there is the formation of chylomicron and this chylomicron in association with hydrophobic drugs helps in the transportation of drugs into the lymphatic circulation. The chylomicrons are not able to permeate into the systemic capillaries because of its large size; hence it is absorbed into the leaky mesenteric lymph vessel and moves within the lymph until it gets shattered into the blood circulation. Thus, enhanced lymphatic transport of drugs can be achieved by increasing the process of association of chylomicron with lipophilic drugs (Subramanian & Ghosal, 2004).

Figure 7. Schematic representation of mechanism of uptake of nanoemulsions after oral administration.

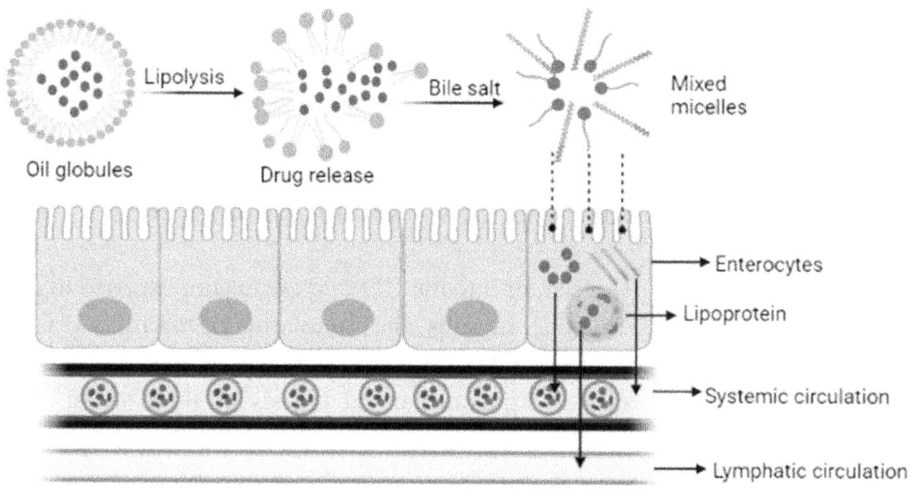

Intestinal lymphatic transport

The lymphatic system is a part of the vascular system that encompasses a convoluted web of channels that hold clear liquid known as lymph. The major role of the lymphatic system is to maintain the water balance of the body by removing excess fluids that ooze out of the blood circulation. It plays a crucial role in protecting the body tissues against infection by filtering the elements from lymph and enhancing the absorption of water-insoluble vitamins, antibiotics, long-chain fatty acids, and cholesterol ester. For drugs that undergo first-pass hepatic metabolism, absorption through the lymphatic system is considered an effective strategy to improve the oral bioavailability of drugs. Lymphatic targeted drug delivery is a fairly new approach for the efficient transportation of drugs, especially used in the delivery of anti-cancer drugs. Lymphatic delivery is shown to improve the mucosal immunity and oral absorption of larger molecules such as protein and peptides (O'Driscoll, 2002). The delivery of drugs via the intestinal lymphatic vasculature occurs because of the nature of lymphatic capillaries which encompass mono-layered, non-perforated endothelial cells with extremely gapped leaky vessels. This permits the entry of higher molecular weight compounds and large water-soluble molecules through the lymphatic system. Another reason for the delivery of drugs via the lymphatic system is the gut-associated lymphoid tissue (GALT). GALT is comprised of both secluded and accumulated lymphoid cavities which form Peyer's patches that act as an entry point for the drug into the lymphatics. Numerous factors such as route of administration, drug delivery system, physicochemical and metabolic properties of the drug mainly influence the lymphatic transport of drug (Kiyasu, Bloom, & Chaikoff, 1952). Various drug delivery systems like nanoemulsion, nanoparticles, liposomes, and micelles were found to deliver the drugs to lymphatic tissues and following oral administration.

Mechanism of Intestinal Drug Absorption

The absorption of the drug administered via the oral route mainly takes place through enterocytes present in intestinal lymphatics, and their entry into the systemic circulation is either by absorption through the portal vein or by trafficking into the pierced lymphatic endothelium (Subramanian & Ghosal, 2004). The actual mechanisms by which lipid or oil component improves the oral absorption of lipophilic drugs are not well known. The possible mechanisms are:

- Lipid assimilated products and bile salts increase the membrane fluidity and thus allow transcellular absorption. Also, paracellular transport is facilitated through the opening of tight junctions.
- Hydrophobic drugs are solubilized in bile salts through the formation of mixed micelles that are absorbed across the intestinal epithelium.
- The lipid component of nanoemulsions causes inhibition of P-glycoprotein and/or cytochrome P450 enzyme and thus increases the concentration and residence time of drugs within the cell.

ORAL NANOEMULSION BASED FORMULATION FOR IMPROVED DELIVERY OF ANTI-HYPERTENSIVE DRUGS

Nanoemulsions are advantageous in the oral delivery of antihypertensive drugs over other nanocarrier systems due to their thermodynamic stability, excellent shelf-life, and enhanced delivery of drugs at the target site. Upon administration via the oral route, the nano-droplets of nanoemulsion and their ability to solubilize lipophilic drugs facilitate the rate of drug dissolution and thus increases the systemic bioavailability of drugs.

In the subsequent sections, published work on oral nanoemulsions for improved delivery of anti-hypertensive drugs is discussed. A few studies are summarized in Table 1.

Lacidipine-loaded Self Nanoemulsifying Drug delivery System (SNEDDS)

Lacidipine is a powerful anti-hypertensive drug of the dihydropyridine class. It is a selective calcium channel blocker and belongs to the BCS class II drug (low water solubility and high permeability). Upon oral administration, lacidipine undergoes hepatic presystemic metabolism via cytochrome p450 3A4 (CYP3A4) enzyme to produce therapeutically inactive metabolites and showed higher variability in absorption (3-59%) due to its lower solubility in the GI tract (Tang, Zhu, Zhao, Cheng, & Peng, 2008). The conventional emulsion formulation of lacidipine showed solubility-dependent release rather than formulation-dependent. Hence, an advanced nanoformulation technique is required to upsurge the bioavailability of lacidipine by avoiding presystemic hepatic first-pass metabolism and to improve the treatment of hypertension. Lacidipine-loaded SNEDDS was prepared by water bath sonication technique using captex 810D, acconon MC8, and Transcutol HP as oil, surfactant, and cosurfactant respectively. *In-vitro* dissolution study revealed that more than 80% of lacidipine is released from SNEDDS within 15 minutes.

The ex-vivo permeation study demonstrated that the permeation of lacidipine from SNEDDS is 4.8 to 9-folds greater in comparison to free drugs. The experimental data depicted that the oral bioavailability of lacidipine-loaded SNEDDS is 2.5 folds greater than the marketed tablet formulation (Subramanian, Sharavanan, Chandrasekar, Balakumar, & Moulik, 2016).

Nisoldipine Loaded Nanoemulsion

Nisoldipine is a potent antihypertensive drug of the dihydropyridine class. It is a calcium channel blocker and belongs to the BCS class II category. Nisoldipine is a substrate of P-gp and CYP3A enzymes and exhibits low oral bioavailability, i.e. only 5% as it is extensively metabolized in the gut wall before reaching the systemic circulation (Wilkinson, 1997). Thus, to enhance the oral bioavailability via intestinal lymphatic uptake nisoldipine nanoemulsion was designed. The nisoldipine-loaded nanoemulsion was prepared by homogenization technique followed by sonication using peceol, cremophor EL, and Transcutol HP as an oil, surfactant, and cosurfactant respectively. The *in-vitro* release study revealed that the nisoldipine-loaded nanoemulsion exhibited higher release of drug (98.51±2.64%) in comparison to the free drug (29.73±2.15%). Cellular uptake study in Caco-2 cells showed higher intestinal absorption and enhanced efficacy of the drug from nanoemulsion as compared to free drug. Further, *in-vitro* and *in-vivo* data have shown significant ($p < 0.05$) enhancement in the bioavailability of nisoldipine when administered in form of nanoemulsion (Mundada, Patel, Mundada, & Sawant, 2020).

Olmesartan Medoxomil (OM) Loaded Nanoemulsion

Olmesartan medoxomil is prodrug (inactive form) of olmesartan. After oral administration, olmesartan medoxomil gets hydrolyzed via intestinal and plasma esterase enzyme and is converted to its active metabolite (olmesartan). It is a powerful, safe, and widely used angiotensin II receptor blocking agent, which exhibits antihypertensive properties by diminishing the release of aldosterone. The low aqueous solubility and abandoned enzymatic conversion of olmesartan medoxomil in the gastrointestinal fluid into its poorly permeable active parent moiety limits its oral bioavailability. Nanoemulsion approaches were used to improve the pharmacokinetic and therapeutic efficacy of olmesartan medoxomil. The OM-loaded nanoemulsion was prepared by phase inversion technique using soyabean oil 700 and sefsol 218 as an oil phase and Solutol HS 15 as surfactant respectively. The pharmacokinetic study data revealed that after oral administration of OM nanoemulsion the area under the curve (AUC_{0-27}) of olmesartan enhanced up to 2.8-fold. A sustained drug release profile was also obtained. Further, *in vivo* studies showed that the OM-loaded nanoemulsion possessed effective and long-term control of hypertension, as well as a 3 times reduction in dose as compared to the marketed formulation (Gorain et al., 2014).

Carvedilol loaded SNEDDS

Carvedilol is a nonselective beta-blocker as well as an alpha-1 receptor blocker. It is a second-line drug for the regulation of high blood pressure. Apart from hypertension treatment, it is used in the management of congestive heart failure and myocardial infarction. The low aqueous solubility, higher log P value (log P = 4.115), and extensive hepatic metabolism of carvedilol result in reduced oral bioavailability i.e., about 20%. The carvedilol-loaded self-nano emulsifying drug delivery system (SNEDDS) was prepared by using a high-speed stirring method using capmul MCM and nikkol HCO 50 as lipid and emulgent respectively. Pharmacokinetic study data showed higher C_{max} (134.2%) and AUC (85.2%) value of drug-loaded SNEDDS as compared to the free drug. The *in-vivo* single-pass intestinal perfusion (SPIP) studies indicated a substantial increase in the transportation of drug-loaded SNEDDS via the lymphatic system. Hence the drug-loaded SNEDDS showed significantly increased oral bioavailability as compared to free drugs. Further, the drug-loaded SNEDDS formulation also possesses an excellent stability profile during storage (B. Singh, Singh, Bandyopadhyay, Kapil, & Garg, 2013).

Candesartan Cilexetil Loaded Nanoemulsion

Candesartan is a powerful, selective angiotensin II type 1 (AT1) receptor antagonist. It exhibits antihypertensive properties by blocking aldosterone-releasing properties of angiotensin II. Candesartan cilexetil (CC) is a prodrug (inactive form) and after oral administration, it gets hydrolyzed through absorption via the GI tract and converts into its active metabolite (candesartan). Lower aqueous solubility of candesartan cilexetil results in only partial intestinal absorption and poor oral bioavailability (Nekkanti, Karatgi, Prabhu, & Pillai, 2010). Thus, to achieve an enhanced intestinal absorption and an improved oral bioavailability candesartan cilexetil nanoemulsion was fabricated by modified emulsification-solvent evaporation technique using soyabean oil, tween 80, and Solutol HS 15 as an oil, surfactant, and co-surfactant respectively.

The study data showed enhancement in the intestinal absorption of candesartan cilexetil loaded nanoemulsion compared to the free drug. Furthermore, the cellular internalization of CC-loaded nanoemulsion into the enterocyte can take place via the clathrin-mediated endocytosis pathway. Subsequently, the drug is delivered to the blood circulatory system through the portal and lymphatic circulation. *In-vivo* study data has shown a 10-fold enhancement in an area under the curve (AUC_{0-t}) of CC-loaded nanoemulsion as compared to free drug. This experimental outcome indicates that a nanoemulsion is an effective approach for the oral delivery of poorly soluble CC (Gao et al., 2011).

Amlodipine Besylate (AB) Loaded Nanoemulsion

Amlodipine, a strong anti-hypertensive drug of dihydropyridine class, is a calcium channel blocker and belongs to the BCS class IV category. The antihypertensive property of amlodipine is due to the inhibition of transmembrane entry of calcium ions into the smooth muscle of blood vessels and myocardium. The low water solubility and poor permeability of amlodipine besylate via the GIT results in low oral bioavailability (60-65%) (McDaid & Deasy, 1996). Moreover, to enhance the solubility, oral bioavailability, and to achieve efficient delivery of drug at the target site, amlodipine besylate nanoemulsion has been prepared by emulsification method using labrafil M, tween 80, and ethanol as an oil, surfactant, and cosurfactant respectively. A significant ($p<0.01$) enhancement in the release of drug from AB nanoemulsion as compared to the marketed formulation has been observed. The pharmacokinetics and biodistribution study results confirmed that the cellular uptake of AB nanoemulsion was substantially greater ($p<0.05$) than the free drug (Chhabra et al., 2011).

Valsartan loaded SNEDDS

Valsartan is an angiotensin II receptor antagonist, having a higher binding tendency for angiotensin type I receptors. It inhibits the activity of angiotensin II resulting in decreased mean systolic blood pressure.

Valsartan is a powerful vasodilator and is widely recommended for the management of congestive heart failure, raised blood pressure, and myocardial infarction. Valsartan belongs to the BCS class II category, possesses poor water solubility (<0.1 mg/mL) and high permeability (log p-value is 5.8). The lower oral bioavailability (i.e., <25%) is due to massive presystemic first-pass metabolism by the liver. Additionally, it exhibits greater P-gp efflux and variable oral bioavailability when administered with food. SNEDDS approach was used to enhance the oral bioavailability of valsartan by avoiding presystemic first-pass metabolism. Valsartan-loaded SNEDDS was prepared by using capmul MCM, labrasol, and tween 20 as an oil, surfactant, and cosurfactant respectively.

I*n-vitro* release study data of valsartan-loaded SNEDDS has shown a 3 to 3.5-fold enhancement in the rate of drug dissolution as compared to the marketed formulation. A noteworthy decrease in systolic blood pressure ($p<0.05$) as the drug-loaded SNEDDS was absorbed via the lymphatic pathway has also been observed. No alteration in physicochemical properties was observed upon storage for 6 months. This showed that the drug-loaded SNEDDS formulation possessed decent stability (Beg, Swain, Singh, Patra, & Rao, 2012).

Ramipril-loaded SNEDDS

Ramipril is an ACE inhibitor. The ACE receptor is responsible for the conversion of angiotensin I into angiotensin II (a vasoconstrictor substance). Hence blockage of ACE receptors causes reduced vasopressor activity resulting in low blood pressure. It is a prodrug (inactive form) of ramiprilat (dicarboxylic acid) and after oral administration, converts into its active metabolite (ramiprilat) in the liver via hydrolytic cleavage of the ester group. It is a very lipophilic drug having a log P value of 3.32 and possesses lower aqueous solubility with a total bioavailability of 28-35%. To achieve an enhanced bioavailability as well as small lipid droplet size, ramipril-loaded SNEDDS was formulated with the use of minimum surfactant concentration. The ramipril-loaded SNEDDS was prepared by using sefsol 218, tween 80, and carbitol as an oil, surfactant, and cosurfactant respectively. The *in-vitro* data showed that the drug release from ramipril-loaded nanoemulsion was significantly ($p<0.01$) greater in comparison to the free drug and marketed formulation. The comparative bioavailability of the drug-loaded nanoemulsion was observed to be 229.62% as compared to the marketed capsule dosage form (Shafiq et al., 2007).

Table 1. A partial list of orally administered anti-hypertensive drug-loaded nanoemulsion formulations.

S. No.	Anti-hypertensive Drug	Formulation Composition	*In-vivo* Study Model	Description	References
1.	Lacidipine	SNEDDS (Capmul and labrafil as oil, Tween 80 and cremophor as surfactant, and Transcutol HP as co-surfactant).	Male Wistar rats	More than 80% lacidipine was released from SNEDDS within 15 minutes. The permeation of lacidipine from SNEDDS is 4.8 to 9-folds greater in comparison to the free drug. The oral bioavailability of lacidipine loaded SNEDDS is 2.5 folds greater than the marketed tablet formulation	(Subramanian et al., 2016)
2.	Nisoldipine	Peceol, Transcutol HP, cremophore EL	Male Sprague-Dawley rats	Nanoemulsion exhibit enhanced release of drug (98.51±2.64%) in comparison to the free drug (29.73±2.15%). The cellular uptake study showed higher intestinal absorption and enhanced efficacy of the drug.	(Mundada et al., 2020)
3.	Olmesartan Medoxomil	SNEDDS	Unisex Wistar rats	The area under the curve (AUC_{0-27}) of olmesartan enhanced up to 2.8-fold and exhibits a sustained drug release profile.	(Gorain et al., 2014)
4.	Carvedilol	SNEDDS (Transcutol HP and Cremophore EL	–	Drug-loaded SNEDDS showed higher Cmax (134.2%) and AUC (85.2%). Substantially enhanced transportation of drug-loaded SNEDDS via the lymphatic system increases the oral bioavailability.	(B. Singh et al., 2013)
5.	Candesartan cilexetil	Nanoemulsion (Tween 80, soybean oil, and Solutol HS-15	Male Sprague-Dawley rats	10-fold enhancement in the area under the curve (AUC_{0-t}) of CC-loaded nanoemulsion as compared to the free drug.	(Gao et al., 2011)
6.	Amlodipine besylate	Dry emulsion (Labrafil CS 1944 CS and Dextrin)	Male Sprague-Dawley rats	Significant ($p<0.01$) enhancement in the release of drug from AB nanoemulsion	(Chhabra et al., 2011)
7.	Valsartan	SNEDDS (Tween 20, capmul MCC, labrasol)	Male Wistar rats	SNEDDS showed a 3 to 3.5-fold enhancement in the rate of drug dissolution and a decrease in the systolic blood pressure ($p<0.05$) was observed.	(Beg et al., 2012)
8.	Ramipril	Tween80, sefsol 218 and carbitol	Wistar male albino rats	Relative bioavailability of drug-loaded nanoemulsion was observed to be 229.62% as compared to marketed capsule	(Shafiq et al., 2007)

CONCLUSION AND FUTURE PERSPECTIVE

Every year hypertension is responsible for a huge number of deaths around the world. Several categories of drugs are in the development stage for the management of hypertension still the consequences of cardiovascular diseases are increasing day by day. The major obstacles in the effective delivery of antihypertensive drugs are the inherent physiochemical properties of drugs like low aqueous solubility, poor permeability, and extensive hepatic first-pass metabolism leading to poor oral bioavailability. The conventional dosage forms do not adequately meet the clinical requirement hence, there is a need to explore newer technologies and strategies to overcome the limitations of conventional therapies. Studies indicate that nanoemulsions offer several advantages in the oral delivery of anti-hypertensive drugs such as improved solubility and enhanced absorption/bioavailability of drugs. The small droplet size of nanoemulsions and large surface area enhances the drug transport and delivery to specific sites. However, still there is scope for enhancement in drug therapy and delivery systems to efficiently regulate blood pressure. Furthermore, an understanding of the nanoemulsion manufacturing process and pharmacological profile is required to accelerate the clinical translation of nanoemulsions.

REFERENCES

Abrams, C. K., Hamosh, M., Lee, T. C., Ansher, A. F., Collen, M. J., Lewis, J. H., Benjamin, S. B., & Hamosh, P. (1988). Gastric lipase: Localization in the human stomach. *Gastroenterology*, *95*(6), 1460–1464. doi:10.1016/S0016-5085(88)80063-5 PMID:3181673

Acierno, M. J., & Labato, M. A. (2005). Hypertension in renal disease: Diagnosis and treatment. *Clinical Techniques in Small Animal Practice*, *20*(1), 23–30. doi:10.1053/j.ctsap.2004.12.004 PMID:15822527

Alam, T., Khan, S., Gaba, B., Haider, M. F., Baboota, S., & Ali, J. (2017). Nanocarriers as treatment modalities for hypertension. *Drug Delivery*, *24*(1), 358–369. doi:10.1080/10717544.2016.1255999 PMID:28165823

Beevers, G., Lip, G. Y., & O'Brien, E. (2001). The pathophysiology of hypertension. *BMJ (Clinical Research Ed.)*, *322*(7291), 912–916. doi:10.1136/bmj.322.7291.912 PMID:11302910

Beg, S., Swain, S., Singh, H. P., Patra, C. N., & Rao, M. B. (2012). Development, optimization, and characterization of solid self-nanoemulsifying drug delivery systems of valsartan using porous carriers. *AAPS PharmSciTech*, *13*(4), 1416–1427. doi:10.120812249-012-9865-5 PMID:23070560

Brenner, B. M., Ballermann, B. J., Gunning, M. E., & Zeidel, M. L. (1990). Diverse biological actions of atrial natriuretic peptide. *Physiological Reviews*, *70*(3), 665–699. doi:10.1152/physrev.1990.70.3.665 PMID:2141944

Cain, A. E., & Khalil, R. A. (2002). *Pathophysiology of essential hypertension: role of the pump, the vessel, and the kidney.* Paper presented at the Seminars in nephrology. 10.1053nep.2002.28639

Cardillo, C., Kilcoyne, C. M., Quyyumi, A. A., Cannon, R. O. III, & Panza, J. A. (1998). Selective defect in nitric oxide synthesis may explain the impaired endothelium-dependent vasodilation in patients with essential hypertension. *Circulation*, *97*(9), 851–856. doi:10.1161/01.CIR.97.9.851 PMID:9521333

Carey, M. C., & Small, D. M. (1970). The characteristics of mixed micellar solutions with particular reference to bile. *The American Journal of Medicine*, *49*(5), 590–608. doi:10.1016/S0002-9343(70)80127-9 PMID:4924587

Carey, M. C., Small, D. M., & Bliss, C. M. (1983). Lipid digestion and absorption. *Annual Review of Physiology*, *45*(1), 651–677. doi:10.1146/annurev.ph.45.030183.003251 PMID:6342528

Carretero, O. A., & Oparil, S. (2000). Essential hypertension: part II: treatment. *Circulation*, *101*(4), 446–453. doi:10.1161/01.CIR.101.4.446 PMID:10653838

Chhabra, G., Chuttani, K., Mishra, A. K., & Pathak, K. (2011). Design and development of nanoemulsion drug delivery system of amlodipine besilate for improvement of oral bioavailability. *Drug Development and Industrial Pharmacy*, *37*(8), 907–916. doi:10.3109/03639045.2010.550050 PMID:21401341

Chime, S., Kenechukwu, F., & Attama, A. (2014). *Nanoemulsions—advances in formulation, characterization and applications in drug delivery* (Vol. 3). Academic Press.

Chircov, C., & Grumezescu, A. M. (2019). *Nanoemulsion preparation, characterization, and application in the field of biomedicine. In Nanoarchitectonics in biomedicine*. Elsevier.

Cody, R. J. (1997). The sympathetic nervous system and the renin–angiotensin–aldosterone system in cardiovascular disease. *The American Journal of Cardiology*, *80*(9), 9J–14J. doi:10.1016/S0002-9149(97)00832-1 PMID:9375943

Delacroix, S., Chokka, R. G., & Worthley, S. G. (2014). Hypertension: Pathophysiology and treatment. *Journal of Neurology & Neurophysiology*, *5*(6), 1–8. doi:10.4172/2155-9562.1000250

DongJin, J., SungTae, K., HwaMi, L., BaeChan, K., & ChongKook, K. (2005). Improvement of bioavailability and photostability of amlodipine using redispersible dry emulsion. 추계총회 및 학술대회, 511-511.

El-Kattan, A., & Varma, M. (2012). Oral absorption, intestinal metabolism and human oral bioavailability. *Topics on Drug Metabolism, 10*, 31087.

Foëx, P., & Sear, J. (2004). Hypertension: Pathophysiology and treatment. *Continuing Education in Anaesthesia, Critical Care & Pain*, *4*(3), 71–75. doi:10.1093/bjaceaccp/mkh020

Ganta, S., Talekar, M., Singh, A., Coleman, T. P., & Amiji, M. M. (2014). Nanoemulsions in translational research—Opportunities and challenges in targeted cancer therapy. *AAPS PharmSciTech*, *15*(3), 694–708. doi:10.120812249-014-0088-9 PMID:24510526

Gao, F., Zhang, Z., Bu, H., Huang, Y., Gao, Z., Shen, J., Zhao, C., & Li, Y. (2011). Nanoemulsion improves the oral absorption of candesartan cilexetil in rats: Performance and mechanism. *Journal of Controlled Release*, *149*(2), 168–174. doi:10.1016/j.jconrel.2010.10.013 PMID:20951749

Gorain, B., Choudhury, H., Kundu, A., Sarkar, L., Karmakar, S., Jaisankar, P., & Pal, T. K. (2014). Nanoemulsion strategy for olmesartan medoxomil improves oral absorption and extended antihypertensive activity in hypertensive rats. *Colloids and Surfaces. B, Biointerfaces*, *115*, 286–294. doi:10.1016/j.colsurfb.2013.12.016 PMID:24388859

Hall, J. E., Granger, J. P., do Carmo, J. M., da Silva, A. A., Dubinion, J., George, E., ... Hall, M. E. (2012). Hypertension: Physiology and pathophysiology. *Comprehensive Physiology*, *2*(4), 2393–2442. doi:10.1002/cphy.c110058 PMID:23720252

Hetal, T., Bindesh, P., & Sneha, T. (2010). A review on techniques for oral bioavailability enhancement of drugs. *International Journal of Pharmaceutical Sciences Review and Research*, *4*(3), 203–223.

Iglarz, M., Binkert, C., Morrison, K., Fischli, W., Gatfield, J., Treiber, A., ... Buchmann, S. (2008). Pharmacology of macitentan, an orally active tissue-targeting dual endothelin receptor antagonist. *The Journal of Pharmacology and Experimental Therapeutics*, *327*(3), 736–745. doi:10.1124/jpet.108.142976 PMID:18780830

Kiyasu, J., Bloom, B., & Chaikoff, I. (1952). The portal transport of absorbed fatty acids. *The Journal of Biological Chemistry*, *199*(1), 415–419. doi:10.1016/S0021-9258(18)44850-8 PMID:12999855

Koroleva, M. Y., & Yurtov, E. V. (2012). Nanoemulsions: The properties, methods of preparation and promising applications. *Russian Chemical Reviews*, *81*(1), 21–43. doi:10.1070/RC2012v081n01ABEH004219

Kotta, S., Khan, A. W., Pramod, K., Ansari, S. H., Sharma, R. K., & Ali, J. (2012). Exploring oral nanoemulsions for bioavailability enhancement of poorly water-soluble drugs. *Expert Opinion on Drug Delivery*, *9*(5), 585–598. doi:10.1517/17425247.2012.668523 PMID:22512597

Kumar, V., Mishra, S., Shrivastava, A., & Zaidi, S. (2020). Prevalence of hypertension and its associated risk factors in rural community of Barabanki. *International Journal of Community Medicine and Public Health*, *7*(3), 893. doi:10.18203/2394-6040.ijcmph20200936

Lu, Y., Qi, J., & Wu, W. (2012). Absorption, disposition and pharmacokinetics of nanoemulsions. *Current Drug Metabolism*, *13*(4), 396–417. doi:10.2174/138920012800166544 PMID:22443535

Mark, A. (1996). The sympathetic nervous system in hypertension: a potential long-term regulator of arterial pressure. *Journal of Hypertension, 14*(5), S159-165.

Mayet, J., & Hughes, A. (2003). Cardiac and vascular pathophysiology in hypertension. *Heart (British Cardiac Society)*, *89*(9), 1104–1109. doi:10.1136/heart.89.9.1104 PMID:12923045

McDaid, D., & Deasy, P. (1996). Formulation development of a transdermal drug delivery system for amlodipine base. *International Journal of Pharmaceutics*, *133*(1-2), 71–83. doi:10.1016/0378-5173(95)04412-4

McMurray, J. J. (2015). Neprilysin inhibition to treat heart failure: A tale of science, serendipity, and second chances. *European Journal of Heart Failure*, *17*(3), 242–247. doi:10.1002/ejhf.250 PMID:25756942

Mundada, V. P., Patel, M. H., Mundada, P. K., & Sawant, K. K. (2020). Enhanced bioavailability and antihypertensive activity of nisoldipine loaded nanoemulsion: Optimization, cytotoxicity and uptake across Caco-2 cell line, pharmacokinetic and pharmacodynamic studies. *Drug Development and Industrial Pharmacy*, *46*(3), 376–387. doi:10.1080/03639045.2020.1724128 PMID:32031412

Nekkanti, V., Karatgi, P., Prabhu, R., & Pillai, R. (2010). Solid self-microemulsifying formulation for candesartan cilexetil. *AAPS PharmSciTech*, *11*(1), 9–17. doi:10.120812249-009-9347-6 PMID:20013081

Niaz, T., Shabbir, S., Manzoor, S., Rehman, A., Rahman, A., Nasir, H., & Imran, M. (2016). Antihypertensive nano-ceuticales based on chitosan biopolymer: Physico-chemical evaluation and release kinetics. *Carbohydrate Polymers*, *142*, 268–274. doi:10.1016/j.carbpol.2016.01.047 PMID:26917399

O'Driscoll, C. M. (2002). Lipid-based formulations for intestinal lymphatic delivery. *European Journal of Pharmaceutical Sciences*, *15*(5), 405–415. doi:10.1016/S0928-0987(02)00051-9 PMID:12036717

Pagore, R., & Biyani, K. (n.d.). *A review on types, treatment, awareness, prevention, pathophysiology and diagnosis of hypertension*. Academic Press.

Panza, J. A., García, C. E., Kilcoyne, C. M., Quyyumi, A. A., & Cannon, R. O. III. (1995). Impaired endothelium-dependent vasodilation in patients with essential hypertension: Evidence that nitric oxide abnormality is not localized to a single signal transduction pathway. *Circulation*, *91*(6), 1732–1738. doi:10.1161/01.CIR.91.6.1732 PMID:7882481

Paulis, L., Rajkovicova, R., & Simko, F. (2015). New developments in the pharmacological treatment of hypertension: Dead-end or a glimmer at the horizon? *Current Hypertension Reports*, *17*(6), 42. doi:10.100711906-015-0557-x PMID:25893478

Paulis, L., & Unger, T. (2010). Novel therapeutic targets for hypertension. *Nature Reviews. Cardiology*, *7*(8), 431–441. doi:10.1038/nrcardio.2010.85 PMID:20567239

Rautureau, M., Bisalli, A., & Rambaud, J. (1981). Bile salts and lipids in aqueous intraluminal phase during the digestion of a standard meal in normal man (author's transl). *Gastroenterologie Clinique et Biologique*, *5*(4), 417–425. PMID:7227749

Sastry, S. V., Nyshadham, J. R., & Fix, J. A. (2000). Recent technological advances in oral drug delivery–a review. *Pharmaceutical Science & Technology Today*, *3*(4), 138–145. doi:10.1016/S1461-5347(00)00247-9 PMID:10754543

Saxena, T., Ali, A. O., & Saxena, M. (2018). Pathophysiology of essential hypertension: An update. *Expert Review of Cardiovascular Therapy*, *16*(12), 879–887. doi:10.1080/14779072.2018.1540301 PMID:30354851

Shafiq, S., Shakeel, F., Talegaonkar, S., Ahmad, F. J., Khar, R. K., & Ali, M. (2007). Development and bioavailability assessment of ramipril nanoemulsion formulation. *European Journal of Pharmaceutics and Biopharmaceutics*, *66*(2), 227–243. doi:10.1016/j.ejpb.2006.10.014 PMID:17127045

Sharma, M., Sharma, R., & Jain, D. K. (2016). Nanotechnology based approaches for enhancing oral bioavailability of poorly water soluble antihypertensive drugs. *Scientifica*, *2016*, 2016. doi:10.1155/2016/8525679 PMID:27239378

Silva, A. C. S., & Flynn, J. T. (2012). The renin–angiotensin–aldosterone system in 2011: Role in hypertension and chronic kidney disease. *Pediatric Nephrology (Berlin, Germany)*, *27*(10), 1835–1845.

Singh, B., Singh, R., Bandyopadhyay, S., Kapil, R., & Garg, B. (2013). Optimized nanoemulsifying systems with enhanced bioavailability of carvedilol. *Colloids and Surfaces. B, Biointerfaces*, *101*, 465–474. doi:10.1016/j.colsurfb.2012.07.017 PMID:23010056

Singh, Y., Meher, J. G., Raval, K., Khan, F. A., Chaurasia, M., Jain, N. K., & Chourasia, M. K. (2017). Nanoemulsion: Concepts, development and applications in drug delivery. *Journal of Controlled Release*, *252*, 28–49. doi:10.1016/j.jconrel.2017.03.008 PMID:28279798

Subramanian, N., & Ghosal, S. K. (2004). *Enhancement of gastrointestinal absorption of poorly water soluble drugs via lipid based systems*. Academic Press.

Subramanian, N., Sharavanan, S. P., Chandrasekar, P., Balakumar, A., & Moulik, S. P. (2016). Lacidipine self-nanoemulsifying drug delivery system for the enhancement of oral bioavailability. *Archives of Pharmacal Research*, *39*(4), 481–491. doi:10.100712272-015-0657-9 PMID:26362165

Sutradhar, K. B., & Amin, M. L. (2013). Nanoemulsions: Increasing possibilities in drug delivery. *European Journal of Nanomedicine*, *5*(2), 97–110. doi:10.1515/ejnm-2013-0001

Tang, J., Zhu, R., Zhao, R., Cheng, G., & Peng, W. (2008). Ultra-performance liquid chromatography–tandem mass spectrometry for the determination of lacidipine in human plasma and its application in a pharmacokinetic study. *Journal of Pharmaceutical and Biomedical Analysis*, *47*(4-5), 923–928. doi:10.1016/j.jpba.2008.04.018 PMID:18534803

Tayeb, H. H., & Sainsbury, F. (2018). Nanoemulsions in drug delivery: Formulation to medical application. *Nanomedicine (London)*, *13*(19), 2507–2525. doi:10.2217/nnm-2018-0088 PMID:30265218

Thakore, S., Patel, R., & Patel, M. (2014). Nanoemulsion or Microemulsion? Understanding the Differences and Similarities. *The Pharma Review*, 136-142.

Vierhapper, H., Wagner, O., Nowotny, P., & Waldhäusl, W. (1990). Effect of endothelin-1 in man. *Circulation*, *81*(4), 1415–1418. doi:10.1161/01.CIR.81.4.1415 PMID:2180594

Wagner, O. F., Christ, G., Wojta, J., Vierhapper, H., Parzer, S., Nowotny, P. J., Schneider, B., Waldhäusl, W., & Binder, B. R. (1992). Polar secretion of endothelin-1 by cultured endothelial cells. *The Journal of Biological Chemistry*, *267*(23), 16066–16068. doi:10.1016/S0021-9258(18)41966-7 PMID:1644793

Weber, M. A., Schiffrin, E. L., White, W. B., Mann, S., Lindholm, L. H., Kenerson, J. G., ... Ram, C. V. S. (2014). Clinical practice guidelines for the management of hypertension in the community: A statement by the American Society of Hypertension and the International Society of Hypertension. *Journal of Hypertension*, *32*(1), 3–15. doi:10.1097/HJH.0000000000000065 PMID:24270181

Wilkinson, G. R. (1997). The effects of diet, aging and disease-states on presystemic elimination and oral drug bioavailability in humans. *Advanced Drug Delivery Reviews*, *27*(2-3), 129–159. doi:10.1016/S0169-409X(97)00040-9 PMID:10837555

Chapter 18
Nanoemulsion and Its Application in Pesticide Formulation

Khemnath Patir
Department of Applied Science and Humanities, Assam University, Silchar, India

ABSTRACT

A nanoemulsion is a colloidal dispersion that contains very small particles with size in the range of 20-200 nm, which may be of the oil-in-water (O/W) or water-in-oil (W/O) types depending on whether the oil is dispersed as droplets in water or vice versa. Pesticides play a critical role in controlling agriculture and ensuring food safety. In the future, the utilization of pesticides will become even more important to solve food security problems associated with providing an adequate food supply for the growing world population. Hence, development of new techniques for nanoemulsions formulation in the field of agriculture and forestry are necessary. In this chapter, the synthesis, characterization, and application of nanoemulsion in pesticides formulation are reviewed. In addition, the effect of antifungal or insecticidal and antimicrobial enhancement activity of nanoemulsions with size, composition, and stability are also discussed.

1. INTRODUCTION

Emulsion is a non-equilibrium heterogeneous colloidal dispersion consisting of two immiscible liquids. Nanoemulsions are isotropic, thermodynamically unstable, kinetically stable transparent colloidal dispersions with particle size in the range, 20-200 nm. It is commonly prepared by using oil, water, surfactant and a co-surfactant as the starting materials. This type of emulsion is thermodynamically unstable due to the effect of large positive interfacial tension between the oil and water phases. The smaller size of the droplets in nanoemulsions typically gives them better stability to gravitational separation and droplet aggregation than other emulsions (Gutierrez et al., 2008, McClements, 2012).

Pesticides play a critical role in reducing yield loss and ensuring food safety (Cooper and Dobson, 2007). In the future, the pesticides utilization will become very important to solve food security problems

DOI: 10.4018/978-1-7998-8378-4.ch018

associated with providing an adequate food supply for the growing world population (Carvalho, 2006). Pesticides are synthetic or natural substances which are used to prevent, destroy or control microorganisms, insects, weeds and other pests that endanger agriculture and forestry. It also helps in regulating the growth of certain plant species. Pesticides are classified based on the biological species that they are used to control: insecticides, fungicides, nematicides, herbicides and plant growth regulators. They can be used for agricultural, forestry, hygiene and nonagricultural purposes (Silva et al., 2012; Weaver et al., 2009, Pavela, 2013, Badawi et al., 2016). Commercially, pesticides are not typically used directly in their pure chemical form. Instead, they are usually incorporated into a suitable delivery system by adding an appropriate dosage of the pesticide into a suitable solvent or colloidal dispersion, which may be used in a solid, liquid, or gaseous form (Rodham, 2000). Thus, commercial pesticide formulations can be divided into solid forms (such as granules or powders), liquid forms (such as solutions or suspensions) and gaseous forms (such as vapours). The liquid forms include solutions, emulsifiable concentrates, oil-in water emulsions and soluble concentrate (Wang et al., 2004). In addition, used of these agrochemical pesticides is also polluting the environment which is hazardous to living being. Hence, the development of suitable formulation technologies is important for efficient delivery systems for hydrophobic pesticides as well as shifting to water based pesticides system is necessary for better results and safety.

In this chapter, synthesis, characterization and application of nanoemulsion in pesticides formulation is reviewed. The effect of antifungal or insecticidal and antimicrobial enhancement activity of nano-emulsion with size, composition and stability are also discussed. This book chapter would be useful as reference for the synthesis of new nanoemulsion based system and its application in pesticides for agriculture and forestry sector.

2. BACKGROUND

There are currently a number of challenges that limit the efficiency of certain types of pesticides. Many pesticides are highly hydrophobic molecules which have low solubility in water and therefore, it is difficult to get them to the required site of action (Wang et al., 2004). Moreover, the problems associated with emulsifiable concentrates, microemulsions and emulsions can be overcome by using nanoemulsions. Like emulsions, nanoemulsions are thermodynamically unstable colloidal dispersions that consist of emulsifier-coated oil droplets dispersed in water, but the mean droplet diameter is smaller (< 200 nm). The smaller size of the droplets in nanoemulsions increases their stability to gravitational separation and droplet aggregation and increases their deposition, diffusion and permeability on plant foliage and insect body surfaces (Kah and Hofmann, 2014). Nanoemulsions provide an effective way to use pesticides efficiently, economically and safely. In view of this, the composition, characteristics, mechanism of formation and stability of pesticide nanoemulsions have important theoretical and practical significances on the promotion and application of pesticides.

2.1 Components of Nanoemulsion

Nanoemulsion may be of the oil-in-water (O/W) or water-in-oil (W/O) types depending on whether the oil is dispersed as droplets in water or vice versa. The structure of the particles in a nanoemulsion consists of non-polar tails of the surfactant molecules protrude into the hydrophobic core formed by the oil phase, while the polar head groups of the surfactant molecules protrude into the surrounding aqueous phase.

2.2 Emulsifier

One of the essential components in nanoemulsions is surfactants or emulsifiers. There are four types of surfactants: cationic, anionic, amphoteric and nonionic. In nanoemulsions formulation for pesticide applications, nonionic surfactants are commonly encapsulated into the nanoemulsion, as they are less affected by the pH and ionic strength. This additional component has the ability to alter the stability and size of the nanoemulsion, as a result of cohesion forces between the anionic surfactant and the solution. The proper surfactants selection is related to its hydrophilic-lipophilic balance (HLB) value. A higher value of HLB indicates an increase in surfactant solubility towards the water which favors O/W system for pesticide formulations. The common value of HLB used to produce good O/W formulation in agriculture is in the range 10-16 while HLB value lower than 10 are oil-soluble surfactants. In producing kinetically stable nanoemulsions, the HLB value of the surfactant is one of the most important parameters to be considered. A varied range of HLB values can be achieved either by single or mixed surfactants. For example, a nanoemulsion against *Aedes aegypti* has been developed separately based on andiroba and copaiba oils which required hydrophilic-lipophilic balance (HLB) values of 11.2 and 14.8 respectively. The nanoemulsions were found to be stable when the HLB value of surfactants was near to the HLB oil in the system (Jesus et al., 2017). This finding has proved that when the HLB value of surfactants is similar to the HLB of the oil used in the system, the formation of the most stable nanoemulsion is possible. The incorporation of surfactant between 1.5-10% range and 5% is the most generally reported as suitable and sufficient amount for nanoemulsion production. The uses of a surfactant alter the electrostatic charge in the nanoemulsion which causes low aggregation (Feng et al., 2016). Some reports have demonstrated that mixed surfactant is capable of producing better hydrophilic-lipophilic balance (HLB), enhances the flexibility of the surfactant layer and ability to partition at high levels into the oil-water interface. In addition, it was observed that a mixture of nonionic surfactants also produced desired HLB value and provides a synergistic effect towards emulsion stability (Du et al., 2016). **Table 1** presents a list of a single (non-ionic) and complex (anionic and non-ionic) surfactants that are used in the preparation of pesticide nanoemulsions. The presence of surfactants has reduced the contact angle and increases the wettability of leaves. The optimum wetted area and contact angle may differ depending on the nature of the surfactants (Lin et al., 2016). The optimum surfactant concentration can give the desired particle size, stability, viscosity properties and antimicrobial activities that are necessary for targeted useful applications. However, at a very high concentration, an excess surfactant may cause toxic effects. For better interaction on negatively charged plant leaves, ionic liquids (ILs) have been introduced to surface-active agents for producing positive-charge nanoemulsion systems. The adsorption mechanism of leaves will adjust the properties of nanoemulsion for better applications (Zhao et al., 2017). In addition, nowadays the nanoemulsions are prepared by first mixing the surfactants with the oily phase which are able to enhance the absorption of poorly water-soluble drugs, like lipid-based nanoemulsion systems. This enhancement of aqueous solubility helps the nanoemulsion carriers in overcoming problems of drug delivery such as absorption, permeation and stability of both orthodox and herbal drugs. It is demonstrated that eco-friendly water based pesticides system, 2.5% Cyhalothrin polydopamine microcapsule suspension has the ability to prolong the retention time on the target plant leaves and reduced the toxicity to aquatic organisms (Wu et al., 2020). It is found that polymeric nanoparticles-metolachlor pesticide formulations improve the water solubility and effective drug delivery. The bioassay test shows that this nanoparticle has higher effect than non-nano forms under relatively low concentrations on *Oryza sativa* and *Digitaria* *sanguinenalis.* The cytotoxicity test involving metolachlor and metolachlor loaded nanoparticles show that the toxic effect is reduced for the latter to the preosteoblast cell line (Tong et al., 2017).

Table 1. Examples of single and mixed surfactants used in the preparation of nanoemulsions.

Type of compound Name Conditions Mean Size (nm) References	
Anionic	Polysorbate 20 (Tween-80) Mixed with organic phase, stirring < 100 (Hazrati et al., 2017) at 2400 rpm for 20 min Polysorbate 20 (Tween-80) Mixed with organic phase and stirred at and sorbitan monooleate 400 rpm for 30 min under constant < 500 (Oliveira et al.,2016) (span 80) temperature (80 °C) Polysorbate 20 Mixed with organic phase and 99-170 (Mossa et al., 2017) performed sonication process Tween-20 Mixed with oil phase at 45 °C 125-134 (Sakulku et al., 2009) Montanov-80 Mixed with aqueous phase 50-150 (Qin et al., 2019) Polyoxyethylene nonyl phenyl Mixed with organic phase 200-400 (Santos et al., 2016) ether (TX-10) Green polyethylene glycerol fatty Mixed with organic phase 41-100 (Jiang et al., 2012) acid ester or glycereth-17cocoate under stirring
Nonionic and anionic	Agnique BL 1754 (AG54) Mixed with organic phase, stirring 250 (Diaz et al., 2016) at 400 rpm for 30 min at 40 °C Palm oil-based, anionic methyl Mixed with oil phase under 1000 rpm ester sulfonate (MES)/Nonionic: for 2 h at room temperature 350-440 (Septiyanti et al., 2019) polyethylene dioleate (PDO)/ temperature (80 °C) Polyethylene glycol monooleate Polyethylene glycoldioleate Mixed with oil phase, vortexes 5-20 (Azim et al., 2016) (non-ionic) and toximol (ionic) at room temperature

2.3 Oil phase

Oils are classified either as essential oils or non-essential oils. Essential oils (EOs) are not true oils; they are concentrated and volatile compound that usually comes from different parts of plant materials and are known for their effectiveness, biodegradable and eco-friendly properties. On the other hand, non-essential oils refer to carrier oil that does not evaporate. The broad use of nanoemulsions as carriers of EOs has been widely reported as they reduce the volatilization, boost the bioactivity and reduce degradation processes such as oxidation, isomerization and polymerization (Pavoni et al., 2020). **Table 2** presents a list of dispersant compositions that have been reported for the production of pesticides nanoemulsion. A proper selection of oil as an oily phase has become one of the most crucial parts for the formation of a nanoemulsion, as it affects the solubility of the active ingredient and enabling of nanoemulsion formulation for the desired purposes. For example, when peanut oil was used as the oil phase, the formation of the nanoemulsion is more difficult compared to long-chain oil, such as hexadecane. The insolubility of oil phase in the system also increases the stability of the nanoemulsion by providing a kinetic barrier to the Ostwald ripening (Amin et al., 2019). Ostwald ripening is the net transport of oil at smaller droplets to larger droplets through continuous phase (Wooster et al., 2008). The dissolution of oil in nanoemulsion improved the cytotoxicity, genotoxicity and antimicrobial activity against pathogens as the oil constituents are rich with biologic properties. Various phytochemical compositions in essential oils will affect the biologic behaviors towards the pathogens (Caputo et al., 2020). Aniba essential oils has the ability to disrupted the cell membrane of all eight phytopathogen types including *Aspergillus flavus*, *Aspergillus Niger* and *Fusarium solani* by triggering more nucleic acid and protein productions (Souza et al., 2020). Vitex Negundo essential oil is introduced to fight against *Avena fatua* and *Echinochloa crus-galli* weeds. Asteriscus graveolens oil inhibits *Fusarium oxysporum* fungi which cause the Bayoud disease in date palm (Chibane et al., 2020). The result proves the V. Negundo EOs is toxic towards weeds by reducing their mitotic index (MI) and chromosomal aberration percentage (Issa et al., 2020). EOs is also used in controlling insects due to its volatility properties, which makes them suitable for fumigation (Nenaah et al., 2014). It is demonstrated that on increasing the jojoba oil concentration causes the rise of death in rice weevil, *Sitophilus oryzae* adults (Shetal et al., 2015). Mosquito larvae (*Culex quinquefasciatus*) mortality rate increases on addition of higher surfactants in neem oil nanoemulsions and *Citrus sinensis* nanoemulsions (Anjali et al., 2012, Azmy et al., 2019). They have found that the mortality rate of pest is dependent on surfactant-oil concentration ratio and droplet size of the nanoemulsions.

Table 2. Examples of oil used in the pesticide nanoemulsions.

Dispersion Composition	Optimal Conditions	Purpose	Reference
Carvone, cinnamaldehye, citral, geraniol, pulegone oils (5 v/v %)	Sonication at 75 kHz, 9 cycle/s for 15 min	Antibacterial activity against *Pectobacterium carotovorum* and *Ralstonia solanacearum*	(Abdelrsoul et al., 2018)
Neem oil (0.5-3w/v)	Sonication at 20 kHz, 9 cycle/min for 10 min	Antifungal activity against *Aspergillus flavus* and *Penicillium citrinum*	(Silva P et al., 2019)
Sweet basil, marjoram, peppermint, spearmint and thyme essential oils (10 v/v %)	Sonication at 700 W for 30 min	Antifungal activity against *Fusarium oxysporum*	(Hassanin et al., 2017)
Thyme essential oils (10 v/v %)	Stirring at 700 rpm for 15-60 min	Antifungal activity against *Sclerotinia sclerotiorum*	(Halawa et al., 2017)
Citronella and lemongrass oil (10 v/v %)	Stirring at 700 rpm for 30 min	Antifungal activity against *Botrytis cinera*	(Hassanin et al., 2019)
Sweet flag oil or Acorus Calamus (6 v/v %)	Stirring at 700 rpm for 2 h	Insecticidal activity against rice *weevil, Sitophilus oryzae*	(Dhivya et al., 2019)
Clove oil (1 w/v %)	Kept under moderate stirring for 10 min	Antifungal/cytotoxicity activity against *Glomerella cingulata* Ecotoxicity and genotoxicity effect against *Folsomia candida*	(Silva S et al., 2019)
Eugenol oil (3.3 v/v %)	Homogenized at room temperature	Antifungal activity against *Glomerella cingulata*	(Velho et al., 2019)
Vegetable oil and fatty acid methyl ester (1.9-7.5 w/w %)	Stirred at 200 rpm for 5 min	Herbicidal activity against *Eleusine indica*	(Jiang et al., 2012)
Clove and lemongrass oil (5 w/w %)	Stirred at 750 rpm for 30 min	Antifungal activity against *Fusarium oxysporum*	(Sharma et al., 2018)
Eucalyptus oil (6-10 w/w %)	Stirred at 750 rpm for 60 min homogenized at 5000-20,000 rpm for 2-20 min	Insecticidal activity against *Tribolium castaneum* and *Sitophilus*	(Pant et al., 2014) (Adak et al., 2020)
Rosmarinus officinalis oil (5w/w %)	The aqueous phase was added drop wise into oil phase at flow rate 3.5 mL/min, stirred at 800 rpm at 60 min	Larvicidal activity against *Aedes aegypti*	(Duarte et al., 2015)
Castor oil (1.38 w/v %)	Preformulation is performed by heating at 80 °C under stirring at 300 rpm and cooling process at room temperature. After neutralization, the final mixture was stirred at 1100 rpm for 40 min	Pesticide applications	(Qin et al., 2019)
Linseed oil (10.5 w/v %)	Preformulation undergoes two heating stages, at 180 °C and 235 °C under stirring speed at 300 rpm for 3 h. The mixture is cooled at a temperature of 150 °C and heated at 175 °C before neutralization process. The mixture was stirred at 1100 rpm for 40 min		
Cinnamon oil, thyme oil manuka oil and tea tree oil (1%-3%)	The aqueous phase is added to oil phase, drop wise with continuous stirring at 500 rpm at room temperature. The mixture is then sonicated 20 W for 1 min.	Antimicrobial activity	(Miastkowska et al., 2020)

2.4 Solvents

Different solvents are used in formulating nanoemulsion systems. The use of solvents can improved the properties of aqueous phase (viscosity, density, interfacial tension) and structural properties of surfactant solutions include optimum curvature and critical micelle concentration (Yuliani et al., 2016). In recent years, the use of green solvent is favour over conventional organic solvents to achieved more environmentally pesticide nanoformulations. Some solvents such as N, N-methyl oleate, dimethyldecanamide (AMD-10) and D-limonene are considered as safety solvents for agrochemical industry leading to low risk for farmer or customers (Solans et al., 2012).

3. PREPARATION METHODS OF PESTICIDE-BASED NANOEMULSIONS

The techniques used to prepare pesticide nanoemulsion can be either high energy or low energy emulsification method which are discuss in detail below.

3.1 Low-energy Method

The low-energy method mainly used the internal energy stored in the system components and therefore requires low energy input, used simple preparation device and hence are relatively low cost. There are many research works in which this technique was demonstrated for the preparation of nanoemulsion (Qin et al., 2019, Jiang et al., 2012, Duarte et al., 2015). This method includes phase inversion and self-emulsifying methods (Solans et al., 2012). The spontaneous or self-emulsification method is a components interaction initiated by the rapid diffusion of solvent or surfactants in the system, **Figure 1**. This method can be accomplished by the aqueous phase or oil phase titration process. The concentration of the oil phase and surfactants are changed during the preparation and ternary phase diagram based on three components: surfactant, water and oil are generated (Anwer et al., 2014). The isotropic region shown by the ternary diagram indicates various combinations of formulation. All nanoemulsion regions in the phase diagram are considered as the optimum formulation with minimum surfactant concentration. Once there are changes in surfactant curvature during the emulsifying process, the method is called the phase inversion method. The phase transitions are induced based on two factors, either temperature or composition, thus known as phase inversion temperature (PIT) and phase inversion composition (PIC) respectively. In the phase inversion composition method, a nanoemulsion is formed by changing the proportion of the different components in the system. An aqueous phase is slowly added dropwise into an organic phase containing a hydrophilic surfactant at low speed. As the water content increases, W/O nanoemulsions or reverse micelles are formed, and then, they are gradually transformed into a double-continuous structure. A further increase in water content results in phase inversion, thereby resulting in the production of O/W nanoemulsions. The phase transition temperature (PIT) method is based on a transition from oil-in-water, to a bicontinuous microemulsion, to a water-in-oil system when certain surfactant-oil-water mixtures are heated. Typically, only nonionic surfactants can be used in this approach since their head group becomes progressively dehydrated as they are heated. Usually, all components are mixed evenly, the temperature is constantly increased at low-speed stirring until the PIT is reached, and then the mixture is transferred to an ice bath for rapid cooling, which results in the formation of a nanoemulsion. Any surfactant can be applied in the PIC method compared to the PIT method, which only accessible for the surfactants which are sensitive to temperature such as polyoxyethylene-type surfactants (Ren et al., 2019). In addition, there is another modified method, called in situ phase inversion. This technique is conducted by mixing oil, surfactant and solvent simultaneously without any equipment needed. Since there is no strong forces applied, adding extra emulsifier into the aqueous phase help to obtain the small nanoemulsion size (Qin et al., 2019).

Figure 1. Formation of nanoemulsions by spontaneous emulsion method.

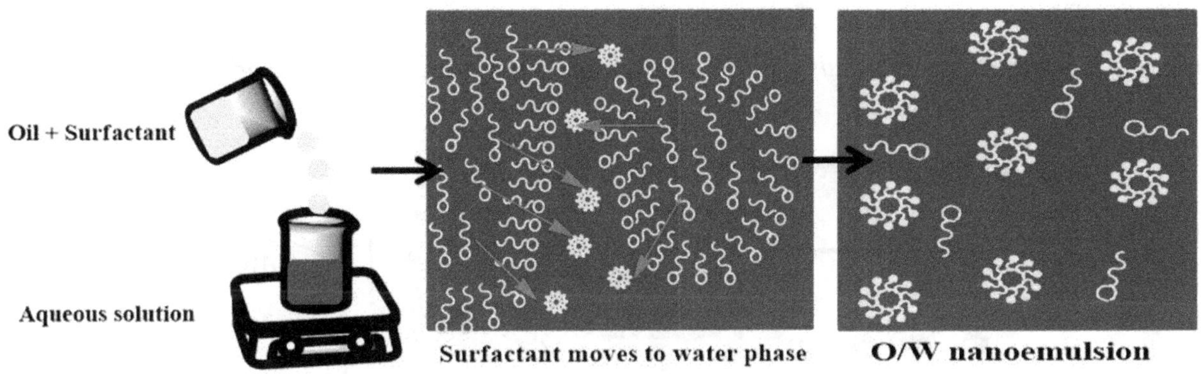

3.2 High-energy Method

This method needs a device to generate intense forces to produced smaller formulations. High shear stirring, ultrasonication and high-pressure homogenization are some techniques adopted within this method. Droplet breakup in high-energy emulsification is mainly due to the strong disruptive forces generated within mechanical homogenizers (Sivastava et al., 2006). These disruptive forces have to be large enough to overcome the Laplace pressure:

$\Delta P = \gamma/2r$

Where γ is the interface tension and r is the droplet radius. The smaller the droplet, the greater the pressure gradient required for the deformation and disruption of the droplets and therefore the greater the mechanical forces and energy expenditure needed to form the small droplets. Hence, the amount of energy input is inversely proportional to the size of the nanoemulsion. Among the high-energy methods, ultrasonication is the simplest method used in producing nanoemulsions. In this method two immiscible liquids are mixed along with surfactants and ultrasonic sound waves are applied which produced strong disruptive forces resulting in formation of cavitation. These shock waves in the surrounding liquid and the formation of liquid jets at the high speed is responsible formation of the nanoemulsions. Higher sonication time creates more kinetic energy to the emulsion and imparts the reduction of particle size (Kumari et al., 2015). Ultra Turrax homogenizer is a mechanical device used in the high-pressure method, **Figure 2**. The applied energy forces a liquid through a specified valve under high pressure to produce high-speed impact leading to droplet rupture before good dispersion is generated. High shear stirring method is referring to the shear forces produced by the velocity of droplet flow at a small gap between rotor and stator. Due to simple operation and low power consumption, the high shear stirring method becomes more favorable than the high-pressure method (Silva P et al., 2019, Hassanin et al., 2017, Halawa et al., 2017). The stability of nanoemulsion-based agrochemicals is also related to the steps involved during the initial preparation. All pre-formulation process involves two phases: organic and aqueous phase. The oil phase consists of oil and dispersant while the aqueous phase is a mixture of surfactants and water. On addition of aqueous phases to organic phases, it contributes to maximal stability of the nanoemulsion (Feng et al., 2016). Emulsification treatment at the final step also plays an important role in the transformation of coarse emulsion to nanoemulsion.

Figure 2. Formation of nanoemulsions by high pressure valve homogenizer method.

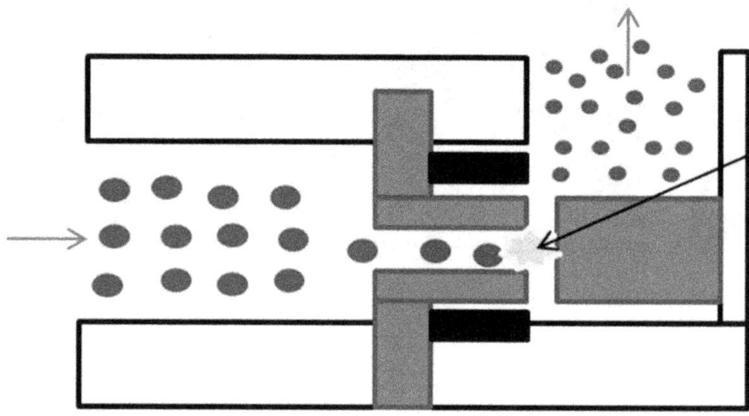

4. PALM OIL DERIVATIVES BASED PESTICIDE NANOEMULSIONS

The fruit part of the palm tree has motivated the production of palm oil. Palm oil is a versatile vegetable oil as it consists of vitamin E and balances the saturated and unsaturated fatty acid components. The promising stability result is obtained when palm oil is used as oil phase in pesticide nanoemulsion. The palm kernel oil ester (PKOE) and parthenium hysterophorus crude extract (PHCE) combination has the ability to control *Diodia ocimifolia* weeds at a fairly lower concentration, 5 g L^{-1} (Zainuddin et al., 2019). Palm oil is utilized as a surfactant in nanoemulsion preparation such as methyl ester sulfonate (MES), polyethylene glycol-monooleate (PMO) or polyethylene glycol-dioleate (PDO) which is quite stable and low environmental risk. It also shows a low concentration value of residue parameter in the fungicide formulations (Septiyanti et al., 2019). A palm-based solvent such as fatty acid methyl ester (FAME) or biodiesel was found to be an eco-friendly alternative solvent due to its lower volatility, toxicity and biodegradability. This solvent is superior for pesticide nanoemulsions application as it has the ability to improve emulsifier, active ingredient solubility, viscosity and spray which is suitably applied in controlling weeds (Purkait et al., 2019) Another interesting carrier solvent from oil palm is palm methyl ester (PME). Due to its availability and cost effective nature, PME-based formulation has the potential for practical application in agricultural field. It was observed that PME can reduce the dose concentration due to its efficient delivery. The PME-based nanoemulsions have shown high efficiency as an insecticide and molluscicides (Mohsin et al., 2017, Massaguni et al., 2016). Due to the versatile properties of palm oil, its demand is raising every year. However, both the quality and yield of the palm oil is affected by basal stem rot (BSR) disease caused by the *Ganoderma boninense* fungus. It is found that *Phenazine* extract and *trichodermin* nanoemulsions have the capability to effectively inhibit the growth of the fungus by 70.74% and 80.74%, respectively (Lee et al., 2016, Hidayah et al., 2018).

5. CHARACTERIZATION OF NANOEMULSIONS

During the preparation of nanoemulsion-based agrochemicals, the samples must undergo two important steps: preformulation and parametric optimization. After the sample has passed the initial screening, the samples will then be chosen for further chemical and biologic characterizations.

5.1 Characterizations for preformulation

During the preformulation process, the value of HLB or critical micelle concentration (CMC) of surfactants is important to be obtained. These parameters are vital to determine the starting point for the optimization process. The HLB value would be an indicator of the types of nanoemulsion produced either O/W or W/O types. CMC value is a point at which self-assembly occurs and the addition of surfactant after this point is no longer reduces the surface tension. Several techniques have been used in revealing the CMC value such as tensiometer, viscometry and electrical conductivity (Wu et al., 2014, Alam et al., 2017). In addition, the fluorescence probe is also one of the convenient and precise methods in obtaining CMC value. The absorbance of solubilized pyrene in systems is measured based on the peak difference method. A graph concentration versus absorbance is plotted in which the intersection point of two line segments is referred to as CMC value (Fu et al., 2015). However, it was found that this technique is not suitable to determine a system that has low CMC value as the minimum concentration of pyrene solution must be at least 6×10^{-6} M to get good pyrene fluorescent spectra. Therefore, a new method called "diluting-concentration" has been introduced as an alternative way for a system with very low CMC value (Lu et al., 2018). During the optimizations process, a range of nanoemulsions with different parameters are developed and undergo initial screening process including centrifugation assay, freeze-thaw cycle, heating cooling test and nanoemulsion stability at room temperature, 25 °C. Nanoemulsion with maximum stability in which the phase separation does not occur along a certain period will be selected for further characterizations. This stage is very important in determining their thermodynamic stability studies

5.2 Particle Size and Its Distribution

Dynamic Light scattering (DLS) is a technique used for determining the particle size distribution of nanoemulsion. The small size of nanoemulsion is essential in achieving maximum efficiency. Samples are first diluted with deionized water before the analysis in order to prevent the multiple scattering caused by aggregation phenomenon which occurs via electrostatic interaction. During the measurement process, the polydispersity index (PDI) value indicates the system stability. The selection of PDI value less than 0.5 is acceptable for agricultural use and is considered as a good uniformity of the particle diameter. The samples with higher PDI will be discarded as they show polydisperse property which is not suitable to be characterized using the DLS measurement (Diaz et al., 2016). The size of nanoemulsion is affected by several factors. Many researchers have reported that the size of nanoemulsion is greatly affected by surfactant concentration and its packing parameters (Zeng et al., 2017). The packing order of the surfactant is highly related to the ratio of hydrophobic and hydrophilic regions. The surfactant arrangement at O/W boundary with low interfacial tension creates a bicontinuous microemulsion which in turn imparts smaller particles.

5.3 Viscosity, Zeta and PH Measurement

Electrophoretic properties or the zeta potential of nanoemulsion are measured using Zetasizer equipment. The surface properties around the particle determine the zeta potential, pH value and indicate the nanoemulsion stability. A negative zeta value induces repulsive forces that are greater than the attraction forces among droplets, thus averting the coagulation and coalescence to occur in disperse emulsion. Increasing the oil concentration in a system may contribute to the decreased stability of nanoemulsion (Heydari et al., 2020). Most pesticide nanoemulsions show acidic properties with a pH value of 5-6 (Al et al., 2019). An Ostwald viscometer has been used to measure the viscosity value of nanoemulsion. The viscosity value may be affected by the nature of surfactants, organic phase components and oil phase. Pesticide nanoemulsion produces low viscosity as it is categorized as O/W type with high water loading. However, the viscosity of nanoemulsion can be altered by surfactant concentration (Alam et al., 2017).

5.4 Morphology and Stability Study

The shape and morphology of nanoemulsions are determined by an atomic force microscope (AFM), transmission electron microscopy (TEM) and cryogenic-field emission scanning electron microscopy (Cryo-FESEM). Commonly reported shapes of pesticide nanoemulsions are spherical or core shell-like structures due to some cluster of nanomicelles that formed during the preparation process (Oman et al., 2017, Surendhiran et al., 2019). Stability tests of nanoemulsions can be conducted by varying the storage time or temperatures. These study are commonly performed at 0, 5, 10 days, but can be over 12 months with the tested temperature at 4-54 °C. During the storage time, the study is accomplished by observing the sample appearance or measuring their physicochemical properties such as zeta potential or particle size at predetermined interval times. Then the sample without any changes in their appearance likes phase's separation, creaming, flocculation, coalescence and sedimentation is considered as a stable system. The zeta potential and particle size of the nanoemulsion is measured and compared before and after the storage. The samples with maintain value are assumed in a stable condition. The study also suggested that high temperature can lead to nanoemulsion instability which may cause by the particle movement and dissolution of emulsifier into water impart to the aggregation of particles in emulsion (Guan et al., 2018). Nanoemulsion stability also influenced by other factors such as Ostwald ripening which usually occurs at the first 5-10 days after the preparation and then flocculation and coalescence will take place at a later stage. Ostwald ripening is dependent on the oil phase fraction in the nanoemulsion system. The coarsening mechanism in nanoemulsion stabilized with Brij 30 or mixture of Span 80 and Tween-80 has explained that Ostwald ripening is the main cause of coarsening phenomenon in nanoemulsion with low oil phase fraction which up to 0.05 (Koroleva et al., 2018). All factors other than Ostwald-ripening can be controlled by the appropriate surfactants and small droplet size. High curvature of nanoemulsion avoids the flocculation and coalescence to occur as Laplace pressure prevents the deformation of larger droplets. Ostwald ripening becomes the main concern as it may occur after long term storage which affects during the application. This phenomenon is associated with the conversion of small droplets into larger droplets with low curvature radius which resulting in larger particle size. However, this occurrence can be avoided by increasing the elasticity of droplet and the addition of surfactant which reduces interfacial free energy forming a mechanical barrier against coalescence (Petal et al., 2012).

5.5 Retention and Contact Angle Measurement

Retention and contact angle of leaves are measured to relate the affinity of the pesticide liquid towards the leaf surfaces. The effectiveness of pesticide nanoemulsion can be improved by increasing the adhesion work of nanoemulsion towards the leaves. Retention can be measured by dipping and micro weighing method while contact angle of leaves is assessed using a precision contact angle measuring instrument equipped with a charged coupled device, CCD camera. It is to be noted that the contact angle of nanoemulsion decreases as the increasing agrochemical content, showing that active ingredient has low interfacial tension which effectively permitting the pesticide diffusion in the plant surface (Qian et al., 2018).

6. BIOLOGICAL STUDY

Pesticide nanoemulsions can be tested for their efficiency towards pathogenic organisms. Researchers applied different approaches in controlling and killing these pathogens-bacteria, fungi and insects, because they damage plants resulting in decreasing the quality and plantation yield. The commonly used pesticide nanoemulsion method for antibacterial and antifungal studies is broth microdilution and agar dilution assay. It is found that preparation of cinnamon essential oil (CEO) nanoemulsion has greater inhibition zone against *Aspergillus Niger, Rhizopus arrhizus, Penicillium sp.* and *Colletotrichum gloeosporioides* when compared to CEO coarse emulsion (Pongsumpun et al., 2020). A peppermint oil-based pesticides nanoemulsion has been reported to reduce biomass of *Alternaria solani* which causes early disease in tomato plants (Pandey et al., 2020). The nanoemulsion with neem and citronella oil composition has decreased the effective dose (ED_{50}) towards *Rhizoctonia solani* and *Sclerotium rolfsii* (Omen et al., 2017). Garlic oil nanoemulsion is highly efficient for destroying proteins in *Penicillin italicum* with lower minimum inhibitory concentration, MIC value of 0.23%, lower than pure garlic oil with value up to 3.7% (Long et al., 2020).

Researchers have developed nanoemulsion containing a microbe for treating crop disease. On incorporation of biologic fungus, *Talaromyces flavus* in nanoemulsion has shown to inhibit *Fusarium oxysporium* species (Naraghi et al., 2020). Encapsulated *Trichoderma* species in nanoemulsion have the ability to control downy mildew diseases (Nandini et al., 2020). The resistance towards microbial diseases such as *Erwinia carotovora* bacteria, *Aspergillus Niger*, *Rhizopus stolonifer fungus* and *Colletotrichum* fungi was observed on addition of polymer in the nanoemulsion (Marei et al., 2018, Solans et al., 2012). In addition, the chitosan nanoemulsion show better results with lower EC_{50} in combating these plant pathogens. Some nanoemulsions were also used for controlling unwanted plants such as weed and grass, as these plants compete with the crops for space and essential nutrients. Since these plants are smaller, it is more susceptible than the larger plants. The inhibition by *Thymus capitatus* and *Majorana hortensis* nanoemulsion towards bindweeds have shown that the nanoemulsion has suppressed the weed growth by reducing more than 50% of the fresh and dry weight of both *Convolvulus arvensis* and *Setaria viridis* than the control (Balah et al., 2016). Palm oil based nanoemulsion used as a pretreatment of slender buttonweed, *Diodia ocimifolia* seeds also indicates that it has toxicity effect towards germination and capable to inhibit at the lowest concentration of 5 g L^{-1} (Zainuddin et al., 2019).

Recent reports have explored the toxicity effect enhancement of pesticide nanoemulsions towards the targeted pests. It is found that *Mentha longifolia* nanoemulsion display slow release of essential oil and longer contact toxicity against *Ephestia kuehniella larva* (Louni et al., 2018). On application of Span 80 and Arabic gum as surfactants in *Tasmanian* blue gum essential oil nanoemulsion has increased toxicity and ovicidal effect towards *Callosobruchus maculatus* adults (Ya et al., 2020). Garlic essential oil nanoemulsion that uses water as a solvent has shown good acaricidal activity towards *eriophyid olive mites* with no toxicity effects towards the albumin and total protein in rats (Mossa et al., 2020). Other investigations have shown the evidence of morphologic and histological damages of insects after treating with nanoemulsions. They have also examined the competence of nanoemulsions to enter the insect cuticle and cause serious injury and irritations to the insect pests. SEM investigation has shown that the *Pimpinella anisum* essential oil nanoemulsions have caused necrosis and blackening on *Tribolium castaneum* (Omen et al., 2017). The cuticle of the insect was severely damaged; resulting in no difference was seen between the exocuticle and endocuticle. The regenerative cells also recorded fewer compared with the control, leading to dilation of internal structure which causes death. A mixture containing soybean oil, tri-n-butyl phosphate and Triton X-100, BCTP nanoemulsion can treat *Bacillus* spore in rats by 98%. After exposure to nanoemulsion, inflammation and inflation of cellular structure in the rat have been observed better results than the untreated ones, showing serious tissue necrosis (Hamouda et al., 1999).

Nanoemulsions were also assessed as growth promoters, as it may have significant effects towards the plant seedling on promoting growth or seed treatment. It was found that the nanoemulsion derived from *thymol* essential oil and *saponin* was able to suppress bacterial growth promoting soybean plant growth (Kumari et al., 2018). Eugenol oil nanoemulsion was found to enhance the germination process and produced high resistance toward the *fusarium wilt* disease on the cottonseed plant (Abd-Elsalam et al., 2015). Methylcellulose nanoemulsion has the ability to improve the growth of maize seedling by increasing the root and shoot length by 18% and 33%, respectively (Surendhiran et al., 2019). Better germination growth of watermelons has resulted on seed priming with turmeric nanoemulsion (Acharya et al., 2020).

Lipid nanomaterials are promising colloidal carriers for controlled drug delivery systems due to non-toxicity, good physiochemical storage stability, high loading capacity and target-orientated release profile. Nanostructured lipid carriers (NLC) and solid lipid nanocarriers (SLC) are commonly used for nanoemulsion formulations. The lipid matrix of nanocarriers has a significant effect on nanoemulsion to penetrate into plant roots and transportation to the stem to target pests entering the plant system. Chitosan coated beeswax NLC was found to have high drug encapsulation efficacy and good protection of deltamethrin against UV photodegradation (Zhang et al., 2013). Three different lipid nanocarrier formulations are fabricated include lipid-based nanoemulsions through a combination of hot homogenization and sonication methods with corn oil (liquid-lipid) and beeswax (solid-lipid). The obtained composite nanoemulsions penetrate at faster rate into the plant roots, 1 day as compared to NLC (3 days) and SLC (6 days) (Pan et al., 2016). In addition, the NLC have no any phytotoxic effects during germination and seeding development in the case of *Zea mays*, *Pisum satium*, and *Brassica rapa* (Nakasato et al., 2017). The encapsulation of *Eucalyptus citriodora* oil into nanoliposomes increases antimicrobial activity against *Staphylococcus aureus*. The preparation of nanolipid carriers by mixing corn oil and beeswax offered high photo-protective encapsulation efficiency (83%) and higher photo-protection efficiency after 12 h exposure of UV light for deltamethrin (Lin et al., 2015).

7. IN VITRO AND IN VIVO APPLICATION OF PESTICIDES NANOEMULSION

Oil in water nanoemulsion was used as a carrier to deliver agrochemicals for agricultural applications. The development of nanoemulsion-based agrochemicals has become a method to overcome the water-insoluble pesticides (Wang et al., 2007). Loading of *Avermectin* and *Cyhalothrin* to biodegradable castor oil polyurethane has been designed and their release property from the carrier has been studied. It has been shown that both acid and alkali medium accelerates the Avermectin release to almost 100% compared to only 72% for the control at the neutral condition (Qian et al., 2018). It was also been found that the release rate of *Cyhalothrin* nanoemulsion is much slower (180 h) compared to commercial product (60 h) (Qin et al., 2017). *Emamectin benzoate* has been synthesized using polymeric nanocapsules and Nanosilicas were used as the carrier. The study demonstrated that a maximum cumulative release of 51% was found in *Emamectin benzoate* loaded with silicon dioxide nanoparticles (Shoaid et al., 2018). The deltamethrin coated using double layers composed of chitosan and lignosulfonate materials absorb UV radiation which reduced the effect of photosensitizers towards *deltamethrin* by half and the thickness of the layer induced slow release of *deltamethrin* from the system (Ngugen et al., 2013). The generation of nanoemulsion-based agrochemicals has enhanced the biologic properties by improving bioavailability and cell uptake of the agrochemicals. Better absorption of active ingredient has created efficient diffusion through the plant cuticle and produce good wettability. A study has reported that *Avermectin* nanoemulsion has increased the spreading ability of cabbage and cucumber leaves with the value of contact angle of 46-58° compared to 79-106° for water (Guan et al. 2018). This innovation has boosted the lipophilicity properties allowing successful impregnation of nanoemulsions through the insect cuticles. This phenomenon causes a deep effect on many parts of body insects thus increase the efficient uptake of agrochemical as an active ingredient. This penetration will disrupt the wax cuticular layer on the insect, impart to rapid water loss and finally causes mortality of insects. *Bifenthrin* nanoemulsions have also increased the mortality rate of cabbage maggots to 100% after 20 h of exposure time (Yan et al., 2019). A microemulsion based on *norcantharidin* and *Emamectin benzoate* has been formulated to control diamondback moth (Shao et al., 2018, Feng et al., 2016). The incorporation of herbicides in nanoemulsion systems is likely to enhance the penetration of herbicides into plant tissue, thus facilitating good weed control with low herbicide dosage. Glyphosate isopropylamine (IPA) based herbicide nanoemulsion was found to promote higher herbicidal efficacy towards weed such as *D. ocimifolia, P. conjugatum, A. gangetica* and *Eleusine indica* (Louni et al., 2018, Lim et al., 2013, Jiang et al., 2012). Fungicide-based nanoemulsions have ability of reaching the fungal cell membrane due to their small size. The formulation of the nanoemulsion increases the water solubility of fungicides, allowing the fusion with phospholipid bilayer of microorganisms and accessing the cell membrane surface leading to breakdown of cells, followed by cell death. *Tebuconazole* nanoemulsion has been fabricated as a potential fungicide with low toxicity compared to its commercial counterparts. The nanoemulsions containing *Mancozeb* have better efficiency on antifungal activity towards *Glomerella cingulate* (Diaz et al., 2016), Velho et al., 2019).

8. ENVIRONMENTAL RISK ASSESSMENT OF PESTICIDES NANOEMULSION

It is highly necessary to first examine the toxicity effects of pesticides nanoemulsion towards environment before introducing them to the market for commercial purposes. The evaluation process involves four steps (i) Microbial community test. (ii) Laboratory and modelling. (iii) Biomagnification and recovery. (iv) Monitoring studies. During each step, comparative predicted environmental concentration and no effect concentration in soil and water are tested. In addition, risk assessment tool are developed by computational method such as quantitative structure-activity-property relationship. This method helps in predicting the physiochemical properties, environmental fate and degradation products (Villaverde et al., 2018).

9. FUTURE RESEARCH DIRECTIONS

Nanoemulsion-based pesticides show many advantages compared to traditional pesticides. These nanoformulations show excellent properties which is seen from the positive impacts on their biological efficiency. The suitable application of nanoformulations may trigger the release of active ingredients to the targeted area in a controlled mode. Water-soluble pesticides are desired in agricultural applications but it may increase solubility causing safety issues. This solubility property may potentially contaminate water sources. Therefore, nanoemulsion-based agrochemicals should be examined further for toxicity studies as well as risk assessment to ensure the level of solubility that may be harmful to living organisms. Most of the reports are concentrated on in vitro studies but in vivo studies are still few. The phytotoxicity studies on the germination of the seedlings should be conducted and proved before introducing in real market applications. For better applications, systematic studies are essential for field investigations using mature plants to examine the effectiveness of pesticide nanoemulsions. Studies on various plant diseases with nanoemulsion-based agrochemicals are based on their physiological parameters. In addition these, relationship of nanoemulsion capabilities in real plantation practices must be identified. Another challenge is the stability of nanoemulsion-based agrochemicals, since pesticides prone to reactive with surrounding groups. The resistance of pesticide nanoemulsions towards environmental factors such as temperature, pH and active ingredients should be investigated. Development of highly specific and selective pesticide nanoemulsions is necessary to reduce the effects on organisms. The cytotoxic and genotoxic effects of pesticide nanoemulsions towards human cells should be further investigated. These would lead to knowledge-generation and contributions which may subsequently help in discovering good nanoemulsion-based agrochemicals for future generation with safety nanopesticides.

10. CONCLUSION

This chapter discusses the physical and chemical properties of nanoemulsions and its application in pesticides formulation. Nanoemulsions are found to be effective for facilitating to develop hydrophilic active-loaded products for pesticide applications. The nanoencapsulation method improves the physicochemical properties and stability by assisting their water solubility, decreasing their volatility and defending them from the external environment. The stability and size distribution of nanoemulsion-based agrochemicals are controlled by the nature of the constituent components and their concentration for the formation of the nanoemulsion systems. It is observed that the modification of nanoemulsions to smaller size enhanced the effectiveness of the antimicrobial activity towards microbial pathogens. The antifungal or insecticidal activities are also found to be more effective in their nanoformulations form due to the smaller size favouring the components or molecules to be entrapped under optimum conditions. Hence, the nanoemulsion-pesticides system with suitable composition and size has the potential to control microorganisms, insects, weeds and other pests for agriculture and forestry.

REFERENCES

Abd El Azim, W. M., & Balah, M. A. (2016). Nanoemulsions formation from essential oil of Thymus capitatus and Majorana hortensis and their use in weed control. *Indian Journal of Weed Science*, *48*(4), 421. doi:10.5958/0974-8164.2016.00108.8

Abd-Elsalam, K. A., & Khokhlov, A. R. (2015). Eugenol oil nanoemulsion: Antifungal activity against Fusarium oxysporum f. sp. vasinfectum and phytotoxicity on cottonseeds. *Applied Nanoscience*, *5*(2), 255–265. doi:10.100713204-014-0398-y

Acharya, P., Jayaprakasha, G. K., Crosby, K. M., Jifon, J. L., & Patil, B. S. (2020). Nanoparticle-Mediated Seed Priming Improves Germination, Growth, Yield, and Quality of Watermelons (Citrullus lanatus) at multi-locations in Texas. *Scientific Reports*, *10*(1), 1–17. doi:10.103841598-020-61696-7 PMID:32193449

Adak, T., Barik, N., Patil, N. B., Govindharaj, G. P. P., Gadratagi, B. G., Annamalai, M., Mukherjee, A. K., & Rath, P. C. (2020). Nanoemulsion of eucalyptus oil: An alternative to synthetic pesticides against two major storage insects Sitophilus oryzae (L.) and Tribolium castaneum (Herbst)) of rice. *Industrial Crops and Products*, *143*, 111849. doi:10.1016/j.indcrop.2019.111849

Al, B. A., Gomah, A., & Ageba, M. E. (2019). Chemical profile, characterization and acaricidal activity of essential oils of three plant species and their nanoemulsions against Tyrophagus putrescentiae, a stored-food mite Basma. *Experimental & Applied Acarology*, *79*(3), 359–376. PMID:31679077

Alam, M. S., Siddiq, A. M., Ragupathy, R., & Mandal, A. B. (2017). Micellization and mixed micellization of cationic gemini (dimeric) surfactants and cationic conventional (monomeric) surfactants: Conductometric, dye solubilization, and surface tension studies. *Journal of Dispersion Science and Technology*, *38*(2), 280–287. doi:10.1080/01932691.2016.1163719

Amin, N., & Das, B. (2019). A Review on Formulation and Characterization of Nanoemulsion. *International Journal of Current Pharmaceutical Research*, *11*, 1–5. doi:10.22159/ijcpr.2019v11i4.34925

Anjali, C., Sharma, Y., Mukherjee, A., & Chandrasekaran, N. (2012). Neem oil nanoemulsion-a potent larvicidal agent against Culex quinquefasciatus. *Pest Management Science*, *68*(2), 158–163. doi:10.1002/ps.2233 PMID:21726037

Anwer, M. K., Jamil, S., Ibnouf, E. O., & Shakeel, F. (2014). Enhanced antibacterial effects of clove essential oil by nanoemulsion. *Journal of Oleo Science*, *63*(4), 13213. doi:10.5650/jos.ess13213 PMID:24599109

Azmy, R. M., Gohary, E., Gohary, E. E., Mahmoud, D. M., Salem, D. A. M., Abdou, M. A., & Salama, M. S. (2019). Assessment of larvicidal activity of nanoemulsion from Citrus sinensis essential oil on Culex pipiens L. (Diptera: Culicidae) Radwa. *Egyptian Journal of Aquatic Biology and Fish*, *23*, 61–67. doi:10.21608/ejabf.2019.35100

Badawi, N., Rosenbom, A. E., Jensen, A. M. D., & Sørensen, S. R. (2016). Degradation and sorption of the fungicide tebuconazole in soils from golf greens. *Environmental Pollution*, *219*, 368–378. doi:10.1016/j.envpol.2016.10.045 PMID:27814554

Balah, M., & Abd El Azim, W. (2016). Emulsions and Nanoemulsions Formation from Wild and Cultivated Thyme and Marjoram Essential Oils for Weeds Control. *Journal of Plant Protection and Pathology*, *7*(10), 641–648. doi:10.21608/jppp.2016.52095

Caputo, L., Smeriglio, A., Trombetta, D., Cornara, L., Trevena, G., Valussi, M., Fratianni, F., De Feo, V., & Nazzaro, F. (2020). Chemical Composition and Biological Activities of the Essential Oils of Leptospermum petersonii and Eucalyptus gunnii. *Frontiers in Microbiology*, *409*(11), 1–15. doi:10.3389/fmicb.2020.00409 PMID:32351456

Carvalho, F. P. (2006). Agriculture, pesticides, food security and food safety. *Environmental Science & Policy*, *9*(7), 685–692. doi:10.1016/j.envsci.2006.08.002

Chibane, E., Essarioui, A., Ouknin, M., Boumezzourh, A., & Bouyanzer, A. (2020). Antifungal activity of Asteriscus graveolens (Forssk) Less essential oil against Fusarium oxysporum f.sp albedinis, the causal agent of "Bayoud" disease on date palm. *Moroccan Journal of Chemistry*, *8*(2), 456–465.

Cooper, J., & Dobson, H. (2007). The benefits of pesticides to mankind and the environment. *Crop Protection (Guildford, Surrey)*, *26*(9), 1337–1348. doi:10.1016/j.cropro.2007.03.022

Da Silva Gündel, S., dos Reis, T. R., Copetti, P. M., Favarin, F. R., Sagrillo, M. R., da Silva, A. S., Segat, J. C., Baretta, D., & Ourique, A. F. (2019). Evaluation of cytotoxicity, genotoxicity and ecotoxicity of nanoemulsions containing Mancozeb and Eugenol. *Ecotoxicology and Environmental Safety*, *169*, 207–215. doi:10.1016/j.ecoenv.2018.11.023 PMID:30448703

Dhivya, V., Nelson, S. J., Subramanian, K. S., Edward, Y. S. J. T., Rajamani, K., Santhanakrishnan, V. P., & Sithanantham, S. (2019). Formulation of sweet flag oil (Acorus calamus) nanoemulsion by spontaneous emulsification method for the management of Sitophilus oryzae. *International Journal of Chemical Studies*, *7*(3), 2072–2076.

Díaz-Blancas, V., Medina, D. I., Padilla-Ortega, E., Bortolini-Zavala, R., Olvera-Romero, M., & Luna-Bárcenas, G. (2016). Nanoemulsion formulations of fungicide tebuconazole for agricultural applications. *Molecules (Basel, Switzerland)*, *21*(10), 1271. doi:10.3390/molecules21101271 PMID:27681721

Du, Z., Wang, C., Tai, X., Wang, G., & Liu, X. (2016). Optimization and Characterization of Biocompatible Oil-in-Water Nanoemulsion for Pesticide Delivery. *ACS Sustainable Chemistry & Engineering*, *4*(3), 983–991. doi:10.1021/acssuschemeng.5b01058

Duarte, J. L., Amado, J. R. R., Oliveira, A. E. M. F. M., Cruz, R. A. S., Ferreira, A. M., Souto, R. N. P., Falcão, D. Q., Carvalho, J. C. T., & Fernandesa, C. P. (2015). Evaluation of larvicidal activity of a nanoemulsion of rosmarinus officinalis essential oil. *Brazilian Journal of Pharmacognosy*, *25*(2), 189–192. doi:10.1016/j.bjp.2015.02.010

Feng, J., Shi, Y., Yu, Q., Sun, C., & Yang, G. (2016). Effect of emulsifying process on stability of pesticide nanoemulsions. *Colloids and Surfaces. A, Physicochemical and Engineering Aspects*, *497*, 286–292. doi:10.1016/j.colsurfa.2016.03.024

Feng, L., Cui, B., Yang, D., Wang, C., Zeng, Z., Wang, Y., Sun, C., Zhao, X., & Cui, H. (2016). Preparation and Evaluation of Emamectin Benzoate Solid Microemulsion. *Journal of Nanomaterials*, *2016*, 2016. doi:10.1155/2016/2386938

Fu, J., Cai, Z., Gong, Y., O'Reilly, S. E., Hao, X., & Zhao, D. (2015). A new technique for determining critical micelle concentrations of surfactants and oil dispersants via UV absorbance of pyrene. *Colloids and Surfaces. A, Physicochemical and Engineering Aspects*, *484*, 1–8. doi:10.1016/j.colsurfa.2015.07.039

Guan, W., Tang, L., Wang, Y., & Cui, H. (2018). Fabrication of an Effective Avermectin Nanoemulsion Using a Cleavable Succinic Ester Emulsifier. *Journal of Agricultural and Food Chemistry*, *66*(29), 7568–7576. doi:10.1021/acs.jafc.8b01388 PMID:29976065

Gutierrez, J. M., Gonzalez, C., Maestro, A., Sole, I., Pey, C. M., & Nolla, J. (2008). Nano-emulsions: New applications and optimization of their preparation. *Current Opinion in Colloid & Interface Science*, *13*(4), 245–251. doi:10.1016/j.cocis.2008.01.005

Halawa, A. E. A., & Ali, A. A. (2017). Nanoemulsion Against Sclerotinia Rot of Fennel. *Egyptian Journal of Agricultural Research*, *95*(3), 1037–1050. doi:10.21608/ejar.2017.149559

Hamouda, T., Hayes, M. M., Cao, Z., Tonda, R., Johnson, K., Wright, D. C., Brisker, J., & Baker, J. R. (1999). A Novel Surfactant Nanoemulsion with Broad-Spectrum Sporicidal Activity against Bacillus Species. *The Journal of Infectious Diseases*, *180*(6), 1939–1949. doi:10.1086/315124 PMID:10558951

Hassanin, M., Abd-El-Sayed, M., & Abdallah, M. (2017). Antifungal Activity of Some Essential Oil Emulsions and Nanoemulsions Against Fusarium Oxysporum Pathogen Affecting Cumin and Geranium Plants. *Scientific Journal of Flowers and Ornamental Plants*, *4*(3), 245–258. doi:10.21608jfop.2017.11326

Hazrati, H., Saharkhiz, M. J., Niakousari, M., & Moein, M. (2017). Natural herbicide activity of Satureja hortensis L. essential oil nanoemulsion on the seed germination and morphophysiological features of two important weed species. *Ecotoxicology and Environmental Safety*, *142*, 423–430. doi:10.1016/j.ecoenv.2017.04.041 PMID:28456128

Heydari, M., Amirjani, A., Bagheri, M., Sharifian, I., & Sabahi, Q. (2020). Eco-friendly pesticide based on peppermint oil nanoemulsion: Preparation, physicochemical properties, and its aphicidal activity against cotton aphid. *Environmental Science and Pollution Research International*, *27*(6), 6667–6679. doi:10.100711356-019-07332-y PMID:31873908

Hidayah, R. E., & Ambarsari, L. (2018). Development of Trichodermin Nanoemulsion Based on Medium Chain Triglycerides as Antifungal of Ganoderma boninense in vitro. *Current Biochemistry*, *4*(2), 12–22.

Issa, M., Chandel, S., Singh, H. P., Batish, D. R., Kohli, R. K., Yadav, S. S., & Kumari, A. (2020). Appraisal of phytotoxic, cytotoxic and genotoxic potential of essential oil of a medicinal plant Vitex negundo. *Industrial Crops and Products*, *145*, 112083. doi:10.1016/j.indcrop.2019.112083

Jesus, F. L. M., De Almeida, F. B., Duarte, J. L., Oliveira, A. E., Cruz, R. A. S., Souto, R. N. P., Ferreira, R. M. A., Kelmann, R. G., Carvalho, J. C. T., Lira-guedes, A. C., Guedes, M., Solans, C., & Fernandes, C. P. (2017). Preparation of a Nanoemulsion with Carapa guianensis Aublet (Meliaceae) Oil by a Low-Energy/Solvent-Free Method and Evaluation of Its Preliminary Residual Larvicidal Activity. *Evidence-Based Complementary and Alternative Medicine*, *2017*, 2017. doi:10.1155/2017/6756793 PMID:28798803

Jiang, C. L., Basri, M., Omar, D., Rahman, M. B. A., Salleh, A. B., Rahman, R. N. Z. R. A., & Selamat, A. (2012). Green nano-emulsion intervention for water-soluble glyphosate isopropylamine (IPA) formulations in controlling Eleusine indica. *Pesticide Biochemistry and Physiology*, *102*(1), 19–29. doi:10.1016/j.pestbp.2011.10.004

Kah, M., & Hofmann, T. (2014). Nanopesticides research: Current trends and future priorities. *Environment International*, *63*, 224–235. doi:10.1016/j.envint.2013.11.015 PMID:24333990

Koroleva, M., Nagovitsina, T., & Yurtov, E. (2018). Nanoemulsions stabilized by non-ionic surfactants: Stability and degradation mechanisms. *Physical Chemistry Chemical Physics*, *20*(15), 10369–10377. doi:10.1039/C7CP07626F PMID:29611566

Kumari, S., Kumaraswamy, R. V., Choudhary, R. C., Sharma, S. S., Pal, A., Raliya, R., Biswas, P., & Saharan, V. (2018). Thymol nanoemulsion exhibits potential antibacterial activity against bacterial pustule disease and growth promotory effect on soybean. *Scientific Reports*, *8*(1), 1–12. doi:10.103841598-018-24871-5 PMID:29703904

Lee, K. W., Omar, D., Abdan, K., & Wong, M. Y. (2016). Physiochemical characterization of nanoemulsion formulation of phenazine and their antifungal efficacy against Ganoderma boninense PER71 in vitro. *Research Journal of Pharmaceutical, Biological and Chemical Sciences*, *7*(6), 3056–3066.

Lim, C. J., Basri, M., Omar, D., Abdul Rahman, M. B., Salleh, A. B., & Raja Abdul Rahman, R. N. Z. (2013). Green nanoemulsion-laden glyphosate isopropylamine formulation in suppressing creeping foxglove (A. gangetica), slender button weed (D. ocimifolia) and buffalo grass (P. conjugatum). *Pest Management Science*, *69*(1), 104–111. doi:10.1002/ps.3371 PMID:22865686

Lin, H., Zhou, H., Xu, L., Zhu, H., & Huang, H. (2016). Effect of surfactant concentration on the spreading properties of pesticide droplets on Eucalyptus leaves. *Biosystems Engineering*, *143*, 42–49. doi:10.1016/j.biosystemseng.2016.01.003

Lin, L., Cui, H., Zhou, H., Zhang, X., Bortolini, C., Chen, M., Liu, L., & Dong, M. (2015). Nanoliposomes containing *Eucalyptus citriodora* as antibiotic with specific antimicrobial activity. *Chemical Communications (Cambridge)*, *51*(13), 2653–2655. doi:10.1039/C4CC09386K PMID:25573466

Long, Y., Huang, W., Wang, Q., & Yang, G. (2020). Green synthesis of garlic oil nanoemulsion using ultrasonication technique and its mechanism of antifungal action against Penicillium italicum. *Ultrasonics Sonochemistry*, *64*, 104970. doi:10.1016/j.ultsonch.2020.104970 PMID:32014757

Louni, M., Shakarami, J., & Negahban, M. (2018). Insecticidal efficiency of nanoemulsion containing mentha longifolia essential oil against ephestia kuehniella (Lepidoptera: Pyralidae). *Journal of Crop Protection*, *7*(2), 171–182.

Lu, Y., Yue, Z., Xie, J., Wang, W., Zhu, H., Zhang, E., & Cao, Z. (2018). Micelles with ultralow critical micelle concentration as carriers for drug delivery. *Nature Biomedical Engineering*, *2*(5), 318–325. doi:10.103841551-018-0234-x PMID:30936455

Marei, G. I. K., Rabea, E. I., & Badawy, M. E. I. (2018). Preparation and characterizations of chitosan/citral nanoemulsions and their antimicrobial activity. *Applied Food Biotechnology*, *5*(2), 69–78.

Massaguni, R., Raman, I. A., Noor, S., & Latip, H. M. (2016). Efficacy and persistence of neem in emulsion-in-water (EW) formulation against the golden apple snail. *International Journal of Scientific Engineering and Applied Science*, *2*(9), 2395–3470.

McClements, D. J. (2012). Nanoemulsions versus microemulsions: Terminology, differences, and similarities. *Soft Matter*, *8*(6), 1719–1729. doi:10.1039/C2SM06903B

Miastkowska, M., Michalczyk, A., Figacz, K., & Sikora, E. (2020). Nanoformulations as a modern form of biofungicide. *Journal of Environmental Health Science & Engineering*, *18*(1), 7–10. doi:10.100740201-020-00445-4 PMID:32399225

Mohsin, S. N. M., Raman, I. A., Hasan, Z. A. A., & Idris, Z. (2017). Palm-based Methyl Esters as Carrier Solvents in Pesticide Formulations. *Palm Oil Developments*, *66*, 32–38.

Mossa, A. T. H., Abdelfattah, N. A. H., & Mohafrash, S. M. M. (2017). Nanoemulsion of camphor (Eucalyptus globulus) essential oil, formulation, characterization and insecticidal activity against wheat weevil, Sitophilus granarius. *Asian Journal of Crop Science*, *9*(3), 50–62. doi:10.3923/ajcs.2017.50.62

Mossa, A. T. H., Afia, S. I., Mohafrash, S. M. M., & Abou-Awad, B. A. (2018). Formulation and characterization of garlic (Allium sativum L.) essential oil nanoemulsion and its acaricidal activity on eriophyid olive mites (Acari: Eriophyidae). *Environmental Science and Pollution Research International*, *25*(11), 10526–10537. doi:10.100711356-017-0752-1 PMID:29181754

Nakasato, D. Y., Pereira, A. E., Oliveira, J. L., Oliveira, H. C., & Fraceto, L. F. (2017). Evaluation of the effects of polymeric chitosan/tripolyphosphate and solid lipid nanoparticles on germination of *Zea mays, Brassica rapa* and *Pisum sativum*. *Ecotoxicology and Environmental Safety*, *142*, 369–374. doi:10.1016/j.ecoenv.2017.04.033 PMID:28437729

Nandini, B., Puttaswamy, H., Prakash, H. S., Adhikari, S., Jogaiah, S., & Nagaraja, G. (2020). Elicitation of novel trichogenic-lipid nanoemulsion signaling resistance against pearl millet downy mildew disease. *Biomolecules*, *10*(1), 25. doi:10.3390/biom10010025 PMID:31878099

Naraghi, L., & Negahban, M. (2020). Efficacy of Talaromyces Flavus coated with nanoparticles in the growth inhibitory of Fusarium Oxysporum F.SP. Cucumerinum. *3C Tecnologia, 9*, 31-45.

Nenaah, G. E. (2014). Chemical composition, toxicity and growth inhibitory activities of essential oils of three Achillea species and their nano-emulsions against Tribolium castaneum. *Industrial Crops and Products*, *53*, 252–260. doi:10.1016/j.indcrop.2013.12.042

Nguyen, H. M., Hwang, I. C., Park, J. W., & Park, H. J. (2012). Photo protection for deltamethrin using chitosan-coated beeswax solid lipid nanoparticles. *Pest Management Science*, *68*(7), 1062–1068. doi:10.1002/ps.3268 PMID:22383397

Nguyen, H. M., Hwang, I. C., Park, J. W., & Park, H. J. (2012). Enhanced payload and photo-protection for pesticides using nanostructured lipid carriers with corn oil as liquid lipid. *Journal of Microcapsules*, *29*(6), 596–604. doi:10.3109/02652048.2012.668960 PMID:22424134

Nguyen, M. H., Hwang, I. C., & Park, H. J. (2013). Enhanced photoprotection for photo-labile compounds using double-layer coated corn oil-nanoemulsions with chitosan and lignosulfonate. *Journal of Photochemical Photobiology B: Biology*, *125*, 194–201. doi:10.1016/j.jphotobiol.2013.06.009 PMID:23851422

Oliveira, A. E. M. F. M., Duarte, J. L., Amado, J. R. R., Cruz, R. A. S., Rocha, C. F., Souto, R. N. P., Ferreira, R. M. A., Santos, K., Da Conceição, E. C., De Oliveira, L. A. R., Kelecom, A., Fernandes, C. P., & Carvalho, J. C. T. (2016). Development of larvicidal nanoemulsion with pterodon emarginatus Vogel oil. *PLoS One*, *11*(1), 1–16. doi:10.1371/journal.pone.0145835 PMID:26742099

Osman, M. A. E., Shakil, N. A., Rana, V. S., Sarkar, D. J., Majumder, S., Kaushik, P., Singh, B. B., & Kumar, J. (2017). Antifungal activity of nano emulsions of neem and citronella oils against phytopathogenic fungi, Rhizoctonia solani and Sclerotium rolfsii. *Industrial Crops and Products*, *108*, 379–387. doi:10.1016/j.indcrop.2017.06.061

Pan, Y., Tikekar, R. V., & Nation, N. (2016). Distribution of a model bioactive within solid lipid nanoparticles and nanostructured lipid carriers influences its loading efficiency and oxidative stability. *International Journal of Pharmaceutics*, *511*(1), 322–330. doi:10.1016/j.ijpharm.2016.07.019 PMID:27418566

Pandey, S., Giri, V. P., Tripathi, A., Kumari, M., Narayan, S., Bhattacharya, A., Srivastava, S., & Mishra, A. (2020). Early blight disease management by herbal nanoemulsion in Solanum lycopersicum with bio-protective manner. *Industrial Crops and Products*, *150*, 112421.

Pant, M., Dubey, S., Patanjali, P. K., Naik, S. N., & Sharma, S. (2014). Insecticidal activity of eucalyptus oil nanoemulsion with karanja and jatropha aqueous filtrates. *International Biodeterioration & Biodegradation*, *91*, 119–127. doi:10.1016/j.ibiod.2013.11.019

Patel, P. R., & Joshi, J. R. (2012). An overview on nanoemulsion: A novel approach. *International Journal of Pharmaceutical Sciences and Research*, *3*(12), 4640–4650.

Pavela, R. (2013). Efficacy of naphthoquinones as insecticides against the house fly, Musca domestica L. *Industrial Crops and Products*, *43*, 745–750. doi:10.1016/j.indcrop.2012.08.025

Pavoni, L., Perinelli, D. R., Bonacucina, G., Cespi, M., & Palmieri, G. F. (2020). An overview of micro- and nanoemulsions as vehicles for essential oils: Formulation, preparation and stability. *Nanomaterials (Basel, Switzerland)*, *10*(1), 135. doi:10.3390/nano10010135 PMID:31940900

Pongsumpun, P., Iwamoto, S., & Siripatrawan, U. (2020). Response surface methodology for optimization of cinnamon essential oil nanoemulsion with improved stability and antifungal activity. *Ultrasonics Sonochemistry*, *60*, 104604. doi:10.1016/j.ultsonch.2019.05.021 PMID:31539730

Purkait, A., & Hazra, D. K. (2019). Biodiesel as a carrier for pesticide formulations: A green chemistry approach. *International Journal of Pest Management*, *66*(4), 341–350. doi:10.1080/09670874.2019.1649740

Qian, K., Zhao, H., & He, L. (2018). Preparation of fenpropathrin nanoemulsion by phase inversion temperature method and its study on performance. *International Journal of Nanoscience*, *17*(5), 1–7. doi:10.1142/S0219581X18500011

Qin, H., Zhang, H., Li, L., Zhou, X., Li, J., & Kan, C. (2017). Preparation and properties of lambda-cyhalothrin/polyurethane drug-loaded nanoemulsions. *RSC Advances*, *7*(83), 52684–52693. doi:10.1039/C7RA10640H

Qin, H., Zhou, X., Gu, D., Li, L., & Kan, C. (2019). Preparation and Characterization of a Novel Water-borne Lambda-Cyhalothrin/Alkyd Nanoemulsion. *Journal of Agricultural and Food Chemistry*, *67*(38), 10587–10594. doi:10.1021/acs.jafc.9b03681 PMID:31497957

Ren, G., Sun, Z., Wang, Z., Zheng, X., Xu, Z., & Sun, D. (2019). Nanoemulsion formation by the phase inversion temperature method using polyoxypropylene surfactants. *Journal of Colloid and Interface Science*, *540*, 177–184. doi:10.1016/j.jcis.2019.01.018 PMID:30640065

Rodham, D. K. (2000). Colloid and interface science in formulation research for crop protection products. *Current Opinion in Colloid & Interface Science*, *5*(6), 280–287. doi:10.1016/S1359-0294(00)00067-4

Rodrigues, C. R., Ferreira, A. M., Vilhena, J. C. E., Almeida, F. B., Cruz, R. A. S., Florentino, A. C., Souto, R. N. P., Carvalho, J. C. T., & Fernandes, C. P. (2014). Development of a larvicidal nanoemulsion with Copaiba (Copaifera duckei) oleoresin. *Revista Brasileira de Farmacognosia*, *24*(6), 699–705. doi:10.1016/j.bjp.2014.10.013

Sakulku, U., Nuchuchua, O., Uawongyart, N., Puttipipatkhachorn, S., Soottitantawat, A., & Ruktanonchai, U. (2009). Characterization and mosquito repellent activity of citronella oil nanoemulsion. *International Journal of Pharmaceutics*, *372*(2), 105–111. doi:10.1016/j.ijpharm.2008.12.029 PMID:19162149

Santos, J., Trujillo-Cayado, L. A., Calero, N., Alfaro, M. C., & Muñoz, J. (2016). Development of eco-friendly emulsions produced by microfluidization technique. *Journal of Industrial and Engineering Chemistry*, *36*, 90–95. doi:10.1016/j.jiec.2016.01.024

Septiyanti, M. (2019). Evaluation of Nanoemulsion Concentrate Botanical Fungicide from Neem, Citronella and Eugenol Oil Using Palm Oil Based Surfactant. *American Journal of Physics and Applications*, *7*(1), 14. doi:10.11648/j.ajpa.20190701.13

Sh, A., Abdelrazeik, A. B., & Rakha, O. M. (2015). Nanoemulsion of Jojoba Oil, Preparation, Characterization and Insecticidal Activity against Sitophilus oryzae (Coleoptera: Curculionidae) on Wheat. *International Journal of Agriculture Innovation and Research*, *4*(1), 72–75.

Shao, H., Xi, N., & Zhang, Y. (2018). Microemulsion formulation of a new biopesticide to control the diamond backmoth (Lepidoptera: Plutellidae). *Scientific Reports*, *8*(1), 1–9. doi:10.103841598-018-28626-0

Sharma, A., Sharma, N. K., Srivastava, A., Kataria, A., Dubey, S., Sharma, S., & Kundu, B. (2018). Clove and lemongrass oil based non-ionic nanoemulsion for suppressing the growth of plant pathogenic Fusarium oxysporum f.sp. lycopersici. *Industrial Crops and Products*, *123*, 353–362. doi:10.1016/j.indcrop.2018.06.077

Shoaib, A., Waqas, M., Elabasy, A., Cheng, X., Zhang, Q., & Shi, Z. (2018). Preparation and characterization of emamectin benzoate nanoformulations based on colloidal delivery systems and use in controlling: Plutella xylostella (L.) (Lepidoptera: Plutellidae). *RSC Advances*, *8*(28), 15687–15697. doi:10.1039/C8RA01913D

Silva, E. (2019). Production and efficacy of neem nanoemulsion in the control of Aspergillus flavus and Penicillium citrinum in soybean seeds. *European Journal of Plant Pathology*, *155*(4), 1105–1116. doi:10.100710658-019-01838-4

Silva, G. N., Faroni, L. R. A., Sousa, A. H., & Freitas, R. S. (2012). Bioactivity of Jatropha curcas L. to insect pests of stored products. *Journal of Stored Products Research*, *48*, 111–113. doi:10.1016/j.jspr.2011.10.009

Sivastava, V. K., Kini, G., & Rout, D. (2006). Detergency in spontaneously formed emulsion. *Journal of Colloid and Interface Science*, *304*(1), 214–221. doi:10.1016/j.jcis.2006.08.035 PMID:16987522

Solans, C., & Solé, I. (2012). Nano-emulsions: Formation by low-energy methods. *Current Opinion in Colloid & Interface Science*, *17*(5), 246–254. doi:10.1016/j.cocis.2012.07.003

Souza, D. P., Pimentel, R. B. Q., Santos, A. S., Albuquerque, P. M., Fernandes, A. V., Junior, S. D., Oliveira, J. T. A., Ramos, M. V., Rathinasabapathi, B., & Gonçalves, J. F. C. (2020). Fungicidal properties and insights on the mechanisms of the action of volatile oils from Amazonian Aniba trees. *Industrial Crops and Products*, *143*, 111914. doi:10.1016/j.indcrop.2019.111914

Surendhiran, M., Raja, K., Jerlin, R., Marimuthu, S., & Srivignesh, S. (2019). Nano Emulsion Seed Invigouration for Improved Germination and Seedling Vigour in Maize. *International Journal of Agricultural Science and Research (IJASR)*, *9*, 333–340. doi:10.24247/ijasrjun201946

Tong, Y., Wu, Y., Zhao, C., Xu, Y., Lu, J., Xiang, S., Zong, F., & Wu, X. (2017). Polymeric Nanoparticles as a Metolachlor Carrier: Water-Based Formulation for Hydrophobic Pesticides and Absorption by Plants. *Journal of Agricultural and Food Chemistry*, *65*(34), 7371–7378. doi:10.1021/acs.jafc.7b02197 PMID:28783335

Velho, M. C., de Oliveira, D. A., da Silva Gündel, S., Favarin, F. R., Santos, R. C. V., & Ourique, A. F. (2019). Nanoemulsions containing mancozeb and eugenol: Development, characterization, and antifungal activity against Glomerella cingulata. *Applied Nanoscience*, *9*(2), 233–241. doi:10.100713204-018-0903-9

Villaverde, J. J., Sevilla-Moran, B., Lopez-Goti, C., Alonso-Prados, J. L., & Sandin-Espana, P. (2018). Contributions of nano-QSAR/QSPR models for nanopesticides risk assessment within the European legislative framework. *The Science of the Total Environment*, *634*, 1530–1539. doi:10.1016/j.scitotenv.2018.04.033 PMID:29710651

Wang, L., Li, X., Zhang, G., Dong, J., & Eastoe, J. (2007). Oil-in-water nanoemulsions for pesticide formulations. *Journal of Colloid and Interface Science*, *314*(1), 230–235. doi:10.1016/j.jcis.2007.04.079 PMID:17612555

Wang, Y. Y., Liu, S. R., Zong, F. L., & Ye, J. M. (2004). Pesticide formulations name and code instructions. *Pesticide Science Admistration*, *25*, 1–5.

Weaver, M. A., Jin, X., Hoagland, R. E., & Boyette, C. D. (2009). Improved bioherbicidal efficacy by Myrothecium verrucaria via spray adjuvants or herbicide mixtures. *Biological Control*, *50*(2), 150–156. doi:10.1016/j.biocontrol.2009.03.007

Wu, C., Li, N. J., Chen, K. C., & Hsu, H. F. (2014). Determination of critical micelle concentrations of ionic and nonionic surfactants based on relative viscosity measurements by capillary electrophoresis. *Research on Chemical Intermediates*, *40*(6), 2371–2379. doi:10.100711164-014-1614-9

Wu, T., Fang, X., Yang, Y., Meng, W., Yao, P., Liu, Q., Zhang, B., Liu, F., Zou, A., & Jiagao Cheng, J. (2020). Eco-friendly Water-Based λ-Cyhalothrin Polydopamine Microcapsule Suspension with High Adhesion on Leaf for Reducing Pesticides Loss. *Journal of Agricultural and Food Chemistry*, *68*(45), 12549–12557. doi:10.1021/acs.jafc.0c02245 PMID:33112140

Ya-Ali, P., Yarahmadi, F., & Mehrnia, M. A. (2020). Efficacies of Two Nano-Formulations of Tasmanian Blue Gum Essential Oil to Control Callosobruchus maculatus. *Journal of Economic Entomology*, *113*(3), 1555–1562. doi:10.1093/jee/toaa069 PMID:32307512

Yan, H., Bao, C., Chen, X., Yu, C., Kong, D., Shi, J., & Lin, Q. (2019). Preparation of biodiesel oil-in-water nanoemulsions by mixed surfactants for bifenthrin formulation. *RSC Advances*, *9*(21), 11649–11658. doi:10.1039/C9RA00591A

Yuliani, S., & Noveriza, R. (2019). Effect of Carrier Oil and Co-Solvent on the Formation of Clove Oil Nanoemulsion by Phase Inversion Technique. *IOP Conference Series. Earth and Environmental Science*, *309*(1), 012036. doi:10.1088/1755-1315/309/1/012036

Yuliani, S., & Noveriza, R. (2019). Effect of Carrier Oil and Co-Solvent on the Formation of Clove Oil Nanoemulsion by Phase Inversion Technique. *IOP Conference Series. Earth and Environmental Science*, *309*(1), 012036. doi:10.1088/1755-1315/309/1/012036

Zainuddin, N. J., Ashari, S. E., Salim, N., Asib, N., Omar, D., & Lian, G. E. C. (2019). Optimization and characterization of palm oil-based nanoemulsion loaded with parthenium hysterophorus crude extract for natural herbicide formulation. *Journal of Oleo Science*, *68*(8), 747–757. doi:10.5650/jos.ess18209 PMID:31292338

Zeng, L., Xin, X., & Zhang, Y. (2017). Development and characterization of promising Cremophor EL stabilized o/w nanoemulsions containing short-chain alcohols as a cosurfactant. *RSC Advances*, *7*(32), 19815–19827. doi:10.1039/C6RA27096D

Zhang, C., Guanming, S., Shen, J., & Hu, R. F. (2015). Productivity effect and overuse of pesticide in crop production in China. *Journal of Integrative Agriculture*, *14*(9), 1903–1910. doi:10.1016/S2095-3119(15)61056-5

Zhao, X., Zhu, Y., Zhang, C., Lei, J., Ma, Y., & Du, F. (2017). Positive charge pesticide nanoemulsions prepared by the phase inversion composition method with ionic liquids. *RSC Advances*, *7*(77), 48586–48596. doi:10.1039/C7RA08653A

Zheng, M., Falkeborg, M., Zheng, Y., Yang, T., & Xu, X. (2013). Formulation an characterization of nanostructured lipid carriers containing a mixed lipids core. *Colloids and Surfaces. A, Physicochemical and Engineering Aspects*, *430*, 76–84. doi:10.1016/j.colsurfa.2013.03.070

ADDITIONAL READING

Carvalho, F. P. (2006). Agriculture, pesticides, food security and food safety. *Environmental Science & Policy*, *9*(7), 685–692. doi:10.1016/j.envsci.2006.08.002

Cooper, J., & Dobson, H. (2007). The benefits of pesticides to mankind and the environment. *Crop Protection (Guildford, Surrey)*, *26*(9), 1337–1348. doi:10.1016/j.cropro.2007.03.022

Gutierrez, J. M., Gonzalez, C., Maestro, A., Sole, I., Pey, C. M., & Nolla, J. (2008). Nano-emulsions: New applications and optimization of their preparation. *Current Opinion in Colloid & Interface Science*, *13*(4), 245–251. doi:10.1016/j.cocis.2008.01.005

Kah, M., & Hofmann, T. (2014). Nanopesticides research: Current trends and future priorities. *Environment International*, *63*, 224–235. doi:10.1016/j.envint.2013.11.015 PMID:24333990

McClements, D. J. (2012). Nanoemulsions versus microemulsions: Terminology, differences, and similarities. *Soft Matter*, *8*(6), 1719–1729. doi:10.1039/C2SM06903B

Mustafa, I. F., & Hussein, M. Z. (2020). Synthesis and Technology of Nanoemulsion-Based Pesticide Formulation. *Nanomaterials (Basel, Switzerland)*, *10*(8), 1608. doi:10.3390/nano10081608 PMID:32824489

Naseema, A., Kovooru, L., Behera, A. K., Pramodh Kumar, K. P., & Srivastava, P. (2020). A critical review of synthesis procedures, applications and future potential of nanoemulsions. *Advances in Colloid and Interface Science*, *287*, 102318. PMID:33242713

Tadros, T., Izquierdo, R., Esquena, J., & Solans, C. (2004). Formation and stability of nano-emulsions. *Advances in Colloid and Interface Science*, *108*, 303–318. doi:10.1016/j.cis.2003.10.023 PMID:15072948

KEY TERMS AND DEFINITIONS

Critical Micelle Concentration (CMC): CMC value is a point at which self-assembly of surfactants occurs and the addition of surfactant after this point is no longer reduces the surface tension which is crucial for the formation of nanoemulsions.

Emulsifiers: Emulsifiers is important components of nanoemulsion which help in controlling its size and stability.

Hydrophilic-Lipophilic Balance (HLB): A higher and lower value of HLB surfactants favours O/W and W/O type pesticide nanoformulations respectively.

Nanoemulsion: A nanoemulsion is a colloidal dispersion that contains very small particles with size in the range, 20-200 nm which may be of the oil-in-water (O/W) or water-in-oil (W/O) types depending on whether the oil is dispersed as droplets in water or vice versa.

Oil Phase: Essential oils are concentrated and volatile compound that usually comes from different parts of plant materials which are biodegradable and eco-friendly while non-essential oils are carrier oil that does not evaporate.

Ostwald Ripening: Ostwald ripening is a physical instability in nanoemulsions that occurs due to a particle size dependence of the local solubility of the oil phase in the surrounding water phase.

Pesticides: Pesticides are synthetic or natural substances used to control microorganisms, insects, weeds, and other pests that endanger agriculture and forestry and to regulate the growth of certain plant species.

Chapter 19
Nanoemulsion Based on Mushroom Bioactive Compounds and Its Application in Food Preservation

Anjali Kosre
Pt. Ravishankar Shukla University, India

Deepali Koreti
Pt. Ravishankar Shukla University, India

Nagendra Kumar Chandrawanshi
https://orcid.org/0000-0002-9093-5365
Pt. Ravishankar Shukla University, India

Ashish Kumar
Sant Gahira Guru Vishwavidyalaya, Sarguja, India

ABSTRACT

Mushrooms belonging to basiodmycetes with their high nutritional value and biologically active compounds of medicinal importance can be developed into potential food products. They have been used as a traditional food, and their medicinal property is also appreciable all over the world. Naturally occurring active compounds such as polysaccharides, proteins, lipids, and glucans, etc. are obtained from various sources including plants, animals, bacteria, algae, and fungi. The efficiency of naturally derived compounds in food industry, as well as factors influencing its effectiveness, has been reported by researchers. Mushrooms produce a diversity of biologically active compounds such as proteoglucans, polysaccharides, phenolic compounds, lectins, steroids β-glucan, chitosan, and terpenoids, etc. The bioactive compounds and their concentration differ from species to species. Thus, these bioactives can be effectively used in the fabrication of fungal (mushroom)-derived nanoemulsions applicable for the food industry.

DOI: 10.4018/978-1-7998-8378-4.ch019

INTRODUCTION

Nanotechnology is the rising technology having enormous potential in every segment from machine to medicine along with the food industry. This technology deals with the manipulation of matter on atomic and molecular level forming it into a nanoparticle. The dimension of nanoparticle in nanometer mostly lies between 1 to 100 nm (Chellaram et al., 2014). It implies with the materials of nanoscale and is capable to create novel devices and techniques. Nanoparticles are supposed to be the small entity that performs functions as a whole unit having unique properties and instigating a new level of activity (Ozimek et al., 2010). Nanoparticles being composed of enhanced surface area and mass transfer rates supposed to hold a greater catalytic behavior, quantum properties that shows chemical, biological activity, enzymatic reactivity and penetrability as compared to the large particles (with same composition) (Avella et al., 2007). The classifications of nanomaterials are based on their structures, size, and characteristics. Nanomaterials with high surface dimensions ratio is capable of exhibiting various exceptional physiochemical characteristics such as diffusivity, solubility, color, bioavailability, strength, in toxicity, optics, magnetism, and thermodynamics (Avella et al., 2005).

NANOTECHNOLOGY IN FOOD PRESERVATION

Nanotechnology has been growing rapidly in global market with the potential application in various sectors such as healthcare, food industry, energy, cosmetics, agriculture and defense. The growth rate of nanotechnology in global market has been estimated to attain 17% for the estimated upcoming year 2017-2024 (Research & Market 2018). The application of nanotechnology in food sector has been principally focused on encapsulation, emulsion, and packaging technologies. Among various nanotechnological food applications, nanoemulsion has found a huge market and application in food industry (Dasgupta, Ranjan & Gandhi, 2019).

In recent world of science and technology, the food market necessitates such techniques, which can be able to manufacture an authentic, convenient, safe and appetizing food product in order to maintain the leadership in food processing and marketing sector. For which, nanotechnology is such a known technique that acquired extensive application in food processing. Nanoparticles are frequently used in the form of food additives thereby shielding the food from contamination, enhancing the taste and lifespan as well. Nanoparticles food additives in form of preservatives, flavoring agent, antimicrobial sensors, encapsulated food components and packaging substances etc. are used to manipulate the nutrient composition and to improve the texture, flavor and its shelf life of food product (Bajpai, Chand, & Chaurasia, 2012). Nanotechnology offers numerous opportunities for the improvement of food products with its expanded applications in food systems such as nutraceuticals, bioactives, functional foods, pharma foods, and so on (Samal et al., 2017). Also it is also used to detect food pathogens to ensure food quality and safety indicators (Bott, Stormer, & Franz, 2014).

In food preservation technology, nanoencapsulation of food components (nano-sized), can be performed to enhance the nutritional value in form of supplements (e.g., proteins vitamins and antioxidants) and additives (e.g., flavor and color) in the form of nanocapsules that can be incorporated in functional foods. This provides diverse taste and off-flavor masking, controlled release, protective barriers, improved bioavailability of many vitamins and their precursors with better delivery and dispensability for water-insoluble ingredients solubilized by a nanoparticle formulation in the form of nanoemulsion (Berekaa, 2015; Prakash et al., 2013) .

Nanomaterials are microstructures made of nano scale building blocks or structures and dimension ranges within a few nanometers (typically 1–10) and the size lies between nanodimension and microdimension that can be developed into various forms (Pathakoti, Manubolu, & Hwang, 2017). Nanomaterials mostly consist of nanorods, nanoparticles, nanowires, and so on and basically form layered films, wire, and atomic structures. NSMs are categorized into three categories based on structural element dimension as (1) zero dimensional for nanoparticles, (2) one dimensional for nanorods or nanotubes, and (3) two dimensional for thin films and three dimensional for nanocomposites and dendrimers. Soft and hard templates are used to form the micro-level structures and nanoparticles are their building blocks in which complicated elementary structures are used to make nanocomposites. The properties of such nanocomposites are dependent on the size and nature of the nanostructure. Food-grade ingredients are used to fabricate nanostructures by means of simple layer by layer techniques with economical approach. Physical, chemical, and biological properties of nanoparticles are distinctive that requires the understanding of biological and physical happening in food system (Pathakoti et al., 2017). A number of methods such as homogenization, milling, ultrasound emulsification, and microfluidization are being used for nanoparticle generation (Degant & Schwechten, 2002). Among which homogenization is an oldest technique used for reducing the fat globules in milk for stability of emulsions (Thiebaud et al., 2003) whereas high-pressure homogenization is more proficient in producing extremely fine emulsions by subjecting to high stress (Swientek, 1990). Texture are enhanced during microfluidization along with size reduction and emulsification (Degant & Schwechten, 2002) and used for production of yoghurts, creams, syrups, malted drinks, icing and fillings, flavored oils, salad dressing (Kentish et al., 2008). Recently, high-intensity ultrasound technique is used for nanoemulsification through cavitation (Li & Chiang, 2012; Singh, 2016).

NANOEMULSIONS

The diffusion of two immiscible liquids composed of sphere-shaped drops results in the formation of the scattered gel phase along with the surrounding liquid forming the continuous phase is known as emulsions (McClements et al., 2007; Acosta, 2009). They are grouped as coarse microemulsions and nanoemulsions on the basis of their droplet size and stability (Komaiko & McClements, 2016).The water and oil medium is widespread medium for the formation of emulsion. The oil mixed in water (aqueous phase) is known as oil-in-water emulsions. This kind of emulsion system is suitable for hydrophobic substances delivery. As well water blended in oil is called the water-in-oil emulsions suitable for the hydrophilic compounds delivery. As well as oil/water/oil and water/oil/water emulsions can also be formulated as multiple emulsion systems. The oil-in-water-in-oil are made of water containing oil are diffused in an oil phase while in water/oil/water large oil droplets, water containing in an aqueous phase (Garti & Benichou, 2004; Weiss et al., 2006).

A heterogeneous system contains at least two non miscible liquids, dispersed in each other in small droplets size from 10 to 1000 nm range forms a nanoemulsion. An ideal nanoemulsion formulation contains an aqueous and oil phase blended with emulsifier. The aqueous phase is generally contains water, which may be also be mixed with other co-solvents including alcohols, minerals, carbohydrates, proteins, bases, acid and other polar solvents. The oil phase can be obtained by formulation of a range of non polar solvents such as acylglycerols (di/mono), triacylglycerols, essential oils, free fatty acids, lipids, waxes, vitamins and lipophilic compounds (McClements, 2011). It can act as delivery systems for functionally essential components like antimicrobials, drugs, antioxidants, flavours and nutraceuticals etc. (Salvia-Trujillo et al., 2017). Nanoemulsion having nanosized droplets, it shows a number of benefits for example they remarkably shows stablility to gravitational separation due to Brownian motion effect dominating the gravitational force. The bioavailability of the encapsulated components can be improved in nanoemulsions owing to small particle size and high surface-to-volume ratio. (Salvia-Trujillo et al., 2015a). It can be manufactured using two different approaches as high and low energy methods. The high-energy techniques exerts extreme disruptive forces due to which macro droplets splits up into micro droplets forming an immiscible fluid via mechanical appliances like fluidisers, high-pressure homogenizers and sonicators (Hasan et al., 2019). On the other hand, the low energy methods are based upon the spontaneous formulation of small droplets in mixed oil–water emulsification methods according to its environmental condition and type of solution involving spontaneous and phase inversion method (Chu et al., 2007). Subsequent to formulation of nanoemulsion, it have different features such as stability, droplet range, composition and texture etc. characterized using techniques for example dynamic and X-ray diffraction or infrared spectroscopy, multiple light scattering technique, differential scanning calorimetry, etc. (Hasan et al., 2019).

Composition of Nanoemulsions

The formulation of nanoemulsion requires two or more immiscible liquids (oil/water) and an emulsifying agent, forming the both phases as dispersed and aqueous respectively. The nanoemulsion formulation contains a core-shell structure of each droplet such as in an oil based formulation an amphipathic covering consists of active surface particles, while the lipid loving core is enclosed by non-polar molecules. Oil phase is made up of triacylglycerols and free fatty acids in a nanoemulsion formulation (Tadros et al., 2004). In contrast, lipid, minerals, waxes, non polar essential oils, fat soluble vitamins and other substances may also be employed in oil phase. The characteristics properties which may influence the stability of oil based nanoemulsion are its viscous nature, refractive index, density and tension between two phases (Wooster et al., 2008). The triacylglycerols polymers are advantageous for the preparation of nanoemulsion due to its ease of availability, cost efficient and nutritional aspects (Witthayapanyanon et al., 2006). The aqueous phase contains polar solvent with a co-solvent after the evaluation of the interfacial tension, phase activities and ionic strength of nano formulation. Generally, water is used due to its polar nature but carbohydrates, protein, polyols and alcohol can also be used as co-solvents (Saxena et al., 2017). Both phases can be fragile due to flocculation, coalescence process and gravitational separation. Addition of stabilizer agent to nano formulation may prevent this breakdown process. The stabilizer spreads over the nanodroplet and can forming a multilayer semi solid cover to nanoemulsion. Various stabilizing agents as emulsifiers, weighting agents, ripening retarders, and texture modifiers are used (McClements & Rao, 2011).

Emulsifiers are surface active molecules acts as stabilizers in nanoemulsion preparation to protect and stabilize small droplets. It reduces the interfacial tension as a result of formulation of stable nanoemulsions with nanocomposites. The usage of emulsification prevents the collision between the nanodroplets increasing the kinetic stability of nanoemulsions (Mason et al., 2006). The nature of emulsifiers can be anionic, cationic, nonionic, and zwitter ionic such as phospholipids, proteins, polysaccharides, polymers (polyvinyl alcohol) (Kralova & Sjoblom, 2009) and small-molecule surfactants etc. (Verma, Jain

& Chandrawanshi, 2020). The nature of surfactants used in nanoemulsion formulation can be ionic or non-ionic. Ionic surfactant can protect the aggregation of nanocomposites via electrostatic repulsion, while non-ionic surfactants decreases the aggregation by means of thermal fluctuation interactions, steric hindrances, and hydration (Silva et al., 2015). More than two emulsifiers may possibly be used together in formulation of nanoemulsion to check their synergistic effect (Qadir et al., 2016). To hinder the gravitational force and decreased sedimentation and creaming of nanoemulsion weighting agents are applied. The substances that shows aquaphobic action assists in hindering the ripening process are called ripening retarders (Schuch et al., 2014). Oils (mineral) and long chain triacyl glycerols are generally used as ripening inhibiting agent by preventing dispersion of small oil molecules throughout the aqueous phase to form large round molecules (Li et al., 2009).

The texture modifiers are the substances used in the formulation of nanoemulsions which interact with the aqueous phase only results in enhancing the viscosity by turning it in the form of gel. The viscosity of nanoemulsion prevents the motion of oil droplets hence giving thickening the texture of aqueous phase. The commonly used texture modifiers are proteins, gums, resins and polysaccharides etc. Recently, increased interest is seen for the usage of food-grade stabilizers to prepare nanoemulsions (Rayner et al., 2014). Pickering stabilizers show greater stability in comparison to typically used surfactant-stabilizer. This pickering particle forms a dense layer at the interface of oil and water thus preventing the droplet coalescence and flocculation via steric mechanism are the three key pre-requisites for proficient pickering stabilization. (Duffus et al., 2016). Several food grade components include flavonoids, proteins, and polysaccharides etc. have been investigated as pickering agent for nanoemulsion preparation whereas colloidal lipid particles have revealed as the most appropriate pickering stabilizers in oil/water nanoemulsions providing excellent physical stability. These are formulated effortlessly with flexible microstructure, and can be used for the nanoemulsions production at high pressure homogenization (Schroder et al., 2018).

Properties of nanoemulsions

On the basis of their droplet size nanoemulsions are optically transparent or slightly turbid. It plays important role in nanoemulsion formulation. The opacity of nanoemulsions are expressed in terms of turbidity (τ) and characterized by measurements of transmission. The distribution of particle size of nanodroplets influences its opacity (Wooster et al., 2008). The rheological properties may cause modification in food texture (Genovese et al., 2007). The evaluation of relative droplet size is important because it may have impacts on the rheological properties e.g. a beverage is a slow viscous liquid so, its nanoemulsions preparation should be incorporate with a suitable droplet size which does not alter the overall viscosity (McClements & Rao, 2011).

Physicochemical stability and Gravitational separation

The nanoemulsions is distinguished as kinetically stable systems based on its physicochemical property as it shatters down with instance as a result of destabilizing mechanisms such as gravitational separation, flocculation and chemical instability etc. Gravitational separation is because of unlike relative densities among the continuous and dispersed phases which results in sedimentation. Formation of small oil droplets and lipid crystals causes sedimentation in an oil/water nanoemulsion. In the same way, creaming takes place because of large size particle and the movement of droplets is effected by gravity in the formulation of nanoemulsion (Walstra, 2003).

Chemical degradation

Chemical degradation occurs due to redox reaction in nanoemulsions. The nanoemulsions have specific surface area makes it susceptible for chemical degradation. The opacity of the nanoemulsion also participates in a playing significant role in its chemical stability. The nanoemulsions with opaque and small droplets cause easy degradation of by UV/visible light due to transparency. Due to their relatively small particle size, droplet aggregation such as flocculation or coalescence is in smaller amount in nanoemulsions (Tadros et al., 2004). A colloidal interaction happens due to the attractive forces like van der Waals and hydrophobic interactions and repulsive interactions (electrostatic and stearic) between two neighboring droplets are related to their droplet size (McClements & Rao, 2011).

Characterizations of nanoemulsions

Nanoemulsions have several characteristics properties for instance opacity, microstructure and stability, requires the characterization for its functioning in food industry as it affects the flavor, taste, final texture, aroma and stability of foods. Structure of the particle and its distribution in nanoemulsions is determined by dynamic light scattering (DLS). This technique measures size allotment of droplets in the range of 3 nm-5μm and the intensity fluctuations which is a result of change in relative spatial location because of Brownian motion of small particles over time when exposed to light (Fryd & Mason, 2012). The size allocation of nanoemulsions is recorded ij form of single/ narrow peak in DLS technique which is appropriate for the evaluation of its stability in external conditions (Yun Zhang, 2003).

Another technique used is scattering technique such as Small-angle X-ray (SAXS) and small-angle neutron (SANS) which is functional for nanostructure characterization of shape and size of the colloidal particles (Gradzielski, 2008). It performs as a function between scattering vector and intensity obtained at point of low angles (typically 0.1–10°) while SANS receives the shorter wavelength ($\lambda < 10$Å) to determine neutron scattering length and density (Mason et al., 2006). Similarly, electron charge is evaluated to determine the zeta potential. Zeta potential is known as the electro-kinetic potential of nanoparticle. The particle charge influences the contact between two adjacent droplets disturbing the nanoemulsions stability. The elevated value for zeta potential (negative or positive) is stabilized for nanoemulsion by means of low zeta potential coagulates (Mohanraj & Chen, 2007).

BIOACTIVE COMPOUNDS IN MUSHROOM

The bioactive compounds from edible mushrooms have become an attractive functional food with related to the beneficial health effects, antioxidants in biological systems. They are a macro-fungus that has a distinctive fruiting body, have been mostly used as human food for centuries and have been famous for texture and flavors as well as having various medicinal properties Mushrooms such as *Scizophyllum commune*, *Pleurotus species* etc. are rich in different biologically active compounds such as phenolics, tocopherol, ß-carotene, polysaccharide, vitamins and secondary metabolites etc. in their fruiting bodies (Chandrawanshi et al., 2017). Some bioactive compounds are as follows:

Polysaccharides

Polysaccharides are the potential substance derived from mushrooms and are responsible for various physiological activities, like antitumor, immunomodulatory, antioxidant, antiviral, anti-inflammatory, anticarcinogenic, and neuroprotective activities (Varghese et al., 2019). Several studies have been documented that some saccharides are produced by mushrooms, such as xylose, rhamnose, arabinose, fucose, glucose, fructose, mannose, sucrose, mannitol, maltose, and trehalose showing bioactivities. Biologically active polysaccharides (glucans derivatives) produced from macrofungi exhibit various structures with different properties (Valverde et al., 2015).

Macrofungal β-glucans

β -glucans are most abundant in mushrooms, and are found mainly in the fungal cell wall. It is the most versatile compound having a wide range of biological activity. Investigative studies of β-glucans confirmed their beneficial impact on human life. These compounds are responsible for many bioactivities, such as immunomodulating, anticholesterolemic, antidiabetic, antioxidant, and neuroprotective activities, as well as lipid balance improvement, and they even have a great impact on the universal feeling of the consumer. It can bind to a membrane receptor and induce the biological responses (Kataoka et al., 2002).

Proteins

Proteins isolated from macrofungi have numerous pharmaceutical applications and possess immunomodulatory properties. Bioactive proteins have great value in terms of pharmaceutical and food potential. Such proteins include lectins, fungal immunomodulatory proteins, ribosome inactivating proteins, ribonucleases and antimicrobial proteins (Xu et al., 2011). Lectins are carbohydrate binding proteins that can aggregate anibodies and may be involved in sugar transport or carbohydrate storage in the cell. These are present in the fruiting bodies of mushroom plays important role in the life cycle of fungi (Liu,et al., 2013). Ergothioneine is an amino acid present in some mushrooms, such as *Pleurotus citrinopileatus* and *A. bisporus*. It is a thiourea derivative of histidine, containing a sulfur atom bonded to the 2-position on the imidazole ring. Various studies showed that human body is unable to synthesize this compound and its presence in the blood is mostly dependent on diet (Weigand-Heller et al., 2012).

Fats

The total amount of fat in mushroom fruiting bodies is comparatively low to other bioactive. The fats present in mushrooms are mostly unsaturated fatty acids. Mushrooms are rich in linolenic acid, which is an essential fatty acid (Heleno et al., 2015). Other lipids, such as tocopherols, are also essential fatty acids that take part in a wide range of physiological functions. They have been reported to show high antioxidant activities which help in protecting the body against degenerative malfunctions, cancer, and cardiovascular diseases (Ferreira et al., 2009). Ergosterol is a steroid precursor of vitamin D2. In human nutrition, exposure to small amounts of ultraviolet light is required for the activation of vitamin D. Among the cultivated mushrooms, the *Pleurotus* species have been shown to have relatively higher concentrations of ergosterol, with better conversion to vitamin D2 (Jasinghe et al., 2007).

Phenolic compounds

Phenolic compounds are secondary metabolites possessing an aromatic ring with one or more hydroxyl groups, and their structure can be that of a simple phenolic molecule or a complex polymer. Phenolic compounds in mushrooms are excellent antioxidants and synergists, while not being mutagenic. Phenolic compounds reported from mushrooms include protocatechuic, p-hydroxybenzoic, p-coumaric, and cinnamic acids (Heleno et al., 2015).

Vitamins

Vitamins are essential nutrients plays vital role in the human body such as metabolism, immunity, and digestion. Mushrooms are a good source of vitamins, especially of group B, namely thiamine (vitamin B1), riboflavin (vitamin B2), pyridoxine (vitamin B6), pantothenic acid (vitamin B5), nicotinic acid/niacin and its amide named nicotinamide (vitamin B3), folic acid (vitamin B9), and cobalamin (vitamin B12) (Mattila et al., 2001). Other vitamins, such as ergosterol, a precursor of vitamin D2, biotin (vitamin B8) and tocopherol (vitamin E) are also present. Mushroom species such as *Boletus edulis* have high group B content. *Pleurotus ostreatus* contains high amounts of folacine (vitamin B9), and vitamins B1 and B3 (Bernas et al., 2006). *Lentinula edodes* and *Boletus edulis* have a high content of vitamin D. The most common vitamin D in mushrooms is vitamin D2, which can also be found in vegetables and, thus, can be used as food supplements for vegetarians. Vitamin D4, chemically known as 22-dihydroergocalciferol, can also be found in some mushrooms (*agaricus*, morel, chanterelle) but in small amounts (Cardwell et al., 2018).

Chitosan

Chitin is an abundant naturally occurring biopolymer and is found in the exoskeleton of crustaceans and in fungal cell walls etc. Structurally, it is poly (β-(1–4)-2-acetamido-D-glucose), which is identical to cellulose exception is that the secondary hydroxyl group present on the second carbon atom of the hexose repeating unit is replaced by an acetamide group. Chitosan is derived from chitin by deacetylation in an alkaline media ((Maher & Abdou, 2013). Actually, chitosan is a copolymer consisting of β-(1–4)-2-acetamido-D-glucose and β-(1–4)-2-amino-D-glucose units with the latter usually exceeding ratio of 60% derived from chitin. It is explained on the basis of degree of deacetylation and average molecular weight with its significance inhabits in its antimicrobial properties in concurrence with its cationicity and film-forming characteristics. Mushroom can also be considered as an alternative non-animal source for chitosan production. It is a natural, biocompatible and biodegradable polycationic polysaccharide that possesses antimicrobial activity and filmogenic properties. Chitin and its derivative, chitosan, are abundant and renewable biopolymers found in nature (Kannan, et al., 2010). Several mushrooms genera including *Agaricus, Pleurotus, Ganoderma* and *Lentinula* have been reported as alternative chitosan sources (Yen & Mau, 2007). Chitosan-based films have been reported to show antimicrobial activity, for example chitosan nanocomposite films loaded with silver and zinc oxide nanoparticles as well as bionanocomposites based on chitosan/poly(vinylalcohol)/titanium nanoparticles and chitosan/calcium silicate nanocomposites mixed with $AgNO_3$ was evaluated to show good antimicrobial activity. Currently, most of the commercial chitosan is derived from crustaceans, such as shrimps and crabs. However, cell walls of mushroom contain high content of chitin, which might be transformed into chitosan through a deacetylation reaction and would be a potential candidate to obtain chitosan (Ospina et al., 2015).

MUSHROOM BASED COMPOUND FOR NANOEMULSION PRODUCTION

Chitosan

Cell wall of mushrooms contains chitin fibers which could potentially be employed as a non-animal source for production of chitosan (Kalac, 2013). Several studies have been described the production of chitosan from various speices of mushrooms (Bilbao-Sainz et al., 2017; Yen & Mau, 2007). Furthermore, chitosan production does not require high quality of mushroom and can also be extracted from waste of cultivated mushroom. Certain mushroom processing units, the cut-off stipe edges are not fully utilized and treated as a waste (Zhang et al., 2012). The potential of using mushroom stipe cutoff, wastage of mushroom industry as a source to extract chitosan for the preparation of active edible coatings in the form of nanoemulsion was reported. The yield and quality of chitosan obtained from both champignon stipe and cap were compared. After complete production, macrofungal chitosan from both sources was compared to a commercial animal-sourced chitosan on the basis of structure and antimicrobial activity. The potential application of chitosan to act as a food preservative from natural origin has been broadly reported on the basis of *in vitro* investigations along with through direct application on real complex matrix of food products (Maher & Abdou, 2013).

Figure 1. Mushroom bioactive compound for the formulation of nanoemusion

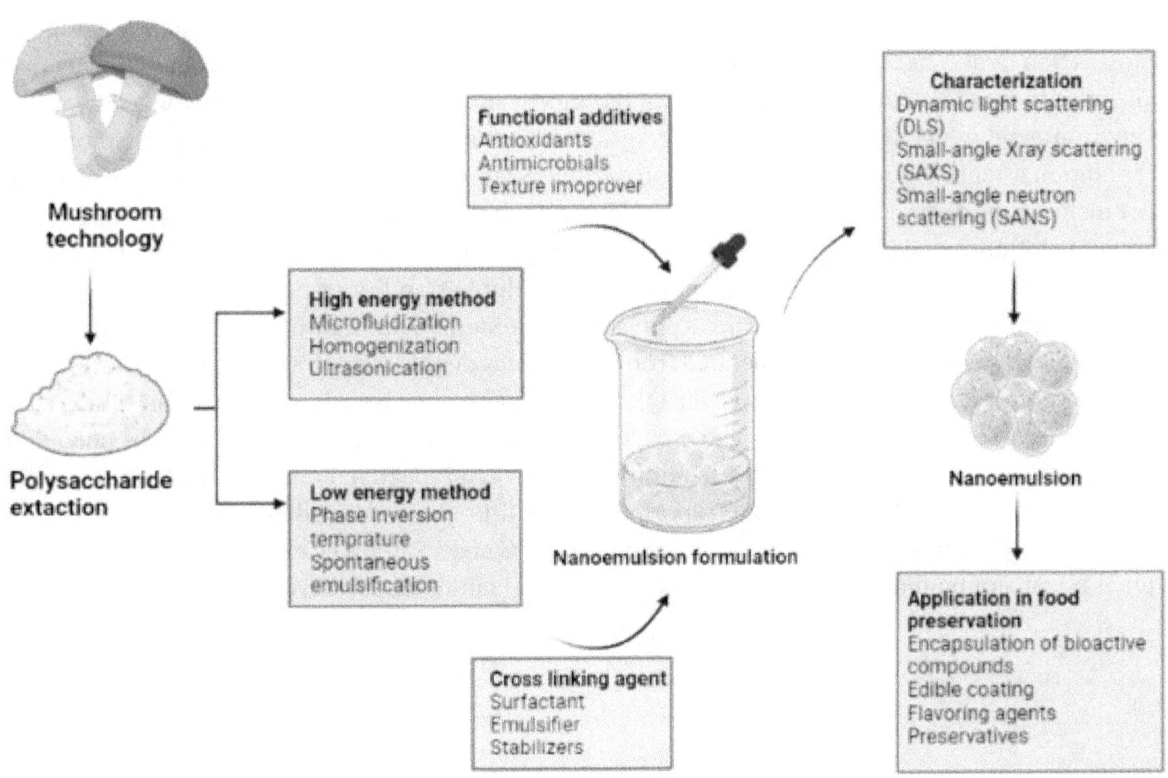

Chitosan films have a selective permeability to gasses (CO_2 and O_2) and good mechanical properties. However, the fact that chitosan films are highly permeable to water vapor limits their use as being an important drawback since an effective control of moisture transfer is a desirable property for most foods, especially in moist environments. Therefore, several strategies have been used to improve the physical properties of biopolymer based films; the functional properties of chitosan-based films can be improved by combining them with hydrocolloids and nanoemulsion technology (Vargas et al., 2009). Another method for the improvement of the physical and mechanical properties of these biofilms is by combining it with proteins (e.g. milk proteins, soy protein, collagen and gelatin) with polysaccharides (e.g. starches, alginates, cellulose and chitosan). Chitosan-gelatin blend films have been shown to be homogeneous due to the good miscibility between both biopolymers (Silva et al., 2007). The possibility of usage wastes generated from mushroom industry as a source of biopolymer chitosan to form active edible coatings in the form of nanoemulsion was studied by several researchers. It was found that the champignon stipe, an underutilized part of the mushroom, gave rise to a higher chitosan yield than caps. Fungal chitosan caused a total growth inhibition of the *Saccharomyces cerevisiae* yeast and *Escherichia coli* bacteria at concentrations of 1% and 2%, respectively. The fungal chitosan-based edible coatings were applied on fresh-cut melons and found to enhance fruit firmness, inhibit off-flavors and reduce the microbial counts. Volatiles profile showed the coated melons have a higher content of esters responsible for fruit flavor. This study demonstrates that waste from the mushroom industry can be utilized for the production of non-animal sourced chitosan to form active edible coatings in the form of nanoemulsions (Poverenov et al., 2018).

Polysaccharides

Polysaccharide is a most important bioactive compound of mushroom. Its biological activities have attracted great attention due to the immunomodulation and antitumor activities. It can protect pancreatic β-cells from STZ-induced damage and suppress human breast cancer growth through inducing autophagy apoptosis (Son et al., 2007). However, the intestinal permeability of SMP is poor, because of its macromolecular properties. Furthermore, as an oral biological substance, SMP will be damaged by gastric acid environment, which will reduce its desired effect (Du et al., 2007). It was easy separated out from water solution after being conserved for a period of time. SMP molecular was also easy to aggregate and further increase the droplet size in water solution. These phenomena result it hardly adsorbed through the intestinal wall thus reduced its bioavailability. Therefore, it is important to find an efficacious delivery system for reducing the particle size and improving the soluble stability of polysaccharides to gain the reasonable biological effects. β-D-glucan as a polysaccharide showed high antioxidant activity against free radicals or reactive oxygen species generated from the process of oxidation (Paterson, 2006). β-D-glucan isolated from *Ganoderma lucidum* was encapsulated with palm olein based nanoemulsions which act as carrier systems to enhance the delivery and bioactivity of these polysaccharides and could be potentially useful for skin care applications. The utilization of mushroom bioactive compounds as nanoemulsion demonstrated in figure 1. In addition, it is reported that the β-D-glucan loaded nano formulation demonstrated good stability below different storage conditions (4°C and 25 °C). The experiments exhibited that, the plam oliean containing β-D-glucan has shown potential antioxidants activity (Alzorqi et al., 2015).

APPLICATION OF NANOEMULSION IN FOOD PRESERVATION

The consumer's demand has risen related to the healthy nutritional diet has promoted the research in the direction of novel studies for the preservation of food products without the requirement of using preservatives. Nanoemulsion-based edible film technology is considered as an alternative option to protect quality of fruit and vegetable. Natural antioxidants (extracted from fruits, vegetables or cereals) in the form of nanoemulsions as food preservative applications have been patented. The encapsulations of natural antioxidants in the form of nanogel have been freeze dried and can be applied for the preserving minimally processed or fresh foods. It is been applied as thin, nanometer-size coating on the raw food which has been reported to prevent fluidic exchange with the external environment results in extending the shelf life food material. It has been also found to improve the sensorial quality of frozen foods on thawing (Aswathanarayan and vital, 2019). The majority of the bioactive compounds of food products are susceptible to oxidative deterioration at the time of processing and storage. The technological limitations to develop functional foods have certain disadvantages such as stability, low solubility, and bioavailability of the bioactive compounds (Shahidi & Zhong, 2010). Different bioactive compounds shows low solubility but have fast metabolism which reduces its bioavailability, whereas some of them are volatile in nature and shows sensitivity against processing methods (Jin et al., 2016). These limitations can be solved by the employment of nanoemulsions derived from mushroom such as chitosan and polysaccharides like b-glucan for the encapsulation of bioactive components to escorting its stability, bioavailability, and controlled rate of release to the food grade nanoemulsion ensuring its function in food industry (McClements et al., 2007).

Nanoencapsulation of Flavor and Coloring Agents

The substances used as flavoring and coloring agent in food contain compounds of aldehyde, ketone, and esters as functional groups which increases its susceptiblility to the degradation (oxidative and photolytic). Encapsulation of these ingredients in the form of nanogel can check these degradative effects and boosts shelf life of the food product (Goindi et al., 2016). Citral, α, β-unsaturated aldehyde is an aromatic compound having one extra double bond and used as flavoring agent in food industry but its deprivation produces off-flavor compounds. Citral based nanoemulsions enhances its stability. Oil water based citral nanoemulsion along with natural antioxidants for example β-carotene, black tea extract and tanshinone showed a higher chemical stability at time of storage. Encapsulation of antioxidants results in low production of off-flavor products for e.g. p-methylacetophenone and α, p-dimethylstyrene (Yang et al., 2011). Usage of emulsifying agents like gelatin and tween 20 enhances the stability shielding it from degradation (Tian et al., 2017). β-Carotene, an antioxidant compound but it is easily degradable by heat, light and oxidative condition and is also used in the food industry. Investigations are being conducted to show the amplified firmness of β-carotene under various condition of production. β-carotene nanoemulsion prepared by emulsification-evaporation technique had shown an increased stability with nanodroplets of 60 to 140 nm diameter range (Tan and Nakajima, 2005). β-carotene stabilized with protein nanoemulsion is formulated using method of emulsification-evaporation enclosing mean particle size of 17 nm. Sodium caseinate acts as an emulsifier at a high concentration reduces particle size with improved polydispersity of the nanodispersions (Yuan et al., 2008). Sonication-assisted with freeze drying method has been applied to for the production of β-carotene based nanoemulsions showing high water dispersion and stable chemical property (Chen et al., 2017). Carotenoids isolated in cantaloupe melon when formulated as nanoemulsion showed improved color stability and water solubility. The oil/water nanoemulsions based on carotenoid encapsulated with porcine gelatin and whey protein is determined to have an average particle size of 70-160 nm. Gelatin has ability to improve the water solubility of carotenoids. The yogurts with nanoemulsion as natural coloring agent have been investigated to show stability for 60 days (Medeiros et al., 2019).

Nanoemulsion Based Food Packaging Materials

Incorporation of nanoemulsions into films and coatings can also be applied for food packaging. The edible nanocoatings formulated using biopolymer contains the continuous phase shows monodispersion giving that stability to nanoemulsion. The increased viscosity lowers blending up of droplets with each other (Artiga-Artigas et al., 2017). The fabrication of nanoemulsion based edible films includes dispersion quality of the bioactive compound in continuous phase forming the matrix. Food grade emulsions are prepared with suitable techniques for homogenization (high or low energy). The homogenized formulation is developed into films with appropriate thickness and then dried. Afterwards, these films are characterized on the basis of properties as mechanical, structural and thermal (Otoni et al., 2014). The biopolymers used for coating also play a significant role in prolongation of the functions of dispersed nanoemulsions. Pectin biopolymer used for the fabrication of edible films via oil based nanoemulsions with enhanced antimicrobial properties with low water and vapor permeability (Sasaki et al., 2016). Cellulosic monomers along with its derivatives are used for the fabrication of edible films via clove and oregano essential oils based nanoemulsions. Chitosan, a chitin derivative has been reported for fabrication of essential oils based nanocoatings showed antimicrobial activity (Severino et al., 2015). Porcine gelatin has been used for the preparation of biofilms with enhanced antioxidant activity from oil based nanoemulsions (Alexandre et al., 2016). Green beans coated with chitosan solution containing oil based nanoemulsions (mandarin oil) irradiated with UV-C, γ-irradiation showed the synergistic effect and inhibits the growth of *Listeria innocua* via UV-C irradiation (Severino et al., 2014).

Nanoemulsion as Antimicrobial Agent

Due to its natural composition, fresh-cut fruits and vegetables are prone to microbial contamination. As these are consumed raw as a results safety problems can come across as there is a risk of infections caused by microbial contaminations (Berger et al., 2010). In view of that, research for innovative methods to retarding microbial contamination which causes spoilage of fruits/vegetables significantly required. Conventionally, spray or dipping methods in antimicrobial solution are applied to degrade microbial growth results in increasing the shelf life of product. Food antimicrobials are those chemical compounds that may hinder the growth of microbes when integrated into the food matrix (Davidson et al., 2013). Therefore, a complex mixture of nonvolatile and volatile compounds from fungal and plants as a source of essential bioactive compounds has been proved as an alternative to synthetic chemical food additives. The insertion of antimicrobials as edible coatings is being used for inhibitory action against microbes (Rojas-Grau et al., 2009). Several researches have been reported regarding the efficiency of edible-coating systems as antimicrobials for the preservation of processed food products. However, hydrophobic property (i.e. solubility, loss of activity) of antimicrobial agent makes their application challenging in food products when used at increased amount (Sanchez Gonzalez et al., 2011).Therefore, application of these natural compounds by means of nanotechnology may conquer these problems improving the efficiency of the antimicrobial agent. So, nanoemulsion could be a novel emerging way to encapsulate the naturally occurring bioactive compounds and control the release of antimicrobials (Salvia-Trujillo et al., 2015b).

Nanoemulsion as Antioxidants/Antibrowning Agent

The quality parameters are change in color and appearance of fresh-cut fruits and vegetables which should remain same before consuming. It was evaluated that the chemical changes of fresh fruit/vegetable is due to oxidation process because of the presence of polyphenol oxidase enzyme, which in combination with oxygen changes phenolic components into dark coloured pigments (Zawistowski et al., 1991). Generally this adverse colour change is protected by immersing fruits and vegetables directly in the solution of antioxidant or antibrowning agents. Ascorbic acid is the most widely used antioxidant compound capable to control the enzymatic changes in texture or colour of cut fruits and vegetables (McEvily et al., 1992). But such surface treatments may less effective and have few drawbacks. Nano coatings based on nanoemulsion technique is an efficient method for encapsulation of naturally available antioxidants like ascorbic acid and a-tocopherol (McClements et al., 2007), which may lessen the color change of food products compared to an individually applied antioxidant agent(Rojas-Grau et al., 2009). Moreover, it was demonstrated that the application of chitosan in edible coatings (nanocoating) as an efficient barricade to detain the activity of polyphenol oxidase on fruit for example apple demonstrated in figure 2. It significantly reduces the browning and preserves the appearance throughout storage (Zambrano-Zaragoza et al., 2014).

Figure 2. Application of nanoemulsion as nanocoating or edible films on food products.

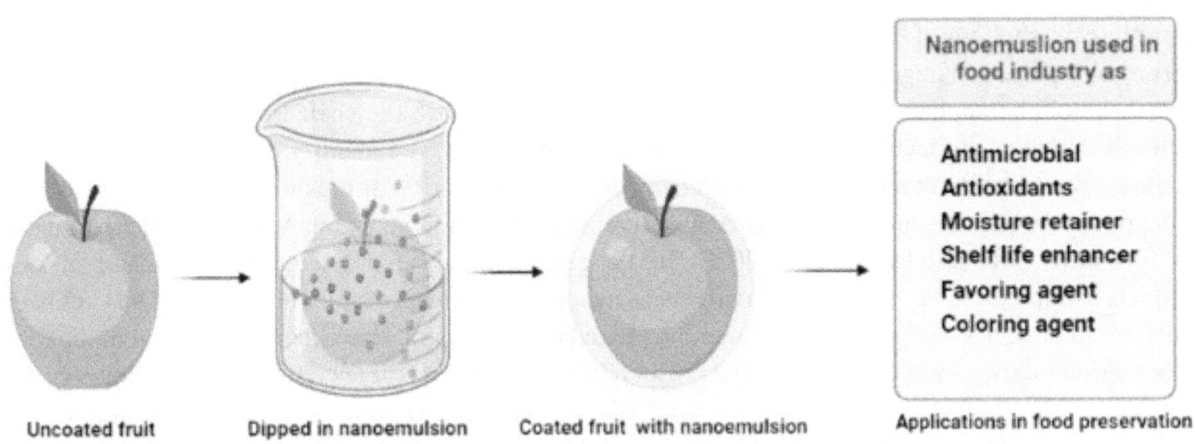

Nanoemulsion as Texture Enhancer Agent

Fresh cut fruits and vegetables losses its texture and cell wall integration at the time of storage due to the enzymatic action on it. Nanoemulsion based edible coatings reported to have potential quality as texture enhancers on fruits and vegetables such as carbohydrate polymer xanthan gum, as a texture enhancer has been evaluated for the formulation of nanoemulsion. It was showed in the results that the nanoemulsion complexed with xanthan gum effectively prevents the spoilage of fresh-cut fruit (Alandes et al., 2006). The study showed that enzymatic activity of polyphenol oxidase and pectin methylesterase was found reduced in fresh-cut guava/apples coated with nanoemulsion containing a-tocopherol and xanthan gum etc. The enzymatic activity is related to the spoilage of fruits. Higher enzymatic activity fastens the spoilage of the fruits (Garca Betanzos et al., 2017). The usage of xanthan gum based nano-emulsion modifies the metabolic process of cell wall deprivation of fruit results in the postponement of the enzymatic activity and retained freshness. However, to modify the characteristics of nanoemulsion the incorporation of different stabilizing substances known as stabilisers are required. These stabilizing agent includes pectin, carrageenan, xanthan, gelatin, alginate, whey protein, esters gum etc. and other compounds. These components significantly protects the texture by preventing moisture, maintaining fluidity and providing the mechanical strength to the food product (Zambrano-Zaragoza et al., 2014).

CHALLENGES AND FUTURE RESEARCH DIRECTION

Nanoemulsification is a technique that enhances the availability of bioactive compounds as reported by numerous *in vitro* studies. But there have been very few studies has been carried out which shows the mushroom as nanoemulsion formulation for food preservation with definite health benefits including it as nanoemulsions in foods and their consumption with public opinion. As well as studies have been carried out to evaluate the approaches applied for the formulation nanoemulsions in recent time. The main target is to optimize the parameters and elements employed for formulation of nanoemulsion. Bioactive compounds from mushroom for nanoemulsion in the form of edible coatings can be an effective method improving the shelf life of fresh fruits and vegetables. Bioactive compounds such as carbohydrates, lipids, and proteins played an important role in sustaining the quality and food preservation. The polysaccharide coatings could be applied to fresh fruits and vegetables to improve the shelf life of fruits, vegetables, shellfish, and meat products by reducing surface darkening, oxidative rancidity, and dehydration of food products. But there is not much research has been done on application of mushroom for the formulation of nano based emulsions in foods industry which requires direction investment and several equipments and further research (Al-Tayyar et al., 2020). In the same way, the risk related to nanoemulsion in food products is unpredictable. The probable affects of nanoparticles and its fate after digestion has not been clarified till now. In contrast, natural compounds based nanoemulsions have a great potential in the food preservation industry. The emulsion-based preservation system and edible coatings in food industry can advances the food functioning and also enhances the quality and shelf life. On the other hand, the food grade nanoemulsions will come across widespread identification and application only if it is economically viable and follows the safety principles of food sector. For that reason it is required for optimizing the conditions for encapsulation of components form natural entities like mushrooms and plants for scaled up production.

CONCLUSION

Nanoemulsions based on natural compounds shows an immense potential for its applications in the food preservation. The improvement of food grade nanoemulsion capable for encapsulation of the active components is an efficient method for a new generation food products. The nanoemulsion-based delivery systems along with edible films could enhance their quality, shelf life and advances the future of food. As well as development of new functional food products, nutraceuticals and pharmaceuticals products using mushroom based nano products which will be commercially feasible including the safety standards also. The formulation of nanoemulsion based on encapsulation of natural compound can be a good alternative for the food industry preventing the food degradation from temperature, light, pH, and oxidative and microbial contamination ensuring its long term storage. The application of mushroom technology along with nanoemulsion based delivery system for food preservation can be an economically feasible technology for industrial scale production of food grade nanoemulsions. Further studies should be focused on the risks associated with the usage of natural compounds from mushroom nanoemulsion and its utility in food applications and ensuring its safety for consumers. Thus it can be concluded that mushroom industry can be a potential alternative for nanoemulsion based delivery system ought to have compatibility with food matrix and have nominal impact on the flavor, appearance, texture and other organoleptic characteristics of the food products.

ACKNOWLEDGMENT

The authors are thankful to the Pt. Ravishankar Shukla University Research scholarship award (797/Fin/Sch./2021), and the Junior Research Fellowship (DBT/JRF/BET-18/I/2018/AL/123), Department of Biotechnology, Biotech Consortium of India Limited for providing funding support. The authors are also are thankful to the Head, School of Studies in Biotechnology, Pt. Ravishankar Shukla University, Raipur.

REFERENCES

Acosta, E. (2009). Bioavailability of nanoparticles in nutrient and nutraceutical delivery. *Current Opinion in Colloid & Interface Science*, *14*(1), 3–15. doi:10.1016/j.cocis.2008.01.002

Al-Tayyar, N. A., Youssef, A. M., & Al-Hindi, R. R. (2020). Edible coatings and antimicrobial nanoemulsions for enhancing shelf life and reducing foodborne pathogens of fruits and vegetables: A review. *Sustainable. Materials Technology*.

Alandes, L., Hernando, I., Quiles, A., Pérez-Munuera, I., & Lluch, M. A. (2006). Cell wall stability of fresh-cut fuji apples treated with calcium lactate. *Journal of Food Science*, *71*(9), 615–620. doi:10.1111/j.1750-3841.2006.00180.x

Alexandre, E. M. C., Lourenco, R. V., Bittante, A. M. Q. B., Moraes, I. C. F., & Sobral, P. J. A. (2016). Gelatine based films reinforced with montmorillonite and activated with nanoemulsion of ginger essential oil for food packaging applications. *Food Packaging and Shelf Life*, *10*, 87–96. doi:10.1016/j.fpsl.2016.10.004

Artiga-Artigas, M., Acevedo-Fani, A., & Martín-Belloso, O. (2017). Effect of sodium alginate incorporation procedure on the physicochemical properties of nanoemulsions. *Food Hydrocolloids*, *70*, 191–200. doi:10.1016/j.foodhyd.2017.04.006

Aswathanarayan, J. B., & Vittal, R. R. (2019). Nanoemulsions and Their Food Applications Nanoemulsions and Their Potential Applications in Food Industry. *Frontiers in Sustainable Food Systems*, *3*, 95. doi:10.3389/fsufs.2019.00095

Avella, M., Bruno, G., Errico, M. E., Gentile, G., Piciocchi, N., Sorrentino, A., & Volpe, M. G. (2007). Innovative packaging for minimally processed fruits. *Packaging Technology & Science*, *20*(5), 325–335. doi:10.1002/pts.761

Avella, M., De Vlieger, J. J., Errico, M. E., Fischer, S., Vacca, P., & Volpe, M. G. (2005). Biodegradable starch/clay nanocomposite films for food packaging applications. *Food Chemistry*, *93*(3), 467–474. doi:10.1016/j.foodchem.2004.10.024

Bajpai, S. K., Chand, N., & Chaurasia, V. (2012). Nano zinc oxide-loaded calcium alginate films with potential antibacterial properties. *Food and Bioprocess Technology*, *5*(5), 1871–1881. doi:10.100711947-011-0587-6

Berekaa, M. M. (2015). Nanotechnology in food industry; advances in food processing, packaging and food safety. *International Journal of Current Microbiology and Applied Sciences*, *4*(5), 345–357.

Berger, C. N., Sodha, S. V., Shaw, R. K., Griffin, P. M., Pink, D., Hand, P., & Frankel, G. (2010). Fresh fruit and vegetables as vehicles for the transmission of human pathogens. *Environmental Microbiology*, *12*(9), 2385–2397. doi:10.1111/j.1462-2920.2010.02297.x PMID:20636374

Bernas, E., Jaworska, G., & Lisiewska, Z. (2006). Edible mushrooms as a source of valuable nutritive constituents. *Acta Scientiarum Polonorum. Technologia Alimentaria*, *5*, 5–20.

Bilbao-Sainz, C., Chiou, B. S., Williams, T., Wood, D., Du, W. X., Sedej, I., & McHugh, T. (2017). Vitamin D-fortified chitosan films from mushroom waste. *Carbohydrate Polymers*, *167*, 97–104. doi:10.1016/j.carbpol.2017.03.010 PMID:28433182

Bott, J., Stormer, A., & Franz, R. (2014). A comprehensive study into the migration potential of nano silver particles from food contact polyolefins. In M. A. Benvenuto, S. Ahuja, T. V. Duncan, G. O. Noonan, & E. S. RobertsKirchhoff (Eds.), Chemistry of food, food supplements, and food contact materials: From production to plate (pp. 51–70). Washington, DC: American Chemical Society. doi:10.1021/bk-2014-1159.ch005

Cardwell, G., Bornman, J., James, A., & Black, L. (2018). A review of mushrooms as a potential source of dietary Vitamin D. *Nutrients*, *10*(10), 1498. doi:10.3390/nu10101498 PMID:30322118

Chandrawanshi, N. K., Tandia, D. K., & Jadhav, S. K. (2017). Nutraceutical Properties Evaluation of *Schizophyllum commune*. *Indian Journal of Scientific Research*, *13*(2), 57–62.

Chellaram, C., Murugaboopathi, G., John, A. A., Sivakumar, R., Ganesan, S., Krithika, S., & Priya, G. (2014). Significance of nanotechnology in food industry. *APCBEE Procedia*, *8*, 109–113. doi:10.1016/j.apcbee.2014.03.010

Chen, J., Li, F., Li, Z., McClements, D. J., & Xiao, H. (2017). Encapsulation of carotenoids in emulsion-based delivery systems: Enhancement of βcarotene water-dispersibility and chemical stability. *Food Hydrocolloids*, *69*, 49–55. doi:10.1016/j.foodhyd.2017.01.024

Chu, B.-S., Ichikawa, S., Kanafusa, S., & Nakajima, M. (2007). Preparation and characterization of b-carotene nanodispersions prepared by solvent displacement technique. *Journal of Agricultural and Food Chemistry*, *55*(16), 6754–6760. doi:10.1021/jf063609d PMID:17630759

Dasgupta, N., Ranjan, S., & Gandhi, M. (2019). Nanoemulsions in food: Market demand. *Environmental Chemistry Letters*, *17*(2), 1003–1009. Advance online publication. doi:10.100710311-019-00856-2

Davidson, P. M., Critzer, F. J., & Taylor, T. M. (2013). Naturally occurring antimicrobials for minimally processed foods. *Annual Review of Food Science and Technology*, *4*(1), 163–190. doi:10.1146/annurev-food-030212-182535 PMID:23244398

Du, B., Bian, Z., & Xu, B. (2014). Skin Health Promotion Effects of Natural Beta-Glucan Derived from Cereals and Microorganisms: A Review. *Phytotherapy Research*, *28*(2), 159–166. doi:10.1002/ptr.4963 PMID:23494974

Duffus, L. J., Norton, J. E., Smith, P., Norton, I. T., & Spyropoulos, F. (2016). A comparative study on the capacity of a range of food-grade particles to form stable O/W and W/O Pickering emulsions. *Journal of Colloid and Interface Science*, *473*, 9–21. doi:10.1016/j.jcis.2016.03.060 PMID:27042820

Ferreira, I., Barros, L., & Abreu, R. (2009). Antioxidants in wild mushrooms. *Current Medicinal Chemistry*, *16*(12), 1543–1560. doi:10.2174/092986709787909587 PMID:19355906

Fryd, M. M., & Mason, T. G. (2012). Advanced nanoemulsions. *Annual Review of Physical Chemistry*, *63*(1), 493–518. doi:10.1146/annurev-physchem-032210-103436 PMID:22475339

Garca-Betanzos, C. I., Hernandez-Sanchez, H., Bernal-Couoh, T. F., Quintanar-Guerrero, D., & Zambrano-Zaragoza, M. (2017). Physicochemical, total phenols and pectin methylesterase changes on quality maintenance on guava fruit (*Psidium guajava* L.) coated with candeuba wax solid lipid nanoparticles-xanthan gum. *Food Research International*, *101*, 218–227. doi:10.1016/j.foodres.2017.08.065 PMID:28941687

Garti, N., & Benichou, A. (2004). Recent developments in double emulsions for food applications. *Food Emulsion*, *132*, 353–412.

Genovese, D. B., Lozano, J. E., & Rao, M. A. (2007). The rheology of colloidal and noncolloidal food dispersions. *Journal of Food Science*, *72*(2), 11–20. doi:10.1111/j.1750-3841.2006.00253.x PMID:17995847

Goindi, S., Kaur, A., Kaur, R., Kalra, A., & Chauhan, P. (2016). Nanoemulsions: An emerging technology in the food industry. *Emulsions*, *3*, 651–688. doi:10.1016/B978-0-12-804306-6.00019-2

Hasan, S. M. K., Ferrentino, G., & Scampicchio, M. (2019). Nanoemulsion as advanced edible coatings to preserve the quality of fresh-cut fruits and vegetables: A review. *International Journal of Food Science & Technology*. Advance online publication. doi:10.1111/ijfs.14273

Heleno, S. A., Martins, A., Queiroz, M. J. R. P., & Ferreira, I. C. F. R. (2015). Bioactivity of phenolic acids: Metabolites versus parent compounds: A review. *Food Chemistry*, *173*, 501–513. doi:10.1016/j.foodchem.2014.10.057 PMID:25466052

Jasinghe, V. J., Perera, C. O., & Sablani, S. S. (2007). Kinetics of the conversion of ergosterol in edible mushrooms. *Journal of Food Engineering*, *79*(3), 864–869. doi:10.1016/j.jfoodeng.2006.01.085

Jin, W., Xu, W., Liang, H., Li, Y., Liu, S., & Li, B. (2016). Nanoemulsions for food: Properties, production, characterization, and applications. *Emulsions*, *3*, 1–36. doi:10.1016/B978-0-12-804306-6.00001-5

Kalac, P. (2013). A review of chemical composition and nutritional value of wild-growing and cultivated mushrooms. *Journal of the Science of Food and Agriculture*, *93*(2), 209–218. doi:10.1002/jsfa.5960 PMID:23172575

Kannan, M., Nesakumari, M., Rajarathinam, K., & Singh, A. (2010). Production and characterization of mushroom chitosan under solid-state fermentation conditions. *Advances in Biological Research (Faisalabad)*, *4*(1), 10–13.

Kataoka, K., Muta, T., Yamazaki, S., & Takeshige, K. (2002). Activation of macrophages by linear (1→3)-β-d-glucans. *The Journal of Biological Chemistry*, *277*(39), 36825–36831. doi:10.1074/jbc.M206756200 PMID:12140296

Kentish, S., Wooster, T. J., Ashokkumar, M., Balachandran, S., Mawson, R., & Simons, L. (2008). The use of ultrasonics for nanoemulsion preparation. *Innovative Food Science & Emerging Technologies*, *9*(2), 170–175. doi:10.1016/j.ifset.2007.07.005

Komaiko, J. S., & McClements, D. J. (2016). Formation of food-grade nanoemulsions using low-energy preparation methods: A review of available methods. *Comprehensive Reviews in Food Science and Food Safety*, *15*(2), 331–352. doi:10.1111/1541-4337.12189 PMID:33371595

Kralova, I., & Sjoblom, J. (2009). Surfactants used in food industry: A review. *Journal of Dispersion Science and Technology*, *30*(9), 1363–1383. doi:10.1080/01932690902735561

Lee, D. K., Wang, C.-F., Yang, Y., & Vaccari, D. (2015). *International Flavors and Fragrances Inc., Flavor Nanoemulsions and Methods of Preparing the Same*. US Patent PCT/US2015/056470, WO2016064883A1. Available online at: https:// patents.google.com/patent/WO2016064883A1/en

Li, P. H., & Chiang, B. H. (2012). Process optimization and stability of Dlimonene-in-water nanoemulsions prepared by ultrasonic emulsification using response surface methodology. *Ultrasonics Sonochemistry*, *19*(1), 192–197. doi:10.1016/j.ultsonch.2011.05.017 PMID:21680223

Li, X., Xing, Y., Jiang, Y., Ding, Y., & Li, W. (2009). Antimicrobial activities of ZnO powder-coated PVC film to inactivate food pathogens. *International Journal of Food Science & Technology*, *44*(11), 2161–2168. doi:10.1111/j.1365-2621.2009.02055.x

Liu, Q., Ng, T., & Wang, H. (2013). Isolation and characterization of a novel lectin from the wild mushroom *Oudemansiella radicata* (Relhan.: Fr.) sing. *Biotechnology and Bioprocess Engineering; BBE*, *18*(3), 465–471. doi:10.100712257-012-0699-5

Maher, Z. E., & Abdou, E. S. (2013). Chitosan based edible films and coatings: A review. *Materials Science and Engineering*, *33*(4), 1819–1841. doi:10.1016/j.msec.2013.01.010 PMID:23498203

Mason, T. G., Wilking, J. N., Meleson, K., Chang, C. B., & Graves, S. M. (2006). Nanoemulsions: Formation, structure, and physical properties. *Journal of Physics Condensed Matter*, *18*(41), 079001. doi:10.1088/0953-8984/18/41/R01

Mattila, P., Konko, K., Eurola, M., Pihlava, J. M., Astola, J., Vahteristo, L., Hietaniemi, V., Kumpulainen, J., Valtonen, M., & Piironen, V. (2001). Contents of vitamins, mineral elements, and some phenolic compounds in cultivated mushrooms. *Journal of Agricultural and Food Chemistry*, *49*(5), 2343–2348. doi:10.1021/jf001525d PMID:11368601

Mcclements, D. J. (2007a). Critical review of techniques and methodologies for characterization of emulsion stability. *Critical Reviews in Food Science and Nutrition*, *47*(7), 611–649. doi:10.1080/10408390701289292 PMID:17943495

McClements, D. J., Decker, E. A., & Weiss, J. (2007b). Emulsion-based delivery systems for lipophilic bioactive components. *Journal of Food Science*, *72*(8), 109–124. doi:10.1111/j.1750-3841.2007.00507.x PMID:17995616

McClements, D. J., & Rao, J. (2011). Food-grade nanoemulsions: Formulation, fabrication, properties, performance, biological fate, and potential toxicity. *Critical Reviews in Food Science and Nutrition*, *51*(4), 285–330. doi:10.1080/10408398.2011.559558 PMID:21432697

McEvily, A. J., Iyengar, R., & Otwell, W. S. (1992). Inhibition of enzymatic browning in foods and beverages. *Critical Reviews in Food Science and Nutrition*, *32*(3), 253–273. doi:10.1080/10408399209527599 PMID:1418602

Medeiros, A. K. O. C., Gomes, C. C., Amaral, M. L. Q. A., Medeiros, L. D. G., Medeiros, I., Porto, D. L., Aragao, C. F. S., Maciel, B. L. L., Morais, A. H. A., & Passos, T. S. (2019). Nanoencapsulation improved water solubility and color stability of carotenoids extracted from *Cantaloupe melon* (Cucumis melo L.). *Food Chemistry*, *270*, 562–572. doi:10.1016/j.foodchem.2018.07.099 PMID:30174087

Mohanraj, V., & Chen, Y. (2007). Nanoparticles-a review. *Tropical Journal of Pharmaceutical Research*, *5*(1), 561–573. doi:10.4314/tjpr.v5i1.14634

Ospina, N. M., Ospina Alvarez, S. P., Escobar Sierra, D. M., Rojas Vahos, D. F., Zapata Ocampo, P. A., & Ossa Orozco, C. P. (2015). Isolation of chitosan from *Ganoderma lucidum* mushroom for biomedical applications. *Journal of Materials Science. Materials in Medicine*, *26*(3), 135. doi:10.100710856-015-5461-z PMID:25716022

Otoni, C. G., de Moura, M. R., Aouada, F. A., Camilloto, G. P., Cruz, R. S., & Lorevice, M. V. (2014a). Antimicrobial and physicalmechanical properties of pectin/papaya puree/cinnamaldehyde nanoemulsion edible composite films. *Food Hydrocolloids*, *41*, 188–194. doi:10.1016/j.foodhyd.2014.04.013

Ozimek, L., Pospiech, E., & Narine, S. (2010). Nanotechnologies in food and meat processing. *Acta Scientiarum Polonorum. Technologia Alimentaria*, *9*(4), 401–412.

Paterson, R. R. M. (2006). *Ganoderma* - A therapeutic fungal biofactory. *Phytochemistry*, *67*(18), 1985–2001. doi:10.1016/j.phytochem.2006.07.004 PMID:16905165

Pathakoti, K., Manubolu, M., & Hwang, H. M. (2017). Nanostructures: Current uses and future applications in food science. *Yao Wu Shi Pin Fen Xi, 25*(2), 245–253. PMID:28911665

Poverenova, E., Ripsa, H. A., Zaitseva, Y., Bara, Y., Danay, O., Horeva, B., Bilbao-Sainz, C., McHugh, T., & Rodov, V. (2018). Potential of chitosan from mushroom waste to enhance quality and storability of fresh-cut melons. *Food Chemistry, 268*, 233–241. doi:10.1016/j.foodchem.2018.06.045 PMID:30064752

Prakash, A., Sen, S., & Dixit, R. (2013). The emerging usage and applications of nanotechnology in food processing industries. The new age of nanofood. *International Journal of Pharmaceutical Sciences and Research, 22*(1), 107–111.

Qadir, A., Faiyazuddin, M. D., Talib Hussain, M. D., Alshammari, T. M., & Shakeel, F. (2016). Critical steps and energetics involved in a successful development of a stable nanoemulsion. *Journal of Molecular Liquids, 214*, 7–18. doi:10.1016/j.molliq.2015.11.050

Rayner, M., Marku, D., Eriksson, M., Sjoo, M., Dejmek, P., & Wahlgren, M. (2014). Biomass-based particles for the formulation of Pickering type emulsions in food and topical applications. *Colloids and Surfaces. A, Physicochemical and Engineering Aspects, 458*, 48–62. doi:10.1016/j.colsurfa.2014.03.053

Research and Market. (2018). *Global nanotechnology market outlook 2024*. Report ID: 3989533. https://www.researchandmarkets.com/reports/4536705/global-nanotechnology-market-outlook-2018-2024

Rojas-Grau, M. A., Soliva-Fortuny, R., & Martın-Belloso, O. (2009). Edible coatings to incorporate active ingredients to fresh-cut fruits: A review. *Trends in Food Science & Technology, 20*(10), 438–447. doi:10.1016/j.tifs.2009.05.002

Salvia-Trujillo, L., Rojas-Grau, A., Soliva-Fortuny, R., & Martın Belloso, O. (2015a). Physicochemical characterization and antimicrobial activity of food-grade emulsions and nanoemulsions incorporating essential oils. *Food Hydrocolloids, 43*, 547–556. doi:10.1016/j.foodhyd.2014.07.012

Salvia-Trujillo, L., Rojas-Grau, M. A., Soliva-Fortuny, R., & Martın Belloso, O. (2015b). Use of antimicrobial nanoemulsions as edible coatings: Impact on safety and quality attributes of fresh-cut Fuji apples. *Postharvest Biology and Technology, 105*, 8–16. doi:10.1016/j.postharvbio.2015.03.009

Samal, D. (2017). Use of nanotechnology in food industry: A review. *International Journal of Environment. Agricultural Biotechnology, 2*(4), 238902.

Sanchez-Gonzalez, L., Vargas, M., Gonzalez-Martınez, C., Chiralt, A., & Chafer, M. (2011). Use of essential oils in bioactive edible coatings: A review. *Food Engineering Reviews, 3*(1), 1–16. doi:10.100712393-010-9031-3

Sasaki, R. S., Mattoso, L. H. C., & de Moura, M. R. (2016). New edible bionanocomposite prepared by pectin and clove essential oil nanoemulsions. *Journal of Nanoscience and Nanotechnology, 16*(6), 6540–6544. doi:10.1166/jnn.2016.11702 PMID:27427749

Saxena, A., Maity, T., Paliwal, A., & Wadhwa, S. (2017). Technological aspects of nanoemulsions and their applications in the food sector. In *Nanotechnology Applications in Food* (pp. 129–152). Academic Press. doi:10.1016/B978-0-12-811942-6.00007-8

Schroder, A., Sprakel, J., Schroen, K., Spaen, N. J., & Berton-Carabin, C. (2018). Coalescence stability of Pickering emulsions produced with lipid particles: A microfluidic study. *Journal of Food Engineering, 234*, 63–72. doi:10.1016/j.jfoodeng.2018.04.007

Schultz, S., Wagner, G., Urban, K., & Ulrich, J. (2004). High-pressure homogenization as a process for emulsion formation. *Chemical Engineering & Technology, 27*(4), 361–368. doi:10.1002/ceat.200406111

Severino, R., Ferrari, G., Vu, K. D., Donsi, F., Salmieri, S., & Lacroix, M. (2015). Antimicrobial effects of modified chitosan based coating containing nanoemulsion of essential oils, modified atmosphere packaging and gamma irradiation against *Escherichia coli* O157:H7 and *Salmonella Typhimurium* on green beans. *Food Control, 50*, 215–222. doi:10.1016/j.foodcont.2014.08.029

Severino, R., Vu, K. D., Donsì, F., Salmieri, S., Ferrari, G., & Lacroix, M. (2014). Antibacterial and physical effects of modified chitosan based-coating containing nanoemulsion of mandarin essential oil and three non-thermal treatments against *Listeria innocua* in green beans. *International Journal of Food Microbiology, 191*, 82–88. doi:10.1016/j.ijfoodmicro.2014.09.007 PMID:25255308

Shahidi, F., & Zhong, Y. (2010). Lipid oxidation and improving the oxidative stability. *Chemical Society Reviews, 39*(11), 4067–4079. doi:10.1039/b922183m PMID:20617249

Silva, H. D., Cerqueira, M. A., & Vicente, A. A. (2015). Influence of surfactant and processing conditions in the stability of oil-in-water nanoemulsions. *Journal of Food Engineering, 167*, 89–98. doi:10.1016/j.jfoodeng.2015.07.037

Silva, S. S., Goodfellow, B. J., Benesch, J., Rocha, J., Mano, J. F., & Reis, R. L. (2007). Morphology and miscibility of chitosan/soy protein blended membranes. *Carbohydrate Polymers, 70*(1), 25–31. doi:10.1016/j.carbpol.2007.02.023

Singh, H. (2016). Nanotechnology applications in functional foods; opportunities and challenges. *Preventive Nutrition and Food Science, 21*(1), 1–8. doi:10.3746/pnf.2016.21.1.1 PMID:27069899

Son, H. J., Bae, H. C., Kim, H. J., Lee, D. H., Han, D. W., & Park, J. C. (2005). Effects of β-glucan on proliferation and migration of fibroblasts. *Current Applied Physics, 5*(5), 468–471. doi:10.1016/j.cap.2005.01.011

Swientek, R. J. (1990). Microfluidizing technology enhances emulsion stability. *Food Processing, 6*, 152–153.

Tadros, T., Izquierdo, P., Esquena, J., & Solans, C. (2004). Formation and stability of nano-emulsions. *Advances in Colloid and Interface Science, 108*, 303–318. doi:10.1016/j.cis.2003.10.023 PMID:15072948

Tan, C. P., & Nakajima, M. (2005a). β-Carotene nanodispersions: Preparation, characterization and stability evaluation. *Food Chemistry, 92*(4), 661–671. doi:10.1016/j.foodchem.2004.08.044

Thiebaud, M., Dumay, E., Picart, L., Guiraud, J. P., & Cheftel, J. C. (2003). Highpressure homogenisation of raw bovine milk. Effects on fat globule size distribution and microbial inactivation. *International Dairy Journal, 13*(6), 427–439. doi:10.1016/S0958-6946(03)00051-7

Tian, H., Li, D., Xu, T., Hu, J., Rong, Y., & Zhao, B. (2017). Citral stabilization and characterization of nanoemulsions stabilized by a mixture of gelatin and Tween 20 in an acidic system. *Journal of the Science of Food and Agriculture*, *97*(9), 2991–2998. doi:10.1002/jsfa.8139 PMID:27859362

Vargas, M., Albors, A., Chiralt, A., & Gonzalez-Martınez, C. (2009). Characterization of chitosan–oleic acid composite films. *Food Hydrocolloids*, *23*(2), 536–547. doi:10.1016/j.foodhyd.2008.02.009

Varghese, R., Dalvi, Y.B., Lamrood, P.Y., Shinde, B.P., & Nair, C.K.K. (2019). Historical and current perspectives on therapeutic potential of higher basidiomycetes: An overview. *3 Biotech*, 9, 362.

Verma, S., Jain, V., & Chandrawanshi, N. K. (2020). *Surfactant Based Anhydrous Nano Carrier System for Poorly Aqueous Soluble Anti-Cancer Drugs*. IGI Global. doi:10.4018/978-1-7998-6530-8.ch014

Walstra, P. (2003). *Physical Chemistry of Foods*. Marcel Decker.

Weigand-Heller, A. J., Kris-Etherton, P. M., & Beelman, R. B. (2012). The bioavailability of ergothioneine from mushrooms (*Agaricus bisporus*) and the acute effects on antioxidant capacity and biomarkers of inflammation. *Preventive Medicine*, *54*, 75–78. doi:10.1016/j.ypmed.2011.12.028 PMID:22230474

Weiss, J., Takhistov, P., & McClements, D. J. (2006). Functional materials in food nanotechnology. *Journal of Food Science*, *71*(9), 107–116. doi:10.1111/j.1750-3841.2006.00195.x

Witthayapanyanon, A., Acosta, E. J., Harwell, J. H., & Sabatini, D. A. (2006). Formulation of ultralow interfacial tension systems using extended surfactants. *Journal of Surfactants and Detergents*, *9*(4), 331–339. doi:10.100711743-006-5011-2

Wooster, T. J., Golding, M., & Sanguansri, P. (2008). Impact of oil type on nanoemulsion formation and Ostwald ripening stability. *Langmuir: ACS Surface and Colloids*, *24*(22), 12758–12765. doi:10.1021/la801685v PMID:18850732

Xu, X., Yan, H., Chen, J., & Zhang, X. (2011). Bioactive proteins from mushrooms. *Biotechnology Advances*, *29*(6), 667–674. doi:10.1016/j.biotechadv.2011.05.003 PMID:21605654

Yang, X., Tian, H., Ho, C. T., & Huang, Q. (2011). Inhibition of citral degradation by oilin-water nanoemulsions combined with antioxidants. *Journal of Agricultural and Food Chemistry*, *59*(11), 6113–6119. doi:10.1021/jf2012375 PMID:21517071

Yen, M. T., & Mau, J. L. (2006). Preparation of fungal chitin and chitosan from shiitake stipes. *Fungal Science*, *21*, 1–11.

Yen, M. T., & Mau, J. L. (2007). Physico-chemical characterization of fungal chitosan from shiitake stipes. *Lebensmittel-Wissenschaft + Technologie*, *40*(3), 472–479. doi:10.1016/j.lwt.2006.01.002

Yuan, Y., Gao, Y., Zhao, J., & Mao, L. (2008). Characterization and stability evaluation of β-carotene nanoemulsions prepared by high pressure homogenization under various emulsifying conditions. *Food Research International*, *41*(1), 61–68. doi:10.1016/j.foodres.2007.09.006

Yun Zhang, P. G. W. (2003). Studying the interaction of gel carbohydrate antigen and protein by quartz-crystal microbalance. *Journal of the American Chemical Society*, *125*(31), 9292–9293. doi:10.1021/ja035350a PMID:12889948

Zambrano-Zaragoza, M. L., Mercado-Silva, E., Del Real, L. A., Gutierrez-Cortez, E., Cornejo-Villegas, M. A., & Quintanar-Guerrero, D. (2014). The effect of nano-coatings with a-tocopherol and xanthan gum on shelf-life and browning index of fresh-cut "Red Delicious" apples. *Innovative Food Science & Emerging Technologies*, *22*, 188–196. doi:10.1016/j.ifset.2013.09.008

Zawistowski, J., Biliaderis, C. G., & Eskin, N. A. M. (1991). *Polyphenol oxidase. In Oxidative enzymes in foods*. Elsevier.

Zhang, Z., Song, H., Peng, Z., Luo, Q., Ming, J., & Zhao, G. (2012). Characterization of stipe and cap powders of mushroom (*Lentinus edodes*) prepared by different grinding methods. *Journal of Food Engineering*, *109*(3), 406–413. doi:10.1016/j.jfoodeng.2011.11.007

KEY TERMS AND DEFINITIONS

β-Glucan: It is carbohydrate derivatives present in the cell of microbes such as fungi, yeasts, lichens, algae, bacteria, and plants having medical advantages.

Carotenoids: A yellow or orange pigment present in plants, algae, and photosynthetic microrganism producing coloration in plants performs photosynthesis also acts as antioxidants.

Chitosan: It is a chitin derivative made up of linear polysaccharide consists of β-linked D-glucosamine and N-acetyl-D-glucosamine. It has numerous commercial and medical applications.

Coalescence: The process of amalgamation of colloidal particles into a larger component due to molecular attraction results in the formation of a film.

Emulsification: The process of diffusion of two immiscible liquid/semisolid in one another is called emulsification without changing the identity of other.

Mushroom: It is a fleshy, spore bearing having distinctive morphology belongs to kingdom fungi, consisting of a cap and stalk which may be edible or inedible.

Polysaccharides: It is composed of two or more monosaccharide unit bound together by glycosidic linkages found most abundantly in food, e.g., starch, cellulose, etc.

Chapter 20
Applications of Nanoemulsions in Food Manufacturing

Tahir Mehmood
The Islamia University of Bahawalpur, Pakistan

Muhammad Kaleem
Pir Mehr Ali Shah Arid Agriculture University, Pakistan

Zulfiqar Ahmad
The Islamia University of Bahawalpur, Pakistan

Muhammad Waseem
The Islamia University of Bahawalpur, Pakistan

Baila Ahmad
The Islamia University of Bahawalpur, Pakistan

Muhammad Adil Rehman
The Islamia University of Bahawalpur, Pakistan

Faiz-ul-Hassan Shah
The Islamia University of Bahawalpur, Pakistan

Ishtiaq Ahmad
Zhejiang University of Technology, Hangzhou, China

ABSTRACT

Nanoemulsions are increasingly used in the food industry for the encapsulation of lipophilic compounds such as nutraceuticals, flavors, vitamins, antioxidants, and colors. The advantages of nanoemulsion-based encapsulation include increased bioavailability, higher solubility, control release, protection from chemical degradation, and incorporation of different ingredients into food products. Food industries are interested in the preparation of natural polymer-based nanoemulsions to prepare clean label products. A variety of polymers such as soya lecithin, WPI, GA, and modified starches have been used to fabricate nanoemulsion-based delivery systems for nutraceuticals and micronutrients. Hence, nanoemulsion-fortified food products have received great attention due to consumer demand for safer and healthier food products. Additionally, nanoemulsions can be used for the texture modification of food products and encapsulation as well as delivery of antimicrobial agents. The main challenge for the preparation of nanoemulsions is the selection of appropriate ingredients.

DOI: 10.4018/978-1-7998-8378-4.ch020

INTRODUCTION

Nanoemulsions are smaller size (<200nm) dispersions consist of two immiscible liquids stabilized through surfactants. These nanoemulsions are thermodynamically unstable and kinetically stable colloidal dispersions (Gupta, Eral, Hatton, & Doyle, 2016; Lima et al., 2021). The nanoemulsions stability against sedimentation and creaming increased due to smaller size of droplets (Solans, Izquierdo, Nolla, Azemar, & Garcia-Celma, 2005). Additionally, the bioavailability of encapsulated bioactive compounds increased in nanoemulsions due to higher surface area to volume ratio (Sotomayor-Gerding et al., 2016). Nanoemulsions are used in the beverages and food industry for the encapsulation of lipophilic components such as colors, flavors, nutraceuticals, preservatives, vitamins and antioxidants (Given, 2009; Graves & Mason, 2008). These components are encapsulated to increase their solubility and bioavailability, protect them from degradation, incorporate them into food products, ease of utilization and to control release rate (McClements & Rao, 2011). We can also introduce lipophilic compounds into clear or slightly turbid food products without changing their appearance. Numerous types of nanoemulsions have been developed to encapsulate a variety of lipophilic compounds such as citral (Mei et al., 2009), β- carotene (Yin, Chu, Kobayashi, & Nakajima, 2009), fat-soluble vitamins (Hatanaka et al., 2010) and co-Enzyme Q (Ozaki et al., 2010).

Lipophilic components are solubilized in oil before emulsification so that these compounds are trapped within lipid phase during nanoemulsion preparation. The location of hydrophobic compounds in nanoemulsions depends on physicochemical and molecular properties, such as surface activity, surface hydrophobicity, melting point, solubility and partition coefficient between oil-water. The location of lipophilic compound exerts a significant effect on physical as well as chemical stability of nanoemulsions. For example, chemical degradation starts in lipophilic compounds when they come in contact with polar compounds. So, it is important for their stability that lipophilic compounds should be trapped into oily phase rather than shell (McClements, 2011). In the previous study, when citral (a component of flavor molecule) come in contact with the proton of water, chemical degradation start in citral molecules which effect on the stability of nanoemulsions (Mei et al., 2009).

The applications of nanoemulsions in the food industry are relatively slower as compared to other industries due to higher homogenization cost, scarcity of food grade surfactants, concerns of regulatory bodies and consumers about nanotechnology and destabilization of nanoemulsions due to Ostwald ripening (Ashaolu, 2021; Donsì, 2018; Tripathi, Sharma, Agarwal, & Haleem, 2021). However, in the future, the cost of emulsification will decrease due to its applications in other fields. Additionally, scientists are working on the naturally-derived food-grade surfactants such as biopolymers, sugar esters, lecithin and modified starches (Dammak, Sobral, Aquino, Neves, & Conte-Junior, 2020; Tan & McClements, 2021). Due to these developments, the application of nanoemulsions in food industry will substantially increase. These nanoemulsions will enable us to develop food products with increased functionality and enhanced sensory properties. Nanoemulsions have many potential applications for the incorporation of lipophilic compounds into food products. In this book chapter, we have summarized the different applications of nanoemulsions in food science such as encapsulation and delivery of micronutrients, nutraceutical compounds, colors, flavors and antimicrobial agents; structuring of food products and control delivery and increased bioavailability of fat-soluble compounds.

PREPARATION OF NANOEMULSIONS

Nanoemulsions can be prepared through high-energy and low-energy methods (Mehmood, 2021a). High energy methods use work-based emulsification approach which requires mechanical devices such as ultrasonic homogenizer, high-pressure homogenizer and microfluidizer (McClements, 2011; Mehmood, Ahmed, & Ahmed, 2021). These mechanical devices generate disruptive forces to break the oil droplets and disperse them into aqueous phase (Laura Salvia-Trujillo, Soliva-Fortuny, Rojas-Graü, McClements, & Martín-Belloso, 2017). In low energy methods, simple process is used for nanoemulsion preparation such as stirring and change in composition and formulation parameters e.g., oil to water ratio, oil type, temperature and surfactant concentration. Phase inversion temperature method and spontaneous emulsification are the examples of low-energy methods (Safaya & Rotliwala, 2020; Laura Salvia-Trujillo et al., 2017). Both approaches are capable to form smaller size droplets. However, high energy methods are preferred in food industry due to ease in preparation of nanoemulsions, possibility of the use of natural surfactant, large scale production and use of lower surfactant concentration (Mehmood & Ahmed, 2020; Mehmood, Ahmed, Ahmad, Ahmad, & Sandhu, 2018).

The nanoemulsions are prepared by emulsifying three major components i.e., emulsifier, water (aqueous phase) and oil (Fig. 1). However, the oil and water may contain one or more other compounds and more than one emulsifier can be used for the stabilization of nanoemulsions (Mehmood, 2015; Mehmood, Ahmad, Ahmed, & Ahmed, 2017). The concentration and properties of these components determine the properties of final nanoemulsions. The nanoemulsions may comprised of bioactive lipid phase (such as flavor oil, essential oil or fish oil), or it may contain bioactive compounds containing lipid phase (Q. Liu, Huang, Chen, Lin, & Wang, 2019). The aqueous phase of nanoemulsions plays a major role in the physical and chemical properties of nanoemulsions. The aqueous phase contains different water-soluble compounds such as acids, bases, minerals, preservatives, vitamins, flavors, polysaccharide, proteins and surfactants (McClements, 2015; McClements & Rao, 2011). The emulsifiers are added to ease the formation of smaller droplets and prevent the aggregation of droplets during storage of nanoemulsions. Additionally, other type of stabilizers can be used to modify the properties of nanoemulsions such as texture modifiers, weighing agent and ripening inhibitors.

ENCAPSULATION AND DELIVERY OF MICRONUTRIENTS

There has been a surge in customer demand for the fortification of functional food with micronutrients to improve their health. These micronutrients include nutraceuticals, vitamins and minerals which have diverse physicochemical properties, biological effects and molecular characteristics (McClements, 2019). Some micronutrients are vital for the proper functioning of our body (vitamins and minerals), whereas other micronutrients give additional health apart from basic nutrition (nutraceuticals) (Joye, Davidov-Pardo, & McClements, 2014). The encapsulation of these micronutrients in functional food products is not easy due to various constraints such as lower solubility in water or oil, degradation of vitamins during processing, storage and distribution, lower bioavailability and reduced acceptability of micronutrient supplemented food products due to distinct off-flavor of some micronutrients (McClements, 2019; Mehmood, 2015; Mehmood et al., 2017; Mehmood & Ahmed, 2020; Mehmood et al., 2018). Hence, there is a surge in interest to develop food grade delivery system for these micronutrients which protect, encapsulate and control release these micronutrients. Nanoemulsions are widely investigated for the encapsulation of micronutrients. Some micronutrients have received particular attention due to health benefits such as polyunsaturated fatty acids, lipophilic vitamins and dietary fibers. Table 1 summarizes the studies which are conducted to encapsulate the micronutrients in nanoemulsions.

The micronutrients are encapsulated in nanoemulsions to improve their solubility in hydrophilic products, protect them from degradation during preparation, processing, storage and distribution of food, prevent their interaction with major food components and control their digestion as well as uptake in gastrointestinal tract (Donsì, 2018; Mehmood et al., 2017; Mehmood & Ahmed, 2020). Relatively, higher concentration of micronutrients is encapsulated in nanoemulsions as compared to bioactive compounds. Moreover, some of these micronutrients do not require carrier lipophilic phase such as omega three and omega six oils (Jafari, Assadpoor, Bhandari, & He, 2008). Masking the unpleasant flavor of micronutrients is another challenge during the preparation of the delivery system for micronutrients. Additionally, efficacy of micronutrients depends on the size and charge of nanoemulsions (García-Márquez, Higuera-Ciapara, & Espinosa-Andrews, 2017). Carrier oil is required for the delivery of some micronutrients such as fat-soluble vitamins. The use of carrier oil containing digestible triglycerides increases the bioavailability of micronutrients (Laura Salvia-Trujillo et al., 2017).

Figure 1. The components of nanoemulsions and bioactive compounds entrapments in nanoemulsions

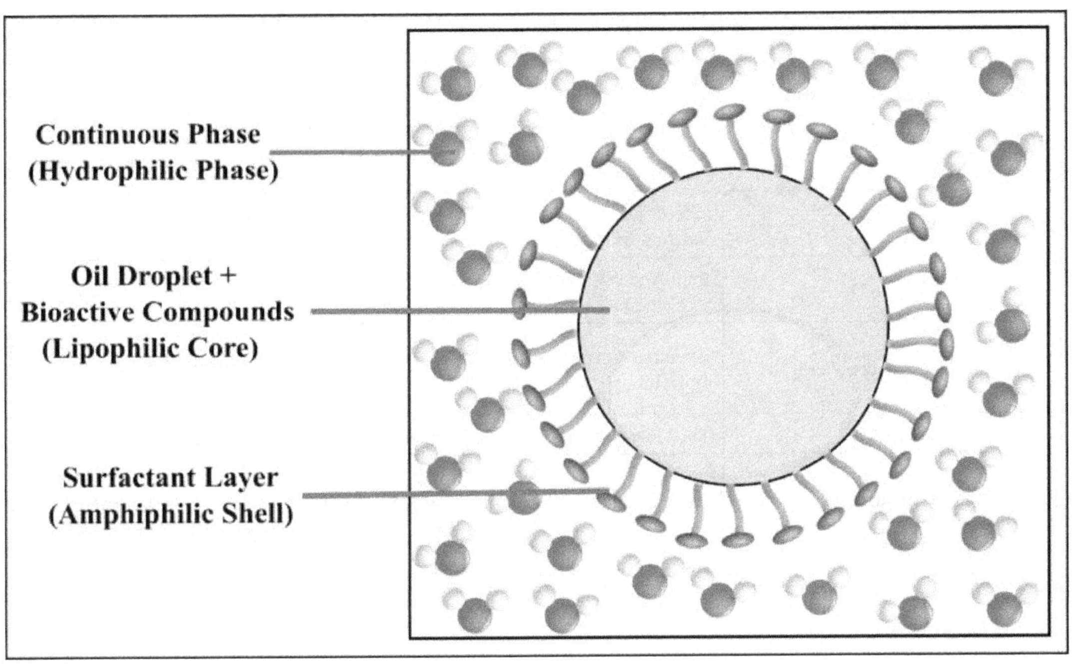

For hydrophilic micronutrients, double emulsions are preferred for the preparation of delivery system for these compounds. In these emulsions, hydrophilic micronutrient is firstly encapsulated in water-in-oil emulsions and then dispersed in double emulsions to increase their protection (Assadpour, Maghsoudlou, Jafari, Ghorbani, & Aalami, 2016). Additionally, single or mixed surfactants are used for stable nanoemulsions preparation (Mehmood, 2021a). Both natural and synthetic surfactants are used for the preparation of nanoemulsions. Recently, natural surfactants are gaining more interest from food industry due to customer demand (Akkam et al., 2021; Mehmood et al., 2018; Mehmood et al., 2021).

ENCAPSULATION AND DELIVERY OF BIOACTIVE COMPOUNDS

Over the last few years, customers are demanding functional food products which can promote their health and protect them from diseases. For this purpose, food industrialists are interested in the incorporation of nutraceutical compounds in food products. The bioactive compounds contain several classes of molecules which exert positive health effect in the human body. The list of bioactive compounds is continuously expanding on the basis of the latest research and clinical evidence.

Table 1. Applications of nanoemulsions for the encapsulation of micronutrients

Class	Compound	Value Addition of Nanoemulsions	Reference
Fatty acids	Omega-3 fatty acids	Physical stability, oxidative stability and bioaccessibility of omega-3 fatty acids improved after their incorporation into nanoemulsions.	(Inapurapu et al., 2020)
	Polyunsaturated fatty acids	The bioavailability of polyunsaturated fatty acids was improved.	(Dey, Koley, Ghosh, Dey, & Dhar, 2019)
	Alpha-linolenic acid	Increased stability against change in pH, ionic strength and light intensity.	(M. Liu et al., 2021)
Vitamins	Beta-carotene	The physicochemical stability of beta-carotene was increased.	(Mehmood et al., 2021)
	Vitamin D	Protect vitamin D from degradation in nanoemulsions based fortified food products	(Akkam et al., 2021)
	Vitamin D	Improved cellular uptake efficiency and bioavailability of vitamin D	(Walia & Chen, 2020)
	Alpha tocopherol	Effective delivery system for encapsulation and protection of alpha tocopherol.	(Sahafi, Goli, Kadivar, Varshosaz, & Shirvani, 2021)
	Vitamin E	The bioaccessibility of encapsulated vitamin E was increased	(Lv et al., 2018)
Dietary Fibers	Mandarin fiber	Preparation of edible coating with desirable properties	(Artiga-Artigas, Acevedo-Fani, & Martín-Belloso, 2017)
Phytosterols	Stigmasterol	Higher storage stability and controlled release of stigmasterol	(Feng et al., 2021)

The use of bioactive compounds in food products is challenging due to their lower solubility and instability in food preparations (Ashaolu, 2021; Mehmood, 2021a). Hence, these compounds are frequently encapsulated in nanoemulsions which increase their stability during storage and release them in specific conditions such as release of flavor in the mouth and nutraceuticals in gastrointestinal tract (Soukoulis et al., 2017). Additionally, the bioaccessibility of encapsulated compounds also increased due to an increase in solubility and incorporation of these compounds into mixed micelles of gastrointestinal tract (Ziani, Fang, & McClements, 2012). Table 2 summarizes the major classes of nutraceuticals that are encapsulated in nanoemulsions. It includes phenolic compounds (both flavonoids and non-flavonoids) and carotenoids. Many studies are conducted to show that nanoemulsions increase the bioavailability of bioactive compounds. But, further research studies are mandatory to understand their interaction with food components in actual food products and the effect of consumption on their properties (Laura Salvia-Trujillo et al., 2017).

Major challenge in the development of functional food products is the encapsulation of nutraceutical compounds through food grade delivery systems which encapsulate, protect and control release nutraceutical compound (McClements, 2011). In this regard, nanoemulsions are an ideal candidate for food-grade delivery system for lipophilic compounds (Mehmood, 2015, 2021a). The encapsulation of bioactive compounds in nanoemulsions is preferred due to the availability of easily scalable homogenizers (i.e., high pressure homogenizer and ultrasonication), lower concentration of emulsifiers, feasibility for using natural ingredients and higher affinity for aqueous food products (Donsì, Wang, & Huang, 2011). In nanoemulsions, lower concentrations of bioactive compounds are required to show health promoting effects. The disperse phase of nanoemulsion is a useful tool to control the biological activity, chemical stability and release characteristics of bioactive compounds (Donsì, 2018). Additionally, nanoemulsions increase the chemical stability of encapsulated compounds by preventing their interactions with external agents such as other food constituents, oxygen, heat and UV light (Donsì, 2018; Mehmood & Ahmed, 2020).

The encapsulated compounds may be localized in the lipophilic core or stabilization layer (shell) depending on their polarity. It had a significant effect on the behavior of nanoemulsions in food products and human body (McClements, 2011). If bioactive molecules have lower solubility in water then the microenvironment around these compounds is changed by using amphiphilic molecules to prevent their crystallization. The selection of appropriate surfactant and cp-surfactant helps the stabilization of hydrophilic molecules at the interface of nanoemulsions (Tsai & Chen, 2016). However, many researchers reported that nanoemulsions can significantly improve the bioavailability of bioactive ingredients in different food products (Mehmood, 2015). Several researchers observed that nanoemulsions increase the bioavailability of carotenoids from tomato juice (L Salvia-Trujillo & McClements, 2016), carrots (Zhang et al., 2016), yellow peppers (X. Liu, Bi, Xiao, & McClements, 2015) and mangoes (X. Liu, Bi, Xiao, & McClements, 2016). Although it is proved that nanoemulsion-based delivery system improved the bioavailability of encapsulated compounds, more research is required to prove that nanoemulsions are associated with improved bioavailability of encapsulated materials in complex food matrices.

Currently, only a few food products are available in the market in which bioactive compounds are incorporated using nanoemulsions based delivery systems. Limited information is available about the potential advantage and disadvantages of this delivery system. Majority of the studies are conducted on simple food systems. The interaction of nanoemulsions based delivery system with other macronutrients in food should be investigated to check their suitability for food products (Laura Salvia-Trujillo et al., 2017).

ENCAPSULATION AND DELIVERY OF COLORS AND FLAVORS

The incorporation of flavors into food products is a tough task, which should bear in mind the protection and control release of the volatile molecules during the shelf life of food products. The release of flavor molecules during the preparation and consumption effect the perception of consumers about food product (Donsì, 2018). Essential oils such as Citral and D-limonene are commonly used to change the sensory properties of food products due to their volatile and pungent taste. But, the incorporation of essential oils in food products is challenging due to lower water solubility, oxidative instability and degradation due to environmental stresses (Yoplac, Vargas, Robert, & Hidalgo, 2021). Hence, the encapsulation of essential oil in nanoemulsions is a viable approach for their protection from degradation. Additionally, the selection of suitable surfactants, biopolymer layers and antioxidants for nanoemulsions are powerful tools to increase the stability of essential oils (Jafari, He, & Bhandari, 2007; Mehmood, 2021b). These essential oils can be incorporated into food products by spray-drying essential oil containing nanoemulsions. Additionally, the use of biopolymers as surfactants did not alter the properties of nanoemulsions upon rehydration (Jafari et al., 2007). However, only limited number of studies are conducted to investigate the effect of essential oil containing nanoemulsions on the sensory properties of food products (Majeed et al., 2016). Hence, more research studies should be carried out to explore the possibility of controlling the organoleptic properties of food products using essential oil based nanoemulsions.

Table 2. Applications of nanoemulsions for the encapsulation of Bioactive Compounds

Class	Category	Value Addition of Nanoemulsions	References
Flavonoids	Isoflavones	Controlled release of isoflavones through adjusting pH	(S. Wang et al., 2020)
	Flavanones	Effective delivery and controlled release of flavanones	(Bustos-Salgado et al., 2021)
	Flavones	Improved oxidative stability and bioavailability of flavones	(Sharma et al., 2020)
	Flavonols	Higher solubility and improved physical and chemical stability	(W. Silva, Torres-Gatica, et al., 2018)
	Flavanols	Enhanced antioxidant activity and good stability against stress conditions	(Ruengdech & Siripatrawan, 2021)
Nonflavonoids	Hydroxybenzoic acids	Higher physical and oxidative stability of encapsulated compounds	(W. Silva, Torres-Gatica, et al., 2018)
	Hydroxycinnamic acid	Extend the shelf life of fresh-cut lettuce through antibiofilm and antimicrobial effects	(Letsididi, Lou, Letsididi, Mohammed, & Maguy, 2018)
	Stilbenes	Increase the shelf life of meat by reducing protein and lipid oxidation and preventing microbial growth	(Xiong, Li, Warner, & Fang, 2020)
	Curcuminoids	Increased bioaccessibility and improved lipid digestibility	(H. D. Silva, Poejo, et al., 2018)
Carotenoids	Carotenes	Improved stability against storage, oxidation and environmental stresses	(Mehmood, 2021a)
	Xanthophylls	Improved solubility and physical stability	(Sedaghat Doost et al., 2020)

The colloidal properties of nanoemulsions such as droplet concentration, droplet size and refractive index affects the optical behaviors of nanoemulsions due to their influence on light absorption and scattering (McClements, 2011; Mehmood & Ahmed, 2020; Mehmood et al., 2021). In general, the lightness of the food samples increased with lower droplet size, increasing droplet concentration and lower concentration of dye, whereas chromophoric material exerts opposite effects (McClements, 2011; Mehmood & Ahmed, 2020). When a beam of light is passed through nanoemulsions samples, some part of this light is reflected or scattered and some part is transmitted or absorbed. Scattering of light determines the lightness, opacity and turbidity of the system, whereas absorption of light determines the chromatics, e.g., redness, greenness and blueness (McClements, 2011; Mehmood et al., 2021). The appearance of nanoemulsions is transparent or slightly turbid due to the presence of smaller size droplets. Hence, nanoemulsions are suitable for use in foods and beverages where a clear look is required (Mehmood et al., 2017). The color of nanoemulsions based food products can be controlled by controlling the physical properties of nanoemulsions such as refractive index, particle size distribution, droplet size, presence of chromophoric compounds and chromophoric components stability. Previously, some studies are conducted to encapsulate natural bioactive pigments such as beta carotene, curcumin and indigo carmine into nanoemulsions due to their effects on the appearance of food products and potential health benefits (Mehmood et al., 2018; Qazi, Ye, Acevedo-Fani, & Singh, 2021). Additionally, we can use a combination of natural pigments to achieve tunable appearance of food products. In a previous study, few researchers incorporated a combination of indigo carmine and curcumin in colloidal dispersions to develop desired coloring system (Patel, Heussen, Dorst, Hazekamp, & Velikov, 2013). There is a need to conduct more studies for understanding the optical properties of nanoemulsions.

ENCAPSULATION AND DELIVERY OF ANTIMICROBIAL AGENTS

Some antimicrobial agents are soluble in water and we cannot incorporate them into lipid-based products. On the other end, some antimicrobial agents are only soluble in oil, so they cannot be incorporated into aqueous products. The usability and effectiveness of these compounds can be increased by encapsulating them in nanoemulsions. These antimicrobial agents can be encapsulated into an amphiphilic exterior or hydrophobic interior of lipid phase or both (Jochen, Sylvia, Michael, & Julian, 2009).

The activity of encapsulated antimicrobial agents depends on the transportation of these materials from nanoemulsions to the bacterial surface. Two different mechanisms may be involved in this process: (1) direct interaction of droplets and microorganisms or (2) diffusion of molecules through aqueous phase. Smaller droplets of nanoemulsions increase their antimicrobial activity against microorganisms in many different ways. Firstly, due to Laplace effect higher concentration of antimicrobial agents are present at droplet surfaces which increases the mass transport into the aqueous phase (McClements, 2011). Secondly, the interaction between droplets and microorganisms is increased due to an increase in the Brownian motion of smaller droplets. Thirdly, nanoemulsions increased the penetration of antimicrobials into bacterial surfaces by facilitating their penetration into biological membranes.

Ozogul and co-workers (2020) determined the antimicrobial properties of purified and non-purified thyme essential oil on food-borne pathogens and bacteria causing fish spoilage. They reported that non-purified thyme oil was more effective against food-borne pathogens. Additionally, the encapsulation of thyme essential oil into nanoemulsions significantly increase their antimicrobial properties. In another study, Manzoor and his colleagues (2021) investigate the antimicrobial effect of nanoemulsions on fresh cut ready to eat slices of kiwi fruit. They reported that antimicrobial agent containing nanoemulsions slowed the growth of mold, bacteria and yeast during storage. In a recent study, Yazgan (2020) evaluated the antimicrobial properties of sage essential oil containing nanoemulsions. The result of this study indicated that sage essential oil based nanoemulsions successfully retarded the growth of food-borne pathogens (*Staphylococcus aureus, Salmonella Paratyphi, Klebsiella pneumoniae and Enterococcus faecalis*) and spoilage microorganism (*Serratia liquefaciens, Pseudomonas luteola, Proteus mirabilis, Enterococcus faecalis, Vibrio vulnificus and Photobacterium damselae*).

Oregano oil based nanoemulsions with droplet size of 150 nm was recently used to retard the growth of food-borne pathogens in lettuce during their storage in the refrigerator (Bhargava, Conti, da Rocha, & Zhang, 2015). Nanoemulsions with cinnamaldehyde (droplet size less than 200 nm) were found effective for the deactivation of bacterial growth in melon juice (Jo et al., 2015). Furthermore, soaking radish, alfalfa and mong bean seeds in nanoemulsions containing carvacrol retarded the growth of Enteritidis, *Escherichia coli* O157: H7 and *Salmonella enterica* without affecting their sprout yield (Landry, Micheli, McClements, & McLandsborough, 2015). Donsi and coworkers investigated the effect of the incorporation of terpene containing nanoemulsion into pear and orange juice during their storage at 32 °C for 16 days. They reported that lower concentration (1.0g/L) of terpene containing nanoemulsions delayed the growth of microorganisms (*Lactobacillus delbrueckii*) while higher concentration (5.0g/L) completely inactivated the bacterial growth without compromising the sensory properties of fruit juice (Donsì, Annunziata, Sessa, & Ferrari, 2011). Joe and coworkers (2012) observed that the shelf life of fish steak can be increased by using sunflower oil based nanoemulsions. They reported a significant reduction in the population of lactic acid, hydrogen sulfide and heterotrophic bacteria as well as extension of shelf life during storage as compared to control samples. Mate and others (2016) observed reduced growth of *Listeria monocytogens* in vegetable cream and chicken broth treated with nisin and D-limonene nanoemulsions. Ma and others reported that eugenol or thymol nanoemulsions exhibit antimicrobial activity in low-fat milk emulsified with LAE and lecithin (Ma, Davidson, & Zhong, 2016).

Despite the potential application of nanoemulsions based delivery systems for the delivery of antimicrobial agents, their benefits over conventional emulsions are not clear due to a lack of consistent data. Some studies reported that the efficiency of nanoemulsions based delivery systems for antimicrobial agents decreased due to smaller particle size due to the adsorption of antimicrobial at droplet surface instead of microorganisms (Laura Salvia-Trujillo et al., 2017).

STRUCTURING OF FOOD PRODUCTS

Nanoemulsions have very smaller size droplets due to which interfacial coating around droplets constitutes an appreciable proportion of the overall volume of droplets. As a result of this, it might be possible that nanoemulsion promotes gelation reactions at lower oil concentrations as compared to conventional emulsions. These properties are desirable in the preparation of the products which required gel-like or viscous appearance such as reduced fat products. Currently, nanoemulsions are not exploited for the texture modifications of food products. However, some studies on non-food grade systems reported that the electrostatic repulsion effect can be used for the preparations of nanoemulsions with a transparent look and gel-like characteristics (Kawada et al., 2010). The present approach may be used for the preparation of viscoelastic and highly viscous nanoemulsions using lower oil concentrate as compared to conventional emulsions (Wilking & Mason, 2007). Polymer-induced and electrostatic interactions between droplets are a powerful tool to control the rheological behavior of the colloidal systems at the molecular scale, through changing the formulation and microstructure of droplets (Helgeson, 2016). For example, higher viscoelastic properties of nanoemulsions gel can be achieved by selecting longer hydrophilic tails containing surfactants (Wulff-Pérez, Martín-Rodríguez, Gálvez-Ruiz, & de Vicente, 2013). Currently, there is no practical application of nanoemulsions gel as a constituent in food products. Further studies should be carried out on the preparation of nanoemulsions gel using food-grade material and their stability during the preparation and storage of food products.

INCREASED BIOAVAILABILITY OF LIPOPHILIC COMPONENTS

Several studies reported that the bioavailability of lipophilic compounds encapsulated into nanoemulsions increased due to the smaller size of nanoemulsions (Acosta, 2009). This increased in bioavailability is due to different reasons. Firstly, nanoemulsion droplets have a larger surface area. Due to this, the digestive enzyme acts more quickly on nanoemulsion as compared to nanoemulsions which led to easy absorption and rapidly release of the encapsulated material. Secondly, smaller droplets of nanoemulsions can penetrate into the mucous layer of epithelium cells in the small intestine which increases residence time and they reached closure to absorption sites. Thirdly, smaller droplets of nanoemulsions can be transported directly through paracellular or transcellular mechanisms across epithelium cells (McClements, 2011). Additionally, partition into the aqueous phase may be greater due to Laplace pressure which results in higher water solubility of lipophilic components. Presently, there is a poor understanding about the significance of these mechanisms for food grade nanoemulsions with different surface characteristics, composition and droplet size.

Recently, some researchers reported that the bioavailability of curcumin nanoemulsions can be increased by encapsulating them into nanoemulsions (Huang, Yu, & Ru, 2010; X. Wang et al., 2008). Various studies confirmed that nanoemulsions are associated with increased bioavailability of lipophilic components in pharmaceuticals and nutraceuticals (Hatanaka et al., 2010; Ozaki et al., 2010; Talegaonkar, Mustafa, Akhter, & Iqbal, 2010). Furthermore, nanoemulsions can be effectively used for the target delivery of bioactive components within the human body which results in improved efficiency (Huang et al., 2010; Laura Salvia-Trujillo et al., 2017).

CONCLUSION AND FUTURE RESEARCH DIRECTIONS

Nanoemulsions are the highly effective for the delivery of lipophilic compounds. These nanoemulsions can be prepared using natural ingredients. Nanoemulsions protect the encapsulated compounds from degradation and control their behavior after ingestion. Additionally, gel formation capability is exploited in many functional food product developments. Currently, nanoemulsions found limited applications in the preparation of commercial food products. However, they have greater potential for functional product development due to their novel functionalities. Particularly, the delivery of lipophilic compounds such as micronutrients, bioactive molecules, antimicrobial agents, colors and flavors, into food products is highly desirable for value addition of food products by enhancing their nutritional profile, health beneficial properties, shelf life, appearance, taste and aroma. The main challenge in the preparation of nanoemulsion is the availability of natural ingredients to control the digestion, release and absorption properties of encapsulated compounds. The functional properties of food products can be controlled by careful designing of nanoemulsions.

Number of research studies are carried out to demonstrate the increased bioavailability of lipophilic compounds which are encapsulated into nanoemulsions based delivery systems, it is the need of time to confirm this evidence after incorporating these compounds into complex food matrices. Hence, a comprehensive approach is required in future research on nanoemulsions to investigate the actual advantages which are associated with nanoemulsions based delivery system. In future research, loss in the functionality of encapsulated material due to processing, storage and effect of food matrix should be considered. Furthermore, higher costs related to nanoemulsions production should be considered due to the requirement of some equipment and higher energy input. Additionally, research studies should be conducted to access the toxicological effect and biological fat of engineered nanoemulsions after digestion.

REFERENCES

Acosta, E. (2009). Bioavailability of nanoparticles in nutrient and nutraceutical delivery. *Current Opinion in Colloid & Interface Science*, *14*(1), 3–15. doi:10.1016/j.cocis.2008.01.002

Akkam, Y., Rababah, T., Costa, R., Almajwal, A., Feng, H., Laborde, J. E. A., Abulmeaty, M. M., & Razak, S. (2021). Pea protein nanoemulsion effectively stabilizes vitamin D in food products: A potential supplementation during the COVID-19 pandemic. *Nanomaterials (Basel, Switzerland)*, *11*(4), 887. doi:10.3390/nano11040887 PMID:33807206

Artiga-Artigas, M., Acevedo-Fani, A., & Martín-Belloso, O. (2017). Improving the shelf life of low-fat cut cheese using nanoemulsion-based edible coatings containing oregano essential oil and mandarin fiber. *Food Control*, *76*, 1–12. doi:10.1016/j.foodcont.2017.01.001

Ashaolu, T. J. (2021). Nanoemulsions for health, food, and cosmetics: A review. *Environmental Chemistry Letters*, *19*(4), 1–15. doi:10.100710311-021-01216-9 PMID:33746662

Assadpour, E., Maghsoudlou, Y., Jafari, S.-M., Ghorbani, M., & Aalami, M. (2016). Evaluation of folic acid nano-encapsulation by double emulsions. *Food and Bioprocess Technology*, *9*(12), 2024–2032. doi:10.100711947-016-1786-y

Bhargava, K., Conti, D. S., da Rocha, S. R., & Zhang, Y. (2015). Application of an oregano oil nanoemulsion to the control of foodborne bacteria on fresh lettuce. *Food Microbiology*, *47*, 69–73. doi:10.1016/j.fm.2014.11.007 PMID:25583339

Bustos-Salgado, P., Andrade-Carrera, B., Domínguez-Villegas, V., Rodríguez-Lagunas, M. J., Boix-Montañes, A., Calpena-Campmany, A., & Garduño-Ramírez, M. L. (2021). Biopharmaceutic study and in vivo efficacy of natural and derivatives flavanones formulations. *Nanomedicine (London)*, *16*(3), 205–220. doi:10.2217/nnm-2020-0368 PMID:33480290

Dammak, I., Sobral, P. J. A., Aquino, A., Neves, M. A., & Conte-Junior, C. A. (2020). Nanoemulsions: Using emulsifiers from natural sources replacing synthetic ones—A review. *Comprehensive Reviews in Food Science and Food Safety*, *19*(5), 2721–2746. doi:10.1111/1541-4337.12606 PMID:33336986

Dey, T. K., Koley, H., Ghosh, M., Dey, S., & Dhar, P. (2019). Effects of nano-sizing on lipid bioaccessibility and ex vivo bioavailability from EPA-DHA rich oil in water nanoemulsion. *Food Chemistry, 275*, 135–142. doi:10.1016/j.foodchem.2018.09.084 PMID:30724179

Donsì, F. (2018). Applications of nanoemulsions in foods. In *Nanoemulsions* (pp. 349–377). Elsevier. doi:10.1016/B978-0-12-811838-2.00011-4

Donsì, F., Annunziata, M., Sessa, M., & Ferrari, G. (2011). Nanoencapsulation of essential oils to enhance their antimicrobial activity in foods. *Lebensmittel-Wissenschaft + Technologie*, *44*(9), 1908–1914. doi:10.1016/j.lwt.2011.03.003

Donsì, F., Wang, Y., & Huang, Q. (2011). Freeze–thaw stability of lecithin and modified starch-based nanoemulsions. *Food Hydrocolloids*, *25*(5), 1327–1336. doi:10.1016/j.foodhyd.2010.12.008

Feng, S., Yan, J., Wang, D., Jiang, L., Sun, P., Xiang, N., & Shao, P. (2021). Preparation and characterization of soybean protein isolate/pectin-based phytosterol nanodispersions and their stability in simulated digestion. *Food Research International*, *143*, 110237. doi:10.1016/j.foodres.2021.110237 PMID:33992350

García-Márquez, E., Higuera-Ciapara, I., & Espinosa-Andrews, H. (2017). Design of fish oil-in-water nanoemulsion by microfluidization. *Innovative Food Science & Emerging Technologies*, *40*, 87–91. doi:10.1016/j.ifset.2016.11.007

Given, P. S. Jr. (2009). Encapsulation of flavors in emulsions for beverages. *Current Opinion in Colloid & Interface Science*, *14*(1), 43–47. doi:10.1016/j.cocis.2008.01.007

Graves, S. M., & Mason, T. G. (2008). Transmission of visible and ultraviolet light through charge-stabilized nanoemulsions. *The Journal of Physical Chemistry C*, *112*(33), 12669–12676. doi:10.1021/jp801019z

Gupta, A., Eral, H. B., Hatton, T. A., & Doyle, P. S. (2016). Nanoemulsions: Formation, properties and applications. *Soft Matter*, *12*(11), 2826–2841. doi:10.1039/C5SM02958A PMID:26924445

Hatanaka, J., Chikamori, H., Sato, H., Uchida, S., Debari, K., Onoue, S., & Yamada, S. (2010). Physicochemical and pharmacological characterization of α-tocopherol-loaded nano-emulsion system. *International Journal of Pharmaceutics*, *396*(1), 188–193. doi:10.1016/j.ijpharm.2010.06.017 PMID:20558261

Helgeson, M. E. (2016). Colloidal behavior of nanoemulsions: Interactions, structure, and rheology. *Current Opinion in Colloid & Interface Science*, *25*, 39–50. doi:10.1016/j.cocis.2016.06.006

Huang, Q., Yu, H., & Ru, Q. (2010). Bioavailability and delivery of nutraceuticals using nanotechnology. *Journal of Food Science*, *75*(1), R50–R57. doi:10.1111/j.1750-3841.2009.01457.x PMID:20492195

Inapurapu, S. P., Ibrahim, A., Kona, S. R., Pawar, S. C., Bodiga, S., & Bodiga, V. L. (2020). Development and characterization of ω-3 fatty acid nanoemulsions with improved physicochemical stability and bioaccessibility. *Colloids and Surfaces. A, Physicochemical and Engineering Aspects*, *606*, 125515. doi:10.1016/j.colsurfa.2020.125515

Jafari, S. M., Assadpoor, E., Bhandari, B., & He, Y. (2008). Nano-particle encapsulation of fish oil by spray drying. *Food Research International*, *41*(2), 172–183. doi:10.1016/j.foodres.2007.11.002

Jafari, S. M., He, Y., & Bhandari, B. (2007). Encapsulation of nanoparticles of d-limonene by spray drying: Role of emulsifiers and emulsifying techniques. *Drying Technology*, *25*(6), 1069–1079. doi:10.1080/07373930701396758

Jo, Y.-J., Chun, J.-Y., Kwon, Y.-J., Min, S.-G., Hong, G.-P., & Choi, M.-J. (2015). Physical and antimicrobial properties of trans-cinnamaldehyde nanoemulsions in water melon juice. *Lebensmittel-Wissenschaft + Technologie*, *60*(1), 444–451. doi:10.1016/j.lwt.2014.09.041

Jochen, W., Sylvia, G., Michael, D., & Julian, M. (2009). Nanostructured Encapsulation Systems: Food Antimicrobials. *Global Issues in Food Science and Technology ISBN*, *483208248*, 425–479.

Joe, M. M., Chauhan, P. S., Bradeeba, K., Shagol, C., Sivakumaar, P. K., & Sa, T. (2012). Influence of sunflower oil based nanoemulsion (AUSN-4) on the shelf life and quality of Indo-Pacific king mackerel (Scomberomorus guttatus) steaks stored at 20 C. *Food Control*, *23*(2), 564–570. doi:10.1016/j.foodcont.2011.08.032

Joye, I. J., Davidov-Pardo, G., & McClements, D. J. (2014). Nanotechnology for increased micronutrient bioavailability. *Trends in Food Science & Technology*, *40*(2), 168–182. doi:10.1016/j.tifs.2014.08.006

Kawada, H., Kume, T., Matsunaga, T., Iwai, H., Sano, T., & Shibayama, M. (2010). Structure and rheology of a self-standing nanoemulsion. *Langmuir*, *26*(4), 2430–2437. doi:10.1021/la902905b PMID:20141204

Landry, K. S., Micheli, S., McClements, D. J., & McLandsborough, L. (2015). Effectiveness of a spontaneous carvacrol nanoemulsion against Salmonella enterica Enteritidis and Escherichia coli O157: H7 on contaminated broccoli and radish seeds. *Food Microbiology*, *51*, 10–17. doi:10.1016/j.fm.2015.04.006 PMID:26187822

Letsididi, K. S., Lou, Z., Letsididi, R., Mohammed, K., & Maguy, B. L. (2018). Antimicrobial and antibiofilm effects of trans-cinnamic acid nanoemulsion and its potential application on lettuce. *Lebensmittel-Wissenschaft + Technologie*, *94*, 25–32. doi:10.1016/j.lwt.2018.04.018

Lima, T. S., Silva, M. F. S., Nunes, X. P., Colombo, A. V., Oliveira, H. P., Goto, P. L., Blanzat, M., Piva, H. L., Tedesco, A. C., & Siqueira-Moura, M. P. (2021). Cineole-containing nanoemulsion: Development, stability, and antibacterial activity. *Chemistry and Physics of Lipids*, *239*, 105113. doi:10.1016/j.chemphyslip.2021.105113 PMID:34216586

Liu, M., Yang, C., Liu, E., Zhang, F., Meng, X., & Liu, B. (2021). Effect of environmental stresses on physicochemical properties of ALA oil-in-water nanoemulsion system prepared by emulsion phase inversion. *Food Chemistry*, *343*, 128475. doi:10.1016/j.foodchem.2020.128475 PMID:33168257

Liu, Q., Huang, H., Chen, H., Lin, J., & Wang, Q. (2019). Food-Grade Nanoemulsions: Preparation, Stability and Application in Encapsulation of Bioactive Compounds. *Molecules (Basel, Switzerland)*, *24*(23), 4242. doi:10.3390/molecules24234242 PMID:31766473

Liu, X., Bi, J., Xiao, H., & McClements, D. J. (2015). Increasing carotenoid bioaccessibility from yellow peppers using excipient emulsions: Impact of lipid type and thermal processing. *Journal of Agricultural and Food Chemistry*, *63*(38), 8534–8543. doi:10.1021/acs.jafc.5b04217 PMID:26357977

Liu, X., Bi, J., Xiao, H., & McClements, D. J. (2016). Enhancement of nutraceutical bioavailability using excipient nanoemulsions: Role of lipid digestion products on bioaccessibility of carotenoids and phenolics from mangoes. *Journal of Food Science*, *81*(3), N754–N761. doi:10.1111/1750-3841.13227 PMID:26816342

Lv, S., Gu, J., Zhang, R., Zhang, Y., Tan, H., & McClements, D. J. (2018). Vitamin E Encapsulation in Plant-Based Nanoemulsions Fabricated Using Dual-Channel Microfluidization: Formation, Stability, and Bioaccessibility. *Journal of Agricultural and Food Chemistry*, *66*(40), 10532–10542. doi:10.1021/acs.jafc.8b03077 PMID:30240207

Ma, Q., Davidson, P. M., & Zhong, Q. (2016). Nanoemulsions of thymol and eugenol co-emulsified by lauric arginate and lecithin. *Food Chemistry*, *206*, 167–173. doi:10.1016/j.foodchem.2016.03.065 PMID:27041312

Majeed, H., Antoniou, J., Hategekimana, J., Sharif, H. R., Haider, J., Liu, F., Ali, B., Rong, L., Ma, J., & Zhong, F. (2016). Influence of carrier oil type, particle size on in vitro lipid digestion and eugenol release in emulsion and nanoemulsions. *Food Hydrocolloids*, *52*, 415–422. doi:10.1016/j.foodhyd.2015.07.009

Manzoor, S., Gull, A., Wani, S. M., Ganaie, T. A., Masoodi, F. A., Bashir, K., Malik, A. R., & Dar, B. N. (2021). Improving the shelf life of fresh cut kiwi using nanoemulsion coatings with antioxidant and antimicrobial agents. *Food Bioscience*, *41*, 101015. doi:10.1016/j.fbio.2021.101015

Mate, J., Periago, P. M., & Palop, A. (2016). Combined effect of a nanoemulsion of D-limonene and nisin on Listeria monocytogenes growth and viability in culture media and foods. *Food Science & Technology International*, *22*(2), 146–152. doi:10.1177/1082013215577034 PMID:25788168

McClements, D. J. (2011). Edible nanoemulsions: Fabrication, properties, and functional performance. *Soft Matter*, *7*(6), 2297–2316. doi:10.1039/C0SM00549E

McClements, D. J. (2015). *Food emulsions: principles, practices, and techniques*. CRC Press. doi:10.1201/b18868

McClements, D. J. (2019). *Nanoparticle-and microparticle-based delivery systems: Encapsulation, protection and release of active compounds*. CRC Press.

McClements, D. J., & Rao, J. (2011). Food-grade nanoemulsions: Formulation, fabrication, properties, performance, biological fate, and potential toxicity. *Critical Reviews in Food Science and Nutrition*, *51*(4), 285–330. doi:10.1080/10408398.2011.559558 PMID:21432697

Mehmood, T. (2015). Optimization of the canola oil based vitamin E nanoemulsions stabilized by food grade mixed surfactants using response surface methodology. *Food Chemistry*, *183*, 1–7. doi:10.1016/j.foodchem.2015.03.021 PMID:25863602

Mehmood, T. (2021). Optimisation of food grade mixed surfactant-based l-ascorbic acid nanoemulsions using response surface methodology. *IET Nanobiotechnology / IET*, *15*(3), 309–317. doi:10.1049/nbt2.12014 PMID:34694664

Mehmood, T., Ahmad, A., Ahmed, A., & Ahmed, Z. (2017). Optimization of olive oil based O/W nanoemulsions prepared through ultrasonic homogenization: A response surface methodology approach. *Food Chemistry*, *229*, 790–796. doi:10.1016/j.foodchem.2017.03.023 PMID:28372245

Mehmood, T., & Ahmed, A. (2020). Tween 80 and Soya-Lecithin-Based Food-Grade Nanoemulsions for the Effective Delivery of Vitamin D. *Langmuir*, *36*(11), 2886–2892. doi:10.1021/acs.langmuir.9b03944 PMID:32118445

Mehmood, T., Ahmed, A., Ahmad, A., Ahmad, M. S., & Sandhu, M. A. (2018). Optimization of mixed surfactants-based β-carotene nanoemulsions using response surface methodology: An ultrasonic homogenization approach. *Food Chemistry*, *253*, 179–184. doi:10.1016/j.foodchem.2018.01.136 PMID:29502819

Mehmood, T., Ahmed, A., & Ahmed, Z. (2021). Food-Grade Nanoemulsions for the Effective Delivery of β-Carotene. *Langmuir*, *37*(10), 3086–3092. doi:10.1021/acs.langmuir.0c03399 PMID:33646002

Mei, L., Choi, S. J., Alamed, J., Henson, L., Popplewell, M., McClements, D. J., & Decker, E. A. (2009). Citral stability in oil-in-water emulsions with solid or liquid octadecane. *Journal of Agricultural and Food Chemistry*, *58*(1), 533–536. doi:10.1021/jf902665b PMID:19911843

Ozaki, A., Muromachi, A., Sumi, M., Sakai, Y., Morishita, K., & Okamoto, T. (2010). Emulsification of coenzyme Q 10 using Gum Arabic increases bioavailability in rats and human and improves food-processing suitability. *Journal of Nutritional Science and Vitaminology*, *56*(1), 41–47. doi:10.3177/jnsv.56.41 PMID:20354345

Ozogul, Y., Kuley Boğa, E., Akyol, I., Durmus, M., Ucar, Y., Regenstein, J. M., & Köşker, A. R. (2020). Antimicrobial activity of thyme essential oil nanoemulsions on spoilage bacteria of fish and food-borne pathogens. *Food Bioscience*, *36*, 100635. doi:10.1016/j.fbio.2020.100635

Patel, A., Heussen, P., Dorst, E., Hazekamp, J., & Velikov, K. P. (2013). Colloidal approach to prepare colour blends from colourants with different solubility profiles. *Food Chemistry*, *141*(2), 1466–1471. doi:10.1016/j.foodchem.2013.03.082 PMID:23790940

Qazi, H. J., Ye, A., Acevedo-Fani, A., & Singh, H. (2021). In vitro digestion of curcumin-nanoemulsion-enriched dairy protein matrices: Impact of the type of gel structure on the bioaccessibility of curcumin. *Food Hydrocolloids*, *117*, 106692. doi:10.1016/j.foodhyd.2021.106692

Ruengdech, A., & Siripatrawan, U. (2021). Application of catechin nanoencapsulation with enhanced antioxidant activity in high pressure processed catechin-fortified coconut milk. *Lebensmittel-Wissenschaft + Technologie*, *140*, 110594. doi:10.1016/j.lwt.2020.110594

Safaya, M., & Rotliwala, Y. (2020). Nanoemulsions: A review on low energy formulation methods, characterization, applications and optimization technique. *Materials Today: Proceedings*, *27*, 454–459. doi:10.1016/j.matpr.2019.11.267

Sahafi, S. M., Goli, S. A. H., Kadivar, M., Varshosaz, J., & Shirvani, A. (2021). Pomegranate seed oil nanoemulsion enriched by α-tocopherol; the effect of environmental stresses and long-term storage on its physicochemical properties and oxidation stability. *Food Chemistry*, *345*, 128759. doi:10.1016/j.foodchem.2020.128759 PMID:33310251

Salvia-Trujillo, L., & McClements, D. (2016). Enhancement of lycopene bioaccessibility from tomato juice using excipient emulsions: Influence of lipid droplet size. *Food Chemistry*, *210*, 295–304. doi:10.1016/j.foodchem.2016.04.125 PMID:27211650

Salvia-Trujillo, L., Soliva-Fortuny, R., Rojas-Graü, M. A., McClements, D. J., & Martín-Belloso, O. (2017). Edible nanoemulsions as carriers of active ingredients: A review. *Annual Review of Food Science and Technology*, *8*(1), 439–466. doi:10.1146/annurev-food-030216-025908 PMID:28125342

Sedaghat Doost, A., Afghari, N., Abbasi, H., Nikbakht Nasrabadi, M., Dewettinck, K., & Van der Meeren, P. (2020). Nano-lipid carriers stabilized by hydrophobically modified starch or sucrose stearate for the delivery of lutein as a nutraceutical beverage model. *Colloids and Surfaces. A, Physicochemical and Engineering Aspects*, *605*, 125349. doi:10.1016/j.colsurfa.2020.125349

Sharma, S., Rabbani, S. A., Narang, J. K., Hyder Pottoo, F., Ali, J., Kumar, S., & Baboota, S. (2020). Role of rutin nanoemulsion in ameliorating oxidative stress: Pharmacokinetic and pharmacodynamics studies. *Chemistry and Physics of Lipids*, *228*, 104890. doi:10.1016/j.chemphyslip.2020.104890 PMID:32032570

Silva, H. D., Poejo, J., Pinheiro, A. C., Donsì, F., Serra, A. T., Duarte, C. M. M., Ferrari, G., Cerqueira, M. A., & Vicente, A. A. (2018). Evaluating the behaviour of curcumin nanoemulsions and multilayer nanoemulsions during dynamic in vitro digestion. *Journal of Functional Foods*, *48*, 605–613. doi:10.1016/j.jff.2018.08.002

Silva, W., Torres-Gatica, M. F., Oyarzun-Ampuero, F., Silva-Weiss, A., Robert, P., Cofrades, S., & Giménez, B. (2018). Double emulsions as potential fat replacers with gallic acid and quercetin nanoemulsions in the aqueous phases. *Food Chemistry*, *253*, 71–78. doi:10.1016/j.foodchem.2018.01.128 PMID:29502846

Solans, C., Izquierdo, P., Nolla, J., Azemar, N., & Garcia-Celma, M. J. (2005). Nano-emulsions. *Current Opinion in Colloid & Interface Science*, *10*(3-4), 102–110. doi:10.1016/j.cocis.2005.06.004

Sotomayor-Gerding, D., Oomah, B. D., Acevedo, F., Morales, E., Bustamante, M., Shene, C., & Rubilar, M. (2016). High carotenoid bioaccessibility through linseed oil nanoemulsions with enhanced physical and oxidative stability. *Food Chemistry*, *199*, 463–470. doi:10.1016/j.foodchem.2015.12.004 PMID:26775996

Soukoulis, C., Tsevdou, M., Andre, C. M., Cambier, S., Yonekura, L., Taoukis, P. S., & Hoffmann, L. (2017). Modulation of chemical stability and in vitro bioaccessibility of beta-carotene loaded in kappa-carrageenan oil-in-gel emulsions. *Food Chemistry*, *220*, 208–218. doi:10.1016/j.foodchem.2016.09.175 PMID:27855891

Talegaonkar, S., Mustafa, G., Akhter, S., & Iqbal, Z. (2010). Design and development of oral oil-in-water nanoemulsion formulation bearing atorvastatin: In vitro assessment. *Journal of Dispersion Science and Technology*, *31*(5), 690–701. doi:10.1080/01932690903120540

Tan, C., & McClements, D. J. (2021). Application of advanced emulsion technology in the food industry: A review and critical evaluation. *Foods*, *10*(4), 812. doi:10.3390/foods10040812 PMID:33918596

Tripathi, A. D., Sharma, R., Agarwal, A., & Haleem, D. R. (2021). Nanoemulsions based edible coatings with potential food applications. *International Journal of Biobased Plastics*, *3*(1), 112–125. doi:10.1080/24759651.2021.1875615

Tsai, Y.-J., & Chen, B.-H. (2016). Preparation of catechin extracts and nanoemulsions from green tea leaf waste and their inhibition effect on prostate cancer cell PC-3. *International Journal of Nanomedicine*, *11*, 1907. PMID:27226712

Walia, N., & Chen, L. (2020). Pea protein based vitamin D nanoemulsions: Fabrication, stability and in vitro study using Caco-2 cells. *Food Chemistry*, *305*, 125475. doi:10.1016/j.foodchem.2019.125475 PMID:31518841

Wang, S., Shao, G., Yang, J., Liu, J., Wang, J., Zhao, H., Yang, L., Liu, H., Zhu, D., Li, Y., & Jiang, L. (2020). The production of gel beads of soybean hull polysaccharides loaded with soy isoflavone and their pH-dependent release. *Food Chemistry*, *313*, 126095. doi:10.1016/j.foodchem.2019.126095 PMID:31923873

Wang, X., Jiang, Y., Wang, Y.-W., Huang, M.-T., Ho, C.-T., & Huang, Q. (2008). Enhancing anti-inflammation activity of curcumin through O/W nanoemulsions. *Food Chemistry*, *108*(2), 419–424. doi:10.1016/j.foodchem.2007.10.086 PMID:26059118

Wilking, J. N., & Mason, T. G. (2007). Irreversible shear-induced vitrification of droplets into elastic nanoemulsions by extreme rupturing. *Physical Review. E*, *75*(4), 041407. doi:10.1103/PhysRevE.75.041407 PMID:17500894

Wulff-Pérez, M., Martín-Rodríguez, A., Gálvez-Ruiz, M. J., & de Vicente, J. (2013). The effect of polymeric surfactants on the rheological properties of nanoemulsions. *Colloid & Polymer Science*, *291*(3), 709–716. doi:10.100700396-012-2780-1

Xiong, Y., Li, S., Warner, R. D., & Fang, Z. (2020). Effect of oregano essential oil and resveratrol nanoemulsion loaded pectin edible coating on the preservation of pork loin in modified atmosphere packaging. *Food Control*, *114*, 107226. doi:10.1016/j.foodcont.2020.107226

Yazgan, H. (2020). Investigation of antimicrobial properties of sage essential oil and its nanoemulsion as antimicrobial agent. *Lebensmittel-Wissenschaft + Technologie*, *130*, 109669. doi:10.1016/j.lwt.2020.109669

Yin, L.-J., Chu, B.-S., Kobayashi, I., & Nakajima, M. (2009). Performance of selected emulsifiers and their combinations in the preparation of β-carotene nanodispersions. *Food Hydrocolloids*, *23*(6), 1617–1622. doi:10.1016/j.foodhyd.2008.12.005

Yoplac, I., Vargas, L., Robert, P., & Hidalgo, A. (2021). Characterization and antimicrobial activity of microencapsulated citral with dextrin by spray drying. *Heliyon*, *7*(4), e06737. doi:10.1016/j.heliyon.2021.e06737 PMID:33898839

Zhang, R., Zhang, Z., Zou, L., Xiao, H., Zhang, G., Decker, E. A., & McClements, D. J. (2016). Enhancement of carotenoid bioaccessibility from carrots using excipient emulsions: Influence of particle size of digestible lipid droplets. *Food & Function*, *7*(1), 93–103. doi:10.1039/C5FO01172H PMID:26583923

Ziani, K., Fang, Y., & McClements, D. J. (2012). Encapsulation of functional lipophilic components in surfactant-based colloidal delivery systems: Vitamin E, vitamin D, and lemon oil. *Food Chemistry*, *134*(2), 1106–1112. doi:10.1016/j.foodchem.2012.03.027 PMID:23107734

KEY TERMS AND DEFINITIONS

Emulsion: Homogeneous mixture of immiscible liquids stabilized by emulsifiers.

High Energy Methods: The methods that use work-based emulsification approach which requires mechanical devices such as ultrasonic homogenizer, high-pressure homogenizer and microfluidizer.

Low Energy Methods: The methods that use simple process for nanoemulsion preparation such as stirring and change in composition and formulation parameters e.g., oil to water ratio, oil type, temperature, and surfactant concentration.

Micronutrients: The essential nutrients that are required in small amount such as vitamins and minerals.

Nanoemulsions: Nanoemulsions are smaller size (<200nm) dispersions consist of two immiscible liquids stabilized through surfactants.

Surfactants: The surface-active agents that reduce the interfacial tension between two liquids.

Vitamins: Essential organic micronutrients that are required in small amount for the proper functioning of human body.

Chapter 21
Applications of Nanoemulsion for the Wound-Healing Process

S. M. Shaheedha
B.S. Abdur Rahman Crescent Institute of Science and Technology, India

ABSTRACT

Attention to nanoemulsions has significantly grown in recent years as a result of their unique features like better stability, special appearance, higher performance, and sensorial merits. Chronic injuries are the consequence of a disturbance in the extremely coordinated cataract of wound healing actions. Nevertheless, correlated with variations in the timescales of various physical methods embroiled in tissue renewal, the aggression of the tumor microenvironment, rich in decaying enzymes, as well as its increased pH, demands the use of efficient drug delivery applications. This chapter summarizes that the various stages of wound healing include four phases: hemostatic stage, inflammation, proliferation, and remodeling process, respectively. Moreover, the major reported classes of lipid-based elements were either vesicular (liposome, permeation increased vesicle, etc.), emulsion-based behavior (nano-emulsion and micro-emulsion), or comprise a solid-based liquid matrix in the wound-healing process.

INTRODUCTION

Nanotechnology (NT), rapidly emerging arena which allows for the creation of ingredients with novel realms, new characteristics, and a broader variety of uses (Achari and Kowshik, 2018; Agarwal et al., 2018). It's considered one of the most important inventions of the twenty-first generation. The National Nanotechnology Initiative of the United States describes nanoparticles (NPs) as particles with a diameter of 1–100 nanometers (Roco, 2007). Both inorganic and organic materials could be used to make NPs and nanoformulations (NF) (Jampilek and Kralova, 2018; Pisárčik et al., 2018). As previously stated, nanoscale materials alter the characteristics and performance of all constituents, allowing for the improvement and innovation of a wider variety of manufacturing, agricultural, medicinal, and medical devices (Patra et al., 2018). Nanoemulsions (NEs) with favorable physico-chemical properties have been widely researched as pesticide delivery vehicles (Hayles et al., 2017; Hazra et al., 2017). NT is, nevertheless, progressively being used in medicine, including theranostic and drug delivery (Prasad et al., 2016;

DOI: 10.4018/978-1-7998-8378-4.ch021

Prasad et al., 2016a). Sustainable drug release and also a decrease in the necessary drug volume could be achieved through encapsulating drugs in NF, and nanomaterials (NMs) can also be used to manage and mitigate infections induced by resistant strains (Jampílek and Kráľová, 2019; Pentak et al., 2016).

NEs with favorable physico-chemical properties have been intensively investigated as pesticide delivery vehicles (Hayles et al., 2017; Hazrati et al., 2017). NT is, nevertheless, gradually being used in medicine and other fields like drug distribution and theranostic. Most natural products, like embedded active compounds as well as floral oils for the NE oil process are used to make bio-based NEs (Raviadaran et al., 2018). The structure of NEs can be used to classify them. Water has been the continuous stage and oil is the dispersal stage in a 'water-based' or oil-in-water (O/W) emulsion (Mason et al., 2006; Singh et al., 2017), while the adjusted state produces an 'oil-based' or water-in-oil (W/O) emulsion. NEs enable for uniform accumulation and incorporation of active compounds via the skin's surface due to their small pore sizes (Sonneville-Aubrun et al., 2004). Because of the greater surface area and low viscosity of the entire emulsifying agent, NEs have greater ingredient diffusion efficacy (Bouchemal et al., 2004), involving just 3–10% surfactants during preparations (Tan et al., 2016).

Surfactant absorptions of 20 percent or developed are needed in microemulsions (Dehghani et al., 2017). As a consequence, NEs become more liquid (at lower oil concentrations), by attractive physical characteristics and skin, particularly when thickeners are not used. Destabilization phenomena like creaming or sedimentation, coagulation, and coalescence, which are common in emulsions, are also avoided, thanks to significant steric stabilization among submicrometric droplet sizes. The influence of the ordering wherein the different substances are combined during formulations is a significant feature to study when making NEs. It is crucial to stress that surfactants must be combined with the oily process before NEs can be prepared. This allows for the production of highly favorable conditions for both the development of NEs. Surfactants combined with water in the early steps of production, on the other hand, would favor the production of 'macroscopic' emulsions (Anton and Vandamme, 2011).

NEs are currently made by means of various methods, which can be considered as either lower- or higher-energy emulsification procedures, or a mixture of both (Ngan et al., 2014). The usage of mechanical devices to produce powerful extreme elements that breakdown the oil and water processes to create oil-based drops characterizes high-energy strategies. Higher-pressure homogenizers, microfluidizers, and sonication techniques are used in this technique (Al-Sabagh et al., 2011). Lower-energy methods, on the other hand, emulsify using the system's own chemical energy. This is done by redirecting the surfactant, co-surfactant, and inert ingredients in the formulation's intrinsic physicochemical characteristics (Saberi et al., 2013).

NP-based distribution has recent times considered as a leading method for treating dermatological disorders (Amer et al., 2019; Hatem et al., 2018), including wounds caused by multiple reasons, in order to expedite wound healing and prevent infection associated with the slower recovery process (Li et al., 2016). NPs have been documented to permeated the stratum corneum and enter the profounder epidermal and dermal coats of the skin via one or more combined processes, including their tiny size, deformable characteristics, lipid behavior, or lipid-fluid nature (Bsieso et al., 2015; Hatem et al., 2018). One among the most widely explored fields is the usage of NPs for wound healing, wherein NPs were either equipped with multiple bioactive agents or demonstrated wound healing properties without even any dosage form. As a consequence, the focus of this chapter was to demonstrate the greatest widely employed NPs, with a focus on various lipid-based schemes as organic NPs and gold, silver, and metal oxide NPs as inorganic NPs.

NE formulations

NEs were bi-phasic suspensions of two im-miscible liquid phases: whether W/O or O/W dewdrops alleviated through an amphiphilic surfactant, in which very minor emulsified drops with really tiny droplet size (usually oil droplets in water) with very tiny emulsion dewdrops (normally oil dewdrops in water dimensions of the direction of 100 nanometers (typically 500 nanometers) and resulting from a thermo-dynamic process. The slower kinetics of a viewpoint in a non-equilibrium system are defined by kinetically balanced due to their lack of destabilization. Because of their small scale, NEs have useful assets like higher surface arena per unit mass, robust stability, visual transparency, and tunable rheology (Singh et al., 2017; Sabry et al., 2018).

Since NEs' tiny size contributed greatly to the protection of droplet coagulation and agglomeration, the destabilizing method was controlled solely through the Ostwald ripening period (Anton and Vandamme, 2011). Surface functionalization of NEs can be utilized to alleviate the O/W interfacer throughout emulsification because of the weakening of NE caused by the shift in drops size caused through Ostwald ripening (Qadir et al., 2016). NEs manufactured from biodegradable and biocompatible components may be utilized for drug/active component encapsulation in medication and the food processing, with the added benefit of continuous and controlled release of embedded substances. The microemulsion (ME), "dispersion composed of water, oil, and surfactant(s) and which an isotropy and chemically steady system by distributed domain width ranging generally from 1 to 100 nm, typically from 10 to 50 nm, in which the word "oil" denotes that any water-insoluble fluid," as per the IUPAC description (Slomkowski et al., 2011). Figure 1 exemplifies the variances among NE and ME, both are comprised of oil, water and etc. These 2 kinds of W/O colloid-based dispersals have a parallel structural assembly: a hydrophobic shell-based oil and surfactant tail and a hydrophilic-based core of water and surfactant headed assemblies.

Figure 1. Representation of water-in-oil micro and NEs comprising of oil, water and surfactant.

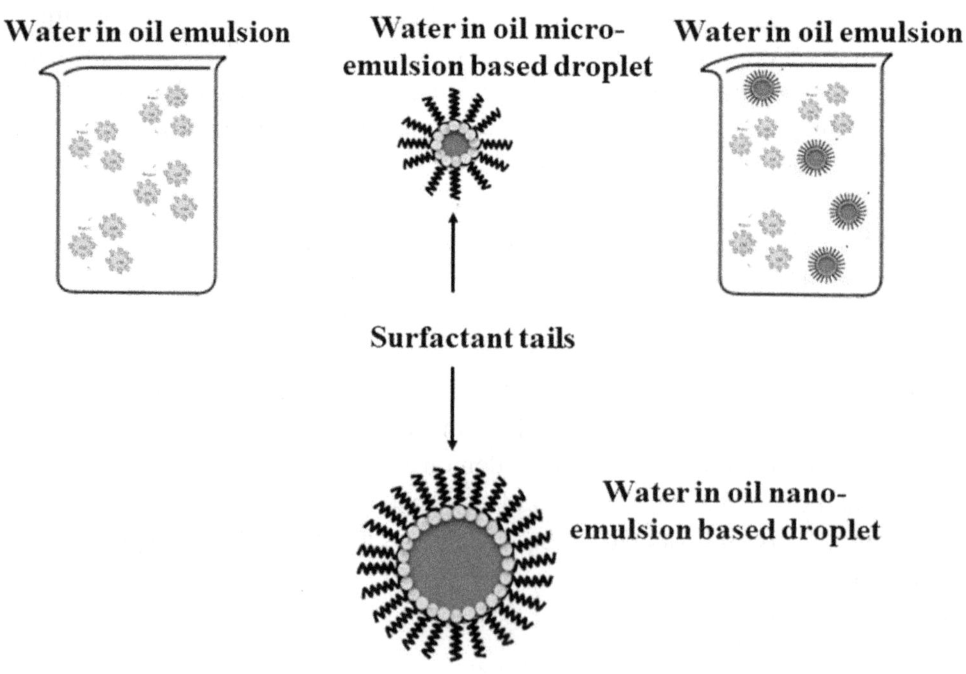

In particular, NEs can be made using "higher-energy" approaches, which rely on individual users to provide sufficient energy to enhance the W/O interface part and produce nano-scale particles, or "lower-energy" techniques, which rely on spontaneous emulsifier without the usage of device or energy. Higher-energy strategies for preparing NEs include higher-pressure homogenization, micro-fluidization (MF), sonication assimilation, jet augmenter elevated ultrasonic process, or membrane emulsifying agent. Elevated approaches are structured to provide the energy needed for emulsifier by exposing it to a destructive hydro-dynamic pressure, such as laminar or turbulent shear/condensation (Jaiswal et al., 2015).

For instance, dual channel MF, is an effective approach for fabricating well NEs with increased oil-based loading stages that might be useful for numerous industrial uses, whereas the single-channel process was proven to be active only at generating NEs at comparatively lower oil absorptions (Bai et al., 2016). With MF stress, number of iterations, and emulsifier concentrations in -carotene NEs generated by MF, droplet extent declined from 416.0 to 97.2 nm, and homogenization stress of 120 MPa and three rotations were calculated as the ideal circumstances for manufacturing such NEs (Jo et al., 2014). Abbas et al., described the concepts and manufacturing machinery of higher-intensity ultrasound, explored the impact of cavity bubbles in the nutrition industry O/W NEs, and addressed the technology's technological challenges, problems, and potential prospects (Abbas et al., 2013).

Applications of NEs in Wound and Others

NEs, which are alleviated heterogeneous schemes made up of 2 immiscible fluids, have a huge amount of possible in biological uses, particularly drug delivery. Enumeration of lipophilic drugs results in medicinal NFs that, include bioavailability adjustment, alternate administration paths, releasing of medicines, and, as a result, a decrease in the necessary drug volume (Pentak et al., 2016; Tayeb et al., 2018).

Transdermal-based NE Formulations

Transdermal drug delivery is a great option for medications with deprived oral biological availability owing to initial-pass digestion. Since the top coating of the epidermis prevents most drugs from getting through, chemical permeability stimulants that can render the stratum corneum more pliable for medications and decrease the initial skin blockade through various mode of actions were employed for drug delivery systems. Physical methods (such as sonophoresis, iontophoresis, electrophoresis, microneedles, and so on) can also be used to improve drug permeation (Jampilek and Brychtova, 2012). Nanoscale drug delivery systems, such as NEs, are commonly used to improve epidermal and dermal drug accumulation (Iqbal et al., 2018). Despite the fact that the skin acts as a natural physical shield towards therapeutic NPs can be delivered via particle diffusion, particularly in infected areas, skin and hair follicle opens, and are ideal candidates because of the first stage, transdermal delivery drugs have a lower oral bioavailability of metabolic rate. The fact that NEs can also serve as transdermal vehicles is beneficial permeation modulators lack the usage of extra penetration relishes (Shakeel et al., 2010).

Rai et al. reviewed the advancement in delivery systems via NE, with a focus on dermal and transdermal administering, as well as the most effective semisolid dosing schedules for the different types of NEs, and also the impact of element diameter and zeta potential on drug distribution via dermal- and trans-dermal routes (Rai et al., 2018). Hydrogel-thickened NEs utilizing oxidizing polymer, such as CS, could resolve the lower thickness of NEs which make them inappropriate for current use. Yukuyama et al. (2017) addressed the problems and useful programs of NEs as a medicine transporter.

De Souza et al. (2018) reviewed the uses of NEs in dermatology, emphasizing the benefits over all other dermatological treatments associated with higher area of contact due to atom size which results in higher drug efficiency. Salim et al. addressed the ability of medicine-loaded NEs for the therapy of psoriasis to improve efficiency and reduce adverse properties, noting that a small droplet size could improve drug release and diffusion via the psoriasis layer of the skin. Nastiti et al. (2017) concentrated on the arrangement and interpretation of MEs and NEs for topical- and transdermal-based distribution, as well as the process of skin distribution through the stratum corneum and through hairy cells.

Nano-emulgels which are essentially O/W NEs that have been gelled through both the addition of a gelling substance, and for that the gel stage in the formulations is nongreasy and stabilizes the preparation by lowering surface and interfacial tension, may be regarded a new transdermal distribution process that can circumvent deprived oral bio-availability of the drug and extra target specific to the location of a wound (Choudhury et al., 2017). The NE-comprising drug was integrated to a gel base of nano-emulgels. Lipophilic-based drugs can be simply integrated, and owing to the finely emulsion droplets of the NE process, the skin permeation of the combined medications might be increased by many folds, resulted in massively enhanced pharmacodynamic and pharmacokinetic profiling of the lipophilic drugs. As a result, nano-emulgel compounds may be suggested as promising and effective prospects for future bioavailability enhancement of lipophilic drugs (Sengupta and Chatterjee, 2017). Numerous lipid substances employed for the formulation of NEs for topical and transdermal route of delivering systems were discussed in Pawar and Babu, (2014).

Because of improved permeation across rat skin, the optimized clove oil-based olmesartan NE demonstrated a 1.23-fold improvement in bioavailability as comparison to the oral medication of the drug. It can be employed as an anti-hypertensive dose type for successful transdermal distribution of Olmesartan (Aqil et al., 2016). O/W NEs comprising the skin permeation improver oleic acid/eucalyptol as oil pronounced improved the skin diffusion of embedded caffeine and naproxen relative to the aqueous control results. The rise in caffeine stratum corneum bioavailability and skin diffusion was attributed to the increased in caffeine maximum flux. Caffeine extreme flux improvements were related to a synergetic rise in caffeine stratum corneum bioavailability and skin diffusivity, with both the stratum corneum solubility being the dominant factor greater naproxen fluxes determining factor (Abd et al., 2016; Mostafa et al., 2015).

Transdermal NE containing 2 percent Foeniculum vulgare Mill. EO, 5.6 percent oleic acid, 68 percent Smix, purified water demonstrated a higher potential for lowering plasma glucose ranges in diabetic rats for seven days after such a single topical use of 120 mg/kg fennel EO, taking glucose levels back to usual in diabetic rats (Mostafa et al., 2015). As compared with the control solutions, O/W NEs comprising minoxidil, an antihypertensive vasodilator, as well as the skin permeation stimulator oleic acid as stages significantly increased medicine penetration via skin. Eucalyptol NEs strengthened minoxidil retention throughout the stratum corneum and bottomless layers of the skin, whereas oleic acid formulas had the highest hair follicle penetration. In both NE processes, rises in minoxidil stratum corneum solubility and skin diffusivity was expressed in increased drug maximum flux, which was attributed to increased fluidity and destruction of stratum corneum lipids (Abd et al., 2018).

Topical formulations comprising NEs usually contains isoflavone-rich soybean extracts, which are deemed attractive skin anti-aging agents because of its antioxidant activity, may be utilized to preserve the skin from UVA/UVB oxidative harm. Topical application of antioxidant hydrogels comprising an Achyroclinesatureioides extract-loaded NEs was found to be successful in protecting the porcine ear skin from oxidative stress caused through UVA or UVB radiation (Balestrin et al., 2016). Brownlow et al. (2015) created vitamin E-containing NE vehicles packed by genistein for chemoprevention of UVB-stimulated skin injury, demonstrating improved drug delivery via the dermis. Nam et al. (2018) mixed surfactant to vitamin E antioxidant body oil and $NaNO_2$ solutions to create W/O NE that produced nitric oxide. The blood NOx level in the penis improved until this NEs with NO was distributed on the penis skin for middle-aged dogs, resulted in penile erection with no significant topical or systemic risk factors, indicating that these NEs with NO could be used to treat penile erection in humans. W/O NEs may be employed as a noninvasive treatment for erectile dysfunction patients caused by a lack of reaction to phosphodiesterase type 5 (PDE5) antagonists including Viagra® and Cialis® are two examples.

According to Bonferoni et al. (2018) NEs of CS oleate containing -tocopherol and having particle sizes of 220 nm are suitable for topical wound healing process (Zimber et al., 2012). As compared to ASX NEs and ASX solution, ASX-loaded carboxymethyl CS functionalized NE preparations by spherical drops with a mean width greater than 100 nm and a -ve overall surface characteristic demonstrated greater ASX thermal strength and skin permeation, and also reduced toxicity (Hong et al., 2017 Argenta et al., 2016).

Argenta et al. (2014) suggested using a factorial design approach to improve the lipid formulations of topical antiherpetic NEs containing the isoflavone genistein. The medication can be directly supplied to the keloid lesion because the 10,11-methylenedioxycamptothecin-loaded hyaluronic acid NEs have a strong skin penetrable potential through human keloid skin. Cell cycle regulation, which stopped cells from reaching mitosis by arrested them at G1/S, was also found to have a growth-inhibitory effect (Gao et al., 2014).

In vitro, NEs with co-enconjugated C6 ceramide and paclitaxel in tributyrin incorporated 2- and 2.4-fold further paclitaxel to feasible skin coatings of porcine skin than MEs at 4 and 8 hours, respectively, and 1.9-fold more C6 ceramide at 8 hours, suggesting that the medications are co-localized largely in the epidermis. Independently expressed paclitaxel and ceramide in NEs had EC_{50} matters pertaining to melanoma treated cells that were 4- and 13-fold lesser than the unconjugated preparations, and synergic action was discovered when both inorganic chemicals were co-encapsulated, with the EC50 being 2.5–4.5-fold lower, and the defined mixture index also suggesting an extremely useful. When NEs were applied to the epidermis of three-dimensional bio-engineered melanoma prototypes for 48 hours, the epidermis was severely damaged, within a limited cells left (Carvalho et al., 2017).

Curcumin (CUR) NE for transdermal delivery by standardized NE considerably increased CUR permeation flux from the hydrophilic matrices gel Viscolam AT 100P, whereas NE configuration increased significantly CUR permeability flux either from the hydrophilic matrices gel Viscolam (Rachmawati et al., 2015). CUR-loaded NE was integrated into gel utilizing Carbopol® 980 and evaluated on Freund's total adjuvant-stimulated arthritic mouse models after topical use of CUR-NE gel in Wistar rats displayed significant reversal of arthritic signs, implying a NF may have medicinal properties (Naz and Ahmad, 2015). The use of CUR-loaded myristic acid MEs to study CUR circulation in neonate pig skin revealed dermal CUR aggregation and transdermal CUR distribution. CUR embodied in ME repressed Staphylococcus epidermidis microbial growth and it was identified as 12-fold further efficient than CUR immersed in dimethyl sulfoxide, implying these MEs would be utilized as a substitute therapy for S. epidermidis-related infections and acne vulgaris (Liu and Huang, 2012).

Zhao et al. (2018) discovered that woody oil-based emulsive-based schemes could effectively distribute unwell water - soluble regular alkaloids, raising lung cancer cell sensitivity. In human alveolar carcinoma cells, O/W cinnamon oil NE and vitamin D embedded cinnamon oil NE induced DNA disruption with an important upsurge in micronucleus frequency, halted cellular propagation within the G0 or G1 stage, and displayed greater production of Bax, capase-3, and caspase-9 proteins and reduced levels of BcL2 proteins alongside important improvement in apoptotic cell populace and lack of mitochondrial membrane. In S180 murine sarcoma cancerous cells, the anticancer capacity of both O/W NE with embedded vitamin K2 and NE combining VK2 by a ligand coupled sialic acid-cholesterol bound on the surface was examined. It was discovered that giving VK2 NE to syngeneic mice by subcutaneous injection developed S180 tumors through i.v. or intragastric administering led to significant tumor growth inhibition, with the surface-modified NE having a stronger effect, and both NEs being highly toxic (Shi et al., 2018).

According to Ahmad et al., (2018) the optimized silymarin NE with a mean particle dimension of 21.24 nm decreased tumor cell feasibility, improved ROS and chromatin condensation were used as an effective transporter for silymarin distribution to human hepatocellular carcinoma without harming human cells. In human umbilical vein endothelial cells, methyl jasmonate full NE with a mean particle diameter of 75.06 nm and a PDI of 0.017 was identified as more successful in destroying tumor cells; it triggered a greater sub-G1 halt than methyl jasmonate and demonstrated highly substantial lack of toxicity (Habibi et al., 2017). The EC50 values for cytotoxicity towards the A549 tumor cell line calculated by pure Casearia sylvestris Sw. extract showed the NF is 4-fold more effective (Pereira et al., 2017).

Paclitaxel NEs were evaluated towards Bx-PC-3 and PANC-1 pancreatic cancer cells where the tocopherol oil center of TocosolTM was replaced with -tocotrienol, and d-tocopheryl polyethylene glycol 1000 succinate by PEGylated -tocotrienol displaying droplet size 300 nm. When NEs loaded for free PTX were used as the heart, the quickest escape was reported. Cytotoxic effect was also enhanced by changing -tocopherol with -tocotrienol, and a dramatic rise in activities were calculated when PTX lipid complexes have been used (Abu-Fayyad et al., 2018; Abu-Fayyad and Nazzal, S. (2017). When comparison to the free drug, confinement of gemcitabine—tocotrienol adducts into NEs significantly enhanced their anticancer effects towards Bx-PC-3 and PNAC-1 pancreatic cancerous cells. Furthermore, as contrasted to free and conjugated gemcitabine in solution, gemcitabine—tocotrienol conjugates were found to be minimally impacted by the deamination revocation reaction *in vitro* (Steuber et al., 2016).

Wound Healing Method

Wound healing, complex method wherein the body's tissues rebuild themselves after being injured (Nguyen et al., 2019). It is a mechanism in which intercellular pathways are activated, tissue integrity is coordinated, and homeostasis is maintained. Wound healing can be categorized based on the extent and deepness of the wound (Zhong et al., 2010). An acute wound occurs when the honesty of the soft tissue cover that surrounds every portion of the body was compromised. The scale, depth, and anatomic structures involved in an acute wound describe it. Healing time and the distinction amongst acute and chronic wounds are difficult to predict and depend on the location and origin of the wound, and the patient's age and physical health.

Tissue repair, modest mathematical method wherein growth factors induce cell growth, resulting in the incorporation of complex changes connecting soluble intermediaries, blood cells, extracellular matrices (ECM) formation, and parenchymal gene expression (Shaw and Martin, 2009). Wound repair was a natural physiological procedure which includes the interaction of dermal and epidermis, the ECM, plasma-based enzymes, growth factors, etc. to promote curative process (Metcalfe and Ferguson, 2008; Troxler et al., 2006). The hemostasis stage, the provocative process, the propagation process, and the remodeling stage are the four stages of healing process. Initial healing, deferred primary healing, and recovery by direct extension are some of the other types of wound healing (Maxson et al., 2012).

Despite the fact that cellular and extracellular components are classified differently, their relations are identical (Turksen, 2018).

Different Stages of Wound Repair

There are 4 levels in the wound healing method as shown in Figure 2.

Stage 1 – hemostasis

This process includes a series of complex reactions, the most important of which are hemostasis and clot development. A fibrin mesh by distributed platelets is incorporated in the clot. The mesh catches crimson red cells, which were the main cause of clot plugging. Fibrin, protein that is formed as a result of coagulation cascades that are caused by vascular injuries. Using different events, an intrinsic and an extrinsic clotting cascade is triggered (Hawkins and Abrahamse, 2006). When blood is introduced to foreign materials, Factor XII is activated, which starts the intrinsic coagulation pathway. The extrinsic coagulation process is started when a tissue element binds component VII. Tissue problem isn't found on vascular endothelial-based cells, but it was found in large quantities on extravascular cell exteriors, particularly adventitial fibroblasts. The component is exposed when the cells are injured. Both pathways lead to the production of thrombin, which catalyzes the breakdown of fibrinogen to fibrin.

Figure 2. The four phases executed in wound healing process.

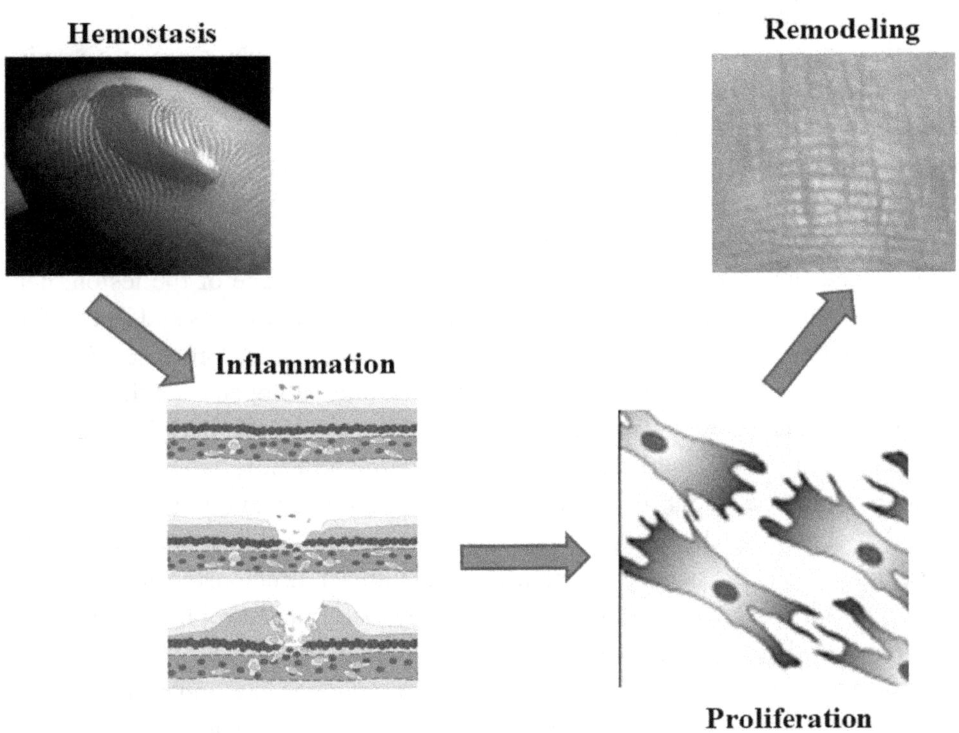

Fibrin, initial component of the provisional matrices which bureaucracy inside the wound during the initial recovery phase, in addition to contributing to hemostasis. Fibronectins are a type of glycoprotein that helps moving cells bind to the fibrin latticework, and they are extremely essential in the earliest matrix as well as the matured dermis (Zimber et al., 2012). Fibroblasts and epithelial cells both contain fibronectin. Normal endothelial cells contain prostacyclin, which prevents platelet aggregation, so activation of the hemostatic pathways is restricted to the site of injury. Antithrombin III associates' thrombin and attaches its last time in uninjured areas, and protein C damages factors V and VII (Ferguson and O'Kane, 2004; Penn et al., 2012). While the causes for clot activation wear off, the blood coagulation and platelet aggregation strategies come to an end. Clot deterioration begins as soon as the masses are formed. (Thom, 2011).

Stage 2 – inflammation

The lesioned blood-vessels agreement and the leaking blood coalesces in a vascular anti-inflammation-based response, helping to protect the integrity of the vessel. Coagulation is the accumulation of thrombocytes and platelets within a fibrin culture, which is based on the stimulation and accumulation of certain cells through particular factors (Martin, 1997). Additionally, reestablishing homeostasis and creating a shield against microbe invasion, the fibrin group organizes the requisite temporary network for cellular migration, restoring the skin's role as a defensive barrier and sustaining the skin's integrity (Medrado et al., 2003). This also allows for mobile movement to the lesion's micro - environmental and fibroblast multiplication enhancement. The influx of leukocytes throughout the wound area characterizes the anti-inflammatory response. In overall, the cell-based response is formed within the initial 24 hrs and can last until 2 days (Buckley, 2011).

Astrocytes and Langerhans cells which secreted chemo- and cyto-kines, can cause a rapid stimulation of immune cells inside the tissue. Infection was a localized and defensive tissue response which was triggered through the lesion, resulting in tissue damage. Inflammatory cells are significant in wound repair because they aid in the release of lysosomal enzymes and ROS and the smoothing of different mobile particles (Gurtner et al., 2008). Buckley claims in a report that the association of leukocytes and stromal cells resolves through the anti-inflammatory awareness at a certain point during an acute anti-inflammatory reaction (Thuraisingam et al., 2010). On the surface of the lesion, neutrophils are responsible for expressing a large number of proanti-inflammatory cytokines and a large number of fantastically effective antibacterial substances, like ROS, cationic peptides, and proteases. Anti-inflammatory cytokines formed by stimulated platelets, endothelial cells, and pathogenic marketers' decomposition products attracts them. Cells that play major part in the easy up were recruited. In addition to contributing to the mortality of the tissue, marketers invading (Rodero and Khosrotehrani, 2010).

Just a few hours after the abscess, a large number of neutrophils transmigrate over the blood capillary walls through endothelial cells, which may be triggered by pro-anti-inflammatory cytokines like interleukin 1 (IL1), TNF-α, and IFN-γ at the site of the lesion. These cytokines stimulate the development of a variety of adhesion molecules like selectins and integrins, are a key factor in neutrophil diapedesis because they associate with that already exist on the membranes layer of endothelial cells (Singer and Clark, 1999). The choice of fibrin and ECM blood clotting, the urging of angiogenesis, and reepithelialization are all factors that referent cells have an impact on during tissue repair. As early as 48 years after the initial of the lesion, monocytes from nearby blood vessels, which also invade the lesion region, accelerate their movement and, with the appearance of brand-new genic expression levels, distinguish into macrophages.

Enabled by chemokine signaling, they can serve as antigen-gifting cells and valuable resource neutrophils in phagocytosis (Li et al., 2003). As either a result, the main population of macrophages within the lesion is activated in relation to resident macrophages. As a result of chemotactic compounds in the blood, as ECM protein fragments can be seen transforming development, TGF and MCP1 are two types of growth factors (Mendonça and Coutinho-Netto, 2009). On the basis of the profile's macrophages can be categorized based on their genic expression M1 and M2 are normally activated.

Macrophages perform phagocytosis of muscular waste, as well as the development and release of cytokines, proangiogenic, anti-inflammatory, and fibrogen, and free radicals. When macrophages secrete chemotactic features, they attract a variety of anti-inflammatory cells to the wound-repair site. Inclusion, they make prostaglandins that are powerful hormones, vasodilators, which affect micro-blood permeability a vessel such components, when combined, aim to activate endothelial cells. Coutinho-Netto also makes PDGF, FGF and VEGF are two of the most significant growth factors, main cytokines that can promote the development of tissue of granulation (Burnouf et al., 2013).

Stage 3 –proliferation

This section is made up of many subphases. Those subphases are no lengthier seen in distinct time edges, but rather as a popular and continuous procedure. "Fibroplasia, matrix deposition, angiogenesis, and reepithelialization" are the subphases (Giusti et al., 2013). Fibroblasts also moved into the wound on days 5–7, setting down new collagen subtypes I and III. Kind III collagen prevails earlier in normal wound healing, although this is modified later with the use of kind I collagen. Tropocollagen was the counterpart of all categories of collagen and altered in the uneven endoplasmic reticulum of the cell, where proline and lysine were hydroxylated. Three tropo-collagen elements create a triple lefthanded triple helix, known as procollagen, thanks to disulfide bonds. Peptases inside the cell wall cleave terminal peptide chains as procollagen is absorbed into the extracellular space, forming genuine collagen fibrils. Peptases inside the cell wall cleave terminal peptide chains as procollagen is metabolized into the extracellular space, allowing it to expand, collagen fibrils that are real. The wound is dripping with blood, GAGs (GLycosAminoGlycans) and fibronectin were developed by way of fibroblasts (Posthauer et al., 2013).

Heparan sulphate, hyaluronic acid, and keratan sulphate are examples of GAGs. GAGs which were covalently bound to a protein center and subsidize to matrix deposition are known as proteoglycans. The process of angiogenesis begins with the identification of vessel offshoots. The development of recent vasculature necessitates the deterioration of the ECM and basement membrane, as well as endothelial cell proliferation, mitosis, and maturation. Angiogenesis is thought to be modulated by primary FGF and vascular endothelial growth factors. The movement of cells from of the wound's periphery and adnexal structures causes reepithelization. The first step in this method is to find a suitable location within 24 hours, cells disperse (Posthauer, 2012).

Stage 4 – remodeling

Reworking starts 2 to 3 weeks after the start of the lesion which may lasting for a year of 1/3 of the healing process. The remodeling stage's main aim is to get the most tensile energy out of the ECM by reorganizing, decaying, and resynthesizing it. In this end process of lesion reconstruction, an effort is made to restore natural tissue assembly, and the granulation tissue was slowly redesigned, resulting in scar tissue which was much low cell and vascular 3 and has an advanced improve in collagen fiber consideration. Collagen is remodeled from type III to I in this chapter. The diversity of blood vessels in the injured region retreats and decline, and cellular interest decreases. After the third week, the wound goes through a series of changes called transforming that can last for years after the first wound has healed. When compared to primary healing, secondary recovery is more successful. The wound's maximum tensile strength was accomplished by the 12th week has passed, and the final mark has appeared. The authentic skin's tensile strength is best 80 percent and it has taken its place (Boyapati and Wang, 2007).

Factors Influencing Wound Healing

Wound healing can be slowed by a variety of factors. The factors that affect reparation can be divided as two categories: local and systemic. Local components were those which have a direct effect on the wound's characteristics, whilst systemic components were the person's overall healthiness or illness condition, which affects his or her capability to reconcile. Numerous elements were interconnected, and systemic variables influence wound healing from nearby influences (Brem and Tomic-Canic, 2007).

Oxygenation

Oxygen is required for cellular digest, particularly vitality generation through adenosine triphosphate, and it was required for closely all wound repair procedures. It protects wounds from contagion, stimulates angiogenesis, improves keratinocyte isolation, motion, and reepithelialization, and boosts fibroblast reproduction and collagen incorporation, and speeds wound solidity (George Broughton et al., 2006). Furthermore, the amount of superoxide (O_2) generated by polymorphonuclear leukocytes (a key component in oxidative killing pathogens) is strongly reliant on oxygen levels. Restoration is hampered in wounds whereby oxygenation has not been preserved. Restoration is hampered in wounds where oxygenation has not been preserved.

Hypoxia may cause macrophages, keratinocytes, and fibroblasts to produce cytokines and growth factors. PDGF, TGF, VEGF, tumor necrosis, and endothelin1 are important regulators of mobile propagation, movement and angiogenic in wound repair and are developed in response to hypoxia (Calabro and Yeh, 2007). ROS, like hydrogen peroxide and oxygen, were assumed to serve as cell messengers in normally healing wounds, enhancing essential wound-healing strategies cellular motility, cytokine production, and wound healing a shift as well as angiogenesis Both hypoxia and hyperoxia increase the creation of reactive oxygen species (ROS) development, but a higher level of ROS goes beyond that. It has a positive impact and causes further tissue damage (Campos et al., 2008).

Inflammation

It's a necessary component of the wound healing procedure in order to remove contaminating microbes. Furthermore, in the lack of successful decontamination, because of microbial resistance, infection will last for a long time the clearance is insufficient. Endotoxins and bacteria may also cause damage result in a long-term increase in seasoned inflammatory markers cytokines like IL1β and TNF-α and lengthen the section that is inflammatory. If this continues, the wound can become infected even, if you're in a chronic country, you won't be able to recover. A higher number of matrix metalloproteases, a separate family of enzymes, is produced as a result of the sustained irritation, proteases that could break down the ECM in conjunction with a rise in protease content, and a reduction in the degree, there are naturally occurring protease inhibitors. This is a good example, a change in protease balance may cause boom essentials, seem to be rapidly deteriorated in continuous wounds Simply put, the microbes in infected people are infected in the same way as other infective systems are infected. Wounds are caused by biofilms, which are a type of bacteria, embedded complex classes of distributed microbe in an extracellular polysaccharide matrix secreted by the cells (Chan et al., 2006).

Age

The elderly populace was increasing faster than many other age groups, and becoming older was a main danger factor for poor wound remedial. Many experimental and animal models have looked at age-associated variations and late in wound repair at the cell and molecular extent. The consequence of becoming older origins a temporal stays in wound repair in steady elder adults, but no significant loss in relations of the quality of recovery (Da Costa, 2003). Wound healing in the elderly is decreased, and this is linked to changes in the inflammatory process, like delayed Tcell penetration to the wound site, differences in chemokine manufacture, and reduced macrophage phagocytic capability (Davis et al., 2008) (). Elderly mice have also been shown to have impaired reepithelialization, collagen synthesis, and angiogenesis as compared to younger mice (Dong et al., 1993).

Hormones

Female hormones show a part in age-associated wound repair delays in the elderly. When comparison to elderly women, ageing men have a lesser degree of acute wound healing. One reason for this is that female estrogens, and its steroid forerunner dehydroepiandrosterone all seem to have important impacts on wound repair (Dvivedi et al., 1997). In recent times, they discovered where the changes in genome expression amid aged male and younger human injuries were virtually entirely controlled by estrogen. Estrogen disturbs wound remedial through controlling a number of genes involved in rejuvenation, matrices development, protease suppression, epidermal characteristics, and inflammatory (Emery et al., 2005).

Stress

Stress has a major effect on human strength and communal behavior. Stress was related to a number of illnesses, like cardiovascular illness, most tumors, slow wound healing, and diabetes. Many other studies have found that stress causes a disturbance of neuroendocrine-immune balance, which has a negative impact on health (Fontana et al., 2007). The pathophysiology of strain is caused by immune system dysregulation, which is mediated primarily by the sympathetic nervous system's hypothalamic–pituitary–and sympathetic–adrenal–medullary axes.

Diabetes

Diabetes causes 100's of thousands of people around the world. A reported weakness in the curing of acute injuries has been observed in diabetics. Furthermore, this demographic is vulnerable to a broader range of diabetic foot ulcers (DFUs) that were expected to affect 15 percent of all genders by diabetes. DFUs were a significant problem of diabetes, and they are the cause of 84 percent of all diabetes-related lesser leg eliminations. In diabetes patients, delayed recovery of each DFU and acute cutaneous injuries is caused by a number of complex pathophysiological mechanisms. Hypoxia is often present in DFUs including venous stasis illness and pressure-associated chronic non-healing injuries. Hypoxia, dysfunction in fibroblasts and epidermal-based cells, damaged angiogenesis and neovascularization, elevated levels of metalloproteases, damage from system examination and a longer period, reduced host immune resist, and neuropathy are all symptoms of impaired recovery in diabetic patients (Galiano et al., 2004).

NEs in Wound Healing Process

Lipid-based NEs

Lipidic nanomaterials are those whose composition includes one or more lipidic/oily substances. As shown in Figure 3, nanoparticles can permeate the skin through a variety of mechanisms. Lipidic nanoparticles have been shown to improve skin hydration and enable skin cell reconfiguration, which could lead to nanoparticle fragmentation or improved permeability across skin layers (Li et al., 2013). Liposomes, diffusion enhancer vesicles, ethosomes, and transfersomes are examples of lipid-based elements that were vesicular in behavior, emulsion-based in nature (NEs and MEs), or comprise a solid-based lipid framework.

Figure 3. Mode of actions or NPs permeation across the skin.

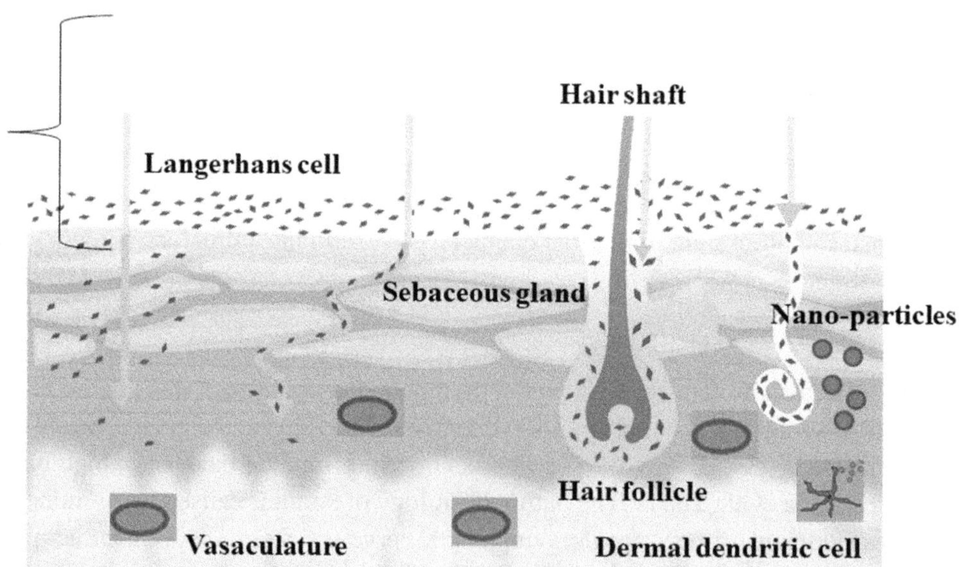

Liposomes

Liposomes were by far the greatest widely mentioned vesicular structures, and name "liposomes" means which were only made up of phospholipids and have no another additive including surfactants relishes, making them first-generation vesicles. Liposomes' popularity stems from their phospholipid material, which would be biocompatible to lipids found in the skin. Liposomes are medication reservoirs that allow for controlled drug delivery, but they are often found in the outermost layer of skin. Despite their rigidity, several studies have found that they are effective as topical delivery systems. The semi occlusive existence of hydrogels, which encourages angiogenesis and enables the development of granulation tissue, was documented to help wound healing by incorporate liposomes within them (Thirumaleshwar et al., 2012). Table 1 recapitulates few of the auspicious applications of liposome-based for wound repair within a designated vesicle were lacking of any dispersion improver constituent.

Permeation increaser vesicle

Table 1. Penetration improver including vehicles utilized for wound healing.

S. No.	Arrangement	Diffusion enhancer utilized	Investigation status	Ingredient used as active	Merits	References
1	PEVs prepared from soybean phospholipids	Polyethylene glycol	In vitro or In vivo	Quercetin and curcumin	PEVs equipped by quercetin and curcumin showed improved antiinflammatory action and more important dermal statement of medicines, by greater consequence met with curcumin in comparison with quercetin vesicle in animal inflammation study	(Manca, 2019)
		Ethylene Glycol	In vitro or In vivo	Polyphenolic phytocomplex from Fraxinus angustifolia	PEVs capturing the removed polyphenols caused in substantial decrease in edema by associated reduction in myeloperoxidase action associated to unconjugated extracts, by normal skin presence in animal provocative study	(Verma & Pathak, 2010)
2	PEVs processed from phospholipids mix	Polyethylene glycol	In vitro or In vivo	Quercetin	PEVs-embodied quercetin Showed vital decrease in tissue injury with important tissue renewal established by enhance in collagen-based fibers in comparison with un-encapsulated drug	(Godin et al., 2005)

Continued on following page

Table 1. Continued

S. No.	Arrangement	Diffusion enhancer utilized	Investigation status	Ingredient used as active	Merits	References
3	PEVs processed from soyabean lecithin and tween-80	Sorbitol	In vitro	Baicalin	The changed sorbitol PEVs preparation exhibited improved skin accretion in overall coatings, provided greater proliferative stimulus of fibroblasts, and providing improved skin defense associated with transfersomes, and was confirmed extra greater in indorsing cellular-based wound closing	(Partoazar et al., 2016)

Penetration enhancer vesicle (PEV) are similar to liposomes, which they comprise extra permeation enhancers like polyethylene glycol, labrasol, and transcutol in their formulation. Since the penetration improvers cause more skin disordering than traditional vesicles (Castangia et al., 2016), it was classified as supple vesicles which can promote dermal diffusion and enable improved fibroblast cellular intake. Table 2 lists few of the confirmed effective efforts to deliver active compounds via PEVs for wound healing.

Ethosomes

Ethosomes are alike to liposomes where it comprises phospholipids as the bi-layer formation substance as well as ethanol that serves as a skin permeation improver (Verma and Pathak, 2010), and therefore they were classified as supple vesicles. Just some analysis mentioned ethanol as a topical vesicular therapy for wounds, most likely due to its ethanolic component, which may irritate open wounds. Godin et al. (2005) found the topical use of erythromycin-loaded hydroalcoholic fluid was less efficient as a wound repair death than topically executed ethosomal erythromycin, emphasizing the role of vesicular assembly in skin permeation.

Due to an excellent reepithelization, collagen formation, and granulation tissue tissue promoting possible, ethosomal curcumin were reported to strongly suppress gram +ve and -ve microbes extracted from injuries comparison to unconjugated curcumin, and to decrease overall wound site in a 2nd degree injury rat models in research shown by Partoazar et al. (2016). Furthermore, when comparison to povidone iodine ointment, an ethosomal gel overloaded for Sesamum indicum seed extracts showed significantly greater wound reduction proportion through the 16th day in a rat removal template, highly correlated with higher collagen synthesis derived from vastly greater hydroxyproline level determined with the ethosomal type (Somwanshi and Hiremath, 2018). Eventually, as contrasted to a silver sulfadiazine gel commercial substance, an ethosomal preparation packed by silver sulfadiazine showed a quicker wound repair phase cataract and a stronger wound contraction percentage.

NEs

NEs are water, oil, and surfactant/cosurfactant-based soft delivery systems made up of nano-lipids (Nasr, 2016). Microemulsions are similar to NEs in that they have a higher surfactant absorption were made with a delicate proportion of surfactant mechanisms, typically and use the water dilution process. In this study, microemulsions will be referred to as NEs because they have a nanometer-size spectrum. The oily step of a NE has been shown to impact wound healing potential by affecting the proclamation of constituents from the oily center (Steinbrenner et al., 2016), and the addition of actives to nano or micro-emulsion preparations has been depicted to improve wound repair. Cinnamon oil NE was identified as better efficient in speeding wound healing in animals than that of the oil alone (Ghosh et al., 2013). Furthermore, NEs can involve multiple active ingredients, such as astaxanthin and alpha tocopherol, that Shanmugapriya et al. (2018) loaded into a NE formulation that improved cell closure in a scratch wound repair cell-based assay in comparison with control, establishing this NE method as a capable therapeutic topical carrier. NE hydrogels are packed by growth factor blends and displayed superior skin permeability than the unencapsulated type, making them promising wound healing systems (Choi et al., 2017).

Solid-based Lipid NPs Lipid Transporters

Solid-based lipid NPs and nanostructured lipid-based carriers are two common lipid delivery schemes. Both structures were made up of a fluid lipid matrix, with the NLCs' matrices also comprising oil (Garcês et al., 2018). Both systems may be used independently or in combination with other dressings. When compared to quercetin alone, quercetin-loaded SLNs improved wound healing (Ma et al., 2018). SLNs also developed a combination therapy containing the peptide LL37 and the elastase suppressor Serpin A1 which intensified curative process in cell-based studies (Fumakia and Ho, 2016), as well as a curcumin-ampicillin standard treatment that accelerated wound healing in an animal study (Ghaffari et al., 2018). Tetrahydrocurcumin embodied in SLNs gel showed quicker wound healing than the unconjugated form (Kakkar et al., 2018).

NLCs made with olive and eucalyptus oil reduced wound areas in rats significantly (Saporito et al., 2018). Gainza et al. (2015) contrasted the efficacy of NLCs-encoded recombinant human epidermal growth factor to the different mode in a porcine full-width repair. The NLCs form of the growth factor exhibited improved repair in case of healing wounds rate, epithelization, collagen production, and reduce irritation than the greater dosage of the free form, due to its defense of the latter towards hydrolytic hydrolysis. Other important study by Alalaiwe et al., (2018) showed that mounting oxacillin antibiotic in cationic NLCs improved the antibiotic's bactericidal effect towards methicillin-resilient S. aureus MRSA, and resulting in nearly absolute abscess care, with substantially less water loss from the diseased wound relative to merely administering antibiotics, followed by much less lesions.

CONCLUSION

Hemostasis, inflammation, multiplication, and remodeling are all part of the wound curative method. There are a huge number of diverse cellular forms, including neutrophils, macrophages, and keratinocytes are all types of white blood cells, etc. were involved in this process. NEs have proven to be an adaptable medium for wound healing. Either organic or synthetic in nature, their distinctive characteristics allow them to transcend the limitations of traditional therapeutic approaches and result in a better treatment outcome. Future research on composite NEs tailored for healing process are expected to rise as functional materials were discovered, ultimately replacing traditional treatments.

REFERENCES

Abbas, S., Hayat, K., Karangwa, E., Bashari, M., & Zhang, X. (2013). An overview of ultrasound-assisted food-grade NEs. *Food Engineering Reviews*, *5*(3), 139–157. doi:10.100712393-013-9066-3

Abd, E., Benson, H., Roberts, M., & Grice, J. (2018). Minoxidil skin delivery from NE formulations containing eucalyptol or oleic acid: Enhanced diffusivity and follicular targeting. *Pharmaceutics*, *10*(1), 19. doi:10.3390/pharmaceutics10010019 PMID:29370122

Abd, E., Namjoshi, S., Mohammed, Y. H., Roberts, M. S., & Grice, J. E. (2016). Synergistic skin penetration enhancer and NE formulations promote the human epidermal permeation of caffeine and naproxen. *Journal of Pharmaceutical Sciences*, *105*(1), 212–220. doi:10.1002/jps.24699 PMID:26554868

Abu-Fayyad, A., Kamal, M. M., Carroll, J. L., Dragoi, A.-M., Cody, R., Cardelli, J., & Nazzal, S. (2018). Development and in-vitro characterization of NEs loaded with paclitaxel/γ-tocotrienol lipid conjugates. *International Journal of Pharmaceutics*, *536*(1), 146–157. doi:10.1016/j.ijpharm.2017.11.062 PMID:29195915

Abu-Fayyad, A., & Nazzal, S. (2017). Gemcitabine-vitamin E conjugates: Synthesis, characterization, entrapment into NEs, and in-vitro deamination and antitumor activity. *International Journal of Pharmaceutics*, *528*(1-2), 463–470. doi:10.1016/j.ijpharm.2017.06.031 PMID:28627455

Achari, G. A., & Kowshik, M. (2018). Recent developments on nanotechnology in agriculture: Plant mineral nutrition, health, and interactions with soil microflora. *Journal of Agricultural and Food Chemistry*, *66*(33), 8647–8661. doi:10.1021/acs.jafc.8b00691 PMID:30036480

Agarwal, V., Bajpai, M., & Sharma, A. (2018). Patented and approval scenario of nanopharmaceuticals with relevancy to biomedical application, manufacturing procedure and safety aspects. *Recent Patents Drug Delivery Formulation, 12*(1), 40-52.

Ahmad, U., Akhtar, J., Singh, S. P., Badruddeen, Ahmad, F. J., Siddiqui, S., & Wahajuddin. (2018). Silymarin NE against human hepatocellular carcinoma: Development and optimization. *Artificial Cells, Nanomedicine, and Biotechnology*, *46*(2), 231–241. doi:10.1080/21691401.2017.1324465 PMID:28503949

Al-Sabagh, A. M., Emara, M. M., Noor El-Din, M. R., & Aly, W. R. (2011). Formation of water-in-diesel oil nano-emulsions using high energy method and studying some of their surface active properties. *Egyptian Journal of Petroleum*, *20*(2), 17–23. doi:10.1016/j.ejpe.2011.06.005

Alalaiwe, A., Wang, P.-W., Lu, P.-L., Chen, Y.-P., Fang, J.-Y., & Yang, S.-C. (2018). Synergistic anti-MRSA activity of cationic nanostructured lipid carriers in combination with oxacillin for cutaneous application. *Frontiers in Microbiology*, *9*, 1493. doi:10.3389/fmicb.2018.01493 PMID:30034381

Amer, S. S., Nasr, M., Mamdouh, W., & Sammour, O. (2019). Insights on the use of nanocarriers for acne alleviation. *Current Drug Delivery*, *16*(1), 18–25. doi:10.2174/1567201815666180913144145 PMID:30210000

Anton, N., & Vandamme, T. F. (2011). Nano-emulsions and micro-emulsions: Clarifications of the critical differences. *Pharmaceutical Research*, *28*(5), 978–985. doi:10.100711095-010-0309-1 PMID:21057856

Aqil, M., Kamran, M., Ahad, A., & Imam, S. S. (2016). Development of clove oil based NE of olmesartan for transdermal delivery: Box–Behnken design optimization and pharmacokinetic evaluation. *Journal of Molecular Liquids*, *214*, 238–248. doi:10.1016/j.molliq.2015.12.077

Argenta, D. F. (2014). Factorial design applied to the optimization of lipid composition of topical antiherpetic NEs containing isoflavone genistein. *International Journal of Nanomedicine*, *9*, 4737. PMID:25336951

Argenta, D. F., Bidone, J., Misturini, F. D., Koester, L. S., Bassani, V. L., Simões, C. M. O., & Teixeira, H. F. (2016). *In vitro* evaluation of mucosa permeation/retention and antiherpes activity of genistein from cationic NEs. *Journal of Nanoscience and Nanotechnology*, *16*(2), 1282–1290. doi:10.1166/jnn.2016.11676 PMID:27433578

Bai, L., & McClements, D. J. (2016). Development of MF methods for efficient production of concentrated NEs: Comparison of single-and dual-channel microfluidizers. *Journal of Colloid and Interface Science*, *466*, 206–212. doi:10.1016/j.jcis.2015.12.039 PMID:26724703

Balestrin, L. A., Bidone, J., Bortolin, R. C., Moresco, K., Moreira, J. C., & Teixeira, H. F. (2016). Protective effect of a hydrogel containing Achyrocline satureioides extract-loaded NE against UV-induced skin damage. *Journal of Photochemistry and Photobiology. B, Biology*, *163*, 269–276. doi:10.1016/j.jphotobiol.2016.08.039 PMID:27599114

Bonferoni, M. C., Riva, F., Invernizzi, A., Dellera, E., Sandri, G., Rossi, S., Marrubini, G., Bruni, G., Vigani, B., Caramella, C., & Ferrari, F. (2018). Alpha tocopherol loaded chitosan oleate NEs for wound healing. Evaluation on cell lines and ex vivo human biopsies, and stabilization in spray dried Trojan microparticles. *European Journal of Pharmaceutics and Biopharmaceutics*, *123*, 31–41. doi:10.1016/j.ejpb.2017.11.008 PMID:29155053

Bouchemal, K., Briançon, S., Perrier, E., & Fessi, H. (2004). Nano-emulsion formulation using spontaneous emulsification: Solvent, oil and surfactant optimisation. *International Journal of Pharmaceutics*, *280*(1-2), 241–251. doi:10.1016/j.ijpharm.2004.05.016 PMID:15265563

Boyapati, L., & Wang, H. L. (2007). The role of stress in periodontal disease and wound healing. *Periodontology 2000*, *44*(1), 195–210. doi:10.1111/j.1600-0757.2007.00211.x PMID:17474934

Brem, H., & Tomic-Canic, M. (2007). Cellular and molecular basis of wound healing in diabetes. *The Journal of Clinical Investigation*, *117*(5), 1219–1222. doi:10.1172/JCI32169 PMID:17476353

Brownlow, B., Nagaraj, V. J., Nayel, A., Joshi, M., & Elbayoumi, T. (2015). Development and in vitro evaluation of vitamin E-enriched NE vehicles loaded with genistein for chemoprevention against UVB-induced skin damage. *Journal of Pharmaceutical Sciences*, *104*(10), 3510–3523. doi:10.1002/jps.24547 PMID:26108889

Bsieso, E. A., Nasr, M., Moftah, N. H., Sammour, O. A., & Abd El Gawad, N. A. (2015). Could nanovesicles containing a penetration enhancer clinically improve the therapeutic outcome in skin fungal diseases? *Nanomedicine (London), 10*(13), 2017–2031. doi:10.2217/nnm.15.49 PMID:26135513

Buckley, C. D. (2011). Why does chronic inflammation persist: An unexpected role for fibroblasts. *Immunology Letters, 138*(1), 12–14. doi:10.1016/j.imlet.2011.02.010 PMID:21333681

Burnouf, T., Goubran, H. A., Chen, T.-M., Ou, K.-L., El-Ekiaby, M., & Radosevic, M. (2013). Blood-derived biomaterials and platelet growth factors in regenerative medicine. *Blood Reviews, 27*(2), 77–89. doi:10.1016/j.blre.2013.02.001 PMID:23434399

Caddeo, C., Díez-Sales, O., Pons, R., Fernàndez-Busquets, X., Fadda, A. M., & Manconi, M. (2014). Topical anti-inflammatory potential of quercetin in lipid-based nanosystems: In vivo and in vitro evaluation. *Pharmaceutical Research, 31*(4), 959–968. doi:10.100711095-013-1215-0 PMID:24297068

Calabro, P., & Yeh, E. T. (2007). Obesity, Inflammation, and Vascular Disease. *Inflammation in the Pathogenesis of Chronic Diseases*, 63-91.

Campos, A. C., Groth, A. K., & Branco, A. B. (2008). Assessment and nutritional aspects of wound healing. *Current Opinion in Clinical Nutrition and Metabolic Care, 11*(3), 281–288. doi:10.1097/MCO.0b013e3282fbd35a PMID:18403925

Carvalho, V. F., Migotto, A., Giacone, D. V., de Lemos, D. P., Zanoni, T. B., Maria-Engler, S. S., Costa-Lotufo, L. V., & Lopes, L. B. (2017). Co-encapsulation of paclitaxel and C6 ceramide in tributyrin-containing nanocarriers improve co-localization in the skin and potentiate cytotoxic effects in 2D and 3D models. *European Journal of Pharmaceutical Sciences, 109*, 131–143. doi:10.1016/j.ejps.2017.07.023 PMID:28735040

Castangia, I., Nácher, A., Caddeo, C., Valenti, D., Fadda, A. M., Díez-Sales, O., Ruiz-Saurí, A., & Manconi, M. (2014). Fabrication of quercetin and curcumin bionanovesicles for the prevention and rapid regeneration of full-thickness skin defects on mice. *Acta Biomaterialia, 10*(3), 1292–1300. doi:10.1016/j.actbio.2013.11.005 PMID:24239901

Chan, L. K., Withey, S., & Butler, P. E. M. (2006). Smoking and wound healing problems in reduction mammaplasty: Is the introduction of urine nicotine testing justified? *Annals of Plastic Surgery, 56*(2), 111–115. doi:10.1097/01.sap.0000197635.26473.a2 PMID:16432315

Choi, S. W., Pangeni, R., & Park, J. W. (2017). NE-Based Hydrogel for Topical Delivery of Highly Skin-Permeable Growth Factor Combinations: Preparation and In Vitro Evaluation. *Journal of Nanoscience and Nanotechnology, 17*(4), 2363–2369. doi:10.1166/jnn.2017.13318 PMID:29641166

Choudhary, V., Shivakumar, H., & Ojha, H. (2019). Curcumin-loaded liposomes for wound healing: Preparation, optimization, in-vivo skin permeation and bioevaluation. *Journal of Drug Delivery Science and Technology, 49*, 683–691. doi:10.1016/j.jddst.2018.12.008

Choudhury, H., Gorain, B., Pandey, M., Chatterjee, L. A., Sengupta, P., Das, A., Molugulu, N., & Kesharwani, P. (2017). Recent update on nanoemulgel as topical drug delivery system. *Journal of Pharmaceutical Sciences, 106*(7), 1736–1751. doi:10.1016/j.xphs.2017.03.042 PMID:28412398

Da Costa, M. A., Campos, A. C., Coelho, J. C., de Barros, A. M., & Matsumoto, H. M. (2003). Oral glutamine and the healing of colonic anastomoses in rats. *J Parent Ent Nutri*, *27*(3), 182–185. doi:10.1177/0148607103027003182 PMID:12757111

Davis, S. C., Ricotti, C., Cazzaniga, A., Welsh, E., Eaglstein, W. H., & Mertz, P. M. (2008). Microscopic and physiologic evidence for biofilm-associated wound colonization in vivo. *Wound Repair and Regeneration*, *16*(1), 23–29. doi:10.1111/j.1524-475X.2007.00303.x PMID:18211576

de Souza, M. L., Oliveira, D. D., Pereira, N. P., & Soares, D. M. (2018). NEs and dermatological diseases: Contributions and therapeutic advances. *International Journal of Dermatology*, *57*(8), 894–900. doi:10.1111/ijd.14028 PMID:29766488

Dehghani, F., Farhadian, N., Golmohammadzadeh, S., Biriaee, A., Ebrahimi, M., & Karimi, M. (2017). Preparation, characterization and in-vivo evaluation of microemulsions containing tamoxifen citrate anticancer drug. *European Journal of Pharmaceutical Sciences*, *96*, 479–489. doi:10.1016/j.ejps.2016.09.033 PMID:27693298

Dong, Y. L., Declan Fleming, R. Y., Yan, T. Z., Herndon, D. N., & Waymack, J. P. (1993). Effect of ibuprofen on the inflammatory response to surgical wounds. *The Journal of Trauma*, *35*(3), 340–343. doi:10.1097/00005373-199309000-00002 PMID:8371289

Dvivedi, S. (1997). Effect of ibuprofen and diclofenac sodium on experimental would healing. *Indian Journal of Experimental Biology*, *35*(11), 1243–1245. PMID:9567757

Emery, C. F., Kiecolt-Glaser, J. K., Glaser, R., Malarkey, W. B., & Frid, D. J. (2005). Exercise accelerates wound healing among healthy older adults: A preliminary investigation. *The Journals of Gerontology. Series A, Biological Sciences and Medical Sciences*, *60*(11), 1432–1436. doi:10.1093/gerona/60.11.1432 PMID:16339330

Ferguson, M. W., & O'Kane, S. (2004). Scar–free healing: From embryonic mechanisms to adult therapeutic intervention. *Philosophical Transactions of the Royal Society of London. Series B, Biological Sciences*, *359*(1445), 839–850. doi:10.1098/rstb.2004.1475 PMID:15293811

Ferreira, H., Matamá, T., Silva, R., Silva, C., Gomes, A. C., & Cavaco-Paulo, A. (2013). Functionalization of gauzes with liposomes entrapping an anti-inflammatory drug: A strategy to improve wound healing. *Reactive & Functional Polymers*, *73*(10), 1328–1334. doi:10.1016/j.reactfunctpolym.2013.05.012

Fontana, L., Eagon, J. C., Colonna, M., & Klein, S. (2007). Impaired mononuclear cell immune function in extreme obesity is corrected by weight loss. *Rejuvenation Research*, *10*(1), 41–46. doi:10.1089/rej.2006.0504 PMID:17378751

Fumakia, M., & Ho, E. A. (2016). Nanoparticles encapsulated with LL37 and serpin A1 promotes wound healing and synergistically enhances antibacterial activity. *Molecular Pharmaceutics*, *13*(7), 2318–2331. doi:10.1021/acs.molpharmaceut.6b00099 PMID:27182713

Gainza, G., Bonafonte, D. C., Moreno, B., Aguirre, J. J., Gutierrez, F. B., Villullas, S., Pedraz, J. L., Igartua, M., & Hernandez, R. M. (2015). The topical administration of rhEGF-loaded nanostructured lipid carriers (rhEGF-NLC) improves healing in a porcine full-thickness excisional wound model. *Journal of Controlled Release*, *197*, 41–47. doi:10.1016/j.jconrel.2014.10.033 PMID:25449803

Galiano, R. D., Tepper, O. M., Pelo, C. R., Bhatt, K. A., Callaghan, M., Bastidas, N., Bunting, S., Steinmetz, H. G., & Gurtner, G. C. (2004). Topical vascular endothelial growth factor accelerates diabetic wound healing through increased angiogenesis and by mobilizing and recruiting bone marrow-derived cells. *American Journal of Pathology*, *164*(6), 1935–1947. doi:10.1016/S0002-9440(10)63754-6 PMID:15161630

Gao, Y., Cheng, X., Wang, Z., Wang, J., Gao, T., Li, P., Kong, M., & Chen, X. (2014). Transdermal delivery of 10, 11-methylenedioxycamptothecin by hyaluronic acid based NE for inhibition of keloid fibroblast. *Carbohydrate Polymers*, *112*, 376–386. doi:10.1016/j.carbpol.2014.05.026 PMID:25129757

Garcês, A., Amaral, M. H., Sousa Lobo, J. M., & Silva, A. C. (2018). Formulations based on solid lipid nanoparticles (SLN) and nanostructured lipid carriers (NLC) for cutaneous use: A review. *European Journal of Pharmaceutical Sciences*, *112*, 159–167. doi:10.1016/j.ejps.2017.11.023 PMID:29183800

George Broughton, I. I. (2006). The basic science of wound healing. *Plastic and Reconstructive Surgery*, *117*(7S), 12S–34S. doi:10.1097/01.prs.0000225430.42531.c2 PMID:16799372

Ghaffari, S., Alihosseini, F., Rezayat Sorkhabadi, S. M., Arbabi Bidgoli, S., Mousavi, S. E., Haghighat, S., Afshar Nasab, A., & Kianvash, N. (2018). Nanotechnology in wound healing; semisolid dosage forms containing curcumin-ampicillin solid lipid nanoparticles, in-vitro, ex-vivo and in-vivo characteristics. *Advanced Pharmaceutical Bulletin*, *8*(3), 395–400. doi:10.15171/apb.2018.046 PMID:30276135

Ghosh, V., Saranya, S., Mukherjee, A., & Chandrasekaran, N. (2013). Antibacterial microemulsion prevents sepsis and triggers healing of wound in wistar rats. *Colloids and Surfaces. B, Biointerfaces*, *105*, 152–157. doi:10.1016/j.colsurfb.2013.01.009 PMID:23357738

Giusti, I., Rughetti, A., D'Ascenzo, S., Di Stefano, G., Nanni, M. R., Millimaggi, D., Dell'Orso, L., & Dolo, V. (2013). The effects of platelet gel–released supernatant on human fibroblasts. *Wound Repair and Regeneration*, *21*(2), 300–308. doi:10.1111/wrr.12025 PMID:23438188

Godin, B., Touitou, E., Rubinstein, E., Athamna, A., & Athamna, M. (2005). A new approach for treatment of deep skin infections by an ethosomal antibiotic preparation: An in vivo study. *The Journal of Antimicrobial Chemotherapy*, *55*(6), 989–994. doi:10.1093/jac/dki125 PMID:15857943

Habibi, K. (2017). Design and production of methyl jasmonate NEs using experimental design technique and evaluation of its anti-cancer efficacy. *Die Pharmazie-An International Journal of Pharmaceutical Sciences*, *72*(11), 652–662. PMID:29442039

Hatem, S., Nasr, M., Elkheshen, S. A., & Geneidi, A. S. (2018). Recent advances in antioxidant cosmeceutical topical delivery. *Current Drug Delivery*, *15*(7), 953–964. doi:10.2174/1567201815666180214143551 PMID:29446743

Hatem, S., Nasr, M., Moftah, N. H., Ragai, M. H., Geneidi, A. S., & Elkheshen, S. A. (2018). Clinical cosmeceutical repurposing of melatonin in androgenic alopecia using nanostructured lipid carriers prepared with antioxidant oils. *Expert Opinion on Drug Delivery*, *15*(10), 927–935. doi:10.1080/17425247.2018.1517740 PMID:30169980

Hawkins, D. H., & Abrahamse, H. (2006). The role of laser fluence in cell viability, proliferation, and membrane integrity of wounded human skin fibroblasts following helium-neon laser irradiation. *Lasers in Surgery and Medicine: The Official Journal of the American Society for Laser Medicine and Surgery*, *38*(1), 74–83. doi:10.1002/lsm.20271 PMID:16444694

Hayles, J., Johnson, L., Worthley, C., & Losic, D. (2017). Nanopesticides: a review of current research and perspectives. *New pesticides and soil sensors*, 193-225.

Hazra, D. K., & Karmakar, R. (2017). Recent advances in pesticide formulations for eco-friendly and sustainable vegetable pest management: A review. *Arch Agri Environ Sci*, *2*(3), 232–237.

Hazrati, H., Saharkhiz, M. J., Niakousari, M., & Moein, M. (2017). Natural herbicide activity of *Satureja hortensis* L. essential oil NE on the seed germination and morphophysiological features of two important weed species. *Ecotoxicology and Environmental Safety*, *142*, 423–430. doi:10.1016/j.ecoenv.2017.04.041 PMID:28456128

Hong, L., Zhou, C.-L., Chen, F.-P., Han, D., Wang, C.-Y., Li, J.-X., Chi, Z., & Liu, C.-G. (2017). Development of a carboxymethyl chitosan functionalized NE formulation for increasing aqueous solubility, stability and skin permeability of astaxanthin using low-energy method. *Journal of Microencapsulation*, *34*(8), 707–721. doi:10.1080/02652048.2017.1373154 PMID:29141479

Iqbal, B., Ali, J., & Baboota, S. (2018). Recent advances and development in epidermal and dermal drug deposition enhancement technology. *International Journal of Dermatology*, *57*(6), 646–660. doi:10.1111/ijd.13902 PMID:29430629

Jaiswal, M., Dudhe, R., & Sharma, P. K. (2015). NE: an advanced mode of drug delivery system. *3 Biotech*, *5*(2), 123-127.

Jampilek, J., & Brychtova, K. (2012). Azone analogues: Classification, design, and transdermal penetration principles. *Medicinal Research Reviews*, *32*(5), 907–947. doi:10.1002/med.20227 PMID:22886628

Jampílek, J., & Kráľová, K. (2019). Recent advances in lipid nanocarriers applicable in the fight against cancer. *Nanoarchitectonics in Biomedicine*, 219-294.

Jampilek, J. O. S. E. F., & Kralova, K. (2018). *Application of nanobioformulations for controlled release and targeted biodistribution of drugs. In Nanobiomaterials: Appl Drug Deliv* (pp. 131–208). doi:10.1201/9781315204918-5

Jangde, R., & Singh, D. (2016). Preparation and optimization of quercetin-loaded liposomes for wound healing, using response surface methodology. *Artificial Cells, Nanomedicine, and Biotechnology*, *44*(2), 635–641. doi:10.3109/21691401.2014.975238 PMID:25375215

Jo, Y. J., & Kwon, Y. J. (2014). Characterization of β-carotene NEs prepared by MF technique. *Food Science and Biotechnology*, *23*(1), 107–113. doi:10.100710068-014-0014-7

Kakkar, V., Kaur, I. P., Kaur, A. P., Saini, K., & Singh, K. K. (2018). Topical delivery of tetrahydrocurcumin lipid nanoparticles effectively inhibits skin inflammation: In vitro and in vivo study. *Drug Development and Industrial Pharmacy*, *44*(10), 1701–1712. doi:10.1080/03639045.2018.1492607 PMID:29938544

Langer, S., Botteck, N. M., Bosse, B., Reimer, K., Vogt, P. M., Steinau, H.-U., & Mueller, S. (2006). Effect of polyvinylpyrrolidone-iodine liposomal hydrogel on wound microcirculation in SKH1-hr hairless mice. *European Surgical Research, 38*(1), 27–34. doi:10.1159/000091524 PMID:16490991

Li, J., Chen, J., & Kirsner, R. (2007). Pathophysiology of acute wound healing. *Clinics in Dermatology, 25*(1), 9–18. doi:10.1016/j.clindermatol.2006.09.007 PMID:17276196

Li, Z. (2016). Increased cutaneous wound healing effect of biodegradable liposomes containing madecassoside: Preparation optimization, in vitro dermal permeation, and in vivo bioevaluation. *International Journal of Nanomedicine, 11*, 2995–3007. doi:10.2147/IJN.S105035 PMID:27486319

Liu, C. H., & Huang, H. Y. (2012). Antimicrobial activity of curcumin-loaded myristic acid microemulsions against Staphylococcus epidermidis. *Chemical & Pharmaceutical Bulletin, 60*(9), 1118–1124. doi:10.1248/cpb.c12-00220 PMID:22976319

Ma, J., Ji, C., Xiao, D., & Fan, Q. (2018). Quercetin-loaded solid lipid nanoparticles-enriched hydrogel prevents the formation of skin scars by inhibiting TGF-β/Smad signaling pathway. *Biomedical Research, 29*(7). Advance online publication. doi:10.4066/biomedicalresearch.29-17-2202

Manca, M. L. (2019). Sorbitol-penetration enhancer containing vesicles loaded with baicalin for the protection and regeneration of skin injured by oxidative stress and UV radiation. *Int J Pharma, 555*, 175-183.

Martin, P. (1997). Wound healing—Aiming for perfect skin regeneration. *Science, 276*(5309), 75–81. doi:10.1126cience.276.5309.75 PMID:9082989

Mason, T. G., Wilking, J. N., Meleson, K., Chang, C. B., & Graves, S. M. (2006). NEs: Formation, structure, and physical properties. *Journal of Physics Condensed Matter, 18*(41), R635–R666. doi:10.1088/0953-8984/18/41/R01

Maxson, S., Lopez, E. A., Yoo, D., Danilkovitch-Miagkova, A., & LeRoux, M. A. (2012). Concise review: Role of mesenchymal stem cells in wound repair. *Stem Cells Translational Medicine, 1*(2), 142–149. doi:10.5966ctm.2011-0018 PMID:23197761

Medrado, A. R., Pugliese, L. S., Reis, S. R. A., & Andrade, Z. A. (2003). Influence of low-level laser therapy on wound healing and its biological action upon myofibroblasts. *Lasers in Surgery and Medicine, 32*(3), 239–244. doi:10.1002/lsm.10126 PMID:12605432

Mendonça, R. J. D., & Coutinho-Netto, J. (2009). Cellular aspects of wound healing. *Anais Brasileiros de Dermatologia, 84*(3), 257–262. PMID:19668939

Mengoni, T., Adrian, M., Pereira, S., Santos-Carballal, B., Kaiser, M., & Goycoolea, F. (2017). A chitosan—Based liposome formulation enhances the in vitro wound healing efficacy of substance P neuropeptide. *Pharmaceutics, 9*(4), 56. doi:10.3390/pharmaceutics9040056 PMID:29211047

Metcalfe, A. D., & Ferguson, M. W. (2008). Skin stem and progenitor cells: Using regeneration as a tissue-engineering strategy. Cellular and molecular life sciences. *Cellular and Molecular Life Sciences, 65*(1), 24–32. doi:10.100700018-007-7427-x PMID:18030423

Mostafa, D. M., Abd El-Alim, S. H., Asfour, M. H., Al-Okbi, S. Y., Mohamed, D. A., & Awad, G. (2015). Transdermal NEs of Foeniculum vulgare Mill. essential oil: Preparation, characterization and evaluation of antidiabetic potential. *Journal of Drug Delivery Science and Technology*, *29*, 99–106. doi:10.1016/j.jddst.2015.06.021

Moulaoui, K., Caddeo, C., Manca, M. L., Castangia, I., Valenti, D., Escribano, E., Atmani, D., Fadda, A. M., & Manconi, M. (2015). Identification and nanoentrapment of polyphenolic phytocomplex from Fraxinus angustifolia: *In vitro* and *in vivo* wound healing potential. *European Journal of Medicinal Chemistry*, *89*, 179–188. doi:10.1016/j.ejmech.2014.10.047 PMID:25462238

Nam, E., Yoo, S., Kim, H.-Y., Kim, Y.-R., & Heo, Y. J. (2018). Transdermal water-in-oil nanocarriers of nitric oxide for triggering penile erection. *Scientific Reports*, *8*(1), 1–8. doi:10.103841598-018-25786-x PMID:29743519

Nasr, M. (2016). Development of an optimized hyaluronic acid-based lipidic NE co-encapsulating two polyphenols for nose to brain delivery. *Drug Delivery*, *23*(4), 1444–1452. doi:10.3109/10717544.2015.1092619 PMID:26401600

Nasr, M., Abdel-Hamid, S., Moftah, N. H., Fadel, M., & Alyoussef, A. A. (2017). Jojoba oil soft colloidal nanocarrier of a synthetic retinoid: Preparation, characterization and clinical efficacy in psoriatic patients. *Current Drug Delivery*, *14*(3), 426–432. doi:10.2174/1567201813666160513132321 PMID:27174314

Nastiti, C. M., Ponto, T., Abd, E., Grice, J., Benson, H., & Roberts, M. (2017). Topical nano and microemulsions for skin delivery. *Pharmaceutics*, *9*(4), 37. doi:10.3390/pharmaceutics9040037 PMID:28934172

Naz, Z., & Ahmad, F. J. (2015). Curcumin-loaded colloidal carrier system: Formulation optimization, mechanistic insight, ex vivo and in vivo evaluation. *International Journal of Nanomedicine*, *10*, 4293. doi:10.2147/IJN.S82788 PMID:26170665

Ngan, C. L., Basri, M., Lye, F. F., Fard Masoumi, H. R., Tripathy, M., Abedi Karjiban, R., & Abdul-Malek, E. (2014). Comparison of Box–Behnken and central composite designs in optimization of fullerene loaded palm-based nano-emulsions for cosmeceutical application. *Industrial Crops and Products*, *59*, 309–317. doi:10.1016/j.indcrop.2014.05.042

Nguyen, D. T. (2009). The pathophysiologic basis for wound healing and cutaneous regeneration. In *Biomaterials for treating skin loss* (pp. 25–57). Woodhead Publishing. doi:10.1533/9781845695545.1.25

Nishitani Yukuyama, M. (2017). Challenges and future prospects of NE as a drug delivery system. *Current Pharmaceutical Design*, *23*(3), 495–508. doi:10.2174/1381612822666161027111957 PMID:27799037

Partoazar, A., Kianvash, N., Darvishi, M., Nasoohi, S., Rezayat, S., & Bahador, A. (2016). Ethosomal curcumin promoted wound healing and reduced bacterial flora in second degree burn in rat. *Drug Research*, *66*(12), 660–665. doi:10.1055-0042-114034 PMID:27626605

Patra, J. K., Das, G., Fraceto, L. F., Campos, E. V. R., Rodriguez-Torres, M. P., Acosta-Torres, L. S., Diaz-Torres, L. A., Grillo, R., Swamy, M. K., Sharma, S., Habtemariam, S., & Shin, H.-S. (2018). Nano based drug delivery systems: Recent developments and future prospects. *Journal of Nanobiotechnology*, *16*(1), 1–33. doi:10.118612951-018-0392-8 PMID:30231877

Pawar, K. R., & Babu, R. J. (2014). Lipid materials for topical and transdermal delivery of NEs. *Critical Reviews™ in Therapeutic Drug Carrier Systems, 31*(5).

Penn, J. W. (2012). The role of the TGF-β family in wound healing, burns and scarring: A review. *International Journal of Burns and Trauma, 2*(1), 18. PMID:22928164

Pentak, D., Kozik, V., Bąk, A., Dybał, P., Sochanik, A., & Jampilek, J. (2016). Methotrexate and cytarabine—Loaded nanocarriers for multidrug cancer therapy. Spectroscopic study. *Molecules (Basel, Switzerland), 21*(12), 1689. doi:10.3390/molecules21121689 PMID:27941655

Pereira, F. G. (2017). Cytotoxic effects of the essential oil from leaves of Casearia sylvestris Sw.(Salicaceae) and its NE on A549 tumor cell line. *Boletín Latinoamericano y del Caribe de Plantas Medicinales y Aromáticas, 16*(5), 506–512.

Pisárčik, M., Lukáč, M., Jampílek, J., Bilka, F., Bilková, A., Pašková, Ľ., Devínsky, F., Horáková, R., & Opravil, T. (2018). Silver nanoparticles stabilised with cationic single-chain surfactants. Structure-physical properties-biological activity relationship study. *Journal of Molecular Liquids, 272*, 60–72. doi:10.1016/j.molliq.2018.09.042

Posthauer, M. E. (2012). The role of nutrition in wound care. *Advances in Skin & Wound Care, 25*(2), 62–63. doi:10.1097/01.ASW.0000411404.19016.ad PMID:22258214

Posthauer, M. E., Collins, N., Dorner, B., & Sloan, C. (2013). Nutritional strategies for frail older adults. *Advances in Skin & Wound Care, 26*(3), 128–140. doi:10.1097/01.ASW.0000427920.74379.8c PMID:23426414

Prasad, R., Pandey, R., & Barman, I. (2016). Engineering tailored nanoparticles with microbes: Quo vadis? *Wiley Interdisciplinary Reviews. Nanomedicine and Nanobiotechnology, 8*(2), 316–330. doi:10.1002/wnan.1363 PMID:26271947

Prasad, R., Pandey, R., Varma, R., & Barman, I. (2016a). *Polymer based nanoparticles for drug delivery systems and cancer therapeutics. Natural polymers for drug delivery.* CABI.

Qadir, A., Faiyazuddin, M. D., Talib Hussain, M. D., Alshammari, T. M., & Shakeel, F. (2016). Critical steps and energetics involved in a successful development of a stable NE. *Journal of Molecular Liquids, 214*, 7–18. doi:10.1016/j.molliq.2015.11.050

Rachmawati, H., Budiputra, D. K., & Mauludin, R. (2015). Curcumin NE for transdermal application: Formulation and evaluation. *Drug Development and Industrial Pharmacy, 41*(4), 560–566. doi:10.3109/03639045.2014.884127 PMID:24502271

Rai, V. K., Mishra, N., Yadav, K. S., & Yadav, N. P. (2018). NE as pharmaceutical carrier for dermal and transdermal drug delivery: Formulation development, stability issues, basic considerations and applications. *Journal of Controlled Release, 270*, 203–225. doi:10.1016/j.jconrel.2017.11.049 PMID:29199062

Raviadaran, R., Chandran, D., Shin, L. H., & Manickam, S. (2018). Optimization of palm oil in water nano-emulsion with curcumin using microfluidizer and response surface methodology. *Lebensmittel-Wissenschaft + Technologie, 96*, 58–65. doi:10.1016/j.lwt.2018.05.022

Roco, M. C. (2007). National nanotechnology initiative-past, present, future. *Handbook on nanoscience, engineering and technology*, 2.

Rodero, M. P., & Khosrotehrani, K. (2010). Skin wound healing modulation by macrophages. *International Journal of Clinical and Experimental Pathology*, *3*(7), 643. PMID:20830235

Roesken, F., Uhl, E., Curri, S. B., Menger, M. D., & Messmer, K. (2000). Acceleration of wound healing by topical drug delivery via liposomes. *Langenbeck's Archives of Surgery*, *385*(1), 42–49. doi:10.1007004230050010 PMID:10664120

Saberi, A. H., Fang, Y., & McClements, D. J. (2013). Fabrication of vitamin E-enriched NEs: Factors affecting particle size using spontaneous emulsification. *Journal of Colloid and Interface Science*, *391*, 95–102. doi:10.1016/j.jcis.2012.08.069 PMID:23116862

Sabry, A. K. H., & Ragaei, M. (2018). Nanotechnology and their applications in insect's pest control. In *Nanobiotechnology Applications in Plant Protection* (pp. 1–28). Springer. doi:10.1007/978-3-319-91161-8_1

Saporito, F. (2018). Essential oil-loaded lipid nanoparticles for wound healing. *Int J Nanomed, 13*, 175.

Sengupta, P., & Chatterjee, B. (2017). Potential and future scope of nanoemulgel formulation for topical delivery of lipophilic drugs. *International Journal of Pharmaceutics*, *526*(1-2), 353–365. doi:10.1016/j.ijpharm.2017.04.068 PMID:28461261

Shakeel, F., Ramadan, W., Faisal, M., Rizwan, M., Faiyazuddin, M., Mustafa, G., & Shafiq, S. (2010). Transdermal and topical delivery of anti-inflammatory agents using NE/microemulsion: An updated review. *Current Nanoscience*, *6*(2), 184–198. doi:10.2174/157341310790945740

Shanmugapriya, K., Kim, H., Saravana, P. S., Chun, B.-S., & Kang, H. W. (2018). Astaxanthin-alpha tocopherol NE formulation by emulsification methods: Investigation on anticancer, wound healing, and antibacterial effects. *Colloids and Surfaces. B, Biointerfaces*, *172*, 170–179. doi:10.1016/j.colsurfb.2018.08.042 PMID:30172200

Shaw, T. J., & Martin, P. (2009). Wound repair at a glance. *Journal of Cell Science*, *122*(18), 3209–3213. doi:10.1242/jcs.031187 PMID:19726630

Shi, J., Zhou, S., Kang, L., Ling, H., Chen, J., Duan, L., Song, Y., & Deng, Y. (2018). Evaluation of the antitumor effects of vitamin K 2 (menaquinone-7) NEs modified with sialic acid-cholesterol conjugate. *Drug Delivery and Translational Research*, *8*(1), 1–11. doi:10.100713346-017-0424-1 PMID:28849577

Singer, A. J., & Clark, R. A. (1999). Cutaneous wound healing. *The New England Journal of Medicine*, *341*(10), 738–746. doi:10.1056/NEJM199909023411006 PMID:10471461

Singh, Y., Meher, J. G., Raval, K., Khan, F. A., Chaurasia, M., Jain, N. K., & Chourasia, M. K. (2017). NE: Concepts, development and applications in drug delivery. *Journal of Controlled Release*, *252*, 28–49. doi:10.1016/j.jconrel.2017.03.008 PMID:28279798

Slomkowski, S., Alemán, J. V., Gilbert, R. G., Hess, M., Horie, K., Jones, R. G., Kubisa, P., Meisel, I., Mormann, W., Penczek, S., & Stepto, R. F. T. (2011). Terminology of polymers and polymerization processes in dispersed systems (IUPAC Recommendations 2011). *Pure and Applied Chemistry*, *83*(12), 2229–2259. doi:10.1351/PAC-REC-10-06-03

Somwanshi, S. B., & Hiremath, S. N. (2018). In-vivo Evaluation of the Wound Healing Activity of the Sesamum Indicum L. Seed Extract in Novel Ethosomal Vesicular System. *Journal of Drug Delivery and Therapeutics*, *8*(5), 411–420. doi:10.22270/jddt.v8i5.1895

Sonneville-Aubrun, O., Simonnet, J.-T., & L'Alloret, F. (2004). NEs: A new vehicle for skincare products. *Advances in Colloid and Interface Science*, *108*, 145–149. doi:10.1016/j.cis.2003.10.026 PMID:15072937

Steinbrenner, I., Houdek, P., Pollok, S., Brandner, J. M., & Daniels, R. (2016). Influence of the oil phase and topical formulation on the wound healing ability of a birch bark dry extract. *PLoS One*, *11*(5), e0155582. doi:10.1371/journal.pone.0155582 PMID:27219110

Steuber, N., Vo, K., Wadhwa, R., Birch, J., Iacoban, P., Chavez, P., & Elbayoumi, T. (2016). Tocotrienol NE platform of curcumin elicit elevated apoptosis and augmentation of anticancer efficacy against breast and ovarian carcinomas. *International Journal of Molecular Sciences*, *17*(11), 1792. doi:10.3390/ijms17111792 PMID:27792193

Tan, S. F., Masoumi, H. R. F., Karjiban, R. A., Stanslas, J., Kirby, B. P., Basri, M., & Basri, H. B. (2016). Ultrasonic emulsification of parenteral valproic acid-loaded NE with response surface methodology and evaluation of its stability. *Ultrasonics Sonochemistry*, *29*, 299–308. doi:10.1016/j.ultsonch.2015.09.015 PMID:26585010

Tayeb, H. H., & Sainsbury, F. (2018). NEs in drug delivery: Formulation to medical application. *Nanomedicine (London)*, *13*(19), 2507–2525. doi:10.2217/nnm-2018-0088 PMID:30265218

Thirumaleshwar, S. K., Kulkarni, P., & Gowda, D. (2012). Liposomal hydrogels: A novel drug delivery system for wound dressing. *Current Drug Therapy*, *7*(3), 212–218. doi:10.2174/157488512803988021

Thom, S. R. (2011). Hyperbaric oxygen–its mechanisms and efficacy. *Plastic and Reconstructive Surgery*, *127*(Suppl 1), 131S–141S. doi:10.1097/PRS.0b013e3181fbe2bf PMID:21200283

Thuraisingam, T., Xu, Y. Z., Eadie, K., Heravi, M., Guiot, M.-C., Greemberg, R., Gaestel, M., & Radzioch, D. (2010). MAPKAPK-2 signaling is critical for cutaneous wound healing. *The Journal of Investigative Dermatology*, *130*(1), 278–286. doi:10.1038/jid.2009.209 PMID:19609314

Troxler, M., Vowden, K., & Vowden, P. (2006). Integrating adjunctive therapy into practice: the importance of recognising 'hard-to-heal' wounds. *World Wide Wounds, 32*, 99-105.

Turksen, K. (Ed.). (2018). *Wound healing: Stem cells repair and restorations, basic and clinical aspects*. John Wiley & Sons.

Verma, P., & Pathak, K. (2010). Therapeutic and cosmeceutical potential of ethosomes: An overview. *Journal of Advanced Pharmaceutical Technology & Research*, *1*(3), 274. doi:10.4103/0110-5558.72415 PMID:22247858

Yin, F. (2014). Preparation of redispersible liposomal dry powder using an ultrasonic spray freeze-drying technique for transdermal delivery of human epithelial growth factor. *International Journal of Nanomedicine*, *9*, 1665. PMID:24729702

Zhao, J., Liu, S., Hu, X., Zhang, Y., Yan, S., Zhao, H., Zeng, M., Li, Y., Yang, L., & Zhang, J. (2018). Improved delivery of natural alkaloids into lung cancer through woody oil-based emulsive nanosystems. *Drug Delivery*, *25*(1), 1426–1437. doi:10.1080/10717544.2018.1474970 PMID:29890855

Zhong, S. P., Zhang, Y. Z., & Lim, C. T. (2010). Tissue scaffolds for skin wound healing and dermal reconstruction. *Nanomed Nanobiotechnol*, *2*(5), 510–525. doi:10.1002/wnan.100 PMID:20607703

Zimber, M. P., Mansbridge, J. N., Taylor, M., Stockton, T., Hubka, M., Baumgartner, M., Rheins, L., Hubka, K., Brandt, E. N., Kellar, R., & Naughton, G. K. (2012). Human cell-conditioned media produced under embryonic-like conditions result in improved healing time after laser resurfacing. *Aesthetic Plastic Surgery*, *36*(2), 431–437. doi:10.100700266-011-9787-8 PMID:21735336

Chapter 22
Application of Nanoemulsions in the Vaccination Process

Sumira Malik
https://orcid.org/0000-0001-5077-1493
Amity University, Ranchi, India

Shristi Kishore
https://orcid.org/0000-0002-8851-5951
Amity University, Ranchi, India

Shradha A. Kumar
Amity University, Ranchi, India

Anjali Kumari
Amity University, Ranchi, India

Manisha Kumari
Amity University, Ranchi, India

Archna Dhasmana
https://orcid.org/0000-0002-1120-2009
School of Applied and Life Sciences, Uttaranchal University, Dehradun, India

ABSTRACT

Vaccination is one of the most effective approaches for the prevention of deadly and highly contagious diseases. One of the important biomedical applications of nanoemulsions is in the process of vaccination. Nanoemulsions are made from nano-sized safe, well-considered ingredients, amalgamated in a protective way to bring forth a stable emulsion. They have provided ways for vaccine delivery using intranasal or mucosal oil-based emulsions rather than using needles. Also, some nanoemulsions have effectively shown anti-pathogenic activities against several germs. Nanoemulsions are also used as vaccine adjuvants and are used to boost the effectiveness of vaccines. Nanoemulsion-based adjuvants put forward the possibility of non-irritating, needle-free vaccines, handed out as nose drops or as a simple nasal sprayer. The chapter aims to discuss the applications of nanoemulsions in the process of vaccination.

DOI: 10.4018/978-1-7998-8378-4.ch022

1. INTRODUCTION

Vaccines are possibly the most paramount interventions developed in the history of mankind to date that have led to a significant reduction in the death rates of the populations since the testing of the world's first vaccine of smallpox by Edward Jenner in 1798 (Riedel, 2005). The aim behind the development of vaccines is to prevent the infections that lead to morbidity, disability, and mortality of individuals. The process of vaccination has improved the quality of life as well as the life expectancy of people all over the world. Although vaccination has been a very successful process, improvement in the development and delivery procedures of vaccines would be of great benefit in this new world in which the unjustifiable anthropogenic actions are leading to the frequent development of new serious infectious diseases. The development of vaccines has seen a technological revolution from the concept of three Is (isolate, inactivate, and inject) to the formulations of rationally designed vaccines having a minimalist composition (Bragazzi et al., 2018; De Gregorio and Rappuoli, 2014). In the past few years, more focus has been given to the development of vaccines that are safe, stable, effective, eassy to administer, and are also cost-effective. An overview of the timeline of vaccine development from 1798 to the present has been depicted in Figure-1 (Gonzalez-Aramundiz et al., 2012).

Figure 1. Timeline of vaccine development

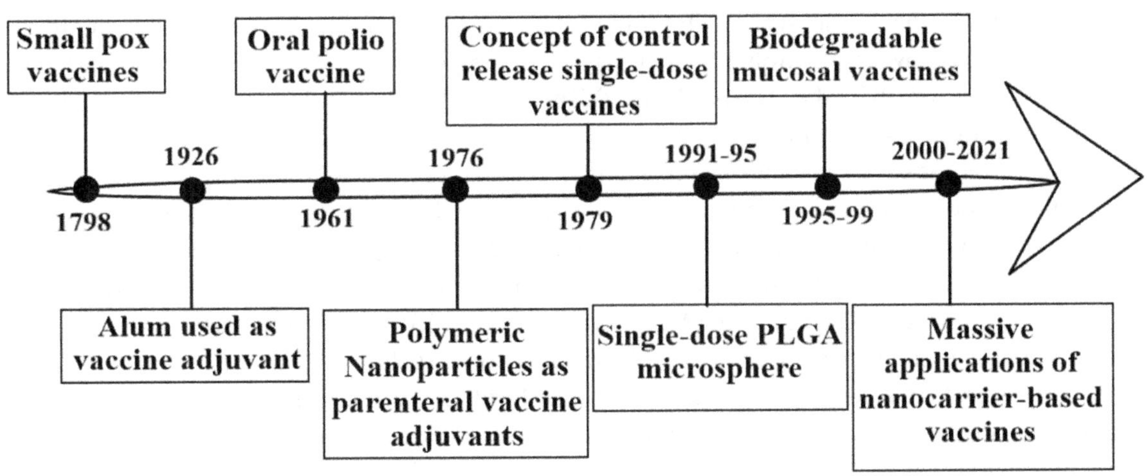

Nowadays, nanocarriers-based vaccines are widely under investigation for needle-free administration. The most common nanocarriers being investigated to be used in the vaccination process are polymeric nanoparticles (NPs), liposomes, virus-like particles, carbon nanotubes, and nanoemulsions (NEs) (Karandikar et al., 2017).

NEs are submicron-sized emulsions that are under broad examination as medicate carriers for moving forward the conveyance of therapeutics such as drugs, vaccines, etc. NEs are thermodynamically steady straightforward (translucent) scatterings of oil and water stabilized by an interfacial film of surfactant and co-surfactant atoms having an average droplet diameter size ranging between 100 nm to 500 nm (Aboofazeli, 2010). They are colloidal scatterings composed of an oil phase, water phase, surfactant, and co-surfactant at suitable proportions. NEs possess varied benefits over other lipid-based carriers. For example, NEs have a way higher extent and free energy than macro emulsions that consign them a good transport system. They do not show the issues of inherent creaming action, coalescence, and deposit that are normally associated in the case of macroemulsions (McClements, 2012). Moreover, they can be developed in a variety of formulations like foams, creams, liquids, and sprays.

Specifically, nanoemulsions (NEs) in the nanometric scale are transparent, translucent, and non-stable dispersion colloidal phase droplets, that is about 100nm, but in some of the bibliography they refer to the limit which is up to 300nm (Comfort et al., 2015, Bahamondez-Canas and Cui 2018). The droplets of 300nm might also be called submicron emulsions, mini emulsions, ultrafine emulsions, and translucent emulsions (Cinar, 2017). In contrast to microemulsions, NEs cannot be manufactured impulsively, thus, in the preparation of NEs high energy methods are required, generally by ultrasound generators, high-pressure homogenizers, and high-shear stirring. During this process, the thermolabile compounds and other proteins, like some nucleic acids and enzymes, may go through deterioration because of the high temperatures and pressure (Çinar, 2017). Although, they are some other alternatives for the preparations of nanoemulsions, that consist of low-energy methods such as emulsion inversion and phase inversion temperature (Bonferoni et al., 2019)

At the mucosal level, under all the physiological conditions NEs might be utilized to solve the problems related to drug stability and/or drug solubility like pH, enzymatic degradation, hydrolysis, and oxidation (Bonferoni et al., 2019). Although, some studies and resources show that even though the NEs have good stability but the droplet size of NEs could follow the Ostwald ripening process and might even breakdown or increase over time. Solutions that have high polydisperse systems or high solubility can follow a passive route or suffer the micellar-based diffusion that forms a droplet of larger size at the cost of the smaller size droplets (Comfort et al., 2015). Kinetic stability is shown by NEs but they do not have thermodynamic stability. NEs because of the small size of the drop have long-term physical stability, thus prevent the process of destabilization such as creaming, sedimentation during the time of storage, flocculation, and coalescence (Comfort et al., 2015, Çinar, 2017). The stability of NEs against cooling, heating, ionic strength, pH, and long-term storage are given by emulsifiers (McClements and Rao, 2011). Surfactants give the ionic surfactants to the NEs and provide the electrical charge whereas steric barriers are created with the bulky groups of molecules by the non-ionic surfactants. Therefore, the emulsifiers that are mainly used for the NEs are surfactants, however lipids and protein are also been reported (Ruiz et al., 2019). Other than these some of the more utilized are sodium deoxycholate (bile salts), sorbitan monolaurate, lecithin (phosphatidylcholine) polyoxyethelene sorbitan monolaurate, sodium dodecyl sulfate, block polymers containing poly-ethylene-glycol (PEG), poloxamers, and some of the amphiphilic proteins such as casein and polysaccharides (gums and starch derivatives) (Bonferoni et al., 2019). NEs are non-toxic and non-irritant, thus can be simply applied to the skin and mucosal membranes. Since NEs are developed with surfactants that are approved for human consumption (GRAS) and thus can be taken by the enteric routes. NEs also have the potency to act as tissue layer immunizing agent adjuvants. NEs adjuvants offer the opportunity of accomplishing immunity with much fewer antigens. Thus, in the process of vaccination, NE-based transport systems can be utilized to a wider extent.

The current chapter aims to discuss the application of NEs in the process of vaccination, focusing mainly on the mucosal and nasal routes of administration.

2. BACKGROUND

The immune system plays a very crucial role in protecting our body from germs and diseases. It also protects our body from various infections that are mainly caused by alien sources such as viruses, bacteria, fungi, and parasites. In human beings, there are mainly three types of immunities that include innate immunity, adaptive immunity, and passive immunity. Among them, two of the immunities, i.e., innate immunity and adaptive immunity are co-related with each other (Kaper et al., 2004). The utilization of the same effector cells leads to their correlation.

The innate immune reaction is non-precise and is the body's first line of defense against contamination from pathogens. The most important role of innate immune reaction is to eliminate or control the infection by foreign pathogens (Kaper et al., 2004). Whenever the human body comes in contact with foreign particles, the innate immune is the first to decide whether or not to react to the particular antigen. Antigens are normally made up of proteins or polysaccharides, which include the protein coat of the virus, cell envelope components, or the DNA or RNA of bacteria, protozoa, viruses, or fungi. These are often any foreign particles or cells that the innate immune cells do not acknowledge as "self". The cells of innate immune response acknowledge foreign particles and also their antigens with the help of a special method known as pattern recognize receptor (PRRs), that acknowledge the molecules present in a wide range of microbes referred to as pathogen-associated molecular patterns (PAMPs) (Medzhitov et al., 2002). Among the PRRs, on the surface of cells within the innate immune system, Toll-like receptors (TLRs) play a very important role in the immune acknowledgement. A total of ten TLRs have been diagnosed in mice and humans (Underhill & Ozinsky, 2002). TLRs just like all PRRs are receptor specific to structurally preserved molecules derived from microbes (Kawai & Akira, 2007). Once PRRs acknowledge PAMPs, cells trigger communication pathways, that cause the activation of phagocytes (that engulf and destroy pathogens) and lead to the initiation of the inflammatory process. This is often accomplished by the secretion of varied cytokines, comparable to interleukins, interferons, and chemokines, which successively trigger the discharge of different cytokines and stimulate cell actions (Kaper et al., 2004). The cells which are operating in innate immune reactions mainly contain macrophages, dendritic cells, mast cells, neutrophils, eosinophils, and natural killer cells (Kumar et al., 1998).

The adaptive immune system (also known as the specific immune system) comes into action when the cells of the innate immune system are unable to stop or end the infection. After the activation of the adaptive immune system, it mainly fights against the infection and protects the body. The adaptive system has the flexibility to recollect and adapt to infections, however, the adaptive immune reaction is additionally stated because of the non-inheritable immune response. As a result of this, cellular memory is not transferred from one generation to another, except for the restricted transfer of antibodies from a mother to a foetus. Thus, each individual must acquire this immunological memory over their lifetime. The main connections between innate and adaptive immune systems are the dendrite cells that are found in almost all the tissues present in a human body. Dendritic cells, one among the classes of antigen-presenting cells, are largely liable for alerting the cells of the adaptive system by the expression of costimulatory molecules (Guermonprez et al., 2002). Dendritic cells perceive unfamiliar microbes with their surface receptors, like TLRs, and go through dendritic cell development, which includes the movement of the dendritic cells to the lymph hubs to alarm antigen-explicit Lymphocytes (Banchereau et al., 1998). The principle cell kinds of versatile resistance must be activated once antigen-introducing cells (like dendritic cells) take-up, interacts, and presents an antigen to gullible lymphocytes in the lymph nodes (Kaper et al., 2004). B-lymphocytes separate into plasma cells, which leads to the discharge of antibodies within the blood, which subsequently flags other cells to start the defense against the restricting antigen or cells bearing them. Some B-lymphocytes do not separate into plasma cells but stay as memory B cells to evoke a quick and expansive discharge of antibodies if the antigen reoccurs by ensuing contaminations taking after the beginning acknowledgment of the antigen. Cytotoxic T lymphocytes that are also alluded to as CD8+ cells, directly assault and slaughter cells bearing the antigen (Harty et al., 2000). Antibody-mediated reactions are moreover called humoral reactions and are the most defense against microbes, infections, organisms, and toxins. Cell-mediated responses are those that do not include antibodies, such as the activity of antigen-specific T lymphocytes, common executioner cells, and other cells that discharge different cytokines in reaction to an antigen. T -helper cells (also known as CD4+ cells) are mainly of two types which include Th-1 and Th-2. Both of them are distinguished i.e., Th1 and Th2 by the discharge of distinctive and particular cytokines (Jankovic et al., 2001; McGuirk et al., 2002). The design of cytokine generation decides the sort of resistant reaction, whether transcendently a humoral or cellular reaction. Understanding what kind of reaction is required to bestow resistance and what instruments lead to a Th1 or Th2 reaction is basic to the advancement of antibodies.

Conventional immunizations comprise of little amounts of living or dead organisms or atoms determined from the remote organism that enacts the adaptive resistant reaction and the acceptance of memory B cells, which can create a quick, effective safe reaction to future disease by that organism. For occurrence, flu immunizations are either live constricted or inactivated immunizations. The inactivated immunizations have three diverse shapes of antigen preparation-inactivated entirety infection, subdivisions where filtered infection particles are disturbed to solubilize the lipid envelope (“split” immunization), or as a subunit antibody (filtered surface proteins, that are commonly known antigens) (Ledet et al., 2013). The outside factors that can affect the adequacy of an antibody incorporate the course of organization, the dosage, the measurement plan, and the immunologic status of the beneficiary.

3. NANOEMULSIONS AS VACCINE ADJUVANTS

Molecules or compounds that act as additives in the formulations of vaccines in order to strengthen, modulate, and boost their immunogenicity, while maintaining the least/negligible toxic effects are known as adjuvants (Wack and Rappuoli, 2005). The use of adjuvants in the formulation of vaccines leads to an increase in the longevity in the response of the immune cells to some specific antigens. Moreover, it has been reported that adjuvants play a key role in increasing the efficiency of the vaccine in neonates, aged, and immunocompromised persons, as well as help in reducing the dosage and number of requisite immunizations (Kenney and Edelman, 2003). Based on their source of components, physical and chemical properties, and mechanisms of action, adjuvants can be classified into two types: immunostimulants and vehicles or delivery systems. Immunostimulants are those adjuvants that help in increasing the response of the immune system to antigens by directly acting on it. TLR ligands, saponins, bacterial toxins, cytokines, etc. are examples of immunostimulant adjuvants. On the other hand, vehicles or delivery systems are those adjuvants that aid in delivering the antigens to the specific immune sites in a controlled and efficient manner. Vehicle adjuvants include mineral salts, liposomes, virosomes, microspheres, and NEs (Reed et al., 2009). NEs due to their wide range of beneficial properties have gained huge attention in the area of vaccination and drug delivery in recent years. They are regarded as appropriate adjuvants in the vaccine formulations due to the following reasons. Firstly, most of the adjuvants such as aluminum salts do not induce a suitable immune response against all the antigens. Secondly, emulsions are considered safe for human use for a long time. Moreover, the stability of NEs confers them an advantage for storage for a long period and manufacturing at a large scale. Lastly, the success rate of use of NEs as adjuvants in the vaccine is quite high, although, before the 1990s, they were mostly used in the formulations of veterinary medicines (Aucouturier, Dupuis, and Ganne, 2001).

3.1. Approved Nanoemulsions as Vaccine Adjuvants

Before the 1990s, the only adjuvants approved for their use in the human vaccine formulations were aluminum salts, sometimes also referred to as alum. No other adjuvant was approved for the safe use in humans until the emergence of MF59®, an oil-in-water NE composed of squalene (5%), Tween 80 (0.5%), and Span 85 (0.5%) in an aqueous buffer of sodium citrate and citric acid having a pH of 6.5 (Podda and Del Giudice, 2003). MF59® containing the influenza vaccine Flaud® was initially developed to deliver muramyl tripeptide phosphatidylethanolamine (MTP-PE) which is an immunostimulant. But due to its toxic effects, MTP-PE was drawn out of this formulation (Schultze et al., 2008). MF59® formulated with inactivated split-virion influenza vaccine has been reported with a good safety profile in the population of elderly individuals, as they were found to increase their immunogenicity (Galli et al., 2009). MF59® adjuvant formulations adjuvant with influenza vaccine Flaud® have been licensed in Europe since 1997. In 2009, they were also approved for use in Europe as the pandemic vaccine Focetria® (Hatz et al., 2012). In a study, MF59® was compared with alum, calcium phosphate, poly-(lactide-co-glycolide), and CpG to test for the induction of better immune response in mice. Among all other adjuvants, MF59® was found to be the most potent adjuvant in terms of CD4 T cells and protective antibodies (Wack et al., 2008). Although the mechanism of action of MF59® is not still clear, it has been found that it stimulates the recognition and uptake of antigens, recruits innate immune cells, and forms an immunocompetent environment at the site of injection. The recruitment of the innate immune cells triggers them to take up the antigen and the adjuvant and transport them to the locally-confined lymph nodes. In this way, the immune response is induced by them (O'Hagan et al., 2012).

AS03 is another oil-in-water NE that is composed of squalene, α-tocopherol, and tween-80 in a phosphate-saline buffer of pH 6.8 (Garçon et al., 2012). A pre-pandemic influenza vaccine that contains AS03 is Prepandrix®. Another vaccine approved as H1N1 influenza A pandemic vaccine with AS03 is Pandemrix®. This vaccine has been approved in Europe in 2009. It has been found that the response of T cells and antibodies against viral proteins in split cells is enhanced by the actions of AS03 (Leroux-Roels et al., 2007). Moreover, the innate immune response triggered by S03 is mostly localized. Moreover, in comparison with other adjuvants such as alum, the antigen-specific adaptive immune response enhanced by AS03 is relatively higher (Morel et al., 2011). The mechanism of action of AS03 vaccine adjuvants is illustrated in Figure-2.

Figure 2. Mechanism of action of AS03.

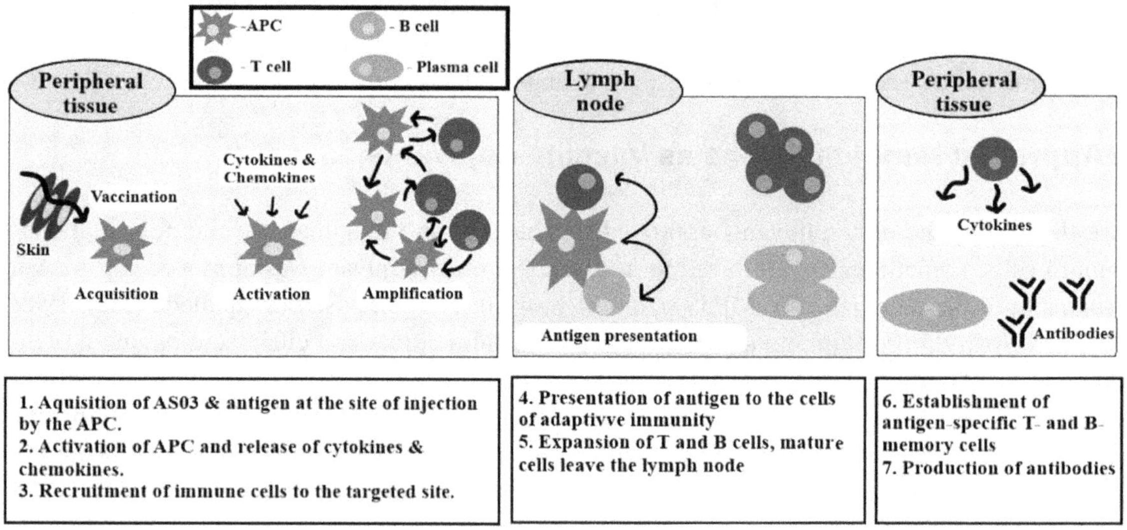

Humenza® is a vaccine that had been approved for pandemic influenza prophylaxis in Europe. One of the components of this vaccine formulation was AF03 that was used as an adjuvant. AS03 is a thermoreversible (emulsion that can interconvert into oil-in-water type to water-in-oil type) NE composed of squalene, polyoxoethylene cetostearyl ether, Span-80, and mannitol in a phosphate/citrate buffer (Klucker et al., 2012). Due to its thermoreversible nature, AF03 is very stable in comparison with other adjuvants. Unlike the previous two adjuvants, it can be stored in high temperatures due to its higher stability. But due to commercial reasons, Humenza® containing AF03 was drawn out of the market in 2011.

3.2. Current State of Development

Currently, numerous research works are being performed by scientists all over the world to test the efficiency and safety of different types of vaccine adjuvants. Presently, a clinical trial (NCT04148118) of BW-1010, a NE adjuvant recombinant protein (rPA) is going on. The aim of the trial is to assess the safety, tolerability, and immunogenicity of BW-1010 NE adjuvant recombinant anthrax vaccine administered through nasal routes in healthy adult individuals. Currently, the trial is in its phase-1 and is active, but not recruiting. Another active clinical trial (NCT02500680) is being done for testing the safety, tolerance, and immunogenicity of MAS-1 emulsion adjuvant seasonal inactivated influenza vaccine in healthy adults and elderly individuals. This trial is also in its phase-1 and is active but not recruiting). In a study, the efficiency of MF59 was tested for delivering a synthetic TLR4 agonist-E6020. When single adjuvant MF59, MF59 with TLR agonist EL020, and MF59 with TLR9 agonist CpG were administered in mice, single adjuvant MF59 was found to induce the hemagglutination inhibition (HI) titres and responses of T-cells the most. However, adjuvanting MF59 with any of TLR agonists did not lead to an increase in the antibody production but still, the response of the immune cells was shifted more towards Th1-based (Baudner et al., 2009). Furthermore, the use of emulsion-based adjuvants such as AS03 and MF59 is also under investigation for vaccination against the coronavirus (Liang et al., 2020).

4. ROLE OF NANOEMULSIONS-BASED VACCINES IN ENHANCING IMMUNE RESPONSE

NEs are being used to supply either inactivated organisms or recombinant proteins to a mucosal surface to produce an immune response. Influenza vaccines and HIV vaccines are their first applications and are proceeded to clinical trials. The proteins applied to the mucosal surface get adjuvant because of NE and it expedites the uptake of antigen-presenting cells and which results in a substantial systemic and mucosal immune response that also encompasses the production of specific IgA & IgG antibodies over and above cellular immunity (Lycke, 2012). Many works have been conducted in influenza for exhibiting that animals can be safeguarded against influenza by just providing a single exposure of mucosal vaccine admixed with NEs, researches have also showcased that when animals are subjected to recombinant gp120 in NE on their nasal mucosa they have developed momentous responses to HIV, consequently offering a foundation to assess the use of NE in HIV vaccine, further researches are still going on to accomplish the validation of this concept in animals trials for such other vaccines like for anthrax and Hepatitis B. Epidemiology and experimental data suggests that both full-bodied neutralizing antibodies and effective cellular responses play important roles in controlling primary HIV-1 infection (Rezende & Weiner, 2017). Studies were conducted in mice and guinea pigs, they were immunized intranasally by the application of recombinant HIV gp120 antigen admixed in NE and they show healthy serum anti-gp120 IgG along with serum anti-gp120 IgA, vaginal and bronchial in mice. The serum of the above animals showcased antibodies when cross-reacted with heterologous serotypes anti-gp120 IgA, they had much necessary neutralizing activity against two Clade-B laboratory strains of HIV and five primary HIV-1 isolates. All the studies advocated that NE be supposed to evaluate as a mucosal adjuvant for multivalent HIV vaccines (Bielinska et al., 2008).

Hepatitis B virus infection is persistently a major health problem globally regardless of the accessibility of effective prophylactic vaccines, it is because of the limitations like a refrigerant requirement and three immunizations thus creates a restriction in their use, especially in a developing country. New nasal hepatitis B vaccine that is made of recombinant hepatitis B surface antigen in NE adjuvant has been found out to be effective with lesser administration. Pre-clinical toxicology evaluation has been conducted to show that the HBsAg-NE vaccine is safe and effective as well as is well tolerated in many animal models; the results obtained suggest that nasal immunization with HBsAg-NE vaccine could be a much safe and more effective vaccine for hepatitis B. It also provides another possible booster administration for the parenteral vaccine for hepatitis B (Makidon et al., 2008). The vaccine can induce a Th1 associated cellular immunity and also offer therapeutic benefit to the patients who have chronic infection of hepatitis B as they lack cellular immune response thus not able to control the replication of the viral DNA. Stability for the long term of this type of formulated vaccine at higher temperature could be beneficial as it shows a direct advantage since potential deviations from cold chain maintenance could be tolerated without any loss in the therapeutic efficiency (Bielinska et al., 2007). NE Vaccine administration without the use of needles has proved to be able to produce a much strong immune response against HIV and smallpox in many studies conducted recently. The studies showed that NE somehow has the capability of developing vaccines from many miscellaneous materials (Wack, & Rappuoli, 2005). The published results show that the HIV NE vaccine when tested in mice represents a different approach in the way it produces immunity and the type of immunity produced, a normal vaccine that was administered through the nose were not able to induce mucosal immunity in the genital mucosa as there are shreds of evidence that HIV can infect the mucosal immune system. For that reason, mucosal immunity has to be developed and it will only play a crucial part in protection against HIV. The NE HIV vaccine has shown that it can induce mucosal immunity, cellular immunity, and neutralizing antibody to various isolates of HIV.

Adjuvants are additives that are included in vaccine formulations to mainly enhance, direct or modulate the immunogenicity of the vaccine while causing minimal toxicity (Reed, 2009), they are generally classified by their mode of action as immunostimulatory agents and as vehicles. Immunostimulatory agents are those which act on the immune system to increase the efficiency of vaccine whereas vehicles are those which help in delivering and deliver the antigen to the immune system (Byars & Allison, 1987), but in some cases, adjuvants can fit into both the categories. Adjuvants can modify the immune response to the vaccine in several ways that specifically include depot creation for long-drawn-out release of antigen, engaging with PRRs to increase the activity of DC, and cytotoxic T lymphocytes induction (Banzhoff et al., 2009). There are many benefits of adjuvants, some of them are increased immunogenic response to those who have relegated receptiveness to normal vaccines or non-adjuvanted vaccines and also give the chance to reduce the amount of needed antigen to provoke the anticipated immune response that somehow set aside for additional doses to be administered with the equal amount of antigen. Another benefit of adjuvants is that it reduces or simply eliminates the numerous doses at the same time as even so attaining passable fortification along with that adjuvants also can alter some type of immune response to get on the whole an apt response to a particular antigen in the query (Boyle et al., 2007). All the above benefits conclude that adjuvants are very important and necessary to enhance the immunogenicity and efficiency of several vaccines especially for those who are more susceptible to a specific disease and cannot fight against that particular disease. As it is known that old age people, in particular, are more susceptible to diseases and it is so because as one gets older their immune system gets weaken and perpetually loses its ability to fight against diseases, similarly new-born baby or infants and children also have more chances to get affected with transmittable disease because their immune system is not fully developed as compared to an adult (Mastelic et al., 2010). Thus, immunocompromised individuals are in more need of an effective and more reliable vaccine to prevent any by chance infections and any other kind of transmittable disease for which their immune system is not well equipped to fight off (Schwarz et al., 2009). Adjuvants are in process of becoming more important as a vaccine because as the technology progresses it has come beyond the use of live attenuated vaccines and is moving forward for the use of isolated antigens like the influenza vaccine. On the whole, live virus vaccine and inactivated viruses or bacteria do not have the requirement for an adjuvant to increase their immunogenicity, and highly purified recombinant subunit antigens have the capability of reducing the risk of toxicity that are allied with live viruses but they are often found out to be less immunogenic than any other traditional live virus vaccines (Petrovsky & Aguilar, 2004). Recombinant vaccines are the only alternative for the diseases for which live or inactivated vaccines are not available, not practical, or are not safe and in such cases the development of appropriate adjuvants becomes necessary. Adjuvants would have to be customized to their specific target antigen because of the complex nature of the immune response, the range of pathogens, and several other routes of entry into the body available to pathogens (Kaper & Cobon, 2004). The supreme objective of enhancing an adjuvant for any specific antigen is to boost the potential of the vaccine. The common criteria that are normally considered while selecting the appropriate type of adjuvant required for a specific vaccine mainly include an understanding of the nature of the antigen, the type of immune response required to provide the best protection, and the delivery route of the vaccine. The most favourable adjuvant also has to be biocompatible, biodegradable, safe, stable for long shelf life, and inexpensive to manufacture (Johnson-Weaver et al., 2020). The risk of adverse side effects of adjuvants must be balanced with the benefits of including it in vaccine formulations. NEs exemplify a promising mucosal vaccine adjuvant system that authorizes persistent assessment in clinical studies because they can activate the immune system employing various pathways.

5. NANOEMULSIONS AS MUCOSAL VACCINE ADJUVANTS

Mucosal vaccine adjuvants have the capability of inducing a protective immune response in both the mucosal immune compartments as well as systematic immune responses, it has many advantages in the therapeutic field and thus it is often characterized as the blue ocean in the vaccine industry. Mucosal vaccines offer lower cost, have better accessibility, are a needle-free delivery and most importantly it has a higher capacity for mass immunizations during pandemics, thus can be very useful in times of crisis during the pandemic. But a very limited number of vaccines have been approved for human use in the market yet. For induction of the immune response through mucosal immunization proper co-administration of adjuvants is required as then only it can commence and provide for the effective collaboration between both the immunities i.e., innate and adaptive immunity. Researches on adjuvants have rather been experiential than scientific, then again during the last several years, scientific achievements in innate immunity have been transformed into the advancement of new mucosal adjuvants.

In view of the fact that many pathogens infect the host via mucosal route like inhalation, ingestion, etc., so the development of a vaccine that can both prevent the invasion of the pathogen through the mucosal route at the infection stage and can also neutralize the pathogen-derived toxin or during later replication stages the replication of pathogen in the body can be inhibited, is necessary then only such diseases can be prevented (Holmgren & Czerkinsky, 2005). Immunization through mucosal routes could meet the above-mentioned requirements as it stimulates the immune response (Chen et al., 2010). Mucosal vaccines have many advantages but it also has some disadvantages, like poor immunogenicity, physical mucosal barriers, heavy dose due to instability of antigens in the mucosal surfaces, etc. (Holmgren, et al., 2003). But most of its disadvantages can be by-passed with effective adjuvants and drug delivery systems. The most potent mucosal adjuvants are descended from bacterial toxins, TLR (Toll-like receptors) ligands, non-TLR immunostimulants (immunostimulants are naturally occurring compounds that amend the immune system by way of increasing the host's resistance to disease), and novel small molecules (Freytag & Clements, 2005). The table-1 below shows the current mucosal vaccine adjuvants and vaccine delivery systems under clinical application or preclinical researches.

The sophistication of vaccine adjuvants design has been improved by getting more understanding about the features and importance of NEs and their immunological function. NEs also shows innate immune-activating adjuvants especially in vaccine administered mucosally. Although NEs are mostly important in drug delivery they are also useful as immune-modulating adjuvants for a mucosal vaccine. One such example of NE formulation with nasal adjuvant activity is W805EC and which contains Cetylpyrimidine chloride, soybean oil, ethanol, and water (Bielinska et al., 2014) Nasal immunization with this W805EC along with an inactivated influenza virus has seemed to enhance the induction of antigen-specific serum IgG and IgA in the nasal ashes of immunized mice (Das et al., 2012). Antigen formulated with phosphate-buffered saline when compared with W805EC adjuvant vaccine, this NE formulated vaccine was found to have more effect in enhancing immune response as it has increased the serum HA-inhibiting antibodies leading to decrease in viral loads mainly in lungs and nasal cavity also morbidity was reduced. NE-adjuvant influenza virus vaccine has induced antibody responses that were cross-reacted with some other strains of influenza that were not present in the vaccine perchance it will be the NE adjuvant that influences the antigenic determinant specificity of antibodies induced by the vaccine (Kallewaard et al., 2016). When Guinea pigs were immunized through nasal administration with hepatitis B antigens brought together with NEs showed to have developed an elevated antigen-specific serum IgG (Makidon et al., 2008), correspondingly mice and guinea pigs were used to ascertain the adjuvant activity of NE when they were co-processed with rPA (Recombinant protective antigen) (Bielinska et al., 2007). In mice when immunization through nasal route or nasal immunization with NE-adjuvanted rPA was done it induced serum anti-rPA IgG responses which were much greater than antigen-rPA IgG induced by MPL (Monophosphoryl lipid A) or CpG oligodeoxynucleotide adjuvant rPA (Stanberry et al., 2012) and similarly guinea pigs were also nasally immunized with NE-adjuvanted rPA they also developed increased serum rPA-specific IgG, antibodies that were anthrax lethal toxin-neutralizing and thus were protected against lethal anthrax infection. Many clinical trials have been conducted to evaluate the safety and immunogenicity of NE adjuvant W805EC after nasal immunization with seasonal influenza virus vaccines and no major or adverse effects were found except for some minor throat irritation (Man & Kanneganti, 2016). When intranasal immunization was compared with intramuscular immunization, the latter was found to develop serum anti-influenza antibody responses and W805EC was found to be more effective nasal adjuvant activity mainly based on its ability to stimulate serum anti-influenza IgG responses and when this was also compared with nasal vaccines containing influenza antigen in nonexistence of added adjuvant, then the former one was significantly showing higher responses (Saito et al., 2016). NEs arbitrate their adjuvant activity as a consequence of effective activation of the innate immune system, given that epithelial cells are the first cell source to come across with NE adjuvants after mucosal delivery afterward their response to them may perhaps initiate the proinflammatory cascade eventually leading to adjuvant activity. NE W805EC in human epithelial cells will induce apoptosis using several mechanisms together with necrosis, pyroptosis, and increased production of caspase enzymes because the release of caspase enzyme from epithelial cells may initiate a cascade that will cause enhancement in proinflammatory cytokine production thus activating the innate immune system to provide adjuvant activity (Makidon et al., 2012). Nasal administration of NEs enhances the inflammatory cytokine production whereas in presence of NEs the epithelial cells swamp antigen in vitro and in vivo and increases the molecule expression of MHC II (Myc et al., 2013), afterward the dendritic cells consume the apoptotic epithelial cells and eventually increasing the expression of costimulatory molecules (Čerpnjak et al., 2013), that may perhaps lead to more increase activation of T cell. Mucosal administration of NEs also increases the ability of dendritic cells to circulating antigens to lymph nodes and there they may activate T cells thus enhancing adaptive immune responses.

Table 1. Current mucosal vaccine adjuvants and vaccine delivery systems under clinical application or preclinical researches

Adjuvants/delivery systems	References
1. Live attenuated bacterial and viral vaccines	(Duthie et al., 2011)
2. Live bacterial and viral vectors	(Duthie et al., 2011)
Nanoparticles and microparticles	
1. Poly(DL-lactide-co-glycolide) microparticles	(Mishra et al., 2010)
2. Polystyrene (latex) microparticles	(Mishra et al., 2010)
Lipid-based or lipid-containing	
1. Liposomes (anionic and cationic)	(Chen et al., 2010)
2. Proteoliposomes	(Chen et al., 2010)
3. Cochleates	(Bracho et al., 2006)
4. Immune-stimulating complexes and Iscomatrix	(Cox et al., 2006)
5. Eurocine (monoglycerides with fatty acids)	(Chen et al., 2010)
6. Archaeal lipid mucosal vaccine adjuvant and delivery	(Chen et al., 2010)
7. Virosomes and virus-like particles	(Yoneyama & Fujita, 2009)
8. Outer membrane vesicles	(Chen et al., 2010)
Bacterial toxins and their derivatives	
1. Cholera toxin	(Holmgren, et al., 2003)
2. Escherichia coli heat labile toxin	(Holmgren & Czerkinsky, 2005)
3. Other AB5 toxins and their derivatives	(Lawson et al., 2011)
TLR ligands	
1. Monophosphoryl lipid A	(Holmgren & Czerkinsky, 2005)
2. DNA/CpG-ODN motifs	(Holmgren & Czerkinsky, 2005)
3. Flagellin	(Lee et al., 2015)
Others	
1. c-di-GMP	(Chen et al., 2010)
2. Supramolecular bio-vectors	(Chen et al., 2010)
3. Transgenic plant "edible" vaccines	(Chen et al., 2010)
4. Saponins	(Guy, 2007)
5. Chitosan	(Simões, et al., 2005)
6. α-Galactosylceramide	(Noda et al., 2010)
7. Endogenous molecules (cytokines, chemokines and defensins)	(Holmgren, et al., 2003)

6. NANOEMULSIONS IN NASAL VACCINE FORMULATIONS

In the mucous of the body large amount of human infections take place, therefore activating immunological defence in this particular area is of great importance. Although efficient systemic protection is induced by the parenteral administration of the antigen, the mucous formed by the body remain unprotected. However various attempts have been made to trigger immunological defence in mucous membranes following distinct routes of application but among them, some of these have had side effects or the effect that has been expected clinically has not been achieved (Luna-Pineda et al., 2019). The nasal mucous route is a type of non-invasive route, mainly for diseases related to respiration (Comfort et al., 2015). This route does not require any type of needles for their application and permits the use of vaccines that are heat resistant, since no special type of personnel is needed for their application or waste disposal they can potentially lower the costs of their use, which would be favourable in mass vaccination campaigns. Furthermore, this nasal mucous route permits the administration of both dry as well as liquid vaccines and in some preclinical studies, it has been shown that to activate immunological defence the more convenient way is to use intranasal dry powder vaccines (DPVs) instead of liquid vaccines because they are more stable (Sandri et al., 2015).

In recent studies, it has been seen that the advantages of mucosal vaccination are more than the intramuscular immunization when it comes to giving mucosal protection, probably because of the most important fact that intranasal administration induces cellular and hormonal immunity, that permit immunization at different sites of mucous as well as in nasal mucous and also provide systemic protection (Comfort et al., 2015).

In the animal model, it has been demonstrated that the use of nasal vaccination cause protection against the antigens from HIV gp120, recombinant anthrax protective antigen, hepatitis B surface antigen (HBsAg), and influenza virus (Makidon et al., 2010). In comparison to the nasal route, the much larger mucosal surface area is offered by the pulmonary route and also has a great vascularization in the tissues of alveoli. This can be helpful in the systemic delivery of the antigens and may enhance systemic and respiratory immunization. For some diseases like influenza, tuberculosis, measles, and hepatitis the studies on dry powder vaccines through inhalation have already been done. As we know that on the surface of the upper ways the particles that are above 5 micrometres get trapped, therefore it is important for the particles to get to the deep lung measure between 1 and 5 micrometres. Therefore, the device and the formulation that is used to direct the formulation are critical, because they can find the particle size that is used for lung deposition. However, in the case of vaccine delivery, it may not totally be a problem because the directed antigens can still manage to reach the tissue of lymphoid in the oropharyngeal region (Bahamondez-Canas and Cui, 2018).

6.1 Immunological Responses Activated Through Nasal Administered Antigen

The nasally administered antigen can obey the following two routes: particle-sized antigens could either be absorbed through M cells (micro fold cell) in NALT (Nasopharynx-Associated Lymphoid Tissue) or could be removed through the mucociliar system, however, soluble antigens have the ability to penetrate the nasal mucosa effectively and also activate the antigen-presenting cells, such as macrophages and dendritic cells (Bahamondez-Canas and Cui 2018). The mucosal response has been assigned to activate the Toll-Like-Receptors (TLR), mainly TLR-4 and TLR-3, and for the internalization of droplets of nanoemulsion (Comfort et al., 2015). The immune response of B and T cells get triggered when the antigen-presenting cells are activated, so they release immunoglobulin A in the lumen and differentiate in the plasma cells, which neutralizes particular antigens and shape the complex antigen-antibodies, thus in the mucus, they get trapped easily and later on the ciliary movement shown by epithelial cells remove them. In some of the specific mucosal surface lymphocytes were activated and possibly reach remote sites of the mucosa by traveling through the lymphatic system, therefore provide protection in this new site also and this process is commonly known as "the immune system of common mucosa" (Ruiz et al., 2019).

Antigens that are captured through NALT drain into the lymphatic nodes, so they generate the production of IgG and IgA serum in the systemic organs of lymphoid. However, the systemic immunological protection induced by parental immunization is usually more effective than that caused by the mucosal vaccination. In nasal vaccines the ciliary clearance can be a problem because antigens might not be able to activate the immune response effectively before they get eliminated, however, the chitosan has been previously proved as a mucoadhesive that can increase the time of antigens that are in contact with descendental immunological response and immune effector sites (Bahamondez-Canas and Cui 2018). Figure 3 illustrates the use of NEs in the nasal vaccines formulations.

Figure 3. Nanoemulsions used as adjuvants in intranasal vaccines

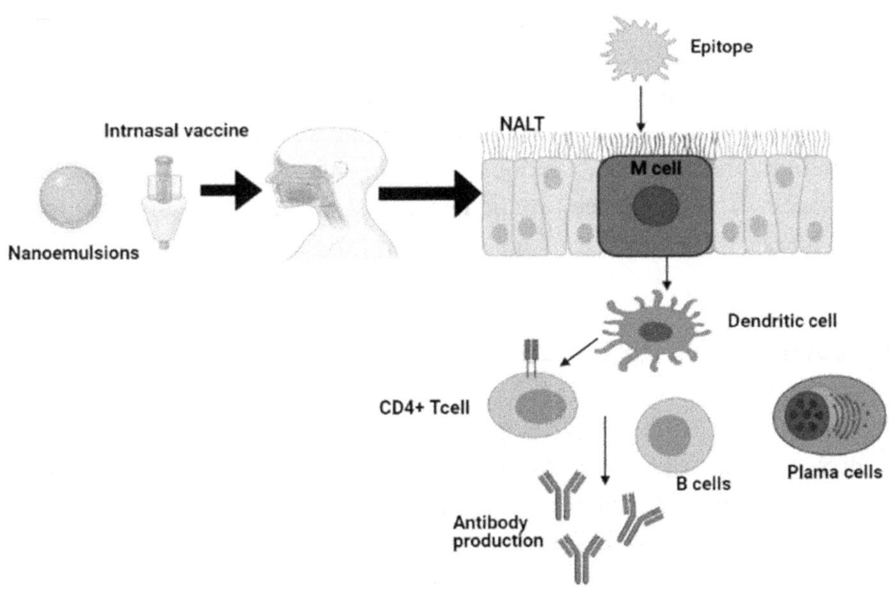

6.2 Nanoemulsions as Nasal Vaccines Adjuvants

In the past few years, in the nasal mucosal vaccination, nanoemulsions (NEs) have been mainly studied as adjuvants, because of their immunomodulatory or/and immunostimulatory effect. As the antigens get established into the nanoemulsion oil droplets, the degradation of epitopes gets prevented, resulting in an increased stability. (Comfort et al., 2015). Moreover, in comparison to other formulations, NEs are characterized through a high surface area when it comes to nasal drug delivery.

In the development of intranasal vaccines nanocarriers are required which are capable of delivering the antigens to the cells and of entering the mucosal cells. In the process of administration of protein, the biodegradable nanoparticles that are based on cationic maltodextrin and porous are the best option among the 5 types of nanoparticles that are studies (Quan Le et al., 2018). For the stabilization of NEs, the co-surfactants such as propylene glycol, glycerine, polyethylene glycol, and ethylene glycol can also be used. In these formulations, no mutagenic effects have been reported to date in the literature and thus they seem to be safe (Bonferoni et al., 2019). Makidon and their colleague proved that if administered antigens go through any functional change it is due to protein folding and they concluded that vaccines that are based on NEs do not require any type of engineered delivery devices to be effective (Makidon et al., 2010).

The immunizations against the influenza B subtype viruses and type A viruses are indicated by FluMist Quadrivalent which is a live attenuated influenza vaccine (LAIV). However, in 2016, 2017, and 2018 the Centers for Disease Control and Prevention (CDC) and the American Academy of Pediatrics (AAP) stop recommending the use of LAIV due to its poor effect against the strains of HINI, since it has been again formulated, in 2019 American Academy of Pediatrics announced that it does not have a preference between the inactivated influenza vaccine and LAIV. The 2019 version of FluMist Quadrivalent protects against the 4 weakened influenza viruses such as A (H3N2), B Victoria lineage, A (HINI), and B Yamagata lineage. The inactivated ingredients in them are monobasic potassium phosphate, gelatine, sucrose, monosodium glutamate, dibasic potassium phosphate, gentamicin, and sucrose (Ruiz, Varela and Cornejo., 2019). Some intranasal vaccines have been already commercialized like FluMist®, which is the first LAIV among the intranasal vaccines that were commercialized and approved in the Europe and US as Fluenz®. In India the influenza vaccine that has been approved is Nasovac®. For intranasal administration, other vaccines have also been studied against some infectious diseases such as tuberculosis, pneumonia, and measles (Bahamondez-Canas and Cui 2018).

7. CONCLUSION AND FUTURE PROSPECTIVE

NEs because of their numerous advantages are being utilized as effective vaccine delivery systems. This effective vaccine delivery system uses nanotechnology in the process of vaccination against some deadly pathogens such as human immune-deficiency viruses (HIV). Recently, some pieces of evidence show HIV are contaminated in the mucosal immune system. Therefore, to increase the further fight with HIV the use of nanoemulsions in the development of mucosal immunity is becoming very important day by day. NEs based on oil are administered and it also helps in the advancement in the characterization, formulation, and applications in drug delivery operated into the nasal route in contrast to the vaccines that are used traditionally. The result of researches and recent trends shows that genital mucosa immunity might be achieved with the help of vaccines that are administered to the nasal mucosa route. NEs are used for the transportation of inactivated organisms to the mucosal surface to produce the immune response. The first application of the nanoemulsions as the influenza vaccine and an HIV vaccine is started to clinical trials for future studies. Due to its appropriate cost benefits and its wide use, the intranasal route of NEs is given as a great option for vaccine administration, mainly for mass vaccination to prevent epidemics. Despite having many advantages there are some disadvantages of NEs such as an increased amount of mild adverse effects at the site of injection applied and they have a very limited role to play in immunity. However, NEs have proven their effectiveness and versatility, still, there are much more growth and researches needed before the establishment of NEs as a favourable vaccine adjuvant.

Today, as the international travel and global population increases, the rate of the global threat of pandemics, biological welfare, and emerging diseases boost up. In the past years more than 30 emerging (AIDS), re-emerging infections (malaria and anthrax), and swine influenza have been reported. The effectiveness and versatility of NEs make them more favourable to encourage researchers to develop effective vaccines for emerging diseases. In addition to this that new vaccines are essential to meet the needs of a population with a different immune response from a larger elderly population, researchers are also assigned the task of finding the possible vaccines for age-related vaccines because of these growing statistics in the developed countries. Despite the challenges and activities that have come to light, we still encourage to continue the investigation of vaccination with the testing and use of nanoemulsions as the adjuvants, that enhance and induce immunological activity and provide the stability to the components of vaccines additionally.

REFERENCES

Aboofazeli, R. (2010). Nanometric-scaled emulsions (nanoemulsions). *Iranian journal of pharmaceutical research. IJPR*, *9*(4), 325–326. PMID:24381596

Aucouturier, J., Dupuis, L., & Ganne, V. (2001). Adjuvants designed for veterinary and human vaccines. *Vaccine*, *19*(17-19), 2666–2672.

Bahamondez-Canas, T. F., & Cui, Z. (2018). Intranasal immunization with dry powder vaccines. *European Journal of Pharmaceutics and Biopharmaceutics*, *122*, 167–175.

Banzhoff, A., Gasparini, R., Laghi-Pasini, F., Staniscia, T., Durando, P., Montomoli, E., Capecchi, P. L., di Giovanni, P., Sticchi, L., Gentile, C., Hilbert, A., Brauer, V., Tilman, S., & Podda, A. (2009). Correction: MF59®-Adjuvanted H5N1 Vaccine Induces Immunologic Memory and Heterotypic Antibody Responses in Non-Elderly and Elderly Adults. *PLoS One*, *4*(3). Advance online publication. doi:10.1371/annotation/5f2ae0fb-53f1-48c2-aa19-9114765ba029 PMID:19197383

Baudner, B. C., Ronconi, V., Casini, D., Tortoli, M., Kazzaz, J., Singh, M., Hawkins, L. D., Wack, A., & O'Hagan, D. T. (2009). MF59 emulsion is an effective delivery system for a synthetic TLR4 agonist (E6020). *Pharmaceutical Research*, *26*(6), 1477–1485. doi:10.100711095-009-9859-5 PMID:19255727

Bielinska, A., Janczak, K., Landers, J., Makidon, P., Sower, L., Peterson, J., & Baker, J. Jr. (2007). Mucosal Immunization with a Novel Nanoemulsion-Based Recombinant Anthrax Protective Antigen Vaccine Protects against Bacillus anthracis Spore Challenge. *Infection and Immunity*, *75*(8), 4020–4029. doi:10.1128/IAI.00070-07 PMID:17502384

Bielinska, A., Makidon, P., Janczak, K., Blanco, L., Swanson, B., Smith, D., Pham, T., Szabo, Z., Kukowska-Latallo, J. F., & Baker, J. R. Jr. (2014). Distinct Pathways of Humoral and Cellular Immunity Induced with the Mucosal Administration of a Nanoemulsion Adjuvant. *Journal of Immunology (Baltimore, Md.: 1950)*, *192*(6), 2722–2733. doi:10.4049/jimmunol.1301424 PMID:24532579

Bielinska, A. U., Janczak, K. W., Landers, J. J., Makidon, P., Sower, L. E., Peterson, J. W., & Baker, J. R. Jr. (2007). Mucosal immunization with a novel nanoemulsion-based recombinant anthrax protective antigen vaccine protects against Bacillus anthracis spore challenge. *Infection and Immunity*, *75*(8), 4020–4029. doi:10.1128/IAI.00070-07 PMID:17502384

Bielinska, A. U., Janczak, K. W., Landers, J. J., Markovitz, D. M., Montefiori, D. C., & Baker, J. R. Jr. (2008). Nasal immunization with a recombinant HIV gp120 and nanoemulsion adjuvant produces Th1 polarized responses and neutralizing antibodies to primary HIV type 1 isolates. *AIDS Research and Human Retroviruses*, *24*(2), 271–281. doi:10.1089/aid.2007.0148 PMID:18260780

Bonferoni, M. C., Rossi, S., Sandri, G., Ferrari, F., Gavini, E., Rassu, G., & Giunchedi, P. (2019). Nanoemulsions for "nose-to-brain" drugdelivery. *Pharmaceutics*, *11*(2), 84. doi:10.3390/pharmaceutics11020084 PMID:30781585

Boyle, J., Eastman, D., Millar, C., Camuglia, S., Cox, J., Pearse, M., Good, J., & Drane, D. (2007). The utility of ISCOMATRIX™ adjuvant for dose reduction of antigen for vaccines requiring antibody responses. *Vaccine*, *25*(14), 2541–2544. doi:10.1016/j.vaccine.2006.12.018 PMID:17240491

Bracho, G., Lastre, M., Campo, J. D., Zayas, C., González, D., Gil, D., Acevedo, R., Taboada, C., Solís, R. L., & Pérez, O. (2006). Proteoliposome derived cochleate as novel adjuvant. *Vaccine*, *24*, S30–S31. doi:10.1016/j.vaccine.2005.01.108 PMID:16823914

Bragazzi, N. L., Gianfredi, V., Villarini, M., Rosselli, R., Nasr, A., Hussein, A., Martini, M., & Behzadifar, M. (2018). Vaccines Meet Big Data: State-of-the-Art and Future Prospects. From the Classical 3Is ("Isolate-Inactivate-Inject") Vaccinology 1.0 to Vaccinology 3.0, Vaccinomics, and Beyond: A Historical Overview. *Frontiers in Public Health*, *6*, 62. doi:10.3389/fpubh.2018.00062 PMID:29556492

Byars, N. E., & Allison, A. C. (1987). Adjuvant formulation for use in vaccines to elicit both cell-mediated and humoral immunity. *Vaccine*, *5*(3), 223–228. doi:10.1016/0264-410X(87)90105-8 PMID:3499713

Čerpnjak, K., Zvonar, A., Gašperlin, M., & Vrečer, F. (2013). Lipid-based systems as a promising approach for enhancing the bioavailability of poorly water-soluble drugs. *Acta Pharmaceutica (Zagreb, Croatia)*, *63*(4), 427–445. doi:10.2478/acph-2013-0040 PMID:24451070

Chen, W., KuoLee, R., & Yan, H. (2010). The potential of 3′, 5′-cyclic diguanylic acid (c-di-GMP) as an effective vaccine adjuvant. *Vaccine*, *28*(18), 3080–3085. doi:10.1016/j.vaccine.2010.02.081 PMID:20197136

Chen, W., Patel, G. B., Yan, H., & Zhang, J. (2010). Recent advances in the development of novel mucosal adjuvants and antigen delivery systems. *Human Vaccines*, *6*(9), 706–714. doi:10.4161/hv.6.9.11561 PMID:20861683

Çinar, K. (2017). A review on nanoemulsions: Preparation methods and stability. *Trakya University Journal of Engineering Sciences*, *18*(1), 73–83.

Comfort, C., Garrastazu, G., Pozzoli, M., & Sonvico, F. (2015). Opportunities and challenges for the nasaladministration of nanoemulsions. *Current Topics in Medicinal Chemistry*, *15*(4), 356–368. doi:10.2174/1568026615666150108144655 PMID:25579345

Cox, E., Verdonck, F., Vanrompay, D., & Goddeeris, B. (2006). Adjuvants modulating mucosal immune responses or directing systemic responses towards the mucosa. *Veterinary Research*, *37*(3), 511–539. doi:10.1051/vetres:2006014 PMID:16611561

Das, S., Hatta, M., Wilker, P., Myc, A., Hamouda, T., Neumann, G., Baker, J. R. Jr, & Kawaoka, Y. (2012). Nanoemulsion W805EC improves immune responses upon intranasal delivery of an inactivated pandemic H1N1 influenza vaccine. *Vaccine*, *30*(48), 6871–6877. doi:10.1016/j.vaccine.2012.09.007 PMID:22989689

De Gregorio, E., & Rappuoli, R. (2014). From empiricism to rational design: A personal perspective of the evolution of vaccine development. *Nature Reviews. Immunology*, *14*(7), 505–514. doi:10.1038/nri3694 PMID:24925139

Duthie, M. S., Windish, H. P., Fox, C. B., & Reed, S. G. (2011). Use of defined TLR ligands as adjuvants within human vaccines. *Immunological Reviews*, *239*(1), 178–196. doi:10.1111/j.1600-065X.2010.00978.x PMID:21198672

Freytag, L. C., & Clements, J. D. (2005). Mucosal adjuvants. *Vaccine*, *23*(15), 1804–1813. doi:10.1016/j.vaccine.2004.11.010 PMID:15734046

Galli, G., Hancock, K., Hoschler, K., DeVos, J., Praus, M., Bardelli, M., Malzone, C., Castellino, F., Gentile, C., McNally, T., Del Giudice, G., Banzhoff, A., Brauer, V., Montomoli, E., Zambon, M., Katz, J., Nicholson, K., & Stephenson, I. (2009). Fast rise of broadly cross-reactive antibodies after boosting long-lived human memory B cells primed by an MF59 adjuvanted prepandemic vaccine. *Proceedings of the National Academy of Sciences of the United States of America*, *106*(19), 7962–7967. doi:10.1073/pnas.0903181106 PMID:19416838

Garçon, N., Vaughn, D. W., & Didierlaurent, A. M. (2012). Development and evaluation of AS03, an Adjuvant System containing α-tocopherol and squalene in an oil-in-water emulsion. *Expert Review of Vaccines*, *11*(3), 349–366. doi:10.1586/erv.11.192 PMID:22380826

Gonzalez-Aramundiz, J. V., Cordeiro, A. S., Csaba, N., de la Fuente, M., & Alonso, M. J. (2012). Nanovaccines: Nanocarriers for antigen delivery. *Biologie Aujourd'hui*, *206*(4), 249–261. doi:10.1051/jbio/2012027 PMID:23419252

Guy, B. (2007). The perfect mix: Recent progress in adjuvant research. *Nature Reviews. Microbiology*, *5*(7), 396–397. doi:10.1038/nrmicro1681 PMID:17558426

Hatz, C., von Sonnenburg, F., Casula, D., Lattanzi, M., & Leroux-Roels, G. (2012). A randomized clinical trial to identify the optimal antigen and MF59(®) adjuvant dose of a monovalent A/H1N1 pandemic influenza vaccine in healthy adult and elderly subjects. *Vaccine*, *30*(23), 3470–3477. doi:10.1016/j.vaccine.2012.03.017 PMID:22446638

Holmgren, J., & Czerkinsky, C. (2005). Mucosal immunity and vaccines. *Nature Medicine*, *11*(S4), S45–S53. doi:10.1038/nm1213 PMID:15812489

Holmgren, J., Czerkinsky, C., Eriksson, K., & Mharandi, A. (2003). Mucosal immunisation and adjuvants: A brief overview of recent advances and challenges. *Vaccine*, *21*, S89–S95. doi:10.1016/S0264-410X(03)00206-8 PMID:12763689

Johnson-Weaver, B. T., Abraham, S. N., & Staats, H. F. (2020). Innate Immunity-Based Mucosal Modulators and Adjuvants. *Mucosal Vaccines*, 167-183.

Kallewaard, N. L., Corti, D., Collins, P. J., Neu, U., Mcauliffe, J. M., Benjamin, E., Wachter-Rosati, L., Palmer-Hill, F. J., Yuan, A. Q., Walker, P. A., Vorlaender, M. K., Bianchi, S., Guarino, B., De Marco, A., Vanzetta, F., Agatic, G., Foglierini, M., Pinna, D., Fernandez-Rodriguez, B., ... Skehel, J. J. (2016). Structure and Function Analysis of an Antibody Recognizing All Influenza A Subtypes. *Cell*, *166*(3), 596–608. doi:10.1016/j.cell.2016.05.073 PMID:27453466

Karandikar, S., Mirani, A., Waybhase, V., Patravale, V., & Patankar, S. (2017). Nanovaccines for oral delivery-formulation strategies and challenges. *Nanostructures for Oral Medicine*, 263-293.

Kenney, R. T., & Edelman, R. (2003). Survey of human-use adjuvants. *Expert Review of Vaccines*, *2*(2), 167–188. doi:10.1586/14760584.2.2.167 PMID:12899569

Klucker, M. F., Dalençon, F., Probeck, P., & Haensler, J. (2012). AF03, an alternative squalene emulsion-based vaccine adjuvant prepared by a phase inversion temperature method. *Journal of Pharmaceutical Sciences*, *101*(12), 4490–4500. doi:10.1002/jps.23311 PMID:22941944

Lawson, L. B., Norton, E. B., & Clements, J. D. (2011). Defending the mucosa: Adjuvant and carrier formulations for mucosal immunity. *Current Opinion in Immunology*, *23*(3), 414–420. doi:10.1016/j.coi.2011.03.009 PMID:21511452

Lee, S. E., Nguyen, C. T., Kim, S. Y., Thi, T. N., & Rhee, J. H. (2015). Tetanus toxin fragment C fused to flagellin makes a potent mucosal vaccine. *Clinical and Experimental Vaccine Research*, *4*(1), 59. doi:10.7774/cevr.2015.4.1.59 PMID:25649002

Leroux-Roels, I., Borkowski, A., Vanwolleghem, T., Dramé, M., Clement, F., Hons, E., Devaster, J. M., & Leroux-Roels, G. (2007). Antigen sparing and cross-reactive immunity with an adjuvanted rH5N1 prototype pandemic influenza vaccine: A randomised controlled trial. *Lancet*, *370*(9587), 580–589. doi:10.1016/S0140-6736(07)61297-5 PMID:17707753

Levine, M. M., & Woodrow, G. C. (Eds.). (2004). New generation vaccines. New York: Marcel Dekker.

Liang, Z., Zhu, H., Wang, X., Jing, B., Li, Z., Xia, X., Sun, H., Yang, Y., Zhang, W., Shi, L., Zeng, H., & Sun, B. (2020). Adjuvants for Coronavirus Vaccines. *Frontiers in Immunology*, *11*, 2896. doi:10.3389/fimmu.2020.589833 PMID:33240278

Luna-Pineda, V. M., Ochoa, S., Cruz-Córdova, A., Cázares-Domínguez, V., Vélez-González, F., Hernández-Castro, R., & Xicohtencatl-Cortes, J. (2019). Infecciones del tracto urinario, inmunidad y vacunación. *Boletín Médico del Hospital Infantil de México*, *75*(2), 67–78. doi:10.24875/BMHIM.M18000011 PMID:29658949

Lycke, N. (2012). Recent progress in mucosal vaccine development: Potential and limitations. *Nature Reviews. Immunology*, *12*(8), 592–605. doi:10.1038/nri3251 PMID:22828912

Makidon, P., Belyakov, I., Blanco, L., Janczak, K., Landers, J., Bielinska, A., Groom, J. V. II, & Baker, J. R. Jr. (2012). Nanoemulsion mucosal adjuvant uniquely activates cytokine production by nasal ciliated epithelium and induces dendritic cell trafficking. *European Journal of Immunology*, *42*(8), 2073–2086. doi:10.1002/eji.201142346 PMID:22653620

Makidon, P. E., Bielinska, A. U., Nigavekar, S. S., Janczak, K. W., Knowlton, J., Scott, A. J., Mank, N., Cao, Z., Rathinavelu, S., Beer, M. R., Wilkinson, J. E., Blanco, L. P., Landers, J. J., & Baker, J. R. (2008). Pre-Clinical Evaluation of a Novel Nanoemulsion-Based Hepatitis B Mucosal Vaccine. *PLoS One*, *3*(8), e2954. doi:10.1371/journal.pone.0002954 PMID:18698426

Makidon, P. E., Nigavekar, S. S., Bielinska, A. U., Mank, N., Shetty, A. M., Suman, J., & Baker, J. R. (2010). Characterization of stability and nasaldeliverysystems for immunization with nanoemulsion-basedvaccines. *Journal of Aerosol Medicine and Pulmonary Drug Delivery*, *23*(2), 77–89. doi:10.1089/jamp.2009.0766 PMID:19778268

Man, S. M., & Kanneganti, T. D. (2016). Converging roles of caspases in inflammasome activation, cell death and innate immunity. *Nature Reviews. Immunology*, *16*(1), 7–21. doi:10.1038/nri.2015.7 PMID:26655628

Mastelic, B., Ahmed, S., Egan, W., Del Giudice, G., Golding, H., Gust, I., Neels, P., Reed, S. G., Sheets, R. L., Siegrist, C.-A., & Lambert, P.-H. (2010). Mode of action of adjuvants: Implications for vaccine safety and design. *Biologicals*, *38*(5), 594–601. doi:10.1016/j.biologicals.2010.06.002 PMID:20659806

McClements, D. (2012). Nanoemulsions versus microemulsions: Terminology, differences, and similarities. *Soft Matter*, *8*(6), 1719–1729. doi:10.1039/C2SM06903B

McClements, D. J., & Rao, J. (2011). Food-gradenanoemulsions: Formulation, fabrication, properties, performance, biologicalfate, and potentialtoxicity. *Critical Reviews in Food Science and Nutrition*, *51*(4), 285–330. doi:10.1080/10408398.2011.559558 PMID:21432697

Mishra, N., Goyal, A. K., Tiwari, S., Paliwal, R., Paliwal, S. R., Vaidya, B., Mangal, S., Gupta, M., Dube, D., Mehta, A., & Vyas, S. P. (2010). Recent advances in mucosal delivery of vaccines: Role of mucoadhesive/biodegradable polymeric carriers. *Expert Opinion on Therapeutic Patents*, *20*(5), 661–679. doi:10.1517/13543771003730425 PMID:20345332

Morel, S., Didierlaurent, A., Bourguignon, P., Delhaye, S., Baras, B., Jacob, V., Planty, C., Elouahabi, A., Harvengt, P., Carlsen, H., Kielland, A., Chomez, P., Garçon, N., & Van Mechelen, M. (2011). Adjuvant System AS03 containing α-tocopherol modulates innate immune response and leads to improved adaptive immunity. *Vaccine*, *29*(13), 2461–2473. doi:10.1016/j.vaccine.2011.01.011 PMID:21256188

Myc, A., Kukowska-Latallo, J., Smith, D., Passmore, C., Pham, T., Wong, P., Bielinska, A. U., & Baker, J. R. Jr. (2013). Nanoemulsion nasal adjuvant W805EC induces dendritic cell engulfment of antigen-primed epithelial cells. *Vaccine*, *31*(7), 1072–1079. doi:10.1016/j.vaccine.2012.12.033 PMID:23273511

Noda, K., Kodama, S., Umemoto, S., Abe, N., Hirano, T., & Suzuki, M. (2010). Nasal vaccination with P6 outer membrane protein and α-galactosylceramide induces nontypeable Haemophilus influenzae-specific protective immunity associated with NKT cell activation and dendritic cell expansion in nasopharynx. *Vaccine*, *28*(31), 5068–5074. doi:10.1016/j.vaccine.2010.05.005 PMID:20478344

O'Hagan, D. T., Ott, G. S., De Gregorio, E., & Seubert, A. (2012). The mechanism of action of MF59 - an innately attractive adjuvant formulation. *Vaccine*, *30*(29), 4341–4348. doi:10.1016/j.vaccine.2011.09.061 PMID:22682289

Petrovsky, N., & Aguilar, J. C. (2004). Vaccine adjuvants: Current state and future trends. *Immunology and Cell Biology*, *82*(5), 488–496. doi:10.1111/j.0818-9641.2004.01272.x PMID:15479434

Podda, A., & Del Giudice, G. (2003). MF59-adjuvanted vaccines: Increased immunogenicity with an optimal safety profile. *Expert Review of Vaccines*, *2*(2), 197–203. doi:10.1586/14760584.2.2.197 PMID:12899571

Quan Le, M., Carpentier, R., Lantier, I., Ducournau, C., Dimier-Poisson, I., & Betbeder, D. (2018). Residence time and uptake of porousand cationic maltodextrin-based nanoparticles in the nasal mucosa: Comparison with cationic or anionic nanoparticles. *International Journal of Pharmaceutics*, *550*(1–2), 316–324. PMID:30171898

Reed, S. G., Bertholet, S., Coler, R. N., & Friede, M. (2009). New horizons in adjuvants for vaccine development. *Trends in Immunology*, *30*(1), 23–32. doi:10.1016/j.it.2008.09.006 PMID:19059004

Rezende, R. M., & Weiner, H. L. (2017). History and mechanisms of oral tolerance. *Seminars in Immunology*, *30*, 1–13. doi:10.1016/j.smim.2017.07.004 PMID:28774470

Riedel, S. (2005). Edward Jenner and the history of smallpox and vaccination. *Proceedings - Baylor University. Medical Center*, *18*(1), 21–25. doi:10.1080/08998280.2005.11928028 PMID:16200144

Ruiz, A., Varela, A., & Cornejo, L. (2019). Nanoemulsions as coadyuvants in intranasal vaccines. *International Conference Series on Multidisciplinary Sciences*, *4*, 5078-2624.

Saito, S., Ainai, A., Suzuki, T., Harada, N., Ami, Y., Yuki, Y., Takeyama, H., Kiyono, H., Tsukada, H., & Hasegawa, H. (2016). The effect of mucoadhesive excipient on the nasal retention time of and the antibody responses induced by an intranasal influenza vaccine. *Vaccine*, *34*(9), 1201–1207. doi:10.1016/j.vaccine.2016.01.020 PMID:26802605

Sandri, G., Bonferoni, M. C., Ferrari, F., Rossi, S., Mori, M., & Caramella, C. (2015). Opportunities offered by chitosan-basednanotechnology in mucosal/skindrugdelivery. *Current Topics in Medicinal Chemistry*, *15*(4), 401–412. doi:10.2174/1568026615666150108124122 PMID:25579349

Schultze, V., D'Agosto, V., Wack, A., Novicki, D., Zorn, J., & Hennig, R. (2008). Safety of MF59 adjuvant. *Vaccine*, *26*(26), 3209–3222. doi:10.1016/j.vaccine.2008.03.093 PMID:18462843

Schwarz, T., Horacek, T., Knuf, M., Damman, H., Roman, F., Dramé, M., Gillard, P., & Jilg, W. (2009). Single dose vaccination with AS03-adjuvanted H5N1 vaccines in a randomized trial induces strong and broad immune responsiveness to booster vaccination in adults. *Vaccine*, *27*(45), 6284–6290. doi:10.1016/j.vaccine.2009.01.040 PMID:19856521

Simões, S., Filipe, A., Faneca, H., Mano, M., Penacho, N., Düzgünes, N., & Pedroso de Lima, M. (2005). Cationic liposomes for gene delivery. *Expert Opinion on Drug Delivery*, *2*(2), 237–254. doi:10.1517/17425247.2.2.237 PMID:16296751

Stanberry, L., Simon, J., Johnson, C., Robinson, P., Morry, J., Flack, M., Gracon, S., Myc, A., Hamouda, T., & Baker, J. R. Jr. (2012). Safety and immunogenicity of a novel nanoemulsion mucosal adjuvant W805EC combined with approved seasonal influenza antigens. *Vaccine*, *30*(2), 307–316. doi:10.1016/j.vaccine.2011.10.094 PMID:22079079

Wack, A., Baudner, B. C., Hilbert, A. K., Manini, I., Nuti, S., Tavarini, S., Scheffczik, H., Ugozzoli, M., Singh, M., Kazzaz, J., Montomoli, E., Del Giudice, G., Rappuoli, R., & O'Hagan, D. T. (2008). Combination adjuvants for the induction of potent, long-lasting antibody and T-cell responses to influenza vaccine in mice. *Vaccine*, *26*(4), 552–561. doi:10.1016/j.vaccine.2007.11.054 PMID:18162266

Wack, A., & Rappuoli, R. (2005). Vaccinology at the beginning of the 21st century. *Current Opinion in Immunology*, *17*(4), 411–418. doi:10.1016/j.coi.2005.05.005 PMID:15950445

Yoneyama, M., & Fujita, T. (2009). RNA recognition and signal transduction by RIG-I-like receptors. *Immunological Reviews*, *227*(1), 54–65. doi:10.1111/j.1600-065X.2008.00727.x PMID:19120475

ADDITIONAL READING

Farazuddin, M., Goel, R. R., Kline, N. J., Landers, J. J., O'Konek, J. J., & Baker, J. R. Jr. (2019). Nanoemulsion Adjuvant Augments Retinaldehyde Dehydrogenase Activity in Dendritic Cells via MyD88 Pathway. *Frontiers in Immunology*, *10*, 916. doi:10.3389/fimmu.2019.00916 PMID:31134057

Huang, C. H., Huang, C. Y., Ho, H. M., Lee, C. H., Lai, P. T., Wu, S. C., Liu, S. J., & Huang, M. H. (2020). Nanoemulsion adjuvantation strategy of tumor-associated antigen therapy rephrases mucosal and immunotherapeutic signatures following intranasal vaccination. *Journal for Immunotherapy of Cancer*, *8*(2), e001022. doi:10.1136/jitc-2020-001022 PMID:33037116

O'Konek, J. J., & Baker, J. R. Jr. (2020). Treatment of allergic disease with nanoemulsion adjuvant vaccines. *Allergy*, *75*(1), 246–249. doi:10.1111/all.13977 PMID:31298741

Shah, R. R., Hassett, K. J., & Brito, L. A. (2017). Overview of Vaccine Adjuvants: Introduction, History, and Current Status. *Methods in Molecular Biology (Clifton, N.J.)*, *1494*, 1–13. doi:10.1007/978-1-4939-6445-1_1 PMID:27718182

Wang, S. H., Chen, J., Smith, D., Cao, Z., Acosta, H., Fan, Y., Ciotti, S., Fattom, A., & Baker, J. Jr. (2020). A novel combination of intramuscular vaccine adjuvants, nanoemulsion and CpG produces an effective immune response against influenza A virus. *Vaccine*, *38*(19), 3537–3544. doi:10.1016/j.vaccine.2020.03.026 PMID:32245642

Yang, Y., Chen, L., Sun, H. W., Guo, H., Song, Z., You, Y., Yang, L. Y., Tong, Y. N., Gao, J. N., Zeng, H., Yang, W. C., & Zou, Q. M. (2019). Epitope-loaded nanoemulsion delivery system with ability of extending antigen release elicits potent Th1 response for intranasal vaccine against Helicobacter pylori. *Journal of Nanobiotechnology*, *17*(1), 6. doi:10.118612951-019-0441-y PMID:30660182

Zhao, L., Zhu, Z., Ma, L., & Li, Y. (2020). O/W Nanoemulsion as an Adjuvant for an Inactivated H3N2 Influenza Vaccine: Based on Particle Properties and Mode of Carrying. *International Journal of Nanomedicine*, *15*, 2071–2083. doi:10.2147/IJN.S232677 PMID:32273703

KEY TERMS AND DEFINITIONS

Adaptive Immune System: The subsystem of the immune system that is made of specialized, systemic cells and have means that eradicate pathogens or avert their growth.

Costimulatory Molecules: A heterogeneous group of cell surface molecules that pretence to intensify or counteract the initial activating signals provided to T cells from the T cell receptor following its interaction with a major histocompatibility complex (MHC), by this means influences T cell differentiation and fate.

Hemagglutination Inhibition: A laboratory test in which the lack of agglutination of red blood cells specifies that antibodies are present in the patient's blood.

Immunization: The process by which an individual's immune system becomes invigorated against an infectious agent.

Immunostimulants: Also known as immunostimulators, are substances that stimulate the immune system by inducing activation or increasing activity of any of its components.

Innate Immune System: The defense system with which one is born, protects against all antigens.

Pathogen-Associated Molecular Patterns: Small molecular patterns that are maintained within a class of microbes.

Toll-Like Receptors: Single-pass membrane-spanning receptors usually expressed on sentinel cells such as macrophages and dendritic cells, that recognize structurally conserved molecules derived from microbes.

Chapter 23
Healthcare Applications of Nanoemulsions

Dr.Sumira Malik
https://orcid.org/0000-0001-5077-1493
Amity University, Ranchi, India

Shristi Kishore
https://orcid.org/0000-0002-8851-5951
Amity University, Ranchi, India

Shradha A. Kumari
Amity University, Ranchi, India

Anjali Kumari
Amity University, Ranchi, India

ABSTRACT

The area of healthcare needs new innovative methods and tools for improvisation and to impart better efficiencies. Nanoemulsions are pharmaceutical formulations containing nanometre-sized particles used for controlled and systemic delivery of bioactive pharmaceuticals. Various advantageous properties of nanoemulsions such as the presence of hydrophobic core region, higher stability, and smaller size have made them useful to a large extent in the biomedical field. They have been employed in transdermal drug delivery, intranasal drug delivery, pulmonary drug delivery, parenteral drug delivery, and improvised delivery of hydrophobic drugs. This chapter aims to discuss various applications of nanoemulsions in healthcare including cosmetics, antimicrobials, vaccine delivery, targeted drug delivery, gene delivery, cancer therapy, and many more in detail.

1. INTRODUCTION

Nanotechnology is a fast-growing field in the arena of science and technology. To date, nanotechnology has proven its advantages in various fields. One of the finest known products of nanotechnology

DOI: 10.4018/978-1-7998-8378-4.ch023

are nanoemulsions (NEs). NEs, sometimes also referred to as mini emulsions, ultrafine emulsions, and submicron emulsions by some researchers, are the biphasic colloidal dispersions in which the oil droplets are dispersed in an aqueous phase (Solans & Solé, 2012; Ganta et al., 2014;McClements, D., 2011). These heterogeneous systems of particles are usually stabilized by an emulsifying agent and their internal phase droplets size generally lies below 1000 nm. On the basis of the dispersed and continuous phase, emulsions can be of four types: water-in-oil (w/o), oil-in-water (o/w), oil-in-water-in-oil (O/W/O), and water-in-oil-in-water (W/O/W) emulsions (Figure 1)(Tayeb & Sainsbury, 2018). The NEs particles can exist either in the form of W/O or O/W emulsions. In W/O NEs, the inner core of the particle is in an aqueous phase whereas, in O/W NEs, the inner core is composed of oil. O/W NEs have more applications in the healthcare industry because of the presence of a hydrophobic core region that aids in the delivery of water-insoluble drugs and bioactive compounds. Although the terms microemulsion and nanoemulsion look similar, there are pronounced differences between the two on the account of stability, particle size, etc. Microemulsions are thermodynamically stable whereas NEs are kinetically stable. Moreover, microemulsions are formed by the process of spontaneous thermodynamic self-assembly whereas NEs are formed by the applications of low or high external shear (McClements D., 2012).

Figure 1. Types of emulsions based on the dispersed and continuous phase.

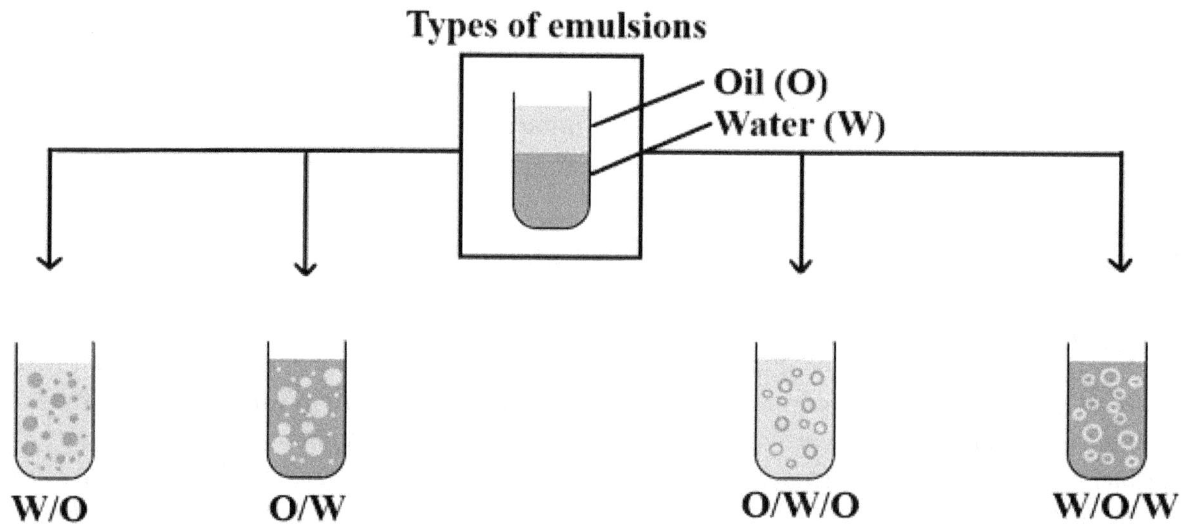

In the healthcare care industry, there is a need for tools and technologies that improvise the existing process and innovate new processes having better efficiencies. Most of the current therapeutic agents show poor aqueous solubility and according to a study, about 40% of them have been eliminated from the drug development lines because of the limitation of having poor solubility in water (Ganta et al., 2014). For the delivery of hydrophobic bioactive compounds,NEs have evolved as effective carriers. Apart from having better solubility, NEs show other beneficial properties. NEs can be produced by easy methods. They have higher and long-term stability, high loading capacity, high biocompatibility, and increased bioavailability. Moreover, they can be used for the controlled release of bioactive compounds

(Fryd & Mason, 2012). These all attractive properties make them advantageous tools to be used in the healthcare industry.

The current chapter attempts to address various applications of NEs in the healthcare industry including preparation of antimicrobials, non-toxic disinfectant cleaners, drug delivery agents, vaccine adjuvants, cancer therapeutics, and many more. Furthermore, the chapter aims to discuss the application of NEs in the cosmetic industry and the treatment of some other disease conditions.

2. APPLICATIONS OF NANOEMULSION IN THE PREPARATION OF ANTIMICROBIALS AND NON-TOXIC DISINFECTANT CLEANERS

Agents that kill the microorganisms or stop their growth are entitled antimicrobials. NEs are stable colloidal systems having small droplet sizes whereas antimicrobial NEs are also oil in water droplets usually of size ranging from 200 to 600nm and the stabilization of oil and water is done by lowering the interfacial tension using surfactant and alcohol. The particles of NEs are driven thermodynamically to fuse with organisms containing lipid. Most bacteria have teichoic acid linked to peptidoglycan and have a negative charge in their cell wall. As most of the NEs have a positive charge, their fusion is enhanced by a strong electrostatic force of attraction. The stable lipid membrane of the pathogen gets destabilized by the release of energy that was trapped within the emulsion when enough particles fuse with pathogens and the active ingredient. This process results in the lysis of the cell (Debnath, Satayanarayana, and Kumar, 2011). For actions against the microbial spores, additional germination enhancers are incorporated into the emulsion and as the germination starts, the spores become more susceptible to the NEs' antimicrobial activity. The selective toxicity of NEs on microbes at safe concentration does not cause irritation to the skin or mucous membrane. This property makes them unique and is because each droplet has a low level of detergents but still having enough surfactant and energy that can destabilize the pathogenic cells without damaging the healthy cells. The activity of NEs against bacteria, viruses, fungi, and spores is broad. For example, they act as antimicrobials against *E. coli*, HIV, *Candida*, and *Anthrax*. These all aspects of antimicrobial NEs can make them topical antimicrobials (Lovelyn and Attama,2011). Some essential oil NE formulations tested against various microorganisms are listed in Table-1.

Table 1. Antimicrobial activities of some essential oils nanoemulsion formulations.

Target microorganism	Name of the essential oil	NE Formulations	References
Salmonella typhimurium	Cinnamaldehyde	Tween 20	(Jo et al., 2015)
Salmonella enteritidis	Thyme oil	Sodium caseinate + lecithin	(Xue, Michael, and Zhong, 2015)
		Soluble soybean polysaccharide	(Wu, Lin, and Zhong, 2014)
	Clove oil	Whey proteins + gum Arabic + lecithin	(Luo et al., 2014)
Fusarium oxysporum	Eugenol	Tween 20	(Abd-Elsalam and Khokhlov, 2015)
Lactobacillus delbrueckii	d-limonene	Palm oil and lecithin; Modified starch	(Donsì et al., 2011);
Staphylococcus aureus	d-limonene	Tween 80	(Zhang et al., 2014)
	Eucalyptus oil	Tween 80	(Sugumar et al., 2014)
	Eugenol	Sesame oil and Tween 20/80	(Ghosh et al., 2014)

EnviroSystems has developed NEs as a nontoxic disinfectant cleaner for use in routine markets. As antimicrobial NEs have been effective to a broad spectrum of bacteria, viruses, and fungi without any damage to the mucous membrane. There are no warning labels present on the products as it is completely harmless. The disinfectant NEs are made up of small spheres of oil droplets of size less than 100μm and are suspended in water to produce parachlorometaxylenol (PCMX) that requires less amount of active ingredient. The charges on the nanospheres are enough to penetrate through the surface charges of the membrane of microorganisms. Due to the formulation, PCMX are able to target and penetrate cell walls easily instead of drowning the cells, therefore are applicable at a concentration lower than the other disinfectants and thus are non-toxic for humans, animals, and also for the environment (Yashpal, Tanuj, and Harsh,2013).

3. APPLICATIONS OF NANOEMULSIONS IN TARGETED DRUG DELIVERY

Drug delivery is considered ideal only when it stands up on the aim of having a maximum therapeutic effect and minimum toxic effect (Wang et al., 2007). NEs have emerged out to be the novel drug delivery system as they can easily be formulated for the targeted delivery of drugs through various routes. Because of this reason, they are used as vehicles for drugs active against infections caused by fungi, bacteria, etc. (Primo et al., 2007).

3.1. Transdermal drug delivery

Drugs can penetrate through the skin by mainly three routes. They are by hair follicles, directly through stratum corneum, or through sweat ducts but they can obstruct their absorption to great extent and thus limiting the bioavailability (Sahu et al., 2017). Thus, for proper and improved pharmacokinetics and targeting of drugs, the primary barriers of skin need to be overcome. In such cases, NEs came into play as they are easily able to penetrate the skin pores and reach the systemic circulation thus direct for effective delivery. For transdermal drug delivery, W/O NE formulation with caffeine has been developed (Thiagarajan,2011). The NE loaded drugs show more prominent significance in terms of the permeability parameters, thus its use in transdermal drug delivery has been emerged out as an important area of research in drug delivery and resulted in eradicating the primary problems and enhancing the bioavailability without any serious effects. It has several advantages like high storage stability, low preparation cost, high thermodynamic stability, and a good feasible rate of production. Many studies have been conducted and it has been shown that NE formulations have improved the dermal and transdermal delivery properties *in vitro* as well as *in vivo*. The emulsions and gels are the conventional topical formulation that has shown improved transdermal permeation of many drugs (Osborne, Ward, and O'NEILL,1991).

The data collected from *in vitro* and *in vivo* have concluded that the NE developed have much greater potential for transdermal drug delivery of aceclofenac, while the system of NEs containing ketoprofen have a high degree of stability and has enhanced the permeation rate in vitro through mouse skins (Shakeel et al., 2007).This was all done to examine and evaluate the potential of NEs mainly for increasing their solubility, the NEs that were prepared were firstly subjected to physical stability tests, and the studies related to the irritation suggested that the optimized NE was acting as a non-irritant transdermal drug delivery agent. For example, Celecoxib is a selective cyclo-oxygenase inhibitor and it is recommended orally for the treatment of arthritis and osteoarthritis but in the long run, it may produce some serious gastrointestinal side effects thus, skin permeation of celecoxib from NE was evaluated by conduction many examinations and the optimized NE was subjected to male rats and the results showed that the NE applied in transdermal and NE gel has increased the bioavailability fold by 3.30 and 2.97 respectively as compared to oral capsule formulation. All the studies in transdermal drug delivery have indicated the NEs have great potential in being a successful vehicle for enhancement of skin permeation and bioavailability of less soluble drugs (Kim et al., 2008).

3.2. Intranasal drug delivery

This type of drug delivery system is now getting recognized as a more reliable route for drug administration as nasal mucosa is a more therapeutically viable channel and has a more favorable way to overcome the barriers for direct entry of drugs to the target site and if taken into consideration it is also painless, well-tolerated and non-invasive (Pires et al., 2009). Nasal cavities have very few enzymatic activities, many immunoreactive sites, and permeable epithelium. Drugs that are hydrophilic and have high molecular weight can lead to several problems when they are targeted to the brain because the endothelium is impervious but the olfactory region of the nasal mucosa provides a direct connection with the brain (PardridgeW. M, 1999). Diseases like Alzheimer's disease, migraine, schizophrenia, meningitis, Parkinson's disease, etc., can be treated with the help of NEs that are loaded with the drugs and delivered to the brain via the nose. Another application of the intranasal drug delivery system is the development of vaccines, by the administration of mucosal antigen immunity can be achieved (Csaba, Garcia-Fuentes and Alonso, 2009). Diseases related to the central nervous system (CNS) can be treated by targeting NEs formulation of drugs to the brain through nasal mucosa or intranasal drug delivery system. The nanocarriers protect the biomolecules and the nano-carrier interaction between mucosa and direct antigen and the lymphoid tissue is promoted thus,it is efficient in treating diseases related to CNS. Insulin and testosterone have also been formulated for nasal deliveries.

3.3. Parenteral drug delivery

Usually, the active agent having low bioavailability and a narrow therapeutic index, the most effective and common route adopted for drug administration is the parenteral drug delivery system. NEs can dissolve large quantities of hydrophobic elements, have mutual compatibility, and great ability to protect the drugs from possible enzymatic degradation and hydrolysis, thus making it an exemplar vehicle for parenteral drug delivery (Thiagarajan P, 2011). NEs assures the release of drugs in a more perpetual and controlled manner for a long period consequently reducing the frequency and dosage of injection throughout the therapy period furthermore sedimentation, lack of flocculation, creaming, large surface area, and free energy add up its advantages for this route of administration. Having a large interfacial area influences it positively for ease in drug transport and delivery and also in targeting drugs to specific sites (ReissH, 1975). Henceforth many clinical and pre-clinical trials have been conducted in NEs parenteral drug delivery system. Some problems were also faced like a decrease in the content of NE when drug formulation was about 0.01% and plasma concentration when thalidomide containing NE has dose as low as 25mg but these problems have been resolved by the addition of polysorbate 80. Chlorambucil has been used against breast cancer and ovarian cancer. It is a lipophilic anticancer agent and its anticancer activity was studied by high energy ultrasonication method (Patel and Patel,2014). Colon adenocarcinoma in the mouse when treated with such type of NE led to a much higher rate of suppression of tumor when put side by side to plain drug solution treatment thus concluding that NE drug formulations are the more applicable carrier for drug deliverance in cancer treatment. Carbamazepine has poor water solubility, therefore; it has non-parenteral treatment but a NE has been developed for intravenous delivery that shows favorable release kinetics in vitro. Diazepam, Propofol, Dexamethasone, Etomidate, Flurbiprofen, and Prostaglandin are some of the drugs documented parenteral NE formulation (Araújo et al., 2011). Prostaglandin E1 is incorporated in lipid NE because it has a short life and high doses are needed thus having adverse side effects. Lipid formulations are used to treat cardiovascular diseases because they generally accumulate in walls that are injured. As a result, the drug is transported to the site of vascular injury thus protecting it from inactivation by the lungs (Brussel et al., 2012).

3.4. Pulmonary drug delivery

For drug delivery lung is considered to be an attractive target because of its non-invasive administration employing inhalation aerosols, the first-pass metabolism can easily be avoided, treatment of respiratory diseases is made easy by its direct delivery to the target site and it provides a huge surface area for drug action and its systematic absorption (Heidi, Yun-Seok and Xiao,2009). Nanocarrier systems in the pulmonary drug have many advantages such as having the capability to achieve a homogenous dispersal of drug dose among alveoli, the solubility of the drug is convalesced from its aqueous solubility, a perpetual release of drugs that eventually reduces dosing frequency. As a result, the compliance of the patient gets better and they have less possibility of getting side effects (Bivas-Benita et al., 2004). Cationic submicron emulsions are also emerging as a potential carrier for DNA vaccines to the lungs given that they can infect the pulmonary epithelial cells with free nucleic acids. This process is also called transfection and it induces the immunogenic cross-presentation of antigen-presenting cells (APCs) and directly activates the dendritic cells thus, as a result, the antigen-specific T cells get stimulated. Further studies are required for the better and successful formulation of submicronemulsions that can be inhaled because they can have side effects on alveoli of the lungs due to surfactant interactions. Trichlorotrifluoroethane is a microemulsion formulation stabilized by lecithin used for the pulmonary delivery of salbutamol. This microemulsion formulation is an aerosol system (Lawrence and Rees, 2000).

3.5. Ocular and Otic Drug Delivery

This type of drug delivery is one of the most challenging and interesting ventures faced by the scientist. It is known that eye drops when used as ophthalmic delivery systems have poor bioavailability and their therapeutic response is also inefficient. It is mainly because of lachrymal secretion and nasolacrimal drainage in the eye and most of it gets drawn off from its precorneal area in a few minutes (Macha, Hughes and Mitra, 2003). Therefore, recurrent installation of concentrated solution is needed to be accomplished to get the anticipated medicinal effects but because of the drainage, the important part of the drug that has administrated is transported by means of the nasolacrimal duct to the gastric intestinal tract where there are chances that it may get absorbed and might cause side effects.To increase the efficiency of the drug, a special form of dosage should be chosen which has the potential to increase the contact timing of the drug in the eye, as a result, it may increase the bioavailability and reduce the need for continuous installation thus compliance of patient can be improved (Patel et al., 2013). NEs that can be diluted is considered the potential for drug delivery vehicles for ophthalmic use because it has several advantages like prolonged effect and high ability of penetration of drugs into the deep layer of ocular structure and the aqueous humor. Dorzolamide hydrochloride is an antiglaucoma drug formulated as an ocular NE. When it was evaluated biologically in a normotensive albino rabbit, it was found that they have higher therapeutic effects, non-irritant, prolonged effect, and faster onset of action. This summarizes that Dorzolamide hydrochloride in NE form offers better intensive treatment of glaucoma and better compliance in the patient compared to normal typical eye drops (Ammar et al., 2009).

4. APPLICATION OF NANOEMULSIONS IN VACCINATION

NE can be used as an efficient vaccine delivery system to deliver weakened viruses and this type of delivery system uses nanotechnology to vaccinate against HIV. Of late shreds of evidence are that HIV can infect the mucosal immune system and for that reason, NE can be used for developing mucosal immunity and there are chances that it may become very important in the future fight in opposition to HIV (Yashpal, Tanuj and Harsh, 2013). Intranasal vaccine delivery system is considered more efficient than the traditional vaccine routes. Studies in recent times have shown that genital mucosa immunity may be attained through vaccine administration via nasal mucosa or intranasal vaccine delivery. Inactivated organisms can be transported to the mucosal surface with the help of NE to produce an immune response. APC uptakes the adjuvants formed due to NE that causes the protein applies to the mucosal surface to be adjuvants thus resulting in systematics and mucosal immune response due to which IgG and IgA antibody is produced as well as cellular immunity is achieved. Single mucosal exposure to virus mixed with NE can prevent animals from getting influenza (Lovelyn and Attama et al., 2011). Research is being done and it has shown that when animals are exposed to recombinant gp120 in NE on their nasal mucosa creates a significant response to HIV as follows gives basis to use it as a vaccine for HIV. NEs have a great impact on vaccine adjuvant research as they are less expensive, easily produced, can stay in storage for long period, and is safe. Still, many kinds of research are being conducted as more growth is needed before they can be established as the preferred vaccine adjuvant.(Rutvij et al., 2011) H5N1 influenza vaccine was enhanced by oil-in-water NE adjuvant as they show good stability in temperature and proper particle size is maintained, it also enhances the production of IL-6 and MCP-1 to employ innate cells, macrophages and enhances the functioning of dendritic cells to induce robust antigen-specific T and B cell immune responses. The NE-adjuvanted H5N1 vaccine has elevated and enduring antibody responses it also has improved fortification against homologous clade 1 accompanied by heterogeneous clade 2 H5N1 virus in young and old mice.

Nucleic acid-based vaccines for instance plasmid DNA and mRNA are technologically advanced as means to look after several unaddressed medical needs that existing vaccine technologies have not been able to address. A Cationic NE delivery system was also developed to deliver a self-amplifying mRNA vaccine, it a type of nonviral delivery system and is based on Novartis's proprietary adjuvant MF59 which has already been examined and evaluated to show a great clinical safety profile among children, adults, and elderly. CNE is well tolerated and immunogenic in several animal models and studies are still going on. For vaccines against Anthrax and Hepatitis B are being developed and research is going on to complete the proof (Rodriguez et al., 1999).

5. APPLICATION OF NANOEMULSION IN COSMETICS

In the past few years, the world is facing numerous problems due to the fast-growing population. With the growth of population and technology, the demands for healthy living are also increasing day by day in almost every field. Among these, one of the most popular demands is from the field of cosmetics (Antignac et al., 2011; Philippe et al., 2012; Barbulova et al., 2015). A disadvantage of using traditional cosmetic formulations is that they show very poor absorption rates in the deep layer of the skin. The advancement in nanotechnology and the discovery of NEs have put forward a step in resolving the problem of poor absorption of cosmetics.

NEs are very small droplets having a diameter in nanoscales that get easily absorbed in any skin type. They have emerged as excellent mobile agents for the managed shipping of cosmetics and the optimized dispersion of energetic substances especially in pores and skin layers. NEs can easily transport lipophilic drugs than liposomes because of the interior of the lipophilic environment. Similar to liposomes, NEs support the penetration of active ingredients and thus increases their concentration in the skin (Debnath et al., 2011). Another important property of NEs is that they have a high surface-to-volume ratio because of their small size that allows effective delivery of bioactive compounds to the skin. Emulsions that are mainly used in cosmetics industries are basically of two types: oil-in-water emulsions (o/w) and water-in-oil emulsions (w/o). Oil-in-water emulsions (o/w) are used for the delivery of general and hydrophilic drugs whereas water-in-oil emulsions (w/o) act as moisture for dry skin type (Eid et al. 2014). NEs have made themselves engrossing for use in cosmetic industries due to their attractive properties which include low viscosity and transparency. Due to their bio-active effects, the demand for NEs is increasing throughout the industrial markets. The bio-active effect also plays a key role in reducing the trans-epidermal water loss (TWEL) suggesting that the carrier function of the skin is strengthened. NEs can integrate lipophilic, hydrophilic, and amphiphilic compounds very easily because all of them contain oil, inter-facial region, and water which make them perfect for the making of cosmetics products.

Numerous varieties of ingredients are used in cosmetics but emulsions are the major ones. The emulsions are spread all over the market. The pharmaceutical creams go deep inside the skin but emulsions are only present on the upper layer of the skin i.e., the epidermis. Cosmetics are products that do not penetrate deep inside the skin, rather they just remain at the upper layer. The formulation of cosmetics is quite risky and demanding at the same time as it demands many features like safety for skin, moisturizing skin, soft skin, eco-friendly to the environment, have anti-aging properties, free from harmful surfactants, and also be cost-effective. NEs are considered far better than macroemulsions as they are nanoscale particles and there is no inherent creaming, sedimentation, flocculation, or coalescence. In cosmetics, NEs are available in different forms like cream, lotion, gel, and moisturizer as the particles are easily absorbable in every layer of skin because of the small-sized droplets and they also help the skin to be moisturized throughout the day. The advantage of using NEs in cosmetics is the presence of small-sized droplets and absenteeism of smoothening and coagulation that ease the absorption by antigen-presenting cells and leading to a filter product formulation (Chellapa et al., 2016; Ribeiro et al., 2015).

Some oil-water prepared NEs have accommodated *Opuntia ficus Indica* (L.) mill hydro glycolic extract and xanthan gum whose droplet size ranges from 92.2 to 233.6 nm. The oil-water nanoemulsion has 1% of *Opuntia ficus Indica* (L.). Mill extract was kept steady for almost 60 days. After all these mixings it was observed that the stratum corneum has increased its water content which proves to be an excellent part of the cosmetics products as due to the increase in water content, its moisturizing effectiveness has increased (Ribeiro et al., 2015). NE formulations were prepared from the hydroalcoholic extracts of *Vellozia Squamata* leaves and stem with high antioxidant activity by using the phase inversion method. Stable formulations have been received from each extract from leaves and stems. Antioxidants are well-recognized as anti-aging agents, hence are suitable for cosmetics formulation (Quintão et al., 2013). Comparative research reveals that after 1 hour the hydration power of the NEs was higher than body milk and body water. Again after 24 hours, the outcome came was that NEs long-lasted from the other two products (Chellapa et al., 2016). Besides the substantive hobby in NEs for their specific texture, other advantages of NEs are valued in lots of beauty applications, especially skin take care and solubilization, and transport of energetic ingredients. In the hair-care domain, they play a crucial position inside the silicon deposition onto hair fiber or scalp treatment. There are a few exemplars of copyright NEs used in the cosmetic sector. L'Oreal (Paris, France) has employed NEs based on phosphoric acid and fatty acid esters in beauty products, dermatological, pharmaceutical, and/or ophthalmological fields as well as NEs dependent on ethylene oxide and propylene oxide block copolymers and their uses in the cosmetics, dermatological, and/or ophthalmological fields (Shah et al., 2010). NanoGel generation presents an easy system and device to create submicron emulsions from an easy-to-use, oil-in-water concentrate. The method is in particular suitable for minimizing trans-epidermal water loss and penetration of lively ingredients in the skin pores. These traits suggest that it would be in particular beneficial for solar care merchandise as well as moisturizing and anti-aging creams—precise regions where nanotechnology is already being included in a bunch of merchandise presently at the market. NEs have also attracted great interest in the latest years for utility in private care merchandise as capability cars for the managed transport of cosmetics and the optimized dispersion of lively elements mainly pores and skin layers (Guglielmini, G., 2008). According to some studies, NEs droplet length is an issue that affects the maintenance of beauty products. To overcome these limitations of NEs, the preservative dose needs to be tailored to NEs to assure appropriate maintenance. Advantages and challenges of using cosmetic NEs are summarized in Figure 2.

Figure 2. Advantages and challenges of using cosmetic NEs.

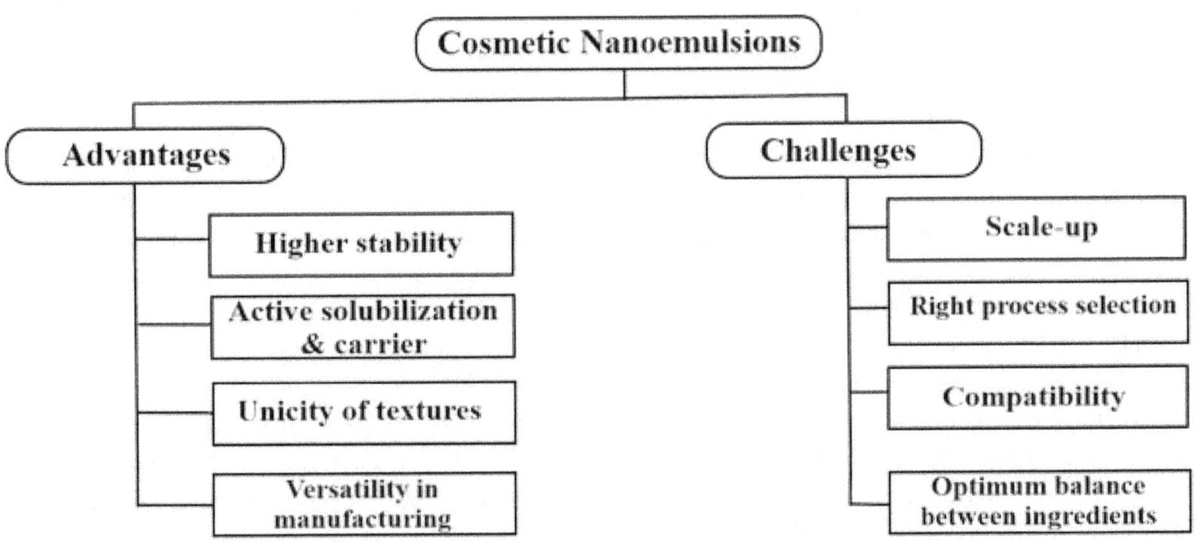

6. APPLICATION OF NANOEMULSIONS IN CANCER THERAPEUTICS

Cancer is the uncontrolled growth of abnormal cells. The treatment of cancer is done by different therapies including radiation therapy, chemotherapy, and surgery. Cancer is divided into sub-stages and each stage requires a different type of treatment. These therapies cause damage to the normal cell and are the causes of different side effects such as hair fall, body pain, nose bleed, and many more. With the growth and development of nanotechnology, different treatments are coming up for the treatment of cancer. Nanotechnology has formulized different products such as nanoparticles, NEs, nanocapsules, liposomes, micelles dendrimers, polymer-drug compounds, and immune compounds. These all together have intensified the therapeutic applications by boosting up circulation duration of the drug, prolonging half-life, controlling and sustaining release, enhancing site-specificity, etc. (Jain, Jain, and Mahajan, 2015). Up till now, there are nearly no shreds of evidence proving appearance of toxicity and instability of nano-based elements that have developed to be applied in human beings. Instead, NEs mainly use oil, water, and surfactant and they have certain safety limits. But it has been shown that NEs used in the treatment of cancer have their safety limits and display very few side effects (Kumar & Divya 2015). NEs are colloidal particulate system with nano-sized particles which act as a drug delivery molecule. NEs are immiscible liquids stabilized by emulsifying agents. They have additional benefits of having nano-sized droplets, optical clarity, good physical stability, improvised bioavailability, non-toxicity, and non- irritability (McClements, D., 2012).

In the last few years, the spice-based NEs demand is increasing day by day in cancer therapy. The phenolic compound of spices is known to have the great power of anticancer activity and they were found to be antagonistic to cancer of the liver, breast, large intestine, stomach, skin, bladder, and colon (Beidokhti, M. N., 2013). However, the efficacy of these oils made from spices and the essential oil is limited due to their low aqueous content and high lipophilicity (Bilia et al., 2014). By keeping the issues at the center of attention, scientists have developed spice oil-based NEs because of their biocompatibility and higher efficiency. In research, the eugenol-loaded NEs have manifest/showed apoptosis of both colons (HTB37) and liver (HB8065) cancer cell lines via reactive oxygen species (ROS) proofed through flow cytometry and microscopy (Majeed et al., 2014). To upgrade the oral bioavailability of curcumin, a solid dosage form of curcumin was also designed. In another research, curcuminoid NEs were developed and showed that mitochondria and death receptor pathways were responsible for A549 and H460 apoptosis (Chang, & Chen, 2015).

Even though NEs are mostly seen as vehicles for directing aqueous insoluble medications, they are currently getting more popular as colloidal carriers for target delivery of the different anticancer drugs, photosensitizers, neutron capture therapy agents, or diagnostic agents. Another advancement in cancer therapy is the development of magnetic NEs. They can mainly carry photosensitizers like Foscan to the underlying tissue layers beyond the skin by activating hyperthermia for a subsequent free radical generation. In the form of photodynamic therapy, treatment of cancer can be efficiently done (Primo et al., 2007). Cancer treatment has recently included a brand new theranostic approach and photodynamic remedy, that may be noninvasively implemented to malignant and nonmalignant sicknesses and to visualize most cancer cells via fluorescent imaging (Leandro et al., 2017). The photodynamic remedy is mainly the combination of red light having a wavelength range of -600-800nm, molecular oxygen (O_2), and photosensitizers components. The illumination excites the photosensitizers and it passes its power to oxygen, which changes normal oxygen into ROS (de Menezes Furtado et al., 2017). The photodynamic technique will result in preferential absorption of the photosensitizer drug through the diseased tissue, inflicting irreversible photo-damage to a selected region (tumor tissues and blood vessels) and activation of antitumor, immunity, and inflammatory responses (Monge-Fuentes et al., 2017). In photodynamic therapy and cancer, imaging photons are used that are near-infrared region because they perforate the most and are less harmful to human tissues (Leandro et al., 2017)

Unfortunately, none of the photosensitizing compounds used so far are ideal for these applications (Monge-Fuentes et al., 2017). Photosensitizer medications are aqua-phobic components and the assimilation into the distinct system to increase their bioavailability are required. Oil-water-based NEs are one of the most important drug delivery systems for not only cancer imaging but also for photodynamic remedy (Castilho-Fernandes et al., 2017). The use of chloro-aluminum phthalocyanine has been broadly researched as a potential theranostic agent (Leandro et al., 2017; Castilho-Fernandes et al., 2017). Chloro-aluminum phthalocyanine is a second-era photosensitizer, with photophysical residences appropriate for photodynamic remedy and most cancers imaging, together with the absorption of some distance and near-infrared light, excessive quantum yields, and feature a protracted lifetime (Leandro et al., 2017; Castilho-Fernandes et al., 2017). The use of chloro-aluminum phthalocyanine-loaded NE systems has been researched as a remedy for glioblastoma (Leandro et al., 2017; Castilho-Fernandes et al., 2017), prostate cancer, lung adenocarcinoma, and cervical adenocarcinoma (Franchi et al., 2016). The technique proved efficient suppression of glioblastoma cells, inducing apoptosis or necrosis. However, when the research was executed in-vitro, then also nanoemulsion was proofed to be the appropriate delivery technique for chloro- aluminum phthalocyanine to glioma cells (Castilho-Fernandes et al., 2017). Research on prostate cancer has demonstrated a better theranostic behavior on chloro-aluminum phthalocyanine-loaded nanocapsules than that through chloro-aluminum phthalocyanine-loaded NEs (Leandro et al., 2017). Comparability among lung and cervical adenocarcinoma demonstrated an enhanced effect on cervical cancer cells, where increased cytotoxicity and apoptosis were observed (Franchi et al., 2016).

Acai oil and extracts from *Tectona grandis* leaves were found as new potential photosensitizer agents. The use of incorporated natural components in NEs showed incontestable potency within the medication of melanoma, with a reduction in vital death and tumors (de Menezes Furtado et al., 2017; Monge-Fuentes et al., 2017). Breast cancer medication using camptothecin-loaded NEs stabilized with chitosan or poloxamer has also shown the potential of NEs as a carrier of cancer remedy (Natesan et al., 2017; Sugumaran et al., 2018). The loading of paclitaxel in NEs was demonstrated in the research for the cure o cancer was observed. High tumor-accumulation and effectiveness demonstrate its ability as a promising candidate drug device for scientific applications (Chen et al., 2017). Some of anticancer NEs are shown in table-2.

Table 2. Some emulsions based anticancer drug formulations and their respective activities.

Name of the anticancer drug	Nanoemulsion based formulation	Observed activities	References
Methotrexate	Soybean oil based microemulsion	*In-vitro* cytotoxicity against MCF-7 cell lines	(Karasulu et al., 2007)
Curcumin	Microemulsion based on lecithin	*In-vitro* cytotoxicity against OSCC-4 & OSCC-25 cell lines	(Lin et al., 2012)
Paclitaxel	Drug loaded NE co-administered with curcumin	An increase in tumor uptake by 3.2 fold	(Ganta, Devalapally, & Amiji, 2010)
Benzyl isothiocyanate	Flax seed oil based NE	In-vitro cytotoxicity against SKOV-3 and A549 cell lines	(Qhattal et al., 2011)

7. OTHER HEALTHCARE APPLICATIONS OF NANOEMULSIONS

7.1. Nanoemulsions as Prophylactic in Bio-Terrorism Attack

The NEs consisting of antimicrobial agents have demonstrated prophylactic activities to fight against bioterrorism attacks. NEs were earlier used as prophylactic medicated forms in the protective treatment of the people who were susceptible to bioterrorism attacks including Anthrax and Ebola. Bio-terrorism was arranged and conscious discharge of viruses, bacteria, or other microbial agents that caused diseases and in severe cases death to humans that led to an extensive loss in the population. The drug which is commonly found in nature can be modified to increase disease resistance, increase current resistance to drugs, and increase its natural ability to spread. For many centuries, biological tools were a peril for human life.

Earlier, there were different ways to pollute water sources like fecal matter, animal carcasses, etc. but at present time there are many biological chemicals and microbes that pollute the water. Few of them are like dried spores and genetically modified items or animals that can infect a large population even in a very small amount (Pinto, V.N., 2013). For a long time, Indian defense and medical experts were engaged in the battle of harmful drugs to the population. There were some doubtful events in old times such as in the 1965 Indo-Pakistan war, scrub typhus also known as bush typhus - a disease outburst in north-eastern India came under doubtful. In 1994, the outburst of pneumonic plague in Surat and Bubonic plague in Beed all caused damage to the population (Sharma, R., 2001). In December 1999, wide-ranging NEs were examined on the surfaces by the US Army (RestOps) for disinfecting the spores of Anthrax. In March 2001, it was again examined by RestOps as a chemical decontamination microbe. NEs as prophylactic have been tested on many pathogenic agents such as gangrene, *Clostridium botulism* spores, and even on contaminated injuries. NEs are available in different forms such as creams, sprays, and foams to disinfect a bulk amount of items which are retailed as NANOSTAT (Nanonio Corp.) (Rutvij et al., 2011; Lovelyn & Attama, 2011; Debnath, 2011).

7.2. Nanoemulsions as Gene Delivery Vectors

Gene therapy is one of the advancements in genetics and biotechnology. Gene therapy is the delivery of nucleic material by vectors. Based on their low cell uptake and biologic uncertainty, vectors are essentially of two types - viral and non-viral. Non-viral vectors have been widely preferred due to their advantages such as low immune response, ease of production, cost-effectiveness, and safety (Teixeira et al., 2017). There are some alternative non-viral potential vectors to deliver nucleic acid. They are nanoparticles, NEs, liposomes, dendrimers, and many others (Silva et al., 2016). For single and double-stranded nucleic acid, oil-water cationic NEs were put forward for the delivery system. Their implementation can address issues such as particle aggregation. There has also been a huge amount of research into the best manufacturing conditions for cationic NEs for gene therapy applications (Teixeira et al., 2017). Cationic NEs have the power to resolve the problems that are faced during nucleic acid gene delivery. The formation of an ideal system includes consideration of the interactions of oil-water surfactants, the toxicity profile, and the transfection quantity (Teixeira et al., 2017). Reports in the literature have shown the high efficiency of a NE prepared by high-pressure homogenization and stabilized with various lipids as an intravenous gene transfer system. Transfection efficiency was detected for the medication of mucopolysaccharidosis Type-I disorder and is measured as Alpha-L-iduronidase enzyme activity (Fraga et al., 2015). The study was based on the ability to load plasmid DNA into the NEs system. Gene delivery requires a positive charge for the system so incorporation of stearylamine was compulsory for the system to work. Due to suitable stability, ultrasonic stearyl amine NEs are used in gene therapy but advanced studies on transfection are also mandatory (Silva et al., 2012, 2016). The same study was carried out on ultrasonic NEs stabilized with arginine, lysine, and histidine. The highest transfection potency was ascertained for lysine and arginine, followed by histidine, demonstrating that positively charged cationic NEs have increased effects (Liu & Yu, 2010). Emulsions are found as the latest possible way of gene transfer vectors to liposomes (Liu et al., 1996). Emulsions analysis for gene delivery (non-pulmonary route) has proved that binding to the emulsion or DNA complex was much stronger than liposome carrier (Yi et al., 2000). This stable emulsion system was found to be more efficient than liposomes for delivered genes (Liu& Yu, 2010; Silva et al., 2016).

8. APPLICATIONS OF NANOEMULSIONS IN THE TREATMENT OF VARIOUS OTHER DISEASES

Due to their broad applications in the drug delivery through various routes like oral, nasal, parental, and ocular,NEs have emerged out as potent alternatives for the traditional treatments of various diseases.

For example, NEs were employed in the treatment of Alzheimer's disease. Alzheimer's disease is a neurodegenerative disease that leads to the loss of neurons, particularly in the cortex area of the brain, and is common among older people. With the progression of the Alzheimer's, it leads to short memory of the patient, personality loss, loss of motor functions, long memory, disorientation, and ultimately death due to infections like pneumonia (Nirale, Paul, and Yadav, 2020). To treat this neurodegenerative disease, the drugs need to cross the Blood Brain Barrier (BBB). The BBB does not allow large molecules to cross the barrier. To overcome this limitation, NEs have gained huge attention due to their small size, high solubility, and presence of a hydrophobic core region. The small size of NEs allows them to cross the BBB and deliver the target drug effectively. In a study, O/W NEs loaded with donepezil were explored for administration through intranasal routes. When the *in vitro* release of the drug was studied, it showed a maximum release (99.22% in four hours) in phosphate buffered saline, followed by good result (98% in two hours) in artificial cerebrospinal fluid, and fair release (96% in two hours) in simulated nasal fluid. This study suggested that NEs loaded with donepezil can serve as an effective tool for the treatment of Alzheimer's via intranasal routes for the drug delivery to the brain (Kaur et al., 2020a). Similar studies showed the potential of memantine loaded NEs for the treatment of Alzheimer's through nose to brain drug delivery. The memantine loaded NEs showed 80% release of the drug in simulated nasal fluid *in vitro*(Kaur et al., 2020b).NE formulations have also been tested for the treatment of some lysosome-related neurodegenerative disorders like Parkinson's disease. Impaired lysosomal pH is mostly responsible for these pathological conditions. A study showed that acidic O/W NEs loaded with poly(dl-lactide-co-glycolide) (PLGA) were able to restore the pH of the lysosome in cellular models of Parkinson's disease. *In vivo* studies in mouse also demonstrated the crossing of these NEs systems through the BBB and restoration of the cells' lysosomal pH (Prévot et al., 2018).

As the current approaches for the delivery of protein therapeutics to the heart are not fully mature, there is an urgent need for efficient delivery agents. In a recent study, NEs were tested for the treatment of cardiac diseases by pulmonary administration. Protein therapeutics loaded Perfluorocarbons (PFC) nanoemulsions (PNEs) having particle size of 140 nm were synthesized. Pulmonary administration of

PNEs in rats showed successful delivery of protein therapeutics and reduced the ischemic myocardial injury brought about by acute myocardial infarction. The study suggested that NEs can act as better delivery agents for the treatment of cardiac diseases (Qin et al., 2021).

NEs have also been tested inthe treatment of tuberculosis. The conventional methodologies being used in the treatment of tuberculosis are not satisfactory for a long run because of lower bioavailability of the therapeutics. NE formulations loaded with anti-tuberculosis drugs have the ability to cross the barriers and reach the systemic circulation. After reaching the systemic circulations, these formulations subsequently target the causative pathogen, *Mycobacterium tuberculosis*. The lipid soluble nature of these NEs help to target the drugs to the lymph nodes, and to reduce the frequency of the dosages (Beg et al., 2017). NEs have also been employed in the treatment of various other diseases like cutaneous leishmaniosis (Kawakami et al., 2021), rheumatoid arthritis (Poonia et al., 2020), dry eye syndrome (Kim et al., 2017), cerebral ischemia (Ahmad et al., 2018), and many more.

9. CONCLUSION

NEs are one of the important products of nanotechnology. These biphasic systems have come out as important tools for the targeted delivery of drugs, therapeutics, genes, vaccines, and other bioactive compounds in the area of healthcare. Because of their small size, high surface-to-volume ratio, high stability, high bioactivity, and lipophilicity, NEs have been utilized as efficient vehicles for the diagnostic and drug delivery through transdermal, intranasal, parenteral, ocular, and optic routes. They are able to cross several physiological barriers and deliver the loaded bioactive compounds. Due to their advantageous properties, NEs have also been explored in the formulation of anticancer drugs and prophylactics for bioterrorism attack. They hold promise for different healthcare areas including the cosmetic industry, pharmaceutical industry, biotechnology industry, etc. It is expected that further research and development works would be carried out on increasing the practical applications of NEs in the area of healthcare, especially in the delivery of drugs and therapeutics to the target sites.

REFERENCES

Abd-Elsalam, K., & Khokhlov, A. (2015). Eugenol oil nanoemulsion: Antifungal activity against Fusarium oxysporum f. sp. vasinfectum and phytotoxicity on cottonseeds. *Applied Nanoscience*, *5*(2), 255–265. doi:10.100713204-014-0398-y

Ahmad, N., Ahmad, R., Naqvi, A. A., Alam, M. A., Ashafaq, M., Abdur Rub, R., & Ahmad, F. J. (2018). Intranasal delivery of quercetin-loaded mucoadhesive nanoemulsion for treatment of cerebral ischaemia. *Artificial Cells, Nanomedicine, and Biotechnology*, *46*(4), 717–729. doi:10.1080/21691401.2017.1337024 PMID:28604104

Ammar, H. O., Salama, H. A., Ghorab, M., & Mahmoud, A. A. (2009). Nanoemulsion as a potential ophthalmic delivery system for dorzolamide hydrochloride. *AAPS PharmSciTech*, *10*(3), 808–819. doi:10.120812249-009-9268-4 PMID:19536653

Antignac, E., Nohynek, G. J., Re, T., Clouzeau, J., & Toutain, H. (2011). Safety of botanical ingredients in personal care products/cosmetics. *Food and Chemical Toxicology*, *49*(2), 324–341. doi:10.1016/j.fct.2010.11.022 PMID:21111022

Araújo, F. A., Kelmann, R. G., Araújo, B. V., Finatto, R. B., Teixeira, H. F., & Koester, L. S. (2011). Development and characterization of parenteral nanoemulsions containing thalidomide. *European Journal of Pharmaceutical Sciences*, *42*(3), 238–245. doi:10.1016/j.ejps.2010.11.014 PMID:21130164

Barbulova, A., Colucci, G., & Apone, F. (2015). New trends in cosmetics: By-products of plant origin and their potential use as cosmetic active ingredients. *Cosmetics*, *2*(2), 82–92. doi:10.3390/cosmetics2020082

Beg, S., Saini, S., Imam, S. S., Rahman, M., Swain, S., & Hasnain, M. S. (2017). Nanoemulsion for the Effective Treatment and Management of Anti-tubercular Drug Therapy. *Recent Patents on Anti-infective Drug Discovery*, *12*(2), 85–94. doi:10.2174/1574891X12666170504094330 PMID:28480833

Beidokhti, M. N. (2013). 55 Anticancer Fruits, Vegetables, Beverages, Oils and Spices. *International Journal of Phytomedicine*, *5*(4), 415.

Bilia, A., Guccione, C., Isacchi, B., Righeschi, C., Firenzuoli, F., & Bergonzi, M. (2014). Essential Oils Loaded in Nanosystems: A Developing Strategy for a Successful Therapeutic Approach. *Evidence-Based Complementary and Alternative Medicine*, *2014*, 1–14. doi:10.1155/2014/651593 PMID:24971152

Bivas-Benita, M., Oudshoorn, M., Romeijn, S., van Meijgaarden, K., Koerten, H., van der Meulen, H., Lambert, G., Ottenhoff, T., Benita, S., Junginger, H., & Borchard, G. (2004). Cationic submicron emulsions for pulmonary DNA immunization. *Journal of Controlled Release*, *100*(1), 145–155. doi:10.1016/j.jconrel.2004.08.008 PMID:15491818

Brussel, F., Manoela, L., Luisa, B. W., Michelle, F., Koester, L. S., & Teixeira, H. F. (2012). Nanoemulsions as parenteral drug delivery systems: A review. *Chem. New*, *35*, 34–39.

Castilho-Fernandes, A., Lopes, T. G., Primo, F. L., Pinto, M. R., & Tedesco, A. C. (2017). Photodynamic process induced by chloro-aluminum phthalocyanine nanoemulsion in glioblastoma. *Photodiagnosis and Photodynamic Therapy*, *19*, 221–228. doi:10.1016/j.pdpdt.2017.05.003 PMID:28599959

Chang, H. B., & Chen, B. H. (2015). Inhibition of lung cancer cells A549 and H460 by curcuminoid extracts and NEs prepared from Curcuma longa Linnaeus. *International Journal of Nanomedicine*, *10*, 5059. PMID:26345201

Chellapa, P., Ariffin, F. D., Eid, A. M., Almahgoubi, A. A., Mohamed, A. T., Issa, Y. S., & Elmarzugi, N. A. (2016). Nanoemulsion for cosmetic application. *European Journal of Biomedical and Pharmaceutical Sciences*, *3*(7), 8–11.

Chen, L., Chen, B., Deng, L., Gao, B., Zhang, Y., Wu, C., Yu, N., Zhou, Q., Yao, J., & Chen, J. (2017). An optimized two-vial formulation lipid nanoemulsion of paclitaxel for targeted delivery to tumor. *International Journal of Pharmaceutics*, *534*(1-2), 308–315. doi:10.1016/j.ijpharm.2017.10.005 PMID:28986321

Csaba, N., Garcia-Fuentes, M., & Alonso, M. J. (2009). Nanoparticles for nasal vaccination. *Advanced Drug Delivery Reviews*, *61*(2), 140–157. doi:10.1016/j.addr.2008.09.005 PMID:19121350 de Menezes Furtado, C., de Faria, F., Azevedo, R., Py-Daniel, K., dos Santos Camara, A., da Silva, J., de Holanda Oliveira, E., Rodriguez, A. F. R., & Degterev, I. A. (2017). Tectona grandis leaf extract, free and associated with nanoemulsions, as a possible photosensitizer of mouse melanoma B16 cell. *Journal of Photochemistry and Photobiology. B, Biology*, *167*, 242–248. doi:10.1016/j.jphotobiol.2017.01.004 PMID:28088105

Debnath, S., Satayanarayana, K. V., & Kumar, G. V. (2011). Nanoemulsion—A method to improve the solubility of lipophilic drugs. *Pharmanest*, *2*(2-3), 72–83.

Donsì, F., Annunziata, M., Sessa, M., & Ferrari, G. (2011). Nanoencapsulation of essential oils to enhance their antimicrobial activity in foods. *Lebensmittel-Wissenschaft + Technologie*, *44*(9), 1908–1914. doi:10.1016/j.lwt.2011.03.003

Eid, A. M., El-Enshasy, H. A., Aziz, R., & Elmarzugi, N. A. (2014). Preparation, characterization, and anti-inflammatory activity of Swietenia macrophylla nanoemulgel. *Journal of Nanomedicine & Nanotechnology*, *5*(2), 1–10. doi:10.4172/2157-7439.1000190

Fraga, M., de Carvalho, T., da Silva Diel, D., Bruxel, F., Filho, N., Teixeira, H., & Matte, U. (2015). Cationic Nanoemulsions as a Gene Delivery System: Proof of Concept in the Mucopolysaccharidosis I Murine Model. *Journal of Nanoscience and Nanotechnology*, *15*(1), 810–816. doi:10.1166/jnn.2015.9179 PMID:26328445

Franchi, L. P., Amantino, C. F., Melo, M. T., de Lima Montaldi, A. P., Primo, F. L., & Tedesco, A. C. (2016). In vitro effects of photodynamic therapy induced by chloroaluminum phthalocyanine nanoemulsion. *Photodiagnosis and Photodynamic Therapy*, *16*, 100–105. doi:10.1016/j.pdpdt.2016.09.003 PMID:27612654

Fryd, M. M., & Mason, T. G. (2012). Advanced nanoemulsions. *Annual Review of Physical Chemistry*, *63*(1), 493–518. doi:10.1146/annurev-physchem-032210-103436 PMID:22475339

Ganta, S., Devalapally, H., & Amiji, M. (2010). Curcumin enhances oral bioavailability and anti-tumor therapeutic efficacy of paclitaxel upon administration in nanoemulsion formulation. *Journal of Pharmaceutical Sciences*, *99*(11), 4630–4641. doi:10.1002/jps.22157 PMID:20845461

Ganta, S., Talekar, M., Singh, A., Coleman, T. P., & Amiji, M. M. (2014). Nanoemulsions in translational research-opportunities and challenges in targeted cancer therapy. *AAPS PharmSciTech*, *15*(3), 694–708. doi:10.120812249-014-0088-9 PMID:24510526

Ghosh, V., Mukherjee, A., & Chandrasekaran, N. (2014). Eugenol-loaded antimicrobial nanoemulsion preserves fruit juice against, microbial spoilage. *Colloids and Surfaces. B, Biointerfaces*, *114*, 392–397. doi:10.1016/j.colsurfb.2013.10.034 PMID:24252231

Guglielmini, G. (2008). Nanostructured novel carrier for topical application. *Clinics in Dermatology*, *26*(4), 341–346. doi:10.1016/j.clindermatol.2008.05.004 PMID:18691513

Heidi, M. M., Yun-Seok, R., & Xiao, W. (2009). Nanomedicine in pulmonary delivery. *International Journal of Nanomedicine*, *4*, 299–319. PMID:20054434

Jain, V., Jain, S., & Mahajan, S. C. (2015). Nanomedicines based drug delivery systems for anti-cancer targeting and treatment. *Current Drug Delivery*, *12*(2), 177–191. doi:10.2174/1567201811666140822112516 PMID:25146439

Jo, Y., Chun, J., Kwon, Y., Min, S., Hong, G., & Choi, M. (2015). Physical and antimicrobial properties of trans-cinnamaldehyde nanoemulsions in water melon juice. *Lebensmittel-Wissenschaft + Technologie*, *60*(1), 444–451. doi:10.1016/j.lwt.2014.09.041

Karasulu, H., Karabulut, B., Göker, E., Güneri, T., & Gabor, F. (2007). Controlled Release of Methotrexate from W/O Microemulsion and Its In Vitro Antitumor Activity. *Drug Delivery*, *14*(4), 225–233. doi:10.1080/10717540601067760 PMID:17497355

Kaur, A., Nigam, K., Bhatnagar, I., Sukhpal, H., Awasthy, S., Shankar, S., Tyagi, A., & Dang, S. (2020 a). Treatment of Alzheimer's diseases using donepezil nanoemulsion: An intranasal approach. *Drug Delivery and Translational Research*, *10*(6), 1862–1875. doi:10.100713346-020-00754-z PMID:32297166

Kaur, A., Nigam, K., Srivastava, S., Tyagi, A., & Dang, S. (2020 b). Memantine nanoemulsion: A new approach to treat Alzheimer's disease. *Journal of Microencapsulation*, *37*(5), 355–365. doi:10.1080/02652048.2020.1756971 PMID:32293915

Kawakami, M., Zamora, L. O., Araújo, R. S., Fernandes, C. P., Ricotta, T., de Oliveira, L. G., Queiroz-Junior, C. M., Fernandes, A. P., da Conceição, E. C., Ferreira, L., Barros, A., Aguiar, M. G., & Oliveira, A. (2021). Efficacy of nanoemulsion with Pterodon emarginatus Vogel oleoresin for topical treatment of cutaneous leishmaniasis. *Biomedicine & pharmacotherapy = Biomedecine & pharmacotherapie, 134*, 111109. doi:10.1016/j.biopha.2020.111109

Kim, B. S., Won, M., Yang, Lee, K. M., & Kim, C. S. (2008). In vitro permeation studies of nanoemulsions containing ketoprofen as a model drug. *Drug Delivery*, *15*(7), 465–469. doi:10.1080/10717540802328599 PMID:18712624

Kim, H. S., Kim, T. I., Kim, J. H., Yoon, K. C., Hyon, J. Y., Shin, K. U., & Choi, C. Y. (2017). Evaluation of Clinical Efficacy and Safety of a Novel Cyclosporin A Nanoemulsion in the Treatment of Dry Eye Syndrome. *Journal of Ocular Pharmacology and Therapeutics*, *33*(7), 530–538. doi:10.1089/jop.2016.0164 PMID:28759302

Kumar, G. P., & Divya, A. (2015). Nanoemulsion based targeting in cancer therapeutics. *Medicinal Chemistry (Shariqah, United Arab Emirates)*, *5*(5), 272–284.

Lawrence, M. J., & Rees, G. D. (2000). Microemulsion-based media as novel drug delivery systems. *Advanced Drug Delivery Reviews*, *45*(1), 89–121. doi:10.1016/S0169-409X(00)00103-4 PMID:11104900

Leandro, F. Z., Martins, J., Fontes, A. M., & Tedesco, A. C. (2017). Evaluation of theranostic nanocarriers for near-infrared imaging and photodynamic therapy on human prostate cancer cells. *Colloids and Surfaces. B, Biointerfaces*, *154*, 341–349. doi:10.1016/j.colsurfb.2017.03.042 PMID:28365423

Lin, H. Y., Thomas, J. L., Chen, H. W., Shen, C. M., Yang, W. J., & Lee, M. H. (2012). In vitro suppression of oral squamous cell carcinoma growth by ultrasound-mediated delivery of curcumin microemulsions. *International Journal of Nanomedicine*, *7*, 941–951. doi:10.2147/IJN.S28510 PMID:22393291

Liu, C. H., & Yu, S. Y. (2010). Cationic NE's as non-viral vectors for plasmid DNA delivery. *Colloids and Surfaces. B, Biointerfaces*, *79*(2), 509–515. doi:10.1016/j.colsurfb.2010.05.026 PMID:20541375

Liu, F., Yang, J., Huang, L., & Liu, D. (1996). Effect of non-ionic surfactants on the formation of DNA/emulsion complexes and emulsion-mediated gene transfer. *Pharmaceutical Research*, *13*(11), 1642–1646. doi:10.1023/A:1016480421204 PMID:8956328

Lovelyn, C., & Attama, A. A. (2011). Current state of nanoemulsions in drug delivery. *Journal of Biomaterials and Nanobiotechnology*, *2*(05), 626–639. doi:10.4236/jbnb.2011.225075

Luo, Y., Zhang, Y., Pan, K., Critzer, F., Davidson, P., & Zhong, Q. (2014). Self-Emulsification of Alkaline-Dissolved Clove Bud Oil by Whey Protein, Gum Arabic, Lecithin, and Their Combinations. *Journal of Agricultural and Food Chemistry*, *62*(19), 4417–4424. doi:10.1021/jf500698k PMID:24758517

Macha, S., Hughes, P. M., & Mitra, A. K. (2003). Overview of ocular drug delivery. *Drugs and the Pharmaceutical Sciences*, *130*, 1–12.

Majeed, H., Antoniou, J., & Fang, Z. (2014). Apoptotic effects of eugenol-loaded NEs in human colon and liver cancer cell lines. *Asian Pacific Journal of Cancer Prevention*, *15*(21), 9159–9164. doi:10.7314/APJCP.2014.15.21.9159 PMID:25422195

McClements, D. (2011). Edible nanoemulsions: Fabrication, properties, and functional performance. *Soft Matter*, *7*(6), 2297–2316. doi:10.1039/C0SM00549E

McClements, D. (2012). Nanoemulsions versus microemulsions: Terminology, differences, and similarities. *Soft Matter*, *8*(6), 1719–1729. doi:10.1039/C2SM06903B

Monge-Fuentes, V., Muehlmann, L., Longo, J., Silva, J., Fascineli, M., de Souza, P., Faria, F., Degterev, I. A., Rodriguez, A., Carneiro, F. P., Lucci, C. M., Escobar, P., Amorim, R. F. B., & Azevedo, R. B. (2017). Photodynamic therapy mediated by acai oil (Euterpe oleracea Martius) in nanoemulsion: A potential treatment for melanoma. *Journal of Photochemistry and Photobiology. B, Biology*, *166*, 301–310. doi:10.1016/j.jphotobiol.2016.12.002 PMID:28024281

Natesan, S., Sugumaran, A., Ponnusamy, C., Thiagarajan, V., Palanichamy, R., & Kandasamy, R. (2017). Chitosan stabilized camptothecin NEs: Development, evaluation and biodistribution in preclinical breast cancer animal mode. *International Journal of Biological Macromolecules*, *104*, 1846–1852. doi:10.1016/j.ijbiomac.2017.05.127 PMID:28545970

Nirale, P., Paul, A., & Yadav, K. S. (2020). Nanoemulsions for targeting the neurodegenerative diseases: Alzheimer's, Parkinson's and Prion's. *Life Sciences*, *245*, 117394. doi:10.1016/j.lfs.2020.117394 PMID:32017870

Osborne, D. W., Ward, A. J. I., & O'Neill, K. J. (1991). Microemulsions as topical drug delivery vehicles: In-vitro transdermal studies of a model hydrophilic drug. *The Journal of Pharmacy and Pharmacology*, *43*(6), 451–454. doi:10.1111/j.2042-7158.1991.tb03511.x PMID:1717675

Pardridge, W. M. (1999). Non-invasive drug delivery to the human brain using endogenous blood–brain barrier transport systems. *Pharmaceutical Science & Technology Today*, *2*(2), 49–59. doi:10.1016/S1461-5347(98)00117-5 PMID:10234207

Patel, A., Cholkar, K., Agrahari, V., & Mitra, A. K. (2013). Ocular drug delivery systems: An overview. *World Journal of Pharmacology*, *2*(2), 47–64. doi:10.5497/wjp.v2.i2.47 PMID:25590022

Patel, R., & Patel, K. P. (2014). Advances in novel parentral drug delivery systems. *Asian Journal of Pharmaceutics (AJP), 4*(3), 193-199.

Philippe, M., Didillon, B., & Gilbert, L. (2012). Industrial commitment to green and sustainable chemistry: Using renewable materials & developing eco-friendly processes and ingredients in cosmetics. *Green Chemistry*, *14*(4), 952–956. doi:10.1039/c2gc16341a

Pinto, V. N. (2013). Bioterrorism: Health sector alertness. *Journal of Natural Science, Biology, and Medicine*, *4*(1), 24. doi:10.4103/0976-9668.107256 PMID:23633831

Pires, A., Fortuna, A., Alves, G., & Falcão, A. (2009). Intranasal drug delivery: How, why and what for? *Journal of Pharmacy & Pharmaceutical Sciences*, *12*(3), 288–311. doi:10.18433/J3NC79 PMID:20067706

Poonia, N., Lather, V., Kaur, B., Kirthanashri, S. V., & Pandita, D. (2020). Optimization and Development of Methotrexate- and Resveratrol-Loaded Nanoemulsion Formulation Using Box-Behnken Design for Rheumatoid Arthritis. *Assay and Drug Development Technologies*, *18*(8), 356–368. doi:10.1089/adt.2020.989 PMID:33052698

Prévot, G., Soria, F. N., Thiolat, M. L., Daniel, J., Verlhac, J. B., Blanchard-Desce, M., Bezard, E., Barthélémy, P., Crauste-Manciet, S., & Dehay, B. (2018). Harnessing Lysosomal pH through PLGA Nanoemulsion as a Treatment of Lysosomal-Related Neurodegenerative Diseases. *Bioconjugate Chemistry*, *29*(12), 4083–4089. doi:10.1021/acs.bioconjchem.8b00697 PMID:30424597

Primo, F. L., Michieleto, L., Rodrigues, M. A., Macaroff, P. P., Morais, P. C., Lacava, Z. G., Bentley, M. V. L. B., & Tedesco, A. C. (2007). Magnetic nanoemulsions as drug delivery system for Foscan®: Skin permeation and retention in vitro assays for topical application in photodynamic therapy (PDT) of skin cancer. *Journal of Magnetism and Magnetic Materials*, *311*(1), 354–357. doi:10.1016/j.jmmm.2006.10.1183

Primo, F. L., Michieleto, L., Rodrigues, M. A., Macaroff, P. P., Morais, P. C., Lacava, Z. G., Bentley, M. V. L. B., & Tedesco, A. C. (2007). Magnetic NEs as drug delivery system for Foscan®: Skin permeation and retention in vitro assays for topical application in photodynamic therapy (PDT) of skin cancer. *Journal of Magnetism and Magnetic Materials*, *311*(1), 354–357. doi:10.1016/j.jmmm.2006.10.1183

Qhattal, H. S., Wang, S., Salihima, T., Srivastava, S. K., & Liu, X. (2011). Nanoemulsions of cancer chemopreventive agent benzyl isothiocyanate display enhanced solubility, dissolution, and permeability. *Journal of Agricultural and Food Chemistry*, *59*(23), 12396–12404. doi:10.1021/jf202612b PMID:22011260

Qin, X., Zhou, Y., Wang, Y., Wang, Z., Wang, Y., Chen, J., Zhu, L., Quan, X., Liu, Z., Zhang, H., Jiang, L., Dong, H., & Zhang, Z. (2021). Preparation and Characterization of Protein-loaded PFC Nanoemulsions for the Treatment of Heart Diseases by Pulmonary Administration. *European Journal of Pharmaceutical Sciences, 158*, 105690. doi:10.1016/j.ejps.2020.105690

Quintão, F. J., Tavares, R. S., Vieira-Filho, S. A., Souza, G. H., & Santos, O. D. (2013). Hydroalcoholic extracts of Vellozia Squamata: A study of its NE's for pharmaceutical or cosmetic applications. *Revista Brasileira de Farmacognosia*, *23*(1), 101–107. doi:10.1590/S0102-695X2013005000001

Reiss, H. (1975). Entropy-induced dispersion of bulk liquids. *Journal of Colloid and Interface Science, 53*(1), 61–70. doi:10.1016/0021-9797(75)90035-1

Ribeiro, R. C. D. A., Barreto, S. M. A. G., Ostrosky, E. A., Rocha-Filho, P. A. D., Veríssimo, L. M., & Ferrari, M. (2015). Production and characterization of cosmetic NE's containing Opuntia ficus-indica (L.) Mill extract as moisturizing agent. *Molecules (Basel, Switzerland), 20*(2), 2492–2509. doi:10.3390/molecules20022492 PMID:25648593

Rodriguez, L., Passerini, N., Cavallari, C., Cini, M., Sancin, P., & Fini, A. (1999). Description and preliminary evaluation of a new ultrasonic atomizer for spray-congealing processes. *International Journal of Pharmaceutics, 183*(2), 133–143. doi:10.1016/S0378-5173(99)00076-9 PMID:10361164

Rutvij, J. P., Gunjan, J. P., Bharadia, P. D., Pandya, V. M., & Modi, D. A. (2011). Nanoemulsion: An advanced concept of dosage form. *Int. J. Pharm. Cosmetol, 1*(5), 122–133.

Sahu, P., Das, D., Mishra, V. K., Kashaw, V., & Kashaw, S. K. (2017). Nanoemulsion: A novel eon in cancer chemotherapy. *Mini-Reviews in Medicinal Chemistry, 17*(18), 1778–1792. doi:10.2174/1389557516666160219122755 PMID:26891931

Shah, P., Bhalodia, D., & Shelat, P. (2010). Nanoemulsion: A pharmaceutical review. *Systematic Reviews In Pharmacy, 1*(1), 24. doi:10.4103/0975-8453.59509

Shakeel, F., Baboota, S., Ahuja, A., Ali, J., Aqil, M., & Shafiq, S. (2007). Nanoemulsions as vehicles for transdermal delivery of aceclofenac. *AAPS PharmSciTech, 8*(4), 191–199. doi:10.1208/pt0804104 PMID:18181525

Sharma, R. (2001). India wakes up to threat of bioterrorism. *BMJ (Clinical Research Ed.), 323*(7315), 714. doi:10.1136/bmj.323.7315.714a PMID:11576976

Silva, A. L., Alexandrino, F., Verissimo, L. M., Agnez-Lima, L. F., Egito, L. C. M., De Oliveira, A. G., & Do Egito, E. S. T. (2012). Physical factors affecting plasmid DNA compaction in stearylamine-containing nanoemulsions intended for gene delivery. *Pharmaceuticals, 5*(6), 643–654. doi:10.3390/ph5060643 PMID:24281666

Silva, A. L., Marcelino, H. R., Veríssimo, L. M., Araujo, I. B., Agnez-Lima, L. F., & do Egito, E. S. (2016). Stearylamine-containing cationic nanoemulsion as a promising carrier for gene delivery. *Journal of Nanoscience and Nanotechnology, 16*(2), 1339–1345. doi:10.1166/jnn.2016.11671 PMID:27433584

Solans, C., & Solé, I. (2012). Nano-emulsions: Formation by low-energy methods. *Current Opinion in Colloid & Interface Science, 17*(5), 246–254. doi:10.1016/j.cocis.2012.07.003

Sugumar, S., Ghosh, V., Nirmala, M., Mukherjee, A., & Chandrasekaran, N. (2014). Ultrasonic emulsification of eucalyptus oil nanoemulsion: Antibacterial activity against Staphylococcus aureus and wound healing activity in Wistar rats. *Ultrasonics Sonochemistry, 21*(3), 1044–1049. doi:10.1016/j.ultsonch.2013.10.021 PMID:24262758

Sugumaran, A., Ponnusamy, C., Kandasamy, P., Krishnaswami, V., Palanichamy, R., Kandasamy, R., Lakshmanan, M., & Natesan, S. (2018). Development and evaluation of camptothecin loaded polymer stabilized nanoemulsion: Targeting potential in 4T1-breast tumour xenograft model. *European Journal of Pharmaceutical Sciences*, *116*, 15–25. doi:10.1016/j.ejps.2017.10.005 PMID:28987538

Tayeb, H. H., & Sainsbury, F. (2018). Nanoemulsions in drug delivery: Formulation to medical application. *Nanomedicine (London, England)*, *13*(19), 2507–2525. doi:10.2217/nnm-2018-0088 PMID:30265218

Teixeira, H. F., Bruxel, F., Fraga, M., Schuh, R. S., Zorzi, G. K., Matte, U., & Fattal, E. (2017). Cationic nanoemulsions as nucleic acids delivery systems. *International Journal of Pharmaceutics*, *534*(1-2), 356–367. doi:10.1016/j.ijpharm.2017.10.030 PMID:29038065

Thiagarajan, P. (2011). Nanoemulsions for drug delivery through different routes. *Research in Biotechnology, 2*(3), 1-13.

Wang, L., Li, X., Zhang, G., Dong, J., & Eastoe, J. (2007). Oil-in-water nanoemulsions for pesticide formulations. *Journal of Colloid and Interface Science*, *314*(1), 230–235. doi:10.1016/j.jcis.2007.04.079 PMID:17612555

Wu, J., Lin, J., & Zhong, Q. (2014). Physical and antimicrobial characteristics of thyme oil emulsified with soluble soybean polysaccharide. *Food Hydrocolloids*, *39*, 144–150. doi:10.1016/j.foodhyd.2013.12.029

Xue, J., Michael, P., & Zhong, Q. (2015). Antimicrobial activity of thyme oil co-nanoemulsified with sodium caseinate and lecithin. *International Journal of Food Microbiology*, *210*, 1–8. doi:10.1016/j.ijfoodmicro.2015.06.003 PMID:26082324

Yashpal, S., Tanuj, H., & Harsh, K. (2013). Nanoemulsions: A pharmaceutical review. *Int. J. Pharma. Prof. Res*, *4*(2), 928–935.

Yi, S. W., Yune, T. Y., Kim, T. W., Chung, H., Choi, Y. W., Kwon, I. C., Lee, E. B., & Jeong, S. Y. (2000). A cationic lipid emulsion/DNA complex as a physically stable and serum-resistant gene delivery system. *Pharmaceutical Research*, *17*(3), 314–320. doi:10.1023/A:1007553106681 PMID:10801220

Zhang, Z., Vriesekoop, F., Yuan, Q., & Liang, H. (2014). Effects of nisin on the antimicrobial activity of d-limonene and its nanoemulsion. *Food Chemistry*, *150*, 307–312. doi:10.1016/j.foodchem.2013.10.160 PMID:24360455

Compilation of References

A, N., Kovooru, L., Behera, A. K., Kumar, K. P. P., & Srivastava, P. (2021). A critical review of synthesis procedures, applications and future potential of nanoemulsions. *Adv Colloid Interface Sci, 287*, 102318. doi:10.1016/j.cis.2020.102318

Aagaard, C., Dietrich, J., Doherty, M., & Andersen, P. (2009). TB vaccines: Current status and future perspectives. *Immunology and Cell Biology*, *87*(4), 279–286. doi:10.1038/icb.2009.14 PMID:19350048

Abbasi, F., Samadi, F., Jafari, S. M., Ramezanpour, S., & Shams Shargh, M. (2019). Ultrasound-assisted preparation of flaxseed oil nanoemulsions coated with alginate-whey protein for targeted delivery of omega-3 fatty acids into the lower sections of gastrointestinal tract to enrich broiler meat. *Ultrasonics Sonochemistry*, *50*, 208–217. doi:10.1016/j.ultsonch.2018.09.014 PMID:30249371

Abbasi, Z., Aminzare, M., Azar, H. H., & Rostamizadeh, K. (2020). Effect of corn starch coating incorporated with nanoemulsion of Zatariamultiflora essential oil fortified with cinnamaldehyde on microbial quality of fresh chicken meat and fate of inoculated Listeria monocytogenes. *Journal of Food Science and Technology*, 1–11. doi:10.100713197-020-04774-y

Abbas, S., Hayat, K., Karangwa, E., Bashari, M., & Zhang, X. (2013). An overview of ultrasound-assisted food-grade nanoemulsions. *Food Engineering Reviews*, *5*(3), 139–157. doi:10.100712393-013-9066-3

Abbas, S., Karangwa, E., Bashari, M., Hayat, K., Hong, X., Sharif, H. R., & Zhang, X. (2015). Fabrication of polymeric nanocapsules from curcumin-loaded nanoemulsion templates by self-assembly. *Ultrasonics Sonochemistry*, *23*, 81–92. doi:10.1016/j.ultsonch.2014.10.006 PMID:25453208

Abd El Azim, W. M., & Balah, M. A. (2016). Nanoemulsions formation from essential oil of Thymus capitatus and Majorana hortensis and their use in weed control. *Indian Journal of Weed Science*, *48*(4), 421. doi:10.5958/0974-8164.2016.00108.8

Abd Ellah, N. H., Tawfeek, H. M., John, J., & Hetta, H. F. (2019). Nanomedicine as a future therapeutic approach for Hepatitis C virus. *Nanomedicine (London)*, *14*(11), 1471–1491. doi:10.2217/nnm-2018-0348 PMID:31166139

Abdelmonem, R., Younis, M. K., Hassan, D. H., & El-Sayed, M. A. E. G. (2019). Formulation and characterization of chlorhexidineHCl nanoemulsion as a promising antibacterial root canal irrigant: In-vitro and ex-vivo studies. *International Journal of Nanomedicine*, *14*, 4697. doi:10.2147/IJN.S204550

Abdelrasoul, M. A., Eid, A. R., & Badawy, M. E. I. (2020). Preparation, characterizations and antibacterial activity of different nanoemulsions incorporating monoterpenes: In vitro and in vivo studies. *Archiv für Phytopathologie und Pflanzenschutz*, *53*(7-8), 310–334. doi:10.1080/03235408.2020.1744977

Abd-Elsalam, K. A., & Khokhlov, A. R. (2015). Eugenol oil nanoemulsion: Antifungal activity against Fusarium oxysporum f. sp. vasinfectum and phytotoxicity on cottonseeds. *Applied Nanoscience*, *5*(2), 255–265. doi:10.100713204-014-0398-y

Abdou, E. M., Kandil, S. M., & Miniawy, H. M. F. E. (2017). Brain targeting efficiency of antimigrain drug loaded mucoadhesive intranasal nanoemulsion. *International Journal of Pharmaceutics*, *529*(1-2), 667–677. doi:10.1016/j.ijpharm.2017.07.030 PMID:28729175

Abdulkareem, I. H., & Zurmi, I. B. (2012). Review of hormonal treatment of breast cancer. *Nigerian Journal of Clinical Practice*, *15*(1), 9. doi:10.4103/1119-3077.94088 PMID:22437080

Aboofazeli, R. (2010). Nanometric-scaled emulsions (nanoemulsions). *IJPR*, *9*(4), 325–326. PMID:24381596

Aboofazeli, R. (2010). Nanometric-scaled emulsions (nanoemulsions). *Iranian Journal of Pharmaceutical Research*, *9*(4), 325–326.

Aboofazeli, R. (2010). Nanometric-scaled emulsions (nanoemulsions). *Iranian journal of pharmaceutical research. IJPR*, *9*(4), 325–326. PMID:24381596

Abrams, C. K., Hamosh, M., Lee, T. C., Ansher, A. F., Collen, M. J., Lewis, J. H., Benjamin, S. B., & Hamosh, P. (1988). Gastric lipase: Localization in the human stomach. *Gastroenterology*, *95*(6), 1460–1464. doi:10.1016/S0016-5085(88)80063-5 PMID:3181673

Abu Lila, A. S., Kiwada, H., & Ishida, T. (2013). The accelerated blood clearance (ABC) phenomenon: Clinical challenge and approaches to manage. *Journal of Controlled Release*, *172*(1), 38–47. doi:10.1016/j.jconrel.2013.07.026 PMID:23933235

Acharya, P., Jayaprakasha, G. K., Crosby, K. M., Jifon, J. L., & Patil, B. S. (2020). Nanoparticle-Mediated Seed Priming Improves Germination, Growth, Yield, and Quality of Watermelons (Citrullus lanatus) at multi-locations in Texas. *Scientific Reports*, *10*(1), 1–17. doi:10.103841598-020-61696-7 PMID:32193449

Acierno, M. J., & Labato, M. A. (2005). Hypertension in renal disease: Diagnosis and treatment. *Clinical Techniques in Small Animal Practice*, *20*(1), 23–30. doi:10.1053/j.ctsap.2004.12.004 PMID:15822527

Acosta, E. (2009). Bioavailability of nanoparticles in nutrient and nutraceutical delivery. *Current Opinion in Colloid & Interface Science*, *14*(1), 3–15. doi:10.1016/j.cocis.2008.01.002

Adak, T., Barik, N., Patil, N. B., Govindharaj, G. P. P., Gadratagi, B. G., Annamalai, M., Mukherjee, A. K., & Rath, P. C. (2020). Nanoemulsion of eucalyptus oil: An alternative to synthetic pesticides against two major storage insects Sitophilus oryzae (L.) and Tribolium castaneum (Herbst)) of rice. *Industrial Crops and Products*, *143*, 111849. doi:10.1016/j.indcrop.2019.111849

Adams, L. S., Phung, S., Wu, X., Ki, L., & Chen, S. (2008). White button mushroom (*Agaricus bisporus*) exhibits antiproliferative and proapoptotic properties and inhibits prostate tumor growth in athymic mice. *Nutrition and Cancer*, *60*(6), 44–56. doi:10.1080/01635580802192866 PMID:19005974

Aguilar, Z. P., Alhashim, E., & Benedict, V. H. R. (2020). COVID-19 pathology and vitamin d nanoemulsion as preventive and alternative remediation for severity: A review. *European Journal of Biomedical*, *7*(6), 488–499.

Aguilera-Correa, J. J., Esteban, J., & Vallet-Regí, M. (2021). Inorganic and Polymeric Nanoparticles for Human Viral and Bacterial Infections Prevention and Treatment. *Nanomaterials (Basel, Switzerland)*, *11*(1), 137. doi:10.3390/nano11010137 PMID:33435597

Ahmad, N., Ahmad, R., Al-Qudaihi, A., Alaseel, S. E., Fita, I. Z., Khalid, M. S., & Pottoo, F. H. (2019). Preparation of a novel curcumin nanoemulsion by ultrasonication and its comparative effects in wound healing and the treatment of inflammation. *RSC Advances*, *9*(35), 20192–20206. doi:10.1039/C9RA03102B

Ahmad, N., Ahmad, R., Naqvi, A. A., Alam, M. A., Ashafaq, M., Abdur Rub, R., & Ahmad, F. J. (2018). Intranasal delivery of quercetin-loaded mucoadhesive nanoemulsion for treatment of cerebral ischaemia. *Artificial Cells, Nanomedicine, and Biotechnology*, *46*(4), 717–729. doi:10.1080/21691401.2017.1337024 PMID:28604104

Ahmad, Z., Pandey, R., Sharma, S., & Khuller, G. K. (2008). Novel chemotherapy for tuberculosis: Chemotherapeutic potential of econazole- and moxifloxacin-loaded PLG nanoparticles. *International Journal of Antimicrobial Agents*, *31*(2), 142–146. doi:10.1016/j.ijantimicag.2007.10.017 PMID:18155883

Ahmed, M., Smith, D. M., Hamouda, T., Rangel-Moreno, J., Fattom, A., & Khader, S. A. (2017). A novel nanoemulsion vaccine induces mucosal Interleukin-17 responses and confers protection upon *Mycobacterium tuberculosis* challenge in mice. *Vaccine*, *35*(37), 4983–4989. doi:10.1016/j.vaccine.2017.07.073 PMID:28774560

Ahmedy, O. (2016). Study of the anticancer potential of celery seed oil against chemically induced hepatocellular carcinoma in rats: A mechanistic approach. *Al-Azhar Journal of Pharmaceutical Sciences*, *53*(1), 14–28. doi:10.21608/ajps.2016.6686

Ahn, H., & Park, J. H. (2016). Liposomal delivery systems for intestinal lymphatic drug transport. *Biomaterials Research*, *20*(1), 36. doi:10.118640824-016-0083-1 PMID:27895934

Akkam, Y., Rababah, T., Costa, R., Almajwal, A., Feng, H., Laborde, J. E. A., Abulmeaty, M. M., & Razak, S. (2021). Pea protein nanoemulsion effectively stabilizes vitamin D in food products: A potential supplementation during the COVID-19 pandemic. *Nanomaterials (Basel, Switzerland)*, *11*(4), 887. doi:10.3390/nano11040887 PMID:33807206

Al Faraj, A., Shaik, A. S., & Al Sayed, B. (2015). Preferential magnetic targeting of carbon nanotubes to cancer sites: Noninvasive tracking using MRI in a murine breast cancer model. *Nanomedicine (London)*, *10*(6), 931–948.

Alam, A., Farooq, U., Singh, R., Dubey, V. P., Kumar, S., Kumari, R., ... Dhar, K. L. (2018). Chemotherapy treatment and strategy schemes: A review. *Open Access J Toxicol*, *2*(5), 555600. doi:10.19080/OAJT.2018.02.555600

Alam, M. S., Siddiq, A. M., Ragupathy, R., & Mandal, A. B. (2017). Micellization and mixed micellization of cationic gemini (dimeric) surfactants and cationic conventional (monomeric) surfactants: Conductometric, dye solubilization, and surface tension studies. *Journal of Dispersion Science and Technology*, *38*(2), 280–287. doi:10.1080/01932691.2016.1163719

Alam, T., Khan, S., Gaba, B., Haider, M. F., Baboota, S., & Ali, J. (2017). Nanocarriers as treatment modalities for hypertension. *Drug Delivery*, *24*(1), 358–369. doi:10.1080/10717544.2016.1255999 PMID:28165823

Alandes, L., Hernando, I., Quiles, A., Pérez-Munuera, I., & Lluch, M. A. (2006). Cell wall stability of fresh-cut fuji apples treated with calcium lactate. *Journal of Food Science*, *71*(9), 615–620. doi:10.1111/j.1750-3841.2006.00180.x

Alany, R. G., Rades, T., Nicoll, J., Tucker, I. G., & Davies, N. M. (2006). W/O microemulsions for ocular delivery: Evaluation of ocular irritation and precorneal retention. *Journal of Controlled Release*, *111*(1-2), 145–152.

Alavian, F., & Shams, N. (2019). Oral and intranasal administration of nanoparticles in the cerebral ischemia treatment; Considering the advantages and disadvantages. *Current Clinical Pharmacology*, *14*, 1–10. doi:10.2174/1574884714666190704115345 PMID:31272358

Al, B. A., Gomah, A., & Ageba, M. E. (2019). Chemical profile, characterization and acaricidal activity of essential oils of three plant species and their nanoemulsions against Tyrophagus putrescentiae, a stored-food mite Basma. *Experimental & Applied Acarology*, *79*(3), 359–376. PMID:31679077

Alexandre, E. M. C., Lourenco, R. V., Bittante, A. M. Q. B., Moraes, I. C. F., & Sobral, P. J. A. (2016). Gelatine based films reinforced with montmorillonite and activated with nanoemulsion of ginger essential oil for food packaging applications. *Food Packaging and Shelf Life*, *10*, 87–96. doi:10.1016/j.fpsl.2016.10.004

Ali, J., Ali, M., Baboota, S., Kaur Sahni, J., Ramassamy, C., Dao, L., & Bhavna. (2010). Potential of Nanoparticulate Drug Delivery Systems by Intranasal Administration. *Current Pharmaceutical Design*, *16*(14), 1644–1653. doi:10.2174/138161210791164108 PMID:20210751

Alka, A. J. A., Baboota, S., Shakeel, F., & Shafiq, S. (2007). Design development and evaluation of novel nanoemulsion formulations for transdermal potential of Celecoxib. *Acta Pharmaceutica (Zagreb, Croatia)*, *57*(3), 315–332. doi:10.2478/v10007-007-0025-5 PMID:17878111

Alkhatib, M. H., Al-Otaibi, W. A., & Wali, A. N. (2018). Antineoplastic activity of mitomycin C formulated in nanoemulsions-based essential oils on HeLa cervical cancer cells. *Chemico-Biological Interactions*, *291*, 72–80.

Alkhatib, M. H., Aly, M. M., Saleh, O. A., & Gashlan, H. M. (2016). Antibacterial activity of a microemulsion loaded with cephalosporin. *Biologia*, *71*(7), 748–756. doi:10.1515/biolog-2016-0105

Alkhatib, M. H., Bawadud, R. S., & Gashlan, H. M. (2020). Incorporation of docetaxel and thymoquinone in borage nanoemulsion potentiates their antineoplastic activity in breast cancer cells. *Scientific Reports*, *10*(1), 18124.

Alkhatib, M. H., Binsiddiq, B. M., & Backer, W. S. (2017). in Vivo Evaluation of the Anticancer Activity of a Water-In-Garlic Oil Nanoemulsion Loaded with Docetaxel. *International Journal of Pharmaceutical Sciences and Research*, *8*(12), 5373–5379. doi:10.13040/IJPSR.0975-8232.8(12).5373-79

Alkreathy, H. M., Damanhouri, Z. A., Ahmed, N., Slevin, M., & Osman, A. M. M. (2012). Mechanisms of cardioprotective effect of aged garlic extract against doxorubicin-induced cardiotoxicity. *Integrative Cancer Therapies*, *11*(4), 364–370. doi:10.1177/1534735411426726 PMID:22172987

Allegranzi, B., Nejad, S. B., Combescure, C., Graafmans, W., Attar, H., Donaldson, L., & Pittet, D. (2011). Burden of endemic health-care-associated infection in developing countries: Systematic review and meta-analysis. *Lancet*, *377*(9761), 228–241. doi:10.1016/S0140-6736(10)61458-4

Allen, T. M., & Martin, F. J. (2004). Advantages of liposomal delivery systems for anthracyclines. *Seminars in Oncology*, *31*(6, Suppl 13), 5–15.

Alliod, O., Messager, L., Fessi, H., Dupin, D., & Charcosset, C. (2019). Influence of viscosity for oil-in-water and water-in-oil nanoemulsions production by SPG premix membrane emulsification. *Chemical Engineering Research & Design*, *142*, 87–99. doi:10.1016/j.cherd.2018.11.027

Almeida Da Silva, P. E., & Palomino, J. C. (2011). Molecular basis and mechanisms of drug resistance in *Mycobacterium tuberculosis*: Classical and new drugs. *The Journal of Antimicrobial Chemotherapy*, *66*(7), 1417–1430. doi:10.1093/jac/dkr173 PMID:21558086

Almeida, C. P., Vital, C. G., Contente, T. C., Maria, D. A., & Maranhao, R. C. (2010). Modification of composition of a nanoemulsion with different cholesteryl ester molecular species: Effects on stability, peroxidation, and cell uptake. *International Journal of Nanomedicine*, *5*, 679–686. doi:10.2147/IJN.S12293 PMID:20957219

Alshehri, S., Altamimi, M. A., Hussain, A., Imam, S. S., Singh, S. K., & Faruk, A. (2020). Morphological transition of *M. tuberculosis* and modulation of intestinal permeation by food grade cationic nanoemulsion: *In vitro-ex vivo-in silico* GastroPlus™ studies. *Journal of Drug Delivery Science and Technology*, *60*, 101971. doi:10.1016/j.jddst.2020.101971

Al-Tayyar, N. A., Youssef, A. M., & Al-Hindi, R. R. (2020). Edible coatings and antimicrobial nanoemulsions for enhancing shelf life and reducing foodborne pathogens of fruits and vegetables: A review. *Sustainable. Materials Technology.*

Alzorqi, I., Ketabchi, M. R., Sudheer, S., & Manickam, S. (2016). Optimization of ultrasound induced emulsification on the formulation of palm-olein based nanoemulsions for the incorporation of antioxidant beta-D-glucan polysaccharides. *Ultrasonics Sonochemistry*, *31*, 71–84. doi:10.1016/j.ultsonch.2015.12.004 PMID:26964925

Amin, N., & Das, B. (2019). A Review on Formulation and Characterization of Nanoemulsion. *International Journal of Current Pharmaceutical Research*, *11*, 1–5. doi:10.22159/ijcpr.2019v11i4.34925

Ammar, H. O., Salama, H. A., Ghorab, M., & Mahmoud, A. A. (2009). Nanoemulsion as a potential ophthalmic delivery system for dorzolamide hydrochloride. *AAPS PharmSciTech*, *10*(3), 808–819. doi:10.120812249-009-9268-4 PMID:19536653

Angelini, P., Venanzoni, R., Flores, G. A., Tirillini, B., Orlando, G., Recinella, L., Chiavaroli, A., Brunetti, L., Leone, S., Di Simone, S. C., Ciferri, M. C., Zengin, G., Gunes, A., Menghini, L., & Ferrante, C. (2020). Evaluation of antioxidant, antimicrobial and tyrosinase inhibitory activities of extracts from *Tricholosporum goniospermum,* an edible wild mushroom. *Antibiotics (Basel, Switzerland)*, *9*(8), 9. doi:10.3390/antibiotics9080513 PMID:32823710

Angsubhakorn, S., Bhamarapravati, N., Sahaphong, S., & Sathiropas, P. (1988). Reducing effects of rodent malaria on hepatic carcinogenesis induced by dietary aflatoxin B1. *International Journal of Cancer*, *41*(1), 69–73. doi:10.1002/ijc.2910410114 PMID:3121525

An, H. Z., Safai, E. R., Eral, H. B., & Doyle, P. S. (2013). Synthesis of biomimetic oxygen-carrying compartmentalized microparticles using flow lithography. *Lab on a Chip*, *13*(24), 4765–4774. doi:10.1039/c3lc50610j PMID:24141406

Anisimova, Y. V., Gelperina, S. I., Peloquin, C. A., & Heifets, L. B. (2000). Nanoparticles as antituberculosis drugs carriers: Effect on activity against *Mycobacterium tuberculosis* in human monocyte-derived macrophages. *Journal of Nanoparticle Research*, *2*(2), 165–171. doi:10.1023/A:1010061013365

Anjali, C., Sharma, Y., Mukherjee, A., & Chandrasekaran, N. (2012). Neem oil nanoemulsion-a potent larvicidal agent against Culex quinquefasciatus. *Pest Management Science*, *68*(2), 158–163. doi:10.1002/ps.2233 PMID:21726037

Anjana, D., Ravichandran, S., Parmar, H., & Upadhyay, R. V. (2014). Development of curcumin based ophthalmic formulation development of Curcumin. *American Journal of Infectious Diseases*, *8*, 41–49.

Anke, T., Kupka, J., Schramm, G., & Steglich, W. (1980). Antibiotics from basidiomycetes. X. Scorodonin, a new antibacterial and antifungal metabolite from Marasmius scorononius (Fr.). *The Journal of Antibiotics*, *33*(5), 463–467. doi:10.7164/antibiotics.33.463 PMID:7191846

Antignac, E., Nohynek, G. J., Re, T., Clouzeau, J., & Toutain, H. (2011). Safety of botanical ingredients in personal care products/cosmetics. *Food and Chemical Toxicology*, *49*(2), 324–341. doi:10.1016/j.fct.2010.11.022 PMID:21111022

Antonarakis, E. S., Heath, E. I., Smith, D. C., Rathkopf, D., Blackford, A. L., Danila, D. C., King, S., Frost, A., Ajiboye, A. S., Zhao, M., Mendonca, J., Kachhap, S. K., Rudek, M. A., & Carducci, M. A. (2013). Repurposing itraconazole as a treatment for advanced prostate cancer: A noncomparative randomized phase II trial in men with metastatic castration-resistant prostate cancer. *The Oncologist*, *18*(2), 163–173. doi:10.1634/theoncologist.2012-314 PMID:23340005

Anton, N., Benoit, J. P., & Saulnier, P. (2008). Design and production of nanoparticles formulated from nano-emulsion templates-a review. *Journal of Controlled Release*, *128*(3), 185–199. doi:10.1016/j.jconrel.2008.02.007 PMID:18374443

Anton, N., Hallouard, F., Attia, M. F., & Vandamme, T. F. (2016). Nano-emulsions for drug delivery and biomedical imaging. In A. Propkop & V. Weissig (Eds.), *Intracellular delivery III* (pp. 273–300). Springer. doi:10.1007/978-3-319-43525-1_11

Anton, N., & Vandamme, T. F. (2009). The universality of low-energy nano-emulsification. *International Journal of Pharmaceutics*, *377*(1-2), 142–147. doi:10.1016/j.ijpharm.2009.05.014 PMID:19454306

Anton, N., & Vandamme, T. F. (2011). Nano-emulsions and micro-emulsions: Clarifications of the critical differences. *Pharmaceutical Research*, *28*(5), 978–985. doi:10.100711095-010-0309-1 PMID:21057856

Anwer, M. K., Jamil, S., Ibnouf, E. O., & Shakeel, F. (2014). Enhanced antibacterial effects of clove essential oil by nanoemulsion. *Journal of Oleo Science*, *63*(4), 13213. doi:10.5650/jos.ess13213 PMID:24599109

Aqil, M., Kamran, M., Ahad, A., & Imam, S. S. (2016). Development of clove oil based nanoemulsion of olmesartan for transdermal delivery: Box–Behnken design optimization and pharmacokinetic evaluation. *Journal of Molecular Liquids*, *214*, 238–248. doi:10.1016/j.molliq.2015.12.077

Araújo, F. A., Kelmann, R. G., Araújo, B. V., Finatto, R. B., Teixeira, H. F., & Koester, L. S. (2011). Development and characterization of parenteral nanoemulsions containing thalidomide. *European Journal of Pharmaceutical Sciences*, *42*(3), 238–245. doi:10.1016/j.ejps.2010.11.014 PMID:21130164

Aravalli, R. N., Steer, C. J., & Cressman, E. N. K. (2008). *Molecular Mechanisms of Hepatocellular Carcinoma.* doi:10.1002/hep.22580

Arbain, N. H., Salim, N., Masoumi, H., Wong, T. W., Basri, M., & Abdul Rahman, M. B. (2019). In vitro evaluation of the inhalable quercetin loaded nanoemulsion for pulmonary delivery. *Drug Delivery and Translational Research*, *9*(2), 497–507.

Ardito, F., Pellegrino, M. R., Perrone, D., Troiano, G., Cocco, A., & Muzio, L. L. (2017). In vitro study on anti-cancer properties of genistein in tongue cancer. *OncoTargets and Therapy*, *10*, 5405–5415. doi:10.2147/OTT.S133632 PMID:29180873

Arguelles-Mondal, W., Cabrera, G., Peniche, C., & Rinaudo, M. (2000). Conductimetric study of the interpolyelectrolyte reaction between chitosan and polygalacturonic acid. *Polymer*, *41*(7), 2373–2378. doi:10.1016/S0032-3861(99)00396-1

Armstrong, C. M., & Gao, A. C. (2015). Drug resistance in castration resistant prostate cancer: Resistance mechanisms and emerging treatment strategies. *American Journal of Clinical and Experimental Urology*, *3*(2), 64–76. PMID:26309896

Arslan, I., & Celik, A. (2008). Chemical composition and antistaphylococcal activity of Salvia chrysophylla Stapf. Naturally distributed in Denizli province (Turkey) and its vicinity. *Pakistan Journal of Botany*, •••, 401799–401803.

Artiga-Artigas, M., Acevedo-Fani, A., & Martín-Belloso, O. (2017). Effect of sodium alginate incorporation procedure on the physicochemical properties of nanoemulsions. *Food Hydrocolloids*, *70*, 191–200. doi:10.1016/j.foodhyd.2017.04.006

Artiga-Artigas, M., Acevedo-Fani, A., & Martín-Belloso, O. (2017). Improving the shelf life of low-fat cut cheese using nanoemulsion-based edible coatings containing oregano essential oil and mandarin fiber. *Food Control*, *76*, 1–12. doi:10.1016/j.foodcont.2017.01.001

Artiga-Artigas, M., Lanjari-Perez, Y., & Martin-Belloso, O. (2018). Curcumin-loaded nanoemulsions stability as affected by the nature and concentration of surfactant. *Food Chemistry*, *266*, 466–474. doi:10.1016/j.foodchem.2018.06.043 PMID:30381213

Asensio, C. M., Quiroga, P. R., Al-Gburi, A., Huang, Q., & Grosso, N. R. (2020). Reological behaviour, Antibacterial and Quorum sensing inhibition study of an *Argentinean oregano* essential oil nanoemulsion. *Frontiers in Nutrition*, *7*, 193. doi:10.3389/fnut.2020.569913

Ashaolu, T. J. (2021). Nanoemulsions for health, food, and cosmetics: A review. *Environmental Chemistry Letters*, *19*(4), 1–15. Advance online publication. doi:10.100710311-021-01216-9 PMID:33746662

Ashrafizadeh, M., Bakhoda, M. R., Bahmanpour, Z., Ilkhani, K., Zarrabi, A., Makvandi, P., Khan, H., Mazaheri, S., Darvish, M., & Mirzaei, H. (2020). Apigenin as Tumor Suppressor in Cancers: Biotherapeutic Activity, Nanodelivery, and Mechanisms With Emphasis on Pancreatic Cancer. *Frontiers in Chemistry*, *8*, 8. doi:10.3389/fchem.2020.00829 PMID:33195038

Assadpour, E., & Jafari, S. M. (2018). A systematic review on nanoencapsulation of food bioactive ingredients and nutraceuticals by various nanocarriers. *Critical Reviews in Food Science and Nutrition*, *59*(19), 3129–3151. doi:10.1080/10408398.2018.1484687 PMID:29883187

Assadpour, E., Maghsoudlou, Y., Jafari, S.-M., Ghorbani, M., & Aalami, M. (2016). Evaluation of folic acid nano-encapsulation by double emulsions. *Food and Bioprocess Technology*, *9*(12), 2024–2032. doi:10.100711947-016-1786-y

Aswathanarayan, J. B., & Vittal, R. R. (2019). Nanoemulsions and Their Potential Applications in Food Industry. *Frontiers in Sustainable Food Systems*, *3*, 95. doi:10.3389/fsufs.2019.00095

Atinderpal, K., Kapoor, N., Gupta, S., Tyag, A., Sharma, R. K., Ali, J., & Dang, S. (2018). Development and characterization of green tea catechins and ciprofloxacin-loaded nanoemulsion for intravaginal delivery to treat urinary tract infection. *Indian Journal of Pharmaceutical Sciences*, *80*(3), 442–452. doi:10.4172/pharmaceutical-sciences.1000377

Attari, F., Hazim, H., Zandi, A., Mazarei, Z., & Rafati, H. (2021). Circumventing paclitaxel resistance in breast cancer cells using a nanoemulsion system and determining its efficacy via an impedance biosensor. *Analyst (London)*, *146*(10), 3225–3233.

Attia, M. F., Anton, N., Chiper, M., Akasov, R., Anton, H., Messaddeq, N., Fournel, S., Klymchenko, A. S., Mély, Y., & Vandamme, T. F. (2014). Biodistribution of X-ray iodinated contrast agent in nano-emulsions is controlled by the chemical nature of the oily core. *ACS Nano*, *8*(10), 10537–10550. doi:10.1021/nn503973z PMID:25284066

Atuma, C., Strugala, V., Allen, A., & Holm, L. (2001). The adherent gastrointestinal mucus gel layer: Thickness and physical state in vivo. *American Journal of Physiology. Gastrointestinal and Liver Physiology*, *280*(5), G922–G929.

Aucouturier, J., Dupuis, L., & Ganne, V. (2001). Adjuvants designed for veterinary and human vaccines. *Vaccine*, *19*(17-19), 2666–2672. doi:10.1016/S0264-410X(00)00498-9 PMID:11257407

Avella, M., Bruno, G., Errico, M. E., Gentile, G., Piciocchi, N., Sorrentino, A., & Volpe, M. G. (2007). Innovative packaging for minimally processed fruits. *Packaging Technology & Science*, *20*(5), 325–335. doi:10.1002/pts.761

Avella, M., De Vlieger, J. J., Errico, M. E., Fischer, S., Vacca, P., & Volpe, M. G. (2005). Biodegradable starch/clay nanocomposite films for food packaging applications. *Food Chemistry*, *93*(3), 467–474. doi:10.1016/j.foodchem.2004.10.024

Avlessi, F., Dangou, J., Wotto, V., Alitonou, G., Sohounhloue, D., & Menut, C. (2004). Antioxidant activities of essential oil from Clausena anisata (Wild) Hook. *Comptes Rendus. Chimie*, *7*, 1057–1061. doi:10.1016/j.crci.2003.11.015

Azarmi, S., Roa, W. H., & Löbenberg, R. (2008). Targeted delivery of nanoparticles for the treatment of lung diseases. *Advanced Drug Delivery Reviews*, *60*(8), 863–875. doi:10.1016/j.addr.2007.11.006 PMID:18308418

Azeem, A., Rizwan, M., Ahmad, F. J., Iqbal, Z., Khar, R. K., Aqil, M., & Talegaonkar, S. (2009). Nanoemulsion components screening and selection: A technical note. *AAPS PharmSciTech*, *10*(1), 69–76. doi:10.120812249-008-9178-x PMID:19148761

Azmi, N. A. N., Elgharbawy, A. A. M., Motlagh, S. R., Samsudin, N., & Salleh, H. M. (2019). Nanoemulsions: Factory for Food, Pharmaceutical and Cosmetics. *Processes (Basel, Switzerland)*, *7*(9), 617. doi:10.3390/pr7090617

Azmy, R. M., Gohary, E., Gohary, E. E., Mahmoud, D. M., Salem, D. A. M., Abdou, M. A., & Salama, M. S. (2019). Assessment of larvicidal activity of nanoemulsion from Citrus sinensis essential oil on Culex pipiens L. (Diptera: Culicidae) Radwa. *Egyptian Journal of Aquatic Biology and Fish*, *23*, 61–67. doi:10.21608/ejabf.2019.35100

Baboota, S., Al-Azaki, A., Kohli, K., Ali, J., Dixit, N., & Shakeel, F. (2007). Development and evaluation of a microemulsion formulation for transdermal delivery of terbinafine. *PDA Journal of Pharmaceutical Science and Technology*, *61*(4), 276–285. PMID:17933209

Badalyan, S. M. (2014). Potential of mushroom bioactive molecules to develop healthcare biotech products. *Proceedings of the 8th International Conference on Mushroom Biology and Mushroom Products*, 373-378.

Badawi, N., Rosenbom, A. E., Jensen, A. M. D., & Sørensen, S. R. (2016). Degradation and sorption of the fungicide tebuconazole in soils from golf greens. *Environmental Pollution*, *219*, 368–378. doi:10.1016/j.envpol.2016.10.045 PMID:27814554

Badi, H., Yazdani, D., Ali, S. M., & Nazari, F. (2004). Effects of spacing and harvesting time on herbage yield and quality/quantity of oil in thyme, Thymus vulgaris L. *Industrial Crops and Products*, *19*(3), 231–236. doi:10.1016/j.indcrop.2003.10.005

Bae, Y. H., & Park, K. K. (2011). Targeted drug delivery to tumours: Myths, reality and possibility. *Journal of Controlled Release*, *153*(3), 198–205. doi:10.1016/j.jconrel.2011.06.001 PMID:21663778

Bahamondez-Canas, T. F., & Cui, Z. (2018). Intranasal immunization with dry powder vaccines. *European Journal of Pharmaceutics and Biopharmaceutics*, *122*, 167–175. doi:10.1016/j.ejpb.2017.11.001 PMID:29122735

Bailly, A. L., Correard, F., Popov, A., Tselikov, G., Chaspoul, F., Appay, R., Al-Kattan, A., Kabashin, A. V., Braguer, D., & Esteve, M. A. (2019). In vivo evaluation of safety, biodistribution and pharmacokinetics of laser-synthesized gold nanoparticles. *Scientific Reports*, *9*(1), 1–12. doi:10.103841598-019-48748-3 PMID:31501470

Bainun, I. N., Alias, N. H., & Syed-Hassan, S. S. A. (2015). Nanoemulsion: Formation, Characterization, Properties and Applications - A Review. *Advanced Materials Research*, *1113*, 147–152. . doi:10.4028/www.scientific.net/AMR.1113.147

Bajpai, S. K., Chand, N., & Chaurasia, V. (2012). Nano zinc oxide-loaded calcium alginate films with potential antibacterial properties. *Food and Bioprocess Technology*, *5*(5), 1871–1881. doi:10.100711947-011-0587-6

Baker, J. R. (2006). *Nanoemulsion compositions having anti-inflammatory activity*. https://patents.google.com/patent/EP1951200A2/en

Baker, J. R., Annis, T. C., Hamouda, T., & Vengroff, L. (2004). *Compositions for inactivating pathogenic microorganisms, methods of making the compositions, and methods of use thereof*. https://patents.google.com/patent/US20150150244

Baker, J. R., Rathinavelu, S., Makidon, P. E., LiPuma, J.J., & Nigavekar, S. (2008). *Nanoemulsion therapeutic compositions and methods of using the same*. https://patents.google.com/patent/US8747872B2/en

Baker, J.R., Bielinska, A., & Myc, A. (2008). *Compositions and methods for human immunodeficiency virus vaccination*. US0026988A1.

Bala, A., Kar, B., Haldar, P. K., Mazumder, U. K., & Bera, S. (2010). Evaluation of anticancer activity of Cleome gynandra on Ehrlich's Ascites Carcinoma treated mice. *Journal of Ethnopharmacology*, *129*(1), 131–134. doi:10.1016/j.jep.2010.03.010 PMID:20307641

Balah, M., & Abd El Azim, W. (2016). Emulsions and Nanoemulsions Formation from Wild and Cultivated Thyme and Marjoram Essential Oils for Weeds Control. *Journal of Plant Protection and Pathology*, *7*(10), 641–648. doi:10.21608/jppp.2016.52095

Balakrishnan, S., Bhat, F. A., Raja Singh, P., Mukherjee, S., Elumalai, P., Das, S., Patra, C. R., & Arunakaran, J. (2016). Gold nanoparticle-conjugated quercetin inhibits epithelial-mesenchymal transition, angiogenesis and invasiveness via EGFR/VEGFR-2-mediated pathway in breast cancer. *Cell Proliferation*, *49*(6), 678–697.

Bali, V., Ali, M., & Ali, J. (2011). Nanocarrier for the enhanced bioavailability of a cardiovascular agent: In vitro, pharmacodynamic, pharmacokinetic and stability assessment. *International Journal of Pharmaceutics*, *403*(1-2), 46–56. doi:10.1016/j.ijpharm.2010.10.018 PMID:20969935

Bamrungsap, S., Zhao, Z., Chen, T., Wang, L., Li, C., Fu, T., & Tan, W. (2012). Nanotechnology in therapeutics: A focus on nanoparticles as a drug delivery system. *Nanomedicine (London)*, *7*(8), 1253–1271. https://doi.org/10.2217/nnm.12.87

Bangaru, S., Marrero, J. A., & Singal, A. G. (2020). Review article: New therapeutic interventions for advanced hepatocellular carcinoma. *Alimentary Pharmacology & Therapeutics*, *51*(1), 78–89. doi:10.1111/apt.15573 PMID:31747082

Banker, G. S., Lieberman, H. A., & Rieger, M. M. (2002). Pharmaceutical dosage forms. *Disperse Systems Marcel Dekker*, *2*(3), 339–340.

Banzhoff, A., Gasparini, R., Laghi-Pasini, F., Staniscia, T., Durando, P., Montomoli, E., Capecchi, P., di Giovanni, P., Sticchi, L., Gentile, C., Hilbert, A., Brauer, V., Tilman, S., & Podda, A. (2009). Correction: MF59®-Adjuvanted H5N1 Vaccine Induces Immunologic Memory and Heterotypic Antibody Responses in Non-Elderly and Elderly Adults. *PLoS One*, *4*(3), e4384. doi:10.1371/journal.pone.0004384 PMID:19197383

Bao, F., Yang, K., Wu, C., Gao, S., Wang, P., Chen, L., & Li, H. (2018). New natural inhibitors of hexokinase 2 (HK2): Steroids from *Ganoderma sinense*. *Fitoterapia*, *125*, 123–129. doi:10.1016/j.fitote.2018.01.001 PMID:29305912

Barbulova, A., Colucci, G., & Apone, F. (2015). New trends in cosmetics: By-products of plant origin and their potential use as cosmetic active ingredients. *Cosmetics*, *2*(2), 82–92. doi:10.3390/cosmetics2020082

Barchini, R., & Saville, D. A. (1996). Electrokinetic Properties of Surfactant-Stabilized Oil Droplets. *Langmuir*, *12*(6), 1442–1445. doi:10.1021/la950693a

Barhoi, D., Giri, S., Upadhaya, P., & Barbhuiya, S. N. (2021b). Chemopreventive and Therapeutic Potential of Phytopharmaceuticals Against Oral Cancer: Evidence-Based Reports From Preclinical Studies in Animal Models. In Handbook of Research on Advancements in Cancer Therapeutics (pp. 541-569). IGI Global.

Barhoi, D., Upadhaya, P., Barbhuiya, S. N., Giri, A., & Giri, S. (2021a). Aqueous extract of Moringa oleifera exhibit potential anticancer activity and can be used as a possible cancer therapeutic agent: A study involving in vitro and in vivo approach. *Journal of the American College of Nutrition*, *40*(1), 70–85. doi:10.1080/07315724.2020.1735572 PMID:32191153

Barres, A. R., Wimmer, M. R., & Mecozzi, S. (2017). Multicompartment Theranostic Nanoemulsions Stabilized by a Triphilic Semifluorinated Block Copolymer. *Molecular Pharmaceutics*, *14*(11), 3916–3926. doi:10.1021/acs.molpharmaceut.7b00624 PMID:28945386

Barros, L., Cruz, T., Baptista, P., Estevinho, L. M., & Ferreira, I. C. F. R. (2008). Wild and commercial mushrooms as source of nutrients and nutraceuticals. *Food and Chemical Toxicology*, *46*(8), 2742–2747. doi:10.1016/j.fct.2008.04.030 PMID:18538460

Barta, J. A., Powell, C. A., & Wisnivesky, J. P. (2019). Global Epidemiology of Lung Cancer. *Annals of Global Health*, *85*(1), 8.

Barth, P. J., Weingärtner, K., Köhler, H. H., & Bittinger, A. (1996). Assessment of the vascularization in prostatic carcinoma: A morphometric investigation. *Human Pathology*, *27*(12), 1306–1310. doi:10.1016/S0046-8177(96)90342-1 PMID:8958303

Basak, S., & Guha, P. (2018). A review on antifungal activity and mode of action of essential oils and their delivery as nano-sized oil droplets in food system. *Journal of Food Science and Technology*, *55*(12), 4701–4710. doi:10.100713197-018-3394-5 PMID:30482966

Basha, S. K., Muzammil, M. S., Dhandayuthabani, R., Kumari, V. S., & Kaviyarasu, K. (2019). Nanoemulsion as Oral Drug Delivery-A Review. *Current Drug Research Reviews*, *12*(1), 4–15. doi:10.2174/2589977511666191024173508 PMID:31774040

Baskar, R., Lee, K. A., Yeo, R., & Yeoh, K. W. (2012). Cancer and radiation therapy: Current advances and future directions. *International Journal of Medical Sciences*, *9*(3), 193–199. doi:10.7150/ijms.3635 PMID:22408567

Bassetti, M., Vena, A., Croxatto, A., Righi, E., & Guery, B. (2018). How to manage Pseudomonas aeruginosa infections. *Drugs in Context*, *7*. Advance online publication. doi:10.7573/dic.212527

Baudino, T. A. (2015). Targeted cancer therapy: The next generation of cancer treatment. *Current Drug Discovery Technologies*, *12*(1), 3–20. doi:10.2174/1570163812666150602144310 PMID:26033233

Baudner, B. C., Ronconi, V., Casini, D., Tortoli, M., Kazzaz, J., Singh, M., Hawkins, L. D., Wack, A., & O'Hagan, D. T. (2009). MF59 emulsion is an effective delivery system for a synthetic TLR4 agonist (E6020). *Pharmaceutical Research*, *26*(6), 1477–1485. doi:10.100711095-009-9859-5 PMID:19255727

Bayón-Cordero, L., Alkorta, I., & Arana, L. (2019). Application of Solid Lipid Nanoparticles to Improve the Efficiency of Anticancer Drugs. *Nanomaterials (Basel, Switzerland)*, *9*(3), 474.

Beattie, K. D., Rouf, R., Gander, L., May, T. W., Ratkowsky, D., Donner, C. D., Gill, M., Grice, I. D., & Tiralongo, E. (2010). Antibacterial metabolites from Australian macrofungi from the genus *Cortinarius*. *Phytochemistry*, *71*(8-9), 948–955. doi:10.1016/j.phytochem.2010.03.016 PMID:20392467

Beaumont, E., & Roingeard, P. (2015). Chimeric hepatitis B virus (HBV)/hepatitis C virus (HCV) subviral envelope particles induce efficient anti-HCV antibody production in animals pre-immunized with HBV vaccine. *Vaccine*, *33*(8), 973–976. doi:10.1016/j.vaccine.2015.01.007 PMID:25596457

Beaumont, M., Portune, K. J., Steuer, N., Lan, A., Cerrudo, V., Audebert, M., Dumont, F., Mancano, G., Khodorova, N., Andriamihaja, M., Airinei, G., Tomé, D., Benamouzig, R., Davila, A.-M., Claus, S. P., Sanz, Y., & Blachier, F. (2017). Quantity and source of dietary protein influence metabolite production by gut microbiota and rectal mucosa gene expression: A randomized, parallel, double-blind trial in overweight humans. *The American Journal of Clinical Nutrition*, *106*(4), 1005–1019. doi:10.3945/ajcn.117.158816 PMID:28903954

Beevers, G., Lip, G. Y., & O'Brien, E. (2001). The pathophysiology of hypertension. *BMJ (Clinical Research Ed.)*, *322*(7291), 912–916. doi:10.1136/bmj.322.7291.912 PMID:11302910

Beg, S., Saini, S., Imam, S. S., Rahman, M., Swain, S., & Hasnain, M. S. (2017). Nanoemulsion for the Effective Treatment and Management of Anti-tubercular Drug Therapy. *Recent Patents on Anti-infective Drug Discovery*, *12*(2), 85–94. doi:10.2174/1574891X12666170504094330 PMID:28480833

Beg, S., Swain, S., Singh, H. P., Patra, C. N., & Rao, M. B. (2012). Development, optimization, and characterization of solid self-nanoemulsifying drug delivery systems of valsartan using porous carriers. *AAPS PharmSciTech*, *13*(4), 1416–1427. doi:10.120812249-012-9865-5 PMID:23070560

Beidokhti, M. N. (2013). 55 Anticancer Fruits, Vegetables, Beverages, Oils and Spices. *International Journal of Phytomedicine*, *5*(4), 415.

Beija, M., Salvayre, R., Lauth-de Viguerie, N., & Marty, J. D. (2012). Colloidal systems for drug delivery: From design to therapy. *Trends in Biotechnology*, *30*(9), 485–496. doi:10.1016/j.tibtech.2012.04.008 PMID:22673692

Belhaj, N., Dupuis, F., Arab-Tehrany, E., Denis, F. M., Paris, C., Lartaud, I., & Linder, M. (2012). Formulation, characterisation and pharmacokinetic studies of coenzyme Q10 PUFA's nanoemulsions. *European Journal of Pharmaceutical Sciences*, *47*(2), 305–312.

Bellini, R. G., Guimarães, A. P., Pacheco, M. A., Dias, D. M., Furtado, V. R., de Alencastro, R. B., & Horta, B. A. (2015). Association of the anti-tuberculosis drug rifampicin with a PAMAM dendrimer. *Journal of Molecular Graphics & Modelling*, *60*, 34–42. doi:10.1016/j.jmgm.2015.05.012 PMID:26093506

Bentayeb, K., Vera, P., Rubio, C., & Nerín, C. (2014). The additive properties of Oxygen Radical Absorbance Capacity (ORAC) assay: The case of essential oils. *Food Chemistry*, *148*, 204–208. doi:10.1016/j.foodchem.2013.10.037 PMID:24262547

Berekaa, M. M. (2015). Nanotechnology in food industry; advances in food processing, packaging and food safety. *International Journal of Current Microbiology and Applied Sciences*, *4*(5), 345–357.

Bergamini, E., Bizzarri, R., Cavallini, G., Cerbai, B., Chiellini, E., Donati, A., Gori, Z., Manfrini, A., Parentini, I., Signori, F., & Tamburini, I. (2004). Ageing and oxidative stress: A role for dolichol in the antioxidant machinery of cell membranes? *Journal of Alzheimer's Disease*, *6*(2), 129–135. doi:10.3233/JAD-2004-6204 PMID:15096696

Berger, C. N., Sodha, S. V., Shaw, R. K., Griffin, P. M., Pink, D., Hand, P., & Frankel, G. (2010). Fresh fruit and vegetables as vehicles for the transmission of human pathogens. *Environmental Microbiology*, *12*(9), 2385–2397. doi:10.1111/j.1462-2920.2010.02297.x PMID:20636374

Bernardi, D. S., Pereira, T. A., Maciel, N. R., Bortoloto, J., Viera, G. S., Oliveira, G. C., & Rocha-Filho, P. A. (2011). Formation and stability of oil-in-water nanoemulsions containing rice bran oil: In vitro and in vivo assessments. *Journal of Nanobiotechnology*, *9*(1), 44. doi:10.1186/1477-3155-9-44 PMID:21952107

Bernas, E., Jaworska, G., & Lisiewska, Z. (2006). Edible mushrooms as a source of valuable nutritive constituents. *Acta Scientiarum Polonorum. Technologia Alimentaria*, *5*, 5–20.

Bernstein, D. I., Cardin, R. D., Bravo, F. J., Hamouda, T., Pullum, D. A., Cohen, G., Bitko, V., & Fattom, A. (2019). Intranasal nanoemulsion-adjuvanted HSV-2 subunit vaccine is effective as a prophylactic and therapeutic vaccine using the guinea pig model of genital herpes. *Vaccine*, *37*(43), 6470–6477. doi:10.1016/j.vaccine.2019.08.077 PMID:31515143

Bhargava, K., Conti, D. S., da Rocha, S. R., & Zhang, Y. (2015). Application of an oregano oil nanoemulsion to the control of foodborne bacteria on fresh lettuce. *Food Microbiology*, *47*, 69–73. doi:10.1016/j.fm.2014.11.007 PMID:25583339

Bharmoria, P., Bisht, M., Gomes, M. C., Martins, M., Neves, M. C., Mano, J. F., Bdikin, I., Coutinho, J., & Ventura, S. (2021). Protein-olive oil-in-water nanoemulsions as encapsulation materials for curcumin acting as anticancer agent towards MDA-MB-231 cells. *Scientific Reports*, *11*(1), 9099.

Bhatia, D., Thoppil, R. J., Mandal, A., Samtani, K. A., Darvesh, A. S., & Bishayee, A. (2013). Pomegranate bioactive constituents suppress cell proliferation and induce apoptosis in an experimental model of hepatocellular carcinoma: Role of Wnt/β-catenin signaling pathway. *Evidence-Based Complementary and Alternative Medicine*, *2013*, 2013. doi:10.1155/2013/371813 PMID:23606879

Bhavana, S. M., & Lakshmi, C. R. (2014). Oral oncoprevention by phytochemicals-A systematic review disclosing the therapeutic dilemma. *Advanced Pharmaceutical Bulletin*, *4*(Suppl 1), 413. PMID:25364656

Bhushani, A., & Anandharamakrishnan, C. (2017). Food-Grade Nanoemulsions for Protection and Delivery of Nutrients. *Food-Grade Nanoemulsions for Protection and Delivery of Nutrients.*, *24*, 99–139. doi:10.1007/978-3-319-53112-0_3

Bidone, J., Argenta, D. F., Kratz, J., Pettenuzzo, L. F., Horn, A. P., Koester, L. S., ... Teixeira, H. F. (2015). Antiherpes activity and skin/mucosa distribution of flavonoids from achyrocline satureioides extract incorporated into topical nanoemulsions. *BioMed Research International*, •••, 2015. PMID:26101767

Bielinska, A. U., Janczak, K. W., Landers, J. J., Markovitz, D. M., Montefiori, D. C., & Baker, J. R. Jr. (2008). Nasal immunization with a recombinant HIV gp120 and nanoemulsion adjuvant produces Th1 polarized responses and neutralizing antibodies to primary HIV type 1 isolates. *AIDS Research and Human Retroviruses*, *24*(2), 271–281. doi:10.1089/aid.2007.0148 PMID:18260780

Bielinska, A., Janczak, K., Landers, J., Makidon, P., Sower, L., Peterson, J., & Baker, J. Jr. (2007). Mucosal Immunization with a Novel Nanoemulsion-Based Recombinant Anthrax Protective Antigen Vaccine Protects against Bacillus anthracis Spore Challenge. *Infection and Immunity*, *75*(8), 4020–4029. doi:10.1128/IAI.00070-07 PMID:17502384

Bielinska, A., Makidon, P., Janczak, K., Blanco, L., Swanson, B., Smith, D., Pham, T., Szabo, Z., Kukowska-Latallo, J. F., & Baker, J. R. Jr. (2014). Distinct Pathways of Humoral and Cellular Immunity Induced with the Mucosal Administration of a Nanoemulsion Adjuvant. *Journal of Immunology (Baltimore, Md.: 1950)*, *192*(6), 2722–2733. doi:10.4049/jimmunol.1301424 PMID:24532579

Bilbao-Sainz, C., Chiou, B. S., Williams, T., Wood, D., Du, W. X., Sedej, I., & McHugh, T. (2017). Vitamin D-fortified chitosan films from mushroom waste. *Carbohydrate Polymers*, *167*, 97–104. doi:10.1016/j.carbpol.2017.03.010 PMID:28433182

Bilia, A. R., Guccione, C., Isacchi, B., Righeschi, C., Firenzuoli, F., & Bergonzi, M. C. (2014). Essential oils loaded in nanosystems: a developing strategy for a successful therapeutic approach. *Evidence-Based Complementary and Alternative Medicine*.

Bilia, A., Guccione, C., Isacchi, B., Righeschi, C., Firenzuoli, F., & Bergonzi, M. (2014). Essential Oils Loaded in Nanosystems: A Developing Strategy for a Successful Therapeutic Approach. *Evidence-Based Complementary and Alternative Medicine*, *2014*, 1–14. doi:10.1155/2014/651593 PMID:24971152

Bisht, D., Sharma, D., Sharma, D., Singh, R., & Gupta, V. K. (2019). Recent insights into *Mycobacterium tuberculosis* through proteomics and implications for the clinic. *Expert Review of Proteomics*, *16*(5), 443–456. doi:10.1080/14789450.2019.1608185 PMID:31032653

Bivas-Benita, M., Oudshoorn, M., Romeijn, S., van Meijgaarden, K., Koerten, H., van der Meulen, H., Lambert, G., Ottenhoff, T., Benita, S., Junginger, H., & Borchard, G. (2004). Cationic submicron emulsions for pulmonary DNA immunization. *Journal of Controlled Release*, *100*(1), 145–155. doi:10.1016/j.jconrel.2004.08.008 PMID:15491818

Björn, A., de La Monja, P. S., Karlsson, A., Ejlertsson, J., & Svensson, B. H. (2012). Rheological characterization. *Biogas, 1*, 63-76. doi:10.5772/32596

Blecher, K., Nasir, A., & Friedman, A. (2011). The growing role of nanotechnology in combating infectious disease. *Virulence*, *2*(5), 395–401. doi:10.4161/viru.2.5.17035 PMID:21921677

Bonferoni, M. C., Riva, F., Invernizzi, A., Dellera, E., Sandri, G., Rossi, S., Marrubini, G., Bruni, G., Vigani, B., Caramella, C., & Ferrari, F. (2017). Alpha-tocopherol loaded chitosan oleate nanoemulsions for wound healing. Evaluation on cell lines and ex vivo human biopsies, and stabilization in spray-dried Trojan microparticles. *European Journal of Pharmaceutics and Biopharmaceutics*. Advance online publication. doi:10.1016/j.ejpb.2017.11.008 PMID:29155053

Bonferoni, M. C., Rossi, S., Sandri, G., Ferrari, F., Gavini, E., Rassu, G., & Giunchedi, P. (2019). Nanoemulsions for "nose-to-brain" drug delivery. *Pharmaceutics*, *11*(2), 84. doi:10.3390/pharmaceutics11020084 PMID:30781585

Boni, C., Vecchi, A., Rossi, M., Laccabue, D., Giuberti, T., Alfieri, A., Lampertico, P., Grossi, G., Facchetti, F., Brunetto, M. R., Coco, B., Cavallone, D., Mangia, A., Santoro, R., Piazzolla, V., Lau, A., Gaggar, A., Subramanian, G. M., & Ferrari, C. (2018). TLR7 agonist increases responses of hepatitis B virus–specific T cells and natural killer cells in patients with chronic hepatitis B treated with nucleos (t) ide analogues. *Gastroenterology*, *154*(6), 1764–1777. doi:10.1053/j.gastro.2018.01.030 PMID:29378197

Borhade, V., Pathak, S., Sharma, V., & Patravale, V. (2012). Patravale, Clotrimazole nanoemulsion for malaria chemotherapy. Part I: Preformulation studies, formulation design and physicochemical evaluation. *International Journal of Pharmaceutics*, *431*(1-2), 138–148. doi:10.1016/j.ijpharm.2011.12.040 PMID:22227344

Borse, V., Konwar, A. N., & Buragohain, P. (2020). Oral cancer diagnosis and perspectives in India. *Sensors International*, 100046.

Borthakur, P., Boruah, P. K., Sharma, B., & Das, M. R. (2016). *Nanoemulsion: Preparation and its application in food industry*. doi:10.1016/B978-0-12-804306-6.00005-2

Borugă, O., Jianu, C., Mişcă, C., Goleţ, I., Gruia, A. T., & Horhat, F. G. (2014). Thymus vulgaris essential oil: chemical composition and antimicrobial activity. *Journal of Medicine and Life*, *3*(Spec Iss 3), 56–60.

Bott, J., Stormer, A., & Franz, R. (2014). A comprehensive study into the migration potential of nano silver particles from food contact polyolefins. In M. A. Benvenuto, S. Ahuja, T. V. Duncan, G. O. Noonan, & E. S. RobertsKirchhoff (Eds.), Chemistry of food, food supplements, and food contact materials: From production to plate (pp. 51–70). Washington, DC: American Chemical Society. doi:10.1021/bk-2014-1159.ch005

Bouchaala, R., Mercier, L., Andreiuk, B., Mély, Y., Vandamme, T., Anton, N., Goetz, J. G., & Klymchenko, A. S. (2016). Integrity of lipid nanocarriers in bloodstream and tumor quantified by near-infrared ratiometric FRET imaging in living mice. *Journal of Controlled Release*, *236*, 57–67. doi:10.1016/j.jconrel.2016.06.027 PMID:27327767

Bouchemal, K., Briançon, S., Perrier, E., & Fessi, H. (2004). Nano-emulsion formulation using spontaneous emulsification: Solvent, oil and surfactant optimisation. *International Journal of Pharmaceutics*, *280*(1–2), 241–251. doi:10.1016/j.ijpharm.2004.05.016 PMID:15265563

Boyle, J., Eastman, D., Millar, C., Camuglia, S., Cox, J., Pearse, M., Good, J., & Drane, D. (2007). The utility of ISCOMATRIX™ adjuvant for dose reduction of antigen for vaccines requiring antibody responses. *Vaccine*, *25*(14), 2541–2544. doi:10.1016/j.vaccine.2006.12.018 PMID:17240491

Bracho, G., Lastre, M., Campo, J. D., Zayas, C., González, D., Gil, D., Acevedo, R., Taboada, C., Solís, R. L., & Pérez, O. (2006). Proteoliposome derived cochleate as novel adjuvant. *Vaccine*, *24*, S30–S31. doi:10.1016/j.vaccine.2005.01.108 PMID:16823914

Bragazzi, N. L., Gianfredi, V., Villarini, M., Rosselli, R., Nasr, A., Hussein, A., Martini, M., & Behzadifar, M. (2018). Vaccines Meet Big Data: State-of-the-Art and Future Prospects. From the Classical 3Is ("Isolate-Inactivate-Inject") Vaccinology 1.0 to Vaccinology 3.0, Vaccinomics, and Beyond: A Historical Overview. *Frontiers in Public Health, 6*, 62. doi:10.3389/fpubh.2018.00062 PMID:29556492

Bratten, J., & Jones, M. P. (2006). New directions in the assessment of gastric function: Clinical applications of physiologic measurements. *Digestive Diseases (Basel, Switzerland), 24*(3-4), 252–259.

Brenner, B. M., Ballermann, B. J., Gunning, M. E., & Zeidel, M. L. (1990). Diverse biological actions of atrial natriuretic peptide. *Physiological Reviews, 70*(3), 665–699. doi:10.1152/physrev.1990.70.3.665 PMID:2141944

Brink, A. J., Coetzee, J., Clay, C. G., Sithole, S., Richards, G. A., Poirel, L., & Nordmann, P. (2012). Emergence of New Delhi metallo-beta-lactamase (NDM-1) and Klebsiellapneumoniaecarbapenemase (KPC-2) in South Africa. *Journal of Clinical Microbiology, 50*(2), 525–527. doi:10.1128/JCM.05956-11

Brito-de la Fuente, E., Turcanu, M., Ekberg, O., & Gallegos, C. (2017). Rheological aspects of swallowing and dysphagia: shear and elongational flows. In *Dysphagia* (pp. 687–716). Springer. org/ doi:10.1007/174_2017_119

Brooker, M. I. H., & Kleinig, D. A. (2004). Field guide to Eucalypts (2nd ed.). Bloomings Books.

Bruneton, J. (1995). Pharmacognosy, phytochemistry, medicinal plants. Andover: Intercept / Paris: Lavoisier.

Brunton, L. L., Hilal-Dandan, R., & Knollmann, B. C. (Eds.). (2018). *Goodman & Gilman's The Pharmacological Basis of Therapeutics* (p. 7). McGraw-Hill Education.

Brussel, F., Manoela, L., Luisa, B. W., Michelle, F., Koester, L. S., & Teixeira, H. F. (2012). Nanoemulsions as parenteral drug delivery systems: A review. *Chem. New, 35*, 34–39.

Bulam, S., Ustun, N.S., & Peksen, A. (2019). Health benefits of *Ganoderma lucidum* as a medicinal mushroom. *TURJAF, 7*(sp 1), 84–93.

Burapapadh, K., Kumpugdee-Vollrath, M., Chantasart, D., & Sriamornsak, P. (2010). Fabrication of pectin-based nanoemulsions loaded with itraconazole for pharmaceutical application. *Carbohydrate Polymers, 82*(2), 384–393. doi:10.1016/j.carbpol.2010.04.071

Burt, S. (2004). Essential oils: Their antibacterial properties and potential applications in foods—a review. *International Journal of Food Microbiology, 94*(3), 223–253. doi:10.1016/j.ijfoodmicro.2004.03.022 PMID:15246235

Burt, S. A., & Reinders, R. D. (2003). Antibacterial activity of selected plant essential oils against Escherichia coli O157:H7. *Letters in Applied Microbiology, 36*(3), 162–167. doi:10.1046/j.1472-765X.2003.01285.x PMID:12581376

Bush, K. (2010). Alarming β-lactamase-mediated resistance in multidrug-resistant Enterobacteriaceae. *Current Opinion in Microbiology, 13*(5), 558–564. doi:10.1016/j.mib.2010.09.006

Bustos-Salgado, P., Andrade-Carrera, B., Domínguez-Villegas, V., Rodríguez-Lagunas, M. J., Boix-Montañes, A., Calpena-Campmany, A., & Garduño-Ramírez, M. L. (2021). Biopharmaceutic study and in vivo efficacy of natural and derivatives flavanones formulations. *Nanomedicine (London), 16*(3), 205–220. doi:10.2217/nnm-2020-0368 PMID:33480290

Buya, A. B., Beloqui, A., Memvanga, P. B., & Préat, V. (2020). Self-Nano-Emulsifying Drug-Delivery Systems: From the Development to the Current Applications and Challenges in Oral Drug Delivery. *Pharmaceutics, 12*(1194), 1–55. doi:10.3390/pharmaceutics12121194 PMID:33317067

Buzdar, A. U., Ibrahim, N. K., Francis, D., Booser, D. J., Thomas, E. S., Theriault, R. L., Pusztai, L., Green, M. C., Arun, B. K., Giordano, S. H., Cristofanilli, M., Frye, D. K., Smith, T. L., Hunt, K. K., Singletary, S. E., Sahin, A. A., Ewer, M. S., Buchholz, T. A., Berry, D., & Hortobagyi, G. N. (2005). Significantly higher pathologic complete remission rate after neoadjuvant therapy with trastuzumab, paclitaxel, and epirubicin chemotherapy: Results of a randomized trial in human epidermal growth factor receptor 2-positive operable breast cancer. *Journal of Clinical Oncology*, *23*(16), 3676–3685.

Byars, N. E., & Allison, A. C. (1987). Adjuvant formulation for use in vaccines to elicit both cell-mediated and humoral immunity. *Vaccine*, *5*(3), 223–228. doi:10.1016/0264-410X(87)90105-8 PMID:3499713

Cain, A. E., & Khalil, R. A. (2002). *Pathophysiology of essential hypertension: role of the pump, the vessel, and the kidney.* Paper presented at the Seminars in nephrology. 10.1053nep.2002.28639

Cai, Z., Wang, Y., Zhu, L. J., & Liu, Z. Q. (2010). Nanocarriers: A general strategy for enhancement of oral bioavailability of poorly absorbed or pre-systemically metabolised drugs. *Current Drug Metabolism*, *11*(2), 197–207.

Calderó, G., Montes, R., Llinàs, M., García-Celma, M. J., Porras, M., & Solans, C. (2016). Studies on the formation of polymeric nano-emulsions obtained via low-energy emulsification and their use as templates for drug delivery nanoparticle dispersions. *Colloids and Surfaces. B, Biointerfaces*, *145*, 922–931. doi:10.1016/j.colsurfb.2016.06.013 PMID:27341306

Calligaris, S., Plazzotta, S., Bot, F., Grasselli, S., Malchiodi, A., & Anese, M. (2016). Nanoemulsion preparation by combining high pressure homogenization and high power ultrasound at low energy densities. *Food Research International*, *83*, 25–30. doi:10.1016/j.foodres.2016.01.033

Calvo, P., Remufin-Ldpez, C., Vila-Jato, J. L., & Alonso, M. J. (1997). Development of positively charged colloidal drug carriers: Chitosan-coated polyester nanocapsules and submicron-emulsions. *Colloid & Polymer Science*, *275*(1), 46–53. doi:10.1007003960050050

Caminero, J. A., Sotgiu, G., Zumla, A., & Migliori, G. B. (2010). Best drug treatment for multidrug-resistant and extensively drug-resistant tuberculosis. *The Lancet. Infectious Diseases*, *10*(9), 621–629. doi:10.1016/S1473-3099(10)70139-0 PMID:20797644

Campani, V., Biondi, M., Mayol, L., Cilurzo, F., Pitaro, M., & de Rosa, G. (2016). Development of nanoemulsions for topical delivery of vitamin K1. *International Journal of Pharmaceutics*, *511*(1), 170–177. doi:10.1016/j.ijpharm.2016.07.004 PMID:27386792

Campos, V. E. B., Ricci-Júnior, E., & Mansur, C. R. E. (2012). Nanoemulsions as Delivery Systems for Lipophilic Drugs. *Journal of Nanoscience and Nanotechnology*, *12*(3), 2881–2890. doi:10.1166/jnn.2012.5690 PMID:22755138

Cao, W., Davis, W. G., Kim, J. H., Juan, A., Taylor, A., Hendrickson, G. R., ... Sambhara, S. (2016). An oil-in-water nanoemulsion enhances immunogenicity of H5N1 vaccine in mice. *Nanomedicine; Nanotechnology, Biology, and Medicine*, *12*(7), 1909–1917. doi:10.1016/j.nano.2016.04.005 PMID:27112307

Cao, X., Luo, J., Gong, T., Zhang, Z. R., Sun, X., & Fu, Y. (2015). Coencapsulated doxorubicin and bromotetrandrine lipid nanoemulsions in reversing multidrug resistance in breast cancer in vitro and in vivo. *Molecular Pharmaceutics*, *12*(1), 274–286.

Cao, Z., Spilker, T., Fan, Y., Kalikin, L. M., & Ciotti, S., LiPuma, J. J., & Wang, S. H. (2017). Nanoemulsion is an effective antimicrobial for methicillin-resistant Staphylococcus aureus in infected wounds. *Nanomedicine (London)*, *12*(10), 1177–1185. doi:10.2217/nnm-2017-0025

Caputo, L., Smeriglio, A., Trombetta, D., Cornara, L., Trevena, G., Valussi, M., Fratianni, F., De Feo, V., & Nazzaro, F. (2020). Chemical Composition and Biological Activities of the Essential Oils of Leptospermum petersonii and Eucalyptus gunnii. *Frontiers in Microbiology*, *409*(11), 1–15. doi:10.3389/fmicb.2020.00409 PMID:32351456

Cardillo, C., Kilcoyne, C. M., Quyyumi, A. A., Cannon, R. O. III, & Panza, J. A. (1998). Selective defect in nitric oxide synthesis may explain the impaired endothelium-dependent vasodilation in patients with essential hypertension. *Circulation, 97*(9), 851–856. doi:10.1161/01.CIR.97.9.851 PMID:9521333

Cardwell, G., Bornman, J., James, A., & Black, L. (2018). A review of mushrooms as a potential source of dietary Vitamin D. *Nutrients, 10*(10), 1498. doi:10.3390/nu10101498 PMID:30322118

Carey, M. C., & Small, D. M. (1970). The characteristics of mixed micellar solutions with particular reference to bile. *The American Journal of Medicine, 49*(5), 590–608. doi:10.1016/S0002-9343(70)80127-9 PMID:4924587

Carey, M. C., Small, D. M., & Bliss, C. M. (1983). Lipid digestion and absorption. *Annual Review of Physiology, 45*(1), 651–677. doi:10.1146/annurev.ph.45.030183.003251 PMID:6342528

Carmeliet, P., & Jain, R. K. (2011). Molecular mechanisms and clinical applications of angiogenesis. *Nature, 473*(7347), 298–307.

Carpenter, J., George, S., & Saharan, V. K. (2020). Ultrasonic cavitation assisted synthesis of multilayer emulsions as encapsulating and delivery systems for bioactive compounds. *Encapsulation of Active Molecules and Their Delivery System, 23.*

Carretero, O. A., & Oparil, S. (2000). Essential hypertension: part II: treatment. *Circulation, 101*(4), 446–453. doi:10.1161/01.CIR.101.4.446 PMID:10653838

Carvalho, F. P. (2006). Agriculture, pesticides, food security and food safety. *Environmental Science & Policy, 9*(7), 685–692. doi:10.1016/j.envsci.2006.08.002

Castaneda, S. A., & Strasser, J. (2017). Updates in the Treatment of Breast Cancer with Radiotherapy. *Surgical Oncology Clinics of North America, 26*(3), 371–382.

Castanheira, M., Deshpande, L. M., Mathai, D., Bell, J. M., Jones, R. N., & Mendes, R. E. (2011). Early dissemination of NDM-1-and OXA-181-producing Enterobacteriaceae in Indian hospitals: Report from the SENTRY Antimicrobial Surveillance Program, 2006-2007. *Antimicrobial Agents and Chemotherapy, 55*(3), 1274–1278. doi:10.1128/AAC.01497-10

Castilho-Fernandes, A., Lopes, T. G., Primo, F. L., Pinto, M. R., & Tedesco, A. C. (2017). Photodynamic process induced by chloro-aluminum phthalocyanine nanoemulsion in glioblastoma. *Photodiagnosis and Photodynamic Therapy, 19*, 221–228. doi:10.1016/j.pdpdt.2017.05.003 PMID:28599959

Cecchini, M. E., Paoloni, C., Campra, N., Picco, N., Grosso, M. C., Perez, M. S., & Bellingeri, R. (2021). Nanoemulsion of Minthostachysverticillata essential oil, In-vitro evaluation of its antibacterial activity. *Heliyon, 7*(1), e05896. doi:10.1016/j.heliyon.2021.e05896

Cenobio-Galindo, A. (2019). Development and incorporation of nanoemulsions in food. *International Journal of Food Studies, 8*(2), 105–124. doi:10.7455/ijfs/8.2.2019.a10

Čerpnjak, K., Zvonar, A., Gašperlin, M., & Vrečer, F. (2013). Lipid-based systems as a promising approach for enhancing the bioavailability of poorly water-soluble drugs. *Acta Pharmaceutica (Zagreb, Croatia), 63*(4), 427–445. doi:10.2478/acph-2013-0040 PMID:24451070

Ceylan, E., & Fung, D. Y. C. (2004). Antimicrobial activity of spices. *Journal of Rapid Methods and Automation in Microbiology, 12*(1), 1–55. doi:10.1111/j.1745-4581.2004.tb00046.x

Chae, C. H., Choi, D. J., Shim, H. Y., Byun, E. S., Hong, S. M., Park, Y. H., & Park, J. W. (2007). treatment of oral soft tissue lesions and wounds with high functional tooth paste made from nanoemulsion gel. *Journal of the Korean Association of Oral and Maxillofacial Surgeons, 33*(6), 694–700.

Chakravarty, M., & Vora, A. (2020). Nanotechnology-based antiviral therapeutics. *Drug Delivery and Translational Research*, 1–40. PMID:32748035

Chanamai, R., Horn, G. R., & Mcclements, D. (2002). Influence of oil polarity on droplet growth in oil-in-water emulsions stabilized by a weakly adsorbing biopolymer or a nonionic surfactant. *Journal of Colloid and Interface Science*, *247*(1), 167–176. doi:10.1006/jcis.2001.8110 PMID:16290453

Chanamai, R., & McClements, D. J. (2000). Dependence of creaming and rheology of monodisperse oil-in-water emulsions on droplet size and concentration. *Colloids and Surfaces. A, Physicochemical and Engineering Aspects*, *172*(1-3), 79–86. doi:10.1016/S0927-7757(00)00551-3

Chandler, D. (2005). Interfaces and the driving force of hydrophobic assembly. *Nature*, *437*(7059), 640–647. doi:10.1038/nature04162 PMID:16193038

Chandrawanshi, N. K., Tandia, D. K. & Jadhav, S. K. (2018) Determination of antioxidant and antidiabetic activities of polar solvent extracts of *Daedaleopsis confragosa* (Bolton) J. Schrot. *Research J. Pharm. and Tech, 11*(12), 5623-5630. . doi:10.5958/0974-360X.2018.01020.X

Chandrawanshi, N.K., Tandia, D.K. & Jadhav, S.K. (2018). Determination of antioxidant and antidiabetic activities of polar solvent extracts of *Daedaleopsis confragosa* (Bolton) J. Schrot. *Research Journal of Pharmacy and Technology, 11,* 5623–5630.

Chandrawanshi, N. K., Jadhav, S. K., Tiwari, K. L., & Quraishi, A. (2015). In vitro tuberization and Colchicine content analysis of *Gloriosa superba L. Biotechnology*, *14*, 142–147. doi:10.3923/biotech.2015.142.147

Chandrawanshi, N. K., Tandia, D. K., & Jadhav, S. K. (2017). Nutraceutical properties evaluation of *Schizophyllum commune. Indian Journal of Scientific Research*, *13*(2), 57–62.

Chandrawanshi, N. K., Tandia, D. K., & Jadhav, S. K. (2017). Nutraceutical Properties Evaluation of *Schizophyllum commune. Indian Journal of Scientific Research*, *13*(2), 57–62.

Chandrawanshi, N. K., Tandia, D., & Jadhav, S. K. (2017). Nutraceutical properties evaluation of *schizophyllum commune. Indian Journal of Scientific Research*, *13*, 57–62.

Chandrawanshi, N. K., & Verma, S. (2020). *Recent research and development in stem cell therapy for cancer treatment: future promising and challenges*. IGI Global. doi:10.4018/978-1-7998-6530-8.ch018

Chan, E. D., Strand, M. J., & Iseman, M. D. (2008). Treatment outcomes in extensively resistant tuberculosis. *The New England Journal of Medicine*, *359*(6), 657–659. doi:10.1056/NEJMc0706556 PMID:18687654

Chang, H.Y., Sheu, M. J. & Yang, C.H. (2011) Analgesic effects and themechanisms of anti-inflammation of hispolon in mice. *Evid.-Based Compl.Alt. Med.*

Chang, H. B., & Chen, B. H. (2015). Inhibition of lung cancer cells A549 and H460 by curcuminoid extracts and nanoemulsions prepared from Curcuma longa Linnaeus. *International Journal of Nanomedicine*, *10*, 5059–5080.

Chang, H. B., & Chen, B. H. (2015). Inhibition of lung cancer cells A549 and H460 by curcuminoid extracts and NEs prepared from Curcuma longa Linnaeus. *International Journal of Nanomedicine*, *10*, 5059. PMID:26345201

Chang, S. T. (1999). Global impact of edible and medicinal mushrooms on human welfare in the 21st century: Nongreen revolution. *International Journal of Medicinal Mushrooms*, *1*(1), 1–7. doi:10.1615/IntJMedMushrooms.v1.i1.10

Chang, Y., McLandsborough, L., & McClements, D. J. (2011). Interactions of a cationic antimicrobial (ε-polylysine) with an anionic biopolymer (pectin): An isothermal titration calorimetry, microelectrophoresis, and turbidity study. *Journal of Agricultural and Food Chemistry*, *59*(10), 5579–5588. doi:10.1021/jf104299q PMID:21462961

Chang, Y., McLandsborough, L., & McClements, D. J. (2011). Physicochemical properties and antimicrobial efficacy of electrostatic complexes based on cationic ε-polylysine and anionic pectin. *Journal of Agricultural and Food Chemistry*, *59*(12), 6776–6782. doi:10.1021/jf201162g PMID:21574554

Chang, Y., McLandsborough, L., & McClements, D. J. (2012). Cationic antimicrobial (ε-polylysine)-anionic polysaccharide (pectin) interactions: Influence of polymer charge on physical stability and antimicrobial efficacy. *Journal of Agricultural and Food Chemistry*, *60*(7), 1837–1844. doi:10.1021/jf204384s PMID:22268773

Chang, Y., McLandsborough, L., & McClements, D. J. (2012). Physical properties and antimicrobial efficacy of thyme oil nanoemulsions: Influence of ripening inhibitors. *Journal of Agricultural and Food Chemistry*, *60*(48), 12056–12063. doi:10.1021/jf304045a PMID:23140446

Chang, Y., McLandsborough, L., & McClements, D. J. (2013). Physicochemical properties and antimicrobial efficacy of carvacrol nanoemulsions formed by spontaneous emulsification. *Journal of Agricultural and Food Chemistry*, *61*(37), 8906–8913. doi:10.1021/jf402147p PMID:23998790

Chang, Y., McLandsborough, L., & McClements, D. J. (2015). Fabrication, stability and efficacy of dual-component antimicrobial nanoemulsions: Essential oil (thyme oil) and cationic surfactant (lauric arginate). *Food Chemistry*, *172*, 298–304. doi:10.1016/j.foodchem.2014.09.081 PMID:25442557

Chang, Y., Zhang, M., Jiang, Y., Liu, Y., Luo, H., Hao, C., Zeng, P., & Zhang, L. (2017). Preclinical and clinical studies of *Coriolus versicolor* polysaccharopeptide as an immunotherapeutic in China. *Discovery Medicine*, *23*, 207–219. PMID:28595034

Chan, P. K. S., Chan, M. Y. M., Li, W. W. H., Chan, D. P. C., Cheung, J. L. K., & Cheng, A. F. (2001). Association of human β-herpesviruses with the development of cervical cancer: Bystanders or cofactors. *Journal of Clinical Pathology*, *54*(1), 48–53. doi:10.1136/jcp.54.1.48 PMID:11271789

Charalambous, A., Mico, V., McVeigh, L. E., Marston, G., Ingram, N., Volpato, M., Peyman, S. A., McLaughlan, J. R., Wierzbicki, A., Loadman, P., Bushby, R. J., Markham, A. F., Evans, S. D., & Coletta, P. L. (2021). Targeted microbubbles carrying lipid-oil-nanodroplets for ultrasound-triggered delivery of the hydrophobic drug, Combretastatin A4. *Nanomedicine; Nanotechnology, Biology, and Medicine*, 102401. Advance online publication.

Charles, L., & Attama, A.A. (2011) Current state of nanoemulsions in drug delivery. *Journal of Biomaterials amd Nanobiotechology, 2*, 626-639.

Chavhan, S. S., Petkar, K. C., & Sawant, K. K. (2013). Simvastatin nanoemulsion for improved oral delivery: Design, characterisation, in vitro and in vivo studies. *Journal of Microencapsulation*, *30*(8), 771–779.

Chellapa, P., Ariffin, F. D., Eid, A. M., Almahgoubi, A. A., Mohamed, A. T., Issa, Y. S., & Elmarzugi, N. A. (2016). Nanoemulsion for cosmetic application. *European Journal of Biomedical and Pharmaceutical Sciences*, *3*(7), 8–11.

Chellaram, C., Murugaboopathi, G., John, A. A., Sivakumar, R., Ganesan, S., Krithika, S., & Priya, G. (2014). Significance of nanotechnology in food industry. *APCBEE Procedia*, *8*, 109–113. doi:10.1016/j.apcbee.2014.03.010

Chen, F. J. (2006). *An antibacterial rifampicin nanoemulsion and preparation method thereof.* https://patents.google.com/patent/CN100528160C/en

Chen, E., Cao, L., McClements, D. J., Liu, S., Li, B., & Li, Y. (2018). Enhancement of physicochemical properties of whey protein-stabilized nanoemulsions by interfacial cross-linking using cinnamaldehyde. *Food Hydrocolloids*, *77*, 976–985. doi:10.1016/j.foodhyd.2017.11.047

Cheng, H., & Olvera de la Cruz, M. (2003). Adsorption of rod-like polyelectrolytes onto weakly charged surfaces. *The Journal of Chemical Physics*, *119*(23), 12635–12644. doi:10.1063/1.1626630

Cheng, L. C., Kuei Vehusheia, S. L., & Doyle, P. S. (2020). Tuning Material Properties of Nanoemulsion Gels by Sequentially Screening Electrostatic Repulsions and Then Thermally Inducing Droplet Bridging. *Langmuir*, *36*(13), 3346–3355. doi:10.1021/acs.langmuir.0c00199 PMID:32216359

Cheng, Y., Liu, M., Hu, H., Liu, D., & Zhou, S. (2015). Development, Optimization, and Characterization of PEGylated Nanoemulsion of Prostaglandin E1 for Long Circulation. *AAPS PharmSciTech*, *17*(2), 409–417. doi:10.120812249-015-0366-1 PMID:26195071

Chen, H., Du, D., Mao, C. M. D., Wan, J., Xu, H., & Yang, X. (2008). Hydrogel-thickened nanoemulsion system for topical delivery of lipophilic drugs. *International Journal of Pharmaceutics*, *353*(1-2), 270–276. doi:10.1016/j.ijpharm.2007.11.051 PMID:18215479

Chen, H., Khemtong, C., Yang, X., Chang, X., & Gao, J. (2011). Nanonization strategies for poorly water-soluble drugs. *Drug Discovery Today*, *16*(7-8), 354–360. doi:10.1016/j.drudis.2010.02.009 PMID:20206289

Chen, J., & Hu, L. (2020). Nanoscale Delivery System for Nutraceuticals: Preparation, Application, Characterization, Safety, and Future Trends. *Food Engineering Reviews*, *12*(1), 14–31. doi:10.100712393-019-09208-w

Chen, J., Li, F., Li, Z., McClements, D. J., & Xiao, H. (2017). Encapsulation of carotenoids in emulsion-based delivery systems: Enhancement of βcarotene water-dispersibility and chemical stability. *Food Hydrocolloids*, *69*, 49–55. doi:10.1016/j.foodhyd.2017.01.024

Chen, L., Chen, B., Deng, L., Gao, B., Zhang, Y., Wu, C., Yu, N., Zhou, Q., Yao, J., & Chen, J. (2017). An optimized two-vial formulation lipid nanoemulsion of paclitaxel for targeted delivery to tumor. *International Journal of Pharmaceutics*, *534*(1-2), 308–315. doi:10.1016/j.ijpharm.2017.10.005 PMID:28986321

Chen, L., Remondetto, G., & Subirade, M. (2006). Food protein-based materials as nutraceutical delivery systems. *Trends in Food Science & Technology*, *17*(5), 272–283. doi:10.1016/j.tifs.2005.12.011

Chen, P. N., Chu, S. C., Kuo, W. H., Chou, M. Y., Lin, J. K., & Hsieh, Y. S. (2011). Epigallocatechin-3 gallate inhibits invasion, epithelial– mesenchymal transition, and tumor growth in oral cancer cells. *Journal of Agricultural and Food Chemistry*, *59*(8), 3836–3844. doi:10.1021/jf1049408 PMID:21388137

Chen, T., Gong, T., Zhao, T., Fu, Y., Zhang, Z., & Gong, T. (2018). A comparison study between lycobetaine-loaded nanoemulsion and liposome using nRGD as therapeutic adjuvant for lung cancer therapy. *European Journal of Pharmaceutical Sciences*, *111*, 293–302. doi:10.1016/j.ejps.2017.09.041 PMID:28966099

Chen, W., KuoLee, R., & Yan, H. (2010). The potential of 3′, 5′-cyclic diguanylic acid (c-di-GMP) as an effective vaccine adjuvant. *Vaccine*, *28*(18), 3080–3085. doi:10.1016/j.vaccine.2010.02.081 PMID:20197136

Chen, W., Patel, G. B., Yan, H., & Zhang, J. (2010). Recent advances in the development of novel mucosal adjuvants and antigen delivery systems. *Human Vaccines*, *6*(9), 706–714. doi:10.4161/hv.6.9.11561 PMID:20861683

Chepurnov, A. A., Bakulina, L. F., Dadaeva, A. A., Ustinova, E. N., Chepurnova, T. S., & Baker, J. R. Jr. (2003). Inactivation of Ebola virus with a surfactant nanoemulsion. *Acta Tropica*, *87*(3), 315–320. doi:10.1016/S0001-706X(03)00120-7 PMID:12875924

Chhabra, G., Chuttani, K., Mishra, A. K., & Pathak, K. (2011). Design and development of nanoemulsion drug delivery system of amlodipine besilate for improvement of oral bioavailability. *Drug Development and Industrial Pharmacy*, *37*(8), 907–916. doi:10.3109/03639045.2010.550050 PMID:21401341

Chhabra, G., Chuttani, K., Mishra, A. K., & Pathak, K. (2011). Design and development of nanoemulsion drug delivery system of amlodipine besylate for improvement of oral bioavailability. *Drug Development and Industrial Pharmacy*, *37*(8), 907–916.

Chibane, E., Essarioui, A., Ouknin, M., Boumezzourh, A., & Bouyanzer, A. (2020). Antifungal activity of Asteriscus graveolens (Forssk) Less essential oil against Fusarium oxysporum f.sp albedinis, the causal agent of "Bayoud" disease on date palm. *Moroccan Journal of Chemistry*, *8*(2), 456–465.

Chime, S., Kenechukwu, F., & Attama, A. (2014). *Nanoemulsions—advances in formulation, characterization and applications in drug delivery* (Vol. 3). Academic Press.

Chime, S. A., Kenechukwu, F. C., & Attama, A. A. (2014). *Nanoemulsions — Advances in Formulation.* In *Characterization and Applications in Drug Delivery, Application of Nanotechnology in Drug Delivery*. IntechOpen.

Chime, S. A., Kenechukwu, F. C., & Attama, A. A. (2014). Nanoemulsions—Advances in formulation, characterization and applications in drug delivery. In A. D. Sezer (Ed.), *Application of Nanotechnology in Drug Delivery* (Vol. 3, pp. 77–126). Intech.

Chime, S. A., Kenechukwu, F. C., & Attama, A. A. (2014). Nanoemulsions-Advances in Formulation. *Characterisation and Applications in Drug Delivery*, *3*, 77–126.

Ching, Y. P., Wong, C. M., Chan, S. F., Leung, T. H. Y., Ng, D. C. H., Jin, D. Y., & Ng, I. O. L. (2003). Deleted liver cancer (DLC) 2 encodes a RhoGAP protein with growth suppressor function and is underexpressed in hepatocellular carcinoma. *The Journal of Biological Chemistry*, *278*(12), 10824–10830. doi:10.1074/jbc.M208310200 PMID:12531887

Chircov, C., & Grumezescu, A. M. (2019). Nanoemulsion preparation, characterization, and application in the field of biomedicine. In *Nanoarchitectonics in biomedicine* (pp. 169–188). William Andrew Publishing. doi:10.1016/B978-0-12-816200-2.00019-0

Chircov, C., & Grumezescu, A. M. (2019). *Nanoemulsion preparation, characterization, and application in the field of biomedicine. In Nanoarchitectonics in biomedicine*. Elsevier.

Choi, A. J., Kim, C. J., Cho, Y. J., Hwang, J. K., & Kim, C. T. (2011). Characterization of capsaicin-loaded nanoemulsions stabilized with alginate and chitosan by self-assembly. *Food and Bioprocess Technology*, *4*(6), 1119–1126. doi:10.100711947-011-0568-9

Choi, S. J., & McClements, D. J. (2020). Nanoemulsions as delivery systems for lipophilic nutraceuticals: Strategies for improving their formulation, stability, functionality and bioavailability. *Food Science and Biotechnology*, *29*(2), 149–168. doi:10.100710068-019-00731-4 PMID:32064124

Cholley, P., Ka, R., Guyeux, C., Thouverez, M., Guessennd, N., Ghebremedhin, B., & Hocquet, D. (2014). Population structure of clinical Pseudomonas aeruginosa from West and Central African countries. *PLoS One*, *9*(9), e107008. doi:10.1371/journal.pone.0107008

Cholo, M. C., Steel, H. C., Fourie, P. B., Germishuizen, W. A., & Anderson, R. (2012). Clofazimine: Current status and future prospects. *The Journal of Antimicrobial Chemotherapy*, *67*(2), 290–298. doi:10.1093/jac/dkr444 PMID:22020137

Chong, P. S., Fung, M., Wong, K. H., & Lim, L. W. (2020). Therapeutic potential of *Hericium erinaceus* for depressive disorders. *International Journal of Molecular Sciences*, *2*, 163.

Choradiya, B., & Patil, S. (2021). A comprehensive review on nanoemulsion as an ophthalmic drug delivery system. *Journal of Molecular Liquids*, *339*, 116751. doi:10.1016/j.molliq.2021.116751

Choudhari, A. S., Mandave, P. C., Deshpande, M., Ranjekar, P., & Prakash, O. (2020). Phytochemicals in cancer treatment: From preclinical studies to clinical practice. *Frontiers in Pharmacology*, *10*, 1614. doi:10.3389/fphar.2019.01614 PMID:32116665

Choudhary, D., Goykar, H., Karanwad, T., Kannaujia, S., Gadekar, V. & Mishra, M. (2020). An understanding of mitochondria and its role in targeting nanocarriers for diagnosis and treatment of cancer. *Asian Journal of Pharmaceutical Sciences*. . doi:10.1016/j.ajps.2020.10.002

Choudhury, H., Gorain, B., Karmakar, S., Biswas, E., Dey, G., Barik, R., & Pal, T. K. (2014). Improvement of cellular uptake, in vitro antitumor activity and sustained release profile with increased bioavailability from a nanoemulsion platform. *International Journal of Pharmaceutics*, *460*(1-2), 131–143.

Choudhury, H., Pandey, M., Grain, B., Chatterjee, B., Madheswaran, T., Md, S., Mak, K.-K., Tambuwala, M., Chourasia, M. K., & Kesharwani, P. (2019). Nanoemulsions as Effective Carriers for the Treatment of Lung Cancer. In *Nanotechnology-Based Targeted Drug Delivery Systems for Lung Cancer*. Elsevier Inc. doi:10.1016/B978-0-12-815720-6.00009-5

Chouhan, S., Sharma, K., & Guleria, S. (2017). Antimicrobial activity of some essential oils present status and future perspectives. *Medicines (Basel, Switzerland)*, *4*(3), 58. doi:10.3390/medicines4030058

Chowdhury, M. M. H., Kubra, K., & Ahmed, S. R. (2015). Screening of antimicrobial, antioxidant properties and bioactive compounds of some edible mushrooms cultivated in Bangladesh. *Annals of Clinical Microbiology and Antimicrobials*, *14*(1), 2–6. doi:10.118612941-015-0067-3 PMID:25858107

Chrastina, A., Baron, V. T., Abedinpour, P., Rondeau, G., Welsh, J., & Borgström, P. (2018). Plumbagin-Loaded Nanoemulsion Drug Delivery Formulation and Evaluation of Antiproliferative Effect on Prostate Cancer Cells. *BioMed Research International*.

Chuan, L. I., Zhang, J., Yu-Jiao, Z. U., Shu-Fang, N. I. E., Jun, C. A. O., Qian, W. A. N. G., ... Shu, W. A. N. G. (2015). Biocompatible and biodegradable nanoparticles for enhancement of anti-cancer activities of phytochemicals. *Chinese Journal of Natural Medicines*, *13*(9), 641–652. doi:10.1016/S1875-5364(15)30061-3 PMID:26412423

Chu, B.-S., Ichikawa, S., Kanafusa, S., & Nakajima, M. (2007). Preparation and characterization of b-carotene nanodispersions prepared by solvent displacement technique. *Journal of Agricultural and Food Chemistry*, *55*(16), 6754–6760. doi:10.1021/jf063609d PMID:17630759

Chuesiang, P., Siripatrawan, U., Sanguandeekul, R., McLandsborough, L., & McClements, D. J. (2018). Optimization of cinnamon oil nanoemulsions using phase inversion temperature method: Impact of oil phase composition and surfactant concentration. *Journal of Colloid and Interface Science*, *514*, 208–216. doi:10.1016/j.jcis.2017.11.084 PMID:29257975

Çinar, K. (2017). A review on nanoemulsions: Preparation methods and stability. *Trakya University Journal of Engineering Sciences*, *18*(1), 73–83.

Cinausero, M., Aprile, G., Ermacora, P., Basile, D., Vitale, M. G., Fanotto, V., Parisi, G., Calvetti, L., & Sonis, S. T. (2017). New frontiers in the pathobiology and treatment of cancer regimen-related mucosal injury. *Frontiers in Pharmacology*, *8*, 354. doi:10.3389/fphar.2017.00354 PMID:28642709

Clemens, D. L., Lee, B. Y., Xue, M., Thomas, C. R., Meng, H., Ferris, D., Nel, A. E., Zink, J. I., & Horwitz, M. A. (2012). Targeted intracellular delivery of antituberculosis drugs to *Mycobacterium tuberculosis*-infected macrophages via functionalized mesoporous silica nanoparticles. *Antimicrobial Agents and Chemotherapy*, *56*(5), 2535–2545. doi:10.1128/AAC.06049-11 PMID:22354311

Codjoe, F. S., & Donkor, E. S. (2018). Carbapenem resistance: A review. *Medical Science*, *6*(1), 1. doi:10.3390/medsci6010001

Cody, R. J. (1997). The sympathetic nervous system and the renin–angiotensin–aldosterone system in cardiovascular disease. *The American Journal of Cardiology*, *80*(9), 9J–14J. doi:10.1016/S0002-9149(97)00832-1 PMID:9375943

Cohen Stuart, M. A., Waajen, F. H. W. H., & Dukhin, S. S. (1984). Electrokinetic effects of adsorbed neutral polymers. *Colloid & Polymer Science*, *262*(5), 423–426. doi:10.1007/BF01410263

Comfort, C., Garrastazu, G., Pozzoli, M., & Sonvico, F. (2015). Opportunities and challenges for the nasaladministration of nanoemulsions. *Current Topics in Medicinal Chemistry*, *15*(4), 356–368. doi:10.2174/156802661566615010814 4655 PMID:25579345

Comstock, G. W., Livesay, V. T., & Woolpert, S. F. (1974). The prognosis of a positive tuberculin reaction in childhood and adolescence. *American Journal of Epidemiology*, *99*(2), 131–138. doi:10.1093/oxfordjournals.aje.a121593 PMID:4810628

Cone, R. A. (2009). Barrier properties of mucus. *Advanced Drug Delivery Reviews*, *61*(2), 75–85.

Consortium, T. H. M. P. (2012). Structure, Function and Diversity of the Healthy Human Microbiome. *Nature*, *486*, 207–214.

Constantinides, P. P., Chaubal, M. V., & Shorr, R. (2008). Advances in lipid nanodispersions for parenteral drug delivery and targeting. *Advanced Drug Delivery Reviews*, *60*(6), 757–767. doi:10.1016/j.addr.2007.10.013 PMID:18096269

Constantinides, P. P., Lambert, K. J., Tustian, A. K., Schneider, B., Lalji, S., Ma, W., & Quay, S. C. (2000).. . *Pharmaceutical Research*, *17*(2), 175–182. doi:10.1023/A:1007565230130 PMID:10751032

Constantinides, P. P., Tustian, A., & Kessler, D. R. (2004). Tocol emulsions for drug solubilisation and parenteral delivery. *Advanced Drug Delivery Reviews*, *56*(9), 1243–1255.

Cooper, J., & Dobson, H. (2007). The benefits of pesticides to mankind and the environment. *Crop Protection (Guildford, Surrey)*, *26*(9), 1337–1348. doi:10.1016/j.cropro.2007.03.022

Cosentino, S., Tuberoso, C. I., Pisano, B., Satta, M., Mascia, V., Arzedi, E., & Palmas, F. (1999). In-vitro antimicrobial activity and chemical composition of Sardinian Thymus essential oils. *Letters in Applied Microbiology*, *29*(2), 130–135. doi:10.1046/j.1472-765X.1999.00605.x PMID:10499301

Costa, C., Medronho, B., Filipe, A., Mira, I., Lindman, B., Edlund, H., & Norgren, M. (2019). Emulsion Formation and Stabilization by Biomolecules: The Leading Role of Cellulose. *Polymers*, *11*(10), 1570. doi:10.3390/polym11101570 PMID:31561633

Cox, E., Verdonck, F., Vanrompay, D., & Goddeeris, B. (2006). Adjuvants modulating mucosal immune responses or directing systemic responses towards the mucosa. *Veterinary Research*, *37*(3), 511–539. doi:10.1051/vetres:2006014 PMID:16611561

Cruz-Valenzuela, M. R., Tapia-Rodríguez, M. R., Vazquez-Armenta, F. J., Silva-Espinoza, B. A., & Ayala-Zavala, J. F. (2017). Lime (Citrus aurantifolia) Oils. *Essential Oils in Food Preservation. Flavor and Safety*, *61*(3), 532–537. doi:10.1016/B978-0-12-416641-7.00061-4

Csaba, N., Garcia-Fuentes, M., & Alonso, M. J. (2009). Nanoparticles for nasal vaccination. *Advanced Drug Delivery Reviews*, *61*(2), 140–157. doi:10.1016/j.addr.2008.09.005 PMID:19121350

Cui, F., Zan, X., Li, Y., Sun, W., Yang, Y., & Ping, L. (2016). *Grifola frondosa* glycoprotein GFG-3a arrests S phase, alters proteome, and induces apoptosis in human gastric cancer cells. *Nutrition and Cancer*, *68*(2), 267–279. doi:10.1080/01635581.2016.1134599 PMID:27040446

Cui, H., Bai, M., Rashed, M., & Lin, L. (2018). The antibacterial activity of clove oil/chitosan nanoparticles embedded gelatin nanofibers against *Escherichia coli* O157:H7 biofilms on cucumber. *International Journal of Food Microbiology*, *266*, 69–78. doi:10.1016/j.ijfoodmicro.2017.11.019 PMID:29179098

Czarnota, A., Offersgaard, A., Pihl, A. F., Prentoe, J., Bukh, J., Gottwein, J. M., Bieńkowska-Szewczyk, K., & Grzyb, K. (2020). Specific Antibodies Induced by Immunization with Hepatitis B Virus-Like Particles Carrying Hepatitis C Virus Envelope Glycoprotein 2 Epitopes Show Differential Neutralization Efficiency. *Vaccines*, *8*(2), 294. doi:10.3390/vaccines8020294 PMID:32532076

D'Ambrosio, L., Centis, R., Sotgiu, G., Pontali, E., Spanevello, A., & Migliori, G. B. (2015). New anti-tuberculosis drugs and regimens: 2015 update. *ERJ Open Research*, *1*(1), 00010–02015. doi:10.1183/23120541.00010-2015 PMID:27730131

D'Archivio, M., Filesi, C., Varì, R., Scazzocchio, B., & Masella, R. (2010). Bioavailability of the polyphenols: Status and controversies. *International Journal of Molecular Sciences*, *11*(4), 1321–1342. doi:10.3390/ijms11041321 PMID:20480022

Da Silva Gündel, S., dos Reis, T. R., Copetti, P. M., Favarin, F. R., Sagrillo, M. R., da Silva, A. S., Segat, J. C., Baretta, D., & Ourique, A. F. (2019). Evaluation of cytotoxicity, genotoxicity and ecotoxicity of nanoemulsions containing Mancozeb and Eugenol. *Ecotoxicology and Environmental Safety*, *169*, 207–215. doi:10.1016/j.ecoenv.2018.11.023 PMID:30448703

Damjanović Vratnica, B., Djakov, T., Sukovic, D., & Damjanovic, J. (2011). Antimicrobial Effect of Essential Oil Isolated from Eucalyptus globulus Labill. from Montenegro. *Czech Journal of Food Sciences*, *29*(3), 277–284. doi:10.17221/114/2009-CJFS

Dammak, I., Sobral, P. J. A., Aquino, A., Neves, M. A., & Conte-Junior, C. A. (2020). Nanoemulsions: Using emulsifiers from natural sources replacing synthetic ones—A review. *Comprehensive Reviews in Food Science and Food Safety*, *19*(5), 2721–2746. doi:10.1111/1541-4337.12606 PMID:33336986

Danaraddi, S., Koneru, A., Hunasgi, S., Ramalu, S., & Vanishree, M. (2014). Natural ways to prevent and treat oral cancer. *Journal of Oral Research and Review*, *6*(1), 34. doi:10.4103/2249-4987.140213

Danhier, F., Feron, O., & Préat, V. (2010). To exploit the tumor microenvironment: Passive and active tumor targeting of nanocarriers for anti-cancer drug delivery. *Journal of Controlled Release*, *148*(2), 135–146. doi:10.1016/j.jconrel.2010.08.027 PMID:20797419

Danilov, L. L., Maltsev, S. D., Deyeva, A. V., Narovlyansky, A. N., Sanin, A. V., Ozherelkov, S. V., & Pronin, A. V. (1996). Phosprenyl: A novel drug with antiviral and immunomodulatory activity. *Archivum Immunologiae et Therapiae Experimentalis*, *44*(5-6), 395–400. PMID:9017158

Darvesh, Aggarwal, & Bishayee. (2011). Curcumin and Liver Cancer: A Review. *Current Pharmaceutical Biotechnology*, *13*(1), 218–228. doi:10.2174/138920112798868791

Das, A. K., Nanda, P. K., Bandyopadhyay, S., Banerjee, R., Biswas, S., & McClements, D. J. (2020). Application of nanoemulsion-based approaches for improving the quality and safety of muscle foods: A comprehensive review. *Comprehensive Reviews in Food Science and Food Safety*, *19*(5), 2677–2700. doi:10.1111/1541-4337.12604 PMID:33336977

Dasgupta, N., & Ranjan, S. (2018). Food Nanoemulsions: Stability, Benefits and Applications. *An Introduction to Food Grade Nanoemulsions*, 19-48.

Dasgupta, N., Ranjan, S., & Gandhi, M. (2019). Nanoemulsion ingredients and components. *Environmental Chemistry Letters*, *17*(2), 917–928. doi:10.100710311-018-00849-7

Dasgupta, N., Ranjan, S., & Gandhi, M. (2019). Nanoemulsions in food: Market demand. *Environmental Chemistry Letters*, *17*(2), 1003–1009. Advance online publication. doi:10.100710311-019-00856-2

Dash, M., Chiellini, F., Ottenbrite, E., & Chiellini, E. (2011). Chitosan-A Versatile Semi-Synthetic Polymer in Biomedical Applications. *Progress in Polymer Science*, *36*(8), 981–1014. doi:10.1016/j.progpolymsci.2011.02.001

Das, S. C., Hatta, M., Wilker, P. R., Myc, A., Hamouda, T., Neumann, G., & Kawaoka, Y. (2012). Nanoemulsion W805EC improves immune responses upon intranasal delivery of an inactivated pandemic H1N1 influenza vaccine. *Vaccine*, *30*(48), 6871–6877. doi:10.1016/j.vaccine.2012.09.007 PMID:22989689

Das, S., Sarmah, S., Lyndem, S., & Singha Roy, A. (2020). An investigation into the identification of potential inhibitors of SARS-CoV-2 main protease using molecular docking study. *Journal of Biomolecular Structure & Dynamics*, 1–11. doi:10.1080/07391102.2020.1763201 PMID:32362245

Date, A. A., Dixit, R., & Nagarsenker, M. (2010). *Self-nanoemulsifying drug delivery systems : formulation insights, applications and advances review.* Academic Press.

Davidov-Pardo, G., & McClements, D. J. (2015). Nutraceutical delivery systems: Resveratrol encapsulation in grape seed oil nanoemulsions formed by spontaneous emulsification. *Food Chemistry*, *167*, 205–212. doi:10.1016/j.foodchem.2014.06.082 PMID:25148980

Davidson, P. M., Critzer, F. J., & Taylor, T. M. (2013). Naturally occurring antimicrobials for minimally processed foods. *Annual Review of Food Science and Technology*, *4*(1), 163–190. doi:10.1146/annurev-food-030212-182535 PMID:23244398

Davies, J. T. (1957). A quantitative kinetic theory of emulsion type. I. Physical chemistry of the emulsifying agent. In *Gas/Liquid and Liquid/Liquid Interface. Proceedings of the International Congress of Surface Activity* (*Vol. 1*, pp. 426-438). Academic Press.

Davin-Regli, A. (2015). Enterobacteraerogenes and Enterobacter cloacae; versatile bacterial pathogens confronting antibiotic treatment. *Frontiers in Microbiology*, *6*, 392. doi:10.3389/fmicb.2015.00392

Davis, T., & Farag, S. S. (2013). Treating relapsed or refractory Philadelphia chromosome-negative acute lymphoblastic leukemia: Liposome-encapsulated vincristine. *International Journal of Nanomedicine*, *8*, 3479–3488. doi:10.2147/IJN.S47037 PMID:24072970

De Aguiar, H. B., De Beer, A. G. F., Strader, M. L., & Roke, S. (2010). The interfacial tension of nanoscopic oil droplets in water is hardly affected by SDS surfactant. *Journal of the American Chemical Society*, *132*(7), 2122–2123. doi:10.1021/ja9095158 PMID:20121125

De Gregorio, E., & Rappuoli, R. (2014). From empiricism to rational design: A personal perspective of the evolution of vaccine development. *Nature Reviews. Immunology*, *14*(7), 505–514. doi:10.1038/nri3694 PMID:24925139

De Matos, R. P. A., Calmon, M. F., Amantino, C. F., Villa, L. L., Primo, F. L., Tedesco, A. C., & Rahal, P. (2018). Effect of curcumin-nanoemulsion associated with photodynamic therapy in cervical carcinoma cell lines. *BioMed Research International*, *2018*, 2018. doi:10.1155/2018/4057959 PMID:29581972

de Matos, R., Calmon, M. F., Amantino, C. F., Villa, L. L., Primo, F. L., Tedesco, A. C., & Rahal, P. (2018). Effect of Curcumin-Nanoemulsion Associated with Photodynamic Therapy in Cervical Carcinoma Cell Lines. *BioMed Research International*, *2018*, 4057959.

de Meneses, A. C., Sayer, C., Puton, B. M. S., Cansian, R. L., Araújo, P. H. H., & de Oliveira, D. (2019). Production of clove oil nanoemulsion with rapid and enhanced antimicrobial activity against gram-positive and gram-negative bacteria. *Journal of Food Process Engineering*, *6*(42),. doi:10.1111/jfpe.13209

de Menezes Furtado, C., de Faria, F., Azevedo, R., Py-Daniel, K., dos Santos Camara, A., da Silva, J., de Holanda Oliveira, E., Rodriguez, A. F. R., & Degterev, I. A. (2017). Tectona grandis leaf extract, free and associated with nanoemulsions, as a possible photosensitizer of mouse melanoma B16 cell. *Journal of Photochemistry and Photobiology. B, Biology*, *167*, 242–248. doi:10.1016/j.jphotobiol.2017.01.004 PMID:28088105

De Oliveira, D. M., Forde, B. M., Kidd, T. J., Harris, P. N., Schembri, M. A., Beatson, S. A., & Walker, M. J. (2020). Antimicrobial resistance in ESKAPE pathogens. *Clinical Microbiology Reviews*, *33*(3), e00181–e19. doi:10.1128/CMR.00181-19

de Souza, E. L., de Barros, J. C., de Oliveira, C. E., & da Conceição, M. L. (2010). Influence of Origanum vulgare L. essential oil on enterotoxin production, membrane permeability and surface characteristics of Staphylococcus aureus. *International Journal of Food Microbiology*, *137*(2-3), 308–311. doi:10.1016/j.ijfoodmicro.2009.11.025 PMID:20015563

de Vos, P., Faas, M. M., Spasojevic, M., & Sikkema, J. (2010). Encapsulation for preservation of functionality and targeted delivery of bioactive food components. *International Dairy Journal*, *20*(4), 292–302. doi:10.1016/j.idairyj.2009.11.008

Dean, M. (2009). ABC transporters, drug resistance, and cancer stem cells. *Journal of Mammary Gland Biology and Neoplasia*, *14*(1), 3–9.

Debnath, S., And, S., & Kumar, G.V., (2010), Nanoemulsion-a method to improve the solubility of lipophilic drugs. *International Journal of Advances in Pharmaceutical Sciences, 2*(2-3), 72-83.

Debnath, S., Satayanarayana, K. V., & Kumar, G. V. (2011). Nanoemulsion—A method to improve the solubility of lipophilic drugs. *Pharmanest*, *2*(2-3), 72–83.

Decher, G., Hong, J. D., & Schmitt, J. (1992). Buildup of ultrathin multilayer films by a self-assembly process: III. Consecutively alternating adsorption of anionic and cationic polyelectrolytes on charged surfaces. *Thin Solid Films*, *210/211*, 831–835. doi:10.1016/0040-6090(92)90417-A

Deeksha, K., & Hari, S. (2019). Synthesis, Characterization and Antibacterial effect of Neem Oil Nanoemulsion. *Research Journal of Pharmacy and Technology*, *12*(9), 4400–4404. doi:10.5958/0974-360X.2019.00757.1

Dehelean, C. A., Feflea, S., Gheorgheosu, D., Ganta, S., Cimpean, A. M., Muntean, D., & Amiji, M. M. (2013). Anti-angiogenic and anti-cancer evaluation of betulin nanoemulsion in chicken chorioallantoic membrane and skin carcinoma in Balb/c mice. *Journal of Biomedical Nanotechnology*, *9*(4), 577–589.

Del Blanco, L. F., Rodriguez, M. S., Schulz, P. C., & Agguló, E. (1999). Influence of the deacetylation degree on chitosan emulsification properties. *Colloid & Polymer Science*, *277*(11), 1087–1092. doi:10.1007003960050495

Delacroix, S., Chokka, R. G., & Worthley, S. G. (2014). Hypertension: Pathophysiology and treatment. *Journal of Neurology & Neurophysiology*, *5*(6), 1–8. doi:10.4172/2155-9562.1000250

Delgado, A. V., Gonzalez-Caballero, F., Hunter, R. J., Koopal, L. K., & Lyklema, J. (2007). Measurement and interpretation of electrokinetic phenomena. *Journal of Colloid and Interface Science*, *309*(2), 194–224. doi:10.1016/j.jcis.2006.12.075 PMID:17368660

Demisli, S., Mitsou, E., Pletsa, V., Xenakis, A., & Papadimitriou, V. (2020). Development and Study of Nanoemulsions and Nanoemulsion-Based Hydrogels for the Encapsulation of Lipophilic Compounds. *Nanomaterials (Basel, Switzerland)*, *10*(12), 2464. doi:10.3390/nano10122464 PMID:33317080

Deng, S., Zhang, G., Kuai, J., Fan, P., Wang, X., Zhou, P., Yang, D., Zheng, X., Liu, X., Wu, Q., & Huang, Y. (2018). Lentinan inhibits tumor angiogenesis via interferon and in a T cell independent manner. *Journal of Experimental & Clinical Cancer Research*, *37*(1), 270. doi:10.118613046-018-0932-y PMID:30373628

Derkach, S. R. (2009). Rheology of emulsions. *Advances in Colloid and Interface Science*, *151*(1-2), 1-23. doi:org/10.1016/j.cis.2009.07.001

Desai, D., Vaka, S., Shah, N., Jain, A., & Phuapradit, W. (2014). *Self-nanoemulsion of poorly soluble drugs*. Academic Press.

Desai, A., Vyas, T., & Amiji, M. (2008). Cytotoxicity and apoptosis enhancement in brain tumor cells upon coadministration of paclitaxel and ceramide in nanoemulsion formulations. *Journal of Pharmaceutical Sciences*, *97*(7), 2745–2756. doi:10.1002/jps.21182 PMID:17854074

Desbois, A. P., & Smith, V. J. (2010). Antibacterial free fatty acids: Activities, mechanisms of action and biotechnological potential. *Applied Microbiology and Biotechnology*, *85*(6), 1629–1642. doi:10.100700253-009-2355-3

Deshpande, D., Kethireddy, S., Gattacceca, F., & Amiji, M. (2014). Comparative pharmacokinetics and tissue distribution analysis of systemically administered 17-β-estradiol and its metabolites in vivo delivered using a cationic nanoemulsion or a peptide-modified nanoemulsion system for targeting atherosclerosis. *Journal of Controlled Release*, *180*, 117–124. doi:10.1016/j.jconrel.2014.02.009 PMID:24556419

Devalapally, H., Silchenko, S., Zhou, F., McDade, J., Goloverda, G., Owen, A., & Hidalgo, I. J. (2013). Evaluation of a Nanoemulsion Formulation Strategy for Oral Bioavailability Enhancement of Danazol in Rats and Dogs. *Journal of Pharmaceutical Sciences*, *102*(10), 3808–3815. doi:10.1002/jps.23667 PMID:23878097

Dey, T. K., Ghosh, S., Ghosh, M., Koley, H., & Dhar, P. (2012). Comparative study of gastrointestinal absorption of EPA & DHA rich fish oil fromnano and conventional emulsion formulation in rats. *Food Research International*, *49*(1), 72–79. doi:10.1016/j.foodres.2012.07.056

Dey, T. K., Koley, H., Ghosh, M., Dey, S., & Dhar, P. (2019). Effects of nano-sizing on lipid bioaccessibility and ex vivo bioavailability from EPA-DHA rich oil in water nanoemulsion. *Food Chemistry*, *275*, 135–142. doi:10.1016/j.foodchem.2018.09.084 PMID:30724179

Dhankhar, R., Vyas, S. P., Jain, A. K., Arora, S., Rath, G., & Goyal, A. K. (2010). Advances in novel drug delivery strategies for breast cancer therapy. *Artificial Cells, Blood Substitutes, and Immobilization Biotechnology*, *38*(5), 230–249.

Dhivya, V., Nelson, S. J., Subramanian, K. S., Edward, Y. S. J. T., Rajamani, K., Santhanakrishnan, V. P., & Sithanantham, S. (2019). Formulation of sweet flag oil (Acorus calamus) nanoemulsion by spontaneous emulsification method for the management of Sitophilus oryzae. *International Journal of Chemical Studies*, *7*(3), 2072–2076.

Di Pasqua, R., Betts, G., Hoskins, N., Edwards, M., Ercolini, D., & Mauriello, G. (2007). Membrane toxicity of antimicrobial compounds from essential oils. *Journal of Agricultural and Food Chemistry*, *55*(12), 4863–4870. doi:10.1021/jf0636465 PMID:17497876

Díaz-Blancas, V., Medina, D. I., Padilla-Ortega, E., Bortolini-Zavala, R., Olvera-Romero, M., & Luna-Bárcenas, G. (2016). Nanoemulsion formulations of fungicide tebuconazole for agricultural applications. *Molecules (Basel, Switzerland)*, *21*(10), 1271. doi:10.3390/molecules21101271 PMID:27681721

Dickinson, E. (1992). *Introduction to food colloids*. Oxford University Press.

Dickinson, E., Semenova, M. G., Antipova, A. S., & Pelan, E. (1998). Effect of high-methoxy pectin on properties of casein-stabilized emulsions. *Food Hydrocolloids*, *12*(4), 425–432. doi:10.1016/S0268-005X(98)00057-5

Diling, C., Chaoqun, Z., Jian, Y., Jian, L., Jiyan, S., Yizhen, X., & Guoxiao, L. (2017). Immunomodulatory activities of a fungal protein extracted from *Hericium erinaceus* through regulating the gut microbiota. *Frontiers in Immunology*, *8*, 666. doi:10.3389/fimmu.2017.00666 PMID:28713364

Dillen, K., Bridts, C., Van der Veken, P., Cos, P., Vandervoort, J., Augustyns, K., Stevens, W., & Ludwig, A. (2008). Adhesion of PLGA or Eudragit/PLGA nanoparticles to Staphylococcus and Pseudomonas. *International Journal of Pharmaceutics*, *349*(1-2), 234–240. doi:10.1016/j.ijpharm.2007.07.041

Dima, C., & Dima, S. (2015). Essential oils in foods: Extraction, stabilization, and toxicity. *Current Opinion in Food Science*, *5*, 29–35. doi:10.1016/j.cofs.2015.07.003

Dinache, A., Tozar, T., Smarandache, A., Andrei, I. R., Nistorescu, S., Nastasa, V., & Romanitan, M. O. (2020). Spectroscopic characterization of emulsions generated with a new laser-assisted device. *Molecules (Basel, Switzerland)*, *25*(7), 1729. doi:10.3390/molecules25071729 PMID:32283754

Ding, M., Liu, L., Zhang, T., Tao, N., Wang, X., & Zhong, J. (2021). Effect of interfacial layer number on the storage stability and in vitro digestion of fish oil-loaded multilayer emulsions consisting of gelatin particle and polysaccharides. *Food Chemistry*, *336*, 127686. doi:10.1016/j.foodchem.2020.127686 PMID:32763735

Donalisio, M., Leone, F., Civra, A., Spagnolo, R., Ozer, O., Lembo, D., & Cavalli, R. (2018). Acyclovir-loaded chitosan nanospheres from nano-emulsion templating for the topical treatment of herpesviruses infections. *Pharmaceutics*, *10*(2), 46. doi:10.3390/pharmaceutics10020046 PMID:29642603

DongJin, J., SungTae, K., HwaMi, L., BaeChan, K., & ChongKook, K. (2005). Improvement of bioavailability and photostability of amlodipine using redispersible dry emulsion. 추계총회 및 학술대회, 511-511.

Donovan, B. W., Reuter, J. D., Cao, Z., Myc, A., Johnson, K. J., & Baker, J. R. Jr. (2000). Prevention of murine influenza A virus pneumonitis by surfactant nano-emulsions. *Antiviral Chemistry & Chemotherapy*, *11*(1), 41–49. doi:10.1177/095632020001100104 PMID:10693653

Donsì, F. (2018). Applications of nanoemulsions in foods. In *Nanoemulsions* (pp. 349–377). Elsevier. doi:10.1016/B978-0-12-811838-2.00011-4

Donsì, F., Annunziata, M., Sessa, M., & Ferrari, G. (2011). Nanoencapsulation of essential oils to enhance their antimicrobial activity in foods. *Lebensmittel-Wissenschaft + Technologie*, *44*(9), 1908–1914. doi:10.1016/j.lwt.2011.03.003

Donsì, F., Annunziata, M., Vincensi, M., & Ferrari, G. (2012). Design of nanoemulsion-based delivery systems of natural antimicrobials: Effect of the emulsifier. *Journal of Biotechnology*, *159*(4), 342–350. doi:10.1016/j.jbiotec.2011.07.001 PMID:21763730

Donsì, F., & Ferrari, G. (2016). Essential oil nanoemulsions as antimicrobial agents in food. *Journal of Biotechnology*, *233*, 106–120. doi:10.1016/j.jbiotec.2016.07.005 PMID:27416793

Donsì, F., Sessa, M., & Ferrari, G. (2012). Effect of Emulsifier Type and Disruption Chamber Geometry on the Fabrication of Food Nanoemulsions by High Pressure Homogenization. *Industrial & Engineering Chemistry Research*, *51*(22), 7606–7618. doi:10.1021/ie2017898

Donsì, F., Wang, Y., & Huang, Q. (2011). Freeze–thaw stability of lecithin and modified starch-based nanoemulsions. *Food Hydrocolloids*, *25*(5), 1327–1336. doi:10.1016/j.foodhyd.2010.12.008

Donsi, F., Wang, Y., Li, J., & Huang, Q. (2010). Preparation of curcumin sub-micrometer dispersions by high-pressure homogenization. *Journal of Agricultural and Food Chemistry*, *58*(5), 2848–2853. doi:10.1021/jf903968x PMID:20146520

Dordevic, S. M., Radulovic, T. S., Cekic, N. D., Randelovic, D. V., Savic, M. M., Krajisnik, D. R., & Savic, S. D. (2013). Experimental design in formulation of diazepam nanoemulsions: Physicochemical and pharmacokinetic performances. *Journal of Pharmaceutical Sciences*, *102*(11), 4159–4172.

Dorman, H. J., & Deans, S. G. (2000). Antimicrobial agents from plants: Antibacterial activity of plant volatile oils. *Journal of Applied Microbiology*, *88*(2), 308–316. doi:10.1046/j.1365-2672.2000.00969.x PMID:10736000

Drew, R. H. (2020). Isoniazid: An overview. *UpToDate*. https://www.uptodate.com/contents/isoniazid-an-overview

du Toit, L. C., Pillay, V., & Danckwerts, M. P. (2006). Tuberculosis chemotherapy: Current drug delivery approaches. *Respiratory Research*, *7*(1), 118–125. doi:10.1186/1465-9921-7-118 PMID:16984627

Duarte, J. L., Amado, J. R. R., Oliveira, A. E. M. F. M., Cruz, R. A. S., Ferreira, A. M., Souto, R. N. P., Falcão, D. Q., Carvalho, J. C. T., & Fernandesa, C. P. (2015). Evaluation of larvicidal activity of a nanoemulsion of rosmarinus officinalis essential oil. *Brazilian Journal of Pharmacognosy*, *25*(2), 189–192. doi:10.1016/j.bjp.2015.02.010

Du, B., Bian, Z., & Xu, B. (2014). Skin Health Promotion Effects of Natural Beta-Glucan Derived from Cereals and Microorganisms: A Review. *Phytotherapy Research*, *28*(2), 159–166. doi:10.1002/ptr.4963 PMID:23494974

Duffus, L. J., Norton, J. E., Smith, P., Norton, I. T., & Spyropoulos, F. (2016). A comparative study on the capacity of a range of food-grade particles to form stable O/W and W/O Pickering emulsions. *Journal of Colloid and Interface Science*, *473*, 9–21. doi:10.1016/j.jcis.2016.03.060 PMID:27042820

Du, R., Wang, X., Ma, L., Larcher, L. M., Tang, H., Zhou, H., Chen, C., & Wang, T. (2021). Adverse reactions of targeted therapy in cancer patients: A retrospective study of hospital medical data in China. *BMC Cancer*, *21*(1), 1–13. doi:10.118612885-021-07946-x PMID:33639888

Durai, R. D. (2015). Drug delivery approaches of an antiviral drug: A comprehensive review. *Asian Journal of Pharmaceutics, 9*(1), 1-12.

Duru, M. E., & Cayan, G. T. (2015). Biologically active terpenoids from mushroom origin:a review. *Records of Natural Products*, *9*, 456–483.

Duskey, J. T., & Rice, K. G. (2014). Nanoparticle ligand presentation for targeting solid tumors. *AAPS PharmSciTech*, *15*(5), 1345–1354. doi:10.120812249-014-0143-6 PMID:24927668

Duthie, M. S., Windish, H. P., Fox, C. B., & Reed, S. G. (2011). Use of defined TLR ligands as adjuvants within human vaccines. *Immunological Reviews*, *239*(1), 178–196. doi:10.1111/j.1600-065X.2010.00978.x PMID:21198672

Dutt, M., & Khuller, G. K. (2001). Chemotherapy of *Mycobacterium tuberculosis* infections in mice with a combination of isoniazid and rifampicin entrapped in Poly (DL-lactide-co-glycolide) microparticles. *The Journal of Antimicrobial Chemotherapy*, *47*(6), 829–835. doi:10.1093/jac/47.6.829 PMID:11389115

Dutt, M., & Khuller, G. K. (2001). Therapeutic efficacy of poly(DL-lactide-co-glycolide)-encapsulated antitubercular drugs against *Mycobacterium tuberculosis* infection induced in mice. *Antimicrobial Agents and Chemotherapy*, *45*(1), 363–366. doi:10.1128/AAC.45.1.363-366.2001 PMID:11121000

Du, Z., Wang, C., Tai, X., Wang, G., & Liu, X. (2016). Optimization and Characterization of Biocompatible Oil-in-Water Nanoemulsion for Pesticide Delivery. *ACS Sustainable Chemistry & Engineering*, *4*(3), 983–991. doi:10.1021/acssuschemeng.5b01058

Early Breast Cancer Trialists' Collaborative Group (EBCTCG). (2005). Effects of chemotherapy and hormonal therapy for early breast cancer on recurrence and 15-year survival: An overview of the randomised trials. *Lancet*, *365*(9472), 1687–1717.

Edefonti, V., Hashibe, M., Parpinel, M., Ferraroni, M., Turati, F., Serraino, D., Matsuo, K., Olshan, A. F., Zevallos, J. P., Winn, D. M., Moysich, K., Zhang, Z.-F., Morgenstern, H., Levi, F., Kelsey, K., McClean, M., Bosetti, C., Schantz, S., Yu, G.-P., ... Decarli, A. (2015). Vitamin E intake from natural sources and head and neck cancer risk: A pooled analysis in the International Head and Neck Cancer Epidemiology consortium. *British Journal of Cancer*, *113*(1), 182–192. doi:10.1038/bjc.2015.149 PMID:25989276

Edris, A. E. (2007). Pharmaceutical and therapeutic potentials of essential oils and their individual volatile constituents: A review. *Phytotherapy Research*, *21*(4), 308–323. doi:10.1002/ptr.2072 PMID:17199238

Efferth, T. (2017). Cancer combination therapies with artemisinin-type drugs. *Biochemical Pharmacology*, *139*(March), 56–70. doi:10.1016/j.bcp.2017.03.019 PMID:28366726

Efferth, T., Li, P. C. H., Konkimalla, V. S. B., & Kaina, B. (2007). From traditional Chinese medicine to rational cancer therapy. *Trends in Molecular Medicine*, *13*(8), 353–361. doi:10.1016/j.molmed.2007.07.001 PMID:17644431

Egan, J.E., Ojima, I., Amiji, M.M., & Botchkina, G.I. (2018). *Nanoemulsion compositions of taxoid drugs, and methods for the use thereof to target cancer cells and cancer stem cells*. Academic Press.

Eid, A. M., El-Enshasy, H. A., Aziz, R., & Elmarzugi, N. A. (2014). Preparation, characterization, and anti-inflammatory activity of Swietenia macrophylla nanoemulgel. *Journal of Nanomedicine & Nanotechnology*, *5*(2), 1–10. doi:10.4172/2157-7439.1000190

El-Kattan, A., & Varma, M. (2012). Oral absorption, intestinal metabolism and human oral bioavailability. *Topics on Drug Metabolism, 10*, 31087.

El-Leithy, E. S., Ibrahim, H. K., & Sorour, R. M. (2013). In vitroandin vivoevaluation of indomethacin nanoemulsion as a transdermal delivery system. *Drug Delivery*, *22*(8), 1010–1017. doi:10.3109/10717544.2013.844742 PMID:24188427

El-readi, M. Z. (2019). *Review Article Cancer Nanomedicine : A New Era of Successful Targeted Therapy*. Academic Press.

El-Sayed, H. S., Chizzola, R., Ramadan, A. A., & Edris, A. E. (2017). Chemical composition and antimicrobial activity of garlic essential oils evaluated in organic solvent, emulsifying, and self-microemulsifying water based delivery systems. *Food Chemistry*, *221*, 196–204. doi:10.1016/j.foodchem.2016.10.052 PMID:27979186

Elshafeey, A. H., Kamel, A. O., & Fathallah, M. M. (2009). Utility of Nanosized Microemulsion for Transdermal Delivery of Tolterodine Tartrate: Ex-Vivo Permeation and In-Vivo Pharmacokinetic Studies. *Pharmaceutical Research*, *26*(11), 2446–2453. doi:10.100711095-009-9956-5 PMID:19697106

Elsheikh, M. A., Elnaggar, Y. S. R., Gohar, E. Y., & Abdallah, O. Y. (2012). Nanoemulsion liquid preconcentrates for raloxifene hydrochloride: Optimization and in vivo appraisal. *International Journal of Nanomedicine*, *3787*. Advance online publication. doi:10.2147/IJN.S33186 PMID:22888234

El-Zahaby, S. A., Elnaggar, Y. S. R., & Abdallah, O. Y. (2019). Reviewing two decades of nanomedicine implementations in targeted treatment and diagnosis of pancreatic cancer: An emphasis on state of art. *Journal of Controlled Release*, *293*, 21–35. doi:10.1016/j.jconrel.2018.11.013 PMID:30445002

Ensign, L. M., Cone, R., & Hanes, J. (2012). Oral drug delivery with polymeric nanoparticles: The gastrointestinal mucus barriers. *Advanced Drug Delivery Reviews*, *64*(6), 557–570.

Epstein, J. I., Amin, M. B., Beltran, H., Lotan, T. L., Mosquera, J. M., Reuter, V. E., Robinson, B. D., Troncoso, P., & Rubin, M. A. (2014). Proposed morphologic classification of prostate cancer with neuroendocrine differentiation. *The American Journal of Surgical Pathology*, *38*(6), 756–767. doi:10.1097/PAS.0000000000000208 PMID:24705311

Epstein, J. I., Egevad, L., Amin, M. B., Delahunt, B., Srigley, J. R., Humphrey, P. A., & Grading, C. (2016). The 2014 International Society of Urological Pathology (ISUP) Consensus Conference on Gleason Grading of Prostatic Carcinoma: Definition of grading patterns and proposal for a new grading system. *The American Journal of Surgical Pathology*, *40*(2), 244–252. doi:10.1097/PAS.0000000000000530 PMID:26492179

Eral, H. B., Mahony, M. O., Shaw, R., Trout, B. L., Myerson, A. S., & Doyle, P. S. (2014). Composite Hydrogels Laden with Crystalline Active Pharmaceutical Ingredients of Controlled Size and Loading. *Chemistry of Materials*, *26*(21), 6213–6220. doi:10.1021/cm502834h

Erdmann, M. E., Lautenschlaeger, R., Schmidt, H., Zeeb, B., Gibis, M., Bruggemann, D. A., & Weiss, J. (2017). Influence of droplet size on the antioxidant efficacy of oil-in-water emulsions loaded with rosemary in rawfermented sausages. *European Food Research and Technology*, *243*(8), 1415–1427. doi:10.100700217-017-2852-z

Ernst, E. (2004). Phytotherapy: A Quick Reference to Herbal Medicine. *Focus on Alternative and Complementary Therapies*, *9*(2), 164–164. doi:10.1111/j.2042-7166.2004.tb04305.x

Erramreddy, V. V., & Ghosh, S. (2014). Influence of emulsifier concentration on nanoemulsion gelation. *Langmuir*, *30*(37), 11062–11074. doi:10.1021/la502733v PMID:25137632

Erramreddy, V. V., Tu, S., & Ghosh, S. (2017). Rheological reversibility and long-term stability of repulsive and attractive nanoemulsion gels. *RSC Advances*, *7*(75), 47818–47832. doi:10.1039/C7RA09605D

Esfahani, K., Roudaia, L., Buhlaiga, N., Del Rincon, S. V., Papneja, N., & Miller, W. H. (2020). A review of cancer immunotherapy: From the past, to the present, to the future. *Current Oncology (Toronto, Ont.)*, *27*(12, s2), 87–97. doi:10.3747/co.27.5223 PMID:32368178

Esmaeili, F., Hosseini-Nasr, M., Rad-Malekshahi, M., Samadi, N., Atyabi, F., & Dinarvand, R. (2007). Preparation and antibacterial activity evaluation of rifampicin-loaded poly lactide-co-glycolide nanoparticles. *Nanomedicine (London)*, *3*(2), 161–167. doi:10.1016/j.nano.2007.03.003 PMID:17468055

Espina, L., Somolinos, M., Lorán, S., Conchello, P., García, D., & Pagán, R. (2011). Chemical composition of commercial citrus fruit essential oils and evaluation of their antimicrobial activity acting alone or in combined processes. *Food Control*, *22*(6), 896–902. doi:10.1016/j.foodcont.2010.11.021

Espinoza, J. A., Zisi, A., Kanellis, D. C., Carreras-Puigvert, J., Henriksson, M., Hühn, D., Watanabe, K., Helleday, T., Lindström, M. S., & Bartek, J. (2020). The antimalarial drug amodiaquine stabilizes p53 through ribosome biogenesis stress, independently of its autophagy-inhibitory activity. *Cell Death and Differentiation*, *27*(2), 773–789. doi:10.103841418-019-0387-5 PMID:31285544

Espitia, P. J. P., Fuenmayor, C. A., & Otoni, C. G. (2019). Nanoemulsions: Synthesis, characterization, and application in bio-based active food packaging. *Comprehensive Reviews in Food Science and Food Safety*, *18*(1), 264–285. doi:10.1111/1541-4337.12405 PMID:33337016

Esquerdo, V. M., Silva, P. P., Dotto, G. L., & Pinto, L. A. A. (2018). Nanoemulsions From Unsaturated Fatty Acids Concentrates of Carp Oil Using Chitosan, Gelatin, and Their Blends as Wall Materials. *European Journal of Lipid Science and Technology*, *120*(2), 1700240. doi:10.1002/ejlt.201700240

Faghmous, N., Bouzid, D., Boumaza, M., Touati, A., & Boyron, O. (2020). Optimization of chitosan-coated W/O/W multiple emulsion stabilized with Span 80 and Tween 80 using Box–Behnken design. *Journal of Dispersion Science and Technology*, 1–13. doi:10.1080/01932691.2020.1774387

Fallingborg, J., Christensen, L. A., Jacobsen, B. A., & Rasmussen, S. N. (1993). Very low intraluminal colonic pH in patients with active ulcerative colitis. *Digestive Diseases and Sciences*, *38*(11), 1989–1993.

Fang, Z. X., & Bhandari, B. (2010). Encapsulation of polyphenols – a review. *Trends in Food Science & Technology*, *21*(10), 510–523. doi:10.1016/j.tifs.2010.08.003

Fardous, J., Omoso, Y., Joshi, A., Yoshida, K., Patwary, M., Ono, F., & Ijima, H. (2021). Development and characterization of gel-in-water nanoemulsion as a novel drug delivery system. *Materials Science and Engineering C*, *124*, 112076.

Farhan, A. J., Mushir, A., Faiyaz, S., Cushman, T., Roop, K. K., & Sheikh, S. (2008). Investigation of nanoemulsion system for transdermal delivery of domperidone: *Ex-vivo* and *in vivo* studies. *Current Nanoscience*, *4*(4), 381–390. doi:10.2174/157341308786306071

Farshbaf-Sadigh, A., Jafarizadeh-Malmiri, H., Anarjan, N., & Najian, Y. (2019). Preparation of Ginger Oil in Water Nanoemulsion Using Phase Inversion Composition Technique: Effects of Stirring and Water Addition Rates on their Physico-Chemical Properties and Stability. *Zeitschrift für Physikalische Chemie*, *0*(0). Advance online publication. doi:10.1515/zpch-2019-1427

Farshi, P., Tabibiazar, M., Ghorbani, M., & Hamishehkar, H. (2017). Evaluation of Antioxidant Activity and Cytotoxicity of Cumin Seed Oil Nanoemulsion Stabilized by Sodium Caseinate- Guar Gum. Tabriz University of Medical Sciences, 23(4), 293–300. doi:10.15171/PS.2017.43

Fathordoobady, F., Sannikova, N., Guo, Y., Singh, A., Kitts, D. D., & Pratap-Singh, A. (2021). Comparing microfluidics and ultrasonication as formulation methods for developing hempseed oil nanoemulsions for oral delivery applications. *Scientific Reports*, *11*(1), 72. doi:10.103841598-020-79161-w PMID:33420136

Fattom, A.I., Simon, J., & Baker, J.R. (2013). *Herpes simplex virus nanoemulsion vaccine*. Academic Press.

Fausto, N., & Campbell, J. S. (2003). The role of hepatocytes and oval cells in liver regeneration and repopulation. *Mechanisms of Development*, *120*(1), 117–130. doi:10.1016/S0925-4773(02)00338-6 PMID:12490302

Fazelifar, P., Tabrizi, M. H., & Rafiee, A. (2021). The *Arachis hypogaea* Essential Oil Nanoemulsion as an Efficient Safe Apoptosis Inducer in Human Lung Cancer Cells (A549). *Nutrition and Cancer*, *73*(6), 1059–1067.

Feng, J., Shi, Y., Yu, Q., Sun, C., & Yang, G. (2016). Effect of emulsifying process on stability of pesticide nanoemulsions. *Colloids and Surfaces. A, Physicochemical and Engineering Aspects*, *497*, 286–292. doi:10.1016/j.colsurfa.2016.03.024

Feng, L., Cui, B., Yang, D., Wang, C., Zeng, Z., Wang, Y., Sun, C., Zhao, X., & Cui, H. (2016). Preparation and Evaluation of Emamectin Benzoate Solid Microemulsion. *Journal of Nanomaterials*, *2016*, 2016. doi:10.1155/2016/2386938

Feng, S., Yan, J., Wang, D., Jiang, L., Sun, P., Xiang, N., & Shao, P. (2021). Preparation and characterization of soybean protein isolate/pectin-based phytosterol nanodispersions and their stability in simulated digestion. *Food Research International*, *143*, 110237. doi:10.1016/j.foodres.2021.110237 PMID:33992350

Ferlay, J., Soerjomataram, I., Dikshit, R., Eser, S., Mathers, C., Rebelo, M., Parkin, D. M., Forman, D., & Bray, F. (2015). *Cancer incidence and mortality worldwide : Sources, methods and major patterns in GLOBOCAN 2012*. doi:10.1002/ijc.29210

Fernández-López, J., & Viuda-Martos, M. (2018). Introduction to the Special Issue: Application of Essential Oils in Food Systems. *Foods (Basel, Switzerland)*, *7*(4), 56. doi:10.3390/foods7040056 PMID:29621143

Ferreira, I., Barros, L., & Abreu, R. (2009). Antioxidants in wild mushrooms. *Current Medicinal Chemistry*, *16*(12), 1543–1560. doi:10.2174/092986709787909587 PMID:19355906

Finimundy, T., Gambato, G., Fontana, R., Camassola, M., Salvador, M., Moura, S., Hess, J., Henriques, J. A. P., Dillon, A. J. P., & Roesch-Ely, M. (2013). Aqueous extracts of *Lentinula edodes* and *Pleurotus sajor-caju* exhibit high antioxidant capability and promising *in vitro* antitumor activity. *Nutrition Research (New York, N.Y.), 33*(1), 76–84. doi:10.1016/j.nutres.2012.11.005 PMID:23351413

Firoozi, M., Rezapour-Jahani, S., Shahvegharasl, Z., & Anarjan, N. (2020). Ginger essential oil nanoemulsions: Preparation and physicochemical characterization and antibacterial activities evaluation. *Journal of Food Process Engineering, 43*(8), 43. doi:10.1111/jfpe.13434

Firoozy, M., & Anarjan, N. (2019). Preparation of maltodextrin stabilized alpha-tocopherol nanoemulsions using solvent-displacement technique. *Food Science & Technology International, 25*(5), 404–413. doi:10.1177/1082013219825893 PMID:30704297

Fisher, B., Costantino, J. P., Wickerham, D. L., Redmond, C. K., Kavanah, M., Cronin, W. M., Vogel, V., Robidoux, A., Dimitrov, N., Atkins, J., Daly, M., Wieand, S., Tan-Chiu, E., Ford, L., & Wolmark, N. (1998). Tamoxifen for prevention of breast cancer: Report of the national surgical adjuvant breast and bowel project P-1 study. *Journal of the National Cancer Institute, 90*(18), 1371–1388. doi:10.1093/jnci/90.18.1371 PMID:9747868

Fisher, K., & Phillips, C. (2009). In vitro inhibition of vancomycin-susceptible and vancomycin-resistant *Enterococcus faecium* and E. faecalis in the presence of citrus essential oils. *British Journal of Biomedical Science, 66*(4), 180–185. doi:10.1080/09674845.2009.11730270

Florence, A. T. (2012). "Targeting" nanoparticles: The constraints of physical laws and physical barriers. *Journal of Controlled Release, 164*(2), 115–124.

Florence, A. T., & Hussain, N. (2001). Transcytosis of nanoparticle and dendrimer delivery systems: Evolving vistas. *Advanced Drug Delivery Reviews, 50*, S69–S89.

Floriano, B. F., Carvalho, T., Lopes, T. Z., Takahashi, L. A. U., Rahal, P., Tedesco, A. C., & Calmon, M. F. (2021). Effect of berberine nanoemulsion Photodynamic therapy on cervical carcinoma cell line. *Photodiagnosis and Photodynamic Therapy, 33*, 102174. doi:10.1016/j.pdpdt.2020.102174 PMID:33401021

Flory, P. J. (1942). Thermodynamics of high polymer solutions. *The Journal of Chemical Physics, 10*(1), 51–61. doi:10.1063/1.1723621

Floss, H. G., & Yu, T. W. (2005). Rifamycin-mode of action, resistance, and biosynthesis. *Chemical Reviews, 105*(2), 621–632. doi:10.1021/cr030112j PMID:15700959

Foëx, P., & Sear, J. (2004). Hypertension: Pathophysiology and treatment. *Continuing Education in Anaesthesia, Critical Care & Pain, 4*(3), 71–75. doi:10.1093/bjaceaccp/mkh020

Fofaria, N. M., Qhattal, H. S. S., Liu, X., & Srivastava, S. K. (2016). Nanoemulsion formulations for anticancer agent piplartine - Characterisation, toxicological, pharmacokinetics and efficacy studies. *International Journal of Pharmaceutics, 498*(1-2), 12–22.

Fofaria, N. M., Qhattal, H. S. S., Liu, X., & Srivastava, S. K. (2016). Nanoemulsion formulations for anti-cancer agent piplartine—Characterization, toxicological, pharmacokinetics and efficacy studies. *International Journal of Pharmaceutics, 498*(1-2), 12–22. doi:10.1016/j.ijpharm.2015.11.045 PMID:26642946

Fontana, S., Flugy, A., Schillaci, O., Cannizzaro, A., Gargano, M. L., Saitta, A., De Leo, G., Venturella, G., & Alessandro, R. (2014). *In vitro* antitumor effects of the cold-water extracts of mediterranean species of genus *Pleurotus* (higher Basidiomycetes) on human colon cancer cells. *International Journal of Medicinal Mushrooms, 16*(1), 49–63. doi:10.1615/IntJMedMushr.v16.i1.50 PMID:24940904

Forgiarini, A., Esquena, J., Gonzalez, C., & Solans, C. (2001). Formation of nanoemulsions by low-energy emulsification methods at constant temperature. *Langmuir, 17*(7), 2076–2083.

Fox, C. B., & Haensler, J. (2013). An update on safety and immunogenicity of vaccines containing emulsion-based adjuvants. *Expert Review of Vaccines, 12*(7), 747–758. doi:10.1586/14760584.2013.811188 PMID:23885820

Fox, C. B., Van Hoeven, N., Granger, B., Lin, S., Guderian, J. A., Hartwig, A., Marlenee, N., Bowen, R. A., Soultanov, V., & Carter, D. (2019). Vaccine adjuvant activity of emulsified oils from species of the Pinaceae family. *Phytomedicine, 64*, 152927. doi:10.1016/j.phymed.2019.152927 PMID:31465981

Fraga, M., de Carvalho, T., da Silva Diel, D., Bruxel, F., Filho, N., Teixeira, H., & Matte, U. (2015). Cationic Nanoemulsions as a Gene Delivery System: Proof of Concept in the Mucopolysaccharidosis I Murine Model. *Journal of Nanoscience and Nanotechnology, 15*(1), 810–816. doi:10.1166/jnn.2015.9179 PMID:26328445

Franchi, L. P., Amantino, C. F., Melo, M. T., de Lima Montaldi, A. P., Primo, F. L., & Tedesco, A. C. (2016). In vitro effects of photodynamic therapy induced by chloroaluminum phthalocyanine nanoemulsion. *Photodiagnosis and Photodynamic Therapy, 16*, 100–105. doi:10.1016/j.pdpdt.2016.09.003 PMID:27612654

Frank, D. N., Amand, A. L. S., Feldman, R. A., Boedeker, E. C., Harpaz, N., & Pace, N. R. (2007). Molecular-phylogenetic characterisation of microbial community imbalances in human inflammatory bowel diseases. *Proceedings of the National Academy of Sciences of the United States of America, 104*(34), 13780–13785.

Franklyne, J. S., Gopinath, P. M., Mukherjee, A., & Chandrasekaran, N. (2021). Nanoemulsions: The rising star of antiviral therapeutics and nano-delivery system-current status and prospects. *Current Opinion in Colloid & Interface Science, 54*, 101458. doi:10.1016/j.cocis.2021.101458 PMID:33814954

Franklyne, J. S., Mukherjee, A., & Chandrasekaran, N. (2016). Essential oil micro-and nanoemulsions: Promising roles in antimicrobial therapy targeting human pathogens. *Letters in Applied Microbiology, 63*(5), 322–334. doi:10.1111/lam.12631

Freires, I. A., Denny, C., Benso, B., de Alencar, S. M., & Rosalen, P. L. (2015). Antibacterial Activity of Essential Oils and Their Isolated Constituents against Cariogenic Bacteria: A Systematic Review. *Molecules (Basel, Switzerland), 20*(4), 7329–7358. doi:10.3390/molecules20047329 PMID:25911964

Freytag, L. C., & Clements, J. D. (2005). Mucosal adjuvants. *Vaccine, 23*(15), 1804–1813. doi:10.1016/j.vaccine.2004.11.010 PMID:15734046

Fryd, M. M., & Mason, T. G. (2012). Advanced nanoemulsions. *Annual Review of Physical Chemistry, 63*(1), 493–518. doi:10.1146/annurev-physchem-032210-103436 PMID:22475339

Fu, J., Cai, Z., Gong, Y., O'Reilly, S. E., Hao, X., & Zhao, D. (2015). A new technique for determining critical micelle concentrations of surfactants and oil dispersants via UV absorbance of pyrene. *Colloids and Surfaces. A, Physicochemical and Engineering Aspects, 484*, 1–8. doi:10.1016/j.colsurfa.2015.07.039

Fukumura, D., Duda, D. G., Munn, L. L., & Jain, R. K. (2010). Tumor microvasculature and microenvironment: Novel insights through intravital imaging in pre-clinical models. *Microcirculation (New York, N.Y.), 17*(3), 206–225. doi:10.1111/j.1549-8719.2010.00029.x PMID:20374484

Fuller, T. L., Calvet, G., Genaro Estevam, C., Rafael Angelo, J., Abiodun, G. J., Halai, U. A., De Santis, B., Carvalho Sequeira, P., Machado Araujo, E., Alves Sampaio, S., Lima de Mendonça, M. C., Fabri, A., Ribeiro, R. M., Harrigan, R., Smith, T. B., Raja Gabaglia, C., Brasil, P., Bispo de Filippis, A. M., & Nielsen-Saines, K. (2017). Behavioral, climatic, and environmental risk factors for Zika and Chikungunya virus infections in Rio de Janeiro, Brazil, 2015-16. *PLoS One, 12*(11), e0188002. doi:10.1371/journal.pone.0188002 PMID:29145452

Gaba, B., Khan, T., Haider, M. F., Alam, T., Baboota, S., Parvez, S., & Ali, J. (2019). Vitamin E Loaded Naringenin Nanoemulsion via Intranasal Delivery for the Management of Oxidative Stress in a 6-OHDA Parkinson's Disease Model. *BioMed Research International*, *2382563*, 1–20. Advance online publication. doi:10.1155/2019/2382563 PMID:31111044

Galli, G., Hancock, K., Hoschler, K., DeVos, J., Praus, M., Bardelli, M., Malzone, C., Castellino, F., Gentile, C., McNally, T., Del Giudice, G., Banzhoff, A., Brauer, V., Montomoli, E., Zambon, M., Katz, J., Nicholson, K., & Stephenson, I. (2009). Fast rise of broadly cross-reactive antibodies after boosting long-lived human memory B cells primed by an MF59 adjuvanted prepandemic vaccine. *Proceedings of the National Academy of Sciences of the United States of America*, *106*(19), 7962–7967. doi:10.1073/pnas.0903181106 PMID:19416838

Ganai, A. A., & Farooqi, H. (2015). Bioactivity of genistein: A review of in vitro and in vivo studies. *Biomedicine and Pharmacotherapy*, *76*, 30–38. doi:10.1016/j.biopha.2015.10.026 PMID:26653547

Ganesan, P., Karthivashan, G., Park, S. Y., Kim, J., & Choi, D. K. (2018). Microfluidization trends in the development of nanodelivery systems and applications in chronic disease treatments. *International Journal of Nanomedicine*, *13*, 6109–6121. doi:10.2147/IJN.S178077 PMID:30349240

Ganesan, P., Ko, H.-M., Kim, I.-S., & Choi, D.-K. (2015). Recent trends in the development of nanophytobioactive compounds and delivery systems for their possible role in reducing oxidative stress in Parkinson's disease models. *International Journal of Nanomedicine*, *10*, 6757–6772. doi:10.2147/IJN.S93918 PMID:26604750

Ganesh, G. N. K., Singh, M. K., Datri, S., & Karri, V. V. S. R. (2019). Design and Development of Curcumin Nanogel for Squamous Cell Carcinoma. *Journal of Pharmaceutical Sciences and Research*, *11*(4), 1683–1683.

Ganta, S., & Amiji, M. (2009). Coadministration of paclitaxel and curcumin in nanoemulsion formulations to overcome multidrug resistance in tumor cells. *Molecular Pharmaceutics*, *6*(3), 928–939. doi:10.1021/mp800240j PMID:19278222

Ganta, S., Devalapally, H., & Amiji, M. (2010). Curcumin enhances oral bioavailability and anti-tumor therapeutic efficacy of paclitaxel upon administration in nanoemulsion formulation. *Journal of Pharmaceutical Sciences*, *99*(11), 4630–4641. doi:10.1002/jps.22157 PMID:20845461

Ganta, S., Devalapally, H., Shahiwala, A., & Amiji, M. (2008). A review of stimuli-responsive nanocarriers for drug and gene delivery. *Journal of Controlled Release*, *126*(3), 187–204.

Ganta, S., Singh, A., Patel, N. R., Cacaccio, J., Rawal, Y. H., Davis, B. J., Amiji, M. M., & Coleman, T. P. (2014). Development of EGFR targeted nanoemulsion for imaging and novel platinum therapy of ovarian cancer. *Pharmaceutical Research*, *31*(9), 2490–2502. doi:10.100711095-014-1345-z PMID:24643932

Ganta, S., Talekar, M., Singh, A., Coleman, T. P., & Amiji, M. M. (2014). Nanoemulsions in translational research-opportunities and challenges in targeted cancer therapy. *AAPS PharmSciTech*, *15*(3), 694–708. doi:10.120812249-014-0088-9 PMID:24510526

Gao, F., Zhang, Z., Bu, H., Huang, Y., Gao, Z., Shen, J., Zhao, C., & Li, Y. (2011). Nanoemulsion improves the oral absorption of candesartan cilexetil in rats: Performance and mechanism. *Journal of Controlled Release*, *149*(2), 168–174. doi:10.1016/j.jconrel.2010.10.013 PMID:20951749

Gao, W., Jiang, Z., Du, X., Zhang, F., Liu, Y., Bai, X., & Sun, G. (2020). Impact of Surfactants on Nanoemulsions based on Fractionated Coconut Oil: Emulsification Stability and in vitro Digestion. *Journal of Oleo Science*, *69*(3), 227–239. doi:10.5650/jos.ess19264 PMID:32051356

Gao, Y., Qi, X., Zheng, Y., Ji, H., Wu, L., Zheng, N., & Tang, J. (2016). Nanoemulsion enhances α-tocopherol succinate bioavailability in rats. *International Journal of Pharmaceutics*, *515*(1-2), 506–514. doi:10.1016/j.ijpharm.2016.10.026 PMID:27746330

Gao, Y., Tang, M., Leung, E., Svirskis, D., Shelling, A., & Wu, Z. (2020). Dual or multiple drug loaded nanoparticles to target breast cancer stem cells. *RSC Advances*, *10*(32), 19089–19105. doi:10.1039/D0RA02801K

Garabagi, F., Morello, G., & Wagner, C. (2021). *Improved cannobinoid bioavailability*. Academic Press.

Garca-Betanzos, C. I., Hernandez-Sanchez, H., Bernal-Couoh, T. F., Quintanar-Guerrero, D., & Zambrano-Zaragoza, M. (2017). Physicochemical, total phenols and pectin methylesterase changes on quality maintenance on guava fruit (*Psidium guajava* L.) coated with candeuba wax solid lipid nanoparticles-xanthan gum. *Food Research International*, *101*, 218–227. doi:10.1016/j.foodres.2017.08.065 PMID:28941687

García-Márquez, E., Higuera-Ciapara, I., & Espinosa-Andrews, H. (2017). Design of fish oil-in-water nanoemulsion by microfluidization. *Innovative Food Science & Emerging Technologies*, *40*, 87–91. doi:10.1016/j.ifset.2016.11.007

Garçon, N., Vaughn, D. W., & Didierlaurent, A. M. (2012). Development and evaluation of AS03, an Adjuvant System containing α-tocopherol and squalene in an oil-in-water emulsion. *Expert Review of Vaccines*, *11*(3), 349–366. doi:10.1586/erv.11.192 PMID:22380826

Garti, N., & Benichou, A. (2003). Recent developments in double emulsions for food applications. *Sematic Scholar*, *132*, 353–412. doi:10.1201/9780203913222.ch10

Garti, N., & Benichou, A. (2004). Recent developments in double emulsions for food applications. *Food Emulsion*, *132*, 353–412.

Garti, N., Yaghmur, A., Leser, M. E., Clement, V., & Watzke, H. J. (2001). Improved oil solubilization in oil/water food grade microemulsions in the presence of polyols and ethanol. *Journal of Agricultural and Food Chemistry*, *49*(5), 2552–2562. doi:10.1021/jf001390b PMID:11368635

Garti, N., Yuli, I. A., Spernath, A., & Hoffman, R. E. (2004). Transition and loci of solubilization of nutraceuticals in U-type nonionic microemulsions studied by self-diffusion NMR. *Physical Chemistry*, *6*, 2968–2976.

Gasco, M. R., Gallarate, M., & Pattarino, F. (1991). *In vitro* permeation of azelaic acid from viscosized microemulsions. *International Journal of Pharmaceutics*, *69*(3), 193–196. doi:10.1016/0378-5173(91)90361-Q

Gaspar, M. M., Bakowsky, U., & Ehrhardt, C. (2008). Inhaled Liposomes–Current Strategies and Future Challenges. *Journal of Biomedical Nanotechnology*, *4*(3), 245–257. doi:10.1166/jbn.2008.334

Gaspar, M. M., Cruz, A., Fraga, A. G., Castro, A. G., Cruz, M. E., & Pedrosa, J. (2008). Developments on drug delivery systems for the treatment of mycobacterial infections. *Current Topics in Medicinal Chemistry*, *8*(7), 579–591. doi:10.2174/156802608783955629 PMID:18473884

Gavin, A., Pham, J. T., Wang, D., Brownlow, B., & Elbayoumi, T. A. (2015). Layered nanoemulsions as mucoadhesive buccal systems for controlled delivery of oral cancer therapeutics. *International Journal of Nanomedicine*, *10*, 1569. PMID:25759580

Gavini, E., Sanna, V., Sharma, R., Juliano, C., Usai, M., Marchetti, M., Karlsen, J., & Giunchedi, P. (2005). Solid lipid microparticles (SLM) containing juniper oil as anti-acne topical carriers: Preliminary studies. *Pharmaceutical Development and Technology*, *10*(4), 479–487. doi:10.1080/10837450500299727 PMID:16370177

Gebreyohannes, G., Nyerere, A., Bii, C., & Sbhatu, D. B. (2019). Determination of antimicrobial activity of extracts of indigenous wild mushrooms against pathogenic organisms. *Evidence-Based Complementary and Alternative Medicine*, *2019*, 1–7. doi:10.1155/2019/6212673 PMID:30906415

Gedaly, R., Galuppo, R., Daily, M. F., Shah, M., Maynard, E., Chen, C., Zhang, X., Esser, K. A., Cohen, D. A., Evers, B. M., Jiang, J., & Spear, B. T. (2014). Targeting the Wnt/β-catenin signalling pathway in liver cancer stem cells and hepatocellular carcinoma cell lines with FH535. *PLoS One*, *9*(6), e99272. Advance online publication. doi:10.1371/journal.pone.0099272 PMID:24940873

Gelband, H., Molly Miller, P., Pant, S., Gandra, S., Levinson, J., Barter, D., & Laxminarayan, R. (2015). The state of the world's antibiotics. *Wound Healing Southern Africa, 8*(2), 30-34. doi.org/hdl.handle.net/10520/EJC180082

Gelderblom, H., Verweij, J., Nooter, K., & Sparreboom, A. (2001). Cremophor EL: The drawbacks and advantages of vehicle selection for drug formulation. *European Journal of Cancer*, *37*(13), 1590–1598. doi:10.1016/S0959-8049(01)00171-X PMID:11527683

Genovese, D. B., Lozano, J. E., & Rao, M. A. (2007). The rheology of colloidal and noncolloidal food dispersions. *Journal of Food Science*, *72*(2), 11–20. doi:10.1111/j.1750-3841.2006.00253.x PMID:17995847

George, J., & Kuttan, R. (1997). CANCER shell liquid. *Group*, *112*, 5–10.

Georghiou, S. B., Magana, M., Garfein, R. S., Catanzaro, D. G., Catanzaro, A., & Rodwell, T. C. (2012). Evaluation of genetic mutations associated with *Mycobacterium tuberculosis* resistance to amikacin, kanamycin and capreomycin: A systematic review. *PLoS One*, *7*(3), e33275. doi:10.1371/journal.pone.0033275 PMID:22479378

Ghaderi, L., Aliahmadi, A., Ebrahimi, S. N., & Rafati, H. (2021). Effective Inhibition and eradication of Pseudomonas aeruginosa biofilms by Saturejakhuzistanica essential oil nanoemulsion. *Journal of Drug Delivery Science and Technology*, *61*, 102260. doi:10.1016/j.jddst.2020.102260

Ghaderi, L., Moghimi, R., Aliahmadi, A., McClements, D. J., & Rafati, H. (2017). Development of antimicrobial nano-emulsion-based delivery systems against selected pathogenic bacteria using a thymol-rich Thymus daenensis essential oil. *Journal of Applied Microbiology*, *123*(4), 832–840. doi:10.1111/jam.13541 PMID:28714250

Ghani, S., Barzegar, H., Noshad, M., & Hojjati, M. (2019). Development of nanoemulsion-based antimicrobial activity of cinnamon prepared with soy protein isolate-lecithin. *Journal of Food and Bioprocess Engineering*, *2*(2), 147–154.

Gharehbeglou, P., Jafari, S. M., Homayouni, A., Hamishekar, H., & Mirzaei, H. (2019). Fabrication of double W1/O/W2 nano-emulsions loaded with oleuropein in the internal phase (W1) and evaluation of their release rate. *Food Hydrocolloids*, *89*, 44–55. doi:10.1016/j.foodhyd.2018.10.020

Ghazy, O. A., Fouad, M. T., Saleh, H. H., Kholif, A. E., & Morsy, T. A. (2021). Ultrasound-assisted preparation of anise extract nanoemulsion and its bioactivity against different pathogenic bacteria. *Food Chemistry*, *341*(Pt 2), 128259. doi:10.1016/j.foodchem.2020.128259 PMID:33068847

Gheibi, S., Ghaleh, H. E. G., Motlagh, B. M., & Azarbayjani, A. F. (2019). Therapeutic effects of curcumin and urso-dexycholic acid on non-alcoholic fatty liver disease. *Biomedicine and Pharmacotherapy*, *115*, 108938. doi:10.1016/j.biopha.2019.108938 PMID:31071511

Ghosh, V., Mukherjee, A., & Chandrasekaran, N. (2014). *Colloids and Surfaces B : Biointerfaces Eugenol-loaded antimicrobial nanoemulsion preserves fruit juice against, microbial spoilage*. Academic Press.

Ghosh, V., Mukherjee, A., & Chandrasekaran, N. (2014). Eugenol-loaded antimicrobial nanoemulsion preserves fruit juice against, microbial spoilage. *Colloids and Surfaces. B, Biointerfaces*, *114*, 392–397. doi:10.1016/j.colsurfb.2013.10.034 PMID:24252231

Gill, A. O., & Holley, R. A. (2006). Disruption of *Escherichia coli, Listeria monocytogenes and Lactobacillus sakei* cellular membranes by plant oil aromatics. *International Journal of Food Microbiology*, *108*(1), 1–9. doi:10.1016/j.ijfoodmicro.2005.10.009 PMID:16417936

Giordano, A., & Tommonaro, G. (2019). Curcumin and cancer. *Nutrients*, *11*(10), 2376. doi:10.3390/nu11102376 PMID:31590362

Giri, S., Barhoi, D., Barbhuiya, S. N., Giri, A., Das, S., Das, A., ... Singh, S. (2021). Consumption pattern and genotoxic potential of various smokeless tobacco products in Assam, India: A public health concern. *Mutation Research/Genetic Toxicology and Environmental Mutagenesis*, *866*, 503349. doi:10.1016/j.mrgentox.2021.503349 PMID:33985693

Given, P. S. Jr. (2009). Encapsulation of flavors in emulsions for beverages. *Current Opinion in Colloid & Interface Science*, *14*(1), 43–47. doi:10.1016/j.cocis.2008.01.007

Glass, B. D., Agatonovic-Kustrin, S., & Wisch, M. H. (2005). Artificial neural networks to optimize formulation components of a fixed-dose combination of rifampicin, isoniazid and pyrazinamide in a microemulsion. *Current Drug Discovery Technologies*, *2*(3), 195–201. doi:10.2174/1570163054866864 PMID:16472228

Glassman, D., Hignett, S., Rehman, S., Linforth, R., & Salhab, M. (2017). Adjuvant Endocrine Therapy for Hormone-positive Breast Cancer, Focusing on Ovarian Suppression and Extended Treatment: An Update. *Anticancer Research*, *37*(10), 5329–5341.

Gleason, D. F., & Mellinger, G. T. (1974). Prediction of prognosis for prostatic adenocarcinoma by combined histological grading and clinical staging. *The Journal of Urology*, *111*(1), 58–64. doi:10.1016/S0022-5347(17)59889-4 PMID:4813554

Goindi, S., Kaur, A., Kaur, R., Kalra, A., & Chauhan, P. (2016). Nanoemulsions: An emerging technology in the food industry. *Emulsions*, *3*, 651–688. doi:10.1016/B978-0-12-804306-6.00019-2

Goldberg, M. A., & Schneider, T. J. (1994). Similarities between the oxygen-sensing mechanisms regulating the expression of vascular endothelial growth factor and erythropoietin. *The Journal of Biological Chemistry*, *269*(6), 4355–4359.

Golding, M., Wooster, T. J., Day, L., Xu, M., Lundin, L., Keogh, J., & Clifton, P. (2011). Impact of gastric structuring on the lipolysis of emulsified lipids. *Soft Matter*, *7*(7), 3513–3523.

Goldstein, D., Gofrit, O., Nyska, A., & Benita, S. (2007). Anti-HER2 cationic immunoemulsion as a potential targeted drug delivery system for the treatment of prostate cancer. *Cancer Research*, *67*(1), 269–275. doi:10.1158/0008-5472.CAN-06-2731 PMID:17210707

Gonnet, M., Lethuaut, L., & Boury, F. (2010). New trends in encapsulation of liposoluble vitamins. *Journal of Controlled Release*, *146*(3), 276–290. doi:10.1016/j.jconrel.2010.01.037 PMID:20600399

Gönüllü, Ü., Üner, M., & Yener, G., Fatma Karaman, E. C. E. M., & Aydoğmuş, Z. (2015). Formulation and characterization of solid lipid nanoparticles, nanostructured lipid carriers and nanoemulsion of lornoxicam for transdermal delivery. *ActaPharmaceutica*, *65*(1), 1–13. PMID:25781700

Gonzalez-Aramundiz, J. V., Cordeiro, A. S., Csaba, N., de la Fuente, M., & Alonso, M. J. (2012). Nanovaccines: Nanocarriers for antigen delivery. *Biologie Aujourd'hui*, *206*(4), 249–261. doi:10.1051/jbio/2012027 PMID:23419252

Goodarzi, F., & Zendehboudi, S. (2018). A Comprehensive Review on Emulsions and Emulsion Stability in Chemical and Energy Industries. *Canadian Journal of Chemical Engineering*, *97*(1), 281–309. doi:10.1002/cjce.23336

Gorain, B., Choudhury, H., Kundu, A., Sarkar, L., Karmakar, S., Jaisankar, P., & Pal, T. K. (2014). Nanoemulsion strategy for olmesartan medoxomil improves oral absorption and extended antihypertensive activity in hypertensive rats. *Colloids and Surfaces. B, Biointerfaces*, *115*, 286–294. doi:10.1016/j.colsurfb.2013.12.016 PMID:24388859

Gorain, B., Choudhury, H., Nair, A. B., Dubey, S. K., & Kesharwani, P. (2020). Theranostic application of nanoemulsions in chemotherapy. *Drug Discovery Today*, *25*(7), 1174–1188. doi:10.1016/j.drudis.2020.04.013 PMID:32344042

Gortzi, O., Lalas, S., Tsaknis, J., & Chinou, I. (2007). Enhanced bioactivity of Citrus limon (Lemon Greek cultivar) extracts, essential oil and isolated compounds before and after encapsulation in liposomes. *Planta Medica*, *73*(09), 184. doi:10.1055-2007-986965

Goyal, A., Sharma, V., Upadhyay, N., Singh, A. K., Arora, S., Lal, D., & Sabikhi, L. (2015). Development of stable flaxseed oil emulsions as a potential delivery system of ω-3 fatty acids. *Journal of Food Science and Technology*, *52*(7), 4256–4265. doi:10.100713197-014-1370-2 PMID:26139890

Goycoolea, F. M., & Milkova, V. (2017). Electrokinetic behaviour of chitosan adsorbed on o/w nanoemulsion droplets. *Colloids and Surfaces. A, Physicochemical and Engineering Aspects*, *519*, 205–211. doi:10.1016/j.colsurfa.2016.05.093

Goycoolea, F. M., Valle-Gallego, A., Stefani, R., Menchicchi, B., David, L., Rochas, C., Santander-Ortega, M. G., & Alonso, M. J. (2012). Chitosan-based nanocapsules: Physical characterization, stability in biological media and capsaicin encapsulation. *Colloid & Polymer Science*, *290*(14), 1423–1434. doi:10.100700396-012-2669-z

Gradzielski, M. (1998). Effect of the cosurfactant structure on the bending elasticity in nonionic oil-in-water microemulsions. *Langmuir*, *14*(21), 6037–6044. doi:10.1021/la980074c

Graves, S. M., & Mason, T. G. (2008). Transmission of visible and ultraviolet light through charge-stabilized nanoemulsions. *The Journal of Physical Chemistry C*, *112*(33), 12669–12676. doi:10.1021/jp801019z

Graves, S., Meleson, K., Wilking, J., Lin, M. Y., & Mason, T. G. (2005). Structure of concentrated nanoemulsions. *The Journal of Chemical Physics*, *122*(13), 134703.

Gresser, U. (2001). Amoxicillin-clavulanic acid therapy may be associated with severe side effects - review of the literature. *European Journal of Medical Research*, *6*, 139–149. PMID:11309226

Grobmyer, S. R., Zhou, G., Gutwein, L. G., Iwakuma, N., Sharma, P., & Hochwald, S. N. (2012). Nanoparticle delivery for metastatic breast cancer. *Nanomedicine; Nanotechnology, Biology, and Medicine*, *8*(Suppl 1), S21–S30.

Groo, A. C., De Pascale, M., Voisin-Chiret, A. S., Corvaisier, S., Since, M., & Malzert-Freon, A. (2017). Comparison of 2 strategies to enhance pyridoclax solubility: Nanoemulsion delivery system versus salt synthesis. *European Journal of Pharmaceutical Sciences*, *97*, 218–226.

Groo, A. C., Hedir, S., Since, M., Brotin, E., Weiswald, L. B., Paysant, H., Nee, G., Coolzaet, M., Goux, D., Delépée, R., Freret, T., Poulain, L., Voisin-Chiret, A. S., & Malzert-Fréon, A. (2020). Pyridoclax-loaded nanoemulsion for enhanced anticancer effect on ovarian cancer. *International Journal of Pharmaceutics*, *587*, 119655.

Guagliardo, R., Pérez-Gil, J., De Smedt, S., & Raemdonck, K. (2018). Pulmonary surfactant and drug delivery: Focusing on the role of surfactant proteins. *Journal of Controlled Release*, *291*, 116–126. doi:10.1016/j.jconrel.2018.10.012 PMID:30321577

Guan, W., Tang, L., Wang, Y., & Cui, H. (2018). Fabrication of an Effective Avermectin Nanoemulsion Using a Cleavable Succinic Ester Emulsifier. *Journal of Agricultural and Food Chemistry*, *66*(29), 7568–7576. doi:10.1021/acs.jafc.8b01388 PMID:29976065

Gué, E., Since, M., Ropars, S., Herbinet, R., Le Pluart, L., & Malzert-Fréon, A. (2016). Evaluation of the versatile character of a nanoemulsion formulation. *International Journal of Pharmaceutics*, *498*(1-2), 49–65. doi:10.1016/j.ijpharm.2015.12.010 PMID:26685727

Guenther, E. (1952). The essential oils. *Journal of the American Pharmaceutical Association, 41*(6), 339–340. doi:10.1002/jps.3030410627

Guerra-Rosas, M. I., Morales-Castro, J., Cubero-Márquez, M. A., Salvia-Trujillo, L., & Martín-Belloso, O. (2017). Antimicrobial activity of nanoemulsions containing essential oils and high methoxyl pectin during long-term storage. *Food Control, 77*, 131–138. doi:10.1016/j.foodcont.2017.02.008

Guglielmini, G. (2008). Nanostructured novel carrier for topical application. *Clinics in Dermatology, 26*(4), 341–346. doi:10.1016/j.clindermatol.2008.05.004 PMID:18691513

Gündel, S., de Souza, M. E., Quatrin, P. M., Klein, B., Wagner, R., Gündel, A., Vaucher, R. A., Santos, R., & Ourique, A. F. (2018). Nanoemulsions containing Cymbopogon flexuosus essential oil: Development, characterization, stability study and evaluation of antimicrobial and antibiofilm activities. *Microbial Pathogenesis, 118*, 268–276. doi:10.1016/j.micpath.2018.03.043 PMID:29581028

Gündel, S., Velho, M. C., Diefenthaler, M. K., Favarin, F. R., Copetti, P. M., Fogaça, A. O., Klein, B., Wagner, R., Gündel, A., Sagrillo, M. R., & Ourique, A. F. (2018). Basil oil-nanoemulsions: Development, cytotoxicity and evaluation of antioxidant and antimicrobial potential. *Journal of Drug Delivery Science and Technology, 46*, 378–383. doi:10.1016/j.jddst.2018.05.038

Guo, Y., Wu, Z., Shen, S., Guo, R., Wang, J., Wang, W., Zhao, K., Kuang, M., & Shuai, X. (2018). Nanomedicines reveal how PBOV1 promotes hepatocellular carcinoma for effective gene therapy. *Nature Communications, 9*(1), 3430. Advance online publication. doi:10.103841467-018-05764-7 PMID:30143633

Gupta, A., Eral, H. B., Hatton, T. A., & Doyle, P. S. (2016). Nanoemulsions: Formation, properties and applications. *Soft Matter, 12*(11), 2826–2841. doi:10.1039/C5SM02958A PMID:26924445

Gupta, K., & Metgud, R. (2013). Evidences suggesting involvement of viruses in oral squamous cell carcinoma. *Pathology Research International, 2013*, 2013. doi:10.1155/2013/642496 PMID:24455418

Gupta, P. K., Bhandari, N. N., Shah, H., Khanchandani, V., Keerthana, R., Nagarajan, V., & Hiremath, L. (2019). *An update on nanoemulsions using nanosized liquid in liquid colloidal systems*. Published by Intechopen. doi:10.5772/intechopen.84442

Gupta, R., Shea, J., Scaife, C., Shurlygina, A., & Rapoport, N. (2015). Polymeric micelles and nanoemulsions as drug carriers: Therapeutic efficacy, toxicity, and drug resistance. *Journal of Controlled Release, 212*, 70–77.

Gurpreet, K., & Singh, S. K. (2017). Review of nanoemulsion formulation and characterization techniques. *Indian Journal of Pharmaceutical Sciences, 80*, 781–789.

Gurpret, K., & Singh, S. K. (2018). Review of Nanoemulsion Formulation and Characterization Techniques. *Indian Journal of Pharmaceutical Sciences, 80*(5). Advance online publication. doi:10.4172/pharmaceutical-sciences.1000422

Gutierrez, J. M., Gonzalez, C., Maestro, A., Sole, I., Pey, C. M., & Nolla, J. (2008). Nano-emulsions: New applications and optimization of their preparation. *Current Opinion in Colloid & Interface Science, 13*(4), 245–251. doi:10.1016/j.cocis.2008.01.005

Guy, B. (2007). The perfect mix: Recent progress in adjuvant research. *Nature Reviews. Microbiology, 5*(7), 396–397. doi:10.1038/nrmicro1681 PMID:17558426

Guzey, D., & McClements, D. J. (2006). Formation, stability and properties of multilayer emulsions for application in the food industry. *Advances in Colloid and Interface Science, 128–130*, 227–248. doi:10.1016/j.cis.2006.11.021 PMID:17223060

Hadziabdi, J., Dzana, O., Alisa, E., Edina, V., & Ognjenka, R. (2017). Preparation of nanoemulsions by high-energy and low- energy emulsification methods. *IFMBE Procesing*, *62*, 317–322. doi:10.1007/978-981-10-4166-2_48

Hadziyannis, S. J., Tassopoulos, N. C., Heathcote, E. J., Chang, T. T., Kitis, G., Rizzetto, M., Marcellin, P., Lim, S. G., Goodman, Z., Wulfsohn, M. S., Xiong, S., Fry, J., & Brosgart, C. L. (2003). Adefovir dipivoxil for the treatment of hepatitis B e antigen–negative chronic hepatitis B. *The New England Journal of Medicine*, *348*(9), 800–807. doi:10.1056/NEJMoa021812 PMID:12606734

Haham, M., Ish-Shalom, S., Nodelman, M., Duek, I., Segal, E., Kustanovich, M., & Livney, Y. D. (2012). Stability and bioavailability of vitamin D nanoencapsulated in casein micelles. *Food & Function*, *3*(7), 737–744. doi:10.1039/c2fo10249h PMID:22569895

Hakemi-Vala, M., Rafati, H., Aliahmadi, A., & Ardalan, A. (2017). Nanoemulsions: A Novel Antimicrobial Delivery System. *Nano- and Microscale Drug Delivery Systems*, 245–266.

Halawa, A. E. A., & Ali, A. A. (2017). Nanoemulsion Against Sclerotinia Rot of Fennel. *Egyptian Journal of Agricultural Research*, *95*(3), 1037–1050. doi:10.21608/ejar.2017.149559

Hall, J. E. (2015). Secretory Functions of the Alimentary Tract. In J. E. Hall & A. C. Guyton (Eds.), *Guyton and Hall Textbook of Medical Physiology* (pp. 817–832). Elsevier, Inc.

Hall, J. E., Granger, J. P., do Carmo, J. M., da Silva, A. A., Dubinion, J., George, E., ... Hall, M. E. (2012). Hypertension: Physiology and pathophysiology. *Comprehensive Physiology*, *2*(4), 2393–2442. doi:10.1002/cphy.c110058 PMID:23720252

Hamedi, H., Moradi, S., Tonelli, A. E., & Hudson, S. M. (2019). Preparation and Characterization of Chitosan–Alginate Polyelectrolyte Complexes Loaded with Antibacterial Thyme Oil Nanoemulsions. *Applied Sciences (Basel, Switzerland)*, *9*(18), 3933. doi:10.3390/app9183933

Hamed, R., Basil, M., AlBaraghthi, T., Sunoqrot, S., & Tarawneh, O. (2015). Nanoemulsion-based gel formulation of diclofenac diethylamine: Design, optimization, rheological behavior and in vitro diffusion studies. *Pharmaceutical Development and Technology*, *21*(8), 980–989. doi:10.3109/10837450.2015.1086372 PMID:26369621

Hamouda, T., & Baker, J. R. Jr. (2000). Antimicrobial mechanism of action of surfactant lipid preparations in enteric Gram-negative *bacilli. Journal of Applied Microbiology*, *89*(3), 397–403. doi:10.1046/j.1365-2672.2000.01127.x

Hamouda, T., Hayes, M. M., Cao, Z., Tonda, R., Johnson, K., Wright, D. C., Brisker, J., & Baker, J. R. (1999). A Novel Surfactant Nanoemulsion with Broad-Spectrum Sporicidal Activity against Bacillus Species. *The Journal of Infectious Diseases*, *180*(6), 1939–1949. doi:10.1086/315124 PMID:10558951

Hamouda, T., Hayes, M. M., Cao, Z., Tonda, R., Johnson, K., Wright, D. C., Brisker, J., & Baker, J. R. Jr. (1999). A novel surfactant nanoemulsion with broad-spectrum sporicidal activity against Bacillus species. *The Journal of Infectious Diseases*, *180*(6), 1939–1949.

Hamouda, T., Myc, A., Donovan, B., Shih, A. Y., Reuter, J. D., & Baker, J. R. (2001). A novel surfactant nanoemulsion with a unique non-irritant topical antimicrobial activity against bacteria, enveloped viruses and fungi. *Microbiological Research*, *156*(1), 1–7. doi:10.1078/0944-5013-00069 PMID:11372645

Handa, M., Ujjwal, R. R., Vasdev, N., Flora, S. J. S., & Shukla, R. (2021). Optimization of Surfactant- And Cosurfactant-Aided Pine Oil Nanoemulsions by Isothermal Low-Energy Methods for Anticholinesterase Activity. *ACS Omega*, *6*(1), 559–568.

Handa, M., Ujjwal, R., Vasdev, N., Flora, S., & Shukla, R. (2021). Optimization of Surfactant- and Cosurfactant-Aided Pine Oil Nanoemulsions by Isothermal Low-Energy Methods for Anticholinesterase Activity. *ACS Omega*, *6*(1), 559–568. doi:10.1021/acsomega.0c05033 PMID:33458508

Han, J., Zhang, S., Liu, X., & Xiao, C. (2020). Fabrication of capsaicin emulsions: Improving the stability of the system and relieving the irritation to the gastrointestinal tract of rats. *Journal of the Science of Food and Agriculture*, *100*(1), 129–138. doi:10.1002/jsfa.10002 PMID:31452202

Han, M., He, C.-X., Fang, Q.-L., Yang, X.-C., Diao, Y.-Y., Xu, D.-H., & Gao, J.-Q. (2008). A Novel Camptothecin Derivative Incorporated in Nano-Carrier Induced Distinguished Improvement in Solubility, Stability and Anti-tumor Activity Both In Vitro and In Vivo. *Pharmaceutical Research*, *26*(4), 926–935. doi:10.100711095-008-9795-9 PMID:19048358

Hare, J. I., Lammers, T., Ashford, M. B., Puri, S., Storm, G., & Barry, S. T. (2017). Challenges and strategies in anticancer nanomedicine development: An industry perspective. *Advanced Drug Delivery Reviews*, *108*, 25–38. doi:10.1016/j.addr.2016.04.025 PMID:27137110

Harris, J. R. (2014). Fifty years of progress in radiation therapy for breast cancer. *American Society of Clinical Oncology educational book. American Society of Clinical Oncology Annual Meeting*, 21–25.

Hartshorn, K. L. (2010). Role of surfactant protein A and D (SP-A and SP-D) in human antiviral host defense. *Frontiers in Bioscience (Scholar Edition)*, *2*(1), 527–546. doi:10.274183 PMID:20036966

Harush-Frenkel, O., Debotton, N., Benita, S., & Altschuler, Y. (2007). Targeting of nanoparticles to the clathirin mediated endocytic pathway. *Biochemical and Biophysical Research Communications*, *353*(1), 26–32. doi:10.1016/j.bbrc.2006.11.135 PMID:17184736

Harwansh, R. K., Mukherjee, P. K., & Biswas, S. (2017). Nanoemulsion as a novel carrier system for improvement of betulinic acid oral bioavailability and hepatoprotective activity. *Journal of Molecular Liquids*, *237*, 361–371.

Hasan, S. M. K., Ferrentino, G., & Scampicchio, M. (2019). Nanoemulsion as advanced edible coatings to preserve the quality of fresh-cut fruits and vegetables: A review. *International Journal of Food Science & Technology*. Advance online publication. doi:10.1111/ijfs.14273

Hashem, A. S., Awadalla, S. S., Zayed, G. M., Maggi, F., & Benelli, G. (2018). Pimpinella anisum essential oil nanoemulsions against Tribolium castaneum-insecticidal activity and mode of action. *Environmental Science and Pollution Research International*, *25*(19), 18802–18812. doi:10.100711356-018-2068-1 PMID:29713977

Hashtjin, A. M., & Abbasi, S. (2015). Optimization of ultrasonic emulsification conditions for the production of orange peel essential oil nanoemulsions. *Journal of Food Science and Technology*, *52*(5), 2679–2689. doi:10.100713197-014-1322-x PMID:25892765

Hassanin, M., Abd-El-Sayed, M., & Abdallah, M. (2017). Antifungal Activity of Some Essential Oil Emulsions and Nanoemulsions Against Fusarium Oxysporum Pathogen Affecting Cumin and Geranium Plants. *Scientific Journal of Flowers and Ornamental Plants*, *4*(3), 245–258. doi:10.21608jfop.2017.11326

Hassan, K. A., & Mujtaba, M. A. (2019). Antibacterial efficacy of garlic oil nano-emulsion. *AIMS Agriculture and Food*, *4*(1), 194–205. doi:10.3934/agrfood.2019.1.194

Hassanshahian, M., Saadatfar, A., & Masoumipour, F. (2020). Formulation and characterization of nanoemulsion from Alhagimaurorum essential oil and study of its antimicrobial, antibiofilm, and plasmid curing activity against antibiotic-resistant pathogenic bacteria. *Journal of Environmental Health Science & Engineering*, *18*(2), 1015–1027. doi:10.100740201-020-00523-7

Hatanaka, J., Chikamori, H., Sato, H., Uchida, S., Debari, K., Onoue, S., & Yamada, S. (2010). Physicochemical and pharmacological characterization of α-tocopherol-loaded nano-emulsion system. *International Journal of Pharmaceutics*, *396*(1), 188–193. doi:10.1016/j.ijpharm.2010.06.017 PMID:20558261

Hatanaka, J., Kimura, Y., Lai-Fu, Z., Onoue, S., & Yamada, S. (2008). Physicochemical and pharmacokinetic characterisation of water-soluble Coenzyme Q10 formulations. *International Journal of Pharmaceutics*, *363*(1-2), 112–117.

Hatz, C., von Sonnenburg, F., Casula, D., Lattanzi, M., & Leroux-Roels, G. (2012). A randomized clinical trial to identify the optimal antigen and MF59(®) adjuvant dose of a monovalent A/H1N1 pandemic influenza vaccine in healthy adult and elderly subjects. *Vaccine*, *30*(23), 3470–3477. doi:10.1016/j.vaccine.2012.03.017 PMID:22446638

Hazrati, H., Saharkhiz, M. J., Niakousari, M., & Moein, M. (2017). Natural herbicide activity of Satureja hortensis L. essential oil nanoemulsion on the seed germination and morphophysiological features of two important weed species. *Ecotoxicology and Environmental Safety*, *142*, 423–430. doi:10.1016/j.ecoenv.2017.04.041 PMID:28456128

Hearst, R., Nelson, D., McCollum, G., Millar, B. C., Maeda, Y., Goldsmith, C. E., Rooney, P. J., Loughrey, A., Rao, J. R., & Moore, J. E. (2009). An examination of antibacterial and antifungal properties of constituents of Shiitake (*Lentinula edodes*) and Oyster (*Pleurotus ostreatus*) mushrooms. *Complementary Therapies in Clinical Practice*, *15*(1), 5–7. doi:10.1016/j.ctcp.2008.10.002 PMID:19161947

Heleno, S. A., Martins, A., Queiroz, M. J. R. P., & Ferreira, I. C. F. R. (2015). Bioactivity of phenolic acids: Metabolites versus parent compounds: A review. *Food Chemistry*, *173*, 501–513. doi:10.1016/j.foodchem.2014.10.057 PMID:25466052

Helgeson, M. E. (2016). Colloidal behavior of nanoemulsions: Interactions, structure, and rheology. *Current Opinion in Colloid & Interface Science*, *25*, 39–50. doi:10.1016/j.cocis.2016.06.006

Helgeson, M. E., Moran, S. E., & Doyle, H. Z. (2012). Mesoporous organohydrogels from thermogelling photocross linkable nanoemulsions. *Nature Materials*, *11*(4), 344–352. doi:10.1038/nmat3248 PMID:22327746

Hemmila, M. R., Mattar, A., Taddonio, M. A., Arbabi, S., Hamouda, T., Ward, P. A., Wang, S. C., & Baker, J. R. Jr. (2010). Topical nanoemulsion therapy reduces bacterial wound infection and inflammation after burn injury. *Surgery*, *148*(3), 499–509. doi:10.1016/j.surg.2010.01.001 PMID:20189619

Hendry, E. R., Worthington, T., Conway, B. R., & Lambert, P. A. (2009). Antimicrobial efficacy of eucalyptus oil and 1,8-cineole alone and in combination with chlorhexidine digluconate against microorganisms grown in planktonic and biofilm cultures. *The Journal of Antimicrobial Chemotherapy*, *64*(6), 1219–1225. doi:10.1093/jac/dkp362 PMID:19837714

Henry, R. Y., & O'mahony, D. (1999). Treatment of prostate cancer. *Journal of Clinical Pharmacy and Therapeutics*, *24*(2), 93–102. doi:10.1046/j.1365-2710.1999.00207.x PMID:10380061

Herbst, R. S., Heymach, J. V., & Lippman, S. M. (2008). Lung cancer. *The New England Journal of Medicine*, *359*(13), 1367–1380.

Hetal, T., Bindesh, P., & Sneha, T. (2010). A review on techniques for oral bioavailability enhancement of drugs. *International Journal of Pharmaceutical Sciences Review and Research*, *4*(3), 203–223.

Hewlings, S. J., & Kalman, D. S. (2017). Curcumin: A review of its effects on human health. *Foods*, *6*(10), 92. doi:10.3390/foods6100092 PMID:29065496

Heydari, M., Amirjani, A., Bagheri, M., Sharifian, I., & Sabahi, Q. (2020). Eco-friendly pesticide based on peppermint oil nanoemulsion: Preparation, physicochemical properties, and its aphicidal activity against cotton aphid. *Environmental Science and Pollution Research International*, *27*(6), 6667–6679. doi:10.100711356-019-07332-y PMID:31873908

Hidajat, M. J., Jo, W., Kim, H., & Noh, J. (2020). Effective Droplet Size Reduction and Excellent Stability of Limonene Nanoemulsion Formed by High-Pressure Homogenizer. *Colloids and Interfaces*, *4*(1), 5. doi:10.3390/colloids4010005

Hidayah, R. E., & Ambarsari, L. (2018). Development of Trichodermin Nanoemulsion Based on Medium Chain Triglycerides as Antifungal of Ganoderma boninense in vitro. *Current Biochemistry*, *4*(2), 12–22.

Higgins, L. H., Withers, H. G., Garbens, A., Love, H. D., Magnoni, L., Hayward, S. W., & Moyes, C. D. (2009). Hypoxia and the metabolic phenotype of prostate cancer cells. *Biochimica et Biophysica Acta*, *1787*(12), 1433–1443. doi:10.1016/j.bbabio.2009.06.003 PMID:19524545

Hill, R. J. (2020). Electrokinetic spectra of dilute surfactant-stabilized nano-emulsions. *Journal of Fluid Mechanics*, *902*, A15. doi:10.1017/jfm.2020.579

Hill, R. J., & Afuwape, G. (2020). Dynamic mobility of surfactant-stabilized nano-drops: Unifying equilibrium thermodynamics, electrokinetics and Marangoni effects. *Journal of Fluid Mechanics*, *895*, A14. doi:10.1017/jfm.2020.256

Hobson, J. J., Edwards, S., Slater, R. A., Martin, P., Owen, A., & Rannard, S. P. (2018). Branched copolymer-stabilised nanoemulsions as new candidate oral drug delivery systems. *RSC Advances*, *8*(23), 12984–12991.

Ho, C. H., Hsu, J. L., Liu, S. P., Hsu, L. C., Chang, W. L., Chao, C. C., & Guh, J. H. (2015). Repurposing of phentolamine as a potential anticancer agent against human castration-resistant prostate cancer: A central role on microtubule stabilization and mitochondrial apoptosis pathway. *The Prostate*, *75*(13), 1454–1466. doi:10.1002/pros.23033 PMID:26180030

Hoffman, A., Domb, A.J., & Cherniakov, A.E. (2020). *Formulation and method for increasing oral bioavailability of drugs*. Academic Press.

Hoiby, N., Bjarnsholt, T., Givskov, M., Molin, S., & Ciofu, O. (2010). Antibiotic resistance of bacterial biofilms. *International Journal of Antimicrobial Agents*, *35*(4), 322–332. doi:10.1016/j.ijantimicag.2009.12.011

Holmgren, J., & Czerkinsky, C. (2005). Mucosal immunity and vaccines. *Nature Medicine*, *11*(S4), S45–S53. doi:10.1038/nm1213 PMID:15812489

Holmgren, J., Czerkinsky, C., Eriksson, K., & Mharandi, A. (2003). Mucosal immunisation and adjuvants: A brief overview of recent advances and challenges. *Vaccine*, *21*, S89–S95. doi:10.1016/S0264-410X(03)00206-8 PMID:12763689

Homayun, B., Lin, X., & Choi, H. J. (2019). Challenges and Recent Progress in Oral Drug Delivery Systems for Biopharmaceuticals. *Pharmaceutics*, *11*(129), 1–29. doi:10.3390/pharmaceutics11030129 PMID:30893852

Homayun, B., Lin, X., & Choi, H. J. (2019). Challenges and recent progress in oral drug delivery systems for biopharmaceuticals. *Pharmaceutics*, *11*(3), 129.

Hong, C., Kim, H. A., Firestone, G. L., & Bjeldanes, L. F. (2002). 3, 3′-Diindolylmethane (DIM) induces a G1 cell cycle arrest in human breast cancer cells that is accompanied by Sp1-mediated activation of p21WAF1/CIP1 expression. *Carcinogenesis*, *23*(8), 1297–1305. doi:10.1093/carcin/23.8.1297 PMID:12151347

Hong, L., Wang, J. L., Geng, J. X., Zhao, Y. H., Zhou, G. X., Zhang, J., Liu, L. W., & Qu, J. L. (2021). Rational design of an oxygen-enriching nanoemulsion for enhanced near-infrared laser activatable photodynamic therapy against hypoxic tumors. *Colloids and Surfaces. B, Biointerfaces*, *198*, 111500.

Horita, Y., Maeda, S., Kazumi, Y., & Doi, N. (2014). *In vitro* susceptibility of *Mycobacterium tuberculosis* isolates to an oral carbapenem alone or in combination with *β*-lactamase inhibitors. *Antimicrobial Agents and Chemotherapy*, *58*(11), 7010–7014. doi:10.1128/AAC.03539-14 PMID:25224000

Hörmann, K., & Zimmer, A. (2016). Drug delivery and drug targeting with parenteral lipid nanoemulsions— A review. *Journal of Controlled Release*, *223*, 85–98. doi:10.1016/j.jconrel.2015.12.016 PMID:26699427

Hornik, C. P., Herring, A. H., Benjamin, D. K., Capparelli, E. V., Kearns, G. L., van den Anker, J., Cohen-Wolkowiez, M., Clark, R. H., & Smith, P. B. (2013). Best pharmaceuticals for children act-pediatric trials network. Adverse events associated with meropenem versus imipenem/cilastatin therapy in a large retrospective cohort of hospitalized infants. *The Pediatric Infectious Disease Journal*, *32*(7), 748–753. doi:10.1097/INF.0b013e31828be70b PMID:23838776

Hortobagyi, G. N., Ueno, N. T., Xia, W., Zhang, S., Wolf, J. K., Putnam, J. B., Weiden, P. L., Willey, J. S., Carey, M., Branham, D. L., Payne, J. Y., Tucker, S. D., Bartholomeusz, C., Kilbourn, R. G., De Jager, R. L., Sneige, N., Katz, R. L., Anklesaria, P., Ibrahim, N. K., ... Hung, M. C. (2001). Cationic liposome-mediated E1A gene transfer to human breast and ovarian cancer cells and its biologic effects: A phase I clinical trial. *Journal of Clinical Oncology*, *19*(14), 3422–3433.

Hosny, K. M., Al Nahyah, K. S., & Alhakamy, N. A. (2020). Self-Nanoemulsion Loaded with a Combination of Isotretinoin, an Anti-Acne Drug, and Quercetin: Preparation, Optimization, and In Vivo Assessment. *Pharmaceutics*, *13*(1), 46. Advance online publication. doi:10.3390/pharmaceutics13010046 PMID:33396942

Hossain, M., Al-Hashmi, R. A., Weli, A. M., Al-Riyami, Q., & Al-Sabahib, J. N. (2012). Constituents of the essential oil from different brands of Syzigium caryophyllatum L by gas chromatography–mass spectrometry. *Asian Pacific Journal of Tropical Biomedicine*, *2*(3), S1446–S1449. doi:10.1016/S2221-1691(12)60435-3

Houssami, N., Macaskill, P., von Minckwitz, G., Marinovich, M. L., & Mamounas, E. (2012). Meta-analysis of the association of breast cancer subtype and pathologic complete response to neoadjuvant chemotherapy. *European Journal of Cancer*, *48*(18), 3342–3354.

Howard, A., O'Donoghue, M., Feeney, A., & Sleator, R. D. (2012). Acinetobacter baumannii: An emerging opportunistic pathogen. *Virulence*, *3*(3), 243–250. doi:10.4161/viru.19700

Hsiao, Y. C., Peng, S. F., Lai, K. C., Liao, C. L., Huang, Y. P., Lin, C. C., Lin, M.-L., Liu, K.-C., Tsai, C.-C., Ma, Y.-S., & Chung, J. G. (2019). Genistein induces apoptosis in vitro and has antitumor activity against human leukemia HL-60 cancer cell xenograft growth in vivo. *Environmental Toxicology*, *34*(4), 443–456. doi:10.1002/tox.22698 PMID:30618158

Hsu, H. J., Huang, R. F., Kao, T. H., Inbaraj, B. S., & Chen, B. H. (2017). Preparation of carotenoid extracts and nanoemulsions from Lycium barbarum L. and their effects on growth of HT-29 colon cancer cells. *Nanotechnology*, *28*(13), 135103.

https://clinicaltrials.gov/

Huang, C. Y., Ju, D. T., Chang, C. F., Reddy, P. M., & Velmurugan, B. K. (2017). A review on the effects of current chemotherapy drugs and natural agents in treating non–small cell lung cancer. *Biomedicine (Taipei)*, *7*(4), 23. doi:10.1051/bmdcn/2017070423 PMID:29130448

Huang, J., Xiong, Y., Zeng, G., Guo, S., Xie, G., Zhang, D., Tang, X., & Liu, Z. (2012). Separation of phenol from various micellar solutions using MEUF. *Separation and Purification Technology*, *98*, 1–6. doi:10.1016/j.seppur.2012.06.037

Huang, P. H., Chuang, H. C., Chou, C. C., Wang, H., Lee, S. L., Yang, H. C., Chiu, H.-C., Kapuriya, N., Wang, D., Kulp, S. K., & Chen, C. S. (2013). Vitamin E facilitates the inactivation of the kinase Akt by the phosphatase PHLPP1. *Science Signaling*, *6*(267), ra19–ra19. doi:10.1126cisignal.2003816 PMID:23512990

Huang, Q., Yu, H., & Ru, Q. (2010). Bioavailability and delivery of nutraceuticals using nanotechnology. *Journal of Food Science*, *75*(1), R50–R57. doi:10.1111/j.1750-3841.2009.01457.x PMID:20492195

Huang, Y., Zhou, B., Luo, H., Mao, J., Huang, Y., Zhang, K., Mei, C., Yan, Y., Jin, H., Gao, J., Su, Z., Pang, P., Li, D., & Shan, H. (2019). ZnAs@SiO2 nanoparticles as a potential anti-tumor drug for targeting stemness and epithelial-mesenchymal transition in hepatocellular carcinoma via SHP-1/JAK2/STAT3 signaling. *Theranostics*, *9*(15), 4391–4408. doi:10.7150/thno.32462 PMID:31285768

Huggins, C., & Hodges, C. V. (1941). Studies on prostatic cancer. I. The effect of castration, of estrogen and of androgen injection on serum phosphatases in metastatic carcinoma of the prostate. *Cancer Research*, *1*, 293–297.

Hughes, K. S., Schnaper, L. A., Bellon, J. R., Cirrincione, C. T., Berry, D. A., McCormick, B., Muss, H. B., Smith, B. L., Hudis, C. A., Winer, E. P., & Wood, W. C. (2013). Lumpectomy plus tamoxifen with or without irradiation in women age 70 years or older with early breast cancer: Long-term follow-up of CALGB 9343. *Journal of Clinical Oncology*, *31*(19), 2382–2387.

Hu, J., Chen, D., Jiang, R., Tan, Q., Zhu, B., & Zhang, J. (2014). Improved absorption and in vivo kinetic characteristics of nanoemulsions containing evodiamine–phospholipid nanocomplex. *International Journal of Nanomedicine*, *9*, 4411.

Hunter, R. J. (2001). *Foundations of colloid science*. Oxford university press.

Hu, Q., Gerhard, H., Upadhyaya, I., Venkitanarayanan, K., & Luo, Y. (2016). Antimicrobial eugenol nanoemulsion prepared by gum arabic and lecithin and evaluation of drying technologies. *International Journal of Biological Macromolecules*, *87*, 130–140. doi:10.1016/j.ijbiomac.2016.02.051 PMID:26902894

Hu, S. H., Liang, Z. C., Chia, Y. C., Lien, J. L., Chen, K. S., Lee, M. Y., & Wang, J. C. (2006). Antihyperlipidemic and antioxidant effects of extracts from *Pleurotus citrinopileatus*. *Journal of Agricultural and Food Chemistry*, *54*(6), 2103–2110. doi:10.1021/jf052890d PMID:16536582

Hussain, A., Altamimi, M. A., Alshehri, S., Imam, S. S., Shakeel, F., & Singh, S. K. (2020). Novel approach for transdermal delivery of rifampicin to induce synergistic antimycobacterial effects against cutaneous and systemic tuberculosis using a cationic nanoemulsion gel. *International Journal of Nanomedicine*, *15*, 1073–1094. doi:10.2147/IJN.S236277 PMID:32103956

Hu, Y., Zhang, K., Zhu, X., Zheng, X., Wang, C., Niu, X., Jiang, T., Ji, X., Zhao, W., Pang, L., Qi, Y., Li, F., Li, L., Xu, Z., Gu, W., & Zou, H. (2021). Synergistic Inhibition of Drug-Resistant Colon Cancer Growth with PI3K/mTOR Dual Inhibitor BEZ235 and Nano-emulsioned Paclitaxel via Reducing Multidrug Resistance and Promoting Apoptosis. *International Journal of Nanomedicine*, *16*, 2173–2186.

Huynh, H., Tuyen Nguyen, T. T., Pierce Chow, K. H., Tan, P. H., Soo, K. C., & Tran, E. (2003). Over-expression of the mitogen-activated protein kinase (MAPK) kinase (MEK)-MAPK in hepatocellular carcinoma: Its role in tumor progression and apoptosis. *BMC Gastroenterology*, *3*(1), 1–21. doi:10.1186/1471-230X-3-19 PMID:12906713

Hu, Z., Mawatari, S., Shibata, N., Takada, K., Yoshikawa, H., Arakawa, A., & Yosida, Y. (2000). Application of a biomagnetic measurement system (BMS) to the evaluation of gastrointestinal transit of intestinal pressure-controlled colon delivery capsules (PCDCs) in human subjects. *Pharmaceutical Research*, *17*(2), 160–167.

Hwang, Y. Y., Ramalingam, K., Bienek, D. R., Lee, V., You, T., & Alvarez, R. (2013). Antimicrobial activity of nanoemulsion in combination with cetylpyridinium chloride in multidrug-resistant Acinetobacter baumannii. *Antimicrobial Agents and Chemotherapy*, *57*(8), 3568–3575. doi:10.1128/AAC.02109-12

IARC. (2020). *Cancer fact sheet*. https://gco.iarc.fr/today/data/factsheets/cancers/1-Lip-oral-cavity-fact-sheet.pdf

Ibekwe, V. C., Fadda, H. M., McConnell, E. L., Khela, M. K., Evans, D. F., & Basit, A. W. (2008). Interplay between intestinal pH, transit time and feed status on the in vivo performance of pH responsive ileo-colonic release systems. *Pharmaceutical Research*, *25*(8), 1828–1835.

Ibrahim, D., Abdelfattah-Hassan, A., Badawi, M., Ismail, T. A., Bendary, M. M., Abdelaziz, A. M., & Abd El-Hamid, M. I. (2021). Thymol nanoemulsion promoted broiler chicken's growth, gastrointestinal barrier and bacterial community and conferred protection against Salmonella Typhimurium. *Scientific Reports*, *11*(1), 1–20. doi:10.103841598-021-86990-w

Iglarz, M., Binkert, C., Morrison, K., Fischli, W., Gatfield, J., Treiber, A., ... Buchmann, S. (2008). Pharmacology of macitentan, an orally active tissue-targeting dual endothelin receptor antagonist. *The Journal of Pharmacology and Experimental Therapeutics*, *327*(3), 736–745. doi:10.1124/jpet.108.142976 PMID:18780830

Imeson, A. (2010). *Food Stabilizers, Thickeners and Gelling Agents*. JohnWiley & Sons.

Inaba, T. A. (1992). Quantitative measurements of prostatic blood flow and blood volume by positron emission tomography. *The Journal of Urology*, *148*(5), 1457–1460. doi:10.1016/S0022-5347(17)36939-2 PMID:1279212

Inapurapu, S. P., Ibrahim, A., Kona, S. R., Pawar, S. C., Bodiga, S., & Bodiga, V. L. (2020). Development and characterization of ω-3 fatty acid nanoemulsions with improved physicochemical stability and bioaccessibility. *Colloids and Surfaces. A, Physicochemical and Engineering Aspects*, *606*, 125515. doi:10.1016/j.colsurfa.2020.125515

Ines, G. R. M., Juliana, M. C., Araceli, O. M., Laura, S. T., & Olga, M. B. (2015). Long-term stability of food-grade nanoemulsions from high methoxyl pectin containing essential oils. *Food Hydrocolloids*, *52*, 438–446. doi:10.1016/j.foodhyd.2015.07.017

Inouye, S., Tsuruoka, T., Uchida, K., & Yamaguchi, H. (2001). Effect of sealing and Tween 80 on the antifungal susceptibility testing of essential oils. *Microbiology and Immunology*, *45*(3), 201–208. doi:10.1111/j.1348-0421.2001.tb02608.x PMID:11345529

Ipuge, Y. A. I., Rieder, H. L., & Enarson, D. A. (1995). Adverse cutaneous reactions to thiacetazone for tuberculosis treatment in Tanzania. *Lancet*, *346*(8976), 657–660. doi:10.1016/S0140-6736(95)92278-4 PMID:7544858

Iqbal, H., Razzaq, A., Uzair, B., Ain, N. U., Sajjad, S., Khan, N. U., & Menaa, F. (2021). *Breast Cancer Inhibition by Biosynthesized Titanium Dioxide Nanoparticles Is Comparable to Free Doxorubicin but Appeared Safer in BALB / c Mice*. Academic Press.

Irby, D., Du, C., & Li, P. (2017). Lipid–drug conjugate for enhancing drug delivery. *Molecular Pharmaceutics*, *14*(5), 1325–1338. doi:10.1021/acs.molpharmaceut.6b01027 PMID:28080053

Ishikawa, N. K., Kasuya, M. C. M., & Vanetti, M. C. D. (2001). Antibacterial activity of *Lentinula edodes* grown in liquid medium. *Brazilian Journal of Microbiology*, *32*(3), 206–210. doi:10.1590/S1517-83822001000300008

Islam, M. T., Streck, L., Vinícius, M., Barros, O., Wanessa, S., Silva, C., Machado, C., Machado, C., Luiz, A., Júnior, G., Fernanda, M., & Jardim, C. (2017). Evaluation of toxic, cytotoxic and genotoxic effects of phytol and its nanoemulsion. *Manuscript*, *177*, 93–101. Advance online publication. doi:10.1016/j.chemosphere.2017.02.145 PMID:28284120

Islam, S. U., Shehzad, A., Ahmed, M. B., & Lee, Y. S. (2020). Intranasal Delivery of Nanoformulations: A Potential Way of Treatment for Neurological Disorders. *Molecules (Basel, Switzerland)*, *25*(8), 1929. doi:10.3390/molecules25081929 PMID:32326318

Issa, M., Chandel, S., Singh, H. P., Batish, D. R., Kohli, R. K., Yadav, S. S., & Kumari, A. (2020). Appraisal of phytotoxic, cytotoxic and genotoxic potential of essential oil of a medicinal plant Vitex negundo. *Industrial Crops and Products*, *145*, 112083. doi:10.1016/j.indcrop.2019.112083

Izadiyan, Z., Basri, M., Fard Masoumi, H.R., Abedi Karjiban, R., Salim, N., & Shameli, K. (2017). Modeling and optimization of nanoemulsion containing Sorafenib for cancer treatment by response surface methodology. *Chemistry Central Journal*, *11*(21), 1-9. doi:10.1186/s13065-017-0248-6

Izquierdo, P., Esquena, J., Tadros, T. F., Dederen, C., Garcia, M. J., Azemar, N. S. C., & Solans, C. (2002). Formation and stability of nano-emulsions prepared using the phase inversion temperature method. *Langmuir, 18*(1), 26–30. doi:10.1021/la010808c

Izquierdo, P., Feng, J., Esquena, J., Tadros, T. F., Dederen, J. C., Garcia, M. J., & Solans, C. (2005). The influence of surfactant mixing ratio on nanoemulsion formation by the pit method. *Journal of Colloid and Interface Science, 285*(1), 388–394.

Jackson, S. P., & Bartek, J. (2009). The DNA-damage response in human biology and disease. *Nature, 461*(7267), 1071–1078. doi:10.1038/nature08467 PMID:19847258

Jacob, J. S., Richards, J., Augustine, J. G., & Milea, J. S. (2010). Liquid Bevirimat Dosage Forms for Oral Administration. Academic Press.

Jafari, S. M. (2017). An overview of nanoencapsulation techniques and their classification. In S. M. Jafari (Ed.), *Nanoencapsulation Technologies for the Food and Nutraceutical Industries* (pp. 1–34). Academic Press. doi:10.1016/B978-0-12-809436-5.00001-X

Jafari, S. M., Assadpoor, E., Bhandari, B., & He, Y. (2008). Nano-particle encapsulation of fish oil by spray drying. *Food Research International, 41*(2), 172–183. doi:10.1016/j.foodres.2007.11.002

Jafari, S. M., He, Y., & Bhandari, B. (2007). Encapsulation of nanoparticles of d-limonene by spray drying: Role of emulsifiers and emulsifying techniques. *Drying Technology, 25*(6), 1069–1079. doi:10.1080/07373930701396758

Jafari, S. M., He, Y., & Bhandari, B. (2007). Optimisation of nano-emulsions production by microfluidization. *European Food Research and Technology, 225*(5), 733–741.

Jafari, S. M., & McClements, D. J. (Eds.). (2018). *Nanoemulsions: Formulation, applications, and characterization.* Academic Press.

Jafari, S., He, Y., & Bhandari, B. (2007). Production of sub-micron emulsions by ultrasound and microfluidization techniques. *Journal of Food Engineering, 82*(4), 478–488. doi:10.1016/j.jfoodeng.2007.03.007

Jain, C. P., & Vyas, S. P. (1995). Preparation and characterization of niosomes containing rifampicin for lung targeting. *Journal of Microencapsulation, 12*(4), 401–407. doi:10.3109/02652049509087252 PMID:8583314

Jain, R., & Patravale, V. B. (2009). Development and Evaluation of Nitrendipine Nanoemulsion for Intranasal Delivery. *Journal of Biomedical Nanotechnology, 5*(1), 62–68. doi:10.1166/jbn.2009.031 PMID:20055107

Jain, S. K., Chourasia, M. K., Masuriha, R., Soni, V., Jain, A., Jain, N. K., & Gupta, Y. (2005). Solid Lipid Nanoparticles Bearing Flurbiprofen for Transdermal Delivery. *Drug Delivery, 12*(4), 207–215. doi:10.1080/10717540590952591 PMID:16036715

Jain, V., Jain, S., & Mahajan, S. C. (2015). Nanomedicines based drug delivery systems for anti-cancer targeting and treatment. *Current Drug Delivery, 12*(2), 177–191. doi:10.2174/1567201811666140822112516 PMID:25146439

Jaiswal, M., Dudhe, R., & Sharma, P. K. (2015). Nanoemulsion: an advanced mode of drug delivery system. *3 Biotech, 5*(2), 123-127.

Jaiswal, M., Dudhe, R., & Sharma, P. K. (2015). Nanoemulsion: An advanced mode of drug delivery system. *Biotech.* doi:org/10.1007/s13205-014-0214-0

Jaiswal, M., Dudhe, R., & Sharma, P.K. (2015). Nanoemulsion: an advanced mode of drug delivery system. *3 Biotech, 5*(2), 123-127. doi:10.1007/s13205-014-0214-0

Jaiswal, M., Dudhe, R., & Sharma, P. K. (2015). Nanoemulsion: An advanced mode of drug delivery system. *Biotech*, *5*(2), 123–127.

Jamieson, F. B., Guthrie, J. L., Neemuchwala, A., Lastovetska, O., Melano, R. G., & Mehaffy, C. (2014). Profiling of rpoB mutations and MICs for rifampin and rifabutin in *Mycobacterium tuberculosis*. *Journal of Clinical Microbiology*, *52*(6), 2157–2162. doi:10.1128/JCM.00691-14 PMID:24740074

Jampilek, J., & Brychtova, K. (2010). Azone analogues: Classification, design, and transdermal penetration principles. *Medicinal Research Reviews*, *32*(5), 907–947.

Jang, J. H., Jeong, S. H., & Lee, Y. B. (2020). Enhanced Lymphatic Delivery of Methotrexate Using W/O/W Nanoemulsion: In Vitro Characterization and Pharmacokinetic Study. *Pharmaceutics*, *12*(10), 978. Advance online publication. doi:10.3390/pharmaceutics12100978 PMID:33081266

Jasinghe, V. J., Perera, C. O., & Sablani, S. S. (2007). Kinetics of the conversion of ergosterol in edible mushrooms. *Journal of Food Engineering*, *79*(3), 864–869. doi:10.1016/j.jfoodeng.2006.01.085

Jasmina, H., Džana, O., Alisa, E., Edina, V., & Ognjenka, R. (2017). *Preparation of nanoemulsions by high-energy and lowenergy emulsification methods CMBEBIH 2017*. Springer.

Jatav, V. S., Singh, S. K., Khatri, P., Sharma, A. K., & Singh, R. (2011). Formulation and *in-vitro* evaluation of rifampicin-loaded niosomes. *Journal of Chemical and Pharmaceutical Research*, *3*(2), 199–203.

Jaworska, M., Sikora, E., & Ogonowski, J. (2015). Rheological Properties of Nanoemulsions Stabilized by Polysorbate 80. *Chemical Engineering & Technology*, *38*(8), 1469–1476. doi:10.1002/ceat.201500069

Jeitler, M., Michalsen, A., Frings, D., Hübner, M., Fischer, M., Koppold-Liebscher, D. A., Murthy, V., & Kessler, C. S. (2020). Significance of medicinal mushrooms in integrative oncology: A narrative review. *Frontiers in Pharmacology*, *11*, 580656. doi:10.3389/fphar.2020.580656 PMID:33424591

Jemaa, M.B., Falleh, H., & Ksouri, R. (2019). *Encapsulation of Natural Bioactive Compounds: Nanoemulsion Formulation to Enhance Essential Oils Activities*. Academic Press.

Jemal, A., Bray, F., Center, M. M., Ferlay, J., Ward, E., & Forman, D. (2011). Global cancer statistics. *CA: a Cancer Journal for Clinicians*, *61*(2), 69–90.

Jensen, L. B., Magnussson, E., Gunnarsson, L., Vermehren, C., Nielsen, H. M., & Petersson, K. (2010). Corticosteroid solubility and lipid polarity control release from solid lipid nanoparticles. *International Journal of Pharmaceutics*, *390*(1), 53–60. doi:10.1016/j.ijpharm.2009.10.022 PMID:19836439

Jeong, Y. J., Kim, H. E., Han, S. J., & Choi, J. S. (2021). Antibacterial and antibiofilm activities of cinnamon essential oil nanoemulsion against multi-species oral biofilms. *Scientific Reports*, *11*(1), 1–8. doi:10.103841598-021-85375-3 PMID:33723345

Jerobin, J., Sureshkumar, R. S., Anjali, C. H., Mukherjee, A., & Chandrasekaran, N. (2012). Biodegradable polymer-based encapsulation of neem oil nanoemulsion for controlled release of Aza-A. *Carbohydrate Polymers*, *90*(4), 1750–1756. doi:10.1016/j.carbpol.2012.07.064 PMID:22944443

Jesser, E., Lorenzetti, A. S., Yeguerman, C., Murray, A. P., Domini, C., & Werdin-Gonzalez, J. O. (2020). Ultrasound assisted formation of essential oil nanoemulsions: Emerging alternative for Culex pipiens pipiens Say (Diptera: Culicidae) and Plodia interpunctella Hubner (Lepidoptera: Pyralidae) management. *Ultrasonics Sonochemistry*, *61*, 104832. doi:10.1016/j.ultsonch.2019.104832 PMID:31675660

Jesus, F. L. M., De Almeida, F. B., Duarte, J. L., Oliveira, A. E., Cruz, R. A. S., Souto, R. N. P., Ferreira, R. M. A., Kelmann, R. G., Carvalho, J. C. T., Lira-guedes, A. C., Guedes, M., Solans, C., & Fernandes, C. P. (2017). Preparation of a Nanoemulsion with Carapa guianensis Aublet (Meliaceae) Oil by a Low-Energy/Solvent-Free Method and Evaluation of Its Preliminary Residual Larvicidal Activity. *Evidence-Based Complementary and Alternative Medicine*, *2017*, 2017. doi:10.1155/2017/6756793 PMID:28798803

Jha, D., & Mazumder, P. M. (2018). Biological, chemical and pharmacological aspects of Madhucalongifolia. *Asian Pacific Journal of Tropical Medicine*, *11*(1), 9. doi:10.4103/1995-7645.223528

Jiang, C. L., Basri, M., Omar, D., Rahman, M. B. A., Salleh, A. B., Rahman, R. N. Z. R. A., & Selamat, A. (2012). Green nano-emulsion intervention for water-soluble glyphosate isopropylamine (IPA) formulations in controlling Eleusine indica. *Pesticide Biochemistry and Physiology*, *102*(1), 19–29. doi:10.1016/j.pestbp.2011.10.004

Jiang, Q., Christen, S., Shigenaga, M. K., & Ames, B. N. (2001). γ-Tocopherol, the major form of vitamin E in the US diet, deserves more attention. *The American Journal of Clinical Nutrition*, *74*(6), 714–722. doi:10.1093/ajcn/74.6.714 PMID:11722951

Jiang, Q., Wong, J., & Ames, B. N. (2004a). γ-Tocopherol Induces Apoptosis in Androgen-Responsive LNCaP Prostate Cancer Cells via Caspase-Dependent and Independent Mechanisms. *Annals of the New York Academy of Sciences*, *1031*(1), 399–400. doi:10.1196/annals.1331.056 PMID:15753180

Jiang, Q., Wong, J., Fyrst, H., Saba, J. D., & Ames, B. N. (2004b). γ-Tocopherol or combinations of vitamin E forms induce cell death in human prostate cancer cells by interrupting sphingolipid synthesis. *Proceedings of the National Academy of Sciences of the United States of America*, *101*(51), 17825–17830. doi:10.1073/pnas.0408340102 PMID:15596715

Jiang, Q., Yin, X., Lill, M. A., Danielson, M. L., Freiser, H., & Huang, J. (2008). Long-chain carboxychromanols, metabolites of vitamin E, are potent inhibitors of cyclooxygenases. *Proceedings of the National Academy of Sciences of the United States of America*, *105*(51), 20464–20469. doi:10.1073/pnas.0810962106 PMID:19074288

Jin, W., Xu, W., Liang, H., Li, Y., Liu, S., & Li, B. (2016). Nanoemulsions for food: Properties, production, characterization, and applications. *Emulsions*, *3*, 1–36. doi:10.1016/B978-0-12-804306-6.00001-5

Jiyu, Z., Jianyong, L., & Xuzheng, Z. (2010). *Ivermectin nanoemulsion drug combination and preparation method thereof*. Academic Press.

Jochen, W., Sylvia, G., Michael, D., & Julian, M. (2009). Nanostructured Encapsulation Systems: Food Antimicrobials. *Global Issues in Food Science and Technology ISBN*, *483208248*, 425–479.

Joe, M. M., Chauhan, P. S., Bradeeba, K., Shagol, C., Sivakumaar, P. K., & Sa, T. (2012). Influence of sunflower oil based nanoemulsion (AUSN-4) on the shelf life and quality of Indo-Pacific king mackerel (Scomberomorus guttatus) steaks stored at 20 C. *Food Control*, *23*(2), 564–570. doi:10.1016/j.foodcont.2011.08.032

Johansson, M. E., Sjövall, H., & Hansson, G. C. (2013). The gastrointestinal mucus system in health and disease. *Nature Reviews. Gastroenterology & Hepatology*, *10*(6), 352.

John, H., Barghorn, A., Funke, G., Sulser, T., Hailemariam, S., Hauri, S., & Joller-Jemelka, H. M. (2000). Noninflammatory chronic pelvic pain syndrome: Immunological study in blood, ejaculate and prostate tissue. *European Urology*, *39*(1), 72–78. doi:10.1159/000052415 PMID:11173942

Johnson-Weaver, B. T., Abraham, S. N., & Staats, H. F. (2020). Innate Immunity-Based Mucosal Modulators and Adjuvants. *Mucosal Vaccines*, 167-183.

Johnson-Weaver, B., Abraham, S., & Staats, H. (2020). Innate Immunity-Based Mucosal Modulators and Adjuvants. In Mucosal Vaccines (2nd ed., pp. 167-183). Academic Press. doi:10.1016/B978-0-12-811924-2.00010-9

Jones, F. A. (1996). Herbs- useful plants. Their role in history and today. *European Journal of Gastroenterology & Hepatology*, *8*(12), 1227–1231. doi:10.1097/00042737-199612000-00018 PMID:8980946

Jores, K., Haberland, A., Wartewig, S., Mäder, K., & Mehnert, W. (2005). Solid lipid nanoparticles (SLN) and oil-loaded SLN studied by spectrofluorometry and Raman spectroscopy. *Pharmaceutical Research*, *22*(11), 1887–1897. doi:10.100711095-005-7148-5 PMID:16132349

Joseph Schwarz, J., & Michael Weisspapir, M. (2017). *Medicinal Composition for enhanced delivery of triterpenes*. Academic Press.

Joseph Schwarz, J., & Michael Weisspapir, M. (2018). *Composition for intraoral delivery of biologically active peptides and proteins*. Academic Press.

Joseph, T. P., Chanda, W., Padhiar, A. A., Batool, S., Li Qun, S., Zhong, M., & Huang, M. (2018). Preclinical evaluation of the antitumor activities of edible and medicinal mushrooms: A molecular insight. *Integrative Cancer Therapies*, *17*(2), 200–209. doi:10.1177/1534735417736861 PMID:29094602

Journal, A., & Issue, S. (2013). *Formulation and Characterization of Plant Essential Oil Based Nanoemulsion : Evaluation of its Larvicidal Activity Against Aedes aegypti*. Academic Press.

Jo, Y.-J., Chun, J.-Y., Kwon, Y.-J., Min, S.-G., Hong, G.-P., & Choi, M.-J. (2015). Physical and antimicrobial properties of trans-cinnamaldehyde nanoemulsions in water melon juice. *Lebensmittel-Wissenschaft + Technologie*, *60*(1), 444–451. doi:10.1016/j.lwt.2014.09.041

Joyce, N., & Ramamurthy, N. (2017). Recent Research Trends in Fabrication and Applications of Plant Essential Oil Based Nanoemulsions. *Journal of Nanomedicine & Nanotechnology*, *08*(02). Advance online publication. doi:10.4172/2157-7439.1000434

Joye, I. J., Davidov-Pardo, G., & McClements, D. J. (2014). Nanotechnology for increased micronutrient bioavailability. *Trends in Food Science & Technology*, *40*(2), 168–182. doi:10.1016/j.tifs.2014.08.006

Ju, J., Picinich, S. C., Yang, Z., Zhao, Y., Suh, N., Kong, A. N., & Yang, C. S. (2010). Cancer-preventive activities of tocopherols and tocotrienols. *Carcinogenesis*, *31*(4), 533–542. doi:10.1093/carcin/bgp205 PMID:19748925

Jumaa, M., & Muller, B. W. (1999). Physicochemical properties of chitosan-lipid emulsions and their stability during autoclaving process. *International Journal of Pharmaceutics*, *183*(2), 175–184. doi:10.1016/S0378-5173(99)00086-1 PMID:10361168

Kabal'nov, A. S., Pertzov, A. V., & Shchukin, E. D. (1987). Ostwald ripening in two-component disperse phase systems: Application to emulsion stability. *Colloids and Surfaces*, *24*(1), 19–32. doi:10.1016/0166-6622(87)80258-5

Kadomatsu, M., Nakajima, S., Kato, H., Gu, L., Chi, Y., Yao, J., & Kitamura, M. (2012). Cordycepin as a sensitizer to tumour necrosis factor (TNF) - α-induced apoptosis through eukaryotic translation initiation factor 2α (eIF2 α) - and mammalian target of rapamycin complex 1 (mTORC1)-mediated inhibition of nuclear factor (NF)-κB. *Clinical and Experimental Immunology*, *168*(3), 325–332. doi:10.1111/j.1365-2249.2012.04580.x PMID:22519596

Kagan, L., & Hoffman, A. (2008). Systems for region selective drug delivery in the gastrointestinal tract: Biopharmaceutical considerations. *Expert Opinion on Drug Delivery*, *5*(6), 681–692.

Kah, M., & Hofmann, T. (2014). Nanopesticides research: Current trends and future priorities. *Environment International*, *63*, 224–235. doi:10.1016/j.envint.2013.11.015 PMID:24333990

Kalac, P. (2013). A review of chemical composition and nutritional value of wild-growing and cultivated mushrooms. *Journal of the Science of Food and Agriculture*, *93*(2), 209–218. doi:10.1002/jsfa.5960 PMID:23172575

Kaladhar, K., & Sharma, C. P. (2006). *Novel albumin self-assembled liposomes for drug delivery applications.* Annual Meeting-Society For Biomaterials In Conjunction With The International Biomaterials Symposium, Annual Meeting: Biomaterials- The enabling Technology, Pittsburgh, PA.

Kallewaard, N. L., Corti, D., Collins, P. J., Neu, U., Mcauliffe, J. M., Benjamin, E., Wachter-Rosati, L., Palmer-Hill, F. J., Yuan, A. Q., Walker, P. A., Vorlaender, M. K., Bianchi, S., Guarino, B., De Marco, A., Vanzetta, F., Agatic, G., Foglierini, M., Pinna, D., Fernandez-Rodriguez, B., ... Skehel, J. J. (2016). Structure and Function Analysis of an Antibody Recognizing All Influenza A Subtypes. *Cell*, *166*(3), 596–608. doi:10.1016/j.cell.2016.05.073 PMID:27453466

Kalyoncu, F., & Oskay, M. (2008). Antimicrobial activities of four wild mushroom species collected from Turkey. *Proceeding of the 6th International Conference on Mushroom Biology and Mushroom Products.*

Kalyoncu, F., Oskay, M., Saglam, H., Erdogan, T. F., & Tamer, A. U. (2010). Antimicrobial and antioxidant activities of mycelia of 10 wild mushroom species. *Journal of Medicinal Food*, *13*(2), 415–419. doi:10.1089/jmf.2009.0090 PMID:20192845

Kamba, T., & McDonald, D. M. (2007). Mechanisms of adverse effects of anti-VEGF therapy for cancer. *British Journal of Cancer*, *96*(12).

Kannan, M., Nesakumari, M., Rajarathinam, K., & Singh, A. (2010). Production and characterization of mushroom chitosan under solid-state fermentation conditions. *Advances in Biological Research (Faisalabad)*, *4*(1), 10–13.

Karandikar, S., Mirani, A., Waybhase, V., Patravale, V., & Patankar, S. (2017). Nanovaccines for oral delivery-formulation strategies and challenges. *Nanostructures for Oral Medicine*, 263-293.

Karandikar, S., Mirani, A., Waybhase, W., Patravale, V. B., & Patankar, S. (2017). Nanovaccines for oral delivery-formulation strategies and challenges. In Nanostructures for Oral Medicine (pp. 263-293). Elsevier. doi:10.1016/B978-0-323-47720-8.00011-0

Karasulu, H., Karabulut, B., Göker, E., Güneri, T., & Gabor, F. (2007). Controlled Release of Methotrexate from W/O Microemulsion and Its In Vitro Antitumor Activity. *Drug Delivery*, *14*(4), 225–233. doi:10.1080/10717540601067760 PMID:17497355

Kataoka, K., Muta, T., Yamazaki, S., & Takeshige, K. (2002). Activation of macrophages by linear (1→3)-β-d-glucans. *The Journal of Biological Chemistry*, *277*(39), 36825–36831. doi:10.1074/jbc.M206756200 PMID:12140296

Kathe, K., & Kathpalia, H. (2017). Film Forming Systems for Topical and Transdermal Drug Delivery. *Asian Journal of Pharmaceutical Sciences*. doi:10.1016/j.ajps.2017.07.004

Kaufmann, S.H., & McMichael, A.J. (2005). Annulling a dangerous liaison: Vaccination strategies against AIDS and tuberculosis. *Nature Medicine, 4*, S33-44. . doi:10.1038/nm1221

Kaufmann, S. H. E. (2001). How can immunology contribute to the control of tuberculosis? *Nature Reviews. Immunology*, *1*(1), 20–30. doi:10.1038/35095558 PMID:11905811

Kaur, A., Nigam, K., Bhatnagar, I., Sukhpal, H., Awasthy, S., Shankar, S., Tyagi, A., & Dang, S. (2020 a). Treatment of Alzheimer's diseases using donepezil nanoemulsion: An intranasal approach. *Drug Delivery and Translational Research*, *10*(6), 1862–1875. doi:10.100713346-020-00754-z PMID:32297166

Kaur, A., Nigam, K., Srivastava, S., Tyagi, A., & Dang, S. (2020 b). Memantine nanoemulsion: A new approach to treat Alzheimer's disease. *Journal of Microencapsulation*, *37*(5), 355–365. doi:10.1080/02652048.2020.1756971 PMID:32293915

Kaur, G., & Mehta, S. K. (2014). Probing location of anti-TB drugs loaded in Brij 96 microemulsions using thermo-analytical and photophysical approach. *Journal of Pharmaceutical Sciences*, *103*(3), 937–944. doi:10.1002/jps.23857 PMID:24425102

Kaur, G., Mehta, S. K., Kumar, S., Bhanjana, G., & Dilbaghi, N. (2015). Coencapsulation of hydrophobic and hydrophilic antituberculosis drugs in synergistic Brij 96 microemulsions: A biophysical characterization. *Journal of Pharmaceutical Sciences*, *104*(7), 2203–2212. doi:10.1002/jps.24469 PMID:25951802

Kawada, H., Kume, T., Matsunaga, T., Iwai, H., Sano, T., & Shibayama, M. (2010). Structure and rheology of a self-standing nanoemulsion. *Langmuir*, *26*(4), 2430–2437. doi:10.1021/la902905b PMID:20141204

Kawakami, M., Zamora, L. O., Araújo, R. S., Fernandes, C. P., Ricotta, T., de Oliveira, L. G., Queiroz-Junior, C. M., Fernandes, A. P., da Conceição, E. C., Ferreira, L., Barros, A., Aguiar, M. G., & Oliveira, A. (2021). Efficacy of nano-emulsion with Pterodon emarginatus Vogel oleoresin for topical treatment of cutaneous leishmaniasis. *Biomedicine & pharmacotherapy = Biomedecine & pharmacotherapie, 134*, 111109. doi:10.1016/j.biopha.2020.111109

Kelmann, R. G., Kuminek, G., Teixeira, H. F., & Koester, L. S. (2007). Carbamazepine parenteral nanoemulsions prepared by spontaneous emulsification process. *International Journal of Pharmaceutics*, *342*(1-2), 231–239.

Kennedy, J. F., & Thorley, M. (2000). Introduction to soft matter: polymers, colloids, amphiphiles and liquid crystals-IW Hamley. Carbohydrate Polymere, 1, 78.

Kenney, R. T., & Edelman, R. (2003). Survey of human-use adjuvants. *Expert Review of Vaccines*, *2*(2), 167–188. doi:10.1586/14760584.2.2.167 PMID:12899569

Kentish, S., Wooster, T. J., Ashokkumar, M., Balachandran, S., Mawson, R., & Simons, L. (2008). *The use of ultrasonics for nanoemulsion preparation.* doi:10.1016/j.ifset.2007.07.005

Kerns, S. L., Ostrer, H., & Rosenstein, B. S. (2014). Radiogenomics: Using genetics to identify cancer patients at risk for development of adverse effects following radiotherapy. *Cancer Discovery*, *4*(2), 155–165. doi:10.1158/2159-8290.CD-13-0197 PMID:24441285

Keykhosravy, K., Khanzadi, S., Hashemi, M., & Azizzadeh, M. (2020). Chitosan-loaded nanoemulsion containing ZatariaMultifloraBoiss and BuniumpersicumBoiss essential oils as edible coatings: Its impact on microbial quality of turkey meat and fate of inoculated pathogens. *International Journal of Biological Macromolecules*, *150*, 904–913. doi:10.1016/j.ijbiomac.2020.02.092

Khaled, A., Soonham, Y., Taha, K., & David, M. (2016). Utilization of nanoemulsions to enhance bioactivity of pharmaceuticals, supplements, and nutraceuticals: Nanoemulsion delivery systems and nanoemulsion excipient systems. *Expert Opinion on Drug Delivery*, *13*(9), 1327–1336. Advance online publication. doi:10.1517/17425247.2016.1162154 PMID:26984045

Khaleque, M., Keya, C. A., Hasan, K. N., Hoque, M., Inatsu, Y., & Bari, M. (2016). Use of cloves and cinnamon essential oil to inactivate Listeria monocytogenes in ground beef at freezing and refrigeration temperatures. *Lebensmittel-Wissenschaft + Technologie*, *74*, 219–223. doi:10.1016/j.lwt.2016.07.042

Khalid, N., Kobayashi, I., Wang, Z., Neves, M. A., Uemura, K., Nakajima, M., & Nabetani, H. (2015). Formulation of monodisperse oil-in-water emulsions loaded with ergocalciferol and cholecalciferol by microchannel emulsification: Insights of production characteristics and stability. *International Journal of Food Science & Technology*, *50*(8), 1807–1814. doi:10.1111/ijfs.12832

Khalil, H. E., Alqahtani, N. K., Darrag, H. M., Ibrahim, H. M., Emeka, P. M., Badger-Emeka, L. I., Matsunami, K., Shehata, T. M., & Elsewedy, H. S. (2021). Date Palm Extract (*Phoenix dactylifera*) PEGylated Nanoemulsion: Development, Optimization and Cytotoxicity Evaluation. *Plants*, *10*(4), 735.

Khan, M., & Nazzal, S. (2012). *Eutectic based self-nanoemulsified drug delivery system*. Academic Press.

Khan, D. H., Bashir, S., Khan, M. I., Figueiredo, P., Santos, H. A., & Peltonen, L. (2020). Formulation optimization and *in vitro* characterization of rifampicin and ceftriaxone dual drug loaded niosomes with high energy probe sonication technique. *Journal of Drug Delivery Science and Technology*, *58*, 101763. doi:10.1016/j.jddst.2020.101763

Khan, F., Sarker, M., Rahman, M., Ming, L. C., Mohamed, I. N., Zhao, C., ... Rashid, M. A. (2019). Comprehensive review on phytochemicals, pharmacological and clinical potentials of Gymnema sylvestre. *Frontiers in Pharmacology*, *10*, 1223. doi:10.3389/fphar.2019.01223 PMID:31736747

Khan, I., Bahuguna, A., Kumar, P., Bajpai, V. K., & Kang, S. C. (2018). *In vitro* and *in vivo* antitumor potential of carvacrol nanoemulsion against human lung adenocarcinoma A549 cells via mitochondrial mediated apoptosis. *Scientific Reports*, *8*(1), 144. doi:10.103841598-017-18644-9 PMID:29317755

Khani, S., Keyhanfar, F., & Amani, A. (2016). Design and evaluation of oral nanoemulsion drug delivery system of mebudipine. *Drug Delivery*, *23*(6), 2035–2043.

Khan, M. H., & Ramalingam, K. (2019). Synthesis of antimicrobial nanoemulsions and its effectuality for the treatment of multi-drug resistant ESKAPE pathogens. *Biocatalysis and Agricultural Biotechnology*, *18*, 101025. doi:10.1016/j.bcab.2019.101025

Khanna, V., Singh, S., Baghel, D., Kaur, B., Nayar, A., & Ghimire, R. (2018). *Nanoemulsion present aand future perspective – An Overview*. Academic Press.

Khatamian, N., Homayouni Tabrizi, M., Ardalan, P., Yadamani, S., & Darchini Maragheh, A. (2019). Synthesis of Carum Carvi essential oil nanoemulsion, the cytotoxic effect, and expression of caspase 3 gene. *Journal of Food Biochemistry*.

Khater, D., Nsairat, H., Odeh, F., Saleh, M., Jaber, A., Alshaer, W., & Mubarak, M. S. (2021). Design, Preparation, and Characterization of Effective Dermal and Transdermal Lipid Nanoparticles: *A Review. Cosmetics*, *8*(2), 39. doi:10.3390/cosmetics8020039

Khatri, P., & Shao, J. (2017). Transport of lipid nano-droplets through MDCK epithelial cell monolayer. *Colloids and Surfaces. B, Biointerfaces*, *153*, 237–243. doi:10.1016/j.colsurfb.2017.02.024 PMID:28267668

Khatri, S., Lohani, P., & Gandhi, S. (2013). Nanoemulsions in cancer therapy. Indo Global. *Journal of Pharmaceutical Sciences*, *3*(2), 124–133.

Khattab, A., Hassanin, L., & Zaki, N. (2017). Self-nanoemulsifying drug delivery system of coenzyme (Q10) with improved dissolution, bioavailability, and protective efficiency on liver fibrosis. *AAPS PharmSciTech*, *18*(5), 1657–1672.

Khurana, S., Jain, N. K., & Bedi, P. M. S. (2013). Nanoemulsion based gel for transdermal delivery of meloxicam: Physico-chemical, mechanistic investigation. *Life Sciences*, *92*(6-7), 383–392. doi:10.1016/j.lfs.2013.01.005 PMID:23353874

Kiho, T., Shiose, Y., Nagai, K., & Ukai, S. (1992). Polysaccharides in fungi. XXX. Antitumor and immunomodulating activities of two polysaccharides from the fruiting bodies of *Armillariella tabescens. Chemical & Pharmaceutical Bulletin, 40*(8), 2110–2114. doi:10.1248/cpb.40.2110 PMID:1423764

Kim, B. S., Won, M., Yang, Lee, K. M., & Kim, C. S. (2008). In vitro permeation studies of nanoemulsions containing ketoprofen as a model drug. *Drug Delivery, 15*(7), 465–469. doi:10.1080/10717540802328599 PMID:18712624

Kim, C. K., Cho, Y. J., & Gao, Z. G. (2001). Preparation and evaluation of biphenyl dimethyl dicarboxylate microemulsions for oral delivery. *Journal of Controlled Release, 70*(1-2), 149–155. doi:10.1016/S0168-3659(00)00343-6 PMID:11166415

Kim, H. S., Kim, T. I., Kim, J. H., Yoon, K. C., Hyon, J. Y., Shin, K. U., & Choi, C. Y. (2017). Evaluation of Clinical Efficacy and Safety of a Novel Cyclosporin A Nanoemulsion in the Treatment of Dry Eye Syndrome. *Journal of Ocular Pharmacology and Therapeutics, 33*(7), 530–538. doi:10.1089/jop.2016.0164 PMID:28759302

Kim, J. E., & Park, Y. J. (2017). Paclitaxel-loaded hyaluronan solid nanoemulsions for enhanced treatment efficacy in ovarian cancer. *International Journal of Nanomedicine, 12*, 645–658.

Kim, J., Kim, H., Kim, J. H., Kwon, O., Son, E., Lee, C., & Park, Y. (2016). Effects of Ganodermanondiol, a new melanogenesis inhibitor from the medicinal mushroom *Ganoderma lucidum. Int. Mol. Sci, 17*(11), 1798. doi:10.3390/ijms17111798 PMID:27801787

Kisich, K.O., Gelperina, S., Higgins, M.P., Wilson, S., Shipulo, E., Oganesyan, E., & Heifets, L. (2007). Encapsulation of moxifloxacin within poly (butyl cyanoacrylate) nanoparticles enhances efficacy against intracellular *Mycobacterium tuberculosis. International Journal of Pharmaceutics, 10*(345), 154-162. . doi:10.1016/j.ijpharm.2007.05.062

Kiyasu, J., Bloom, B., & Chaikoff, I. (1952). The portal transport of absorbed fatty acids. *The Journal of Biological Chemistry, 199*(1), 415–419. doi:10.1016/S0021-9258(18)44850-8 PMID:12999855

Klang, V., Matsko, N. B., Valenta, C., & Hofer, F. (2012). Electron microscopy of nanoemulsions: An essential tool for characterisation and stability assessment. *Micron (Oxford, England), 43*(2-3), 85–103. doi:10.1016/j.micron.2011.07.014 PMID:21839644

Klaus, A., Kozarski, M., & Vunduk, J. (2017). Antibacterial and antifungal potential of wild basidiomycete mushroom *Ganoderma applanatum. Lek. Sirovine, 36*, 37–46.

Klinkesorn, U. (2013). The role of chitosan in emulsion formation and stabilization. Food Reviews International. *Taylor & Francis Group, 29*(4), 371–393. doi:10.1080/87559129.2013.818013

Klucker, M. F., Dalençon, F., Probeck, P., & Haensler, J. (2012). AF03, an alternative squalene emulsion-based vaccine adjuvant prepared by a phase inversion temperature method. *Journal of Pharmaceutical Sciences, 101*(12), 4490–4500. doi:10.1002/jps.23311 PMID:22941944

Kohli, K., Chopra, S., Khar, R.K., Pillai, K.K., & Arora, S. (2011). *Self emulsifying drug delivery system for a curcuminoid based composition*. Academic Press.

Kohli, A. G., Kierstead, P. H., Venditto, V. J., Walsh, C. L., & Szoka, F. C. (2014). Designer lipids for drug delivery: From heads to tails. *Journal of Controlled Release, 190*, 274–287. doi:10.1016/j.jconrel.2014.04.047 PMID:24816069

Komaiko, J. (2016). *Optimization of the fabrication, stability, and performance of food grade nanoemulsions with low and high energy methods*. Academic Press.

Komaiko, J., & Mcclements, D. (2016). Formation of Food-Grade Nanoemulsions Using Low-Energy Preparation Methods: A Review of Available Methods. *Comprehensive Reviews in Food Science and Food Safety, 15*(2), 331–352. doi:10.1111/1541-4337.12189 PMID:33371595

Kong, T., Hao, L., Wei, Y., Cai, X., & Zhu, B. (2018). Doxorubicin conjugated carbon dots as a drug delivery system for human breast cancer therapy. *Cell Proliferation*, *51*(5), e12488.

Koroleva, M. Y., & Yurtov, E. V. (2012). Nanoemulsions: The properties, methods of preparation and promising applications. *Russian Chemical Reviews*, *81*(1), 21.

Koroleva, M., Nagovitsina, T., & Yurtov, E. (2018). Nanoemulsions stabilized by non-ionic surfactants: Stability and degradation mechanisms. *Physical Chemistry Chemical Physics*, *20*(15), 10369–10377. doi:10.1039/C7CP07626F PMID:29611566

Ko, S. Y., Chang, K. W., Lin, S. C., Hsu, H. C., & Liu, T. Y. (2007). The repressive effect of green tea ingredients on amyloid precursor protein (APP) expression in oral carcinoma cells in vitro and in vivo. *Cancer Letters*, *245*(1-2), 81–89. doi:10.1016/j.canlet.2005.12.029 PMID:16458426

Kotta, S., Khan, A. W., Ansari, S. H., Sharma, R. K., & Ali, J. (2014). Anti HIV nanoemulsion formulation: Optimisation and in vitro–in vivo evaluation. *International Journal of Pharmaceutics*, *462*(1-2), 129–134.

Kotta, S., Khan, A. W., Pramod, K., Ansari, S. H., Sharma, R. K., & Ali, J. (2012). Exploring oral nanoemulsions for bioavailability enhancement of poorly water-soluble drugs. *Expert Opinion on Drug Delivery*, *9*(5), 585–598.

Kotta, S., Khan, A., Ansari, S., Sharma, R., & Ali, J. (2013). Formulation of nanoemulsion: A comparison between phase inversion composition method and high-pressure homogenization method. *Drug Delivery*, *22*(4), 455–466. doi:10.3109/10717544.2013.866992 PMID:24329559

Koubaa, M., Roohinejad, S., Sharma, P., Nikmaram, N., Hashemi, S. S., Abbaspourrad, A., & Greiner, R. (2018). Multiple Emulsions. *Emulsion-based Systems for Delivery of Food Active Compounds: Formation, Application, Health and Safety*, 69.

Kováčik, A., Kopečná, M., & Vávrová, K. (2020). Permeation enhancers in transdermal drug delivery: Benefits and limitations. *Expert Opinion on Drug Delivery*, *17*(2), 145–155. https://doi.org/10.1080/17425247.2020.1713087

Králová, I., & Sjöblom, J. (2009). Surfactants Used in Food Industry: A Review. *Journal of Dispersion Science and Technology*, *30*(9), 1363–1383. doi:10.1080/01932690902735561

Kramer, R. M., Archer, M. C., Orr, M. T., Dubois, C. N., Beebe, E. A., Huang, P. D., Dowling, Q. M., Schwartz, A. M., Fedor, D. M., Vedvick, T. S., & Fox, C. B. (2018). Development of a thermostable nanoemulsion adjuvanted vaccine against tuberculosis using a design-of-experiments approach. *International Journal of Nanomedicine*, *26*(13), 3689–3711. doi:10.2147/IJN.S159839 PMID:29983563

Kreutz, F.T. (2019). *Nanoemulsions and methods for cancer therapy*. Academic Press.

Krishnamoorthy, R., Athinarayanan, J., Periasamy, V. S., Adisa, A. R., Al-Shuniaber, M. A., Gassem, M. A., & Alshatwi, A. A. (2018). Antimicrobial activity of nanoemulsion on drug-resistant bacterial pathogens. *Microbial Pathogenesis*, *120*, 85–96. doi:10.1016/j.micpath.2018.04.035

Krishnamoorthy, R., Gassem, M. A., Athinarayanan, J., Periyasamy, V. S., Prasad, S., & Alshatwi, A. A. (2021). Antifungal activity of nanoemulsion from Cleome viscosa essential oil against food-borne pathogenic Candida albicans. *Saudi Journal of Biological Sciences*, *28*(1), 286–293. doi:10.1016/j.sjbs.2020.10.001 PMID:33424308

Kriwet, K., & Müller-Goymann, C. (1995). Diclofenac release from phospholipid drug systems and permeation through excised human stratum corneum. *International Journal of Pharmaceutics*, *125*(2), 231–242. doi:10.1016/0378-5173(95)00130-B

Kruger, S., Ilmer, M., Kobold, S., Cadilha, B. L., Endres, S., Ormanns, S., Schuebbe, G., Renz, B. W., D'Haese, J. G., Schloesser, H., Heinemann, V., Subklewe, M., Boeck, S., Werner, J., & von Bergwelt-Baildon, M. (2019). Advances in cancer immunotherapy 2019–latest trends. *Journal of Experimental & Clinical Cancer Research*, *38*(1), 1–11. doi:10.118613046-019-1266-0 PMID:30606223

Kullberg, M., Owens, J. L., & Mann, K. (2010). Listeriolysin O enhances cytoplasmic delivery by Her-2 targeting liposomes. *Journal of Drug Targeting*, *18*(4), 313–320.

Kumar, M., & Jain, C. P. (2020). Possible benefits of reformulating antiviral drugs with nanoemulsion system in the treatment of novel coronavirus infection. *Current Drug Therapy, 2*, 3.

Kumar, S., Ali, J., & Baboota, S. (2016). Design Expert®supported optimization and predictive analysis of selegiline nanoemulsion via the olfactory region with enhanced behavioural performance in Parkinson's disease. *Nanotechnology*, *27*(43). doi:10.1088/0957-4484/27/43/435101

Kumar, D. D., Mann, B., Pothuraju, R., Sharma, R., Bajaj, R., & Minaxi, M. (2016). Formulation and characterization of nanoencapsulated curcumin using sodium caseinate and its incorporation in ice cream. *Food Functinal*, *7*(1), 417–424. doi:10.1039/C5FO00924C PMID:26501983

Kumar, D., Aqil, M., Rizwan, M., Sultana, Y., & Ali, M. (2009). Investigation of a nanoemulsion as vehicle for transdermal delivery of amlodipine. *Die Pharmazie*, *64*, 80–85. doi:10.1691/ph.2009.8200 PMID:19320278

Kumar, G. P., & Divya, A. (2015). Nanoemulsion based targeting in cancer therapeutics. *Medicinal Chemistry (Shariqah, United Arab Emirates)*, *5*(5), 272–284.

Kumari, S., Kumaraswamy, R. V., Choudhary, R. C., Sharma, S. S., Pal, A., Raliya, R., Biswas, P., & Saharan, V. (2018). Thymol nanoemulsion exhibits potential antibacterial activity against bacterial pustule disease and growth promotory effect on soybean. *Scientific Reports*, *8*(1), 1–12. doi:10.103841598-018-24871-5 PMID:29703904

Kumar, M., Bishnoi, R. S., Shukla, A. K., & Jain, C. P. (2019). Techniques for Formulation of Nanoemulsion Drug Delivery System: A Review. *Preventive Nutrition and Food Science*, *24*(3), 225–234. doi:10.3746/pnf.2019.24.3.225 PMID:31608247

Kumar, M., Misra, A., Babbar, A. K., Mishra, A. K., Mishra, P., & Pathak, K. (2008). Intranasal nanoemulsion based brain targeting drug delivery system of risperidone. *International Journal of Pharmaceutics*, *358*(1-2), 285–291. doi:10.1016/j.ijpharm.2008.03.029 PMID:18455333

Kumar, P. V., Asthana, A., Dutta, T., & Jain, N. K. (2006). Intracellular macrophage uptake of rifampicin loaded mannosylated dendrimers. *Journal of Drug Targeting*, *14*(8), 546–556. doi:10.1080/10611860600825159 PMID:17050121

Kumar, S. H., & Singh, V. (2012). Nanoemulsification-a novel targeted drug delivery tool. *Journal of Drug Delivery and Therapeutics*, *2*(4). Advance online publication. doi:10.22270/jddt.v2i4.185

Kumar, V., Mishra, S., Shrivastava, A., & Zaidi, S. (2020). Prevalence of hypertension and its associated risk factors in rural community of Barabanki. *International Journal of Community Medicine and Public Health*, *7*(3), 893. doi:10.18203/2394-6040.ijcmph20200936

Kummar, S., Gutierrez, M., Doroshow, J. H., & Murgo, A. J. (2006). Drug development in oncology: Classical cytotoxics and molecularly targeted agents. *British Journal of Clinical Pharmacology*, *62*(1), 15–26. doi:10.1111/j.1365-2125.2006.02713.x PMID:16842375

Kümmel, S., Holtschmidt, J., & Loibl, S. (2014). Surgical treatment of primary breast cancer in the neoadjuvant setting. *British Journal of Surgery*, *101*(8), 912–924.

Kundranda, M. N., & Niu, J. (2015). Albumin-bound paclitaxel in solid tumors: Clinical development and future directions. *Drug Design, Development and Therapy*, *9*, 3767–3777. doi:10.2147/DDDT.S88023 PMID:26244011

Kunin, C. M. (1996). Antimicrobial activity of rifabutin. *Clinical Infectious Diseases*, *22*(1, Supplement_1), S3–S13. doi:10.1093/clinids/22.Supplement_1.S3 PMID:8785253

Lai, L. K., Abidin, N. Z., Abdullah, N., & Sabaratnam, V. (2010). Anti-human papillomavirus (HPV) 16E6 activity of Ling Zhi or reishi medicinal mushroom, *Ganoderma lucidum* (W. Curt.: Fr.) P. Karst.(Aphyllophoromycetideae) extracts. *International Journal of Medicinal Mushrooms*, *12*(3), 279–286. doi:10.1615/IntJMedMushr.v12.i3.70

Lamaallam, S., Bataller, H., Dicharry, C., & Lachaise, J. (2005). Formation and stability of miniemulsions produced by dispersion of water/oil/surfactants concentrates in a large amount of water. *Colloids and Surfaces. A, Physicochemical and Engineering Aspects*, *270*, 44–51.

Lambert, L. K., Balneaves, L. G., Howard, A. F., Chia, S. K., & Gotay, C. C. (2018). Understanding adjuvant endocrine therapy persistence in breast Cancer survivors. *BMC Cancer*, *18*(1), 732.

Lambert, R. J., Skandamis, P. N., Coote, P. J., & Nychas, G. J. (2001). A study of the minimum inhibitory concentration and mode of action of oregano essential oil, thymol and carvacrol. *Journal of Applied Microbiology*, *91*(3), 453–462. doi:10.1046/j.1365-2672.2001.01428.x PMID:11556910

Lammers, T., Kiessling, F., Hennink, W. E., & Storm, G. (2012). Drug targeting to tumors: Principles, pitfalls and (pre-) clinical progress. *Journal of Controlled Release*, *161*(2), 175–187.

Landry, K. S., Micheli, S., McClements, D. J., & McLandsborough, L. (2015). Effectiveness of a spontaneous carvacrol nanoemulsion against Salmonella enterica Enteritidis and Escherichia coli O157: H7 on contaminated broccoli and radish seeds. *Food Microbiology*, *51*, 10–17. doi:10.1016/j.fm.2015.04.006 PMID:26187822

Lante, A., & Friso, D. (2013). Oxidative stability and rheological properties of nanoemulsions with ultrasonic extracted green tea infusion. *Food Research International*, *54*(1), 269–276. doi:10.1016/j.foodres.2013.07.009

Larson, R. G. (1999). *The structure and rheology of complex fluids* (Vol. 150). Oxford university press.

Lawler, J. (2009). Introduction to the tumour microenvironment review series. *Journal of Cellular and Molecular Medicine*, *13*(8A), 1403–1404.

Lawrence, M. J., & Rees, G. D. (2000). Microemulsion-based media as novel drug delivery systems. *Advanced Drug Delivery Reviews*, *45*(1), 89–121. doi:10.1016/S0169-409X(00)00103-4 PMID:11104900

Lawson, L. B., Norton, E. B., & Clements, J. D. (2011). Defending the mucosa: Adjuvant and carrier formulations for mucosal immunity. *Current Opinion in Immunology*, *23*(3), 414–420. doi:10.1016/j.coi.2011.03.009 PMID:21511452

Laxmi, M., Bhardwaj, A., Mehta, S., & Mehta, A. (2015). Development and characterisation of nanoemulsion as carrier for the enhancement of bioavailability of artemether. *Artificial Cells, Nanomedicine, and Biotechnology*, *43*(5), 334–344.

Leandro, F. Z., Martins, J., Fontes, A. M., & Tedesco, A. C. (2017). Evaluation of theranostic nanocarriers for near-infrared imaging and photodynamic therapy on human prostate cancer cells. *Colloids and Surfaces. B, Biointerfaces*, *154*, 341–349. doi:10.1016/j.colsurfb.2017.03.042 PMID:28365423

LeCluyse, E. L., Alexandre, E., Hamilton, G. A., Viollon-Abadie, C., Coon, D. J., Jolley, S., & Richert, L. (2005). Isolation and culture of primary human hepatocytes. *Methods in Molecular Biology (Clifton, N.J.)*, *290*, 207–229. doi:10.1385/1-59259-838-2:207 PMID:15361665

Lee, D. K., Wang, C.-F., Yang, Y., & Vaccari, D. (2015). *International Flavors and Fragrances Inc., Flavor Nanoemulsions and Methods of Preparing the Same.* US Patent PCT/US20 15/056470, WO2016064883A1. Available online at: https:// patents.google.com/patent/WO2016064883A1/en

Lee, I., Ahn, B., Choi, J., Hattori, M., Min, B. S., & Bae, K. H. (2011). Selective cholinesterase inhibition by lanostane-triterpenes from fruiting bodies of *Ganoderma lucidum. Bioorganic & Medicinal Chemistry Letters*, *21*(21), 6603–6607. doi:10.1016/j.bmcl.2011.04.042 PMID:21924611

Lee, I., Kim, J., Ryio, I., Kim, Y., Choo, S., Yoo, I., Min, B., Na, M., Hattori, M., & Bae, K. (2010). Lanostane triterpenes from *Ganoderma lucidum* suppress the adipogenesis in 3T3-L1 cells through down-regulation of SREBP-1c. *Bioorganic & Medicinal Chemistry Letters*, *20*(18), 5577–5581. doi:10.1016/j.bmcl.2010.06.093 PMID:20702092

Lee, J. H., & Nan, A. (2012). Combination drug delivery approaches in metastatic breast cancer. *Journal of Drug Delivery*, *2012*.

Lee, J. S., Heo, J., Libbrecht, L., Chu, I. S., Kaposi-Novak, P., Calvisi, D. F., Mikaelyan, A., Roberts, L. R., Demetris, A. J., Sun, Z., Nevens, F., Roskams, T., & Thorgeirsson, S. S. (2006). A novel prognostic subtype of human hepatocellular carcinoma derived from hepatic progenitor cells. *Nature Medicine*, *12*(4), 410–416. doi:10.1038/nm1377 PMID:16532004

Lee, K. W., Omar, D., Abdan, K., & Wong, M. Y. (2016). Physiochemical characterization of nanoemulsion formulation of phenazine and their antifungal efficacy against Ganoderma boninense PER71 in vitro. *Research Journal of Pharmaceutical, Biological and Chemical Sciences*, *7*(6), 3056–3066.

Lee, S. E., Nguyen, C. T., Kim, S. Y., Thi, T. N., & Rhee, J. H. (2015). Tetanus toxin fragment C fused to flagellin makes a potent mucosal vaccine. *Clinical and Experimental Vaccine Research*, *4*(1), 59. doi:10.7774/cevr.2015.4.1.59 PMID:25649002

Lee, S. J., Yeo, W. H., Yun, B. S., & Yoo, I. D. (1999). Isolation and sequence analysis of new Peptaibol, Boletusin, from Boletus spp. *Journal of Peptide Science*, *5*(8), 374–378. doi:10.1002/(SICI)1099-1387(199908)5:8<374::AID-PSC211>3.0.CO;2-X PMID:10507687

Lee, S., Lee, D., Jang, T. S., Kang, K., Nam, J.-W., Lee, H.-J., & Kim, K. (2017). Anti-inflammatory phenolic metabolites fromthe edible fungus *Phellinus* baumii in LPS-stimulated RAW 264.7 cells. *Molecules (Basel, Switzerland)*, *22*(10), 1583. doi:10.3390/molecules22101583

Lee, V. A., Karthikeyan, R., Rawls, H. R., & Amaechi, B. T. (2010). Anti-cariogenic effect of a cetylpyridinium chloride-containing nanoemulsion. *Journal of Dentistry*, *38*(9), 742–749. doi:10.1016/j.jdent.2010.06.001 PMID:20600554

Lehtola & Anttila. (2005). *Novel oral formulations of ospemifene.* Academic Press.

Leite, C. B. S., Coelho, J. M., Muehlmann, L. A., Azevedo, R. B., & Sousa, M. H. (2018). Microemulsions as Platforms for Transdermal Delivery of Hydrophilic Drugs - A Review. *Current Nanoscience*, *14*(3), 170–178. Advance online publication. doi:10.2174/1573413714666171218145416

Leone, A., Di Gennaro, E., Bruzzese, F., Avallone, A., & Budillon, A. (2014). New perspective for an old antidiabetic drug: Metformin as anticancer agent. *Cancer Treatment and Research*, *159*, 355–376. doi:10.1007/978-3-642-38007-5_21 PMID:24114491

Leong, T. S. H., Zhou, M., Kukan, N., Ashokkumar, M., & Martin, G. J. O. (2017). Preparation of water-in-oil-in-water emulsions by low frequency ultrasound using skim milk and sunflower oil. *Food Hydrocolloids*, *63*, 685–695. doi:10.1016/j.foodhyd.2016.10.017

Leroux-Roels, I., Borkowski, A., Vanwolleghem, T., Dramé, M., Clement, F., Hons, E., Devaster, J. M., & Leroux-Roels, G. (2007). Antigen sparing and cross-reactive immunity with an adjuvanted rH5N1 prototype pandemic influenza vaccine: A randomised controlled trial. *Lancet*, *370*(9587), 580–589. doi:10.1016/S0140-6736(07)61297-5 PMID:17707753

Leth-Larsen, R., Zhong, F., Chow, V. T., Holmskov, U., & Lu, J. (2007). The SARS coronavirus spike glycoprotein is selectively recognized by lung surfactant protein D and activates macrophages. *Immunobiology*, *212*(3), 201–211. doi:10.1016/j.imbio.2006.12.001 PMID:17412287

Letsididi, K. S., Lou, Z., Letsididi, R., Mohammed, K., & Maguy, B. L. (2018). Antimicrobial and antibiofilm effects of trans-cinnamic acid nanoemulsion and its potential application on lettuce. *Lebensmittel-Wissenschaft + Technologie*, *94*, 25–32. doi:10.1016/j.lwt.2018.04.018

Levine, M. M., & Woodrow, G. C. (Eds.). (2004). New generation vaccines. New York: Marcel Dekker.

Liang, J., Yan, H., Yang, H.-J., Kim, H. W., Wan, X., Lee, J., & Ko, S. (2016). Synthesis and controlled-release properties of chi-tosan/ Lactoglobulin nanoparticles as carriers for oral administration of epigallocatechin gallate. *Food Science and Biotechnology*, *25*(6), 1583–1590. doi:10.100710068-016-0244-y PMID:30263448

Liang, R., Xu, S., Shoemaker, C. F., Li, Y., Zhong, F., & Huang, Q. (2012). Physical and antimicrobial properties of peppermint oil nanoemulsions. *Journal of Agricultural and Food Chemistry*, *60*(30), 7548–7555. doi:10.1021/jf301129k PMID:22746096

Liang, Z., Zhu, H., Wang, X., Jing, B., Li, Z., Xia, X., Sun, H., Yang, Y., Zhang, W., Shi, L., Zeng, H., & Sun, B. (2020). Adjuvants for Coronavirus Vaccines. *Frontiers in Immunology*, *11*, 2896. doi:10.3389/fimmu.2020.589833 PMID:33240278

Liao, F., Liu, L., Luo, E., & Hu, J. (2018). Curcumin enhances anti-tumor immune response in tongue squamous cell carcinoma. *Archives of Oral Biology*, *92*, 32–37. doi:10.1016/j.archoralbio.2018.04.015 PMID:29751146

Liew, S., Utra, U., Alias, A. K., Tan, T. B., Tan, C. P., & Yussof, N. S. (2020). Physical, morphological and antibacterial properties of lime essential oil nanoemulsions prepared via spontaneous emulsification method. *Lebensmittel-Wissenschaft + Technologie*, *128*, 109388. Advance online publication. doi:10.1016/j.lwt.2020.109388

Li, F., Wan, L., Xiao, T., Liu, H., Jiang, Y., Zhao, X., Wang, R., & Wan, K. (2018). *In vitro* activity of *β*-Lactams in Combination with *β*-Lactamase Inhibitors against *Mycobacterium tuberculosis* Clinical Isolates, *BioMed Research International. Article ID*, *3579832*, 1–8. Advance online publication. doi:10.1155/2018/3579832 PMID:30065936

Li, G., Zhang, Z., Liu, H., & Hu, L. (2021). Nanoemulsion-based delivery approaches for nutraceuticals: Fabrication, application, characterization, biological fate, potential toxicity and future trends. *Food & Function*, *12*(5), 1933–1953. doi:10.1039/D0FO02686G PMID:33596279

Li, J., & Bluth, M. H. (2011). Pharmacogenomics of drug metabolizing enzymes and transporters: Implications for cancer therapy. *Pharmacogenomics and Personalized Medicine*, *4*, 11–33. doi:10.2147/PGPM.S18861 PMID:23226051

Li, M., & Fogler, H. (1978). Acoustic emulsification- II Break-up of the large primary oil droplets in a water medium. *Journal of Fluid Mechanics*, *88*(3), 513–528. doi:10.1017/S0022112078002244

Li, M., McClements, D. J., Liu, X., & Liu, F. (2020). Design principles of oil-in-water emulsions with functionalized interfaces: Mixed, multilayer, and covalent complex structures. *Comprehensive Reviews in Food Science and Food Safety*, *19*(6), 3159–3190. doi:10.1111/1541-4337.12622 PMID:33337043

Lima, T. S., Silva, M. F. S., Nunes, X. P., Colombo, A. V., Oliveira, H. P., Goto, P. L., Blanzat, M., Piva, H. L., Tedesco, A. C., & Siqueira-Moura, M. P. (2021). Cineole-containing nanoemulsion: Development, stability, and antibacterial activity. *Chemistry and Physics of Lipids*, *239*, 105113. doi:10.1016/j.chemphyslip.2021.105113 PMID:34216586

Lim, C. J., Basri, M., Omar, D., Abdul Rahman, M. B., Salleh, A. B., & Raja Abdul Rahman, R. N. Z. (2013). Green nanoemulsion-laden glyphosate isopropylamine formulation in suppressing creeping foxglove (A. gangetica), slender button weed (D. ocimifolia) and buffalo grass (P. conjugatum). *Pest Management Science*, *69*(1), 104–111. doi:10.1002/ps.3371 PMID:22865686

Li, N., Chen, Y., Sun, H., Huang, T., Chen, T., Jiang, Y., Yang, Q., Yan, X., & Wu, M. (2020). Decreasing acute toxicity and suppressing colorectal carcinoma using Sorafenib-loaded nanoparticles. *Pharmaceutical Development and Technology*, *25*(5), 556–565. doi:10.1080/10837450.2020.1718704 PMID:31958240

Li, N., Nguyen, H. H., Byrom, M., & Ellington, A. D. (2011). Inhibition of cell proliferation by an anti-EGFR aptamer. *PLoS One*, *6*(6), e20299. https://doi.org/10.1371/journal.pone.0020299

Lin, C.-H., Sheu, G.-T., Lin, Y.-W., Yeh, C.-S., Huang, Y.-H., Lai, Y.-C., Chang, J.-G., & Ko, J.-L. (2010). A new immunomodulatory protein from *Ganoderma microsporum* inhibits epidermal growth factor mediated migration and invasion in A549 lung cancer cells. *Process Biochemistry (Barking, London, England)*, *45*(9), 1537–1542. doi:10.1016/j.procbio.2010.06.006

Lin, H. Y., Thomas, J. L., Chen, H. W., Shen, C. M., Yang, W. J., & Lee, M. H. (2012). In vitro suppression of oral squamous cell carcinoma growth by ultrasound-mediated delivery of curcumin microemulsions. *International Journal of Nanomedicine*, *7*, 941–951. doi:10.2147/IJN.S28510 PMID:22393291

Lin, H., Zhou, H., Xu, L., Zhu, H., & Huang, H. (2016). Effect of surfactant concentration on the spreading properties of pesticide droplets on Eucalyptus leaves. *Biosystems Engineering*, *143*, 42–49. doi:10.1016/j.biosystemseng.2016.01.003

Lin, L., Cui, H., Zhou, H., Zhang, X., Bortolini, C., Chen, M., Liu, L., & Dong, M. (2015). Nanoliposomes containing *Eucalyptus citriodora* as antibiotic with specific antimicrobial activity. *Chemical Communications (Cambridge)*, *51*(13), 2653–2655. doi:10.1039/C4CC09386K PMID:25573466

Liolios, C. C., Gortzi, O., Lalas, S., Tsaknis, J., & Chinou, I. (2009). Liposomal incorporation of carvacrol and thymol isolated from the essential oil of Origanum dictamnus L. and in vitro antimicrobial activity. *Food Chemistry*, *112*(1), 77–83. doi:10.1016/j.foodchem.2008.05.060

Li, P. H., & Chiang, B. H. (2012). Process optimization and stability of Dlimonene-in-water nanoemulsions prepared by ultrasonic emulsification using response surface methodology. *Ultrasonics Sonochemistry*, *19*(1), 192–197. doi:10.1016/j.ultsonch.2011.05.017 PMID:21680223

Li, S., Sun, J., Yan, J., Zhang, S., Shi, C., McClements, D. J., Liu, X., & Liu, F. (2021). Development of antibacterial nanoemulsions incorporating thyme oil: Layer-by-layer self-assembly of whey protein isolate and chitosan hydrochloride. *Food Chemistry*, *339*, 128016. doi:10.1016/j.foodchem.2020.128016 PMID:33152858

Lis-Balchin, M., & Deans, S. G. (1997). Bioactivity of selected plant essential oils against Listeria monocytogenes. *Journal of Applied Microbiology*, *82*(6), 759–762. doi:10.1046/j.1365-2672.1997.00153.x PMID:9202441

Li, T., Han, X., Bao, R., Hao, Y., & Li, S. (2019). Preparation and properties of water-in-oil shiitake mushroom polysaccharide nanoemulsion. *International Journal of Biological Macromolecules*, *140*, 343–349. doi:10.1016/j.ijbiomac.2019.08.134 PMID:31425759

Liu, C. H., & Yu, S. Y. (2010). Cationic NE's as non-viral vectors for plasmid DNA delivery. *Colloids and Surfaces. B, Biointerfaces*, *79*(2), 509–515. doi:10.1016/j.colsurfb.2010.05.026 PMID:20541375

Liu, F., Lin, S., Zhang, Z., Hu, J., Liu, G., Tu, Y., Yang, Y., Zou, H., Mo, Y., & Miao, L. (2014). pH responsive nanoemulsions for controlled drug release. *Biomacromolecules*, *15*(3), 968–977. doi:10.1021/bm4018484 PMID:24527876

Liu, F., Yang, J., Huang, L., & Liu, D. (1996). Effect of non-ionic surfactants on the formation of DNA/emulsion complexes and emulsion-mediated gene transfer. *Pharmaceutical Research, 13*(11), 1642–1646. doi:10.1023/A:1016480421204 PMID:8956328

Liu, K., Xiao, X., Wang, J., Chen, C.-Y. O., & Hu, H. (2017). Polyphenolic composition and antioxidant, antiproliferative, and antimicrobial activities of mushroom *Inonotus sanghuang. Lebensmittel-Wissenschaft + Technologie, 82*, 154–161. doi:10.1016/j.lwt.2017.04.041

Liu, M., Yang, C., Liu, E., Zhang, F., Meng, X., & Liu, B. (2021). Effect of environmental stresses on physicochemical properties of ALA oil-in-water nanoemulsion system prepared by emulsion phase inversion. *Food Chemistry, 343*, 128475. doi:10.1016/j.foodchem.2020.128475 PMID:33168257

Liu, Q., Huang, H., Chen, H., Lin, J., & Wang, Q. (2019). Food-Grade Nanoemulsions: Preparation, Stability and Application in Encapsulation of Bioactive Compounds. *Molecules (Basel, Switzerland), 24*(23), 4242. doi:10.3390/molecules24234242 PMID:31766473

Liu, Q., Ng, T., & Wang, H. (2013). Isolation and characterization of a novel lectin from the wild mushroom *Oudemansiella radicata* (Relhan.: Fr.) sing. *Biotechnology and Bioprocess Engineering; BBE, 18*(3), 465–471. doi:10.100712257-012-0699-5

Liu, R. H. (2013). Dietary bioactive compounds and their health implications. *Journal of Food Science, 78*(s1, Suppl), A18–A25. doi:10.1111/1750-3841.12101 PMID:23789932

Liu, T., & Liu, L. (2020). Fabrication and characterization of chitosan nanoemulsions loading thymol or thyme essential oil for the preservation of refrigerated pork. *International Journal of Biological Macromolecules, 162*, 1509–1515. doi:10.1016/j.ijbiomac.2020.07.207 PMID:32738320

Liu, X. T., Winkler, A. L., Schwan, W. R., Volk, T. J., Rott, M. A., & Monte, A. (2010). Antibacterial compounds from mushrooms I: A Lanostane-type triterpene and prenylphenol derivates from *Jahnoporus hiritus* and *Albatrellus flettii* and their activities against *Bacillus cereus* and *Enterococcus faecalis. Planta Medica, 76*(02), 182–185. doi:10.1055-0029-1186001 PMID:19644795

Liu, X., Bi, J., Xiao, H., & McClements, D. J. (2015). Increasing carotenoid bioaccessibility from yellow peppers using excipient emulsions: Impact of lipid type and thermal processing. *Journal of Agricultural and Food Chemistry, 63*(38), 8534–8543. doi:10.1021/acs.jafc.5b04217 PMID:26357977

Liu, X., Bi, J., Xiao, H., & McClements, D. J. (2016). Enhancement of nutraceutical bioavailability using excipient nanoemulsions: Role of lipid digestion products on bioaccessibility of carotenoids and phenolics from mangoes. *Journal of Food Science, 81*(3), N754–N761. doi:10.1111/1750-3841.13227 PMID:26816342

Livermore, D. M. (2002). Multiple mechanisms of antimicrobial resistance in Pseudomonas aeruginosa: Our worst nightmare. *Clinical Infectious Diseases, 34*(5), 634–640. doi:10.1086/338782

Li, W., Chen, H., He, Z., Han, C., Liu, S., & Li, Y. (2015). Influence of surfactant and oil composition on the stability and antibacterial activity of eugenol nanoemulsions. *Lebensmittel-Wissenschaft + Technologie, 62*(1), 39–47. doi:10.1016/j.lwt.2015.01.012

Li, X., Anton, N., Zuber, G., Zhao, M., Messaddeq, N., Hallouard, F., Fessi, H., & Vandamme, T. F. (2013). Iodinated α-tocopherol nano-emulsions as non-toxic contrast agents for preclinical X-ray imaging. *Biomaterials, 34*(2), 481–491. doi:10.1016/j.biomaterials.2012.09.026 PMID:23083930

Li, X., Müller, R. H., Keck, C. K., & Bou-Chacra, N. A. (2016). Mucoadhesive dexamethasone acetate-polymyxin B sulfate cationic ocular nanoemulsion – novel combinatorial formulation concept. *Die Pharmazie*, *71*, 327–333. doi:10.1691/ph.2016.5190 PMID:27455551

Li, X., Wu, G., Qi, X., Zhang, H., Wang, L., & Qian, H. (2019). Physicochemical properties of stable multilayer nanoemulsion prepared via the spontaneously-ordered adsorption of short and long chains. *Food Chemistry*, *274*, 620–628. doi:10.1016/j.foodchem.2018.09.002 PMID:30372987

Li, X., & Xia, W. (2011). Effect of concentration, degree of deacetylation and molecular weigh on emulsifying properties of chitosan. *International Journal of Biological Macromolecules*, *48*(5), 768–772. doi:10.1016/j.ijbiomac.2011.02.016 PMID:21382402

Li, X., Xing, Y., Jiang, Y., Ding, Y., & Li, W. (2009). Antimicrobial activities of ZnO powder-coated PVC film to inactivate food pathogens. *International Journal of Food Science & Technology*, *44*(11), 2161–2168. doi:10.1111/j.1365-2621.2009.02055.x

Li, Y., Hu, M., & McClements, D. J. (2011). Factors affecting lipase digestibility of emulsified lipids using an in vitro digestion model: Proposal for a standardised pH-stat method. *Food Chemistry*, *126*(2), 498–505.

Li, Y., Jin, H., Sun, X., Sun, J., Liu, C., Liu, C., & Xu, J. (2018). Physicochemical Properties and Storage Stability of Food Protein-Stabilized Nanoemulsions. *Nanomaterials (Basel, Switzerland)*, *9*(1), 25. doi:10.3390/nano9010025 PMID:30585224

Lococo, D., Mora-Huertas, C. E., Fessi, H., Zaanoun, I., & Elaissari, A. (2012). Argan oil nanoemulsions as new hydrophobic drug-loaded delivery system for transdermal application. *Journal of Biomedical Nanotechnology*, *8*(5), 843–848. doi:10.1166/jbn.2012.1445 PMID:22888756

Lohith Kumar, D. H., & Sarkar, P. (2017). Encapsulation of bioactive compounds using nanoemulsions. *Environmental Chemistry Letters*, *16*(1), 59–70. doi:10.100710311-017-0663-x

Long, Y., Huang, W., Wang, Q., & Yang, G. (2020). Green synthesis of garlic oil nanoemulsion using ultrasonication technique and its mechanism of antifungal action against Penicillium italicum. *Ultrasonics Sonochemistry*, *64*, 104970. doi:10.1016/j.ultsonch.2020.104970 PMID:32014757

Lopes, E., Pohlmann, A. R., Bassani, V., & Guterres, S. S. (2000). Polymeric colloidal systems containing ethionamide: Preparation and physico-chemical characterization. *Die Pharmazie*, *55*(7), 527–530. PMID:10944782

Lopez-Diez, E., Perez, S., Carballo, M., Iñarrea, A., de la Orden, A., Castro, M., Rodríguez, M., Almuster, S., Montero, R., Perez, M., Sanchez, J., & Ojea, A. (2017). Lifestyle factors and oncogenic papillomavirus infection in a high-risk male population. *PLoS One*, *12*(9), e0184492. doi:10.1371/journal.pone.0184492 PMID:28898279

Louni, M., Shakarami, J., & Negahban, M. (2018). Insecticidal efficiency of nanoemulsion containing mentha longifolia essential oil against ephestia kuehniella (Lepidoptera: Pyralidae). *Journal of Crop Protection*, *7*(2), 171–182.

Lou, Z., Chen, J., Yu, F., Wang, H., Kou, X., Ma, C., & Zhu, S. (2017). The antioxidant, antibacterial, antibiofilm activity of essential oil from Citrus medica L. var. sarcodactylis and its nanoemulsion. *Lebensmittel-Wissenschaft + Technologie*, *80*, 371–377. doi:10.1016/j.lwt.2017.02.037

Lou, Z., Letsididi, K. S., Yu, F., Pei, Z., Wang, H., & Letsididi, R. (2019). Inhibitive effect of eugenol and its nanoemulsion on quorum sensing–mediated virulence factors and biofilm formation by Pseudomonas aeruginosa. *Journal of Food Protection*, *82*(3), 379–389. doi:10.4315/0362-028X.JFP-18-196

Lovelyn, C., & Attama, A. A. (2011). Current state of nanoemulsions in drug delivery. *Journal of Biomaterials and Nanobiotechnology*, *2*(05), 626–639. doi:10.4236/jbnb.2011.225075

Lowel, G., Amselem, S., Friedman, D., & Aviv, H. (1999). *Submicron emulsions as vaccine adjuvants*. Academic Press.

Luengo, J., Weiss, B., Schneider, M., Ehlers, A., Stracke, F., König, K., & Schaefer, U. F. (2006). Influence of Nanoencapsulation on Human Skin Transport of Flufenamic Acid. *Skin Pharmacology and Physiology*, *19*(4), 190–197. doi:10.1159/000093114 PMID:16679821

Lu, H. L., Syu, W. J., Nishiyama, N., Kataoka, K., & Lai, P. S. (2011). Dendrimer phthalocyanine-encapsulated polymeric micelle-mediated photochemical internalization extends the efficacy of photodynamic therapy and overcomes drug-resistance in vivo. *Journal of Controlled Release*, *155*(3), 458–464.

Luís, Â., Duarte, A., Gominho, J., Domingues, F., & Duarte, A. P. (2016). Chemical composition, antioxidant, antibacterial and anti-quorum sensing activities of Eucalyptus globulus and Eucalyptus radiata essential oils. *Industrial Crops and Products*, *79*, 274–282. doi:10.1016/j.indcrop.2015.10.055

Luke, M. C., & Coffey, D. S. (1994). Human androgen receptor binding to the androgen response element of prostate specific antigen. *Journal of Andrology*, *15*(1), 41–51. PMID:7514587

Luna-Pineda, V. M., Ochoa, S., Cruz-Córdova, A., Cázares-Domínguez, V., Vélez-González, F., Hernández-Castro, R., & Xicohtencatl-Cortes, J. (2019). Infecciones del tracto urinario, inmunidad y vacunación. *Boletín Médico del Hospital Infantil de México*, *75*(2), 67–78. doi:10.24875/BMHIM.M18000011 PMID:29658949

Luo, Y., Zhang, Y., Pan, K., Critzer, F., Davidson, P., & Zhong, Q. (2014). Self-Emulsification of Alkaline-Dissolved Clove Bud Oil by Whey Protein, Gum Arabic, Lecithin, and Their Combinations. *Journal of Agricultural and Food Chemistry*, *62*(19), 4417–4424. doi:10.1021/jf500698k PMID:24758517

Lu, R., Liu, S., Wang, Q., & Li, X. (2015). Nanoemulsions as novel oral carriers of stiripentol: Insights into the protective effect and absorption enhancement. *International Journal of Nanomedicine*, *10*, 4937–4946. PMID:26261418

Lu, W. C., Huang, D. W., Wang, C. R., Yeh, C. H., Tsai, J. C., Huang, Y. T., & Li, P. H. (2018). Preparation, characterization, and antimicrobial activity of nanoemulsions incorporating citral essential oil. *Yao Wu Shi Pin Fen Xi*, *26*(1), 82–89. doi:10.1016/j.jfda.2016.12.018 PMID:29389592

Lu, Y., Qi, J., & Wu, W. (2012). Absorption, disposition and pharmacokinetics of nanoemulsions. *Current Drug Metabolism*, *13*(4), 396–417. doi:10.2174/138920012800166544 PMID:22443535

Lu, Y., Wang, Y., & Tang, X. (2008). Formulation and thermal sterile stability of a less painful intravenous clarithromycin emulsion containing vitamin E. *International Journal of Pharmaceutics*, *346*(1-2), 47–56. doi:10.1016/j.ijpharm.2007.06.003 PMID:17640833

Lu, Y., Yue, Z., Xie, J., Wang, W., Zhu, H., Zhang, E., & Cao, Z. (2018). Micelles with ultralow critical micelle concentration as carriers for drug delivery. *Nature Biomedical Engineering*, *2*(5), 318–325. doi:10.103841551-018-0234-x PMID:30936455

Lv, S., Gu, J., Zhang, R., Zhang, Y., Tan, H., & McClements, D. J. (2018). Vitamin E encapsulation in plant-Based nanoemulsions fabricated using dual-channel microfluidization: Formation, stability, and bioaccessibility. *Journal of Agricultural and Food Chemistry*, *66*(40), 10532–10542. doi:10.1021/acs.jafc.8b03077 PMID:30240207

Lycke, N. (2012). Recent progress in mucosal vaccine development: Potential and limitations. *Nature Reviews. Immunology*, *12*(8), 592–605. doi:10.1038/nri3251 PMID:22828912

Maccelli, A., Vitanza, L., Imbriano, A., Fraschetti, C., Filippi, A., Goldoni, P., Maurizi, L., Ammendolia, M. G., Crestoni, M. E., Fornarini, S., Menghini, L., Carafa, M., Marianecci, C., Longhi, C., & Rinaldi, F. (2019). Satureja montana L. Essential Oils: Chemical Profiles/Phytochemical Screening, Antimicrobial Activity and O/W NanoEmulsion Formulations. *Pharmaceutics*, *12*(1), 7. doi:10.3390/pharmaceutics12010007 PMID:31861717

MacDougall, D. B. (2010). Colour measurement of food: principles and practice. In *Colour Measurement* (pp. 312–342). Woodhead Publishing. doi:10.1533/9780857090195.2.312

Macfarlane, G. T., & Macfarlane, S. (2011). Fermentation in the human large intestine: Its physiologic consequences and the potential contribution of prebiotics. *Journal of Clinical Gastroenterology*, *45*, S120–S127.

Machado, F. C., Adum de Matos, R. P., Primo, F. L., Tedesco, A. C., Rahal, P., & Calmon, M. F. (2019). Effect of curcumin-nanoemulsion associated with photodynamic therapy in breast adenocarcinoma cell line. *Bioorganic & Medicinal Chemistry*, *27*(9), 1882–1890.

Macha, S., Hughes, P. M., & Mitra, A. K. (2003). Overview of ocular drug delivery. *Drugs and the Pharmaceutical Sciences*, *130*, 1–12.

Ma, D., Tu, Z. C., Wang, H., Zhang, Z., & McClements, D. J. (2017). Fabrication and characterization of nanoemulsion-coated microgels: Electrostatic deposition of lipid droplets on alginate beads. *Food Hydrocolloids*, *71*, 149–157. doi:10.1016/j.foodhyd.2017.05.015

Maeda, H. (2012). Vascular permeability in cancer and infection as related to macromolecular drug delivery, with emphasis on the EPR effect for tumor-selective drug targeting. *Proceedings of the Japan Academy. Series B, Physical and Biological Sciences*, *88*(3), 53–71. Advance online publication. doi:10.2183/pjab.88.53 PMID:22450535

Maeda, H., Wu, J., Sawa, T., Matsumura, Y., & Hori, K. (2000). Tumor vascular permeability and the EPR effect in macromolecular therapeutics: A review. *Journal of Controlled Release*, *65*(1-2), 271–284.

Ma, G., Yang, W., Zhao, L., Pei, F., Fang, D., & Hu, Q. (2018). A critical review on the health promoting effects of mushrooms nutraceuticals. *Food Science and Human Wellness*, *7*(2), 125–133. doi:10.1016/j.fshw.2018.05.002

Mahabir, S., Schendel, K., Dong, Y. Q., Barrera, S. L., Spitz, M. R., & Forman, M. R. (2008). Dietary α-, β-, γ-and δ-tocopherols in lung cancer risk. *International Journal of Cancer*, *123*(5), 1173–1180. doi:10.1002/ijc.23649 PMID:18546288

Mahajan, H. S., Mahajan, M. S., Nerkar, P. P., & Agrawal, A. (2013). Nanoemulsion-based intranasal drug delivery system of saquinavir mesylate for brain targeting. *Drug Delivery*, *21*(2), 148–154. doi:10.3109/10717544.2013.838014 PMID:24128122

Mahato, R. (2017). Nanoemulsion as Targeted Drug Delivery System for Cancer Therapeutics. *Journal of Pharmaceutical Sciences and Pharmacology*, *3*(2), 83–97. doi:10.1166/jpsp.2017.1082

Mahato, R. (2017). Nanoemulsion as targeted drug delivery system for cancer therapeutics. *Journal of Pharmaceutical Sciences and Pharmacology*, *3*, 82–97.

Mahato, R., Tai, W., & Cheng, K. (2011). Prodrugs for improving tumor targetability and efficiency. *Advanced Drug Delivery Reviews*, *63*(8), 659–670.

Mahboobian, M. M., Seyfoddin, A., Aboofazeli, R., Foroutan, S. M., & Rupenthal, I. D. (2019). Brinzolamide-loaded nanoemulsions: Ex vivo transcorneal permeation, cell viability and ocular irritation tests. *Pharmaceutical Development and Technology*, *24*(5), 600–606.

Mahdi Jafari, S., He, Y., & Bhandari, B. (2006). Nano-Emulsion production by sonication and microfluidization - A aomparison. *International Journal of Food Properties*, *9*(3), 475–485. doi:10.1080/10942910600596464

Maher, Z. E., & Abdou, E. S. (2013). Chitosan based edible films and coatings: A review. *Materials Science and Engineering*, *33*(4), 1819–1841. doi:10.1016/j.msec.2013.01.010 PMID:23498203

Mahour, R., Sahni, J. K., Sharma, S., Kumar, S., Ali, J., & Baboota, S. (2015). Nanoemulsion as a tool for improvement of cilostazol oral bioavailability. *Journal of Molecular Liquids*, *212*, 792–798.

Mahour, R., Sahni, J. K., Sharma, S., Kumar, S., Ali, J., & Baboota, S. (2015). Nanoemulsion as a tool for improvement of Cilostazol oral bioavailability. *Journal of Molecular Liquids*, *212*, 792–798. doi:10.1016/j.molliq.2015.10.027

Majeed, H., Liu, F., Hategekimana, J., Sharif, H. R., Qi, J., Ali, B., Bian, Y.Y., Ma, J., Yokoyama, W., & Zhong, F. (2016). Bactericidal action mechanism of negatively charged food grade clove oil nanoemulsions. *Food Chemistry*, *197*(Prt A), 75-83. . doi:10.1016/j.foodchem.2015.10.015

Majeed, H., Antoniou, J., & Fang, Z. (2014). Apoptotic effects of eugenol-loaded NEs in human colon and liver cancer cell lines. *Asian Pacific Journal of Cancer Prevention*, *15*(21), 9159–9164. doi:10.7314/APJCP.2014.15.21.9159 PMID:25422195

Majeed, H., Antoniou, J., Hategekimana, J., Sharif, H. R., Haider, J., Liu, F., Ali, B., Rong, L., Ma, J., & Zhong, F. (2016). Influence of carrier oil type, particle size on in vitro lipid digestion and eugenol release in emulsion and nanoemulsions. *Food Hydrocolloids*, *52*, 415–422. doi:10.1016/j.foodhyd.2015.07.009

Makidon, P. E. (2011). *Moxifloxacin nanoemulsion and its preparation method.* https://patents.google.com/patent/CN102920656A/en

Makidon, P. E., Bielinska, A. U., Nigavekar, S. S., Janczak, K. W., Knowlton, J., Scott, A. J., Mank, N., Cao, Z., Rathinavelu, S., Beer, M. R., Wilkinson, J. E., Blanco, L. P., Landers, J. J., & Baker, J. R. Jr. (2008). Pre-clinical evaluation of a novel nanoemulsion-based hepatitis B mucosal vaccine. *PLoS One*, *3*(8), e2954. doi:10.1371/journal.pone.0002954 PMID:18698426

Makidon, P. E., Nigavekar, S. S., Bielinska, A. U., Mank, N., Shetty, A. M., Suman, J., & Baker, J. R. (2010). Characterization of stability and nasal delivery systems for immunization with nanoemulsion-based vaccines. *Journal of Aerosol Medicine and Pulmonary Drug Delivery*, *23*(2), 77–89. doi:10.1089/jamp.2009.0766 PMID:19778268

Makidon, P., Belyakov, I., Blanco, L., Janczak, K., Landers, J., Bielinska, A., Groom, J. V. II, & Baker, J. R. Jr. (2012). Nanoemulsion mucosal adjuvant uniquely activates cytokine production by nasal ciliated epithelium and induces dendritic cell trafficking. *European Journal of Immunology*, *42*(8), 2073–2086. doi:10.1002/eji.201142346 PMID:22653620

Makvandi, P., Josic, U., Delfi, M., Pinelli, F., Jahed, V., Kaya, E., Ashrafizadeh, M., Zarepour, A., Rossi, F., Zarrabi, A., Agarwal, T., Zare, T. N., Ghomi, M., Maiti, T. K., Breschi, L., & Tay, F. R. (2021). Drug Delivery (Nano)Platforms for Oral and Dental Applications: Tissue Regeneration, Infection Control, and Cancer Management. *Advanced Science (Weinheim, Baden-Wurttemberg, Germany)*, *8*(8), 2004014. doi:10.1002/advs.202004014 PMID:33898183

Malacara, D. (2002). Color vision and colorimetry: Theory and applications. *SPIE Press*, *7*, 164.

Malarvizhi, G. L., Retnakumari, A. P., Nair, S., & Koyakutty, M. (2014). Transferrin targeted core-shell nanomedicine for combinatorial delivery of doxorubicin and sorafenib against hepatocellular carcinoma. *Nanomedicine (London)*, *10*(8), 1649–1659. doi:10.1016/j.nano.2014.05.011 PMID:24905399

Mallick, S., Pattnaik, S., Swain, K., & De, P. K. (2007). Current Perspectives of Solubilization: Potential for Improved Bioavailability. *Drug Development and Industrial Pharmacy*, *33*(8), 865–873. doi:10.1080/03639040701429333 PMID:17729104

Manchanda, V., Sanchaita, S., & Singh, N. P. (2010). Multidrug resistant acinetobacter. *Journal of Global Infectious Diseases*, *2*(3), 291. doi:10.4103/0974-777X.68538

Mann, B., Sonu, K. S., Sharma, R., Kumar, R., & Singh, R. (2020). Formulation of peppermint oil nanoemulsion using conjugates of whey proteins with maltodextrin and its characterization. *Indian Journal of Traditional Knowledge*, *19*(2), 394–400.

Manning, G. S. (1978). Limiting laws and counterion condensation in polyelectrolyte solutions V. Further development of the chemical model. *Biophysical Chemistry*, *9*(1), 65–70. doi:10.1016/0301-4622(78)87016-1 PMID:753405

Manoharan, Y., Haridas, V., Vasanthakumar, K. C., Muthu, S., Thavoorullah, F. F., & Shetty, P. (2020). Curcumin: A wonder drug as a preventive measure for COVID19 management. *Indian Journal of Clinical Biochemistry*, *35*(3), 373–375. doi:10.100712291-020-00902-9 PMID:32641876

Manosroi, J., Dhumtanom, P., & Manosroi, A. (2006). Anti-proliferative activity of essential oil extracted from Thai medicinal plants on KB and P388 cell lines. *Cancer Letters*, *235*(1), 114–120. doi:10.1016/j.canlet.2005.04.021 PMID:15979235

Man, S. M., & Kanneganti, T. D. (2016). Converging roles of caspases in inflammasome activation, cell death and innate immunity. *Nature Reviews. Immunology*, *16*(1), 7–21. doi:10.1038/nri.2015.7 PMID:26655628

Mansour, H. M., Rhee, Y. S., & Wu, X. (2009). Nanomedicine in pulmonary delivery. *International Journal of Nanomedicine*, *4*, 299–319.

Manyarara, T. E., Khoza, S., Dube, A., & Maponga, C. C. (2018). Formulation and characterization of a paediatric nanoemulsion dosage form with modified oral drug delivery system for improved dissolution rate of nevirapine. *MRS Advances*, *3*(37), 2203–2219. doi:10.1557/adv.2018.320

Manzoor, S., Gull, A., Wani, S. M., Ganaie, T. A., Masoodi, F. A., Bashir, K., Malik, A. R., & Dar, B. N. (2021). Improving the shelf life of fresh cut kiwi using nanoemulsion coatings with antioxidant and antimicrobial agents. *Food Bioscience*, *41*, 101015. doi:10.1016/j.fbio.2021.101015

Mao, Y., & McClements, D. J. (2013). Modulation of food texture using controlled heteroaggregation of lipid droplets: Principles and applications. *Journal of Applied Polymer Science*, *130*(6), 3833–3841. doi:10.1002/app.39631

Maphosa, Y., & Jideani, V. A. (2018). Factors affecting the stability of emulsions stabilised by biopolymers. *Science and Technology Behind Nanoemulsions*, *65*. Advance online publication. doi:10.5772/intechopen.75308

Ma, Q., Davidson, P. M., & Zhong, Q. (2016). Antimicrobial properties of microemulsions formulated with essential oils, soybean oil, and Tween 80. *International Journal of Food Microbiology*, *226*, 20–25. doi:10.1016/j.ijfoodmicro.2016.03.011 PMID:27016636

Ma, Q., Davidson, P. M., & Zhong, Q. (2016). Nanoemulsions of thymol and eugenol co-emulsified by lauric arginate and lecithin. *Food Chemistry*, *206*, 167–173. doi:10.1016/j.foodchem.2016.03.065 PMID:27041312

Maragheh, A. D., Tabrizi, M. H., Karimi, E., Seyedi, S., & Khatamian, N. (2019). Producing the sour cherry pit oil nanoemulsion and evaluation of its anti-cancer effects on both breast cancer murine model and MCF-7 cell line. *Journal of Microencapsulation*, *36*(4), 399–409.

Marchiori, M. L., Lubini, G., Dalla Nora, G., Friedrich, R. B., Fontana, M. C., Ourique, A. F., Bastos, M. O., Rigo, L. A., Silva, C. B., Tedesco, S. B., & Beck, R. C. (2010). Hydrogel containing dexamethasone-loaded nanocapsules for cutaneous administration: Preparation, characterization, and in vitro drug release study. *Drug Development and Industrial Pharmacy*, *36*(8), 962–971. doi:10.3109/03639041003598960 PMID:20590450

Marcu, A., Pop, S., Dumitrache, F., Mocanu, M., Niculite, C. M., Gherghiceanu, M., Lungu, C. P., Fleaca, C., Ianchis, R., Barbut, A., Grigoriu, C., & Morjan, I. (2013). Magnetic iron oxide nanoparticles as drug delivery system in breast cancer. *Applied Surface Science*, 281.

Marei, G. I. K., Rabea, E. I., & Badawy, M. E. I. (2018). Preparation and characterizations of chitosan/citral nanoemulsions and their antimicrobial activity. *Applied Food Biotechnology*, *5*(2), 69–78.

Marei, G. I. K., Rabea, E. I., & Badawy, M. E. I. (2018). Preparation and Characterizations of Chitosan/Citral Nanoemulsions and their Antimicrobial Activity. *Applied Food Biotechnology*, *5*(2), 69–78. doi:10.22037/afb.v5i2.19005

Marieb, E. N., & Hoehn, K. (2007). *Human Anatomy & Physiology*. Pearson education.

Mark, A. (1996). The sympathetic nervous system in hypertension: a potential long-term regulator of arterial pressure. *Journal of Hypertension, 14*(5), S159-165.

Mark, S., Valentina, K., Yulia, M., Alexander, G., Andrey, P., Andrey, S., Nikolay, P., & Nina, D. (2004). Immunomodulating and anti-tumor action of extracts of several mushrooms. *Journal of Biotechnology*, *113*(1-3), 77–83. doi:10.1016/j.jbiotec.2004.04.034 PMID:15380649

Marsup, P., Yeerong, K., Neimkhum, W., Sirithunyalug, J., Anuchapreeda, S., To-Anun, C., & Chaiyana, W. (2020). Enhancement of chemical stability and dermal delivery of *Cordyceps militaris* Extracts by Nanoemulsion. *Nanomaterials (Basel, Switzerland), 10*(8), 1565. doi:10.3390/nano10081565 PMID:32784892

Martinez, M. N., & Amidon, G. L. (2002). A mechanistic approach to understanding the factors affecting drug absorption: A review of fundamentals. *Journal of Clinical Pharmacology*, *42*(6), 620–643.

Martins, C., Higaki, N. T. F., Montrucchio, D. P., Oliveira, C. F., Gomes, M. L. S., Miguel, M. D., Miguel, O. G., Zanin, S. M. W., & Dias, J. F. G. (2020). Development of W1/O/W2 emulsion with gallic acid in the internal aqueous phase. *Food Chemistry*, *314*, 126174. doi:10.1016/j.foodchem.2020.126174 PMID:31981885

Mason, T. G., Graves, S. M., Wilking, J. N., & Lin, M. Y. (2006). Extreme emulsification: Formation and structure of nanoemulsions. *Condensed Matter Physics*, *9*, 193–199.

Mason, T. G., Wilking, J., Meleson, K., Chang, C. B., & Graves, S. (2006). Nanoemulsions: Formation, structure, and physical properties. *Journal of Physics Condensed Matter*, *18*(41), 079001. doi:10.1088/0953-8984/18/41/R01

Massaguni, R., Raman, I. A., Noor, S., & Latip, H. M. (2016). Efficacy and persistence of neem in emulsion-in-water (EW) formulation against the golden apple snail. *International Journal of Scientific Engineering and Applied Science*, *2*(9), 2395–3470.

Mastelic, B., Ahmed, S., Egan, W., Del Giudice, G., Golding, H., Gust, I., Neels, P., Reed, S. G., Sheets, R. L., Siegrist, C.-A., & Lambert, P.-H. (2010). Mode of action of adjuvants: Implications for vaccine safety and design. *Biologicals*, *38*(5), 594–601. doi:10.1016/j.biologicals.2010.06.002 PMID:20659806

Mate, J., Periago, P. M., & Palop, A. (2016). Combined effect of a nanoemulsion of D-limonene and nisin on Listeria monocytogenes growth and viability in culture media and foods. *Food Science & Technology International*, *22*(2), 146–152. doi:10.1177/1082013215577034 PMID:25788168

Mattila, P., Konko, K., Eurọla, M., Pihlava, J. M., Astola, J., Vahteristo, L., Hietaniemi, V., Kumpulainen, J., Valtonen, M., & Piironen, V. (2001). Contents of vitamins, mineral elements, and some phenolic compounds in cultivated mushrooms. *Journal of Agricultural and Food Chemistry*, *49*(5), 2343–2348. doi:10.1021/jf001525d PMID:11368601

Matuszewska, A., Karp, M., Jaszek, M., Janusz, G., Osińska-Jaroszuk, M., Sulej, J., Stefaniuk, D., Tomczak, W., & Giannopoulos, K. (2016). Laccase purified from *Cerrena unicolor* exerts antitumor activity against leukemic cells. *Oncology Letters*, *11*(3), 2009–2018. doi:10.3892/ol.2016.4220 PMID:26998114

Mau, J. L., Lin, H. C., & Chen, C. C. (2002). Antioxidant properties of several medicinal mushrooms. *Journal of Agricultural and Food Chemistry*, *50*(21), 6072–6077. doi:10.1021/jf0201273 PMID:12358482

Maurya, V. K., & Aggarwal, M. (2019). A phase inversion based nanoemulsion fabrication process to encapsulate vitamin D3 for food applications. *The Journal of Steroid Biochemistry and Molecular Biology*, *190*, 88–98. doi:10.1016/j.jsbmb.2019.03.021 PMID:30923018

Ma, Y., Li, H., & Guan, S. (2015). Enhancement of the oral bioavailability of breviscapine by nanoemulsions drug delivery system. *Drug Development and Industrial Pharmacy*, *41*(2), 177–182.

Ma, Y., Liu, D., Wang, D., Wang, Y., Fu, Q., Fallon, J. K., Yang, X., He, Z., & Liu, F. (2014). Combinational delivery of hydrophobic and hydrophilic anticancer drugs in single nanoemulsions to treat MDR in cancer. *Molecular Pharmaceutics*, *11*(8), 2623–2630.

Mayet, J., & Hughes, A. (2003). Cardiac and vascular pathophysiology in hypertension. *Heart (British Cardiac Society)*, *89*(9), 1104–1109. doi:10.1136/heart.89.9.1104 PMID:12923045

McCarron, P. A., & Faheem, A. M. (2010). Nanomedicine-based cancer targeting: A new weapon in an old war. *Nanomedicine (London)*, *5*(1), 3–5. doi:10.2217/nnm.09.89 PMID:20025458

McClements, D. J., Saliva-Trujillo, L., Zhang, R., Zhang, Z., Zou, L., Yao, M., & Xiao, H. (2016). Boosting the bioavailability of hydrophobic nutrients, vitamins, and nutraceuticals in natural products using excipient emulsions. *Food Research International*, *88*(Pt A), 140–152. doi:10.1016/j.foodres.2015.11.017

McClements, D. J. (1994). Ultrasonic determination of depletion flocculation in oil-in-water emulsions containing a non-ionic surfactant. *Colloids and Surfaces. A, Physicochemical and Engineering Aspects*, *90*(1), 25–35. doi:10.1016/0927-7757(94)02881-8

McClements, D. J. (2002). Theoretical prediction of emulsion color. *Advances in Colloid and Interface Science*, *97*(1-3), 63–89. doi:10.1016/S0001-8686(01)00047-1 PMID:12027025

McClements, D. J. (2004). *Food Emulsions: Principles, Practices, and Techniques* (2nd ed.). CRC Press., org/ doi:10.1201/9781420039436

McClements, D. J. (2007). *Critical review of techniques and methodologies for characterization of emulsion stability*. Critical Reviews in Food Science and Nutrition. doi:10.1080/10408390701289292

McClements, D. J. (2010). Emulsion design to improve the delivery of functional lipophilic components. *Annual Review of Food Science and Technology*, *1*(1), 241–269. doi:10.1146/annurev.food.080708.100722 PMID:22129337

McClements, D. J. (2011). Edible nanoemulsions: Fabrication, properties, and functional performance. *Soft Matter*, *7*(6), 2297–2316. doi:10.1039/C0SM00549E

Mcclements, D. J. (2012). Nanoemulsions versus microemulsions: Terminology, differences, and similarities. *Soft Matter*, *8*(6), 1719–1729. doi:10.1039/C2SM06903B

McClements, D. J. (2013). Nanoemulsion-based oral delivery systems for lipophilic bioactive components: Neutraceuticals and pharmaceuticals. *Therapeutic Delivery*, *4*(7), 841–857.

McClements, D. J. (2013). Nanoemulsion-based oral delivery systems for lipophilic bioactive components: Nutraceuticals and pharmaceuticals. *Therapeutic Delivery*, *4*(7), 841–857. doi:10.4155/tde.13.46 PMID:23883127

McClements, D. J. (2014). Advances in fabrication of emulsions with enhanced functionality using structural design principles. *Current Opinion in Colloid & Interface Science*, *17*(5), 235–245. doi:10.1016/j.cocis.2012.06.002

McClements, D. J. (2015). *Food emulsions: Principles, practices, and techniques*. CRC Press. doi:10.1201/b18868

McClements, D. J. (2019). *Nanoparticle-and microparticle-based delivery systems: Encapsulation, protection and release of active compounds*. CRC Press.

McClements, D. J., Decker, E. A., & Weiss, J. (2007). Emulsion-based delivery systems for lipophilic bioactive components. *Journal of Food Science*, *72*(8), R109–R124. doi:10.1111/j.1750-3841.2007.00507.x PMID:17995616

McClements, D. J., & Gumus, C. E. (2016). Natural emulsifiers - Biosurfactants, phospholipids, biopolymers, and colloidal particles: Molecular and physicochemical basis of functional performance. *Advances in Colloid and Interface Science*, *234*, 3–26. doi:10.1016/j.cis.2016.03.002 PMID:27181392

McClements, D. J., Henson, L., Popplewell, L. M., Decker, E. A., & Choi, S. J. (2012). Inhibition of Ostwald ripening in model beverage emulsions by addition of poorly water soluble triglyceride oils. *Journal of Food Science*, *77*(1), C33–C38. doi:10.1111/j.1750-3841.2011.02484.x PMID:22133014

Mcclements, D. J., & Jafari, S. M. (2018). General Aspects of Nanoemulsions and Their Formulation. In *Nanoemulsions*. Elsevier Inc. doi:10.1016/B978-0-12-811838-2.00001-1

McClements, D. J., & Rao, J. (2011). Food-grade nanoemulsions: Formulation, fabrication, properties, performance, biological fate, and potential toxicity. *Critical Reviews in Food Science and Nutrition*, *51*(4), 285–330.

McClements, D. J., & Rao, J. (2011). Food-Grade Nanoemulsions: Formulation, Fabrication, Properties, Performance, Biological Fate, and Potential Toxicity. *Critical Reviews in Food Science and Nutrition*, *51*(4), 285–330. doi:10.1080/10408398.2011.559558 PMID:21432697

McClements, D. J., & Xiao, H. (2012). Potential biological fate of ingested nanoemulsions: Influence of particle characteristics. *Food & Function*, *3*(3), 202–220.

McDaid, D., & Deasy, P. (1996). Formulation development of a transdermal drug delivery system for amlodipine base. *International Journal of Pharmaceutics*, *133*(1-2), 71–83. doi:10.1016/0378-5173(95)04412-4

McEvily, A. J., Iyengar, R., & Otwell, W. S. (1992). Inhibition of enzymatic browning in foods and beverages. *Critical Reviews in Food Science and Nutrition*, *32*(3), 253–273. doi:10.1080/10408399209527599 PMID:1418602

McMurray, J. J. (2015). Neprilysin inhibition to treat heart failure: A tale of science, serendipity, and second chances. *European Journal of Heart Failure*, *17*(3), 242–247. doi:10.1002/ejhf.250 PMID:25756942

Md Othman, S., Hassan, M. A., Nahar, L., Basar, N., Jamil, S., & Sarker, S. D. (2016). Essential Oils from the Malaysian Citrus (Rutaceae) Medicinal Plants. *Medicines (Basel, Switzerland)*, *3*(2), 13. doi:10.3390/medicines3020013 PMID:28930124

Medeiros, A. K. O. C., Gomes, C. C., Amaral, M. L. Q. A., Medeiros, L. D. G., Medeiros, I., Porto, D. L., Aragao, C. F. S., Maciel, B. L. L., Morais, A. H. A., & Passos, T. S. (2019). Nanoencapsulation improved water solubility and color stability of carotenoids extracted from *Cantaloupe melon* (Cucumis melo L.). *Food Chemistry*, *270*, 562–572. doi:10.1016/j.foodchem.2018.07.099 PMID:30174087

Mehanna, M.M., Mohyeldin, S.M., & Elgindy, N.A. (2019). Rifampicin-carbohydrate spray-dried nanocomposite: A futuristic multiparticulate platform for pulmonary delivery. *International Journal of Nanomedicine, 14*, 9089-9112. Doi: org/ doi:10.2147/IJN.S211182

Mehmood, T. (2015). Optimization of the canola oil based vitamin E nanoemulsions stabilized by food grade mixed surfactants using response surface methodology. *Food Chemistry*, *183*, 1–7. doi:10.1016/j.foodchem.2015.03.021 PMID:25863602

Mehmood, T. (2021). Optimisation of food grade mixed surfactant-based l-ascorbic acid nanoemulsions using response surface methodology. *IET Nanobiotechnology / IET*, *15*(3), 309–317. doi:10.1049/nbt2.12014 PMID:34694664

Mehmood, T., Ahmad, A., Ahmed, A., & Ahmed, Z. (2017). Optimization of olive oil based O/W nanoemulsions prepared through ultrasonic homogenization: A response surface methodology approach. *Food Chemistry*, *229*, 790–796. doi:10.1016/j.foodchem.2017.03.023 PMID:28372245

Mehmood, T., & Ahmed, A. (2020). Tween 80 and Soya-Lecithin-Based Food-Grade Nanoemulsions for the Effective Delivery of Vitamin D. *Langmuir*, *36*(11), 2886–2892. doi:10.1021/acs.langmuir.9b03944 PMID:32118445

Mehmood, T., Ahmed, A., Ahmad, A., Ahmad, M. S., & Sandhu, M. A. (2018). Optimization of mixed surfactants-based β-carotene nanoemulsions using response surface methodology: An ultrasonic homogenization approach. *Food Chemistry*, *253*, 179–184. doi:10.1016/j.foodchem.2018.01.136 PMID:29502819

Mehmood, T., Ahmed, A., & Ahmed, Z. (2021). Food-Grade Nanoemulsions for the Effective Delivery of β-Carotene. *Langmuir*, *37*(10), 3086–3092. doi:10.1021/acs.langmuir.0c03399 PMID:33646002

Mehta, P., Bothiraja, C., Kadam, S., & Pawar, A. (2018). Potential of dry powder inhalers for tuberculosis therapy: Facts, fidelity and future. *ArtifCells Nanomedicine and Biotechnology, 46*(3), 791-806. Doi: org/ doi:10.1080/21691401.2018.1513938.3

Mehta, S. K., Kaur, G., & Bhasin, K. K. (2007). Analysis of Tween based microemulsion in the presence of TB drug rifampicin. *Colloids and Surfaces. B, Biointerfaces*, *60*(1), 95–104. doi:10.1016/j.colsurfb.2007.06.012 PMID:17646089

Mehta, S. K., Kaur, G., & Bhasin, K. K. (2008). Incorporation of antitubercular drug isoniazid in pharmaceutically accepted microemulsion: Effect on microstructure and physical parameters. *Pharmaceutical Research*, *25*(1), 227–236. doi:10.100711095-007-9355-8 PMID:17577642

Mei, L., Choi, S. J., Alamed, J., Henson, L., Popplewell, M., McClements, D. J., & Decker, E. A. (2009). Citral stability in oil-in-water emulsions with solid or liquid octadecane. *Journal of Agricultural and Food Chemistry*, *58*(1), 533–536. doi:10.1021/jf902665b PMID:19911843

Mei, L., Zhang, Z., Zhao, L., Huang, L., Yang, X.-L., Tang, J., & Feng, S.-S. (2013). Pharmaceutical nanotechnology for oral delivery of anticancer drugs. *Advanced Drug Delivery Reviews*, *65*(6), 880–890. doi:10.1016/j.addr.2012.11.005 PMID:23220325

Melancon, M. P., Yevich, S., Avritscher, R., Swigost, A., Lu, L., Tian, L., Damasco, J. A., Dixon, K., Cortes, A. C., Munoz, N. M., Liang, D., Liu, D., & Tam, A. L. (2021). A novel irinotecan-lipiodol nanoemulsion for intravascular administration: Pharmacokinetics and biodistribution in the normal and tumor bearing rat liver. *Drug Delivery*, *28*(1), 240–251.

Melo, K. J., Souza, N., Ferraz, H., Fotaki, N., Lobenberg, R., & Bou-Chacra, N. (2015). Formulation with Povacoat® to Prevent Aggregation of Rifampin Nanocrystal Using High-Pressure Homogenization. *AAPS Annual Meeting*.

Melo, K. J. C., Henostroza, M. A. B., Löbenberg, R., & Bou-Chacra, N. A. (2020). Rifampicin nanocrystals: Towards an innovative approach to treat tuberculosis. *Materials Science and Engineering C*, *112*, 110895. doi:10.1016/j.msec.2020.110895 PMID:32409052

Mendes Miranda, S. E., Alcântara Lemos, J., Fernandes, R. S., Silva, J. O., Ottoni, F. M., Townsend, D. M., Rubello, D., Alves, R. J., Cassali, G. D., Ferreira, L., & de Barros, A. (2021). Enhanced antitumor efficacy of lapachol-loaded nanoemulsion in breast cancer tumor model. *Biomedicine and Pharmacotherapy*, *133*, 110936.

Meng, R., Wang, C., & Shen, Z. (2020). Optimization and Characterization of Highly Stable Nanoemulsion for Effective Oil-Based Drilling Fluid Removal. *SPE Journal*, *25*(03), 1259–1271. doi:10.2118/199904-PA

Metcalfe, P. D., & Thomas, S. (2010). Challenges in the prediction and modeling of oral absorption and bioavailability. *Current Opinion in Drug Discovery & Development*, *13*(1), 104–110.

Meynier, A., Lecoq, C., & Genot, C. (2005). Emulsification enhances the retention of esters and aldehydes to a greater extent than changes in the droplet size distribution of the emulsion. *Food Chemistry*, *93*(1), 153–159. doi:10.1016/j.foodchem.2004.10.018

Miastkowska, M., Michalczyk, A., Figacz, K., & Sikora, E. (2020). Nanoformulations as a modern form of biofungicide. *Journal of Environmental Health Science & Engineering*, *18*(1), 7–10. doi:10.100740201-020-00445-4 PMID:32399225

Michiels, J. E., Van den Bergh, B., Verstraeten, N., Fauvart, M., & Michiels, J. (2016). In vitro emergence of high persistence upon periodic aminoglycoside challenge in the ESKAPE pathogens. *Antimicrobial Agents and Chemotherapy*, *60*(8), 4630–4637. doi:10.1128/AAC.00757-16

Milkova, V. (2021). Electrosteric stabilization of oil/water emulsions by adsorption of chitosan oligosaccharides – An electrokinetic study. *Carbohydrate Polymers*, *265*, 118072–1180820. doi:10.1016/j.carbpol.2021.118072 PMID:33966836

Milkova, V., & Goycoolea, F. M. (2020). Encapsulation of caffeine in polysaccharide oil-core nanocapsules. *Colloid & Polymer Science*, *298*(8), 1035–1041. doi:10.100700396-020-04653-0

Milkova, V., & Radeva, Ts. (2006). Counterion release from adsorbed highly charged polyelectrolyte – an electrooptical study. *Journal of Colloid and Interface Science*, *298*(2), 550–555. doi:10.1016/j.jcis.2006.01.020 PMID:16480999

Millard, J., & Moore, D. A. J. (2015). Multidrug resistant tuberculosis. *BMJ (Clinical Research Ed.)*, *350*(feb26 10), h882. doi:10.1136/bmj.h882 PMID:25721508

Min, J.-Y., Ahn, S.-I., Lee, Y.-K., Kwak, H.-S., & Chang, Y. H. (2018). Optimized conditions to produce water-in-oil-in-water nanoemulsion and spray-dried nanocapsule of red ginseng extract. *Food Science and Technology (Campinas)*, *38*(3), 485–492. doi:10.1590/fst.09517

Minz, S., & Pandey, R. S. (2018). Lipid a adjuvanted chylomicron mimicking solid fat nanoemulsions for immunization against hepatitis B. *AAPS PharmSciTech*, *19*(3), 1168–1181. doi:10.120812249-017-0932-9 PMID:29243216

Mishra, P., Balaji, A., Mukherjee, A., & Chandrasekaran, N. (2018). Bio-Based Nanoemulsions: An Eco-Safe Approach Towards the Eco-Toxicity Problem. Handbook of Ecomaterials, 1-23. doi:org/10.1007/978-3-319-48281-1_94-1

Mishra, N., Goyal, A. K., Tiwari, S., Paliwal, R., Paliwal, S. R., Vaidya, B., Mangal, S., Gupta, M., Dube, D., Mehta, A., & Vyas, S. P. (2010). Recent advances in mucosal delivery of vaccines: Role of mucoadhesive/biodegradable polymeric carriers. *Expert Opinion on Therapeutic Patents*, *20*(5), 661–679. doi:10.1517/13543771003730425 PMID:20345332

Mishra, P., R. S, S. K., Jerobin, J., Thomas, J., Mukherjee, A., & Chandrasekaran, N. (2014). Study on antimicrobial potential of neem oil nanoemulsion against P seudomonasaeruginosa infection in L abeorohita. *Biotechnology and Applied Biochemistry*, *61*(5), 611–619. doi:10.1002/bab.1213

Mitnick, C. D., Appleton, S. C., & Shin, S. S. (2008). Epidemiology and treatment of multidrug resistant tuberculosis. *Seminars in Respiratory and Critical Care Medicine*, *29*(5), 499–524. doi:10.1055-0028-1085702 PMID:18810684

Mkandawire, M., & Aryee, A. N. (2018). Resurfacing and modernization of edible packaging material technology. *Current Opinion in Colloid & Interface Science*, *19*, 104–112.

Modi, S., & Anderson, B. D. (2013). Determination of drug release kinetics from nanoparticles: Overcoming pitfalls of the dynamic dialysis method. *Molecular Pharmaceutics*, *10*(8), 3076–3089.

Moghimi, R., Aliahmadi, A., Mcclements, D., & Rafati, H. (2016). Investigations of the effectiveness of nanoemulsions from sage oil as antibacterial agents on some food borne pathogens. *Lebensmittel-Wissenschaft + Technologie*, *71*, 69–76. doi:10.1016/j.lwt.2016.03.018

Moghimi, R., Aliahmadi, A., Rafati, H., Abtahi, H. R., Amini, S., & Feizabadi, M. M. (2018). Antibacterial and anti-biofilm activity of nanoemulsion of Thymus daenensis oil against multi-drug resistant Acinetobacterbaumannii. *Journal of Molecular Liquids*, *265*, 765–770. doi:10.1016/j.molliq.2018.07.023

Moghimi, R., Ghaderi, L., Rafati, H., Aliahmadi, A., & McClements, D. J. (2016). Superior antibacterial activity of nanoemulsion of Thymus daenensis essential oil against E. coli. *Food Chemistry*, *194*, 410–415. doi:10.1016/j.foodchem.2015.07.139 PMID:26471573

Mohammad, I. S., He, W., & Yin, L. (2018). Understanding of human ATP binding cassette superfamily and novel multidrug resistance modulators to overcome MDR. *Biomedicine and Pharmacotherapy*, *100*, 335–348.

Mohanraj, V., & Chen, Y. (2007). Nanoparticles-a review. *Tropical Journal of Pharmaceutical Research*, *5*(1), 561–573. doi:10.4314/tjpr.v5i1.14634

Mohsin, S. N. M., Raman, I. A., Hasan, Z. A. A., & Idris, Z. (2017). Palm-based Methyl Esters as Carrier Solvents in Pesticide Formulations. *Palm Oil Developments*, *66*, 32–38.

Monga, S. P. S., Pediaditakis, P., Mule, K., Stolz, D. B., & Michalopoulos, G. K. (2001). Changes in wnt/β-catenin pathway during regulated growth in rat liver regeneration. *Hepatology (Baltimore, Md.)*, *33*(5), 1098–1109. doi:10.1053/jhep.2001.23786 PMID:11343237

Monge-Fuentes, V., Muehlmann, L., Longo, J., Silva, J., Fascineli, M., de Souza, P., Faria, F., Degterev, I. A., Rodriguez, A., Carneiro, F. P., Lucci, C. M., Escobar, P., Amorim, R. F. B., & Azevedo, R. B. (2017). Photodynamic therapy mediated by acai oil (Euterpe oleracea Martius) in nanoemulsion: A potential treatment for melanoma. *Journal of Photochemistry and Photobiology. B, Biology*, *166*, 301–310. doi:10.1016/j.jphotobiol.2016.12.002 PMID:28024281

Moraes-Lovison, M., Marostegan, L., Peres, M., Menezes, I., Ghiraldi, M., Rodrigues, R., Fernandes, A., & Pinho, S. (2016). Nanoemulsions encapsulating oregano essential oil: Production, stability, antibacterial activity and incorporation in chicken pâté. *Lebensmittel-Wissenschaft + Technologie*, *77*, 233–240. Advance online publication. doi:10.1016/j.lwt.2016.11.061

Morel, S., Didierlaurent, A., Bourguignon, P., Delhaye, S., Baras, B., Jacob, V., Planty, C., Elouahabi, A., Harvengt, P., Carlsen, H., Kielland, A., Chomez, P., Garçon, N., & Van Mechelen, M. (2011). Adjuvant System AS03 containing α-tocopherol modulates innate immune response and leads to improved adaptive immunity. *Vaccine*, *29*(13), 2461–2473. doi:10.1016/j.vaccine.2011.01.011 PMID:21256188

Moreno, M. A., Ballesteros, M. P., & Frutos, P. (2003). Lecithin-based oil-in-water microemulsions for parenteral use: Pseudoternary phase diagrams, characterisation and toxicity studies. *Journal of Pharmaceutical Sciences*, *92*(7), 1428–1437.

Moretton, M. A., Chiappetta, D. A., & Sosnik, A. (2012). Cryoprotection-lyophilization and physical stabilization of rifampicin-loaded flower-like polymeric micelles. *Journal of the Royal Society, Interface*, *9*(68), 487–502. doi:10.1098/rsif.2011.0414 PMID:21865255

Moretton, M. A., Glisoni, R. J., Chiappetta, D. A., & Sosnik, A. (2010). Molecular implications in the nanoencapsulation of the anti-tuberculosis drug rifampicin within flower-like polymeric micelles. *Colloids and Surfaces. B, Biointerfaces*, *79*(2), 467–479. doi:10.1016/j.colsurfb.2010.05.016 PMID:20627665

Moretton, M. A., Hocht, C., Taira, C., & Sosnik, A. (2014). Rifampicin-loaded 'flower-like' polymeric micelles for enhanced oral bioavailability in an extemporaneous liquid fixed-dose combination with isoniazid. *Nanomedicine (London)*, *9*(11), 25–30. doi:10.2217/nnm.13.154 PMID:24410279

Morlock, G. P., Metchock, B., Sikes, D., Crawford, J. T., & Cooksey, R. C. (2003). ethA, inhA, and katG loci of ethionamide-resistant clinical *Mycobacterium tuberculosis* isolates. *Antimicrobial Agents and Chemotherapy*, *47*(12), 3799–3805. doi:10.1128/AAC.47.12.3799-3805.2003 PMID:14638486

Mossa, A. T. H., Abdelfattah, N. A. H., & Mohafrash, S. M. M. (2017). Nanoemulsion of camphor (Eucalyptus globulus) essential oil, formulation, characterization and insecticidal activity against wheat weevil, Sitophilus granarius. *Asian Journal of Crop Science*, *9*(3), 50–62. doi:10.3923/ajcs.2017.50.62

Mossa, A. T. H., Afia, S. I., Mohafrash, S. M. M., & Abou-Awad, B. A. (2018). Formulation and characterization of garlic (Allium sativum L.) essential oil nanoemulsion and its acaricidal activity on eriophyid olive mites (Acari: Eriophyidae*). Environmental Science and Pollution Research International*, *25*(11), 10526–10537. doi:10.100711356-017-0752-1 PMID:29181754

Mostafa, D. M., Abd El-Alim, S. H., Asfour, M. H., Al-Okbi, S. Y., Mohamed, D. A., & Awad, G. (2015). Transdermal nanoemulsions of Foeniculum vulgare Mill. essential oil: Preparation, characterization and evaluation of antidiabetic potential. *Journal of Drug Delivery Science and Technology*, *29*, 99–106. doi:10.1016/j.jddst.2015.06.021

Mothana, R. A. A., Jansen, R., Julich, W.-D., & Lindequist, U. (2000). Ganomycins A and B, new antimicrobial Farnesyl hydroquinones from the Basidiomycete *Ganoderma pfeifferi. Journal of Natural Products*, *63*(3), 416–418. doi:10.1021/np990381y PMID:10757736

Mountain, G. A., Jelier, B. J., Bagia, C., Friesen, C. M., & Janjic, J. M. (2014). Design and formulation of nanoemulsions using 2-(poly(hexafluoropropylene oxide)) perfluoropropyl benzene in combination with linear perfluoro(polyethylene glycol dimethyl ether). *Journal of Fluorine Chemistry*, *162*, 38–44. doi:10.1016/j.jfluchem.2014.03.007 PMID:24976645

Muheem, A., Shakeel, F., Jahangir, M. A., Anwar, M., Mallick, N., Jain, G. K., & Ahmad, F. J. (2016). A review on the strategies for oral delivery of proteins and peptides and their clinical perspectives. *Saudi Pharmaceutical Journal*, *24*(4), 413–428.

Mullaicharam, A., & Murthy, R. (2004). Lung accumulation of niosome-entrapped rifampicin following intravenous and intratracheal administration in the rat. *Journal of Drug Delivery Science and Technology*, *14*(2), 99–104. doi:10.1016/S1773-2247(04)50020-5

Muller, R. H., Gohla, S., & Keck, C. M. (2011). State of the art of nanocrystals—Special features, production, nanotoxicology aspects and intracellular delivery. *European Journal of Pharmaceutics and Biopharmaceutics*, *78*(1), 1–9. doi:10.1016/j.ejpb.2011.01.007 PMID:21266197

Mundada, V. P., Patel, M. H., Mundada, P. K., & Sawant, K. K. (2020). Enhanced bioavailability and antihypertensive activity of nisoldipine loaded nanoemulsion: Optimization, cytotoxicity and uptake across Caco-2 cell line, pharmacokinetic and pharmacodynamic studies. *Drug Development and Industrial Pharmacy*, *46*(3), 376–387. doi:10.1080/03639045.2020.1724128 PMID:32031412

Mun, S., Decker, E. A., & McClements, D. J. (2005). Influence of droplet characteristics on the formation of oil-in water emulsions stabilized by surfactant-chitosan layers. *Langmuir*, *12*(14), 6228–6234. doi:10.1021/la050502w PMID:15982024

Mu, Q., Wang, H., & Zhang, M. (2017). Nanoparticles for imaging and treatment of metastatic breast cancer. *Expert Opinion on Drug Delivery*, *14*(1), 123–136.

Murakawa, K., Fukunaga, K., Tanouchi, M., Hosokawa, M., Hossain, Z., & Takahashi, K. (2007). Therapy of myeloma *in vivo* using marine phospholip in combination with *Agaricus blazei* Murill as an immune respond activator. *Journal of Oleo Science*, *56*(4), 179–188. doi:10.5650/jos.56.179 PMID:17898480

Mura, S., Nicolas, J., & Couvreur, P. (2013). Stimuli-responsive nanocarriers for drug delivery. *Nature Materials*, *12*(11), 991–1003. doi:10.1038/nmat3776 PMID:24150417

Murthy, K. N. C., Monika, P., Jayaprakasha, G. K., & Patil, B. S. (2018) Nanoencapsulation: An advanced nanotechnological approach to enhance the biological efficacy of curcumin. *Proceedings of the ACS Symposium Series*, 383-405. 10.1021/bk-2018-1286.ch021

Musa, S. H., Basri, M., Masoumi, H. R. F., Karjiban, R. A., Abd Malek, E., Basri, H., & Shamsuddin, A. F. (2013). Formulation optimisation of palm kernel oil esters nanoemulsion-loaded with chloramphenicol suitable for meningitis treatment. *Colloids and Surfaces. B, Biointerfaces*, *112*, 113–119.

Myc, A., Kukowska-Latallo, J. F., Bielinska, A. U., Cao, P., Myc, P. P., Janczak, K., Sturm, T. R., Grabinski, M. S., Landers, J. J., Young, K. S., Chang, J., Hamouda, T., Olszewski, M. A., & Baker, J. R. Jr. (2003). Development of immune response that protects mice from viral pneumonitis after a single intranasal immunization with influenza A virus and nanoemulsion. *Vaccine*, *21*(25-26), 3801–3814. doi:10.1016/S0264-410X(03)00381-5 PMID:12922114

Myc, A., Kukowska-Latallo, J., Smith, D., Passmore, C., Pham, T., Wong, P., Bielinska, A. U., & Baker, J. R. Jr. (2013). Nanoemulsion nasal adjuvant W805EC induces dendritic cell engulfment of antigen-primed epithelial cells. *Vaccine*, *31*(7), 1072–1079. doi:10.1016/j.vaccine.2012.12.033 PMID:23273511

Nabavi, S. M., Marchese, A., Izadi, M., Curti, V., Daglia, M., & Nabavi, S. F. (2015). Plants belonging to the genus Thymus as antibacterial agents: From farm to pharmacy. *Food Chemistry*, *173*, 339–347. doi:10.1016/j.foodchem.2014.10.042 PMID:25466031

Nabila, N., Suada, N. K., Denis, D., Yohan, B., Adi, A. C., Veterini, A. S., Anindya, A. L., Sasmono, R. T., & Rachmawati, H. (2020). Antiviral action of curcumin encapsulated in nanoemulsion against four serotypes of dengue virus. *Pharmaceutical Nanotechnology*, *8*(1), 54–62. doi:10.2174/2211738507666191210163408 PMID:31858909

Nacey, J. N., Meffan, P. J., & Delahunt, B. (1995). The effect of finasteride on prostate volume, urinary flow rate and symptom score in men with benign prostatic hyperplasia. *The Australian and New Zealand Journal of Surgery*, *65*(5), 35–39. doi:10.1111/j.1445-2197.1995.tb01744.x PMID:7529489

Nácher, M. P. A., Jesús, M., Ofelia, H. M. A., & Busó, V. (2019). A novel lidocaine hydrochloride mucoadhesive films for periodontal diseases. *Journal of Materials Science. Materials in Medicine*, *30*(1), 14. Advance online publication. doi:10.100710856-018-6213-7 PMID:30635738

Nagi, A., Iqbal, B., Kumar, S., Sharma, S., Ali, J., & Baboota, S. (2017). Quality by design based silymarin nanoemulsion for enhancement of oral bioavailability. *Journal of Drug Delivery Science and Technology*, *40*, 35–44.

Naithani, V., Kumar, S., Srivastava, S. M., Pathak, N., & Chaudhary, M. (2010). An evaluation of antimicrobial efficacy of acano against some acne causing microorganisms. *Int. J. Drug Dev. Res*, *2*(1), 134–140.

Nakalembe, I., Kabasa, J. D., & Olila, D. (2015). Comparative nutrient composition of selected wild edible mushrooms from two agro-ecological zones, Uganda. *SpringerPlus*, *4*(1), 433. doi:10.118640064-015-1188-z PMID:26306295

Nakasato, D. Y., Pereira, A. E., Oliveira, J. L., Oliveira, H. C., & Fraceto, L. F. (2017). Evaluation of the effects of polymeric chitosan/tripolyphosphate and solid lipid nanoparticles on germination of *Zea mays, Brassica rapa* and *Pisum sativum*. *Ecotoxicology and Environmental Safety*, *142*, 369–374. doi:10.1016/j.ecoenv.2017.04.033 PMID:28437729

Namratha Vinaya, K., Mary John, A., Mangsatabam, M., & Anna Philip, M. (2021). Ultrasound-assisted synthesis and characterization of sesame oil based nanoemulsion. *IOP Conference Series. Materials Science and Engineering*, *1114*(1), 012085. doi:10.1088/1757-899X/1114/1/012085

Nam, Y. S., Kim, J. W., Park, J., Shim, J., Lee, J. S., & Han, S. H. (2012). Tocopheryl acetate nanoemulsions stabilized with lipid-polymer hybrid emulsifiers for effective skin delivery. *Colloids and Surfaces. B, Biointerfaces*, *94*, 51–57. doi:10.1016/j.colsurfb.2012.01.016 PMID:22326341

Nandini, B., Puttaswamy, H., Prakash, H. S., Adhikari, S., Jogaiah, S., & Nagaraja, G. (2020). Elicitation of novel trichogenic-lipid nanoemulsion signaling resistance against pearl millet downy mildew disease. *Biomolecules*, *10*(1), 25. doi:10.3390/biom10010025 PMID:31878099

Nantarat, T., Chansakaow, S., & Leelapornpisid, P. (2015). Optimization, characterization and stability of essential oils blend loaded nanoemulsions by PIC technique for anti-tyrosinase activity. *International Journal of Pharmacy and Pharmaceutical Sciences*, *7*, 308–312.

Naraghi, L., & Negahban, M. (2020). Efficacy of Talaromyces Flavus coated with nanoparticles in the growth inhibitory of Fusarium Oxysporum F.SP. Cucumerinum. *3C Tecnologia, 9*, 31-45.

Narang, A. S., Delmarre, D., & Gao, D. (2007). Stable drug encapsulation in micelles and microemulsions. *International Journal of Pharmaceutics*, *345*(1-2), 9–25.

Narita, M., Tsuji, B. T., & Yu, V. L. (2007). Linezolid-associated peripheral and optic neuropathy, lactic acidosis, and serotonin syndrome. *Pharmacotherapy*, *27*(8), 1189–1197. doi:10.1592/phco.27.8.1189 PMID:17655517

Naseri, N., Valizadeh, H., & Zakeri-Milani, P. (2015). Solid Lipid Nanoparticles and Nanostructured Lipid Carriers: Structure, Preparation and Application. *Advanced Pharmaceutical Bulletin*, *5*(3), 305–313. doi:10.15171/apb.2015.043 PMID:26504751

Nasiri, M. J., Haeili, M., Ghazi, M., Goudarzi, H., Pormohammad, A., Fooladi, A. A. I., & Feizabadi, M. M. (2017). New insights in to the intrinsic and acquired drug resistance mechanisms in mycobacteria. *Frontiers in Microbiology*, *8*, 681. Advance online publication. doi:10.3389/fmicb.2017.00681 PMID:28487675

Natesan, S., Sugumaran, A., Ponnusamy, C., Thiagarajan, V., Palanichamy, R., & Kandasamy, R. (2017). Chitosan stabilized camptothecin NEs: Development, evaluation and biodistribution in preclinical breast cancer animal mode. *International Journal of Biological Macromolecules*, *104*, 1846–1852. doi:10.1016/j.ijbiomac.2017.05.127 PMID:28545970

National Cancer Institute. (2020). *Prostate cancer treatment.* Retrieved January 16, 2020, from http://www.cancer.gov/types/prostate/patient/prostate-treatment-pdq#section/_120

National Library of Medicine (U.S.). (2015). *The Safety, Tolerance, and Immunogenicity of MAS-1-Adjuvanted Seasonal Inactivated Influenza Vaccine (MER4101).* Identifier NCT02500680. https://clinicaltrials.gov/ct2/show/NCT02500680

National Library of Medicine (U.S.). (2019). *A Safety and Immunogenicity of Intranasal Nanoemulsion Adjuvant Recombinant Anthrax Vaccine in Healthy Adults*. Identifier NCT04148118. https://clinicaltrials.gov/ct2/show/NCT04148118)

Naylor, N. R., Atun, R., Zhu, N., Kulasabanathan, K., Silva, S., Chatterjee, A., & Robotham, J. V. (2018). Estimating the burden of antimicrobial resistance: A systematic literature review. *Antimicrobial Resistance and Infection Control, 7*(1), 1–17. doi:10.118613756-018-0336-y

Nazzaro, F., Fratianni, F., De Martino, L., Coppola, R., & De Feo, V. (2013). Effect of essential oils on pathogenic bacteria. *Pharmaceuticals (Basel, Switzerland), 6*(12), 1451–1474. doi:10.3390/ph6121451 PMID:24287491

Nejak-Bowen, K. N., Zeng, G., Tan, X., Cieply, B., & Monga, S. P. (2009). β-catenin regulates vitamin C biosynthesis and cell survival in murine liver. *The Journal of Biological Chemistry, 284*(41), 28115–28127. doi:10.1074/jbc.M109.047258 PMID:19690176

Nekkanti, V., Karatgi, P., Prabhu, R., & Pillai, R. (2010). Solid self-microemulsifying formulation for candesartan cilexetil. *AAPS PharmSciTech, 11*(1), 9–17. doi:10.120812249-009-9347-6 PMID:20013081

Nenaah, G. E. (2014). Chemical composition, toxicity and growth inhibitory activities of essential oils of three Achillea species and their nano-emulsions against Tribolium castaneum. *Industrial Crops and Products, 53*, 252–260. doi:10.1016/j.indcrop.2013.12.042

Nepal, P. R., Han, H. K., & Choi, H. K. (2010). Preparation and in vitro–in vivo evaluation of Witepsol® H35 based self-nanoemulsifying drug delivery systems (SNEDDS) of coenzyme Q10. *European Journal of Pharmaceutical Sciences, 39*(4), 224–232.

Newling, D. W. (1997). The palliative therapy of advanced prostate cancer, with particular reference to the results of recent European clinical trials. *British Journal of Urology, 79*(S1), 72–81. doi:10.1111/j.1464-410X.1997.tb00805.x PMID:9088277

Newman, J. W., Floyd, R. V., & Fothergill, J. L. (2017). The contribution of Pseudomonas aeruginosa virulence factors and host factors in the establishment of urinary tract infections. *FEMS Microbiology Letters, 364*(15). Advance online publication. doi:10.1093/femsle/fnx124

Ngai, P. H. K., & Ng, T. B. (2004). A ribonuclease with antimicrobial, antimitogenic and antiproliferative activities from the edible mushroom *Pleurotus sajor-caju. Peptides, 25*(1), 11–17. doi:10.1016/j.peptides.2003.11.012 PMID:15003351

Nguyen, H. M., Hwang, I. C., Park, J. W., & Park, H. J. (2012). Enhanced payload and photo-protection for pesticides using nanostructured lipid carriers with corn oil as liquid lipid. *Journal of Microcapsules, 29*(6), 596–604. doi:10.3109/02652048.2012.668960 PMID:22424134

Nguyen, H. M., Hwang, I. C., Park, J. W., & Park, H. J. (2012). Photo protection for deltamethrin using chitosan-coated beeswax solid lipid nanoparticles. *Pest Management Science, 68*(7), 1062–1068. doi:10.1002/ps.3268 PMID:22383397

Nguyen, M. H., Hwang, I. C., & Park, H. J. (2013). Enhanced photoprotection for photo-labile compounds using double-layer coated corn oil-nanoemulsions with chitosan and lignosulfonate. *Journal of Photochemical Photobiology B: Biology, 125*, 194–201. doi:10.1016/j.jphotobiol.2013.06.009 PMID:23851422

Niaz, T., Shabbir, S., Manzoor, S., Rehman, A., Rahman, A., Nasir, H., & Imran, M. (2016). Antihypertensive nanoceuticales based on chitosan biopolymer: Physico-chemical evaluation and release kinetics. *Carbohydrate Polymers, 142*, 268–274. doi:10.1016/j.carbpol.2016.01.047 PMID:26917399

Nicolosi R & Jean T. (2008). *Compositions and methods for treating cancer with dacarbazine nanoemulsions*. Academic Press.

Nicolosi R & Shea T. (2007). *Antioxidant synergy formulation nanoemulsions to treat cancer.* Academic Press.

Nicolosi, R., & Shea, T. (2015). *Anti-oxidant synergy formulation nanoemulsions to treat cancer*. Academic Press.

Nicolosi, R., & Tagne, J. (2010). *Compositions and methods for treating cancer with dacarbazine nanoemulsions*. Academic Press.

Nik, A. M., Corredig, M., & Wright, A. J. (2011). Release of lipophilic molecules during in vitro digestion of soy protein-stabilised emulsions. *Molecular Nutrition & Food Research*, *55*(S2), S278–S289.

Nikaido, H. (2003). Molecular basis of bacterial outer membrane permeability revisited. *Microbiology and Molecular Biology Reviews*, *67*(4), 593–656. doi:10.1128/MMBR.67.4.593-656.2003

Nikaido, H. (2009). Multidrug resistance in bacteria. *Annual Review of Biochemistry*, *78*, 119–146. doi:10.1146/annurev.biochem.78.082907.145923

Niknam, S. M., Escudero, I., & Benito, J. M. (2020). Formulation and Preparation of Water-In-Oil-In-Water Emulsions Loaded with a Phenolic-Rich Inner Aqueous Phase by Application of High Energy Emulsification Methods. *Foods (Basel, Switzerland)*, *9*(10), 1411. doi:10.3390/foods9101411 PMID:33027911

Nirale, P., Paul, A., & Yadav, K. S. (2020). Nanoemulsions for targeting the neurodegenerative diseases: Alzheimer's, Parkinson's and Prion's. *Life Sciences*, *245*, 117394. doi:10.1016/j.lfs.2020.117394 PMID:32017870

Nirmala, M. J., Durai, L., Gopakumar, V., & Nagarajan, R. (2020). Preparation of Celery Essential Oil-Based Nanoemulsion by Ultrasonication and Evaluation of Its Potential Anticancer and Antibacterial Activity. *International Journal of Nanomedicine*, *15*, 7651–7666. doi:10.2147/IJN.S252640 PMID:33116493

Nirmal, N. P., Meredy, R., Li, L., & Sultanbawa, Y. (2018). Formulation, characterisation and antibacterial activity of lemon myrtle and anise myrtle essential oil in water nanoemulsion. *Food Chemistry*, *254*, 1–7. doi:10.1016/j.foodchem.2018.01.173 PMID:29548427

Niu, Z., Acevedo-Fani, A., McDowell, A., Barnett, A., Loveday, S. M., & Singh, H. (2020). Nanoemulsion structure and food matrix determine the gastrointestinal fate and in vivo bioavailability of coenzyme Q10. *Journal of Controlled Release*, *327*, 444–455.

Noda, K., Kodama, S., Umemoto, S., Abe, N., Hirano, T., & Suzuki, M. (2010). Nasal vaccination with P6 outer membrane protein and α-galactosylceramide induces nontypeable Haemophilus influenzae-specific protective immunity associated with NKT cell activation and dendritic cell expansion in nasopharynx. *Vaccine*, *28*(31), 5068–5074. doi:10.1016/j.vaccine.2010.05.005 PMID:20478344

Noguchi, M., Stamey, T. A., Mc Neal, J. E., & Yemoto, C. E. (2000). An analysis of 148 consecutive transition zone cancers: Clinical and histological characteristics. *The Journal of Urology*, *163*(6), 1751–1752. doi:10.1016/S0022-5347(05)67535-0 PMID:10799175

Nurgali, K., Jagoe, R. T., & Abalo, R. (2018). Adverse effects of cancer chemotherapy: Anything new to improve tolerance and reduce sequelae? *Frontiers in Pharmacology*, *9*, 245. doi:10.3389/fphar.2018.00245 PMID:29623040

O'Brien, R., & White, L. R. (1978). Electrophoretic mobility of a spherical colloidal particle. *Journal of the Chemical Society. Faraday Transactions II*, *43*, 1607–1626. doi:10.1039/f29787401607

O'Driscoll, C. M. (2002). Lipid-based formulations for intestinal lymphatic delivery. *European Journal of Pharmaceutical Sciences*, *15*(5), 405–415. doi:10.1016/S0928-0987(02)00051-9 PMID:12036717

O'Hagan, D. T., Ott, G. S., De Gregorio, E., & Seubert, A. (2012). The mechanism of action of MF59 - an innately attractive adjuvant formulation. *Vaccine*, *30*(29), 4341–4348. doi:10.1016/j.vaccine.2011.09.061 PMID:22682289

O'Konek, J. J., Makidon, P. E., Landers, J. J., Cao, Z., Malinczak, C. A., Pannu, J., Sun, J., Bitko, V., Ciotti, S., Hamouda, T., Wojcinski, Z. W., Lukacs, N. W., Fattom, A., & Baker, J. R. Jr. (2015). Intranasal nanoemulsion-based inactivated respiratory syncytial virus vaccines protect against viral challenge in cotton rats. *Human Vaccines & Immunotherapeutics*, *11*(12), 2904–2912. doi:10.1080/21645515.2015.1075680 PMID:26307915

Odriozola-Serrano, I., Oms-Oliu, G., & Martín-Belloso, O. (2014). Nanoemulsion-based delivery systems to improve functionality of lipophilic components. *Frontiers in Nutrition*, *1*, 24. doi:10.3389/fnut.2014.00024 PMID:25988126

Ogawa, S., Decker, E. A., & McClements, D. J. (2004). Production and characterization of O/W emulsions containing droplets stabilized by lechithin-chitosan-pectin multilayered membranes. *Journal of Agricultural and Food Chemistry*, *52*(11), 3595–3600. doi:10.1021/jf034436k PMID:15161236

Oliveira, A. E. M. F. M., Duarte, J. L., Amado, J. R. R., Cruz, R. A. S., Rocha, C. F., Souto, R. N. P., Ferreira, R. M. A., Santos, K., Da Conceição, E. C., De Oliveira, L. A. R., Kelecom, A., Fernandes, C. P., & Carvalho, J. C. T. (2016). Development of larvicidal nanoemulsion with pterodon emarginatus Vogel oil. *PLoS One*, *11*(1), 1–16. doi:10.1371/journal.pone.0145835 PMID:26742099

Oraki Kohshour, M., Mirzaie, S., Zeinali, M., Amin, M., Said Hakhamaneshi, M., Jalili, A., Mosaveri, N., & Jamalan, M. (2014). Ablation of breast cancer cells using trastuzumab-functionalized multi-walled carbon nanotubes and trastuzumab-diphtheria toxin conjugate. *Chemical Biology & Drug Design*, *83*(3), 259–265.

Orr, M. T., Kramer, R. M., Barnes, L., Dowling, Q. M., Desbien, A. L., Beebe, E. A., Laurance, J. D., Fox, C. B., Reed, S. G., Coler, R. N., & Vedvick, T. S. (2014). Elimination of the cold-chain dependence of a nanoemulsion adjuvanted vaccine against tuberculosis by lyophilization. *Journal of Controlled Release*, *177*, 20–26. doi:10.1016/j.jconrel.2013.12.025 PMID:24382398

Osborne, D. W., Ward, A. J. I., & O'Neill, K. J. (1991). Microemulsions as topical drug delivery vehicles: In-vitro transdermal studies of a model hydrophilic drug. *The Journal of Pharmacy and Pharmacology*, *43*(6), 451–454. doi:10.1111/j.2042-7158.1991.tb03511.x PMID:1717675

Osman, M. A. E., Shakil, N. A., Rana, V. S., Sarkar, D. J., Majumder, S., Kaushik, P., Singh, B. B., & Kumar, J. (2017). Antifungal activity of nano emulsions of neem and citronella oils against phytopathogenic fungi, Rhizoctonia solani and Sclerotium rolfsii. *Industrial Crops and Products*, *108*, 379–387. doi:10.1016/j.indcrop.2017.06.061

Ospina, N. M., Ospina Alvarez, S. P., Escobar Sierra, D. M., Rojas Vahos, D. F., Zapata Ocampo, P. A., & Ossa Orozco, C. P. (2015). Isolation of chitosan from *Ganoderma lucidum* mushroom for biomedical applications. *Journal of Materials Science. Materials in Medicine*, *26*(3), 135. doi:10.100710856-015-5461-z PMID:25716022

Otoni, C. G., de Moura, M. R., Aouada, F. A., Camilloto, G. P., Cruz, R. S., & Lorevice, M. V. (2014a). Antimicrobial and physicalmechanical properties of pectin/papaya puree/cinnamaldehyde nanoemulsion edible composite films. *Food Hydrocolloids*, *41*, 188–194. doi:10.1016/j.foodhyd.2014.04.013

Ozaki, A., Muromachi, A., Sumi, M., Sakai, Y., Morishita, K., & Okamoto, T. (2010). Emulsification of coenzyme Q 10 using Gum Arabic increases bioavailability in rats and human and improves food-processing suitability. *Journal of Nutritional Science and Vitaminology*, *56*(1), 41–47. doi:10.3177/jnsv.56.41 PMID:20354345

Ozen, T., Darcan, C., Aktop, O., & Turkekul, I. (2011). Screening of antioxidant, antimicrobial activities and chemical contents of edible mushrooms wildly grown in the Black Sea region of Turkey. *Combinatorial Chemistry & High Throughput Screening*, *14*(2), 72–84. doi:10.2174/138620711794474079 PMID:20958253

Ozimek, L., Pospiech, E., & Narine, S. (2010). Nanotechnologies in food and meat processing. *Acta Scientiarum Polonorum. Technologia Alimentaria, 9*(4), 401–412.

Özoğul, Y., Boğa, E. K., Akyol, I., Durmuş, M., Uçar, Y., Regenstein, J., & Kosker, A. R. (2020). Antimicrobial activity of thyme essential oil nanoemulsions on spoilage bacteria of fish and food-borne pathogens. *Food Bioscience, 36*, 100635.

Ozturk, B., & McClements, D. J. (2016). Progress in natural emulsifiers for utilization in food emulsions. *Current Opinion in Food Science, 7*, 1–6. https://doi.org/10.1016/j.cofs.2015.07.008

Ozturk, M., Duru, M. E., Kivrak, S., Mercan-Dogan, N., Turkoglu, A., & Ozler, M. A. (2011). In vitro antioxidant, anticholinesterase and antimicrobial activity studies on three Agaricus species with fatty acid compositions and iron contents: A comparative study on the three most edible mushrooms. *Food and Chemical Toxicology, 49*(6), 1353–1360. doi:10.1016/j.fct.2011.03.019 PMID:21419821

Padayatchi, N., & Friedland, G. (2008). Decentralised management of drug-resistant tuberculosis (MDR- and XDR-TB) in South Africa: An alternative model of care. *The International Journal of Tuberculosis and Lung Disease, 12*, 978–980. PMID:18647461

Pagar, K.R., & Darekar, A.B. (2019). Nanoemulsion: A new concept of Delivery System. *Nanoemulsion: A New Concept of Delivery System, 9*(1), 39-46. doi:org/10.5958/2231-5659.2019.00006.7

Pagore, R., & Biyani, K. (n.d.). *A review on types, treatment, awareness, prevention, pathophysiology and diagnosis of hypertension*. Academic Press.

Pal, K., Banerjee, I., Sarkar, P., Kim, D., Deng, W. P., Dubey, N. K., & Majumder, K. (2020). *Biopolymer-Based Formulations: Biomedical and Food Applications*. Academic Press.

Palacios, I., Lozano, M., Moro, C., D'Arrigo, M., Rostagno, M. A., Martínez, J. A., García-Lafuente, A., Guillamón, E., & Villares, A. (2011). Antioxidant properties of phenolic compounds occurring in edible mushrooms. *Food Chemistry, 128*(3), 674–678. doi:10.1016/j.foodchem.2011.03.085

Palafox-Carlos, H., Ayala-Zavala, J. F., & Gonzalez-Aguilar, G. A. (2011). The role of dietary fiber in the bioaccessibility and bioavailability of fruit and vegetable antioxidants. *Journal of Food Science, 76*(1), 6–15. doi:10.1111/j.1750-3841.2010.01957.x PMID:21535705

Pal, R. (2016). Modeling the viscosity of concentrated nanoemulsions and nanosuspensions. *Fluids, 1*(2), 11. doi:10.3390/fluids1020011

Pandey, M., Choudhury, H., Yeun, O. C., Yin, H. M., Lynn, T. W., Tine, C., Wi, N. S., Yen, K., Phing, C. S., Kesharwani, P., Bhattamisra, S. K., & Gorain, B. (2018). Perspectives of Nanoemulsion Strategies in the Improvement of Oral, Parenteral and Transdermal Chemotherapy. *Current Pharmaceutical Biotechnology, 19*(4), 276–292. doi:10.2174/1389201019666180605125234 PMID:29874994

Pandey, R., & Khuller, G. K. (2005). Solid lipid particle-based inhalable sustained drug delivery system against experimental tuberculosis. *Tuberculosis (Edinburgh, Scotland), 85*(4), 227–234. doi:10.1016/j.tube.2004.11.003 PMID:15922668

Pandey, R., Sharma, A., Zahoor, A., Sharma, S., Khuller, G. K., & Prasad, B. (2003). Poly (DL-lactide-co-glycolide) nanoparticle-based inhalable sustained drug delivery system for experimental tuberculosis. *The Journal of Antimicrobial Chemotherapy, 52*(6), 981–986. doi:10.1093/jac/dkg477 PMID:14613962

Pandey, R., Zahoor, A., Sharma, S., & Khuller, G. K. (2003). Nanoparticle encapsulated antitubercular drugs as a potential oral drug delivery system against murine tuberculosis. *Tuberculosis (Edinburgh, Scotland), 83*(6), 373–378. doi:10.1016/j.tube.2003.07.001 PMID:14623168

Pandey, S., Giri, V. P., Tripathi, A., Kumari, M., Narayan, S., Bhattacharya, A., Srivastava, S., & Mishra, A. (2020). Early blight disease management by herbal nanoemulsion in Solanum lycopersicum with bio-protective manner. *Industrial Crops and Products*, *150*, 112421.

Pang, X. F., Zhang, L. H., Bai, F., Wang, N. P., Garner, R. E., McKallip, R. J., & Zhao, Z. Q. (2015). Attenuation of myocardial fibrosis with curcumin is mediated by modulating expression of angiotensin II AT1/AT2 receptors and ACE2 in rats. *Drug Design, Development and Therapy*, *9*, 6043. PMID:26648693

Panigrahi, S. S., Syed, I., Sivabalan, S., & Sarkar, P. (2018). Nanoencapsulation strategies for lipid-soluble vitamins. *Chemical Papers*, *73*(1), 1–16. doi:10.100711696-018-0559-7

Pan, K., Chen, H., Davidson, P. M., & Zhong, Q. (2014). Thymol nanoencapsulated by sodium caseinate: Physical and antilisterial properties. *Journal of Agricultural and Food Chemistry*, *62*(7), 1649–1657. https://doi.org/10.1021/jf4055402

Pannu, J., Mccarthy, A., Martin, A., Hamouda, T., Ciotti, S., Fothergill, A., & Sutcliffe, J. (2009). NB-002, a novel nanoemulsion with broad antifungal activ-ity against dermatophytes, other filamentous fungi, and Candida albicans. *Antimicrobial Agents and Chemotherapy*, *53*, 3273–3279.

Pant, M., Dubey, S., Patanjali, P. K., Naik, S. N., & Sharma, S. (2014). Insecticidal activity of eucalyptus oil nanoemulsion with karanja and jatropha aqueous filtrates. *International Biodeterioration & Biodegradation*, *91*, 119–127. doi:10.1016/j.ibiod.2013.11.019

Pan, Y., Tikekar, R. V., & Nation, N. (2016). Distribution of a model bioactive within solid lipid nanoparticles and nanostructured lipid carriers influences its loading efficiency and oxidative stability. *International Journal of Pharmaceutics*, *511*(1), 322–330. doi:10.1016/j.ijpharm.2016.07.019 PMID:27418566

Panza, J. A., García, C. E., Kilcoyne, C. M., Quyyumi, A. A., & Cannon, R. O. III. (1995). Impaired endothelium-dependent vasodilation in patients with essential hypertension: Evidence that nitric oxide abnormality is not localized to a single signal transduction pathway. *Circulation*, *91*(6), 1732–1738. doi:10.1161/01.CIR.91.6.1732 PMID:7882481

Pardridge, W. M. (1999). Non-invasive drug delivery to the human brain using endogenous blood–brain barrier transport systems. *Pharmaceutical Science & Technology Today*, *2*(2), 49–59. doi:10.1016/S1461-5347(98)00117-5 PMID:10234207

Parhi, P., Mohanty, C., & Sahoo, S. K. (2012). Nanotechnology-based combinational drug delivery: An emerging approach for cancer therapy. *Drug Discovery Today*, *17*(17-18), 1044–1052. doi:10.1016/j.drudis.2012.05.010 PMID:22652342

Park, J.W., Byun, Y.R., Choi, Y.K., Chang, K.Y., & Lee, J.B. (2021). *Current Oral drug delivery composition containing oxaliplatin and method for preparing same*. Academic Press.

Park, H., Mun, S., & Kim, Y. R. (2019). UV and storage stability of retinol contained in oil-in-water nanoemulsions. *Food Chemistry*, *272*, 404–410. doi:10.1016/j.foodchem.2018.08.076 PMID:30309562

Parra, E., & Pérez-Gil, J. (2015). Composition, structure and mechanical properties define performance of pulmonary surfactant membranes and films. *Chemistry and Physics of Lipids*, *185*, 153–175. doi:10.1016/j.chemphyslip.2014.09.002 PMID:25260665

Pateiro, M., Gomez, B., Munekata, P. E. S., Barba, F. J., Putnik, P., Bursa, D., Cevi, K., & Lorenzo, J. M. (2021). Nanoencapsulation of promising bioactive compounds to improve their absorption, stability, functionality and the appearance of the final food products. *Molecules (Basel, Switzerland)*, *26*(6), 1547. doi:10.3390/molecules26061547 PMID:33799855

Patel, M.R., Patel, R.B., & Thakore, S.D. (2018). Nanoemulsion in drug delivery. Applications of Nanocomposite. *Materials in Drug Delivery*, 667–700. doi:10.1016/B978-0-12-813741-3.00030-3

Patel, R., & Patel, K. P. (2014). Advances in novel parentral drug delivery systems. *Asian Journal of Pharmaceutics (AJP), 4*(3), 193-199.

Patel, A., Cholkar, K., Agrahari, V., & Mitra, A. K. (2013). Ocular drug delivery systems: An overview. *World Journal of Pharmacology, 2*(2), 47–64. doi:10.5497/wjp.v2.i2.47 PMID:25590022

Patel, A., Heussen, P., Dorst, E., Hazekamp, J., & Velikov, K. P. (2013). Colloidal approach to prepare colour blends from colourants with different solubility profiles. *Food Chemistry, 141*(2), 1466–1471. doi:10.1016/j.foodchem.2013.03.082 PMID:23790940

Patel, P. R., & Joshi, J. R. (2012). An overview on nanoemulsion: A novel approach. *International Journal of Pharmaceutical Sciences and Research, 3*(12), 4640–4650.

Patel, R. B., Patel, M. R., Thakore, S. D., & Patel, B. G. (2017). Nanoemulsion as a valuable nanostructure platform for pharmaceutical drug delivery. In *Nano-and Microscale Drug Delivery Systems* (pp. 321–341). Elsevier.

Patel, R. B., Thakore, S. D., & Patel, M. R. (2016). Recent Survey on Patents of Nanoemulsions. *Current Drug Delivery, 13*(6), 857–881. doi:10.2174/1567201812666150901111930 PMID:26324230

Patel, R. P., & Joshi, J. R. (2012). An overview on nanoemulsion: A Novel Approach. *International Journal of Pharmaceutical Sciences and Research, 3*(12), 4640–4650. org/ doi:10.13040/IJPSR.0975-8232.3(12).4640-50

Patel, S., & Goyal, A. (2012). Recent developments in mushrooms as anti-cancer therapeutics: A review. *Biotechnology, 2*, 1–15. PMID:22582152

Paterson, R. R. M. (2006). *Ganoderma* - A therapeutic fungal biofactory. *Phytochemistry, 67*(18), 1985–2001. doi:10.1016/j.phytochem.2006.07.004 PMID:16905165

Pathak, M. (2017). *Nanoemulsions and Their Stability for Enhancing Functional Properties of Food Ingredients*. doi:10.1016/B978-0-12-811942-6.00005-4

Pathak, M. (2017). Nanoemulsions and Their Stability for Enhancing Functional Properties of Food Ingredients. Nanotechnology Applications in Food, 87-106. doi:org/10.1016/B978-0-12-811942-6.00005-4

Pathakoti, K., Manubolu, M., & Hwang, H. M. (2017). Nanostructures: Current uses and future applications in food science. *Yao Wu Shi Pin Fen Xi, 25*(2), 245–253. PMID:28911665

Patil, K., Bagade, S., Bonde, S., Sharma, S., & Saraogi, G. (2018). Recent therapeutic approaches for the management of tuberculosis: Challenges and opportunities. *Biomedicine and Pharmacotherapy, 99*, 735–745. doi:10.1016/j.biopha.2018.01.115 PMID:29710471

Paudel, S., Bhargava, K., & Kotturi, H. (2019). Antimicrobial activity of cinnamon oil nanoemulsion against Listeria monocytogenes and Salmonella spp. on melons. *LWT- Food Science and Technology*. doi:. doi:10.1016/j.lwt.2019.05.087

Paulis, L., Rajkovicova, R., & Simko, F. (2015). New developments in the pharmacological treatment of hypertension: Dead-end or a glimmer at the horizon? *Current Hypertension Reports, 17*(6), 42. doi:10.100711906-015-0557-x PMID:25893478

Paulis, L., & Unger, T. (2010). Novel therapeutic targets for hypertension. *Nature Reviews. Cardiology, 7*(8), 431–441. doi:10.1038/nrcardio.2010.85 PMID:20567239

Pavela, R. (2013). Efficacy of naphthoquinones as insecticides against the house fly, Musca domestica L. *Industrial Crops and Products, 43*, 745–750. doi:10.1016/j.indcrop.2012.08.025

Pavoni, L., Perinelli, D. R., Bonacucina, G., Cespi, M., & Palmieri, G. F. (2020). An Overview of Micro- and Nanoemulsions as Vehicles for Essential Oils: Formulation, Preparation and Stability. *Nanomaterials (Basel, Switzerland)*, *10*(1), 135. doi:10.3390/nano10010135 PMID:31940900

Pawar, V. K., Panchal, S. B., Singh, Y., Meher, J. G., Sharma, K., Singh, P., Bora, H. K., Singh, A., Datta, D., & Chourasia, M. K. (2014). Immunotherapeutic vitamin E nanoemulsion synergies the antiproliferative activity of paclitaxel in breast cancer cells via modulating Th1 and Th2 immune response. *Journal of Controlled Release*, *196*, 295–306.

Pedro, A. S., Cabral-Albuquerque, E., Ferreira, D., & Sarmento, B. (2009). Chitosan: An option for development of essential oil delivery systems for oral cavity care? *Carbohydrate Polymers*, *76*(4), 501–508. doi:10.1016/j.carbpol.2008.12.016

Peer, D., Karp, J. M., Hong, S., Farokhzad, O. C., Margalit, R., & Langer, R. (2007). Nanocarriers as an emerging platform for cancer therapy. *Nature Nanotechnology*, *2*(12), 751–760.

Peña, A. A., & Miller, C. A. (2006). Solubilization rates of oils in surfactant solutions and their relationship to mass transport in emulsions. *Advances in Colloid and Interface Science*, *123-126*, 241–257. https://doi.org/10.1016/j.cis.2006.05.005

Pendleton, J. N., Gorman, S. P., & Gilmore, B. F. (2013). Clinical relevance of the ESKAPE pathogens. *Expert Review of Anti-Infective Therapy*, *11*(3), 297–308. doi:10.1586/eri.13.12

Perez-Cacho, P. R., & Rouseff, R. (2008). Processing and storage effects on orange juice aroma: A review. *Journal of Agricultural and Food Chemistry*, *56*(21), 9785–9796. https://doi.org/10.1021/jf801244j

Perez-Herrero, E., & Fernandez-Medarde, A. (2015). Advanced targeted therapies in cancer: Drug nanocarriers, the future of chemotherapy. *European Journal of Pharmaceutics and Biopharmaceutics*, *93*, 52–79. Advance online publication. doi:10.1016/j.ejpb.2015.03.018 PMID:25813885

Periasamy, K. (2005). Novel antibacterial compounds obtained from some edible mushrooms. *International Journal of Medicinal Mushrooms*, *7*, 443–444. doi:10.1615/IntJMedMushrooms.v7.i3.820

Periasamy, V. S., Athinarayanan, J., & Alshatwi, A. A. (2016). Anticancer activity of an ultrasonic nanoemulsion formulation of Nigella sativa L. essential oil on human breast cancer cells. *Ultrasonics Sonochemistry*, *31*, 449–455.

Peters, K., Leitzke, S., Diederichs, J. E., Borner, K., Hahn, H., Müller, R. H., & Ehlers, S. (2000). Preparation of a clofazimine nanosuspension for intravenous use and evaluation of its therapeutic efficacy in murine *Mycobacterium avium* infection. *The Journal of Antimicrobial Chemotherapy*, *45*(1), 77–83. doi:10.1093/jac/45.1.77 PMID:10629016

Petkov, P. M., Zavadil, J., Goetz, D., Chu, T., Carver, R., Rogler, C. E., Bottinger, E. P., Shafritz, D. A., & Dabeva, M. D. (2004). Gene Expression Pattern in Hepatic Stem/Progenitor Cells during Rat Fetal Development Using Complementary DNA Microarrays. *Hepatology (Baltimore, Md.)*, *39*(3), 617–627. doi:10.1002/hep.20088 PMID:14999680

Peto, R., Davies, C., Godwin, J., Gray, R., Pan, H. C., Clarke, M., Cutter, D., Darby, S., McGale, P., Taylor, C., Wang, Y. C., Bergh, J., Di Leo, A., Albain, K., Swain, S., Piccart, M., & Pritchard, K. (2012). Comparisons between different polychemotherapy regimens for early breast cancer: Meta-analyses of long-term outcome among 100,000 women in 123 randomised trials. *Lancet*, *379*(9814), 432–444.

Petrovska, B. B. (2001). Protein fraction in edible Macedonian mushrooms. *European Food Research and Technology*, *212*(4), 469–472. doi:10.1007002170000285

Petrovsky, N., & Aguilar, J. C. (2004). Vaccine adjuvants: Current state and future trends. *Immunology and Cell Biology*, *82*(5), 488–496. doi:10.1111/j.0818-9641.2004.01272.x PMID:15479434

Pham, D. H., & Nguyen, T. T. (2020). Preparation of Tamanu Oil Nanoemulsions by Phase Inversion Temperature. *IOP Conference Series. Materials Science and Engineering*, *991*, 012116. doi:10.1088/1757-899X/991/1/012116

Pham, J., Grundmann, O., & Elbayoumi, T. (2015). Mitochondriotropic nanoemulsified genistein-loaded vehicles for cancer therapy. In *Mitochondrial Medicine* (pp. 85–101). Humana Press. doi:10.1007/978-1-4939-2288-8_7

Philippe, M., Didillon, B., & Gilbert, L. (2012). Industrial commitment to green and sustainable chemistry: Using renewable materials & developing eco-friendly processes and ingredients in cosmetics. *Green Chemistry*, *14*(4), 952–956. doi:10.1039/c2gc16341a

Pianetti, S., Arsura, M., Romieu-Mourez, R., Coffey, R. J., & Sonenshein, G. E. (2001). Her-2/neu overexpression induces NF-κB via a PI3-kinase/Akt pathway involving calpain-mediated degradation of IκB-α that can be inhibited by the tumor suppressor PTEN. *Oncogene*, *20*(11), 1287–1299. doi:10.1038j.onc.1204257 PMID:11313873

Pidaparthi, K., & Suares, D. (2016). Comparison of Nanoemulsion and Aqueous Micelle Systems of Paliperidone for Intranasal Delivery. *AAPS PharmSciTech*, *18*(5), 1710–1719. doi:10.120812249-016-0640-x PMID:27714701

Pilcer, G., & Amighi, K. (2010). Formulation strategy and use of excipients in pulmonary drug delivery. *International Journal of Pharmaceutics*, *392*(1-2), 1–19. doi:10.1016/j.ijpharm.2010.03.017 PMID:20223286

Pinto, V. N. (2013). Bioterrorism: Health sector alertness. *Journal of Natural Science, Biology, and Medicine*, *4*(1), 24. doi:10.4103/0976-9668.107256 PMID:23633831

Pires, A., Fortuna, A., Alves, G., & Falcão, A. (2009). Intranasal drug delivery: How, why and what for? *Journal of Pharmacy & Pharmaceutical Sciences*, *12*(3), 288–311.

Pires, L. A., Hegg, R., Valduga, C. J., Graziani, S. R., Rodrigues, D. G., & Maranhão, R. C. (2009). Use of cholesterol-rich nanoparticles that bind to lipoprotein receptors as a vehicle to paclitaxel in the treatment of breast cancer: Pharmacokinetics, tumor uptake and a pilot clinical study. *Cancer Chemotherapy and Pharmacology*, *63*(2), 281–287.

Plaza-Oliver, M., Santander-Ortega, M. J., & Lozano, M. V. (2021). Current approaches in lipid-based nanocarriers for oral drug delivery. *Drug Delivery and Translational Research*, *11*(2), 471–497. doi:10.100713346-021-00908-7 PMID:33528830

Podda, A., & Del Giudice, G. (2003). MF59-adjuvanted vaccines: Increased immunogenicity with an optimal safety profile. *Expert Review of Vaccines*, *2*(2), 197–203. doi:10.1586/14760584.2.2.197 PMID:12899571

Poller, J. M., Zaloga, J., Schreiber, E., Unterweger, H., Janko, C., Radon, P., Eberbeck, D., Trahms, L., Alexiou, C., & Friedrich, R. P. (2017). Selection of potential iron oxide nanoparticles for breast cancer treatment based on in vitro cytotoxicity and cellular uptake. *International Journal of Nanomedicine*, *12*, 3207–3220.

Pongsumpun, P., Iwamoto, S., & Siripatrawan, U. (2020). Response surface methodology for optimization of cinnamon essential oil nanoemulsion with improved stability and antifungal activity. *Ultrasonics Sonochemistry*, *60*, 104604. doi:10.1016/j.ultsonch.2019.05.021 PMID:31539730

Poonia, N., Lather, V., Kaur, B., Kirthanashri, S. V., & Pandita, D. (2020). Optimization and Development of Methotrexate- and Resveratrol-Loaded Nanoemulsion Formulation Using Box-Behnken Design for Rheumatoid Arthritis. *Assay and Drug Development Technologies*, *18*(8), 356–368. doi:10.1089/adt.2020.989 PMID:33052698

Porras, M., Solans, C., Gonzalez, C., & Gutierrez, J. M. (2008). Properties of water-in-oil (W/O) nanoemulsions prepared by a low-energy emulsification method. *Colloids and Surfaces. A, Physicochemical and Engineering Aspects*, *324*(1-3), 181–188.

Porter, C. J., Pouton, C. W., Cuine, J. F., & Charman, W. N. (2008). Enhancing intestinal drug solubilisation using lipid-based delivery systems. *Advanced Drug Delivery Reviews*, *60*(6), 673–691.

Porter, C. J., Trevaskis, N. L., & Charman, W. N. (2007). Lipids and lipid-based formulations: Optimising the oral delivery of lipophilic drugs. *Nature Reviews. Drug Discovery*, *6*(3), 231–248.

Pouton, C. W. (1997). Formulation of selfemulsifying drug delivery systems. *Advanced Drug Delivery Reviews*, *25*(1), 47–58. doi:10.1016/S0169-409X(96)00490-5

Pouton, C. W., & Porter, C. J. (2008). Formulation of lipid-based delivery systems for oral administration: Materials, methods9 and strategies. *Advanced Drug Delivery Reviews*, *60*(6), 625–637.

Poverenova, E., Ripsa, H. A., Zaitseva, Y., Bara, Y., Danay, O., Horeva, B., Bilbao-Sainz, C., McHugh, T., & Rodov, V. (2018). Potential of chitosan from mushroom waste to enhance quality and storability of fresh-cut melons. *Food Chemistry*, *268*, 233–241. doi:10.1016/j.foodchem.2018.06.045 PMID:30064752

Prabhakar, K., Afzal, S. M., Surender, G., & Kishan, V. (2013). Tween 80 containing lipid nanoemulsions for delivery of indinavir to brain. *Acta Pharmaceutica Sinica. B*, *3*(5), 345–353. doi:10.1016/j.apsb.2013.08.001

Praditya, D., Kirchhoff, L., Brüning, J., Rachmawati, H., Steinmann, J., & Steinmann, E. (2019). Anti-infective properties of the golden spice curcumin. *Frontiers in Microbiology*, *10*, 912. doi:10.3389/fmicb.2019.00912 PMID:31130924

Prakash, R.T., & Thiagarajan, P. (2011). Nanoemulsions for drug delivery through different routes. *Research in Biotechnology*, *2*(3), 1-13.

Prakash, A., Baskaran, R., Nithyanand, P., & Vadivel, V. (2020). Effect of nanoemulsification on the antibacterial and anti-biofilm activities of selected spice essential oils and their major constituents against *Salmonella enterica Typhimurium. Journal of Cluster Science*, *31*(5), 1123–1135. doi:10.100710876-019-01720-7

Prakash, A., Sen, S., & Dixit, R. (2013). The emerging usage and applications of nanotechnology in food processing industries. The new age of nanofood. *International Journal of Pharmaceutical Sciences and Research*, *22*(1), 107–111.

Prasad, C., Bhatia, E., & Banerjee, R. (2020). curcumin encapsulated Lecithin nanoemulsions: An oral platform for Ultrasound Mediated Spatiotemporal Delivery of curcumin to the tumor. *Scientific Reports*, *10*(1), 1–15. doi:10.103841598-020-65468-1 PMID:32444829

Praveen-Kumar, G., & Divya, A. (2015). Nanoemulsion based targeting in cancer therapeutics. *Medicinal Chemistry (Sunnyvale, Calif.)*, *5*(6), 272–284. doi:10.4172/2161-0444.1000275

Press, D. (2015a). *Antibacterial activity of neem nanoemulsion and its toxicity assessment on human lymphocytes in vitro*. Academic Press.

Press, D. (2015b). *Eucalyptus oil nanoemulsion-impregnated chitosan film : antibacterial effects against a clinical pathogen, Staphylococcus aureus, in vitro*. Academic Press.

Prestinaci, F., Pezzotti, P., & Pantosti, A. (2015). Antimicrobial resistance: A global multifaceted phenomenon. *Pathogens and Global Health*, *109*(7), 309–318. https://doi.org/10.1179/2047773215Y.0000000030

Prévot, G., Soria, F. N., Thiolat, M. L., Daniel, J., Verlhac, J. B., Blanchard-Desce, M., Bezard, E., Barthélémy, P., Crauste-Manciet, S., & Dehay, B. (2018). Harnessing Lysosomal pH through PLGA Nanoemulsion as a Treatment of Lysosomal-Related Neurodegenerative Diseases. *Bioconjugate Chemistry*, *29*(12), 4083–4089. doi:10.1021/acs.bioconjchem.8b00697 PMID:30424597

Primo, F. L., Bentley, M. V., & Tedesco, A. C. (2008). Photophysical studies and in vitro skin permeation/retention of foscan/ nanoemulsion (NE) applicable to photodynamic therapy skin cancer treatment. *Journal of Nanoscience and Nanotechnology*, *8*, 340–347.

Primo, F. L., Michieleto, L., Rodrigues, M. A., Macaroff, P. P., Morais, P. C., Lacava, Z. G., Bentley, M. V. L. B., & Tedesco, A. C. (2007). Magnetic nanoemulsions as drug delivery system for Foscan®: Skin permeation and retention in vitro assays for topical application in photodynamic therapy (PDT) of skin cancer. *Journal of Magnetism and Magnetic Materials*, *311*(1), 354–357. doi:10.1016/j.jmmm.2006.10.1183

Prins, G. S., & Lindgren, M. (2014). Accessory sex glands in the male. In Physiology of Reproduction (pp. 773–804). Elsevier Inc.

Prokop, A., & Davidson, J. M. (2008). Nanovehicular intracellular delivery systems. *Journal of Pharmaceutical Sciences*, *97*(9), 3518–3590. doi:10.1002/jps.21270 PMID:18200527

Prow, T. W., Grice, J. E., Lin, L. L., Faye, R., Butler, M., Becker, W., Wurm, E. M., Yoong, C., Robertson, T. A., Soyer, H. P., & Roberts, M. S. (2011). Nanoparticles and microparticles for skin drug delivery. *Advanced Drug Delivery Reviews*, *63*(6), 470–491.

Puhalla, S., Bhattacharya, S., & Davidson, N. E. (2012). Hormonal therapy in breast cancer: A model disease for the personalization of cancer care. *Molecular Oncology*, *6*(2), 222–236. doi:10.1016/j.molonc.2012.02.003 PMID:22406404

Puri, M., Kaur, I., Perugini, M. A., & Gupta, R. C. (2012). Ribosome-inactivating proteins: Currentstatus and biomedical applications. *Drug Discovery Today*, *17*(13-14), 774–783. doi:10.1016/j.drudis.2012.03.007 PMID:22484096

Purkait, A., & Hazra, D. K. (2019). Biodiesel as a carrier for pesticide formulations: A green chemistry approach. *International Journal of Pest Management*, *66*(4), 341–350. doi:10.1080/09670874.2019.1649740

Qadir, A., Faiyazuddin, M. D., Talib Hussain, M. D., Alshammari, T. M., & Shakeel, F. (2016). Critical steps and energetics involved in a successful development of a stable nanoemulsion. *Journal of Molecular Liquids*, *214*, 7–18. doi:10.1016/j.molliq.2015.11.050

Qadir, A., Faiyazuddin, M. D., Talib Hussain, M. D., Alshammari, T. M., & Shakeel, F. (2016). Critical steps and energetics involved in a successful development of a stable nanoemulsion. *Journal of Molecular Liquids*, *214*, 7–18. https://doi.org/10.1016/j.molliq.2015.11.050

Qazi, H. J., Ye, A., Acevedo-Fani, A., & Singh, H. (2021). In vitro digestion of curcumin-nanoemulsion-enriched dairy protein matrices: Impact of the type of gel structure on the bioaccessibility of curcumin. *Food Hydrocolloids*, *117*, 106692. doi:10.1016/j.foodhyd.2021.106692

Qhattal, H. S., Wang, S., Salihima, T., Srivastava, S. K., & Liu, X. (2011). Nanoemulsions of cancer chemopreventive agent benzyl isothiocyanate display enhanced solubility, dissolution, and permeability. *Journal of Agricultural and Food Chemistry*, *59*(23), 12396–12404. doi:10.1021/jf202612b PMID:22011260

Qian, C., Decker, E. A., Xiao, H., & McClements, D. J. (2012). Solid lipid nanoparticles: Effect of carrier oil and emulsifier type on phase behavior and physical stability. *Journal of the American Oil Chemists' Society, 89*(1), 17-28. doi:10.1007/s11746-011-1882-0

Qian, C., & McClements, D. J. (2011). Formation of nanoemulsions stabilised by model food-grade emulsifiers using high-pressure homogenisation: Factors affecting particle size. *Food Hydrocolloids*, *25*(5), 1000–1008.

Qian, C., & McClements, D. J. (2011). Formation of nanoemulsions stabilized by model food-grade emulsifiers using high-pressure homogenization: Factors affecting particle size. *Food Hydrocolloids*, *25*, 1000–1008.

Qian, K., Zhao, H., & He, L. (2018). Preparation of fenpropathrin nanoemulsion by phase inversion temperature method and its study on performance. *International Journal of Nanoscience*, *17*(5), 1–7. doi:10.1142/S0219581X18500011

Qin, X., Zhou, Y., Wang, Y., Wang, Z., Wang, Y., Chen, J., Zhu, L., Quan, X., Liu, Z., Zhang, H., Jiang, L., Dong, H., & Zhang, Z. (2021). Preparation and Characterization of Protein-loaded PFC Nanoemulsions for the Treatment of Heart Diseases by Pulmonary Administration. *European Journal of Pharmaceutical Sciences, 158*, 105690. doi:10.1016/j.ejps.2020.105690

Qin, H., Zhang, H., Li, L., Zhou, X., Li, J., & Kan, C. (2017). Preparation and properties of lambda-cyhalothrin/polyurethane drug-loaded nanoemulsions. *RSC Advances*, *7*(83), 52684–52693. doi:10.1039/C7RA10640H

Qin, H., Zhou, X., Gu, D., Li, L., & Kan, C. (2019). Preparation and Characterization of a Novel Waterborne Lambda-Cyhalothrin/Alkyd Nanoemulsion. *Journal of Agricultural and Food Chemistry*, *67*(38), 10587–10594. doi:10.1021/acs.jafc.9b03681 PMID:31497957

Qiu, N., Liu, Y., Liu, Q., Chen, Y., Shen, L., Hu, M., Zhou, X., Shen, Y., Gao, J., & Huang, L. (2021). Celastrol nanoemulsion induces immunogenicity and downregulates PD-L1 to boost abscopal effect in melanoma therapy. *Biomaterials*, *269*, 120604.

Quan Le, M., Carpentier, R., Lantier, I., Ducournau, C., Dimier-Poisson, I., & Betbeder, D. (2018). Residence time and uptake of porousand cationic maltodextrin-based nanoparticles in the nasal mucosa: Comparison with cationic or anionic nanoparticles. *International Journal of Pharmaceutics*, *550*(1–2), 316–324. PMID:30171898

Quatrin, P. M., Verdi, C. M., de Souza, M. E., de Godoi, S. N., Klein, B., Gundel, A., Wagner, R., de Almeida Vaucher, R., Ourique, A. F., & Santos, R. (2017). Antimicrobial and antibiofilm activities of nanoemulsions containing Eucalyptus globulus oil against Pseudomonas aeruginosa and Candida spp. *Microbial Pathogenesis*, *112*, 230–242. https://doi.org/10.1016/j.micpath.2017.09.062

Quemada, D., & Berli, C. (2002). Energy of interaction in colloids and its implications in rheological modeling. *Advances in Colloid and Interface Science*, *98*(1), 51–85. doi:10.1016/S0001-8686(01)00093-8 PMID:12061712

Quemeneur, F., Rammal, A., Rinaudo, M., & Pepin-Donat, B. (2007). Large and giant vesicles "decorated" with chitosan: Effects of pH, salt or glucose stress, and surface adhesion. *Biomacromolecules*, *8*(8), 2512–2519. doi:10.1021/bm061227a PMID:17658883

Quemeneur, F., Rinaudo, M., Maret, G., & Pepin-Donat, B. (2010). Decoration of lipid vesicles by polyelectrolytes: Mechanism and structure. *Soft Matter*, *6*(18), 4471–4481. doi:10.1039/c0sm00154f

Quereshi, S., Pandey, A. K., & Sandhu, S. S. (2010). Evaluation of antibacterial activity of different *Ganoderma lucidum* extracts. *Journal of Science and Research*, *3*, 9–13.

Quintão, F. J., Tavares, R. S., Vieira-Filho, S. A., Souza, G. H., & Santos, O. D. (2013). Hydroalcoholic extracts of Vellozia Squamata: A study of its NE's for pharmaceutical or cosmetic applications. *Revista Brasileira de Farmacognosia*, *23*(1), 101–107. doi:10.1590/S0102-695X2013005000001

Qu, Y., Li, A., Ma, L., Iqbal, S., Sun, X., Ma, W., Li, C., Zheng, D., Xu, Z., Zhao, Z., & Ma, D. (2021). Nose-to-brain delivery of disulfiram nanoemulsion in situ gel formulation for glioblastoma targeting therapy. *International Journal of Pharmaceutics*, *597*, 120250.

Rabelo, A. S., Sutili, F. K., Meneses, J. O., Severino, P., Souto, E. B., Fujimoto, R. Y., & Cardoso, J. C. (2021). 23 central composite rotatable design for the production of neem oil nanoemulsion for antifungal and antiparasitic applications. *Journal of Chemical Technology and Biotechnology (Oxford, Oxfordshire)*, *96*, 2159–2167. https://doi.org/10.1002/jctb.6623

Ragelle, H., Crauste-Manciet, S., Seguin, J., Brossard, D., Scherman, D., Arnaud, P., & Chabot, G. G. (2012). Nanoemulsion formulation of fisetin improves bioavailability and antitumour activity in mice. *International Journal of Pharmaceutics*, *427*(2), 452–459. doi:10.1016/j.ijpharm.2012.02.025 PMID:22387278

Raghavan, P. R. (2016). Inhibition of Dengue and other enveloped viruses by Metadichol®, a novel Nano emulsion Lipid. *Journal of the science of Healing Outcomes, 8*, 19-25.

Rai, V. K., Mishra, N., Yadav, K. S., & Yadav, N. P. (2018). Nanoemulsion as pharmaceutical carrier for dermal and transdermal drug delivery: Formulation development, stability issues, basic considerations and applications. *Journal of Controlled Release, 270*, 203–225. doi:10.1016/j.jconrel.2017.11.049

Rai, V. K., Sharma, A., & Thakur, A. (2021). Quality Control of Nanoemulsion: By PDCA Cycle and 7QC Tools. *Current Drug Delivery, 18*(9), 1–12. doi:10.2174/1567201818666210203180516 PMID:33538674

Ramalingam, K. & Lee, V. A. (2018). Antibiofilm activity of an EDTA-containing nanoemulsion on multidrug-resistant Acinetobacterbaumannii. *Artificial Cells, Nanomedicine, and Biotechnology, 46*(sup2), 737-743. doi:10.1080/21691401.2018.1468771

Ramalingam, K., & Amaechi, B. T. (2019). Phytochemicals: Their Therapeutic Potential Against Dental Caries. In Computational Techniques for Dental Image Analysis (pp. 238-275). IGI Global.

Ramalingam, K., Amaechi, B. T., Ralph, R. H., & Lee, V. A. (2012). Antimicrobial activity of nanoemulsion on cariogenic planktonic and biofilm organisms. *Archives of Oral Biology, 57*(1), 15–22. doi:10.1016/j.archoralbio.2011.07.001 PMID:21807359

Ramalingam, K., & Lee, V. (2019). Synergistic Effects of Nanoemulsion and Deferiprone (1, 2 Dimethyl-3-Hydroxypyrid-4-One) on Multi-Drug Resistant Acinetobacterbaumannii. *Nano Biomedicine and Engineering, 11*(3), 226–237. doi:10.5101/nbe.v11i3.p226-237

Rani, D., Saxena, R., Nayak, B., & Srivastava, S. (2018). Cloning and expression of truncated ORF2 as a vaccine candidate against hepatitis E virus. *3 Biotech, 8*(10), 1-5.

Rani, N. P., Suriyaprakash, T. N. K., & Senthamarai, R. (2010). Formulation and evaluation of rifampicin and gatifloxacin niosomes on logarithmic-phase cultures of *Mycobacterium tuberculosis. International Journal of Pharma and Bio Sciences, 1*(4), 379–387.

Rao, J., & Mcclements, D. (2012). Impact of lemon oil composition on formation and stability of model food and beverage emulsions. *Food Chemistry, 134*(2), 749–757.

Rapoport, N. Y., Nam, K. H., Gupta, R., Gao, Z., Mohan, P., Payne, A., Todd, N., Liu, X., Kim, T., Shea, J., Scaife, C., Parker, D. L., Jeong, E. K., & Kennedy, A. M. (2011). Ultrasound mediated tumor imaging and nanotherapy using drug loaded, block copolymer stabilized perfluorocarbon nanoemulsions. *Journal of Controlled Release, 153*(1), 4–15. doi:10.1016/j.jconrel.2011.01.022 PMID:21277919

Rathore, H., Prasad, S., & Sharma, S. (2017). Mushroom nutraceuticals forimproved nutrition and better human health: A review. *Pharma Nutr, 5*, 35–46.

Rautureau, M., Bisalli, A., & Rambaud, J. (1981). Bile salts and lipids in aqueous intraluminal phase during the digestion of a standard meal in normal man (author's transl). *Gastroenterologie Clinique et Biologique, 5*(4), 417–425. PMID:7227749

Raviadaran, R., Chandran, D., Shin, L. H., & Manickam, S. (2018). Optimization of palm oil in water nano-emulsion with curcumin using microfluidizer and response surface methodology. *Lebensmittel-Wissenschaft + Technologie, 96*, 58–65. doi:10.1016/j.lwt.2018.05.022

Raviadaran, R., Ng, M. H., Chandran, D., Ooi, K. K., & Manickam, S. (2021). Stable W/O/W multiple nanoemulsion encapsulating natural tocotrienols and caffeic acid with cisplatin synergistically treated cancer cell lines (A549 and HEP G2) and reduced toxicity on normal cell line (HEK 293). *Materials Science and Engineering C, 121*, 111808.

Rayner, M., Marku, D., Eriksson, M., Sjoo, M., Dejmek, P., & Wahlgren, M. (2014). Biomass-based particles for the formulation of Pickering type emulsions in food and topical applications. *Colloids and Surfaces. A, Physicochemical and Engineering Aspects*, *458*, 48–62. doi:10.1016/j.colsurfa.2014.03.053

Reddy, V. M., Bogatcheva, E., Einck, L., & Nacy, C. A. (2011). Nanoemulsion formulation enhances intracellular activity of capuramycin analogues against *Mycobacterium tuberculosis*. *Drug Delivery Letters*, *1*(2), 150–158. doi:10.2174/2210304x11101020150

Reed, S. G., Bertholet, S., Coler, R. N., & Friede, M. (2009). New horizons in adjuvants for vaccine development. *Trends in Immunology*, *30*(1), 23–32. doi:10.1016/j.it.2008.09.006 PMID:19059004

Rehman, A. U., Akram, S., Seralin, A., Vandamme, T., & Anton, N. (2020). *Lipid nanocarriers: Formulation, properties, and applications.* doi:10.1016/B978-0-12-816770-0.00021-6

Rehman, F. U., Shah, K. U., Shah, S. U., Khan, I. U., Khan, G. M., & Khan, A. (2017). From nanoemulsions to self-nanoemulsions, with recent advances in self-nanoemulsifying drug delivery systems (SNEDDS). *Expert Opinion on Drug Delivery*, *14*(11), 1325–1340. doi:10.1080/17425247.2016.1218462 PMID:27485144

Reinbolt, R. E., Mangini, N., Hill, J. L., Levine, L. B., Dempsey, J. L., Singaravelu, J., Koehler, K. A., Talley, A., & Lustberg, M. B. (2015). Endocrine therapy in breast cancer: The neoadjuvant, adjuvant, and metastatic approach. *Seminars in Oncology Nursing*, *31*(2), 146–155.

Reinus, J., & Simon, D. (2014). *Gastrointestinal Anatomy and Physiology: The essentials*. Wiley-Blackwell Publishers.

Reiss, H. (1975). Entropy-induced dispersion of bulk liquids. *Journal of Colloid and Interface Science*, *53*(1), 61–70. doi:10.1016/0021-9797(75)90035-1

Ren, G., Sun, Z., Wang, Z., Zheng, X., Xu, Z., & Sun, D. (2019). Nanoemulsion formation by the phase inversion temperature method using polyoxypropylene surfactants. *Journal of Colloid and Interface Science*, *540*, 177–184. doi:10.1016/j.jcis.2019.01.018 PMID:30640065

Ren, J. N., Dong, M., Hou, Y. Y., Fan, G., & Pan, S. Y. (2018). Effect of olive oil on the preparation of nanoemulsions and its effect on aroma release. *Journal of Food Science and Technology*, *55*(10), 4223–4231. doi:10.100713197-018-3358-9 PMID:30228421

Research and Market. (2018). *Global nanotechnology market outlook 2024*. Report ID: 3989533. https://www.researchandmarkets.com/reports/4536705/global-nanotechnology-market-outlook-2018-2024

Reverchon, E., De Marco, I., & Della Porta, G. (2002). Rifampicin microparticles production by supercritical antisolvent precipitation. *International Journal of Pharmaceutics*, *243*(1-2), 83–91. doi:10.1016/S0378-5173(02)00261-2 PMID:12176297

Rezende, R. M., & Weiner, H. L. (2017). History and mechanisms of oral tolerance. *Seminars in Immunology*, *30*, 1–13. doi:10.1016/j.smim.2017.07.004 PMID:28774470

Ribeiro, B., de Pinho, P. G., Andrade, P. B., Oliveira, C., Silva Ferreira, A. C., Baptista, P., & Valentao, P. (2011). Do bioactive carotenoids contribute to the color of edible mushrooms? *The Open Chemical and Biomedical Methods Journal*, *4*(1), 14–18. doi:10.2174/1875038901104010014

Ribeiro, R. C. D. A., Barreto, S. M. A. G., Ostrosky, E. A., Rocha-Filho, P. A. D., Veríssimo, L. M., & Ferrari, M. (2015). Production and characterization of cosmetic NE's containing Opuntia ficus-indica (L.) Mill extract as moisturizing agent. *Molecules (Basel, Switzerland)*, *20*(2), 2492–2509. doi:10.3390/molecules20022492 PMID:25648593

Ribes, S., Fuentes, A., Talens, P., Barat, J., Ferrari, G., & Donsì, F. (2017). Influence of emulsifier type on the antifungal activity of cinnamon leaf, lemon and bergamot oil nanoemulsions against Aspergillus niger. *Food Control*, *73*, 784–795.

Rice, L. B. (2008). Federal funding for the study of antimicrobial resistance in nosocomial pathogens, no ESKAPE. doi.org/ doi:10.1086/533452

Riedel, S. (2005). Edward Jenner and the history of smallpox and vaccination. *Proceedings - Baylor University. Medical Center*, *18*(1), 21–25. doi:10.1080/08998280.2005.11928028 PMID:16200144

Rinaldi, F., Oliva, A., Sabatino, M., Imbriano, A., Hanieh, P. N., Garzoli, S., & Carafa, M. (2020). Antimicrobial Essential Oil Formulation: Chitosan Coated Nanoemulsions for Nose to Brain Delivery. *Pharmaceutics*, *12*(7), 678. doi:10.3390/pharmaceutics12070678

Rinaudo, M., Milas, M., & Le Dung, P. (1993). Characterization of chitosan. Influence of ionic strength and degree of acetylation on chain expansion. *International Journal of Biological Macromolecules*, *15*(5), 281–285. doi:10.1016/0141-8130(93)90027-J PMID:8251442

Risso, N. H., Stopiglia, C. D. O., Oliveira, M. T., Haas, S. E., Maciel, T. R., Lazzari, N. R., & Beckmann, D. V. (2020). Chlorhexidine Nanoemulsion: A New Antiseptic Formulation. *International Journal of Nanomedicine*, *15*, 6935. doi:10.2147/IJN.S228280

Roco, M. C. (1999). Nanoparticles and nanotechnology research. *Journal of Nanoparticle Research*, *1*(1), 1–6. doi:10.1023/A:1010093308079

Rodham, D. K. (2000). Colloid and interface science in formulation research for crop protection products. *Current Opinion in Colloid & Interface Science*, *5*(6), 280–287. doi:10.1016/S1359-0294(00)00067-4

Rodrigues, C. R., Ferreira, A. M., Vilhena, J. C. E., Almeida, F. B., Cruz, R. A. S., Florentino, A. C., Souto, R. N. P., Carvalho, J. C. T., & Fernandes, C. P. (2014). Development of a larvicidal nanoemulsion with Copaiba (Copaifera duckei) oleoresin. *Revista Brasileira de Farmacognosia*, *24*(6), 699–705. doi:10.1016/j.bjp.2014.10.013

Rodriguez-Burneo, N., Busquets, M. A., & Estelrich, J. (2017). Magnetic Nanoemulsions: Comparison between Nano-emulsions Formed by Ultrasonication and by Spontaneous Emulsification. *Nanomaterials (Basel, Switzerland)*, *7*(7), 190. Advance online publication. doi:10.3390/nano7070190 PMID:28737673

Rodríguez, J., Martín, M. J., Ruiz, M. A., & Clares, B. (2016). Current encapsulation strategies for bioactive oils: From alimentary to pharmaceutical perspectives. *Food Research International*, *83*, 41–59. doi:10.1016/j.foodres.2016.01.032

Rodriguez, L., Passerini, N., Cavallari, C., Cini, M., Sancin, P., & Fini, A. (1999). Description and preliminary evaluation of a new ultrasonic atomizer for spray-congealing processes. *International Journal of Pharmaceutics*, *183*(2), 133–143. doi:10.1016/S0378-5173(99)00076-9 PMID:10361164

Roehrborn, C. (2006). Insights into the relationships between prostatic disorders and their potential impact on future urologic practice. *European Urology Supplements*, *5*(12), 698–703. doi:10.1016/j.eursup.2006.06.007

Roger, K. (2016). Nanoemulsification in the vicinity of phase inversion: Disruption of bicontinuous structures in oil / surfactant / water systems. *Current Opinion in Colloid & Interface Science*, *25*, 120–128. doi:10.1016/j.cocis.2016.09.015

Rojas-Grau, M. A., Soliva-Fortuny, R., & Martın-Belloso, O. (2009). Edible coatings to incorporate active ingredients to fresh-cut fruits: A review. *Trends in Food Science & Technology*, *20*(10), 438–447. doi:10.1016/j.tifs.2009.05.002

Romberg, B., Hennink, W. E., & Storm, G. (2008). Sheddable coatings for long-circulating nanoparticles. *Pharmaceutical Research*, *25*(1), 55–71.

Rosa, L. H., Machado, K. M. G., Jacob, C. C., Capelari, M., Rosa, C. A., & Zani, C. L. (2003). Screening of Brazilian Basidiomycetes for antimicrobial activity. *Memorias do Instituto Oswaldo Cruz*, *98*(7), 967–974. doi:10.1590/S0074-02762003000700019 PMID:14762527

Rosas-Durazo, A., Lizardi, J., Higuera-Ciapara, I., Arduelles-Monal, W., & Goycoolea, F. M. (2011). Development and characterization of nanocapsules comprising dodecyltrimethylammonium chloride and κ-carrageenan. *Colloids and Surfaces. B, Biointerfaces*, *86*(1), 242–246. doi:10.1016/j.colsurfb.2011.03.020 PMID:21497069

Roskams, T., De Vos, R., Van Eyken, P., Myazaki, H., Van Damme, B., & Desmet, V. (1998). Hepatic OV-6 expression in human liver disease and rat experiments: Evidence for hepatic progenitor cells in man. *Journal of Hepatology*, *29*(3), 455–463. doi:10.1016/S0168-8278(98)80065-2 PMID:9764994

Rowe, R. C., Sheskey, P., & Quinn, M. (2009). *Handbook of Pharmaceutical Excipients*. Pharmaceutical Press.

Ruberto, G. (2002). Analysis of Volatile Components of Citrus Fruit Essential Oils. In J. F. Jackson & H. F. Linskens (Eds.), Analysis of Taste and Aroma. Molecular Methods of Plant Analysis, 21. Springer. https://doi.org/10.1007/978-3-662-04857-3_7

Ruengdech, A., & Siripatrawan, U. (2021). Application of catechin nanoencapsulation with enhanced antioxidant activity in high pressure processed catechin-fortified coconut milk. *Lebensmittel-Wissenschaft + Technologie*, *140*, 110594. doi:10.1016/j.lwt.2020.110594

Ruiz, A., Varela, A., & Cornejo, L. (2019). Nanoemulsions as coadyuvants in intranasal vaccines. *International Conference Series on Multidisciplinary Sciences*, *4*, 5078-2624.

Ruiz, M. E., & Montoto, S. S. (2018). Routes of drug administration. In *ADME Processes in Pharmaceutical Sciences* (pp. 97–133). Springer.

Rupa, E. J., Li, J. F., Arif, M. H., Yaxi, H., Puja, A. M., Chan, A. J., Hoang, V., Kaliraj, L., Yang, D. C., & Kang, S. C. (2020). *Cordyceps militaris* fungus extracts-mediated nanoemulsion for improvement antioxidant, antimicrobial, and anti-inflammatory activities. *Molecules (Basel, Switzerland)*, *25*(23), 5733. doi:10.3390/molecules25235733 PMID:33291776

Russo, M., Russo, G. L., Daglia, M., Kasi, P. D., Ravi, S., Nabavi, S. F., & Nabavi, S. M. (2016). Understanding genistein in cancer: The "good" and the "bad" effects: A review. *Food Chemistry*, *196*, 589–600. doi:10.1016/j.foodchem.2015.09.085 PMID:26593532

Rutvij, J. P., Gunjan, J. P., Bharadia, P. D., Pandya, V. M., & Modi, D. A. (2011). Nanoemulsion: An advanced concept of dosage form. *Int. J. Pharm. Cosmetol*, *1*(5), 122–133.

Ryu, V. (2017). *Use Of Different Ripening Inhibitors To Enhance Antimicrobial Activity Of Essential Oil Nanoemulsion* (Thesis). University of Massachusetts Amherst.

Ryu, V., Corradini, M. G., McClements, D. J., & McLandsborough, L. (2019). Impact of ripening inhibitors on molecular transport of antimicrobial components from essential oil nanoemulsions. *Journal of Colloid and Interface Science*, *556*, 568–576. https://doi.org/10.1016/j.jcis.2019.08.059

Ryu, V., McClements, D. J., Corradini, M. G., & McLandsborough, L. (2018). Effect of ripening inhibitor type on formation, stability, and antimicrobial activity of thyme oil nanoemulsion. *Food Chemistry*, *245*, 104–111. https://doi.org/10.1016/j.foodchem.2017.10.084

Ryu, V., McClements, D. J., Corradini, M. G., Yang, J. S., & McLandsborough, L.National Center for Biotechnology Information. (2021). *PubChem* Natural antimicrobial delivery systems: Formulation, antimicrobial activity, and mechanism of action of quillaja saponin-stabilized carvacrol nanoemulsions. *Food Hydrocolloids, 82*, 442–450. doi:10.1016/j.foodhyd.2018.04.017

Sabir, D. K., & Sidiq, K. R. (2020). Antimicrobial Activity of Combined Cinnamon Nano Emulsions-Antibiotics against AcinetobacterBaumannii. *Passer Journal, 1*(2), 2. doi:10.24271/psr.03

Sabir, F., Ismail, R., & Csoka, I. (2020). Nose-to-brain delivery of antiglioblastoma drugs embedded into lipid nanocarrier systems: Status quo and Outlook. *Drug Discovery Today, 25*(1), 185–194. doi:10.1016/j.drudis.2019.10.005 PMID:31629966

Sable, S., Plikaytis, B. B., Shinnick, T. M., Amara, R. R., & Cheruvu, M. (2009). *Polypeptide vaccine and vaccination strategy against Mycobacterium.* https://patents.google.com/patent/EP2818178A1/en

Sachdeva, M.S., Patel, K., & Godugu, C. (2017). *Composition for improving boavailibility and efficcacy of taxane.* Academic Press.

Sadeghpour Galooyak, S., & Dabir, B. (2015). Three-factor response surface optimization of nano-emulsion formation using a microfluidizer. *Journal of Food Science and Technology, 52*(5), 2558–2571. doi:10.100713197-014-1363-1 PMID:25892755

Safaya, M., & Rotliwala, Y. (2020). Nanoemulsions: A review on low energy formulation methods, characterization, applications and optimization technique. *Materials Today: Proceedings, 27*, 454–459. doi:10.1016/j.matpr.2019.11.267

Saffarionpour, S. (2019). Preparation of Food Flavor Nanoemulsions by High- and Low-Energy Emulsification Approaches. *Food Engineering Reviews, 11*(4), 259–289. doi:10.100712393-019-09201-3

Sahafi, S. M., Goli, S. A. H., Kadivar, M., Varshosaz, J., & Shirvani, A. (2021). Pomegranate seed oil nanoemulsion enriched by α-tocopherol; the effect of environmental stresses and long-term storage on its physicochemical properties and oxidation stability. *Food Chemistry, 345*, 128759. doi:10.1016/j.foodchem.2020.128759 PMID:33310251

Sahay, G., Alakhova, D. Y., & Kabanov, A. V. (2010). Endocytosis of nanomedicines. *Journal of Controlled Release, 145*(3), 182–195. doi:10.1016/j.jconrel.2010.01.036 PMID:20226220

Sahu, P., Das, D., Mishra, V. K., Kashaw, V., & Kashaw, S. K. (2017). Nanoemulsion: A novel eon in cancer chemotherapy. *Mini-Reviews in Medicinal Chemistry, 17*(18), 1778–1792. doi:10.2174/1389557516666160219122755 PMID:26891931

Saini, R., Saini, S., & Sharma, S. (2010). Nanotechnology: The future medicine. *Journal of Cutaneous and Aesthetic Surgery, 3*(1), 32–33. doi:10.4103/0974-2077.63301 PMID:20606992

Sainsbury, J., Grypa, R., Ellingworth, J., Duodu, K. G., & De Kock, H. L. (2016). The effects of antioxidants and shelf life conditions on oxidation markers in a sunflower oil salad dressing emulsion (SOSDE). *Food Chemistry, 213*, 230–237. doi:10.1016/j.foodchem.2016.06.081 PMID:27451176

Saito, S., Ainai, A., Suzuki, T., Harada, N., Ami, Y., Yuki, Y., Takeyama, H., Kiyono, H., Tsukada, H., & Hasegawa, H. (2016). The effect of mucoadhesive excipient on the nasal retention time of and the antibody responses induced by an intranasal influenza vaccine. *Vaccine, 34*(9), 1201–1207. doi:10.1016/j.vaccine.2016.01.020 PMID:26802605

Sakulku, U., Nuchuchua, O., Uawongyart, N., Puttipipatkhachorn, S., Soottitantawat, A., & Ruktanonchai, U. (2009). Characterization and mosquito repellent activity of citronella oil nanoemulsion. *International Journal of Pharmaceutics, 372*(2), 105–111. doi:10.1016/j.ijpharm.2008.12.029 PMID:19162149

Salehi, F., Jamali, T., Kavoosi, G., Ardestani, S. K., & Vahdati, S. N. (2020). Stabilization of Zataria essential oil with pectin-based nanoemulsion for enhanced cytotoxicity in monolayer and spheroid drug-resistant breast cancer cell cultures and deciphering its binding mode with gDNA. *International Journal of Biological Macromolecules*, *164*, 3645–3655.

Saleh, M. H., Rashedi, I., & Keating, A. (2017). Immunonodulatory properties of *Coriuolus versicolor*: The role of Polysaccharopeptide. *Frontiers in Immunology*, *8*, 1087. doi:10.3389/fimmu.2017.01087 PMID:28932226

Saleh, M. M., Hashem, F. A., & Glombitza, K. W. (1998). Cytotoxicity and in vitro effects on human cancer cell lines of volatiles of Apium graveolens var filicinum. *Pharmaceutical and Pharmacological Letters*, *8*(2), 97–99.

Salem, M., & Ezzat, S. (2019). Nanoemulsions in Food Industry. *Some New Aspects of Colloidal Systems in Foods.* doi:org/10.5772/intechopen.79447

Salminen, H., Herrmann, K., & Weiss, J. (2013). Oil-in-water emulsions as delivery systemfor n-3 fatty acids in meat products. *Meat Science*, *93*(3), 659–667. doi:10.1016/j.meatsci.2012.11.035 PMID:23273478

Salvia-trujillo, L., Rojas-graü, A., Soliva-fortuny, R., & Martín-belloso, O. (2013). *Physicochemical Characterization of Lemongrass Essential Oil – Alginate Nanoemulsions : Effect of Ultrasound Processing Parameters*. doi:10.1007/s11947-012-0881-y

Salvia-Trujillo, L., & McClements, D. (2016). Enhancement of lycopene bioaccessibility from tomato juice using excipient emulsions: Influence of lipid droplet size. *Food Chemistry*, *210*, 295–304. doi:10.1016/j.foodchem.2016.04.125 PMID:27211650

Salvia-Trujillo, L., Rojas-Graü, A., Soliva-Fortuny, R., & Martín-Belloso, O. (2015). Physicochemical characterization and antimicrobial activity of food-grade emulsions and nanoemulsions incorporating essential oils. *Food Hydrocolloids*, *43*, 547–556. https://dx.doi.org/10.1016/j.foodhyd.2014.07.012

Salvia-Trujillo, L., Rojas-Grau, M. A., Soliva-Fortuny, R., & Martın Belloso, O. (2015b). Use of antimicrobial nanoemulsions as edible coatings: Impact on safety and quality attributes of fresh-cut Fuji apples. *Postharvest Biology and Technology*, *105*, 8–16. doi:10.1016/j.postharvbio.2015.03.009

Salvia-Trujillo, L., Soliva-Fortuny, R., Rojas-Grau, M. A., McClements, D. J., & Martin-Belloso, O. (2017). Edible Nanoemulsions as carriers of active ingredients: A Review. *Annual Review of Food Science and Technology*, *8*(1), 439–466. doi:10.1146/annurev-food-030216-025908 PMID:28125342

Samadi, A., Pourmadadi, M., Yazdian, F., Rashedi, H., Navaei-Nigjeh, M., & Eufrasio-da-Silva, T. (2021). Ameliorating quercetin constraints in cancer therapy with pH-responsive agarose-polyvinylpyrrolidone -hydroxyapatite nanocomposite encapsulated in double nanoemulsion. *International Journal of Biological Macromolecules*, *182*, 11–25. Advance online publication.

Samal, D. (2017). Use of nanotechnology in food industry: A review. *International Journal of Environment. Agricultural Biotechnology*, *2*(4), 238902.

Samiun, W. S., Ashari, S. E., Salim, N., & Ahmad, S. (2020). Optimization of Processing Parameters of Nanoemulsion Containing Aripiprazole Using Response Surface Methodology. *International Journal of Nanomedicine*, *15*, 1585–1594. doi:10.2147/IJN.S198914 PMID:32210553

Sanchez-Gonzalez, L., Vargas, M., Gonzalez-Martınez, C., Chiralt, A., & Chafer, M. (2011). Use of essential oils in bioactive edible coatings: A review. *Food Engineering Reviews*, *3*(1), 1–16. doi:10.100712393-010-9031-3

Sánchez-López, E., Guerra, M., Dias-Ferreira, J., Lopez-Machado, A., Ettcheto, M., Cano, A., Espina, M., Camins, A., Garcia, M. L., & Souto, E. B. (2019). Current Applications of Nanoemulsions in Cancer Therapeutics. *Nanomaterials (Basel, Switzerland)*, *9*(6), 821. doi:10.3390/nano9060821 PMID:31159219

Sandri, G., Bonferoni, M. C., Ferrari, F., Rossi, S., Mori, M., & Caramella, C. (2015). Opportunities offered by chitosan-basednanotechnology in mucosal/skindrugdelivery. *Current Topics in Medicinal Chemistry*, *15*(4), 401–412. doi:10.2174/1568026615666150108124122 PMID:25579349

Santajit, S., & Indrawattana, N. (2016). Mechanisms of antimicrobial resistance in ESKAPE pathogens. *BioMed Research International*. Advance online publication. doi:10.1155/2016/2475067

Santander-Ortega, M. J., Peula-Garcia, J. M., Goycoolea, F. M., & Ortega-Vinuesa, J. L. (2011). Chitosan nanocapsules: Effect of chitosan molecular weight and acetylation degree on electrokinetic behavior and colloidal stability. *Colloids Surf. B. Biointerphases*, *82*(2), 571–580. doi:10.1016/j.colsurfb.2010.10.019 PMID:21071187

Santander-Ortega, M. J., Stauner, T., Loretz, B., Ortega-Vinuesa, J. L., Bostos-Gonzalez, D., Wenz, G., Schaefer, U. F., & Lehr, C. M. (2010). Nanoparticles made from novel starch derives for transdermal drug delivery. *Journal of Controlled Release*, *141*(1), 85–92. doi:10.1016/j.jconrel.2009.08.012 PMID:19699771

Santos, D. T., & Meireles, M. A. A. (2010). Carotenoid pigments encapsulation: Fundamentals, techniques and recent trends. *The Open Chemical Engineering Journal*, *4*(1), 42–50. doi:10.2174/1874123101004010042

Santos, J., Trujillo-Cayado, L. A., Calero, N., Alfaro, M. C., & Muñoz, J. (2016). Development of eco-friendly emulsions produced by microfluidization technique. *Journal of Industrial and Engineering Chemistry*, *36*, 90–95. doi:10.1016/j.jiec.2016.01.024

Sapra, B., Thatai, P., Bhandari, S., Sood, J., Jindal, M., & Tiwary, A. (2013). A critical appraisal of microemulsions for drug delivery: Part I. *Therapeutic Delivery*, *4*(12), 1547–1564.

Saraiva, S. M., Gutiérrez-Lovera, C., Martínez-Val, J., Lores, S., Bouzo, B. L., Díez-Villares, S., Alijas, S., Pensado-López, A., Vázquez-Ríos, A. J., Sánchez, L., & de la Fuente, M. (2021). Edelfosine nanoemulsions inhibit tumor growth of triple negative breast cancer in zebrafish xenograft model. *Scientific Reports*, *11*(1), 9873.

Sarangi, I., Ghosh, D., Bhutia, S. K., Mallick, S. K., & Maiti, T. K. (2006). Anti-tumor and immunomodulating effects of *Pleurotus ostreatus* mycelia-derived proteoglycans. *International Immunopharmacology*, *6*(8), 1287–1297. doi:10.1016/j.intimp.2006.04.002 PMID:16782541

Saraogi, G. K., Gupta, P., Gupta, U. D., Jain, N. K., & Agrawal, G. P. (2010). Gelatin nanocarriers as potential vectors for effective management of tuberculosis. *International Journal of Pharmaceutics*, *385*(1-2), 143–149. doi:10.1016/j.ijpharm.2009.10.004 PMID:19819315

Sarathchandraprakash, N. K., Mahendra, C., Prashanth, S. J., Manral, K., Babu, U. V., & Gowda, D. (2013). Emulsions and emulsifiers. *Asian Journal of Experimental Chemistry*, *8*, 30–45.

Saravana, P. S., Shanmugapriya, K., Gereniu, C. R. N., Chae, S.-J., Kang, H. W., Woo, H.-C., & Chun, B.-S. (2019). Ultrasound-mediated fucoxanthin rich oil nanoemulsions stabilized by κ -carrageenan: Process optimization, bio-accessibility and cytotoxicity. *Ultrasonics Sonochemistry*, *55*, 105–116. doi:10.1016/j.ultsonch.2019.03.014 PMID:31084784

Sareen, S., Mathew, G., & Joseph, L. (2012). Improvement in solubility of poor water-soluble drugs by solid dispersion. *International Journal of Pharmaceutical Investigation*, *2*(1), 12–17.

Sarheed, O., Shouqair, D., Ramesh, K., Khaleel, T., Amin, M., Boateng, J., & Drechsler, M. (2020). Formation of stable nanoemulsions by ultrasound-assisted two-step emulsification process for topical drug delivery: Effect of oil phase composition and surfactant concentration and loratadine as ripening inhibitor. *International Journal of Pharmaceutics*, *576*, 118952. doi:10.1016/j.ijpharm.2019.118952 PMID:31843549

Sari, T. P., Mann, B., Kumar, R., Singh, R. R. B., Sharma, R., Bhardwaj, M., & Athira, S. (2015). Preparation and characterization of nanoemulsion encapsulating curcumin. *Food Hydrocolloids*, *43*, 540–546. doi:10.1016/j.foodhyd.2014.07.011

Sarkar, F. H., Banerjee, S., & Li, Y. (2007). Pancreatic cancer: Pathogenesis, prevention and treatment. *Toxicology and Applied Pharmacology*, *224*(3), 326–336.

Sarker, D. K. (2005). Engineering of Nanoemulsions for Drug Delivery. *Current Drug Delivery*, *2*(4), 297–310. doi:10.2174/156720105774370267 PMID:16305433

Sartor, R. B. (2008). Microbial influences in inflammatory bowel diseases. *Gastroenterology*, *134*(2), 577–594.

Sartor, R. B. (2010). Genetics and environmental interactions shape the intestinal microbiome to promote inflammatory bowel disease versus mucosal homeostasis. *Gastroenterology*, *139*(6), 1816–1819.

Sasaki, R. S., Mattoso, L. H. C., & de Moura, M. R. (2016). New edible bionanocomposite prepared by pectin and clove essential oil nanoemulsions. *Journal of Nanoscience and Nanotechnology*, *16*(6), 6540–6544. doi:10.1166/jnn.2016.11702 PMID:27427749

Sasikumar, A., & Kamalasanan, K. (2017). Nanomedicine for prostate cancer using nanoemulsion: A review. *Journal of Controlled Release*, *260*, 111–123. doi:10.1016/j.jconrel.2017.06.001 PMID:28583444

Sastry, S. V., Nyshadham, J. R., & Fix, J. A. (2000). Recent technological advances in oral drug delivery–a review. *Pharmaceutical Science & Technology Today*, *3*(4), 138–145. doi:10.1016/S1461-5347(00)00247-9 PMID:10754543

Satyal, P., Craft, J. D., Dosoky, N. S., & Setzer, W. N. (2017). The chemical compositions of the volatile oils of garlic (Allium sativum) and wild garlic (Allium vineale). *Foods*, *6*(8), 1–10. doi:10.3390/foods6080063 PMID:28783070

Saxena, A., Maity, T., Paliwal, A., & Wadhwa, S. (2017). Technological aspects of nanoemulsions and their applications in the food sector. In *Nanotechnology Applications in Food* (pp. 129–152). Academic Press. doi:10.1016/B978-0-12-811942-6.00007-8

Saxena, T., Ali, A. O., & Saxena, M. (2018). Pathophysiology of essential hypertension: An update. *Expert Review of Cardiovascular Therapy*, *16*(12), 879–887. doi:10.1080/14779072.2018.1540301 PMID:30354851

Schatten, H. (2018). Brief overview of prostate cancer statistics, grading, diagnosis and treatment strategies. *Advances in Experimental Medicine and Biology*, *1095*, 1–14. doi:10.1007/978-3-319-95693-0_1 PMID:30229546

Schatz, A., Bugie, E., Waksman, S. A., Hanssen, A. D., Patel, R., & Osmon, D. R. (2005). Streptomycin, a substance exhibiting antibiotic activity against gram-positive and gram-negative bacteria. *Clinical Orthopaedics and Related Research*, (437), 3–6. doi:10.1097/01.blo.0000175887.98112.fe PMID:16056018

Schatz, Ch., Lucas, J.-M., Viton, Ch., Domard, A., Pichot, Ch., & Delair, Th. (2004). Formation and properties of positively charged colloids based on polyelectrolyte complexes of biopolymers. *Langmuir*, *20*(18), 7766–7778. doi:10.1021/la049460m PMID:15323530

Scherze, I., Marzilger, K., & Muschiolik, G. (1999). Emulsification using micro porous glass (MPG): Surface behaviour of milk protein. *Colloids and Surfaces. B, Biointerfaces*, *12*(3-6), 213–221. doi:10.1016/S0927-7765(98)00076-9

Schianti, J. N., Cerize, N. N. P., Oliveira, A. M., Derenzo, S., & Gingora-Rubio, M. R. (2011). Rifampicin nanosuspension production using microfluidic device. *NSTI-Nanotech*, *1*, 229–232.

Schreiner, T. B., Santamaria-Echart, A., Ribeiro, A., Peres, A. M., Dias, M. M., Pinho, S. P., & Barreiro, M. F. (2020). Formulation and Optimization of Nanoemulsions Using the Natural Surfactant Saponin from Quillaja Bark. *Molecules (Basel, Switzerland)*, *25*(7), 1538. doi:10.3390/molecules25071538

Schroder, A., Sprakel, J., Schroen, K., Spaen, N. J., & Berton-Carabin, C. (2018). Coalescence stability of Pickering emulsions produced with lipid particles: A microfluidic study. *Journal of Food Engineering*, *234*, 63–72. doi:10.1016/j.jfoodeng.2018.04.007

Schubert, H., Ax, K., & Behrend, O. (2003). Product engineering of dispersed systems. *Trends in Food Science & Technology*, *14*(1-2), 9–16. doi:10.1016/S0924-2244(02)00245-5

Schultze, V., D'Agosto, V., Wack, A., Novicki, D., Zorn, J., & Hennig, R. (2008). Safety of MF59 adjuvant. *Vaccine*, *26*(26), 3209–3222. doi:10.1016/j.vaccine.2008.03.093 PMID:18462843

Schultz, S., Wagner, G., Urban, K., & Ulrich, J. (2004). High-pressure homogenization as a process for emulsion formation. *Chemical Engineering & Technology*, *27*(4), 361–368. doi:10.1002/ceat.200406111

Schulz, P. C., Rodriguez, M. S., Del Blanco, L. F., Pistonesi, M., & Agullo, E. (1998). Emulsification properties of chitosan. *Colloid & Polymer Science*, *267*(12), 1159–1165. doi:10.1007003960050359

Schwarz, T., Horacek, T., Knuf, M., Damman, H., Roman, F., Dramé, M., Gillard, P., & Jilg, W. (2009). Single dose vaccination with AS03-adjuvanted H5N1 vaccines in a randomized trial induces strong and broad immune responsiveness to booster vaccination in adults. *Vaccine*, *27*(45), 6284–6290. doi:10.1016/j.vaccine.2009.01.040 PMID:19856521

Scrobota, I., Bolfa, P., Filip, A. G., Catoi, C., Alb, C., Pop, O., ... Baciut, G. (2016). Natural chemopreventive alternatives in oral cancer chemoprevention. *Journal of Physiology and Pharmacology*, *67*(1), 161–172. PMID:27010905

Sedaghat Doost, A., Afghari, N., Abbasi, H., Nikbakht Nasrabadi, M., Dewettinck, K., & Van der Meeren, P. (2020). Nano-lipid carriers stabilized by hydrophobically modified starch or sucrose stearate for the delivery of lutein as a nutraceutical beverage model. *Colloids and Surfaces. A, Physicochemical and Engineering Aspects*, *605*, 125349. doi:10.1016/j.colsurfa.2020.125349

Segvić Klarić, M., Kosalec, I., Mastelić, J., Piecková, E., & Pepeljnak, S. (2007). Antifungal activity of thyme (Thymus vulgaris L.) essential oil and thymol against moulds from damp dwellings. *Letters in Applied Microbiology*, *44*(1), 36–42. https://doi.org/10.1111/j.1472-765X.2006.02032.x

Seibert, J. B., Bautista-Silva, J. P., Amparo, T. R., Petit, A., Pervier, P., Almeida, J. C. D. S., Azevedo, M. C., Silveira, B. M., Brandao, G. C., de Souza, G. H. B., & ... . (2019). Development of propolis nanoemulsion with antioxidant and antimicrobial activity for use as a potential natural preservative. *Food Chemistry*, *287*, 61–67. doi:10.1016/j.foodchem.2019.02.078 PMID:30857719

Sens, P., & Joanny, J.-F. (2000). Counterion release and electrostatic adsorption. *Physical Review Letters*, *84*(21), 4862–4265. doi:10.1103/PhysRevLett.84.4862 PMID:10990817

Şenyiğit, T., Sonvico, F., Barbieri, S., Özer, Ö., Santi, P., & Colombo, P. (2010). Lecithin/chitosan nanoparticles of clobetasol-17-propionate capable of accumulation in pig skin. *Journal of Controlled Release*, *142*(3), 368–373. doi:10.1016/j.jconrel.2009.11.013 PMID:19932722

Septiyanti, M. (2019). Evaluation of Nanoemulsion Concentrate Botanical Fungicide from Neem, Citronella and Eugenol Oil Using Palm Oil Based Surfactant. *American Journal of Physics and Applications*, *7*(1), 14. doi:10.11648/j.ajpa.20190701.13

Serdaroglu, M., Ozturk, B., & Kara, A. (2015). An overview of food emulsions: Description, classification and recent potential applications. *Turkish Journal of Agriculture -. Food Science and Technology (Campinas)*, *3*, 430–438.

Sessa, M., Balestrieri, M. L., Ferrari, G., Servillo, L., Castaldo, D., D'Onofrio, N., & Tsao, R. (2014). Bioavailability of encapsulated resveratrol into nanoemulsion-based delivery systems. *Food Chemistry*, *147*, 42–50.

Sessa, M., Tsao, R., Liu, R., Ferrari, G., & Donsì, F. (2011). Evaluation of the stability and antioxidant activity of nanoencapsulated resveratrol during in vitro digestion. *Journal of Agricultural and Food Chemistry*, *59*(23), 12352–12360.

Severino, R., Ferrari, G., Vu, K. D., Donsi, F., Salmieri, S., & Lacroix, M. (2015). Antimicrobial effects of modified chitosan based coating containing nanoemulsion of essential oils, modified atmosphere packaging and gamma irradiation against *Escherichia coli* O157:H7 and *Salmonella Typhimurium* on green beans. *Food Control*, *50*, 215–222. doi:10.1016/j.foodcont.2014.08.029

Severino, R., Vu, K. D., Donsì, F., Salmieri, S., Ferrari, G., & Lacroix, M. (2014). Antibacterial and physical effects of modified chitosan based-coating containing nanoemulsion of mandarin essential oil and three non-thermal treatments against *Listeria innocua* in green beans. *International Journal of Food Microbiology*, *191*, 82–88. doi:10.1016/j.ijfoodmicro.2014.09.007 PMID:25255308

Sfanos, S., & De Marzo, A. M. (2012). Prostate cancer and inflammation: The evidence. *Histopathology*, *60*(1), 199–215. doi:10.1111/j.1365-2559.2011.04033.x PMID:22212087

Sha, L., & Rongmin, Y. (2010). *Method for preparing oral chondroitin sulfate nanoemulsion*. Academic Press.

Sh, A., Abdelrazeik, A. B., & Rakha, O. M. (2015). Nanoemulsion of Jojoba Oil, Preparation, Characterization and Insecticidal Activity against Sitophilus oryzae (Coleoptera: Curculionidae) on Wheat. *International Journal of Agriculture Innovation and Research*, *4*(1), 72–75.

Shafiee, G., Saidijam, M., Tavilani, H., Ghasemkhani, N., & Khodadadi, I. (2016). Genistein induces apoptosis and inhibits proliferation of HT29 colon cancer cells. *International Journal of Molecular and Cellular Medicine*, *5*(3), 178. PMID:27942504

Shafiq, S., Shakeel, F., Talegaonkar, S., Ahmad, F. J., Khar, R. K., & Ali, M. (2007). Development and bioavailability assessment of ramipril nanoemulsion formulation. *European Journal of Pharmaceutics and Biopharmaceutics*, *66*(2), 227–243. doi:10.1016/j.ejpb.2006.10.014 PMID:17127045

Shafiq-un-Nabi, S., Shakeel, F., Talegaonkar, S., Ali, J., Baboota, S., Ahuja, A., Khar, R. K., & Ali, M. (2007). Formulation development and optimization using nanoemulsion technique: A technical note. *AAPS PharmSciTech*, *8*(2), E12–E28. https://doi.org/10.1208/pt0802028

Shahabi, N., Tajik, H., Moradi, M., Forough, M., & Ezati, P. (2017). Physical, antimicrobial and antibiofilm properties of Zataria multiflora Boiss essential oil nanoemulsion. *International Journal of Food Science & Technology*, *52*, 1645–1652.

Shahavi, M. H., Hosseini, M., Jahanshahi, M., Meyer, R. L., & Darzi, G. N. (2015). Clove oil nanoemulsion as an effective antibacterial agent: Taguchi optimization method. *Desalination and Water Treatment*, *57*, 18379–18390. doi:10.1080/19443994.2015.1092893

Shahavi, M. H., Hosseini, M., Jahanshahi, M., Meyer, R. L., & Darzi, G. N. (2019). Evaluation of critical parameters for preparation of stable clove oil nanoemulsion. *Arabian Journal of Chemistry*, *12*(8), 3225–3230. doi:10.1016/j.arabjc.2015.08.024

Shahbazi, Y. (2019). Antioxidant, antibacterial, and antifungal properties of nanoemulsion of clove essential oil. *Nanomedicine Research Journal*, *4*(4), 204–208. doi:10.22034/nmrj.2019.04.001

Shahidi, F., & Zhong, Y. (2010). Lipid oxidation and improving the oxidative stability. *Chemical Society Reviews*, *39*(11), 4067–4079. doi:10.1039/b922183m PMID:20617249

Shahi, S. K., Shukla, A. C., Bajaj, A. K., Banerjee, U., Rimek, D., Midgely, G., & Dikshit, A. (2000). Broad spectrum herbal therapy against superficial fungal infections. *Skin Pharmacology and Applied Skin Physiology*, *13*(1), 60–64. https://doi.org/10.1159/000029909

Shah, P., Bhalodia, D., & Shelat, P. (2010). Nanoemulsion: A pharmaceutical review. *Systematic Review in Pharmacy*, *1*, 24–32.

Shah, P., Bhalodia, D., & Shelat, P. (2010). Nanoemulsion: A pharmaceutical review. *Systematic Reviews in Pharmacy*, *1*(1), 24–32. doi:10.4103/0975-8453.59509

Shakeel, F., Baboota, S., Ahuja, A., Ali, J., Aqil, M., & Shafiq, S. (2007). Nanoemulsions as vehicles for transdermal delivery of aceclofenac. *AAPS PharmSciTech*, *8*(4), 191–199. doi:10.1208/pt0804104 PMID:18181525

Shakeel, F., & Ramadan, W. (2010). Transdermal delivery of anticancer drug caffeine from water-in-oil nanoemulsions. *Colloids and Surfaces. B, Biointerfaces*, *75*(1), 356–362.

Shakeel, F., Ramadan, W., Faisal, M., Rizwan, M., Faiyazuddin, M., Mustafa, G., & Shafiq, S. (2010). Transdermal and Topical Delivery of Anti-inflammatory Agents Using Nanoemulsion/Microemulsion: An Updated Review. *Current Nanoscience*, *6*(2), 184–198.

Shakeell, F., Baboota, S., Ahuja, A., Ali, J., Faisal, M. S., & Shafiq, S. (2008). Stability evaluation of celecoxib nanoemulsion containing Tween 80. *Thaiphesatchasan*, *32*, 4–9.

Shaker, D. S., Ishak, R. A., Ghoneim, A., & Elhuoni, M. A. (2019). Nanoemulsion: A review on mechanisms for the transdermal delivery of hydrophobic and hydrophilic drugs. *Scientia Pharmaceutica*, *87*(3), 17. doi:10.3390cipharm87030017

Shamseddine, A., Saleh, A., Charafeddine, M., Seoud, M., Mukherji, D., Temraz, S., & Sibai, A. M. (2014). Cancer trends in Lebanon: A review of incidence rates for the period of 2003-2008 and projections until 2018. *Population Health Metrics*, *12*(1), 2–9. doi:10.1186/1478-7954-12-4 PMID:24593777

Shao, H., Xi, N., & Zhang, Y. (2018). Microemulsion formulation of a new biopesticide to control the diamond backmoth (Lepidoptera: Plutellidae). *Scientific Reports*, *8*(1), 1–9. doi:10.103841598-018-28626-0

Sharafi, S. M., Rasooli, I., Owlia, P., Taghizadeh, M., & Astaneh, S. D. A. (2010). Protective effects of bioactive phytochemicals from Mentha piperita with multiple health potentials. *Pharmacognosy Magazine*, *6*(23), 147. doi:10.4103/0973-1296.66926 PMID:20931070

Sharif, H. R., Abbas, S., Majeed, H., Safdar, W., Shamoon, M., Khan, M. A., Shoaib, M., Raza, H., & Haider, J. (2017). Formulation, characterization and antimicrobial properties of black cumin essential oil nanoemulsions stabilized by OSA starch. *Journal of Food Science and Technology*, *54*(10), 3358–3365. https://doi.org/10.1007/s13197-017-2800-8

Sharma, A., Sharma, N. K., Srivastava, A., Kataria, A., Dubey, S., Sharma, S., & Kundu, B. (2018). Clove and lemongrass oil based non-ionic nanoemulsion for suppressing the growth of plant pathogenic *Fusarium oxysporum f.sp. lycopersici*. *Industrial Crops and Products*, *123*, 353–362. doi:10.1016/j.indcrop.2018.06.077

Sharma, M., Sharma, R., & Jain, D. K. (2016). Nanotechnology based approaches for enhancing oral bioavailability of poorly water soluble antihypertensive drugs. *Scientifica, 2016*, 2016. doi:10.1155/2016/8525679 PMID:27239378

Sharma, N., Bansal, M., Visht, S., Sharma, P., & Kulkarni, G. (2010). Nanoemulsion: A new concept of delivery system. *Chronicles of Young Scientists*, *1*(2), 2.

Sharma, N., Kaur, G., & Khatkar, S. K. (2021). Optimization of emulsification conditions for designing ultrasound assisted curcumin loaded nanoemulsion: Characterization, antioxidant assay and release kinetics. *Lebensmittel-Wissenschaft + Technologie*, *141*, 110962. doi:10.1016/j.lwt.2021.110962

Sharma, N., Mishra, S., Sharma, S., Desphande, R. D., & Sharma, R. K. (2013). Preparation and Optimization of Nanoemulsions for targeting Drug Delivery. *Journal of Drug Development and Research*, *5*(4), 37–48.

Sharma, N., Thomas, S., Golden, E. B., Hofman, F. M., Chen, T. C., Petasis, N. A., Schönthal, A. H., & Louie, S. G. (2012). Inhibition of autophagy and induction of breast cancer cell death by mefloquine, an antimalarial agent. *Cancer Letters*, *326*(2), 143–154. doi:10.1016/j.canlet.2012.07.029 PMID:22863539

Sharma, N., & Tripathi, A. (2008). Effects of Citrus sinensis (L.) Osbeck epicarp essential oil on growth and morphogenesis of Aspergillus niger (L.) Van Tieghem. *Microbiological Research*, *163*(3), 337–344. https://doi.org/10.1016/j.micres.2006.06.009

Sharma, R. (2001). India wakes up to threat of bioterrorism. *BMJ (Clinical Research Ed.)*, *323*(7315), 714. doi:10.1136/bmj.323.7315.714a PMID:11576976

Sharma, S., & SaranSgdevot, K. (2012). Nanoemulsions for cosmetics. *International Journal of Advanced Research in Pharmaceuticals Bio Sciences*, *1*(4), 408.

Sharma, S., Rabbani, S. A., Narang, J. K., Hyder Pottoo, F., Ali, J., Kumar, S., & Baboota, S. (2020). Role of rutin nanoemulsion in ameliorating oxidative stress: Pharmacokinetic and pharmacodynamics studies. *Chemistry and Physics of Lipids*, *228*, 104890. doi:10.1016/j.chemphyslip.2020.104890 PMID:32032570

Sharma, S., Sahni, J. K., Ali, J., & Baboota, S. (2014). Effect of high-pressure homogenization on formulation of TPGS loaded nanoemulsion of rutin – pharmacodynamic and antioxidant studies. *Drug Delivery*, *22*(4), 541–551. doi:10.3109/10717544.2014.893382 PMID:24625264

Shchukina, E. M., & Shchukin, D. (2012). Layer-by-layer coated emulsion microparticles as storage and delivery tool. *Current Opinion in Colloid & Interface Science*, *17*(5), 281–289. doi:10.1016/j.cocis.2012.06.003

Sheth, T., Seshadri, S., Prileszky, T., & Helgeson, M. E. (2020). Multiple nanoemulsions. *Nature Reviews. Materials*, *5*(3), 214–228. doi:10.103841578-019-0161-9

Shin, J. W., & Chung, Y. H. (2013). Molecular targeted therapy for hepatocellular carcinoma: Current and future. *World Journal of Gastroenterology*, *19*(37), 6144–6155. doi:10.3748/wjg.v19.i37.6144 PMID:24115810

Shinoda, K., & Saito, H. (1969). The stability of O / W Type emulsions as functions of temperature and the HLB of emulsifiers: The emulsification by PIT method. *Journal of Colloid and Interface Science*, *30*(2), 258–263. doi:10.1016/S0021-9797(69)80012-3

Shipulo, E. V., Lyubimov, I. I., Maksimenko, O. O., Vanchugova, L. V., Oganesyan, E. A., Sveshnikov, P. G., Biketov, S. F., Severin, E. S., Heifets, L. B., & Gel'perina, S. E. (2008). Development of a nanosomal formulation of moxifloxacin based on poly(butyl-2-cyanoacrylate). *Pharmaceutical Chemistry Journal*, *42*(3), 43–47. doi:10.100711094-008-0073-2

Shivangi, & Meena, L.S. (2018). A novel approach in treatment of tuberculosis by targeting drugs to infected macrophages using biodegradable nanoparticles. *Applied Biochemistry and Biotechnology, 185,* 815–821. Doi:org/10.1007/s12010-018-2695-5

Shoaib, A., Waqas, M., Elabasy, A., Cheng, X., Zhang, Q., & Shi, Z. (2018). Preparation and characterization of emamectin benzoate nanoformulations based on colloidal delivery systems and use in controlling: Plutella xylostella (L.) (Lepidoptera: Plutellidae). *RSC Advances*, *8*(28), 15687–15697. doi:10.1039/C8RA01913D

Shobo, A., Pamreddy, A., Kruger, H. G., Makatini, M. M., Naicker, T., Govender, T., & Baijnath, S. (2018). Enhanced brain penetration of pretomanid by intranasal administration of an oil-in-water nanoemulsion. *Nanomedicine (London)*, *13*(9), 997–1008. doi:10.2217/nnm-2017-0365 PMID:29790418

Shojaee-Aliabadi, S., Hosseini, H., Mohammadifar, M. A., Mohammadi, A., Ghasemlou, M., Hosseini, S. M., & Khaksar, R. (2014). Characterization of κ-carrageenan films incorporated plant essential oils with improved antimicrobial activity. *Carbohydrate Polymers*, *101*, 582–591. https://doi.org/10.1016/j.carbpol.2013.09.070

Siddappa, G., Kulsum, S., Ravindra, D. R., Kumar, V. V., Raju, N., Raghavan, N., Sudheendra, H. V., Sharma, A., Sunny, S. P., Jacob, T., Kuruvilla, B. T., Benny, M., Antony, B., Seshadri, M., Lakshminarayan, P., Hicks, W. Jr, Suresh, A., & Kuriakose, M. A. (2017). Curcumin and metformin-mediated chemoprevention of oral cancer is associated with inhibition of cancer stem cells. *Molecular Carcinogenesis*, *56*(11), 2446–2460. doi:10.1002/mc.22692 PMID:28618017

Siegel, R. L., Jemal, A., Wender, R. C., Gansler, T., Ma, J., & Brawley, O. W. (2018). An assessment of progress in cancer control. *CA: a Cancer Journal for Clinicians*, *68*(5), 329–339. doi:10.3322/caac.21460 PMID:30191964

Signoretto, C., Marchi, A., Bertoncelli, A., Burlacchini, G., Tessarolo, F., Caola, I., Pezzati, E., Zaura, E., Papetti, A., Lingstrom, P., Pratten, J., Spratt, D. A., Wilson, M., & Canepari, P. (2011). Effects of mushroom and chicory extracts on the physiology and shape of *Prevotella intermedia*, a Periodontopathogenic bacterium. *Journal of Biomedicine & Biotechnology*, *2011*, 1–8. Advance online publication. doi:10.1155/2011/635348 PMID:21918594

Silva, A. C. S., & Flynn, J. T. (2012). The renin–angiotensin–aldosterone system in 2011: Role in hypertension and chronic kidney disease. *Pediatric Nephrology (Berlin, Germany)*, *27*(10), 1835–1845.

Silva, A. L., Alexandrino, F., Verissimo, L. M., Agnez-Lima, L. F., Egito, L. C. M., De Oliveira, A. G., & Do Egito, E. S. T. (2012). Physical factors affecting plasmid DNA compaction in stearylamine-containing nanoemulsions intended for gene delivery. *Pharmaceuticals*, *5*(6), 643–654. doi:10.3390/ph5060643 PMID:24281666

Silva, A. L., Marcelino, H. R., Veríssimo, L. M., Araujo, I. B., Agnez-Lima, L. F., & do Egito, E. S. (2016). Stearylamine-containing cationic nanoemulsion as a promising carrier for gene delivery. *Journal of Nanoscience and Nanotechnology*, *16*(2), 1339–1345. doi:10.1166/jnn.2016.11671 PMID:27433584

Silva, E. (2019). Production and efficacy of neem nanoemulsion in the control of Aspergillus flavus and Penicillium citrinum in soybean seeds. *European Journal of Plant Pathology*, *155*(4), 1105–1116. doi:10.100710658-019-01838-4

Silva, G. N., Faroni, L. R. A., Sousa, A. H., & Freitas, R. S. (2012). Bioactivity of Jatropha curcas L. to insect pests of stored products. *Journal of Stored Products Research*, *48*, 111–113. doi:10.1016/j.jspr.2011.10.009

Silva, H. D., Beldíková, E., Poejo, J., Abrunhosa, L., Serra, A. T., Duarte, C. M. M., Brányik, T., Cerqueira, M. A., Pinheiro, A. C., & Vicente, A. A. (2019). Evaluating the effect of chitosan layer on bioaccessibility and cellular uptake of curcumin nanoemulsions. *Journal of Food Engineering*, *243*, 89–100. doi:10.1016/j.jfoodeng.2018.09.007

Silva, H. D., Cerqueira, M. A., & Vicente, A. A. (2015). Influence of surfactant and processing conditions in the stability of oil-in-water nanoemulsions. *Journal of Food Engineering*, *167*, 89–98. doi:10.1016/j.jfoodeng.2015.07.037

Silva, H. D., Poejo, J., Pinheiro, A. C., Donsì, F., Serra, A. T., Duarte, C. M. M., Ferrari, G., Cerqueira, M. A., & Vicente, A. A. (2018). Evaluating the behaviour of curcumin nanoemulsions and multilayer nanoemulsions during dynamic in vitro digestion. *Journal of Functional Foods*, *48*, 605–613. doi:10.1016/j.jff.2018.08.002

Silva, M., Anh Bui, T. H., Dharmadana, D., Zisu, B., & Chandrapala, J. (2020). Ultrasound-assisted formation of double emulsions stabilized by casein-whey protein mixtures. *Food Hydrocolloids*, *109*(June), 106143. doi:10.1016/j.foodhyd.2020.106143

Silva, M., Ricelli, N. L., El Seoud, O., Valentim, C. S., Ferreira, A. G., Sato, D. N., Leite, C. Q., & Ferreira, E. I. (2006). Potential tuberculostatic agent: Micelle-forming pyrazinamide prodrug. *Archiv der Pharmazie*, *339*(6), 283–290. doi:10.1002/ardp.200500039 PMID:16688684

Silva, S. S., Goodfellow, B. J., Benesch, J., Rocha, J., Mano, J. F., & Reis, R. L. (2007). Morphology and miscibility of chitosan/soy protein blended membranes. *Carbohydrate Polymers*, *70*(1), 25–31. doi:10.1016/j.carbpol.2007.02.023

Silva, W., Torres-Gatica, M. F., Oyarzun-Ampuero, F., Silva-Weiss, A., Robert, P., Cofrades, S., & Gimenez, B. (2018). Double emulsions as potential fat replacers with gallic acid and quercetin nanoemulsions in the aqueous phases. *Food Chemistry*, *253*, 71–78. doi:10.1016/j.foodchem.2018.01.128 PMID:29502846

Simões, M., Simões, L. C., Cleto, S., Pereira, M. O., & Vieira, M. J. (2008). The effects of a biocide and a surfactant on the detachment of Pseudomonas fluorescens from glass surfaces. *International Journal of Food Microbiology*, *121*(3), 335–341. https://doi.org/10.1016/j.ijfoodmicro.2007.11.041

Simões, S., Filipe, A., Faneca, H., Mano, M., Penacho, N., Düzgünes, N., & Pedroso de Lima, M. (2005). Cationic liposomes for gene delivery. *Expert Opinion on Drug Delivery*, *2*(2), 237–254. doi:10.1517/17425247.2.2.237 PMID:16296751

Simonsen, L., Kane, A., Lloyd, J., Zaffran, M., & Kane, M. (1999). Unsafe injections in the developing world and transmission of bloodborne pathogens: A review. *Bulletin of the World Health Organization*, *77*(10), 789. PMID:10593026

Singh, B., Singh, R., Bandyopadhyay, S., Kapil, R., & Garg, B. (2013). Optimized nanoemulsifying systems with enhanced bioavailability of carvedilol. *Colloids and Surfaces. B, Biointerfaces*, *101*, 465–474. doi:10.1016/j.colsurfb.2012.07.017 PMID:23010056

Singh, G., Raghuvanshi, H. K., Anand, A., Pundir, R., & Dwivedi, H. (2010). Targeted delivery of rifampicin by niosomal drug delivery system. *Journal of Pharmacy Research*, *3*(5), 1152–1154.

Singh, H. (2016). Nanotechnology applications in functional foods; opportunities and challenges. *Preventive Nutrition and Food Science*, *21*(1), 1–8. doi:10.3746/pnf.2016.21.1.1 PMID:27069899

Singh, J. A., Upshur, R., & Padayatchi, N. (2007). XDR-TB in South Africa: No time for denial or complacency. *PLoS Medicine*, *4*(1), e50. doi:10.1371/journal.pmed.0040050 PMID:17253901

Singh, K. K., & Vingkar, S. K. (2008). Formulation, antimalarial activity and biodistribution of oral lipid nanoemulsion of primaquine. *International Journal of Pharmaceutics*, *347*(1&2), 136–143. doi:10.1016/j.ijpharm.2007.06.035 PMID:17709216

Singh, K. K., & Vingkar, S. K. (2008). Formulation, antimalarial activity and biodistribution of oral lipid nanoemulsion of primaquine. *International Journal of Pharmacology*, *347*, 136–143.

Singh, N., Verma, S. M., Singh, S. K., & Verma, P. R. P. (2015). Evidence for bactericidal activities of lipidic nanoemulsions against Pseudomonas aeruginosa. *Antonie van Leeuwenhoek*, *107*(6), 1555–1568. doi:10.100710482-015-0449-8

Singh, S. (2004). Tuberculosis. *Current Anaesthesia and Critical Care*, *15*(3), 165–171. doi:10.1016/j.cacc.2004.05.007

Singh, S. S., Wang, H., Chan, Y. S., Pan, W., Dan, X., Yin, C., Akkouh, O., & Ng, T. (2014). Lectins from edible mushrooms. *Molecules (Basel, Switzerland)*, *20*(1), 446–469. doi:10.3390/molecules20010446 PMID:25558856

Singh, S., Hussain, A., Shakeel, F., Ahsan, M. J., Alshehri, S., Webster, T. J., & Lal, U. R. (2019). Recent insights on nanomedicine for augmented infection control. *International Journal of Nanomedicine*, *14*, 2301–2325. doi:10.2147/IJN.S170280

Singh, Y., Meher, J. G., Raval, K., Khan, F. A., Chaurasia, M., Jain, N. K., & Chourasia, M. K. (2017). Nanoemulsion: Concepts, development and applications in drug delivery. *Journal of Controlled Release*, *252*, 28–49. doi:10.1016/j.jconrel.2017.03.008 PMID:28279798

Singh, Y., Meher, J. G., Raval, K., Khan, F. A., Chaurasia, M., Jain, N. K., & Chourasia, M. K. (2017). Nanoemulsion: Concepts, development, and applications in drug delivery. *Journal of Controlled Release*, *252*, 28–49.

Sinico, C., De Logu, A., Lai, F., Valenti, D., Manconi, M., Loy, G., Bonsignore, L., & Fadda, A. M. (2005). Liposomal incorporation of Artemisia arborescens L. essential oil and in vitro antiviral activity. *European Journal of Pharmaceutics and Biopharmaceutics*, *59*(1), 161–168. doi:10.1016/j.ejpb.2004.06.005 PMID:15567314

Sivakumar, M., Tang, S. Y., & Tan, K. W. (2014). Cavitation technology – A greener processing technique for the generation of pharmaceutical nanoemulsions. *Ultrasonics Sonochemistry*, *21*(6), 2069–2083. doi:10.1016/j.ultsonch.2014.03.025 PMID:24755340

Sivakumar, M., Tang, S. Y., & Tan, K. W. (2014). Cavitation technology - A greener processing technique for the generation of pharmaceutical nanoemulsions. *Ultrasonics Sonochemistry*, *21*, 2069–2083.

Sivaramakrishnan, R., Nakamura, C., Mehnert, W., Korting, H. C., Kramer, K. D., & Schäfer-Korting, M. (2004). Glucocorticoid entrapment into lipid carriers—Characterisation by parelectric spectroscopy and influence on dermal uptake. *Journal of Controlled Release*, *97*(3), 493–502. doi:10.1016/S0168-3659(04)00169-5 PMID:15212881

Sivastava, V. K., Kini, G., & Rout, D. (2006). Detergency in spontaneously formed emulsion. *Journal of Colloid and Interface Science*, *304*(1), 214–221. doi:10.1016/j.jcis.2006.08.035 PMID:16987522

Skeiky, Y. A., & Sadoff, J. C. (2006). Advances in tuberculosis vaccine strategies. *Nature Reviews. Microbiology*, *4*(6), 469–476. doi:10.1038/nrmicro1419 PMID:16710326

Słotwiński, R., Lech, G., & Słotwińska, S. M. (2018). MicroRNAs in pancreatic cancer diagnosis and therapy. *Central European Journal of Immunology*, *43*(3), 314–324. doi:10.5114/ceji.2018.80051 PMID:30588176

SMigotto, A., Carvalho, V., Salata, G. C., da Silva, F., Yan, C., Ishida, K., Costa-Lotufo, L. V., Steiner, A. A., & Lopes, L. B. (2018). Multifunctional nanoemulsions for intraductal delivery as a new platform for local treatment of breast cancer. *Drug Delivery*, *25*(1), 654–667.

Söderlind, E., Wollbratt, M., & Corswant, C. (2003). The usefulness of sugar surfactants as solubilizing agents in parenteral formulations. *International Journal of Pharmaceutics*, *252*(1-2), 61–71. doi:10.1016/S0378-5173(02)00599-9 PMID:12550781

Sohretoglu, D., & Huang, S. (2018). *Ganoderma lucidum* Polysaccharides as an anti-cancer agent. *Anti-cancer Agents in Medicinal Chemistry*, *18*(5), 667–674. doi:10.2174/1871520617666171113121246 PMID:29141563

Solans, C., Esquena, J., Forgiarini, A. M., Uson, N., & Izquierdo, P. (2002). Nanoemulsions: formation and properties. In K. L. Mittal & D. O. Shah (Eds.), *Surfactants in Solution: Fundamentals and Applications*. Marcel Dekker.

Solans, C., Izquierdo, P., Nolla, J., Azemar, N., & García-Celma, M. J. (2005). Nano-emulsion. *Current Opinion in Colloid & Interface Science*, *10*(3-4), 102–110. doi:10.1016/j.cocis.2005.06.004

Solans, C., Izquierdo, P., Nolla, J., Azemar, N., & Garcia-Celma, M. J. (2005). Nano-emulsions. *Current Opinion in Colloid & Interface Science*, *10*(3-4), 102–110.

Solans, C., & Solé, I. (2012). Nano-emulsions: Formation by low-energy methods. *Current Opinion in Colloid & Interface Science*, *17*(5), 246–254. doi:10.1016/j.cocis.2012.07.003

Sole, I., Maestro, A., Pey, C. M., Gonzalez, C., Solans, C., & Gutierrez, J. M. (2006). Nano-emulsions preparation by low energy methods in an ionic surfactant system. *Colloids and Surfaces. A, Physicochemical and Engineering Aspects*, *288*(1-3), 138–143.

Sole, I., Pey, C. M., Maestro, A., Gonzalez, C., Porras, M., Solans, C., & Gutierrez, J. M. (2010). Nano-emulsions prepared by the phase inversion composition method: Preparation variables and scale up. *Journal of Colloid and Interface Science*, *344*(2), 417–423.

Somoskovi, A., Parsons, L. M., & Salfinger, M. (2001). The molecular basis of resistance to isoniazid, rifampin, and pyrazinamide in *Mycobacterium tuberculosis*. *Respiratory Research*, *2*(3), 164–168. doi:10.1186/rr54 PMID:11686881

Sonawane, S. J., Kalhapure, R. S., Jadhav, M., Rambharose, S., Mocktar, C., & Govender, T. (2015). Transforming linoleic acid into a nanoemulsion for enhanced activity against methicillin susceptible and resistant Staphylococcus aureus. *RSC Advances*, *5*(110), 90482–90492.

Sonawane, S., Bhanvase, B. A., & Sivakumar, M. (2020). *Encapsulation of Active Molecules and Their Delivery System*. Elsevier.

Song, F.-Q., Liu, Y., Kong, X.-S., & Chang, S. (2013). Progress on understanding the anticancer mechanisms of medicinal mushroom: *Inonotus obliquus*. *AsianPac. Journal of Cancer Prevention*, *14*, 1571–1578. PMID:23679238

Song, R., Yan, F., Cheng, M., Dong, F., Lin, Y., Wang, Y., & Song, B. (2020). Ultrasound-Assisted Preparation of Exopolysaccharide/Nystatin Nanoemulsion for Treatment of Vulvovaginal Candidiasis. *International Journal of Nanomedicine*, *15*, 2027–2044. doi:10.2147/IJN.S241134 PMID:32273700

Son, H. J., Bae, H. C., Kim, H. J., Lee, D. H., Han, D. W., & Park, J. C. (2005). Effects of β-glucan on proliferation and migration of fibroblasts. *Current Applied Physics*, *5*(5), 468–471. doi:10.1016/j.cap.2005.01.011

Soni, H., & Sharma, S. (2021). Current Update on Nanoemulsion: A Review. *Sch Int J Anat Physiol*, *4*(1), 6–13.

Sonneville-Aubrun, O., Yukuyama, M. N., & Pizzino, A. (2018). Application of Nanoemulsions in Cosmetics. Nanoemulsions, 435-475. doi:org/10.1016/B978-0-12-811838-2.00014-X

Sonneville-Aubrun, O., Simonnet, J. T., & L'Alloret, F. (2004). Nanoemulsions: A new vehicle for skincare products. *Advances in Colloid and Interface Science*, *108-109*, 145–149. https://doi.org/10.1016/j.cis.2003.10.026

Sonu, K. S., Mann, B., Sharma, R., Kumar, R., & Singh, R. (2018). Physico-chemical and antimicrobial properties of d-limonene oil nanoemulsion stabilized by whey protein-maltodextrin conjugates. *Journal of Food Science and Technology*, *55*(7), 2749–2757. doi:10.100713197-018-3198-7 PMID:30042591

Sonvico, F., Cagnani, A., Rossi, A., Motta, S., Di Bari, T. T., Cavatorta, F., Alonso, M. J., Deriu, A., & Colombo, P. (2006). Formation of self-organized nanoparticles by lecithin/chitosan ionic interaction. *International Journal of Pharmaceutics*, *324*(1), 67–73. doi:10.1016/j.ijpharm.2006.06.036 PMID:16973314

Soottitantawat, A., Yoshii, H., Furuta, T., Ohkawara, M., & Linko, P. (2003). Microencapsulation by Spray Drying: Influence of Emulsion Size on the Retention of Volatile Compounds. *Journal of Food Science*, *68*, 2256–2262. https://doi.org/10.1111/j.1365-2621.2003.tb05756.x

Sotomayor-Gerding, D., Oomah, B. D., Acevedo, F., Morales, E., Bustamante, M., Shene, C., & Rubilar, M. (2016). High carotenoid bioaccessibility through linseed oil nanoemulsions with enhanced physical and oxidative stability. *Food Chemistry*, *199*, 463–470. doi:10.1016/j.foodchem.2015.12.004 PMID:26775996

Soukoulis, C., Tsevdou, M., Andre, C. M., Cambier, S., Yonekura, L., Taoukis, P. S., & Homann, L. (2017). Modulation of chemical stability and in vitro bioaccessibility of beta-carotene loaded in kappa-carrageenan oil-in-gel emulsions. *Food Chemistry*, *220*, 208–218. doi:10.1016/j.foodchem.2016.09.175 PMID:27855891

Souza, D. P., Pimentel, R. B. Q., Santos, A. S., Albuquerque, P. M., Fernandes, A. V., Junior, S. D., Oliveira, J. T. A., Ramos, M. V., Rathinasabapathi, B., & Gonçalves, J. F. C. (2020). Fungicidal properties and insights on the mechanisms of the action of volatile oils from Amazonian Aniba trees. *Industrial Crops and Products*, *143*, 111914. doi:10.1016/j.indcrop.2019.111914

Stanberry, L., Simon, J., Johnson, C., Robinson, P., Morry, J., Flack, M., Gracon, S., Myc, A., Hamouda, T., & Baker, J. R. Jr. (2012). Safety and immunogenicity of a novel nanoemulsion mucosal adjuvant W805EC combined with approved seasonal influenza antigens. *Vaccine*, *30*(2), 307–316. doi:10.1016/j.vaccine.2011.10.094 PMID:22079079

Stanculeanu, D. L., Daniela, Z., Lazescu, A., Bunghez, R., & Anghel, R. (2016). Development of new immunotherapy treatments in different cancer types. *Journal of Medicine and Life*, *9*(3), 240. PMID:27974927

Streck, L., Sarmento, V. H., Machado, P. R., Farias, K. J., Fernandes-Pedrosa, M. F., & da Silva-Júnior, A. A. (2016). Phase Transitions of Isotropic to Anisotropic Biocompatible Lipid-Based Drug Delivery Systems Overcoming Insoluble Benznidazole Loading. *International Journal of Molecular Sciences*, *17*(7), 981. doi:10.3390/ijms17070981 PMID:27376278

Subhashis, D., Satayanarayana, J., & Gampa, V. K. (2011). Nanoemulsion-a method to improve the solubility of lipophilic drugs. *International Journal of Advanced Pharmacology and Science*, *2*(2-3), 72–83.

Subramanian, N., & Ghosal, S. K. (2004). *Enhancement of gastrointestinal absorption of poorly water soluble drugs via lipid based systems*. Academic Press.

Subramanian, N., Sharavanan, S. P., Chandrasekar, P., Balakumar, A., & Moulik, S. P. (2016). Lacidipine self-nanoemulsifying drug delivery system for the enhancement of oral bioavailability. *Archives of Pharmacal Research*, *39*(4), 481–491. doi:10.100712272-015-0657-9 PMID:26362165

Sudirman, L. I. (2010). Partial purification of antimicrobial compounds isolated from mycelia of tropical *lentinus cladopus* lc4. *Hayati Journal of Biosciences*, *2*(2), 63–67. doi:10.4308/hjb.17.2.63

Sugumar, S, Clarke, S. K., Nirmala, M. J., Tyagi, B. K., Mukherjee, A., & Chandrasekaran, N. (2014). *Nanoemulsion of eucalyptus oil and its larvicidal activity against Culex quinquefasciatus*. doi:10.1017/S0007485313000710

Sugumar, S., & Singh, S. (2015). *Nanoemulsion of orange oil with non ionic surfactant produced emulsion using ultrasonication technique : Evaluating against food spoilage yeast*. doi:10.1007/s13204-015-0412-z

Sugumar, S., Nirmala, J., Anjali, H., Mukherjee, A., & Chandrasekaran, N. (2012). *Bio-based nanoemulsion formulation, characterization and antibacterial activity against food-borne pathogens*. doi:10.1002/jobm.201200060

Sugumaran, A., Ponnusamy, C., Kandasamy, P., Krishnaswami, V., Palanichamy, R., Kandasamy, R., Lakshmanan, M., & Natesan, S. (2018). Development and evaluation of camptothecin loaded polymer stabilized nanoemulsion: Targeting potential in 4T1-breast tumour xenograft model. *European Journal of Pharmaceutical Sciences*, *116*, 15–25.

Sugumar, S., Ghosh, V., Nirmala, M. J., Mukherjee, A., & Chandrasekaran, N. (2014). Ultrasonic emulsification of eucalyptus oil nanoemulsion: Antibacterial activity against Staphylococcus aureus and wound healing activity in Wistar rats. *Ultrasonics Sonochemistry*, *21*(3), 1044–1049. doi:10.1016/j.ultsonch.2013.10.021

Sugumar, S., Mukherjee, A., & Chandrasekaran, N. (2015). Eucalyptus oil nanoemulsion-impregnated chitosan film: Antibacterial effects against a clinical pathogen, Staphylococcus aureus, in vitro. *International Journal of Nanomedicine*, *10*(Suppl 1), 67–75. https://doi.org/10.2147/IJN.S79982

Sugumar, S., Nirmala, J., Ghosh, V., Anjali, H., Mukherjee, A., & Chandrasekaran, N. (2013). Bio-based nanoemulsion formulation, characterization and antibacterial activity against food-borne pathogens. *Journal of Basic Microbiology*, *53*(8), 677–685. https://doi.org/10.1002/jobm.201200060

Sugumar, S., Singh, S., Mukherjee, A., & Chandrasekaran, N. (2016). Nanoemulsion of orange oil with non ionic surfactant produced emulsion using ultrasonication technique: Evaluating against food spoilage yeast. *Applied Nanoscience*, *6*, 113–120. https://doi.org/10.1007/s13204-015-0412-z

Sundararajan, B., Moola, A. K., Vivek, K., & Kumari, B. R. (2018). Formulation of nanoemulsion from leaves essential oil of Ocimumbasilicum L. and its antibacterial, antioxidant and larvicidal activities (Culexquinquefasciatus). *Microbial Pathogenesis*, *125*, 475–485. doi:10.1016/j.micpath.2018.10.017

Sung, H., Ferlay, J., Siegel, R. L., Laversanne, M., Soerjomataram, I., Jemal, A., & Bray, F. (2021). Global cancer statistics 2020: GLOBOCAN estimates of incidence and mortality worldwide for 36 cancers in 185 countries. *CA: a Cancer Journal for Clinicians*, *71*(3), 209–249. doi:10.3322/caac.21660 PMID:33538338

Sun, Z., Zhang, J., Song, H., Zhang, X., Li, Y., Tian, M., Liu, Y., Zhao, Y., & Li, C. (2010). Concomitant increases in spectrum and level of drug resistance in *Mycobacterium tuberculosis* isolates. *The International Journal of Tuberculosis and Lung Disease*, *14*(11), 1436–1441. PMID:20937184

Surendhiran, M., Raja, K., Jerlin, R., Marimuthu, S., & Srivignesh, S. (2019). Nano Emulsion Seed Invigouration for Improved Germination and Seedling Vigour in Maize. *International Journal of Agricultural Science and Research (IJASR)*, *9*, 333–340. doi:10.24247/ijasrjun201946

Suslick, K. S., Didenko, Y., Fang, M. M., Hyeon, T., Kolbeck, K. J., McNamara III, W. B., ... Wong, M. (1999). Acoustic cavitation and its chemical consequences. *Philosophical Transactions of the Royal Society of London. Series A: Mathematical, Physical and Engineering Sciences, 357*(1751), 335-353.

Sutradhar, K., & Amin, M. (2013). Nanoemulsions: Increasing possibilities in drug delivery. *European Journal of Nanomedicine*, *5*(2), 97–110. doi:10.1515/ejnm-2013-0001

Swathy, J. S., Mishra, P., Thomas, J., Mukherjee, A., & Chandrasekaran, N. (2018). Antimicrobial potency of high-energy emulsified black pepper oil nanoemulsion against aquaculture pathogen. *Aquaculture (Amsterdam, Netherlands)*, *491*, 210–220. doi:10.1016/j.aquaculture.2018.03.045

Swientek, R. J. (1990). Microfluidizing technology enhances emulsion stability. *Food Processing*, *6*, 152–153.

Syed, I., Banerjee, P., & Sarkar, P. (2020). Oil-in-water emulsions of geraniol and carvacrol improve the antibacterial activity of these compounds on raw goat meat surface during extended storage at 4°C. *Food Control*, *107*, 106757. doi:10.1016/j.foodcont.2019.106757

Syed, Y. Y. (2020). Ramucirumab : A Review in Hepatocellular Carcinoma. *Drugs*, (3), 315–322. Advance online publication. doi:10.100740265-020-01263-6 PMID:32034692

Symonds, R. P. (2001). Radiotherapy. *BMJ (Clinical Research Ed.)*, *323*(7321), 1107–1110. doi:10.1136/bmj.323.7321.1107 PMID:11701579

Szczepanowicz, K., Bazylinska, U., Pietkiewicz, J., Szyk-Warszynska, L., Wilk, K. A., & Warszynski, P. (2015). Biocompatible long-sustained release oil-core polyelectrolyte nanocarriers: From controlling physical state and stability to biological impact. *Advances in Colloid and Interface Science*, *222*, 678–691. doi:10.1016/j.cis.2014.10.005 PMID:25453660

Tadros, T. F. (2011). *Rheology of dispersions: principles and applications*. John Wiley & Sons.

Tadros, T., Izquierdo, P., Esquena, J., & Solans, C. (2004). Formation and stability of nano-emulsions. *Advances in Colloid and Interface Science*, *108–109*, 303–318. doi:10.1016/j.cis.2003.10.023 PMID:15072948

Tagne, J. B., Kakumanu, S., Ortiz, D., Shea, T., & Nicolosi, R. J. (2008). A nanoemulsion formulation of tamoxifen increases its efficacy in a breast cancer cell line. *Molecular Pharmaceutics*, *5*(2), 280–286.

Taha, A., Ahmed, E., Hu, T., Xu, X., Pan, S., & Hu, H. (2019). Effects of different ionic strengths on the physicochemical properties of plant and animal proteins-stabilized emulsions fabricated using ultrasound emulsification. *Ultrasonics Sonochemistry*, *58*, 104627. doi:10.1016/j.ultsonch.2019.104627 PMID:31450289

Taha, A., Ahmed, E., Ismaiel, A., Ashokkumar, M., Xu, X., Pan, S., & Hu, H. (2020). Ultrasonic emulsification: An overview on the preparation of different emulsifiers-stabilized emulsions. *Trends in Food Science & Technology*, *105*, 363–377. doi:10.1016/j.tifs.2020.09.024

Tai, W., Shukla, R. S., Qin, B., Li, B., & Cheng, K. (2011). Development of a peptide-drug conjugate for prostate cancer therapy. *Molecular Pharmaceutics*, *8*(3), 901–912. doi:10.1021/mp200007b PMID:21510670

Takala, P., Vu, K., Salmieri, S., & Lacroix, M. (2011). Effect of antimicrobial coatings on the radiosensitization of *Escherichia coli, Salmonella Typhimurium,* and *Listeria monocytogenes* in fresh broccoli. *Journal of Food Protection*, *74*(7), 1065–1069. doi:10.4315/0362-028X.JFP-10-520 PMID:21740707

Talegaonkar, S., Mustafa, G., Akhter, S., & Iqbal, Z. (2010). Design and development of oral oil-in-water nanoemulsion formulation bearing atorvastatin: In vitro assessment. *Journal of Dispersion Science and Technology*, *31*(5), 690–701. doi:10.1080/01932690903120540

Talegaonkar, S., & Negi, L. M. (2015). Nanoemulsion in drug targeting. In P. V. Devarajan & S. Jain (Eds.), *Targeted Drug Delivery: Concepts and Design* (pp. 433–459). Springer. doi:10.1007/978-3-319-11355-5_14

Tambekar, D. H., Sonar, T. P., Khodke, M. V., & Khante, B. S. (2006). The Novel antibacterials from two edible mushrooms: *Agaricus bisporus and Pleurotus sajor caju. International Journal of Pharmacology*, *2*(5), 584–587. doi:10.3923/ijp.2006.584.587

Tan, C. P., & Nakajima, M. (2005a). β-Carotene nanodispersions: Preparation, characterization and stability evaluation. *Food Chemistry*, *92*(4), 661–671. doi:10.1016/j.foodchem.2004.08.044

Tan, C., & McClements, D. J. (2021). Application of Advanced Emulsion Technology in the Food Industry: A Review and Critical Evaluation. *Foods (Basel, Switzerland)*, *10*(4), 812. doi:10.3390/foods10040812 PMID:33918596

Tang, S.Y., Sivakumar, M., Ng, A.M., & Shridharan, P. (2012). Anti-inflammatory and analgesic activity of novel oral aspirin-loaded nanoemulsion and nano multiple emulsion formulations generated using ultrasound cavitation. *International Journal of Pharmaceutics, 430*(1-2), 299-306.

Tang, J., Zhu, R., Zhao, R., Cheng, G., & Peng, W. (2008). Ultra-performance liquid chromatography–tandem mass spectrometry for the determination of lacidipine in human plasma and its application in a pharmacokinetic study. *Journal of Pharmaceutical and Biomedical Analysis*, *47*(4-5), 923–928. doi:10.1016/j.jpba.2008.04.018 PMID:18534803

Tang, K., Du, S., Wang, Q., Zhang, Y., & Song, H. (2020). Traditional Chinese medicine targeting cancer stem cells as an alternative treatment for hepatocellular carcinoma. *Journal of Integrative Medicine*, *18*(3), 196–202. doi:10.1016/j.joim.2020.02.002 PMID:32067923

Tang, S. Y., Manickam, S., Wei, T. K., & Nashiru, B. (2012). Formulation development and optimisation of a novel cremophor EL-based nanoemulsion using ultrasound cavitation. *Ultrasonics Sonochemistry*, *19*, 330–345.

Tan, S. F., Masoumi, H. R., Karjiban, R. A., Stanslas, J., Kirby, B. P., Basri, M., & Basri, H. B. (2016). Ultrasonic emulsification of parenteral valproic acid-loaded nanoemulsion with response surface methodology and evaluation of its stability. *Ultrasonics Sonochemistry*, *29*, 299–308. doi:10.1016/j.ultsonch.2015.09.015 PMID:26585010

Tan, T. B., Chu, W. C., Yussof, N. S., Abas, F., Mirhosseini, H., Cheah, Y. K., Nehdi, I. A., & Tan, C. P. (2016). Physicochemical, morphological and cellular uptake properties of lutein nanodispersions prepared by using surfactants with different stabilizing mechanisms. *Food & Function*, *7*(4), 2043–2051. doi:10.1039/C5FO01621E PMID:27010495

Tayeb, H. H., & Sainsbury, F. (2018). Nanoemulsions in drug delivery: Formulation to medical application. *Nanomedicine (London)*, *13*(19), 2507–2525. doi:10.2217/nnm-2018-0088 PMID:30265218

Taylor, P. (1998). Ostwald ripening in emulsions. *Advances in Colloid and Interface Science*, *75*(2), 107–163. https://doi.org/10.1016/S0001-8686(98)00035-9

Taylor, P. R., Qiao, Y. L., Abnet, C. C., Dawsey, S. M., Yang, C. S., Gunter, E. W., Wang, W., Blot, W. J., Dong, Z.-W., & Mark, S. D. (2003). Prospective study of serum vitamin E levels and esophageal and gastric cancers. *Journal of the National Cancer Institute*, *95*(18), 1414–1416. doi:10.1093/jnci/djg044 PMID:13130117

Tedesco, A. C., Silva, E., Jayme, C. C., Piva, H. L., & Franchi, L. P. (2021). Cholesterol-rich nanoemulsion (LDE) as a novel drug delivery system to diagnose, delineate, and treat human glioblastoma. *Materials Science and Engineering C*, *123*, 111984.

Teixeira, B., Marques, A., Pires, C., Ramos, C., Batista, I., Saraiva, J., & Nunes, M. (2014). Characterization of fish protein films incorporated with essential oils of clove, garlic and origanum: Physical, antioxidant and antibacterial properties. *Lebensmittel-Wissenschaft + Technologie*, *59*, 533–539.

Teixeira, B., Marques, A., Ramos, C., Serrano, C., Matos, O., Neng, N. R., Nogueira, J. M. F., Saraiva, J. A., & Nunes, M. L. (2013). Chemical composition and bioactivity of different oregano (Origanum vulgare) extracts and essential oil. *Journal of the Science of Food and Agriculture*, *93*, 2707–2714. https://doi.org/10.1002/jsfa.6089

Teixeira, H. F., Bruxel, F., Fraga, M., Schuh, R. S., Zorzi, G. K., Matte, U., & Fattal, E. (2017). Cationic nanoemulsions as nucleic acids delivery systems. *International Journal of Pharmaceutics*, *534*(1-2), 356–367. doi:10.1016/j.ijpharm.2017.10.030 PMID:29038065

Teo, A., Goh, K., & Lee, S. J. (2014). *Nanoparticles and nanoemulsions. Functional foods and dietary supplements: Processing effects and health benefits*. John Wiley & Sons, Ltd. Publication.

Teo, A., Lee, S. J., & Goh, K. K. T. (2017). Formation and stability of single and bi-layer nanoemulsions using WPI and lactoferrin as interfacial coatings under different environmental conditions. *Food Structure*, *14*, 60–67. doi:10.1016/j.foostr.2017.06.005

Tepe, B., Akpulat, H. A., Sokmen, M., Daferera, D., Yumrutas, O., Aydin, E., Polissiou, M., & Sokmen, A. (2006). Screening of the antioxidative and antimicrobial properties of the essential oils of Pimpinella anisetum and Pimpinella flabellifolia from Turkey. *Food Chemistry*, *97*(4), 719–724. https://doi.org/10.1016/j.foodchem.2005.05.045

Tesch, S., Gerhards, C., & Schubert, H. (2002). Stabilization of emulsions by OSA starches. *Journal of Food Engineering, 54*(2), 167–174. doi:10.1016/S0260-8774(01)00206-0

Thakore, S., Patel, R., & Patel, M. (2014). Nanoemulsion or Microemulsion? Understanding the Differences and Similarities. *The Pharma Review*, 136-142.

Thatoi, H., & Singdevsachan, S. K. (2014). Diversity, nutritional composition andmedicinal potential of Indian mushrooms: A review. *African Journal of Biotechnology, 13*(4), 523–545. doi:10.5897/AJB2013.13446

The Future of the Global Nanoemulsion Market to 2024: Rising Adoption of Targeted Therapeutics & Image-Guided Therapies. (2019). https://www.prnewswire.com/news-releases/the-future-of-the-global-nanoemulsion-market-to-2024-rising-adoption-of-targeted-therapeutics--image-guided-therapies-300873615.html

Thiagarajan, P. (2011). Nanoemulsions for drug delivery through different routes. *Research in Biotechnology, 2*(3), 1-13.

Thiagarajan, P. (2011). Nanoemulsions for drug delivery through different routes. *Research in Biotechnology, 2*(3), 1–13.

Thiebaud, M., Dumay, E., Picart, L., Guiraud, J. P., & Cheftel, J. C. (2003). Highpressure homogenisation of raw bovine milk. Effects on fat globule size distribution and microbial inactivation. *International Dairy Journal, 13*(6), 427–439. doi:10.1016/S0958-6946(03)00051-7

Thomas, M. B., & Abbruzzese, J. L. (2005). Opportunities for targeted therapies in hepatocellular carcinoma. *Journal of Clinical Oncology, 23*(31), 8093–8108. doi:10.1200/JCO.2004.00.1537 PMID:16258107

Thormar, H. (2010). *Lipids and Essential Oils as Antimicrobial Agents*. John Wiley & Sons.

Tian, H., Li, D., Xu, T., Hu, J., Rong, Y., & Zhao, B. (2017). Citral stabilization and characterization of nanoemulsions stabilized by a mixture of gelatin and Tween 20 in an acidic system. *Journal of the Science of Food and Agriculture, 97*(9), 2991–2998. doi:10.1002/jsfa.8139 PMID:27859362

Tian, W.-L., Lei, L.-L., Zhang, Q., & Li, Y. (2016). Physical stability and antimicrobial activity of encapsulated Cinnamaldehyde by self-emulsifying nanoemulsion. *Journal of Food Process Engineering, 39*(5), 462–471. doi:10.1111/jfpe.12237

Tiwari, S. B., & Amiji, M. M. (2006) Nanoemulsion formulations for tumor-targeted delivery. In Nanotech Cancer Therapy (pp. 723-739). Taylor and Francis Group.

Tiwari, S. B., & Amiji, M. M. (2006). Improved Oral Delivery of Paclitaxel Following Administration in Nanoemulsion Formulations. *Journal of Nanoscience and Nanotechnology, 6*(9), 3215–3221. doi:10.1166/jnn.2006.440 PMID:17048539

Tiwari, S., Tan, Y. M., & Amiji, M. (2006). Preparation and *in vitro* characterization of multifunctional Nanoemulsions for simultaneous MR imaging and targeted drug delivery. *Journal of Biomedical Nanotechnology, 2*(3), 217–224. doi:10.1166/jbn.2006.038

Tiwari, S., Tan, Y. M., & Amiji, M. (2006). Preparation and In Vitro Characterization of Multifunctional Nanoemulsions for Simultaneous MR Imaging and Targeted Drug Delivery. *Journal of Biomedical Nanotechnology, 2*, 217–224.

Toledano-Medina, M. A., Pérez-Aparicio, J., Moreno-Rojas, R., & Merinas-Amo, T. (2016). Evolution of some physicochemical and antioxidant properties of black garlic whole bulbs and peeled cloves. *Food Chemistry, 199*, 135–139. https://doi.org/10.1016/j.foodchem.2015.11.128

Tomeh, M. A., Hadianamrei, R., & Zhao, X. (2019). A review of curcumin and its derivatives as anticancer agents. *International Journal of Molecular Sciences, 20*(5), 1033. doi:10.3390/ijms20051033 PMID:30818786

Tong, Y., Wu, Y., Zhao, C., Xu, Y., Lu, J., Xiang, S., Zong, F., & Wu, X. (2017). Polymeric Nanoparticles as a Metolachlor Carrier: Water-Based Formulation for Hydrophobic Pesticides and Absorption by Plants. *Journal of Agricultural and Food Chemistry*, *65*(34), 7371–7378. doi:10.1021/acs.jafc.7b02197 PMID:28783335

Topuz, O. K., Özvural, E. B., Zhao, Q., Huang, Q., Chikindas, M., & Gölükçü, M. (2016). Physical and antimicrobial properties of anise oil loaded nanoemulsions on the survival of foodborne pathogens. *Food Chemistry*, *203*, 117–123. https://doi.org/10.1016/j.foodchem.2016.02.051

Torchilin, V. P. (2000). Drug targeting. *European Journal of Pharmaceutical Sciences*, *11*(Suppl 2), S81–S91. doi:10.1016/S0928-0987(00)00166-4 PMID:11033430

Torre, L. A., Bray, F., Siegel, R. L., Ferlay, J., Lortet-Tieulent, J., & Jemal, A. (2015). Global cancer statistics, 2012. *CA: a Cancer Journal for Clinicians*, *65*(2), 87–108. doi:10.3322/caac.21262 PMID:25651787

Toti, U. S., Guru, B. R., Hali, M., McPharlin, C. M., Wykes, S. M., Panyam, J., & Whittum-Hudson, J. A. (2011). Targeted delivery of antibiotics to intracellular chlamydial infections using PLGA nanoparticles. *Biomaterials*, *32*(27), 6606–6613. doi:10.1016/j.biomaterials.2011.05.038 PMID:21652065

Toub, N., Malvy, C., Fattal, E., & Couvreur, P. (2006). Innovative nanotechnologies for the delivery of oligonucleotides and siRNA. *Biomedicine and Pharmacotherapy*, *60*(9), 607–620.

Trauner, A., Borrell, S., Reither, K., & Gagneux, S. (2014). Evolution of drug resistance in tuberculosis: Recent progress and implications for diagnosis and therapy. *Drugs*, *74*(10), 1063–1072. doi:10.100740265-014-0248-y PMID:24962424

Trevaskis, N. L., Charman, W. N., & Porter, C. J. (2008). Lipid-based delivery systems and intestinal lymphatic drug transport: A mechanistic update. *Advanced Drug Delivery Reviews*, *60*(6), 702–716.

Tricker, A. R., & Preussmann, R. (1991). Carcinogenic N-nitrosamines in the diet: occurrence, formation, mechanisms and carcinogenic potential. *Mutation Research/Genetic Toxicology*, *259*(3–4), 277–289. doi:10.1016/0165-1218(91)90123-4

Tripathi, A. D., Sharma, R., Agarwal, A., & Haleem, D. R. (2021). Nanoemulsions based edible coatings with potential food applications. *International Journal of Biobased Plastics*, *3*(1), 112–125. doi:10.1080/24759651.2021.1875615

Tripathi, A., Gupta, R., & Saraf, S. A. (2010). PLGA nanoparticles of antitubercular drug: Drug loading and release studies of a water-in-soluble drug. *International Journal of Pharm Tech Research*, *2*(3), 2116–2123.

Trotta, M. (1999). Influence of phase transformation on indomethacin release from microemulsions. *Journal of Controlled Release*, *60*(2-3), 399–405. doi:10.1016/S0168-3659(99)00094-2 PMID:10425344

Trujillo-Cayado, L. A., Santos, J., Calero, N., Alfaro-Rodríguez, M. C., & Muñoz, J. (2020). Strategies for reducing Ostwald ripening phenomenon in nanoemulsions based on thyme essential oil. *Journal of the Science of Food and Agriculture*, *100*(4), 1671–1677. https://doi.org/10.1002/jsfa.10181

Trujillo, L. S., Belloso, O. M., & McClements, D. J. (2016). Excipient Nanoemulsions for Improving Oral Bioavailability of Bioactives. *Nanomaterials (Basel, Switzerland)*, *6*(1), 17. doi:10.3390/nano6010017 PMID:28344274

Trumbeckas, D., Milonas, D., Jievaltas, M., Matjosaitis, A. J., Kincius, M., Grybas, A., & Kopustinskas, V. (2011). Importance of prostate volume and urinary flow rate in prediction of bladder outlet obstruction in men with symptomatic benign prostatic hyperplasia. *Central European Journal of Urology*, *64*(2), 75–79. doi:10.5173/ceju.2011.02.art5 PMID:24578868

Tsai, Y. J., & Chen, B. H. (2016). Preparation of catechin extracts and nanoemulsions from green tea leaf waste and their inhibition effect on prostate cancer cell PC-3. *International Journal of Nanomedicine*, *11*, 1907–1926.

Tsukihara, H., Nakagawa, F., Sakamoto, K., Ishida, K., Tanaka, N., Okabe, H., Uchida, J., Matsuo, K., & Takechi, T. (2015). Efficacy of combination chemotherapy using a novel oral chemotherapeutic agent, TAS-102, together with bevacizumab, cetuximab, or panitumumab on human colorectal cancer xenografts. *Oncology Reports*, *33*(5), 2135–2142. doi:10.3892/or.2015.3876 PMID:25812794

Tubtimsri, S., Limmatvapirat, C., Limsirichaikul, S., Akkaramongkolporn, P., Inoue, Y., & Limmatvapirat, S. (2018). Fabrication and characterization of spearmint oil loaded nanoemulsions as cytotoxic agents against oral cancer cell. *Asian Journal of Pharmaceutical Sciences, 13*(5), 425-437.

Tuli, H. S., Tuorkey, M. J., Thakral, F., Sak, K., Kumar, M., Sharma, A. K., Sharma, U., Jain, A., Aggarwal, V., & Bishayee, A. (2019). Molecular mechanisms of action of genistein in cancer: Recent advances. *Frontiers in Pharmacology*, *10*, 1336. doi:10.3389/fphar.2019.01336 PMID:31866857

Turek, C., & Stintzing, F. C. (2013). Stability of Essential Oils: A Review. *Comprehensive Reviews in Food Science and Food Safety*, *12*, 40–53. https://doi.org/10.1111/1541-4337.12006

Ultee, A., Bennik, M. H., & Moezelaar, R. (2002). The phenolic hydroxyl group of carvacrol is essential for action against the food-borne pathogen Bacillus cereus. *Applied and Environmental Microbiology*, *68*(4), 1561–1568. https://doi.org/10.1128/AEM.68.4.1561-1568.2002

Uner, M., Wissing, S. A., Yener, G., & Müller, R. H. (2005). Solid lipid nanoparticles (SLN) and nanostructured lipid carriers (NLC) for application of ascorbylpalmitate. *Die Pharmazie*, *60*(8), 577–582. PMID:16124399

Upadhaya, P., Barhoi, D., Giri, A., Bhattacharjee, A., & Giri, S. (2019). Joint detection of claudin-1 and junctional adhesion molecule-A as a therapeutic target in oral epithelial dysplasia and oral squamous cell carcinoma. *Journal of Cellular Biochemistry*, *120*(10), 18117–18127. doi:10.1002/jcb.29115 PMID:31161679

Upadhyay, T., Ansari, V. A., Ahmad, U., Sultana, N., & Akhtar, J. (2020). Exploring Nanoemulsion for Liver Cancer Therapy. *Current Cancer Therapy Reviews*, *16*(4), 260–268. doi:10.2174/1573394716666200302123336

Vacher, S., Pellegrin, J. L., Leblanc, F., Fourche, J., & Maugein, J. (1999). Comparative antimycobacterial activities of ofloxacin, ciprofloxacin and grepafloxacin. *The Journal of Antimicrobial Chemotherapy*, *44*(5), 647–652. doi:10.1093/jac/44.5.647 PMID:10552981

Valente, S. A., & Shah, C. (2020). The Landmark Series: Adjuvant Radiation Therapy for Breast Cancer. *Annals of Surgical Oncology*, *27*(7), 2203–2211.

Valverde, M. E., Hernandez-Perez, T., & Paredes-Lopez, O. (2015). Edible mushrooms: Improving human health and promoting quality life. *International Journal of Microbiology*, *2015*, 1–14. doi:10.1155/2015/376387 PMID:25685150

van der Meel, R., Lammers, T., & Hennink, W. E. (2017). Cancer nanomedicines: Oversold or underappreciated? *Expert Opinion on Drug Delivery*, *14*(1), 1–5. doi:10.1080/17425247.2017.1262346 PMID:27852113

Van der Paardt, A. L., Akkerman, O. W., Gualano, G., Palmieri, F., Davies Forsman, L., Aleksa, A., Tiberi, S., de Lange, W. C., Bolhuis, M. S., Skrahina, A., van Soolingen, D., Kosterink, J. G., Migliori, G. B., van der Werf, T. S., & Alffenaar, J. C. (2017). Safety and tolerability of clarithromycin in the treatment of multidrug-resistant tuberculosis. *The European Respiratory Journal*, *49*(3), 1601612. doi:10.1183/13993003.01612-2016 PMID:28331034

van Ginckel, R., De Coster, R., Wouters, W., Vanherck, W., Veer, V., Der, D., Goeminne, N., Jagers, E., Van Cauteren, H., Wouters, L., & Distelmans, W. (1990). Antitumoral effects of R 75251 on the growth of transplantable R3327 prostatic adenocarcinoma in rats. *The Prostate*, *16*(4), 313–323. doi:10.1002/pros.2990160406 PMID:2371176

Vandamme, T. F. (2002). Microemulsions as ocular drug delivery systems: Recent developments and future challenges. *Progress in Retinal and Eye Research*, *21*(1), 15–34.

Vargas, M., Albors, A., Chiralt, A., & Gonzalez-Martınez, C. (2009). Characterization of chitosan–oleic acid composite films. *Food Hydrocolloids*, *23*(2), 536–547. doi:10.1016/j.foodhyd.2008.02.009

Varghese, R., Dalvi, Y.B., Lamrood, P.Y., Shinde, B.P., & Nair, C.K.K. (2019). Historical and current perspectives on therapeutic potential of higher basidiomycetes: An overview. *3 Biotech*, 9, 362.

Vasir, J. K., & Labhasetwar, V. (2005). Targeted drug delivery in cancer therapy. *Technology in Cancer Research & Treatment*, *5*(5), 363–374.

Vatanparast, H., Shahabi, F., Bahramian, A., Javadi, A., & Miller, R. (2018). The role of electrostatic repulsion on increasing surface activity of anionic surfactants in the presence of hydrophilic silica nanoparticles. *Scientific Reports*, *8*(1), 1–11. doi:10.103841598-018-25493-7 PMID:29740036

Vatsraj, S., Chauhan, K., & Pathak, H. (2014). Formulation of a novel nanoemulsion system for enhanced solubility of a sparingly water soluble antibiotic, clarithromycin. *Journal of Nanoscience*. Advance online publication. doi:10.1155/2014/268293

Vaupel, P., & Kelleher, D. K. (2013). Blood flow and oxygenation status of prostate cancers. *Advances in Experimental Medicine and Biology*, *765*, 299–305. doi:10.1007/978-1-4614-4989-8_42 PMID:22879048

Veerareddy, P.R., Poluri, K., Sistla, R., Chaganty, S., & Venishetty, V.K. (2012). Formulation development and comparative pharmacokinetic evaluation of felodipine nanoemulsion in SD rats. *American Journal of PharmTech Research*, *2*(3), 931–45.

Velderrain-Rodríguez, G. R., Acevedo-Fani, A., González-Aguilar, G. A., & Martín-Belloso, O. (2019). Encapsulation and stability of a phenolic-rich extract from mango peel within water-in-oil-in-water emulsions. *Journal of Functional Foods*, *56*, 65–73. doi:10.1016/j.jff.2019.02.045

Velho, M. C., de Oliveira, D. A., da Silva Gündel, S., Favarin, F. R., Santos, R. C. V., & Ourique, A. F. (2019). Nanoemulsions containing mancozeb and eugenol: Development, characterization, and antifungal activity against Glomerella cingulata. *Applied Nanoscience*, *9*(2), 233–241. doi:10.100713204-018-0903-9

Velikov, K. P., & Pelan, E. (2008). Colloidal delivery systems for micronutrients and nutraceuticals. *Soft Matter*, *4*(10), 1964–1980. doi:10.1039/b804863k

Venkateshwarlu, I., Prabhakar, K., Ali, M., & Kishan, V. (2010). Development and In Vitro Cytotoxic Evaluation of Parenteral Docetaxel Lipid Nanoemulsions for Application in Cancer Treatment. *PDA Journal of Pharmaceutical Science and Technology*, *64*(3), 233–241. PMID:21502023

Venkateswaran, S., Dos Santos, O. D. H., Scholefield, E., Lilienkampf, A., Gwynne, P. J., Swann, D. G., & Bradley, M. (2016). Fortified interpenetrating polymers–bacteria resistant coatings for medical devices. *Journal of Materials Chemistry. B, Materials for Biology and Medicine*, *4*(32), 5405–5411. doi:10.1039/c6tb01110a

Venturella, G., Ferraro, V., Cirlincione, F., & Gargano, M. L. (2021). Medicinal mushrooms: Bioactive compounds, use, and clinical trials. *International Journal of Molecular Sciences*, *22*(2), 634. doi:10.3390/ijms22020634 PMID:33435246

Verma, P., Meher, J. G., Asthana, S., Pawar, V. K., Chaurasia, M., & Chourasia, M. K. (2014). Perspectives of nanoemulsion assisted oral delivery of docetaxel for improved chemotherapy of cancer. *Drug Delivery*, *23*(2), 479–488. doi:10.3109/10717544.2014.920430 PMID:24901205

Verma, S., Jain, V., & Chandrawanshi, N. K. (2020). *Surfactant Based Anhydrous Nano Carrier System for Poorly Aqueous Soluble Anti-Cancer Drugs*. IGI Global. doi:10.4018/978-1-7998-6530-8.ch014

Vibe, C. B., Fenaroli, F., Pires, D., Wilson, S. R., Bogoeva, V., Kalluru, R., Speth, M., Anes, E., Griffiths, G., & Hildahl, J. (2016). Thioridazine in PLGA nanoparticles reduces toxicity and improves rifampicin therapy against mycobacterial infection in zebrafish. *Nanotoxicology*, *10*(6), 680–688. doi:10.3109/17435390.2015.1107146 PMID:26573343

Vierhapper, H., Wagner, O., Nowotny, P., & Waldhäusl, W. (1990). Effect of endothelin-1 in man. *Circulation*, *81*(4), 1415–1418. doi:10.1161/01.CIR.81.4.1415 PMID:2180594

Vila, J., Martí, S., & Sanchez-Céspedes, J. (2007). Porins, efflux pumps and multidrug resistance in Acinetobacterbaumannii. *The Journal of Antimicrobial Chemotherapy*, *59*(6), 1210–1215. doi:10.1093/jac/dkl509

Villares, A., Mateo-Vivaracho, L., & Guillamon, E. (2012). Structural features and healthy properties of polysaccharides occurring in mushrooms. *Agriculture*, *2*(4), 452–471. doi:10.3390/agriculture2040452

Villaverde, J. J., Sevilla-Moran, B., Lopez-Goti, C., Alonso-Prados, J. L., & Sandin-Espana, P. (2018). Contributions of nano-QSAR/QSPR models for nanopesticides risk assessment within the European legislative framework. *The Science of the Total Environment*, *634*, 1530–1539. doi:10.1016/j.scitotenv.2018.04.033 PMID:29710651

Vinarov, Z., Abrahamsson, B., Artursson, P., Batchelor, H., Berben, P., Bernkop-Schnürch, A., Butler, J., Ceulemans, J., Davies, N., Dupont, D., Flaten, G. E., Fotaki, N., Griffin, B. T., Jannin, V., Keemink, J., Kesisoglou, F., Koziolek, M., Kuentz, M., Mackie, A., ... Augustijns, P. (2021). Current challenges and future perspectives in oral absorption research: An opinion of the UNGAP network. *Advanced Drug Delivery Reviews*, *171*, 289–331. doi:10.1016/j.addr.2021.02.001 PMID:33610694

Viswanathan, P., Muralidaran, Y., & Ragavan, G. (2017). Challenges in oral drug delivery: a nano-based strategy to overcome. *Nanostructures for Oral Medicine*, 173–201. doi:10.1016/B978-0-323-47720-8.00008-0

Vlachopoulos, J., Polychronopoulos, N., & Kontopoulou, M. (2011). Basic concepts in polymer melt rheology and their importance in processing. *Applied Polymer Rheology: Polymeric Fluids with Industrial Applications,* 1-27. doi:org/10.1002/9781118140611.ch1

Von Groll, A., Martin, A., Jureen, P., Hoffner, S., Vandamme, P., Portaels, F., Palomino, J. C., & da Silva, P. A. (2009). Fluoroquinolone resistance in *Mycobacterium tuberculosis* and mutations in gyrA and gyrB. *Antimicrobial Agents and Chemotherapy*, *53*(10), 4498–4500. doi:10.1128/AAC.00287-09 PMID:19687244

Vyas, S. P., Kannan, M. E., Jain, S., Mishra, V., & Singh, P. (2004). Design of liposomal aerosols for improved delivery of rifampicin to alveolar macrophages. *International Journal of Pharmaceutics*, *269*(1), 37–49. doi:10.1016/j.ijpharm.2003.08.017 PMID:14698575

Vyas, T. K., Shahiwala, A., & Amiji, M. M. (2008). Improved oral bioavailability and brain transport of Saquinavir upon administration in novel nanoemulsion formulations. *International Journal of Pharmaceutics*, *347*(1-2), 93–101. doi:10.1016/j.ijpharm.2007.06.016 PMID:17651927

Wack, A., Baudner, B. C., Hilbert, A. K., Manini, I., Nuti, S., Tavarini, S., Scheffczik, H., Ugozzoli, M., Singh, M., Kazzaz, J., Montomoli, E., Del Giudice, G., Rappuoli, R., & O'Hagan, D. T. (2008). Combination adjuvants for the induction of potent, long-lasting antibody and T-cell responses to influenza vaccine in mice. *Vaccine*, *26*(4), 552–561. doi:10.1016/j.vaccine.2007.11.054 PMID:18162266

Wack, A., & Rappuoli, R. (2005). Vaccinology at the beginning of the 21st century. *Current Opinion in Immunology*, *17*(4), 411–418. doi:10.1016/j.coi.2005.05.005 PMID:15950445

Wagner, J. G., Gerrard, E. S., & Kaiser, D. G. (1996). The effect of the dosage form on serum levels of indoxole. *Clinical Pharmacology and Therapeutics*, *7*(5), 610–619. doi:10.1002/cpt196675610 PMID:5957166

Wagner, O. F., Christ, G., Wojta, J., Vierhapper, H., Parzer, S., Nowotny, P. J., Schneider, B., Waldhäusl, W., & Binder, B. R. (1992). Polar secretion of endothelin-1 by cultured endothelial cells. *The Journal of Biological Chemistry*, *267*(23), 16066–16068. doi:10.1016/S0021-9258(18)41966-7 PMID:1644793

Wahgiman, N. A., Salim, N., Abdul Rahman, M. B., & Ashari, S. E. (2019). Optimization of nanoemulsion containing gemcitabine and evaluation of its cytotoxicity towards human fetal lung fibroblast (MRC5) and human lung carcinoma (A549) cells. *International Journal of Nanomedicine*, *14*, 7323–7338. doi:10.2147/IJN.S212635 PMID:31686809

Walia, N., & Chen, L. (2020). Pea protein based vitamin D nanoemulsions: Fabrication, stability and in vitro study using Caco-2 cells. *Food Chemistry*, *305*, 125475. doi:10.1016/j.foodchem.2019.125475 PMID:31518841

Walinska, K. (2004). Comparison of the influence of the polyprenol structure on model membranes. *Desalination*, *163*(1-3), 239–245. doi:10.1016/S0011-9164(04)90195-6

Walinska, K., & Janas, T. (2001). The effect of phosphate ester derivatives of polyprenol on model membrane stability. *Cellular & Molecular Biology Letters*, *6*(2A).

Walstra, P. (1993). Principles of emulsion formation. *Chemical Engineering Science*, *48*(2), 333–349. https://doi.org/10.1016/0009-2509(93)80021-H

Walstra, P. (2003). *Physical Chemistry of Foods*. Marcel Decker.

Wang, C. Y., & Huang, L. (1987). pH-sensitive immunoliposomes mediate target-cell-specific delivery and controlled expression of a foreign gene in mouse. *Proceedings of the National Academy of Sciences of the United States of America*, *84*(22), 7851–7855.

Wang, C. Z., Li, W. J., Tao, R., Ye, J. Z., & Zhang, H. Y. (2015). Antiviral activity of a nanoemulsion of polyprenols from ginkgo leaves against influenza A H3N2 and hepatitis B virus in vitro. *Molecules (Basel, Switzerland)*, *20*(3), 5137–5151. doi:10.3390/molecules20035137 PMID:25808155

Wang, D., Veena, M. S., Stevenson, K., Tang, C., Ho, B., Suh, J. D., Duarte, V. M., Faull, K. F., Mehta, K., Srivatsan, E. S., & Wang, M. B. (2008). Liposome-encapsulated curcumin suppresses growth of head and neck squamous cell carcinoma in vitro and in xenografts through the inhibition of nuclear factor κB by an AKT-independent pathway. *Clinical Cancer Research*, *14*(19), 6228–6236. doi:10.1158/1078-0432.CCR-07-5177 PMID:18829502

Wang, F., Wang, Y. C., Dou, S., Xiong, M. H., Sun, T. M., & Wang, J. (2011). Doxorubicin-tethered responsive gold nanoparticles facilitate intracellular drug delivery for overcoming multidrug resistance in cancer cells. *ACS Nano*, *5*(5), 3679–3692. doi:10.1021/nn200007z PMID:21462992

Wang, H., & Ng, T. (2006). A laccase from the medicinal mushroom *Ganoderma lucidum. Applied Microbiology and Biotechnology*, *72*(3), 508–513. doi:10.100700253-006-0314-9 PMID:16636832

Wang, J., Shi, A., Agyei, D., & Wang, Q. (2017). Formulation of water-in-oil-in-water (W/O/W) emulsions containing trans-resveratrol. *RSC Advances*, *7*(57), 35917–35927. doi:10.1039/C7RA05945K

Wang, L., Li, X., Zhang, G., Dong, J., & Eastoe, J. (2007). Oil-in-water nanoemulsions for pesticide formulations. *Journal of Colloid and Interface Science*, *314*(1), 230–235.

Wang, L., Tabor, R., Eastoe, J., Li, X., Heenan, R. K., & Dong, J. (2009). Formation and stability of nanoemulsions with mixed ionic–nonionic surfactants. *Physical Chemistry Chemical Physics*, *11*(42), 9772–9778. doi:10.1039/b912460h PMID:19851556

Wang, P., Liu, H. J., Mei, X. Y., Nakajima, M., & Yin, L. J. (2012). Preliminary study into the factors modulating β-carotene micelle formation in dispersions using an in vitro digestion model. *Food Hydrocolloids*, *26*(2), 427–433.

Wang, Q., Gong, J., Huang, X., Yu, H., & Xue, F. (2009). In vitro evaluation of the activity of microencapsulated carvacrol against Escherichia coli with K88 pili. *Journal of Applied Microbiology*, *107*(6), 1781–1788. doi:10.1111/j.1365-2672.2009.04374.x PMID:19645769

Wang, S., Shao, G., Yang, J., Liu, J., Wang, J., Zhao, H., Yang, L., Liu, H., Zhu, D., Li, Y., & Jiang, L. (2020). The production of gel beads of soybean hull polysaccharides loaded with soy isoflavone and their pH-dependent release. *Food Chemistry*, *313*, 126095. doi:10.1016/j.foodchem.2019.126095 PMID:31923873

Wang, S., Wang, X., Liu, M., Zhang, L., Ge, Z., Zhao, G., & Zong, W. (2020). Preparation and characterization of Eucommia ulmoides seed oil O/W nanoemulsion by dynamic high-pressure microfluidization. *Lebensmittel-Wissenschaft + Technologie*, *121*, 108960. doi:10.1016/j.lwt.2019.108960

Wang, W., Chen, K., Liu, Q., Johnston, N., Ma, Z., Zhang, F., & Zheng, X. (2014). Suppression of tumor growth by *Pleurotus ferulae* ethanol extract through induction of cell apoptosis, and inhibition of cell proliferation and migration. *PLoS One*, *9*(7), e102673. doi:10.1371/journal.pone.0102673 PMID:25029345

Wang, X. Y., & Heuzey, M. C. (2016). Chitosan-based conventional and pickering emulsions with long-term stability. *Langmuir*, *32*(4), 924–936. doi:10.1021/acs.langmuir.5b03556 PMID:26743171

Wang, X., Jiang, Y., Wang, Y. W., Huang, M. T., Ho, C. T., & Huang, Q. (2008). Enhancing anti-inflammation activity of curcumin through O/W nanoemulsions. *Food Chemistry*, *108*(2), 419–424. doi:10.1016/j.foodchem.2007.10.086 PMID:26059118

Wang, X., Yang, L., Chen, Z. G., & Shin, D. M. (2008). Application of nanotechnology in cancer therapy and imaging. *CA: a Cancer Journal for Clinicians*, *58*(2), 97–110.

Wang, Y. Y., Liu, S. R., Zong, F. L., & Ye, J. M. (2004). Pesticide formulations name and code instructions. *Pesticide Science Admistration*, *25*, 1–5.

Wang, Y., Bao, L., Yang, X., Li, L., Li, S., Gao, H., Yao, X.-S., Wen, H., & Liu, H.-W. (2012). Bioactive sesquiterpenoids from the solidculture of the edible mushroom *Flammulina velutipes* growing on cookedrice. *Food Chemistry*, *132*(3), 1346–1353. doi:10.1016/j.foodchem.2011.11.117 PMID:29243621

Wang, Y., Li, P., Chen, F., Jia, L., Xu, Q., Gai, X., Yu, X., Di, Y., Zhu, Z., Liang, Y., Liu, M., Pan, W., & Yang, X. (2017). A novel pH-sensitive carrier for the delivery of antitumor drugs: Histidine-modified *Auricularia auricular* polysaccharide nanomicelles. *Scientific Reports*, *7*(1), 4751. doi:10.103841598-017-04428-8 PMID:28684737

Wani, T.A., & Masoodi, F.A., MahdiJafari, S., & McClements, D.J. (2018). *Safety of Nanoemulsions and Their Regulatory Status, Nanoemulsions, Formulation, Applications, and Characterization.* Academic Press.

Wan, J., Jin, Z., Zhong, S., Schwarz, P., Chen, B., & Rao, J. (2020). Clove oil-in-water nanoemulsion: Mitigates growth of Fusarium graminearum and trichothecene mycotoxin production during the malting of Fusarium infected barley. *Food Chemistry*, *312*, 126120. https://doi.org/10.1016/j.foodchem.2019.126120

Wan, J., Zhong, S., Schwarz, P., Chen, B., & Rao, J. (2019). Enhancement of antifungal and mycotoxin inhibitory activities of food-grade thyme oil nanoemulsions with natural emulsifiers. *Food Control*, *106*, 106709. https://doi.org/10.1016/j.foodcont.2019.106709

Wan, J., Zhong, S., Schwarz, P., Chen, B., & Rao, J. (2019). Physical properties, antifungal and mycotoxin inhibitory activities of five essential oil nanoemulsions: Impact of oil compositions and processing parameters. *Food Chemistry*, *291*, 199–206. https://doi.org/10.1016/j.foodchem.2019.04.032

Ward, M. C., Vicini, F., Chadha, M., Pierce, L., Recht, A., Hayman, J., Thaker, N. G., Khan, A., Keisch, M., & Shah, C. (2019). Radiation Therapy Without Hormone Therapy for Women Age 70 or Above with Low-Risk Early Breast Cancer: A Microsimulation. *International Journal of Radiation Oncology, Biology, Physics*, *105*(2), 296–306.

Warisnoicharoen, W., Lansley, A. B., & Lawrence, M. J. (2000). Nonionic oil-in-water microemulsions: The effect of oil type on phase behaviour. *International Journal of Pharmaceutics*, *198*(1), 7–27.

Washington, C., Chawla, A., Christy, N., & Davis, S. S. (1989). Electrokinetic properties of phospholipid-stabilized fat emulsions. *Intertational J. of Pharmaceutics*, *54*(3), 191–197. doi:10.1016/0378-5173(89)90096-3

Wasser, S. P. (2002). Medicinal mushrooms as a source of antitumor and immunomodulating polysaccharides. *Applied Microbiology and Biotechnology*, *60*(3), 258–274. doi:10.100700253-002-1076-7 PMID:12436306

Wasser, S. P. (2014). Medicinal mushroom science: Current perspectives, advances, evidences, and challenges. *Biomedical Journal*, *37*(6), 345–356. doi:10.4103/2319-4170.138318 PMID:25179726

Weaver, M. A., Jin, X., Hoagland, R. E., & Boyette, C. D. (2009). Improved bioherbicidal efficacy by Myrothecium verrucaria via spray adjuvants or herbicide mixtures. *Biological Control*, *50*(2), 150–156. doi:10.1016/j.biocontrol.2009.03.007

Weber, M. A., Schiffrin, E. L., White, W. B., Mann, S., Lindholm, L. H., Kenerson, J. G., ... Ram, C. V. S. (2014). Clinical practice guidelines for the management of hypertension in the community: A statement by the American Society of Hypertension and the International Society of Hypertension. *Journal of Hypertension*, *32*(1), 3–15. doi:10.1097/HJH.0000000000000065 PMID:24270181

Wehrli, W. (1983). Rifampin: Mechanisms of action and resistance. *Reviews of Infectious Diseases*, *5*(3), S407–S411. doi:10.1093/clinids/5.Supplement_3.S407 PMID:6356275

Weidner, N., Carroll, P. R., Flax, J., Blumenfeld, W., & Folkman, J. (1993). Tumor angiogenesis correlates with metastasis in invasive prostate carcinoma. *American Journal of Pathology*, *143*(2), 401–409. PMID:7688183

Weigand-Heller, A. J., Kris-Etherton, P. M., & Beelman, R. B. (2012). The bioavailability of ergothioneine from mushrooms (*Agaricus bisporus*) and the acute effects on antioxidant capacity and biomarkers of inflammation. *Preventive Medicine*, *54*, 75–78. doi:10.1016/j.ypmed.2011.12.028 PMID:22230474

Weiner, L. M., Webb, A. K., Limbago, B., Dudeck, M. A., Patel, J., Kallen, A. J., & Sievert, D. M. (2016). Antimicrobial-resistant pathogens associated with healthcare-associated infections: Summary of data reported to the National Healthcare Safety Network at the Centers for Disease Control and Prevention, 2011–2014. *Infection Control and Hospital Epidemiology*, *37*(11), 1288–1301. doi:10.1017/ice.2016.174

Weiss, J., Gaysinksy, S., Davidson, M., & Mcclements, J. (2009). Nanostructured encapsulation systems: food antimicrobials. In *IUFoST World Congress Book: Global Issues in Food Science and Technology*. Elsevier Inc.

Weiss, J., Cancelliere, C., & Mcclements, D. J. (2000). Mass Transport Phenomena in Oil-in-Water Emulsions Containing Surfactant Micelles: Ostwald Ripening. *Langmuir*, *16*, 6833–6838. https://doi.org/10.1021/la991477v

Weiss, J., Decker, E. A., McClements, D. J., Kristbergsson, K., Helgason, T., & Awad, T. (2008). Solid lipid nanoparticles as delivery systems for bioactive food components. *Food Biophysics*, *3*(2), 146–154. doi:10.100711483-008-9065-8

Weiss, J., Herrmann, N., & McClements, D. J. (1999). Ostwald ripening of hydrocarbon emulsion droplets in surfactant solutions. *Langmuir*, *15*(20), 6652–6657. doi:10.1021/la981739d

Weiss, J., & McClements, D. J. (2000). Influence of Ostwald ripening on rheology of oil-in-water emulsions containing electrostatically stabilized droplets. *Langmuir*, *16*(5), 2145–2150. doi:10.1021/la9909392

Weiss, J., Takhistov, P., & McClements, D. J. (2006). Functional materials in food nanotechnology. *Journal of Food Science*, *71*(9), 107–116. doi:10.1111/j.1750-3841.2006.00195.x

Wenzel, E., & Somoza, V. (2005). Metabolism and bioavailability of trans-resveratrol. *Molecular Nutrition & Food Research*, *49*(5), 472–481. doi:10.1002/mnfr.200500010 PMID:15779070

Whitmore, S. E., & Lamont, R. J. (2014). Oral bacteria and cancer. *PLoS Pathogens*, *10*(3), e1003933. doi:10.1371/journal.ppat.1003933 PMID:24676390

Widakowich, C., de Castro, G. Jr, De Azambuja, E., Dinh, P., & Awada, A. (2007). Side effects of approved molecular targeted therapies in solid cancers. *The Oncologist*, *12*(12), 1443–1455. doi:10.1634/theoncologist.12-12-1443 PMID:18165622

Wilhelm, S., Tavares, A. J., Dai, Q., Ohta, S., Audet, J., Dvorak, H. F., & Chan, W. C. W. (2016). Analysis of nanoparticle delivery to tumours. *Nature Reviews. Materials*, *1*(5), 16014. doi:10.1038/natrevmats.2016.14

Wilking, J. N., & Mason, T. G. (2007). Irreversible shear-induced vitrification of droplets into elastic nanoemulsions by extreme rupturing. *Physical Review. E*, *75*(4), 041407. doi:10.1103/PhysRevE.75.041407 PMID:17500894

Wilkinson, G. R. (1997). The effects of diet, aging and disease-states on presystemic elimination and oral drug bioavailability in humans. *Advanced Drug Delivery Reviews*, *27*(2-3), 129–159. doi:10.1016/S0169-409X(97)00040-9 PMID:10837555

Willby, M., Sikes, R. D., Malik, S., Metchock, B., & Posey, J. E. (2015). Correlation between GyrA substitutions and ofloxacin, levofloxacin, and moxifloxacin cross-resistance in Mycobacterium tuberculosis. *Antimicrobial Agents and Chemotherapy*, *59*(9), 5427–5434. doi:10.1128/AAC.00662-15 PMID:26100699

Winter, P. M., Schmieder, A. H., Caruthers, S. D., Keene, J. L., Zhang, H., Wickline, S. A., & Lanza, G. M. (2008). Minute dosages of alpha(nu)beta3-targeted fumagillin nanoparticles impair Vx-2 tumor angiogenesis and development in rabbits. *The FASEB Journal*, *22*(8), 2758–2767.

Wiradharma, N., Zhang, Y., Venkataraman, S., Hedrick, J. L., & Yang, Y. Y. (2009). Self-assembled polymer nanostructures for delivery of anticancer therapeutics. *Nano Today*, *4*(4), 302–317.

Witthayapanyanon, A., Acosta, E. J., Harwell, J. H., & Sabatini, D. A. (2006). Formulation of ultralow interfacial tension systems using extended surfactants. *Journal of Surfactants and Detergents*, *9*(4), 331–339. doi:10.100711743-006-5011-2

Włodarska, K., Pawlak-Lemańska, K., Górecki, T., & Sikorska, E. (2019). Factors influencing consumers' perceptions of food: A study of apple juice using sensory and visual attention methods. *Foods*, *8*(11), 545. doi:10.3390/foods8110545 PMID:31684206

Wochna Loerzel, V. (2015). Symptom Experience in Older Adults Undergoing Treatment for Cancer. *Oncology Nursing Forum*, *42*(3), E269–E278. doi:10.1188/15.ONF.E269-E278 PMID:25901389

Wolucka, B. A., McNeil, M. R., de Hoffmann, E., Chojnacki, T., & Brennan, P. J. (1994). Recognition of the lipid intermediate for arabinogalactan/arabinomannan biosynthesis and its relation to the mode of action of ethambutol on mycobacteria. *The Journal of Biological Chemistry*, *269*(37), 23328–23335. doi:10.1016/S0021-9258(17)31657-5 PMID:8083238

Wong, M. Y., & Chiu, G. N. (2010). Simultaneous liposomal delivery of quercetin and vincristine for enhanced estrogen-receptor-negative breast cancer treatment. *Anti-Cancer Drugs*, *21*(4), 401–410.

Wong, P. T., & Choi, S. K. (2015). Mechanisms of drug release in nanotherapeutic delivery systems. *Chemical Reviews*, *115*(9), 3388–3432. doi:10.1021/cr5004634 PMID:25914945

Wong, P. T., Goff, P. H., Sun, R. J., Ruge, M. J., Ermler, M. E., Sebring, A., ... Baker, J. R. Jr. (2020). Combined Intranasal Nanoemulsion and RIG-I Activating RNA Adjuvants Enhance Mucosal, Humoral, and Cellular Immunity to Influenza Virus. *Molecular Pharmaceutics*. PMID:32491861

Woodward, W. A., Barlow, W. E., Jagsi, R., Buchholz, T. A., Shak, S., Baehner, F., Whelan, T. J., Davidson, N. E., Ingle, J. N., King, T. A., Ravdin, P. M., Osborne, C. K., Tripathy, D., Livingston, R. B., Gralow, J. R., Hortobagyi, G. N., Hayes, D. F., & Albain, K. S. (2020). Association between 21-Gene Assay Recurrence Score and Locoregional Recurrence Rates in Patients With Node-Positive Breast Cancer. *JAMA Oncology*, *6*(4), 505–511.

Wooster, T. J., Golding, M., & Sanguansri, P. (2008). Impact of oil type on nanoemulsion formation and Ostwald ripening stability. *Langmuir ACS Journal of Surface Colloids*, *24*(22), 12758–12765. doi:10.1021/la801685v PMID:18850732

Wooster, T. J., Golding, M., & Sanguansri, P. (2008). Impact of oil type on nanoemulsion formation and Ostwald ripening stability. *Langmuir*, *24*(22), 12758–12765. https://doi.org/10.1021/la801685v

World Health Organization. (2014). *World health statistics 2014*. World Health Organization. https://apps.who.int/iris/handle/10665/112738

Wu, C., Li, N. J., Chen, K. C., & Hsu, H. F. (2014). Determination of critical micelle concentrations of ionic and nonionic surfactants based on relative viscosity measurements by capillary electrophoresis. *Research on Chemical Intermediates*, *40*(6), 2371–2379. doi:10.100711164-014-1614-9

Wu, D., Lu, J., Zhong, S., Schwarz, P., Chen, B., & Rao, J. (2019). Influence of nonionic and ionic surfactants on the antifungal and mycotoxin inhibitory efficacy of cinnamon oil nanoemulsions. *Food & Function*, *10*(5), 2817–2827. doi:10.1039/C9FO00470J PMID:31049507

Wu, J., Lin, J., & Zhong, Q. (2014). Physical and antimicrobial characteristics of thyme oil emulsified with soluble soybean polysaccharide. *Food Hydrocolloids*, *39*, 144–150. doi:10.1016/j.foodhyd.2013.12.029

Wulff-Pérez, M., Martín-Rodríguez, A., Gálvez-Ruiz, M. J., & de Vicente, J. (2013). The effect of polymeric surfactants on the rheological properties of nanoemulsions. *Colloid & Polymer Science*, *291*(3), 709–716. doi:10.100700396-012-2780-1

Wulff-Perez, M., Pavon, F. J., Martin-Rodriguez, A., De Vicente, J., Alen, F., De Fonseca, F. R., & Serrano, A. (2014). Preparation, characterisation and in vivo evaluation of nanoemulsions for the controlled delivery of the antiobesity agent N-oleoylethanolamine. *Nanomedicine (London)*, *9*(18), 2761–2772.

Wu, S. H., Hung, Y., & Mou, C. Y. (2013). Compartmentalized Hollow Silica Nanospheres Templated from Nanoemulsions. *Chemistry of Materials*, *25*(3), 352–364. doi:10.1021/cm303116u

Wu, T., Fang, X., Yang, Y., Meng, W., Yao, P., Liu, Q., Zhang, B., Liu, F., Zou, A., & Jiagao Cheng, J. (2020). Eco-friendly Water-Based λ-Cyhalothrin Polydopamine Microcapsule Suspension with High Adhesion on Leaf for Reducing Pesticides Loss. *Journal of Agricultural and Food Chemistry*, *68*(45), 12549–12557. doi:10.1021/acs.jafc.0c02245 PMID:33112140

Wu, X., & Li, Y. (2012). *Signaling Pathways in Liver Cancer*. Liver Tumors., doi:10.5772/31381

Wyres, K. L., Hawkey, J., Hetland, M. A., Fostervold, A., Wick, R. R., Judd, L. M., & Holt, K. E. (2019). Emergence and rapid global dissemination of CTX-M-15-associated *Klebsiella pneumoniae* strain ST307. *The Journal of Antimicrobial Chemotherapy*, *74*(3), 577–581. doi:10.1093/jac/dky492

Xie, G., Zhao, A., Zhao, L., Chen, T., Chen, H., Qi, X., & Jia, W. (2012). Metabolic fate of tea polyphenols in humans. *Journal of Proteome Research*, *11*(6), 3449–3457.

Xiong, Y., Li, S., Warner, R. D., & Fang, Z. (2020). Effect of oregano essential oil and resveratrol nanoemulsion loaded pectin edible coating on the preservation of pork loin in modified atmosphere packaging. *Food Control*, *114*, 107226. doi:10.1016/j.foodcont.2020.107226

Xu, D., Wang, H., Zheng, W., Gao, Y., Wang, M., Zhang, Y., & Gao, Q. (2016). Charaterization and immunomodulatory activities of polysaccharide isolated from Pleurotus eryngii. *International Journal of Biological Macromolecules*, *92*, 30–36. doi:10.1016/j.ijbiomac.2016.07.016 PMID:27392770

Xue, J., Davidson, P. M., & Zhong, Q. (2015). Antimicrobial activity of thyme oil co-nanoemulsified with sodium caseinate and lecithin. *International Journal of Food Microbiology*, *210*, 1–8. doi:10.1016/j.ijfoodmicro.2015.06.003 PMID:26082324

Xu, Q., Zhou, A., Wu, H., & Bi, Y. (2019). Development and in vivo evaluation of baicalin-loaded W/O nanoemulsion for lymphatic absorption. *Pharmaceutical Development and Technology*, *24*(9), 1155–1163. doi:10.1080/10837450.2019.1646757 PMID:31342830

Xu, T., Gao, C., Feng, X., Yang, Y., Shen, X., & Tang, X. (2019). PT SC College of Food Science and Engineering / Collaborative Innovation Center for. *International Journal of Biological Macromolecules*. Advance online publication. doi:10.1016/j.ijbiomac.2019.04.189 PMID:31047924

Xu, W., Huang, J. J., & Cheung, P. C. (2012). Extract of *Pleurotus pulmonarius* suppresses liver cancer development and progression through inhibition of VEGF-induced PI3K/AKT signalling pathway. *PLoS One*, *7*(3), e34406. doi:10.1371/journal.pone.0034406 PMID:22470568

Xu, W., Zhu, D., Xu, B., Huang, L., Xiong, Y., Ge, P., & Wang, Z. (2019). Preparation, antibacterial and antioxidant properties of green tea seed oil nanoemulsions by self-emulsification method. *Micro & Nano Letters*, *14*(12), 1219–1222. doi:10.1049/mnl.2019.0185

Xu, X., Yan, H., Chen, J., & Zhang, X. (2011). Bioactive proteins from mushrooms. *Biotechnology Advances*, *29*(6), 667–674. doi:10.1016/j.biotechadv.2011.05.003 PMID:21605654

Ya-Ali, P., Yarahmadi, F., & Mehrnia, M. A. (2020). Efficacies of Two Nano-Formulations of Tasmanian Blue Gum Essential Oil to Control Callosobruchus maculatus. *Journal of Economic Entomology*, *113*(3), 1555–1562. doi:10.1093/jee/toaa069 PMID:32307512

Yadav, S., & Gupta, S. (2015). Development and in vitro characterization of docetaxel-loaded ligand appended solid fat nanoemulsions for potential use in breast cancer therapy. *Artificial Cells, Nanomedicine, and Biotechnology*, *43*(2), 93–102.

Yamamoto-Ribeiro, M. M., Grespan, R., Kohiyama, C. Y., Ferreira, F. D., Mossini, S. A., Silva, E. L., Filho, B. A., Mikcha, J. M., & Machinski, M. Jr. (2013). Effect of Zingiber officinale essential oil on Fusarium verticillioides and fumonisin production. *Food Chemistry*, *141*(3), 3147–3152. https://doi.org/10.1016/j.foodchem.2013.05.144

Yang, X., Tian, H., Ho, C. T., & Huang, Q. (2011). Inhibition of citral degradation by oilin-water nanoemulsions combined with antioxidants. *Journal of Agricultural and Food Chemistry*, *59*(11), 6113–6119. doi:10.1021/jf2012375 PMID:21517071

Yang, Y. M., Yang, Y., Dai, W. W., Li, X. M., Ma, J. Q., & Tang, L. P. (2016). Genistein-induced apoptosis is mediated by endoplasmic reticulum stress in cervical cancer cells. *European Review for Medical and Pharmacological Sciences*, *20*(15), 3292–3296. PMID:27467006

Yang, Y., Zhou, Z. T., & Ge, J. P. (2006). Effect of genistein on DMBA-induced oral carcinogenesis in hamster. *Carcinogenesis*, *27*(3), 578–583. doi:10.1093/carcin/bgi234 PMID:16195236

Yan, H., Bao, C., Chen, X., Yu, C., Kong, D., Shi, J., & Lin, Q. (2019). Preparation of biodiesel oil-in-water nanoemulsions by mixed surfactants for bifenthrin formulation. *RSC Advances*, *9*(21), 11649–11658. doi:10.1039/C9RA00591A

Yashpal, S., Tanuj, H., & Harsh, K. (2013). Nanoemulsions: A pharmaceutical review. *Int. J. Pharma. Prof. Res*, *4*(2), 928–935.

Yayan, J., Ghebremedhin, B., & Rasche, K. (2015). Antibiotic resistance of Pseudomonas aeruginosa in pneumonia at a single university hospital center in Germany over a 10-year period. *PLoS One*, *10*(10), e0139836. doi:10.1371/journal.pone.0139836

Yazgan, H. (2020). Investigation of antimicrobial properties of sage essential oil and its nanoemulsion as antimicrobial agent. *Lebensmittel-Wissenschaft + Technologie*, *130*, 109669. doi:10.1016/j.lwt.2020.109669

Yazgan, H., Ozogul, Y., & Kuley, E. (2019). Antimicrobial influence of nanoemulsified lemon essential oil and pure lemon essential oil on food-borne pathogens and fish spoilage bacteria. *International Journal of Food Microbiology*, *306*, 108266. https://doi.org/10.1016/j.ijfoodmicro.2019.108266

Ye, A., Zhu, X., & Singh, H. (2013). Oil-in-water emulsion system stabilized by protein-coated nanoemulsion droplets. *Langmuir*, *29*(47), 14403–14410. doi:10.1021/la403493y PMID:24175702

Yen, M. T., & Mau, J. L. (2006). Preparation of fungal chitin and chitosan from shiitake stipes. *Fungal Science*, *21*, 1–11.

Yen, M. T., & Mau, J. L. (2007). Physico-chemical characterization of fungal chitosan from shiitake stipes. *Lebensmittel-Wissenschaft + Technologie*, *40*(3), 472–479. doi:10.1016/j.lwt.2006.01.002

Yildirim, S. T., Oztop, M. H., & Soyer, Y. (2017). Cinnamon oil nanoemulsions by spontaneous emulsification: Formulation, characterization and antimicrobial activity. *Lebensmittel-Wissenschaft + Technologie*, *84*, 122–128. https://doi.org/10.1016/j.lwt.2017.05.041

Yin, L.-J., Chu, B.-S., Kobayashi, I., & Nakajima, M. (2009). Performance of selected emulsifiers and their combinations in the preparation of β-carotene nanodispersions. *Food Hydrocolloids*, *23*(6), 1617–1622. doi:10.1016/j.foodhyd.2008.12.005

Yi, S. W., Yune, T. Y., Kim, T. W., Chung, H., Choi, Y. W., Kwon, I. C., Lee, E. B., & Jeong, S. Y. (2000). A cationic lipid emulsion/DNA complex as a physically stable and serum-resistant gene delivery system. *Pharmaceutical Research*, *17*(3), 314–320. doi:10.1023/A:1007553106681 PMID:10801220

Yoneyama, M., & Fujita, T. (2009). RNA recognition and signal transduction by RIG-I-like receptors. *Immunological Reviews*, *227*(1), 54–65. doi:10.1111/j.1600-065X.2008.00727.x PMID:19120475

Yoo, M. K., Kang, S. K., Choi, J. H., Park, I. K., Na, H. S., Lee, H. C., & Cho, C. S. (2010). Targeted delivery of chitosan nanoparticles to Peyer's patch using M cell-homing peptide selected by phage display technique. *Biomaterials*, *31*(30), 7738–7747. doi:10.1016/j.biomaterials.2010.06.059 PMID:20656343

Yoplac, I., Vargas, L., Robert, P., & Hidalgo, A. (2021). Characterization and antimicrobial activity of microencapsulated citral with dextrin by spray drying. *Heliyon*, *7*(4), e06737. doi:10.1016/j.heliyon.2021.e06737 PMID:33898839

Yousefian Rad, E., Homayouni Tabrizi, M., Ardalan, P., Seyedi, S. M. R., Yadamani, S., Zamani-Esmati, P., & Haghani Sereshkeh, N. (2020). Citrus Lemon essential oil nanoemulsion (CLEO-NE), a safe cell-depended apoptosis inducer in human A549 lung cancer cells with anti-angiogenic activity. *Journal of Microencapsulation*, 1–24.

Yousefian Rad, E., Homayouni Tabrizi, M., Ardalan, P., Seyedi, S. M. R., Yadamani, S., Zamani-Esmati, P., & Haghani Sereshkeh, N. (2020). Citrus lemon essential oil nanoemulsion (CLEO-NE), a safe cell-depended apoptosis inducer in human A549 lung cancer cells with anti-angiogenic activity. *Journal of Microencapsulation*, *37*(5), 394–402. doi:10.1080/02652048.2020.1767223 PMID:32400238

Yuan, Y., Gao, Y., Zhao, J., & Mao, L. (2008). Characterization and stability evaluation of β-carotene nanoemulsions prepared by high pressure homogenization under various emulsifying conditions. *Food Research International*, *41*(1), 61–68. doi:10.1016/j.foodres.2007.09.006

Yu, H., & Huang, Q. (2012). Improving the oral bioavailability of curcumin using novel organogel-based nanoemulsion. *Journal of Agricultural and Food Chemistry*, *60*(21), 5373–5379.

Yu, H., Xu, Z., Chen, X., Xu, L., Yin, Q., Zhang, Z., & Li, Y. (2014). Reversal of lung cancer multidrug resistance by pH-responsive micelleplexes mediating co-delivery of siRNA and paclitaxel. *Macromolecular Bioscience*, *14*(1), 100–109.

Yu, L., Li, C., Xu, J., Hao, J., & Sun, D. (2012). Highly Stable Concentrated Nanoemulsions by the Phase Inversion Composition Method at Elevated Temperature. *Langmuir*, *28*(41), 14547–14552. doi:10.1021/la302995a PMID:22985401

Yuliani, S., & Noveriza, R. (2019). Effect of Carrier Oil and Co-Solvent on the Formation of Clove Oil Nanoemulsion by Phase Inversion Technique. *IOP Conference Series. Earth and Environmental Science*, *309*(1), 012036. doi:10.1088/1755-1315/309/1/012036

Yu, M. K., Park, J., & Jon, S. (2012). Targeting strategies for multifunctional nanoparticles in cancer imaging and therapy. *Theranostics*, *2*(1), 3–44. https://doi.org/10.7150/thno.3463

Yun Zhang, P. G. W. (2003). Studying the interaction of gel carbohydrate antigen and protein by quartz-crystal microbalance. *Journal of the American Chemical Society*, *125*(31), 9292–9293. doi:10.1021/ja035350a PMID:12889948

Yu, S.-H., Chen, S.-Y. T., Li, W.-S., Dubey, N. K., Chen, W.-H., Chuu, J.-J., Leu, S.-J., & Deng, W.-P. (2015). Hypoglycemic activity through a novel combination of fruiting body and mycelia of *Cordyceps militaris* in high-fat diet-induced type 2 diabetes mellitus mice. *Journal of Diabetes Research*. PMID:26258146

Yu, W., & Zhang, N. (2009). Surface Modification of Nanocarriers for Cancer Therapy. *Current Nanoscience*, *5*(2), 123–134.

Zainuddin, N. J., Ashari, S. E., Salim, N., Asib, N., Omar, D., & Lian, G. E. C. (2019). Optimization and characterization of palm oil-based nanoemulsion loaded with parthenium hysterophorus crude extract for natural herbicide formulation. *Journal of Oleo Science*, *68*(8), 747–757. doi:10.5650/jos.ess18209 PMID:31292338

Zakaria, M. K. (2015). Hepatic Cancer Stem Cells and Signaling Pathways. *International Journal of Cancer and Clinical Research*, *2*(2). Advance online publication. doi:10.23937/2378-3419/2/2/1017

Zambrano-Zaragoza, M. L., Mercado-Silva, E., Del Real, L. A., Gutierrez-Cortez, E., Cornejo-Villegas, M. A., & Quintanar-Guerrero, D. (2014). The effect of nano-coatings with a-tocopherol and xanthan gum on shelf-life and browning index of fresh-cut "Red Delicious" apples. *Innovative Food Science & Emerging Technologies*, *22*, 188–196. doi:10.1016/j.ifset.2013.09.008

Zanesco-Fontes, I., Silva, A., da Silva, P. B., Duarte, J. L., Di Filippo, L. D., Chorilli, M., Cominetti, M. R., & Martin, A. (2021). [10]-Gingerol-Loaded Nanoemulsion and its Biological Effects on Triple-Negative Breast Cancer Cells. *AAPS PharmSciTech*, *22*(5), 157.

Zarrilli, R., Pournaras, S., Giannouli, M., & Tsakris, A. (2013). Global evolution of multidrug-resistant Acinetobacterbaumannii clonal lineages. *International Journal of Antimicrobial Agents*, *41*(1), 11–19. doi:10.1016/j.ijantimicag.2012.09.008

Zawistowski, J., Biliaderis, C. G., & Eskin, N. A. M. (1991). *Polyphenol oxidase. In Oxidative enzymes in foods*. Elsevier.

Ždralević, M., Vučetić, M., Daher, B., Marchiq, I., Parks, S. K., & Pouysségur, J. (2018). Disrupting the 'Warburg effect' re-routes cancer cells to OXPHOS offering a vulnerability point via 'ferroptosis'-induced cell death. *Advances in Biological Regulation*, *68*(December), 55–63. doi:10.1016/j.jbior.2017.12.002 PMID:29306548

Zdrali, E., Etienne, G., Smolentsev, N., Amstad, E., & Roke, S. (2019). The interfacial structure of nano- and micron-sized oil and water droplets stabilized with SDS and Span80. *The Journal of Chemical Physics*, *150*(20), 204704. doi:10.1063/1.5083844 PMID:31153210

Zelnak, A. B., & O'Regan, R. M. (2015). Optimizing Endocrine Therapy for Breast Cancer. *Journal of the National Comprehensive Cancer Network: JNCCN*, *13*(8), e56–e64.

Zeng, L., Xin, X., & Zhang, Y. (2017). Development and characterization of promising Cremophor EL stabilized o/w nanoemulsions containing short-chain alcohols as a cosurfactant. *RSC Advances*, *7*(32), 19815–19827. doi:10.1039/C6RA27096D

Zhang, C., Guanming, S., Shen, J., & Hu, R. F. (2015). Productivity effect and overuse of pesticide in crop production in China. *Journal of Integrative Agriculture*, *14*(9), 1903–1910. doi:10.1016/S2095-3119(15)61056-5

Zhang, J., Kinoh, H., Hespel, L., Liu, X., Quader, S., Martin, J., Chida, T., Cabral, H., & Kataoka, K. (2017). Effective treatment of drug resistant recurrent breast tumors harboring cancer stem-like cells by staurosporine/epirubicin co-loaded polymeric micelles. *Journal of Controlled Release*, *264*, 127–135.

Zhang, J., Wang, L., You, X., Xian, T., Wu, J., & Pang, J. (2019). Nanoparticle therapy for prostate cancer: Overview and perspectives. *Current Topics in Medicinal Chemistry*, *19*(1), 57–73. doi:10.2174/1568026619666190125145836 PMID:30686255

Zhang, L. W., Al-Suwayeh, S. A., Hung, C. F., Chen, C. C., & Fang, J. Y. (2011). Oil components modulate the skin delivery of 5-aminolevulinic acid and its ester pro drug from oil-in-water and water-in-oil nanoemulsions. *International Journal of Nanomedicine*, *6*, 693–704.

Zhang, L. W., Al-Suwayeh, S. A., Hung, C. F., Chen, C. C., & Fang, J. Y. (2011). Oil components modulate the skin delivery of 5-aminolevulinic acid and its ester prodrug from oil-in-water and water-in-oil nanoemulsions. *International Journal of Nanomedicine*, *6*, 693–704. doi:10.2147/IJN.S17524 PMID:21556344

Zhang, P., Kopechek, J. A., & Porter, T. M. (2013). The impact of vaporized nanoemulsions on ultrasound-mediated ablation. *Journal of Therapeutic Ultrasound*, *1*(1), 2. doi:10.1186/2050-5736-1-2 PMID:24761223

Zhang, R., Tian, G., Zhao, Y., Zhao, L., Wang, H., Gong, Z., & Ng, T. B. (2015). A novel ribonuclease with HIV-1 reverse transcriptase inhibitory activity purified from the fungus *Ramaria Formosa. Journal of Basic Microbiology*, *55*(2), 269–275. doi:10.1002/jobm.201300876 PMID:24523258

Zhang, R., Zhang, Z., Zou, L., Xiao, H., Zhang, G., Decker, E. A., & McClements, D. J. (2016). Enhancement of carotenoid bioaccessibility from carrots using excipient emulsions: Influence of particle size of digestible lipid droplets. *Food & Function*, *7*(1), 93–103. doi:10.1039/C5FO01172H PMID:26583923

Zhang, S., Chen, J., Cui, P., Shi, W., Zhang, W., & Zhang, Y. (2015). Identification of novel mutations associated with clofazimine resistance in *Mycobacterium tuberculosis. The Journal of Antimicrobial Chemotherapy*, *70*(9), 2507–2510. doi:10.1093/jac/dkv150 PMID:26045528

Zhang, X., Liu, L., Zhang, Y., Dai, G., Huang, H., & Jin, Q. (2015). Genetic determinants involved in *p*-aminosalicylic acid resistance in clinical isolates from tuberculosis patients in northern China from 2006 to 2012. *Antimicrobial Agents and Chemotherapy*, *59*(2), 1320–1324. doi:10.1128/AAC.03695-14 PMID:25421465

Zhang, Y., Liao, Y., Tang, Q., Lin, J., & Huang, P. (2021). Biomimetic Nanoemulsion for Synergistic Photodynamic-Immunotherapy Against Hypoxic Breast Tumor. *Angewandte Chemie*, *60*(19), 10647–10653.

Zhang, Y., Lv, C., Dong, Y., & Yang, Q. (2020). Aspirin-targeted PD-L1 in lung cancer growth inhibition. *Thoracic Cancer, 11*(6), 1587–1593. doi:10.1111/1759-7714.13433 PMID:32297484

Zhang, Y., Shang, Z., Gao, C., Du, M., Xu, S., Song, H., & Liu, T. (2014). Nanoemulsion for solubilization, stabilization, and in vitro release of pterostilbene for oral delivery. *AAPS PharmSciTech, 15*(4), 1000–1008. doi:10.120812249-014-0129-4 PMID:24831090

Zhang, Y., Wade, M. M., Scorpio, A., Zhang, H., & Sun, Z. (2003). Mode of action of pyrazinamide: Disruption of *Mycobacterium tuberculosis* membrane transport and energetics by pyrazinoic acid. *The Journal of Antimicrobial Chemotherapy, 52*(5), 790–795. doi:10.1093/jac/dkg446 PMID:14563891

Zhang, Y., Xin, C., Cheng, C., & Wang, Z. (2020). Antitumor activity of nanoemulsion based on essential oil of *Pinus koraiensis* pinecones in MGC-803 tumor-bearing nude mice. *Arabian Journal of Chemistry, 13*(11), 8226–8238. doi:10.1016/j.arabjc.2020.09.058

Zhang, Z., & McClements, D. J. (2018). Overview of nanoemulsion properties: stability, rheology, and appearance. In *Nanoemulsions* (pp. 21–49). Academic Press. org/ doi:10.1016/B978-0-12-811838-2.00002-3

Zhang, Z., Song, H., Peng, Z., Luo, Q., Ming, J., & Zhao, G. (2012). Characterization of stipe and cap powders of mushroom (*Lentinus edodes*) prepared by different grinding methods. *Journal of Food Engineering, 109*(3), 406–413. doi:10.1016/j.jfoodeng.2011.11.007

Zhang, Z., Vriesekoop, F., Yuan, Q., & Liang, H. (2014). Effects of nisin on the antimicrobial activity of d-limonene and its nanoemulsion. *Food Chemistry, 150*, 307–312. doi:10.1016/j.foodchem.2013.10.160 PMID:24360455

Zhao, A., Lee, J., & Emelianov, S. (2021). Formulation and Acoustic Modulation of Optically Vaporized Perfluorocarbon Nanodroplets. *Journal of Visualized Experiments, 62814*(173). doi:10.3791/62814 PMID:34338676

Zhao, F., Wang, X. D., Erber, L. N., Luo, M., Guo, A. Z., Yang, S. S., Gu, J., Turman, B. J., Gao, Y. R., Li, D. F., Cui, Z. Q., Zhang, Z. P., Bi, L. J., Baughn, A. D., Zhang, X. E., & Deng, J. Y. (2014). Binding pocket alterations in dihydrofolate synthase confer resistance to para-aminosalicylic acid in clinical isolates of *Mycobacterium tuberculosis. Antimicrobial Agents and Chemotherapy, 58*(3), 1479–1487. doi:10.1128/AAC.01775-13 PMID:24366731

Zhao, H., Lu, Gong, T., & zhang. (2013). Nanoemulsion loaded with lycobetaine–oleic acid ionic complex: Physicochemical characteristics, in vitro, in vivo evaluation, and antitumor activity. *International Journal of Nanomedicine, 8*, 1959–1973. doi:10.2147/IJN.S43892 PMID:23723698

Zhao, S., Rong, C.-B., Kong, C., Liu, Y., Xu, F., Miao, Q.-J., Wang, S.-X., Wang, H.-X., & Zhang, G.-Q. (2014). A novel laccase with potentantil proliferative and HIV-1 reverse transcriptase inhibitory activities from mycelia of mushroom *Coprinus comatus. BioMed Research International, 2014*, 1–8. doi:10.1155/2014/417461

Zhao, X., Zhu, Y., Zhang, C., Lei, J., Ma, Y., & Du, F. (2017). Positive charge pesticide nanoemulsions prepared by the phase inversion composition method with ionic liquids. *RSC Advances, 7*(77), 48586–48596. doi:10.1039/C7RA08653A

Zhao, Y., Moddaresi, M., Jones, S. A., & Brown, M. B. (2009). A dynamic topical hydrofluoroalkane foam to induce nanoparticle modification and drug release in situ. *European Journal of Pharmaceutics and Biopharmaceutics, 72*(3), 521–528. doi:10.1016/j.ejpb.2009.03.002 PMID:19303438

Zheng, M., Falkeborg, M., Zheng, Y., Yang, T., & Xu, X. (2013). Formulation an characterization of nanostructured lipid carriers containing a mixed lipids core. *Colloids and Surfaces. A, Physicochemical and Engineering Aspects, 430*, 76–84. doi:10.1016/j.colsurfa.2013.03.070

Zhong, J., Huang, P., Wen, M., Liang, L., Zhang, X., Tan, S., & Zhu, X. (2012). Antigenic epitope peptides of influenza H3N2 virus neuraminidase gene based on experiments. *Chinese Science Bulletin*, *57*(22), 2908–2913. doi:10.100711434-012-5266-0

Zhong, L., Li, Y., Xiong, L., Wang, W., Wu, M., Yuan, T., Yang, W., Tian, C., Miao, Z., Wang, T., & Yang, S. (2021). Small molecules in targeted cancer therapy: Advances, challenges, and future perspectives. *Signal Transduction and Targeted Therapy*, *6*(1), 1–48. doi:10.103841392-021-00572-w PMID:34054126

Zhou, J., Mulshine, J. L., Unsworth, E. J., Scott, F. M., Avis, I. M., Vos, M. D., & Treston, A. M. (1996). Purification and characterization of a protein that permits early detection of lung cancer: Identification of heterogeneous nuclear ribonucleoprotein-A2/B1 as the antigen for monoclonal antibody 703D4. *The Journal of Biological Chemistry*, *271*(18), 10760–10766. doi:10.1074/jbc.271.18.10760 PMID:8631886

Zhu, Y., Gao, H., Liu, W., Zou, L., & McClements, D. J. (2020). A review of the rheological properties of dilute and concentrated food emulsions. *Journal of Texture Studies*, *51*(1), 45–55. doi:10.1111/jtxs.12444 PMID:31107991

Ziani, K., Chang, Y., McLandsborough, L., & McClements, D. J. (2011). Influence of surfactant charge on antimicrobial efficacy of surfactant-stabilized thyme oil nanoemulsions. *Journal of Agricultural and Food Chemistry*, *59*(11), 6247–6255. https://doi.org/10.1021/jf200450m

Ziani, K., Fang, Y., & McClements, D. J. (2012). Encapsulation of functional lipophilic components in surfactant-based colloidal delivery systems: Vitamin E, vitamin D, and lemon oil. *Food Chemistry*, *134*(2), 1106–1112. doi:10.1016/j.foodchem.2012.03.027 PMID:23107734

Zou, M., Toivanen, R., Mitrofanova, A., Floch, N., Hayati, S., Sun, Y., Le Magnen, C., Chester, D., Mostaghel, E. A., Califano, A., Rubin, M. A., Shen, M. M., & Abate-Shen, C. (2017). Transdifferentiation as a mechanism of treatment resistance in a mouse model of castration-resistant prostate cancer. *Cancer Discovery*, *7*(7), 736–749. doi:10.1158/2159-8290.CD-16-1174 PMID:28411207

About the Contributors

Karthikeyan Ramalingam is Dean, Student Affairs and Associate professor, BSA Crescent Institute of Science and Technology with over 20 years of laboratory knowledge, including academic, government, and industrial projects, handling experience with excellent supervisory skills and a strong record of scientific accomplishments and publications. In nanoemulsion sciences experienced from the University of Texas Health Science Center at San Antonio, Texas, USA, US Naval Medical Research Unit, San Antonio, Texas, USA, New York University, USA, Nottingham University, UK and received funds for nanoemulsions research from Department of Defense, Navy, USA ($105,900), DST-SERB (Rs. 23 lakhs) and ICMR (15 lakhs) and USA-NSF-SBIR ($475,766) as principal investigator.

* * *

Nitin Agrawal is accomplished teacher & researcher in the field of Pharmaceutical Sciences. He has more than eleven years of teaching as well as research experience. He authored more than 15 research articles in several national and international journals, one book chapter, delivered invited lecture at RIPANS, Mizoram. He also attended UGC Sponsored workshop and given presentation in more than 15 national & international conferences in India and abroad. He has completed his B. Pharm in 2005, M. Pharm in 2008 in Pharmacognosy & phytochemistry. He has awarded Ph.D from Jadavpur University, Kolkata in 2015. He got Junior Research Fellowship in DBT funded project. Presently working as Head, Faculty of Pharmacy, Raja Balwant Singh Engineering Technical Campus, Bichpuri, Agra.

Baila Ahmad is currently serving as Associate Lecturer at the Islamia University of Bahawalpur, Pakistan.

Dharmeswar Barhoi is working as a Ph.D. research scholar at the Department of Life Science and Bioinformatics, Assam University, Silchar. His research work is mainly screening of medicinal plants against cancer development. Mr. Barhoi is working under the supervision of Prof. Sarbani Giri, Head & Dean, Department of Life Science and Bioinformatics, Assam University, Silchar.

Nagendra Chandrawanshi is working as an Assistant Professor at School of Studies in Biotechnology, Pt. Ravishankar Shukla University, Raipur (Chhattisgarh) India. He has done his graduation from Govt. Digvijay Postgraduate College, Rajnandgaon (Chhattisgarh), and done his post-graduation and Ph.D. from School of Studies in Biotechnology, Pt. Ravishankar Shukla University, Raipur. Dr. Nagendra Kumar Chandrawanshi has vast experience in academic, research, and administrative work. Dr. Nagendra Kumar

Chandrawanshi enthusiastically involved in research in the field of mushroom and medicinal plants for the Nutraceuticals, Drug discovery, Mechanistic study of Anticancerous, Antimicrobials, Herbal immune modulators, and Drug delivery models, etc. Presently he is working on part of a few funded research projects. Dr. Chandrawashi is also dealt with in the post-graduate curriculum of Biotechnology, such as Genetics, Molecular Biology, and Immunology & Animal Biotechnology in M.Sc. theory as well as a laboratory in regularly. Dr. Chandrawanshi published his research work in many standard national and international journals and also presented his work in different national and international conferences and seminars. He also participated and presented his research paper in international conferences in aboard. Dr. Chandrawanshi is a member of many editorial boards of national and international Journals, and continues performing the reviewer task. Dr. Chandrawanshi is a life member of Professional bodies like The Indian Science Congress Association, Kolkata, and The Indian Aerobiological Society (Regd.), India. Presently, he has a secretary of the Alumni Association of Biotechnology, Amanaka Raipur (registered by CG Govt.). Moreover, he is members of the Board of Studies in Biotechnology, Pt. RSU, Raipur, and Board of Central Studies in Biotechnology, Higher Education Department, CG, India.

Sindhu Priya Dhas obtained her Ph.D. in Nanobiotechnology from VIT University, Vellore, India. She is working as an Assistant Professor at Jaya College of Arts and Science, Chennai, India. She has a teaching experience of 3 years. She has also been associated with various research works. Her main areas of research are Nanofabrication, Nanoemulsions, and Cancer treatment.

Archna Dhasmana is an Assistant Professor, Department of Biotechnology, School of Applied and Life Sciences, Uttaranchal University Dehradun.

Elaine Elaine embarked on her postgraduate studies in Doctorate of Philosophy (Science) at UCSI University in 2021. Her research interest is mainly on encapsulation technologies with nanoemulsion delivery systems. She started to be passionate about scientific research when she was in her final year of undergraduate studies. She holds a bachelor's degree in Food Science with Nutrition in her current institution. By improving scientific writing, lab skills, and human relations, she believed that she can contribute to the research society, institutions with her projects.

Omotayo Fagbuaro was born over fifty years ago in Ado-Ekiti. He had his Elementary School in the old Ondo State Nigeria. His Secondary School studies was in MayHill Comprehensive High School, Ado-Ekiti, Nigeria. He obtained his first degree from University of Ado-Ekiti (Now Ekiti State University, Ado-Ekiti) in 1992. He bagged his M.Sc certificate in Cell Biology and Genetics from University of Ibadan in 1996. He had his Ph. D certificate in fish breeding and Genetics from Ekiti State University Ado-Ekiti in 2006. He started Lecturing in Ekiti State University Ado-Ekiti in 1995 as a graduate Assistant. He rose through the ranks to become Professor of Zoology (fish breeding and Genetics) in 2016. He has so many research articles to his credit. He has served as External Examiner to some Universities among which are Obafemi Awolowo University, Ile- Ife, Nigeria, The Federal University of Technology, Akure, Nigeria.

Sarbani Giri is the Principle investigator of Laboratory of Molecular and Cell Biology with more than 25 years of teaching experience and currently serving as Head of the Department of Life Science and Bioinformatics and Dean of the School of Life Sciences, Assam University, Silchar. She is the recipient of DBT Long term Overseas Associateship and joined the Laboratory of Metabolism, National Cancer Institute, National Institute of Health, Maryland USA (2005- 2006). She was awarded with the Indo-US Science and Technology Forum fellowship (2008). She joined Department of Molecular and Cell Biology, University of California, Berkeley, USA, as long term visiting scholar. Her research focus is mainly Life style factors and its impact on human health, xenobiotic / drug metabolism, role of tight junction integrity in oral cancer, downstream signalling pathways involved and environmental health assessment.

Victor Hmingthansanga is currently pursuing PhD in the Department of Pharmaceutics at National Institute of Pharmaceutical Education and Research (NIPER-Kolkata) and have experienced on topical application of nanoemulsion. Further would like to explore this versatile delivery system for oral application.

Ihsana Banu Ishthiaq is a second-year student of biotechnology at B.S. Abdur Rahman Crescent Institute of Science and Technology. She received her bachelor's degree in biotechnology from Kalasalingam Academy of Research and Education. Her research focuses specifically on autophagy in cancer with special focus on fasting, starvation induced autophagy and nanoemulsions in drug delivery.

Muhammad Kaleem is a Ph.D in Food Technology from Arid Agriculture University, Rawalpindi, Punjab, Pakistan working in Research and Development wing of Punjab Food Authority.

Mohd Hashim Khan completed his B. Sc. (Hons) Chemistry, M. Sc. (Microbiology) and PDTC (Tissue Culture) from Aligarh Muslim University, Aligarh- UP, India. Ph. D. (Microbiology) from B.S. Abdur Rahman Crescent Institute of Science and Technology, Vandalur, Chennai- Tamil Nadu- India. He is passionate towards indulging in research work with the experience of guiding more than 30 students including B.Tech., M.Tech. and B.Sc., M.Sc. final projects in various fields of Nanotechnology, Microbiology, Biotechnology and Environmental Science. He presented his research work in national and International Conferences around the country (India) and got First and best research awards in various Universities including: 1- He got First prize for the presentation of research work in National conference on "WONDERS OF THE SMALL 2.0 A National Conference on Recent Trends in Microbiome Research Exploring the Microbial Diversity" for presenting a poster entitled as "Synthesis of Bio-Based AgNPs for the Treatment of Drug Resistant ESKAPE Pathogens", on 20 & 21 March, 2019, Organized by Department of Microbiology Pondicherry University, Puducherry. 2- He got First prize for Best Research Paper Presentation in National conference on "Impact of Biotechnology in Improving Human Health" for presenting a poster entitled as "Anti-biofilm potential of biogenic silver nanoparticles against multi-drug resistant ESKAPE pathogens " on 27 September , 2019, Organized by Department of Biotechnology , SRM Arts and Science College, SRM University, Kattankulathur, Chennai, Tamil Nadu. India. 3- He got Best Research Scholar award-SLS in "International Conference on Recent trends in Bionanotechnology, Energy and Environment(ICBEE 2020)" held on February 18 & 19, 2020, Organized by School of Life Sciences, B.S. Abdur Rahman Crescent Institute of Science and Technology, Chennai, India. He also selected for the Senior Research Fellowship (SRF) from Indian Council of Medical Research (ICMR) New Delhi, for the proposed project entitled as "Synergistic effect of double nano materials (nanoemulsions and nanoparticles) against multi-drug resistant ESKAPE pathogens" and got fellowship for the tenure of two years (2019-2021). File No : AMR/Fellowship/6/2019/ECD-II. He has successfully completed the project entitled as "Synthesis of antimicrobial nanoemulsions for the treatment of multi-drug resistant

ESKAPE pathogens" under the supervision of Dr. Karthikeyan Ramalingam (Dean- Student Affairs) sponsored by DST-SERB, Govt. of India. Currently, he is working on formulations of Dual nanomaterials to combat drug resistance in human ESKAPE pathogens.

Uma Kizhuveetil has done her PhD. in Biotechnology from the Indian Institute of Technology Madras. The focus of her research was on redox alterations induced by various drugs on cancer cells. Currently, she is working as an Assistant Professor in the Department of Food Biotechnology at Jayoti Vidyapeeth Women's University Jaipur.

Deepali Koreti is currently working as a junior research fellow at the School of Studies in Biotechnology, Pandit Ravishankar Shukla University, Raipur, under the guidance of Dr. Nagendra Kumar Chandrawanshi. Her research work is going on edible mushrooms and their cultivation technology. She is also working on mushroom metabolites and the characterization of bioactive properties such as antioxidant and antibacterial.

Anjali Kosre received her M.Phil degree in Biotechnology. She is currently pursuing her Ph.D degree under assistant professor Dr.Nagendra Kumar Chandrawanshi in Pandit Ravishankar Shukla University Raipur, Chhattisgarh. Her topic of research is in edible mushroom.

Ashish Kumar is currently working as an Assistant Professor & Head in the Department of Biotechnology, Sant Gahira Guru University Sarguja, Ambikapur Chhattisgarh. He has received his PhD degree in Biotechnology with an emphasis on Plant Molecular Genetics from Pt. Ravishankar Shukla University, Raipur, Chhattisgarh, India. He has completed his post-doctorate from the Institute of Plant Sciences, the Volcani Center, Ministry of Agriculture and Rural Development Govt. of Israel. He is awarded a doctoral and post-doctoral fellowship from University Grants Commission, New Delhi. He has received accolades for excellent professional achievements such as the Vice Chancellor's Best Employee Award and the Senior Scientist Award by the Microbiologist Society of India among many others. Dr Ashish has worked as Senior Research Fellow at Indian Institute of Vegetable Research (IIVR), Varanasi, and as Research Associate at Central Research Institute for Dryland Agriculture (CRIDA), Hyderabad. Dr. Ashish has published several book chapters, reviews, and research papers in reputed national and international journals. He has completed one research project funded by the Chhattisgarh Council of Science & Technology, Raipur, India. He has reviewed research proposals for several funding agencies such as C-COST, DBT and DST, New Delhi India. Dr Ashish has more than eight years of teaching and fifteen years of research experience in various reputed national and international research institutes. Dr Ashish has successfully mentored more than twenty-five postgraduate students during their final year dissertation project with publications in peer-reviewed journals. His field of research interests includes plant-microbe interaction, symbiotic fungi for the management of biotic and abiotic stresses in crop plants.

Pramod Mahish is currently serving as Assistant Professor of Biotechnology in Govt. Digvijay Autonomous PG College Rajnandgaon Chhattisgarh India. He had completed his M.Sc. in Medical Biotechnology from Pt. JNM Medical College Raipur and M.Phil. and Ph.D. from Pt. Ravishankar Shukla University Raipur. He has 09 Yr of teaching and 12 Yr of research experience. His area of research is bioprocessing. He is affiliated as Research Supervisor in Hemchand Yadav Vishwavidyalaya Durg and currently guiding 02 Ph.D. Scholars. He is also serving as Chairman Board of Studies Biotechnology at same University. He had published 17 research papers, 02 Books, 06 Book chapters, and participated in more than 25 national and international seminars. He had completed 02 Research projects funded by Government agencies. He is a fellow member of World Researchers Association and Life member of Indian Aerobiology Society. He is also serving as Assistant Editor of "Research Fronts" a peer review Journal (ISBN 2250-2653).

Sivakumar Manickam is a Chemical Engineer specializing in the areas of process engineering of nanomaterials, especially nanopharmaceuticals. Currently, he is attached with the Petroleum and Chemical Engineering Department, UTB. Previously he was working with the University of Nottingham, Malaysia Campus. His research focuses on the process development of cavitation-based reactors towards technologically important nanomaterials, greener extraction of natural products, water treatment, development of pharmaceutical nanoemulsions and utilizing novel carbon nanomaterials to design biosensor for the earlier detection of cancer and diabetes. He took various leadership roles at the University of Nottingham. He was the founding director of the Centre for Nanotechnology and Advanced Materials (CENTAM) and Head of Manufacturing and Industrial Processes Research Division, and Director of Research. Later he was the Associate Dean for Research and Knowledge Exchange for the Faculty of Science and Engineering at Nottingham Univ. He was also the recipient of the prestigious JSPS fellowship, Japan. He has published ~250 peer-reviewed journal and conference papers. His h-index is 46 (scholar google). He is also the Executive Editor of Ultrasonics Sonochemistry (Elsevier, Q1, IF 6.5) Journal. He is a board member of the Asia Oceania Sonochemical Society (AOSS). He is the Fellow of Higher Education Academy (UK) and Fellow of the Royal Society of Chemistry (FRSC).

Tahir Mehmood is serving as Assistant Professor at Islamia University of Bahawalpur, Pakistan.

Viktoria Milkova is an Assoc. Prof. in Department of Interfaces and Colloids at the Institute of Physical Chemistry, Bulgarian Academy of Sciences. Her expertise is the field of design and characterization of the surface properties and stability of nanoparticles or advanced chitosan-based nanostructures for encapsulation of drugs or (bio)active substances by using mainly the classical theories of colloid science and electrokinetic phenomena in dispersion systems.

Amitava Mukherjee is a Senior Professor and Director at Centre for Nanobiotechnology, Vellore Institute of Technology (VIT), Vellore, India. He has 370 publications in peer-reviewed journals (citation index 13300; h-index 57; i 10- index 228). He has four granted and ten filed patents in the areas relating to Environmental Technology. Prof. Mukherjee has so far received sixteen funded projects as Principal Investigator from several federal agencies in India. He has been admitted as the Fellow of Royal Society of Chemistry and Royal Society of Biology, UK in 2016. He currently serves as Associate Editor of Frontiers in Nanotechnology, Academic Editor of Plos One, and is the Member of Editorial board in the following journals-Environmental Pollution, Science of the Total Environment, Toxicology Reports, and Proceedings of the National Academy of Sciences (Biological Sciences).

Yogesh Murti is presently working as an Assistant Professor, (Pharmaceutical Chemistry) at the Institute of Pharmaceutical Research, GLA University, Mathura (UP). He has published 30 Research and 30 Review Articles in International and National repute journals. He has co-supervised 10 M Pharm students. He is also life members of many professional bodies like IPA, APTI, IPGA, ISTE, CSI, etc. and

Nominee of IAEC nominated by CPCSEA. He has received an award as "Most Promising Teacher in Pharma Education" in Golden Global Health & Education Awards in the year 2019, as "Star Performer" in AICTE sponsored Two Weeks Online Faculty Development Program (FDP) in the year 2020. Recently he was awarded as "Outstanding Oration Award (Uttar Pradesh)" in the One Day Virtual 13th INDO-US Conference organized by Association of Pharmacy Professionals in the year 2021. His area of expertise is Phytochemistry and Synthetic chemistry.

Subramanian Natesan, Professor, Department of Pharmaceutics, National Institute of Pharmaceutical Education Research – Kolkata, got his B. Pharmacy from Dr. MGR Medical University, M. Pharmacy from Annamalai University, Chidambaram, India and Ph.D. from Jadavpur University, Kolkatta. He has 22 years of teaching & research experience. He served as Dean i/c, University College of Engineering, Ariyalur and Zonal officer/Zonal coordinator in the Anna University Examination System. He has been awarded a BOYSCAST Fellowship by DST, Government of India for post-doctoral research at the School of Pharmacy, University of Missouri - Kansas City, USA. He has received a research grant of more than Rs. 1.5 crores from various funding agencies such as DST, DBT, NASF and ICMR, New Delhi. Dr. Subramanian has won Best Innovation Award, 2013, Anna University, Chennai and Young Scientist Fellowship Award- Tamilnadu State Council for Science and Technology, Chennai, India. Two granted US patent, one PCT patent (granted in USA, Europe, Japan, China, Spain, Mexico, and Australia) and one granted Indian patent to his credit. He has 75 peer-reviewed publications and 15 book chapters to his credit. Dr. Subramanian is the reviewer of many peer reviewed journals and invited speaker at various conferences, seminars and symposiums.

Sweety Nath Barbhuiya is a Ph.D. research scholar in the Department of Life Science and Bio-informatics at Assam University, Silchar, India. Her research work involved the toxicity assessment of smokeless tobacco and environmental pollutants on reproductive health. She is working under the supervision of Prof. Sarbani Giri, Head & Dean, at the Department of Life Science and Bioinformatics, Assam University, Silchar.

Ayodeji Oteyola studied Zoology and Environmental Biology in Ekiti State University, Nigeria, and then joined the research group of Prof. Bakare in Cell Biology and Genetics in the University of Ibadan, Nigeria, where he obtained his M.Sc degree in 2018. He is presently a doctoral (PhD) student in Morphology Department, Institute of Biological Sciences, Federal University of Minas Gerais, Belo-Horizonte, Brazil.

Raffaele Pilla, Pharm.D., Ph.D., Doctor Europaeus, received his Master's degree in Pharmacy at G. d'Annunzio University in Chieti-Pescara, Italy in 2005, where he also served internships at the Cell Physiology Laboratory and Molecular Biology Laboratory. Prior to his degree, he was an Erasmus Student at Faculté de Pharmacie de Reims in Reims, France. He received his Doctor Europaeus in 2010 from Pitié-Salpétrière Institute in Paris, France. Also in 2010, he received his Ph.D. in Biochemistry, Physiology, and Pathology of Muscle at G. d'Annunzio University in Chieti-Pescara, Italy. He was hired as a Postdoctoral Scholar in the Department of Pharmacology and Physiology at the University of South Florida in Tampa, on two research grants respectively funded by the Office of Naval Research (US Navy) and Divers' Alert Network. He has written and lectured widely on his experiences. Dr. Pilla has been involved in ongoing research at the University of South Florida with the use of ketone esters. The initial work, funded by the Office of Naval Research, focused on using ketone esters to prevent CNS oxygen toxicity. Dr. Pilla is working to coordinate efforts in the United States and Italy to expand the ketone ester research to other conditions such as epilepsy and cancer and its use as an alternative metabolic fuel. Dr. Pilla has also served extensively in various international medical missions, orphan and homeless programs, emergency medicine rescue and training, and animal rescue and fostering. He is fluent in Italian, English, French, and Spanish. He has a special interest in journalism and photography, and he enjoys sailing, SCUBA diving, ultralight airplane piloting, soccer, volleyball, jogging, and motorcycles.

Mohamed Akif S. was born on 15 October 1998 in Tiruchirapalli, Tamil Nadu. He did his schooling in Tiruchirapalli at Campion Anglo Indian Higher Secondary School. He is currently pursuing his final year B. Tech (Biotechnology) in B.S. Abdur Rahman Crescent Institute of Science and Technology, Vandalur, Chennai- Tamil Nadu. His research interests include Cancer Biology, Microbiology and Food Biotechnology.

Nidhi Singh is a Ph.D. Scholar, Department of Pharmaceutics, National Institute of Pharmaceutical Education and Research, Kolkata, West Bengal.

Sonia Singh has experience of about 12 years of teaching in the GLA University, Mathura. She has written about more than 20 papers, including review and research in the field of herbs and pharmacognosy. She also attended many national workshops. She has completed her graduation in pharmacy in the year of 2005 and post-graduation in 2008 from Rajiv Gandhi Institute of Health Science, Karnataka. Now she is pursuing her PhD from GLA University, Mathura U.P.

Ali Zeytunluoglu is Assistant Professor at Pamukkale University. The areas of her research interest include biochemistry, protein chemistry, protein purification, enzyme inhibitors, plant seconder metabolite extraction, chromatographic techniques, biomaterials, proteomics, and other omics technologies.

Index

D

E

F

G

H

I

L

M

N

O

P

S

T

V

Z

www.ingramcontent.com/pod-product-compliance
Lightning Source LLC
LaVergne TN
LVHW080357011026
849161LV00009B/2328